Advances in Experimental Medicine and Biology

Volume 1065

More information about this series at http://www.springer.com/series/5584

Peter L. M. Kerkhof • Virginia M. Miller
Editors

Sex-Specific Analysis of Cardiovascular Function

Springer

Editors
Peter L. M. Kerkhof
Department of Radiology
and Nuclear Medicine
Amsterdam Cardiovascular Sciences
VU University Medical Center
Amsterdam, The Netherlands

Virginia M. Miller
Mayo Clinic
Rochester, MN, USA

ISSN 0065-2598 ISSN 2214-8019 (eBook)
Advances in Experimental Medicine and Biology
ISBN 978-3-030-08578-0 ISBN 978-3-319-77932-4 (eBook)
https://doi.org/10.1007/978-3-319-77932-4

Printed on acid-free paper

This Springer imprint is published by the registered company Springer International Publishing AG part of Springer Nature.
The registered company address is: Gewerbestrasse 11, 6330 Cham, Switzerland

Foreword

Sex-Specific Anaysis of Cardiovascular Function

Cardiovascular disease is the leading cause of morbidity and mortality for women in the Western world, and a major contributor to their illness spectrum in developing countries. Despite these facts, until recent decades, cardiovascular disease was considered a man's problem. The consequence was that the major evidence base, randomized controlled trials of cardiovascular diseases and their therapies, were conducted predominantly or exclusively in middle-aged Caucasian men, with results extrapolated (albeit inappropriately) to multiple other populations. Clinical research related to women's health involved largely what I have termed "bikini medicine," addressing the body systems covered by the bikini bathing suit, i.e., the breast and the reproductive system. Only in recent decades have we seen attention to broader aspects of sex-specific medicine, alternately termed gender medicine, and its consequences have been far reaching. Sex/gender differences entail both those characteristics specific to one sex and those that differ in prevalence, manifestations, and/or outcomes between the sexes.

In their innovative and comprehensive textbook, "Sex-Specific Analysis of Cardiovascular Function," Professors Peter Kerkhof and Virginia Miller provide an amazing resource for the research and clinical adult and pediatric cardiovascular communities. In chapters ranging from basic cardiovascular physiology, genetics, hormonal status, genomics, proteomics, metabolomics, biomarkers, biomedical engineering, biostatistics, and the like; to epidemiology, translational issues, and clinical aspects of a broad spectrum of cardiovascular illnesses, the recently identified sex/gender differences (as well as similarities) are explored. The widespread sex differences in cardiovascular structure and function highlight the importance of consideration of sex as a biologic variable and of sex influences on cardiovascular disease both in preclinical research and in the planning and analysis of cardiovascular clinical trials. Although the coronary circulation and coronary heart disease have been the most intensively studied in the arena of cardiovascular sex-specific medicine, the editors and their authors appropriately explore the domains of the cerebral circulation, pulmonary arterial hypertension, etc.

As recently as 2015, the Heart Research for All Act of the US Congress mandated that there be adequate representation by sex, race/ethnicity, and age for federally funded clinical research studies, with the results disaggregated by these parameters. The same mandate applies to the basic science arena.

Scientists are required to provide the provenance of their cells and tissues, i.e., whether they derived from male or female animals, as well as the sex of the animals studied, again disaggregating the results by sex. This legislation thus further highlights the importance of sex as a biological variable in preclinical research. The same mandates are in effect for clinical research studies submitted to the US Food and Drug Administration for drug and device approval.

Given the major sex-specific aspects of cardiovascular function both in health and in major cardiovascular diseases; and the substantial sex-based variation in the prevalence, expression, and costs of multiple cardiovascular problems, several chapter authors suggest to their readers the use of gender innovations and the mandate for development of new technologies, new pharmacotherapies, and new diagnostic modalities. The knowledge gaps highlighted in multiple chapters define a sex-based agenda for basic, epidemiologic, translational, and clinical research, with the outcomes likely to benefit both women and men. The terms "personalized medicine" and "precision medicine" are recurrently cited as advances in patient care; sex/gender considerations are obvious pivotal components and contributors.

A woman's health relates to the health of her family, the health of her community, and thereby to the health of the nation. Thus, efforts to improve women's heart health, evidence-based application of sex-specific preventive, diagnostic, and therapeutic interventions, have the potential to improve national health status. Advances in the USA offer support to this thesis. In the early years of sex-specific cardiovascular medicine, and prior to intensive educational and advocacy campaigns regarding heart disease in women, 1 of 2 US women died of cardiovascular illness. In 2018, this has decreased to 1 in 4 US women, a stunning accomplishment.

Nanette K. Wenger, MD, MACC, MACP, FAHA
Professor of Medicine (Cardiology)
Emory University School of Medicine
Consultant, Emory Heart and Vascular Center
Founding Consultant, Emory Women's Heart Center
Atlanta, Georgia

Preface

One of the most fascinating aspects to explore in living systems concerns a comparison between females and males. In humans, various findings regarding anatomy and physiology differ considerably for both sexes, while cultural and social factors may also have distinct impact on the course of life, including aspects of health and disease. Every parallel description of health and disease in women and men is best interpreted against the background of the age considered. The present book addresses all these elements as far as the cardiovascular system is concerned.

Deciphering underlying properties and mechanisms of cardiac and vascular functioning will contribute to the formulation of effective guidelines to promote health, to timely take preventive measures, to more precisely establish a diagnosis in case of disease, and to undertake curative action on a personalized basis. In fact, application of precision medicine begins with considering sex and age.

The book presents an updated survey of sex-specific characteristics of the cardiovascular system, based on (epi)genetic, molecular, cellular, anatomical, (electro)physiologic, and epidemiologic studies written by eminent experts in those fields. The contents cover important clinical data, spanning from the fetus to newborn, child, adult, and up to the elderly. This range also includes pregnancy and associated cardiovascular adaptations, as well as related disorders. In fact, four chapters are devoted to cardiovascular disorders typically or predominantly encountered in women.

The biomedical engineering spectrum of sex-specific items starts with basic disciplines such as a refreshing presentation of elementary cardiophysiology, to continue with insightful outlines of modern approaches including wave intensity analysis, robot-assisted rehabilitation, big data analysis, and heart rate variability. Furthermore, due attention is given to circulatory system–related effects of exercise and aging in both women and men.

Starting with a short historical introduction, the book ends with annotated reference values, along with guidelines for their prudent interpretation.

This carefully selected anthology is intended both as an eye-opener for students and as a solid resource for clinicians dealing with sex-specific aspects of cardiovascular disease, either at the first encounter level of general practitioners or for the medical specialist dealing with comorbidities associated with cardiovascular disease. The editors expect that the present comprehensive collection will stimulate sex-specific cardiovascular research and will be of distinctive importance in diagnosis and patient management. The extraordinary contributions by all clinicians and basic scientists involved, often delivered during evening and weekend hours, are very much appreciated. They originate from 17 countries around the world, often forming multinational teams, which fact provides evidence that diversity in thinking leads to innovation in science and medicine with a global impact. We are proud to mention that 60% of all authors are women.

Amsterdam, The Netherlands — Peter L. M. Kerkhof
Rochester, MN, USA — Virginia M. Miller
March 7, 2018

Contents

1 **Women and Men in the History of Western Cardiology: Some Notes on Their Position as Patients, Role as Investigational Study Subjects, and Impact as Professionals** 1
Peter L. M. Kerkhof and Elena Osto

2 **An Introduction to Epigenetics in Cardiovascular Development, Disease, and Sexualization** 31
Christine M. Cunningham and Mansoureh Eghbali

3 **Epidemiology of Congenital Heart Disease with Emphasis on Sex-Related Aspects** 49
Byung Won Yoo

4 **Sex Differences in Epidemiology of Cardiac and Vascular Disease** 61
Tatiana Kuznetsova

5 **Biostatistics in Cardiovascular Research with Emphasis on Sex-Related Aspects** 71
Marieke M. ter Wee and Birgit I. Lissenberg-Witte

6 **The Influence of Age and Sex on the Electrocardiogram** . . 93
Peter W. Macfarlane

7 **Safety and Effectiveness of Medical Device Therapy** 107
Robbert Zusterzeel

8 **Cardiophysiology Illustrated by Comparing Ventricular Volumes in Healthy Adult Males and Females** 123
Peter L. M. Kerkhof, Tatiana Kuznetsova, J. Yasha Kresh, and Neal Handly

9 **Sex Differences in Regulation of Blood Pressure** 139
Jane F. Reckelhoff

10 **Arterial Flow, Pulse Pressure and Pulse Wave Velocity in Men and Women at Various Ages** 153
Alberto P. Avolio, Tatiana Kuznetsova, Guy R. Heyndrickx, Peter L. M. Kerkhof, and John K.-J. Li

11 **Wave-Intensity Analysis: Sex-Specific Differences in Hemodynamic and Ventilatory Responses to Graded Exercise—Echocardiographic Measurements** 169
Midori Tanaka, Motoaki Sugawara, Yasuo Ogasawara, and Kiyomi Niki

12 **Sex Differences in Autonomic Response to Exercise Testing in Patients with Brugada Syndrome** 181
Mireia Calvo, Virginie Le Rolle, Daniel Romero, Nathalie Béhar, Pedro Gomis, Philippe Mabo, and Alfredo Hernández

13 **Benchmarking Heart Rate Variability to Overcome Sex-Related Bias** 191
Massimo Pagani, Roberto Sala, Mara Malacarne, and Daniela Lucini

14 **Cardiovascular Allometry: Analysis, Methodology, and Clinical Applications** 207
John K.-J. Li

15 **The Heart as a Psychoneuroendocrine and Immunoregulatory Organ** 225
Carlo Dal Lin, Francesco Tona, and Elena Osto

16 **Mitochondria and Sex-Specific Cardiac Function** 241
Rosa Vona, Barbara Ascione, Walter Malorni, and Elisabetta Straface

17 **Sex Differences in the Coronary System** 257
Viviany R. Taqueti

18 **Cerebral Circulation in Men and Women** 279
Aurangzeb Memon and Louise D. McCullough

19 **Arterial Wall Properties in Men and Women: Hemodynamic Analysis and Clinical Implications** 291
John K.-J. Li

20 **Sex-Specific Characteristics of the Microcirculation** 307
Virginia H. Huxley and Scott S. Kemp

21 **Sex-Specific Ventricular and Vascular Adaptations to Exercise** 329
Jill N. Barnes and Qi Fu

22 **Growth of Cardiovascular Structures from the Fetus to the Young Adult** 347
Frederic Dallaire and Taisto Sarkola

23 **Heart Function Analysis in Cardiac Patients with Focus on Sex-Specific Aspects** 361
Peter L. M. Kerkhof, Richard A. Peace, Guy R. Heyndrickx, Lilian J. Meijboom, Ralf W. Sprengers, and Neal Handly

24 **Heart Transplantation Survival and Sex-Related Differences** 379
Martina Previato, Elena Osto, Peter L. M. Kerkhof, Gareth Parry, and Francesco Tona

25 **Differences in Cardiovascular Aging in Men and Women** 389
Alice E. Kane and Susan E. Howlett

26 **Hemodynamic and Electrocardiographic Aspects of Uncomplicated Singleton Pregnancy** 413
Qi Fu

27 **Sex-Specific Physiology and Cardiovascular Disease** 433
Chrisandra L. Shufelt, Christine Pacheco, Marysia S. Tweet, and Virginia M. Miller

28 **Cardiovascular Sequels During and After Preeclampsia** . . . 455
Nina D. Paauw and A. Titia Lely

29 **Cardiovascular Implications of Diabetes, Metabolic Syndrome, Thyroid Disease, and Cardio-Oncology in Women** 471
Marijana Tadic, Cesare Cuspidi, Dragan Vasic, and Peter L. M. Kerkhof

30 **Sex-Specific Cardiovascular Comorbidities with Associations in Dermatologic and Rheumatic Disorders** 489
Peter L. M. Kerkhof and Irina Khamaganova

31 **The Role of Sex in the Pathophysiology of Pulmonary Hypertension** 511
Craig K. Docherty, Katie Yates Harvey, Kirsty M. Mair, Sinead Griffin, Nina Denver, and Margaret R. MacLean

32 **Sex Differences in Heart Failure** 529
Gianluigi Savarese and Domenico D'Amario

33 **Sex-Related Aspects of Biomarkers in Cardiac Disease** 545
Alma M. A. Mingels and Dorien M. Kimenai

34 **Cardiac Rehabilitation for Women** 565
Vera Bittner

35 **Stroke Rehabilitation: Therapy Robots and Assistive Devices** 579
Verena Klamroth-Marganska

36 **Precision Medicine and Personalized Medicine in Cardiovascular Disease** 589
Gemma Currie and Christian Delles

37 Big-Data Analysis, Cluster Analysis, and Machine-Learning Approaches 607
Amparo Alonso-Betanzos and Verónica Bolón-Canedo

38 Genome-Wide Association Studies and Risk Scores for Coronary Artery Disease: Sex Biases 627
Sean G. Byars and Mike Inouye

39 Gendered Innovation in Health and Medicine 643
Londa Schiebinger and Ineke Klinge

40 Gendered Innovations in the Study of Cardiovascular Diseases 655
Evgeny Shlyakhto

41 Sex- and Age-Related Reference Values in Cardiology, with Annotations and Guidelines for Interpretation 677
Peter L. M. Kerkhof, Richard A. Peace, and Peter W. Macfarlane

Index 707

List of Reviewers

Alberto Avolio Faculty of Medicine and Health Sciences, Macquarie University, Sydney, Australia

Claire Brough Liverpool Heart and Chest Hospital NHS Foundation Trust, Liverpool, UK

Hans-Peter Brunner-LaRocca Maastricht University Medical Center, Maastricht, The Netherlands

Paul D. Chantler West Virginia University – School of Medicine, Morgantown, USA

Christian Delles Institute of Cardiovascular and Medical Sciences, Glasgow, UK

Mansoureh Eghbali Department of Anesthesiology and Perioperative Medicine, Division of Molecular Medicine, David Geffen School of Medicine at University of California Los Angeles, Los Angeles, CA, USA

Theo J. C. Faes VU University Medical Center, Amsterdam, The Netherlands

Qi Fu University of Texas Southwestern Medical Center, Dallas, USA

Chahinda Ghossein-Doha Maastricht University Medical Center, Maastricht, The Netherlands

K. Lance Gould Dept. Medicine, McGovern Medical School, UTHealth Science Center, Houston, TX

Melanie R. Gubbels Bupp Randolph-Macon College, Ashland VA, USA

Neal Handly Drexel University College of Medicine, Philadelphia, USA

Guy R. Heyndrickx Cardiovascular Center, Aalst, Belgium

Peter L. M. Kerkhof Department of Radiology and Nuclear Medicine, VU University Medical Center, Amsterdam, The Netherlands

Vessela Tz. Krasteva Bulgarian Academy of Sciences, Sofia, Bulgaria

Arnoud van der Laarse Leiden University Medical Center, Leiden, The Netherlands

Titia Lely Department of Obstetrics, Utrecht University Medical Center, Utrecht, The Netherlands

Michelle A. Lowes Rockefeller University, New York, USA

Daniela Lucini University of Milan, Milan, Italy

Christoph Maack Heart Failure Center, University Clinic Würzburg, Würzburg, Germany

Angela H. E. M. Maas Radboud Universitair Medisch Centrum, Nijmegen, The Netherlands

Edward H. Morrow School of Life Sciences, University of Sussex, Brighton, UK

Alma M. A. Mingels Maastricht University Medical Center, Maastricht, The Netherlands

Bohuslav Ošťádal Institute of Physiology, Czech Academy of Sciences, Prague, Czech Republic

Massimo Pagani University of Milan, Milan, Italy

Ronney B. Panerai Dept. Cardiovascular Sciences, Leicester Royal Infirmary, Leicester, UK

Volker Perlitz Simplana GmbH, Aachen, Germany

Leticia Prates Roma Universität des Saarlandes, Homburg, Germany

Merry Stuber Springer US, New York, USA

Lutgarde Thijs Hypertension and Cardiovascular Epidemiology, University of Leuven, Leuven, Belgium

John Tyberg University of Calgary Faculty of Medicine, Calgary, Canada

Jessica Wagenseil Mechanical Engineering and Materials Science, Washington University, St. Louis, USA

Robbert Zusterzeel Division of Applied Regulatory Science, U.S. Food and Drug Administration, Silver Spring, MD, USA

Abbreviations and Acronyms

5mC	5-methylcytosine
aa	Amino acid
AAA	Abdominal aortic aneurysm; cf. EVAR, snAAA, TAA
ABPM	Ambulatory blood pressure monitoring; cf. HBPM, HTN, WCH
1-ABT	1-Aminobenzotrizole; eicosanoid synthesis inhibitor
ACC	American College of Cardiology; cf. AHA
ACE(I)	Angiotensin-converting enzyme (inhibitor)
ACHD	Adult congenital heart disease; cf. HF
ACR	American College of Rheumatology
ACS	Acute coronary syndrome; cf. DAPT, DES
ACTH	Adrenocorticotropic hormone
AD	Atopic dermatitis
ADL	Activity of daily living
ADT	Appropriate device therapy
AGG	Advisory Group for Gender; cf. EC
AHA	American Heart Association; cf. ACC, ASA, ESC, NYHA
AI	Artificial intelligence; cf. ML, RF, SVM
AIF	Apoptosis inducing factor
AIx	Augmentation index; cf. PP
AMA	American Medical Association; cf. JAMA
AMI	Acute myocardial infarction; cf. MI
ANA	Antinuclear antibody
AngII	Angiotensin II
ANP	Atrial natriuretic peptide; cf. BNP, CNP
ANS	Autonomic nervous system
ANSI	Autonomic nervous system index
AO	Abdominal obesity; cf. MetS
AoA	Aortic annulus
AP	Action potential; cf. APD (1); angina pectoris (2)
APD	Action potential duration
AR	Androgen receptor; cf. ER, PR (1); autoregressive (2); acute rejection (3)
ARB	Angiotensin receptor blocker
ARD	Aortic root diameter
AS	Aortic (valve) stenosis (1); ankylosing spondylitis (2)
ASA	American Stroke Association; cf. AHA
ASCVD	Atherosclerotic cardiovascular disease; cf. CAD, CVD
ASD	Atrial septum defect; cf. AVSD, VSD

AT	Anaerobic threshold; cf. RC
ATP	Anti-tachycardia pacing
ATP3	Adult treatment panel 3; cf. LDL
AV	Atrio-ventricular; cf. SA
AVSD	Atrioventricular septal defect; cf. ASD, VSD
β-AS	β-Adrenergic stimulation
BAT	Baroreflex activation therapy
BBB	Blood brain barrier
BMI	Body mass index; cf. BSA
BMP	Bone morphogenic protein
BMPR-II	Bone morphogenic protein receptor type II
BNP	B-type natriuretic peptide; cf. ANP, CNP, NT-proBNP
BP	Blood pressure; cf. DBP, SBP
bpm	Beats per minute; cf. HR
BPV	Blood pressure variability; cf. HRV
BS	Brugada syndrome
BSA	Body surface area; cf. BMI, BW, COi, LBM
CABG	Coronary artery bypass graft
CAC	Coronary artery calcium
CACS	Coronary artery calcium scoring
CAD	Coronary artery disease; = CHD; cf. CCTA, CMD
CAL	Cytokine (chemokine)-activated leukocyte
cAMP	Cyclic adenosine monophosphate; cf. cGMP
CARL	Cardiac apoptosis-related lncRNA
CAV	Cardiac allograft vasculopathy; cf. HTX
CAVI	Cardio-ankle vascular index; cf. PWV
CBF	Coronary blood flow; cf. CFR, CMD, MBF
CBP	CREB binding protein
cDBP	central diastolic blood pressure
CE	Common Era
CEE	Conjugated equine estrogen
CIMT	Constrained induced movement therapy
cDNA	Complementary DNA; a DNA molecule that is complementary to a specific messenger RNA
CFR	Coronary flow reserve; cf. CBF (3), CMD, FFR, MFR
CFS	Correlation-based feature selection
CFTR	Cystic fibrosis transmembrane conductance regulator
CFVR	Coronary flow velocity reserve; cf. FFR
cGMP	Cyclic guanosine monophosphates; cf. cAMP
CGRP	Calcitonin gene-related peptide
Chaer	Cardiac-hypertrophy-associated epigenetic regulator
CH	Chronic hypoxia; cf. eNOS
CHA2DS2-VASc	Congestive heart failure, hypertension, age $\geq$75 years, diabetes mellitus, prior stroke or transient ischemic attack, vascular disease, age 65–74 years, sex category
CHD	Coronary heart disease (1); congenital heart disease (2)
CHF	Congestive heart failure; cf. HF
CHRF	Cardiac hypertrophy-related factor
CI	Confidence interval; cf. COi
CIHR	Canadian Institutes of Health Research
CIMT	Carotid intimal medial thickness

CK	Creatine kinase
CKD	Chronic kidney disease; cf. ESRD
CKMB	Creatine kinase muscle-brain type
CMD	Coronary microvascular dysfunction; cf. CAD, CBF (3), CFR, MVA, SVD
CMR(I)	Cardiovascular magnetic resonance (imaging); cf. LGE, MRI
CNP	Type C natriuretic peptide; cf. ANP, BNP
COA	Coarctation of aorta
CO(i)	Cardiac output (index); cf. BSA, SV, HR, VTI
COPD	Chronic obstructive pulmonary disease
COXIB	Cyclooxygenase-2 selective inhibitor; COX-2 inhibitor
CPAP	Continuous positive airway pressure; cf. OSA
CREB	cAMP response element binding protein; cf. CBP
CRF	Coronary risk factor; cf. FRS
CRH	Corticotropin-releasing hormone
CRP	C-reactive protein; hsCRP
CRT	Cardiac resynchronization therapy (1); = coronary reactivity test; cf. CFR, CMR (2)
CRT-D	Cardiac resynchronization therapy defibrillator
CRT-P	Cardiac resynchronization therapy pacemaker
cSBP	Central systolic blood pressure
CTA	Computed tomography angiography
CTEPH	Chronic thromboembolic pulmonary hypertension; cf. PAH
cTn	Cardiac troponin
cTnI	Cardiac troponin I
cTnT	Cardiac troponin T
CV	Cardiovascular (1); coefficient of variation (2)
CVA	Cardiovascular accident, aka "stroke"; cf. FMD, TIA
CVD	Cardiovascular disease; cf. CAD, CHD, CVRF, HF, IHD, MI
CVE	Cardiovascular event; cf. MI, CVA
CVH	Cardiovascular health
CVRF	Cardiovascular risk factor; cf. CVD
CYP	Cytochrome P450
DAMP	Damage-associated molecular pattern; cf. PAMP
DBP	Diastolic blood pressure; cf. BP, HTN, MAP, PP, SBP
DHEA	Dehydroepiandrosterone
DHEA-S	Dehydroepiandrosterone-sulphate
DIC	Disseminated intravascular coagulation
DM	Diabetes mellitus; cf. GDM, T2DM
DMARD	Disease-modifying anti-rheumatic drug
DNMT	DNA methyltransferase
DNP	D-type natriuretic peptide; cf. ANP, BNP, CNP
DORV	Double outlet of right ventricle
DRG	Dorsal root ganglion
DRP1	Dynamin related protein 1
DSS	Disease-specific survival
E2	17β Estradiol
EC	Endothelial cell (1); European Commission (2); excitation–contraction (3)
ECG	Electrocardiogram
EDR	ECG-derived respiration

EET	Epoxyeicosatrienoic acid
EF	Ejection fraction (1); emptying fraction (2); cf. HFpEF, SV(i)
eGFR	Estimated glomerular filtration rate; cf. GFR
EHR	Electronic health record; cf. NLP
EM	Expectation minimization
Emax	End-systolic elastance; cf. ESPVR, Vo
EMI	Electrical myocardial instability; cf. HRT
eNOS	Endothelial nitric oxide synthase
EPC	Endothelial progenitor cell
EPS	Electrophysiological study
ER	Estrogen receptor; cf. PR
ERα	Estrogen receptor alpha
ERE	Estrogen response element
ERK	Extra-cellular regulated kinase
ESC	European Society of Cardiology; cf. AHA
ESKD	End-stage kidney disease; cf. ESRD, CKD
ESRD	End-stage renal disease; cf. CKD
ESPVR	End-systolic pressure-volume relationship; cf. Emax, LV, Vo
ET	Endothelin
ET-1	Endothelin-1
ETA	A subtype of endothelin for vasoconstriction
ETB1	A subtype of endothelin for vasodilation
FAC	Fractional area change; cf. EF, FS
FCG	Four core genotype
FD	Female donor
FDA	Federal Department of Agriculture; cf. NIH
FDG	^{18}F-fluorodeoxyglucose; cf. PET
FFM	Fat free mass
FFR	Force-frequency relation (1); fractional flow reserve (2)
FHS	Framingham Heart Study; cf. MESA
FLS	Fibroblast-like synoviocyte
fQRS	Fragmentation of QRS; cf. ECG
FRS	Framingham risk score; cf. CRF, FHS
FPTP	Fluoropentyl-triphenylphosphonium; cf. PET
FS	Fractional shortening; cf. EF, FAC
FSH	Follicle-stimulating hormone
FU	Follow-up
FVC	Forced vital capacity
GABA	Gamma-aminobutyric acid
GCS	Global circumferential strain; cf. GLS
GDF-15	Growth differentiation factor-15
GDMT	Guideline-directed medical therapy
GEMM	Genetically engineered mouse model
GFR	Glomerular filtration rate; cf. eGFR
GH	Growth hormone
GHES	German health examination surveys
GLP-1	Glucagon-like protein-1
GLS	Global longitudinal strain; cf. LS
GPER	G-protein-coupled estrogen receptor; cf. G1
GRACE	Global Registry of Acute Coronary Events
GRS	Genetic risk score; cf. IDI, NRI

GWAS	Genome-wide association studies; cf. GSEA, WGHS
GWTG	Get with the guidelines; cf. HF
H-FABP	Heart fatty acid binding protein
HAS-BLED	Hypertension, abnormal renal and liver function, stroke, bleeding, labile INR, elderly, drugs or alcohol
HAT	Histone acetyltransferase
HCM	Hypertrophic cardiomyopathy
HDAC	Histone deacetylase
HDL	High-density lipoprotein; cf. LDL
HELLP	Hemolysis, elevated liver enzymes, and low platelets (syndrome)
20-HETE	20-Hydroxyeicosatetraenoic acid
HF	Heart failure; cf. ACHD, CHF; high frequency; cf. LF (2)
HFmrEF	Heart failure with mid-range ejection fraction
HFpEF	Heart failure with preserved ejection fraction
HFrEF	Heart failure with reduced ejection fraction
HLHS	Hypoplastic left heart syndrome
HMT	Histone methyltransferase; cf. DMT
HPAA	Hypothalamic–pituitary–adrenal axis
HPAH	Heritable pulmonary arterial hypertension; cf. IPAH, PAH
HR	Heart rate (1); cf. bpm, HRV, MHR; hazard ratio (2)
HR-PRO	Health-related patient-reported outcome
HRT	Hormone replacement therapy (1); cf. HT, TRT; heart rate turbulence (2)
HRV	Heart rate variability
hs-CRP	High-sensitivity C-reactive protein; cf. CRP
hs-cTnI	High-sensitivity cardiac troponin I
hs-cTnT	High-sensitivity cardiac troponin T
HTN	(Arterial) hypertension; cf. PAH
I/R	Ischemia/reperfusion; cf. injury
IBD	Inflammatory bowel disease
ICD	Implantable cardioverter-defibrillator; cf. S-ICD
ICV	Intracerebroventricular
IFN	Interferon
IGF-1	Insulin-like growth factor 1
IHD	Ischemic heart disease; cf. ACS, CAD, CHD
IJD	Inflammatory joint disease; cf. RA
IL	Interleukin
IMID	Immune-mediated inflammatory disease
INOCA	Ischemia and no obstructive coronary arteries; cf. MINOCA
INR	International normalized ratio
IPAH	Idiopathic pulmonary arterial hypertension; cf. HPAH, PAH
IPC	Ischemic preconditioning
IPF	Idiopathic pulmonary fibrosis
ISH	Isolated systolic hypertension
ISHLT	International Society of Heart and Lung Transplantation
IUGR	Intrauterine growth restriction
JAMA	Journal of the American Medical Association; cf. AMA
JNC	Joint National Committee (on the prevention, detection, evaluation, and treatment of high blood pressure)
KNDy	Kisspeptin/neurokin B/dinorphin

L-NAME	Nitro-L-arginine methyl esther; non-selective NOS inhibitor
LA	Left atrium/atrial; cf. RA
LAD	(Proximal) left anterior descending (coronary artery); cf. LCX, RCX
LAVI	Left atrial volume index
LBBB	Left bundle branch block; cf. RBBB
LBNP	Lower body negative pressure
LDL	Low-density lipoprotein; cf. ATP3, HDL
LF	Low frequency; cf. HF
LI	Leak index; cf. venular permeability
LLN	Lower limit of normal; cf. ULN
LMS	Lambda-mu-sigma; cf. BIC, z-score
lncRNA	Long ncRNAs
LPS	Lipopolysaccharide
LS	Longitudinal strain; cf. GLS
LV	Left ventricle/ventricular; cf. EDP, EDV, ESPVR, ESV
LVAD	Left ventricular assist device; cf. VAD
LVEF	Left ventricular ejection fraction
LVH	Left ventricular hypertrophy
LVM(i)	Left ventricular mass (index)
LVOT(O)	Left ventricular outflow tract (obstruction)
6MWD	Six-minute walk distance
MACE	Major adverse cardiovascular event; cf. AE, SAE
MAP	Mean arterial blood pressure; cf. DBP, PP, SBP
MAPK	Mitogen-activated protein kinase
MAPSE	Mitral annular plane systolic excursion; cf. TAPSE
MAST	Military antishock trousers
MBF	Myocardial blood flow; cf. CBF
MDCT	Multidetector computed tomography
2ME1	2-Methoxyestrone
2ME2	2-Methoxyestradiol
MESA	Multi-ethnic study of atherosclerosis; cf. FHS
MetS(y)	Metabolic syndrome; cf. AO
MHR	Maximum heart rate
Mhrt	Myosin heavy chain-associated RNA transcript
MHT	Menopausal hormone treatment/therapy
MI	Myocardial infarction; cf. AMI, INOCA, NIMI, NSTEMI, STEMI
MINOCA	Myocardial infarction with nonobstructive coronary arteries
miRNA	microRNA
ML	Machine learning; cf. AI, SVM
MMF	Mycophenolate mofetil
MPI	Myocardial perfusion imaging
MR	Male recipient
MRA	Magnetic resonance angiography
MRI	Magnetic resonance imaging; cf. CMR
mRMR	Minimum redundancy maximum relevance
mRS	Modified Rankin scale
MSC	Mesenchymal stem cell
MSIMI	Mental stress induced myocardial ischemia, cf. MI
MSNA	Muscle sympathetic nerve activity

mtDNA	Mitochondrial DNA
mTWA	Microvolt T-wave alternans
NAMS	North American Menopause Society
NBSR	National Bariatric Surgery Registry
NCDR	National Cardiovascular Data Registry
ncRNA	Noncoding RNA; cf. lncRNA
NE	Norepinephrine
NET	Neutrophil extracellular trap
NHANES	National Health and Nutrition Examination Survey
NHS	National Health Service (in the UK)
NIH	National Institutes of Health; cf. FDA, PMI
NO	Nitric oxide
NOAC	New oral anticoagulant; cf. AF, CVA, DAPT
NOS	Oxide synthase; cf. eNOS
NRI	Net reclassification improvement; cf. GRS, IDI
NRVM	Neonatal rat ventricular myocyte
NSTEMI	Non-ST-elevation myocardial infarction; cf. STEMI
NT-proBNP	N-terminal pro brain-type natriuretic peptide; cf. BNP
nu	Normalized unit
NYHA	New York Heart Association; cf. AHA
OHT	Orthotopic heart transplant
OR	Odds ratio
OSA	Obstructive sleep apnea; cf. CPAP
OSSD	Organization for the study of sex differences
PAD	Peripheral arterial disease
PAEC	Pulmonary arterial endothelial cell; cf. PAH, PASMC
PA	Pulmonary artery
PAD	Peripheral artery disease
PAEC	Pulmonary artery endothelial cell
PAH	Pulmonary arterial hypertension; cf. HPAH, HTN, PH
PAMP	Pathogen-associated molecular pattern; cf. DAMP
PASMC	Pulmonary artery smooth muscle cell; cf. PAEC, SMC
PCI	Percutaneous coronary intervention
PCOS	Polycystic ovary syndrome; cf. AE, POI
PDA	Patent/persistent ductus arteriosus
PE	Peak effort; cf. HRV
PEA	Pulseless electrical activity
PDE	Phosphodiesterase
PDE3	Phosphodiesterase isoform 3
PE	Preeclampsia
PET	Positron emission tomography; cf. FDG, MPI, SPECT
PH	Pulmonary hypertension; cf. experimental model of PAH
PI	Principal investigator
PICO(t)	Patient, intervention, comparison, outcome measure (time)
PINK1	PTEN-induced putative kinase 1
PKA	Protein kinase A
PKG	Protein kinase G
PLB	Phospholamban; a small inhibitory phosphoprotein that in humans is encoded by the PLN gene
PLM	Phospholemman; phosphoprotein, in humans encoded by the FXYD1 gene

PMI	Precision Medicine Initiative; cf. NIH
PNEI	Psycho-neuro-endocrine-immunology
PP	Pulse pressure; cf. DBP, HTN, MAP, SBP
PPAR	Peroxisome proliferator-activated receptor
PPHN	Persistent pulmonary hypertension of the newborn
PPHTN	Portal hypertension; cf. HTN, PAH
PROCAM	Prospective Cardiovascular Muenster (Heart Study); cf. SCORE
PRR	Pattern recognition receptor
PSD	Power spectral density
PSP	Peak systolic pressure
PTEN	Phosphatase and tensin homologue deleted in chromosome 20; cf. PINK1
PTH	Parathyroid hormone
PTP	Permeability transition pore
PTT	Pulse transit time; cf. PWV
PVC	Premature ventricular complex
PWV	Pulse wave velocity; cf. PTT
QoL	Quality of life; cf. CR, HRQOL
RA	Right atrium/atrial (1); cf. LA; = rheumatoid arthritis (2); cf. IJD
RAAS	Renin-angiotensin-aldosterone system
RAS	Renin-angiotensin system
RC	Respiratory compensation; cf. AT
RCA	(Proximal) right coronary artery; cf. LAD, RCX
RCT	Randomized controlled trial
RDN	Renal artery denervation; cf. SNS
RF	Random forest; cf. ML, VSM
RFE	Recursive feature elimination; cf. SVM
RNS	Reactive nitrogen species; cf. ROS
RNV	Radionuclide ventriculography
ROC	Receiver operating curve; cf. AUC
ROS	Reactive oxygen species; cf. RNS
RP	Real picture; cf. EF
RR	Relative risk
RRS	Reynolds Risk Score; cf. CRF, FRS, GRS
RSA	Respiratory sinus arrhythmia
RUPP	Reduced uterine perfusion pressure
RV	Right ventricle/ventricular; cf. DORV
RVIT	Right ventricular inflow tract
RVOT	Right ventricular outflow tract
RVSP	Right ventricular systolic pressure
RVH	Right ventricular hypertrophy
S-ICD	Subcutaneous implantable cardioverter-defibrillator; cf. ICD
SA	Sino-atrial; cf. AV
SAN	Sino-atrial node
SBP	Systolic blood pressure; cf. BP, DBP, HTN, MAP, PP
SCAD	Spontaneous coronary artery dissection; cf. ACS, OCT
SCD	Sudden cardiac death
SCORE	Systematic coronary risk evaluation; European scoring system; cf. PROCAM
SD	Standard deviation
sEng	Soluble endoglin

SERCA	Sarcoendoplasmic reticulum Ca^{2+}-ATPase
SERM	Selective estrogen receptor modulator
SGBA	Sex- and gender-based analysis
sGC	Soluble guanylate cyclase
SHBG	Sex hormone binding globulin
SHEP	Systolic hypertension in the elderly program; cf. HTN, ISH
SHR	Spontaneously hypertensive rat; cf. SSH
SI	Stiffness index
sIB	Sequential information bottleneck
SKM	Skeletal muscle
SLE	Systemic lupus erythematosus
SMAD	Homolog of *sma* gene and Drosophila
SMC	Smooth muscle cell; cf. PASMC
SNP	Single-nucleotide polymorphism
SPECT	Single-photon emission computed tomography; cf. MRI, PET
SPRINT	Systolic blood pressure intervention trial
SPWVD	Smoothed pseudo Wigner-Ville distribution
SR	Sarcoplasmic reticulum (1); strain rate (2)
Sry	Sex-determining region on the Y-chromosome
SSH	Salt-sensitive hypertension; cf. SHR
SSRI	Selective serotonin reuptake inhibitor
STE	Speckle tracking echocardiography
STEMI	ST-elevation myocardial infarction; cf. NSTEMI
STI	Speckle tracking imaging
SV	Single ventricle; cf. SV(i)
SV(i)	Stroke volume (index); cf. CO(i), EDV(i), ESV(i)
SVM	Support vector machine; cf. ML, AI, RFE
SySc	Systemic sclerosis, scleroderma
T2D(M)	Type 2 diabetes (mellitus); cf. DM, GDM
TAC	Transaortic constriction
TAPSE	Tricuspid annular plane systolic excursion
TCR	Torsion to circumference ratio
TDI	Tissue Doppler imaging
TET	Ten-eleven translocation
TGA	Transposition of great arteries
TGF-beta	Beta-transforming growth factor
TLR	Toll-like receptor; cf. PRR
TMS	2,3′,4, 5′-tetramethoxystilbene
TnC	Troponin C
TNFa	Tumor necrosis factor alpha
TnI	Troponin I
TnT	Troponin T
TOF	Tetralogy of Fallot
TP	Total power; cf. HF, HRV, LF
tPA	Tissue Plasminogen Activator
TPH	Tryptophan hydroxylase
TPR	True positive rate
TSH	Thyroid stimulating hormone
TTE	Transthoracic echocardiography
TTh	Testosterone therapy
TTR	Transthyretin; cf. amyloidosis

UA	Unstable angina
ULN	Upper limit of normal; cf. LLN
UNOS	United Network for Organ Sharing [USA]
Vo	Volume axis intercept of end-systolic elastance; cf. Emax, ESPVR
VA	Ventricular arrhythmia; cf. VT
VAC	Ventriculo-arterial coupling
VAR	Variance accounted for
VDAC	Voltage-dependent anion channel
VEGF	Vascular endothelial growth factor
VF	Ventricular fibrillation
VIP	Vasoactive intestinal peptide
VSD	Ventricular septum defect
VSM	Vascular smooth muscle
VT	Ventricular tachycardia; cf. VA
WBC	While blood cell (count)
WGHS	Women's Genome Health Study; cf. GWAS
WHI	Women's Health Initiative; cf. MHT
WI(A)	Wave intensity (analysis)
WPB	Weibel-Palade body
WSDM	Women and sex differences in medicine
wt	Wild type
XBcy	Cross-bridge cycling rate
Xic	X-inactivation center
Xist	X-inactive-specific transcript
XWAS	Software toolset for genetic data analysis and association studies of the X chromosome

1 Women and Men in the History of Western Cardiology: Some Notes on Their Position as Patients, Role as Investigational Study Subjects, and Impact as Professionals

Peter L. M. Kerkhof and Elena Osto

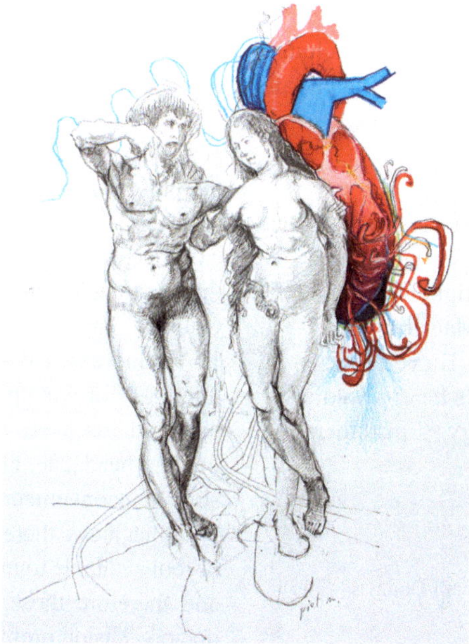

Adam and Eve (after Jean Gossaert). Illustration by Piet Michiels, Leuven, Belgium.

P. L. M. Kerkhof (✉)
Department of Radiology and Nuclear Medicine, Amsterdam Cardiovascular Sciences, VU University Medical Center, Amsterdam, The Netherlands
e-mail: plm.kerkhof@vumc.nl

E. Osto
Laboratory of Translational Nutrition Biology, Federal Institute of Technology Zurich ETHZ, Zurich, Switzerland

Center for Molecular Cardiology, University of Zurich, Zurich, Switzerland

University Heart Center, Cardiology, University Hospital Zurich, Zurich, Switzerland

Abstract

Nowadays, it is generally appreciated that studies in the medical field should not only include sex-related aspects but also consider age. In the past, taking the era of Hippocrates as a starting point for the Western medical sciences, such aspects were less urgent and barely relevant. However, considering such details during daily life became increasingly important as the traditional roles of men and

P. L. M. Kerkhof, V. M. Miller (eds.), *Sex-Specific Analysis of Cardiovascular Function*, Advances in Experimental Medicine and Biology 1065, https://doi.org/10.1007/978-3-319-77932-4_1

women in society and household converged. In the Western world, this fundamental transition process started recently and is advancing at an accelerated pace. Research about the role of women has also evolved, starting from plain history about the lives of women to a description of the relation between men and women, resulting in the gender concept. The present survey highlights a historical selection of observations referring to the impact of men and women on the medical sciences, as patient, study object, and professional. Whenever relevant, focus will be on the field of cardiovascular investigations as documented in the Western world. Rather than being exhaustive, we focus on a few remarkable icons, including Trota of Salerno, Hildegard von Bingen, and Miguel Serveto.

Keywords
History of cardiology · Heart research · Heart as symbol · Merit Ptah · Trota of Salerno · Hildegard von Bingen · Miguel Serveto · Theresa of Avila · Dario Maestrini · Bloodletting · Midwifery · Bicycle face · Anorexia religiosa · Lady's heart · Soldiers heart · E-patient · Pulsology · Epistotherapy · Cardiomythology

Naturae morborum curatrices

(William Douglass 1755)

Introduction

The historical description regarding attention for the heart should not only be viewed against the general developments observed in medicine in the broadest sense but also incorporate the exclusive symbolic meaning of the heart during various epochs and in a variety of cultures [117]. In Mesopotamia [98] and likewise in most ancient cultures, the concept of disease, the mediation by healers, and the practice of healing were very similar. A disease was considered a divine punishment for disobedience or resulting from an external malign influence. Therefore, medicine started by being preventive, by the use of appropriate amulets, and by offerings or sacrifices intending to pacify those malign forces. No organ besides the heart has received permanent and outspoken attention in religion, art, and the fundamentals of living systems, not to forget the numerous emotional connotations [137].

The basic reason for a delayed clinical interest in cardiology likely stems from the absence of direct applications in view of the relative short life span, even more so than in other directions of medicine. In his monumental book on the history of cardiology, Snellen [129] describes the dilemma of pointing to the origin of cardiology. When going back thousands of years and taking feeling the pulse as first indicator, there is no connection with the heart as a pump. Therefore, this author considered Harvey as a "very good" starting point, thereby neglecting contributions by Servetus, Colombo, and other even earlier candidates.

Wound care (following battlefield activities or hunting), for example, was a relevant field to be explored, as well as removing decayed teeth, treating headache, eliminating kidney stones, and similar inconveniences of daily life and old age. For other areas, there was often simply no rationale or tool available to make any appreciable progress, and therefore those attempts were left to quack doctors. Vision problems may have existed similar as in our era, but with the absence of knowledge about optics little remains to be done. Therefore, we follow another path and look into the nature of favorable circumstances and bright or devoted people who have promoted the advance of medicine. Herbal medicine has certainly contributed to the study of diseases, with respect to cause, course, and treatment. Later, convents, monasteries, and prototypes of clinics have facilitated more systematic analysis and the introduction of bedside teaching [5, 57].

How to Define the Origin of Western Cardiology?

- *Thousands of years ago (when considering pulsology)*
- *Around the year 1500 (if pointing to the cardiac pump, as conceived by Borelli, Colombo, Servetus, and later by Harvey)*

In terms of the cardiovascular system, it is not so much the heart that was initially investigated (apart from all sorts of symbolic connotations), but rather the blood as a fluid that has fascinated people, possibly regarded as the nourishing juice of the living organism [82, 93]. Apart from pulsology, the primary fascination may be related to the ostentatious confrontation while blood leaves the body, both in men and in animals, such as during penetrating injury, and in women during their periods and at the time of childbirth. Interventions based on phlebotomy (including the use of leeches and cupping glasses) as well as transfusion were (understood to be) among the lifesaving procedures applied in relation to the cardiovascular system. Notably, massive blood loss such as during accidents, fighting, or delivery was unmistakably associated with fatal outcome. After all, bleeding by severing the carotid vessels was known as an effective means to introduce rapid death with early loss of brain function, already described in the bible for sacrificing animals with minimal discomfort.

The Antiques in the Western World and Dealing with Illness

The splendor and superficial anatomical richness of sculptures representing gods, heroes and statesmen as created by antique Greek and Roman artists is considered unparalleled. In contrast, the internal anatomical structures and workings of the human body remained mysterious in classical times [130], apart from some dissection findings based on animals at the butcher's shop after hunting or observed following entertainment involving wild animal fights in the arena. Mythological inhabitants of Mount Olympus were considered to rule unexplained events in daily life. Natural disasters were attributed to anger of the gods. Philosophers still argued whether the soul is seated in the heart or in the brain. Against this sober and naïve background, we place the origins of Western clinical medicine in the fifth century BC. Greek practitioners commonly referred to themselves as τεχνιται (craftsmen), mastering various skills such as cauterization or extracting antidotes from plant roots [130]. Longevity mainly depended on the number of slaves available. However, the plague attacked the strong and the weak alike, as described by Thucydides (431 BC).

Initially, care for the sick was provided by family members and (if they were lucky) under the guidance of an acknowledged healer. Privileged groups such as gladiators received advanced assistance, and Galen learned a lot by observing and treating the wounded heroes in the Colosseum. Besides, there were court physicians, often with famous reputation, for the kings and emperors. Until a few hundred years ago, patients with nonfatal ailments were often "treated" by communication based on writing letters, termed epistotherapy [52, 119]. The route was particularly suitable for women patients, as the (typically male) doctor was not supposed to touch a woman. The great invention of the stethoscope, inspired when Laennec watched a boy using a tube to listen to a remote sound, was instrumental to bridge the desired distance between the female patient and the investigating physician.

The bodies of both men and women have served to advance medical sciences. Historically, there has been a keen strategic interest in curing warriors and maximum prevention of inflicting wounds especially at vulnerable anatomical structures. Galen had direct access to a number of organs in wounded but still living gladiators, permitting him to assemble a model about bodily functions which survived for nearly one and a half millennium. Within this scheme, the medical care of women was of less interest, unless this concerned childbearing and delivery. The situation changed with the introduction of autopsies in Padua, Leyden, and a few other medical centers, either for investigational purposes or for forensic medicine.

> *Time is nature's way of keeping everything from happening at once.*
>
> (Dossey [43])

Historical Perspective

The traditional portrait of the classical gallery of woman icons features Penelope and Cleopatra, besides the brave amazons. It remains unclear if besides beauty and fighting spirit, qualities such as mastering the healing arts were refined by gifted women in those times. During recent decades, the study of women in Greek and Roman times has been advanced with the repeated analysis, reevaluation of sources, and the introduction of new translations of older documents, culminating in reports facilitated by feministic movements. The ensuing masterpiece by Sarah Pomeroy [114] is nowadays considered a turning point in the study of women in ancient history. Foxhall [58], writing about the historiography of gender in classical antiquity, has described her book as "revolutionary" and "a major step forward" compared to existing scholarship [45].

The notion that women bleed and breed [114] has certain relevance for the circulation of blood and physiological adaptation of the heart. Apart from the extreme biological simplicity of the formulation, a more respectful understanding can be obtained from studying historical material available to analyze daily life. Part of the relative roles of women and men, including their social and hierarchical interaction, may be derived from classical plays where tragedy and comedy addressed contemporary issues [110], as well as from extant recordings about possessions, gifts, rituals, clothing, and even the mythology concerning the Olympian gods which likely reflected many aspects of (sometimes idealized) society [23, 45]. Some studies have been criticized as disappointing pointillism, floating in the direction of what is in vogue and being short on historical perspective [45]. Hence, the term "gender," while often referring to "sexual asymmetry" [16], tended to appear in titles and subtitles of publications in order to be taken more seriously [45]. After all, it is nearly impossible to compare the intrinsic "value" of female and male lives. There is one exceptional character who from experience can talk about both sides. Beyond the own will (and thus in contrast to transgender cases), the blind clairvoyant Tiresias from Thebes, son of the nymph Chariclo, had been woman for 7 years and gave birth to several children. When asked, he reported that certain pleasures were clearly superior, while he experienced the female phase [23, 113].

Monogamy seems to have been the rule, if we discount gods, whose norms often seem to be slightly different from humans because of their divine immortality [3]. If offspring is lacking or insufficient, however, we find reproduction "by proxy": men turn to co-wives, concubines, or slaves [16]. A study about the physiology of the marriage provides some mathematical insight: physically a man is a man for a longer time than a woman is a woman [10], or "men not only cycle, but cycle faster, for longer than women" [78].

Leaving the sunny side, we must admit that little is known about human suffering due to disease in antiquity. A bit more is known about the heroic art of healing. In the ancient Greek world, the field of medicine [110] belonged to the category of τεχνη (skill or craft, i.e., the purposeful application of human intelligence to some aspect of the real world as observed around) in contrast to τυχη (fate, i.e., a factor controlled by external elements, sometimes simply referring to human shortcomings). Hippocrates of Kos (c. 460–377 Larissa) in his Περι Αρχαιες Ιατρικης was the first to explicitly define the foundation of systematic medicine, implying general applicability, being based on reproducible experience, having a high level of precision (ακριβεια), and clearly directed toward providing an explanation [110]. Hippocrates categorized all foods and herbs by fundamental qualities – hot, cold, dry, or damp. In his view, good health was maintained by keeping these qualities in balance, as well as by taking plenty of exercise and fresh air. Remarkably, current research on obesity and fine particulate matter arrives at similar recommendations. When Hippocrates had reduced medicine to a system, observation was abandoned and philosophy was introduced into medicine [57].

Medical treatment serves to help the ones suffering from disease. As we gradually learned to understand, diseases do not result from the anger of the gods but often originate from zoonoses, with cattle transmitting tuberculosis, mosquitos being responsible for malaria, and rats for rapidly spreading plague, to name a few examples. The scale of some disasters was beyond the casualty size of atomic bombs thus far, with nearly one-third of the European population wiped out in the 1340 black death episode. Nowadays, we face a potential newcomer, (tick-borne) Lyme disease, which during an advanced stage may severely affect various organs including the heart.

Central Discoveries in Medicine with Relevance for Cardiology

Under different aspects of space and time, all phases of folk medicine and ancient medicine have been essentially alike in tendency, differing only in unimportant details [59]. Ancient physicians did not receive scholarly, scientific training but rather steered to philosophy and rhetoric [74]. This mixture of individual clinical experience and higher-order insights has dominated medical practice for centuries. However, a few exceptional minds provided incentives for future directions. As far as known from antique literature, Rufus of Ephesus (c. 80–c. 150) composed the first treatise devoted to anamnesis. His oeuvre *Medical Questions* was rediscovered and translated from Greek into French in the nineteenth century. Rufus attached great importance to the interview with the patient and in particular to questions concerning the patient's lifestyle prior to the illness. In this respect, modern views closely resemble his approach. The term "anamnesis" was not used in this sense by other physicians in antiquity. Actually, the expression for "history taking" only came into use since the mid-nineteenth century in German-speaking countries and in the Netherlands [68, 69]. For a wider perspective, we refer to the section on pulsology. Albrecht von Haller (1708–1777) refers to Rufus as *illustris medicus*. In 1806 a collection of previous unpublished fragments was printed in Moscow [69]. Rufus recommended "it is eminently necessary for those who would learn the art of medicine to be instructed on the name that should be given to each part of the body" [73]. It took about seven centuries before this task was realized by Meletius.

The Greek medico-philosopher Hippocrates and his followers contributed to the treatment of patients and the ethical code of physicians, partly by introducing an uncoupling of medicine and religion. They also composed a medical dictionary, called *Corpus Hippocraticum* [136], which is actually a collection of some 60 documents composed over a longer period, but mostly between 430 and 350, with further additions during the first century BC [68]. *De Medicina* is a first-century medical treatise by Aulus Cornelius Celsus, a Roman encyclopedist [108]. It is the only surviving section of a much larger encyclopedia (now available from the Perseus Collection, Tufts University, Boston). The work draws upon knowledge from ancient Greek documents, probably being the best surviving treatise on Alexandrian medicine [7]. The prolific Greek-Roman physician Galen (130–201) wrote a text which served as the standard during 13 centuries and contributed to knowledge about the cardiovascular system. However, his erroneous theory of blood circulation blocked any new ideas in this field for a long period [85].

The Byzantine monk Meletius Monachus (eighth to ninth century), reportedly practicing cautery and bloodletting, is the author of *De Natura Hominis* [73]. This much copied work describes in detail physiological components and functions of all regions of the body, from head to feet, concluding with a chapter on the soul. Like many ancient medical compendia, the collection forms a deliberate pastiche of earlier patristic, philosophical, and medical texts, often apparently preserving fragments otherwise lost. His combination of gendered concepts does not occur in any other ancient source [73].

Paracelsus (1494–1541), born Bombastus von Hohenheim, was a Swiss physician, alchemist, and astrologer, also known as the *Hippocrates of the Renaissance*. He directed therapy to causes rather than to symptoms. His sober spirit of scientific observation and critical reason was accompanied by emotional turbulence and a volcanic temperament. He remains both a mystery and an object of nostalgia. He treated the poor for free and is today celebrated as the first modern medical scientist [60, 128]. Also from Switzerland was Caspar Bauhin (1560–1624), anatomist, physician, and botanist. He commenced the study of medicine in Basel but had to move to Padua after an outbreak of plague in his native city. In Padua, he was privileged to study under Fabricius. His treatise, *Theatrum anatomicum* (1605), forms a major contribution to descriptive terminology in anatomy [62].

Between the thirteenth and eighteenth century, eight sages discovered that the air we breathe contains something that we need. Severinghaus [124] provides an excellent account which is summarized here. Ibn al-Nafis (1210–1288) in Cairo published on the pulmonary circulation [1, 138]. The Spaniard Michael Servetus, while in France, accurately described the pulmonary circulation and its effect on blood color. The long and fascinating trajectory leading to the identification of oxygen is revealing [124].

The French barber-surgeon Ambroise Paré (1509–1590) worked for the army where he treated wounded soldiers, for example, by tying off lacerated blood vessels. Based upon his wide experience he stated: "There are five duties of surgery: to remove what is superfluous, to restore what has been dislocated, to separate what has grown together, to reunite what has been divided, and to redress the defects of nature." [44, 60]. His motto was: "I treated the patient, and God cured him."

Measurement, Diagnosis, Treatment, and Cure

Arterial pressure: Stephen Hales, Scipione Riva-Rocci, Nikolai Korotkoff
Ventricular volume: Stephen Hales, Allan Cormack, Godfrey Hounsfield
Bioelectricity: Luigi Galvani, Augustus Waller, Willem Einthoven

Ambroise Paré (1509–1590): "I treated the patient, and God cured him."

Giovanni Alfonso Borelli was an Italian scientist (Naples 1608–1679 Rome). He was the first to demonstrate that cardiac muscle contraction is associated with the heartbeat and that the circulatory system resembles a hydraulic system [60]. The Austrian physician Leopold Auenbrugger (1722–1809) introduced the percussion technique as described in his *Inventum Novum ex Percussione Thoracis Humani* (1761) [15, 60] in 1816 followed by the introduction of the stethoscope by the French physician René Laennec (1781–1826) [60, 119]. In 1887, the Paris-born British physiologist Augustus Waller (1856–1922) recorded the first electrocardiogram using surface electrodes on his dog Jimmy [26], which followed the contemporary fascination about mesmerism, the experimental studies on bioelectricity described by the Italian scientist Luigi Galvani (1737–1798) in *De viribus electricitatis in motu musculari* (1791). This type of studies was continued by the Dutch Nobel laureate Willem Einthoven who at Leyden University in 1903 designed the galvanometer-based ECG recorder.

The first successful operation on the heart was carried out by the African American surgeon Daniel Williams (1856–1931) in the Provident Hospital (Chicago) in 1893 [49]. In 1896 the Italian Scipione Riva-Rocci (1863–1937) developed a cuff-based version of the mercury sphygmomanometer for the measurement of blood pressure, providing a practical instrument after the initial studies by Stephen Hales (1677–1761) in horse [60]. The Russian physician Nikolai Korotkoff (1874–1920) related vessel sounds to systole and diastole. Further technological advancements followed rapidly, such as the heart-lung apparatus in 1934 [63]; the implantable artificial pacemaker by Greatbatch and Chardack in 1958 [14, 33], culminating in heart transplantation [11, 39]; and the development of artificial hearts [81, 102].

Ratio of Finger Lengths and In Utero Exposure to Sex Hormones

When considering various anatomical aspects and physiological process, it is universally noted that sex and age matter. Size is a key feature of any organism since it influences the rate at which resources are consumed and thus affects metabolic rates. In the 1930s, size-dependent relationships were codified as "allometry," with most of these being quantifiable using the slopes of log-log plots of any relevant pair of variables. Physiologists explored how animal respiration rates varied as a function of body size, expecting rates to scale as the 2/3 power of body size, thus reflecting the ratio of surface area to volume. However, Max Kleiber (1893–1976) in his book *The Fire of Life* [90] revealed that animal respiration rates apparently scale more closely as the 3/4 power of body size [109].

The sex-specific ratio of finger lengths has perplexed scientists and travelers alike. Any book on sex-related differences in the field of medicine is not complete if without a detailed treatise on the ratio of index (D2) to ring finger (D4) length. The issue has been controversial since the days of Giacomo Casanova [27], who criticizes digital details of a portrait by Anton Raphael Mengs [95] and states in replying to Mengs: "It is certain that you are not of my species." Studies of D2:D4 as a function of sexual orientation have produced discrepant results, while support for the prenatal hormonal hypothesis is also lacking [92]. Thus far, no study has considered a connection between D2:D4 and the observed sex-specific differences in cardiac size and vessel diameter. Yet, the fewer the scientific foundations, the more room for discussions. While the capacities of nonverbal communication using the index finger [132] are explored in detail, and *The Finger Handbook* [135] discusses *Hox* genes, the latter does not really touch the issue of finger length, with the exception of the right D4 of Franz Liszt. This omission is largely compensated in the work by Manning [95], explaining the success of soccer players and attractive women on the basis of hormonal effects during their prenatal life. Other accomplishments important during life can be predicted, such as the first heart attack, simply by estimating left- and right-sided D2:D4. The *Hoxa* and *Hoxd* genes are reported to be essential for growth of digits and also strongly expressed in the genital bud [91]. Removal of the posterior *Hox* gene function in mice leads to concomitant loss of digits and bud derivatives. Apart from horrifying pathology, the consequences of a "normal" pattern have been studied. Women of reproductive age often experience a variety of unpleasant symptoms prior to the onset of menstruation. While genetic factors may influence symptom variability and severity, the exact causes remain unknown. A recent study [84] hypothesized that symptom variability originates from exposure to sex hormone differences in the embryonic environment. The authors measured D2:D4 in 402 young women and investigated the potential relationships of this ratio and premenstrual symptoms. They concluded that the embryonic environment, notably the relative concentration of sex hormones an embryo is exposed to, is associated with the severity of premenstrual symptoms.

Bloodletting as Therapeutic Intervention

The practice of bloodletting began around 3000 years ago with the Egyptians; then continued with the Greeks and Romans, the Arabs, and Asians; and spread through Europe during the Middle Ages and the Renaissance

[87]. It reached its peak in Europe in the nineteenth century but subsequently declined and today in Western medicine is used only for a few select conditions [67]. To appreciate the rationale for bloodletting, one must first understand the paradigm of disease in the time of Hippocrates. He believed that existence was represented by the four basic elements – earth, air, fire, and water – which in humans were related to the four basic humors, blood, phlegm, black bile, and yellow bile, each corresponding with a particular personality type (sanguine, phlegmatic, melancholic, and choleric). Being ill meant having an imbalance of the four humors and the need to remove excessive humor by bloodletting, purging, catharsis, diuresis, and so on. When Galen declared blood as the most dominant humor, the practice of venesection gained even greater importance. Thus, the phlebotomy therapy was equivalent to "one strategy (bene)fits all." Not only do Rabbis argue whether or not bloodletting is a therapeutic remedy, but many physicians in antiquity from Hippocrates and on ask the question as to where and when to perform phlebotomy. The necessary information is to be found in Coelius Aurelianus [116].

Bloodletting is divided into a systemic method performed by venesection (usually the median cubital vein) using a lancet and a localized method employing scarification combined with cupping. The latter route involves scraping the skin and then placing a dome-shaped glass while creating vacuum by suction or heating. Certain women were often specialized to perform the intervention (Fig. 1.1). An alternative service is provided by a leech (*Hirudo medicinalis*) which

Fig. 1.1 The practice of cupping was often taken care of by specially trained women. (Painting by Q. Gerritzn van Brekelenkam (Zwammerdam c.1620–1668 Leyden). Mauritshuis collection, The Hague)

ingests 5–10 mL of blood, almost ten times its weight. Their use was promoted by the Parisian physician François Broussais (1772–1838).

Famous therapeutic bleedings have been described in the medical literature [120]. One well-documented case concerns the retired president George Washington (1732–1799) who after riding in snowy weather developed a fever and respiratory distress. His wife asked her husband's secretary, Col. Tobias Lear. Seeing how ill the general was, Lear immediately sent for Dr. Craik, who had been Washington's physician for more than 40 years, and Rawlins, who was well practiced in the art of bloodletting. Later Craik ordered another bleeding, even though Dick, the second physician to arrive at Mount Vernon, objected to such a heroic measure. A third doctor, Brown, made it to the mansion at 4 p.m. Thus, under the care of his three physicians, Washington had copious amounts of blood drawn (a total of >2 L), followed by emetics and laxatives (Fig. 1.2). Ever the gentleman, even in extremis, general Washington made a point of thanking all three doctors for their help. He died the next night (December 14) of what has been diagnosed retrospectively as epiglottitis and shock. His medical treatment aroused significant controversy, particularly the repeated bloodletting procedure. Among others, Dr. Morens [107] wrote about these harrowing last hours.

Human Beings as Investigated Subjects

Anatomical studies: Tryn van Hamburch and many others who remained anonymous

Phlebotomy: George Washington and many less known (e.g., promoted by the Parisian physician Broussais)

Partial ventriculectomy: patients investigated by Batista and other centers around the world

Heart transplantation: Louis Washkansky and Denise Darvall and all those with nondisclosed names

Heart failure phenotyping based on ejection fraction: millions of "hidden" patients giving informed consent

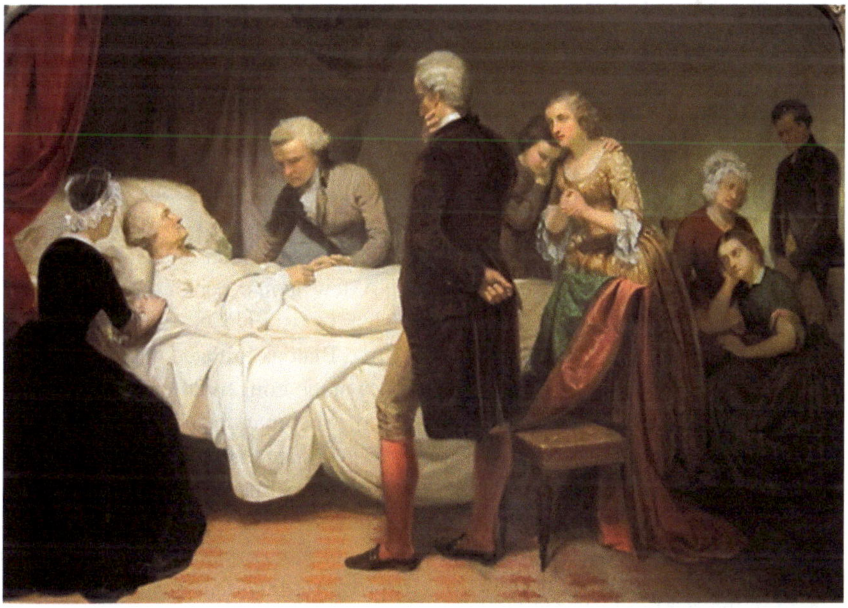

Fig. 1.2 Bloodletting therapy and George Washington (https://www.pbs.org/newshour/show/bloodletting-blisters-solving-medical-mystery-george-washingtons-death)

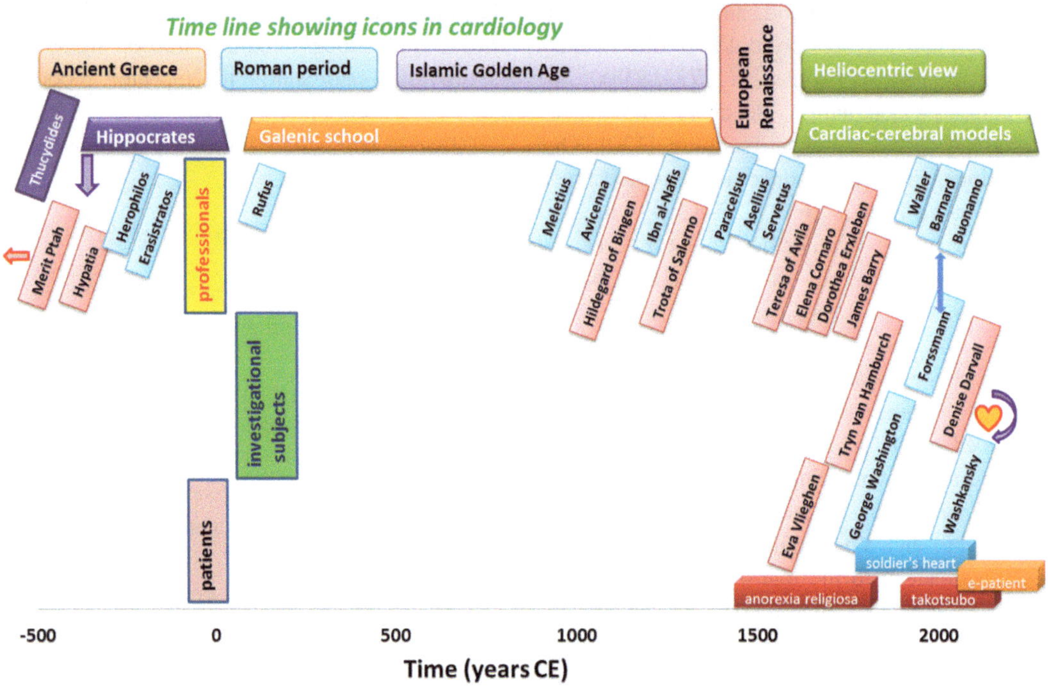

Fig. 1.3 Time line identifying remarkable individuals and developments starting from early Western medicophilosophy to modern cardiology. Not all icons of the last century fit the space available and are omitted because of this limitation. There are three levels as discussed in the text: from bottom to top we see patients, subjects used for investigation, and professionals, all stratified by sex. Ibn Sina (980–1037) is in the West known as Avicenna. Herophilos, Erasistratus, and Asellius are crucial for the microcirculation and lymph system [79]

We Stand on the Shoulders of the Giants of Cardiology

If we are to have a clear view of the panorama of medical history (Fig. 1.3), we must necessarily stand upon the shoulders of our predecessors [59]. The history of medicine is, in fact, the history of humanity itself, with its ups and downs, its brave aspirations after truth and finality, and its pathetic failures [59]. The first institute on the history of medicine was founded in Leipzig in 1906 [86] and is now called *Karl-Sudhoff-Institut*.

In a semi-biographical *Festschrift*, Hurst [80] lists the giants of cardiology which he calls "pilgrims" in analogy with Chaucer's tales: among them at least 24 men with no single woman. Looking at a wider horizon, landmark publications of the last five centuries have been nicely presented in the richly illustrated book by Gedeon [60]. The author identified 99 scientists (including one woman, Maria Curie, winning the Nobel Prize twice) who pioneered advances in the biomedical sciences. Gedeon elucidates their background and motivation, while documenting the significance of their contribution(s). Alternatively, Table 1.1 lists Noble prize winners who contributed to the field of cardiology. Here Gertrude Elion is the only woman. As expected, there is a long list of overlooked candidates, including A. Waller and C. Chagas [104].

A few iconic persons will be listed. In 1944, Helen Taussig (1898–1986) in collaboration with Alfred Blalock initiated the first surgical treatment of cyanotic congenital malformations of the heart [133]. She may be regarded as the *founder of pediatric cardiology* [48], with the British equivalent Jane Somerville. Also in 1944, Dr. Paul Dudley White in Boston, often referred to as the *father of American cardiology*, pioneered the concept of cardiovascular prevention. Inge Edler, a Swedish cardiologist, and Helmuth Hertz, a Swedish physicist (1952), launched noninvasive imaging by echocardiography. In 1958 Dr. F. Mason

Table 1.1 Cardiology-related discoveries that received the Nobel Prize in chemistry/physiology or medicine

Year	Winner(s)	Contribution
1912	Alexis Carrel	Work on vascular suture/transplant of blood vessels and organs
1924	Willem Einthoven	Electrocardiogram
1938	Corneille Heymans	Carotid sinus and aortic receptors and blood pressure and respiration regulation
1949	Walter Hess	Central coordination of the regulation of respiration and heart function
1953	Hans Adolf Krebs	Citric acid cycle (Krebs cycle)
1956	Werner Forssmann	
	Andre Cournand	
	Dickinson W. Richards	Cardiac catheterization
1960	Frank Burnet	
	Peter Medawar	Discovery of the immunological tolerance mechanism
1964	Konrad Bloch	
	Feodor Lynen	Understanding of cholesterol metabolism
1979	Allan Cormack	
	Godfrey Hounsfield	Computed tomography techniques
1982	Bengt Samuelsson	
	Sune Bergström	
	John Vane	Identification of angiotensin-converting enzyme inhibitors
1985	Michael Brown	
	Joseph Goldstein	Discovery of low-density lipoprotein/cholesterol receptors
1988	James Black	
	Gertrude Elion	
	George Hitchings	Development of beta-blockers
1990	Joseph Edward Murray	
	Edward Thomas	Development of organ and tissue transplant
1998	Robert Furchgott	
	Ferid Murad	
	Louis Ignarro	Discoveries about nitric oxide
2003	Paul Lauterbur	
	Peter Mansfield	Magnetic resonance
2012	Robert Lefkowitz	
	Brian Kobilka	G protein-coupled receptors
2012	Shinya Yamanaka	Conversion of skin fibroblasts into induced pluripotent stem cells

Modified but based on Mesquita et al. [104]

Sones Jr. began working on methods to visualize the coronary anatomy via sequential radiographic films [121]. Andreas Grüntzig was a German radiologist and cardiologist (1939–1985), who was the first to develop successful balloon angioplasty by expanding the lumen of narrowed arteries. Edmund Sonnenblick (1932–2007) was one of the greatest cardiovascular physiologists, originally forming a triad with Eugene Braunwald and John Ross.

Giant Teams in Cardiovascular Research

Braunwald, Ross, and Sonnenblick
Married couples with 300 or more publications
Investigators in epidemiology and hypertension

Finally, we mention Wolfgang and Jutta Schaper, Stephen and Dorothy Vatner, and Mark Eisenberg and Louise Pilote, all indefatigably devoted to cardiovascular science with 300 and more publications per married couple. A list of highly influential biomedical researchers based on Scopus citation data from the period 1996–2011 has been generated [21], based on the Hirsch (h)-index. This number is an author-level metric that attempts to measure both the productivity and citation impact of the publications of a scientist. The researchers selected the top 400 living core biomedical researchers. Based on their findings, we derived a sub-list on cardiology, which includes 54 candidates, among them 8 women (namely, Julie Buring, Stefanie Dimmeler, Barbara Howard, Alice Jacobs, Suzanne Oparil, Silvia Priori, Renu Virmani, and Jacqueline Witteman). Best represented specialty areas for the whole group were epidemiology and hypertension.

The Pulmonary Circulation

The Galen school dominated thinking about anatomy and physiology for nearly 13 centuries. Better insight into the design of the circulatory system required clarification of the function of the lung circulation. Several candidates qualify to receive the honor of discovering the blood circulation through the lungs (Fig. 1.4). Additional names start circulating as time progresses and more ancient manuscripts are delved up, analyzed, and critically commented on [1, 4, 31, 97, 124, 131, 136]. The "golden age of Arabic medicine" [105] came to its peak with the *Great Canon* by Ibn Sina (Avicenna) and remained under his influence [143]. He was a pioneer in pulsology [32], although the art of feeling the pulse (*Ars Sphygmica*) was ubiquitous in China already in 2500 BC [12]. Using this art, experienced physicians were said to be able to even determine the sex of an unborn child. Three hundred years before Paracelsus, a Cairo-based medical scientist by the name of Alauddin Ibn al-Nafis (1210–1288) dares to touch the authority of Ibn Sina. He described the pulmonary circulation, and in a separate work on the anatomy, he challenges Galen's teachings [100, 141]. The list now also features Miguel Serveto (Fig. 1.15), Realdo Colombo (1520–1654), Nicolai Massa (1485–1569), Juan Valverde de Amusco

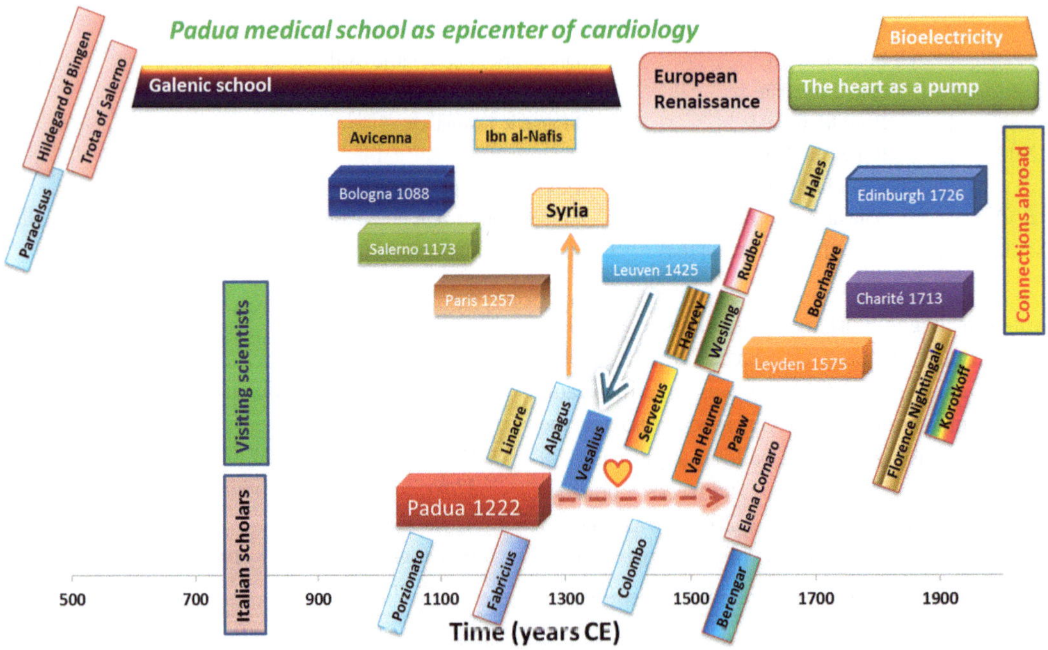

Fig. 1.4 The Padua medical school attracted many scientists, including those from Spain, Germany, Sweden, Flanders, and the Netherlands. Some scholars were involved in the study of the pulmonary circulation, including Ibn al-Nafis, Colombo, Servetus, and later Harvey

(1508–1565), Andrea Cesalpino (1519–1603), Fabrici d'Acquapendente (1533–1619), and William Harvey (1578–1657), while Andreas Vesalius (1514–1654) largely rewrote human anatomy. Furthermore, the Persian scholar Al-Akhawayni Bukhari (?–983 AD) investigated both the anatomy and the physiology of the human body. He reportedly describes the mechanism of pulmonary circulation in his only extant book *Hidayat al-Muta'llemin fi al-Tibb* (A Scholar's Guide to Medicine) [139]. A particular role is played by Servetus [65], who formulated antitrinitarianism (i.e., the denial of the theological concept of the triune Christian God). The doctrine soon won disciples in Northern Italy (Padua, Pisa, and Rome) and in some countries of East-Central Europe (Poland, Moravia, and Transylvania). Physicians who believed in or sympathized with antitrinitarianism revolutionized the theory of blood flow [22]. Sir William Osler became an active member of a committee to erect an expiatory monument to Servetus in Vienne [30], where one of the two remaining copies of his book is secured.

The University of Padua, founded in 1222, started mainly as a law school [115] and soon became the great Italian medical mecca [83]. The medical school has created an established reputation with respect to anatomy and studies on the pulmonary circulation. Andrea Alpagus (c. 1450–1522) was a professor of medicine at the University of Padua and spent 30 years in Syria studying and translating Arabic medical manuscripts. He translated sections of Ibn Al-Nafis' book *Sharh Tashreeh Al-Qanun* into Latin including his views on the pulmonary circulation. This translation, printed in Venice in the year 1547, helped to spread the work of Ibn Nafis to medieval European scholars. It is also interesting to note that Alpagus, Servetus, and Vesalius were fluent in the Arabic language [2] and thus may have had access to primary sources.

The medical school of Padua became the *alma mater* for celebrities including Jacopo Dondi (1293–1359) who constructed the clock in the tower of the Signori in the Palace of the Capitaniato and his son Giovanni Dondi (c.1330–1388) who moved in 1356 from the chair of astronomy to that of medicine, Thomas Linacre (c.1460–1524) from England, Andreas Vesalius (1514–1564) from Flanders, Johann van Heurne (1543–1601) and Pieter Paaw (1564–1617) as well as Joannes van Horne (1621–1670) all from the Netherlands, Johann Wesling (1598–1649) from Germany, and Olof Rudbeck (1630–1702) from Sweden. Paaw was influenced by the humanism of Erasmus from Rotterdam [76]. He arrived in Padua while Fabricius ab Aquapendente occupied the chair of anatomy, who in 1585 had erected the anatomical theater which was modernized in 1594 (in Palazzo Bo).

In those days, knowledge of anatomy was considered an important means to explore the universe created by God [76]. For that reason, most scientific discoveries (be it anatomy or astronomy) were interpreted as being connected to the supreme works of God. Figure 1.5 illustrates this duality by showing explorations with reference to the fragility of life on earth (e.g., with *vanitas* emblems and the banner with *homo bulla*, man is nothing more than a bubble). Therefore, it is not surprising that religion(s) and science often met at the crossroads. Scientists were not supposed to unravel details that belonged to the territory of the Almighty. On this planet, authorities such as the Inquisition and Calvin were present to safeguard that the boundaries should be respected. Renaissance medicine formed an important part of the radical reformation in the sixteenth century, and Michael Servetus, a Spanish physician, played a pivotal role. Physicians who sympathized with his doctrine revolutionized the theory of blood flow. They developed new concepts leading to a theory on the circulation of the blood [22]. Harvey was likely aware of the publications of his predecessors [34], including Colombo [126].

Historical Survey of Women in Medicine

Women have been esteemed as comforting healers for a long period [94], although not always accepted as standard physicians relative to their male counterparts. Initially, even for men, licensure was not required to perform medical

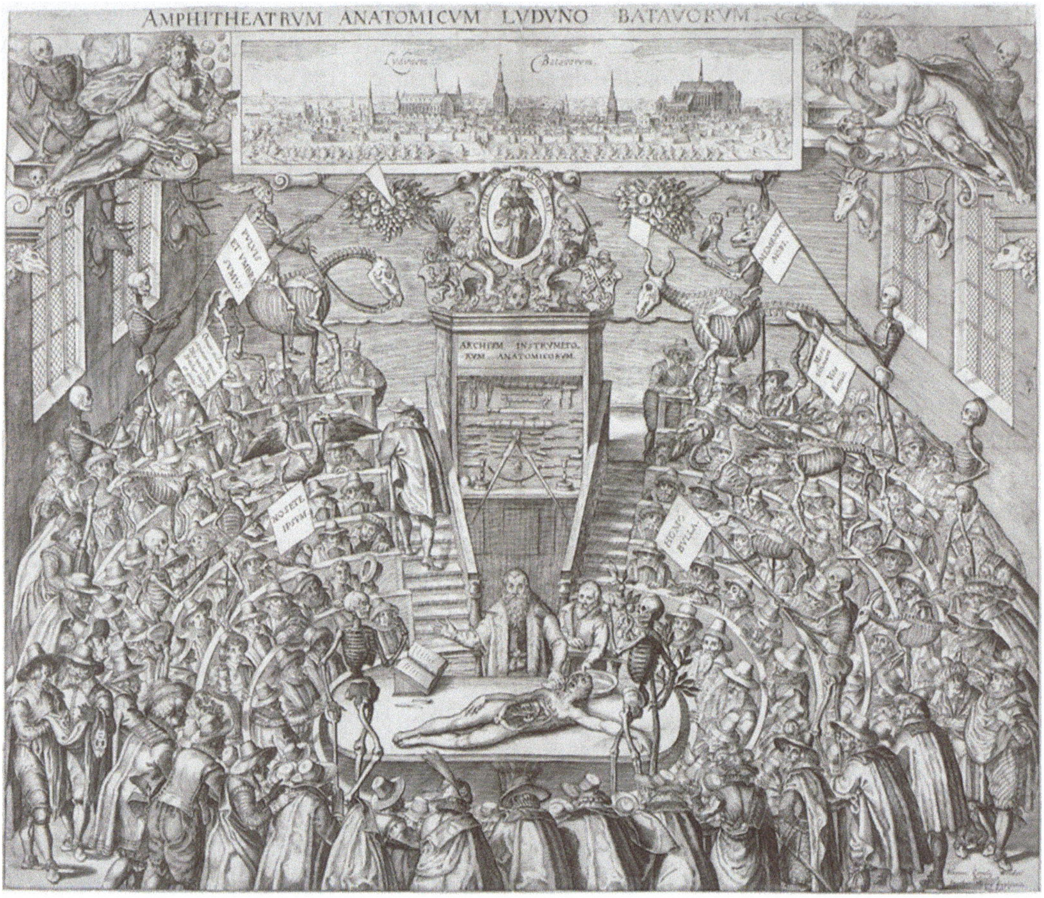

Fig. 1.5 Anatomical theater (founded in 1594) at Leyden university. Engraving by Bartholomeus Dolendo after Jan Cornelisz van't Woud (1609). In the center, we likely see professor Paaw who returned from Padua. Object number RP-P-1887-A-12041, Rijksmuseum Amsterdam (Public domain, http://hdl.handle.net/10934/RM0001.COLLECT.381385)

functions, but all just depended on public acceptance. Since the fourteenth century, examinations were required to practice medicine. As an exception, in Italy women had received medical education for centuries and occupied prestigious university positions [94]. By the end of the fourteenth century, there were 15 licensed female practitioners in Germany. Women involved in medical care often were wives or daughters of wound surgeons. Their chief medical activities concerned midwifery, with famous heroines including Louise Bourgeois (1563–1636), the royal midwife of France [46], and Friesian Catherina Schrader (1656–1746) [47]. After a long period of struggle, occasional secret practices, and slow progress, finally in the nineteenth century women were gradually accepted as medical practitioners in an increasing number of countries.

Women's Medical Schools and Initiatives

When in the past women were routinely forbidden from medical school, they ultimately sought to form their own medical schools. For example, the New England Female Medical College, Boston, was founded in 1848 by Dr. Samuel Gregory with the purpose of offering modern medical training in female-related fields (Fig. 1.6).

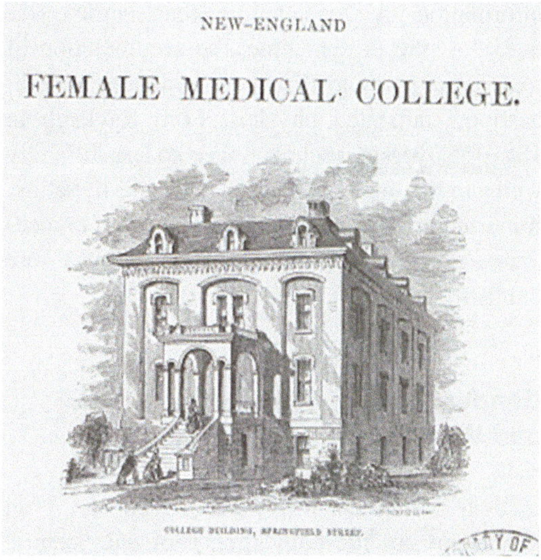

Fig. 1.6 Annual announcement of the New England Female Medical College, Boston [New England Female Medical College]. Persistent Link http://nrs.harvard.edu/urn-3:HMS.COUNT:682467?n=1 Accessed 10 December 2017. Description New England Female Medical College. Page (seq. 1), Repository Countway Library of Medicine, Harvard University

Other countries soon followed: London School of Medicine for Women (founded 1874 by Sophia Jex-Blake), Edinburgh School of Medicine for Women (founded 1886, also by Sophia Jex-Blake), Saint Petersburg State Medical University (founded 1897 as Female Medical University), and Tokyo Women's Medical University (founded 1900 by Yoshioka Yayoi).

The founding of the Medical Women's Federation in the UK in 1917 was a statement that women had established their rights to train and practice [54].

Concerning specific developments in the USA, More (2008) recently edited a book entitled *Women Physicians and the Cultures of Medicine*, which examines the wide-ranging careers and diverse lives of American women physicians, shedding light on their struggles for equality, professional accomplishment, and personal happiness over the past 150 years. Covered are the trials and triumphs of such extraordinary women such as Marie Zakrzewska (founder of the New England Hospital for Women and Children), Mary S. Calderone (medical director of Planned Parenthood in the mid-twentieth century), and Esther Pohl Lovejoy (risking her life to bring medical aid and supplies to countries experiencing war, famine, and other catastrophes) [106]. Ann Preston (1813–1872), from Pennsylvania, became the first female dean of any medical school. In 1985 Margaret Allen became the first female heart transplant surgeon in the USA. Two publications were critical in establishing the women's health movement and scholarship about women in medicine: first, the publication of *Our Bodies, Ourselves* (1973) by the Boston Women's Health Collective [20] and second, *Witches, Midwives, and Nurses: A History of Female Healers*, a short paper by Barbara Ehrenreich and Deirdre English, also in 1973.

Bedside Teaching and Medical Journals

Although not much involved in the primary acts of diagnostics and healing, women played an important role in the development of hospitals since 1450 [94]. Such well-organized centers formed the sources for advances in medicine, where intercollegiate communication flourished and experiences were shared and transferred to students via bedside teaching [5, 94]. The new system of clinical teaching was introduced in Padua in 1539 by Da Monte (1498–1551), who took his students to the local hospital of San Francesco to discuss whatever could be observed while facing the patients. This method became very popular, and foreign students trained in Padua introduced the approach in their home countries. In particular, Paaw and Heurnius advocated the new educational strategy in Leyden, but to no avail. Later, in 1636, Otho, the son of Heurnius, was more successful and launched a plan to take students around the public hospitals within the city and inspect the sick, examine the nature of the internal diseases as well as all external afflictions, and discuss the cures and surgical operations while also showing how to prescribe medication [76]. Importantly, keeping hospitals operational was made possible by the devoted care offered by women who day and night served as nurses. This specific role of women in hospitals, taking care of the sick, actually permitted bedside teaching because patients were permanently housed at a single location. In 1658 Francois de la Boe (also known as Sylvius) opened a clinical school in the hospital at Leyden [57].

Similar to the impact of bedside teaching, the introduction of medical journals has been of instrumental importance to distribute information about clinical findings and to promote scientific discussion. Monthly editions of *Nouvelles Découvertes* were published in Paris, after the journal was founded by the French surgeon Nicolas De Blegny in 1679.

The "Medical Women Question"

"The medical women question is perennial," an anonymous contributor to *The Lancet*'s editorial pages wearily opined in a November issue of 1877. "It knows no limits; we encounter it at every turn ... its appeals to periodical literature, instead of awakening a spirit of conciliation, have usually aroused a feeling of resentment." Alison Moulds, a student at the University of Oxford, is involved in the project, examining the construction of professional identities in nineteenth-century medical writing and fiction. She is particularly interested in representations of the medical-woman movement in the professional press and in fiction by early women doctors. She found that opinion on the medical training of women was far from uniform. One *Lancet* contributor ventured that women might treat female patients and children, whereas another harrumphed that "women hate one another, often at first sight." Moulds' study of the "medical women question" points to the dangers of leaping to conclusions about past opinions or practices without exploring the historical context [55]. Medical journals offer an unparalleled source of historical data not only on doctors, diseases, and medical practice but also on social transformations such as the participation of women in professional life. By the end of the nineteenth century, some hundreds of medical periodicals had been launched. Scientific commentators were already worrying about information overload before the century was over. "At the present time, the accumulation of material is so rapid that there is danger of indigestion," lamented physicist Lord Rayleigh in 1877 [55]. Modern scholars face no less difficulty while retrieving information from the literature. Importantly, various books about women in medicine do exist [77], and new ones are published [42].

Gender Models, Female Physicians, and Women Writers

In their monumental work, Børresen and Vogt [19] point out the convergence of androcentric gender models in the Christian and Islamic traditions. This book presents a pioneering investigation of correlated Christian and Islamic gender models which hitherto were not compared by women's studies in religion. Women writers of Christian tradition are included: Heloise, Hildegard von Bingen, Birgitta of Sweden, Caterina da Siena, Christine de Pisan, Hypatia of Alexandria, Birgitta of Sweden as a model of theological inculturation, Julian of Norwich as a model of feminist theology, and the poet Vittoria Colonna (1490–1547) who was befriended with Michelangelo [19]. Merit Ptah (2700 BC) from Egypt has been identified as the earliest cited female physician [88], while Agnodice reportedly was the first female physician to practice legally in fourth century BC in Athens.

Historians and poets, Victorian novelists, and contemporary feminists have adopted Hypatia (c. 370–415) as a symbol of the waning of classical culture and freedom of inquiry, of the rise of fanatical Christianity, or of sexual freedom [19, 50]. How does gender give meaning to the organization and perception of historical knowledge? The answers depend on gender as an analytic category. For the most part, the attempts of historians to theorize about gender have remained within traditional social scientific frameworks, using long-standing formulations that provide universal causal explanations. These theories have been limited at best because they tend to contain reductive or overly simple generalizations that undercut not only history's disciplinary sense of the complexity of social causation but also feminist commitments to analyses that will lead to change [123].

The Southern Italian coastal town of Salerno was an important center of medical learning and practice in the twelfth century. Here the physician Trota of Salerno (?–1097) collected many of her empirical practices in writings. One work on women's medicine that was associated with her, the *De curis mulierum* ("On Treatments for Women"), formed the core of what came to be known as the *Trotula*, a compendium of three texts that circulated throughout medieval Europe [66]. Trota herself gained a reputation that spreads as far as France and England [13].

The abbess Hildegard of Bingen (Bermersheim 1098–1179 Rupertsberg) is considered Germany's first female physician (Fig. 1.7). She conducted and published comprehensive studies of medicine and natural science [18]. During the Middle Ages, convents were an important place of education for women, and some of these communities provided opportunities for women to contribute to scholarly research. Her prolific writings (c.1151–58) include various scientific subjects, including medicine, botany, and natural history.

Christine de Pisan (1364–1430) was a medieval writer and historiographer who advocated for women's equality. Her works, considered to be some of the earliest feminist writings, include poetry, novels, biography, and autobiography, as well as literary, political, and religious commentary.

Elena Cornaro Piscopia (1646–1684) received an academic degree, and in Padua (1678) she became the first woman in the world to receive a PhD degree, well before Morgagni (1682–1771) founded pathological anatomy [140]. Dorothea Erxleben (1715–1762) was the first woman granted an MD in Germany. James (Miranda) Barry (1797–1865), a renowned female doctor who passed as a man to gain a medical education and practice medicine, was in 1858 appointed to inspector general of the British Army and worked successfully as surgeon for 50 years. At autopsy, it was discovered that James was a woman [38].

Fig. 1.7 Hildegard of Bingen, a medieval German abbess who wrote *Causae et Curae*, a medical text. (Artwork by Piet Michiels, Leuven, Belgium)

Not relying on a disguise, Harriot Hunt (1805–1875) was lucky to be admitted by the dean of Harvard Medical School. However, she was sent home by her (male) fellow students. She ended up as professor of obstetrics in Rochester, although she was not a qualified physician. With medical training in New York and Paris, Elizabeth Blackwell (1821–1910) similarly was appointed professor of obstetrics in London in 1874.

Johann Jakob Bachofen, a Swiss antiquarian, lawyer, philologist, and anthropologist (1815–1887), was a professor of Roman law at the University of Basel. He is often cited in relation to his theories concerning *Das Mutterrecht* (about prehistoric matriarchy), the title of his seminal 1861 book, being an investigation of the religious and juridical character of matriarchy in the Ancient World. He assembled documentation demonstrating that motherhood is the source of human society, religion, morality, and decorum. He postulated an archaic "mother-right" within the context of a primeval matriarchal religion or *Urreligion*. The first form of kinship follows no other rule than to trace descent uniquely through women. Matrilineal systems were therefore the first to develop and with them *Mutterrecht*, which corresponds with a matriarchal *lunar* phase in cultural evolution, based on agriculture, characterized by the emergence of mystery cults and law [8].

By the time diseases were considered curable by some type of intervention, and at least worth the attempt, doctors had mixed ideas about distinctions between male and female patients. The situation has succinctly been characterized in 1986 by the Bostonian physician Mark Altschule, with some excerpts selected here:

> For the most part, physicians think of gender-related illnesses as pertaining to obstetrics and gynecology and genitourinary diseases—and perhaps psychiatry also. However, cardiologists have long had similar concerns, albeit to a limited degree. [. . .] The fact that some women with mitral stenosis might have paroxysmal dyspnea in association with their menstrual periods occasioned interest but no surprise.
>
> Leaving to one side the anatomic and physiologic differences between left ventricles of women and men reviewed by Buonanno of Verona and concentrating only on coronary artery disease, we find marked differences between the manifestations in the two sexes. Normal coronary arteriograms are common in women with angina pectoris and even with myocardial infarction. [. . .] when symptoms occur, including those of infarction, they are likely to occur at younger ages in women than in men. The reasons for these differences are totally unknown. The clinical differences in coronary atherosclerosis in women and in men may make the diagnosis more difficult to establish in women.

As highlighted by Altschule, the publications by Dr. C. Buonanno [24, 25] marked the beginning of a new era, which was, very unfortunately, not immediately recognized by the cardiology community. This is what Dr. Buonanno (Fig. 1.8) reported in 1983:

> In females, then, with smaller ventricular volume, the antegrade flow seems to be maintained by more complete and rapid systolic emptying, which implies a constant hyperdynamic condition. The higher heart rate often reported for women, associated with the more complete and faster ventricular contraction, may well represent another effect of a physiologically present sympathetic hyperactivity. [. . .] As a consequence, when studies are performed in which the left ventricular function of a disease group is evaluated in comparison with normals, the ratio of the two sexes in the groups should be considered since the mean values for the parameters of left ventricular function may be strongly influenced by their sex composition.

In an earlier publication [24], his team reported:

> In a group of 70 patients, 29 women and 41 men, with atypical chest pain and normal findings at coronary arteriography, some hemodynamic and angiographic parameters of left ventricular function were measured for the purpose of determining the values of normality and to document possible differences between sexes. [. . .] the left ventricular ejection fraction of 74 ± 7% in women was significantly higher than that of 67 ± 7% in men ($P < 0.0005$).

It must be emphasized that this comparison is based on volumetric data which are adjusted for differences in body height and weight, while

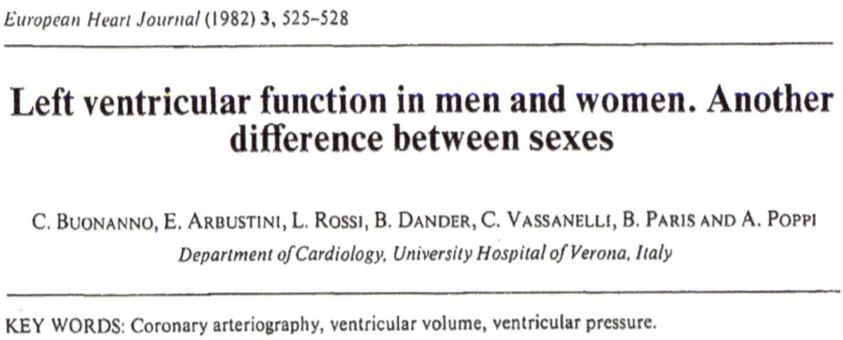
European Heart Journal (1982) **3**, 525–528

Left ventricular function in men and women. Another difference between sexes

C. Buonanno, E. Arbustini, L. Rossi, B. Dander, C. Vassanelli, B. Paris and A. Poppi

Department of Cardiology, University Hospital of Verona, Italy

KEY WORDS: Coronary arteriography, ventricular volume, ventricular pressure.

Fig. 1.8 Title page of the 1982 publication by Dr. C. Buonanno et al. from Verona (Italy)

ejection fraction (EF), being a ratio, is independent of body size. The implications of the findings reported in 1982 are enormous. Ventricular dimensions have been measured in millions of cardiac patients, while the EF metric has been the subject of more than 54,140 scientific papers listed in PubMed (accessed December 12, 2017). As far as known, no publication describes a connection between the ratio EF and the other sex-specific ratio mentioned before, namely, D2: D4. What is well known is that EF forms the central criterion to distinguish heart failure phenotypes and to design clinical trials in cardiology.

Hospitals, Infirmaries, and Clinics

Spanning 2000 years of Christian religious women's quest for spiritual and vocational fulfillment, *Sisters in Arms* is the first definitive history of Catholic nuns in the Western world. Unfolding century by century, this epic story encompasses every period from the dawn of Christianity to the present [103]. Until recently the role of nuns over the centuries has been minimized. These nuns were women scholars, mystics, artists, political activists, healers, and teachers – individuals whose religious vocation enabled them to pursue goals beyond traditional gender roles. They range from Thecla, the legendary companion of Paul, who baptized herself in preparation for facing the lions in the Roman arena, via Hrotsvitha (Hroswitha) of Gandersheim (c. 935–1002) as writer [61] to Hildegard of Bingen, whose visions unlocked her extraordinary talents for music, medicine, and moral teaching in the twelfth century [112]. By entering the convent, nuns gained a community that allowed them to evolve spiritually, intellectually, and emotionally; but the convent was never a perfect refuge. Women's struggles continued against the male church hierarchy, the broader lay community, and the larger cultural and historical forces of change. The history of nuns is an important part of the larger story of Western women whose gender provoked resistance to their claims to autonomy and power. The book is a tribute to the sisters who have labored with prayer and service for 2000 years, struggled to achieve greater recognition and authority, and forged *opportunities for all women* while holding true to the teachings of the Gospel [103].

Traditionally, in continental Europe, institutions offering care for the sick and providing social welfare were supervised and staffed by Catholic ecclesiastical orders (Fig. 1.9). These facilities included hospitals and were maintained by convents and monasteries. For example, Teresa of Avila (Avila 1515–1582 Alba de Tormes) was a prominent Spanish mystic, Carmelite nun, and author. Although she suffered many complaints (including fever, anemia, convulsions, headache, palpitations, painful attacks, tinnitus) connected to a variety of diseases, with tuberculosis likely being fatal, she devoted her life to helping the sick and neglected [41]. Her books include an autobiography, *El Castillo Interior* (The Interior Castle), and *Camino de Perfección* (The Way of Perfection).

Many of these units were taken over by the city administration following the Protestant revolution during the sixteenth century, such as *Cecilia Hospital* in Leyden founded in 1636 [76], Berlin's *Charité* (1768), and Saint Petersburg's *Obukhovkaya* (1779, while the city was founded

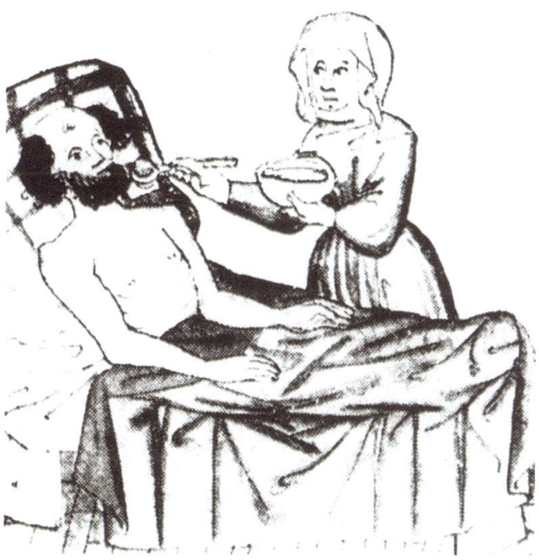

Fig. 1.9 Nurse feeding a patient (c.1450). The role of Christian charity in motivating women to do "good works" was important in the development of hospitals

not earlier than in 1703). Britain was behind the rest of Europe [71], with Westminster (1720), Guy's (1724), and St George's (1733). Edinburgh Royal Infirmary opened in 1729. The USA followed with Philadelphia (1751) and Massachusetts General (1811) in Boston. Before 1800 hospitals offered mostly food, shelter, and rest. They were crowded, with infections thriving [71].

The first (public) dissection by Paaw on a woman (called Jannetgen Jorisdochter van Deventer) at a temporary anatomical theater in the Faliedebagijne church (Leyden) took place in 1589 [76]. The official theater was founded in 1594 and required the appointment of an anatomy professor. As part of a competition to acquire such a professorship, another dissection on a woman (called Tryn van Hamburch, executed in Amsterdam) was carried out by the candidate Henricus Florentius on December 23, 1617 [76]. Tryn reportedly had been condemned several times because of theft. Eight times she escaped punishment because she was pregnant. As additional penalty, both her ears were cut. Finally she died on the gallows, and her body was donated to the Leyden anatomical theater to instruct students [122].

The Heart and the Heliocentric Paradigm

Since cardiovascular disease (CVD)-related problems (as we observe them nowadays) are mostly an issue of older age and often a consequence of affluent society, it is likely that with the exception of congenital heart disease, typical CVD was of relatively minor concern in the older times [17]. Jetter [82] mentions clinical cardiology only in the twentieth century. In the past, threats endangering life were of another order, with epidemics and famine often just around the corner. Serveto, Colombo, and Fludd [64] devoted attention to the heart, not solely because of medical interest or to primarily treat CVD but rather as a logical result of the recently developed heliocentric paradigm and associated interest in the capabilities of mankind, irrespective (or at least less dependent) of the divine authority. For Serveto and Fludd, the intention was to integrate the heart within the theological concepts of that time. Harvey even did not consider publishing his findings until his friend Fludd persuaded him to do so.

As a matter of fact, even scientists from outside the medical field, such as Copernicus and Galileo, created a new view on the cosmos by integrating astronomy and unifying all knowledge about nature, reserving in the higher region a place for God. Insights in the medical sciences at that time were not isolated items, but part of an all-embracing picture, as exemplified by new attempts to create an encyclopedia. The French prototype was *Dictionnaire raisonné des sciences, des arts et des métiers* (1751–1777) with 35 volumes edited by Diderot.

Medicine and the Military System

Special medical care has always been offered to certain "privileged" groups including political and religious leaders, the national military (Fig. 1.10), and heroes (such as favorite gladiators treated by Galen himself). It is said that the city of Cordoba at one time had over 50 hospitals, mostly military. Similarly, pregnant women of the higher economical class received special care at the time of delivery. Court and army physicians, as well as those professing midwifery, were specialized to perform these tasks. Similarly, the Veterans Health Administration is home to the USA's largest integrated healthcare system consisting of 170 medical centers, providing care at 1243 healthcare facilities, including 1063 outpatient sites of care of varying complexity, and serving nine million enrolled veterans each year (Fig. 1.11). As far as the discipline of cardiology is concerned, we see specialized centers, often directed to heart transplantation.

There is a remarkable resemblance between medical and military terminology: (heart) attack, fighting (a disease) [35], combat, withdrawal, aggressive (therapy), collateral damage [99], and the current "precision" strategies. Interestingly, Karen Hagemann, editor of *The Oxford Handbook "Gender, War and the Western World since 1600"* investigated how ideas and practices of gender differences contributed to the shaping

Fig. 1.10 The north front of the Invalides in Paris: Mansart's dome above Bruant's pedimented central block. The military hospital and home for retired soldiers were built in the period 1671–1679. (Copyright by Daniel Vorndran/DXR, CC BY-SA 3.0, https://commons.wikimedia.org/w/index.php?curid=31962731)

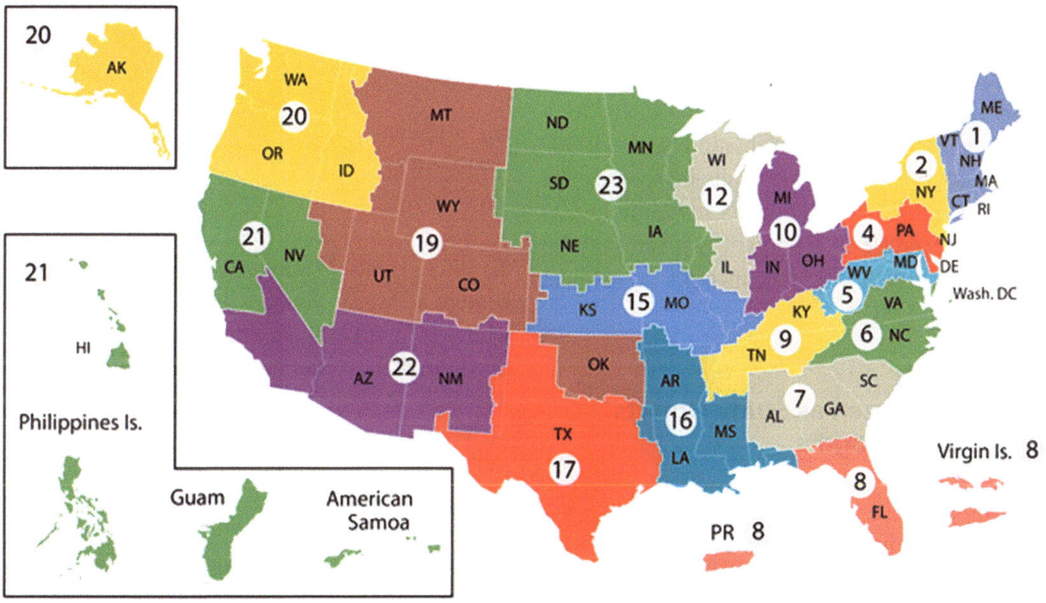

Fig. 1.11 Distributions of Veterans Health Administration locations. (Source: https://www.va.gov/directory/guide/division.asp?dnum=1 (accessed October 29, 2017))

of warfare and military culture and were at the same time transformed by them.

> *Women were introduced in the medical profession during times of war.*
> *The resemblance between medical and military terminology is remarkable.*

As we have seen, it took a long time before women were admitted to the medical profession. A major breakthrough was discernable since the First World War, when women were active as physicians to treat victims in their home country and cared for wounded soldiers abroad. Thus, the initial midwifery specialty reserved for the well-being of mother and newborn was rapidly extended to most disciplines of medicine. Half a century before, in 1854 Florence Nightingale (1820–1910), also called "The lady with the lamp," contributed by founding the modern concept of nursing (schools) [36]. Her accomplishments during the disastrous years the British Army experienced in the Crimea were largely the result of her concerns.

Medical Care in More Recent Times

Medicine in modern times is not limited to treatment of disease but also covers prevention and health coaching. Exercise has been a fascinating area of what is called cardiomythology.

In the period 1880 to 1980, three distinct changes in biomedical attitudes toward vigorous exercise had been detected: the mid-Victorian interpretation of pathological hypertrophy of the heart and the reinvention of hypertrophy as a beneficial physiological adaptation in the 1940s and 1950s, followed by medicalizing sport by the leisure revolution where doctors and cardiologists prescribe exercise as a polypill that can only be safely used with the guidance of a medical professional [72].

Men and Women as Study Objects and Their Strengths

Cardiology as a clinical discipline arose only recently, and there are not many heroic patients similar to, for example, Anna O (in psychiatry, real name Bertha Pappenheim, later referred to as a benefactor of mankind) or Henrietta Lacks (her name perpetuates in the HeLa cell line). An exception may be Werner Forssmann (1904–1979), being a German urologist and surgeon who rendered himself a (potential) patient by demonstrating in 1929 auto-catheterization of the right heart [17, 56]. He shared the Nobel prize in 1956 (Fig. 1.12).

A few patients appeared on front pages in newspapers. Among them was the first heart transplant recipient, but the glory period lasted longer for his cardiac surgeon. On December

Fig. 1.12 Werner T.O. Forssmann M.D., Nobel Prize winner. Artwork by Piet Michiels, Leuven, Belgium

3, 1967, Louis Washkansky (53) received the heart of a 25-year-young woman (Denise Darvall, who was involved in a traffic accident), and he lived for 18 days after the operation. Interestingly, this was also the first case of sex mismatch between donor and recipient, in which "asymmetry" became the subject of extensive research in subsequent years and still continues.

One other surgeon made history and was featured in *Newsweek*, but after a decade his name (Batista) disappeared from the scene, and the fate of some of his patients remained obscure (Fig. 1.13). Adam Rogers reported on June 23, 1996, in *Newsweek* as follows:

> The annals of heart disease are full of failed treatments and faddish operations that gave desperate patients lots of hope but little help.... A new operation devised by a heart surgeon in rural Brazil seems to have just the credentials to join that dubious-achievement list. It flouts a central tenet of cardiology. ... Almost a dozen American physicians have trekked to Brazil to observe Batista's work, and they have been sufficiently impressed that four leading hospitals are now performing the experimental surgery. Called

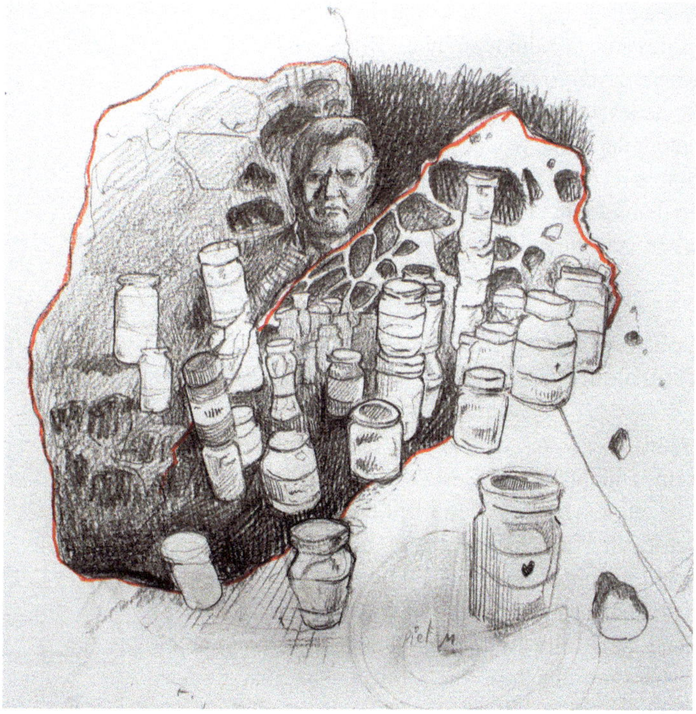

Fig. 1.13 Dr. Randas Batista showing his collection of specimens resulting from partial ventriculectomy. (Artwork by Piet Michiels, Leuven, Belgium)

> ventricular remodeling, the operation is meant for patients with enlarged hearts due to end-stage congestive heart failure, which directly or indirectly kills almost 300,000 Americans each year.... .. One reason for optimism is numbers: 70,000 Americans need heart transplants every year, but only 2,300 get them. Even so, a clear verdict on the operation is no sure bet. Unlike drugs, procedures do not have to be approved by the Food and Drug Administration. Cardiac patients will have to decide for themselves whether to let their surgeon take a little piece of their heart out.

Norman Cousins (1915–1990) wrote a brilliant book *Anatomy of an Illness* (1979) [37]. When the author was diagnosed with a crippling and irreversible disease, he forged an unusual collaboration with his physician, and together they were able to beat the odds. The doctor's genius was in helping his patient to use his own powers: laughter, courage, and tenacity. The patient's talent was in mobilizing his body's own natural resources, proving what an effective healing tool the mind can be. This remarkable story concerns the triumph of the human spirit.

In 1978 the said author joined the University of California, Los Angeles, faculty and became an adjunct professor and taught ethics and medical literature. He died of heart failure, 36 years after his heart disease was first diagnosed.

Then there is the remarkable *e-Patient Dave*: he survived kidney cancer and nobody knows why. Cancer survivor Richard Davies deBronkart Jr. (1950) is an international keynote speaker and academic lecturer who consistently earns extraordinary ratings by understanding each audience to deliver the client's unique objective. In January 2007, a routine shoulder X-ray incidentally disclosed a shadow in the lung, which turned out to be metastasized kidney cancer (stage IV, grade 4 renal cell carcinoma). His median survival time at diagnosis was 24 weeks. Now, a decade later, at conferences he acts as a speaker about the "e-Patient" movement, also referred to as "patient engagement" and "participatory medicine."

Men and women alike have been featuring as patients, a few of them as study objects in medical research, as well as a subgroup in the role of professionals, although not always and everywhere at an equal pace or with similar benefits. Societies which in the past subordinated members belonging to the female sex often made no exception to medical treatment. In that respect, selected epidemiologic studies may have been biased when inadequately considering this confounder.

Soldier's Heart and Other Cardiac Troubles

Heart disease in soldiers was known by several names, including Da Costa's syndrome (referring to a set of symptoms described during the American Civil War), irritable heart, and DAH (disordered action of the heart). When German troops marched into Belgium in the summer of 1914, there was already a significant literature on diseases of the heart in soldiers. Not that this seemed a particularly urgent problem at first. *Home by Christmas* was the cry. But, as the war dragged on, devastating new entities like poison gas, machine guns, and barbed wire presented physicians with hitherto unknown medical concerns. Heart disease first attracted serious attention following the August 1914 retreat from Mons, which saw many soldiers sent back to England with chest pain, dyspnea, palpitations on exertion, and tachycardia. It eventually became the third leading cause of discharge from the British Army during the war [75]. There is a long list of ailments which have been the fashion during certain episodes, including railway spine/neurosis (morbus Erichsen) associated with train accident, broken heart/takotsubo [53], anorexia religiosa (as in Teresa of Avila), and bicycle face which is about the dangers of young women riding bicycles, thus trading their femininity for fitness [55]. Another case concerns Eva Vlieghen (c. 1575–after 1628), who reportedly did not eat for 17 years but remained in blooming health thanks to her faith in God. Her portrait (Fig. 1.14) belonged to the collection of Otho Heurnius and decorated the local anatomical theater. Eva probably also suffered from anorexia religiosa.

Fig. 1.14 Eva Vlieghen (nickname Besje van Meurs), mentioned in Huisman [76] (http://resources.huygens.knaw.nl/vrouwenlexicon/lemmata/data/Vlieg(h)en)

Cardiomythology and Other Inflictions

Bicycle face: about the dangers of young women riding bicycles, trading femininity for fitness [55]

Anorexia religiosa: blooming health thanks to faith in God [41]

Lady's heart: Altschule [6] and Buonanno [24] on real (fe)male differences regarding the heart

(continued)

Soldiers heart (Da Costa's syndrome): on irritable heart [75]
Polypill: Heggie [72] on exercise and the heart
e-Patient (deBronkart Jr): patient engagement [37] and participatory medicine
Takotsubo (a.k.a. broken heart or happy heart syndrome): Engel [53]

The Doctor as a Patient

Dr. Walter Randall (1916–1993) is a famous cardiophysiologist, president (1982–1983) of the American Physiological Society, and editor, e.g., of the book *Neural Regulation of the Heart* [118]. In a book Randall states: I had undergone quadruple coronary bypass surgery [127]. The book does not tell us that this intervention was followed by a serious complication which he survived because it was auto-diagnosed.

Douglas Martin (in the June 21, 2014, edition of *The New York Times*) reported: Dr. Arnold S. Relman rose to the top of the medical profession as a researcher, administrator, and longtime editor of *The New England Journal of Medicine*, which became a platform for his early and influential attacks on the profit-driven healthcare system. Relman and his wife Dr. Marcia Angell filled top editorial posts at the journal for almost a quarter-century, becoming *American medicine's royal couple*, as the physician and journalist Abigail Zuger wrote earlier in the same newspaper. June 2014, Relman fell down a flight of stairs and broke bones including three vertebrae. When he reached the emergency room, his heart had stopped three times.

> "Technically, I died", he told *The Boston Globe*. He went on to write an article about his experience for *The New York Review of Books*, offering the unusual perspective of both a patient and a doctor. The health care system, he said, was in need of a more aggressive solution to fundamental problems, which he had discussed, somewhat philosophically, in an interview with *Technology Review* in 1989. "Many people think that doctors make their recommendations from a basis of scientific certainty, that the facts are very clear and there's only one way to diagnose or treat an illness. In reality, that's not always the case. Many things are a matter of conjecture, tradition, convenience, habit. In this gray area, where the facts are not clear and one has to make certain assumptions, it is unfortunately very easy to do things primarily because they are economically attractive".

Consolatory Reflections

> From the remotest times there exists abundant record that in each and all of these fields [as social workers, hospital consultants, moral reformers, as peacemakers] women were always active and always without ostentation. But to men for countless centuries the stirring events of war rather than those of peace were paramount. (Kate C. Hurd-Mead. *A History of Women in Medicine* [77])

There are real and imagined sex differences, sometimes with a documented biological or psychological basis [134]. One trouble has not confronted women scientists as much as it has done harm to men: the battle about who discovered what first. For example, Otto Frank and Ernest Starling vs. Carl Ludwig [144] and Dario Maestrini [101] or the candidates Harvey, Serveto, Colombo, and Ibn Nafis. Actually, a fierce debate about the discovery of the pulmonary circulation started recently anew [4, 142]. Clinical firsts often trigger sensational media reporting, both celebratory and critical [111]. Such moments highlight the promissory nature of medicine and surgery, the boldness of physicians and patients who dare to go first, the assessment of risks and benefits, and the decision to transition a procedure from the bench to the bedside [102].

Fig. 1.15 Statue of Michael Servetus in Champel, near Geneva. (Illustration by Piet Michiels, Leuven, Belgium)
Vignette: Miguel Servet (also known as Michael Servetus or Serveto) was a natural philosopher, theologian, and physician (Villanueva de Sijena, Aragon 1511–1553 Geneva). His father was a local notary. At the age of 14, Miguel entered the service of the Franciscan Juan de Quintana, who arranged for him to study law at Toulouse (1528/1529). Following stays in Basel and Strasburg (1530/1531), he published two works on *The Trinity* that met rejection from both Catholics and Reformers. Then he studied medicine in Paris and, under the pseudonym Michel de Villeneuve, earned his doctorate. After 1534 he worked as an editor, physician, and author of his own works on the natural sciences, until in 1540 Archbishop Palmier invited him to Vienne. There he spent 12 quiet years as a respected physician and wrote his major work, *Christianismi restitutio*, describing the nature and work of God in a theory influenced by Platonism. Here he describes the circulation of the blood through the lungs. His primary purpose, however, was to present an alternative to Calvin's *Institutio*. The book was published anonymously in Vienne early in 1553. Calvin brought him to a trial that lasted 2 months. He was condemned and on October 17, 1553, burned at the stake in the Champel quarter of the city. Many recent publications are devoted to this hero [22, 29–31, 65, 124, 125, 131], with a biography by Bainton [9]

While it may remain uncertain who was first when it comes to eminent discoveries, it must be noted that those brilliant men who came up with groundbreaking news occasionally ended up at the stakes which fate did not strike women in science (as far as we know). Joan of Arc was burnt, as well as a number of "witches," but not primarily because of scientific disputes, with the possible exception of Hypatia (who was murdered first).

Cases of fraud have been reported, and unfortunately the field of cardiology has not been free from scientific misconduct. It seems that men are relatively more often involved in cases where fraud has been detected, with only one woman in the top 30 list (ranking candidates with 16 or more retractions) (http://retractionwatch.com/the-retraction-watch-leaderboard/). However, another woman – from the high-profile Baker Institute in Melbourne – made it to (a retraction from) the *New England Journal of Medicine* in 2008. Sources of error have been traced [28], and focus has been on specialized areas such as the surgical literature [89].

Punishment can be fierce, implying expulsion from the scientific community in case of plain fraud. If self-declared authorities or assigned peers do not like the message, burning at the stake was followed in the older times. Nowadays, rejection of a manuscript may simply be based on "priority reasons."

Conclusions

We have seen that the organization of prototype hospitals and the subsequent establishment of bedside teaching were only possible with the permanent availability of nurses (often nuns) to care for the sick at selected locations. Admittedly, ambitious women aiming to enter the medical profession were mostly redirected to midwifery and thus laid the basis for a next generation of pioneers to excel. Newly recovered documents, wider availability combined with translations of source material, and not to forget the impetus generated by gender studies have widened the historical landscape to explore the course of medicine. Considering established facts and insights not only on the basis of reported activities of men and women but also guided by analysis of the enforced legal practices, cultural structures, and opportunities within the prevailing social context may open new horizons [40].

Besides the reported influential *Mutterrecht* [8, 96], starting around 1200 CE and uniquely in Europe west of the line Saint Petersburg to Trieste [70], we observe legal arrangements about (*consensus*) marriage and property (with spouses having equal rights) that have shaped society in these countries differently from traditional patterns [40]. In particular, the role of women as patients, study objects, and healthcare workers may become better documented and understood as these investigations advance. All histories, to be of value, must show people in their environment and, after the lapse of many generations, that environment must often be reconstructed from all sorts of scraps and traces, from bits of recorded history, and from the tales told by wandering minstrels and found in literary traditions. And this reconstruction of history, in so far as it concerns women, will probably have to be done by women [77].

References

1. Abdel-Halim RE. Contributions of Ibn Al-Nafis (1210–1288 AD) to the progress of medicine and urology. A study and translations from his medical works. Saudi Med J. 2008;29:13–22.
2. Abdel-Halim RE. The missing link in the history of circulation. Neonatology. 2011;99(4):311.
3. Adkins AWH. Homeric Gods and the values of Homeric society. J Hellenic Stud. 1972;92:1–9.
4. Akmal M, Zulkifle M, Ansari A. Ibn Nafis – a forgotten genius in the discovery of pulmonary blood circulation. Heart Views. 2010;11:26–30.
5. Altschule MD. Essays on the rise and decline of bedside medicine. Philadelphia: Lea & Febiger; 1989.
6. Altschule MD. Lady's heart. Chest. 1986;89:751.
7. Androutsos G, Karamanou M, Stefanadis C. The contribution of Alexandrian physicians to cardiology. Hell J Cardiol. 2013;54:15–7.
8. Bachofen JJ. Das Mutterrecht: eine Untersuchung über die Gynaikokratie der alten Welt nach ihrer religiösen und rechtlichen Natur. Stuttgart: Verlag von Krais und Hoffmann; 1861.
9. Bainton RH. Hunted heretic: the life and death of Michael Servetus, 1511–1553. Providence: Blackstone Editions; 2005.
10. de Balzac H. The physiology of marriage. Baltimore: Johns Hopkins University Press; 1997.
11. Barnard C, Pepper CB. One life. New York: Macmillan; 1969.
12. Bedford DE. The ancient art of feeling the pulse. Br Heart J. 1951;13:423–37.
13. Benton JF. Trotula, women's problems, and the professionalization of medicine in the Middle Ages. Bull Hist Med. 1985;59(1):30–53.
14. Bigelow WG, Callaghan JC, Hopps JA. General hypothermia for experimental intracardiac surgery. The use of electrophrenic respirations, an artificial pacemaker for cardiac standstill and the radio-frequency rewarming in general hypothermia. Ann Surg. 1950;132:531–9.
15. Bishop PJ. A list of papers etc., on Leopold Auenbrugger (1722–1809) and the history of percussion. Med Hist. 1961;5:192–6.
16. Blok J, Mason P. Sexual asymmetry; studies in ancient society. Amsterdam: JC Gieben; 1987.
17. Blömer H. When cardiology became a separate matter. Eur J Med Res. 2006;27:415–7.
18. Bollmann S. Women who write. New York: Merrell Publishers; 2007.
19. Børresen KE, Vogt K. Women's studies of the Christian and Islamic traditions. Ancient, medieval and renaissance foremothers. New York: Springer; 1993.
20. Boston Women's Health Book Collective Staff. Our bodies, a book by and for women. Boston: Simon and Schuster Trade; 1976.
21. Boyack KW, Klavans R, Sorensen AA, Ioannidis JP. A list of highly influential biomedical researchers, 1996–2011. Eur J Clin Investig. 2013;43:1339–65.
22. Bröer R. Blutkreislauf und Dreieinigkeit Medizinischer Antitrinitarismus von Michael Servet (1511–1553) bis Giorgio Biandrata (1515–1588). Ber Wiss. 2006;29:21–37.
23. Bull M. The mirror of the gods; how renaissance artists rediscovered the pagan gods. New York: Oxford University Press; 2005.
24. Buonanno C, Arbustini E, Rossi L, Dander B, Vassanelli C, Paris B, Poppi A. Left ventricular function in men and women. Another difference between sexes. Eur Heart J. 1982;3:525–8.
25. Buonanno C. The female left ventricle: a pathophysiological entity? Int J Cardiol. 1983;4:382–6.
26. Burchell HB. A centennial note on Waller and the first human electrocardiogram. Am J Cardiol. 1987;59:979–83.

27. Casanova G. Histoire de ma vie jusqu'à l'an. 1797 (published posthumously in an expurgated version).
28. Casadevall A, Steen RG, Fang FC. Sources of error in the retracted scientific literature. FASEB J. 2014;28:3847–55.
29. Cattermole GN. Michael Servetus: physician, Socinian and victim. R Soc Med. 1997;90:640–4.
30. Cattermole GN. Hero and heretic: William Osler's interest in Michael Servetus. J Med Biogr. 2000;8:187–90.
31. Cattermole GN. Al-Nafis and Servetus. Saudi Med J. 2008;29:1359.
32. Chamsi-Pasha MA, Chamsi-Pasha H. Avicenna's contribution to cardiology. Avicenna J Med. 2014;4:9–12.
33. Chardack WM, Gage AA, Greatbatch W. A transistorized, self-contained, implantable pacemaker for the long-term correction of complete heart block. Surgery. 1960;48:643–54.
34. Cohen H. The germ of an idea or what put Harvey on the scent? J Hist Med Allied Sci. 1957;12:102–5.
35. Cohn PF, Cohn JK. Fighting the silent killer. Wellesley: A.K. Peters Ltd; 1993.
36. Cope Z. John Shaw Billings, Florence Nightingale and the Johns Hopkins Hospital. Med Hist. 1957;1:367–8.
37. Cousins N. Anatomy of an illness as perceived by the patient: reflections on healing and regeneration. New York: WW Norton and Company Inc.; 1979.
38. Cule J. The timechart history of medicine. Hertfordshire: Worth Press Ltd; 1999.
39. Davis GP, Park E. The heart; the living pump. New York: Torstar Books Inc.; 1984.
40. De Moor T, van Zanden JL. Vrouwen en de geboorte van het kapitalisme in West-Europa. Amsterdam: Boom; 2006.
41. Dequeker J. Bij de dokter; beroemdheden en hun ziektes. Leuven: Davidsfonds Uitgeverij; 2011.
42. Deruyttere M. Markante vrouwen in de geneeskunst. Antwerpen: VBK – Houtekiet; 2015.
43. Dossey L. Space, time and medicine. Boulder: Shambhala Publications Inc.; 1982.
44. Drucker CB. Ambroise Paré and the birth of the gentle art of surgery. Yale J Biol Med. 2008;81:199–202.
45. Duby G, Perrot M. Storia delle donne in occidente: L'Antichità. Rome: Editori Laterza; 1990.
46. Dunn PM. Louise Bourgeois (1563–1636): royal midwife of France. Arch Dis Child Fetal Neonatal Ed. 2004a;89:F185–7.
47. Dunn PM. Catherina Schrader (1656–1746): the memoirs of a Friesian midwife. Arch Dis Child Fetal Neonatal Ed. 2004b;89:F560–2.
48. Dunn PM. Dr Helen Taussig (1898–1986): pioneering American pediatric cardiologist. Arch Dis Child Fetal Neonatal Ed. 2008;93:F74–6.
49. Dunn RR. The man who touched his own heart. New York: Little, Brown and Company; 2015.
50. Dzielska M. Hypatia of Alexandria. Boston: Harvard University Press; 1996.
51. Ehrenreich B, English D. Witches, midwives, and nurses. New York: The Feminist Press; 1973.
52. Ellis C. Epistotherapy. Br Med J. 1998;299:1230.
53. Engel GL. Sudden and rapid death during psychological stress. Ann Int Med. 1971;74:771–82.
54. Ferry G. Inspirational women in medicine. Lancet. 2017;390:1825.
55. Ferry G. Medical periodicals: mining the past. The Lancet. 2015;385(9987):2569–70.
56. Forssmann W. Experiments on myself. Memoirs of a surgeon in Germany. New York: St. Martin's Press; 1974.
57. Foucault J. The birth of the clinic. New York: Vintage books; 1975.
58. Foxhall L. Studying gender in classical antiquity. Cambridge: Cambridge University Press; 2013.
59. Garrison FH. An introduction to the history of medicine. Philadelphia: W. B. Saunders Co; 1913.
60. Gedeon A. Science and technology in medicine. New York: Springer; 2006.
61. Gerchow J von, Marti S. Malereien, Versorgungsanstalten und Frauenbewegungen. In: Krone und Schleier. München: Hirmer Verlag GmbH; 2005.
62. Ghosh SK. Caspar Bauhin (1560–1624): Swiss anatomist and reformer of anatomical nomenclature. Ital J Anat Embryol. 2016;121:159–64.
63. Gibbon JH Jr. The development of the heart-lung apparatus. Am J Surg. 1978;135:608–19.
64. Godwin J. Robert Fludd. Grand Rapids: Phanes Press; 1979.
65. Grafton A, Most GW, Settis S. The classical tradition. Cambridge, MA: Belknap Press of Harvard University Press; 2010.
66. Green MH, editor. The 'Trotula': a medieval compendium of women's medicine. Philadelphia: University of Pennsylvania Press; 2001.
67. Greenstone G. The history of bloodletting. BCMJ. 2010;52:12–4.
68. Haak HL, Horstmanshoff HFJ. The anamnesis in antiquity; medical questions by Rufus Ephesius (1st to 2nd century AD). Ned Tijdschr Geneeskd. 2006;150:2825–9.
69. Haak HL. Rufus Ephesius – Medicus gratiosus (PhD dissertation). Leiden: Leiden University Press; 2013.
70. Hajnal J. European marriage in perspective, in Glass DV and Eversley DEC: population in history. London: Edward Arnold Ltd; 1965.
71. Hall T. History of medicine. London: Hodder & Sloughton Ltd; 2013.
72. Heggie V. A century of cardiomythology: exercise and the heart c.1880–1980. Soc Hist Med. 2010;23:280–98.
73. Holman SR. On Phoenix and eunuchs: sources for Meletius the Monk's anatomy of gender. J Early Christ Stud. 2008;16:79–101.
74. Horstmanshoff HFJ. The ancient physician craftsman or scientist? J Hist Med Allied Sci. 1990;45(2):176–97.
75. Howell JD. "Soldier's heart": the redefinition of heart disease and specialty formation in early twentieth-century Great Britain. Med Hist Suppl. 1985;5:34–52.
76. Huisman T. The finger of god; anatomical practice in 17th century Leiden. Leyden: Primavera Press; 2009.
77. Hurd-Mead KC. A history of women in medicine, from the earliest of times to the beginning of the nineteenth century. Haddam: The Haddam Press; 1938.
78. Huxley VH. Sex and the cardiovascular system: the intriguing tale of how women and men regulate cardiovascular function differently. Adv Physiol Educ. 2007;31:17–22.
79. Huxley VH, Kemp SS. Sex-specific characteristics of the microcirculation. In: Kerkhof PLM, Miller VM, editors. Sex-specific analysis of cardiovascular function. Cham: Springer; 2018. p. 307–28.
80. Hurst JW. The Canterbury tales and cardiology. Circulation. 1982;65:4–6.
81. Jarvis P, Tarbell JM, Frangos JA. An in vitro evaluation of an artificial heart. ASAIO Trans. 1991;37(1):27–32.
82. Jetter D. Geschichte der Medizin. Stuttgart: Georg Thieme Verlag; 1992.
83. Joy M. Harvey, Pasteur, and the truth. Am J Med. 1986;80:917–24.
84. Kaneoke Y, Donishi T, Iwahara A, Shimokawa T. Severity of premenstrual symptoms predicted by second to fourth

digit ratio. Front Med (Lausanne). 2017;4:144. https://doi.org/10.3389/fmed.2017.00144. eCollection 2017
85. Karamanou M, Stefanadis C, Tsoucalas G, Laios K, Androutsos G. Galen's (130–201 AD) conceptions of the heart. Hell J Cardiol. 2015;56:197–200.
86. Kästner I. Das Leipziger Karl-Sudhoff-Institut und das Fach Geschichte der Medizin in der DDR. Med J. 2014;49 (1–2):118–58.
87. Kerridge IH, Lowe M. Bloodletting: the story of a therapeutic technique. Med J Aust. 1995;163:631–3.
88. Khalil R, Moustafa AA, Moftah MZ, Karim AA. How knowledge of ancient Egyptian women can influence today's gender role: does history matter in gender psychology? Front Psychol. 2016;7:2053. https://doi.org/10.3389/fpsyg.2016.02053.
89. King EG, Oransky I, Sachs TE et al. Analysis of retracted articles in the surgical literature. Am J Surg. 2017. pii: S0002-9610(17)31228-X.
90. Kleiber M. The fire of life. 2nd ed. New York: Wiley; 1975.
91. Kondo T, Zákány J, Innis JW, Duboule D. Of fingers, toes, and penises. Nature. 1997;390(6655):29.
92. Legato M, editor. Principles of gender-specific medicine. 2nd ed. Amsterdam: Elsevier; 2010.
93. Lindeboom GA. Inleiding tot de Geschiedenis der Geneeskunde. 3rd ed. Amsterdam: Rodopi; 1979.
94. Lyons AS, Petrucelli RJ. Medicine, an illustrated history. New York: Harry N. Abrams Inc.; 1987.
95. Manning J. The finger book. London: Faber & Faber Ltd; 2007.
96. Margalit Y. The Jewish family – between family law and contract law. Cambridge: Cambridge University Press; 2017.
97. Martins e Silva J. From the discovery of the circulation of the blood to the first steps in hemorheology: part 1. Rev Port Cardiol. 2009;28:1245–68.
98. Martins e Silva J. Medicine in ancient Mesopotamia – part 2. Acta Medica Port. 2010;23:125–40.
99. Marx JL, Kolata GB. Combating the #1 killer; the SCIENCE report on heart research. Washington, DC: AAAS; 1978.
100. Masic I. On occasion of 800th anniversary of birth of Ibn al-Nafis – discoverer of cardiac and pulmonary circulation. Med Arh. 2010;64:309–13.
101. Mazzoni M. Dario Maestrini e la legge del cuore Storia di un mancato Premio Nobel. Notizie dalla Delfico. 2005:1–2.
102. McKellar S. Clinical firsts – Christiaan Barnard's heart transplantations. N Engl J Med. 2017;377:2211–3.
103. McNamara JAK. Sisters in arms – Catholic nuns through two millennia. Boston: Harvard University Press; 1998.
104. Mesquita ET, de Decco Marchese L, Warol Dias D, Brasil Barbeito A, Costa Gomes J, Soares Muradas MC, et al. Nobel Prizes: contributions to cardiology. Arq Bras Cardiol. 2015;105:188–96.
105. Meyerhof M. La découverte de la circulation pulmonaire par Ibn al-Nafis, médecin arabe du Caire (XIIIe siècle). Bull l'Institut d'Egypte (Cairo). 1934;16:33–46.
106. More ES, Fee E, Parry M (Eds). Women physicians and the cultures of medicine. Baltimore: Johns Hopkins University Press; 2008.
107. Morens DM. Death of a president. N Engl J Med. 1999;341:1845–9.
108. Murray C. Human accomplishments – the pursuit of excellence in the arts and sciences, 800 BC to 1950. New York: Harper Collins Publishers Inc.; 2003.
109. Niklas KJ, Kutschera U. Kleiber's Law: how the fire of life ignited debate, fueled theory, and neglected plants as model organisms. Plant Signal Behav. 2015;10(7):e1036216.
110. Nussbaum M. The fragility of goodness. Cambridge: Cambridge University Press; 1986.
111. Onfray M. Féeries anatomiques: Généalogie du corps faustien. Paris: Editions Grasset & Fasquelle; 2003.
112. Pernoud R. Hildegarde de Bingen. Conscience inspirée du XIIe siècle. Monaco: Editions du Rocher; 1995.
113. Ovid. Metamorphoses. Translated by FJ Miller. Revised by G. P. Goold. Loeb Classical Library 42. Cambridge, MA: Harvard University Press; 1916.
114. Pomeroy SB. Goddesses, whores, wives, and slaves: women in classical antiquity. 2nd ed. London: Pimlico; 1994.
115. Porzionato A, Macchi V, Stecco C, Parenti A, De Caro R. The anatomical school of Padua. Anat Rec (Hoboken). 2012;295:902–16.
116. Preuss J. Biblisch Talmudische Medizin. Berlin: Karger; 1911.
117. Putscher M. Geschichte der medizinischen Abbilding. Muenchen: Heinz Moos Verlag; 1972.
118. Randall WC, editor. Neural regulation of the heart. New York: Oxford University Press; 1977.
119. Reiser SJ. Medicine and the reign of technology. London: Cambridge University Press; 1978.
120. Rutkow I. Seeking the cure: a history of medicine in America. New York: Scribner; 2010.
121. Ryan TJ. The coronary angiogram and its seminal contributions to cardiovascular medicine over five decades. Circulation. 2002;106(6):752.
122. Saelemaekers M. The lives of 1001 women, Digitaal Vrouwenlexicon van Nederland, Royal Dutch Academy of Sciences. Huygens Resources. 2017. http://resources.huygens.knaw.nl/vrouwenlexicon/lemmata/data/Pieters,%20Trijn. Accessed 30 Oct 2017.
123. Scott JW. Gender: a useful category of historical analysis. Am Hist Rev. 1986;91:1053–75.
124. Severinghaus JW. Eight sages over five centuries share oxygen's discovery. Adv Physiol Educ. 2016a;40:370–6.
125. Severinghaus JW. The most important discovery of science. Adv Exp Med Biol. 2016b;876:1–16. https://doi.org/10.1007/978-1-4939-3023-4_1.
126. Shackelford J. William Harvey and the mechanics of the heart. Oxford: Oxford University Press; 2003.
127. Sherry CJ. Animal rights: A reference handbook. Santa Barbara: ABC-CLIO; 1994.
128. Siddiqui MA, Mehta NJ, Khan IA. Paracelsus: the Hippocrates of the renaissance. J Med Biogr. 2003;11:78–80.
129. Snellen HA. History of cardiology. Rotterdam: Donker Academic Publications; 1984.
130. Spivey N, Squire M. Panorama of the classical world: London, Thames & Hudson Ltd; 2004.
131. Stefanadis C, Karamanou M, Androutsos G, Servetus M. (1511–1553) and the discovery of pulmonary circulation. Hell J Cardiol. 2009;50:373–8.
132. Tallis R. Michelangelo's finger. London: Atlantic Books; 2010.
133. Taussig HB. Congenital malformations of the heart. 2nd ed. Cambridge, MA: Harvard University Press; 1960.
134. Tavris C, Wade C. The longest war: sex differences in perspective. New York: CG Thomas; 1984.
135. Trumble A. The finger – a handbook. London: Yale University Press; 2010.
136. Tsoucalas G, Sgantzos M. The pulmonary circulation, it all started in the Hippocratic era. Eur Heart J. 2017;38:851.
137. Vinken PJ. The shape of the heart. Amsterdam: Elsevier Science; 2000. ISBN: 9780444829870.
138. West JB. Ibn al-Nafis, the pulmonary circulation and the islamic golden age. J Appl Physiol. 2008;105:1877–80.

139. Yarmohammadi H, Dalfardi B, Rezaian J, Ghanizadeh A. Al-Akhawayni's description of pulmonary circulation. Int J Cardiol. 2013;168:1819–21.
140. Zampieri F, Zanatta A, Thiene G. An etymological "autopsy" of Morgagni's title: De sedibus et causis morborum per anatomen indagatis (1761). Hum Pathol. 2014;45:12–6.
141. Zareba KM. Circulation over the centuries: William Harvey (1578–1657). Cardiol J. 2007;14:214–5.
142. Zargaran A, Arezaei H. Discovery of the pulmonary circulation. Eur Heart J. 2016;37:3494.
143. Zarshenas MM, Zargaran A. A review on the Avicenna's contribution to the field of cardiology. Int J Cardiol. 2015;182:237–41.
144. Zimmer HG. Who discovered the Frank-Starling mechanism? News Physiol Sci. 2002;17:181–4.

An Introduction to Epigenetics in Cardiovascular Development, Disease, and Sexualization

2

Christine M. Cunningham and Mansoureh Eghbali

Art work by Piet Michiels, Leuven, Belgium

C. M. Cunningham · M. Eghbali (✉)
Department of Anesthesiology and Perioperative Medicine, Division of Molecular Medicine, David Geffen School of Medicine at University of California Los Angeles, Los Angeles, CA, USA
e-mail: meghbali@ucla.edu

Abstract

Epigenetic regulation of gene expression is integral to cell differentiation, development, and disease. Modes of epigenetic regulation—including DNA methylation, histone modifications, and ncRNA-based regulation—

P. L. M. Kerkhof, V. M. Miller (eds.), *Sex-Specific Analysis of Cardiovascular Function*,
Advances in Experimental Medicine and Biology 1065, https://doi.org/10.1007/978-3-319-77932-4_2

alter chromatin structure, promotor accessibility, and contribute to posttranscriptional modifications. In the cardiovascular system, epigenetic regulation is necessary for proper cardiovascular development and homeostasis, while epigenetic dysfunction is associated with improper cardiac development and disease.

Early sexualization of tissues, including X-inactivation in females and maternal and paternal imprinting, is also orchestrated through epigenetic mechanisms. Furthermore, sex chromosomes encode various sex-specific genes involved in epigenetic regulation, while sex hormones can act as regulatory cofactors that may predispose or protect males and females against developing diseases with a marked sex bias.

The following book chapter summarizes the field of epigenetics in the context of cardiovascular development and disease while also highlighting the role of epigenetic regulation as a powerful source of sex differences within the cardiovascular system.

Keywords
Epigenetics · Cardiovascular disease · Cardiovascular development · Sex differences · Sexualization · Sex chromosomes · Sex hormones

Introduction

The study of epigenetics has evolved and expanded quite rapidly in recent years. Originally coined by Waddington in 1942, the term "epigenetics" was first used to describe phenotypic phenomena that failed to be understood by genetics alone (Fig. 2.1) [82]. More recently, the field of epigenetics refers to stably heritable changes in phenotype that are not a result of direct changes in the genetic code [8, 33]. Epigenetics primarily influence phenotype by regulating the transcription of genes via DNA methylation, histone modification, and noncoding RNA-dependent mechanisms. DNA and histone modifications influence transcription by altering DNA accessibility while noncoding RNAs can also impart posttranscriptional modifications [2].

Epigenetic modifications are powerful regulators throughout the life span that drive both developmental and pathological processes including the differentiation of embryonic stem cells into cardiomyocytes, the pathological hypertrophy of adult cardiomyocytes in response to physiological and environmental stressors, and much in between [1]. Epigenetic regulation may also help explain the heritability and pathophysiology of complex diseases, including cardiovascular disease (CVD), where there is a wide range of heritability that fails to be accounted for by genomic variation alone [54]. Early sexualization

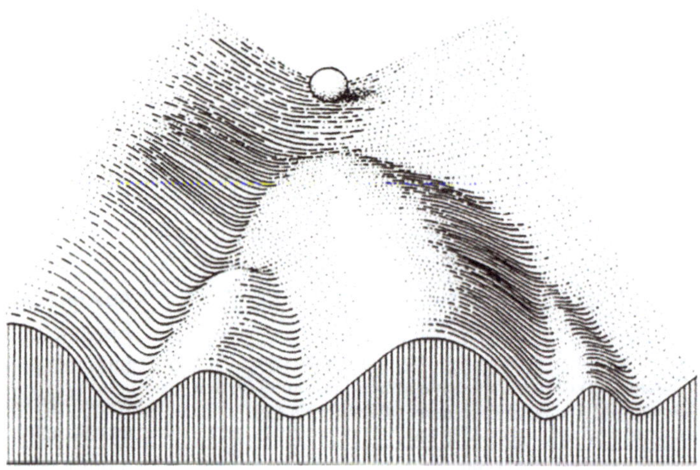

Fig. 2.1 *Waddington's classic rendition of the epigenetic landscape.* Waddington depicts one application of epigenetics as genes tethered to a landscape laden with valleys, or different cell fates, in which an embryonic cell may fall. (Reprinted with permission from Waddington [83])

of tissues, including X-inactivation in females and maternal and paternal imprinting, is also orchestrated through epigenetic mechanisms [55]. Furthermore, sex chromosomes encode various sex-specific genes involved in epigenetic regulation, while sex hormones can act as regulatory cofactors that may predispose or protect males and females against developing diseases with a marked sex bias [13]. Taken together, epigenetic mechanisms are powerful purveyors of phenotype that contribute to development, disease, and sex differences within the cardiovascular system.

Epigenetic Mechanisms and Cardiovascular Disease

Epigenetic regulation is achieved through a variety of dynamic and reversible mechanisms including DNA methylation, histone modification, and alterations in gene expression via noncoding RNAs (Fig. 2.2).

DNA Methylation

Overview

DNA methylation, the most common form of DNA modification, contributes to widespread and reversible changes in gene expression that influence both embryonic development and risk of CVD. DNA methylation is an active process that occurs when a methyl group is added to the C5 position of cytosine, one of the four DNA nucleotides, forming 5-methylcytosine (5mC). DNA methyltransferase (DNMT) enzymes dynamically mediate and maintain this process. Regions of DNA enriched with methylated cytosine residues adjacent to guanine nucleotides (known as CpG islands) serve as docking sites for methyl-binding proteins that, along with other protein complexes, remodel and condense chromatin (Fig. 2.2) [60]. Thus, CpG methylation inhibits the transcription of the nearby genes by condensing chromatin and blocking its accessibility to various transcription factors [77, 80].

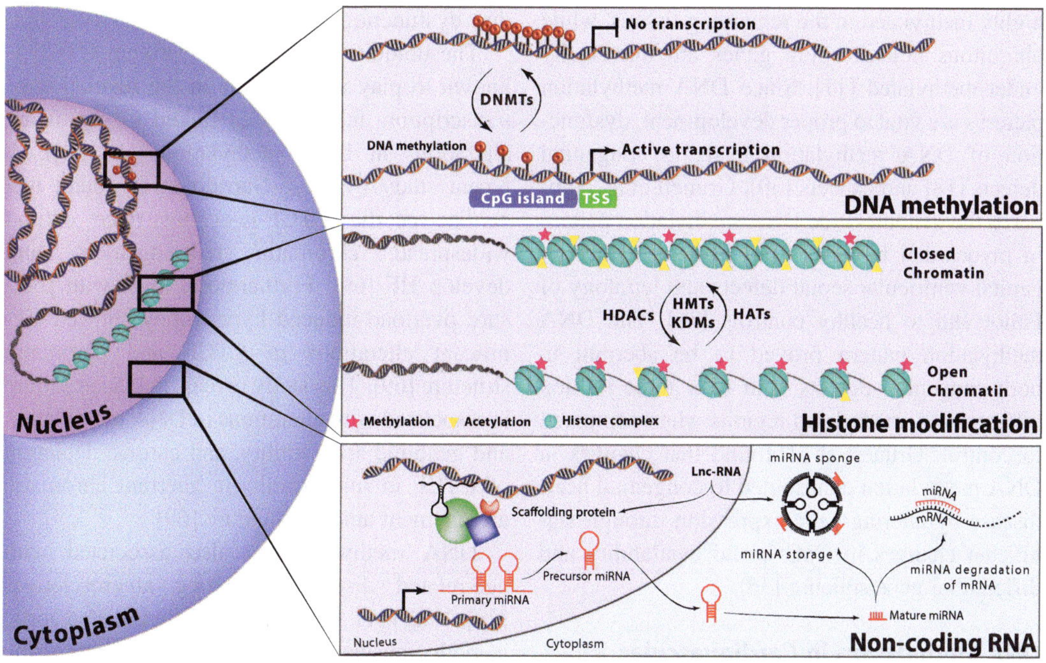

Fig. 2.2 *Main mechanisms of epigenetic regulation*. Epigenetic regulation is largely mediated by dynamic and reversible mechanisms including DNA methylation, histone acetylation and methylation, and alterations in gene expression via lncRNAs and miRNAs

DNA methylation does not permanently alter the transcriptome, as methyl groups can be removed from cytosine residues through active demethylation. TET enzymes initiate the DNA demethylation process by converting 5mC to 5-hydroxymethylcytosine, 5-formylcytosine, and 5-carboxylcytosine via oxidation. These modified 5mC derivatives are replaced by unmodified cytosines during rounds of DNA replication or directly via thymine DNA glycosylase enzymes [88]. Demethylation of DNA results in chromatin decondensation allowing for transcription of downstream genes to occur.

DNA Methylation in Cardiovascular Development

During development, DNA methylation patterns regulate cell differentiation, maturation, and specific gene expression. Genome-wide demethylation during the blastula stage allows for rapid cell division; while later during development, cell-specific methylation patterns allow for tissue-specific gene expression [17]. Tissue-specific genes tend to have unmethylated CpG promoter regions in tissues where they are abundant and are highly methylated in the remaining tissues, while ubiquitous housekeeping genes are universally under-methylated [17]. Since DNA methylation patterns are vital to proper development, dysfunction of DNA methylation leads to congenital defects [18] and cancers [30]. Grunert et al. compared the genome-wide DNA methylation pattern in myocardial biopsies from patients with congenital ventricular septal defects and tetralogy of Fallot and to healthy controls [36]. The DNA methylation pattern proved to be aberrant in both congenital defects with over three million differentially methylated regions when compared to control. Grunert et al. found that changes in DNA methylation contributed to congenital heart disease by altering gene expression through significant changes in CpG island availability and differential gene splicing [36].

DNA Methylation in Cardiovascular Disease

DNA methylation is not only necessary for proper cardiac development, but alterations in DNA methylation are also implicated in CVD development and risk [97]. Movassagh et al. studied the DNA methylation profile of left ventricular samples from male patients with ischemic and idiopathic cardiomyopathy undergoing heart transplantation [59]. Overall, patients undergoing heart transplantation exhibited a similar methylation profile to healthy controls; however, the methylation signature of certain individual genes and CpG islands was distinct [59]. Notably, Movassagh et al. found a significant correlation between reduced DNA methylation in the promoter sites of genes that are upregulated in cardiomyopathy [59].

A later study by Haas et al. also found that, while most CpG islands were similarly methylated in diseased patient samples when compared to healthy controls, many of the altered CpG regions were upstream of genes known to play a role in heart failure (HF), while the remaining altered genes potentially encode for novel HF genes [39]. Using a zebrafish model, Haas et al. found that altering the expression of these novel genes in the direction found altered in dilated cardiomyopathy patients resulted in cardiac dysfunction or failure [39].

The ubiquitously expressed protein CTCF is known to play a major role in the regulation of transcription and has been linked to significant alterations in DNA methylation (H. [84]). A recent study by Rosa-Garrido et al. found that cardiac-specific CTCF-knockout mice exhibit widespread chromatin rearrangement and develop HF [69]. Furthermore, mice with pressure overload-induced hypertrophy showed significant alterations in CTCF and chromatin structure [69]. This study reveals that heart failure is associated with alterations in DNA methylation and genomic accessibility, and cardiac-depletion of CTCF in mice results in aberrant chromatin arrangement and heart failure [69].

DNA methylation is also associated with age-related diseases including atherosclerosis [28]. Ying et al. found that de-differentiated aortic smooth muscle cells (SMCs), exhibiting the proliferative and noncontractile phenotype indicative of atherosclerosis, had significantly different methylation profiles when compared to differentiated, contractile aortic SMCs [91]. Interestingly, the CpG promoter region of ERalpha was significantly more methylated in the de-differentiated phenotype [91]. ERalpha has been linked to cell proliferation, NO production, and angiogenesis [56].

More recently, the association between DNA methylation and atherosclerosis has been further elucidated. Valencia-Morales et al. compared the DNA methylation profile of tissue from human aortic atherosclerotic samples with donor-matched controls [81]. This study found a significant correlation between atherosclerosis severity and change in CpG methylation with high-grade atherosclerosis being associated with hypermethylation [81].

Summary

DNA methylation is an integral part of embryogenesis and aging. The addition of methyl groups to promoter CpG sites, through the dynamic and reversible process of DNA methylation, directly inhibits the transcription of downstream genes. Congenital heart disease, HF, and atherosclerosis exhibit significant alterations in methylation profiles when compared to healthy tissue, and a reduction in DNA methylation activity is sufficient to induce HF.

Histone Modifications

Overview

In the nucleus, DNA is wrapped around nucleosomes to form highly regulated, tightly packed chromatin. Nucleosomes are comprised of eight core histone proteins, dimers of H2-A, H2-B, H-3, and H-4, that interact directly with DNA to facilitate packaging. Each histone has an amino acid tail that can be modified to influence histone-DNA interactions and subsequent DNA packaging. Common histone modifications include lysine acetylation and methylation, arginine methylation, and serine phosphorylation [7], and it is hypothesized that these modifications act sequentially to influence one another to contribute to the overall transcriptional state [76]. Lysine acetylation, carried out by histone acetyltransferases (HATs), is largely activating as it induces chromatin decondensation leaving DNA accessible for transcription [32]. Conversely, lysine deacetylation, carried out by histone deacetylases (HDACs), results in chromatin condensation and transcriptional suppression [32]. Histone methylation is regulated by two classes of enzymes: histone methyltransferases (HMTs) and histone demethylases [32]. Along histone tails, the addition or removal of methyl groups influences transcription in a dose- and site-specific manner (Fig. 2.2).

Histone Modifications in Cardiovascular Disease

Histone modifications impart powerful changes in gene expression, and alterations in histone acetylation and methylation have been reported in disease development. In the context of CVD, histone modifications, as well as the expression of enzymes that facilitate them, are associated with cardiomyocyte hypertrophy, HF, and ischemia/reperfusion (I/R) injury.

The result of histone modifications on gene expression is largely site- and dose-specific; therefore, the combination of different lysine residues along histone amino acid tails acts as a signature for various diseases. A genome-wide histone methylation profile by Kaneda et al. found that cardiomyocyte tri-methylation of H-3 at lysine residue four (H3K4) or nine (H3K9) is affected by the administration of high-sodium diet in Dahl salt-sensitive rats [47]. Dahl rats administered a high-sodium diet are a widely used animal model of congestive HF (CHF). Keneda et al. further analyzed the H3K4 and H3K9 methylation sites in human left ventricle (LV) biopsies from patients with congestive HF and patients with preserved LV ejection fraction. Again, they discovered altered methylation profiles in both sites [47]. Tri-methylation of the lysine residue H3K36 was also found to be associated with end-stage HF in patients [59].

By generating genome-wide maps of chromatin states, Papait et al. further identified the presence of a unique epigenetic signature in LV-derived cardiomyocytes from hypertrophic mice that is indicative of HF [64]. This epigenetic signature includes an overall reduction in both activating and repressive histone modifications which modulate promotor activity and result in altered gene expression [64]. Histone modifications that comprise the epigenetic signature of HF are orchestrated by a number of enzymes that influence lysine residues along histone tails.

Zhang et al. found that JMJD2A, a histone demethylase that removes methyl groups from tri-methylated H3K9 and H3K36 lysine residues, promotes cardiomyocyte hypertrophy in mice subjected to cardiac pressure overload [94]. In this study, JMJD2A-knockout mice subjected to transaortic constriction exhibited less cardiac hypertrophy and pathological remodeling when compared to mice overexpressing JMJD2A [94]. Rosales et al. further found that an increase in JMJD2A is concomitant with cardiac hypertrophy in cultured rat cardiomyocytes treated with angiotensin II (AngII) and endothelin-1 (ET-1) [70]. Treatment also resulted in the reactivation of the fetal gene profile as observed in pathological hypertrophy. Overexpression of JMJD2A and treatment with AngII and ET-1 resulted in an increase in the fetal gene profile, while inhibition of JMJD2A expression resulted in downregulation [70]. Taken together, these studies show that the histone residues H3K4, H3K9, and H3K36 are all altered in HF; and histone demethylase JMJD2A, which removes methylation marks at H3K9 and H3K36, is implicated in increasing cardiac hypertrophy potentially through activating the fetal gene profile that is characteristic of pathological hypertrophy. Interestingly, JMJD2A may contribute to certain sex differences as it has been shown to work along with JMJD2D as a cofactor to activate androgen signaling [73].

Papait et al. recently identified G9a, a HMT associated with a largely repressive epigenetic state, as an important modulator of both cardiac homeostasis and hypertrophy in the adult heart [65]. Like JMJD2a, G9a demethylates the H3K9 residue resulting in the inhibition of anti-hypertrophic genes [65]. Conversely, G9a is necessary in silencing developmental genes in the adult heart by interacting with the MEF2C complex [65]. In this way, the HMT G9a contributes to both necessary cardiac homeostasis and maladaptive cardiac hypertrophy.

HAT activity has also been linked to pathological cardiac hypertrophy. Two of the most studied HATs include the transcriptional co-activators CREB-binding protein (CBP) and p300—both of which are necessary for proper cardiomyocyte development and growth. CBP and p300 HAT activity mediate chromatin decondensation and transcriptional activation. Gusterson et al. found that enhancing CBP and p300 HAT activity is sufficient to induce cardiac hypertrophy in neonatal rat ventricular myocytes (NRVMs), while suppressing HAT activity prevented phenylephrine-induced cardiac hypertrophy [38].

Unlike HATs, HDACs, which enzymatically remove acetyl groups from lysine residues along the histone tails, confer chromatin condensation and transcriptional repression. In this way, HDACs are powerful epigenetic regulators that play a role in the pathogenesis and progression of a variety of diseases. Most notably, HDAC activity is increased in tumorigenic cancer cells, and the FDA has recently approved multiple HDAC inhibitors for use as cancer treatments [92]. The role of HDACs in CVD is complicated as studies show HDACs are anti-hypertrophic, while drugs inhibiting HDAC activity have also been associated with anti-hypertrophic and cardioprotective effects [92]. HDACs are separated into four different classes based on their functional and structural characteristics which could help explain the paradoxical findings associated with HDACs and CVD. Class IIa HDACs demonstrate anti-hypertrophic qualities as deletion of Class IIa proteins, HDAC5 or HDAC9, results in a more aggressive hypertrophic phenotype likely mediated through the reactivation of the fetal genes ([19], p. 5; [93]). Conversely, Class I HDACs have been associated with cardiac hypertrophy as overexpression of the Class I protein HDAC2 is sufficient to induce cardiac hypertrophy, while inhibition results in resistance to hypertrophy ([78], p. 2).

Further supporting the negative role of HDAC activity in the heart, suppression of HDAC activity by pharmacological HDAC inhibitors also demonstrated anti-hypertrophic and cardioprotective effects. Ooi et al. found that mice treated with an FDA-approved Class I/II HDAC inhibitor had less severe pathological cardiac remodeling and hypertrophic gene expression in the transaortic constriction (TAC) model of pressure overload [62]. Genome-wide chromatin immunoprecipitation coupled with massive parallel sequencing (ChIPseq) experiments found that treatment with HDAC inhibitors results in dynamic changes in the acetylation of promoter regions at H3K9 and H3K14 residues in whole hearts subjected to in vivo pressure overload and cultured vascular endothelial cells [62, 68]. Interestingly, HDAC inhibition did not solely result in increased acetylation at these residues as expected [34, 68]. Instead, HDAC inhibition resulted in acetylation and also deacetylation of hyper-acetylated residues [62, 68]. These dynamic changes in H3K9 and H3K14 residues induced downstream changes in gene expression including reduced expression of inflammatory and hypertrophic markers and suppression of autophagy [14, 62, 68].

Hypoxia has been shown to induce HDAC activity; as such, inhibition of HDAC activity has been demonstrated as cardioprotective against I/R injury [34]. Granger et al. found that treatment with pharmacological HDAC inhibitors significantly reduced the area of infarct in a murine in vivo model of I/R injury [34]. Follow-up in vitro experiments using isolated cardiomyocytes revealed that HDAC inhibition reduced cell death and the expression of genes typically upregulated in response to I/R injury including hypoxia inducible factor-1α and genes affecting vascular permeability [34]. Taken together, while HDAC activity has been shown to attenuate cardiac hypertrophy, it is largely viewed as maladaptive. FDA-approved HDAC inhibitors suppress cardiac hypertrophy and ischemic death by regulating acetylation of lysine residues and modulating the expression of genes relating to hypertrophy, inflammation, autophagy, and vascular function.

Summary

Histone modifications largely refer to the addition and removal of methyl and acetyl groups from lysine residues along amino acid histone tails. These modifications, as well as changes in the expression or activity of the enzymes that impart them, induce changes in chromatin conformation, promoter accessibility, and subsequent gene expression. In this way, dynamic histone modifications can drastically alter CVD risk and the development and progression of pathological hypertrophy and HF.

Noncoding RNA

Overview

More recently, noncoding RNAs (ncRNAs) have been identified as powerful epigenetic regulators. While never translated into a protein, ncRNAs influence gene expression by aiding in DNA methylation, histone modifications, and gene silencing. ncRNAs are separated into two groups based on size: long ncRNAs (lncRNAs) are over 200 nucleotides, and short ncRNAs are less than 30 nucleotides. LncRNAs typically inhibit gene expression inside the nucleus where they interact with chromatin-modifying proteins to influence chromatin structure and DNA accessibility. MicroRNAs (miRNAs), a subset of short ncRNA, reduce gene expression mainly through posttranscriptional degradation of target RNA strands (Fig. 2.2). Both lncRNAs and miRNAs are involved in regulating gene expression during development and are altered in various disease states and between sexes. Additionally, they have been identified in cardiomyocytes and vascular cells and have been associated with normal cardiac development and CVD [40, 63].

MicroRNAs in Cardiovascular Disease

The relatively recent discovery of miRNAs as powerful epigenetic regulars has sparked much investigation into their role in CVD and their potential as biomarkers or therapeutic strategies. The effect of miRNAs on CVD varies between individual miRNAs as each 21-24 nucleotide miRNA targets a specific RNA sequence that could result in the altered expression of up to a hundred different genes [66]. As such, modifying the expression of miRNAs can produce varying results. In vitro experiments using neonatal cardiomyocytes reveal that the overexpression of miRNA-23a, miRNA-23b, miRNA-24, miRNA-195, or miRNA-214 is sufficient to induce cardiac hypertrophy, while overexpression of miRNA-133 resulted in the attenuation of the hypertrophic phenotype [15, 50]. MiRNA expression is also associated with HF, myocardial infarction (MI), and coronary artery disease since miRNA expression levels are altered in cardiac tissue or plasma of murine models and human patients with these pathologies [45, 50, 63]. MiRNAs implicated in CVD have been shown to regulate genes in pathways associated with inflammation, autophagy, hypertrophy, and fetal gene reactivation.

LncRNAs in Cardiovascular Development

LncRNAs have also been demonstrated to play a role in CV development and disease by influencing gene expression through interactions with RNA or chromatin [40]. During embryogenesis, the presence of lncRNA Fendrr in the lateral mesoderm is necessary to ensure proper heart development [35]. Fendrr modulates the transcription of important developmental genes by altering chromatin structure as Fendrr-null mice exhibit changes in H3K4 and H3K27 methylation marks in the promoter region of several genes linked to CV development [35]. Another heart-associated lncRNA, Braveheart, is integral in driving cells toward a cardiac fate during embryogenesis and maintaining their cardiac fate by activating CV gene networks and mediating epigenetic regulation of cardiac genes [49].

LncRNAs in Cardiovascular Disease

Following development, lncRNAs are also associated with cardioprotection and cardiac hypertrophy. Han et al. identified a particular lncRNA-chromatin feedback loop that contributes to cardioprotection against CVD [41]. Han et al. found that alternative splicing at the myosin heavy chain 7 loci resulted in the transcription of a cluster of cardio-specific lncRNAs that they aptly named the myosin heavy chain-associated RNA transcript (Mhrt) [41]. Mhrt acts as a competitive inhibitor to Brg1 which functions as a member of the chromatin repressor complex that, when bound to DNA, induces cardiac hypertrophy [41, 42]. During periods of pathological stress, Brg1 can inhibit Mhrt expression thus promoting cardiac hypertrophy in a Brg1-Mhrt feedback circuit [41]. Han et al. found that restoring levels of Mhrt during pathological stress can protect the heart from hypertrophy and failure [41].

Another novel lncRNA has recently been discovered to epigenetically induce cardiac hypertrophy in response to cardiac stress. Cardiac-hypertrophy-associated epigenetic regulator (Chaer) was found to be upregulated in the hearts of pressure-overloaded mice [87]. Wang et al. determined that Chaer promotes the expression of hypertrophic genes by interfering with Polycomb repressor complex 2 to block the methylation of H3K27 [87]. Downregulation of Chaer expression prior to pressure overload was found to prevent pressure overload-induced pathological hypertrophy ([87]).

LncRNAs have also been shown to impart epigenetic regulation through the modulation of other ncRNA expression [40]. Cardiac-hypertrophy-related factor (CHRF) and cardiac apoptosis-related lncRNA (CARL) act as miRNA sponges for miRNA-489 and miRNA-539, respectively [85, 86]. In this way, CHRF and CARL block the function of their respective miRNA resulting in the increased expression of the miRNA-target mRNA. CHRF was found to be increased in cardiac tissue from HF patients, while CARL may regulate cardiac remodeling following I/R and MI injury [85, 86].

Summary

NcRNAs, including lncRNAs and miRNAs, contribute to the risk, pathogenesis, and progression of CVD through epigenetic mechanisms. More specifically, miRNAs are small ncRNAs that target complementary mRNA sequences causing their degradation and reduced expression. miRNAs have been associated with increased and decreased progression and/or risk of CVD. Furthermore, miRNAs are being investigated for their potential use as biomarkers and therapies in CVD. LncRNAs regulate cardiac gene expression by interacting with chromatin-modifying proteins to activate or inhibit gene expression or by blocking the effects of miRNAs.

Epigenetics and Sex Differences

Beginning very early in mammalian development, epigenetic modifications regulate the sexualization of cells, tissues, and entire organisms. These modifications contribute to the differential expression of protein-coding and non-coding genes between sexes, which influence sex differences in disease—including CVD. Additionally, sex hormones have been demonstrated to act as epigenetic cofactors, modifying the epigenetic landscape between men and women.

X-Inactivation and Genomic Imprinting

Mammalian sex is genetically determined by the combination of two sex chromosomes: X and Y. Each female cell contains two X-chromosomes (XX), while each male cell contains one X-chromosome and one Y-chromosome (XY). While the Y-chromosome encodes for the sex-determining Sry gene, it is much less rich in genomic data when compared to the X-chromosome or autosomes. To reduce the genetic density in female cells to match that of males, one X-chromosome in females becomes largely inactivated in a finely tuned process orchestrated by a series of epigenetic modifiers (Fig. 2.3). The process of X-inactivation begins as the X-inactivation center (Xic), a locus on the X-chromosome, dictates which X-chromosome will be silenced. On the soon-to-be inactive X-chromosome, X-inactive-specific transcript (Xist) is transcribed from the Xic locus [11, 24]. Xist is a lncRNA that interacts with PCR2 to methylate H3K27 residues to initiate widespread chromatin condensation and gene silencing—thus effectively inactivating the genes on the X-chromosome [23, 26, 96]. The condensed, X-inactivated state is maintained throughout all subsequent cell divisions [12]. Since both X-chromosomes encode the Xic locus, they are both capable of expressing the Xist lncRNA and undergoing X-inactivation. In fact, active expression of Xist will cause chromosomal inactivation even in autosomes if expressed as a transgene [52]. To prevent inactivation of the active X-chromosome, the 5′ end of the Xist gene is fully methylated to repress Xist transcription, while the Xist gene on the silenced X-chromosome is fully unmethylated and transcriptionally active [61]. DNA methylation also contributes to maternal and paternal imprinting in somatic cells by mediating the expression of Xist [27, 61]. Imprinting dictates whether the actively

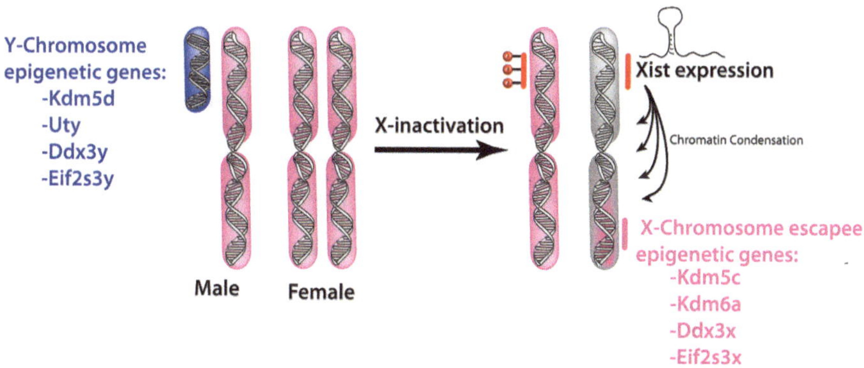

Fig. 2.3 *X-inactivation and sex chromosome epigenetic genes.* X-inactivation in females is initiated by the lncRNA Xist, which is transcribed from the inactivated X-chromosome. Xist renders the chromosome inactive by triggering hypermethylation and condensation of DNA with the exception of the few X-escapee genes that are expressed. The X and Y-chromosomes both encode genes involved in epigenetic regulation that are expressed in the heart

expressed X-chromosome is of maternal or paternal heritage.

X-Escapee Genes

While X-inactivation silences one X-chromosome in female cells to match the X-chromosome gene expression levels in male cells, this inactivation does not confer silencing of all genes. In fact, it is hypothesized that between 12% and 20% of the protein-coding genes on the inactivated X-chromosome are actually expressed in tissues [5]. Genes that escape X-inactivation are commonly referred to as X-escapee genes. The expression of X-escapee genes may be due to altered epigenetic regulation as X-escapee genes have different chromatin marks when compared to X-inactivated genes. Notably, X-escapee genes are void of the chromatin marks characteristic of chromatin condensation and gene repression that are indicative of X-inactivated genes (reviewed in [4]).

While it is not entirely clear why certain X-chromosome genes escape activation, studies have shown that X-chromosome dosage (one X-chromosome versus two) imparts sex differences in disease progression and severity. Most notably, the four core genotype (FCG) mouse model allows researchers to study sex differences imparted by sex chromosomes independent of gonads or gonadal hormones (reviewed in [13]). In the FCG mouse model, the testis-determining Sry gene is removed from the Y-chromosome and placed on an autosome. The now autosomal Sry gene segregates independently from the Y-chromosome allowing for the generation of gonadal males and females with both XX and XY genotypes. Thus, mice with four genotypes are created: XY males, XX males, XY females, and XX females (Fig. 2.4) [13].

Studies using FCG mice reveal that differences in sex chromosomes are implicated in various pathologies including cardiovascular disease [22, 29, 53]. The majority of these studies demonstrate that having one X-chromosome is better than having two [3, 22, 29, 53]. In the context of I/R injury, our lab found that gonadectomized male and female mice with two X-chromosomes developed a larger area of infarct, lower hemodynamic functional recovery post reperfusion, and greater mitochondrial dysfunction than male and female mice with one X-chromosome [53].

X-escapee genes are hypothesized to play a role in X-chromosome dosage-dependent sex differences. The genes that have been regularly demonstrated to escape X-inactivation in various tissues, including the heart, include Kdm5c, Kdm6a, Ddx3x, and Eif2s3x [3]. Interestingly, these genes affect protein expression through direct or indirect epigenetic mechanisms as Kdm5c and Kdm6a are histone demethylases; Ddx3y is an RNA helicase involved in RNA splicing, transcription, and translation; and Eif2s3x is a translation initiation factor. All of

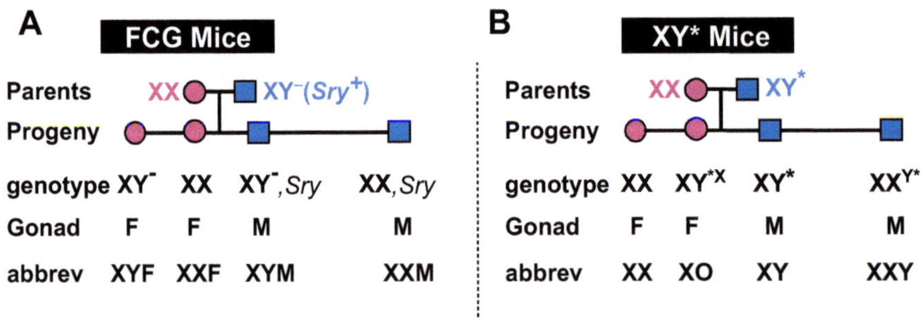

Fig. 2.4 *FCG and XY* mouse models.* (**a**) The four core genotype (FCG) mouse model produces gonadal males and females with both XX and XY genotypes to test the effects of sex chromosomes and gonadal sex hormones separately. (**b**) The XY* mouse model produces near-equivalent XX and XO gonadal females and XY and XXY gonadal males to compare the effect of the number of X chromosomes or the presence of a Y-chromosome. (Adapted from Li et al. [53])

these genes are capable of affecting widespread changes to normal homeostasis as they have been implicated in a disease or developmental processes [6, 71, 75, 89, 90].

Protein-coding genes are not the only possible source of X-chromosome-determined sex differences. In fact, the X-chromosome is densely packed with miRNAs as it contains approximately double the number of miRNAs found on autosomes in both mice and humans [37]. The degree to which X-chromosome-linked miRNAs escape X-inactivation continues to be investigated, although it is known that X-chromosome-encoded miRNAs undergo evolution much more rapidly than autosomal miRNAs. These X-encoded miRNAs offer exciting potential as drivers of sex differences in tissues and diseases [37, 58]. In stark contrast to the miRNA-enriched X-chromosome, the Y-chromosome encodes little to no miRNAs [72].

Y-Chromosome Epigenetic Modifiers

In the FCG mouse model, the number of X-chromosomes is not the sole sex chromosomal difference that could influence disease risk and/or pathogenesis. In fact, differences identified between XX and XY mice of both sexes could be due to the number of X-chromosomes or the presence of a Y-chromosome. As mentioned earlier, the Y-chromosome is much less dense than autosomes and its X-chromosome counterpart. Unlike the X-chromosome, the Y-chromosome is not necessary for proper development; however, it contains the genes necessary for proper male gonad formation and spermatogenesis. It also encodes a subset of ancestral genes leftover from the Y-chromosome's rapid divergence from an autosome beginning millions of years ago [44]. These ancestral genes, expressed only on the Y-chromosome, provide another source of sex differences between males who have a Y-chromosome and females who do not.

Y-chromosome effects in CVD, while difficult to examine in human populations, have been documented. A large study examining a cohort of Polish and Scottish men found that the Y-chromosome polymorphic HindIII biallelic marker was associated with systolic and diastolic blood pressure (BP). Men with HindIII(+) genotype had significantly higher BP than those with HindIII(−) genotype [21]. This study could help explain some of the sex differences associated with BP as females, who do not possess a HindIII gene, have significantly lower blood pressure than age-matched men [48]. This association was not found in a later study examining a male Japanese cohort indicating that the association may also be linked to ethnicity [43].

Single nucleotide polymorphisms within the Y-chromosome may also be associated with an increased risk of CVD in men. A large study found British men inheriting the Y-chromosome haplogroup I had a 50% greater risk of developing coronary artery disease (CAD) than British men with differing haplogroups [20]. Interestingly, a separate study found that macrophages with Y-chromosome haplogroup I had decreased expression of the Y-chromosome genes UTY and PRKY [10]. As macrophages play an integral role in the development of CAD and atherosclerosis, these Y-chromosome genes could be play a role in mediating inflammation in response to CAD [57].

To investigate the role of the Y-chromosome in CVD experimentally, the FCG mouse model and the XY* model have been employed. While the FCG mouse model allows researchers to test the effects of sex chromosomes independent from gonadal sex (see section "*X-escapee genes*"), the XY* mouse model tests whether the sex differences found are due to the number of X-chromosomes or the presence of a Y-chromosome. The XY* mouse model produces near-equivalent XX and XO gonadal females and XY and XXY gonadal males (Fig. 2.4). Comparing mice with one X-chromosome (XO, XY) to those with two X-chromosomes (XX, XXY) distinguishes the effect of X-chromosome number, whereas comparing mice with a Y-chromosome (XY, XXY) to those without a Y-chromosome (XO, XX) distinguishes the effect of the presence/absence of the Y-chromosome.

Our lab recently used the FCG and XY* mouse models to investigate sex differences in pulmonary hypertension (PH), a right ventricular

CVD that is up to four times more common in females. In this study, we found that the presence of a Y-chromosome was protective against the development of PH as male and female mice with a Y-chromosome had lower right ventricular developed pressure and less severe vascular remodeling following experimental PH, regardless of the number of X-chromosomes [79]. We hypothesize that genes encoded by the Y-chromosome that are expressed in the heart and lung, Kdm5d, Uty, Ddx3y, and Eif2s3y, may confer protection against the development of PH. Interestingly, all of these genes impart direct or indirect epigenetic regulation. While each of these genes has a similar partner gene encoded by the X-chromosome, the Y-chromosome is subjected to different evolutionary pressures. As such, Y-chromosome paralogs have demonstrated different functional properties than their X-chromosome counterparts which could influence the diseased state [25, 74].

Sex Hormones and Receptors in Epigenetic Regulation

Sex chromosomes are not the only purveyor of sex differences in epigenetic regulation. Sex steroid hormones and their receptors also influence the epigenetic landscape and, in turn, are epigenetically regulated as well. The androgen and estrogen receptors (AR and ER) are nuclear receptors that complex with other cofactors and regulators and act as transcription factors. AR and ER activity is mediated by androgen and estrogen ligand biding, respectively; and, like other transcription factors, their efficiency is dependent on chromatin accessibility per epigenetic modifications (Fig. 2.5a).

As transcription factors, ERs and ARs modulate the transcription of genes involved in epigenetic regulation including ncRNAs. In this way, sex hormones and their receptors influence epigenetic regulation by modifying the expression of

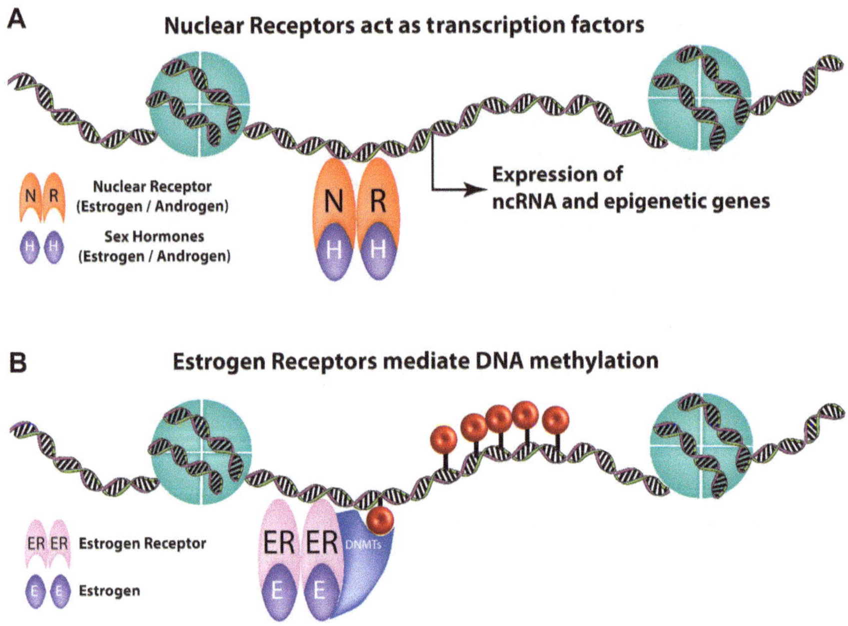

Fig. 2.5 *Sex hormones and receptors mediate epigenetic regulation.* (**a**) Estrogen and androgen receptors can act as transcription factors regulating the expression of various genes including those involved in epigenetic regulation. The availability and efficiency of nuclear receptors are dependent on sex hormone binding and DNA accessibility. (**b**) Estrogen receptors influence epigenetic regulation of genes by aiding in DNA methylation

proteins and ncRNAs that confer epigenetic changes. A study by Quierós et al. found that estrogen mediated a sex-specific response to pressure overload in mice by regulating a set of miRNAs associated with attenuating cardiac fibrosis [67]. Their results indicate that estrogen induces this miRNA expression through ERbeta as the sex-specific miRNA profile was not found in female ERbeta KO mice [67]. These findings could help explain sex differences present in fibrotic deposition in pressure overload injury as female mice, which express estrogen- and ERbeta-induced anti-fibrotic miRNAs, exhibit less cardiac fibrosis than male mice when exposed to TAC.

In breast cancer cells, estrogen was also found to induce the expression of 21 miRNAs and repress the expression of seven miRNAs through activation of ERs [9]. Some of the induced miRNAs were directly linked to breast cancer disease progression [9]. Other studies confirm that estrogen regulates miRNA expression through ERalpha and even contributes to the sophisticated process of miRNA maturation [16, 31]. While controversial, estrogen and ERs are considered to be largely protective against CVD and CVD risk [56].

Estrogen has also been shown to mediate epigenetic regulation by altering DNA methylation (Fig. 2.5b). A genome-wide study by Jadhav et al. found that estrogen treatment in a breast cancer cell line resulted in DNA hypermethylation and epigenetic suppression of a large cluster of genes associated with breast cancer survival [46]. In this way, sex hormones can shift the epigenetic landscape to influence disease risk or progression in a sex-specific manner.

While ER and AR activity can directly influence the expression of epigenetic modifiers including ncRNAs and DNA methylation, the activity of ERs and ARs is also epigenetically regulated. As transcription factors, both receptors directly interact with DNA to initiate transcription, and their efficiency is dependent on DNA accessibility via epigenetic chromatin packaging [95]. ER and AR also interact with other proteins that, in turn, regulate their activity. The activity of ERalpha has been shown to be silenced and/or regulated by DNA methylation, HDAC and HAT activity, and interactions with histone modifier complexes (reviewed in [51]). Likewise, AR activity is regulated by DNA methylation as well as histone acetylation and methylation (reviewed in [51]). Together, sex hormones and their receptors influence gene expression by acting as transcription factors modulating the expression of many genes, including epigenetic regulators, which can elicit widespread genomic effects.

Conclusions and Future Directions

As outlined throughout this chapter, epigenetic regulation of gene expression is integral to cell differentiation, development, and disease. Modes of epigenetic regulation, including DNA methylation, histone modification, and ncRNA-based regulation, alter chromatin structure, promoter accessibility, and contribute to posttranscriptional modifications. In this way, epigenetic regulation allows for the same, steadfast genetic code to be interpreted and expressed in countless ways leading to seemingly limitless variation and widespread differences in potential phenotype.

While epigenetic regulation is a necessary part of development and growth, it is also implicated in the diseased state as well. In the earliest of days, epigenetic dysfunction is associated with improper cardiac development and, later in life, is associated with diseases of aging like atherosclerosis. Epigenetics also serve as a link between the environment and physiology by translating various external stressors into phenotype. High-fat or high-sodium diets can progress to pathological hypertrophy and HF by modulating epigenetic marks along the way. Other stimuli, including pressure overload and I/R injury, also induce alterations in epigenetic markers leading to HF. As such, the inhibition of pathological chromatin rearrangement and gene expression could serve as a novel mechanism by which to prevent CVD.

Epigenetic regulation also orchestrates sex differences during development and disease. Altered gene expression between men and women, as a result of sex chromosomes and/or sex hormones, is mediated through epigenetic control and can contribute to the marked sex differences found in CVD. Understanding how the epigenetic landscape is cultivated by sex could help shed insight into sex differences in CVD risk, pathogenesis, and treatment.

References

1. Abi Khalil C. The emerging role of epigenetics in cardiovascular disease. Ther Adv Chronic Dis. 2014;5(4):178–87. https://doi.org/10.1177/2040622314529325.
2. Allis CD, Jenuwein T. The molecular hallmarks of epigenetic control. Nat Rev Genet. 2016;17(8):487–500. https://doi.org/10.1038/nrg.2016.59.
3. Arnold AP, Reue K, Eghbali M, Vilain E, Chen X, Ghahramani N, et al. The importance of having two X chromosomes. Philos Trans R Soc Lond Ser B Biol Sci. 2016;371(1688):20150113. https://doi.org/10.1098/rstb.2015.0113.
4. Balaton BP, Brown CJ. Escape artists of the X chromosome. Trends Genet. 2016;32(6):348–59. https://doi.org/10.1016/j.tig.2016.03.007.
5. Balaton BP, Cotton AM, Brown CJ. Derivation of consensus inactivation status for X-linked genes from genome-wide studies. Biol Sex Differ. 2015;6:35. https://doi.org/10.1186/s13293-015-0053-7.
6. Banka S, Lederer D, Benoit V, Jenkins E, Howard E, Bunstone S, et al. Novel KDM6A (UTX) mutations and a clinical and molecular review of the X-linked kabuki syndrome (KS2). Clin Genet. 2015;87(3):252–8. https://doi.org/10.1111/cge.12363.
7. Bannister AJ, Kouzarides T. Regulation of chromatin by histone modifications. Cell Res. 2011;21(3):381–95. https://doi.org/10.1038/cr.2011.22.
8. Berger SL, Kouzarides T, Shiekhattar R, Shilatifard A. An operational definition of epigenetics. Genes Dev. 2009;23(7):781–3. https://doi.org/10.1101/gad.1787609.
9. Bhat-Nakshatri P, Wang G, Collins NR, Thomson MJ, Geistlinger TR, Carroll JS, et al. Estradiol-regulated microRNAs control estradiol response in breast cancer cells. Nucleic Acids Res. 2009;37(14):4850–61. https://doi.org/10.1093/nar/gkp500.
10. Bloomer LDS, Nelson CP, Eales J, Denniff M, Christofidou P, Debiec R, et al. Male-specific region of the Y chromosome and cardiovascular risk: phylogenetic analysis and gene expression studies. Arterioscler Thromb Vasc Biol. 2013;33(7):1722–7. https://doi.org/10.1161/ATVBAHA.113.301608.
11. Brown CJ, Ballabio A, Rupert JL, Lafreniere RG, Grompe M, Tonlorenzi R, Willard HF. A gene from the region of the human X inactivation centre is expressed exclusively from the inactive X chromosome. Nature. 1991;349(6304):38–44. https://doi.org/10.1038/349038a0.
12. Brown CJ, Willard HF. The human X-inactivation centre is not required for maintenance of X-chromosome inactivation. Nature. 1994;368(6467):154–6. https://doi.org/10.1038/368154a0.
13. Burgoyne PS, Arnold AP. A primer on the use of mouse models for identifying direct sex chromosome effects that cause sex differences in non-gonadal tissues. Biol Sex Differ. 2016;7:68. https://doi.org/10.1186/s13293-016-0115-5.
14. Cao DJ, Wang ZV, Battiprolu PK, Jiang N, Morales CR, Kong Y, et al. Histone deacetylase (HDAC) inhibitors attenuate cardiac hypertrophy by suppressing autophagy. Proc Natl Acad Sci. 2011;108(10):4123–8. https://doi.org/10.1073/pnas.1015081108.
15. Carè A, Catalucci D, Felicetti F, Bonci D, Addario A, Gallo P, et al. MicroRNA-133 controls cardiac hypertrophy. Nat Med. 2007;13(5):613–8. https://doi.org/10.1038/nm1582.
16. Castellano L, Giamas G, Jacob J, Coombes RC, Lucchesi W, Thiruchelvam P, et al. The estrogen receptor-α-induced microRNA signature regulates itself and its transcriptional response. Proc Natl Acad Sci. 2009;106(37):15732–7. https://doi.org/10.1073/pnas.0906947106.
17. Cedar H, Bergman Y. Programming of DNA methylation patterns. Annu Rev Biochem. 2012;81:97–117. https://doi.org/10.1146/annurev-biochem-052610-091920.
18. Chamberlain AA, Lin M, Lister RL, Maslov AA, Wang Y, Suzuki M, et al. DNA methylation is developmentally regulated for genes essential for cardiogenesis. J Am Heart Assoc. 2014;3(3):e000976. https://doi.org/10.1161/JAHA.114.000976.
19. Chang S, McKinsey TA, Zhang CL, Richardson JA, Hill JA, Olson EN. Histone deacetylases 5 and 9 govern responsiveness of the heart to a subset of stress signals and play redundant roles in heart development. Mol Cell Biol. 2004;24(19):8467–76. https://doi.org/10.1128/MCB.24.19.8467-8476.2004.
20. Charchar FJ, Bloomer LD, Barnes TA, Cowley MJ, Nelson CP, Wang Y, et al. Inheritance of coronary artery disease in men: an analysis of the role of the Y chromosome. Lancet. 2012;379(9819):915–22. https://doi.org/10.1016/S0140-6736(11)61453-0.
21. Charchar FJ, Tomaszewski M, Padmanabhan S, Lacka B, Upton MN, Inglis GC, et al. The Y chromosome effect on blood pressure in two European populations. Hypertension (Dallas, Tex.: 1979). 2002;39(2 Pt 2):353–6.
22. Chen X, McClusky R, Chen J, Beaven SW, Tontonoz P, Arnold AP, Reue K. The number of x chromosomes causes sex differences in adiposity in mice. PLoS Genet. 2012;8(5):e1002709. https://doi.org/10.1371/journal.pgen.1002709.
23. Cifuentes-Rojas C, Hernandez AJ, Sarma K, Lee JT. Regulatory interactions between RNA and polycomb repressive complex 2. Mol Cell. 2014;55(2):171–85. https://doi.org/10.1016/j.molcel.2014.05.009.
24. Clemson CM, McNeil JA, Willard HF, Lawrence JB. XIST RNA paints the inactive X chromosome at interphase: evidence for a novel RNA involved in nuclear/chromosome structure. J Cell Biol. 1996;132(3):259–75. https://doi.org/10.1083/jcb.132.3.259.
25. Cortez D, Marin R, Toledo-Flores D, Froidevaux L, Liechti A, Waters PD, et al. Origins and functional evolution of Y chromosomes across mammals. Nature. 2014;508(7497):488–93. https://doi.org/10.1038/nature13151.
26. da Rocha ST, Boeva V, Escamilla-Del-Arenal M, Ancelin K, Granier C, Matias NR, et al. Jarid2 is implicated in the initial Xist-induced targeting of PRC2 to the inactive X chromosome. Mol Cell. 2014;53(2):301–16. https://doi.org/10.1016/j.molcel.2014.01.002.
27. Dan J, Chen T. Genetic studies on mammalian DNA methyltransferases. In: DNA methyltransferases – role and

function. Cham: Springer; 2016. p. 123–50. https://doi.org/10.1007/978-3-319-43624-1_6.
28. Dong C, Yoon W, Goldschmidt-Clermont PJ. DNA methylation and atherosclerosis. J Nutr. 2002;132(8 Suppl):2406S–9S.
29. Du S, Itoh N, Askarinam S, Hill H, Arnold AP, Voskuhl RR. XY sex chromosome complement, compared with XX, in the CNS confers greater neurodegeneration during experimental autoimmune encephalomyelitis. Proc Natl Acad Sci U S A. 2014;111(7):2806–11. https://doi.org/10.1073/pnas.1307091111.
30. Feinberg AP, Vogelstein B. Hypomethylation distinguishes genes of some human cancers from their normal counterparts. Nature. 1983;301(5895):89–92. https://doi.org/10.1038/301089a0.
31. Fletcher CE, Dart DA, Bevan CL. Interplay between steroid signalling and microRNAs: implications for hormone-dependent cancers. Endocr Relat Cancer. 2014;21(5):R409–29. https://doi.org/10.1530/ERC-14-0208.
32. Gillette TG, Hill JA. Readers, writers and erasers: chromatin as the whiteboard of heart disease. Circ Res. 2015;116(7):1245–53. https://doi.org/10.1161/CIRCRESAHA.116.303630.
33. Goldberg AD, Allis CD, Bernstein E. Epigenetics: a landscape takes shape. Cell. 2007;128(4):635–8. https://doi.org/10.1016/j.cell.2007.02.006.
34. Granger A, Abdullah I, Huebner F, Stout A, Wang T, Huebner T, et al. Histone deacetylase inhibition reduces myocardial ischemia-reperfusion injury in mice. FASEB J. 2008;22(10):3549–60. https://doi.org/10.1096/fj.08-108548.
35. Grote P, Wittler L, Hendrix D, Koch F, Währisch S, Beisaw A, et al. The tissue-specific lncRNA Fendrr is an essential regulator of heart and body wall development in the mouse. Dev Cell. 2013;24(2):206–14. https://doi.org/10.1016/j.devcel.2012.12.012.
36. Grunert M, Dorn C, Cui H, Dunkel I, Schulz K, Schoenhals S, et al. Comparative DNA methylation and gene expression analysis identifies novel genes for structural congenital heart diseases. Cardiovasc Res. 2016;112(1):464–77. https://doi.org/10.1093/cvr/cvw195.
37. Guo X, Su B, Zhou Z, Sha J. Rapid evolution of mammalian X-linked testis microRNAs. BMC Genomics. 2009;10:97. https://doi.org/10.1186/1471-2164-10-97.
38. Gusterson RJ, Jazrawi E, Adcock IM, Latchman DS. The transcriptional co-activators CREB-binding protein (CBP) and p300 play a critical role in cardiac hypertrophy that is dependent on their histone acetyltransferase activity. J Biol Chem. 2003;278(9):6838–47. https://doi.org/10.1074/jbc.M211762200.
39. Haas J, Frese KS, Park YJ, Keller A, Vogel B, Lindroth AM, et al. Alterations in cardiac DNA methylation in human dilated cardiomyopathy. EMBO Mol Med. 2013;5(3):413–29. https://doi.org/10.1002/emmm.201201553.
40. Haemmig S, Simion V, Yang D, Deng Y, Feinberg MW. Long noncoding Rnas in cardiovascular disease, diagnosis, and therapy. Curr Opin Cardiol. 2017;32:776. https://doi.org/10.1097/HCO.0000000000000454.
41. Han P, Li W, Lin C-H, Yang J, Shang C, Nurnberg ST, et al. A long noncoding RNA protects the heart from pathological hypertrophy. Nature. 2014;514(7520):102–6. https://doi.org/10.1038/nature13596.
42. Hang CT, Yang J, Han P, Cheng H-L, Shang C, Ashley E, et al. Chromatin regulation by Brg1 underlies heart muscle development and disease. Nature. 2010;466(7302):62–7. https://doi.org/10.1038/nature09130.
43. Hiura Y, Fukushima Y, Kokubo Y, Okamura T, Goto Y, Nonogi H, et al. Effects of the Y chromosome on cardiovascular risk factors in Japanese men. Hypertens Res. 2008;31(9):1687–94. https://doi.org/10.1291/hypres.31.1687.
44. Hughes JF, Skaletsky H, Brown LG, Pyntikova T, Graves T, Fulton RS, et al. Strict evolutionary conservation followed rapid gene loss on human and rhesus Y chromosomes. Nature. 2012;483(7387):82–6. https://doi.org/10.1038/nature10843.
45. Ikeda S, Kong SW, Lu J, Bisping E, Zhang H, Allen PD, et al. Altered microRNA expression in human heart disease. Physiol Genomics. 2007;31(3):367–73. https://doi.org/10.1152/physiolgenomics.00144.2007.
46. Jadhav RR, Ye Z, Huang R-L, Liu J, Hsu P-Y, Huang Y-W, et al. Genome-wide DNA methylation analysis reveals estrogen-mediated epigenetic repression of metallothionein-1 gene cluster in breast cancer. Clin Epigenetics. 2015;7:13. https://doi.org/10.1186/s13148-015-0045-9.
47. Kaneda R, Takada S, Yamashita Y, Choi YL, Nonaka-Sarukawa M, Soda M, et al. Genome-wide histone methylation profile for heart failure. Genes Cells. 2009;14(1):69–77. https://doi.org/10.1111/j.1365-2443.2008.01252.x.
48. Khoury S, Yarows SA, O'Brien TK, Sowers JR. Ambulatory blood pressure monitoring in a nonacademic setting. Effects of age and sex. Am J Hypertens. 1992;5(9):616–23.
49. Klattenhoff CA, Scheuermann JC, Surface LE, Bradley RK, Fields PA, Steinhauser ML, et al. Braveheart, a long noncoding RNA required for cardiovascular lineage commitment. Cell. 2013;152(3):570–83. https://doi.org/10.1016/j.cell.2013.01.003.
50. Latronico MVG, Condorelli G. MicroRNAs and cardiac pathology. Nat Rev Cardiol. 2009;6(6):419–29. https://doi.org/10.1038/nrcardio.2009.56.
51. Leader JE, Wang C, Fu M, Pestell RG. Epigenetic regulation of nuclear steroid receptors. Biochem Pharmacol. 2006;72(11):1589–96. https://doi.org/10.1016/j.bcp.2006.05.024.
52. Lee JT, Strauss WM, Dausman JA, Jaenisch R. A 450 kb transgene displays properties of the mammalian X-inactivation center. Cell. 1996;86(1):83–94.
53. Li J, Chen X, McClusky R, Ruiz-Sundstrom M, Itoh Y, Umar S, et al. The number of X chromosomes influences protection from cardiac ischaemia/reperfusion injury in mice: one X is better than two. Cardiovasc Res. 2014;102(3):375–84. https://doi.org/10.1093/cvr/cvu064.
54. McCarthy MI, Hirschhorn JN. Genome-wide association studies: potential next steps on a genetic journey. Hum Mol Genet. 2008;17(R2):R156–65. https://doi.org/10.1093/hmg/ddn289.
55. McCarthy MM, Auger AP, Bale TL, De Vries GJ, Dunn GA, Forger NG, et al. The epigenetics of sex differences in the brain. J Neurosci. 2009;29(41):12815–23. https://doi.org/10.1523/JNEUROSCI.3331-09.2009.
56. Menazza S, Murphy E. The expanding complexity of estrogen receptor signaling in the cardiovascular system. Circ Res. 2016;118(6):994–1007. https://doi.org/10.1161/CIRCRESAHA.115.305376.
57. Moore KJ, Sheedy FJ, Fisher EA. Macrophages in atherosclerosis: a dynamic balance. Nat Rev Immunol. 2013;13(10):709–21. https://doi.org/10.1038/nri3520.
58. Morgan CP, Bale TL. Sex differences in microRNA regulation of gene expression: no smoke, just miRs. Biol Sex Differ. 2012;3(1):22. https://doi.org/10.1186/2042-6410-3-22.
59. Movassagh M, Choy M-K, Knowles DA, Cordeddu L, Haider S, Down T, et al. Distinct epigenomic features in end-stage failing human hearts. Circulation. 2011;124(22):2411–22. https://doi.org/10.1161/CIRCULATIONAHA.111.040071.

60. Nikitina T, Shi X, Ghosh RP, Horowitz-Scherer RA, Hansen JC, Woodcock CL. Multiple modes of interaction between the methylated DNA binding protein MeCP2 and chromatin. Mol Cell Biol. 2007;27(3):864–77. https://doi.org/10.1128/MCB.01593-06.
61. Norris DP, Patel D, Kay GF, Penny GD, Brockdorff N, Sheardown SA, Rastan S. Evidence that random and imprinted Xist expression is controlled by preemptive methylation. Cell. 1994;77(1):41–51.
62. Ooi JYY, Tuano NK, Rafehi H, Gao X-M, Ziemann M, Du X-J, El-Osta A. HDAC inhibition attenuates cardiac hypertrophy by acetylation and deacetylation of target genes. Epigenetics. 2015;10(5):418–30. https://doi.org/10.1080/15592294.2015.1024406.
63. Papageorgiou N, Tousoulis D, Androulakis E, Siasos G, Briasoulis A, Vogiatzi G, et al. The role of microRNAs in cardiovascular disease. Curr Med Chem. 2012;19(16):2605–10.
64. Papait R, Cattaneo P, Kunderfranco P, Greco C, Carullo P, Guffanti A, et al. Genome-wide analysis of histone marks identifying an epigenetic signature of promoters and enhancers underlying cardiac hypertrophy. Proc Natl Acad Sci U S A. 2013;110(50):20164–9. https://doi.org/10.1073/pnas.1315155110.
65. Papait R, Serio S, Pagiatakis C, Rusconi F, Carullo P, Mazzola M, et al. Histone methyltransferase G9a is required for cardiomyocyte homeostasis and hypertrophy. Circulation. 2017;136(13):1233–46. https://doi.org/10.1161/CIRCULATIONAHA.117.028561.
66. Pasquinelli AE. MicroRNAs and their targets: recognition, regulation and an emerging reciprocal relationship. Nat Rev Genet. 2012;13(4):271–82. https://doi.org/10.1038/nrg3162.
67. Queirós AM, Eschen C, Fliegner D, Kararigas G, Dworatzek E, Westphal C, et al. Sex- and estrogen-dependent regulation of a miRNA network in the healthy and hypertrophied heart. Int J Cardiol. 2013;169(5):331–8. https://doi.org/10.1016/j.ijcard.2013.09.002.
68. Rafehi H, Balcerczyk A, Lunke S, Kaspi A, Ziemann M, Kn H, et al. Vascular histone deacetylation by pharmacological HDAC inhibition. Genome Res. 2014;24(8):1271–84. https://doi.org/10.1101/gr.168781.113.
69. Rosa-Garrido M, Chapski DJ, Schmitt AD, Kimball TH, Karbassi E, Monte E, et al. High resolution mapping of chromatin conformation in cardiac myocytes reveals structural remodeling of the epigenome in heart failure. Circulation. 2017;136:1613. https://doi.org/10.1161/CIRCULATIONAHA.117.029430.
70. Rosales W, Carulla J, García J, Vargas D, Lizcano F. Role of histone demethylases in cardiomyocytes induced to hypertrophy. Biomed Res Int. 2016;2016(2634976):1. https://doi.org/10.1155/2016/2634976.
71. Santos-Rebouças CB, Fintelman-Rodrigues N, Jensen LR, Kuss AW, Ribeiro MG, Campos M, et al. A novel nonsense mutation in KDM5C/JARID1C gene causing intellectual disability, short stature and speech delay. Neurosci Lett. 2011;498(1):67–71. https://doi.org/10.1016/j.neulet.2011.04.065.
72. Sharma S, Eghbali M. Influence of sex differences on microRNA gene regulation in disease. Biol Sex Differ. 2014;5(1):3. https://doi.org/10.1186/2042-6410-5-3.
73. Shin S, Janknecht R. Activation of androgen receptor by histone demethylases JMJD2A and JMJD2D. Biochem Biophys Res Commun. 2007;359(3):742–6. https://doi.org/10.1016/j.bbrc.2007.05.179.
74. Shpargel KB, Sengoku T, Yokoyama S, Magnuson T. UTX and UTY demonstrate histone demethylase-independent function in mouse embryonic development. PLoS Genet. 2012;8(9):e1002964. https://doi.org/10.1371/journal.pgen.1002964.
75. Snijders Blok L, Madsen E, Juusola J, Gilissen C, Baralle D, Reijnders MRF, et al. Mutations in DDX3X are a common cause of unexplained intellectual disability with gender-specific effects on Wnt signaling. Am J Hum Genet. 2015;97(2):343–52. https://doi.org/10.1016/j.ajhg.2015.07.004.
76. Strahl BD, Allis CD. The language of covalent histone modifications. Nature. 2000;403(6765):41–5. https://doi.org/10.1038/47412.
77. Suzuki MM, Bird A. DNA methylation landscapes: provocative insights from epigenomics. Nat Rev Genet. 2008;9(6):465–76. https://doi.org/10.1038/nrg2341.
78. Trivedi CM, Luo Y, Yin Z, Zhang M, Zhu W, Wang T, et al. Hdac2 regulates the cardiac hypertrophic response by modulating Gsk3 beta activity. Nat Med. 2007;13(3):324–31. https://doi.org/10.1038/nm1552.
79. Umar S, Cunningham CM, Itoh Y, Moazeni S, Vaillancourt M, Sarji S, et al. The Y chromosome plays a protective role in experimental hypoxic pulmonary hypertension. Am J Respir Crit Care Med. 2017. https://doi.org/10.1164/rccm.201707-1345LE.
80. Umer M, Herceg Z. Deciphering the epigenetic code: an overview of DNA methylation analysis methods. Antioxid Redox Signal. 2013;18(15):1972–86. https://doi.org/10.1089/ars.2012.4923.
81. Valencia-Morales M d P, Zaina S, Heyn H, Carmona FJ, Varol N, Sayols S, et al. The DNA methylation drift of the atherosclerotic aorta increases with lesion progression. BMC Med Genet. 2015;8:7. https://doi.org/10.1186/s12920-015-0085-1.
82. Waddington CH. The epigenotype. Endeavour. 1942;1:18–20.
83. Waddington CH. The strategy of the genes, Routledge library editions: 20th century science. London: Routledge; 2014.
84. Wang H, Maurano MT, Qu H, Varley KE, Gertz J, Pauli F, et al. Widespread plasticity in CTCF occupancy linked to DNA methylation. Genome Res. 2012;22(9):1680–8. https://doi.org/10.1101/gr.136101.111.
85. Wang K, Liu F, Zhou L-Y, Long B, Yuan S-M, Wang Y, et al. The long noncoding RNA CHRF regulates cardiac hypertrophy by targeting miR-489. Circ Res. 2014a;114(9):1377–88. https://doi.org/10.1161/CIRCRESAHA.114.302476.
86. Wang K, Long B, Zhou L-Y, Liu F, Zhou Q-Y, Liu C-Y, et al. CARL lncRNA inhibits anoxia-induced mitochondrial fission and apoptosis in cardiomyocytes by impairing miR-539-dependent PHB2 downregulation. Nat Commun. 2014b;5:3596. https://doi.org/10.1038/ncomms4596.
87. Wang Z, Zhang X-J, Ji Y-X, Zhang P, Deng K-Q, Gong J, et al. The long noncoding RNA Chaer defines an epigenetic checkpoint in cardiac hypertrophy. Nat Med. 2016;22(10):1131–9. https://doi.org/10.1038/nm.4179.
88. Wu X, Zhang Y. TET-mediated active DNA demethylation: mechanism, function and beyond. Nat Rev Genet. 2017;18(9):517–34. https://doi.org/10.1038/nrg.2017.33.
89. Xiao ZG, Shen J, Zhang L, Li LF, Li MX, Hu W, et al. The roles of histone demethylase UTX and JMJD3 (KDM6B) in cancers: current progress and future perspectives. Curr Med Chem. 2016;23:3687–96.
90. Yamauchi Y, Riel JM, Stoytcheva Z, Ward MA. Two Y genes can replace the entire Y chromosome for assisted reproduction in the mouse. Science. 2014;343(6166):69–72. https://doi.org/10.1126/science.1242544.
91. Ying AK, Hassanain HH, Roos CM, Smiraglia DJ, Issa JJ, Michler RE, et al. Methylation of the estrogen receptor-alpha

gene promoter is selectively increased in proliferating human aortic smooth muscle cells. Cardiovasc Res. 2000;46 (1):172–9.

92. Yoon S, Eom GH. HDAC and HDAC inhibitor: from Cancer to cardiovascular diseases. Chonnam Med J. 2016;52(1):1–11. https://doi.org/10.4068/cmj.2016.52.1.1.
93. Zhang CL, McKinsey TA, Chang S, Antos CL, Hill JA, Olson EN. Class II histone deacetylases act as signal-responsive repressors of cardiac hypertrophy. Cell. 2002;110(4):479–88.
94. Zhang Q-J, Chen H-Z, Wang L, Liu D-P, Hill JA, Liu Z-P. The histone trimethyllysine demethylase JMJD2A promotes cardiac hypertrophy in response to hypertrophic stimuli in mice. J Clin Invest. 2011;121(6):2447–56. https://doi.org/10.1172/JCI46277.
95. Zhang X, Ho S-M. Epigenetics meets endocrinology. J Mol Endocrinol. 2011;46(1):R11–32. https://doi.org/10.1677/JME-10-0053.
96. Zhao J, Sun BK, Erwin JA, Song J-J, Lee JT. Polycomb proteins targeted by a short repeat RNA to the mouse X chromosome. Science (New York, N.Y.). 2008;322 (5902):750–6. https://doi.org/10.1126/science.1163045.
97. Zhong J, Agha G, Baccarelli AA. The role of DNA methylation in cardiovascular risk and disease: methodological aspects, study design, and data analysis for epidemiological studies. Circ Res. 2016;118(1):119–31. https://doi.org/10.1161/CIRCRESAHA.115.305206.

Epidemiology of Congenital Heart Disease with Emphasis on Sex-Related Aspects 3

Byung Won Yoo

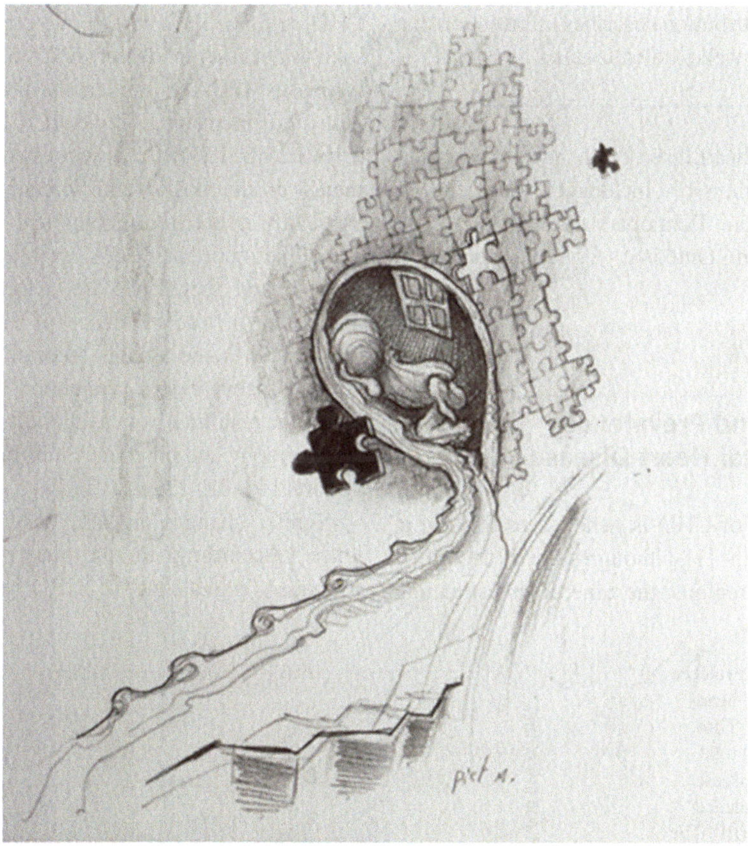

Epidemiology of congenital heart disease. Art work by Piet Michiels, Leuven, Belgium

B. W. Yoo (✉)
Severance Hospital, Yonsei University College of Medicine, Seoul, Republic of Korea
e-mail: BWYOO@yuhs.ac

Abstract

Gender differences in prevalence, manifestation, treatment outcomes, and prognosis have been well known for acquired heart disease

P. L. M. Kerkhof, V. M. Miller (eds.), *Sex-Specific Analysis of Cardiovascular Function*,
Advances in Experimental Medicine and Biology 1065, https://doi.org/10.1007/978-3-319-77932-4_3

such as coronary artery disease. Regarding congenital heart disease (CHD), it is recognized that the incidence of each congenital heart defect varies according to sex observed during a time span of more than 40 years. As diagnostic and surgical methods for CHD have achieved dramatic advances for the past decades, more newborns with CHD were able to survive and reach adulthood. Thereafter gender differences have begun to be reported on mortality, progress to pulmonary arterial hypertension, treatment outcomes, and prognosis in patients with CHD. However, it has been less known in the field of CHD yet, and this contribution describes information that is relatively well studied to date.

Keywords
Congenital heart disease · Gender · Sex ratio · Women · Ethnicity · Incidence · Prevalence · Manifestation · Pulmonary arterial hypertension · Outcome · Complication · Prognosis

Incidence and Prevalence of Congenital Heart Disease (CHD)

The incidence of CHD is generally estimated at 8/1000 live births [1], although it varies according to geographic regions, the time of investigation, target population, and case definition (e.g., exclusion of functionless abnormalities, unspecified anomalies, and spontaneously closed defect or not). However, the incidence of CHD has been rising to 10–14/1000 live births in recent studies [2–5], and this change is more evident in the reports from registries with continuous monitoring of the same populations over time [6–9]. Also a meta-analysis study, including worldwide 114 papers, reported that birth prevalence (<5 years of age) increased substantially over time from 0.6/1000 live births in 1930 to 9.1/1000 live births after 1995 (Fig. 3.1) [10].

When analyzed according to the severity of CHD, simple CHD with subtle physical findings such as atrial septal defect (ASD) or patent ductus arteriosus (PDA) significantly increased, along with the increase of overall CHD incidence (Fig. 3.2) [3, 10–13]. It is suggested that improvements in diagnostic and screening modalities, especially echocardiography and color-Doppler technique since the 1980s, enabled diagnosis of lesions, and therefore case detection increases rather than a true incidence of disease. On the other hand, some studies reported that the incidence of severe and complex CHD decreased over time resulting from availability of fetal echocardiography and pregnancy termination by intervention [10, 12, 13].

Similar changes are also observed in prevalence. According to the data from Quebec (Canada), prevalence of CHD was 6.88/1000

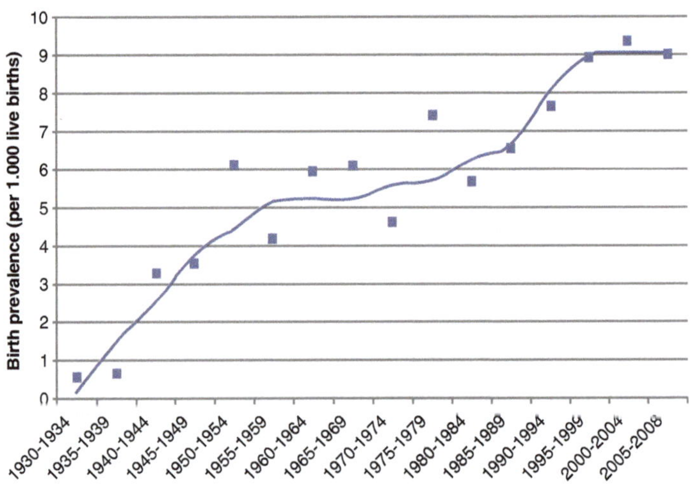

Fig. 3.1 Birth prevalence of total congenital heart disease over time. Time course of reported total congenital heart disease (CHD) birth prevalence from 1930 until 2010. The blue line shows the time trend, and the squares represent the calculated birth prevalence values for each time period. (Reprinted from Ref. [10] with permission from Elsevier)

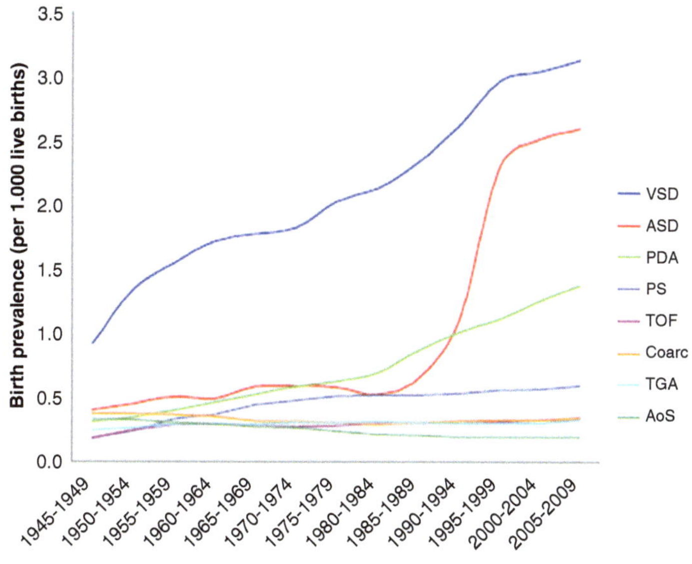

Fig. 3.2 Birth prevalence of specific congenital heart defects over time. Time course of birth prevalence of the eight most common CHD subtypes from 1945 until 2010. *AoS* aortic stenosis, *ASD* atrial septal defect, *Coarc* coarctation of aorta, *PDA* patent ductus arteriosus, *PS* pulmonary valve stenosis, *TGA* transposition of great arteries, *TOF* tetralogy of Fallot, *VSD* ventricular septal defect. (Reprinted from Ref. [10], with permission from Elsevier)

children (<18 years of age) in 1985 and abruptly rose to 11.89/1000 children in 2000 [4]. In addition, the prevalence of all CHD was consistently higher in females over time for both adult and pediatric populations (4.83/1000 females versus 3.94/1000 males in 1985; 4.55/1000 females versus 3.61/1000 males in 2000). While many prevalence studies reported that the sex ratio for overall CHD was greater than 1 [4, 11, 12], some recent studies have reported that total CHD is more prevalent in females [3, 5, 14]. This result could be attributed by that the proportion of simple CHD such as shunt lesions which show female preponderance is recently getting larger in the CHD population. Thus, we need to monitor if this change will continue in the years ahead.

Preponderance of Specific Defects Between Boys and Girls

It is well known that the incidences of CHD are different between male and female, according to each diagnosis, and this observation has been relatively consistent over time. ASD and atrioventricular septal defect (AVSD) have shown a female preponderance, and aortic stenosis (AS), coarctation of aorta (COA), transposition of great arteries (TGA), tetralogy of Fallot (TOF), and double outlet right ventricle (DORV) have shown a male preponderance (Fig. 3.3) [4, 5, 11–21]. Although we have not found the answers to explain these differences yet, it is suggested that a genetic factor is responsible for the male preponderance of aortic valve disease (e.g., AS, COA). The absence of a normal second X chromosome is then thought to be associated with aortopathy [22]. This explanation is based on the speculation that a genetic factor that modulates the development of the aorta and aortic valve is located on the X chromosome, as evidenced by the Turner syndrome (a sex aneuploidy syndrome), in which aortic valve disease occurs 146 times more frequently compared to the general population [23].

Table 3.1 summarizes results from previous studies for gender preponderance of specific congenital heart defects [4, 5, 11–21]. It is generally recognized that male predominance is associated with more complex and severe CHD, whereas female with more simple CHD. However, we should consider that gender preponderance could be different according to the race/ethnicity or geographic regions. In the comparative study from three large birth defect registries in

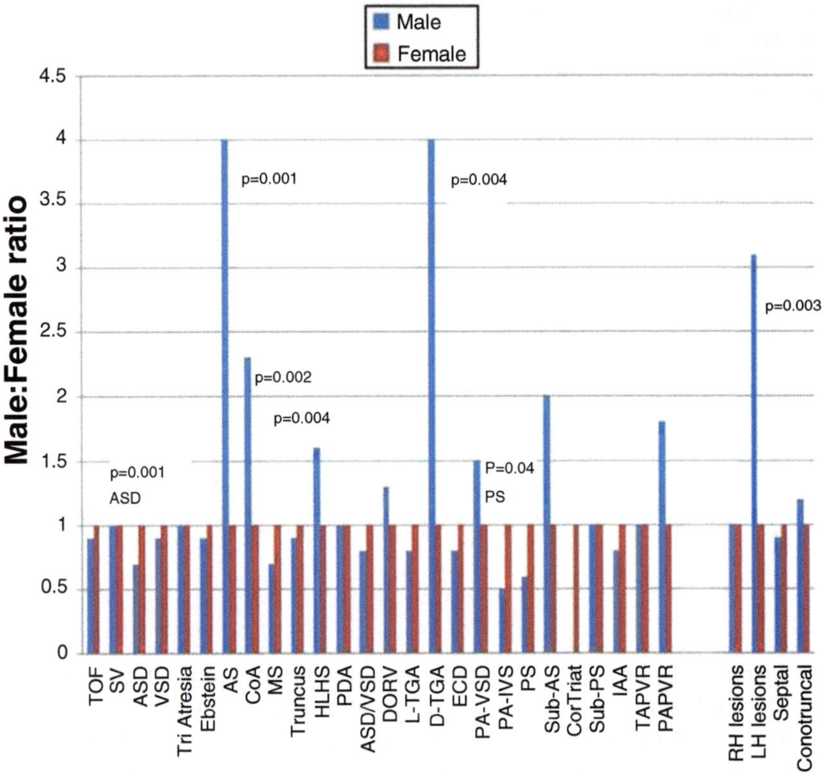

Fig. 3.3 Gender preponderance in congenital heart disease. *AS* aortic stenosis, *CoA* coarctation of aorta, *Cor Triat* cor triatriatum, *d-TGA* complete transposition of great arteries, *ECD* endocardial cushion defect, *DORV* double outlet right ventricle, *HLHS* hypoplastic left heart syndrome, *IAA* interrupted aortic arch, *LH* left heart, *L-TGA* congenitally corrected transposition of great arteries, *MS* mitral stenosis, *PA-IVS* pulmonary atresia with intact ventricular septum, *PA-VSD* pulmonary atresia with ventricular septal defect, *PS* pulmonic stenosis, *RH* right heart, *SV* single ventricle, *TAPVR* total anomalous pulmonary venous return, *TOF* tetralogy of Fallot. (Reprinted from Ref. [4], with permission from Springer)

California, Sweden, and France, the sex ratio of ASD showed a significant heterogeneity between registries (0.54 in French, 0.74 in Swedish, and 1.06 in California registry) [11]. Also, differences in sex ratios between ethnic groups were revealed in some studies [15, 17, 18]. An epidemiologic study of left ventricular outflow tract obstruction lesions, including AS, COA, and hypoplastic left heart syndrome (HLHS) from the Texas Birth Defect Registry, demonstrated racial/ethnic differences not only in prevalence but also in sex ratio of these defects [17]. In these defects which commonly have male preponderance, black males showed lower prevalence than white or Hispanic males and even lower than black females (Fig. 3.4).

Manifestation of CHD

Although atypical symptoms without chest pain occur more frequently in women with acute coronary syndromes resulting in higher mortality, gender difference in the manifestation of CHD has not been researched well. The study from a nationwide registry of adult patients with congenital heart disease in the Netherlands (CONCOR registry) showed a significant gender difference in functional class of patients with pulmonary arterial hypertension (PAH) associated with CHD [24]. More females than males were symptomatic even though mean systolic pulmonary arterial pressure was not different between males and females (Fig. 3.5). Female sex (odds ratio = 1.5) and increased systolic pulmonary arterial pressure (odds ratio = 0.04) were independently associated with a worse NYHA class. Also, in database of the European Heart Survey on adult congenital heart disease (ACHD), females were more likely to have functional limitation than males (OR 1.27; 95% CI 1.09–1.48) [25]. Hormonal fluid retention and predisposition to thrombosis might have contributed to this gender difference in manifestation. However, there has been no study to demonstrate this speculation [26].

Table 3.1 Preponderance of specific congenital heart defects between male and female

Male	Equivocal (or controversial)	Female
AS	VSD[a]	ASD secundum
COA	PS[a]	AVSD
TGA	PA[b]	PDA
TOF	TA[c]	MV anomalies
DORV	Truncus arteriosus[c]	
HLHS		
SV		
Anomalous PV return		

AS aortic stenosis, *ASD* atrial septal defect, *AVSD* atrioventricular septal defect, *COA* coarctation of aorta, *DORV* double outlet right ventricle, *HLHS* hypoplastic left heart syndrome, *MV* mitral valve, *PA* pulmonary atresia, *PDA* patent ductus arteriosus, *PS* pulmonary valve stenosis, *PV* pulmonary venous, *SV* single ventricle, *TA* tricuspid atresia, *TGA* transposition of great arteies, *TOF* tetralogy of Fallot, *VSD* ventricular septal defect

[a]Defects show equivocal sex ratio or female preponderance (female preponderance of VSD in Refs. [5, 12, 14]; of PS in Refs. [4, 5])

[b]Defects show equivocal sex ratio or male preponderance (male preponderance in Refs. [11, 18])

[c]Gender preponderance of defects is controversial [11, 12, 14, 18]

Progression to PAH

It has been recognized that females have a predisposition to PAH. The Registry to Evaluate Early and Long-Term Pulmonary Arterial Hypertension Disease Management (REVEAL registry), the largest ongoing study of PAH in the United States, demonstrated that more women had PAH associated with CHD and connective tissue disease [27]. In the study from the Dutch registry (CONCOR), the prevalence of PAH among 5970 registered patients with ACHD was 4.2% (6.1% among 1824 patients with septal defect), and 60% of these patients were female [24]. It was also noted that women had a 33% higher risk of PAH (OR 1.33; 95% CI 1.07–1.65) than men in the same registry [28]. The sex steroid hormone is considered as a related factor. Estrogen affects angiogenesis, vasculogenesis, and remodeling in response to shear stress via proliferation and migration of both endothelial cells and vascular smooth muscle cells [29]. As another potential influence of estrogen on PAH, it is suggested

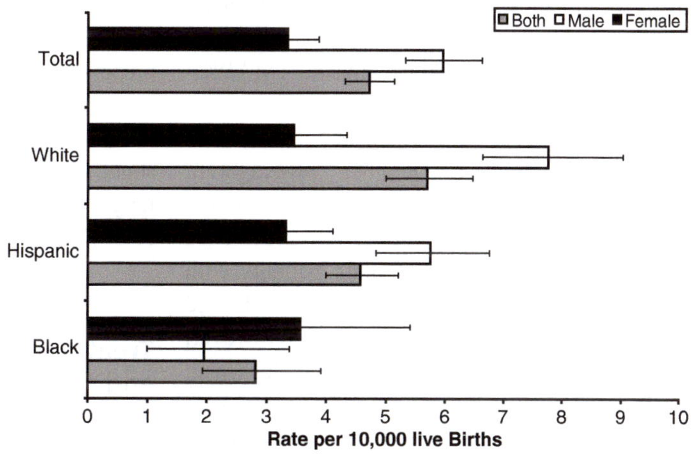

Fig. 3.4 Prevalence of left ventricular outflow tract obstruction malformations by sex and race/ethnicity, Texas, 1999–2001. (From Ref. [17], with permission from John Wiley and Sons)

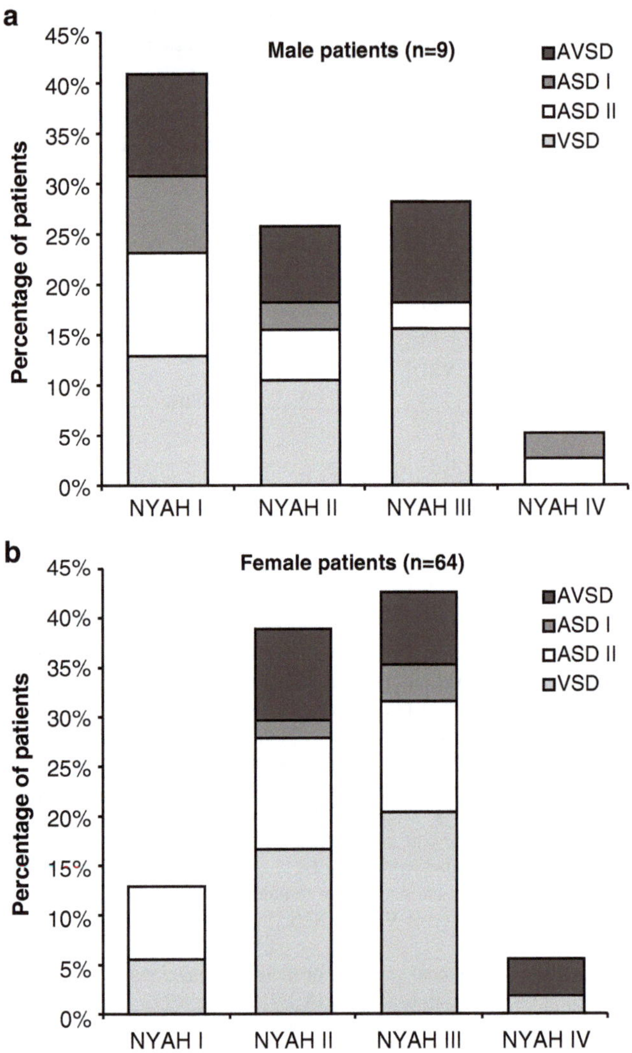

Fig. 3.5 Functional class of patients with pulmonary arterial hypertension due to CHD according to sex. Bars represented percentages. The whole bars sum to 100%. Subdivisions of the bars represent the proportions taken up by the individual defects within each NYHA class. (Reprinted from Ref. [24], with permission from Elsevier)

that estrogen enhances the proliferative capacity of cardiac fibroblasts via estrogen receptor- and mitogen-activated protein kinase (MAPK)-dependent mechanisms [30].

However, we should consider the gender distribution among CHD when investigating the incidence of CHD-associated PAH. In the Dutch study mentioned above, the prevalence of PAH among male and female patients with a septal defect was similar, both overall and per defect (7.8 vs. 7.6%, Fig. 3.6). However, there were more females than males with CHD-associated PAH since the patients with septal defect (especially ASD secundum) were more female.

Taking these two factors into consideration, a "two-hit hypothesis" is suggested for female preponderance in PAH. That is, the female has vulnerability (i.e., genetic susceptibility) to the development of PAH, and then the trigger of a shunt lesion initiates the vascular injury in the lungs resulting in higher prevalence of female in PAH [22].

Treatment Outcomes and Complications

Sex difference in surgical mortality related to CHD is still controversial. The prevalence study

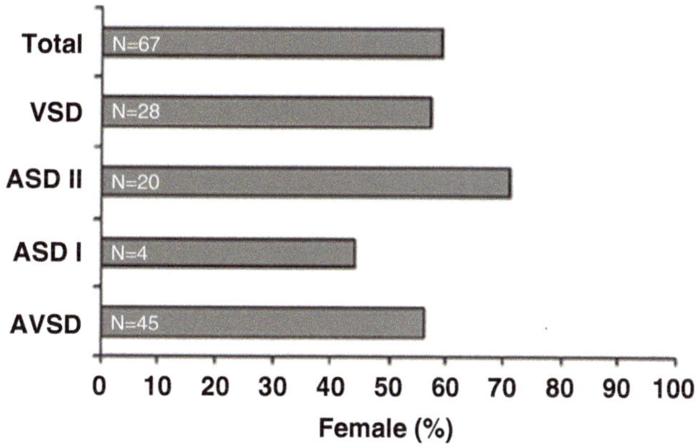

Fig. 3.6 Percentage of females among the patients with pulmonary arterial hypertension and a septal defect. *VSD* ventricular septal defect, *ASD II* atrial septal defect secundum, *ASD I* atrial septal defect primum, *AVSD* complete atrioventricular septal defect. (Reprinted from Ref. [24], with permission from Elsevier)

from Quebec (Canada) revealed the female predominance in ACHD patients with severe disease, but it was not observed in children. On the other hand, for simple disease (e.g., shunt lesions), female predominance was observed in both children and adults [3]. It was suggested that the rising prevalence of severe CHD in the female population is caused by gender difference in mortality, based on that mortality among males has been reported to be 5% greater than in females in a population of high-risk CHD infants [31]. However, some studies have revealed conflicting results [25, 28]. Of the 33,848 hospitalizations for CHD surgery, female infants who had high-risk procedures were at higher risk for death than male (OR, 1.21; 95% CI 1.08–1.36), although males underwent high-risk procedures and CHD surgery more frequently than females [28]. Also, in a recent study, surgical mortality was not different between male and female among 20,399 young patients (<18 years of age) [32]. Based on the results to date, we could infer that the females are affected by severe congenital heart diseases less frequently, but when affected, females could confront with a higher surgical mortality rate.

Regarding the complications of CHD, women had a 33% higher risk of PAH (OR, 1.33), a 33% lower risk of aortic outcomes (OR, 0.67; 95% CI, 0.50–0.90), a 47% lower risk of endocarditis (OR, 0.53; 95% CI, 0.40–0.70), a 55% lower risk of an implantable cardioverter defibrillator (OR, 0.45; 95% CI, 0.26–0.80), and a borderline significant 12% lower risk of arrhythmias (OR, 0.88; 95% CI, 0.77–1.02) in the CONCOR registry (Fig. 3.7) [33].

Prognosis of CHD

Gender differences in the long-term prognosis of CHD have not been studied widely yet. A published lecture provided some information about this issue as follows: In patients with repaired TOF (165 male, 104 female), symptomatic ventricular tachycardia and sudden unexpected death are frequent causes of death and morbidity in males, while the pulmonary vascular disease is the main cause of death in females. In AS, females have less severe manifestation. Females have a greater longevity of the pulmonary artery homograft than males [26]. It was reported that surgery for ACHD is at higher risk in males, and overall mortality in adulthood is greater in male patients with CHD, so the long-term survival rate is higher in females (Fig. 3.8) [34, 35].

As far as TOF is concerned, there are a few studies for gender differences in the long-term prognosis. The study from a cohort of 272 patients with repaired TOF (158 male, 114 female) demonstrated that females with repaired TOF had larger right ventricular (RV) end-systolic volumes (standard deviation scores: women, 4.35; men, 3.25), lower RV ejection fraction (women, −2.83; men, −2.12), lower RV muscle mass (women, 1.58; men, 2.45), and poor exercise capacity relative to sex-matched controls [36]. The other study of the effects of pregnancy

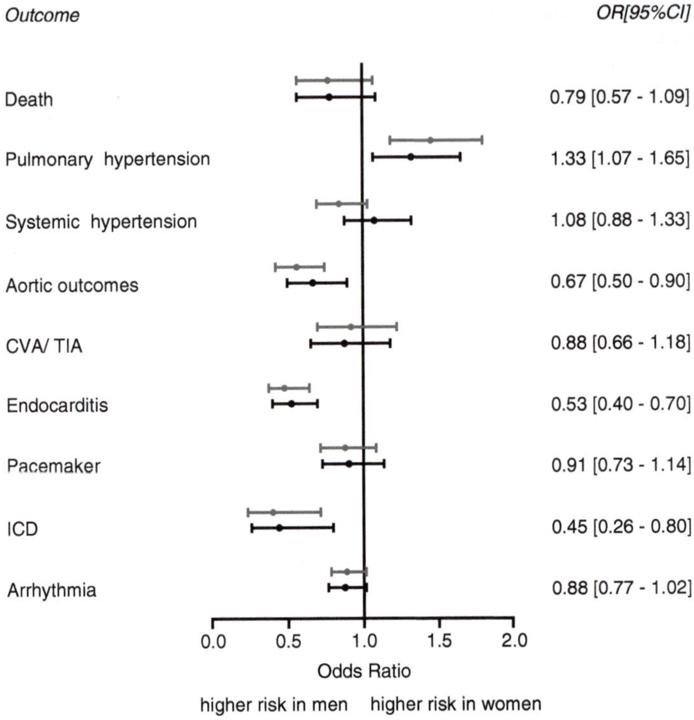

Fig. 3.7 Treatment outcomes and complications of adult congenital heart disease in women compared with men. ORs of outcomes in women compared with men. The gray lines represent ORs with 95% CIs adjusted for age only. The black lines represent ORs with 95% CIs additionally adjusted for underlying congenital heart defects. The numbers adjacent to the figure correspond to the black lines. *CVA* cerebrovascular accident, *TIA* transient ischemic attack. (From Ref. [28])

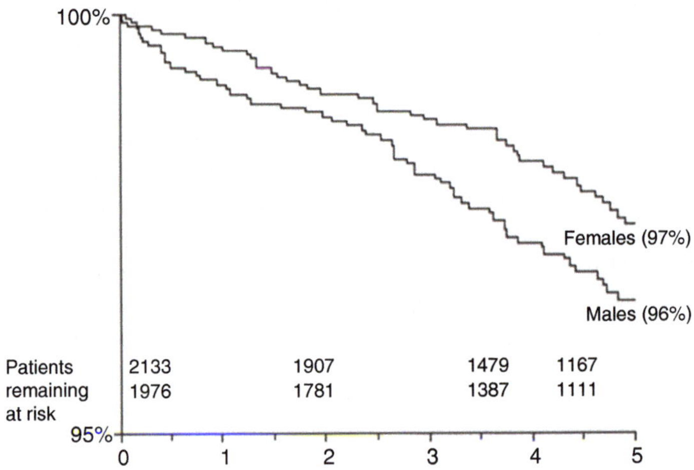

Fig. 3.8 Five-year cumulative survival curves for men and women with adult congenital heart disease, aggregated over all defects. Using Cox regression, cumulative mortality was greater in the male population (hazard ratio, 1.63; 95% CI 1.12–2.38). (Reproduced from Ref. [34], with permission from Bohn Stafleu van Loghum)

on RV remodeling in women with repaired TOF revealed that women with completed pregnancy showed accelerated RV remodeling (an increase in RV end-diastolic volume) compared with nulliparous women with repaired TOF, whereas RV systolic function does not deteriorate [37]. Gender difference in volumetric assessment of ventricles in patients with repaired TOF has been reported, showing that volumes for LV and RV in women are on average smaller, even after indexation for body surface area (Fig. 3.9) [38, 39]. Thus, even if the RV volume overload deteriorates more rapidly in women, it may be less noticeable than in men when evaluated with the same ventricular

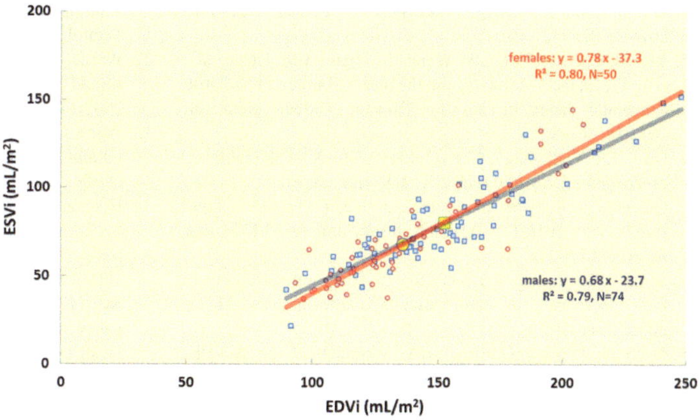

Fig. 3.9 Volume regulation graph. This representation shows end-systolic volume index (ESVi) versus end-diastolic volume index (EDVi) for the right ventricle in 124 patients with repaired tetralogy of Fallot, stratified for sex. Both ESVi ($P = 0.012$) and EDVi ($P = 0.006$) are smaller in females (see yellow circle and square), while ejection fraction is similar. (Data from Ref. [41])

volume criteria as men. Therefore, it is needed to apply sex-specific criteria during long-term follow-up of these patients.

Regarding ASD secundum, a study from CONCOR registry with 2207 adult patients with ASD closed or not revealed that male patients had a lower survival rate compared with the age- and sex-matched general population, although females had equal survival rate with controls. Moreover, males had a higher risk of conduction disturbances (OR, 1.63), supraventricular dysrhythmias (OR 1.41), cerebrovascular thromboembolic events (OR 1.53), and heart failure (OR 1.91) than females [40].

Summary

In addition to the sex-specific preponderance in the incidence of CHD, gender differences have been reported in manifestations, complications, progress to PAH, treatment outcome, and prognosis of CHD. For the gender preponderance of specific congenital heart defects, male predominance is associated with more complex and severe CHD such as AS, COA, TGA, and TOF, whereas female with more simple CHD such as ASD, PDA. Female patients with ACHD were more likely to have functional limitation than male patients and may be attributed by hormonal fluid retention and predisposition to thrombosis. It has been recognized that females have a predisposition to PAH, because female has vulnerability (i.e., genetic and hormonal susceptibility) to the development of PAH, and then the trigger of a shunt lesion initiates the vascular injury in the lungs. Sex differences in surgical mortality related to CHD are still controversial. Regarding the complications of CHD, females show higher risk of PAH and lower risk of aortic outcomes, endocarditis, and an implantable cardioverter defibrillator, while males show generally a higher prevalence of rhythm disorders. Gender differences in the long-term prognosis of CHD have not been studied widely yet, but studies on some diseases have emerged recently. Although there is little information to date, our understanding on the gender differences in CHD will be gradually improved, because the gender-specific prevention and management are essential to devise the optimal strategies for the individual characteristics of CHD.

References

1. Rothman KJ, Fyler DC. Sex, birth order, and maternal age characteristics of infants with congenital heart defects. Am J Epidemiol. 1976;104:527–34.
2. Hoffman JI, Kaplan S. The incidence of congenital heart disease. J Am Coll Cardiol. 2002;39:1890–900.
3. Marelli AJ, Mackie AS, Ionescu-Ittu R, Rahme E, Pilote L. Congenital heart disease in the general population: changing prevalence and age distribution. Circulation. 2007;115:163–72.
4. Egbe A, Uppu S, Stroustrup A, Lee S, Ho D, Srivastava S. Incidences and sociodemographics of specific congenital heart diseases in the United States of America: an evaluation

of hospital discharge diagnoses. Pediatr Cardiol. 2014;35:975–82.
5. Yeh SJ, Chen HC, Lu CW, Wang JK, Huang LM, Huang SC, et al. Prevalence, mortality, and the disease burden of pediatric congenital heart disease in Taiwan. Pediatr Neonatol. 2013;54:113–8.
6. Qu Y, Liu X, Zhuang J, Chen G, Mai J, Guo X, et al. Incidence of congenital heart disease: The 9-year experience of the Guangdong Registry of Congenital Heart Disease, China. PLoS One. 2016;11:e0159257. https://doi.org/10.1371/journal.pone.0159257.
7. Wren C, Richmond S, Donaldson L. Temporal variability in birth prevalence of cardiovascular malformations. Heart. 2000;83:414–9.
8. Oyen N, Poulsen G, Boyd HA, Wohlfahrt J, Jensen PK, Melbye M. National time trends in congenital heart defects, Denmark, 1977–2005. Am Heart J. 2009;157:467–73.
9. Fixler DE, Pastor P, Chamberlin M, Sigman E, Eifler CW. Trends in congenital heart disease in Dallas County births. 1971–1984. Circulation. 1990;81:137–42.
10. van der Linde D, Konings EE, Slager MA, Witsenburg M, Helbing WA, Takkenberg JJ, et al. Birth prevalence of congenital heart disease worldwide: a systematic review and meta-analysis. J Am Coll Cardiol. 2011;58:2241–7.
11. Pradat P, Francannet C, Harris JA, Robert E. The epidemiology of cardiovascular defects, part I: a study based on data from three large registries of congenital malformations. Pediatr Cardiol. 2003;24:195–221.
12. Calzolari E, Garani G, Cocchi G, Magnani C, Rivieri F, Neville A, et al. Congenital heart defects: 15 years of experience of the Emilia-Romagna Registry (Italy). Eur J Epidemiol. 2003;18:773–80.
13. Egbe A, Uppu S, Lee S, Stroustrup A, Ho D, Srivastava S. Temporal variation of birth prevalence of congenital heart disease in the United States. Congenit Heart Dis. 2015;10:43–50.
14. Moons P, Sluysmans T, De Wolf D, Massin M, Suys B, Benatar A, et al. Congenital heart disease in 111 225 births in Belgium: birth prevalence, treatment and survival in the 21st century. Acta Paediatr. 2009;98:472–7.
15. Liebson PR. Cardiovascular risk in special populations IV: congenital heart defects. Prev Cardiol. 2010;13:49–55.
16. Aubry P, Demian H. Sex differences in congenital heart disease. Ann Cardiol Angeiol (Paris). 2016;65:440–5.
17. McBride KL, Marengo L, Canfield M, Langlois P, Fixler D, Belmont JW. Epidemiology of noncomplex left ventricular outflow tract obstruction malformations (aortic valve stenosis, coarctation of the aorta, hypoplastic left heart syndrome) in Texas, 1999–2001. Birth Defects Res A Clin Mol Teratol. 2005;73:555–61.
18. Michalski AM, Richardson SD, Browne ML, Carmichael SL, Canfield MA, Van Zutphen AR, et al. Sex ratios among infants with birth defects, National Birth Defects Prevention Study, 1997–2009. Am J Med Genet. 2015;167A:1071–81.
19. Shaw GM, Carmichael SL, Kaidarova Z, Harris JA. Differential risks to males and females for congenital malformations among 2.5 million California births, 1989–1997. Birth Defects Res A Clin Mol Teratol. 2003;67:953–8.
20. Sokal R, Tata LJ, Fleming KM. Sex prevalence of major congenital anomalies in the United Kingdom: a national population-based study and international comparison meta-analysis. Birth Defects Res A Clin Mol Teratol. 2014;100:79–91.
21. Tennant PWG, Samarasekera SD, Pless-Mulloli T, Rankin J. Sex differences in the prevalence of congenital anomalies: a population-based study. Birth Defects Res A Clin Mol Teratol. 2011;91:894–901.
22. Warnes CA. Sex differences in congenital heart disease: should a woman be more like a man? Circulation. 2008;118:3–5.
23. Warnes CA. Bicuspid aortic valve and coarctation: two villains part of a diffuse problem. Heart. 2003;89:965–6.
24. Duffels MGJ, Engelfriet PM, Berger RMF, van Loon RLE, Hoendermis E, Vriend JWJ, et al. Pulmonary arterial hypertension in congenital heart disease: an epidemiologic perspective from a Dutch registry. Int J Cardiol. 2007;120:198–204.
25. Klitzner TS, Lee M, Rodriguez S, R-KR C. Sex-related disparity in surgical mortality among pediatric patients. Congenit Heart Dis. 2006;1:77–88.
26. Somerville J. The Denolin Lecture: the woman with congenital heart disease. Eur Heart J. 1998;19:1766–75.
27. Shapiro S, Traiger GL, Turner M, McGoon MD, Wason P, Barst RJ. Sex differences in the diagnosis, treatment and outcome of patients with pulmonary arterial hypertension enrolled in the registry to evaluate early and long-term pulmonary arterial hypertension disease management. Chest. 2012;141:363–73.
28. Marelli A, Gauvreau K, Landzberg M, Jenkins K. Sex differences in mortality in children undergoing congenital heart disease surgery. A United States population-based study. Circulation. 2010;122(Suppl. 1):S234–40.
29. Harvey RE, Coffman KE, Miller VM. Women-specific factors to consider in risk, diagnosis and treatment of cardiovascular disease. Womens Health (Lond). 2015;11:239–57.
30. Lee HW, Eghbali-Webb M. Estrogen enhances proliferative capacity of cardiac fibroblasts by estrogen receptor- and mitogen-activated protein kinase-dependent pathways. J Mol Cell Cardiol. 1998;30:1359–68.
31. Fyler DC. Report of the New England regional infant cardiac program. Pediatrics. 1980;65:375–461.
32. DiBardino DJ, Pasquali SK, Hirsch JC, Benjamin DK, Kleeman KC, Sala-zar JD, et al. Effect of sex and race on outcome in patients undergoing congenital heart surgery: an analysis of the Society of Thoracic Surgeons Congenital Heart database. Ann Thorac Surg. 2012;94:2054–9.
33. Verheugt CL, Uiterwaal CS, van der Velde ET, Meijboom FJ, Pieper PG, Vliegen HW, et al. Gender and outcome in adult congenital heart disease. Circulation. 2008;118:26–32.
34. Engelfriet P, Mulder BJM. Gender differences in adult congenital heart disease. Neth Heart J. 2009;17:414–7.
35. Zomer CA, Ionescu-Ittu R, Vaartjes I, Pilote L, Mackie AS, Therrien J, et al. Sex differences in hospital mortality in adults with congenital heart disease. The impact of reproductive health. J Am Coll Cardiol. 2013;62:58–67.
36. Sarikouch S, Boethig D, Peters B, Kropf S, Dubowy KO, Lange P, et al. Poorer right ventricular systolic function and exercise capacity in women after repair of tetralogy of Fallot: a sex comparison of standard deviation scores based on sex-specific reference values in healthy control subjects. Circ Cardiovasc Imaging. 2013;6:924–33.
37. Egidy Assenza G, Cassater D, Landzberg M, Geva T, Schreier J, Graham D, et al. The effects of pregnancy on right ventricular remodeling in women with repaired tetralogy of Fallot. Int J Cardiol. 2013;168:1847–52.
38. Sarikouch S, Koerperich H, Dubowy KO, Boethig D, Boettler P, Mir TS, et al. Impact of gender and age on cardiovascular function late after repair of tetralogy of Fallot: percentiles based on cardiac magnetic resonance. Circ Cardiovasc Imaging. 2011;4:703–11.

39. Kerkhof PLM, van de Ven PM, Yoo BW, Peace RA, Heyndrickx GR, Handly N. Ejection fraction as related to basic components in the left and right ventricular volume domains. Int J Cardiol. 2018;255:105–10.
40. Kuipers JM, van der Bom T, van Riel ACMJ, Meijboom FJ, van Dijk APJ, Pieper PG, et al. Secundum atrial septal defect is associated with reducedsurvival in adult men. Eur Heart J. 2015;36:2079–86.
41. Yoo BW, Kim JO, Kim YJ, Choi JY, Park HK, Park YH, et al. Impact of pressure load caused by right ventricular outflow tract obstruction on right ventricular volume overload in patients with repaired tetralogy of Fallot. J Thorac Cardiovasc Surg. 2012;143:1299–304.

Sex Differences in Epidemiology of Cardiac and Vascular Disease

4

Tatiana Kuznetsova

Epidemiology of cardiovascular diseases. Art work by Piet Michiels, Leuven, Belgium

Abstract

In spite of a remarkable decline in death rates from cardiovascular disease (CVD) observed over the last decades, CVD still remains the leading cause of mortality in both men and women worldwide. Overall the age-adjusted CVD mortality and morbidity rates are highest in men than in women. However, the risk of CVD in women should not be underestimated given that approximately one of two women in developed countries will die of mostly preventable heart diseases or stroke. Although

T. Kuznetsova (✉)
Research Unit Hypertension and Cardiovascular Epidemiology, KU Leuven Department of Cardiovascular Sciences, University of Leuven, Leuven, Belgium
e-mail: tatiana.kouznetsova@med.kuleuven.be

P. L. M. Kerkhof, V. M. Miller (eds.), *Sex-Specific Analysis of Cardiovascular Function*, Advances in Experimental Medicine and Biology 1065, https://doi.org/10.1007/978-3-319-77932-4_4

men and women share the same cardiovascular risk factors, there are substantial sex differences in the first manifestation and clinical presentation of CVD. In this part of the chapter, we will discuss the recent epidemiological data on sex discrepancies in the prevalence and burden of different CVDs.

Keywords
Cardiovascular disease · Coronary heart disease · Stroke · Peripheral arterial disease · Mortality · Morbidity · Microcirculatory dysfunction · Sex differences · Misdiagnosis · Risk profile

Epidemiology of CV Disease in Adults

CVD Mortality

Cardiovascular disease (CVD) is the leading cause of mortality in both men and women; in the world almost 35% of all deaths in women and 32% in men in 2013 are attributable to CVD according to the Global Burden of Disease Study [10]. In Europe, diseases of the heart and circulatory system are responsible for over 3.9 million deaths a year, or 45% of all deaths. In men, CVD accounts for 1.8 million deaths (40% of all deaths), while in women it is responsible for 2.1 million deaths (49% of all deaths; Fig. 4.1) [37]. The observed differences in the percentage of CVD death between men and women might be explained by sex differences in demographics. Women have a higher life expectancy than men do, and therefore, they form a larger proportion of the elderly population in which the prevalence of CVD is highest [24]. Therefore, when adjusted for differences in age distribution, women die from CVD at a lower rate than men do. For instance, in the USA in 2007, the age-adjusted CVD mortality rate in men was 300 per 100,000 compared with 212 per 100,000 women [24]. Similarly, in European countries for which data are lately released by the World Health Organization, age-standardized death rates for coronary heart disease (CHD) and stroke are higher in males than females [6]. As illustrated in Fig. 4.2, men-to-women CHD mortality ratios was above 4 in middle age (30–64 years old) but reduced to 2 in the older age group (65–85 years old) [6]. Men-to-women stroke mortality ratio was around 1.5–2 until old age [6]. Although mortality rate is lower in women than in men, it should therefore not be considered an excuse for CVD prevention and therapeutic remissness in women. The risk of CVD disease in women should not be underestimated taking into consideration the fact that approximately 1 of 2 women in Western countries will die of heart disease or stroke, compared to 1 in 30 of breast cancer (Fig. 4.1; [37]).

First Manifestation of CVD

In addition to its large mortality burden, CVD makes a substantial contribution to overall morbidity. There are also a number of notable sex differences in CVD incidence and risk factor profiles. For further developing primary prevention strategies, it is important to understand the first presentation of CVD in men and women.

Recent follow-up data of 8419 participants recruited within the community-based Rotterdam study demonstrated that the overall lifetime risk of developing CVD in the middle age men and women is similar [18]. Two out of three men and women aged 55 will develop CVD during their remaining lifespan. However, the authors observed large differences in the first manifestation of CVD: men are more likely to develop CHD as first event, while women are more likely to have cerebrovascular disease or heart failure (HF) as their first event (Fig. 4.3; [18]).

Another recent report including almost two million patient electronic health registries from the UK with more than 100,000 fatal and nonfatal CVD end points over 6 years of follow-up demonstrated that initial presentation of CVD was neither myocardial infarction nor ischemic stroke but rather HF, angina, transient ischemic attack, or peripheral arterial diseases (PAD) [11]. Overall, male sex was significantly

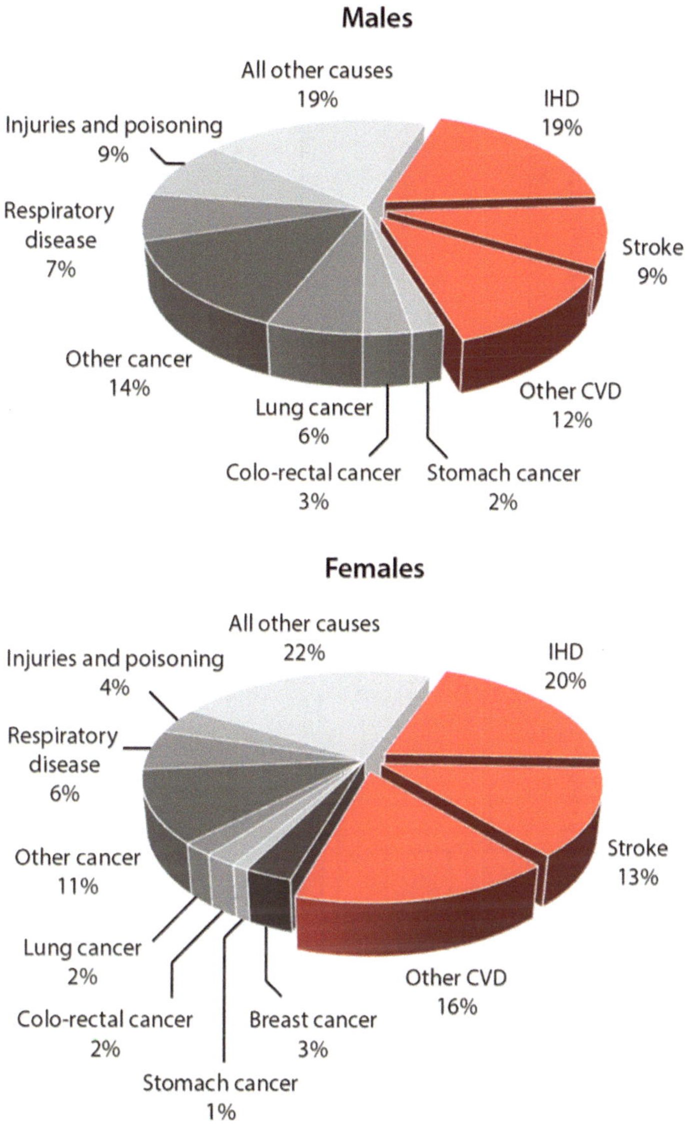

Fig. 4.1 *Deaths by cause, latest available year, Europe.* (Reproduced from [37] with permission)

associated with initial diagnosis of CVD, but the strength of this association is highly variable depending on event. As illustrated in Fig. 4.4, hazard ratios (HR) of men in comparison with women for initial CVD diagnosis varied from strong (HR, 3.6–5.0) for abdominal aortic aneurysm, myocardial infraction, and coronary death, though modest (HR, 1.5–2.0) for stable angina, ischemic stroke, PAD, HF, and cardiac arrest, to weak (HR, <1.5) for transient ischemic attack, intracerebral hemorrhage, and unstable angina [11]. The authors also reported that these associations were modified by age. The larger differences between men and women were observed in the younger and middle age groups for coronary end points, ischemic stroke, PAD, and aortic aneurysm. In all cases, sex differences diminished with higher age. These observations might have an important implication for individual risk prediction and management strategies.

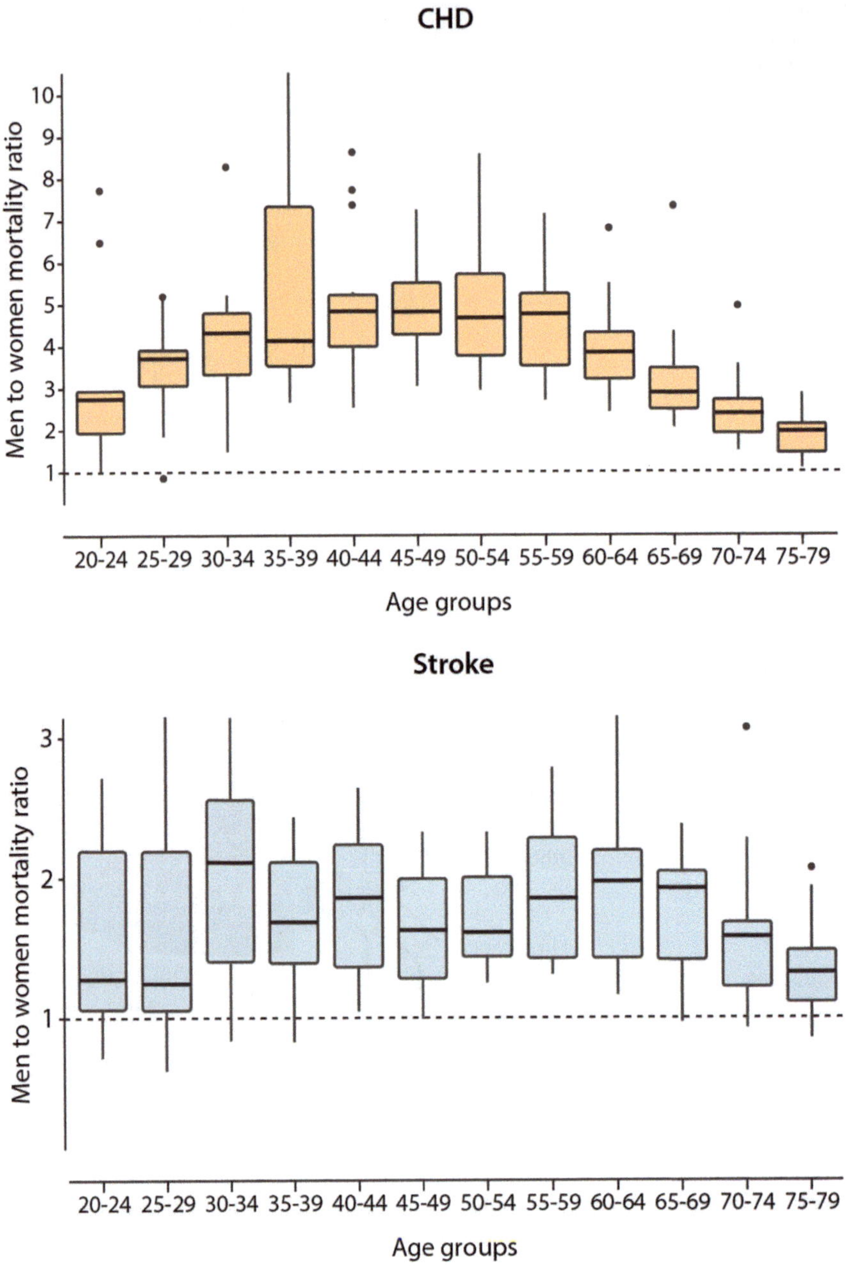

Fig. 4.2 *Men-to-women mortality rate ratios for CHD and stroke across countries in 2010, by age*. The band inside the box is the median mortality rate ratio, the bottom and top of the box are the first and third quartiles, and the ends of the whiskers are placed 1.5 IQR distant from the lower and upper quartile. The dots represent observations outside that range. *CHD*, coronary heart disease. (Reproduced from [6] with permission)

CHD

CHD or ischemic heart disease is the most common form of CVD which accounts for approximately 40% deaths in men and 30% deaths in women of the total cases of CVD mortality [37]. The finding that men have a higher risk of developing CHD during the lifespan than women was initially well illustrated by the Framingham investigators [15, 20]. In their initial study

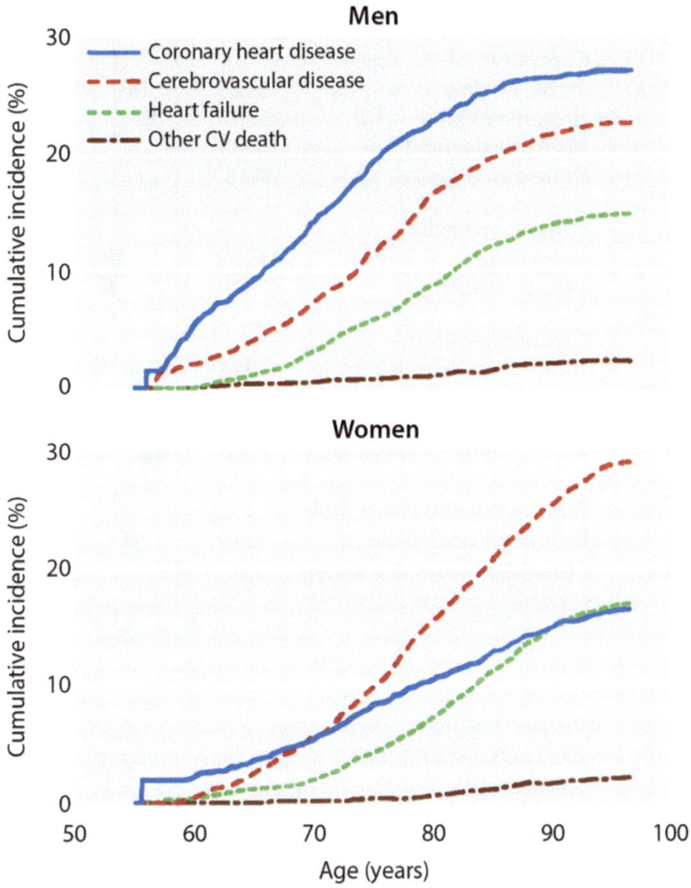

Fig. 4.3 *Cumulative incidence of first CVD manifestations adjusted for competing non-cardiovascular death for men (left) and women (right) aged 55.* Coronary heart disease was defined as myocardial infarction, coronary revascularization, or death from coronary heart disease. Cerebrovascular disease was defined as stroke, transient ischemic attack, or carotid revascularization. Other cardiovascular death included all cardiovascular mortality other than fatal coronary heart disease or stroke. (Reproduced from [18] with permission)

published in 1969, after 14 years of follow-up, the incidence of fatal and nonfatal CHD was higher in men than in women (14.2% vs 5.9%) [15]. Lifetime risk of CHD at age 40 years was 48.6% (95 CI 45.8%–51.3%) for men and 31.7% (29.2%–38.7%) for women [20].

The observed sex differences in the prevalence and presentation of CHD in population might be partially explained by distinct pathophysiological processes leading to myocardial ischemia [8, 35]. Indeed, women have less obstructive and extensive epicardial artery disease than men do but more prone to have an impaired coronary vasomotor function and microcirculatory dysfunction. For instance, the population-based studies reported a lower calcium score [16] and atheroma volume [25] in women when compared with men. On the other hand, based on available experimental and clinical data, "microvascular angina" is considered as a major etiological factor for CHD in the absence of significant coronary obstruction particularly in women [17]. Further clarification of the pathophysiological processes underlining CHD may help with tailoring sex-specific strategies for the prevention, detection, and management of CHD. Due to sex differences in pathophysiology and clinical manifestation of CHD, clinical symptoms of myocardial ischemia in women are often regarded as "atypical" and likely to be ignored or misdiagnosed [1]. As a consequence, women with overt CHD have suboptimal access to healthcare services which leads to delay in diagnosis and treatment and to worse prognosis and outcomes [22].

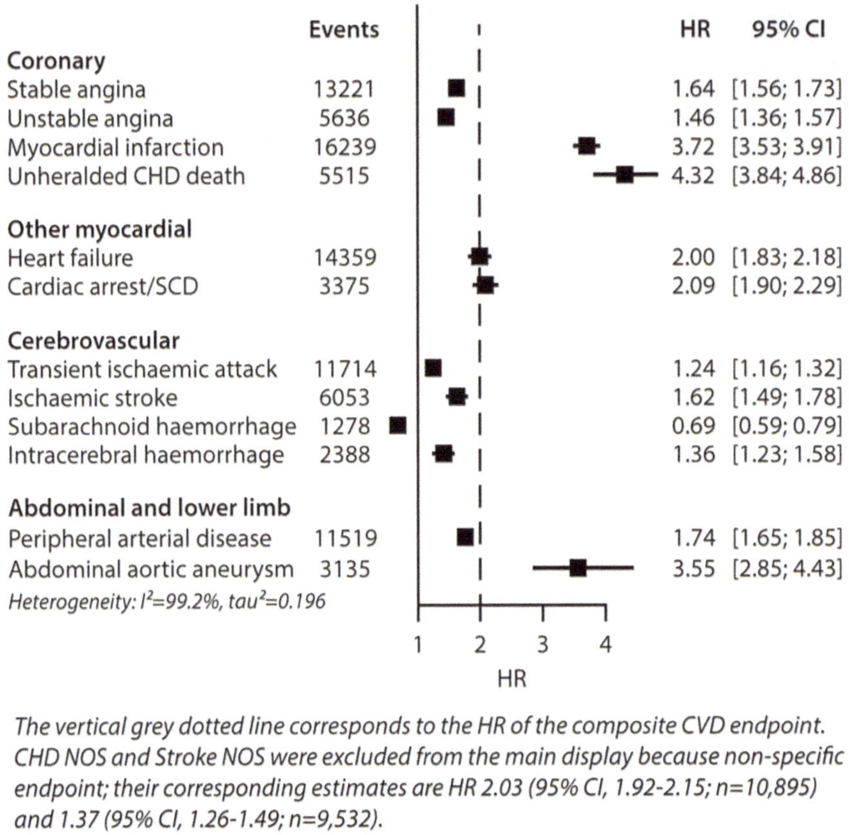

Fig. 4.4 *Hazard ratios of men in comparison with women for initial presentation of 12 different cardiovascular diseases among a population of 1.93 million adults. CHD* indicates coronary heart disease; *CI*, confidence interval; *CVD*, cardiovascular disease; *HR*, hazard ratio; *NOS*, not otherwise specified; and *SCD*, sudden cardiac death. (Reproduced from [11] with permission)

Stroke

Stroke is the second leading cause of CVD mortality which accounts for approximately 22.5% deaths in men and 26.5% deaths in women of the total cases of CVD mortality (Fig. 4.1; [37]). The observation that more women died from stroke than men could be explained by women's higher longevity and the fact that stroke incidence increases in the elderly age group [31]. Sex differences in stroke incidence are smaller as compared to CHD, yet age-adjusted stroke mortality rates remained approximately twice as high in men as compared with women [6]. Similarly, in a systemic review including 59 studies, male stroke incidence and prevalence rates were 33% and 41% higher than the female, but these numbers reversed dramatically over the age of 85 years [2].

With regard to poststroke outcome and prognosis, several studies reported a higher crude case fatality in women when compared with men [26, 27]. For instance, in the International Stroke Trial, 14-day and 6-month crude mortality in women was 11.0% and 24.5%, respectively, compared with 8.7% and 19.3% in men [26]. However, after adjustment for baseline age, severity of stroke, and blood pressure, stroke case fatality was actually 10% lower in women than in men (odds ratio, 0.90; 95% CI, 0.83–0.98) [26].

An embolus originating from the heart caused approximately one-third of ischemic strokes. The most important cause of cardioembolism is atrial fibrillation (AF) which is the common cardiac arrhythmia especially in the aging populations. Although lifetime risk of AF was similar in men and women (>24% by the age of 90), men developed AF a decade earlier than women [21]. Women with sustained AF developed more often cerebrovascular event when compared with men. Indeed, in 15,000 patients with AF recruited within the ATRIA study, Fang et al. [9] reported that the risk of ischemic stroke was greater in women than in men (age-adjusted HR, 1.6; 95% CI, 1.3–1.9).

Heart Failure

Another major burden of modern society is the progressive increase of age-associated disorders such as HF [33]. The overall HF prevalence is similar in men and women [13, 33] and constitutes about a quarter of the first manifestation of cardiovascular disease in both sexes [18]. However, there are important sex differences in the type of HF. While traditionally associated with the concept of pump failure or reduced left ventricular (LV) - ejection fraction (EF), it has become widely recognized that HF can occur even when EF is preserved, constituting the syndrome of HF with preserved EF (HFpEF) [28]. Whereas men develop more frequently HF with reduced ejection fraction (HFrEF) as a complication of CHD, HFpEF is especially frequent in women with risk factors such as hypertension, obesity, and diabetes. Of notice, symptomatic HF with or without reduced EF has a poor prognosis [5, 23, 29]. For instance, in the recent meta-analysis which included 10,347 patients with HFpEF and 31,625 patients with HFrEF, the authors compared survival in these groups using individual patient data [23]. Overall, there were 121 (95% CI, 117–126) and 141 (95% CI, 138–144) deaths per 1000 patient-years in patients with HFpEF and HFrEF, respectively. HFpEF is often underdiagnosed in elderly women even it is already symptomatic. Moreover, due to diagnostic difficulties, women have been often underrepresented in randomized clinical trials testing the effect of treatments on CVD even though these diseases highly affect elderly women [36].

CV Risk Factors

The observed trend in a reduction in CVD mortality over the last 30 years [6] in both men and women is attributable in part to the identification of risk factors and their control via preventive programs. The recent guidelines emphasized the importance of risk assessment to further improve quality of preventive care and highlighted underestimation of CVD risk in women [30].

The classic risk factors for CVD are the same in men and women, but there are sex differences in the prevalence and magnitude of association of these risk factors with CVD events. Age is a major CVD risk factor for both sexes, but women are on average 10 years older than men when they develop CVD [4]. This observation in part might be related to the protective role of female sex hormones in the premenopausal years [32].

Several large-scale population studies reported that ideal cardiovascular health is more frequent in women than in men [12, 14, 34]. To define cardiovascular health in population, seven metrics are often used such as body mass index, smoking status, blood pressure level, total cholesterol and blood glucose levels, physical activity, and diet. For instance, in the cross-sectional population study of 9012 French men and women aged 50–75 years old, women were four times more often in ideal cardiovascular health than men although they were more often depressed and less educated [34].

Blood pressure control is the main target for prevention of CVD such as stroke and HF particularly in women. For instance, the Framingham investigators demonstrated that the hazard for developing HF in hypertensive patients, compared to normotensive subjects, was approximately *twofold* in men and *threefold* in women after adjustment for age and other HF risk factors [19]. The 5-year survival rate for hypertensive symptomatic HF was 24% for men and 31% for women [19]. Along these lines, in the recent meta-analysis including 9357 subjects from 11 populations, Boggia et al. [7] demonstrated that the relation of fatal and nonfatal

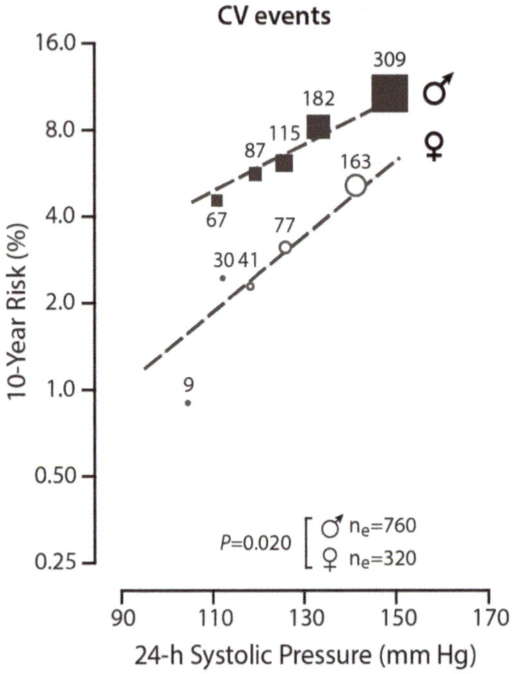

Fig. 4.5 *Absolute 10-year risk of a composite cardiovascular (CV) end point in relation to the 24-hour systolic blood pressure (BP).* The continuous risk functions cover the 5th to 95th percentile interval of the 24-hour systolic BP and were fitted by Cox regression with adjustment for cohort, age, body mass index, smoking and drinking, serum total cholesterol, history of cardiovascular disease, presence of diabetes mellitus, and antihypertensive drug treatment at baseline. Circles (women) and squares (men) represent the multivariable-adjusted HRs in quintiles of the distribution of the 24-hour systolic BP and have a size proportional to the inverse of the variance of the HR. The number of events in each quintile is given next to each circle or square; ne is the total number of events by disease category and sex. The probability values for interaction were derived from multivariable-adjusted Cox models. (Reproduced from [7] with permission)

cardiovascular events with ambulatory 24-hour blood pressure (BP) and with nighttime BP was steeper in women than in men (Fig. 4.5). Consequently, as per 1-SD decrease in ambulatory systolic 24-h and nighttime BP (about 14 mmHg), the proportion of potentially preventable events was higher in women than in men for all cardiovascular events (35.9% vs 24.2% and 35.1% vs 19.4%, respectively). The authors concluded that there is still a vast and largely unused potential for cardiovascular prevention by BP-lowering treatment in women.

Also lower socioeconomic status is independently associated with the greater risk of CVD at least in high-income countries. The recent meta-analysis over 22 million individuals and over 1 million CVD events demonstrated that the excess risk of overall CVD and in particularly CHD associated with the lowest, compared with the highest, level of education was 18% and 24% greater in women when compared with men [3]. The results of epidemiological studies on CVD risk factors suggested that the tailored risk assessments and interventions strategies are required to reduce the burden of CVD in men and women.

Summary

CVD remains the leading cause of mortality and morbidity in both men and women. As shown by numerous epidemiological studies, there are substantial sex differences in the prevalence, clinical presentation, and outcomes of different CVDs. Recently, considerable progress has been made in awareness and prevention of CVD particularly in women in whom the risk of CVD has been underestimated for a long time. Due to women's higher longevity and the fact that CVD incidence increases in the older age group, more women than men die from CVD each year. However, the age-adjusted CVD mortality and morbidity rates are highest in men. There are also a number of notable sex differences in CVD first manifestation and risk factor profiles. Although overall lifetime risk of developing CVD in the middle age men and women is similar, women are on average 10 years older when compared with men when they experience CVD. Men are more likely to develop CHD as first event, while women are more likely to have cerebrovascular disease or HF as their first event. Several population studies reported that overall cardiovascular risk factor profile is more favorable in women than men. On the other hand, the associations of CVD events with blood pressure level and socioeconomic status were greater in women than in men. The reported epidemiological observations have an important implication for tailored risk prediction and management strategies which are required to reduce the burden of CVD in men and women.

References

1. Anderson RD, Pepine CJ. Gender differences in the treatment for acute myocardial infarction: bias or biology? Circulation. 2007;115:823–6.
2. Appelros P, Stegmayr B, Terént A. Sex differences in stroke epidemiology: a systematic review. Stroke. 2009;40:1082–90.
3. Backholer K, Peters SAE, Bots SH, Peeters A, Huxley RR, Woodward M. Sex differences in the relationship between socioeconomic status and cardiovascular disease: a systematic review and meta-analysis. J Epidemiol Community Health. 2017;71:550–7.
4. Bello N, Mosca L. Epidemiology of coronary heart disease in women. Prog Cardiovasc Dis. 2004;46:287–95.
5. Bhatia RS, Tu JV, Lee DS, Austin PC, Fang J, Haouzi A, Gong Y, Liu PP. Outcome of heart failure with preserved ejection fraction in a population-based study. N Engl J Med. 2006;355:260–9.
6. Bots SH, Peters SAE, Woodward M. Sex differences in coronary heart disease and stroke mortality: a global assessment of the effect of ageing between 1980 and 2010. BMJ Glob Health. 2017;2:e000298.
7. Boggia J, Thijs L, Hansen TW, Li Y, Kikuya M, Björklund-Bodegård K, Richart T, Ohkubo T, Jeppesen J, Torp-Pedersen C, Dolan E, Kuznetsova T, Olszanecka A, Tikhonoff V, Malyutina S, Casiglia E, Nikitin Y, Lind L, Maestre G, Sandoya E, Kawecka-Jaszcz K, Imai Y, Wang J, Ibsen H, O'Brien E, Staessen JA, International Database on Ambulatory blood pressure in relation to Cardiovascular Outcomes Investigators. Ambulatory blood pressure monitoring in 9357 subjects from 11 populations highlights missed opportunities for cardiovascular prevention in women. Hypertension. 2011;57:397–405.
8. Crea F, Battipaglia I, Andreotti F. Sex differences in mechanisms, presentation and management of ischaemic heart disease. Atherosclerosis. 2015;241:157–68.
9. Fang MC, Singer DE, Chang Y, Hylek EM, Henault LE, Jensvold NG, Go AS. Gender differences in the risk of ischemic stroke and peripheral embolism in atrial fibrillation: the AnTicoagulation and Risk factors In Atrial fibrillation (ATRIA) study. Circulation. 2005;112:1687–91.
10. GBD 2013 Mortality and Causes of Death Collaborators. Global, regional, and national age-sex specific all-cause and cause-specific mortality for 240 causes of death, 1990–2013: a systematic analysis for the Global Burden of Disease Study 2013. Lancet. 2015;385:117–71.
11. George J, Rapsomaniki E, Pujades-Rodriguez M, Shah AD, Denaxas S, Herrett E, Smeeth L, Timmis A, Hemingway H. How does cardiovascular disease first present in women and men? Incidence of 12 cardiovascular diseases in a contemporary cohort of 1,937,360 people. Circulation. 2015;132:1320–8.
12. Graciani A, León-Muñoz LM, Guallar-Castillón P, Rodríguez-Artalejo F, Banegas JR. Cardiovascular health in a Southern Mediterranean European country: a Nationwide population-based study. Circ Cardiovasc Qual Outcomes. 2013;6:90–8.
13. Huffman MD, Berry JD, Ning H, Dyer AR, Garside DB, Cai X, Daviglus ML, Lloyd-Jones DM. Lifetime risk for heart failure among white and black Americans cardiovascular lifetime risk pooling project. J Am Coll Cardiol. 2013;61:1510–7.
14. Janković J, Marinković J, Stojisavljević D, Erić M, Vasiljević N, Janković S. Sex inequalities in cardiovascular health: a cross-sectional study. Eur J Pub Health. 2016;26:152–8.
15. Kannel WB, Schwartz MJ, McNamara PM. Blood pressure and risk of cardiovascular disease: the Framingham Heart Study. Chest. 1969;56:43–51.
16. Kardys I, Vliegenthart R, Oudkerk M, Hofman A, Witteman JC. The female advantage in cardiovascular disease: do vascular beds contribute equally? Am J Epidemiol. 2007;166:403–12.
17. Lanza GA, Crea F. Primary coronary microvascular dysfunction: clinical presentation, pathophysiology, and management. Circulation. 2010;121:2317–25.
18. Leening MJ, Ferket BS, Steyerberg EW, Kavousi M, Deckers JW, Nieboer D, Heeringa J, Portegies ML, Hofman A, Ikram MA, Hunink MG, Franco OH, Stricker BH, Witteman JC, Roos-Hesselink JW. Sex differences in lifetime risk and first manifestation of cardiovascular disease: prospective population based cohort study. BMJ. 2014;349:g5992.
19. Levy D, Larson MG, Vasan RS, Kannel WB, Ho KK. The progression from hypertension to congestive heart failure. JAMA. 1996;275:1557–62.
20. Lloyd-Jones DM, Larson MG, Beiser A, Levy D. Lifetime risk of developing coronary heart disease. Lancet. 1999;353:89–92.
21. Magnussen C, Niiranen TJ, Ojeda FM, Gianfagna F, Blankenberg S, Njølstad I, Vartiainen E, Sans S, Pasterkamp G, Hughes M, Costanzo S, Donati MB, Jousilahti P, Linneberg A, Palosaari T, de Gaetano G, Bobak M, den Ruijter HM, Mathiesen E, Jørgensen T, Söderberg S, Kuulasmaa K, Zeller T, Iacoviello L, Salomaa V, Schnabel RB, BiomarCaRE Consortium. Sex differences and similarities in atrial fibrillation epidemiology, risk factors, and mortality in community cohorts: results from the BiomarCaRE consortium (Biomarker for Cardiovascular Risk Assessment in Europe). Circulation. 2017;136:1588–97.
22. Manteuffel M, Williams S, Chen W, et al. Influence of patient sex and gender on medication use, adherence, and prescribing alignment with guidelines. J Womens Health (Larchmt). 2014;23:112–9.
23. Meta-analysis Global Group in Chronic Heart Failure (MAGGIC). The survival of patients with heart failure with preserved or reduced left ventricular ejection fraction: an individual patient data meta-analysis. Eur Heart J. 2012;33:1750–7.
24. Mosca L, Barrett-Connor E, Wenger NK. Sex/gender differences in cardiovascular disease prevention: what a difference a decade makes. Circulation. 2011;124:2145–54.
25. Nicholls SJ, Wolski K, Sipahi I, Schoenhagen P, Crowe T, Kapadia SR, Hazen SL, Tuzcu EM, Nissen SE. Rate of progression of coronary atherosclerotic plaque in women. J Am Coll Cardiol. 2007;49:1546–51.
26. Niewada M, Kobayashi A, Sandercock PA, Kamiński B, Członkowska A, International Stroke Trial Collaborative Group. Influence of gender on baseline features and clinical outcomes among 17,370 patients with confirmed ischemic stroke in the international stroke trial. Neuroepidemiology. 2005;24:123–8.
27. Olsen TS, Dehlendorff C, Andersen KK. Sex-related time-dependent variations in post-stroke survival: evidence of a female stroke survival advantage. Neuroepidemiology. 2007;29:218–25.
28. Owan TE, Redfield MM. Epidemiology of diastolic heart failure. Prog Cardiovasc Dis. 2005;47:320–32.

29. Owan TE, Hodge DO, Herges RM, Jacobsen SJ, Roger VL, Redfield MM. Trends in prevalence and outcome of heart failure with preserved ejection fraction. N Engl J Med. 2006;355:251–9.
30. Piepoli MF, Hoes AW, Agewall S, Albus C, Brotons C, Catapano AL, Cooney MT, Corrà U, Cosyns B, Deaton C, Graham I, Hall MS, Hobbs FD, Løchen ML, Löllgen H, Marques-Vidal P, Perk J, Prescott E, Redon J, Richter DJ, Sattar N, Smulders Y, Tiberi M, van der Worp HB, van Dis I, Verschuren WM, Authors/Task Force Members. European Guidelines on cardiovascular disease prevention in clinical practice: The Sixth Joint Task Force of the European Society of Cardiology and Other Societies on Cardiovascular Disease Prevention in Clinical Practice. Developed with the special contribution of the European Association for Cardiovascular Prevention and Rehabilitation (EACPR). Eur Heart J. 2016;37:2315–81.
31. Reeves MJ, et al. Sex differences in stroke: epidemiology, clinical presentation, medical care, and outcomes. Lancet Neurol. 2008;7:915–26.
32. Regitz-Zagrosek V, Kararigas G. Mechanistic pathways of sex differences in cardiovascular disease. Physiol Rev. 2017;97:1–37.
33. Remme WJ, McMurray JJ, Rauch B, Zannad F, Keukelaar K, Cohen-Solal A, Lopez-Sendon J, Hobbs FD, Grobbee DE, Boccanelli A, Cline C, Macarie C, Dietz R, Ruzyllo W. Public awareness of heart failure in Europe: first results from SHAPE. Eur Heart J. 2005;26:2413–21.
34. Simon M, Boutouyrie P, Narayanan K, Gaye B, Tafflet M, Thomas F, Guibout C, Périer MC, Pannier B, Jouven X, Empana JP. Sex disparities in ideal cardiovascular health. Heart. 2017;103:1595–601.
35. Vaccarino V, Badimon L, Corti R, de Wit C, Dorobantu M, Hall A, Koller A, Marzilli M, Pries A, Bugiardini R, Working Group on Coronary Pathophysiology and Microcirculation. Ischaemic heart disease in women: are there sex differences in pathophysiology and risk factors? Position paper from the working group on coronary pathophysiology and microcirculation of the European Society of Cardiology. Cardiovasc Res. 2011;90:9–17.
36. Vitale C, Fini M, Spoletini I, Lainscak M, Seferovic P, Rosano GM. Under-representation of elderly and women in clinical trials. Int J Cardiol. 2017;232:216–21.
37. Wilkins E, Wilson L, Wickramasinghe K, Bhatnagar P, Leal J, Luengo-Fernandez R, Burns R, Rayner M, Townsend N. European cardiovascular disease statistics 2017. Brussels: European Heart Network; 2017.

Biostatistics in Cardiovascular Research with Emphasis on Sex-Related Aspects

5

Marieke M. ter Wee and Birgit I. Lissenberg-Witte

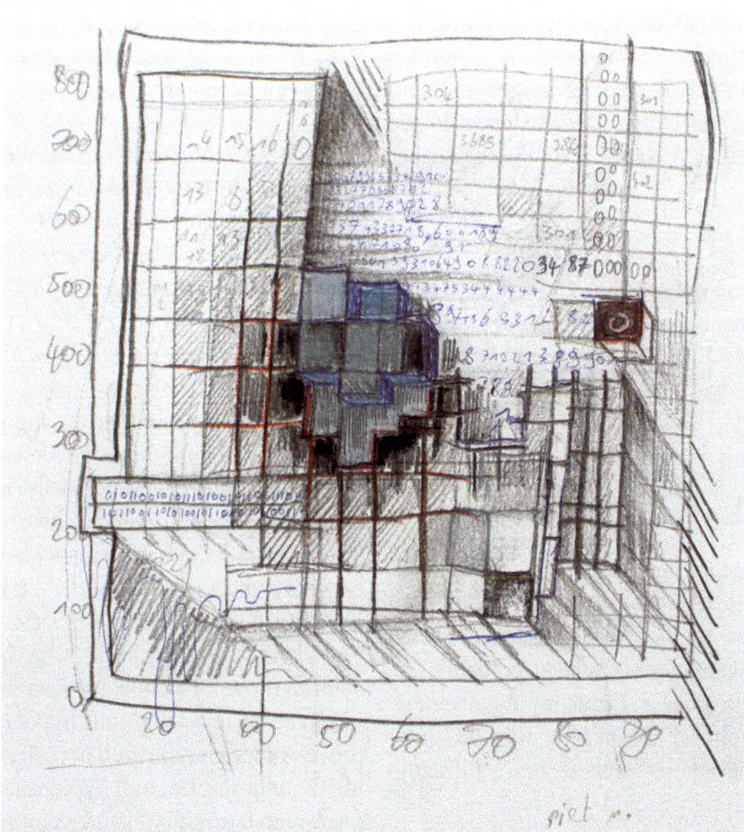

Biostatistics in cardiology. Artwork by Piet Michiels, Leuven, Belgium.

M. M. ter Wee (✉) · B. I. Lissenberg-Witte
Department of Epidemiology and Biostatistics, VU University Medical Center, Amsterdam, The Netherlands
e-mail: m.terwee@vumc.nl

Abstract

Research on sex differences related to cardiovascular dysfunction has become a topic of interest in the last decade. Although scientific

P. L. M. Kerkhof, V. M. Miller (eds.), *Sex-Specific Analysis of Cardiovascular Function*, Advances in Experimental Medicine and Biology 1065, https://doi.org/10.1007/978-3-319-77932-4_5

research has been carried out since ancient times, we still may struggle with performing scientific research in the best way to achieve the highest quality data and solid conclusions. In this chapter, every step of scientific research is explained: from formulating the research question and hypotheses to analyzing the collected data to interpreting and reporting the results. Several fundamental biostatistical techniques—such as the independent samples *t*-test, the chi-square test, the log-rank test, and different regression models—are presented. In addition, methods that can deal with variables influencing the association of interest are discussed. All examples are focused on investigating sex differences in cardiac outcomes, but this chapter is written in such a way that it easily translates to other fields of medical research on every disease or health state.

Keywords
Independent samples *t*-test · Chi-square test · Log-rank test · Regression models · Confounding · Effect modification

Introduction

The term *epidemiology* comes from the Greek words Λογος (logos, or "study"), επι (epi, or "on"), and δεμος (demos, or "people"). A study starts with a research idea, which is probably the easiest part of the process. But then? What are the next steps to follow? Converting the idea into effective and strong research is more difficult. Which study design would be most suitable? How should the participants be selected? If all data are collected, what is the best statistical technique to analyze them? Finally, how should the results be interpreted? Many questions arise when conducting scientific research, so let us take one step at a time.

Step 1: Formulate the Research Question and Hypothesis

All studies start with a research idea that needs to be investigated—for example, whether sex differences are present in treatment responses for left-ventricular ejection fraction (LVEF). This idea is translated into a research question, which can be developed according to the *PICO(t)* model:

P = The patient, research population, or research problem.
I = The intervention, exposure, or prognostic factor.
C = The comparison.
O = The outcome measure.
(t) = The time period, which is sometimes added but not necessary.

The research question to the idea above could be as follows: *What is the difference in the effect of therapy A (intervention [I]) compared with placebo/standard care (comparison [C]) between male and female patients with a first myocardial infarction (MI) (patient [P]) on the LVEF (outcome [O]) after 6 weeks of treatment (time [t])?*

Based on the research question, the null and the alternative hypotheses are defined. In general, the *null hypothesis* (denoted by H0) states that there is no (true) difference between the groups of interest or that there is no association between the exposure and the outcome, while the *alternative hypothesis* (denoted by H1) states the opposite. The alternative hypothesis is the hypothesis you want to prove; the collected data will provide the evidence to reject the null hypothesis. But until proven otherwise, the null hypothesis is the working hypothesis. The null hypothesis for the example research question is that there is no difference of effect of therapy A compared with the placebo/standard care between males and females, and the alternative hypothesis is that there is a difference between males and females in terms of the effect of therapy A.

Step 2: Choose the Research Design
To obtain an answer to the research question, a study must be designed. Many *research designs* exist, all with advantages and disadvantages. To investigate a sex difference in treatment response, four possible designs could be chosen from:

1. A randomized controlled trial (RCT), in which allocation of the treatment (e.g., therapy A or placebo/standard care) given to the patients is defined by randomization. This type of study is highly controlled and therefore leads to less bias in the population, the exposure and the outcome of the measurements.
2. A prospective cohort study, in which, for example, LVEF is monitored over time in a group of patients with a history of MI and including patients receiving therapy A as well as patients receiving standard care. As treatment is not randomly assigned, the characteristics of the patients receiving therapy A or standard care can differ.
3. A retrospective cohort study, in which, for example, treatment response in a group of patients who received therapy A is assessed retrospectively (e.g., by chart review) and compared with the treatment response in a group of patients who received standard care. In a retrospective cohort study, the outcome (i.e., LVEF at 6 months) is already known at the time the investigator initiates the study, while in a prospective cohort study, the participants still must be followed over time to assess their outcome.
4. A systematic review study, in which treatment response in males and females to therapy A compared with the placebo/standard care is retrieved from selected publications and combined to obtain one overall estimate of the difference between males and females.

The RCT or cohort studies all provide information on the causality of sex differences regarding the effect of therapy A and placebo/standard care. However, a cohort study is more prone to indication bias, while an RCT is highly controlled. A retrospective cohort study costs less than a prospective cohort study, but the exposure data cannot always be retrieved reliably from the patient charts, and information on confounding variables usually is not complete. A systematic review, by means of a meta-analysis, requires homogeneity of all publications included, that is, they all must consider the same placebo/control therapy, the same time period, etc.

Step 3: Prepare the Database
During the study, a lot of data are collected that need to be entered into a database. Before analyzing the data, their validity and correctness must be checked. Can very large/small values be considered outliers or are these just typos? Are missing data really missing or just not applicable? Have all privacy data (surname, date of birth, hospital identification number, etc.) been deleted? Although this step is time consuming, if data are not screened carefully, then highly misleading results could be found leading to incorrect conclusions as all analyses of the data in the study depend on the quality of the prepared database.

Step 4: Analyze and Interpret the Data
Finally, it is time to statistically analyze the data to obtain an answer to the research question. After basic statistical terminology is introduced in section "Medical statistics... the basics", various methods are described in more detail in sections "Sex differences in a continuous outcome measure", "Sex differences in a dichotomous outcome measure", "Sex differences in a survival outcome measure", and "Variables influencing the association" for a continuous outcome, a dichotomous outcome, and a survival outcome measure.

All examples in this chapter are illustrated using the Statistical Package for the Social Sciences (SPSS) version 22 (IBM Corp., Armonk, NY). Other statistical software—such as SAS, Stata, and R—can be used equally well, but they are not illustrated here.

Step 5: Report the Results
After all statistical work is completed, the results must be reported in a manuscript, leading to a

scientific publication in an appropriate journal. Some tips and tricks for reporting are discussed in section "Reporting the results".

Medical Statistics... The Basics

An *independent variable*, or *determinant*, is a variable that can be manipulated or that is fixed, for example, being male or female. The *dependent variable*, or *outcome variable*, is the variable that is influenced by the independent variable. For our example, LVEF is the dependent variable and treatment (therapy A or placebo/standard care) is the independent variable. Throughout this chapter, we will consider three types of dependent variables:

- A *continuous* variable, which is measured along a continuum, such as cholesterol level.
- A *dichotomous* variable, which can only consist of two values, such as having diabetes mellitus or not.
- A *time-to-event (survival)* variable, which is a continuous variable with a time component, such as time to recurrent MI.

The independent variable can be dichotomous, categorical (i.e., with more than two groups) or continuous. The type of both the dependent and the independent variables in the research question determines which statistical method needs to be used.

Recall that the research question always requires a certain population, in other words, the P in the PICO(t) model. The ideal situation would be to collect the outcome measure and other variables on every single subject of the whole population. Unfortunately, that is not feasible in practice, so only a small sample of subjects is selected and measured in a study. Due to chance, different samples from the same population will provide different results. Therefore, to generalize the results to the population, the sampling variation (or *uncertainty*) must be taken into account. This area of statistics is called *inferential statistics*, as opposed to *descriptive statistics*, which is used to summarize and describe the outcome measure and other important data, numerically as well as graphically.

The *standard error* (SE) of the effect size is one way to measure the uncertainty, and this is equal to the *standard deviation* (SD) of the sampling distribution. The smaller the SE, the more certain we are that the effect size is a good estimate of the true population effect. The SE is used to construct a *confidence interval* (CI) with a certain precision, denoted by α and usually set to 0.05, which is equal to [effect size $- z_{\alpha/2} \times$SE; effect size + $z_{\alpha/2} \times$SE] for a certain value of $z_{\alpha/2}$. The smaller the α, the wider the CI: $z_{\alpha/2}$=1.96 for $\alpha = 0.05$, while $z_{\alpha/2}$=1.64 for $\alpha = 0.10$ and $z_{\alpha/2}$=2.59 for $\alpha = 0.01$. Of note, this general formula holds for most effect sizes, for example, the mean difference between two groups, but not for all. The CI indicates that there is a $(1-\alpha) \times 100\%$ probability (e.g., 95%) that the interval contains the (unknown!) population effect. This is not the same as saying that with probability $(1-\alpha) \times 100\%$ the population effect lies within the CI, because the population effect is fixed and the CI will vary between samples. In other words, if several independent random samples from the same population are drawn, and the 95% CI for each of these samples is calculated, on average 19 of every 20 samples (i.e., 95%) would contain the true population effect and one of every 20 samples (5%) would not.

Another way to make an inference is to look at the probability that the observed effect is a result of chance under the assumption that the null hypothesis is true. This probability is called the *p-value* and it provides the evidence against the null hypothesis: the smaller the *p*-value, the stronger the evidence against the null-hypothesis. The null hypothesis is rejected in favor of the alternative hypothesis, if the *p*-value is smaller than a predefined *significance level*, which is again denoted by α and usually set to 0.05. If the *p*-value is greater than the significance level, the data does not provide evidence against the null hypothesis. This does not mean the null hypothesis is true! The *p*-value is largely based on the sample size: the larger the sample size, the smaller the *p*-value. If the included sample size in a study is small, statistical significance might not be reached, which could be the result of an underpowered sample.

Depending on the sign or direction of the alternative hypothesis, the hypothesis test is either one-sided or two-sided. In case the direction of the comparison in the alternative hypothesis is specified, a *one-sided* p-value must be reported; in the case of an unspecified direction, a *two-sided p*-value must be reported. For example, to test the alternative hypothesis that males have a smaller LVEF than females, a one-sided *p*-value is used; to test the alternative hypothesis that there is a difference in the LVEF between males and females, a two-sided *p*-value is used.

Both the 95% CI and the significance level in hypothesis testing are based on the same arbitrarily chosen 0.05 threshold. They are related: if the 95% CI does not contain the null value, the *p*-value is below 0.05, and vice versa; if the *p*-value is below 0.05, the 95% CI does not contain the null value. Although this relation exists, we recommend reporting the exact *p*-value (and not just whether it was smaller than α or not) as well as the 95% CI. Large samples are more likely to provide strong evidence against a null hypothesis in the form of a small *p*-value and a narrow 95% CI, but this does not provide any clinical relevance of the observed effect. All hypotheses in the following sections are tested at a significance level $\alpha = 0.05$.

Sex Differences in a Continuous Outcome Measure

In a fictional study, the differences between males and females regarding cholesterol level—a continuous outcome—are assessed with the following research question: *Do cholesterol levels differ between males and females?*

The null and alternative hypotheses for this research question are:

- H0: mean cholesterol level is equal for males and females.
- H1: mean cholesterol level is not equal for males and females.

Put differently, H0 says that the mean difference in cholesterol levels between males and females is equal to 0, while H1 says that the mean difference in cholesterol levels is unequal to 0. The mean difference is referred to as the *effect size*.

Continuous outcomes can be either normally or non-normally distributed, and this distribution is of importance to determine which statistical test is needed. Thus, if the outcome measure is continuous, the first step is to check, for each group separately, whether it is normally distributed. Several tests exist to assess normality; however, for large samples, these tests will almost always reject the null hypothesis that the variable is normally distributed, even for very small deviations from normality. Conversely, it can be impossible to detect non-normality in small samples. Therefore, one should not solely rely on formal normality tests but also examine the distribution visually through a histogram or QQ-plot. In addition, normality of a non-negative variable can be checked using the rule of thumb that approximately 97.5% of the observed values will be greater than the mean minus twice the SD (mean $- 2 \times$ SD). Is the mean minus twice the SD negative? Then it is highly likely that the variable is not normally distributed. Note that this rule of thumb does not hold for variables skewed to the left. Moreover, the mean and median can be compared because for a normally distributed variable the mean and median would be approximately the same. Because many statistical tests and estimators are robust against small to moderate deviations from normality, one should not worry about it too much unless the data are clearly skewed or exhibit extreme outliers.

Independent Samples *t*-Test

In this fictional study, males ($n = 608$) had a mean cholesterol level of 5.42 mmol/L (SD 0.65), while females ($n = 282$) had a mean cholesterol level of 6.06 mmol/L (SD 1.0), leading to a mean difference of 0.64 mmol/L between males and females. First, the distribution of cholesterol level must be checked. The distribution is approximately normal,

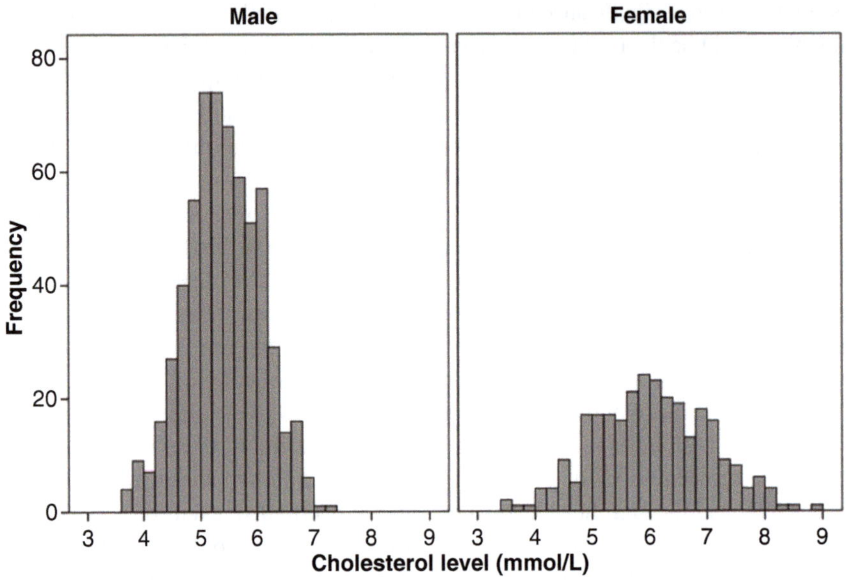

Fig. 5.1 Histogram of cholesterol level (mmol/L) in a fictional study of 608 men (left panel) and 282 women (right panel)

Table 5.1 SPSS output of the independent samples *t*-test comparing cholesterol level (mmol/L) between 608 males and 282 females

		Levene's test for equality of variances		*t*-test for equality of means						
									95% CI of the difference	
		F	Sig.	*t*	*df*	Sig. (2-tailed)	Mean difference	SE difference	Lower	Upper
Cholesterol level (mmol/L)	Equal variances assumed	69.236	0.000	−11.359	888	0.000	−0.639	0.056	−0.749	−0.528
	Equal variances not assumed			−9.770	394.362	0.000	−0.639	0.065	−0.767	−0.510

as can be visually judged from Fig. 5.1, as well as through the fact that the mean and median are nearly the same (median in males = 5.39 mmol/L, median in females = 6.02 mmol/L) and that the mean minus twice the SD is positive both for males (SD = 0.65; mean − 2 × SD = 5.12) and for females (SD = 1.00; mean − 2 × SD = 5.06). Since cholesterol levels are fairly normally distributed, the *independent samples t-test* can be used to test whether the mean difference of 0.64 mmol/L between the sexes is significant. The SPSS output of the independent samples *t*-test is given in Table 5.1.

In Table 5.1, two different rows are displayed: the first is labeled "Equal variances assumed", and the second is labeled "Equal variances not assumed." In addition to determining normality of the outcome in each group separately, the independent samples *t*-test also assumes that the variances (i.e., the squared SD) in both groups are equal. The *Levene's Test for Equality of the Variances*, the first two columns of the table, tests this assumption. The null hypothesis states that the variances are equal versus the alternative hypothesis that they are not. The p-value of the Levene's test is smaller than the significance level of 0.05, thus favors the alternative hypothesis. The p-value for the equality of means should therefore be read from the second row. SPSS always reports p-values rounded at 3 decimals, but a p-value can never be exactly equal to 0. Thus, whenever SPSS reports 0.000 as a p-value, the p-value is reported as being smaller than 0.001. The mean difference, its SE, and the corresponding 95% CI are also reported in the SPSS output. Note that the mean difference and its 95% CI, as given by SPSS, have opposite signs as to what has been manually calculated (i.e., 0.64 mmol/L). This is because SPSS always calculates the mean difference as the mean in group 1 (males) minus the mean in group 2 (females). Since the p-value is smaller than the significance level, and as a consequence the value 0 (the mean difference under the null hypothesis) is not included in the 95% CI, we can conclude with 95% certainty that mean cholesterol levels differ between males and females, females having, on average, a 0.64 mmol/L greater cholesterol level than males (95% CI [0.53; 0.75], $p < 0.001$).

Non-normal Outcome

The independent samples t-test can only be used when the outcome variable is normally distributed in each group separately. If this assumption is violated, there are two possibilities to compare the outcome variable. First, try to transform the outcome variable with a function hoping that the transformed outcome variable is normally distributed. A widely used transformation for variables skewed to the right is logarithmic transformation. Note that in this case, the mean difference reported in the SPSS output is the mean difference in the transformed outcome. To retrieve the difference on the original scale, the difference must be back-transformed by the reciprocal transformation (e.g., the exponential after using the ln-transformation). Second, a non-parametric test can be used. A non-parametric test does not compare the mean of both groups but more generally the distribution of the outcome variable in both groups. The non-parametric version of the independent sample t-test is the *Mann-Whitney U test*, also called the *Wilcoxon rank-sum test*. The Mann-Whitney U test does not compare the mean of the observed values but rather the mean rank of the observed values in each group. The smallest observed value is assigned rank 1, and the second smallest rank 2 up to the largest value, which is assigned rank n (the total sample size). However, the mean rank has no clinical interpretation, and it gives no information about the values and spread. Therefore, for non-normally distributed variables, the median and range (or interquartile range)—instead of the mean and SD—are generally reported to describe the outcome measure in each group.

Linear Regression Model

As will become more evident later, the mean difference in cholesterol levels between males and females can also be investigated by way of a *linear regression* model. For any independent variable x and dependent variable Y, in a linear regression model the relation between x and Y is modeled as follows:

$$Y = b_0 + b_1 x + \varepsilon.$$

The parameters b_0 and b_1 are called the *regression coefficients*, and ε (epsilon) is called the *error term*, which captures all factors that influence Y other than x. The regression coefficients are estimated by the *least squares estimation method*. This method minimizes the squared difference between the observed values of Y (generally denoted by $Y_1, Y_2, \ldots, Y_n$ for a sample of size n) and a straight line with *intercept* b_0 and *slope* b_1. The interpretation of the regression coefficients is derived from the linear function as follows:

- b_0 is the mean outcome for a subject with $x = 0$ [i.e., the intercept of the line].
- b_1 is the mean difference in outcome between two subjects who differ by one unit in the independent variable [i.e., the slope of the line].

The linear regression model assumes that the error-variable ε is normally distributed with mean 0 and that its variance does not depend on x (called *homoscedasticity*). From the first assumption, it follows that the linear regression model can be translated to

$$\text{mean } Y = b_0 + b_1 x.$$

Hence, the linear regression model to assess the difference in cholesterol levels between males and females is equal to

$$\text{mean cholesterol} = b_0 + b_1 \times \text{sex}.$$

If males are coded as 0 and females as 1 in the database, then, by the general interpretation of the regression coefficients:

- b_0 is the mean cholesterol level of males.
- b_1 is the mean difference in cholesterol levels between males and females.

The null and alternative hypotheses for this research question can be translated to H0: $b_1 = 0$ and H1: $b_1 \neq 0$. Table 5.2 gives the results of the linear regression analysis: the unstandardized coefficient is equal to the mean difference in Table 5.1, apart from the minus sign, resulting from the fact that in linear regression males are set as the reference category, while in the independent samples *t*-test females are set as the reference category. In SPSS, regression coefficients can be found in the column labeled "B": $b_1$in the row labeled "Sex" and b_0 in the row labeled "(Constant)". From Table 5.2 we can also conclude (with 95% certainty) that mean cholesterol level differs between males and females, with females having, on average, a 0.64 mmol/L greater cholesterol level than males (95% CI [0.53; 0.75], $p < 0.001$).

The assumptions regarding the error-variable ε still must be tested. However, this variable is unobserved and can only be estimated. For each subject i, the value of the outcome variable Y can be "predicted" by $\hat{Y}_i = \hat{b}_0 + \hat{b}_1 x_i$, where x_i is the value of the independent variable of subject i. The difference between the predicted value $\widehat{Y_i}$ and the observed value Y_i, is called the *residual* (for subject i) and denoted by r_i. Then, the assumptions of the linear regression model can be translated to assumptions regarding the residuals $r_1, r_2, \ldots, r_n$. Thus, it must be checked whether they are normally distributed with mean 0 and whether their variance does not depend on the independent variables $x_1, x_2, \ldots, x_n$. The latter assumption implies that the variance of the residuals is equal in males and females. In SPSS, the residuals can be saved after a linear regression analysis, which are then added to the database as a new variable. Normality of the residuals can then be investigated through a histogram (Fig. 5.2), while equality of the variances can be tested by way of the Levene's test (as part of an independent samples *t*-test on the residuals [Table 5.3]). If normality of the residuals does not hold, transforming the outcome variable might be a possible solution. If homoscedasticity is rejected, more complex techniques exist, but these lie beyond the scope of this chapter.

Standardization

Normally distributed variables are sometimes *standardized* so that they have a mean 0 and an SD of 1. For each subject i, the z-score of their

Table 5.2 SPSS output of a linear regression model comparing cholesterol level (mmol/L) between males and females

Model		Unstandardized coefficients		Standardized coefficients			95% CI for B	
		B	SE	Beta	t	Sig.	Lower bound	Upper bound
1	(Constant)	5.417	0.032		171.133	0.000	5.355	5.479
	Sex	0.639	0.056	0.356	11.359	0.000	0.528	0.749

Dependent variable: Cholesterol level (mmol/L)

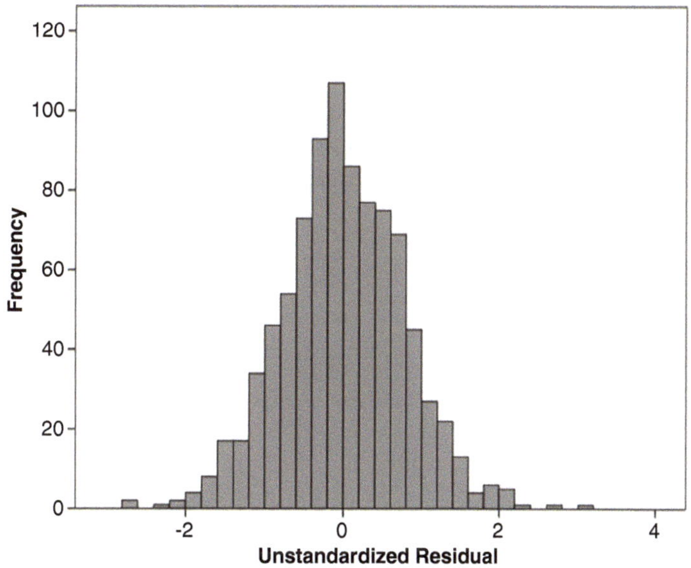

Fig. 5.2 Histogram of residuals in linear regression model comparing cholesterol level (mmol/L) between males and females

Table 5.3 Levene's test for equality of the variances

Levene's test for equality of variances	
F	Sig.
69.236	0.000

observed value y_i is computed via $z_i = (y_i - \mu)/\sigma$, where μ is the sample mean and σ is the sample SD. *Z*-scores denote the deviation from the mean in terms of the SD and lack a dimension, that is, a *z*-score of 0.5 means that the original value was 0.5 times the SD greater than the mean, and a *z*-score of −2 means that the original value was 2 times the SD smaller than the mean. They are frequently used to compare the outcome measure with reference values. More information on how, why, and when to use *z*-scores can be found in Mawad et al. [1].

Sex Differences in a Dichotomous Outcome Measure

An outcome, such as the prevalence of congestive heart failure (CHF), could also be dichotomous. When analyzing sex differences with respect to CHF, the following research question could be tested: *Does the prevalence of congestive heart failure (CHF) differ between males and females?*

Table 5.4 SPSS output of a 2 × 2 contingency table comparing the prevalence of CHF between males and females

		CHF					
		No		Yes		Total	
		n	%	*n*	%	*n*	%
Sex	Male	1192	93.1	147	6.9	2139	100.0
	Female	2148	94.0	136	6.0	2284	100.0
Total		4140	93.6	283	6.4	4423	100.0

The null and alternative hypotheses for this research question would then be as follows:

- H0: prevalence of CHF is equal for males and females.
- H1: prevalence of CHF is not equal for males and females.

A summary of the data is obtained with a *contingency table* (Table 5.4): the prevalence of CHF was 6.9% in males and 6.0% in females.

Several effect sizes can be calculated to compare the prevalence of CHF between males and females based on either the prevalence (or more generally the risk) itself in males and females (denoted by p_0 and p_1) or on the odds for CHF (defined as $p_0/(1 - p_0)$ and $p_1/(1 - p_1)$, respectively):

- *Risk difference* (RD) = $p_1 - p_0 = 0.060 - 0.069 = -0.009$, that is, 0.9% fewer females have CHF than males.
- *Relative risk* (RR) = $p_1/p_0 = 0.060/0.069 = 0.870$, that is, the prevalence of CHF in females is 0.87 times the prevalence in males.
- *Odds ratio* (OR) $= \frac{p_1/(1-p_1)}{p_0/(1-p_0)} = \frac{(0.060/0.940)}{(0.069/0.931)} = 0.861$, that is, the odds for CHF in females is 0.861 times the odds in males.

The RD and RR are easier to interpret than the OR. However, some study designs, such as a case-control study, prevent us from computing the actual risk. In addition, ORs are easy to adjust for confounders, while the adjustment for an RR is much trickier. Note that for rare diseases (i.e., with a prevalence below 1%), the OR is approximately equal to the RR.

Both the RR and the OR and their corresponding 95% CI are obtained as additional output in SPSS (Table 5.5). Note that SPSS computes the RR for the outcome CHF = yes by dividing the risk (or prevalence) in males by the risk in females. The 95% CI for the RR with males as reference equals [1/1.446; 1/0.921] = [0.69; 1.1].

Table 5.5 SPSS output of the estimated RR and OR for CHF of males compared to females

		95% CI	
	Value	Lower	Upper
Odds Ratio for sex (male/female)	0.858	0.674	1.092
For cohort CHF = no	0.990	0.975	1.006
For cohort CHF = yes	1.154	0.921	1.446
N of Valid Cases	4423		

Pearson's Chi-Square Test

From the 95% CI, it can be concluded that the null hypothesis cannot be rejected. Equal risks between males and females indicate equal odds, so under H0, the RR and the OR are equal to 1 (and unequal to 1 under H1). Since the 95% CI for both the RR [0.69; 1.1] and the OR [0.67; 1.1] contain the value 1, it can be concluded that there is not enough evidence to prove a difference in CHF prevalence between males and females.

Note that the 95% CIs for the RR and OR are not symmetric around the effect size because the RR and OR are ratios and therefore do not follow a normal distribution. The ln(RR) and ln(OR) do, and their 95% CIs are based on the standard formula for CIs: [ln(RR) − 1.96 × $SE_{\ln(RR)}$; ln(RR) + 1.96 × $SE_{\ln(RR)}$] and [ln(OR) − 1.96 × $SE_{\ln(OR)}$; OR + 1.96 × $SE_{\ln(OR)}$]. The *p*-value is obtained by the *Pearson's chi-square test* (or χ^2-test; Table 5.6), which compares the number of observed subjects in each of the four cells of the contingency table with the expected number of subjects in these cells under the null hypothesis. The Pearson's chi-square test is only valid when the expected number of subjects is at least 5 in all cells (footnote a in Table 5.6). In the case of lower numbers, the (two-sided) *Fisher's exact test* must be used to analyze the association.

The conclusion with respect to the current research question is that there is not enough evidence to prove that the odds for, and consequently

Table 5.6 SPSS output of the Pearson's chi-square test comparing the prevalence of CHF between males and females

	Value	*df*	Asymp. sig. (2-sided)	Exact sig. (2-sided)	Exact sig. (1-sided)
Pearson Chi-square	1.554[a]	1	0.231		
Continuity correction[b]	1.404	1	0.236		
Likelihood ratio	1.554	1	0.213		
Fisher's exact test				0.219	0.118
Linear-by-linear association	1.554	1	0.213		
N of Valid Cases	14,423				

a. 0 cells (0.0%) have an expected count less than 5. The minimum expected count is 20.50.
b. Computed only for a 2 × 2 table.

the prevalence of, CHF differs between males and females (OR 0.86, 95% CI [0.67; 1.1], $p = 0.23$).

Logistic Regression Model

The equivalent for a linear regression model in case of a dichotomous outcome is the logistic regression model. It models the relationship between CHF and sex as follows:

$$\ln(\text{odds}) = b_0 + b_1 \times \text{sex},$$

where ln(odds) is the natural logarithm of the odds, called the *logit*.

The regression coefficients have a similar interpretation as in a linear regression model:

- b_0 is the ln(odds) for males.
- b_1 is the difference in ln(odds) between males and females.

Both b_0 and b_1 are "measured" at the level of the ln(odds), so that $\exp(b_0)$ is the odds within males, and $\exp(b_1)$ is the odds ratio for females compared with males. After all

$$b_1 = \ln(\text{odds}_F) - \ln(\text{odds}_M) = \ln(\text{odds}_F/\text{odds}_M) = \ln(\text{OR}).$$

Table 5.7 gives the SPSS output for the logistic regression model, from which again the conclusion can be drawn that there is not enough evidence to prove that the odds of CHF differs between males and females. As for linear regression, the regression coefficients b_0 and b_1 can be found in the column labeled "B" and the OR can be found in the column labeled "Exp(B)" and row labeled "Sex".

Sex Differences in a Survival Outcome Measure

In this section, the difference between males and females, with respect to their overall survival after being diagnosed with CHF, is studied. Patients that are alive at the end of the study are then *censored*; it is only known they survived until that date, but it remains uncertain when they would have died. Overall survival (OS) is defined as the time since diagnosis until death (of any cause). Sometimes the cause of death is related to the disease of interest; in that case, one speaks about disease-specific survival (DSS). In this case, patients who died of a non-disease-related cause are also censored at their date of death. For this fictional study, the research question would be: *Does time to death after CHF diagnosis differ between males and females?*

In case the outcome is a time-to-event outcome (or survival outcome), the so-called *survival function*, denoted by $S(t)$, or the *cumulative incidence function*, denoted by $F(t)$ and equal to $1 - S(t)$, is estimated. The null and alternative hypotheses for the research question are therefore:

- H0: survival function is equal for males and females.
- H1: survival is not equal for males and females.

The survival function is estimated by the *Kaplan-Meier curve* for males and females separately (Fig. 5.3). The 5- and 10-year OS in males was 15.3% (95% CI [6.3%; 22.3%]) and 5.1% (95% CI [0.0%; 11.1%]) and in females was 29.4% (95% CI [19.6%; 39.2%]) and 15.3% (95% CI [5.5%; 25.1%]), respectively, as can be

Table 5.7 SPSS output of a logistic regression model for the association between CHF and sex

								95% CI for EXP(B)	
		B	SE	Wald	*df*	Sig.	Exp(B)	Lower	Upper
Step 1	Sex	−0.153	0.123	1.551	1	0.213	0.858	0.674	01.092
	Constant	−2.606	0.085	930.034	1	0.000	0.074		

Variable(s) entered on step 1: Sex

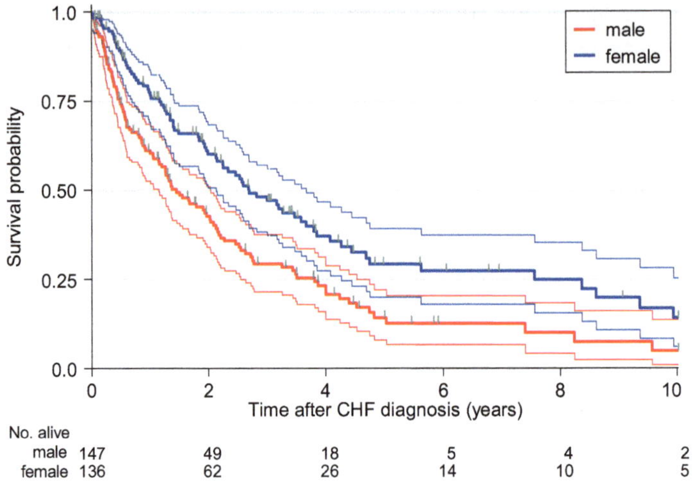

Fig. 5.3 Kaplan-Meier estimate of the survival probability after CHF diagnosis separately for males and females

read from the *survival table* (relevant parts are listed in Table 5.8). That is, 15.5% of the males and 29.4% of the females are still alive 5 years after their CHF diagnosis. Moreover, the cumulative incidence of death after 5 years was 85.7% (95% CI [77.7%; 95.7%] and 70.6% (95% CI [60.8%; 80.4%]) in males and females, respectively. The 95% CI is computed using the standard formula for CIs: [estimated probability − 1.96 × SE; estimated probability +1.96 × SE]. Note that the lower bound of the 95% of the survival probability can never be below 0.0% and the upper bound can never be above 100.0%.

Log-Rank Test

Equality of the survival curves is tested with the *log-rank test*. Since the log-rank test does not have any assumptions with respect to the distribution of the time-to-event outcome (or, equivalently, the shape of the survival/cumulative incidence function), it is also a non-parametric test. Table 5.9 gives the SPSS output of the log-rank test, from which it can be concluded that time to death after CHF diagnosis between males and females differs significantly: males die quicker after a CHF diagnosis than females ($p < 0.001$).

Cox Proportional Hazards Model

The Kaplan-Meier curve estimates the survival function directly in one glance. However, although the survival probabilities (or cumulative incidence) can be reported at one or more time points for males and females separately, it does not provide a single measure that can be considered as the effect size. To quantify the overall difference in survival between the two sexes, the *Cox proportional hazards model* (usually shortened to *Cox regression model*) can be used. This regression model models the *hazard function* $h(t)$ over time. This function is mathematically related to the survival function $S(t)$; it is the derivative of the cumulative hazard function $H(t) = -\ln(S(t))$. If the hazard function of males is denoted by $h_0(t)$ and that of females by $h_1(t)$, the Cox regression model states that:

$$\ln(h_1(t)/h_0(t)) = b_1 \times \text{sex},$$

or equivalently that

$$\ln(h_1(t)) = \ln(h_0(t)) + b_1 \times \text{sex}.$$

The ratio of the two hazard functions, $h_1(t)/h_0(t)$, is called the *hazard ratio* (HR) and is assumed to be constant over time. In other words, the hazard functions $h_0(t)$ and $h_1(t)$ are

Table 5.8 Part of the SPSS survival table stratified for males and females

Sex		Time	Status	Cumulative proportion surviving at the time		N of Cumulative events	N of Remaining cases
				Estimate	SE		
Male	65	0.918	Died	0.612	0.041	55	82
	66	0.951	Censored			55	81
	67	1.006	Died	0.605	0.041	56	80
	137	4.816	Died	0.143	0.036	106	10
	138	5.003	Censored			106	9
	145	9.580	Died	0.051	0.031	110	2
	146	10.000	Censored			110	1
	147	10.000	Censored			110	0
Female	44	0.977	Died	0.776	0.037	28	92
	45	1.001	Died	0.767	0.038	29	91
	118	4.738	Died	0.294	0.050	73	18
	119	4.810	Censored			73	17
	120	4.954	Censored			73	16
	121	5.599	Censored			73	15
	122	5.611	Died	0.275	0.050	74	14
	131	9.938	Died	0.143	0.050	79	5
	132	10.000	Censored			79	4
	133	10.000	Censored			79	3
	134	10.000	Censored			79	2
	135	10.000	Censored			79	1
	136	10.000	Censored			79	0

Table 5.9 SPSS output of the log-rank test comparing time to death after CHF diagnosis between males and females

	Chi-square	*df*	Sig.
Log rank (Mantel-Cox)	12.426	1	0.000

assumed to be proportional to each other, hence the proportionality in the name of the model. Table 5.10 gives the SPSS output of the Cox regression model; the regression coefficient b_1 can be found in the column labeled "B" and the HR in the column labeled "Exp(B)". From this table it can be concluded that time to death after CHF diagnosis between males and females differs significantly: males have worse OS after CHF diagnosis than females (HR = 0.60, 95% CI [0.45; 0.803], $p < 0.001$). The HR of 0.60 indicates that, at each time point, the probability of dying at that moment is estimated to be 40% lower in females than in males.

Similar to the RR and OR for a dichotomous outcome, the null hypothesis in the Cox regression model states that there is no difference between males and females in time to death after CHF diagnosis (i.e., the HR is equal to 1 under the null hypothesis). Since the 95% CI does not contain the value 1, it can be concluded that there was a significant difference based solely on the 95% CI. An HR greater than 1 means that the reference group (in our example males) has a shorter time to event (i.e., performs worst), while an HR smaller than 1 means that the reference group has a longer time to event (i.e., performs best).

The proportionality assumption of the Cox regression model must be checked to determine whether the Cox regression model can be applied. A first a way to investigate this assumption is by looking at the Kaplan-Meier curves. *Proportionality of the curves* means that the difference between the two curves increases at a constant rate; therefore, curves that cross are an example of violation of the proportionality assumption. The survival curve (or the cumulative

Table 5.10 SPSS output of Cox regression comparing time to death after CHF diagnosis between males and females

							95% CI for Exp(B)	
	B	SE	Wald	*df*	Sig.	Exp(B)	Lower	Upper
Sex	−0.518	0.148	12.171	1	0.000	0.596	0.446	0.797

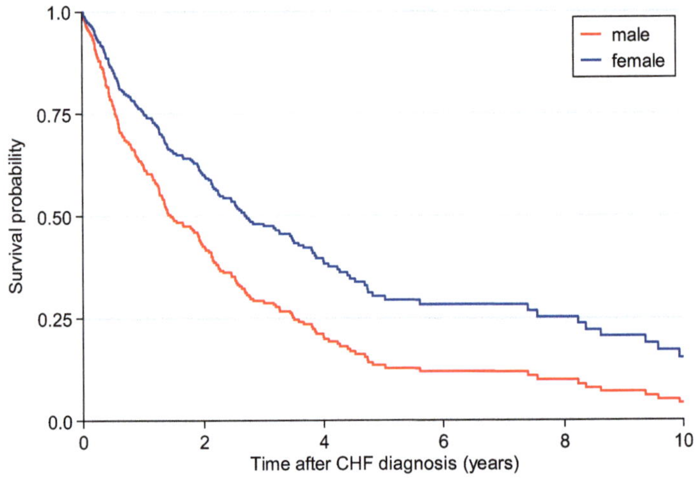

Fig. 5.4 Estimated survival function estimated by Cox regression model comparing time to death after CHF diagnosis between males and females

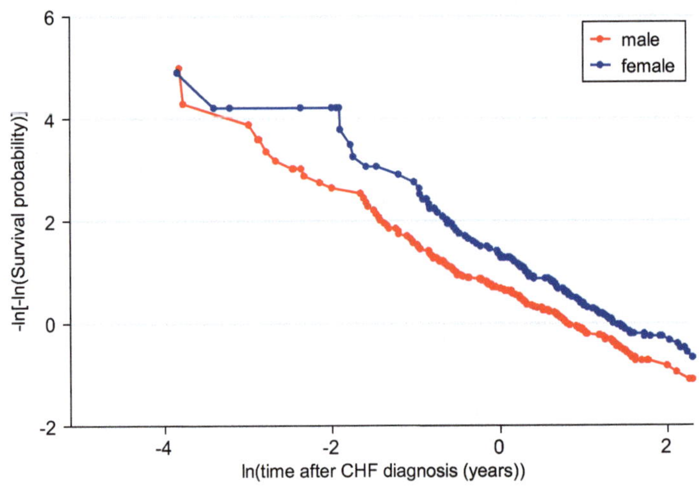

Fig. 5.5 Log–log plot to check the proportionality assumption in a Cox regression model with sex as covariate

incidence curve), as estimated by the Cox regression model, gives a good indication of the meaning of this assumption (Fig. 5.4).

Note that the Kaplan-Meier curves will never be exactly equal to the estimated curves of the Cox regression model because the Cox regression model forces the curves to be proportional, while the Kaplan-Meier curves does not impose any constraints or assumptions. The difference in Kaplan-Meier curves becomes increasingly greater for the first 5 years, and thereafter it seems to remain constant. Thus, proportionality might hold for the first 5 years but not thereafter.

Another way to visually inspect the proportionality assumption is by way of *log–log plot*. In this plot, ln(t) is plotted on the x-axis and $-\ln[-\ln(S(t))]$ is plotted on the y-axis for males and females separately (Fig. 5.5). Parallel curves are indicative for proportionality and crossing curves or converging or diverging curves are not. As the curves are parallel, we can conclude from the log–log plot that proportionality might hold.

To test the proportionality assumption, the regression coefficient b_1 is modeled to change over time. In SPSS, this is referred to as a "Cox regression model with a time-dependent determinant." There are several ways to let b_1 change over time. The easiest method is to set $b_1(t) = c_0 + c_1 d_T(t)$, where d is 0 for all $t \leq T$ and d is 1 for all $t > T$. We could choose $T = 5$ because after 5 years there might not be a difference between males and females, while in the first 5 years the survival probabilities decrease more rapidly in males than in females. Table 5.11 lists the SPSS output of this analysis. The first row in the output gives the estimate for c_0 and the second row for c_1. The hypotheses tested for c_1 are as follows:

- H0: $c_1 = 0$, that is, b_1 does not depend on time.
- H1: $c_1 \neq 0$, that is, b_1 does depend on time.

H0 states that the proportionality assumption holds, while H1 states it does not hold. Since the p-value in the second row ($p = 0.966$) is greater than 0.05, it can be concluded that proportionality seems to be a valid assumption.

If the null hypothesis would be rejected, that is, proportionality does not hold, $\exp(c_0)$ is the HR for the first 5 years (because $T = 5$), and $\exp(c_0 + c_1)$ is the HR for after 5 years. The 95% CI for the later period can be obtained by recoding the time-dependent covariate `T_COV_` in SPSS and setting d to 0 for all $t > T$ and d to 1 for all $t \leq T$.

For a quick reference to help decide which type of statistical analyses needs to be performed, Table 5.12 can be checked.

Table 5.11 SPSS output to test the proportionality assumption of the Cox regression model

	B	SE	Wald	*df*	Sig.	Exp(B)
Sex	−0.490	0.648	0.572	1	0.449	0.612
T_COV_*Sex	−0.029	0.666	0.002	1	0.966	0.972

Table 5.12 Overview of statistical analyses

Independent variable	Outcome		
	Continuous	Dichotomous	Time-to-event
Dichotomous	*Independent samples t-test*	*Chi-square test*	*Log-rank test/Cox regression*
Effect size	Difference in means	RR/RD/OR	Kaplan-Meier curve stratified by group (log-rank)/HR (Cox regression)
Assumptions	Normality of outcome variable in each group separately	All cells should have an expected count of at least 5	Proportional hazards assumption (Cox regression)
Continuous	*Linear regression*	*Logistic regression*	*Cox regression*
Effect size	Regression coefficient (difference in means)	Exponential of regression coefficient (OR)	Exponential of regression coefficient (HR)
Assumptions	1. Normal distribution of the residuals 2. Homoscedasticity 3. Linearity of the relation 4. Rule of thumb: one variable per ten *patients*	1. Linearity of the relation 2. Rule of thumb: one variable per ten patients *with* and ten *without* the outcome	1. Linearity of the relation 2. Rule of thumb: one variable per ten patients *with* an event and ten *without* an event

OR odds ratio, *RD* risk difference, *RR* relative risk, *HR* hazard ratio

Variables Influencing the Association

In the previous section, it was illustrated how to assess sex differences in a continuous, dichotomous and time-to-event dependent variable. However, sex differences can also play an important role in the association between two other variables. For example, males have greater cholesterol levels than females, which needs to be corrected for when analyzing the association between cholesterol levels and age. In addition, other variables influence differences between the sexes. For example, suppose that in the study sample females were on average older than males. In this case, the observed difference in cholesterol levels between the sexes might well be explained by the age difference. Ignoring this influencing effect leads to incorrect estimates of the effect of the independent variable. In this section, two types of influencing effects are considered: confounding and effect modification.

Confounding

A variable is called a *confounder* when part of the association between the outcome and determinant is explained by that variable (Fig. 5.6).

For a variable to be a true confounder, it must fulfill the following three assumptions or rules of thumb:

1. The confounder is significantly associated with the independent variable.
2. The confounder is significantly associated with the outcome variable.
3. The regression coefficient of the independent variable changes by more than 10% after adding the confounder.

Note that assumption no. 3 refers to the regression coefficient of the independent variable of interest, not to that of the potential confounder. Moreover, this rule of thumb is arbitrarily chosen: a change of 9.8% in a large regression coefficient might be clinically relevant, while a change of 12.0% in a small regression coefficient might not be clinically relevant. A confounder can be either dichotomous, categorical, or continuous.

To illustrate how to check for confounding, and to interpret the results after correcting for it, the association between cholesterol levels (outcome variable) and age (independent variable) is considered, and it is investigated whether sex is a confounder.

The uncorrected association between cholesterol levels and age can be assessed with a linear regression model:

$$\text{mean cholesterol} = b_0 + b_1 \times \text{age}.$$

By the general interpretation of the regression coefficients:

- b_0 is the mean cholesterol level of a subject of age 0.
- b_1 is the mean difference in cholesterol levels between two subjects who differ in age by 1 year.

Table 5.13 (model 1) shows that the cholesterol level increases significantly with age by

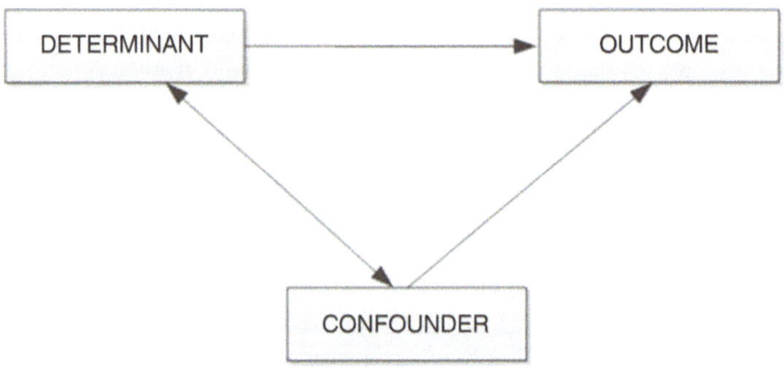

Fig. 5.6 Visualization of the concept of confounding

Table 5.13 SPSS output of a linear regression analysis relating cholesterol level (mmol/L) and age (years)

Model		Unstandardized coefficients B	Unstandardized coefficients SE	Standardized coefficients Beta	*t*	Sig.	95% CI for B Lower bound	95% CI for B Upper bound
1	(Constant)	4.183	0.186		22.528	0.000	3.818	4.547
	Age (yr)	0.022	0.003	0.254	7.819	0.000	0.017	0.028
2	(Constant)	4.344	0.176		24.620	0.000	3.988	4.690
	Age (yr)	0.017	0.003	0.193	6.177	0.000	0.012	0.022
	Sex	0.573	0.056	0.320	10.214	0.000	0.463	0.683

Dependent variable: Cholesterol level (mmol/L)

Table 5.14 SPSS output of the independent samples *t*-test comparing cholesterol level (mmol/L) between males and females

		Levene's test for equality of variances		*t*-test for equality of means						
		F	Sig.	*t*	*df*	Sig. two-tailed)	Mean difference	SE difference	95% CI of the difference Lower	95% CI of the difference Upper
Cholesterol level (mmol/L)	Equal variances assumed	5.297	0.022	−5.749	888	0.000	−3.857	0.671	−5.174	−2.541
	Equal variances not assumed			−5.927	591.187	0.000	−3.857	0.651	−5.136	−2.579

0.022 mmol/L/ year (95% CI [0.017; 0.028], $p < 0.001$).

For sex to be a confounder, it must be associated with age (first assumption): this is the case (Table 5.14) because the *p*-value is below 0.05. For the second assumption, it was already shown that sex is associated with cholesterol level (Table 5.1). To check the last assumption, the *simple* (i.e., only one independent variable; in medical research also referred to as a *univariate analysis*) linear regression model relating cholesterol level to age is extended to a *multiple* (i.e., multiple independent variables; in medical research also referred to as *multivariate analysis*, although this is statistically incorrect) linear regression model by adding sex to this model:

$$\text{mean cholesterol} = b_0 + b_1 \times \text{age} + b_2 \times \text{sex}.$$

After adding sex to the model (model 2 in Table 5.13), the regression coefficient for age changes from 0.022 to 0.017 mmol/L, a decrease of 22.7%. In other words, sex fulfills all three assumptions and hence is a confounder in the association between age and cholesterol level. In other words, the adjusted effect size and corresponding 95% CI and *p*-value must be reported.

The regression coefficient of the corrected (adjusted) model can be interpreted as follows:

- b_0 is the (theoretical) mean cholesterol level for a male of age 0 years.
- b_1 is the mean difference in cholesterol levels between two subjects of the same sex who differ in age by 1 year (e.g., the mean difference in cholesterol levels between two males, one aged 48 and one aged 49 years).
- b_2 is the mean difference in cholesterol levels between males and females of the same age (e.g., the mean difference in cholesterol levels between females and males, both aged 51 years).

Since the association between age and cholesterol level is studied, it is reported that cholesterol level is associated with age, with a mean increase of 0.017 mmol/L/ year (95% CI [0.012; 0.022], $p < 0.001$) after correcting for sex. The procedure for checking confounding in a logistic or Cox regression model is the same as with linear regression.

Effect Modification

With effect modification, the association between an independent variable and the outcome is not partly explained by another variable but is different for different strata of the effect modifier. Effect modification is different from confounding: confounding is bias that must be removed from the dataset, while effect modification is a result on its own and must be reported.

For example, when analyzing the association between cholesterol level (dependent variable) and age (independent variable), sex might be an *effect modifier*. This means that the association between cholesterol level and age might be different in males and females. To analyze whether sex is indeed an effect modifier, the variable sex and the *interaction* between sex and age are added to the simple regression model including only age:

$$\text{mean cholesterol} = b_0 + b_1 \times \text{age} + b_2 \times \text{sex} + b_3 \times \text{age} \times \text{sex}.$$

A variable is considered an effect modifier if its interaction with the independent variable is significant. The interaction must be computed manually in SPSS, in case of a continuous outcome, by multiplying the variables age and sex. In logistic and Cox regression models, SPSS does this automatically. Since the *p*-value of the interaction variable is less than 0.001 (Table 5.15), sex modifies the effect of age on cholesterol level. Consequently, the results are also stratified by sex (Table 5.16). In conclusion, there is a significant association between cholesterol level and age: 0.022 mmol/L/year (95% CI [0.017; 0.028], $p < 0.001$). Males have a significantly lower ($p < 0.001$) increase than females: 0.010 mmol/L/year (95% CI [0.005; 0.016],

Table 5.15 SPSS output of a linear regression model to test effect modification of sex in the association between cholesterol level (mmol/L) and age (years)

	Unstandardized coefficients		Standardized coefficients			95% CI for B	
Model	B	SE	Beta	*t*	Sig.	Lower bound	Upper bound
(Constant)	4.762	0.206		23.160	0.000	4.359	5.166
Age (yr)	0.010	0.003	0.118	3.219	0.001	0.004	0.017
Sex	−0.978	0.404	−0.546	−2.423	0.016	−1.771	−0.186
Interaction	0.024	0.006	0.890	3.878	0.000	0.012	0.036

Dependent variable: Cholesterol level (mmol/L)

Table 5.16 SPSS output of stratified analyses of the association between cholesterol level (mmol/L) and age (years)

		Unstandardized coefficients		Standardized coefficients			95% CI for B	
Stratum	Model	B	SE	Beta	*t*	Sig.	Lower bound	Upper bound
Male	(Constant)	4.762	0.175		27.274	0.000	4.419	5.105
	Age (yr)	0.010	0.003	0.152	3.791	0.000	0.005	0.016
Female	(Constant)	3.784	0.440		8.595	0.000	2.917	4.650
	Age (yr)	0.034	0.007	0.297	5.205	0.000	0.021	0.047

Dependent variable: Cholesterol level (mmol/L)

$p < 0.001$) and 0.034 mmol/L/year (95% CI [0.021; 0.047], $p < 0.001$), respectively.

Although the interaction model is built to investigate effect modification, the regression coefficients do have an interpretation (where males are coded as 0 and females as 1):

- b_0 is the (theoretical) mean cholesterol level for a male of age 0 years.
- b_1 is the mean difference in cholesterol levels between two males who differ in age by 1 year (e.g., the mean difference in cholesterol levels between two males, one aged 48 and one aged 49 years).
- b_2 is the mean difference in cholesterol levels between males and females of the same age (e.g., the mean difference in cholesterol levels between a 51-year-old female and a 51-year-old male).
- $b_1 + b_3$ is the mean difference in cholesterol levels between two females who differ in age by 1 year (e.g., the mean difference in cholesterol levels between two females, one aged 48 and one aged 49 years).

The difference between a confounder and effect modifier can be seen in Fig. 5.7. The blue line is the uncorrected regression line relating cholesterol level to age; the red line is the regression line for males; and the green line represents females. After correcting for sex as a confounder, the slope changes slightly compared with the *crude analysis*, but the slope is the same for males and females (only their intercepts differ): the red and green lines are parallel to each other but not to the blue line. After stratifying for age as an effect modifier, the slopes of males and females differ (as well as their intercepts), that is, they are not parallel.

The procedure for checking effect modification in a logistic or Cox regression model is the same as with linear regression. If a variable is an effect modifier, the analysis must be stratified, and the variable cannot be a confounder.

Multiple Confounders and/or Effect Modifiers

In every study, many variables are measured, and, by definition, all these variables could be possible confounders or effect modifiers. There are several strategies to choose from in case of correcting for multiple variables. None of these strategies is the best: just use what is often done in the particular field of expertise, and keep in mind that these procedures can lead to different results.

The first strategy is to analyze several confounders in groups, and ignore effect modification. For example, model 1 would be the crude, uncorrected model. Model 2 would be the crude model, to which all demographic variables (e.g., age, sex, education, race, etc.) are added. Model 3 would extend model 2 by adding, for example,

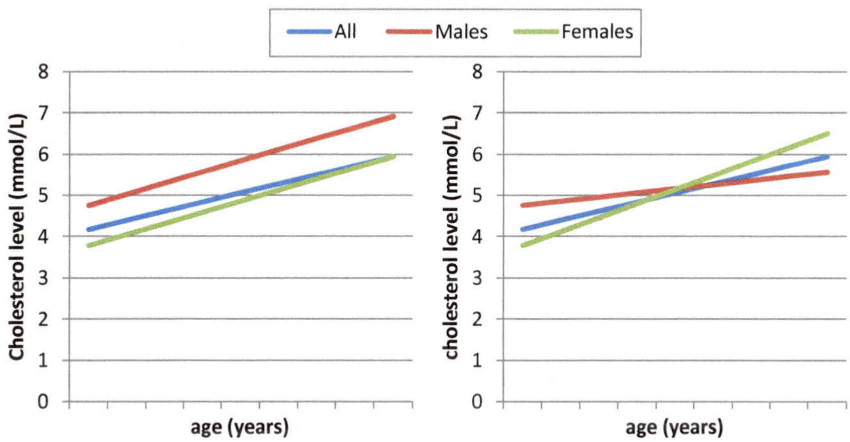

Fig. 5.7 Visualization of the differences between confounding (left panel) and effect modification (right panel)

lifestyle factors (e.g., alcohol use, smoking, physical activity, etc.). When reporting the results, only the effect size for the association of interest for the different (crude and corrected) models is reported. A disadvantage of this strategy is that a correction for these variables is being made a priori without checking if these variables fulfill the assumptions for a confounder. It therefore might be possible that some of the variables that are corrected for are in fact not confounders. The maximum number of confounders that can be included depends on the sample size. The rule of thumb is not to include more confounders than one per 10 subjects.

The second strategy is to determine, one by one, the strongest confounder and then add, step by step, only the strongest confounder to the model. Thus, in the first step every possible confounder is added individually to the uncorrected model. Based on the *p*-value of the confounder, and the change in the regression coefficient of the determinant, the variable with the strongest influence is selected. Then, the model with this particular confounder included is the new "base case" model, to which all remaining confounders are added, one by one, to determine the next strongest confounder. This step is repeated until none of the remaining variables change the regression coefficient by more than 10%. Note that now only the variables fulfilling the three confounder assumptions are corrected for. A disadvantage of this procedure is that analyzing which confounders to correct for can be time consuming if many possible confounders are involved.

These two strategies ignore effect modification. Therefore, another possible procedure is to start with investigating the effect modification, similarly as confounding was investigated in the second procedure. That is, analyze possible effect modifiers one by one; select the strongest effect modifier (with the lowest *p*-value for the interaction with the determinant); and stratify for that variable. Then, again, investigate possible effect modification of the remaining variables, one by one, for each stratum separately, and proceed with these steps until none of the variables modify the effect of interest. Then, add possible confounders to the stratified models by either the first or the second strategy mentioned previously. In theory, many effect modifiers could be found, resulting in many stratified groups containing only very few patients, which is a major disadvantage. It is advised not to stratify any further in case a stratum contains fewer than 10 patients (for a continuous outcome) or fewer than 10 patients with and 10 patients without an "event" (for both a dichotomous and a time-to-event variable). Professional advice is warranted when this procedure is applied.

A last possibility is correcting only for confounders and/or effect modifiers that are known from the literature or that are biological plausible. It does not matter which procedure to choose as long as it is motivated and well described in the final manuscript.

Reporting the Results

After analyzing the data, the results must be published, hopefully in a high-impact journal. In the introduction to the study, the importance of the topic is established by reviewing existing research and describing the gap to be addressed with this study. The applied methods—including study design, patient selection, and statistical analyses—must be specified thoroughly and the results reported. In the last step of the research process, the results of the study are discussed, which often follows a certain structure:

1. Interpreting all results.
2. Comparing the results with those in the existing literature and describing how the findings differ from or are similar to other studies (to strengthen the outcome).
3. Describing the strengths and limitations of the study.
4. Drawing final conclusions.

Writing a publishable article is complex. Nowadays, many journals only consider articles for publication if they have been written according to a specified guideline matching the study design. In these guidelines, several requirements are worked out, which must be taken into consideration and described. The following guidelines have been developed:

1. The CONsolidated Standards Of Reporting Trials (CONSORT) statement is a guideline used to describe the results of experimental studies, such as RCTs [2, 3]. It has several extensions, and all of them can be found at http://www.consort-statement.org.
2. The Transparent Reporting of a multivariable prediction model for Individual Prognosis Or Diagnosis (TRIPOD) statement is a guideline to describe the results of prediction modeling studies [4, 5]. This statement can be found at http://www.tripod-statement.org.
3. The Strengthening The Reporting of OBservational studies in Epidemiology (STROBE) statement is a guideline to report on cohort, case-control, and cross-sectional studies [6, 7]. This statement can be found at http://www.strobe-statement.org.
4. The Preferred Reporting Items for Systematic Reviews and Meta-Analyses (PRISMA) statement is a guideline for writing systematic reviews with or without meta-analyses [8, 9]. This statement can be found at http://www.prisma-statement.org.
5. The STandards for Reporting of Diagnostic Accuracy (STARD) statement is a guideline for reporting studies of diagnostic accuracy [10, 11]. This statement can be found at http://www.equator-network.org/reporting-guidelines/stard/.
6. The COnsensus-based Standards for selection of health Measurement INstruments (COSMIN) standard is a checklist that focuses on standards for studies on the measurement properties of Health-Related Patient-Reported Outcomes (HR-PROs). It helps to define the psychometric properties of existing PROs, and in developing a new PRO [12, 13]. The COSMIN checklist can be found at http://www.cosmin.nl/cosmin_checklist.html.

No guideline has yet been developed for cost-effectiveness studies. The Panel on Cost-Effectiveness in Health and Medicine has developed recommendations for the conduct of cost-effectiveness analysis (CEA) [14–17].

References

1. Mawad W, Drolet C, Dahdah N, Dallaire FA. Review and critique of the statistical methods used to generate reference values in pediatric echocardiography. J Am Soc Echocardiogr. 2013;26(1):29–37.
2. Hopewell S, Clarke M, Moher D, Wager E, Middleton P, Altman DG, et al. CONSORT for reporting randomized trials in journal and conference abstracts. Lancet. 2008;371 (9609):281–3.
3. Hopewell S, Clarke M, Moher D, Wager E, Middleton P, Altman DG, et al. CONSORT for reporting randomized controlled trials in journal and conference abstracts: explanation and elaboration. PLoS Med. 2008;5(1):e20.
4. Collins GS, Reitsma JB, Altman DG, Moons KG. Transparent reporting of a multivariable prediction model for individual prognosis or diagnosis (TRIPOD): the TRIPOD statement. Ann Intern Med. 2015;162(1):55–63.
5. Moons KG, Altman DG, Reitsma JB, Ioannidis JP, Macaskill P, Steyerberg EW, et al. Transparent reporting of a multivariable prediction model for individual prognosis or diagnosis (TRIPOD): explanation and elaboration. Ann Intern Med. 2015;162(1):W1–73.
6. Vandenbroucke JP, von Elm E, Altman DG, Gotzsche PC, Mulrow CD, Pocock SJ, et al. Strengthening the reporting of observational studies in epidemiology (STROBE): explanation and elaboration. Ann Intern Med. 2007;147(8):W163–95.
7. Vandenbroucke JP, von Elm E, Altman DG, Gotzsche PC, Mulrow CD, Pocock SJ, et al. Strengthening the reporting of observational studies in epidemiology (STROBE): explanation and elaboration. Int J Surg. 2014;12(12):1500–25.
8. Moher D, Liberati A, Tetzlaff J, Altman DG, Group P. Preferred reporting items for systematic reviews and meta-analyses: the PRISMA Statement. Open Med. 2009;3(3): e123–30.
9. Shamseer L, Moher D, Clarke M, Ghersi D, Liberati A, Petticrew M, et al. Preferred reporting items for systematic review and meta-analysis protocols (PRISMA-P) 2015: elaboration and explanation. BMJ. 2015;350:g7647.
10. Bossuyt PM, Reitsma JB, Bruns DE, Gatsonis CA, Glasziou PP, Irwig LM, et al. The STARD statement for reporting studies of diagnostic accuracy: explanation and elaboration. Ann Intern Med. 2003;138(1):W1–12.
11. Cohen JF, Korevaar DA, Altman DG, Bruns DE, Gatsonis CA, Hooft L, et al. STARD 2015 guidelines for reporting diagnostic accuracy studies: explanation and elaboration. BMJ Open. 2016;6(11):e012799.
12. Mokkink LB, Prinsen CA, Bouter LM, Vet HC, Terwee CB. The COnsensus-based standards for the selection of health measurement INstruments (COSMIN) and how to select an outcome measurement instrument. Braz J Phys Ther. 2016;20(2):105–13.
13. Mokkink LB, Terwee CB, Knol DL, Stratford PW, Alonso J, Patrick DL, et al. The COSMIN checklist for evaluating the methodological quality of studies on measurement properties: a clarification of its content. BMC Med Res Methodol. 2010;10:22.
14. Mandelblatt JS, Fryback DG, Weinstein MC, Russell LB, Gold MR. Assessing the effectiveness of health interventions

for cost-effectiveness analysis. Panel on cost-effectiveness in health and medicine. J Gen Intern Med. 1997;12(9):551–8.
15. Siegel JE, Torrance GW, Russell LB, Luce BR, Weinstein MC, Gold MR. Guidelines for pharmacoeconomic studies. Recommendations from the panel on cost effectiveness in health and medicine. Panel on cost effectiveness in health and medicine. PharmacoEconomics. 1997;11(2):159–68.
16. Siegel JE, Weinstein MC, Russell LB, Gold MR. Recommendations for reporting cost-effectiveness analyses. Panel on cost-effectiveness in health and medicine. JAMA. 1996;276(16):1339–41.
17. Weinstein MC, Siegel JE, Gold MR, Kamlet MS, Russell LB. Recommendations of the panel on cost-effectiveness in health and medicine. JAMA. 1996;276(15):1253–8.

The Influence of Age and Sex on the Electrocardiogram

6

Peter W. Macfarlane

Electrocardiography and Waller's dog Jimmie. Art work by Piet Michiels, Leuven, Belgium

Abstract

The electrocardiogram (ECG) remains the most commonly used test in medical practice and as such requires to be interpreted with due care and attention to detail. The ECG changes rapidly from birth through childhood with age differences clearly related to increasing QRS

P. W. Macfarlane (✉)
Institute of Health and Wellbeing, University of Glasgow, Glasgow, Scotland, UK
e-mail: peter.macfarlane@glasgow.ac.uk

P. L. M. Kerkhof, V. M. Miller (eds.), *Sex-Specific Analysis of Cardiovascular Function*,
Advances in Experimental Medicine and Biology 1065, https://doi.org/10.1007/978-3-319-77932-4_6

voltages and a widening QRS complex. The only sex difference at this age is a slightly longer QRS duration in boys than girls.

In adulthood, sex differences in QRS voltage are maximum in the under 40 age group and tend to minimise with advancing age. QRS duration is longer in males than in females, but little difference is made of this in diagnostic criteria. In a similar vein, ST amplitudes are higher in young males compared to young females with the difference diminishing as age increases. Corrected QT interval is longer in females than males.

In summary, age and gender differences in the ECG are important and have been incorporated into a variety of criteria for ECG interpretation. Physicians should be aware of the main sex differences in the ECG.

Keywords
ECG · Sex · Age · Reference values · Normal ranges · Databases · Ethnicity · Automated ECG analysis · Diagnostic criteria · QT interval · JT interval

Introduction

It would be remiss to discuss the influence of age and sex on the electrocardiogram (ECG) without reference to the seminal work of Simonson, who published a book on differentiation between the normal and abnormal electrocardiogram in 1961 [1]. Even before the advent of computer methods for measuring the components of the ECG, Simonson and his team drew up extensive tables of references ranges, more often called normal limits, of ECG measurements based on epidemiological studies. While there may be subtle differences between the 1961 measurements and those published more recently, Simonson nevertheless paved the way in pointing out differences in ECG measures between males and females and indeed between different races, as he had access to Japanese ECG databases in particular in addition to his own data from North American-based studies.

Pipberger and his team in Washington, DC, led the way in automating ECG measurements [2]. He used the three orthogonal lead ECGs for his studies. In the same part of the world, Caceres and colleagues also initiated methods for analysing the ECG using computers [3], but they chose to use the standard 12 lead ECG. Initially, the 12 lead ECG was recorded in four groups of three leads [3, 4], but nowadays the eight independent leads of the 12 lead ECG, namely, I, II and V1–V6, are recorded simultaneously, greatly facilitating analysis through having time alignment of ECG appearances in all leads [5]. The other limb leads III, aVR, aVL and aVF are easily calculated from leads I and II [6].

In Glasgow, the development of automated ECG interpretation or as it was initially called locally, computer-assisted reporting of electrocardiograms (CARE), followed a roughly similar route in that initially the three orthogonal lead ECG was used to develop data acquisition and measurement routines before the 12 lead ECG was analysed in groups of three leads. Ultimately, all 12 leads effectively recorded simultaneously were analysed [5].

This chapter documents sex- and age-specific differences in ECG appearances for neonates, children and adults. ECG changes observed in pregnant women are summarised elsewhere in this book [7].

Much of the data in this chapter is drawn from the databases acquired in Glasgow. These databases consist of the following five cohorts.

Glasgow Adult Normal Database

Throughout the late 1980s, a database of ECGs from apparently healthy adults was acquired in Glasgow [8]. These volunteers essentially were recruited from local government and therefore consisted of persons from a variety of occupations from sedentary work in offices through teaching in schools to more manual work. There was therefore a good cross section of the population involved. Each individual was screened by a physician to ensure that there were no known clinical problems and to ensure blood pressure was normal. Blood tests and a chest X-ray were obtained in the first batch of volunteers but proved to be of no value, and these tests were later removed from the protocol. Ultimately, a database of 1496 individuals aged from 18 to 82 years was compiled, and extensive tables of normal limits of various ECG parameters have already been published [9].

Glasgow Paediatric Normal Database

Following the collection of the adult ECGs, it was also thought valuable to acquire 12 lead ECGs in healthy neonates, infants and children as there was no substantial database of ECGs from this age group available with all leads recorded simultaneously. This was achieved through recording ECGs in the Royal Hospital for Sick Children in Glasgow, the Queen Mother's Hospital (Maternity Unit), and also through recording in day nurseries and schools. In all cases, parental permission had to be obtained for the ECG recording.

This remains a rather unique database through having over 500 ECGs recorded from neonates aged from birth to 6 days included in the 1784 entries from which normal limits were derived [10–12]. Improvements in postnatal care mean that nowadays most new mothers leave hospital within 24–48 h of giving birth, and hence, it is no longer feasible to accumulate ECGs from children over the first few days of life. Yet as will be shown, there are significant changes in the ECG at that period of development of a neonate [10]. The availability of the database led to the extension of the Glasgow program to analyse and interpret ECGs from neonates, infants and children [13].

Chinese Database

In order to compare the ECG of Caucasians versus Chinese, 503 ECGs were recorded from healthy individuals in Taipei Veterans General Hospital in Taiwan. All volunteers had a normal cardiovascular system and were hospitalised for a variety of reasons. The 12 lead ECGs were recorded on digital tape and sent to Glasgow for analysis [14]. There were approximately equal numbers of men and women in this study, and the age range was 19–81 years.

Nigerian Database

As part of an MD thesis [15], 1261 ECGs were recorded in and around Ilorin in central Nigeria from apparently healthy volunteers. All ECGs were transferred to a database locally (Burdick, Wisconsin) and then sent to Glasgow for analysis using the same program as for the other databases. A full set of results was published [16]. Ages of the volunteers ranged from 20 to 87.

Indian Database

ECGs for this database were recorded in three separate centres in India, namely, at the Indian Institute of Technology Roorkee, NIT Jalandhar, and SGGS Nanded. In all cases, the same model of electrocardiograph, namely, a Burdick Atria 6100, was used. As for the other studies, all volunteers were apparently healthy with normal blood pressure and no history of cardiovascular disease. The resulting reference ranges (normal limits) of the ECG in this population of 963 individuals aged 18–83 years were previously published [17].

Database Summary

All ECGs in the various databases were analysed using the same version of the Glasgow program, and selected aspects of a comparison among the different groups have been published [18]. It is important to stress that the same software was used for analysis in all databases. Normal ranges were derived by excluding 2% of values at either end of a distribution in order to give a 96 percentile reference range. For comparative studies [18], normal limits were computed from a knowledge of the mean regression line or directly with quantile regression. Additionally, the sample quantiles, which incorporated smoothing with a window of 10 years, were computed and displayed. All statistical analyses were undertaken using the SAS v9.2 package.

P Wave

P Wave Amplitude

Given that cardiac muscle activation is essentially in a base to apex direction from the sinoatrial (SA) node with a mean direction which is roughly parallel to the lead II axis, the highest P wave amplitudes tend to be found in inferiorly directed leads such as II and aVF. In general, there are no sex differences in terms of normal limits of P wave amplitudes. The normal upper limit of P wave amplitude in the limb leads is generally regarded as 0.25 mV for both males and females. Similarly, in children there are no sex differences in P wave amplitudes and no significant difference between paediatric and adult P wave upper limits of normal in the limb leads.

In the precordial leads, the P wave configuration in V1 may be biphasic with an initial positive component and a terminal inversion. It has been suggested that the width of the terminal component multiplied by its amplitude, i.e. the Morris index [19], provides a measure which increases in the presence of left atrial enlargement. A limit of 40 ms duration and 0.1 mV negativity, i.e. the width and depth of one small box on a conventional ECG display, gives a Morris index of 4 mVms, which is generally regarded as the upper limit of normal although the authors of the original paper [19] used 3 mVms.

In right atrial enlargement, P waves in V1 and V2 can be upright. The upper limits of normal of the order of 0.15 mV in V1 and 0.175 in V2 were found in the Glasgow database for both males and females. In children, slightly higher limits of normal are used, being 0.2 mV and 0.25 mV in V1 and V2, and so there is an age dependence on upper limits of normal P wave amplitude in these leads. There are no sex differences in these P wave amplitudes.

P waves in other leads are generally not used for diagnostic purposes, although they can be significantly abnormal in certain forms of congenital heart disease.

P Wave Duration

From the point of view of diagnosing an abnormal ECG, there are no sex differences in normal limits of P wave duration. Most textbooks will quote 120 ms as the upper limit of normal. From the point of view of automated interpretation, where all leads are recorded simultaneously, a slightly wider normal limit of 140 ms is often preferred. Essentially, this relates to the fact that P wave onset does not occur simultaneously in all leads, while the same is true of P wave termination. Thus, an interval from the earliest onset to the latest termination will be slightly longer than P wave durations measured in individual leads.

In the paediatric age range, it tends to be that 100 ms is also used as an upper limit of normal overall P wave duration with 80 ms in children under 1 year [20]. There are no sex differences in P wave durations in this age group.

QRS Complex

Amplitudes

Perhaps the one component of the electrocardiogram where the greatest changes can be seen according to age and sex is the QRS complex. Age-based changes can be seen, even in the first week of life, and so it makes sense to commence a review from that point.

The foetal circulation is decidedly different from adult circulation with blood short circuited from the right ventricle to the aorta via the ductus arteriosus. Blood reaches the left side of the heart from the right atrium via the foramen ovale. The main effect is that, in the newborn, there is a so-called right ventricular preponderance which diminishes over the first days and weeks of life. In summary, there are significant QRS amplitude

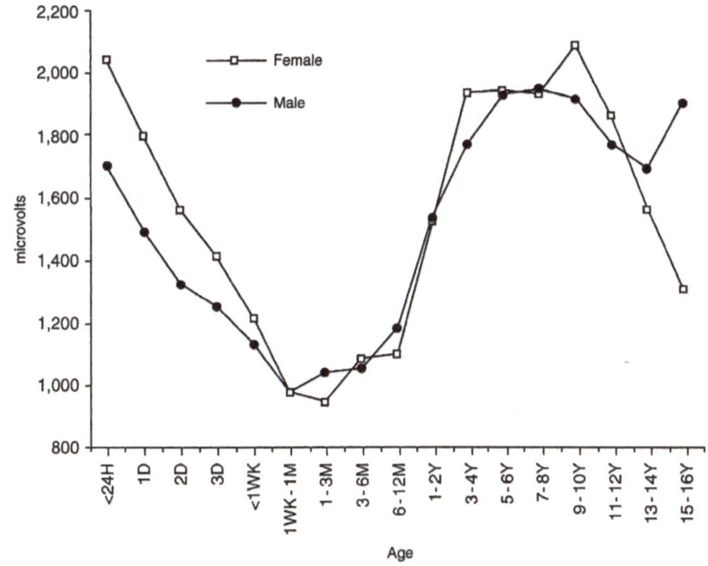

Fig. 6.1 The change in S V2 amplitude from birth to adolescence derived from the Glasgow paediatric database. Note that the scale of the abscissa is non-linear. (Reproduced from [10] with permission)

changes, although there are no sex differences in QRS amplitudes in infants and children.

Figure 6.1 shows the change in S wave amplitude in V2 over the first few days and weeks of life. These data suggest that it is important to use the age of a neonate in days when interpreting an ECG. In practical terms, this is essentially not done because healthy babies will be discharged from the hospital within a day or so after being born, whereas those with more significant problems such as congenital heart disease may have grossly abnormal ECGs not requiring fine adjustment of QRS voltages to determine abnormality.

It can be seen from the figure that over the first years of life, S wave amplitude gradually increases as might be expected due to the natural growth of the child and, hence, increase in heart size. Thus, age differences do play a role in ECG appearances in childhood. This does have relevance when the question of screening children, particularly for participation in sport, is considered.

In the adult situation, Fig. 6.2 shows how the mean S wave amplitude in V2 varies with sex and age. Similar differences can be seen in other amplitudes such as the R wave amplitude in V5. Thus, normal limits of R and S amplitudes are sex dependent.

The reason why ECG amplitudes are sex and age dependent is linked to heart weight and body mass. There are many publications showing that heart weight is higher in males than in females. A recent study [21] provides interesting data. This research, using computed tomography angiography, reported that the mean left ventricular (LV) mass in males was 116 ± 20 g and was 85 ± 14 g in females. This in itself might suggest that QRS voltages in males would be higher than in females due to increased LV mass generating larger electrical signals. Furthermore, the study also showed (Fig. 6.3) that the indexed LV mass, i.e. LV mass indexed by body surface area (LV_{MI}), was higher in males at 60 ± 9 g/m^2 as opposed to 49 ± 7 g/m^2 in females. However, it can be seen in Fig. 6.3 that LV_{MI} decreased with age in males but stayed level in females. This is in keeping with the data in Fig. 6.2 showing that the difference in mean S wave amplitude in V2 between males and females lessens with increasing age.

Sokolow-Lyon Index for LVH

One of the most commonly used indices for reporting left ventricular hypertrophy (LVH) is the Sokolow-Lyon (S-L) index [22]. This index was developed on 147 patients, 90% of whom

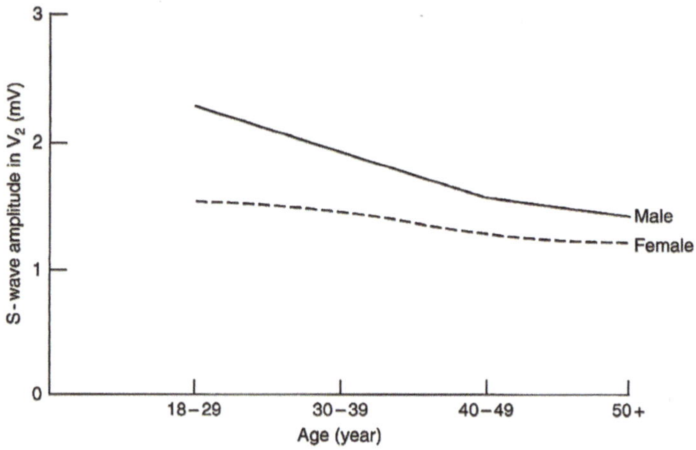

Fig. 6.2 Mean amplitude of S V2 in males and females in the Glasgow database. A significant difference between the two sexes is apparent. (Reproduced from [8] with permission)

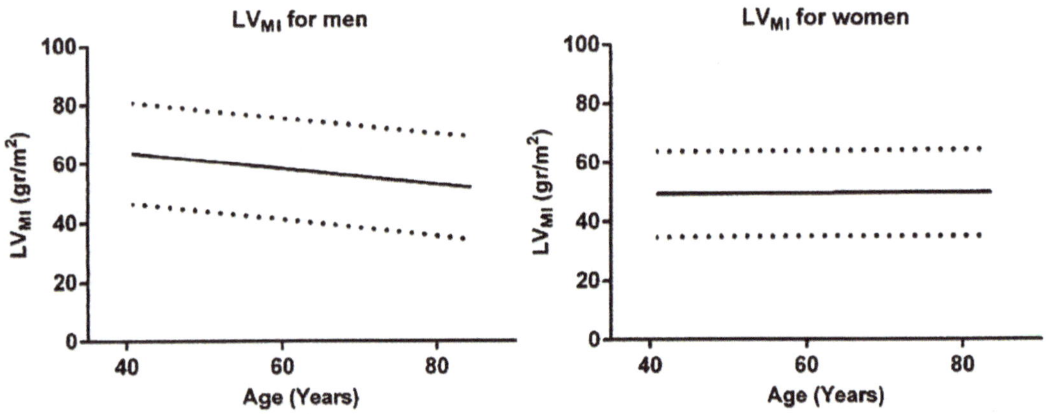

Fig. 6.3 The variation in LV mass index (LV_{MI}) with respect to age is shown for males and females. g = gram, m = metre. (Reproduced from [21] with permission)

were hypertensive with a mean blood pressure of 197/117 and 151 controls. The authors did not detail the percentage of males and females in either cohort, and so sex differences in voltage were not assessed. The controls had a mean age of 35.1 years, but the age of the patients was not stated. The maximum R amplitude in V5 in the controls was 2.6 mV. This has to be compared with a maximum value of 3.5 mV for males and 2.4 mV in females in the Glasgow normal population [12] and 3.5 mV for males and 3.8 mV for females in the Nigerian population [15].

Somehow or other, the index has remained embedded in electrocardiographic criteria for left ventricular hypertrophy since its inception. The basic index is the sum of SV1 + RV5 amplitudes, although there are variations such as SV1 + max (RV5, RV6) amplitudes, because Sokolow and Lyon mixed the two in their article without giving separate criteria. The more commonly used SV1 + RV5 exceeded 3.5 mV in the hypertensive group in 48/147 individuals, i.e. its sensitivity in a training environment was 32.7%. However, the important point to note here is that the index, as originally published, is neither age nor sex dependent. This is a very significant shortcoming.

Figure 6.4 shows the upper and lower limits of the original S-L index for males and females based on the four different adult populations described in Section 1. It can be seen very clearly

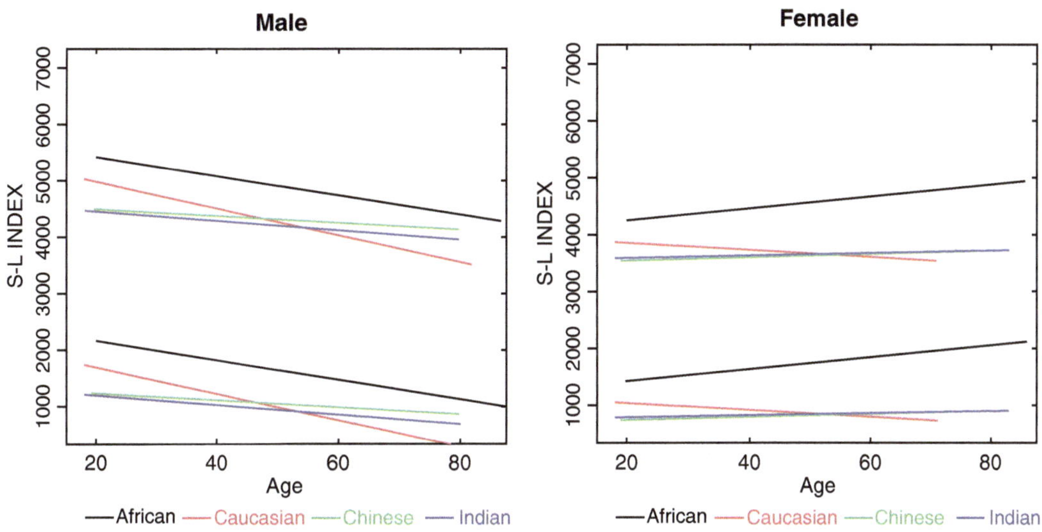

Fig. 6.4 The upper and lower limits of the reference ranges of SV1 + RV5 for the four different adult populations described in the text (Reproduced from [18] with permission)

that the index decreases with increasing age in males and is much higher in young males compared to young females, e.g. has an upper limit of 5.5 mV in black males aged 20 years compared to 4.2 mV in black females of the same age with this difference narrowing with increasing age. Paradoxically, in the Nigerian ladies, an increase in the index was found with increasing age. Therefore, any diagnostic criteria based on this index have to be age and sex dependent. This has been a feature of the Glasgow ECG interpretation program almost since its inception [5, 23]. A newer feature is to use race-, age-, and sex-dependent criteria.

In a hospital environment, most adult individuals with cardiovascular disease will be over 50 years of age, and in this age range, the index is more specific than at younger ages. On the other hand, with the increasing use of the ECG for screening younger individuals with a view to their participation in sport, there are different opinions [24, 25]. In Italy, pre-participation screening, including ECG recording, is obligatory [26], whereas in the USA, while screening may be undertaken in certain situations, ECG recording is not recommended [24]. Where the S-L index is used, it should be age and sex corrected in the author's opinion. This will avoid false-positive reports of LVH and a suspicion of cardiomyopathy in young sportspersons. On the other hand, the most recent version [27] of what were known as the "Seattle criteria" [28] for ECG interpretation in athletes but now termed the International Criteria [27] referenced the S-L index defined as SV1 + max (RV5, RV6) and regarded a value >3.5 mV in athletes, male or female, as "increased QRS voltage" which came under the heading of a "normal ECG finding" in this group of individuals. However, ST depression and T wave inversion, if present in certain leads, are regarded as abnormal in these criteria for athletes and could be an indication of LVH possibly due to cardiomyopathy. The criteria for ST-T segment abnormalities were not sex dependent, and indeed the only such criterion for athletes was the QTc interval (vide infra).

Cornell Criteria for LVH

A group of cardiologists in Weill Medical College, New York, introduced an index for the detection of LVH from the ECG. This was modified from the first publication in 1985 [29] to a later variation published in 1987 [30]. In a more recent study [31], the threshold for females was raised to 2.2 mV by reducing the difference between the male and female thresholds to 0.6 mV.

The Cornell criteria most often used are:

$$\mathrm{RaVL} + \mathrm{SV3} > 2.8\,\mathrm{mV}\ \text{in males}$$

$$\mathrm{RaVL} + \mathrm{SV3} > 2.2\,\mathrm{mV}\ \text{in females}$$

These sex-based criteria have proved to be effective at all ages. In a recent assessment using the Glasgow database, it was shown that the Cornell criteria were essentially ≥94% specific in all age- and sex-based groups with the exception of females aged 40–49 where they were 91% specific [32]. In the same study, the Sokolow and Lyon criterion specificity ranged from 64% in young males to 99% in older females.

The clear message is that criteria for LVH simply must be sex specific. If physicians wish to persist using the Sokolow and Lyon criterion, then it should be both an age- and sex-specific version that is used.

Romhilt-Estes Point Score System for LVH

Romhilt and Estes introduced a point score system for the ECG diagnosis of LVH [33]. These criteria made use not only of voltage changes but also ST-T segment changes, left axis deviation, left atrial abnormalities and a broadened QRS duration. They were not sex specific, but within our own laboratory, we modified voltage criteria so that they were indeed age and sex dependent. This resulted in improved sensitivity as documented elsewhere [34].

Right Ventricular Hypertrophy

There are no criteria for right ventricular hypertrophy (RVH) which are exceptionally sensitive and specific. Indeed, there are few, if any, criteria for RVH which are sex dependent. Previous reference to the neonatal ECG indicates that criteria for RVH are very definitely age dependent. For the adult, an upper limit of 0.5 mV in V1 and/or R/S amplitude ratio > 1 in V1 are commonly cited, whereas for neonates and children, much higher threshold values are required to be used.

QRS Duration

In all of the adult populations studied, the mean QRS duration in males has been higher than in adult females. In the Glasgow database, the QRS mean duration was 96.4 ms in young males and 87.7 ms in young females. Similar changes were seen in different age groups, while in those aged over 50 years, the mean QRS duration was 92.7 ms in males and 87.1 ms in females. Table 6.1 shows how mean QRS duration varies with age and sex in the four different adult populations described earlier but with the much bigger influence being sex. On the other hand, it has to be admitted that if a cardiologist were to be asked what the upper limit of normal QRS duration is, the answer would almost certainly be 120 ms for males and females. There are few diagnostic criteria involving QRS duration which are sex dependent, but recently Strauss et al. introduced the concept of strict left bundle branch block (LBBB) which used a QRS duration >140 ms in men and ≥ 130 ms in women [35]. The need for such differentiation was related to deciding on which patients were most suited to cardiac resynchronisation therapy.

ST-T Segment

ST Amplitude

Although T wave amplitudes are generally not sex dependent, ST amplitudes are certainly age and sex dependent. Young males have higher limits of normal ST amplitude than young women, and older males have lower ST amplitudes than younger males [36]. These changes can be seen in Fig. 6.5 for lead V2. In addition, Fig. 6.6 shows how ST amplitude varies with the precordial lead under consideration as well as age and sex. Figure 6.7 shows the sex-dependent changes for the upper limit of

Table 6.1 Normal reference ranges for QRS duration in the four populations described in the text

	18–29 years		30–39 years		40–49 years		50–59 years		$\geq$ 60 years	
	M	F	M	F	M	F	M	F	M	F
Caucasians	96.4 ± 8.6	87.7 ± 7.8	95.4 ± 9.8	88.6 ± 7.3	94.4 ± 9.9	89.4 ± 7.9	92.7 ± 9.3	87.1 ± 8.7		
	n = 265	n = 317	n = 218	n = 115	n = 119	n = 72	n = 123	n = 79		
Chinese	95.8 ± 8.2	87.0 ± 6.9	94.4 ± 9.8	87.9 ± 10.4	96.0 ± 10.7	88.0 ± 9.6	92.4 ± 10.2	88.9 ± 10.1	94.8 ± 11.2	88.6 ± 8.2
	n = 56	n = 47	n = 50	n = 59	n = 50	n = 50	n = 50	n = 48	n = 49	n = 44
Indians	87.1 ± 8.4	82.2 ± 6.8	87.9 ± 7.2	81.8 ± 7.8	86.5 ± 7.5	81.4 ± 6.7	86.8 ± 7.9	81.9 ± 7.5	87.0 ± 11.3	82.9 ± 8.8
	n = 266	n = 122	n = 130	n = 64	n = 149	n = 39	n = 87	n = 37	n = 38	n = 31
Nigerians	76.7 ± 13.0	66.3 ± 11.8	79.5 ± 11.2	71.2 ± 11.7	76.7 ± 11.6	72.0 ± 10.8	78.6 ± 11.0	72.9 ± 10.3	74.6 ± 12.1	75.5 ± 10.1
	n = 123	n = 97	n = 221	n = 110	n = 199	n = 107	n = 134	n = 91	n = 105	n = 74

Note that for Nigerians, measurements refer to one lead only (V5) and hence are shorter than the overall measurements for the other populations. Sex differences are clear in all cases. The overall mean QRS duration in Nigerians based on 12 simultaneously recorded leads was 86.3 ± 9.0 ms. For the Caucasians, the age group 50–59 years contains data from individuals aged 50 years and above. The same computer program [5] was used to analyse all the ECGs

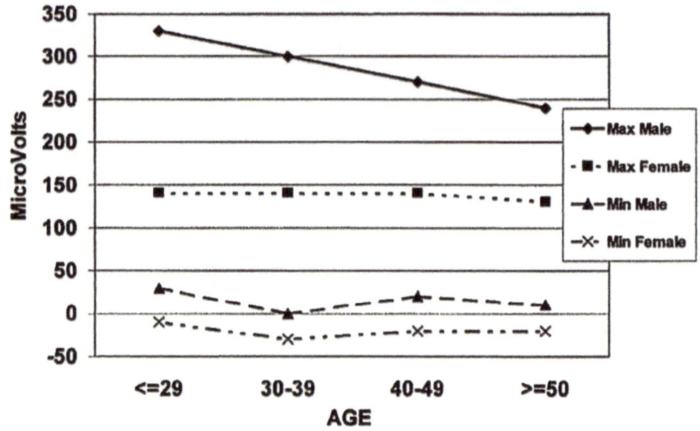

Fig. 6.5 Upper and lower normal limits of ST amplitude (STj) in V2 for males and females in the Glasgow database. (Reproduced from [36] with permission)

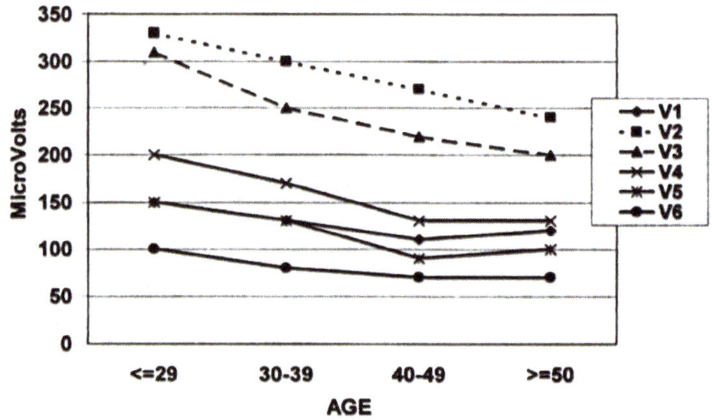

Fig. 6.6 The upper limits of the normal reference range for ST amplitude (STj) for V1–V6 from males in the Glasgow database. Note how V1 and V5 have similar values. (Reproduced from [36] with permission)

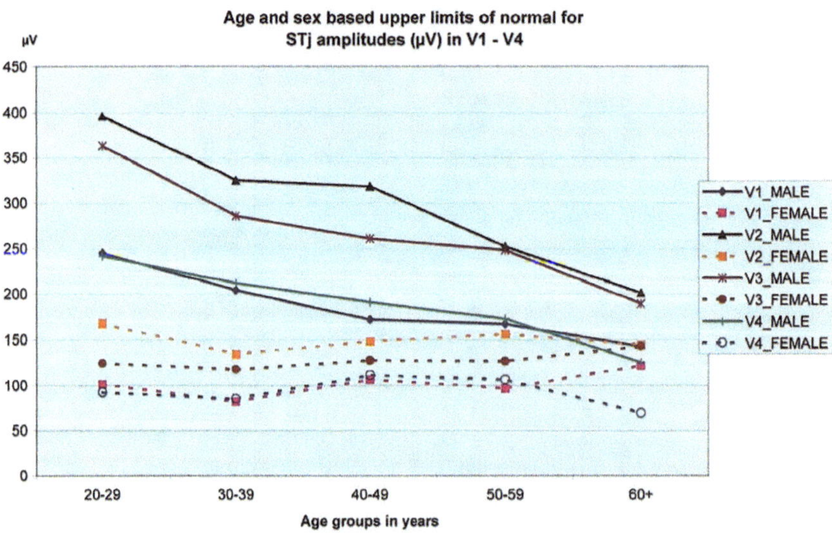

Fig. 6.7 Upper limits of the normal reference ranges in V2–V4 for males and females in the Nigerian population. (Reproduced from [16] with permission)

normal ST amplitude of V2–V4 in the Nigerian population described earlier.

Some of the above work in the author's laboratory led to the introduction of age- and sex-based criteria for ST elevation myocardial infarction (STEMI) as evidenced by the third universal definition of myocardial infarction published in 2012 [37]. This had been preceded by similar criteria in the recommendations of a working group on ST-T changes in acute ischaemia and myocardial infarction [38]. Table 6.2 shows the age- and sex-specific criteria for ST change in diagnosing STEMI as listed in the third universal definition of myocardial infarction [37].

Table 6.2 ST elevation criteria for acute myocardial ischaemia in the absence of LVH and LBBB [37]

New ST elevation at the J point in two contiguous leads with the cut points: J≥0.1mV in all leads other than leads V2-V3 where the following cut points apply: ≥0.2mV in men ≥ 40 years; ≥0.25mV in men < 40 years, or ≥ 015mV in women.

Although ST amplitudes are sex dependent, in clinical practice they are not necessarily among the first thoughts of the cardiac electrophysiologist considering whether or not to undertake a percutaneous coronary intervention (PCI) in a patient with chest pain. These more refined criteria are best applied via automated techniques where improvements in sensitivity can be noted [39, 40].

T Amplitude

There are clear sex differences in the upper limits of the normal precordial T waves as shown in Fig. 6.8. Diagnostic criteria acknowledge such differences when reporting tall T waves in

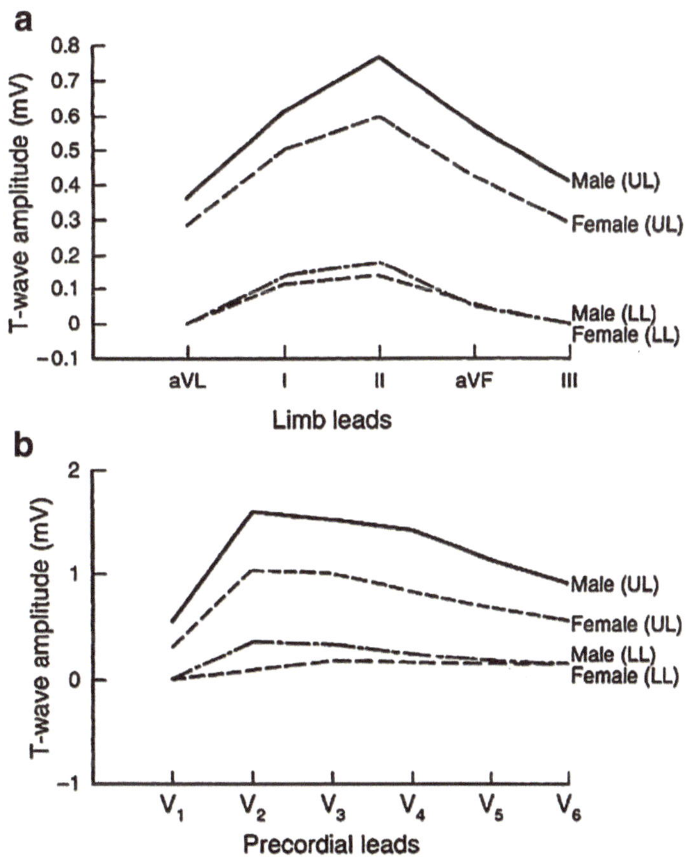

Fig. 6.8 Upper and lower limits of the normal reference range of T waves in the Glasgow database. Sex-dependent differences are clearly seen in the upper limits. (Reproduced from [8] with permission)

precordial leads suggestive of hyperkalaemia or acute myocardial ischaemia. The upper limit of normal in males is of the order of 1.4 mV and 1.0 mV in females.

The only area where sex differences in T wave morphology are considered to be due to normal variation is in the right precordial leads where T wave inversion in V2 would be more frequently regarded as normal in females than in males. The Glasgow data found 1.1% of healthy women had a negative component in the T wave in V2.

There is also said to be racial variation in that a small percentage of healthy young black males tend to have T wave inversion in right precordial leads, which is a normal variant [40]. There have, however, been conflicting results in this connection, and our own studies involving an indigenous Nigerian cohort did not show this feature [16]. On the other hand, the prevalence of T wave changes in black athletes is much higher than in black controls [41].

QT Interval

An important interval in the ECG is the QT interval because a lengthened QT can be due to a number of different causes and may result in death in some cases. Hypokalaemia and hypocalcaemia, for example, are both known to increase the QT interval. Drugs such as moxifloxacin can also increase the QT interval, and a recent comparative study of various computer programs has shown how each can detect even small changes in QT interval in a modest number of individuals after administration of the drug [42].

Various congenital cardiac abnormalities can also result in a long QT interval, and in general terms, such a lengthened QT is associated with life-threatening cardiac arrhythmias.

QT interval varies with heart rate, i.e. it shortens at increased heart rates and lengthens at decreased heart rates. This has led to the concept of correcting the QT interval whereby an attempt is made to predict what the QT interval would be at a heart rate of 60. The most commonly used formula for correcting QT is that of Bazett [43]. This equation is

$$\mathrm{QTc} = \mathrm{QT}\sqrt{(\mathrm{rate}/60)}.$$

It has been suggested recently that the Bazett formula is probably the least accurate of all formulae, and its use is not to be recommended [44].

Other formulae include that of Hodges [45], which is based on a linear correction. The equation used is

$$\mathrm{QTc} = \mathrm{QT} + 175(\mathrm{rate} - 60).$$

Another formula that has increased in popularity recently is that of Fridericia [46]. This is similar to Bazett but uses a cube root rather than a square root, i.e.

$$\mathrm{QTc} = \mathrm{QT}^3\sqrt{(\mathrm{rate}/60)}$$

In a paper using data from this laboratory, Luo et al. [47] compared the different QT formulae and concluded that the Bazett correction was most often out of line with the other formulae.

A small sex-based difference in QT interval has been shown to occur with a mean difference between females and males being the order of 10 ms. The upper limit of normal QTc according to the 2009 guideline [48] is 460 ms for females and 450 ms for males. On the other hand, the international criteria for abnormal ECGs in athletes suggests using 480 ms for females and 470 ms for males [27].

JT/J-Tpeak/Tpeak-Tend Intervals

Other intervals which have been considered recently in relation to drug effects on the ECG [49] include the JT interval (end QRS to end T), J-Tpeak (end QRS to the peak of the T wave) and Tpeak-Tend (peak of the T wave to end T). The JT interval corrected for heart rate (HR) has been shown to be sex dependent in a study of 11,739 adult men and women [50]. The equation derived was

$$\mathrm{JT_{RR}} = \mathrm{JT} - 176\,(60/\mathrm{HR} - 1) \text{ for women and } + 14 \text{ ms for men}$$

In the presence of a conduction defect, the QRS duration was shown to have an effect on the QT interval and a correspondingly weaker effect on the JT interval. A separate sex-dependent equation for JT_{RR} was therefore also derived for use in patients with an interventricular conduction defect, namely,

$$JT_{RR} = JT - 155(60/HR - 1) + k$$

where $k = 34$ ms for men and $k = 22$ ms for women.

Papers from a symposium entitled "The J-Tpeak initiative" were recently published [51].

Conclusion

It should be clear that using age- and sex-based criteria optimises the sensitivity and specificity in reporting an ECG abnormality. For example, the upper limit of normal Sokolow and Lyon index for young males exceeds 5 mV, and so it does not make sense to use the much quoted threshold of 3.5 mV to report left ventricular hypertrophy on the ECG in young males. One implication could be that a young person is prevented from participating in sport, although an echocardiogram should be recorded to look for evidence of increased LV mass if the ECG finding is suspicious.

Similarly, a long QT interval may on occasions be indicative of drug therapy affecting the ECG or sufficient to raise a suspicion of a "long QT interval". It is known that the QT interval is longer in females than males, and so again it is optimum to minimise false-positive diagnoses by using sex-dependent upper limits of normal before suggesting, perhaps wrongly, that drug therapy needs to be altered.

Thus, it is of significant importance to use age- and sex-corrected upper limits of normal in reporting the ECG.

This succinct survey has shown that significant sex differences in ECG measures exist. They may not always be acknowledged in clinical practice, but they can very definitely be incorporated into automated techniques for ECG interpretation. In this way, a second opinion is provided to the cardiologist in respect of all of the measures derived automatically. The ECG is nowadays said not to contain the same appeal to younger physicians as do other cardiological investigations such as echocardiography, MRI, etc., but it still provides a wealth of information. ECG interpretation could and should be optimised using age-, sex- and race-based criteria.

References

1. Simonson E, Harris R. Differentiation between normal and abnormal in electrocardiography. St Louis: Mosby; 1961.
2. Stallman EW, Pipberger HV. Automatic recognition of electrocardiographic waves by digital computer. Circ Res. 1961;9:1138–43.
3. Caceres CA, Steinberg CA, Abrahams S. Computer extraction of electrocardiographic parameters. Circulation. 1962;25:356–62.
4. Bonner RE, et al. A new computer program for analysis of scalar electrocardiograms. Comput Biomed Res. 1972;5:629–53.
5. Macfarlane PW, Devine B, Clark E. The University of Glasgow (Uni-G) ECG analysis program. Comput Cardiol. 2005;32:451–4.
6. Macfarlane PW, Systems L. In: Macfarlane PW, et al., editors. Comprehensive electrocardiology. London: Springer-Verlag; 2011. p. 375–425.
7. Fu Q. Hemodynamic and electrocardiographic aspects of uncomplicated singleton pregnancy. In: Kerkhof PLM, Miller VM, editors. Sex-specific analysis of cardiovascular function. Cham: Springer; 2018. pp. 413–31.
8. Macfarlane PW, Veitch Lawrie TD. The normal electrocardiogram and vectorcardiogram. In: Macfarlane PW, et al., editors. Comprehensive electrocardiology. London: Springer; 2011. p. 483–546.
9. Macfarlane PW. Adult normal limits. Appendix 1. In: Macfarlane PW, et al., editors. Comprehensive electrocardiology. London: Springer; 2011. p. 2057–83.
10. Macfarlane PW, et al. Normal limits of the high fidelity pediatric ECG: preliminary observations. J Electrocardiol. 1990;22 (Suppl):162–8.
11. Macfarlane PW, McLaughlin SC, Rodger JC. Influence of lead selection and population on automated measurement of QT. Circulation. 1998;98:2160–7.
12. Macfarlane PW. Paediatric normal limits. Appendix 2. In: Macfarlane P, et al., editors. Comprehensive electrocardiology. London: Springer; 2011. p. 2128–95.
13. Macfarlane PW, et al. A new 12 lead pediatric ECG interpretation program. J Electrocardiol. 1990;23(Suppl):76–81.
14. Chen CY, Chiang BN, Macfarlane PW. Normal limits of the electrocardiogram in a Chinese population. J Electrocardiol. 1989;22:1–15.
15. Katibi I. Establishment of normal limits of the electrocardiogram in healthy Nigerians using automated methods. Glasgow: Institute of Cardiovascular and Medical Sciences, University of Glasgow; 2011.
16. Katibi I, et al. Normal limits of the electrocardiogram in Nigerians. J Electrocardiol. 2013;46:289–95.
17. Macfarlane PW, et al. Normal limits of the electrocardiogram in Indians. J Electrocardiol. 2015;48:652–68.

18. Macfarlane PW, et al. Racial differences in the ECG – selected aspects. J Electrocardiol. 2014;47(6):809–8014.
19. Morris JJ Jr, et al. P wave analysis in valvular heart disease. Circulation. 1964;29:242–52.
20. Park MK, Guntheroth WG. How to read pediatric ECGs. St Louis: Mosby; 1992.
21. Fuchs A, et al. Normal values of left ventricular mass and cardiac chamber volumes assessed by 320-detector computed tomography angiography in the Copenhagen General Population Study. Eur Heart J Cardiovasc Imag. 2016;17:1009–17.
22. Sokolow M, Lyon TP. The ventricular complex in left ventricular hypertrophy as obtained by unipolar precordial and limb leads. Am Heart J. 1949;37:161–86.
23. Macfarlane PW, et al. Methodology of ECG interpretation in the Glasgow program. Methods Inf Med. 1990;29:354–61.
24. Maron B, et al. Eligibility and disqualification recommendations for athletes with cardiovascular abnormalities: task force 2: Preparticipation screening for cardiovascular disease in competitive athletes. Circulation. 2015;132:e267–72.
25. Corrado D, et al. Cardiovascular pre-participation screening of young competitive athletes for prevention of sudden death: proposal for a common European protocol: consensus statement of the Study Group of Sport Cardiology of the Working Group of Cardiac Rehabilitation and Exercise Physiology and the Working Group of Myocardial and Pericardial Diseases of the European Society of Cardiology. Eur Heart J. 2005;26:516–24.
26. Corrado D, et al. Italian preparticipation screening for prevention of sudden cardiac death: it works! Br Med J. 2016;353: i1156.
27. Drezner JA, et al. International criteria for electrocardiographic interpretation in athletes: consensus statement. Br J Sports Med. 2017;51:704–31.
28. Drezner JA, Ackerman MI, Anderson J. Electrocardiographic interpretation in athletes: the Seattle criteria. Br J Sports Med. 2013;47:122–4.
29. Casale PN, et al. Electrocardiographic detection of left ventricular hypertrophy: development and prospective validation of improved criteria. J Am Coll Cardiol. 1985;6:572–80.
30. Casale PN, et al. Improved sex-specific criteria of left ventricular hypertrophy for clinical and computer interpretation of electrocardiograms: validation with autopsy findings. Circulation. 1987;75:565–72.
31. Dahlof B, et al. Characteristics of 9194 patients with left ventricular hypertrophy: the LIFE Study. Hypertension. 1998;32:989–97.
32. Macfarlane PW, Clark E, Cleland JGF. New criteria for LVH should be evaluated against age. J Am Coll Cardiol. 2017;70:2206–7.
33. Romhilt DW, Estes EH Jr. A point-score system for the ECG diagnosis of left ventricular hypertrophy. Am Heart J. 1968;75:752–8.
34. Morrison I, Clark E, Macfarlane P. Evaluation of the electrocardiographic criteria for left ventricular hypertrophy. Anatolian J Cardiol. 2007;7(Suppl 1):159–63.
35. Strauss DG, et al. Defining left bundle branch block in the era of cardiac resynchronization therapy. Am J Cardiol. 2011;107:927–34.
36. Macfarlane PW. Age, sex and the ST amplitude in health and disease. J Electrocardiol. 2001;34:235–41.
37. Thygesen K, et al. Third Universal Definition of myocardial infarction. ESC/ACCF/AHA/WHF Expert Consensus Document. Circulation. 2009;53(11):1003–11.
38. Wagner GS, et al. AHA/ACCF/HRS recommendations for the standardization and interpretation of the electrocardiogram. Part VI: acute ischemia/infarction. A scientific statement from the American Heart Association Electrocardiography and Arrhythmias Committee, Council on Clinical Cardiology; The American College of Cardiology Foundation and the Heart Rhythm Society. J Am Coll Cardiol. 2009;53 (11):1003–11.
39. Macfarlane PW, et al. Evaluation of age and sex dependent criteria for ST elevation myocardial infarction. Comput Cardiol. 2007;34:293–6.
40. Clark EN, et al. Automated electrocardiogram interpretation programs versus cardiologists' triage decision making based on Teletransmitted data in patients with suspected acute coronary syndrome. Am J Cardiol. 2010;106:1696–702.
41. Papadakis M, et al. The prevalence, distribution and clinical outcomes of electrocardiographic repolarization patterns in male athletes of African/Afro-Caribbean origin. Eur Heart J. 2011;32:2304–13.
42. Kligfield P, et al. Comparison of automated measurements of electrocardiographic intervals and durations by computer based algorithms of digital electrocardiographs. Am Heart J. 2014;167(2):150–159.e1.
43. Bazett HC. An analysis of the time relations of electrocardiograms. Heart. 1920;20:353–70.
44. Vandenberg B, et al. Which QT correction formulae to use for QT monitoring? J Am Heart Assoc. 2016;5:e003264.
45. Hodges M, Salerno D, Erlien D. Bazett's QT correction reviewed. Evidence that a linear QT correction for heart is better. J Am Coll Cardiol. 1993;1:694. (Abstract)
46. Fridericia LS. Die Systolendauer im Elektrokardiogram bei normalen Menschen und bei Herzkranken. Acta Med Scand. 1920;53:469–86.
47. Luo S, et al. A comparison of commonly used QT correction formulae: the effect of heart rate of the QTc of normal ECGs. J Electrocardiol. 2004;37(Suppl):81–90.
48. Rautaharju PM, Surawicz B, Gettes LS. AHA/ACCF/HRS recommendations for the standardization and interpretation of the electrocardiogram. Part IV: the ST segement and U waves, and the QT interval. A scientific statement from the American Heart Association Electrocardiography and Arrhythmias Committee, Council on Clinical Cardiology; the American College of Cardiology Foundation; and the Heart Rhythm Society. J Am Coll Cardiol. 2009;53(11):982–91.
49. Johannesen L, et al. Differentiating drug-induced multichannel block on the electrocardiogram: randomized study of dofetilide, quinidine, ranolazine and verapamil. Clin Pharmacol Ther. 2014;96:549–58.
50. Rautaharju PM, et al. Assessment of prolonged QT and JT intervals in ventricular conduction defects. Am J Cardiol. 2004;93:1017–21.
51. Symposium, The J-Tpeak Initiative. J Electrocardiol. 2017;50:748–75.

Safety and Effectiveness of Medical Device Therapy

7

Robbert Zusterzeel

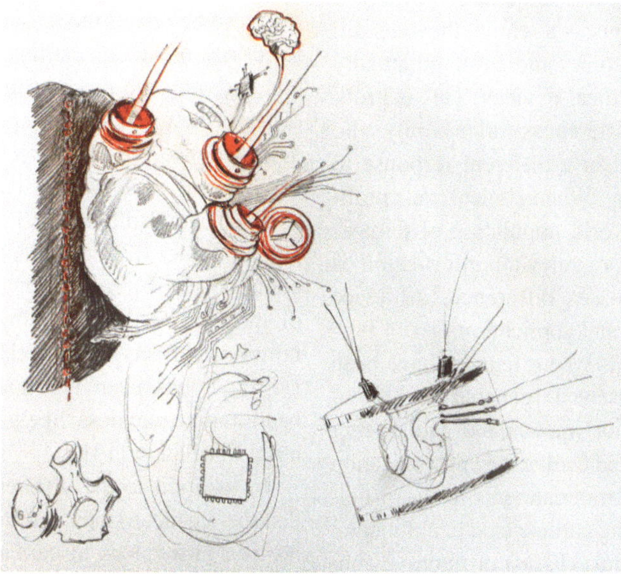

Medical device therapy in cardiology. Art work by Piet Michiels, Leuven, Belgium

Abstract

When enough females and males are enrolled in clinical trials, much more relevant information is available on potential sex differences in device safety and effectiveness. Unfortunately, females have largely been underrepresented in clinical studies of cardiac medical device therapies for heart failure. In this chapter, sex differences in heart failure characteristics and cardiac electrophysiology and their influence on the safety, effectiveness, and application of implantable cardioverter defibrillators (ICDs), subcutaneous ICDs (S-ICDs), and cardiac resynchronization therapy (CRT) will be discussed. In this way, the research community will hopefully become more appreciative of the potential differences in device effects between females and males.

R. Zusterzeel (✉)
Division of Applied Regulatory Science, U.S. Food and Drug Administration, Silver Spring, MD, USA
e-mail: robbert.zusterzeel@fda.hhs.gov

P. L. M. Kerkhof, V. M. Miller (eds.), *Sex-Specific Analysis of Cardiovascular Function*,
Advances in Experimental Medicine and Biology 1065, https://doi.org/10.1007/978-3-319-77932-4_7

Keywords
Medical devices · Sex differences · Mortality · Demographic subgroups · Defibrillator · Pacemaker · Female

Introduction

Professional society guidelines are used by physicians to apply evidence-based treatment and are largely based on well-designed randomized controlled clinical trials. Even though females now comprise a large portion of clinical trial participants as a whole, they are still underrepresented in some studies of potentially lifesaving cardiac medical devices. This is problematic since sex differences in anatomy and physiology can result in a different response to medical device therapy. When clinical trial enrollment reflects a real-world population of females and males, much more relevant information is available on potential sex differences in device safety, effectiveness, and application.

Cardiac device trials where females have been especially underrepresented (generally <30% of enrollees) are those for implantable cardioverter defibrillators (ICDs) and cardiac resynchronization therapy (CRT). Both are treatments for heart failure based on modifying cardiac electrical function to either treat arrhythmia (ICDs) or improve contraction (CRT). It is, however, important to note that there are significant differences in underlying heart failure characteristics and normal electrical function between females and males that can influence the safety and effectiveness of these devices.

Sex Differences in Heart Failure and Cardiac Electrophysiology

To properly understand sex differences in cardiac device therapy, it is best to start with a general overview of heart failure characteristics, cardiac electrophysiology, and mechanisms of cardiac death in females versus males. We will then come back to these factors and how they tie into device response in the specific sections on ICDs and CRT.

In terms of heart failure, females may present with different underlying characteristics than males. Even though the prevalence of heart failure is similar in both sexes, females comprise only ~35% of the patients with left ventricular systolic dysfunction indicated by a low left ventricular ejection fraction (LVEF, <35%), the rest having LVEF ≥35–40% [13]. In addition, females present with lesser scar tissue after myocardial infarction which contributes to the higher incidence of nonischemic causes of heart failure in females compared to males [13]. This may also be one of the reasons why females have lower risk of sudden cardiac death (SCD) in general and why the mechanism of cardiac death is different between sexes; females are more likely to experience pulseless electrical activity (PEA) or asystole, while ventricular tachycardia (VT)/ventricular fibrillation (VF) is more common among males [39]. An additional explanation could be sex differences in cellular electrophysiological properties, autonomic modulation, and hormonal effects on ion channels in the heart. Lastly, it has been reported that females with heart failure are less likely to have underlying atrial fibrillation [13].

In addition to the differences in heart failure disease characteristics and modes of death, there are sex differences in cardiac electrophysiology that can potentially influence device response. Females have ~5 ms shorter QRS duration than males on the electrocardiogram (ECG) at baseline [29], likely because of their smaller heart size. With heart failure, the left ventricle in both females and males dilates, and conduction becomes less rapid causing dyssynchrony in septal and left ventricular lateral wall contraction and therefore lower LVEF. The most severe form of this is a left bundle branch block (LBBB) (Fig. 7.1) where the two walls are completely uncoupled and the lateral wall contracts ~100 ms later than the septum [47]. It is postulated that since females have shorter QRS duration at baseline, they may require less QRS duration prolongation, and are therefore more

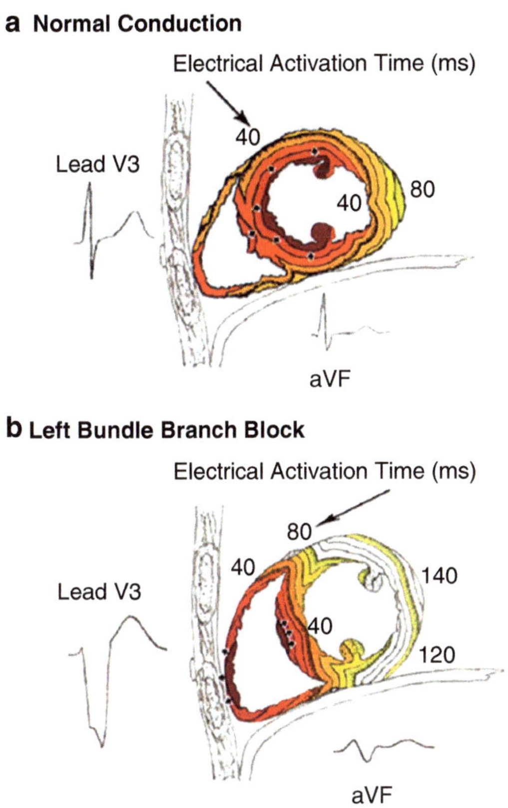

Fig. 7.1 Schematic representation of normal and LBBB activation. Sagittal view of the heart's ventricles in normal electrical activation (**a**) and left bundle branch block (LBBB, **b**); activation starts at the small arrows and spreads with each line representing 10 ms. The delay between septum and left ventricular lateral wall activation is ~40 ms in normal activation but ~100 ms in LBBB. (Adapted from Strauss et al. [46] (with permission))

susceptible, to develop LBBB compared with males [47].

Sex-Specific Safety and Effectiveness of Implantable Defibrillators

Implantable defibrillators have been around since the 1980s, and, even though the technology has significantly advanced since then, their main therapeutic goal of terminating ventricular arrhythmias is still the same. ICDs generally have one lead in the right ventricle (single chamber) or additionally in the right atrium (dual chamber) (Fig. 7.2) and can provide three types of therapy: regular pacing as with a normal pacemaker, antitachycardia pacing (ATP), or defibrillation. ATP aims to terminate ventricular tachycardia by sending pacing pulses to the myocardium at a higher frequency than those caused by the arrhythmia itself, thereby terminating tachycardia and restoring normal sinus rhythm. If ATP is not successful or there is ventricular fibrillation, a defibrillation shock will be applied attempting to terminate the arrhythmia by creating an electrical current between the RV lead and the ICD pulse generator.

Clinical indications for use of ICDs developed by professional societies in the United States (US) (American Heart Association (AHA)/American College of Cardiology (ACC)/Heart Rhythm Society (HRS)) [18] and Europe (European

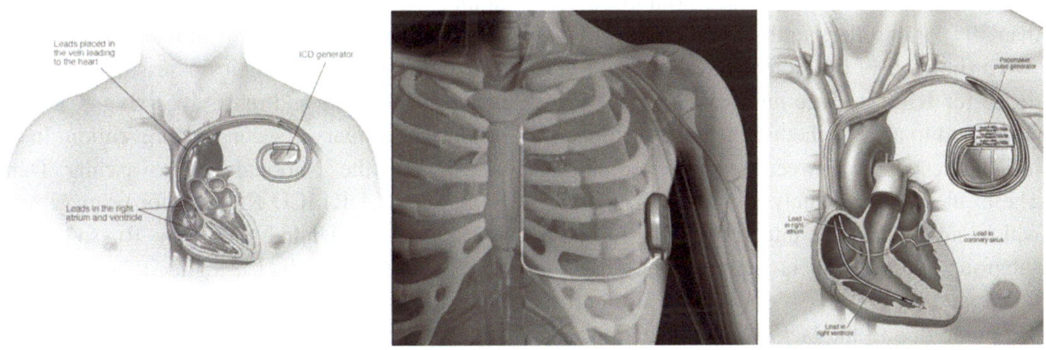

Fig. 7.2 Placement of ICD (left), S-ICD (middle), and CRT(-D) (right) pulse generators and leads. *ICD* implantable cardioverter defibrillator, *S-ICD* subcutaneous ICD, *CRT(-D)* cardiac resynchronization therapy (-defibrillator). Adapted from mayoclinic.org (with permission)

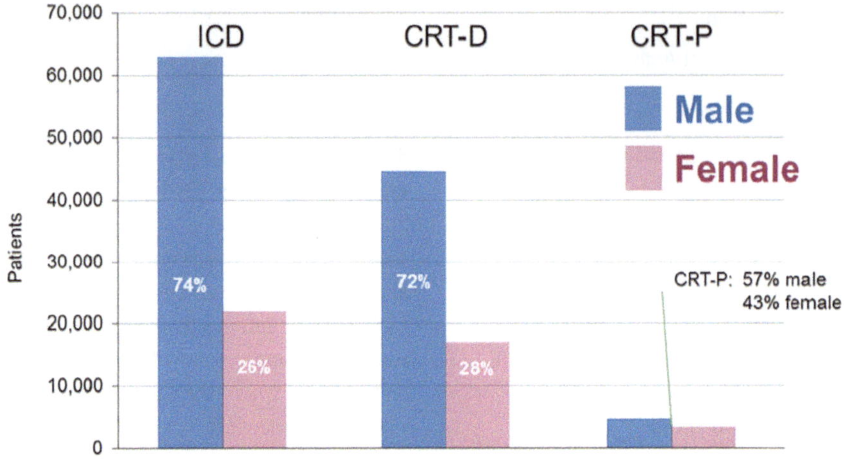

Fig. 7.3 Distribution of male and female patients with implantable devices. Distribution of implantable cardioverter defibrillators (ICD), cardiac resynchronization therapy (CRT)-defibrillator (CRT-D), and CRT-pacemaker (CRT-P) implantations by sex in a nationwide US cohort between 2008 and 2011. (Adapted from Varma et al. [50] (with permission))

Society of Cardiology (ESC)) [36] are largely based on the New York Heart Association (NYHA) heart failure class and LVEF. At the time of writing, ICD therapy is generally applied in patients who survived a previous cardiac arrest (secondary prevention) or those with LVEF ≤30–40% (primary prevention), depending on the specific indication. The overall profound survival benefit of ICDs used for primary prevention demonstrated in the large pivotal trials in the 1990s and 2000s, such as MADIT-II, DEFINITE, and SCD-HeFT, has been the basis for these recommendations. Unfortunately, however, the evidence of safety and effectiveness is much stronger in males since females comprised less than 30% of enrollees in the listed clinical trials.

To (hopefully) overcome the sex bias in evidence for ICDs, a large number of studies were conducted to assess whether there are differences in ICD response between females and males. These included subanalyses of clinical trials, which are notoriously underpowered to answer subgroup questions, meta-analyses, and observational studies. However, causality is difficult to determine due to the lack of randomization, and analyses have demonstrated conflicting results. In addition, after the original clinical trials were conducted and the guidelines were established, real-world use of ICDs has still been significantly more prevalent among males [33, 45, 50] (Fig. 7.3).

Sex-specific analysis of the pivotal trials MUSTT [9] and MADIT-II [32] for primary prevention have shown no difference in mortality outcomes between females and males, while a substudy of the SCD-HeFT trial [2] demonstrated lower mortality in females. A subsequent meta-analysis of all the above trials with additional inclusion of DEFINITE and COMPANION found that there were no sex differences in mortality [42]. However, most of these trials were conducted almost two decades ago, and heart failure treatment in general has changed significantly since then. Starting around 2013, there have been numerous studies assessing mortality benefit in females and males receiving ICDs for primary or primary + secondary prevention. In an analysis of the National Cardiovascular Data Registry (NCDR) ICD Registry for primary prevention in the United States by Russo et al., females with ICDs had a higher mortality than males with ICDs; however, it was also shown that females had greater comorbidity and more advanced heart failure at baseline [41]. A later analysis in the same Registry demonstrated that when matching females and males with ICDs

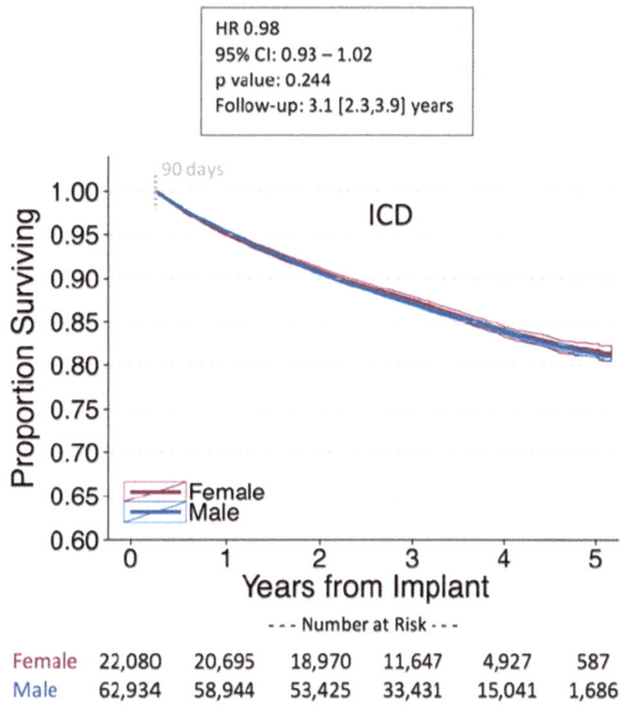

Fig. 7.4 Post-implant survival in females and males with ICD Unadjusted Kaplan-Meier survival rates for implantable cardioverter defibrillator (ICD) patients by sex (bottom) and female-to-male hazard ratios (HR) for all-cause mortality (top) in a nationwide US cohort between 2008 and 2011. (Adapted from Varma et al. [50] (with permission))

based on baseline characteristics, thereby creating a more equal distribution of comorbidity and heart failure, mortality differences between sexes disappeared [52]. A third analysis of the Registry again showed that females had better survival than males [31]. These discrepancies in mortality benefit are, however, not limited to studies of US Registries and can be seen in international registries as well. Bhavnani et al. (in a community cohort) [5], Providencia et al. (in a French registry) [37], Varma et al. (in a remote monitoring database – Fig. 7.4) [50], and Zeitler et al. (in the Get With The Guidelines-Heart Failure [GWTG-HF] study) [53] demonstrated similar mortality between sexes, while a meta-analysis of 20 studies by Conen et al. [14] and a more recent analysis by Sticherling et al. in a combined registry of 11 European countries [45] showed a more favorable survival benefit among females. Some of these studies were conducted just among primary prevention populations, while others included subjects with ICDs for both primary and secondary prevention purposes.

Another part of the balance between benefit and risk for ICDs includes procedural complications and appropriate and inappropriate ICD therapy. Masoudi et al. [31] and Gold et al. [23] demonstrated that females had a higher risk of complications than males including pneumothorax, hemothorax, tamponade, and pericardiocentesis, likely related to smaller heart size and vasculature. Similarly, female sex was an independent predictor of complications associated with lead extractions [17], which are nowadays being performed more often. In addition, in the PAIDLESS study, there were more lead failures in females indicated by an average lead survival time of 13.4 vs. 14.7 years in males; however, this did not result in a mortality difference between sexes [21]. At the same time, females had lower risk of appropriate therapy and generally no difference in inappropriate therapies with ICD; only one substudy of the MADIT-CRT trial demonstrated a lower risk of inappropriate therapy in females compared to males [48].

Finally, three studies investigated social health factors between females and males receiving ICDs. Christensen et al. noted lower quality of life (QoL) and mental health questionnaire scores in females [11], while Pedersen et al. demonstrated that females were more likely to experience anxiety, depression, and fear of ICD shocks compared with males [34]. Hess et al. in addition showed that females receive less ICD counseling than males [24].

Overall, the current evidence for ICD safety and effectiveness generally points toward underuse in females compared to males but also higher complication rates, lower social health, lower appropriate therapy, and an unsure mortality benefit in females. Potential reasons for lower utilization rates in females include these higher risk factors or the plain fact that females do not meet current clinical guideline criteria because of relatively preserved LVEF compared to males, as discussed earlier. However, it has been shown that even when females meet the guideline indications for ICD, they still do not receive it as often as males. The lower appropriate ICD therapies could be explained by females having a lower risk of SCD in general and by the fact that they less often have VT/VF as their mode of death. These two characteristics also go together with females more often having nonischemic cardiomyopathy underlying heart failure with lower amounts of scar than males [28] potentially leading to less of an arrhythmic substrate for ICDs to be beneficial. This is further corroborated by Kwon et al. who showed that males who have larger scars have lower mortality with ICDs [26], likely because this leads to more VT/VF treated by ICDs. Interestingly, however, in females with ICDs, larger scars were shown to be associated with higher mortality. This could indicate that females with larger myocardial scars do not actually develop more VT/VF but instead die from worsening heart failure. Finally, it should be mentioned that LVEF used in clinical guidelines is an inaccurate predictor of risk of SCD, especially in females [15].

More recent developments in the defibrillator space include the approval of the subcutaneous ICD (S-ICD) in the early 2010s. The S-ICD is different from the transvenous ICDs discussed above as it only has one subcutaneous lead implanted next to the sternum and a pulse generator that is implanted on the patients left rib cage, thereby preventing implantation of any transvenous leads (Fig. 7.2). The S-ICD was mainly developed for patients with underlying congenital abnormalities, limited or difficult vascular access precluding placement of transvenous leads and to reduce lead-related complications and infection. Because of its configuration, it cannot apply pacing or ATP like the transvenous ICD but provides defibrillation when necessary. The S-ICD was shown in the pivotal study for approval in the United States (which included younger and healthier subjects compared to the population receiving transvenous ICDs) to be equally safe and effective as transvenous ICDs for treatment of life-threatening ventricular arrhythmias [51].

One would think that the experience with the transvenous ICDs in terms of potential sex differences would lead to the inclusion of more females in clinical trials for S-ICDs, especially given the fact that females suffer from more lead-related complications. Unfortunately, females still comprised only ~26% of patients in the pivotal IDE trial for approval in the United States [23]. Post-approval real-world use in females does not seem to be much better; the subsequent post-approval study [23] and the EFFORTLESS S-ICD Registry [8] have included ~30% females. This low enrollment could potentially be explained by the fact that patients in the S-ICD trials had to have general indications for transvenous ICDs, which suffer from the same sex biases as discussed earlier. Furthermore, even though an analysis of the S-ICD post-approval study showed a higher absolute rate of complications in females versus males (5.8 vs. 2.8%) [23], it is too early to definitively conclude that females benefit less from S-ICDs than males, also given the fact that no sex-specific outcome studies have been performed to date.

The sad conclusion to all of this is that after more than 30 years of technological advancement in the implantable defibrillator space, we still do not know whether they are truly as beneficial in females as in males. There is a need for more inclusion of females in contemporary ICD practices in order to perform larger (randomized) studies that can fully elucidate their benefits and risks and to develop better, possibly sex-specific, predictors of SCD.

Sex-Specific Safety and Effectiveness of Cardiac Resynchronization Therapy

The first large clinical trials on CRT, which is more commonly called biventricular pacing, were conducted during the early 2000s, but CRT treatment has only become more mainstream since the early 2010s with the expanded indications for patients with less severe heart failure. CRT devices are very similar to regular pacemakers and ICDs with one lead in the right atrium and one in the right ventricle, but in addition they have a lead on the epicardium of the left ventricular lateral wall (Fig. 7.2). CRT can either be used on its own (CRT-pacemaker [CRT-P]) or in combination with a defibrillator (CRT-defibrillator [CRT-D]). The right and left ventricular leads together conduct pacing impulses to their respective areas with the goal of restoring physiological mechanical contraction of the left ventricle, thereby increasing pump function (Fig. 7.5). As discussed earlier, in heart failure, the left ventricle dilates leading to dyssynchronous contraction of the septum and the left ventricular free wall, and this dyssynchrony is most severe, and likely most amenable to CRT, with a complete LBBB.

Current clinical indications for CRT developed by the same US and European authorities as for ICDs are mainly based on clinical trials and meta-analyses of clinical trials and use NYHA heart failure class, LVEF, QRS morphology, and QRS duration on the ECG to determine specific indications [18, 35]. The US and European guidelines are somewhat different but mainly recommend CRT treatment in patients with a LVEF ≤35%, LBBB morphology, and QRS duration ≥150 ms (≥130 ms in current European guidelines), while patients without LBBB generally receive lower indications. Unfortunately, as with ICDs, the clinical trials for CRT included a low number of females (only ~20%), and the guidelines may therefore not fully reflect the best treatment options for females.

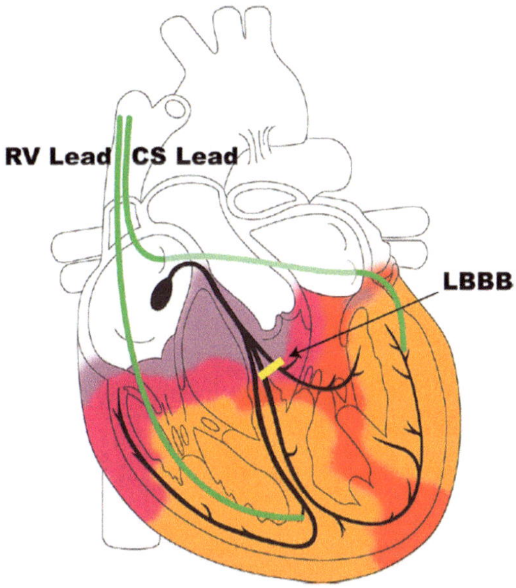

Fig. 7.5 Electrical resynchronization of LBBB with CRT Concurrent pacing of the septum with the right ventricular lead (RV lead) and left ventricular free wall with the coronary sinus lead (CS lead) leads to electrical resynchronization of the left ventricle; earliest activation in orange. LBBB, left bundle branch block. (Adapted from Rao and Faddis [38] (with permission))

A wealth of information on sex-specific CRT effects is available in the literature. They include subgroup analyses from clinical trials, meta-analyses, large registry, and claims analysis and single-center studies conducted mainly in patients with NYHA Class II or III heart failure symptoms. As mentioned before, causality is difficult to determine due to a lack of randomization; however, it is encouraging that almost all of the studies that assessed hard endpoints such as death or heart failure events have pointed in the same direction, a more beneficial effect of CRT in females compared to males (Fig. 7.6). This effect was detected regardless of the fact that females are less likely to receive CRT and are underrepresented in real-world use (Fig. 7.3). Analogous to the clinical guidelines, most sex-specific analyses have made a distinction between patients with NYHA Class II and III heart failure symptoms even though it is questionable whether this is the best measure to determine benefit since heart failure symptoms are mostly subjective.

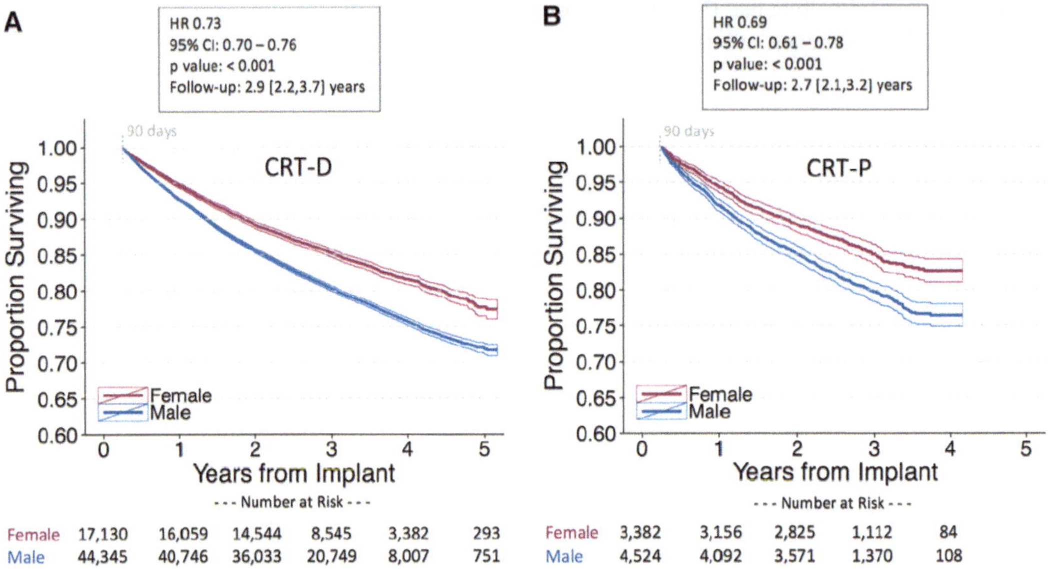

Fig. 7.6 Post-implant survival in females and males with CRT-D and CRT-P. Unadjusted Kaplan-Meier survival rates for cardiac resynchronization therapy defibrillator (CRT-D) (**a**) and CRT-pacemaker (CRT-P) (**b**) patients by sex (bottom) and female-to-male hazard ratios (HR) for all-cause mortality (top) in a nationwide US cohort between 2008 and 2011. (Adapted from Varma et al. [50] (with permission))

The MADIT-CRT, RAFT, and REVERSE trials, which compared patients with CRT-D to those with an ICD, were responsible for the expanded indications in patients with less severe heart failure (NYHA Class II). Most sex-specific information in this category therefore comes from subgroup or meta-analyses of these trials. Two separate substudies of the MADIT-CRT trial showed that females with CRT-D had lower risk of death or heart failure hospitalization than males with CRT-D [1, 6]. The second study further stratified females and males by the presence of LBBB. It was shown that females with LBBB had lower risk of death or heart failure than males with LBBB, even more so than in the first study that compared all females and males together regardless of QRS morphology [6]. In addition, while the first study demonstrated that only females benefited with a QRS duration <150 ms and both sexes benefited with QRS duration ≥150 ms [1], the second study added that this was mainly seen in patients with LBBB [6]. Meanwhile, neither females nor males with non-LBBB morphology benefited from CRT. A - patient-level meta-analysis of all three clinical trials corroborated all these findings but was also able to further stratify patients by 10 ms increments in QRS duration. Patients with LBBB and QRS duration 120–129 ms did not benefit, regardless of sex, while both sexes benefited with LBBB and QRS duration ≥150 ms. Interestingly, in the middle part of the LBBB QRS duration spectrum (130–149 ms), only females demonstrated lower risk of death or heart failure, while males did not [55] (Fig. 7.7, left).

Multiple clinical trials enrolled patients with NYHA Class III heart failure symptoms. A study-level meta-analysis of COMPANION, CARE-HF, and MADIT-CRT along with 42 other observational studies found that females had lower mortality risk and lower composite heart failure or death risk than males [10]. Meanwhile, a subsequent patient-level meta-analysis of the MIRACLE, MIRACLE-ICD, CARE-HF, REVERSE, and RAFT trials did not demonstrate any sex differences for the same endpoints; however, this comparison included a mixed control group of patients with ICDs, medical treatment, or standard pacing [12]. Just as in patients with

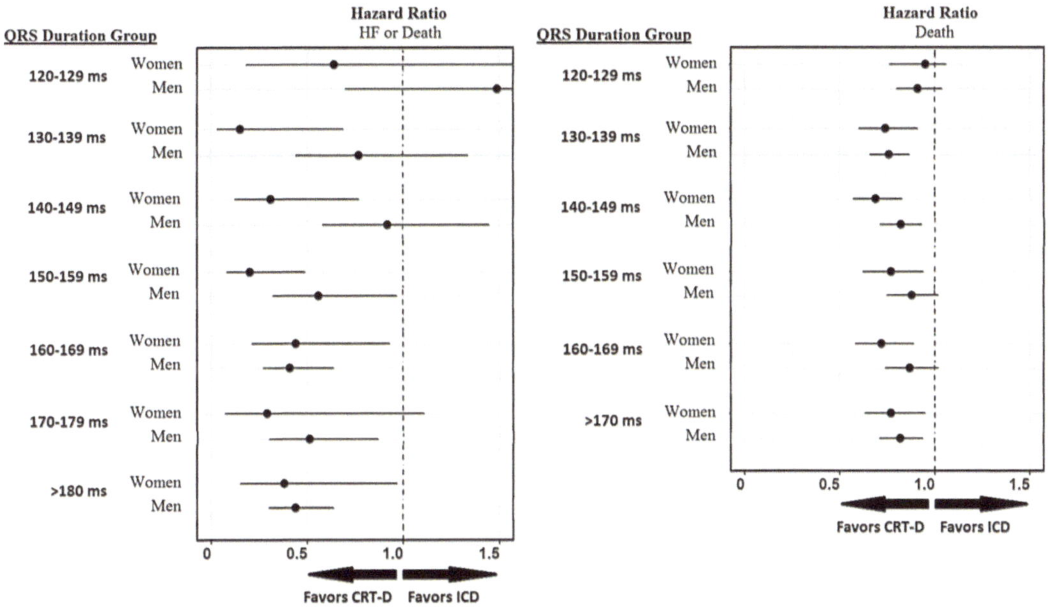

Fig. 7.7 CRT-D benefit across 10 ms QRS duration groups in females and males with LBBB. Left, CRT-D-to-ICD hazard ratios in patients with left bundle branch block (LBBB) for the outcome of heart failure (HF) event or death in patients with predominantly New York Heart Association (NYHA) Class II heart failure symptoms; right, CRT-D-to-ICD hazard ratios in patients with LBBB for the outcome of death in patients with predominantly NYHA Class III heart failure symptoms. Adapted from [55, 56] (with permission)

NYHA Class II heart failure symptoms, LBBB was an important predictor of benefit; an insurance claims data analysis of CRT-D patients revealed that females with LBBB had a lower mortality risk as well as a lower risk of heart failure hospitalization or death than males with LBBB [27]. This was later confirmed by two large studies in the NCDR ICD Registry. These Registry analyses also showed that in both females and males, CRT-D resulted in lower mortality risk with LBBB and QRS duration ≥140 ms (but the effect was larger among females with LBBB and QRS duration 140–149 ms) [54], while the second study detected a CRT-D benefit for both sexes with LBBB and QRS duration ≥130 ms [56] (Fig. 7.7, right). There was no sex difference in CRT-D benefit in patients with non-LBBB morphology. A subanalysis of the ECHO-CRT trial also confirmed that in those with QRS duration <130 ms, there was no benefit of CRT in either sex and may even cause harm [44]. Finally, echocardiographic CRT response in females is also generally better than in males, but these are not discussed here since these measures do not necessarily correlate well with harder endpoints such as heart failure events or mortality and their role in determining benefit is unclear [22].

The above clinical studies have exclusively assessed benefit from CRT-D; however, not much is known about the potential differences in effects of CRT-D versus CRT-P, let alone by sex. This is likely explained by a large overlap in clinical indications for CRT and ICD devices which causes them to often be applied together. In addition, CRT-P devices are not captured in registry data. Only two studies assessed sex differences in CRT-P versus CRT-D. Varma et al., in a large remote monitoring database, showed that CRT-P accounts for ~14% of all CRT device implantations and that there was a somewhat better distribution of CRT-P implantations between females and males compared to CRT-D (43% in females versus 57% in males) [50] (Fig. 7.3). They also demonstrated that, next to females having superior survival with CRT-D compared to males with CRT-D, females also had better survival with CRT-P (Fig. 7.6). Another study by Barra et al. indicated that there was equal survival benefit of CRT-Ps compared to CRT-Ds, regardless of sex [3]. Coming back to the discussion on ICDs, these findings may suggest that complication rates of ICDs among females as discussed above are not a reason for lower defibrillator use. In addition, this could further explain the finding that females receive less appropriate ICD therapy since females who are CRT responders also had a reduced incidence of VT/VF likely caused by increased reverse remodeling creating less of an arrhythmic substrate compared to males. However, more studies are needed to confirm these findings.

Overall, the current evidence points toward females receiving CRT less often than males even though they have a lower risk of mortality and heart failure events. This sex-specific benefit seems largest in the presence of a LBBB. Some differences were noted with regard to QRS duration in patients with NYHA Class II or III heart failure; females with LBBB and QRS duration ≥130 ms benefited from CRT, while males with LBBB only benefited with QRS duration ≥150 ms in NYHA Class II. In NYHA Class III both females and males with LBBB benefited from CRT with QRS duration ≥130 ms. Non-LBBB patients generally did not benefit from CRT, regardless of sex. As with ICDs, the underrepresentation of females may be caused by the fact that they do not meet clinical indications for CRT implantation since they generally have heart failure with relatively preserved LVEF, although it was shown that females receive CRT less often than males even when they do meet professional society guideline indications. Potential explanations for the higher benefit in females could be that females more often have nonischemic causes of heart failure and less atrial fibrillation, where CRT has proven to be more effective, or more often have LBBB and therefore more dyssynchrony amenable to CRT. In a CRT cohort, females were indeed shown to have less myocardial scar compared to males especially in the area where the left ventricular lead is placed [28]. In terms of LBBB, as discussed earlier, females may more often have a true LBBB at lower QRS durations than males due to smaller heart size and shorter QRS duration at baseline. This difference between females and males was captured in newly proposed "strict" ECG criteria for LBBB which require a shorter QRS duration in females than in males [47]. The new sex-specific criteria indeed predicted CRT benefit better than other ECG criteria, both in terms of left ventricular mechanical dyssynchrony [40] and long-term mortality and heart failure outcomes [30]. However, they could benefit from further improvement as it was found that left ventricular electrical activation in some patients with strict LBBB was still heterogeneous [25]. This may also be the reason why a recent study found no differences in echocardiographic benefit from CRT in patients meeting AHA versus strict LBBB criteria [4]. In addition, Varma et al. demonstrated that the sex difference in QRS duration response to CRT was not fully explained by application of the new LBBB criteria or by additional correction for body surface area (BSA) as a measure for differences in ventricular size between females and males [49]. Instead, it was found that sex differences by QRS duration were only resolved by normalization for ventricular size using left ventricular mass. This could indicate that, in addition to QRS duration, differences in ventricular mass need to be taken into account when defining true LBBB; however, this may be difficult when using the ECG alone since criteria for determining ventricular mass have not proven to be accurate [43].

In summary, there is evidence to support that females benefit more from CRT than males even though they are less likely to receive it. As also outlined in a recent paper, the current clinical guidelines for CRT could be updated to reflect this information [57]. Still, more females should be included in trials of cardiac resynchronization to be able to fully understand sex differences in device benefit, especially since application of true resynchronization therapy has been continuously improving to fit individual needs.

Developments and Recommendations to Enhance Sex-Specific Information of Medical Device Therapies

From the above discussion, it is clear that sex differences exist in the application, effectiveness, and safety of some cardiac medical devices, while there is not enough information for others. Without proper representation of both sexes in clinical studies, it is difficult to assess whether devices are equally effective and safe in females and males and, when there are differences, what potential explanations for this could be.

There can be multiple reasons why females are underrepresented in cardiac medical device studies, one of them being that females do not meet guideline indications as discussed earlier. This is, however, somewhat of a vicious cycle when guidelines are mainly based on male characteristics. Other reasons may include a lack of understanding about main obstacles to female participation or about sex differences in disease pathophysiology, fear of potential effects on females of childbearing age, family responsibilities, or avoidance of female patients because of the perception that it takes more time and money to recruit them [19]. Interestingly, in 1 study where 20 pregnancies were conceived by 12 females carrying ICD devices, pregnancy had no effect on ICD operation, and there was no evidence that could link the presence of the device to adverse maternal or fetal outcomes; however, one miscarriage was observed that could have been associated with an ICD shock [7].

Regulatory agencies, such as the US Food and Drug Administration (FDA), have been looking into possibilities to enhance inclusion of females in clinical studies for medical product approval. In 2014, the FDA published an action plan that contains recommendations for improving the quality, transparency, and diversity of clinical data on females (and other demographic subgroups) and to increase enrollment according to sex-specific disease prevalence [20]. In addition, FDA published a guidance document that provides a framework for how to analyze and communicate data on females and males in clinical studies for medical devices [19]. However, these measures are generally only considered recommendations and not requirements, something which is proven by an analysis by Dhruva et al. [16]. They examined 77 studies supporting 42 original device approvals in 2015 that included both females and males and found that sex composition was reported in a little less than half of the device labels. In addition, only 17% of the studies investigated results by sex of which one found less safety in males, one found less safety in females, and one found less effectiveness in females.

In order to truly overcome sex-specific bias in clinical studies, investigators need to become more aware of potential differences in the safety and effectiveness of medical products between females and males. This may only be achieved when levels of enrollment for females are adequate to determine safety and effectiveness by sex.

Summary

Females are underrepresented in some clinical trials of cardiac medical devices, especially for ICDs, S-ICDs, and CRT indicated for the treatment of heart failure, and current clinical guidelines for their applications are based on these clinical trials. This is problematic since the information on device safety and effectiveness primarily reflects the effects in males, while there are sex differences in underlying disease characteristics, anatomy, and (patho)physiology that can influence device performance. Females with heart failure usually have relatively preserved LVEF, less scar leading to more nonischemic heart failure, and a lower risk of SCD (with less underlying VT/VF) compared with males. In addition, they have smaller heart size and shorter QRS duration at baseline. Some of these factors could explain why the current, even though somewhat conflicting, evidence for ICDs points toward an underuse in females. However, a higher complication rate, lower appropriate ICD therapies, and unsure mortality benefit compared to males may be related to this observation as well. More recently, S-ICDs were developed to overcome some of the lead-related complications with regular ICDs, which are more common among females. However, studies have suffered from the same underrepresentation of females, likely because of similar indications for use of regular ICDs and S-ICDs. In addition, large sex-specific outcome studies for S-ICDs have not yet been performed. For CRT, a similar sex-specific bias was noted even though females were shown to benefit more from therapy than males. This may, again, be related to sex differences in underlying disease characteristics, anatomy, and physiology.

There may be multiple reasons why females are underrepresented in some cardiac medical device clinical studies, and regulatory agencies, such as the FDA, have been trying to address this by publishing action plans and guidance documents. Unfortunately, these have not been very effective so far. There is a need for more inclusion of females in clinical studies of cardiac devices in order to perform larger (randomized) studies that can fully elucidate their benefits and risks, especially since device therapies are continuously improving and are being applied in larger populations. To achieve this, a collective effort by researchers, patient advocacy groups, federal agencies, and industry is required.

Acknowledgments The author gratefully acknowledges the insightful comments of Dr. Marjorie Jenkins and Dr. Pamela Scott (FDA) as well as those from the editors and peer reviewers.

Disclaimer The mention of commercial products, their sources, or their use in connection with material reported herein is not to be construed as either an actual or implied endorsement of such products by the Department of Health and Human Services. *The opinions expressed are those of the author and do not necessarily reflect the views or policies of the US Food and Drug Administration.*

References

1. Arshad A, Moss AJ, Foster E, Padeletti L, Barsheshet A, Goldenberg I, Greenberg H, Hall WJ, McNitt S, Zareba W, Solomon S, Steinberg JS, Committee M-CE. Cardiac resynchronization therapy is more effective in women than in men: the MADIT-CRT (multicenter automatic defibrillator implantation trial with cardiac resynchronization therapy) trial. J Am Coll Cardiol. 2011;57(7):813–20. https://doi.org/10.1016/j.jacc.2010.06.061.
2. Bardy GH, Lee KL, Mark DB, Poole JE, Packer DL, Boineau R, Domanski M, Troutman C, Anderson J, Johnson G, McNulty SE, Clapp-Channing N, Davidson-Ray LD, Fraulo ES, Fishbein DP, Luceri RM, Ip JH, Sudden Cardiac Death In Heart Failure Trial I. Amiodarone or an implantable cardioverter-defibrillator for congestive heart failure. N Engl J Med. 2005;352(3):225–37. https://doi.org/10.1056/NEJMoa043399.
3. Barra S, Providencia R, Duehmke R, Boveda S, Marijon E, Reitan C, Borgquist R, Klug D, Defaye P, Sadoul N, Deharo JC, Sadien I, Patel K, Looi KL, Begley D, Chow AW, Le Heuzey JY, Agarwal S, French UKSCRTN. Sex-specific outcomes with addition of defibrillation to resynchronisation therapy in patients with heart failure. Heart. 2017;103(10):753–60. https://doi.org/10.1136/heartjnl-2016-310677.
4. Bertaglia E, Migliore F, Baritussio A, De Simone A, Reggiani A, Pecora D, D'Onofrio A, Rapacciuolo A, Savarese G, Pierantozzi A, Marenna B, Ruffa F, Campari M, Malacrida M, Stabile G. Stricter criteria for left bundle branch block diagnosis do not improve response to CRT. Pacing Clin Electrophysiol. 2017;40(7):850–6. https://doi.org/10.1111/pace.13104.
5. Bhavnani SP, Pavuluri V, Coleman CI, Guertin D, Yarlagadda RK, Clyne CA, Kluger J. The gender-paradox among patients with implantable cardioverter-defibrillators: a propensity-matched study. Pacing Clin Electrophysiol. 2013;36(7):878–84. https://doi.org/10.1111/pace.12141.
6. Biton Y, Zareba W, Goldenberg I, Klein H, McNitt S, Polonsky B, Moss AJ, Kutyifa V, Committee M-CE. Sex differences in long-term outcomes with cardiac resynchronization therapy in mild heart failure patients with left bundle branch block. J Am Heart Assoc. 2015;4(7). https://doi.org/10.1161/JAHA.115.002013.
7. Boule S, Ovart L, Marquie C, Botcherby E, Klug D, Kouakam C, Brigadeau F, Guedon-Moreau L, Wissocque L, Meurice J, Lacroix D, Kacet S. Pregnancy in women with an implantable cardioverter-defibrillator: is it safe? Europace. 2014;16(11):1587–94. https://doi.org/10.1093/europace/euu036.
8. Burke MC, Gold MR, Knight BP, Barr CS, Theuns DA, Boersma LV, Knops RE, Weiss R, Leon AR, Herre JM, Husby M, Stein KM, Lambiase PD. Safety and efficacy of the totally subcutaneous implantable defibrillator: 2-year results from a pooled analysis of the IDE study and EFFORTLESS registry. J Am Coll Cardiol. 2015;65(16):1605–15. https://doi.org/10.1016/j.jacc.2015.02.047.
9. Buxton AE, Lee KL, Fisher JD, Josephson ME, Prystowsky EN, Hafley G. A randomized study of the prevention of sudden death in patients with coronary artery disease. Multicenter Unsustained tachycardia trial investigators. N Engl J Med. 1999;341(25):1882–90. https://doi.org/10.1056/NEJM199912163412503.
10. Cheng YJ, Zhang J, Li WJ, Lin XX, Zeng WT, Tang K, Tang AL, He JG, Xu Q, Mei MY, Zheng DD, Dong YG, Ma H, Wu SH. More favorable response to cardiac resynchronization therapy in women than in men. Circ Arrhythm Electrophysiol. 2014;7(5):807–15. https://doi.org/10.1161/CIRCEP.113.001786.
11. Christensen AV, Zwisler AD, Svendsen JH, Pedersen PU, Blunk L, Thygesen LC, Berg SK. Effect of cardiac rehabilitation in patients with ICD: are gender differences present?

Results from the COPE-ICD trial. Pacing Clin Electrophysiol. 2015;38(1):18–27. https://doi.org/10.1111/pace.12507.
12. Cleland JG, Abraham WT, Linde C, Gold MR, Young JB, Claude Daubert J, Sherfesee L, Wells GA, Tang AS. An individual patient meta-analysis of five randomized trials assessing the effects of cardiac resynchronization therapy on morbidity and mortality in patients with symptomatic heart failure. Eur Heart J. 2013;34(46):3547–56. https://doi.org/10.1093/eurheartj/eht290.
13. Colvin M, Sweitzer NK, Albert NM, Krishnamani R, Rich MW, Stough WG, Walsh MN, Westlake Canary CA, Allen LA, Bonnell MR, Carson PE, Chan MC, Dickinson MG, Dries DL, Ewald GA, Fang JC, Hernandez AF, Hershberger RE, Katz SD, Moore S, Rodgers JE, Rogers JG, Vest AR, Whellan DJ, Givertz MM. Heart failure in non-Caucasians, women, and older adults: a white paper on special populations from the Heart Failure Society of America guideline committee. J Card Fail. 2015;21(8):674–93. https://doi.org/10.1016/j.cardfail.2015.05.013.
14. Conen D, Arendacka B, Rover C, Bergau L, Munoz P, Wijers S, Sticherling C, Zabel M, Friede T. Gender differences in appropriate shocks and mortality among patients with primary prophylactic implantable cardioverter-defibrillators: systematic review and meta-analysis. PLoS One. 2016;11(9): e0162756. https://doi.org/10.1371/journal.pone.0162756.
15. Deo R, Vittinghoff E, Lin F, Tseng ZH, Hulley SB, Shlipak MG. Risk factor and prediction modeling for sudden cardiac death in women with coronary artery disease. Arch Intern Med. 2011;171(19):1703–9. https://doi.org/10.1001/archinternmed.2011.328.
16. Dhruva SS, Mazure CM, Ross JS, Redberg RF. Inclusion of demographic-specific information in studies supporting US Food & Drug Administration Approval of high-risk medical devices. JAMA Intern Med. 2017. https://doi.org/10.1001/jamainternmed.2017.3148.
17. Diemberger I, Mazzotti A, Giulia MB, Cristian M, Matteo M, Letizia ZM, Reggiani B, Battistini P, Boriani G. From lead management to implanted patient management: systematic review and meta-analysis of the last 15 years of experience in lead extraction. Expert Rev Med Devices. 2013;10 (4):551–73. https://doi.org/10.1586/17434440.2013.811837.
18. Epstein AE, JP DM, Ellenbogen KA, Estes NA 3rd, Freedman RA, Gettes LS, Gillinov AM, Gregoratos G, Hammill SC, Hayes DL, Hlatky MA, Newby LK, Page RL, Schoenfeld MH, Silka MJ, Stevenson LW, Sweeney MO, Tracy CM, Epstein AE, Darbar D, JP DM, Dunbar SB, Estes NA 3rd, Ferguson TB Jr, Hammill SC, Karasik PE, Link MS, Marine JE, Schoenfeld MH, Shanker AJ, Silka MJ, Stevenson LW, Stevenson WG, Varosy PD, American College of Cardiology F, American Heart Association Task Force on Practice G, Heart Rhythm S. 2012 ACCF/AHA/HRS focused update incorporated into the ACCF/AHA/HRS 2008 guidelines for device-based therapy of cardiac rhythm abnormalities: a report of the American College of Cardiology Foundation/American Heart Association task force on practice guidelines and the Heart Rhythm Society. J Am Coll Cardiol. 2013;61(3):e6–75. https://doi.org/10.1016/j.jacc.2012.11.007.
19. FDA. Evaluation of sex-specific data in medical device clinical studies: guidance for industry and food and drug administration staff. 2014a. https://www.fdagov/downloads/MedicalDevices/DeviceRegulationandGuidance/GuidanceDocuments/UCM283707.pdf
20. FDA. FDA action plan to enhance the collection and availability of demographic subgroup data. 2014b. https://www.fdagov/downloads/regulatoryinformation/legislation/significantamendmentstothefdcact/fdasia/ucm410474.pdf
21. Feldman AM, Kersten DJ, Chung JA, Asheld WJ, Germano J, Islam S, Cohen TJ. Gender-related and age-related differences in implantable defibrillator recipients: results from the Pacemaker and Implantable Defibrillator Leads Survival Study ("PAIDLESS"). J Invasive Cardiol. 2015;27(12):530–4.
22. Foley PW, Leyva F, Frenneaux MP. What is treatment success in cardiac resynchronization therapy? Europace. 2009;11 (Suppl 5):v58–65. https://doi.org/10.1093/europace/eup308.
23. Gold MR, Aasbo JD, El-Chami MF, Niebauer M, Herre J, Prutkin JM, Knight BP, Kutalek S, Hsu K, Weiss R, Bass E, Husby M, Stivland TM, Burke MC. Subcutaneous implantable cardioverter-defibrillator post-approval study: clinical characteristics and perioperative results. Heart Rhythm. 2017. https://doi.org/10.1016/j.hrthm.2017.05.016.
24. Hess PL, Hernandez AF, Bhatt DL, Hellkamp AS, Yancy CW, Schwamm LH, Peterson ED, Schulte PJ, Fonarow GC, Al-Khatib SM. Sex and race/ethnicity differences in implantable cardioverter-defibrillator counseling and use among patients hospitalized with heart failure: findings from the get with the guidelines-heart failure program. Circulation. 2016;134(7):517–26. https://doi.org/10.1161/CIRCULATIONAHA.115.021048.
25. Jackson T, Sohal M, Chen Z, Child N, Sammut E, Behar J, Claridge S, Carr-White G, Razavi R, Rinaldi CA. A U-shaped type II contraction pattern in patients with strict left bundle branch block predicts super-response to cardiac resynchronization therapy. Heart Rhythm. 2014;11(10):1790–7. https://doi.org/10.1016/j.hrthm.2014.06.005.
26. Kwon DH, Hachamovitch R, Adeniyi A, Nutter B, Popovic ZB, Wilkoff BL, Desai MY, Flamm SD, Marwick T. Myocardial scar burden predicts survival benefit with implantable cardioverter defibrillator implantation in patients with severe ischaemic cardiomyopathy: influence of gender. Heart. 2014;100(3):206–13. https://doi.org/10.1136/heartjnl-2013-304261.
27. Loring Z, Canos DA, Selzman K, Herz ND, Silverman H, MaCurdy TE, Worrall CM, Kelman J, Ritchey ME, Pina IL, Strauss DG. Left bundle branch block predicts better survival in women than men receiving cardiac resynchronization therapy: long-term follow-up of approximately 145,000 patients. JACC Heart Fail. 2013a;1(3):237–44. https://doi.org/10.1016/j.jchf.2013.03.005.
28. Loring Z, Strauss DG, Gerstenblith G, Tomaselli GF, Weiss RG, Wu KC. Cardiac MRI scar patterns differ by sex in an implantable cardioverter-defibrillator and cardiac resynchronization therapy cohort. Heart Rhythm. 2013b;10(5):659–65. https://doi.org/10.1016/j.hrthm.2013.01.003.
29. Macfarlane P, Van Oosterom A, Pahlm O, Kligfield P, Janse M, Camm J (2011) Appendix 1: adult normal limits. In: Macfarlane P, Van Oosterom A, Pahlm O, Kligfield P, Janse MJ, Camm J, editors. Comprehensive electrocardiology, Vol 4, 2nd ed. London: Springer; 2057–2126.
30. Mascioli G, Padeletti L, Sassone B, Zecchin M, Lucca E, Sacchi S, Boggian G, Tondo AL, Belvito C, Bakhtadze N, Borrelli A, Sinagra G. Electrocardiographic criteria of true left bundle branch block: a simple sign to predict a better clinical and instrumental response to CRT. Pacing Clin Electrophysiol. 2012;35(8):927–34. https://doi.org/10.1111/j.1540-8159.2012.03427.x.
31. Masoudi FA, Go AS, Magid DJ, Cassidy-Bushrow AE, Gurwitz JH, Liu TI, Reynolds K, Smith DH, Reifler LM, Glenn KA, Fiocchi F, Goldberg R, Gupta N, Peterson PN, Schuger C, Vidaillet H, Hammill SC, Greenlee RT. Age and sex differences in long-term outcomes following implantable cardioverter-defibrillator placement in contemporary clinical practice: findings from the cardiovascular research network. J

Am Heart Assoc. 2015;4(6):e002005. https://doi.org/10.1161/JAHA.115.002005.
32. Moss AJ, Zareba W, Hall WJ, Klein H, Wilber DJ, Cannom DS, Daubert JP, Higgins SL, Brown MW, Andrews ML, Multicenter Automatic Defibrillator Implantation Trial III. Prophylactic implantation of a defibrillator in patients with myocardial infarction and reduced ejection fraction. N Engl J Med. 2002;346(12):877–83. https://doi.org/10.1056/NEJMoa013474.
33. Patel NJ, Edla S, Deshmukh A, Nalluri N, Patel N, Agnihotri K, Patel A, Savani C, Patel N, Bhimani R, Thakkar B, Arora S, Asti D, Badheka AO, Parikh V, Mitrani RD, Noseworthy P, Paydak H, Viles-Gonzalez J, Friedman PA, Kowalski M. Gender, racial, and health insurance differences in the trend of Implantable Cardioverter-Defibrillator (ICD) utilization: a United States experience over the last decade. Clin Cardiol. 2016;39(2):63–71. https://doi.org/10.1002/clc.22496.
34. Pedersen SS, Nielsen JC, Riahi S, Haarbo J, Videb AR, Larsen ML, Skov O, Knudsen C, Johansen JB. Study design and cohort description of DEFIB-WOMEN: a National Danish Study in patients with an ICD. Pacing Clin Electrophysiol. 2016;39(11):1261–8. https://doi.org/10.1111/pace.12942.
35. Ponikowski P, Voors AA, Anker SD, Bueno H, Cleland JG, Coats AJ, Falk V, Gonzalez-Juanatey JR, Harjola VP, Jankowska EA, Jessup M, Linde C, Nihoyannopoulos P, Parissis JT, Pieske B, Riley JP, Rosano GM, Ruilope LM, Ruschitzka F, Rutten FH, van der Meer P, Task Force M. 2016 ESC guidelines for the diagnosis and treatment of acute and chronic heart failure: the task force for the diagnosis and treatment of acute and chronic heart failure of the European Society of Cardiology (ESC) developed with the special contribution of the Heart Failure Association (HFA) of the ESC. Eur Heart J. 2016;37(27):2129–200. https://doi.org/10.1093/eurheartj/ehw128.
36. Priori SG, Blomstrom-Lundqvist C, Mazzanti A, Blom N, Borggrefe M, Camm J, Elliott PM, Fitzsimons D, Hatala R, Hindricks G, Kirchhof P, Kjeldsen K, Kuck KH, Hernandez-Madrid A, Nikolaou N, Norekval TM, Spaulding C, Van Veldhuisen DJ. 2015 ESC guidelines for the management of patients with ventricular arrhythmias and the prevention of sudden cardiac death: the task force for the management of patients with ventricular arrhythmias and the prevention of sudden cardiac death of the European Society of Cardiology (ESC). Endorsed by: Association for European Paediatric and Congenital Cardiology (AEPC). Eur Heart J. 2015;36(41):2793–867. https://doi.org/10.1093/eurheartj/ehv316.
37. Providencia R, Marijon E, Lambiase PD, Bouzeman A, Defaye P, Klug D, Amet D, Perier MC, Gras D, Algalarrondo V, Deharo JC, Leclercq C, Fauchier L, Babuty D, Bordachar P, Sadoul N, Piot O, Boveda S. Primary prevention Implantable Cardioverter Defibrillator (ICD) therapy in women-data from a multicenter French registry. J Am Heart Assoc. 2016;5(2). https://doi.org/10.1161/JAHA.115.002756.
38. Rao P, Faddis M. Cardiac resynchronisation therapy: current indications, management and basic troubleshooting. Heart. 2017. https://doi.org/10.1136/heartjnl-2016-310656.
39. Rho RW, Patton KK, Poole JE, Cleland JG, Shadman R, Anand I, Maggioni AP, Carson PE, Swedberg K, Levy WC. Important differences in mode of death between men and women with heart failure who would qualify for a primary prevention implantable cardioverter-defibrillator. Circulation. 2012;126(20):2402–7. https://doi.org/10.1161/CIRCULATIONAHA.111.069245.
40. Risum N, Strauss D, Sogaard P, Loring Z, Hansen TF, Bruun NE, Wagner G, Kisslo J. Left bundle-branch block: the relationship between electrocardiogram electrical activation and echocardiography mechanical contraction. Am Heart J. 2013;166(2):340–8. https://doi.org/10.1016/j.ahj.2013.04.005.
41. Russo AM, Daugherty SL, Masoudi FA, Wang Y, Curtis J, Lampert R. Gender and outcomes after primary prevention implantable cardioverter-defibrillator implantation: findings from the National Cardiovascular Data Registry (NCDR). Am Heart J. 2015;170(2):330–8. https://doi.org/10.1016/j.ahj.2015.02.025.
42. Santangeli P, Di Biase L, Basile E, Al-Ahmad A, Natale A. Outcomes in women undergoing electrophysiological procedures. Arrhythm Electrophysiol Rev. 2013;2(1):41–4. https://doi.org/10.15420/aer.2013.2.1.41.
43. Sohaib SM, Payne JR, Shukla R, World M, Pennell DJ, Montgomery HE. Electrocardiographic (ECG) criteria for determining left ventricular mass in young healthy men; data from the LARGE heart study. J Cardiovasc Magn Reson. 2009;11:2. https://doi.org/10.1186/1532-429X-11-2.
44. Steffel J, Varma N, Robertson M, Singh JP, Bax JJ, Borer JS, Dickstein K, Ford I, Gorcsan J 3rd, Gras D, Krum H, Sogaard P, Holzmeister J, Brugada J, Abraham WT, Ruschitzka F. Effect of gender on outcomes after cardiac resynchronization therapy in patients with a narrow QRS complex: a subgroup analysis of the EchoCRT trial. Circ Arrhythm Electrophysiol. 2016;9(6). https://doi.org/10.1161/CIRCEP.115.003924.
45. Sticherling C, Arendacka B, Svendsen J, Wijers S, Friede T, Stockinger J, Dommasch M, Merkely B, Willems R, Lubinski A, Scharfe M, Braunschweig F, Svetlosak M, Zurn C, Huikuri H, Flevari P, Lund-Andersen C, Schaer B, Tuinenburg A, Bergau L, Schmidt G, Szeplaki G, Vanderberk B, Kowalczyk E, Eick C, Juntilla J, Conen D, Zabel M. Sex differences in outcomes of primary prevention implantable cardioverter defibrillator therapy: combined registry data from eleven European countries. Europace. 2017. https://doi.org/10.1093/europace/eux176.
46. Strauss DG, Selvester RH, Lima JA, Arheden H, Miller JM, Gerstenblith G, Marban E, Weiss RG, Tomaselli GF, Wagner GS, Wu KC. ECG quantification of myocardial scar in cardiomyopathy patients with or without conduction defects: correlation with cardiac magnetic resonance and arrhythmogenesis. Circ Arrhythm Electrophysiol. 2008;1(5):327–36. https://doi.org/10.1161/CIRCEP.108.798660.
47. Strauss DG, Selvester RH, Wagner GS. Defining left bundle branch block in the era of cardiac resynchronization therapy. Am J Cardiol. 2011;107(6):927–34. https://doi.org/10.1016/j.amjcard.2010.11.010.
48. Tompkins CM, McNitt S, Polonsky B, Daubert JP, Wang PJ, Moss AJ, Zareba W, Kutyifa V. Sex differences in inappropriate ICD device therapies: MADIT-II and MADIT-CRT. J Cardiovasc Electrophysiol. 2017;28(1):94–102. https://doi.org/10.1111/jce.13102.
49. Varma N, Lappe J, He J, Niebauer M, Manne M, Tchou P. Sex-specific response to cardiac resynchronization therapy: effect of left ventricular size and QRS duration in left bundle branch block. JACC Clin Electrophysiol. 2017a. https://doi.org/10.1016/j.jacep.2017.02.021.
50. Varma N, Mittal S, Prillinger JB, Snell J, Dalal N, Piccini JP. Survival in women versus men following implantation of pacemakers, defibrillators, and cardiac resynchronization therapy devices in a large, Nationwide cohort. J Am Heart Assoc. 2017b;6(5). https://doi.org/10.1161/JAHA.116.005031.

51. Weiss R, Knight BP, Gold MR, Leon AR, Herre JM, Hood M, Rashtian M, Kremers M, Crozier I, Lee KL, Smith W, Burke MC. Safety and efficacy of a totally subcutaneous implantable-cardioverter defibrillator. Circulation. 2013;128 (9):944–53. https://doi.org/10.1161/CIRCULATIONAHA.113.003042.
52. Zeitler EP, Hellkamp AS, Fonarow GC, Hammill SC, Curtis LH, Hernandez AF, Al-Khalidi HR, Curtis JP, Heidenreich PA, Anstrom KJ, Peterson ED, Mark DB, Hammill BG, Sanders GD, Al-Khatib SM. Primary prevention implantable cardioverter-defibrillators and survival in older women. JACC Heart Fail. 2015;3(2):159–67. https://doi.org/10.1016/j.jchf.2014.09.006.
53. Zeitler EP, Hellkamp AS, Schulte PJ, Fonarow GC, Hernandez AF, Peterson ED, Sanders GD, Yancy CW, Al-Khatib SM. Comparative effectiveness of implantable cardioverter defibrillators for primary prevention in women. Circ Heart Fail. 2016;9(1):e002630. https://doi.org/10.1161/CIRCHEARTFAILURE.115.002630.
54. Zusterzeel R, Curtis JP, Canos DA, Sanders WE, Selzman KA, Pina IL, Spatz ES, Bao H, Ponirakis A, Varosy PD, Masoudi FA, Strauss DG. Sex-specific mortality risk by QRS morphology and duration in patients receiving CRT: results from the NCDR. J Am Coll Cardiol. 2014a;64 (9):887–94. https://doi.org/10.1016/j.jacc.2014.06.1162.
55. Zusterzeel R, Selzman KA, Sanders WE, Canos DA, O'Callaghan KM, Carpenter JL, Pina IL, Strauss DG. Cardiac resynchronization therapy in women: US Food and Drug Administration meta-analysis of patient-level data. JAMA Intern Med. 2014b;174(8):1340–8. https://doi.org/10.1001/jamainternmed.2014.2717.
56. Zusterzeel R, Spatz ES, Curtis JP, Sanders WE, Selzman KA, Pina IL, Bao H, Ponirakis A, Varosy PD, Masoudi FA, Canos DA, Strauss DG. Cardiac resynchronization therapy in women versus men: observational comparative effectiveness study from the National Cardiovascular Data Registry. Circ Cardiovasc Qual Outcomes. 2015;8(2 Suppl 1):S4–11. https://doi.org/10.1161/CIRCOUTCOMES.114.001548.
57. Zusterzeel R, Selzman KA, Sanders WE, O'Callaghan KM, Canos DA, Vernooy K, Prinzen FW, Gorgels AP, Strauss DG. Toward sex-specific guidelines for cardiac resynchronization therapy? J Cardiovasc Transl Res. 2016;9(1):12–22. https://doi.org/10.1007/s12265-015-9663-z.

Cardiophysiology Illustrated by Comparing Ventricular Volumes in Healthy Adult Males and Females

8

Peter L. M. Kerkhof, Tatiana Kuznetsova, J. Yasha Kresh, and Neal Handly

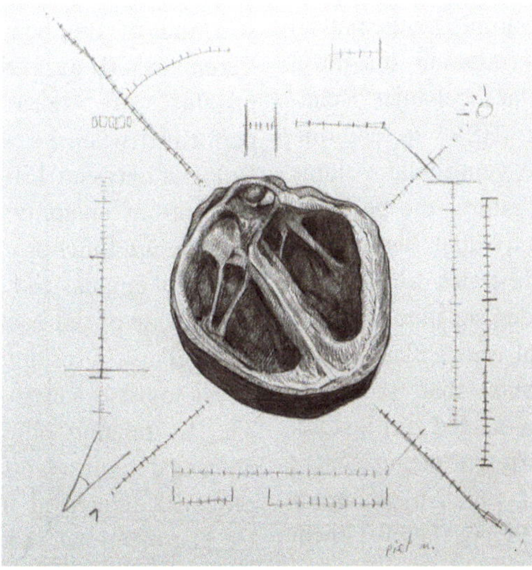

Assessment of cardiac size. Art work by Piet Michiels, Leuven, Belgium.

P. L. M. Kerkhof (✉)
Department of Radiology and Nuclear Medicine, Amsterdam Cardiovascular Sciences, VU University Medical Center, Amsterdam, The Netherlands
e-mail: plm.kerkhof@vumc.nl

T. Kuznetsova
Department of Cardiovascular Sciences, University of Leuven, Leuven, Belgium

J. Yasha Kresh
Departments of Cardiothoracic Surgery and Medicine (Cardiology), Drexel University College of Medicine, IME, University of Pennsylvania, Philadelphia, PA, USA

N. Handly
Department of Emergency Medicine, Drexel University College of Medicine, Philadelphia, PA, USA

Abstract

Recent advances in cardiac imaging techniques have substantially contributed to a growing interest in the analysis of global cardiac chamber dimensions and regional myocardial deformation. During the cardiac cycle, ventricular luminal volume varies due to the contraction process, which also confers a shape change including substantial alteration of long axis length, as well as rotation of the base compared to the apex. Local deformation can be assessed by strain (rate) analysis. Reviewing the present literature, it must be concluded that there is no single metric

P. L. M. Kerkhof, V. M. Miller (eds.), *Sex-Specific Analysis of Cardiovascular Function*,
Advances in Experimental Medicine and Biology 1065, https://doi.org/10.1007/978-3-319-77932-4_8

available to comprehensively characterize ventricular function. Every candidate advanced thus far has been found to incompletely reflect ventricular performance. This observation is not surprising in view of the complexity of the cardiac pump system. Additionally, sex-specific modifiers may play a role. More than three decades ago, it was shown that on average the ventricular volume is smaller in healthy women compared to matched males. Therefore, the present contribution concerns the interpretation of data derived from the healthy heart in both men and women. Starting from the classical Starling concept, we apply a simple mathematical transformation which permits an insightful representation of ventricular mechanics. Relating end-systolic volume (ESV) to end-diastolic volume creates the ventricular volume regulation graph which features the pertinent working point of an individual heart. This fundamental approach illustrates why certain proposed performance indexes cannot individually reveal the essence of ventricular systolic function. We demonstrate that particular metrics are highly interconnected and just tell us the same story in a different disguise. It is imperative to understand which associations exist and if they expectedly are (nearly) linear or frankly nonlinear. Notably, ejection fraction (EF) is primarily determined by ESV, while in turn EF is not much different from ventriculo-arterial coupling (VAC). Insight into cardiac function is promoted by identification of the paramount/essential components involved. The smaller ESV ($p < 0.0001$) implies that EF is higher in women and may also have consequences for VAC.

Keywords

Ventricular volume · Sex-specific analysis · Starling concept · Ejection fraction · Cardiac output · Heart rate · End-systolic volume · Systolic elastance · Ventriculo-arterial coupling · Pressure–volume loop · History of cardiology · Titin isoforms · Soltis–Saucerman model · Cardiomyogenesis · Review

Introduction

Gross anatomical contours of the heart have been depicted already by prehistoric hunters more than 17,000 years ago (the Upper Paleolithic period). Examples of such primitive art have been found in the caves of Lascaux, Southern France. One view holds that these instructional paintings facilitated the work of hunt; another refers to an expiatory ritual [4]. Until the late Middle Ages, most anatomical illustrations concern the complete body, without attention for individual organs [45]. The first illustrated anatomical textbook that includes a detailed description of the structure of the heart was written by Jacopo Berengario (Barigazzi) of Carpi [23, 47]. His insights were based on the many dissections performed when he was professor of surgery at Bologna between 1502 and 1527 [53]. The transition from anatomy (i.e. form) to physiology (*viz.* pump function) took several centuries. It must be emphasized that determination of the phasic size of the beating heart was not relevant until relatively recently (i.e., around 1900), when physiologists started studying pump function, e.g., by relating output to input (see Frank & Starling). Even William Harvey [14] was not primarily interested in ventricular size as such. He just mentioned an estimate of the amount of circulating blood by stating that everyone who crosses streets in the butchers' quarter knows that only a few liters of blood leave the animal once the major neck vessels are severed. It was more than a century later that interest developed in quantifying the size of the (non-beating) heart. Investigations regarding left ventricular (LV) size and shape were carried out by the Reverend Stephen Hales (1677–1761) from Middlesex, who made casts of the lumen of animal heart chambers using the beeswax cast technique [13]. A recently created example based on the same technique is shown in a photographic illustration elsewhere [26].

Cardiac function has often been analyzed by evaluating cardiac output (CO), as well as by considering metrics that directly employ information on end-systolic volume (ESV) and end-diastolic volume (EDV). An indication about the magnitude of CO in animals was noted by Harvey, as mentioned. Following experiments on skeletal muscle, cardiac physiologists commenced investigating the isolated animal heart, while manipulating filling characteristics as in the experiments by O. Frank, E. Starling, and M. Blix [15, 49]. Around the beginning of the twentieth century, physiologists analyzed the relationship between filling (preload) and output, resulting in the famous Starling curve. The Italian scientist Dario Maestrini (Corciano 1886–1975 Arezzo) is said to have arrived at similar statements [36, 38]. It took somewhat longer to devise methods that can actually estimate LV chamber volume in the beating intact (innervated) heart. The introduction of indicator dilution techniques permitted calculation of ejection fraction (EF), being the rate of washout during successive heart beats, i.e., ejected stroke volume (SV) relative to filling volume. With the advent of x-ray imaging, biplane angiography, and radionuclide tracers, besides noninvasive methods such as advanced 3D echocardiography, MRI, and CT scan, accurate estimations of ESV and EDV became within reach [31]. Thus, EF became "redefined" as (1-ESV/EDV). The distinction between ventricular size in men and women was established 35 years ago [7, 8] and is increasingly being appreciated.

Cardiomyocyte and Contraction Process

The fundamental building block of cardiac muscle contraction refers to the cardiomyocyte which contains myofibrils consisting of contractile proteins arranged in units which are called sarcomeres. Activation takes place via an electrical stimulus which in the healthy heart spreads from the sinus node via the AV node and the ventricular septum in the direction of the apex, to continue along the free wall in the direction of the base of the heart [24]. The electrical trigger initiates contraction, and the complete sequence can be recorded as the typical electrocardiogram (ECG). The precise mechanisms of contraction at the cellular level are complex and still an area of intense investigation. The contractile machinery is summarized in Fig. 8.1 [39]. The intrinsic capability to contract and generate force is referred to as contractility (or inotropy). Similarly the property describing the ease of relaxation during diastole is referred to as lusitropy. Both processes are modulated by β-adrenergic stimulation (β-AS), which increases (the rate of) peak force development, in addition to enhancing lusitropy [44]. Notably, β-AS raises intracellular cAMP levels, activating protein kinase A (PKA), which in turn phosphorylates key proteins involved in excitation–contraction (EC) coupling. These include the L-type Ca^{2+} channel and phospholamban (PLB), respectively, increasing Ca^{2+} entry into the cytoplasm and sarcoplasmic reticulum (SR) Ca^{2+} reuptake by the SR Ca^{2+}-ATPase2a (SERCA2a). These effects contribute to increased SR Ca^{2+} content, inotropy, and lusitropy. High SR Ca^{2+} load also enhances Ca^{2+} leak and fractional Ca^{2+} release during EC coupling. Interestingly, ß-AS has been shown to increase the gain of cardiac EC coupling. Mechanisms of cardiac EC coupling have been found to differ between the sexes [10]. Contractions and Ca^{2+} transients in rats are significantly smaller in myocytes from females when compared to males. Ca^{2+} spark frequency does not differ between the sexes, but amplitudes are smaller in cells from females when compared to males. Thus, the gain of EC coupling (i.e., SR Ca^{2+} release per unit Ca^{2+} current) is much lower in cells from females [10]. As the amount of Ca^{2+} released during a spark is regulated by the intrinsic gating of the RyR2 (type 2 release) channels, sex differences in the opening of individual channels could be responsible. Alternatively, as RyR2 levels are actually higher in females, it is possible that there are posttranslational modifications in males that could contribute to increased SR Ca^{2+} release by increasing channel opening [40]. Additionally, estrogen receptor activation was found to influence cardiac myofilament function [33]. At the myofilaments, PKA phosphorylates troponin I (TnI) and myosin binding protein-C (MyBP-C). Cross-bridge (XB) cycling rate (XBcy) is still debated in the literature. Most studies on ion channels and Ca^{2+} transport

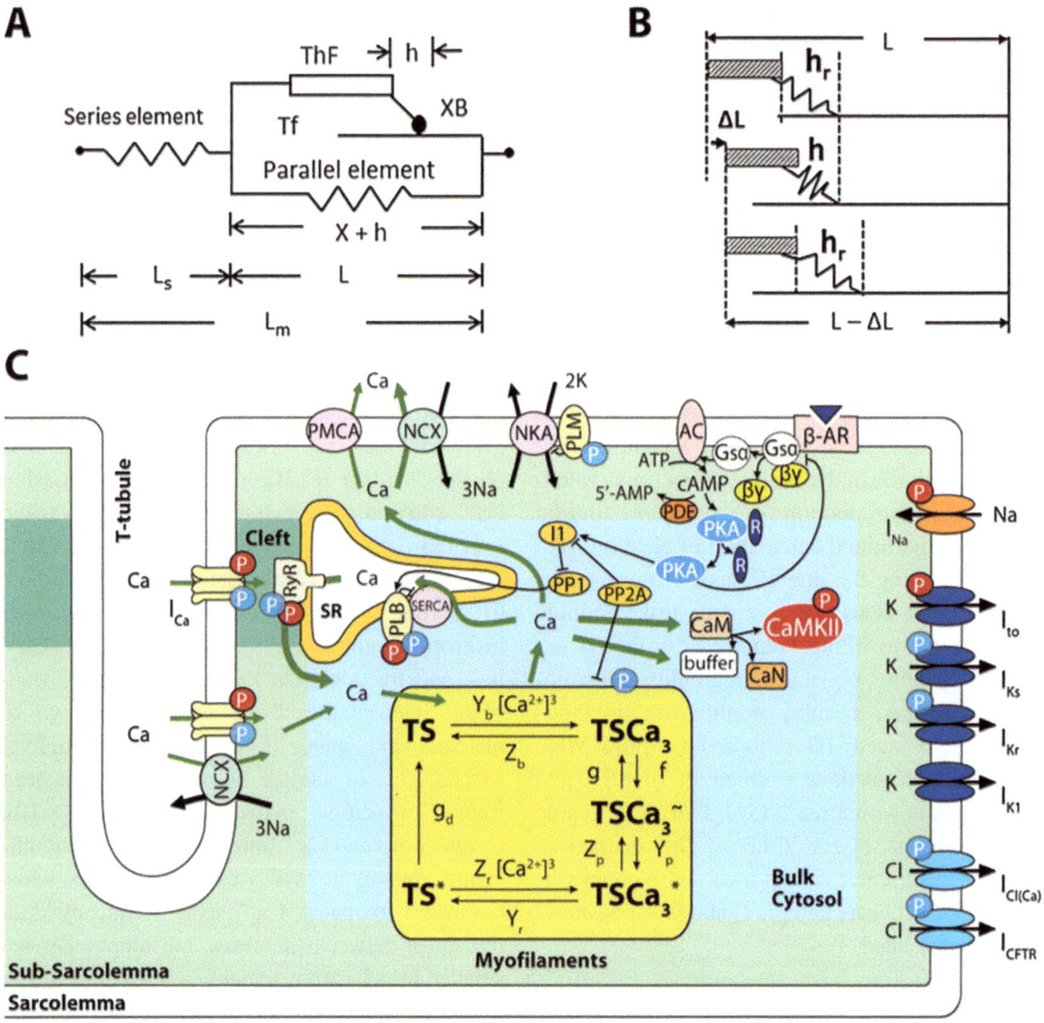

Fig. 8.1 Model summarizing major components and processes involved in cardiac contraction at the cellular level (**A**) Constitutive muscle unit, consisting of half-sarcomere length (L) composed of thick (ThF) and thin (Tf) filaments in parallel with an elastic element. The equivalent cross-bridge (XB) representing all attached cross-bridges is part of the ThF. It attaches to the Tf by the mobile end of its elastic structure, with h elongation, defining an inextensible half-sarcomere length X = L-h. A series elastic element with length L_s accounts for compliant muscle ends and together with L make up total muscle length (L_m). (**B**) Cross-bridge dynamics showing steady-state XB elongation (h_r), its decrease due to ΔL, and later return to h_r at shorter half-sarcomere length (L-ΔL). (**C**) Four-compartment myocyte model: bulk cytosol, cleft, sarcoplasmic reticulum (SR), and subsarcolemmal space (Sub-SL) [50, 54] with ion currents as described by various authors. This model is coupled to the myofilament force development model consisting of 5-state troponin systems (TS) with Ca^{2+} binding. Each TS is composed of three adjacent troponin–tropomyosin regulatory units able to act cooperatively to bind Ca^{2+} in three successive steps. Troponin systems are free TS, Ca^{2+} bound to TS without attached XBs (TSCa3), Ca^{2+} bound to TS with attached XBs in the weak state (TSCa3∼), Ca^{2+} bound to TS with attached XBs in the power state (TSCa3*), and TS without Ca^{2+} with attached XBs in the power state (TS*). Baseline mechanical parameters and all changes used for ISO are in Table 8.1 of the original publication. Reproduced from Negroni et al. [39], with permission. Other abbreviations: *AP(D)* action potential (duration), *CaMKII* Ca^{2+}/calmodulin-dependent protein kinase II, *CFTR* cystic fibrosis transmembrane conductance regulator, *EC* excitation–contraction, *Fm* muscle force, *ICaL* L-type Ca^{2+} channel current, I_{CFTR} CFTR-mediated Cl^- current, *ICl(Ca)* Ca^{2+}-dependent Cl^- current, I_{Kr} rapid delayed rectifier K^+ current, I_{Ks} slow delayed rectifier K+ current, I_{Na} fast Na^+ current, *INCX* Na^+–Ca^{2+} exchanger current, *INKA* Na^+–K^+-ATPase current, *ISO* isoproterenol, *ISO-Cytofl* ISO effect without

have been limited to myocyte ion handling and electrophysiological descriptions. However, an advanced approach by Negroni et al. [39] clarifies how altered Ca^{2+} sensitivity and XBcy interact during β-AS in ventricular myocytes. Thus, they provide a quantitative framework to study β-AS dynamic effects on Ca^{2+} transients, action potential, and contractile properties while incorporating an updated Soltis–Saucerman model [54]. The latter integrates the Shannon–Bers Ca^{2+} electrophysiological myocyte model [50] with dynamic descriptions of CaMKII and PKA signaling pathways. The half-sarcomere is composed of inextensible thick and thin filaments in parallel with an internal elastic load (Fig. 8.1a). The thick (ThF) and thin filaments (Tf) can slide past each other defining a zone of overlap where XBs can attach to the thin filament. Attached XBs act as independent force generators, occupying different states in the XB cycle [19]. It was demonstrated [15] that as the sarcomere shortens below slack length, titin-based restoring forces act to desensitize the myofilaments. That study showed for the first time the importance of titin-based restoring force in length-dependent deactivation during the early phase of diastole. Titin is synthesized as either the more compliant (fetal) N2BA or stiffer (adult) N2B form. Signaling by thyroid hormone, insulin, and Gq-protein-coupled receptors to the phosphoinositol 3 kinase–Akt–mammalian target of rapamycin pathway enhances N2B expression. The N2BA:N2B ratio generally increases in humans with heart failure (HF) and reduced EF, but changes in patients with preserved EF remain less certain [51]. Recently, a validation was realized of an LV finite element model driven by a cell-based cross-bridge cycling descriptor and coupled to a closed-loop lumped parameter circulatory model to simulate different ventricular loading conditions (preload and afterload) and contractility levels. This approach permitted reproduction of a linear end-systolic pressure–volume relationship, a curvilinear end-diastolic pressure–volume relationship, and a linear relationship between myocardial oxygen consumption and pressure–volume area [52].

Sex-Specific Differences in Cardiomyocytes

The observation that receptors for all major sex steroid hormones, including testosterone, are present on individual cardiomyocytes suggests that these hormones may influence the heart at the cellular level. Male/female differences in intracellular Ca^{2+} release and contraction in isolated ventricular myocytes have already been mentioned. Growing evidence suggests that these differences arise from effects of sex steroid hormones on processes involved in intracellular Ca^{2+} homeostasis [2]. Also in adult mammals, sex-based differences in cardiac structure and function have been reported, based on main molecular mechanisms involved in the response of the heart to pathological situations. These differences should be dealt with, not only in basic science or clinical research but also with

Fig. 8.1 (continued) β-AS on parameters involved in cytosolic fluxes, *ISO-titin* ISO effect without β-AS on parameters involved in titin, *ISO-XBCa* ISO effect without β-AS on parameters involved in XBCa, *ISO-XBCa–XBcy* ISO effect without β-AS on parameters in XBCa and XBcy, *ISO-XBcy* ISO effect without β-AS on parameters involved in XBcy, *kdf* rate of force redevelopment (as response to a length step), *krel* rate of force decay, *ktr* rate of force redevelopment (response to a length pulse), *Lm* muscle length, *MyBP-C* myosin binding protein-C, *NCX* Na^+–Ca^{2+} exchanger, *NKA* Na^+–K^+-ATPase, *PKA* protein kinase A, *PLB* phospholamban, *PLM* phospholemman, *PMCA* plasma membrane Ca^{2+} ATPase, a transport protein in the plasma membrane, *RyR* ryanodine receptor, *SERCA* sarcoendoplasmic reticulum Ca^{2+}-ATPase, *SR* sarcoplasmic reticulum, *TnC* troponin C, *TnI* troponin I, *TSCa3* Ca^{2+} bound to troponin system without attached XBs, *TSCa3* ~ Ca^{2+} bound to troponin system with attached XBs in the weak state, *TSCa3** troponin system without Ca^{2+} with attached XBs in the power state, *XB* (myofilament) cross-bridge, *XBCa* XB Ca^{2+} sensitivity, *XBcy* cross-bridge cycling rate, *WT* wild type, *β-AS* β-adrenergic stimulation regard to therapeutic approaches [11]. Recently, in rats contractility in the whole heart, adult ventricular myocytes, as well as myofibrils from both sexes were analyzed. Functional sex differences were observed at all levels. Hearts and cardiomyocytes from female rats displayed greater fractional shortening (FS) than males, and female cells and myofibrils took longer to relax. RNA sequencing experiments on these cardiomyocytes were performed, resulting in the identification of around 600 genes which were expressed in a sexually dimorphic manner. Further analysis revealed sex-specific enrichment of signaling pathways and key regulators. At the protein level, female cardiomyocytes exhibited higher PKA activity, consistent with pathway enrichment identified through RNA sequencing [56]. In contrast, Farrell et al. [10] found FS to be significantly smaller in female rats, and a similar finding for EF was attributed to the load dependence of the metric. However, one would rather expect an increased EF level [32] in women in view of the lower arterial blood pressure levels seen in both healthy humans (Table 8.1) and animals [57]. Additionally, it is important to point out that sex differences in genes and gene families often reverse by sex in diseased versus control, healthy tissues [9].

It remains unclear how these differences as observed in rodents translate to macroscopic distinctions in humans. Reportedly, EF is higher in women ([8], and also this study), but the observation is likely associated with their smaller ESV values [30]. Differences also occur during aging. A progressive loss of telomeric DNA in human cardiac stem cells (hCSCs) occurs with aging, and the newly formed cardiomyocytes inherit short telomeres and rapidly reach the senescent phenotype. One study including female (N = 32) and male (N = 42) hearts from individuals without cardiovascular disease (age range 19–104 years) demonstrated that at all ages, the female heart is

Table 8.1 Hemodynamic variables in healthy adults (N = 410) based on Kuznetsova et al. [34]

Variable	Women		Men	*p*-value	Unit
Persons involved	215		195		number
Age	43.4	≈	42.7	0.62	years
Systolic blood pressure	116.5	<	121.8	<0.0001	mmHg
Diastolic blood pressure	74.6	<	77.8	<0.0001	mmHg
Mean arterial pressure	88.6	<	92.5	<0.0001	mmHg
End-systolic volume index	17.7	<	22.5	<0.0001	mL/m^2
End-diastolic volume index	48.7	<	59.3	<0.0001	mL/m^2
Stroke volume index	31.0	<	36.8	<0.0001	mL/m^2
Cardiac output index	1.93	<	2.13	<0.0001	$L/min.m^2$
Ejection fraction	63.7	>	61.9	0.0026	%
Effective arterial elastance	2.97	>	2.62	<0.0001	$mmHg.m^2/mL$
Emax (assuming Vo = 0)	5.3	>	4.3	<0.0001	$mmHg.m^2/mL$
Ventriculo-arterial coupling	1.83	>	1.69	0.003	dimensionless
Arterial compliance	0.77	<	0.86	<0.0001	$mL/(mmHg.m^2)$
Pulse pressure	41.9	<	44.0	0.009	mmHg
Peripheral resistance	48.1	>	45.4	0.014	$mmHg.min.m^2/L$
Body surface area	1.74	<	1.97	<0.0001	m^2
Heart rate	62.5	>	58.6	<0.0001	beats/minute

Volume data obtained by 2D echocardiography, collected in a north-eastern region of Belgium with mostly Caucasian inhabitants (including 215 women). The subgroup listed here excludes individuals with hypertension or diabetes mellitus. Possible age-dependent variation has not been considered in this survey. Note that all variables (with the exception of age) are significantly different in females when compared to males. In summary: pressures are lower in healthy women, while their ventricular volumes are smaller. Emax and Vo are defined in Fig. 8.3, assuming that Vo vanishes. Details on standard deviations have been omitted for simplicity.

equipped with a larger pool of functionally competent hCSCs and younger myocytes than the male myocardium. The replicative potential is higher, and telomeres are longer in female hCSCs than in male hCSCs. In the female heart, myocyte turnover occurs at a rate of 10%, 14%, and 40% per year at 20, 60, and 100 years of age, respectively. Corresponding values in the male heart are 7%, 12%, and 32% per year, documenting that cardiomyogenesis involves a large and progressively increasing number of parenchymal cells with aging. From 20 to 100 years of age, the myocyte compartment is replaced 15 times in women and 11 times in men [20].

Ventricular Pressure and Volume in Males and Females

So much for the difficult part. All those ion currents and sliding filament activities serve to get something like a compression pump fully operational on a 24-h basis. Now the more straightforward pump function description deserves attention, referring to a macroscopic view. Stroke volume (SV) and heart rate (HR) are important determinants of ventricular pumping capacity. SV is the difference between EDV and ESV. Cardiac output (CO) is obtained by multiplying SV and HR. Figure 8.2 illustrates positions of SV and HR combinations as measured in a population consisting of 410 healthy individuals. Iso-CO curves (i.e., trajectories with fixed CO levels) are inscribed. Obviously, similar values for SV can be generated by various suitable combinations of EDV and ESV. Therefore, we need a framework in which we can depict not just the difference between EDV and ESV (as shown on the ordinate of Fig. 8.2), but rather the particular volumes that actually generate a certain SV, either in a healthy individual or in a cardiac disease patient. In other words, it is required to define a so-called working point, which exactly characterizes the conditions which are responsible for a particular SV. This concept has been presented in detail elsewhere [32].

The Pressure–Volume Loop in Healthy Males and Females

Table 8.1 shows a comparison of various hemodynamically relevant variables routinely explored in clinical investigations. Apart from the slight

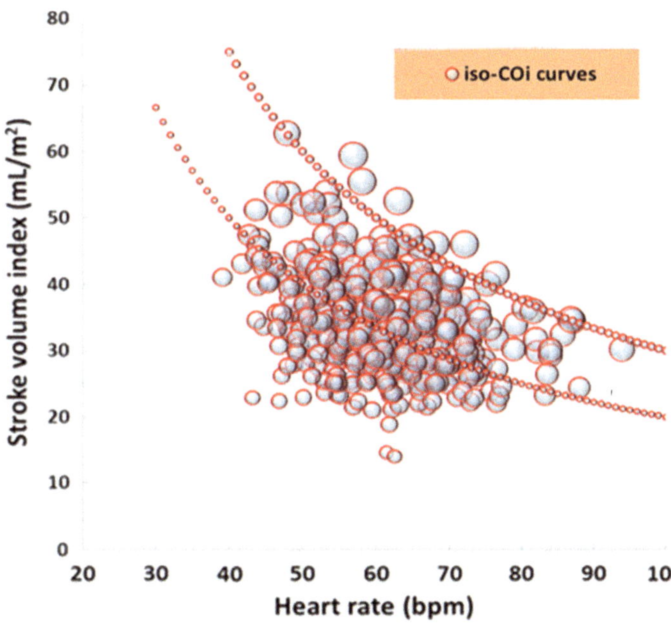

Fig. 8.2 Diagram showing combinations of stroke volume index and heart rate as measured in 410 healthy adults. Multiplication of both components yields cardiac output index (COi), as reflected by the relative bubble size. Two levels of constant COi are indicated by the dotted curves, namely 3 (upper tracing) and 2 L/min.m^2. Average values are different for males and females (see Table 8.1). Abbreviation: bpm, beats per minute

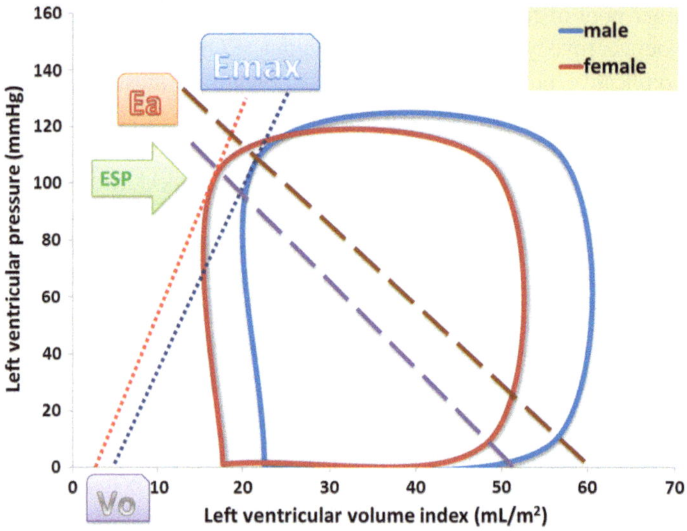

Fig. 8.3 Schematic sketch of the left ventricular pressure–volume loop, reflecting the dynamic behavior during a single cardiac cycle. Although not emphasized in most textbooks, there are considerable differences between the healthy adult male and female. The loops refer to a single cardiac cycle. Filling is complete when reaching end-diastolic volume (EDV). Then pressure builds up, which eventually opens the aortic valve, while starting ejection until end-systolic volume (ESV) with end-systolic pressure (ESP) is reached. Subsequently pressure falls and filling resumes. Essentially the diagram is similar for men and women, except for the fact that ESV and EDV are significantly smaller in healthy adult women, while ESP is slightly lower. End-systolic elastance Emax is defined as ESP / (ESV-Vo), where Vo is an extrapolated value, here based on a reasonable estimate. The effective arterial elastance (Ea) is given by ESP/(EDV–ESV). The ratio of Emax and Ea is called ventriculo-arterial coupling. Based on average data from [34] see Table 8.1

disparity of systolic pressure, it is remarkable to observe the significant differences regarding LV volumes when comparing men and women. This notable fact was first published by Buonanno et al. [8] and unfortunately had not received the attention which it deserves. Figure 8.3 combines LV pressure and volume data to construct average PV loops, typical for healthy males and females. Again, the contours of the loops depict remarkable sex-specific differences which generally are not discussed in standard textbooks. The LV peak systolic pressure (PSP) vs ESV relationship was first described by Holt [17]. Nowadays, end-systolic pressure (ESP) is often approximated as 0.9* PSP, yielding the familiar maximal systolic elastance (Emax), shown in Fig. 8.3.

It is well-known that the size of the healthy heart varies, depending on age and size of the animal or human under consideration. In contrast, in the healthy organism, it is found that arterial blood pressure at resting conditions is nearly constant across all species, although being slightly lower in adult healthy females [48, 57]. A popular index to indicate systolic function (i.e., pump performance) is EF, being the ratio of SV and EDV and often expressed as a percentage. Since average ESV is significantly smaller in healthy adult women, it follows that EF tends to be higher in females [30], and this is confirmed (Table 8.1) based on data described elsewhere [34]. Although EF depends on both ESV and EDV, it appears that ESV is the dominant component [30]. Indeed, the relationship between EF and ESV is based on the boundary conditions of ventricular function, where ESV cannot be greater than EDV, while values for SV occur within a certain range. When these restrictions are applied to a Monte Carlo model, we find a distribution of theoretical working points similar to what we encounter in the actual VRG [27]. In Fig. 8.3 the SV can be

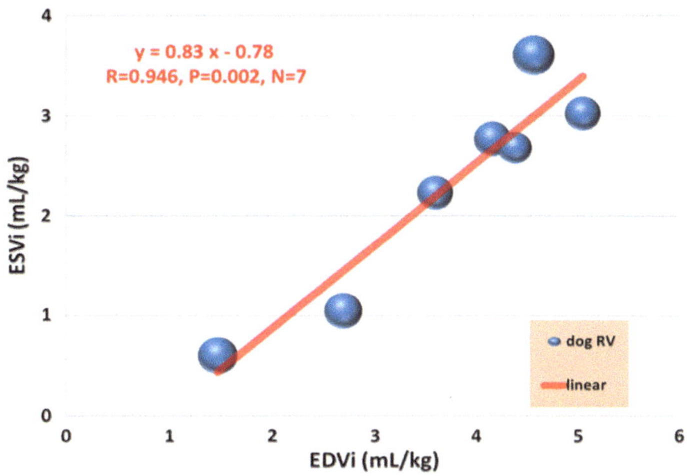

Fig. 8.4 Volume regulation graph, constructed on the basis of right ventricular (RV) volume data (normalized by body weight) for seven anesthetized dogs as reported by Holt [18], using an electrical conductance method and saline injection to obtain indicator dilution curves. Bubble size refers to heart rate (range 70–108 beats per minute)

identified as the width of the loop, since this is the difference between EDV and ESV. The latter two variables determine the working point as illustrated in Fig. 8.4 which is based on the pioneering work by Holt on obtaining LV and right ventricular (RV) volume in dogs [16–18]. The electric conductivity method employed by those authors to measure volumes was refined [3] and adapted for use in humans, horses [22], and small animals (survey in [31]).

The Volume Regulation Graph for the Right Ventricle

Similar to that for the LV, the VRG can be derived for the RV. Figure 8.4 shows the results based on a study by Holt [18], where heart rate (range 70–108 beats per minute) is reflected by "bubble" size. Average fractional emptying (an expression coined by Holt, nowadays called EF) amounted to 39% for the RV in seven anesthetized dogs, while in a preceding study [16], 46% was found for the LV. Although the authors recognized the fractional behavior of the dilution curves, their primary interest was not EF or any equivalent thereof but rather the estimation of residual volume, i.e., ESV. The findings for the RV VRG shown in Fig. 8.4 have been confirmed in numerous other studies, including, e.g., those reporting on post-Fallot repair studies [28].

The Case of Sharing a Pizza

Imagine that someone offers you a slice of pizza. She or he cuts a piece that is 3/5 (60%) of the total size. Now, how many bites do you get? Is this going to be a complete meal or just a snack? Well, apart from the toppings, it all depends on whether we are talking about a mini-pizza or a supersized one. So, given the fraction offered, you need to know the circumference (or diameter), because traditional pizzas, homemade or not, are reported to be round. A similar unspecified attribute is involved when discussing EF, which is often magically endowed with the prestigious title: indicator of systolic function.

The metric EF is also often referred to as an index. Wheelan [58] has nicely described how to view such an indicator: "The advantage of any index is that it consolidates lots of complex information into a single number. We can then rank things that otherwise defy simple comparison – anything from quarterbacks to colleges to beauty pageant contestants. In the Miss America pageant, the overall winner is a combination of five separate competitions: personal interview, swimsuit, evening wear, talent, and onstage question." He continues providing a similar example on ranking sports cars, emphasizing that each classification system has the potential to produce a different outcome. The same line of reasoning applies to the heart when we consider EF.

Robotham et al. [46] concluded: "Thus, EF, although a relatively simple measure that is intuitively easily comprehended, is an extremely complex parameter describing the entire cardiovascular system and requires additional study." Strictly speaking, the outcome for EF only depends on ESV and EDV when employing the more insightful variant of the definition formula:

$$EF = 100^* (1 - ESV/EDV) \quad (8.1)$$

However, both ESV and EDV themselves depend on many underlying factors. Ventricular function can be described via the volume regulation graph (VRG) approach or based on the elastance concept [27]. ESV and EDV can be related [21] to each other via

$$ESV = \alpha + \beta \, EDV \quad (8.2)$$

and graphically represented in the VRG; see Fig. 8.4. In the past, we derived an analytical expression by combining Eq. (8.1) with the definition formula for EF:

$$EF = 1 + C_1 \{ESV/(C_2 - ESV)\} \quad (8.3)$$

with constants C_1 and C_2 calculated for the population under consideration (Kerkhof 1984) [21]. This expression underscores the importance of measuring ESV [29, 30]. From the PV loop (Fig. 8.3), we can define (assuming linearity when considering a sufficiently small ESV range) the Emax using ESP, end-systolic volume (ESV), and the intercept Vo [6]:

$$Emax = ESP/(ESV - Vo) \text{ or,}$$
$$ESV = ESP/Emax + Vo.$$

The ventricular diastolic stiffness is traditionally defined as an exponential curve [27], by relating end-diastolic pressure (EDP) to two constants (i.e., P_o and m), $EDP = P_o\, e^{(m^*EDV)}$, where P_o is a constant and m is the modulus referring to LV distensibility. Thus, EDV can be expressed as EDV = (log EDP − log Po)/m. As a consequence, EF can be described either as a function of {ESV, ESP, Vo, EDP, Po, and m} or as dependent on {ESV, α, β, R^2, EDV_{ave}}.

Importantly, the VRG can be used as a starting point to derive various connections with clinically relevant metrics, such as EF and Emax. This view is summarized in Fig. 8.5, where (1-EF) is shown by the line connecting the origin with any particular operating point of {EDV, ESV}, for example, for the average values in males and females. In dogs of various breeds, admitted to the clinic because of heart disease, age seems to affect ventricular volume regulation (as reflected by the slope difference in the VRG), but this effect is only discernible in female dogs [42].

Another application of the VRG concept refers to EF versus ESVi (Fig. 8.6). The male/female regression curves based on Eq. (8.3) show overlap, but average values for males and females are significantly different (Table 8.1).

These findings can be summarized as shown in Fig. 8.7 where EF is presented in the VRG as a box which contains several surprises when we look at the true contents. The example is based on data (N = 12) obtained in healthy individuals [1], illustrating that EF is inversely associated with both ESVi, and with the VAC index, the latter only when assuming that Vo = 0. It is important to note that EF and VAC are "uncoupled" when the true Vo is taken into account. Otherwise, the relation between EF and VAC is trivial, and one metric does not offer any additional information to that of the other.

Associations Between Clinically Relevant Variables

The insightful and telling Fig. 8.8 obtained from Ky et al. [35] nicely summarizes the associations between major variables of relevance in cardiovascular studies. These authors did not stratify for sex but report that their analysis is based on 466 patients (37% females) with systolic HF with an average EF of 27%. In an earlier study concerning HF patients, we found that both average EF and VAC are significantly higher in women [26]. Table 8.1 documents that this is also the case in

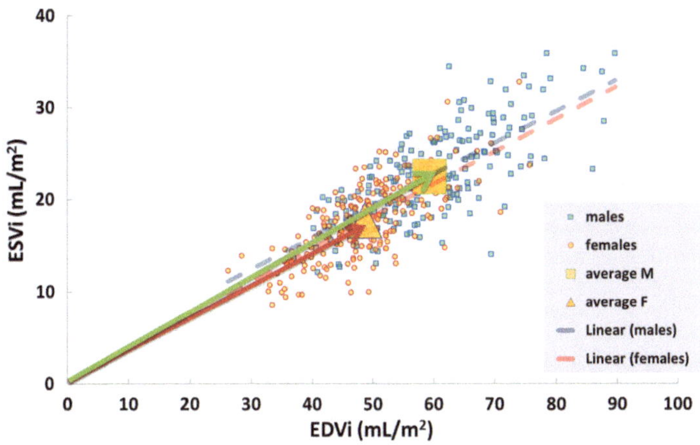

Fig. 8.5 Volume regulation graph, showing relationship between end-systolic volume index (ESVi) and end-diastolic volume index (EDVi) in healthy adult individuals (N = 410), stratified for sex (215 women). The broken lines refer to linear regression analysis. These lines should be distinguished from the solid lines with arrow heads that connect the origin with each group average, illustrating that average ejection fraction (EF) is higher in women (see Table 8.1). A steeper slope indicates a lower EF. Data from Kuznetsova et al. [34], measured by 2D echocardiography

Fig. 8.6 Ejection fraction (EF) versus end-systolic volume index (ESVi) for healthy individuals, stratified for sex, as also shown in Fig. 8.1. The yellow triangle refers to the average value for males, while the yellow star reflects average value for females. Both EF and ESVi are significantly different (see Table 8.1). The regression curve is based on Eq. (8.3)

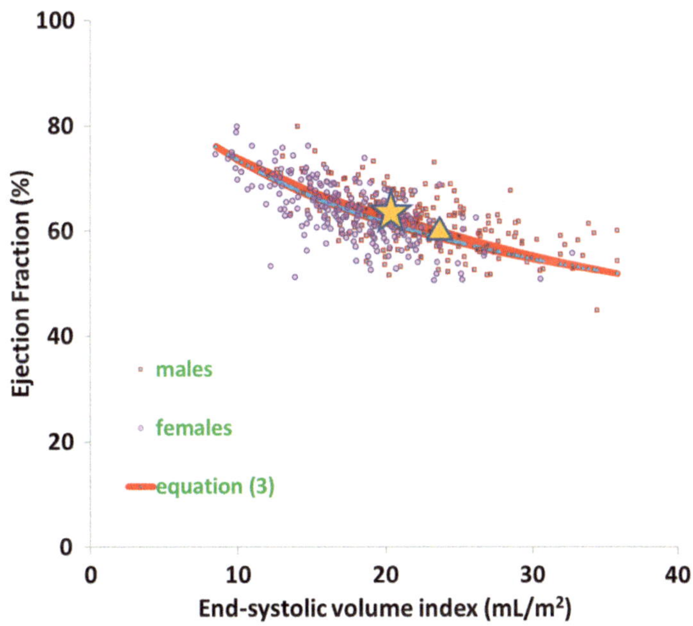

healthy adults. Not all variables listed in Fig. 8.8 were discussed in the present survey, but clearly the correlation reported between ESV and EDV is the highest found among all combinations. This outcome supports our starting point reflected by the VRG (Figs. 8.4 and 8.5), which has also been discussed elsewhere [5, 25].

Discussion

Strictly speaking, any sex-specific analysis is of limited meaning if not interpreted against the context of the potential role which age plays. The data we discussed here refer to healthy adults,

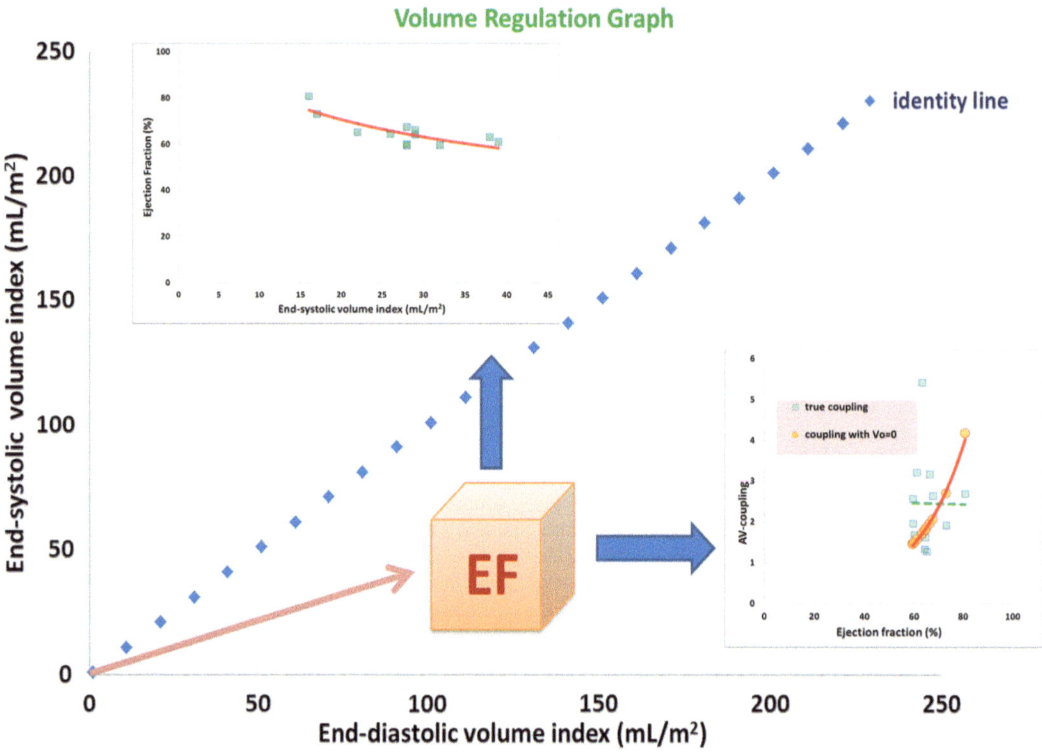

Fig. 8.7 Volume regulation graph illustrating the derivation of ejection fraction (EF) by considering the slope of the line which connects the origin with the pertaining working point defined by {EDV,ESV}. The inset (upper panel) shows that EF is nonlinearly related to ESVi, while an exponential relationship describes ventriculo-arterial coupling versus EF (lower right inset). For explanation see text. Data (N = 12) based on healthy individuals described by Asanoi et al. [1]

with age ranging from 15 to 90 years. From our observations on healthy adults, we conclude that there are significant sex differences regarding average LV size when considering the nearly complete adult lifespan. These findings fully support earlier findings reported in the literature [8], which unfortunately did not receive the attention it deserves within the clinical discipline. The results on the LV have been extended to the RV, e.g., by [37].

Using gated myocardial perfusion SPECT for the LV, Peace et al. [41] studied the effect of sex, age (range), and weight (range) on LV EF and ESV reference limits in adults. These authors fully confirm the findings obtained by Buonanno et al. [8]. Using coronary CT angiography, Gebhard et al. [12] analyzed 1155 individuals (54.5% males) free from overt heart disease and found that average EF was 2.0% points higher in females ($p < 0.001$). This male/female difference increased with advancing age (range 18–92 years). Slight age-dependent variation has been observed in some studies, and this potential modifying effect requires further study, because conflicting results have been reported, e.g., no correlation between EF and age [41].

Why ventricular size is generally smaller in healthy females compared to males is not entirely clear. Various factors have been implicated, including differences in neurohumoral signaling, age-dependent changes in estrogen/progesterone status, gender-related life style factors, and progressive cardiomyocyte loss with age in males (but not observed in women).

The embryologic origins of the RV and the LV differ. The RV exhibits dominant pump function

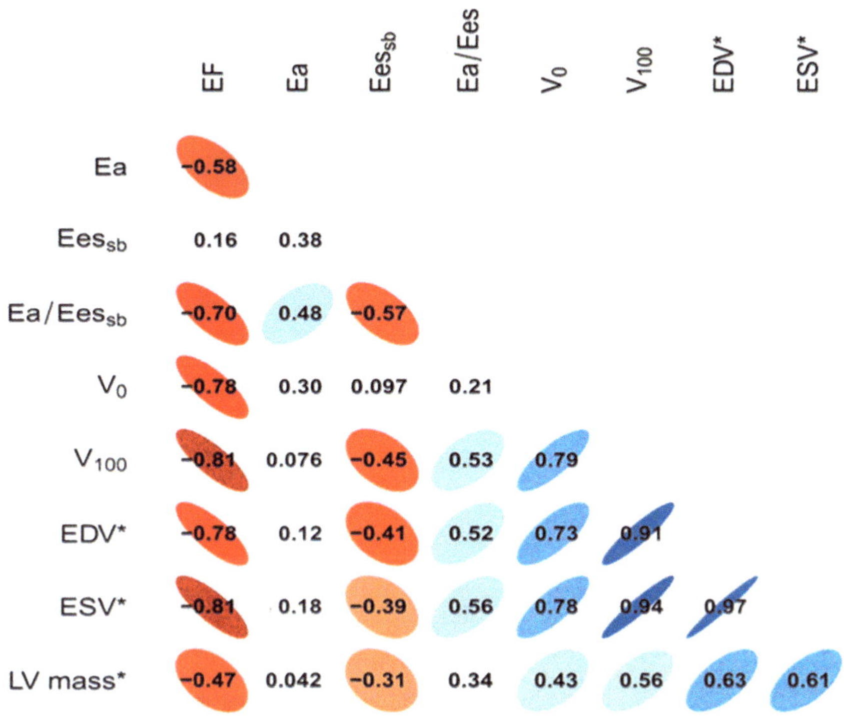

Fig. 8.8 Cross table of volume-related variables of the left ventricle, measured by echocardiography in 466 patients with chronic systolic heart failure, showing correlation coefficients for all combinations of two components. Ees is obtained by the single beat (sb) method and more or less comparable to Emax in our text. Note that the coupling index is here defined as our (k^{-1}). The blue ellipse signifies a positive correlation, while a red contour refers to negative correlation. The intensity of the color reflects the strength of the association. Interestingly, the two variables which we employ for depicting the volume regulation graph, namely, ESV and EDV, yield the highest correlation. V_{100} is the volume which corresponds with a pressure level of 100 mmHg on the linearized Emax curve. Abbreviations as in our text. Reproduced from Ky et al. [35], with permission

during fetal development. At birth, separation of the pulmonary and the systemic circulation causes that the LV assumes primary pump function generating much higher needed pressures. Concomitant with morphological and functional modifications in both the LV and RV, the neonatal cardiomyocytes undergo changes in morphology, respiration, metabolism, and contractile function. Distinct key transcription factors drive the chamber-specific gene networks during development. Existing evidence reveals a critical role of noncanonical Wnt11 in orchestrating early development. Expression of Wnt11 is dynamically regulated in a temporal–spatial manner, playing important roles in myocardial development [55]. However, these researchers only looked at male mice, and further investigations are needed. Another intriguing approach for elucidating male/female phenotypic differences in cardiac size follows the pattern of ventricular volume adaptation for sex-mismatch cases in heart transplant patients, both in children and adults [43].

Conclusions

In healthy adults we observe striking (significant) differences when blood pressure and left ventricular volumes are compared in males and females. These disparities have serious consequences for particular metrics commonly used to evaluate cardiovascular function. Pulse pressure (which is related to vascular stiffness) is higher in males. In contrast EF (often used to assess systolic

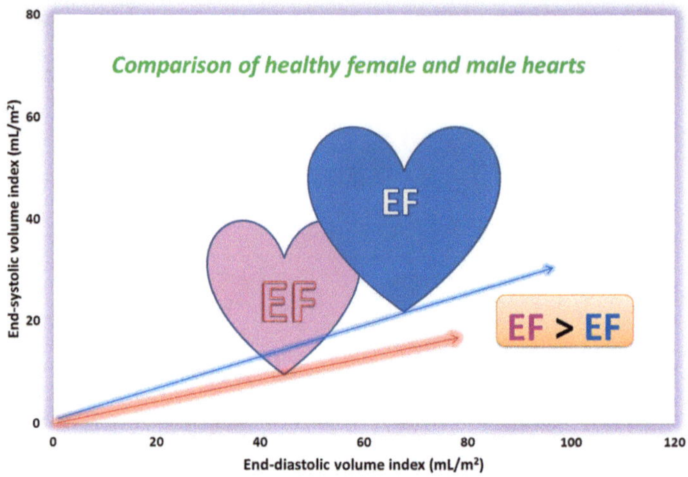

Fig. 8.9 Summary of dimensional findings regarding left ventricular size in healthy adult women and men. Ejection fraction (EF) is generally higher in smaller hearts (as in women), because end-systolic volume index (ESVi) tends to be smaller, while end-diastolic volume index "follows" ESVi. Schematic dimensions of the symbolic heart are somewhat exaggerated to make this point clearly visible

function), Emax (reflecting contractility), and VAC are higher in women (assuming that the intercept Vo can be neglected). However, mean blood pressure (an indicator of afterload) is lower in women. As illustrated, ESV and EDV are linearly correlated, clarifying why EF is slightly higher in women (Figs. 8.4 and 8.9), and permitting the robust description of various relationships, including EF vs ESV(i). It is often emphasized that sex matters when studying cardiovascular properties, and this notion has been confirmed in the present analysis based on healthy individuals. Further studies in cardiac disease patients are required to investigate if these differences persist in selected diagnostic cohort categories and what ultimately will be the implications for more precise diagnosis and treatment.

References

1. Asanoi H, Sasayama S, Kameyama T. Ventriculoarterial coupling in normal and failing heart in humans. Circ Res. 1989;65:483–93.
2. Ayaz O, Howlett SE. Testosterone modulates cardiac contraction and calcium homeostasis: cellular and molecular mechanisms. Biol Sex Differ. 2015;6:9. https://doi.org/10.1186/s13293-015-0027-9.
3. Baan J, Jong TT, Kerkhof PL, Moene RJ, van Dijk AD, van der Velde ET, Koops J. Continuous stroke volume and cardiac output from intra-ventricular dimensions obtained with impedance catheter. Cardiovasc Res. 1981;15:328–34.
4. Bataille G. The cradle of humanity, prehistoric art and culture. New York: Zone Books; 2009.
5. Beringer JY, Kerkhof PLM. A unifying representation of ventricular volumetric indexes. IEEE Trans BME. 1998;45:365–71.
6. Blaudszun G, Morel DR. Relevance of the volume-axis intercept, Vo, compared with the slope of end-systolic pressure-volume relationship in response to large variations in inotropy and afterload in rats. Exp Physiol. 2011;96:1179–95.
7. Buonanno C. The female left ventricle: a pathophysiological entity? Int J Cardiol. 1983;4:382–6.
8. Buonanno C, Arbustini E, Rossi L, Dander B, Vassanelli C, Paris B, Poppi A. Left ventricular function in men and women. Another difference between sexes. Eur Heart J. 1982;3:525–8.
9. Coronado MJ, Fairweather D, Bruno KA. Sex determines cardiac myocyte stretch and relaxation. Circ Cardiovasc Genet. 2017;10(5):e001950.
10. Farrell SR, Ross JL, Howlett SE. Sex differences in mechanisms of cardiac excitation-contraction coupling in rat ventricular myocytes. Am J Physiol Heart Circ Physiol. 2010;299:H36–45. https://doi.org/10.1152/ajpheart.00299.2010.
11. Fazal L, Azibani F, Vodovar N, Cohen Solal A, Delcayre C, Samuel JL. Effects of biological sex on the pathophysiology of the heart. Br J Pharmacol. 2014;171:555–66. https://doi.org/10.1111/bph.12279.
12. Gebhard C, Buechel RR, Stähli BE, Gransar H, Achenbach S, Berman DS, et al. Impact of age and sex on left ventricular function determined by coronary computed tomographic angiography: results from the prospective multicentre CONFIRM study. Eur Heart J Imag. 2017;18:990–1000.
13. Hales S. Statical essays. London: Innys & Co; 1733. See also http://www.bg.ic.ac.uk/research/k.parker/wave_intensity_web/history_hales.htm
14. Harvey W. Exercitatio anatomica de motu cordis et sanguinis in animalibus. Frankfurt a Main: W. Fitzer; 1628.
15. Helmes M, Lim CC, Liao R, Bharti A, Cui L, Sawyer DB. Titin determines the Frank-Starling relation in early diastole. J Gen Physiol. 2003;121:97–110.
16. Holt JP. Estimation of the residual volume of the ventricle of the dog's heart by two indicator dilution technics. Circ Res. 1956;4:187–95.

17. Holt JP. Regulation of the degree of emptying of the left ventricle by the force of ventricular contraction. Circ Res. 1957;5:281–7.
18. Holt JP, Allensworth J. Estimation of the residual volume of the right ventricle of the dog's heart. Circ Res. 1957;5:323–6.
19. Huxley AF. Muscle structure and theories of contraction. Prog Biophys Biophys Chem. 1957;7:255–318.
20. Kajstura J, Gurusamy N, Ogórek B, et al. Myocyte turnover in the aging human heart. Circ Res. 2010;107:1374–86. https://doi.org/10.1161/CIRCRESAHA.110.231498.
21. Kerkhof PLM. Importance of end-systolic volume for the evaluation of cardiac pump performance. In: Chazov EI, Smirnov VN, Oganov RG, editors. Cardiology, an international perspective. New York: Plenum Press; 1984. p. 1339–52.
22. Kerkhof PLM. Combination of Millar and conductance catheter to assess equine left ventricular function. In: Conference Proceedings: annual International Conference of the IEEE Engineering in Medicine and Biology Society, Atlanta. 1999.
23. Kerkhof PLM. Fundamentals of left ventricular volume representation. Conf Proc IEEE Eng Med Biol Soc. 2016Aug;2016:3294–3297. doi: https://doi.org/10.1109/EMBC.2016.7591432. PMID: 28227221.
24. Kerkhof PLM. Physiological modeling survey. In: Webster J, editor. Encyclopedia of electrical and electronics engineering. 3rd revised ed. New York: Wiley; 2018.
25. Kerkhof PLM, Kresh JY, Li JK-J, Heyndrickx GR. Left ventricular volume regulation in heart failure with preserved ejection fraction. Physiol Rep. 2013;1:e00007.
26. Kerkhof PLM, Heyndrickx GR, Li JK. Hemodynamic determinants and ventriculo-arterial coupling are sex-associated in heart failure patients. Conf Proc IEEE Eng Med Biol Soc. 2016 Aug;2016:3286–3289. doi: https://doi.org/10.1109/EMBC.2016.7591430. PMID: 28227219.
27. Kerkhof PLM, Yoo BW, Merillon JP, Peace RA, Handly N. Monte Carlo method applied to the evaluation of the relationship between ejection fraction and its constituent components. Conf Proc IEEE Eng Med Biol Soc. 2017a;2017:1295–8. https://doi.org/10.1109/EMBC.2017.8037069.
28. Kerkhof PLM, Yoo BW, van de Ven PM, Handly N. Sex-specific aspects of left and right ventricular volume regulation in patients following tetralogy of Fallot repair. Conf Proc IEEE Eng Med Biol Soc. 2017b;2017:1303–6. https://doi.org/10.1109/EMBC.2017.8037071.
29. Kerkhof PLM, Peace RA, Handly N. The importance of (measuring) the end-systolic volume index in predicting survival. Heart. 2018a;104:180. https://doi.org/10.1136/heartjnl-2017-312051.
30. Kerkhof PLM, van de Ven PM, Yoo BW, Peace RA, Heyndrickx GR, Handly N. Ejection fraction as related to basic components in the left and right ventricular volume domains. Intern J Cardiol. 2018b;255:105–10.
31. Kerkhof PLM, Heethaar RM, Pearce JA. Bioimpedance: principles, measurements, and applications, In: Webster J, editor. Encyclopedia of electrical and electronics engineering. 3rd revised ed. New York Wiley; 2018c.
32. Kerkhof PLM, Kuznetsova T, Ali R, Handly N. Left ventricular volume analysis as a basic tool to describe cardiac function. Adv Physiol Educ. 2018d;42(1):130–139. doi: https://doi.org/10.1152/advan.00140.2017.
33. Kulpa J, Chinnappareddy N, Pyle WG. Rapid changes in cardiac myofilament function following the acute activation of estrogen receptor-alpha. PLoS One. 2012;7(7):e41076. https://doi.org/10.1371/journal.pone.0041076.
34. Kuznetsova T, Haddad F, Tikhonoff V, et al. Impact and pitfalls of scaling of left ventricular and atrial structure in population-based studies. J Hypertens. 2016;34:1186–94.
35. Ky B, French B, May Khan A, Plappert T, Wang A, Chirinos JA, et al. Ventricular-arterial coupling, remodeling, and prognosis in chronic heart failure. J Am Coll Cardiol. 2013;62:1165–72.
36. Lancet October 5, 1974.
37. Maffessanti F, Muraru D, Esposito R, Gripari P, Ermacora D, Santoro C, et al. Age-, body size-, and sex-specific reference values for right ventricular volumes and ejection fraction by three-dimensional echocardiography: a multicenter echocardiographic study in 507 healthy volunteers. Circ Cardiovasc Imaging. 2013;6:700–10.
38. Mazzoni M. Dario Maestrini e la legge del cuore Storia di un mancato Premio Nobel. Notizie dalla Delfico. 2005; 1–2. See also http://www.treccani.it/enciclopedia/dario-maestrini_(Dizionario-Biografico)/
39. Negroni JA, Morotti S, Lascano EC, Gomes AV, Grandi E, Puglisi JL, Bers DM. β-adrenergic effects on cardiac myofilaments and contraction in an integrated rabbit ventricular myocyte model. Mol Cell Cardiol. 2015;81:162–75.
40. Parks RJ, Howlett SE. Sex differences in mechanisms of cardiac excitation-contraction coupling. Pflugers Arch. 2013 May;465(5):747–63. doi: https://doi.org/10.1007/s00424-013-1233-0
41. Peace RA, Adams PC, Lloyd JJ. Effect of sex, age, and weight on ejection fraction and end-systolic volume reference limits in gated myocardial perfusion SPECT. J Nucl Cardiol. 2008;15:86–93.
42. Pijnenburg HLM, Kerkhof PLM, Kocsis S, Voorhout G. Ventricular volume regulation in dogs as a function of gender, age and breed. FASEB J. 1996;10:A346.
43. Previato M, Kerkhof PLM, Osto E, Parry G, Tona F. Heart transplantation survival and sex-related differences. In: Kerkhof PLM, Miller VM, editors. Sex-specific analysis of cardiovascular function. New York: Springer; 2018.
44. Puglisi JL, Negroni JA, Chen-Izu Y, Bers DM. The force-frequency relationship: insights from mathematical modeling. Adv Physiol Educ. 2013;37:28–34.
45. Putscher M. Geschichte der medizinischen Abbildung. Heinz Moos Verlag: Muenchen; 1972.
46. Robotham JL, Takata M, Berman M, Harasawa Y. Ejection fraction revisited. Anesthesiology. 1991;74(1):172–83.
47. Sakai S. Jacob Berengario. Igaku Toshokan. 1986;33:29–31. https://doi.org/10.7142/igakutoshokan.33.29 [The Japan Medical Library Association].
48. Sandberg K, Ji H. Sex differences in primary hypertension. Biol Sex Differ. 2012;3:7. https://doi.org/10.1186/2042-6410-3-7.
49. Sequeira V, van der Velden J. Historical perspective on heart function: the Frank–Starling law. Biophys Rev. 2015;7:421–47.
50. Shannon TR, Wang F, Puglisi J, Weber C, Bers DM. A mathematical treatment of integrated ca dynamics within the ventricular myocyte. Biophys J. 2004;87:3351–71.
51. Sharma K, Kass DA. Heart failure with preserved ejection fraction: mechanisms, clinical features, and therapies. Circ Res. 2014;115:79–96.
52. Shavik SM, Wall ST, Sundnes J, Burkhoff D, Lee LC. Organ-level validation of a cross-bridge cycling descriptor in a left ventricular finite element model: effects of ventricular loading on myocardial strains. Physiol Rep. 2017;5(21):e13392. https://doi.org/10.14814/phy2.13392.
53. Singer C. A short history of anatomy and physiology from the Greeks to Harvey. New York: Dover Publications; 1957.

54. Soltis AR, Saucerman JJ. Synergy between CaMKII substrates and β-adrenergic signaling in regulation of cardiac myocyte Ca(2+) handling. Biophys J. 2010;99:2038–47.
55. Touma M, Kang X, Gao F, Zhao Y, Cass AA, Biniwale R, Xiao X, Eghbali M, Coppola G, Reemtsen B, Wang Y. Wnt11 regulates cardiac chamber development and disease during perinatal maturation. JCI Insight. 2017;2(17):94904.
56. Trexler CL, Odell AT, Jeong MY, Dowell RD, Leinwand LA. Transcriptome and functional profile of cardiac myocytes is influenced by biological sex. Circ Cardiovasc Genet. 2017;10(5):e001770. https://doi.org/10.1161/CIRCGENETICS.117.001770.
57. Van Liere EJ, Stickney JC, Marsh DF. Sex differences in blood pressure of dogs. Science. 1949;109:489.
58. Wheelan C. Naked statistics. New York: WW Norton; 2013.

Sex Differences in Regulation of Blood Pressure

9

Jane F. Reckelhoff

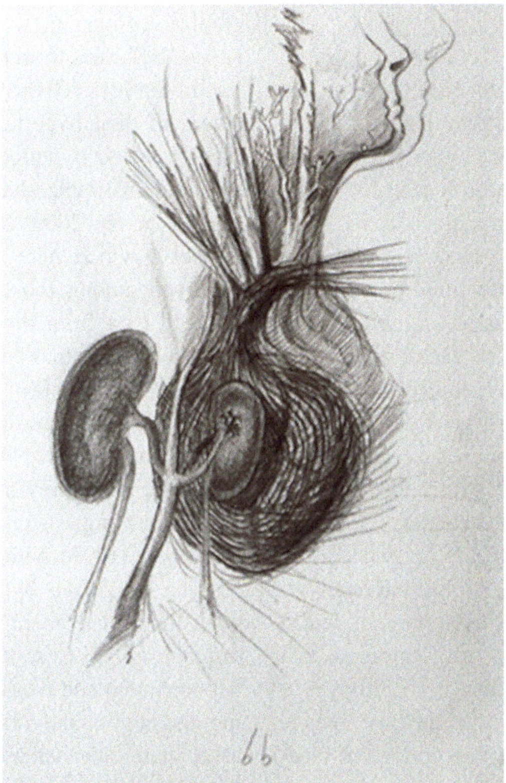

Art work by Piet Michiels, Leuven, Belgium.

J. F. Reckelhoff (✉)
Department of Cell and Molecular Biology and Women's Health Research Center and The Mississippi Center of Excellence in Perinatal Research, University of Mississippi Medical Center, Jackson, MS, USA
e-mail: jreckelhoff@umc.edu

Abstract

Hypertension is one of the leading risk factors for cardiovascular disease, myocardial infarction, and stroke. There are gender differences in the prevalence of hypertension and in the mechanisms responsible for hypertension in

P. L. M. Kerkhof, V. M. Miller (eds.), *Sex-Specific Analysis of Cardiovascular Function*,
Advances in Experimental Medicine and Biology 1065, https://doi.org/10.1007/978-3-319-77932-4_9

humans. This review will discuss the mechanisms for regulation of blood pressure, sex differences that have been identified in animal studies, and the gender differences that have been identified in humans.

Keywords
Hypertension · Obesity · Metabolic syndrome · Postmenopausal women · Hypogonadism · Immune system-mediated hypertension · Androgens · Estrogens · Endothelin

Introduction

Hypertension affects approximately 72 million US adults, with an overall prevalence of 29.3% [6, 24]. Hypertension is also a major public health problem worldwide both because of its high prevalence and also its role as a major risk factor for cardiovascular and kidney disease. The Global Burden of Disease Study identified hypertension as the leading global risk factor for mortality and the third leading global risk factor for disease burden in both men and women [16, 41]. Hypertension is not only a key risk factor for cardiovascular disease but the leading cause of death in both women and men. Moreover, more than half of all deaths in the United States (USA) are related to diseases which are aggravated by hypertension: heart disease, stroke, and renal failure [16]. Despite increasing awareness of cardiovascular disease and treatment of cardiovascular disease risk factors, the incidence of cardiovascular disease has increased over the last decade, especially in women. Thus, the control of blood pressure (BP) is paramount to protect the quality of life in men and women.

Prevalence of hypertension is significantly and independently associated with increasing age, increasing body mass index (BMI, kg/m^2), being African-American, and having less education [43]. Younger women have a lower prevalence of hypertension than men. In fact, the prevalence of hypertension is higher in men than women at young and middle ages and lower at elderly ages [24]. As such, women older than 60 years tend to have a higher prevalence of hypertension compared to men of similar ages. The highest prevalence of hypertension occurs among black women, older than 75 years, in whom prevalence rates exceed 70% [43].

Around the world, women with hypertension are more likely to be treated than men ([34, 54]; US). However, there are society differences as to whether men or women are better controlled for their hypertension. In the US National Health and Nutrition Examination Survey (1999–2004), 61.4% of women with hypertension were treated compared to 56.8% of hypertensive men. However, only 44.8% of treated women achieved BP control compared to 51.1% of treated men [21, 43]. Studies from the German Health Examination Surveys (GHES) showed that men are less aware of their hypertension, received less treatment for their hypertension, and had lower control of their BP than did women, which was a new finding in the 2008–2011 compared with 1998 studies in which there were no gender differences observed among those treated for hypertension [54]. In data from the US National Health and Nutrition Examination Survey (NHANES) IV (1999–2004), 50.8% of men and 55.9% of women had uncontrolled BP, despite the fact that women more frequently had their BP measured in the previous 6 months [30]. Furthermore, in a comparison of the NHANES III cohort (ending in 1999) with the NHANES IV cohort (ending in 2004), hypertension was less well-controlled in women than men, although the drugs used to treat hypertension were similar between men and women. BP normally falls during the night, and non-dipping BP at night is associated with increased target organ damage in both men and women [53, 61, 68]. Non-dipping BP in women appears associated with greater target organ damage than in men [4, 61], and postmenopausal women are more likely than premenopausal women to exhibit nocturnal non-dipping of BP [68].

Drug treatment regimens are also different in men and women often depending on the culture. In the GHES, younger women were more often prescribed β-adrenergic receptor blockers [54] and less angiotensin-converting enzyme inhibitors (ACEIs) than men. The Swedish Primary Care Cardiovascular Database (SPCCD) studies recently reported that women were also more often treated than men, more often treated with diuretics, and men were treated with ACEIs or angiotensin receptor blockers (ARBs) [34]. Men also interrupted their treatment more often than women. There are no gender specific guidelines for treatment of hypertension. In addition, whether certain classes of antihypertensive medications are more efficacious in men or women is virtually unstudied or has been studied but not evaluated for gender differences. The fact that animal studies have shown that there are sex differences in the mechanisms responsible for hypertension suggest that there may also be different mechanisms responsible for the hypertension in men and women, and further human studies are warranted.

Increased BP is independently and continuously associated with cardiovascular disease (5). Hypertension is defined as a systolic BP of 140 mmHg and higher or a diastolic BP of 90 mmHg and higher; prehypertension is defined as a BP between 120 and 39 systolic or 80–89 mmHg diastolic [47]. The higher the BP, the greater the risk for target organ injury, including myocardial infarction, heart failure, renal injury, and stroke [26], and an increase in cardiovascular disease risk begins with blood pressures as low as 115/75 mmHg [32]. Previously, hypertension was considered controlled if BP was less than 140/90 mmHg or, for those with diabetes or chronic kidney disease, less than 130/80 mmHg. However, the recent multicenter Systolic Blood Pressure Intervention Trial (SPRINT) showed that intensive reduction of BP with medications to 120 mmHg in individuals with hypertension and elevated cardiovascular disease (CVD) risk caused such a significant reduction in event rates, including myocardial infarction, acute coronary syndrome, stroke, decompensated heart failure, and cardiovascular disease-related death, that the trail was halted early [66].

SPRINT was performed in men and women average age 70 years > 50 years old [59], and the results were consistent across pre-specified clinical subgroups including age, sex, race, tertiles of SBP, history of cardiovascular disease, and any history of chronic kidney disease. Interestingly, only 2.8 medications were needed to reduce BP to 120 mmHg in subjects versus 1.8 medications in the standard treatment group [66]. Only 35–36% of the subjects in the SPRINT were women, and since the trial was stopped early, the data were not statistically significant that treating lower levels of BP was actually successful in reducing events in women [70]. The new American College of Cardiology/American Heart Association guidelines recommended stricter blood pressure control, and treatment in women was relegated to "Other Groups", despite that >50% of women over the age of 70 years are hypertensive.

Mechanisms Responsible for Hypertension

BP control is mediated via the kidney since studies have shown that the way the body manages salt and water is central to the level of BP [22, 23]. A normal kidney, when given a salt load, will cause an increase in BP such that there is a shift in the pressure-natriuresis relationship, whereby at higher salt content, the BP increases. As the salt is excreted with water by the kidney, the BP will return to normal levels in a few hours. In abnormal kidneys, when a salt load is taken, the BP increases, but this time the kidney is unable to excrete the salt and water, and the BP remains elevated causing a permanent shift in the pressure-natriuresis relationship to higher blood pressures. Thus during the steady state, elevated BP becomes necessary for salt excretion to be maintained at "normal" levels. This is why sodium excretion levels are similar for animals with similar salt intake regardless of one that is hypertensive and one that is normotensive.

Many homeostatic and hormonal systems contribute to BP regulation, including the renin-angiotensin system, the sympathetic nervous system via the renal nerves, the eicosanoid system, the oxidative stress/nitric oxide system, obesity and metabolic syndrome, the endothelin system, and sex steroids. In this review, the mechanisms and the sex differences in them will be discussed in terms of studies done in common animal models.

The Spontaneously Hypertensive Rat (SHR) as a Model of Hypertension

The SHR is a model that develops elevated BP as they progress through puberty [48]. When young, the BP is similar between males and females, but after puberty, the BP is significantly higher in males than females. Castration of male SHR reduces BP to levels found in females. Ovariectomy of the females has no effect on the BP, but testosterone treatment after puberty increases BP as in males. Thus the hypertension in the males is androgen-mediated, but the hypertension in the females is independent of estrogens. With aging, the BP remains static after approximately 9 months of age in the males, but in females, BP increases after cessation of estrous cycling, between 10 and 12 months of age, such that by 16 months of age, BP is similar to or higher in females than males [46]. This model has been used as a model of postmenopausal hypertension.

The mechanisms regulating BP are not only different between male and female SHR but are also different between aging and young animals. For example, in young SHR, the hypertension is mediated via the renin-angiotensin system (RAS) [49] and the sympathetic nervous system (SNS) [28]. This is not the case in aging animals as will be described below.

Role of the Renin-Angiotensin System (RAS)

Angiotensin II is an important regulator of BP [5]. Angiotensin (Ang) II is produced by the conversion of angiotensinogen to Ang I by renin and subsequent conversion of Ang I to Ang II by Ang I-converting enzyme (ACE). Ang II has its biological activity to increase BP through the Ang II AT1 receptor (AT1R). This makes up the vasoconstrictor arm of the RAS. Androgens have been shown to increase angiotensinogen synthesis in SHR [49], whereas estrogens have been shown to reduce expression of AT1Rs [5]. In addition to it's vascular effects, in the kidney, Ang II causes sodium reabsorption via the proximal tubule, also leading to increase in BP. Ang II can also be converted by ACE2 to Ang(1–7) [5], a vasodilator that has biological activity via the Mas receptor. Estrogens have activity to increase ACE2 activity and promote the vasodilatory effects of Ang(1–7). Thus there are sex differences in the RAS that can affect BP.

Essential hypertension in humans is associated with activation of the RAS in many cases. Even salt-sensitive hypertension, which is considered a low-renin hypertension and is prevalent in African-American individuals [63], is associated with abnormally elevated plasma renin activity compared to levels that should be accomplished when salt levels are elevated. Both men and women are prescribed ACEI and AT1R antagonists, although in some societies as mentioned above, men receive RAS blockers more frequently than women [34]. Both ACEIs and ARBs reduce BP to similar normotensive levels in young male and female SHR [49, 80], showing the RAS plays a greater role in the hypertension in male SHR than females since BP at baseline is higher in males. Furthermore, in the presence of male-level androgens in females, enalapril also lowers the BP [49], suggesting that androgens are working through the RAS to increase BP in male SHR.

The hypertension developed with Ang II infusion in rodents is also sex dependent. If the endogenous production of Ang II is blocked by enalapril, female rats exhibit a greater pressor response than males, but the hypertension is not salt sensitive, whereas it is in males [55]. Male mice, regardless of whether the endogenous RAS is blocked or not, have a greater pressor response to Ang II than females [67, 71]. Zimmerman and colleagues also reported that candesartan, an ARB, does not prevent the pressor response to Ang II infusion in males, but does prevent the response in female SHR. In addition they showed that depressor response to candesartan in the females is mediated via production of Ang(1–7) [80].

The contribution of the RAS of the hypertension in aging SHR is different, however. Enalapril reduces the BP to 100 mmHg in male SHR, aged 16–18 months, but to only 130 mmHg in age-matched females [74]. Thus the RAS is a major mediator of the hypertension in aging males, but in aging females, there are other mediators of the hypertension.

The Sympathetic Nervous System

As noted above, β-adrenergic receptor antagonists are common antihypertensive medications given to men and women. As noted above, in some studies women are prescribed them more frequently than men [54], although there is little evidence that they are better at controlling BP in women.

Studies in young SHR show that renal denervation reduces BP by 8–10% in both males and females. β-blockers, such as terazosin and propranolol, reduce BP in both male and female SHR [28, 37, 38]. β-Blockers had a greater effect in reducing BP in old females than in young females [38], suggesting that aging may be associated with increased sympathetic activation in female SHR. The RAS is thought to be stimulated by activation of the sympathetic nervous system. In aging female SHR, renal denervation reduced BP, but concomitant renal denervation followed by chronic losartan (ARB) treatment caused a further reduction in BP, but the rats still remained significantly hypertensive (approximately 140 mmHg) [37]. These data suggest that the hypertension in aging female SHR is mediated by both the sympathetic nervous system and the RAS independently and that other mechanisms contribute to the hypertension in aging females. These data have significant implications for hypertension in postmenopausal women and suggest that inappropriate, inefficacious medications may contribute to their resistant hypertension.

The mechanisms responsible for sympathetic activation in SHR are not clear. Activation of the melanocortin-4 receptor (MC4R) in the hypothalamus has been shown to increase sympathetic activation, and leptin is one of the mediators of MC4R activation. This is one of the mechanisms thought to play a role mediating obesity-induced sympathetic activation and hypertension since leptin is produced in adipose tissue. Da Silva and colleagues reported that intracerebroventricular (ICV) blockade of the MC4R reduces BP in young male SHR [8]. However, similar ICV infusion of MC4R antagonist in young and old female SHR failed to reduce BP, whereas the antagonist reduced BP in aging male SHR [39]. Thus the data suggest that different mechanisms may be responsible for activation of the sympathetic nervous system in males and females. The mechanism(s) that mediate sympathetic activation and hypertension in humans is not clear. In morbidly obese humans with MC4R deficiency, BP, heart rate, and urinary catecholamine excretion, all indicators of sympathetic activity, are reduced [14]. Thus whether the MC4R plays a role in mediating hypertension by causing activation of the sympathetic nervous system in men and women remains to be determined.

The Role of 20-HETE

Arachidonic acid can be converted by cytochrome P450 (CYP) ω-hydroxylases to 20-hydroxyeicosatetraenoic acids (20-HETE), respectively [51]. 20-HETE is a vasoconstrictor produced in endothelial cells. In the kidney, 20-HETE is also produced in renal tubules and reduces sodium reabsorption. Thus the location of 20-HETE production in the kidney (vascular vs tubular) determines whether it is prohypertensive or antihypertensive, respectively. An increase in plasma 20-HETE has been shown to occur in humans with acute ischemic stroke [77, 78] and acute coronary syndrome [81]. Thus 20-HETE could contribute to BP control in humans.

With regard to SHR, Zhang et al. [79] reported that adenoviral vector delivery of a CYP4A1 cDNA caused a decrease in BP in young male SHR, whereas the antisense CYP4A1 cDNA caused an increase in BP in control Sprague-Dawley rats. The investigators interpreted the data to mean that 20-HETE is important in both maintaining normotension and in contributing to hypertension. In old female SHR, blockade of 20-HETE reduces BP, but there is no effect of the inhibitor on BP in young females [75, 76]. To our knowledge, there have been no studies in which inhibitors of 20-HETE synthesis have been given to old male SHR.

Interestingly, the combination of enalapril and the eicosanoid synthesis inhibitor, 1-aminobenzotriazole (1-ABT), has different effects on BP in old female SHR depending on which drug is given first [33]. Treatment of old female SHR with either 1-ABT or enalapril reduces BP to similar levels. Addition of enalapril to the 1-ABT causes a significantly greater reduction in BP than does the addition of 1-ABT to enalapril. However, the BP remained approximately 125–130 mmHg. The data suggest that both the RAS and 20-HETE contribute independently to a part of the hypertension in old female SHR but that 20-HETE may also have activity via the RAS mechanism.

Whether 20-HETE plays a role in mediating the BP in men and women or whether 20-HETE blockade would cause a reduction in BP, and thus could be a novel antihypertensive agent in individuals with resistant hypertension, is unknown and is a topic for further discovery.

Role of Oxidative Stress/Nitric Oxide System

Oxidative stress is defined as the production of superoxide or other reactive oxygen and/or nitrogen species (ROS/RNS) [52]. The NADPH oxidase pathway, the xanthine oxidase pathway, and mitochondria are thought to contribute to the production of ROS. ROS are naturally produced as a consequence of mitochondrial energy production and play important roles in mediating intracellular signaling mechanisms. Superoxide that is produced in the vasculature can bind nitric oxide (NO), thus binding NO and causing vasoconstriction. The combination of superoxide and NO also produces peroxynitrite, a strong oxidant that is a vasodilator that with time causes tachyphylaxis resulting in vasoconstriction. Peroxynitrite can also oxidize other factors, such as the vasodilator, prostacyclin, which causes vasoconstriction.

Oxidative stress has been implicated as playing a role in mediating hypertension, since treatment of adult male SHR with tempol, the superoxide dismutase mimetic, causes a reduction in their BP [56]. Adult female SHR do not respond with a reduction in their BP when tempol is given [19]. However, if SHR are given tempol from weaning, there is an attenuation of their BP in both males and females, suggesting that oxidative stress may be playing a role in development of hypertension in males and females but that maintenance of hypertension in male, but not female SHR, is mediated by oxidative stress. Male SHR also have a depressor response to an NADPH oxidase inhibitor, apocynin, whereas females do not. Sullivan et al. [62] reported that male SHR exhibit greater excretion of hydrogen peroxide than females. Interestingly, the levels of F2-isoprostanes, an indicator of oxidative stress, is similar in plasma of male and female SHR, are higher in kidney tissue of males than females, but excretion rate of F2-isoprostanes is tenfold higher in females than males. Plasma total antioxidant capacity measured in serum is similar in males and females, whereas basal and NADPH-stimulated lucigenin chemiluminescence, an indicator of NADPH oxidase activity, is not different in kidneys of males and females, but higher in aortae of male SHR. Thus the lack of a depressor response to tempol is not due to lack of oxidative stress in female SHR. Female mREN2 rats are hypertensive and also do not respond to tempol [42].

In order to evaluate the mechanisms responsible for the pressor response to oxidative stress, we gave molsidomine, a drug that causes an increase in both superoxide and nitric oxide. Molsidomine caused an increase in BP in male SHR but not females or male WKY controls [18]. The increase in BP in males was accompanied by an increase in lucigenin chemiluminescence, an indicator of increased oxidative stress, an increase in nitrate/nitrite excretion, an index of NO production, and an increase in renal expression of catalase and glutathione peroxidase in WKY males, but not SHR males [18, 19]. Because it was hypothesized that the lack of a response in females was mediated by the increased endogenous NO levels, rats were given nitro-L-arginine methyl ester (L-NAME), the nonselective NOS inhibitor, after molsidomine treatment [35]. While L-NAME increased the BP in females, there was no protective effect with molsidomine.

Endothelial NOS requires cofactors, such as calcium and tetrahydrobiopterin, for activity. In oxidative stress situations, tetrahydrobiopterin is converted to dihydrobiopterin causing the "uncoupling" of eNOS, such that the enzyme produces superoxide rather than NO. In male SHR we tested the hypothesis that infusion of tetrahydrobiopterin would circumvent the increased production of superoxide thus causing a reduction in BP. Studies in male SHR showed that, indeed, tetrahydrobiopterin infusion did reduce BP in male SHR; however, the mechanism was independent of NO and was due instead to a reduction in the synthesis of androgens [17]. Since the BP in male SHR is androgen dependent, a reduction in androgens mediated the reduction in BP.

Unfortunately, there are no studies in humans that suggest that oxidative stress contributes to hypertension. Clinical trials using various antioxidants, such as vitamin E or C, either have shown no benefit to BP, or actually increased BP. Although most studies said the data were "factored for sex," the data were evaluated separately for men and women. Since in our studies and others hypertension in female animals is independent of oxidative stress, it is possible that the antioxidants were effective in men, but not women. In addition, the studies were performed in individuals who had been hypertensive for years. In our studies we found that if the NO system is blocked, tempol and apocynin are incapable of reducing BP in male SHR [73]. These data suggest that an active NO system is necessary for antioxidants to reduce the BP. If the individuals in the clinical trials had significant endothelial dysfunction and thus reductions in synthesis of NO, then the antioxidant therapy would not be expected to be effective. Perhaps antioxidant therapy would be more beneficial in younger men who have little endothelial dysfunction.

Role of the Immune System

Although there have been numerous animal studies showing that T cells contribute to hypertension, the data in humans is less striking (see review [50]). ACEI does reduce the number of circulating CD4+ T cells. Data from the Multicenter AIDS Cohort Study indicate that untreated HIV-positive patients with chronically low numbers of CD4+ T cells have a lower prevalence of systolic hypertension than treated HIV patients and uninfected control subjects [57]. In addition, Herrera et al. [25] showed in a small cohort of essential hypertensive individuals receiving mycophenolate mofetil (MMF), an inhibitor of T cell production, for rheumatoid arthritis or psoriasis, that their BP fell. When MMF was discontinued, BP returned to previously high levels, suggesting that amelioration of hypertension was the result of immune suppression. Whether there are sex differences in the immune system-mediated increase in BP in humans is unknown, however.

Sex differences have been shown to be present in the hypertension in SHR. Tipton et al. [65] reported that there are sex differences in the type of T cells that infiltrate the kidney in SHR. They found that the circulating levels of anti-inflammatory CD3+ and CD4+ and pro-inflammatory CD3 + CD4 + RORγTh17 cells were higher in female SHR than males and that males had more immune-suppressive circulating CD3 + CD4 + Foxp3+ T regulatory cells. The kidneys of females also had higher levels of CD8+ and T regulatory cells than males, whereas kidneys of male SHR had higher levels of CD4+ and Th17 cells. MMF decreased BP in both male and female SHR, but the reduction was greater in the females, suggesting an immune component in the hypertension in both males and females.

Interestingly, unlike discussed above, experimental hypertension in females that has an immune component is responsive to antioxidants, such as tempol or apocynin. For example, tempol/apocynin [40] and MMF [64] reduce BP in a mouse model of lupus erythematosus. In addition, the reduced uterine perfusion pressure (RUPP) model of preeclampsia, which mimics conditions in women, is associated with increases in T(H)17 cells that secrete IL-17, autoantibodies activating the AT1R (AT1-AA), and placental oxidative stress [12]. Tempol not only reduced the BP in the model [58] but also reduced urinary excretion of F2-isoprostanes and AT1-AAs [12]. Why immune system-mediated hypertension in females is susceptible to antioxidants but not SHR hypertension when the immune system is activated in SHR females is not clear and requires further study. Studies into the role played by the immune system in hypertension in aging men and women or animals, and whether or not there are sex differences, are limited and will also require additional research.

Role of Obesity and Metabolic Syndrome

Obesity is a common risk factor for hypertension in both men and women of all ages. Over 70% of adults in the United States with hypertension are currently overweight or obese [47]. Epidemiological and longitudinal studies clearly demonstrate an association between increasing body weight and increasing BP, and the increase in prevalence of hypertension increases with increasing BMI. Whether there are gender differences in the increase in BP with increasing body weight in humans has not been studied to our knowledge.

Sex differences in BP have not been well studied in animal models of obesity. BP measurements in males and females exist but in most cases, have not been done at the same time or by the same groups. For example, telemetry studies in the obese, male MC4R knockout rat show that their BP is not different than control Wistar rats [60], whereas telemetry BP is significantly higher in obese female MC4R knockout rats than WT controls [39]. In addition, no studies have been done in which the increase in body weight has been correlated with BP level. Future studies will need to be done to address this question.

Role of Sex Steroids

Numerous studies have shown that sex steroids affect BP. Acutely, both androgens and estrogens cause vasodilation. In experimental animals, chronic estrogens cause an increase in eNOS expression [69] and thus cause vasodilation which should reduce BP. Androgens have different effects on BP whether they are given to males or females. Postmenopausal women often have elevated androgen levels. Women with polycystic ovary syndrome (PCOS), the most common endocrine abnormality in young reproductive-aged women, develop elevated BP with chronic androgen excess [2, 13].

BP increases in men and women as they age and endogenous sex hormone levels drop [47]. As mentioned previously, the prevalence of hypertension increases more in women after menopause when estrogen levels drop than in age-matched men. However, the role that lack of estrogens play in hypertension in postmenopausal women is controversial. There have been no studies to our knowledge in which ambulatory BP has been serially measured over the perimenopausal transition. Hormone replacement therapy (HRT), even with estradiol rather than conjugated equine estrogens (CEE), has not been shown to consistently lower BP in postmenopausal women. The method of drug delivery, whether oral or transdermal, whether estradiol or CEE, likely contributes to the variable BP effect. In addition, many studies have been short term (e.g., < 1 year). Ichikawa et al. [27] reported lower diastolic and mean BP with transdermal HRT for 12 and 24 months in normotensive postmenopausal women. In another study by Prelevic et al. [45], healthy postmenopausal women who had been taking HRT for at least 5 years had blood pressures similar to untreated age-matched controls or the BP was even higher than controls.

A recent meta-analysis supports these previous findings, that oral HRT had a neutral effect on BP in both normotensive and hypertensive postmenopausal women [29]. Transdermal estrogen and micronized progesterone had a beneficial effect on normotensive women, but had only a neutral effect on hypertensive women. These data do not support estrogens as being the major modulator of BP control in postmenopausal women.

Obesity is associated with reductions in androgen synthesis in men and male animals. For example, in the male obese Zucker rat, androgen levels by 22 weeks of age are reduced by almost 70% compared to their lean littermates [10]. By the time these rats are 32 weeks of age, the obese rats have elevated BP, whereas the lean rats do not. Testosterone supplements in the obese Zucker rats cause a further increase in BP despite the fact that the supplements reverse the inflammation, insulin resistance, and hyperlipidemia. BP does not increase with testosterone supplements in lean Zucker rats. These data suggest that androgen supplements in obese men may reduce some of the cardiovascular risk factors, but care should be taken to monitor their BP carefully.

Alternatively, as shown in Table 9.1, the effects of androgens in women and female rats are very different than in men or male rats. Androgen supplements in normotensive female Sprague-Dawley rats have the opposite effect as in males and cause an increase in food intake, an increase in body weight, increased inflammation and hyperlipidemia, and an increase in BP [76]. This is a model of polycystic ovary syndrome (PCOS) in women [36]. Women with PCOS have elevated BP and many of them are overweight or obese [15]. Because women with PCOS are young when diagnosed, their elevated BP often does not meet the current guidelines for treatment. It remains to be seen how the SPRINT may change those guidelines.

Table 9.1 Sex differences in response to androgens

Parameter	Obese Zucker males	Female Sprague Dawley rats
Body weight	Decrease	Increase
Inflammation	Decrease	Increase
Insulin resistance	Decrease	Increase
Cholesterol	Decrease	Increase
Blood pressure	Increase	Increase

Female-to-male transsexual individuals who take high doses of androgens for virilization also develop PCOS-type symptoms [3], elevated BP, and elevated endothelin levels [20]. In contrast, men with various forms of hypogonadism display significantly higher endothelin levels in comparison with age-matched healthy males, and testosterone therapy decreased ET-1 levels in these individuals [7].

The mechanisms by which androgens can increase BP in females are not clear. Androgen supplements do not increase BP in female rats that are null for CYP4A2 [9], one of the ω-hydroxylases that produce 20-HETE, suggesting that activation of the 20-HETE pathway may contribute. Also, if the MC4R is blocked or if the rats are MC4R null, androgens fail to increase their BP [9], suggesting that an active MC4R system is necessary to increase BP in females with androgens.

Role of Endothelin

Endothelin is known to be one of the most potent vasoconstrictors. A thorough review of sex differences in endothelin (ET1), the receptors, ET_A and ET_B, and the mechanisms responsible is available elsewhere [20]. In postmenopausal women, plasma endothelin levels are increased [31], suggesting that endothelin may contribute to the increased BP following menopause.

The mechanism by which endothelin increases in postmenopausal women is not clear. However, Ang II increases preproendothelin synthesis [1], and thus activation of the RAS after menopause could contribute to the elevated ET-1. While untreated postmenopausal women have been shown to have elevated levels of endothelin, hormone replacement therapy (HRT) with either micronized 17β-estradiol and dydrogesterone or CEE and medroxyprogesterone results in further increases in endothelin levels [11, 44]. Thus the role that the endothelin system plays in mediating gender differences in BP in humans remains to be elucidated.

In young and old male and female SHR, endothelin likely plays little role in their BP. In aging females ET_A receptor antagonism reduces the BP by approximately 10 mmHg [72, 73, 75], but the rats remain very hypertensive. If the three systems, the RAS (with enalapril), 20-HETE (with aminobenzotriazole (1-ABT)), and the ET_A receptor (with ABT-627 – Abbott Labs), are all blocked together, the BP in old female SHR is reduced to approximately 120 mmHg [33]. These data suggest that in aging women, especially those who have essential hypertension that is hard to control, several drugs from different categories of antihypertensives may be beneficial in managing their BP.

Summary

Sex differences in BP control employ a myriad of mechanisms and the mechanisms are different whether the subjects are young or aging. As shown in Fig. 9.1, in young male SHR, the hypertension is mediated via androgens to activate the MC4R and subsequently activate the sympathetic nervous system. Androgens also activate the RAS by increasing expression of angiotensinogen and thus increasing renin activity leading to increases in Ang II. In addition, androgens activate the oxidative stress pathway and could in turn increase RAS acitvity. In young female SHR, sex steroids and the MC4R do not contribute to the hypertension, and thus only activation of sympathetic nervous system and the RAS control their BP.

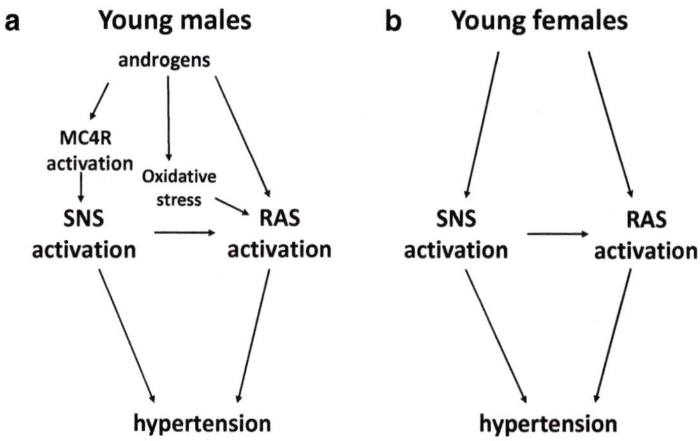

Fig. 9.1 Mechanisms responsible for hypertension in young male (**a**) and female (**b**) SHR. Abbreviations: *MC4R* melanocortin-4 receptor, *SNS* sympathetic nervous system, *RAS* renin-angiotensin system

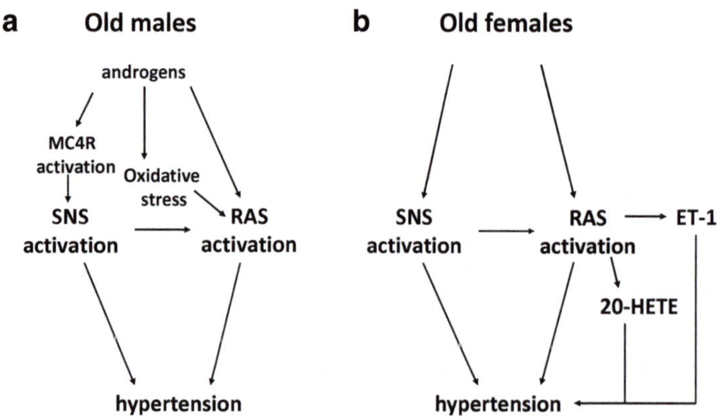

Fig. 9.2 Mechanisms responsible for hypertension in old male (**a**) and female (**b**) SHR. Abbreviations: *MC4R* melanocortin-4 receptor, *SNS* sympathetic nervous system, *RAS* renin-angiotensin system, *ET-1* endothelin

As shown in Fig. 9.2, the aging males do not change their mechanisms responsible for the hypertension. In aging females, however, there is a role for RAS activation of both endothelin (ET-1) and 20-HETE, both of which are vasoconstrictors and increase the BP.

Future studies are needed to clarify the mechanisms responsible for their hypertension, whether they are similar or different than in SHR, and then how these mechanisms can impact the way hypertension in men and women is addressed therapeutically, especially in this age of "precision medicine." The problem of non-compliance with antihypertensive medications needs to be addressed to determine why it is that the medications are not taken or are dropped. To improve compliance, more education is needed to convince individuals that despite their current lack of symptoms without antihypertensive medication, treatment of hypertension reduces the probability of them developing major debilitating cardiovascular diseases, such as myocardial infarction and stroke, later. Finally, future studies are needed to develop new pharmacological tools to specifically address the various mechanisms responsible for hypertension in men and women individually in order to develop more "precision medicine" approach.

Acknowledgments This work is supported by NIH grants, R01HL66072, R01HL67028, P01HL51971, and P20GM121334.

References

1. Alexander BT, Cockrell KL, Rinewalt AN, Herrington JN, Granger JP. Enhanced renal expression of preproendothelin mRNA during chronic angiotensin II hypertension. Am J Physiol Regul Integr Comp Physiol. 2001;280:R1388–92.
2. Azziz R, Carmina E, Dewailly D, Diamanti-Kandarakis E, Escobar-Morreale HF, Futterweit W, Janssen O, Legro R, Normal R, Taylor A, Witchel S. Position statement: criteria for defining polycystic ovary syndrome: an androgen excess society guideline. J Clin Endocrinol Metab. 2006;91:4237–5.
3. Baba T, Endo T, Honnma H, et al. Association between polycystic ovary syndrome and female-to male transsexuality. Hum Reprod. 2006;22:1011–6.
4. Ben-Dov IZ, Kark JD, Ben-Ishay D, Mekler J, Ben-Arie L, Bursztyn M. Predictors of all-cause mortality in clinical ambulatory monitoring. Unique aspects of BP during sleep. Hypertension. 2007;49:235–1241.
5. Chappell MC. Biochemical evaluation of the renin-angiotensin system: the good, bad, and absolute? Am J Physiol Heart Circ Physiol. 2016;310:H137–52.
6. Chobanian AV, Bakris GL, Black HR, Cushman WC, Green LA, Izzo Jr JL, et al. National Heart, Lung, and Blood Institute joint National Committee on prevention, detection, evaluation, and treatment of high blood pressure: the JNC 7 report. JAMA. 2003;289:256072.
7. Christakou C, Economou F, Livadas S, Piperi C, Adamopoulos C, Marinakis E, Jdiamanti-Kandarakis E. Strong and Positive association of endothelin-1 with AGEs in PCOS: a causal relationship or a bystander? Hormones (Athens). 2011;10:292–7.
8. da Silva AA, do Carmo JM, Kanyicska B, Dubinion J, Brandon E, Hall JE. Endogenous melanocortin system activity contributes to the elevated arterial pressure in spontaneously hypertensive rats. Hypertension. 2008;51:884–90.
9. Dalmasso C, Maranon R, Patil C, Moulana M, Romero DG, Reckelhoff JF. 20-HETE and CYP4A2 ω-hydroxylase contribute to the elevated blood pressure in hyperandrogenemic female rats. Am J Physiol Renal Physiol. 2016;311:F71–7.
10. Davis DD, Ruiz AL, Yanes LL, Iliescu R, Yuan K, Moulana M, Racusen LL, Reckelhoff JF. Testosterone supplementation in male obese Zucker rats reduces body weight and improves insulin sensitivity but increases blood pressure. Hypertension. 2012;59:726–31.
11. de Kraker AT, Kenemans P, Smolders RG, Kroeks MV, van der Mooren MJ. Short-term effects of continuous combined oestrogenprogestogen therapy on several cardiovascular risk markers in healthy postmenopausal women; a randomized control trial. Eur J Obstet Gynecol Reprod Biol. 2009;142:139–44.
12. Dhillion P, Wallace K, Herse F, Scott J, Wallukat G, Heath J, Mosely J, Martin JN Jr, Dechend R, La Marca B. IL-17-mediated oxidative stress is an important stimulator of AT1-AA and hypertension during pregnancy. Am J Physiol Regul Integr Comp Physiol. 2012;303:R353–8.
13. Diamanti-Kandarakis E. Polycystic ovarian syndrome: pathophysiology, molecular aspects and clinical implications. Exp Rev Molec Med. 2008;10:1–21.
14. do Carmo JM, da Silva AA, Wang Z, Fang T, Aberdein N, Perez de Lara CE, Hall JE. Role of the brain melanocortins in blood pressure regulation. Biochim Biophys Acta. 2017;1863 (10 Pt A):2508–14.
15. Escobar-Morreale HF, San Millan JL. Abdominal adiposity and the polycystic ovary syndrome. Trends in Endocrinol Metab. 2007;18:266–72.
16. Ezzati M, Lopez AD, Rodgers A, Vander Hoorn S, Murray CJ. Comparative risk assessment collaborating group. Selected major risk factors and global and regional burden of disease. Lancet. 2002;360:134760.
17. Fortepiani LA, Reckelhoff JF. Treatment with tetrahydrobiopterin reduces blood pressure in male SHR by reducing testosterone synthesis. Am J Physiol Regul Integr Comp Physiol. 2005a;R288:R733–6.
18. Fortepiani LA, Reckelhoff JF. Increasing oxidative stress with molsidomine increases blood pressure in genetically hypertensive rats but not normotensive controls. Am J Physiol Regul Integr Comp Physiol. 2005b;289:R763–70.
19. Fortepiani LA, Reckelhoff JF. Role of oxidative stress in the sex differences in blood pressure in spontaneously hypertensive rats. J Hypertens. 2005c;23:801–5.
20. Gohar EY, Giachini FR, Pollock DM, Tostes RC. Role of endothelin system in sexual dimorphism in cardiovasular and renal diseaes. Life Sci. 2016;159:20–9.

21. Gu A, Burt VL, Paulose-Ram R, Dillon CF. Gender differences in hypertension treatment, drug utilization patterns, and blood pressure control am19 US adults with hypertension: data from the national health and examination survey 19992004. Am J Hypertens. 2008;21:78998.
22. Hall JE, Guyton AC, Brands MW. Control of sodium excretion and arterial pressure by intrarenal mechanisms and the renin-angiotensin system. In: Laragh JH, Brenner BM, editors. Hypertension, pathophysiology, diagnosis, and management. 2nd ed. New York: Raven Press; 1995. p. 1452–75.
23. Hall JE, Guyton AC, Brands MW. Pressure-volume regulation in hypertension. Kidney Int Suppl. 1996;55:S35–41.
24. Health, United States. National Center for Health Statistics. With special feature on death abd dying. Hyattsville, MD 2011. Table 67 Hypertension and high blood pressure among persons 20 years of age and over, by selected characteristics: US, selected years 1988–1994 through 2005-2008, 2010. p. 250–51.
25. Herrera J, Ferrebuz A, MacGregor EG, Rodriguez-Iturbe B. Mycophenolate mofetil treatment improves hypertension in patients with psoriasis and rheumatoid arthritis. J Am Soc Nephrol. 2006;17(Suppl 3):S218–25.
26. Hsia J, Margolis KL, Eaton CB, et al. Prehypertension and cardiovascular disease risk in the women's health initiative. Circulation. 2007;115:85560.
27. Ichikawa A, Sumino H, Ogawa T, Ichikawa S, Nitta K. Effects of long term transdermal hormone replacement therapy on the renin-angiotensin-aldosterone system, plasma bradykinin levels and BP in normotensive postmenopausal women. Geriatr Gerontol Int. 2008;8:259–64.
28. Iliescu R, Yanes LL, Bell W, Dwyer T, Baltatu OC, Reckelhoff JF. Role of the renal nerves in blood pressure in male and female SHR. Am J Physiol Regul Integr Comp Physiol. 2006;290:R341–4.
29. Issa Z, Seely EW, Rahme M, El-Hajj Fuleihan G. Effects of hormone therapy on blood pressure. Menopause. 2015;22:456–68.
30. Kim JK, Alley D, Seeman T, Karlamangla A, Crimmins E. Recent changes in cardiovascular risk factors among women and men. J Women's Health (Larchmont). 2006;15:73446.
31. Komatsumoto S, Nara M. Changes in the level of endothelin-1 with aging. Nippon Ronen Igakkai Zasshi. 1995;32:664–9.
32. Lewington S, Clarke R, Qizilbash N, et al. Age-specific relevance of usual blood pressure to vascular mortality. Lancet. 2002;360:190313.
33. Lima R, Yanes LL, Davis DD, Reckelhoff JF. Roles played by 20-HETE, angiotensin II and endothelin in mediating the hypertension in aging female spontaneously hypertensive rats. Am J Physiol Regul Integr Comp Physiol. 2013;304: R248–51.
34. Ljungman C, Kahan T, Schiöler L, Hjerpe P, Hasselström J, Wettermark B, Boström KB, Manhem K. Gender differences in antihypertensive drug treatment: results from the Swedish Primary Care Cardiovascular Database (SPCCD). J Am Soc Hypertens. 2014;8:882–90.
35. Lopez-Ruiz AF, Iliescu R, Reckelhoff JF. Refractory blood pressure in female SHR to increased oxidative stress is not mediated by NO or by upregulation of renal antioxidant enzymes. Am J Physiol Regul Integr Comp Physiol. 2010;298:R266–71.
36. Mannerås L, Cajander S, Holmäng A, Seleskovic Z, Lystig T, Lönn M, Stener-Victorin E. A new rat model exhibiting both ovarian and metabolic characteristics of polycystic ovary syndrome. Endocrinology. 2007;148:3781–91.
37. Maranon RO, Reckelhoff JF. Mechanisms responsible for postmenopausal hypertension in a rat model: Roles of the renal sympathetic nervous system and the renin-angiotensin system. Physiol Rep. 2016;4:e12669.
38. Maranon RO, Lima R, Mathbout M, Do Carmo JM, Hall JE, Roman RJ, Reckelhoff JF. Postmenopausal hypertension: role of the sympathetic nervous system in an animal model. Am J Physiol Regul Integr Comp Physiol. 2014;306:R248–56.
39. Maranon R, Lima R, Spradley FT, do Carmo JM, Zhang H, Smith AD, Bui E, Thomas RL, Moulana M, Hall JE, Granger JP, Reckelhoff JF. Roles for the sympathetic nervous system, renal nerves, and CNS melanocortin-4 receptor in the elevated blood pressure in hyperandrogenemic female rats. Am J Physiol Regul Integr Comp Physiol. 2015;308:R708–13.
40. Mathis KW, Venegas-Pont M, Masterson CW, Stewart NJ, Wasson KL, Ryan MJ. Oxidative stress promotes hypertension and albuminuria during the autoimmune disease systemic lupus erythematosus. Hypertension. 2012;59:673–9.
41. Mosca L, Banka CL, Benjamin EJ, Berr K, Bushnell C, Ganiats T, et al. Update: American Heart Association evidence-based guidelines for cardiovascular disease 2007 update. Circulation. 2007;115:1481–501.
42. Nautiyal M, Shaltout HA, de Lima DC, do Nascimento K, Chappell MC, Diz DI. Central angiotensin-(1-7) improves vagal function independent of blood pressure in hypertensive (mRen2)27 rats. Hypertension. 2012;60:1257–65.
43. Ong KL, Cheung BM, Man YB, Lau CP, Lam KS. Prevalence, awareness, treatment and control of hypertension among United States adults 1999–2004. Hypertension. 2007;49:6975.
44. Polderman KH, Stehouwer CD, van Kamp GJ, Dekker GA, Verheugt FW, Gooren LJ. Influence of sex hormones on plasma endothelin levels. Ann Intern Med. 1993;118:429–32.
45. Prelevic GM, Kwong P, Byrne DJ, Jagroop IA, Ginsburg J, Mikhailidis DPA. Cross-sectional study of the effect of hormone replacement therapy on cardiovascular disease risk profile in healthy postmenopausal women. Fertil Steril. 2002;77:945–51.
46. Reckelhoff JF, Fortepiani LA. Novel mechanisms responsible for postmenopausal hypertension. Hypertension. 2004;43:918–23.
47. Reckelhoff JF, Wofford M. Chapter 70: Hypertension in women. In: Rexrode K, Wickline M, editors. Women and health. 2nd ed. New York: Elsevier; 2012.
48. Reckelhoff JF, Zhang H, Granger JP. Testosterone exacerbates hypertension and reduces pressure-natriuresis in male spontaneously hypertensive rats. Hypertension. 1998;31:435–9.
49. Reckelhoff JF, Zhang H, Srivastava K. Gender differences in development of hypertension in spontaneously hypertensive rats: role of the renin-angiotensin system. Hypertension. 2000;35:480–3.
50. Rodriguez-Iturbe B, Pons H, Johsnon RJ. Role of the immune system in hypertension. Physiol Rev. 2017;97:1127–64.
51. Roman RJ. P-450 metabolites of arachidonic acid in the control of cardiovascular function. Physiol Rev. 2002;82:131–85.
52. Romero JC, Reckelhoff JF. State-of-the-Art lecture. Role of angiotensin and oxidative stress in essential hypertension. Hypertension. 1999;34:943–9.
53. Routledge FS, McFetridge-Durdle JA, Stress DCR. Menopausal status and nocturnal BP dipping patterns among hypertensive women. Can J Cardiol. 2009;25:e157–63.
54. Sarganas G, Neuhauser HK. The persisting gender gap in hypertension management and control in Germany: 1998 and 2008-2011. Hypertens Res. 2016;39:457–66.
55. Sartori-Valinotti JC, Iliescu R, Yanes LL, Dorsett Martin W, Reckelhoff JF. Sex differences in the pressor response to

angiotensin II when the endogenous renin-angiotensin system is blocked. Hypertension. 2008;51:1170–6.
56. Schnackenberg CG, Welch WJ, Wilcox CS. Normalization of blood pressure and renal vascular resistance in SHR with a membrane-permeable superoxide dismutase mimetic: role of nitric oxide. Hypertension. 1998;32:59–64.
57. Seaberg EC, Muñoz A, Lu M, Detels R, Margolick JB, Riddler SA, Williams CM, Phair JP. Multicenter AIDS cohort study. Association between highly active antiretroviral therapy and hypertension in a large cohort of men followed from 1984 to 2003. AIDS. 2005;19:953–60.
58. Sedeek M, Gilbert JS, LaMarca BB, Sholook M, Chandler DL, Wang Y, Granger JP. Role of reactive oxygen species in hypertension produced by reduced uterine perfusion in pregnant rats. Am J Hypertens. 2008;21:1152–6.
59. SPRINT Research Group, Wright JT Jr, Williamson JD, Whelton PK, Snyder JK, Sink KM, Rocco MV, Reboussin DM, Rahman M, Oparil S, Lewis CE, Kimmel PL, Johnson KC, Goff DC Jr, Fine LJ, Cutler JA, Cushman WC, Cheung AK, Ambrosius WT. A randomized trial of intensive versus standard blood pressure control. N Engl J Med. 2015;373:2103–16.
60. Stepp DW, Osakwe CC, Belin de Chantemele EJ, Mintz JD. Vascular effects of deletion of melanocortin-4 receptors in rats. Physiol Rep. 2013;1:e00146.
61. Stolarz K, Staessen JA, O'Brien ET, Night-time BP. Dipping into the future? J Hypertension. 2002;21:2131–3.
62. Sullivan JC, Sasser JM, Pollock JS. Sexual dimorphism in oxidant status in spontaneously hypertensive rats. Am J Physiol Regul Integr Comp Physiol. 2007;292:R764–8.
63. Swart R, van Rooyen JM, Mels CMC. Change in renin, cardiovascular and inflammatory markers over three years in black and white population: the SABPA Study. BMC Cardiovasc Disord. 2017;17:104.
64. Taylor EB, Ryan MJ. Immunosuppression with mycophenolate mofetil attenuates hypertension in an experimental model of autoimmune disease. J Am Heart Assoc. 2017;27:6.
65. Tipton AJ, Baban B, Sullivan JC. Female spontaneously hypertensive rats have greater renal anti-inflammatory T lymphocyte infiltration than males. Am J Physiol Regul Integ Comp Physiol. 2012;303:R359–67.
66. Tyson CC, Coffman TM. In the wake of systolic blood pressure intervention trial: new targets for improving hypertension management in chronic kidney disease? Nephron. 2017;135:287–90.
67. Venegas-Pont M, Sartori-Valinotti JC, Glover PH, Reckelhoff JF, Ryan MJ. Sexual dimorphism in the blood pressure response to angiotensin II in mice after angiotensin converting enzyme blockade. Am J Hypertens. 2010;23:92–6.
68. Verdecchia P, Porcellati C, Schillaci G, Borgioni C, Ciucci A, Battistelli M, et al. An independent predictor of prognosis in essential hypertension. Hypertension. 1994;24:793–801.
69. Weiner CP, Lizasoain I, Baylis SA, Knowles RG, Charles IG, Moncada S. Induction of calcium-dependent nitric oxide synthases by sex hormones. Proc Natl Acad Sci USA. 1994;91:5212–6.
70. Wenger NK, Ferdinand KC, Bairey Merz CN, Walsh MN, Gulati M, Pepine CJ. American College of Cardiology Cardiovascular Disease in women committee. Women, hypertension, and the systolic blood pressure intervention trial. Am J Med. 2016;29:1030–6.
71. Xue B, Pamidimukkala J, Hay M. Sex differences in the development of angiotensin II induced hypertension in conscious mice. Am J Physiol Heart Circ Physiol. 2005;288: H2177–84.
72. Yanes LL, Romero DG, Cucchiarelli VE, Fortepiani LA, Gomez-Sanchez CE, Santacruz F, Reckelhoff JF. Role of endothelin in mediating postmenopausal hypertension in a rat model. Am J Physiol Regul Integr Comp Physiol. 2005a;288:R229–33.
73. Yanes L, Romero D, Iliescu R, Cucchiarelli VE, Fortepiani LA, Santacruz F, Bell W, Zhang H, Reckelhoff JF. Systemic arterial pressure response to two weeks of Tempol therapy in SHR: involvement of NO, the RAS, and oxidative stress. Am J Physiol Regul Integr Comp Physiol. 2005b;288:R903–8.
74. Yanes LL, Romero DG, Iles JW, Iliescu R, Gomez-Sanchez C, Reckelhoff JF. Sexual dimorphism in the renin-angiotensin system in aging spontaneously hypertensive rats. Am J Physiol Regul Integr Comp Physiol. 2006;291:R383–90.
75. Yanes LL, Lima R, Moulana M, Romero DG, Yuan K, Ryan MJ, Baker R, Zhang H, Fan F, Davis DD, Roman RJ, Reckelhoff JF. Postmenopausal hypertension: role of 20-HETE. Am J Physiol Regul Integr Comp Physiol. 2011a;300:R1543–8.
76. Yanes LL, Romero DG, Moulana M, Lima R, Davis DD, Zhang H, Lockhart R, Racusen LC, Reckelhoff JF. Cardiovascular-renal and metabolic characterization of a rat model of polycystic ovary syndrome. Gend Med. 2011b;8:103–15.
77. Yi X, Han Z, Zhou Q, Lin J, Liu P. 20-Hydroxyeicosatetraenoic acid as a predictor of neurological deterioration in acute minor ischemic stroke. Stroke. 2016a;47:3045–7.
78. Yi X, Wu L, Liao D, Wang C, Zhang B. Interactions among CYP2C8, EPHX2, and CYP4A11 variants and CYP plasma metabolite levels in ischemic stroke. J Atheroscler Thromb. 2016b;23:1286–93.
79. Zhang F, Chen CL, Qian JQ, Yan JT, Cianflone K, Xiao X, Wang DW. Long-term modifications of blood pressure in normotensive and spontaneously hypertensive rats by gene delivery of rAAV-mediated cytochrome P450 arachidonic acid hydroxylase. Cell Res. 2005;15:717–24.
80. Zimmerman MA, Harris RA, Sullivan JC. Female spontaneously hypertensive rats are more dependent on ANG (1-7) to mediate effects of low-dose AT1 receptor blockade than males. Am J Physiol Renal Physiol. 2014;306:F1136–42.
81. Zu L, Guo G, Zhou B, Gao W. Relationship between metabolites of arachidonic acid and prognosis in patients with acute coronary syndrome. Thromb Res. 2016;144:192–201.

Arterial Flow, Pulse Pressure and Pulse Wave Velocity in Men and Women at Various Ages

10

Alberto P. Avolio, Tatiana Kuznetsova, Guy R. Heyndrickx, Peter L. M. Kerkhof, and John K.-J. Li

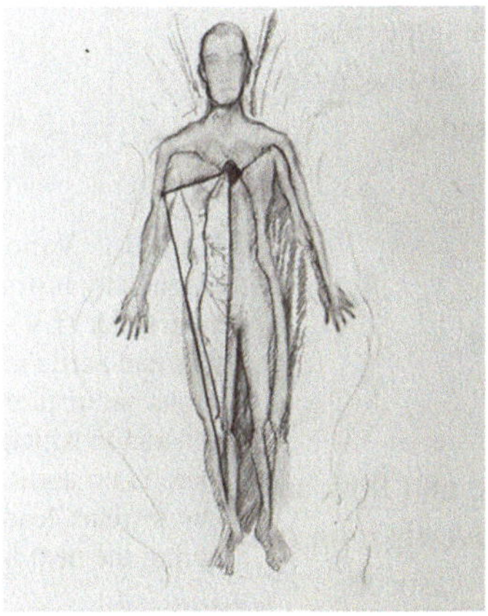

Art work by Piet Michiels, Leuven, Belgium

Abstract

The increase in pulse pressure (PP) that occurs with advancing age is predominantly due to reduced arterial distensibility leading to decreased aortic compliance, particularly in the elderly, in whom high blood pressure mainly manifests as isolated systolic hypertension. Since age-related changes in stroke volume are minimal compared with changes in PP, PP is often considered a surrogate measure of arterial stiffness. However, since PP is determined by both cardiac and arterial function, a more precise and reliable means of assessment of arterial stiffness is arterial pulse wave velocity (PWV), a parameter that is only dependent on arterial properties. Arterial stiffness as measured by PWV has been

A. P. Avolio (✉)
Department of Biomedical Sciences, Faculty of Medicine and Health Sciences, Macquarie University, Sydney, NSW, Australia
e-mail: alberto.avolio@mq.edu.au

T. Kuznetsova
Department of Cardiovascular Sciences, University of Leuven, Leuven, Belgium

G. R. Heyndrickx
Cardiovascular Center, Aalst, Belgium

P. L. M. Kerkhof
Department of Radiology and Nuclear Medicine, Amsterdam Cardiovascular Sciences, VU University Medical Center, Amsterdam, The Netherlands

J. K.-J. Li
Department of Biomedical Engineering, Rutgers University, Piscataway, NJ, USA

P. L. M. Kerkhof, V. M. Miller (eds.), *Sex-Specific Analysis of Cardiovascular Function*,
Advances in Experimental Medicine and Biology 1065, https://doi.org/10.1007/978-3-319-77932-4_10

found to be a powerful pressure-related indicator for cardiovascular morbidity and mortality. We analyzed PP and PWV in men and women of various age groups in healthy volunteers as well as cardiac patients with different types of diseases. The findings identified several striking sex-specific differences which demand consideration in guidelines for diagnostic procedures, for epidemiological analysis, and in evaluation of therapeutic interventions.

Keywords

Pulse pressure · Blood pressure · Aging arterial system · Arterial compliance · Sex-specific analysis · Pulse wave velocity · Cardio-ankle vascular index · Augmentation index

Basic Components: Pressure, Diameter, Flow

Characteristics of pulsatile pressure and flow are important physical attributes for investigation of vascular physiology. Physical aspects of relationships of pressure and flow are most pronounced in the systemic arterial vasculature because of the relatively high pressure levels and the strong pulsatile nature of these phenomena [1]. While it is important to continuously record the frequency-dependent characteristics of pressure and flow waveforms, in humans such accurate measurements are only feasible during catheterization. Much progress in the field has been based on early investigations performed in chronically instrumented animals quantifying pulsatile phenomena in arteries [2] and vascular and ventricular function [3, 4]. An example is presented in Fig.10.1, showing aortic diameter, flow, and left ventricular (LV) pressure and ECG, all sampled at 250 Hz. Using similar techniques, simultaneous aortic pressure and diameter have been measured to calculate viscoelastic properties of the aorta in fetal sheep, exercising lambs and adult animals [5]. Pulsatile

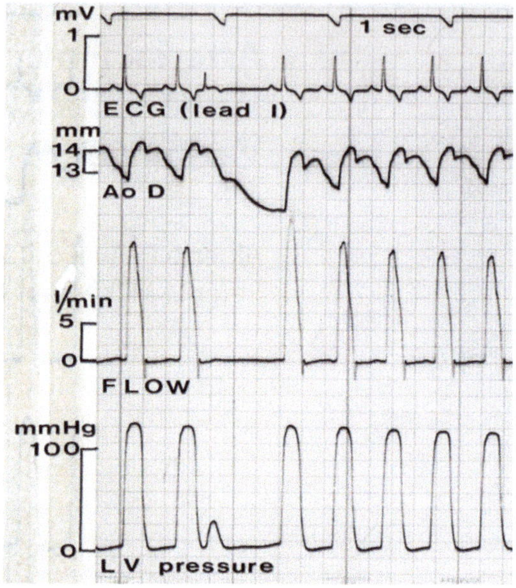

Fig. 10.1 Various calibrated recordings obtained from a chronically instrumented conscious dog. They include left ventricular (LV) pressure, proximal aortic (Ao) diameter (D), and aortic root flow. The registration shows a spontaneous premature beat with reduced ventricular pressure generation which fails to open the aortic valve. As a result there is no aortic flow, while the aortic diameter continuous to relax leading to a transient increase of peak flow during the next beat. (Data from [4])

properties of the arterial system interact with the pumping behavior of the ventricle, implying that the study of ventricular-arterial coupling attracted considerable interest, also by clinicians. A popular approach to study coupling refers to the ratio of ventricular and effective arterial elastance [6]. However, if simply defined as the ratio of SV and ESV, then this index does not reveal anything that is superior to the (problematic) metric of ejection fraction. In the present survey, we will not address these aspects and limit ourselves to arterial pressure and its derived metrics.

In clinical practice, one often has to rely on noninvasive measurements yielding systolic pressure (Ps) and diastolic pressure (Pd) and their difference, Ps-Pd, the pulse pressure (PP). An elevated PP is a powerful independent predictor of cardiovascular end points in the elderly

[7]. This has been confirmed in large cohorts such as the Framingham Heart Study [8]. The important observation is that whereas systolic pressure continues to rise with age in both men and women, diastolic pressure increases up to the age of 55 and then decreases. It is at this age that the systolic pressure for men and women diverge, with women showing a steeper increase with age [9] (Fig. 10.2). In addition, males and females also show different degrees of pressure pulse amplification between the aorta and brachial artery as observed in a large global survey of central aortic blood pressure which aimed to established reference values in a healthy population [10].

From youth to middle age, men consistently have significantly higher systolic and diastolic pressures than women. The significantly increased PP in women after age 55 is clearly observed in these healthy subjects, due to a greater increase in systolic pressure and similar decrease in diastolic pressure as compared with men.

In a recent study [11], four-dimensional (4D) flow MRI at 1.5T and 3T was used for the assessment of three-dimensional (3D) blood flow in the thoracic aorta in 98 healthy subjects (aged 9–78 years, 41 women). Subjects were categorized into age groups with divisions at 15, 20, 40, and 60 years. Data analysis included the 3D segmentation of the aorta, aortic valve peak velocity, mid-ascending aortic diameter (normalized by BSA), and calculation of flow velocity distribution descriptors (mean, median, standard deviation, incidence of velocities >1 m/s, skewness, and kurtosis of aortic velocity magnitude). Men and women revealed significant

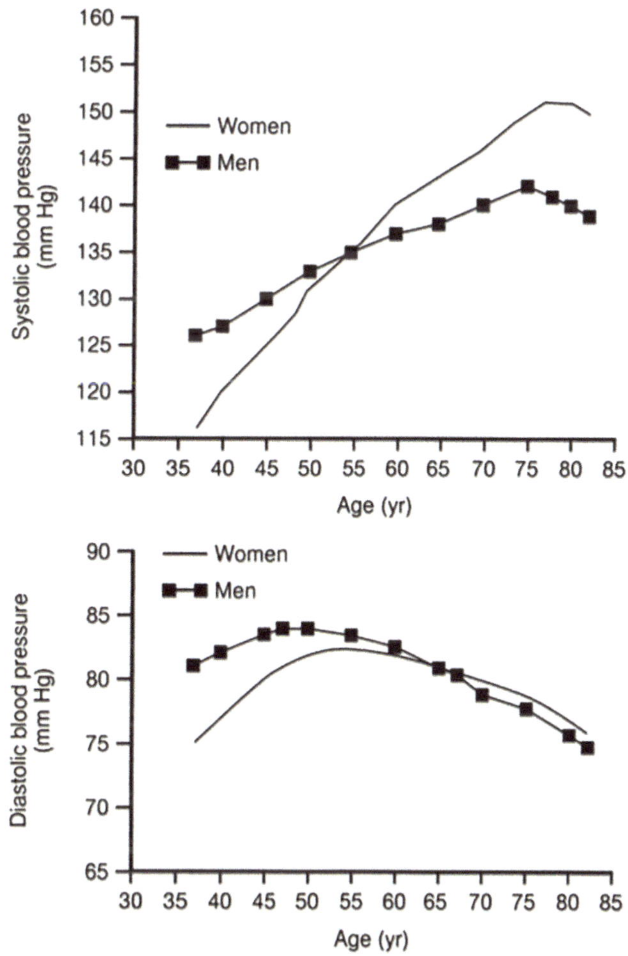

Fig. 10.2 The significant increase in pulse pressure in women compared to men after age 55 in healthy subjects is due to a greater increase in systolic pressure and similar decrease in diastolic pressure. (Data From [9], reproduced with permission)

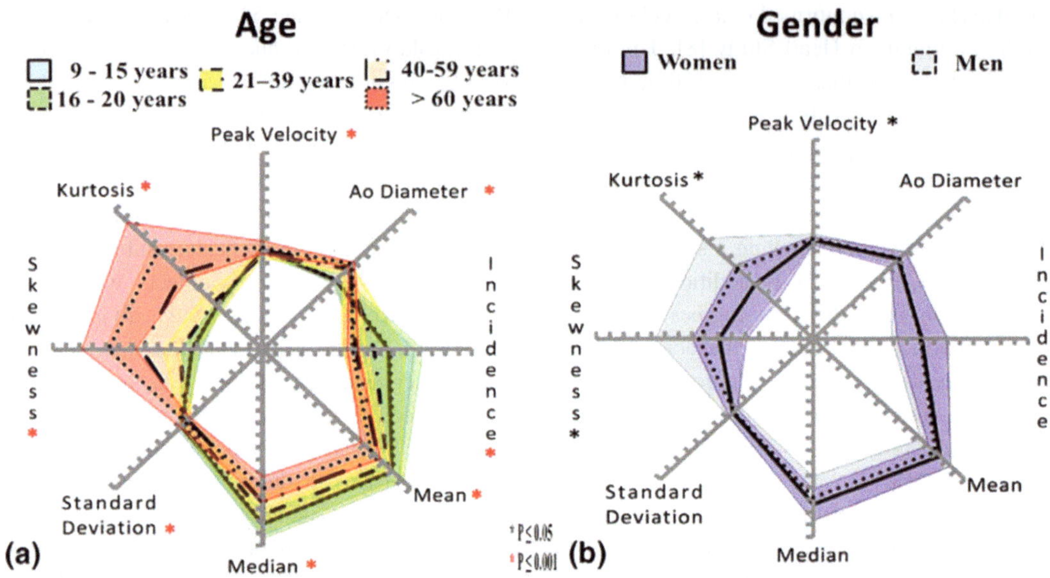

Fig. 10.3 Spider plots showing that the velocity magnitude distribution analysis of values for aortic blood flow depends on age and sex in healthy humans. Individuals (N = 98) were subdivided into five age groups (panel **a**), while analysis was carried out separately for men and women (N = 41) in (panel **b**). (Data from [11], reproduced with permission)

differences ($p \leq 0.05$) for peak velocity, incidence, mean, median, standard deviation, and skewness, all adjusted by heart rate. Results are graphically summarized in Fig. 10.3. The authors conclude that age and sex should be considered for assessment of the impact of cardiovascular disease on aortic blood flow.

Methods to Study Vascular Properties

Arterial Pressure and PWV in Humans

Population Studies

In population studies, blood pressure measurements were performed with conventional brachial cuff sphygmomanometry using auscultatory or oscillometric methods. PWV measurements were performed as a means of noninvasive assessment of arterial stiffness. Increased arterial stiffness and reduced large artery compliance have been found in large population studies with different prevalence of hypertension [12]. PWV is readily obtained during measurement of arterial pulses; it has become a most commonly measured index of vascular stiffness. In experimental studies, PWV has been found to increase with increasing pressure and also increase in wave reflections. In this context, PWV is most commonly measured from foot-to-foot pulse transit time (PTT) over a given distance (d) and computed as PWV = d/PTT. When the pressure wave is available, pulse wave analysis is performed to obtain a central aortic pressure waveform using mathematical transformations [13] from which wave indices are obtained such as augmentation index (AIx) to quantify the effect of increased systolic pressure augmentation due to wave reflections. AIx is calculated from the ratio of augmented pressure from the first systolic shoulder and PP. Similar analysis using central aortic wave components has been shown to discriminate the potential for predisposition to myocardial ischemia in women in a study of 1628 cardiology outpatients (590 females) [14].

The population studies reported are from a cohort of healthy volunteers (N = 987) with no hypertension or diabetes, nor clinically detectable heart disease, and not taking medication that would be interfering with cardiac function.

Pressure-Independent Measure of Arterial Stiffness: Pulse Wave Velocity and Cardio-Ankle Vascular Index

Since PWV as a measure of arterial stiffness is pressure dependent, a pressure-independent index, the cardio-ankle vascular index (CAVI), has been developed [15] and provides a measure of arterial stiffness based on ECG, heart sounds, and PTT of brachial and ankle pulses (indicated as T in Fig. 10.4). However, due to the inherent assumptions involving underlying relationship between pressure and diameter and measurement of reference pressure, the measurement as proposed still contains some residual pressure dependency. Methods have been recently proposed to correct for this form conventional measurements of CAVI [16].

Studies in Children

In children studies reported here, similar techniques in adults were used for measurement of

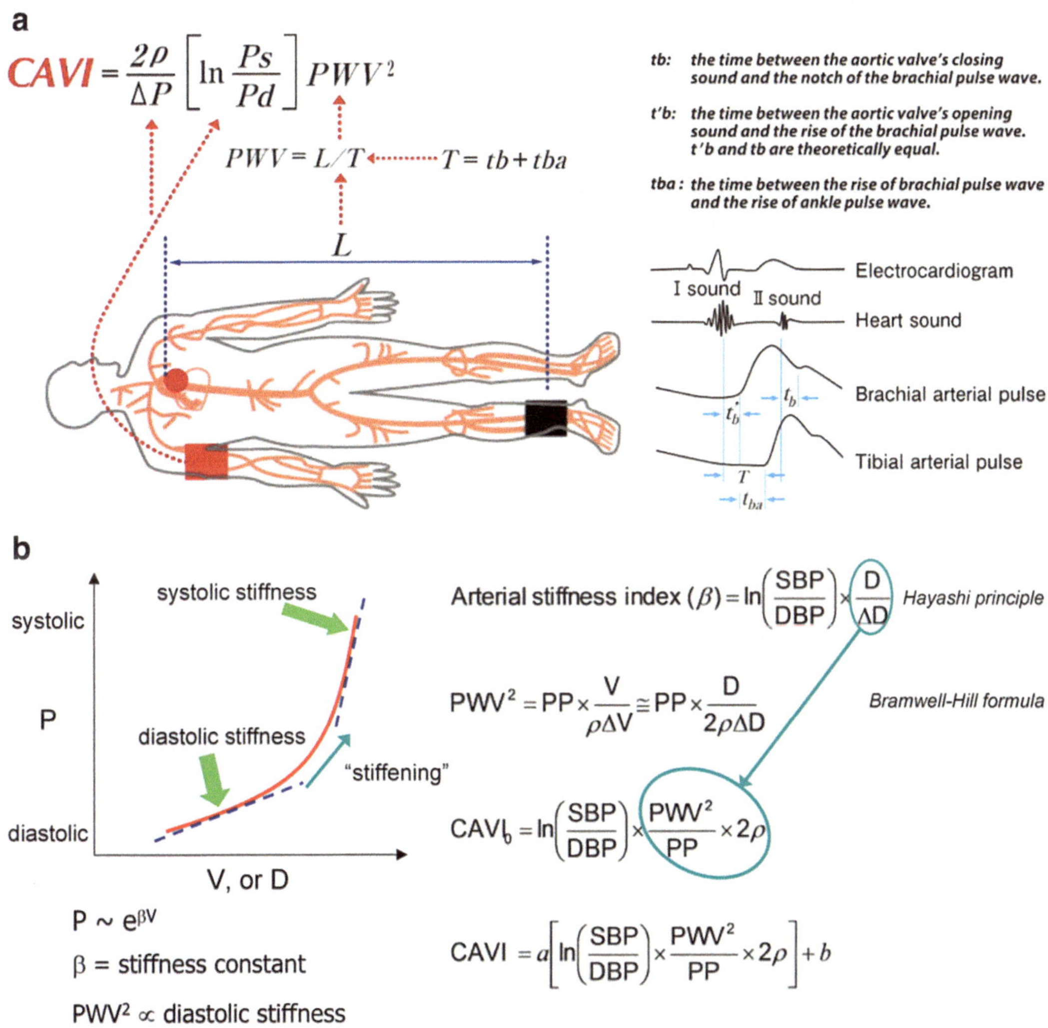

$$CAVI = \frac{2\rho}{\Delta P}\left[\ln\frac{Ps}{Pd}\right]PWV^2$$

$$PWV = L/T \quad T = tb + tba$$

$$P \sim e^{\beta V}$$

$$\text{Arterial stiffness index }(\beta) = \ln\left(\frac{SBP}{DBP}\right)\times\frac{D}{\Delta D}$$

$$PWV^2 = PP\times\frac{V}{\rho\Delta V} \cong PP\times\frac{D}{2\rho\Delta D}$$

$$CAVI_0 = \ln\left(\frac{SBP}{DBP}\right)\times\frac{PWV^2}{PP}\times 2\rho$$

$$CAVI = a\left[\ln\left(\frac{SBP}{DBP}\right)\times\frac{PWV^2}{PP}\times 2\rho\right]+b$$

Fig. 10.4 (**a**) Method for measurement of cardio-ankle vascular index (CAVI). Principles and calculations used to define CAVI are based on capture of the aortic valve closure using a microphone and wave analysis using cuffs on the arm and leg. [Data from [17], with permission]. (**b**) Theoretical association of quantities associated with CAVI in relation to the nonlinear relationship between pressure and vessel diameter. (Data from [18], reproduced with permission) blood pressure and PWV. However, since the transfer function techniques are validated only for adult subjects for determination of central aortic pressure [13], some studies have used the carotid waveform as a surrogate of the central aortic pressure wave [19, 20].

Catheterization Studies

Blood pressure was measured invasively in patients admitted to the catheterization laboratory for evaluation of chest pain and dyspnea. Prior to biplane angiography, blood pressure was recorded in the proximal aorta by an intra-arterial fluid line connected to an external pressure transducer (Statham P23Db). With lines thoroughly flushed, the manometer system has an adequate frequency response for reliable measurement of peak pressures and detection of pulse waveform features. LV volumes were determined by the area-length method.

Results

Populations Studies

The clinical and hemodynamic characteristics for the healthy cohort (N = 987) are described in Table 10.1. Men have higher brachial systolic and diastolic pressures compared to women, with both having similar brachial PP. However, for similar brachial PP and lower aortic PWV, women have higher central aortic PP. Relationships between components of blood pressure, PWV, and age are given in the various panels of Figs. 10.5 and 10.6.

Figure 10.7 shows the blood pressure component associated with CAVI, which slightly but significantly declines with age, although it is generally higher in women for the younger range. However, CAVI value is lower in women at all ages from a study in a large urban Japanese cohort

Table 10.1 Clinical and haemodynamic characteristics of 987 healthy subjects

Characteristics	Men (n = 475)	Women (n = 512)
Anthropometrics		
Age, y	45.4 ± 14.4	46.0 ± 14.1
Body weight, kg	82.9 ± 13.0	68.7 ± 13.4‡
Body height, cm	177.7 ± 6.7	164.4 ± 6.7‡
Body mass index, kg/m^2	26.2 ± 3.80	25.4 ± 4.49‡
Brachial systolic BP, mm Hg	129.1 ± 13.9	122.8 ± 15.2‡
Brachial diastolic BP, mm Hg	83.0 ± 9.8	77.7 ± 9.0‡
Brachial PP, mm Hg	46.2 ± 10.8	45.0 ± 11.5
MAP, mm Hg	101.4 ± 10.3	95.7 ± 10.5‡
Heart rate, beats/minute	62.6 ± 8.6	65.9 ± 9.1‡
Questionnaire data		
Current smoking, n (%)	94(19.8)	90 (17.6)
Drinking alcohol, n (%)	414 (87.2)	316 (61.7)‡
Hypertensive, n (%)	139 (29.3)	79 (15.4)‡
Central haemodynamics and stiffness		
Central PP, mm Hg	34.6 ± 10.8	37.0±12.9†
Pulse wave velocity, m/s	7.34 ± 1.57	6.91±1.53‡

Values are mean (±SD) or number of subjects (%). †$P \leq 0.01$; ‡$P \leq 0.001$; *BP* blood pressure, *MAP* mean arterial pressure, *PP* pulse pressure
Data in table from T. Kuznetzova

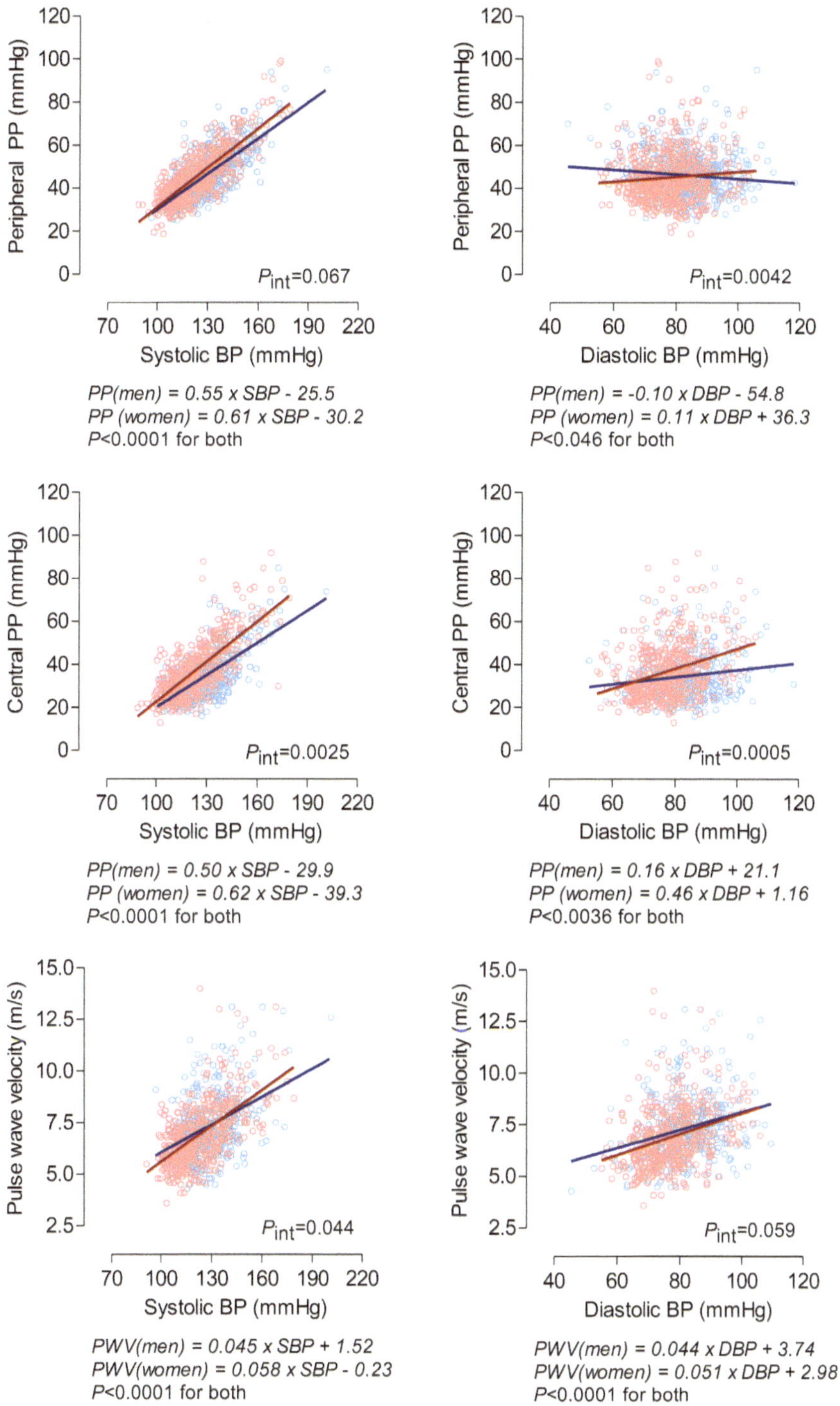

Fig. 10.5 Univariate correlations between central hemodynamics and arterial stiffness with peripheral BP components in men (blue) and women (red)

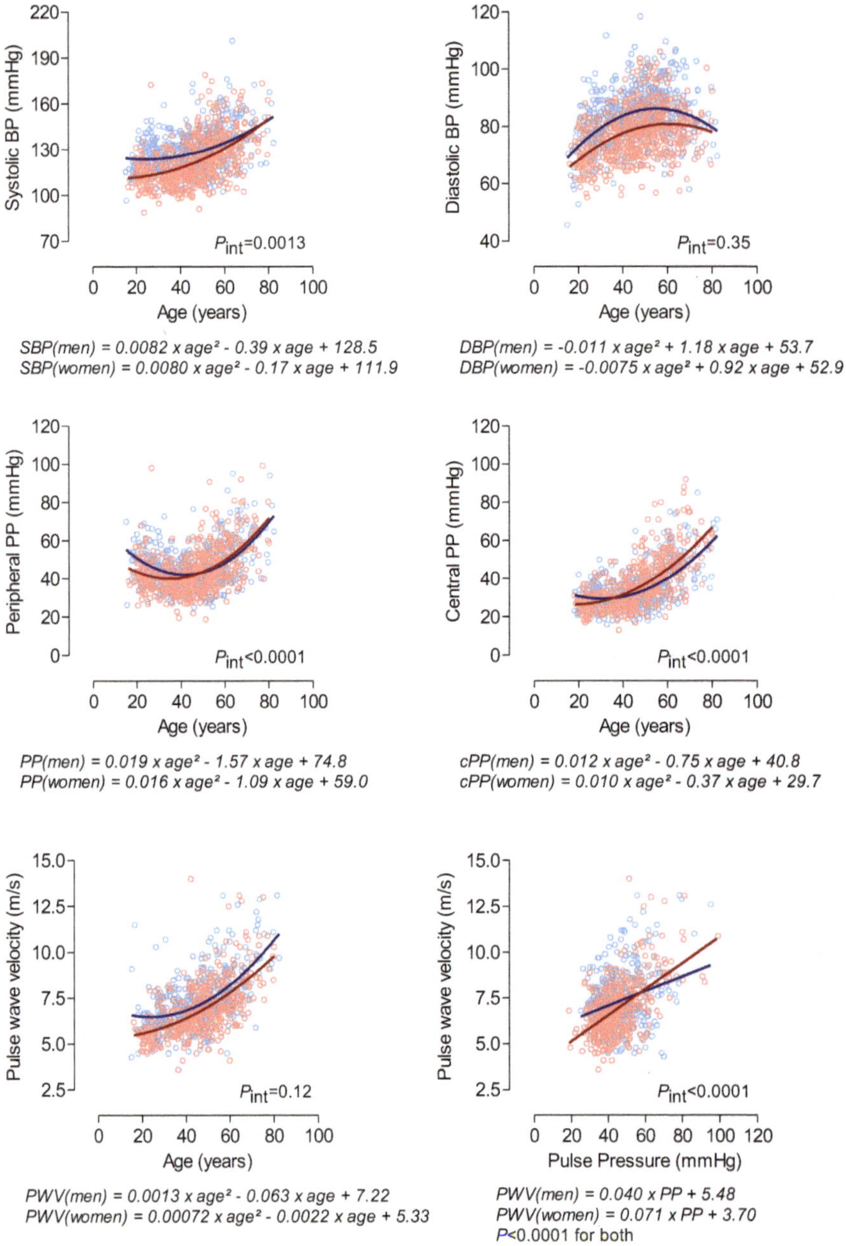

Fig. 10.6 Univariate correlations of peripheral BP components, central hemodynamics, and arterial stiffness with age in healthy men (blue) and women (red)

(N = 32,627) [15] (Fig. 10.8). (Note: Figs. 10.7 and 10.8 show different variables on the ordinate). Other studies show that a high CAVI in women is associated with a greater left ventricular mass index [21].

As seen in Table 10.1, women have similar brachial PP compared to men but greater central aortic PP. This can be explained by a relatively higher systolic augmentation in women as shown in Fig. 10.9 between second and seventh decades.

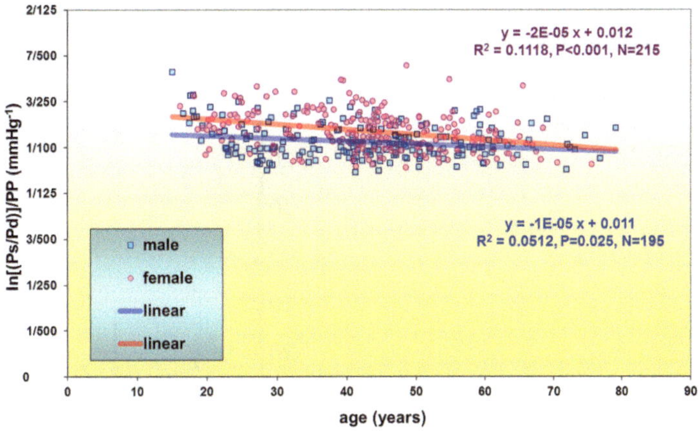

Fig. 10.7 Distribution of the pressure-associated component of the cardio-ankle vascular index (CAVI) versus age in healthy individuals shown for males and females separately. CAVI is obtained by multiplication with pulse wave velocity squared. (Data from T. Kuznetsova)

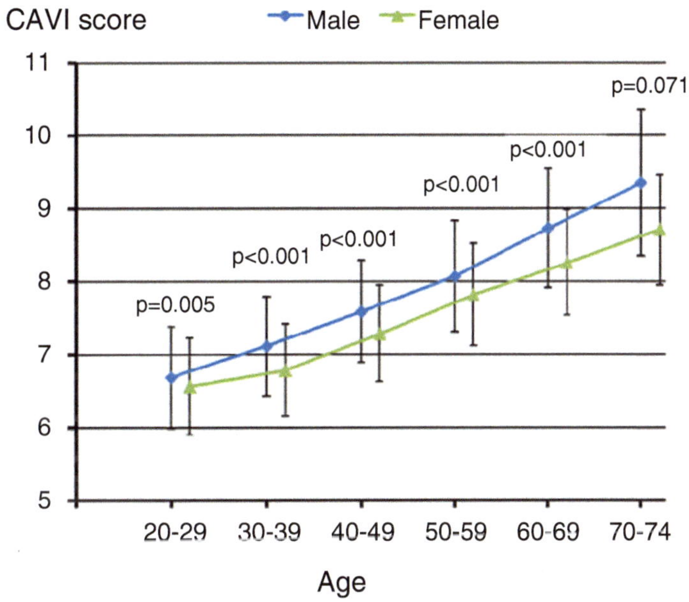

Fig. 10.8 CAVI scores by age between males and females in subjects free from cardiovascular risk. Data from a large urban Japanese population (N = 32,672) [15]

Studies in Children

Sex Differences in Blood Pressure in Children and Adolescents

Blood pressure in children and young adults tends to increase in both males and females. However, changes are more varied during puberty given the different rates of growth in both sexes. A study in over 32,000 normal-weight Chinese adolescents (aged 12–17 years) [23] shows marked differences in systolic and diastolic pressure between males and females when controlling for age and height (Fig. 10.10).

A study in Australian children assessed the differences in blood pressure and components of PP (augmentation index) at age 8 [19]) and then again at age 14 [20]. At age 8, there was no difference in systolic or diastolic pressure and height and weight similar for boys and girls. However, girls had a relatively higher systolic augmentation than boys. This was similar at age 14, with a statically different lower height in girls.

The relatively higher systolic augmentation of central aortic pressure seen in prepubescent girls compared to boys is continued through the growth phase and into adulthood (Fig. 10.9)

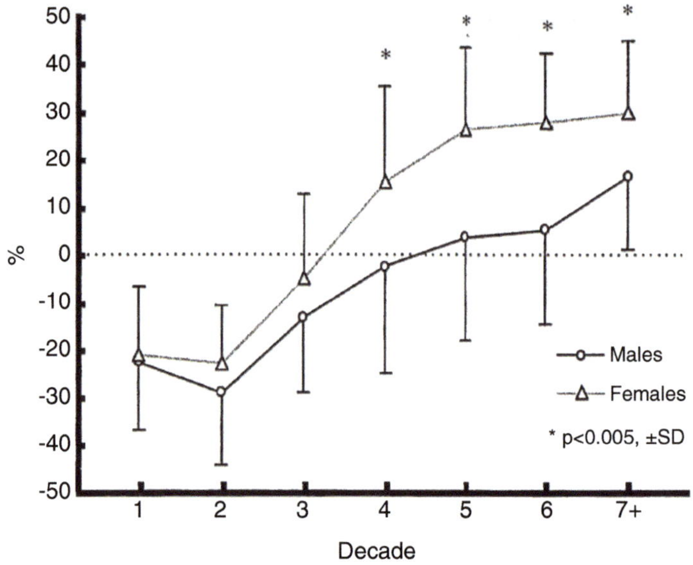

Fig. 10.9 Central aortic augmentation index in male and females in each age decade. (Data from [22], reproduced with permission)

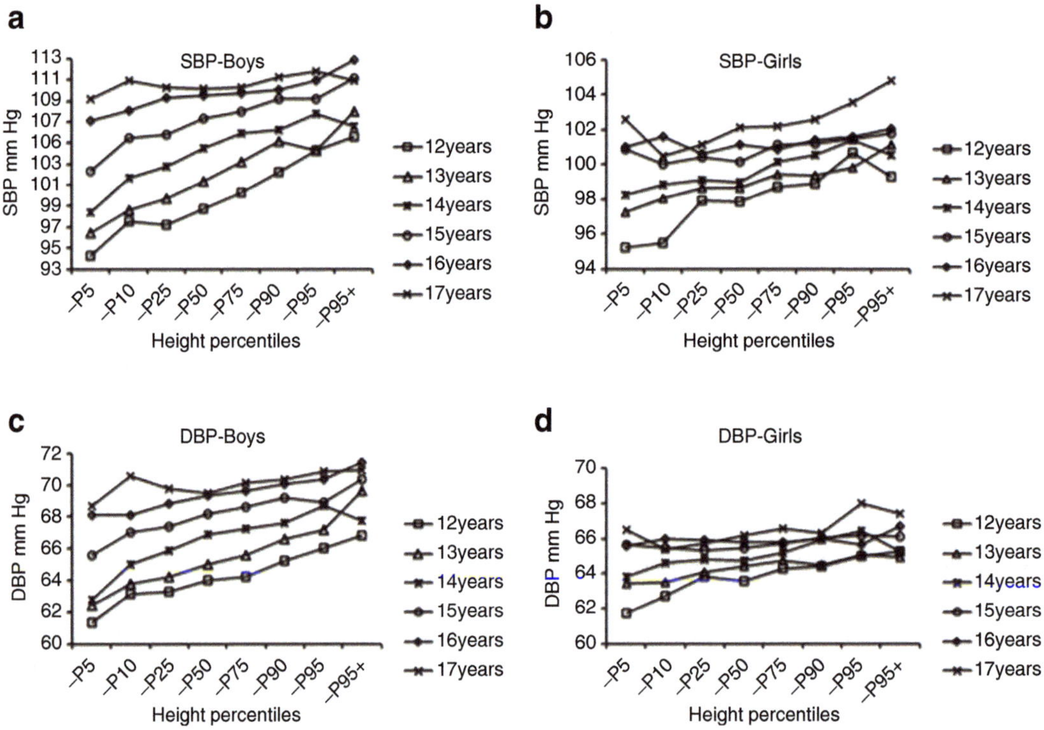

Fig. 10.10 Height percentiles and mean systolic (SBP) and diastolic (DBP) blood pressure for boys and girls for 12–17 years. SBP and DBP increased with height in each age group among boys and girls. Height percentiles in each age group were closely associated with SBP and DBP ($p < 0.01$ or $p < 0.05$) in both boys and girls. (Data from [23])

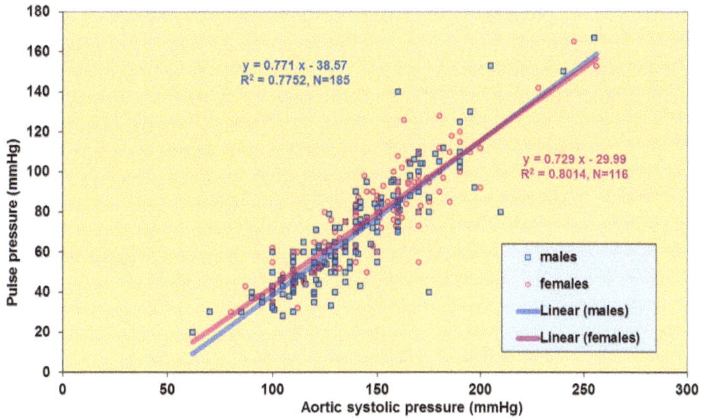

Fig. 10.11 The pulse pressure (PP) as related to systolic arterial pressure in cardiac patients (M vs F)

Table 10.2 Blood pressure components in a similar cohort of boys and girls at age 8 and followed up at age 14

	Age 8			Age 14		
	Boys	Girls	*p*	Boys	Girls	*p*
Height (m)	1.29 ± 0.06	1.28 ± 0.06	ns	1.66 ± 0.80	1.60 ± 0.0.06	<0.001
Weight (kg)	29.4 ± 6.7	28.9 ± 6.7	ns	59.2 ± 13.88	58.4 ± 12.32	ns
bSBP mmHg	100 ± 7	101 ± 7	ns	115 ± 10.3	113 ± 9.5	ns
bDBP mmHg	59 ± 6	60 ± 5	ns	64 ± 6.0	66 ± 7.1	ns
cSBP mmHg	93 ± 8	95 ± 8	ns	102.8 ± 12.53	100.1 ± 10.52	ns
AIx@75 (%)	−18 ± 11.0	−13.1 ± 9.0	<0.001	−32.3 ± 12.37	24.5 ± 12.14	<0.001
cAP mmHg	−6.1 ± 3.8	−4.2 ± 3.2	<0.001	−11.4 ± 6.54	−8.5 ± 5.36	<0.01

Data from [19, 20]

bSBP Brachial systolic pressure, *bDBP* brachial diastolic pressure, *cSBP* central (carotid) systolic pressure, *AIx* augmentation index at heart rate of 75 bpm, *AP* augmentation pressure

[22]. In this study significant differences are found after the fourth decade, suggestive of the time when systolic pressure in females overtakes than in males (Fig. 10.2, Table 10.2).

Catheterization Studies

Figures 10.11, 10.12, 10.13, 10.14, and 10.15 present relationships of invasive pressure measurements in a cohort of cardiac patients (n = 301) evaluated during catheterization procedures. Table 10.3 summarizes pressure data (systolic, diastolic, mean, and pulse pressure) obtained in the proximal aorta, stratified for sex. All measures, apart from diastolic pressure, are significantly different when males and females are compared. Despite the higher afterload (as reflected by mean aortic pressure), women have a higher ejection fraction (EF). Figure 10.16 shows how peripheral resistance (Rs, calculated as mean pressure divided by cardiac output) relates to arterial compliance (C, based on PP and stroke volume). The curves for men and women almost coincide, with no significant differences for Rs and C.

Discussion

Systolic Pressure, Diastolic Pressure, and Pulse Pressure

The chapter has presented data on arterial function parameters such as blood pressure and arterial stiffness measures for men and women over the adult lifespan and also in children and adolescents. Measurements were performed non-invasively in population cohorts and invasively in patients during cardiac catheterization.

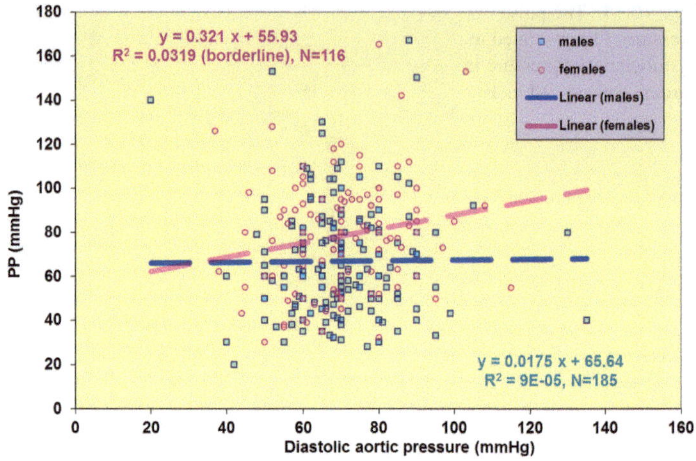

Fig. 10.12 The pulse pressure (PP) as related to diastolic arterial pressure in cardiac patients (M vs F)

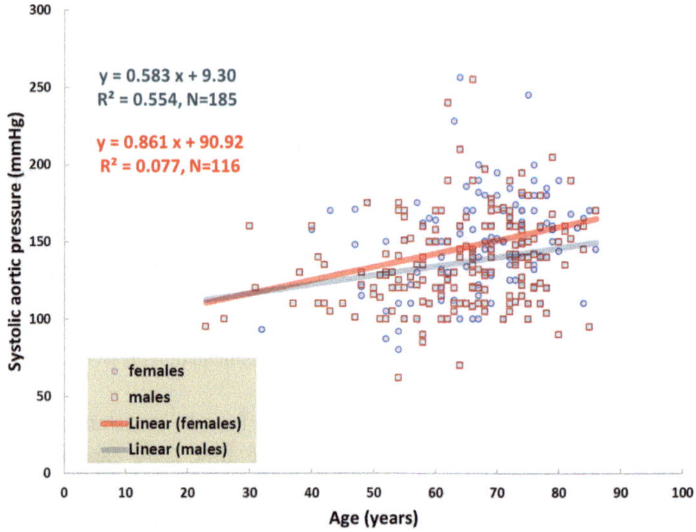

Fig. 10.13 The systolic pressure, Ps, as related to age in cardiac patients (M vs F)

The heart and the arterial system interact dynamically on a beat-to-beat basis. No attempt is made here to separate the components of pulsatile pressure and flow due to the contraction of the left ventricle from those of the arterial system. We addressed how systolic pressure, diastolic pressure, and PP are affected by age and sex in healthy and cardiovascular disease states; PWV is selected as an index of vascular stiffness, as it is solely dependent on the properties of the arterial system.

The principal reason many consider the importance of systolic pressure over diastolic pressure is that systolic pressure is associated with ventricular ejection and is the principal driving pressure for perfusion [1]. Additionally, it has long been thought that systolic time index is critical in determining the energy expenditure of the heart. Thus, a higher systolic arterial pressure signals a greater afterload to ventricular ejection.

The increase in PP is also predominantly due to an increase in systolic pressure, as we have shown here (Fig. 10.6). Note that when systolic pressures are mostly within the normal range, there is little observed difference between the sexes, but the differences become striking at higher pressure

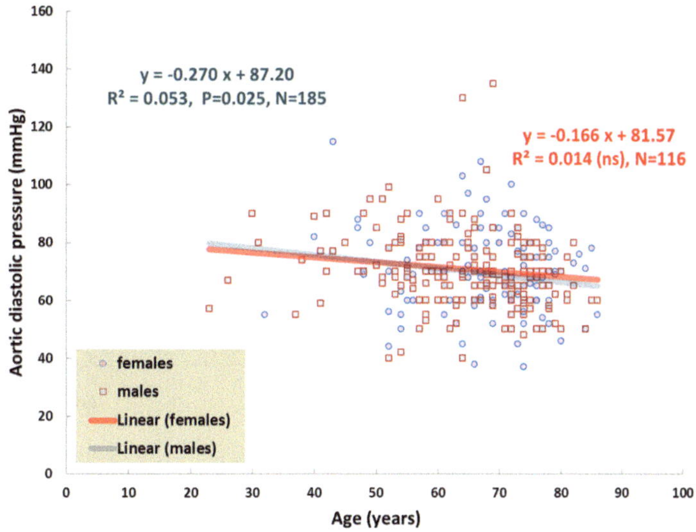

Fig. 10.14 The diastolic pressure, Pd, as related to age in cardiac patients (M vs F)

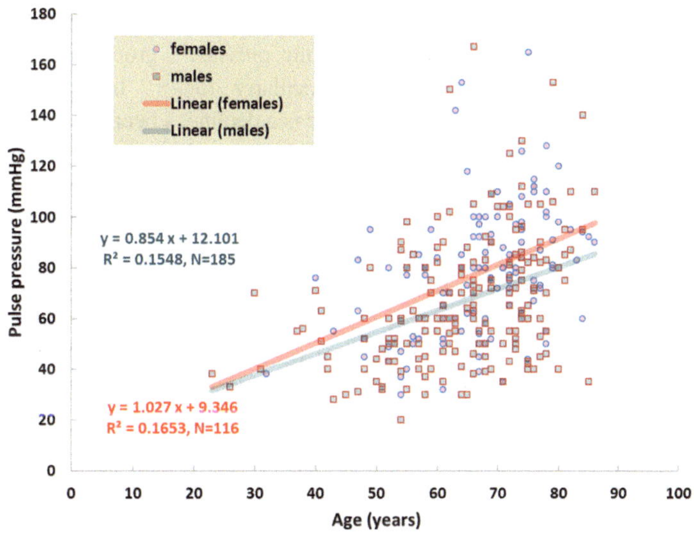

Fig. 10.15 The pulse pressure as related to age in cardiac patients (M vs F)

Table 10.3 PP statistics for the 301 cardiac patients (185 males)

	Ao-meanP mmHg	LVEF %	Ao-SBP mmHg	Ao-DBP mmHg	Ao-PP mmHg
Males					
Mean	95.1	61.7	136.7	69.9	66.9
SD	16.4	17.9	29.2	13.8	25.6
Females					
Mean	100.4	70.3	148.9	70.4	78.5
SD	18.3	16.7	31.4	14.2	25.6
p	0.011	<0.001	<0.001	0.76	<0.001

SD Standard deviation, *Ao-SBP* Aortic systolic pressure, *Ao-DBP* Aortic diastolic pressure, *Ao-PP* Aortic pulse pressure, p values designates level of significance for difference between males and females

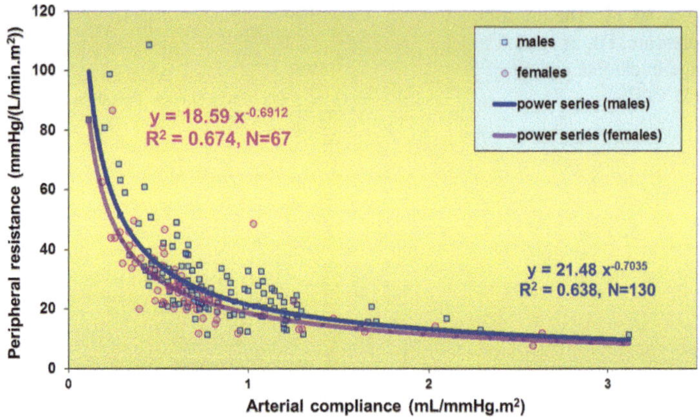

Fig. 10.16 The asymptotic inverse relationship between peripheral resistance, Rs, and arterial compliance, C, in male and female heart failure patients. Calculations based on mean arterial pressure, cardiac output, stroke volume, and pulse pressure. Average values for Rs and C do not differ for males and females

levels. Of course, these relations may be altered in differing pathological conditions.

The American Heart Association defines hypertension as a systolic/diastolic pressure greater than 140/90 mmHg (recently [late 2017] revised to 130/80 mmHg). With age, the tendency toward hypertension is only related to systolic pressure from the Framingham Study, particularly for elderly women (Fig. 10.2). Since there is a decline in diastolic pressure, PP is greater with age and much more significantly so in elderly women than men (Fig. 10.2). Our findings support this view (Figs. 10.5 and 10.6). The increase in PP is mostly due to significantly increased systolic pressure. Such increase in PP is exaggerated further with a greater increase in systolic pressure (Fig. 10.5), rather than the decrease in diastolic pressure. It is this large increase in systolic pressure with increasing age that isolated systolic hypertension (ISH) has been found to be more prevalent in the elderly (also from the Systolic Hypertension in the Elderly Program or (SHEP) study [24]. And the rate of ISH is higher in women than men. ISH is generally defined as a blood pressure greater than 160/90 mmHg.

In cardiac patients, the increase in PP due to the increase in systolic pressure is accentuated (greater slope; Fig. 10.11 vs Fig.10.5), although the differences between sexes are small. This is due to compounded arterial hypertension in many such cardiac patients (Fig. 10.11). The somewhat muted increase in diastolic pressure (Fig. 10.12) showed significant differences among the sexes. Since aortic diastolic pressure serves as the perfusion pressure to the coronary arteries, cardiac patients with hypertension do not receive the benefit of increased coronary perfusion pressure. This aspect is also clear from our data (Fig. 10.14). The higher overall PP in female cardiac patients with associated hypertension may signal a higher overall mortality in this group of elderly women.

Increased PP and PWV, both of which have been shown to be directly related to increased vascular stiffness, are major risk factors of coronary heart disease and stroke. It should be noted here that PWV measured over a long distance is less characteristic of the individual or regional arterial stiffness [1, 25]. In this regard, the CAVI can be said to provide a measure of "average" arterial stiffness over the measurement distances or pulse transmission path.

Central and Peripheral Arterial Function

Overall arterial function is essentially determined by the compliance of the large arteries and resistance of the peripheral vessels, as characterized by the Windkessel model. Peripheral muscular arteries are stiffer than central aorta. As such, the increased stiffness found in hypertensive patients is mostly due to reduced compliance in the aorta rather than that in peripheral arteries, (e.g., radial or femoral arteries). The elastic modulus is much higher in these peripheral arteries than in the central aorta, for both sexes [26]. Thus, a change in increased PWV is less if measured in the radial artery of hypertensive patients than that in the central aorta.

Characterization of the arterial system by the Windkessel model is prescribed by the diastolic aortic pressure decay time constant τ, i.e., the product of total peripheral resistance (Rs) and total arterial system compliance. Since the aorta contributes largely to the total arterial compliance, (C) has been used to reflect the compliance of the aorta. Since $\tau = \text{Rs C}$, the relation between Rs and C is curvilinear or, more precisely, inverse asymptotic, as seen in Fig. 10.16. Rs is defined as the ratio of mean pressure to mean flow in this context. Thus, for any given mean pressure or peripheral resistance, total arterial compliance is lower in females than in males. This sex difference is significant at increased vascular stiffness or greatly reduced compliance. Reduced aortic compliance has been found in coronary artery disease patients [27], thus it is not surprising that Fig. 10.16 reflects greatly reduced compliance values as compared with normal in both male and female patients with heart failure. Systolic hypertension, which is prevalent in the elderly, has been found to be associated with a much greater reduction in arterial compliance than a moderate increase in peripheral resistance [28, 29].

Increased pulse wave reflections are associated with a reduced arterial compliance and exhibit a greater influence on the pressure pulse than that by peripheral resistance [30]. Wave reflection also has considerable impact on the propagating pulse wave [31]. Although not equivalent, AIx has been more popularly used in the clinical setting to describe the effect of wave reflections.

AIx measured in central arteries, such as the carotid artery [22], showed increasing values of AIx with age. The difference in AIx between male and female is more striking with each increasing decade (Fig. 10.9). Again, this reflects the increased systolic pressure augmentation in women, especially beyond age 55, as we also found here.

References

1. Li JK-J. The arterial circulation: physical principles and clinical applications. Totowa: Humana Press; 2000.
2. O'Rourke MF, Taylor MG. Input impedance of the systemic circulation. Circ Res. 1967;20(4):365–80.
3. Vatner SF, et al. Effects of isoproterenol on regional myocardial function, electrogram, and blood flow in conscious dogs with myocardial ischemia. J Clin Invest. 1976;57(5):1261–71.
4. Kerkhof PLM. Beat-to-beat analysis of high-fidelity signals obtained from the left ventricle and aorta in chronically instrumented dogs. Automedica. 1986;7:83–90.
5. Pagani M, et al. Effects of age on aortic pressure-diameter and elastic stiffness-stress relationships in unanesthetized sheep. Circ Res. 1979;44(3):420–9.
6. Hayashi K, et al. Single-beat estimation of ventricular end-systolic elastance-effective arterial elastance as an index of ventricular mechanoenergetic performance. Anesthesiology. 2000;92(6):1769–76.
7. Vaccarino V, Holford TR, Krumholz HM. Pulse pressure and risk for myocardial infarction and heart failure in the elderly. J Am Coll Cardiol. 2000;36(1):130–8.
8. Franklin SS, et al. Is pulse pressure useful in predicting risk for coronary heart disease? The Framingham heart study. Circulation. 1999;100(4):354–60.
9. Kannel WB. Historic perspectives on the relative contributions of diastolic and systolic blood pressure elevation to cardiovascular risk profile. Am Heart J. 1999;138(3 Pt 2):S205–10.
10. Herbert A, et al. Establishing reference values for central blood pressure and its amplification in a general healthy population and according to cardiovascular risk factors. Eur Heart J. 2014;35(44):3122–33.
11. Garcia J, et al. Distribution of blood flow velocity in the normal aorta: effect of age and gender. J Magn Reson Imaging. 2017;47:487–98.
12. Avolio AP, et al. Effects of aging on arterial distensibility in populations with high and low prevalence of hypertension: comparison between urban and rural communities in China. Circulation. 1985;71(2):202–10.
13. Chen CH, et al. Estimation of central aortic pressure waveform by mathematical transformation of radial tonometry pressure. Validation of generalized transfer function. Circulation. 1997;95(7):1827–36.
14. Namasivayam M, Adji A, O'Rourke MF. Influence of aortic pressure wave components determined noninvasively on myocardial oxygen demand in men and women. Hypertension. 2011;57(2):193–200.
15. Namekata T, et al. Establishing baseline criteria of cardio-ankle vascular index as a new indicator of arteriosclerosis: a cross-sectional study. BMC Cardiovasc Disord. 2011;11:51.
16. Spronck B, et al. Direct means of obtaining CAVI0-a corrected cardio-ankle vascular stiffness index (CAVI)-from conventional CAVI measurements or their underlying variables. Physiol Meas. 2017;38(10):N128–37.
17. Leeson P. Hypertension and cardiovascular risk in young adult life: insights from CAVI. Eur Heart J. 2017;19(Suppl B):B24–9.
18. Schillaci G, et al. The impact of the cardio-ankle vascular index on left ventricular structure and function. Eur Heart J Suppl. 2017;19(Suppl B):B30–4.
19. Ayer JG, et al. Central arterial pulse wave augmentation is greater in girls than boys, independent of height. J Hypertens. 2010;28(2):306–13.
20. Barraclough JY, et al. Sex differences in aortic augmentation index in adolescents. J Hypertens. 2017;35(10):2016–24.
21. Ye Z, Pellikka PA, Kullo IJ. Sex differences in associations of cardio-ankle vascular index with left ventricular function and geometry. Vasc Med. 2017;22(6):465–72.

22. Hayward CS, Kelly RP. Gender-related differences in the central arterial pressure waveform. J Am Coll Cardiol. 1997;30(7):1863–71.
23. Jiang X, et al. Blood pressure tables for Chinese adolescents: justification for incorporation of important influencing factors of height, age and sex in the tables. BMC Pediatr. 2014;14:10.
24. Staessen J, Fagard R, Amery A. Isolated systolic hypertension in the elderly: implications of Systolic Hypertension in the Elderly Program (SHEP) for clinical practice and for the ongoing trials. J Hum Hypertens. 1991;5(6):469–74.
25. Li JK, et al. Pulse wave propagation. Circ Res. 1981;49 (2):442–52.
26. Li JK-J. Dynamics of the vascular system. Singapore: World Scientific; 2004.
27. Mohiaddin RH, et al. Regional aortic compliance studied by magnetic resonance imaging: the effects of age, training, and coronary artery disease. Br Heart J. 1989;62(2):90–6.
28. Berger DS, Li JK. Concurrent compliance reduction and increased peripheral resistance in the manifestation of isolated systolic hypertension. Am J Cardiol. 1990;65(1):67–71.
29. Li JK, et al. Allometric hemodynamic analysis of isolated systolic hypertension and aging. Cardiovasc Eng. 2007;7(4):135–9.
30. Berger DS, Li JK, Noordergraaf A. Differential effects of wave reflections and peripheral resistance on aortic blood pressure: a model-based study. Am J Phys. 1994;266(4 Pt 2): H1626–42.
31. Li JK-J, et al. Noninvasive pulse wave velocity and apparent phase velocity in normal and hypertensive subjects. J Cardiovas Diagn Procedures. 1996;13:31–6.

Wave Intensity Analysis: Sex-Specific Differences in Hemodynamic and Ventilatory Responses to Graded Exercise—Echocardiographic Measurements

11

Midori Tanaka, Motoaki Sugawara, Yasuo Ogasawara, and Kiyomi Niki

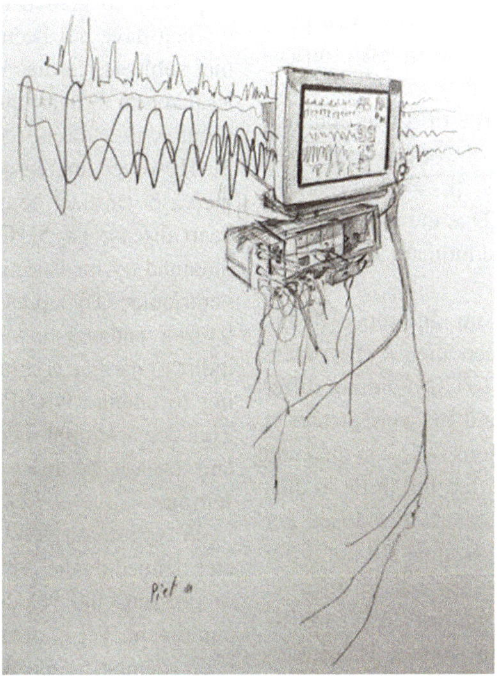

Wave intensity analysis. Artwork by Piet Michiels, Leuven, Belgium

M. Tanaka (✉) · M. Sugawara
Faculty of Health Care Sciences, Himeji Dokkyo University, Himeji, Japan
e-mail: mikko17@gm.himeji-du.ac.jp

Y. Ogasawara
Department of Medical Engineering, Kawasaki University of Medical Welfare, Okayama, Japan

K. Niki
Department of Medical Engineering, Tokyo City University, Tokyo, Japan

Abstract

Background: The differences in hemodynamic and ventilatory responses to graded exercise between men and women have not been well documented. Using wave intensity (WI) analysis, which is useful for analyzing ventriculo-arterial interaction, we aimed to elucidate the sex-specific differences.

P. L. M. Kerkhof, V. M. Miller (eds.), *Sex-Specific Analysis of Cardiovascular Function*, Advances in Experimental Medicine and Biology 1065, https://doi.org/10.1007/978-3-319-77932-4_11

Methods: We enrolled 48 healthy subjects (24 men and 24 women, age 21.3 ± 1.6 and 20.5 ± 0.9 years, n.s. [not significant]). Using ultrasonic diagnostic equipment, we measured WI, arterial stiffness parameter (β), force-frequency relation (FFR) and other hemodynamic parameters in the carotid artery before and during graded bicycle exercise. We also analyzed expired gas volume (VE) during the exercise. The workload was increased stepwise by 20 W at 1-min intervals up to respiratory compensation (RC) point through the anaerobic threshold (AT). WI is defined as WI = (dP/dt) (dU/dt), where P is blood pressure, U is velocity, and t is time. The peak value of WI (W_1) increases with left ventricular (LV) peak dP/dt, in other words, an index of cardiac contractility. The FFR was obtained as the linear regression line of W_1 on heart rate. β is defined as $\beta = \ln (Ps/Pd)/[(Ds - Dd)/Dd]$, where D is the arterial diameter, and suffixes s and d indicate systolic and diastolic, respectively.

Results: There was no difference in the body mass index between men and women. Carbon dioxide outputs (VCO_2) did not differ at rest, but those at AT and RC were greater in men. Oxygen consumptions (VO_2) in men and women at rest did not differ, but those in men at AT and RC were greater. Workloads per body weight in men and women did not differ at AT, but they were greater in men at RC. Systolic pressures at rest, AT and RC were greater in men than women. Heart rates in men and women did not differ at any stage of the graded exercise. W_1 did not differ at rest and AT, but it was greater in men than women at RC. The slope of the FFR during the period from rest to AT did not differ between men and women. However, the slope of the FFR during the period from AT to RC was greater in men.

Conclusions: The reached values of workload/weight at RC, VCO_2 at AT and RC, VO_2 at AT and RC, W_1 at RC, and the slope of the FFR during the period from AT to RC were greater in men than women.

Keywords

Graded exercise · Ultrasonic measurements · Cardiopulmonary test · Wave intensity · Force-frequency relation

Introduction

The differences in hemodynamic and ventilatory responses to graded exercise between men and women have not been well documented. In normal subjects, cardiac contractility increases with an increase in heart rate (HR). This phenomenon is called the "force-frequency relation" (FFR) [1–3]. However, increased HR does not necessarily increase cardiac contractility in patients with heart disease [4, 5]. Conventionally, FFRs were obtained by measuring the maximum rate of left ventricular (LV) pressure increase (peak dP/dt) with a catheter-tipped micromanometer as an index of cardiac contractility and using atrial pacing to change HR (Peak dP/dt – HR relation). This conventional method is an invasive method and cannot be used repeatedly in the clinical setting.

Wave intensity (WI) is a hemodynamic parameter defined as the product of the time-derivatives of pressure and velocity [6, 7]. Using ultrasonic equipment, WI is obtained non-invasively. It has been demonstrated that the magnitude of the first peak of WI increases with an increase in LV peak dP/dt [7]. During exercise, HR increases with an increase in workload; therefore, atrial pacing is not needed for changing HR. Moreover, force-frequency effects on the inotropic state of the intact left ventricle are markedly enhanced by exercise [8]. By measuring WI during graded exercise, we can obtain the FFR non-invasively [9].

We measured WI, FFR and other hemodynamic parameters and ventilatory parameters to investigate sex-specific differences in the responses of these parameters to the graded exercise among healthy young subjects.

Methods

Subjects

We enrolled 24 healthy men and 24 age-matched healthy women (age 21.3 ± 1.6 vs. 20.5 ± 0.9 years, $P = 0.11$). All subjects provided informed consent, and the Ethics Committee of Himeji Dokkyo University approved the study protocol.

Using ultrasonic diagnostic equipment, we measured wave intensity (WI), arterial stiffness parameter (β), force-frequency relation (FFR) and other hemodynamic parameters in the carotid artery before and during graded bicycle exercise.

We also recorded cardiopulmonary data with a breath-by-breath cardiopulmonary testing system (AE300S, MINATO Co LTD, Japan).

Theoretical Consideration of Wave Intensity

Parker et al. [10] proposed a novel index related to wave transmission and named it "wave intensity." They derived the features of WI using a mathematically sophisticated method called "the method of characteristics." However, the same features and further physiological meanings can be obtained by elementary methods in the following way.

Let P_f and P_b be the forward and the backward component of pressure waveform and U_f and U_b be those of velocity waveform, respectively. Then we have

$$P = P_0 + P_f + P_b, \tag{11.1}$$

$$U = U_f + U_b, \tag{11.2}$$

where P and U are measured instantaneous pressure and velocity, and P_0 is the end-diastolic pressure in the artery. On condition that the pulse-wave velocity (PWV) in the artery, c, is available and is constant during each cardiac cycle (partly linear approximation), we have

$$P_f = \rho c\, U_f, \tag{11.3}$$

$$P_b = -\rho c U_b, \tag{11.4}$$

where ρ is the blood density. Eqs. (11.3) and (11.4) are often referred to as the "water hammer formula." From Eqs. (11.1) and (11.2), we obtain

$$P_f = P - P_0 - P_b, \tag{11.5}$$

$$U_f = U - U_b. \tag{11.6}$$

Substituting Eq. (11.4) in Eq. (11.5) and using Eqs. (11.3) and (11.6), we have

$$P_f = P - P_0 + \rho c U_b, \tag{11.7}$$

$$P_f = \rho c U_f = \rho c U - \rho c U_b. \tag{11.8}$$

Adding Eq. (11.7) and Eq. (11.8) and dividing the result by 2, we obtain

$$P_f = (P - P_0 + \rho c U)/2. \tag{11.9}$$

Similarly, we obtain

$$P_b = (P - P_0 - \rho c \mathrm{U})/2. \tag{11.10}$$

WI is defined as

$$\mathrm{WI} = (\mathrm{d}P/\mathrm{d}t)(\mathrm{d}U/\mathrm{d}t). \tag{11.11}$$

Substituting Eqs. (11.1) and (11.2) in Eq. (11.11), and bearing in mind that P_0 is constant, we have

$$\begin{aligned}\mathrm{WI} &= (\mathrm{d}P_f/\mathrm{dt} + \mathrm{d}P_b/\mathrm{dt})(\mathrm{d}U_f/\mathrm{d}t + \mathrm{d}U_b/\mathrm{d}t)\\ &= (\mathrm{d}P_f/\mathrm{d}t)(\mathrm{d}U_f/\mathrm{dt}) + (\mathrm{d}P_b/\mathrm{dt})(\mathrm{d}U_f/\mathrm{d}t)\\ &\quad + (\mathrm{d}P_f/\mathrm{d}t)(\mathrm{d}U_b/\mathrm{d}t) + (\mathrm{d}P_b/\mathrm{d}t)(\mathrm{d}U_b/\mathrm{d}t)\end{aligned}$$

Because

$$(\mathrm{d}P_b/\mathrm{d}t)(\mathrm{d}U_f/\mathrm{d}t) = -\rho c(\mathrm{d}U_b/\mathrm{d}t)(\mathrm{d}U_f/\mathrm{d}t)$$

$$\text{and } (\mathrm{d}P_f/\mathrm{d}t)(\mathrm{d}U_b/\mathrm{d}t) = \rho c(\mathrm{d}U_f/\mathrm{d}t)(\mathrm{d}U_b/\mathrm{d}t),$$

we have

$$\begin{aligned}\mathrm{WI} &= (\mathrm{d}P_f/\mathrm{d}t)(\mathrm{d}U_f/\mathrm{d}t) + (\mathrm{d}P_b/\mathrm{d}t)(\mathrm{d}U_b/\mathrm{d}t)\\ &= (\mathrm{d}P_f/\mathrm{d}t)^2/\rho c - (\mathrm{d}P_b/\mathrm{d}t)^2/\rho c\end{aligned} \tag{11.12}$$

Here,

$$WI_f = (dP_f/dt)(dU_f/dt) = (dP_f/dt)^2/\rho c$$

and $WI_b = (dP_b/dt)(dU_b/dt) = -(dP_b/dt)^2/\rho c$

are the forward and backward components of WI, respectively.

The physiologically important feature that follows from the previous relationship is that forward components are shown always to contribute positively to WI $[(dP_f/dt)^2/\rho c]$, whereas backward components always contribute negatively $[-(dP_b/dt)^2/\rho c]$. This is true whether dP_f/dt or dP_b/dt is positive or negative. Thus, we reach the assertion that a positive value of WI indicates that forward-traveling waves predominate, whereas a negative value indicates that backward-traveling waves predominate.

WI has another important feature. During the period of the first positive peak, which appears in early ejection (Fig. 11.1), P_b and U_b are practically zero, and hence WI_b is also practically zero. Thus, we have

$$WI = WI_f = (dP_f/dt)^2/\rho c = (dP/dt)^2/\rho c, \qquad (11.13)$$

where P is actually measured pressure. Let W_1 be the height of the first peak (Fig. 11.1); then we have

$$W_1 = (\text{Peak } dP/dt)^2/\rho c.$$

When peak dP/dt is measured in the ascending aorta, aortic peak dP/dt is approximately equal to LV peak dP/dt unless there is aortic stenosis. The strong correlation between W_1 in the aorta and LV peak dP/dt ($r = 0.84$, $P < 0.0001$) was confirmed by animal experiments [11].

Noninvasive Measurements of Wave Intensity

In our method of obtaining carotid arterial WI [12], the carotid diameter-change waveform was used as a surrogate for the carotid-pressure

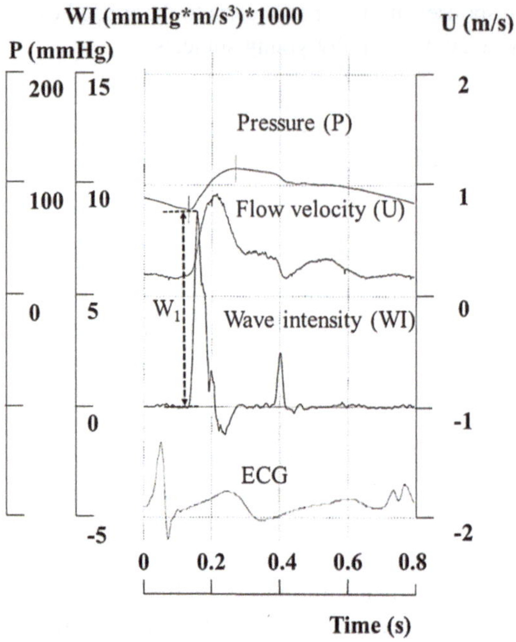

Fig. 11.1 Representative recordings of carotid arterial pressure-change waveform (P), blood flow–velocity waveform (U), calculated WI, and ECG in a healthy human. WI is defined as WI = (dP/dt) (dU/dt)

waveform. The blood-flow velocity averaged along the Doppler beam (Fig. 11.2, line B) crossing the carotid artery was measured using range-gated color-Doppler signals. Because the carotid artery is apart from the ascending aorta, carotid arterial W_1 may not represent LV peak dP/dt. Nevertheless, the correlation between carotid arterial W_1 and LV peak dP/dt was confirmed by clinical measurements (Fig. 11.3) [13]. The FFR was obtained as the linear regression line of W_1 on HR. Then we analyzed the differences in the slope of W_1–HR relations.

We also measured stiffness parameter (ß) in the carotid artery. The definition of the stiffness parameter β is

$$\beta = \ln(Ps/Pd)/[(Ds-Dd)/Dd], \qquad (11.14)$$

where D is the arterial diameter, and suffixes s and d indicate systolic and diastolic, respectively. ß is considered to be independent of pressure [14].

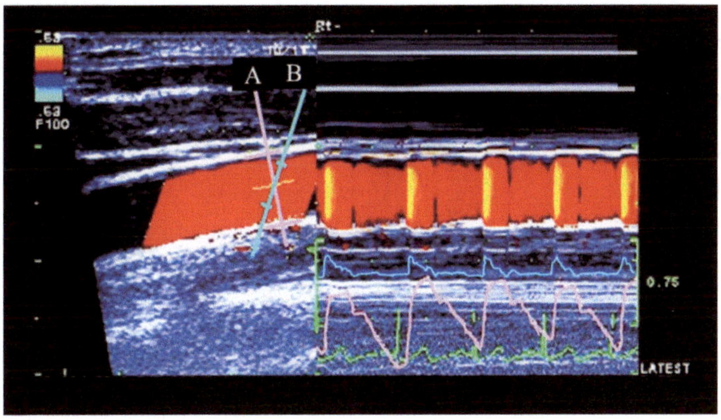

Fig. 11.2 Measurements of pressure-change waveform and blood velocity. Left: Long-axis view of the common carotid artery and ultrasound beams. By setting the tracking positions, displayed as small pink bars on the echo-tracking beam (line A), to arterial walls, echo tracking automatically starts. The blood flow–velocity averaged along the Doppler beam (line B) crossing the artery was measured using range-gated color-Doppler signals. Right: The diameter-change waveform is calculated by subtracting the distance to the near wall from that to the far wall. The maximum and minimum values of the diameter-change waveform are calibrated by Ps and Pd. The pressure-change waveform is displayed on the M-mode view. The blood flow–velocity waveform is also displayed on the M-mode view

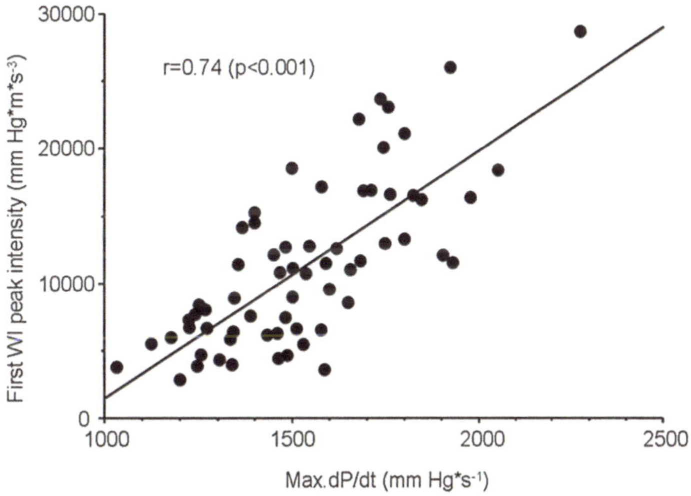

Fig. 11.3 Relationship between the magnitude of the first peak (W_1) of carotid arterial WI obtained by ultrasonic methods and the maximum rate of LV-pressure increase (max dP/dt) (from Ohte N et al. [13] with investigators' permission)

Determination of Anaerobic Threshold and Respiratory Compensation Points

The AT was determined as the point of deflection of VCO_2 versus the VO_2 plot obtained by computerized regression analysis (V-slope method) (Fig. 11.4). The RC was determined by the following: (1) the point of deflection of VE versus VCO_2; (2) the minimal value or nonlinear increase of VE/VCO_2; and (3) the point at which end-tidal CO_2 pressure starts to decrease.

Protocol

First, before the measurements, we had the subjects lie down in semi-supine position for 10 min on the recumbent type ergometer. The location to be measured was the common carotid artery at approximately 2 cm proximal to the carotid bulb. We used scanning in the long-axis view and obtained a B-mode image of a longitudinal section of the artery (Fig. 11.2, left). With the B- and M-mode scans displayed simultaneously on a split screen, the

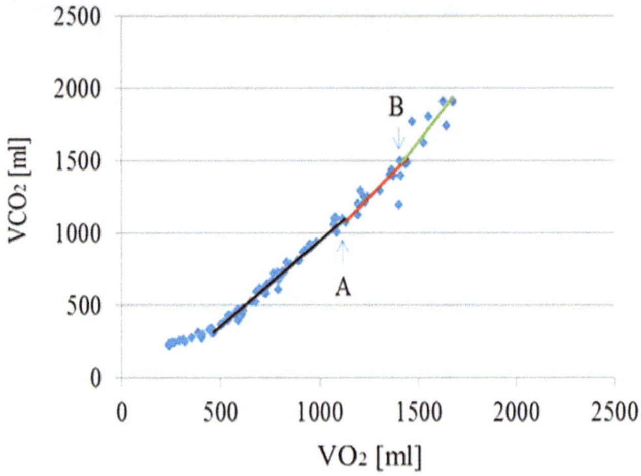

Fig. 11.4 Representative recordings of the breath-by-breath cardiopulmonary test during the graded exercise. $VCO_2 = CO_2$ output; $VO_2 = O_2$ consumption; point A = AT; and point B = RC

echo-tracking system tracked the vessel-wall movements to produce displacement waveforms of the anterior and posterior artery walls (Fig. 11.2, right). This gave the diameter-change waveforms.

Next, after taking the measurements at rest, graded bicycle exercise was performed starting at an initial workload of 20 W and lasting for 2 min; thereafter, the workload was increased stepwise by 20 W at 1-min intervals up to the respiratory compensation point (RC) through the anaerobic threshold (AT). Electrocardiogram was continuously monitored. We measured W_1 and other hemodynamic parameters during the exercise. We also analyzed VE during the exercise.

Data Analysis

The obtained data are expressed as mean ± SD. The scatter diagram of the points (HR, W_1) for the data during exercise from each subject was analyzed by the linear regression method, and the regression line was considered the FFR. Comparisons between men and women were performed by two-way analysis of variance followed by Bonferroni test when necessary.

Results

There were differences in height and weight between men and women (height 172.5 ± 6.4 vs. 159.5 ± 5.0 cm, P <0.0001; weight 63.1 ± 11.8 vs. 53.4 ± 7.7 kg, P <0.0001). However, there was no difference in body mass index (BMI) between men and women (BMI 21.1 ± 2.9 vs. 21.0 ± 2.6 kg/m^2, n.s.). There were no differences in VCO_2 between men and women at rest, but those at AT and RC were greater in men than women (Fig. 11.5, left). VO_2 in men and women at rest did not differ, but VO_2 measurements in men at AT and RC were greater than those in women (Fig. 11.5, right). Workloads/body weight in men and women did not differ at AT, but they were greater in men at RC (Fig. 11.6). Systolic pressures (Ps) at rest, AT and RC were greater in men than in women (Fig. 11.7, left). Diastolic pressure (Pd) at AT was greater in men than women, but there were no differences in Pd at rest and RC (Fig. 11.7, right). HR measurements in men and women did not differ at any stage of the graded exercise (Fig. 11.8). Ps/Pd in men did not differ from that in women throughout the period of the graded exercise (Fig. 11.9, left). (Ds – Dd)/Dd increased in men, but it did not change in women during the exercise (Fig. 11.9, middle). As a result, β in men did not change during the exercise, but β in women significantly increased at AT and RC (Fig. 11.9, right).

At any stage of the graded exercise, carotid arterial blood flow volume during a cardiac cycle (carotid arterial stroke volume) and carotid arterial blood flow volume per minute (carotid arterial output) in men and women did not differ (Fig. 11.10, left and right). W_1 did not differ at

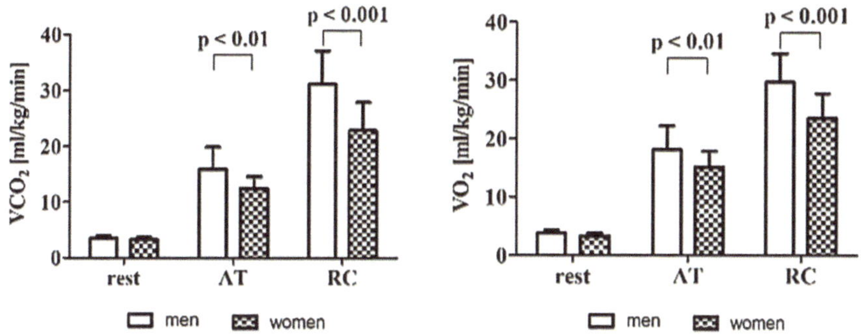

Fig. 11.5 Comparison of CO_2 outputs (VCO_2) (left) and O_2 consumptions (VO_2) (right) at rest, AT and RC point during graduated exercise in men and women

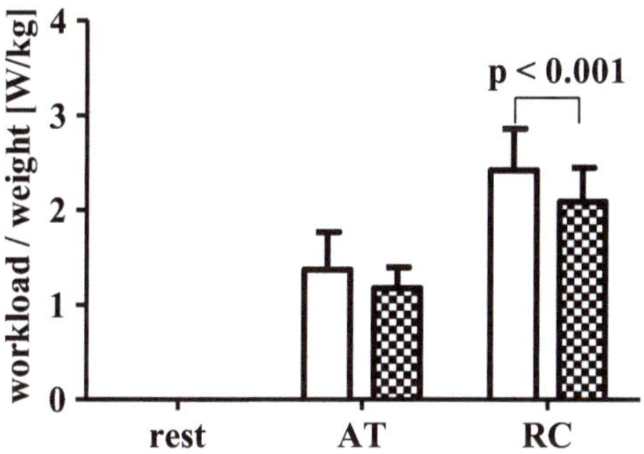

Fig. 11.6 Comparison of W_1 and WR (workload/body weight) at AT and RC point during graded exercise between men and women

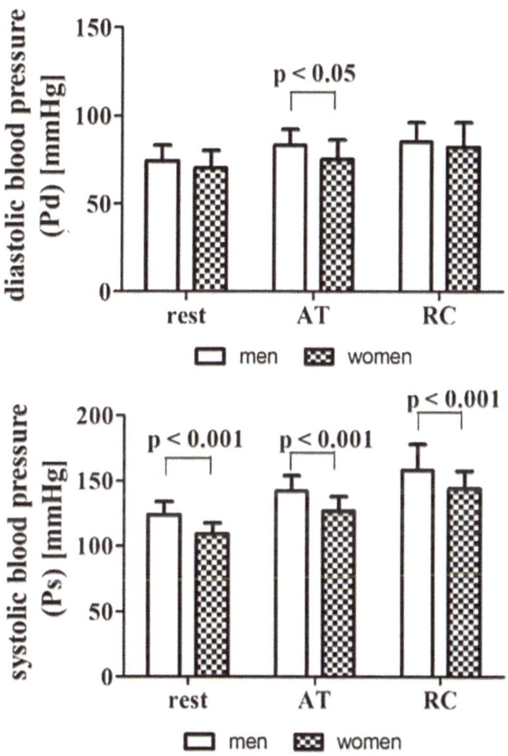

Fig. 11.7 Comparison of Ps and Pd at rest, AT and RC point during graded exercise between men and women

rest and AT, but it was greater in men than in women at RC (Fig. 11.11). In men, FFR slope increased during the period from AT to RC compared with the period from rest to AT (Fig. 11.12, left). However, in women, the slope did not change significantly throughout the period from rest to RC (Fig. 11.12, right). FFR slope during the period from rest to AT did not differ between men and women. However, the slope of the FFR during the period from AT to RC was greater in men than in women (Fig. 11.13).

Discussion

There were no significant differences in VCO_2 and VO_2 between the men and women groups at rest. However, men had greater VCO_2 and VO_2 at AT and RC than did women. The workload/body weight observed at AT in men did not differ significantly from that in women, but men showed greater workload/body weight at RC than that of women. The male and female study groups were age- and BMI-matched, but men had greater heights and weights than women. Because the lung volume and forced vital capacity (FVC) mainly depend on height, men are considered to have greater lung volumes and FVC measurements. Murias [15] reported that in young women, during ramp incremental exercise, the matching of O_2 delivery to O_2 use within the exercising limb is less effective compared with that in men, likely reflecting impairments in blood flow. Although the mechanisms responsible for the discrepancies between sexes are currently unclear, these may account for the above-mentioned differences.

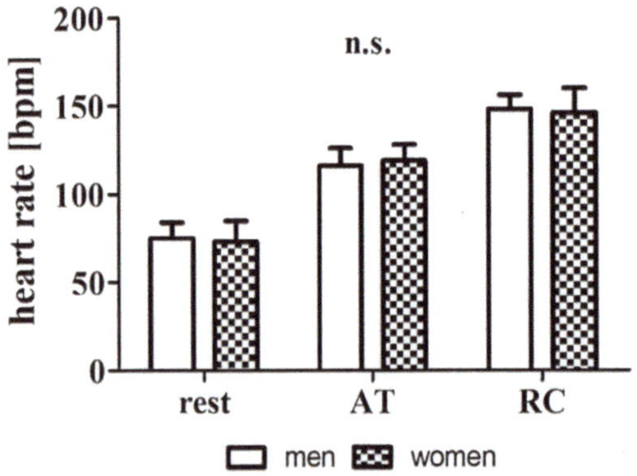

Fig. 11.8 Comparison of HR at rest, AT and RC point during graded exercise between men and women

Studies using the technique of 24-h ambulatory blood-pressure monitoring have shown that blood pressure is greater in men than in women at similar age [16, 17]. In this study, Ps was greater in men at rest, AT and RC. Diastolic pressure did not differ between men and women at rest and RC, but it differed at AT. Women have been noted to have greater HRs at rest [18]. However, in this study, there were no significant differences in HR at rest, AT and RC between men and women. During the exercise, carotid arterial stroke volume decreased with an increase in HR. There were no differences in carotid arterial stroke volume and carotid arterial output between men and women during the exercise.

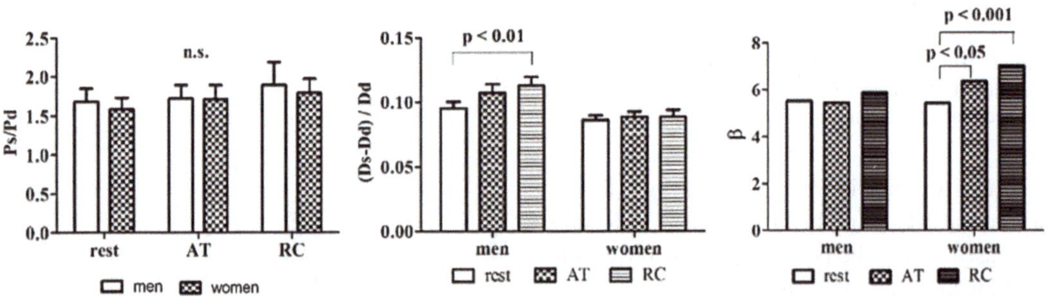

Fig. 11.9 Comparison of Ps/Pd (left), (Ds − Dd)/Dd (middle), andβ (right) at rest, AT and RC point during graded exercise between men and women

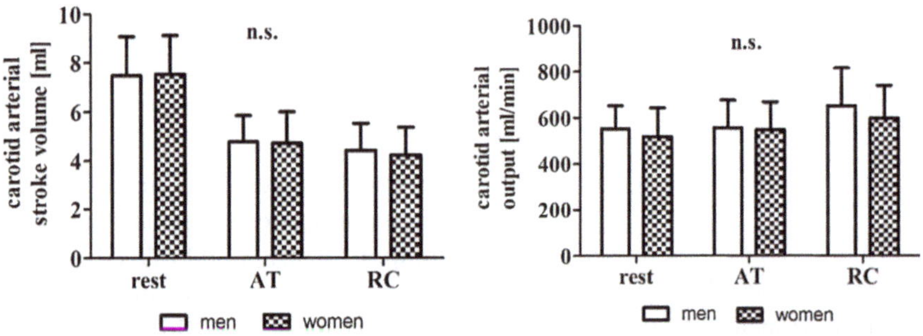

Fig. 11.10 Comparison of carotid arterial stroke volumes and carotid arterial outputs at rest, AT and RC point during graded exercise between men and women

Various theories have been offered to explain the sex-specific differences in change in ejection performance during exercise. Among them, one article recently suggested that central artery stiffness increases during exercise [19], and increased central artery stiffness has a direct impact on myocardial energetics: There is an increased energetic cost to maintain a given ejection/outflow due to a less-compliant central vasculature. This may account, at least in part, for the sex-specific differences in ejection performance during exercise.

Theoretically, the above-mentioned inference is reinforced in the following way. WI analyses assert that during the period of the first positive peak of WI (Fig. 11.1), effects of reflected waves are negligible and the contours of pressure and velocity waves are dominated by forward waves only [7, 10], in other words, $P_b = 0$ and $U_b = 0$. Under such conditions, Eqs. (11.1) and (11.2) give

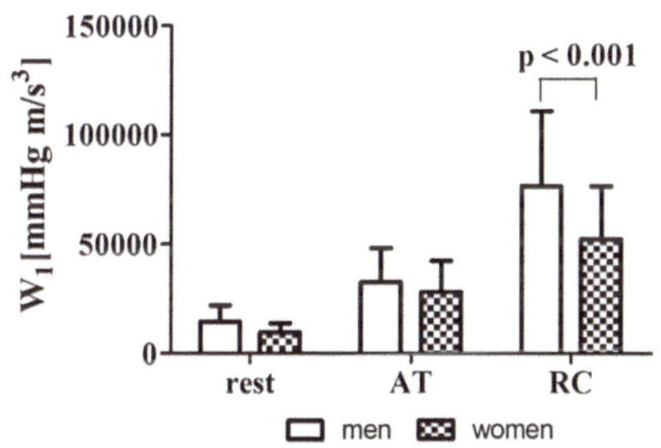

Fig. 11.11 Comparison of W_1 at rest, AT and RC point during graded exercise between men and women

$$P - P_0 = P_f, \text{and } U = U_f.$$

Substituting the previous relations in Eq. (11.3), we obtain

$$P - P_0 = \rho c U. \qquad (11.15)$$

Let us consider the WI in the ascending aorta. P and U are aortic pressure and velocity at any time during the period of the first positive peak of WI; P_0 is aortic end-diastolic pressure; and ρ and c are the blood density and the PWV, respectively [20]. Although ρc looks like resistance to flow, U, and is sometimes called "characteristic impedance" [21], it has no relation to the viscous systemic resistance. Nevertheless, ρc plays the role of afterload encountered by the left ventricle during initial ejection. Most of the studies evaluated arterial stiffness by measuring PWV. Conventional methods of measuring c have been based on two-point measurements, in other words, measurement of the time of travel of the wave over a known distance. Therefore, c measured over a relatively long distance is the integration of regional c in each artery involved within that range. In contrast, carotid arterial stiffness parameter β was measured at one point. Thus, changes in β may not represent overall changes in arterial stiffness. Nevertheless, the Swiss SAPALDIA cohort study reported a strong correlation (r = 0.7, P < 0.001) between brachial–ankle c and the carotid arterial distensibility coefficient, which is approximately the inverse of β [22]. Consequently, we consider that changes in carotid arterial β are associated with changes in c. Thus, the increase in β suppresses the indices of LV-ejection performance.

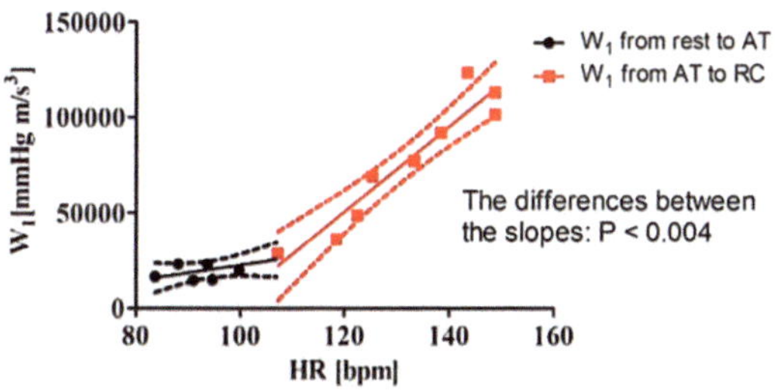

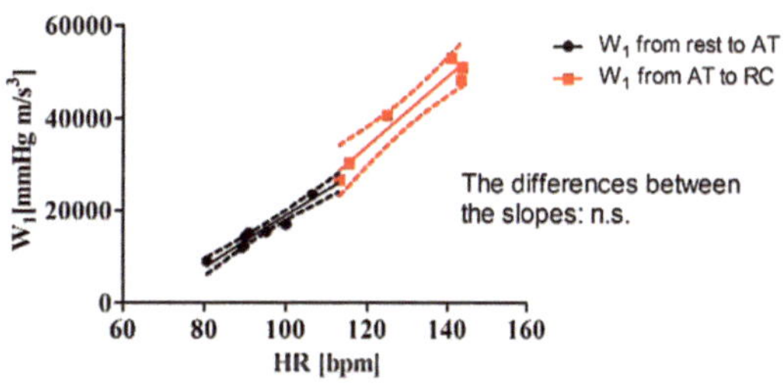

Fig. 11.12 Change in FFR slope during the period from rest to AT and the period from AT to RC in a representative man (left) and a representative woman (right)

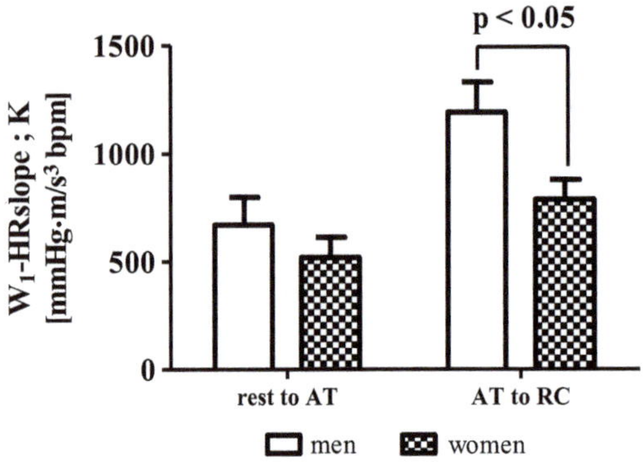

Fig. 11.13 Comparison of FFR (K) slopes during the period from rest to AT and the period from AT to RC between men and women

In this study, *Ps*/*Pd* did not differ between men and women during exercise. However, (*D*s – *D*d)/*D*d in men increased during exercise, but it did not change in women. As a result, β did not change in men, but it increased in women during exercise according to Eq. (11.14). This may have caused an impairment of LV-ejection performance as discussed previously. There were no differences in W_1, the contractility index, between men and women at rest and AT. However, women had lower W_1 at RC. The slope of FFR (W_1–HR relation) is another manifestation of LV-ejection performance. In men, the slope of FFR suddenly increased at AT, but it did not change significantly up to RC in women. There were no differences in FFR slope between men and women during the period from rest to AT, but FFR slope was greater in men than women during the period from AT to RC.

Conclusions

The reached values of workload/weight at RC, VCO_2 and VO_2 at AT and at RC, W_1 at RC, and the slopes of the FFR during the period from AT to RC were greater in men than women.

The mechanism of these differences is still the target of further investigations.

References

1. Ross J Jr, Miura T, Kambayashi M, Eising GP, Ryu KH. Adrenergic control of the force-frequency relation. Circulation. 1995;92(8):2327–32.
2. Lemaire S, Piot C, Leclercq F, Leuranguer V, Nargeot J, Richard S. Heart rate as a determinant of L-type Ca2+ channel activity: mechanisms and implication in force-frequency relation. Basic Res Cardiol. 1998;93(Suppl 1):51–9.
3. Schillinger W, Lehnart SE, Prestle J, Preuss M, Pieske B, Maier LS, et al. Influence of SR Ca(2+)-ATPase and Na(+)-Ca(2+)-exchanger on the force-frequency relation. Basic Res Cardiol. 1998;93(Suppl 1):38–45.
4. Schwinger RH, Bohm M, Koch A, Uhlmann R, Uberfuhr P, Kreuzer E, et al. Force-frequency-relation in human atrial and ventricular myocardium. Mol Cell Biochem. 1993;119 (1–2):73–8.
5. Schmidt U, Schwinger RH, Bohm M, Erdmann E. Alterations of the force-frequency relation depending on stages of heart failure in humans. Am J Cardiol. 1994;74(10):1066–8.
6. Parker KH. An introduction to wave intensity analysis. Med Biol Eng Comput. 2009;47(2):175–88.
7. Sugawara M, Niki K, Ohte N, Okada T, Harada A. Clinical usefulness of wave intensity analysis. Med Biol Eng Comput. 2009;47(2):197–206. Epub 2008/09/03
8. Miura T, Miyazaki S, Guth BD, Kambayashi M, Ross J Jr. Influence of the force-frequency relation on left ventricular function during exercise in conscious dogs. Circulation. 1992;86(2):563–71.
9. Tanaka M, Sugawara M, Ogasawara Y, Suminoe I, Izumi T, Niki K, et al. Noninvasive evaluation of left ventricular force-frequency relationships by measuring carotid arterial wave intensity during exercise stress. J Med Ultrason. 2015;42 (1):65–70. Epub 2015/01/27
10. Parker KH, Jones CJ. Forward and backward running waves in the arteries: analysis using the method of characteristics. J Biomech Eng. 1990;112(3):322–6. Epub 1990/08/01
11. Jones CJ, Sugawara M, Kondoh Y, Uchida K, Parker KH. Compression and expansion wavefront travel in canine ascending aortic flow: wave intensity analysis. Heart Vessel. 2002;16(3):91–8. Epub 2002/05/25
12. Niki K, Sugawara M, Chang D, Harada A, Okada T, Sakai R, et al. A new noninvasive measurement system for wave intensity: evaluation of carotid arterial wave intensity and reproducibility. Heart Vessel. 2002;17(1):12–21. Epub 2002/11/16
13. Ohte N, Narita H, Sugawara M, Niki K, Okada T, Harada A, et al. Clinical usefulness of carotid arterial wave intensity in assessing left ventricular systolic and early diastolic performance. Heart Vessel. 2003;18(3):107–11. Epub 2003/09/05
14. Hayashi K, Nagasawa S, Naruo Y, Okamura A, Moritake K, Handa H. Mechanical properties of human cerebral arteries. Biorheology. 1980;17:211–8.
15. Murias JM, Keir DA, Spencer MD, Paterson DH. Sex-related differences in muscle deoxygenation during ramp incremental exercise. Respir Physiol Neurobiol. 2013;189(3):530–6. Epub 2013/09/03
16. Reckelhoff JF. Gender differences in the regulation of blood pressure. Hypertension. 2001;37(5):1199–208. Epub 2001/05/23
17. Wiinberg N, Hoegholm A, Christensen HR, Bang LE, Mikkelsen KL, Nielsen PE, et al. 24-h ambulatory blood pressure in 352 normal Danish subjects, related to age and gender. Am J Hypertens. 1995;8(10 Pt 1):978–86. Epub 1995/10/01

18. Hanley PC, Zinsmeister AR, Clements IP, Bove AA, Brown ML, Gibbons RJ. Gender-related differences in cardiac response to supine exercise assessed by radionuclide angiography. J Am Coll Cardiol. 1989;13(3):624–9. Epub 1989/03/01
19. Babcock MC, Lefferts WK, Heffernan KS. Relation between exercise central haemodynamic response and resting cardiac structure and function in young healthy men. Clin Physiol Funct Imaging. 2017;37:372–8.
20. Niki K, Sugawara M, Uchida K, Tanaka R, Tanimoto K, Imamura H, et al. A noninvasive method of measuring wave intensity, a new hemodynamic index: application to the carotid artery in patients with mitral regurgitation before and after surgery. Heart Vessel. 1999;14(6):263–71. Epub 2000/07/20
21. Nichols W, O'Rourke M. McDonald's blood flow in arteries: theoretical, experimental and clinical principles. 5th ed. London: Hodder Arnold; 2005. p. 264–5.
22. Endes S, Caviezel S, Schaffner E, Dratva J, Schindler C, Künzli N, et al. Associations of novel and traditional vascular biomarkers of arterial stiffness: results of the SAPALDIA 3 cohort study. PLoS One. 2016;11(9):e0163844.

Sex Differences in Autonomic Response to Exercise Testing in Patients with Brugada Syndrome

12

Mireia Calvo, Virginie Le Rolle, Daniel Romero, Nathalie Béhar, Pedro Gomis, Philippe Mabo, and Alfredo Hernández

Artwork by Piet Michiels, Leuven, Belgium

M. Calvo
Univ Rennes, CHU Rennes, Inserm, LTSI – UMR 1099, Rennes, France

ESAII Department, EUETIB, CREB, Universitat Politècnica de Catalunya, Barcelona, Spain

V. Le Rolle (✉) · D. Romero · N. Béhar · P. Mabo
A. Hernández
Univ Rennes, CHU Rennes, Inserm, LTSI – UMR 1099, Rennes, France
e-mail: virginie.lerolle@univ-rennes1.fr

P. Gomis
ESAII Department, EUETIB, CREB, Universitat Politècnica de Catalunya, Barcelona, Spain

CIBER of Bioengineering, Biomaterials and Nanomedicine (CIBER-BBN), Málaga, Spain

P. L. M. Kerkhof, V. M. Miller (eds.), *Sex-Specific Analysis of Cardiovascular Function*, Advances in Experimental Medicine and Biology 1065, https://doi.org/10.1007/978-3-319-77932-4_12

Abstract

Introduction: Cardiac events in patients with Brugada syndrome (BS) typically occur at rest and mainly during sleep, suggesting that changes in autonomic modulation play an important role in the arrhythmogenesis of the disease. Moreover, sex differences in clinical manifestations of BS have been reported, identifying male patients with worse prognosis. The aim of our work was to assess and compare, according to sex, autonomic response to exercise in a clinical series including 105 BS patients.

Method: Standard 12-lead electrocardiogram recordings were collected during a physical stress test divided into four phases: warm-up, incremental exercise, active recovery, and passive recovery. Spectral non-stationary heart rate variability indicators were extracted by means of a smoothed pseudo Wigner-Ville distribution approach that adapts frequency bands to respiratory information. These indicators were then averaged in non-overlapped windows of 1 min for each patient to compare groups at each minute of the physical stress test.

Results: From the last minute of warm-up and until the third minute of incremental exercise, asymptomatic male patients presented significantly greater low-frequency (LF) values ($\overline{\mathrm{LF}^{WU2}}$: $p = 0.015$; $\overline{\mathrm{LF}^{EX1}}$: $p = 0.024$; $\overline{\mathrm{LF}^{EX2}}$: $p = 0.011$; $\overline{\mathrm{LF}^{EX3}}$: $p = 0.002$) than asymptomatic females. Conversely, asymptomatic women showed increased vagal modulation during the first minutes of incremental exercise ($\overline{\mathrm{HF}^{EX1}}$: $p = 0.031$; $\overline{\mathrm{HF}^{EX2}}$: $p = 0.001$). However, no significant differences were observed between symptomatic male and female patients.

Conclusion: As previously reported in healthy subjects, enhanced parasympathetic and decreased sympathetic tones appear to be not only greater in women but also defensive during cardiac stress. Based on the results, asymptomatic patients presented same-sex tendencies. However, we observed that symptomatic males developed a more female-like autonomic modulation, probably related to a more protective autonomic response to exercise. These results could be a step forward toward the understanding of the autonomic function in BS along with a potential impact on risk stratification.

Keywords

Heart rate variability · Time-frequency analysis · Brugada syndrome · Autonomic function · Sex differences · Exercise · Arrhythmogenesis · Parasympathetic system · Cardioprotection

Introduction

Brugada syndrome (BS) is an inherited disease presenting a typical electrocardiographic pattern characterized by a distinct ST-segment elevation in right precordial leads (V1-V3). It is associated with a high risk for unexpected sudden cardiac death secondary to ventricular fibrillation in the absence of any apparent structural cardiopathy. Since its description in 1992 as a new cardiac disorder by Pedro and Josep Brugada [1], BS has attracted great interest because of its high incidence in some parts of the world (especially in Eastern countries), as well as its association with sudden death in young adults and, less frequently, in infants and children. It has been estimated that BS is responsible for 4%–12% of the total amount of sudden cardiac death (SCD) and for 20% of SCD in the absence of structural cardiopathy [2, 3].

Differences in phenotype and prognosis between men and women in a large population of patients with BS were reported in [4]. The analysis was based on electrocardiographic parameters extracted before and after a pharmacological test, and events during follow-up were recorded for all patients. The results clearly identify men having a greater risk clinical profile than women, although conduction disturbances could be a marker of risk in the female population.

Nevertheless, the multifactorial etiology of BS requires complex approaches capable of capturing multiple mechanisms underlying the pathology. The autonomic nervous system (ANS) plays a determinant role in the pathophysiology, arrhythmogenesis and prognosis of the disease. Major cardiac events in BS most commonly occur during parasympathetic dominance at rest or during sleep [5, 6] and studies based on positron emission tomography have confirmed autonomic dysfunction in this population [5, 7, 8]. Thus, indicators capturing changes in ANS modulation may provide useful knowledge for the analysis of sex differences in patients with BS.

Evaluation of the autonomic response can be better characterized by exciting the ANS in a controlled and repeatable fashion—by applying standardized maneuvers, such as physical-stress testing, which originates an increase in sympathetic activity and a parasympathetic withdrawal—thus resulting in higher heart rates (HR). Conversely, post-exercise cardio-deceleration is mediated by a progressive increase in parasympathetic activity [9] and a continued sympathetic recession [10].

Heart rate variability (HRV) analysis is widely employed in clinical practice to characterize ANS function [11]. However, classic approaches require the signals to be stationary, which is inappropriate for studying non-stationary processes induced by physical exercise. To overcome these limitations, a time-frequency analysis could be used to evaluate the autonomic function. Thus, the aim of this study was to analyze sex differences concerning the temporal variation of different HRV features in BS patients during each phase of a standardized physical stress test by means of a time-frequency approach.

Introduction to HRV

HRV is used to describe variations of HR and respiratory rate (RR) intervals as they result from the complex mechanisms involved in HR regulation, notably the combined action of the two subsystems of the ANS: the sympathetic and parasympathetic nervous systems.

On one hand, parasympathetic neurons release acetylcholine, a cholinergic hormone that activates muscarinic M2-receptors in the heart, causing HR to decrease by slowing down nodal conduction. Because parasympathetic innervation is mainly provided by the vagus nerve, parasympathetic modulation is commonly referred to as "vagal tone." At rest, although both sympathetic and parasympathetic tones are exerted on the heart, the latter predominates. In contrast, sympathetic neurons release norepinephrine, a catecholamine that activates the β_1 receptors on the heart, causing HR to increase by increasing sinoatrial-node discharge.

Although several methods have been suggested for the study of HRV, including a recently proposed approach based on a multivariate unitary index to represent cardiac autonomic regulation [12], frequency-domain analyses are particularly popular because they correlate the autonomic mechanisms involved in RR interval fluctuations on an HR record with the analysis of two main spectral indices: the low-frequency (LF) and the high-frequency (HF) components (Fig. 12.1). These components are usually measured in absolute units of power (ms^2), although they may also be calculated in normalized units (n.u.) to emphasize their balanced behavior while minimizing the effect of changes in total power (TP) under stimulated conditions.

Vagal activity is the major contributor to the HF component, whereas the interpretation of the LF component is controversial. Although some studies suggest LF as a marker for sympathetic modulation, both sympathetic and parasympathetic influences have been reported in other studies. Consequently, the LF/HF ratio has been considered to capture sympathovagal balance as well as only sympathetic modulations.

In healthy patients, spectral analyses performed during 24 h have shown that LF and HF (expressed in n.u.) present a circadian pattern and reciprocal fluctuations with greater values of LF at daytime and of HF at nighttime. Moreover, LF has been proven to increase during tilting, mental stress and exercise testing; however, increases in HF have been induced by controlled respiration tests and cold stimulations [13].

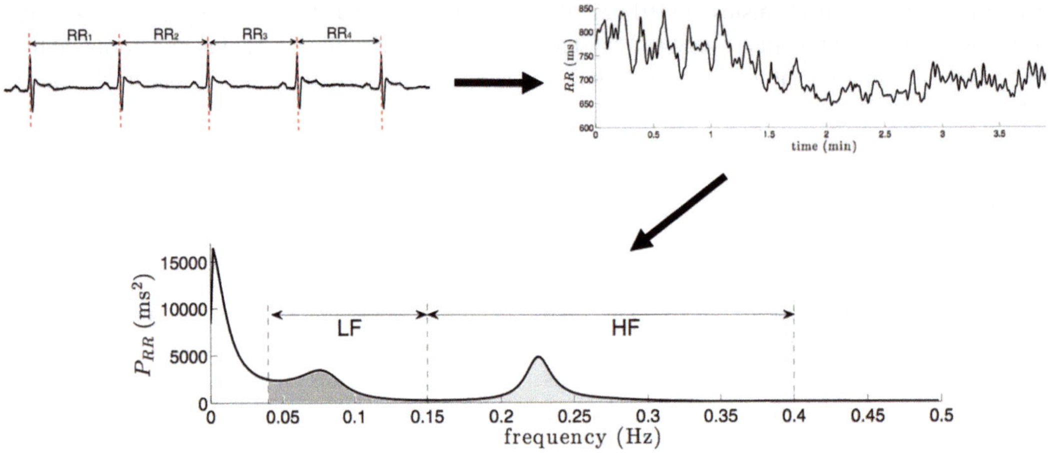

Fig. 12.1 Representative R-wave peak detections from an electrocardiographic signal displaying a BS-like saddleback pattern, extraction of the RR signal and spectral analysis of the RR interval variability

Table 12.1 Patients' baseline characteristics

	Symptomatic (n = 24)		Asymptomatic (n = 81)	
	Males (n = 20)	Females (n = 4)	Males (n = 60)	Females (n = 21)
Age (y)	46.15 ± 15.18	46.75 ± 17.80	43.35 ± 12.69	49.14 ± 13.98
No. (%) ICD implantation	20 (100)	4 (100)	15 (25)	3 (14.3)
No./total/% SCN5A mutation	5/18 (27.8)	1/1 (100)	12/43 (27.9)	9/14 (64.3)

total: Number of patients in whom genetic tests were performed.

Methods

Study Population

Patients diagnosed with BS (n = 105) took part in a physical exercise stress test while continuous ECG recordings were collected. They were enrolled in eight French hospitals located in Rennes, Angers, Brest, Poitiers, Nantes, Bordeaux, La Rochelle and Tours. After approval by the ethics committee of each center, all participants provided informed consent to participate in the study, which was performed in accordance with clinical research and ethics recommendations. Participants' ages ranged from 19 to 74 (45.17 ± 13.62) years, and 76.2% were male. Twenty-four patients had the following documented symptoms: syncope (50%), cardiac arrest (41.7%), dizziness (12.5%) and, less frequently, palpitations and nocturnal convulsions (4.2%).

Implantable cardioverter defibrillator (ICD) implantation had been performed in 18 of 81 (22.2%) asymptomatic patients based on a positive electrophysiological study test, whereas all symptomatic patients were ICD carriers. Among 76 patients in whom genetic analysis was performed (19 were symptomatic), an SCN5A mutation was found in 27 (35.5%), of whom 6 were symptomatic patients. Patients' baseline characteristics in the symptomatic and asymptomatic groups are listed in Table 12.1.

Signal Acquisition During Physical Stress Test

Patients underwent a clinical protocol recommended by the American Heart Association [14] described as a triangular stress test because the load was increased until it reached the patient's submaximal potential. It was performed on a

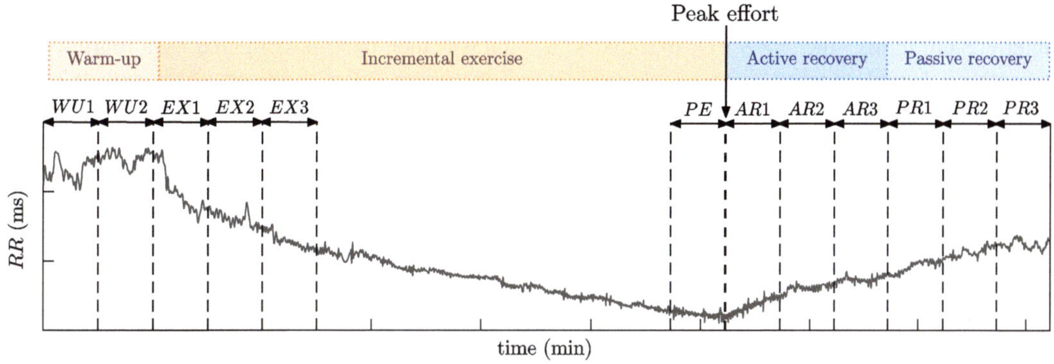

Fig. 12.2 Exercise testing was divided into four phases: warm-up, incremental exercise, active recovery and passive recovery. HRV time series estimated from the RR series were averaged in the following 1-min windows: both mins of warm-up (*WU1* and *WU2*), first 3 min of incremental exercise (*EX1, EX2, EX3*), last min of exercise before peak effort (*PE*), 3 min of active recovery (AR1, AR2, AR3) and 3 min of passive recovery (*PR1, PR2, PR3*)

cyclo-ergometer (Ergoline 900 Egamed, Piestany, Slovakia) and it was divided in the following phases, as represented in Fig. 12.2, where the non-stationarity of the RR series can be noted:

- Warm-up phase: for men, initial load of 50 watts (W); for women, initial load of 30 W; both for 2 min
- Incremental exercise phase: for men, initial load of 80 W for 2 min and then incrementing load by 20 W every 2 min; for women, initial load of 50 W and then incrementing load 20 W every 2 min
- Active recovery phase: for men, fixed load of 50 W; for women, fixed load of 30 W; both for 3 min
- Passive recovery phase: total cessation of effort for 3 min

The purpose of the test was to reach at least the 80% of the theoretical maximum HR of each patient as defined by the formula $MHR = 220 - age$ [15]. The 12-lead ECG recordings were collected and analyzed by the central board (Centre Hospitalier Universitaire de Rennes) using a Holter monitor (ELA Medical, Sorin Group, Le Plessis Robinsson, France) at a sampling frequency of 1000 Hz. Signals were continuously recorded during the warm-up (2 min), incremental exercise (11.86 $\pm$ 3.24 min), and recovery (6 min) phases.

As illustrated in Fig. 12.1, RR interval and R-peak amplitude series were obtained using a noise-robust wavelet-based algorithm for QRS complex detection and subsequent R-wave peak location [16]. After performing manual corrections to the obtained series when necessary, HRV parameters were obtained for the different test phases.

Time–Frequency Signal Analysis

Given that signals on a physical stress test are typically non-stationary, spectral characteristics associated with HRV were analyzed using a time–frequency approach. After cubic splines interpolation and regular resampling at 10 Hz, RR series were high-pass filtered at 0.03 Hz with a fourth-order Butterworth filter applied in both forward and backward directions to remove phase distortion.

Then, a smoothed pseudo Wigner-Ville distribution (SPWVD) transform was employed because it has proved its usefulness for the analysis of cardiovascular signals [17]. The Wigner-Ville distribution is a quadratic time–frequency technique defined as the Fourier transform of the instantaneous auto-correlation/cross-correlation function [18]. Because it is affected by significant interference terms, the SPWVD introduces a smoothing kernel function, $\psi(\tau, \nu)$, defined in Costa and Boudreau-Bartels GF [19], that attenuates interferences while maintaining a suitable time–frequency resolution. Because $A_{RR}(\tau, \nu)$, the ambiguity function of the RR series, was $x_{RR}(t)$, the SPWVD is defined as follows:

$$A_{RR}(\tau,\nu) = \int_{-\infty}^{\infty} x_{RR}\left(t+\frac{\tau}{2}\right)x_{RR}^{*}\left(t-\frac{\tau}{2}\right)e^{-j2\pi\nu t}dt$$

$$\psi(\tau,\nu) = \exp\left\{-\pi\left[\left(\frac{\nu}{\nu_o}\right)^2+\left(\frac{\tau}{\tau_o}\right)^2\right]^{2\lambda}\right\}$$

$$C_{RR}(t,f) = \iint \psi(\tau,\nu)A_{RR}(\tau,\nu)e^{j2\pi(t\nu-\tau f)}d\nu d\tau$$

The parameters ν_o and τ_o, associated with time and frequency resolution, were defined from [19]. HRV was measured as the TP in the LF and HF bands (noted as LF_b and HF_b), obtained from the SPWVD:

$$\mathrm{LF}(t) = \int_{LF_b} C_{RR}(t,f)df$$

$$\mathrm{HF}(t) = \int_{HF_b} C_{RR}(t,f)df$$

Assuming that sympathetic activity always lies within the standard LF band, this band was fixed between 0.04 and 0.15 Hz for the whole stress test. However, the TP in the HF band captures parasympathetic activity, which is closely related to respiratory sinus arrhythmia. Because respiratory frequency during ANS stimulation, especially during exertion, may not be restricted to the classic HF band (0.15–0.4 Hz) and can increase up to 0.7 Hz, HRV analysis within the standard frequency band would lead to unreliable measures of the parasympathetic activity. To overcome this limitation, a time-varying HF band, based on an estimation of the respiration activity captured by the ECG-derived respiration (EDR) series, has been defined [20].

The applied EDR method estimates respiratory information from the amplitude modulation of R-wave peaks [21]. The estimated respiration signal was then band-pass filtered by a fourth-order Butterworth filter between 0.15 and 0.7 Hz applied in both forward and backward directions to remove frequencies from the respiratory range. The same SPWVD transform used for the RR series was then applied to the EDR-filtered signals to estimate the instantaneous respiratory frequency, $f_r(t)$. Once the estimated respiratory frequency series $f_r(t)$ was obtained, the time-varying HF band for HRV analysis was defined as $HF_b(t) = [f_r(t) - 0.125, f_r(t) + 0.125]$ Hz with t covering the whole test.

Unlike classical HRV parameters, SPWVD leads to time–frequency HRV estimators that are indeed time series that vary during the application of the stress test. These markers, accounting for the sympathetic and parasympathetic influences of the ANS on HR, were normalized by the TP, which was defined as the sum of both spectral bands ($TP(t) = LF(t) + HF(t)$), leading to the time series $LF_{nu}(t)$ and $HF_{nu}(t)$. $LF/HF(t)$ was also obtained from dividing $LF(t)$ by $HF(t)$ in the time–frequency domain to obtain the global sympathovagal balance.

Analysis of Effort Testing

HRV time series were extracted from the time–frequency analysis based on the SPWVD of the RR series, which adapts frequency bands to respiratory information resulting from EDR signals. These HRV time series were averaged in temporal non-overlapped windows of 1 min for each patient, leading to $\overline{\mathrm{LF}^i}$, $\overline{\mathrm{LF}^i_{nu}}$, $\overline{\mathrm{HF}^i}$, $\overline{\mathrm{HF}^i_{nu}}$, $\overline{\mathrm{LF/HF}^i}$, which stand for each time series intra-patient mean for the following time periods: $i \in \{WU1, WU2, EX1, EX2, EX3, PE, AR1, AR2, AR3, PR1, PR2, PR3\}$.

Because each test differed in the duration of the incremental exercise phase and the shortest case in our clinical series lasted less than 5 min, only the first 3 min of incremental exertion (*EX1, EX2 and EX3*) and the last minute before peak effort (*PE*) were assessed. In addition, the entire warm-up (*WU1* and *WU2*) and active (*AR1, AR2* and *AR3*) and passive recovery (*PR1, PR2* and

PR3) phases were compared between male and female patients. Figure 12.2 displays the analyzed periods for different phases of the exercise test, indicating the peak effort instant.

Finally, HRV features extracted during exercise and recovery were compared between male and female groups using Mann-Whitney U non-parametric tests. To compare the last minute of exertion and recovery, all patients had to be synchronized with respect to the peak-effort instant. The analysis was made using the commercially available software MatLab (Mathworks Inc., MI, USA) and setting the level of significance at $p < 0.05$.

Results

Figure 12.3 displays an exemplifying RR series and its associated SPWVD spectral power for the same patient; the LF and respiration-centered HF bands are represented by dashed white lines (second panel). The third panel shows the time series *LF(t)* and *HF(t)* extracted from time–frequency analysis, where the 1-min windows used to calculate the mean value of HF during the first minute of active recovery ($\overline{\mathrm{HF}^{AR1}}$) and the mean value of LF at the peak effort ($\overline{\mathrm{LF}^{PE}}$) are indicated.

Table 12.2 lists the evolution of significant HRV features and their associated *p*-values obtained when comparing male and female asymptomatic patients for different segments of physical stress and recovery.

From the last minute of warm-up and until the third minute of incremental exercise, asymptomatic male patients presented significantly greater LF values ($\overline{\mathrm{LF}^{WU2}}$: $p = 0.015$; $\overline{\mathrm{LF}^{EX1}}$: $p = 0.024$; $\overline{\mathrm{LF}^{EX2}}$: $p = 0.011$; $\overline{\mathrm{LF}^{EX3}}$: $p = 0.002$) compared with asymptomatic females. Conversely, based on HF values, asymptomatic women showed increased vagal modulation during the first minutes of incremental exercise ($\overline{\mathrm{HF}^{EX1}}$:

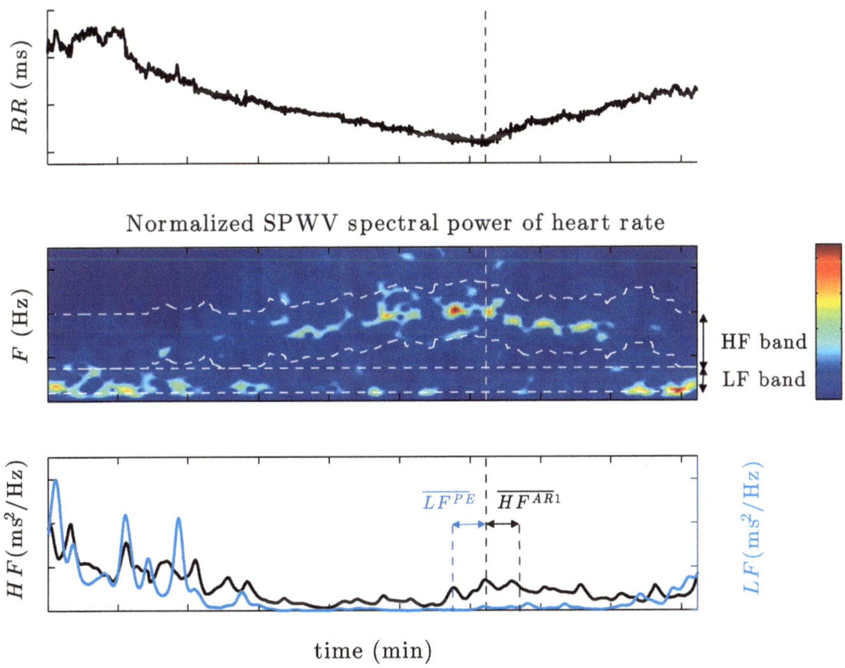

Fig. 12.3 From the RR series, the normalized SPWVD spectral power is calculated. The second panel indicates the LF and HF bands in dashed white lines. In the third panel, the TP in LF (blue) and HF (black) bands at each time instant are represented. Then, two examples of averaged time series ($\overline{\mathrm{LF}^{PE}}$ and $\overline{\mathrm{HF}^{AR1}}$) account for the sympathetic and parasympathetic contributions along the whole exercise test. A vertical dashed line in all panels refers to the peak effort instant

Table 12.2 Mean ± SD for asymptomatic men and women, and associated *p*-values, in significant HRV features

	Asymptomatic males	Asymptomatic females	*p*-value
$\overline{\mathrm{LF}^{WU2}}$	1405.3 ± 1455.6	583.5 ± 470.7	0.015
$\overline{\mathrm{LF}^{EX1}}$	711.0 ± 718.6	404.1 ± 522.2	0.024
$\overline{\mathrm{LF}^{EX2}}$	483.2 ± 761.1	320.0 ± 812.3	0.011
$\overline{\mathrm{LF}^{EX3}}$	307.4 ± 574.0	75.09 ± 92.2	0.002
$\overline{\mathrm{HF}^{EX1}}$	165.0 ± 242.2	228.0 ± 613.6	0.031
$\overline{\mathrm{HF}^{EX2}}$	127.8 ± 251.2	149.7 ± 400.6	0.001

$p = 0.031$; $\overline{\mathrm{HF}^{EX2}}$: $p = 0.001$). However, no significant differences were observed regarding normalized variables or between symptomatic male and female patients.

Discussion

HRV is recognized as a relevant measure of autonomic function. The major findings of this study are as follows: (1) females showed greater HRV-HF values after exercise; (2) asymptomatic male patients presented significantly greater LF values than asymptomatic females at the end of warm-up and during the beginning of incremental exercise; and (3) sex differences seem to disappear for symptomatic male and female patients. Overall, the results suggest an evolution of sex differences with symptomatic status. Asymptomatic females presented greater vagal modulation after exercise, whereas males displayed increased sympathetic dominance. These results are supported by the results of previous works [22] showing that females exhibit a cardio-protective autonomic profile compared with males and thus show a different autonomic response to exercise. In Koenig and Thayer [23], women showed a greater mean HR while presenting a greater vagal contribution captured by HF power of the HRV.

Main differences in autonomic responses were found during the end of warm-up and the beginning of exercise, which is associated with rapid parasympathetic decrease, gradual sympathetic activation and circulating catecholamines [24]. Asymptomatic women presented increased vagal modulation during the first minutes of incremental exercise compared with men. This difference in response to exercise may be due to ovarian hormones [25], which have been shown to have a clear influence on autonomic response [26] through the preservation of sympathovagal balance [27] and baroreflex sensitivity [28]. Moreover, estrogen was shown to improve vasomotor tone, vascular integrity, lipid profiles and cholesterol metabolism and decrease blood pressure [29].

Concerning BS, Benito et al. [4] reported that men were predominant and displayed a higher risk profile compared with women, which could be explained by sex-related intrinsic differences in ionic currents and hormonal influences. In fact, Di Diego et al. [30] proved that transient outward potassium currents were significantly greater in male than in female right-ventricle epicardia of arterially perfused canine heart preparations. Hormones also have important influences in the phenotypic manifestations of BS [31], which could explain the significant levels of testosterone reported in male BS patients compared with control subjects [32]. Moreover, results of experimental studies [33, 34] have demonstrated that hormones could induce ionic membrane current variations, which might explain sex-related ionic current differences in patients with BS.

According to the current results, asymptomatic males and females presented significant differences in the autonomic response to exercise, which concurs with observations obtained in healthy subjects [22]. Increased parasympathetic and decreased sympathetic modulation were found in asymptomatic women because it is associated with defensive behavior during cardiac stress. No significant differences were observed between symptomatic male and female patients.

Consequently, the results suggest that the ANS response evolves with symptomatology because symptomatic males seem to develop a female-like autonomic modulation, probably related to the emergence of cardio-protective behavior.

Conclusion

As previously reported in healthy subjects, increased parasympathetic and decreased sympathetic modulation appears to be not only greater in women but also defensive during cardiac stress. Based on the present results, asymptomatic patients presented the same-sex tendencies. However, we observed that symptomatic males developed a more female-like autonomic modulation, probably related to a more protective autonomic response to exercise. Although the results are based on approximations of ANS modulation and conclusions on neural activity cannot be inferred, this work presents a step forward toward the understanding of the autonomic function in BS patients and thus a potential impact on risk stratification. The differences found between male and female patients suggest that risk stratification should consider sex when evaluating indicators of risk based on autonomic information.

References

1. Brugada P, Brugada J. Right bundle branch block, persistent ST segment elevation and sudden cardiac death: a distinct clinical and electrocardiographic syndrome: a multicenter report. J Am Coll Cardiol. 1992;20(6):1391–6.
2. Brugada J, Brugada R, Brugada P. Right bundle-branch block and ST-segment elevation in leads V1 through V3 : a marker for sudden death in patients without demonstrable structural heart disease. Circulation. 1998;97(5):457–60.
3. Antzelevitch C. Heart Rhythm Society; European Heart Rhythm Association. Brugada syndrome: report of the second consensus conference: endorsed by the Heart Rhythm Society and the European Heart Rhythm Association. Circulation. 2005;111(5):659–70.
4. Benito B, Sarkozy A, Mont L, Henkens S, Berruezo A, Tamborero D, et al. Gender differences in clinical manifestations of Brugada syndrome. J Am Coll Cardiol. 2008;52(19):1567–73.
5. Kies P, Wichter T, Schäfers M, Paul M, Schäfers KP, Eckardt L, et al. Abnormal myocardial presynaptic norepinephrine recycling in patients with Brugada syndrome. Circulation. 2004;110(19):3017–22.
6. Matsuo K, Kurita T, Inagaki M, Kakishita M, Aihara N, Shimizu W, et al. The circadian pattern of the development of ventricular fibrillation in patients with Brugada syndrome. Eur Heart J. 1999;20(6):465–70.
7. Paul M, Meyborg M, Boknik P, Gergs U, Schmitz W, Breithardt G, et al. Autonomic dysfunction in patients with Brugada syndrome: further biochemical evidence of altered signaling pathways. Pacing Clin Electrophysiol. 2011;34(9):1147–53.
8. Wichter T, Matheja P, Eckardt L, Kies P, Schäfers K, Schulze-Bahr E, et al. Cardiac autonomic dysfunction in Brugada syndrome. Circulation. 2002;105(6):702–6.
9. Imai K, Sato H, Hori M, Kusuoka H, Ozaki H, Yokoyama H, et al. Vagally mediated heart rate recovery after exercise is accelerated in athletes but blunted in patients with chronic heart failure. J Am Coll Cardiol. 1994;24(6):1529–35.
10. Savin WM, Davidson DM, Haskell WL. Autonomic contribution to heart rate recovery from exercise in humans. J Appl Physiol. 1982;53(6):1572–5.
11. Camm AJ, Malik M, Bigger JT, Breithardt G, Cerutti S, Cohen RJ, et al. Heart rate variability: standards of measurement, physiological interpretation and clinical use. Task Force of the European Society of Cardiology and the North American Society of Pacing and Electrophysiology. Circulation. 1996;93(5):1043–65.
12. Sala R, Malacarne M, Solaro N, Pagani M, Lucini D. A composite autonomic index as unitary metric for heart rate variability: a proof of concept. Eur J Clin Investig. 2017;47(3):241–9.
13. Malliani A, Pagani M, Lombardi F, Cerutti S. Cardiovascular neural regulation explored in the frequency domain. Circulation [Internet]. 1991;84(2):482–92. Available from: http://circ.ahajournals.org/content/84/2/482.abstract
14. Gibbons RJ, Balady GJ, Bricker JT, Chaitman BR, Fletcher GF, Froelicher VF, et al. ACC/AHA 2002 guideline update for exercise testing: summary article: a report of the American College of Cardiology/American Heart Association Task Force on Practice Guidelines (Committee to Update the 1997 Exercise Testing Guidelines). J Am Coll Cardiol. 2002;40(8):1531–40.
15. Fox SM 3rd, Haskell WL. Physical activity and the prevention of coronary heart disease. Bull N Y Acad Med. 1968;44(8):950–65.
16. Dumont J, Hernandez AI, Carrault G. Improving ECG beats delineation with an evolutionary optimization process. IEEE Trans Biomed Eng. 2010;57(3):607–15.
17. Orini M, Mainardi LT, Gil E, Laguna P, Bailón R. Dynamic assessment of spontaneous baroreflex sensitivity by means of time-frequency analysis using either RR or pulse interval variability. In: 2010 Annual international conference of the IEEE engineering in medicine and biology. 2010. pp 1630–1633.
18. Hlawatsch F, Boudreaux-Bartels GF. Linear and quadratic time-frequency signal representations. IEEE Signal Process Mag. 1992;9(2):21–67.
19. Costa AH, Boudreau-Bartels GF. Design of time-frequency representations using a multiform, tiltable exponential kernel. IEEE Trans Signal Process. 1995;43(10):2283–301.
20. Bailón R, Laguna P, Mainardi L, Sornmo L. Analysis of heart rate variability using time-varying frequency bands based on respiratory frequency. In: Engineering in medicine and biology society, 2007 EMBS 2007 29th Annual international conference of the IEEE. 2007. pp 6674–6677.
21. Moody GB, Mark RG, Bump MA, Weinstein JS, Berman AD, Mietus JE, et al. Clinical validation of the ECG-derived respiration (EDR) technique. Comput Cardiol. 1986;13:507–10.

22. Kappus RM, Ranadive SM, Yan H, Lane-Cordova AD, Cook MD, Sun P, et al. Sex differences in autonomic function following maximal exercise. Biol Sex Differ [Internet]. 2015;6(1):28. Available from: http://www.bsd-journal.com/content/6/1/28
23. Koenig J, Thayer JF. Sex differences in healthy human heart rate variability: a meta-analysis. Neurosci Biobehav Rev. 2016;64:288–310.
24. Breuer HW, Skyschally A, Schulz R, Martin C, Wehr M, Heusch G. Heart rate variability and circulating catecholamine concentrations during steady state exercise in healthy volunteers. Heart [Internet]. 1993;70(2):144–9. Available from: http://heart.bmj.com/cgi/doi/10.1136/hrt.70.2.144
25. Mozaffarian D, Benjamin EJ, Go AS, Arnett DK, Blaha MJ, Cushman M, et al. Heart disease and stroke statistics-2016 update a report from the American Heart Association. Circulation. 2016;133:e38–48.
26. Minson CT, Halliwill JR, Young TM, Joyner MJ. Influence of the menstrual cycle on sympathetic activity, baroreflex sensitivity, and vascular transduction in young women. Circulation. 2000;101(8):862.
27. Saleh TM, Connell BJ. Estrogen-induced autonomic effects are mediated by NMDA and GABAA receptors in the parabrachial nucleus. Brain Res. 2003;973(2):161–70.
28. Mohamed MK, El-Mas MM. Abdel-Rahman AA. Estrogen enhancement of baroreflex sensitivity is centrally mediated. Am J Phys. 1999;276(4 Pt 2):R1030–7.
29. Benjamin IJ, Christians E. Exercise, estrogen, and ischemic cardioprotection by heat shock protein 70. Circ Res. 2002;90(8):833–5.
30. Di Diego JM, Cordeiro JM, Goodrow RJ, Fish JM, Zygmunt AC, Pérez GJ, et al. Ionic and cellular basis for the predominance of the Brugada syndrome phenotype in males. Circulation. 2002;106(15):2004–11.
31. Antzelevitch C. Androgens and male predominance of the Brugada syndrome phenotype. Pacing Clin Electrophysiol. 2003;26(7p1):1429–31.
32. Shimizu W, Matsuo K, Kokubo Y, Satomi K, Kurita T, Noda T, et al. Sex hormone and gender difference – role of testosterone on male predominance in Brugada syndrome. J Cardiovasc Electrophysiol. 2007;18(4):415–21.
33. Song M, Helguera G, Eghbali M, Zhu N, Zarei MM, Olcese R, et al. Remodeling of Kv4.3 Potassium Channel gene expression under the control of sex hormones. J Biol Chem. 2001;276(34):31883–90.
34. Bai CX, Kurokawa J, Tamagawa M, Nakaya H, Furukawa T. Nontranscriptional regulation of cardiac repolarization currents by testosterone. Circulation. 2005;112(12):1701–10.

Benchmarking Heart Rate Variability to Overcome Sex-Related Bias

13

Massimo Pagani, Roberto Sala, Mara Malacarne, and Daniela Lucini

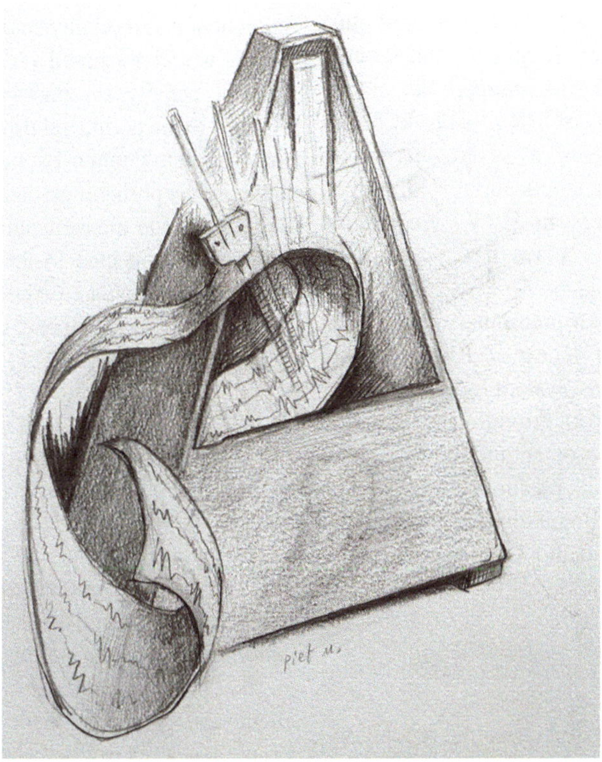

Heart rate variability. Illustration by Piet Michiels, Leuven, Belgium

Abstract

Since the seminal studies by Sayers (Ergonomics 16:17–32, 1973) and Akselrod et al. (Science 213:220–222, 1981) a few decades ago, it became clear that beat-by-beat oscillations in RR interval length (i.e. heart-rate variability [HRV]) contain information on underlying neural-control mechanisms based on the instantaneous balance between parasympathetic and sympathetic innervation.

M. Pagani (✉) · R. Sala · M. Malacarne · D. Lucini
BIOMETRA, University of Milan, Milan, Italy

P. L. M. Kerkhof, V. M. Miller (eds.), *Sex-Specific Analysis of Cardiovascular Function*,
Advances in Experimental Medicine and Biology 1065, https://doi.org/10.1007/978-3-319-77932-4_13

Over the years, the number of studies addressing HRV has increased markedly and now outnumbers 23,000. Despite such a large interest, there is still a continuing debate about interpretation of indices produced by computer analysis of HRV.

The main part of studies relies on spectral techniques to extract parameters that are linked to hidden information. The general idea is that these proxies of autonomic regulation can be useful to clinical applications in various conditions in which autonomic dysregulation may play a role. There are, however, serious shortcomings related to algorithms, interpretation, and the hidden value of individual indices. In particular, it appears that specific training is necessary to interpret the hidden informational value of HRV. This technical complexity represents a severe barrier to large-scale clinical applications. Moreover, important differences in HRV separate the sexes, and age plays an additional confounding role.

We present here a preliminary application of a novel unitary index of RR variability (Autonomic Nervous System Index of cardiac regulation) capable of providing information on the *performance* of autonomic regulation using a percentile rank position as projected on a large benchmark population. A summary of the underlying sympatho-vagal model is also presented.

Keywords
Heart-rate variability · Autonomic regulation · Beat-by-beat oscillations · Parasympathetic innervation · Sympathetic innervation · Sympatho-vagal model · Excitatory–inhibitory balance · Sex-related bias

All affections of the soul are associated with the body

Aristotle, De Anima Book I

Conflicts are frequently over semantics, not substance

Dan Brown, Origin

Introduction

Novelties often arise from the fruitful combination of multiple epistemologies. Accordingly, the same word (say, "anchor") may carry a different meaning according to the context (nautical, construction, or even TV), or may take multiple meanings, thus potentially generating funny, at times comical, effects (*mistaking own wife for a hat*) [1]. Likewise, heart rate variability (HRV) may morph according to the context: In bioengineering it would lean on algorithms and computer programs; in information science it would refer to patterns and meaning; in cardiology it would be associated with arrhythmias, infarction, and mortality statistics; in neurophysiology it would be based on vagal or sympathetic efferent activity; in pharmacology it would be directed to the peripheral flows of autonomic transmitters (acetylcholine for parasympathetic control and norepinephrine [noradrenaline] for sympathetic control); and in medicine and psychology it would mostly be connected to the behavioral dynamics of arousal. Thus, to properly address HRV we must consider both the hard and soft sciences [2].

A Unitary Aim

In 1949, R. W. Hess was awarded the Nobel Prize for Physiology and Medicine for his studies on neural control of the activity of internal organs [3]. Here, again, semantics were implicated. At variance with English usage, dealing with the *autonomic* nervous system, Hess was interested in the "paired antagonistic innervation" (sympathetic and parasympathetic) of the *visceral* nervous system, grouped by various functional regions, "linked to the central nervous system" and therefore seen as a component of an integrated regulatory organization whereby multiple organs aim at a unitary function (*e pluribus unum*). He also conjectured that experimental studies of the neurovisceral system were rendered difficult by the "direct contiguity of functionally multivalent pathways and nuclei [confusing] the . . . elucidation of related symptoms." It was clear, however, that "the parts of the brain communicating directly with the spinal cord at the upper end –

the medulla oblongata, and the segment lying directly beneath the cerebrum, the so-called diencephalon – exert a decisive influence on the vegetative controlling mechanisms"—hence the major contradiction of an *autonomic* section of the nervous system that is (paradoxically) directly controlled by higher structures and communicates with them toward a unitary goal.

From "Autonomy" of Pharmacology to Innervated Medicine

Pharmacological experiments with catecholamines mimicking sympathetic stimulation, and physostigmine-simulating parasympathetic excitation, supported a monolithic view of "autonomic" innervation basically functioning as an overall efferent structure, consisting of two "fundamentally different systems" [4]. It was therefore an obligatory consequence to state that afferent fibers from visceral organs do not have a physiological function because "all autonomic nerves [are] motor" [4]. However, a physiological function was later attributed to visceral neural reflexes, organized like simple reflexes [5], of both negative and positive feedback sign [6] considering that "all parts of the nervous system are connected together" [5]. More specifically, the model underlying neural cardiac regulation considers a complex structure in which both efferent and afferent information travels through visceral nerves: Cardiac innervation would therefore be characterized by a dual innervation (sympathetic and parasympathetic) made up of mixed (afferent and efferent) nerves.

The neural innervation of the cardiovascular system may remain a laboratory curiosity until new experimental needs suggest the appropriate technique of investigation [3] or new users' needs might suggest innovative applications (e.g., electroceuticals).

The Emergence of a New Paradigm: Bioengineering and Information

Importantly, a change of paradigm followed the introduction of bioengineering principles, shifting attention from pharmacology to biomathematics and devising electrophysiological techniques to investigate the complex dynamics of the (antagonistic) heart-rate response to electrical stimulation of the vagal and sympathetic nerves, whereby the vagus influence dominates the control of heart rate [7]. However, even if this model demonstrated an obligatory interaction between sympathetic and parasympathetic regulatory activity (Fig. 13.1), cardiovascular neural regulation is frequently (and simplistically) schematized as "autonomic" and either sympathetic or parasympathetic. The nonlinear interaction between these two components is frequently left out of the picture.

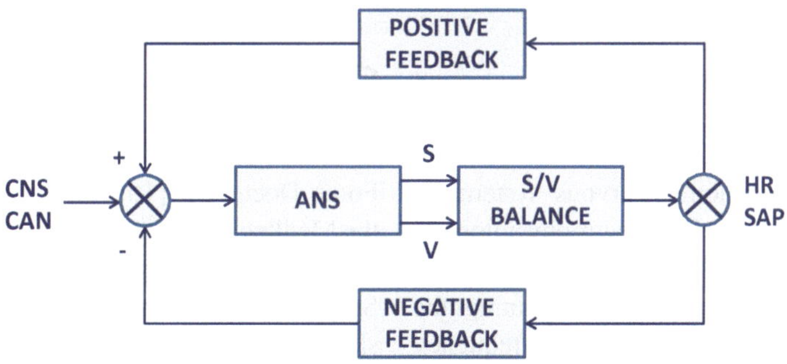

Fig. 13.1 Schematic representation of opposing feedback mechanisms that, in addition to central integration, sub-serve neural control of the cardiovascular system. Baroreceptive and vagal afferent fibers from the cardiopulmonary region mediate negative feedback mechanisms (exciting the vagal outflow and inhibiting the sympathetic outflow), whereas positive feedback mechanisms are mediated by sympathetic afferent fibers (exciting the sympathetic outflow and inhibiting the vagal outflow. (Redrawn from Ref. [6])

The progressive availability of growing computing power employed to study cardiovascular variables on a beat-by-beat basis, initially off line [8] then in real time, has opened the way to assess by proxy the *information* [9] about the dynamics of the balance [6] between sympathetic and parasympathetic regulatory activity.

Initial applications focused on HRV and on ergonomics [8] according to the hypothesis that information on physiological regulatory mechanisms could be coded in the oscillations hidden in the HRV (or rather RR variability) signal. The implications were therefore that a continuous series of symbols (RR intervals from the electrocardiogram [ECG]) could contain *information* [9] on cardiovascular regulation. What remained to be done was to crack the hidden code: We will not delve into the informational properties of scale (amplitude) and pattern because this is beyond the goal of this paper. However, allow us a brief detour to explore what information [9] might bring to our understanding of the physiology of autonomic regulation and HRV.

A Short Biased History of Findings

A seminal study by Akselrod et al. [10] formalized the idea that "sympathetic and parasympathetic nervous activity make frequency-specific contributions to the heart rate power spectrum," thus proposing a parallel between physiology and information. Accordingly, HR fluctuations could furnish a probe (i.e., proxy) of short-term neural cardiac regulation.

From our end, we reasoned that a key element in neural cardiac regulation was related to the obligatory neural interaction between the two branches of the autonomic nervous system, as shown, for example, by electrophysiological experiments on single cardiac vagal efferent fibers [11]. We thus suggested examining the relative powers of low- (LF) and high-frequency (HF) oscillations by shifting the attention beyond raw spectral data and computing normalized units ([nu] essentially focusing on spectral *patterns* as roughly synthesized by the LF/HF ratio) [12]. More importantly, we suggested to assess the excitatory responsiveness to upright stimulation (Fig. 13.2) as a key element of a dynamic protocol [12]. For clinical applications, this test can be simply performed by having subjects stand up for a few minutes.

The general underlying idea was that the key properties of neural structures revolve around dynamic activity, as epitomized by the time-varying spike sequence of nerve firing, which implicitly negates stationarity and implies a large repertoire of coding modalities [13]. Neural information can be hidden in various codes (such as digital or analog [14]), for example, *amplitude* (i.e., average number of spikes per unit time), *frequency* (i.e., instantaneous number of spikes as function of time), *gain* (i.e., a relationship between input and output), *phase* (i.e., a time relation between oscillations; relations between oscillators that we simplified with the ratio between LF and HF frequency oscillations of RR V), and so on, with increasing formality and complexity (such as the nonlinear properties) [15].

The model behind the LF/HF ratio can easily be considered inspired by the historical proposal of the unitary integration of two antagonistic control elements [2]. Numerically it could easily be obtained with a simple mathematical ratio of the LF and HF oscillations extracted from the variability signal. This approach has the advantage of describing changes in pattern [9], such as a power shifts toward the LF region (or vice versa) with a numerical increase (or decrease) of the LF/HF value.

Clinical Applications

Over the years, after a slow beginning and a Task Force Document [16], there was fast growth in the Medline database for HRV, now amounting to >23,000 hits and growing at >1000 hits/year. Surprisingly, however, there is still a need for shared standards of use in terms of underlying neural model and coding, data acquisition, algorithms of analysis, importance of given variables (time or frequency domain), and normative values for health or disease conditions. More importantly, reverse engineering of RR variability

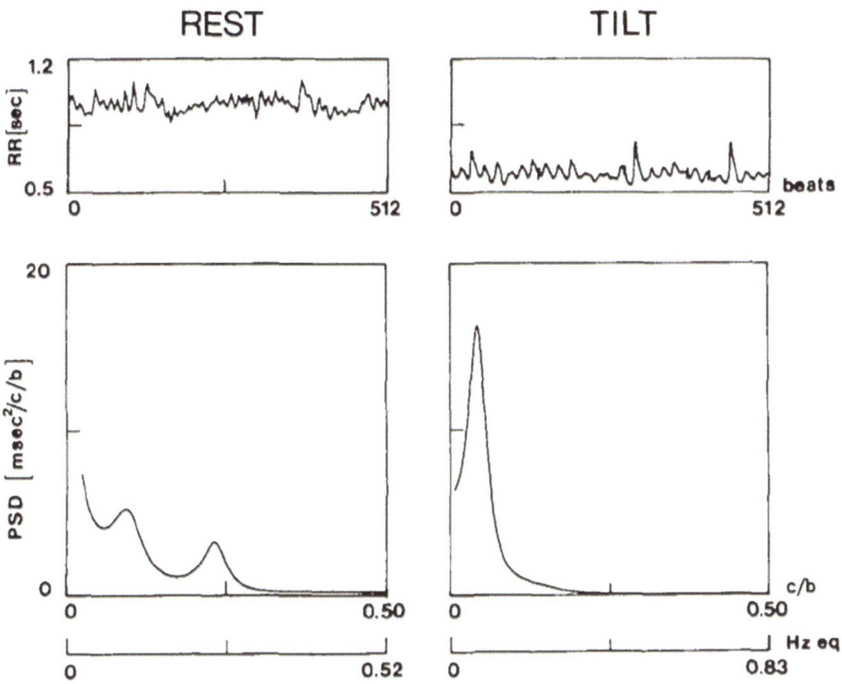

Fig. 13.2 RR interval series, that is, tachogram at rest and during passive upright 90° tilt. In the auto-spectra (bottom panels), two clearly separated LF and HF components are present at rest. During tilt, the low-frequency component becomes preponderant. Notice the change of *pattern*. (Taken from Ref. [12] with permission)

(RR V) should consider all elements together, aiming at reconstituting the unitary function that was broken into several indices by the process of analysis. Said otherwise: Does HRV provide a measure *of* physiology (hard science) or one of information *about* physiology (soft science) [2]?

To substantiate the hypothesis that (LF and HF) rhythms were the key elements carrying the information about a set level of the system, we performed a series of investigations in which we simultaneously recorded cardiovascular variables and electrophysiological signals of efferent sympathetic nerve fibers in human volunteers [17]. The level of sympathetic (and, by inference, parasympathetic) activity was increased or decreased by small infusions of vasoactive drugs, thus eliciting baroreflex-mediated changes (Fig. 13.3). We reported that during sympathetic activation in normal humans, there is a predominance in the LF oscillation of blood pressure, RR interval, and sympathetic nerve activity. During sympathetic inhibition, the HF component of cardiovascular variability predominates. This relationship is best seen when power spectral components are normalized for total power. The use of normalized units accounts in fact for the potentially diverging changes in total power (diminishing) and LF frequency oscillations increasing in relative but not absolute power, for example, with the volunteer standing up or performing light exercise. In any case, synchronous changes in the LF and HF rhythms of both the RR interval and muscle sympathetic nerve activity (MSNA) during different levels of sympathetic drive are suggestive of common central mechanisms governing both parasympathetic and sympathetic cardiovascular modulation. There is a similarity of patterns across different domains (activity of the central and peripheral nervous systems and cardiac rhythm [see Fig. 13.4]). Consequently we proposed [13] that RR V should be interpreted considering at least two different coding modalities: average amplitude (RR and RR variance) and dynamic oscillation (best

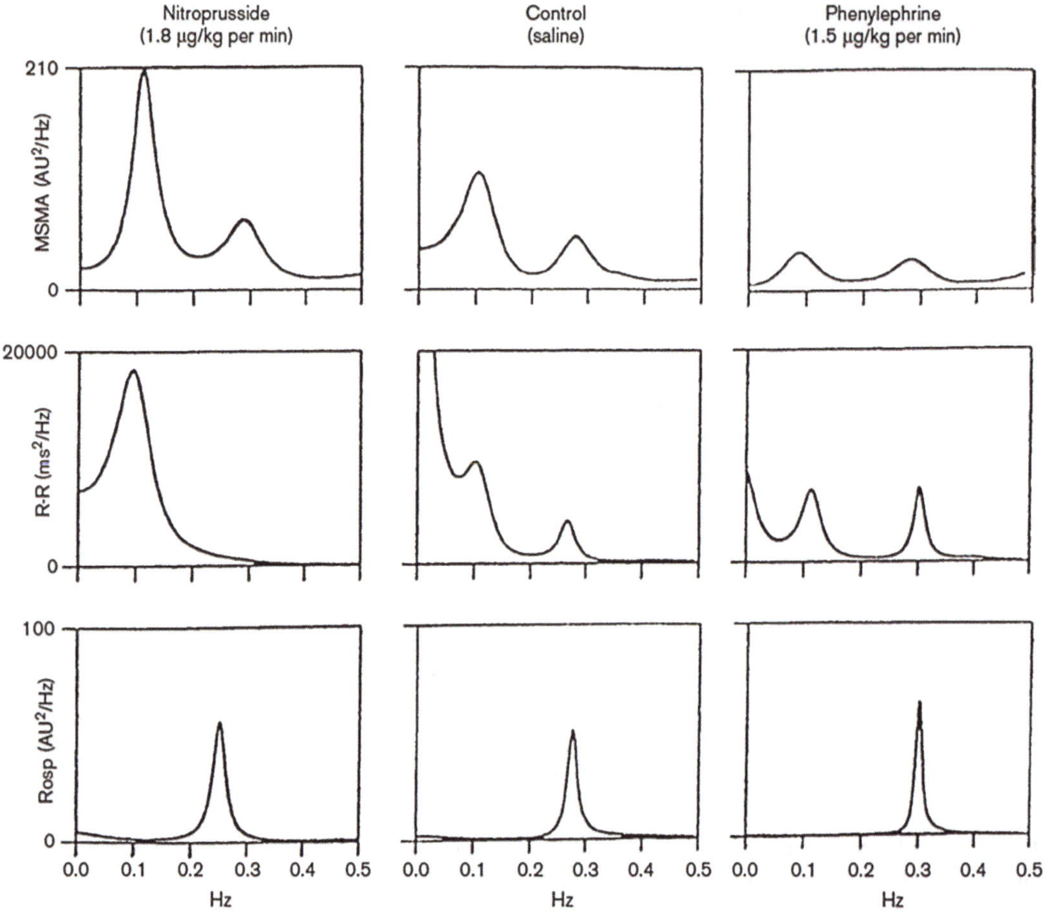

Fig. 13.3 Power spectra of MSNA, RR interval, and respiration (Resp) in a single subject during infusions of saline (Control), nitroprusside, and phenylephrine. During sympathetic activation induced by nitroprusside (left), the LF component of neural and cardiovascular variability signals predominates relative to the HF component. Conversely, during sympathetic inhibition and vagal activation induced by phenylephrine (right), there is an increase of the HF component relative to the LF component. a.u. indicates arbitrary units. Notice the change of patterns. (Taken from Ref. [17] with permission)

appreciated with LF and HF in normalized units). In this way, we avoid the implication of scaling and may easily focus on the change of pattern. As an example, Fig. 13.2 shows the change of pattern from a balanced LF and HF occurring at horizontal rest to a prevailing LF power of RR variability that follows a shift of autonomic balance (toward excitatory prevalence) accompanying the attainment of an upright posture. Similar shifts can be obtained by increasing (or decreasing) the excitatory (sympathetic) set level with manipulation of baroreflex activity using infusions of vasoactive substances. Importantly we should never forget that we are dealing with a complex integrated multi-domain structure.

Information about neurovisceral performance under various conditions might be useful to both physiological and pathophysiological applications. Initially one of the major applications regarded cardiac diseases, in particular, sudden coronary death and arrhythmias [16]. An additional area of potential bias for practical applications, even if not recognized, regards the difference related to sex [20]. It is in fact well recognized that men and women behave differently in terms of cardiovascular pathophysiology.

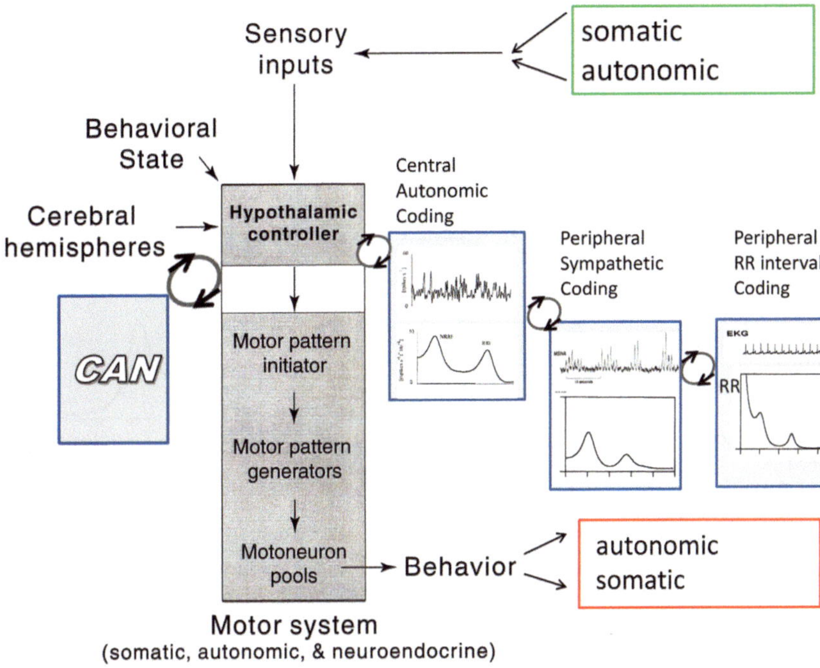

Fig. 13.4 Idealized, schematic representation of the circuitry responsible for generating simultaneous autonomic and somatic behavior as derived from motoneuron pools' activity. This activity is the outcome of the input from sensors (somatic and autonomic) after it has been processed by various controllers. The overall organization maintains an integrated performance of the motor system subdivided into somatic, autonomic and neuroendocrine. In parallel, it is possible to extract central, peripheral sympathetic and peripheral RR interval coding from related variability signals. Notice the similarity of patterns across different domains. (Inspired by Ref. [18], and data taken from Refs. [17, 19])

As an example, let us focus on the higher heart rate and lower arterial pressure observed in women as well as the different profile of cardiac diseases [21], such as the peculiar profile of coronary disease, the different hypertension history, and the emergence of conditions that appear easier to occur in women, such as the Takotsubo syndrome [22]. Approximately 10% of papers stored in the Medline database refer to "sex" or "gender" as a keyword. We will focus on some of the sex-related aspects of HRV, and we will present a novel unitary approach capable of superseding the sex (and age) bias of current autonomic evaluation [23].

Methodology in Practice

The practical value of HRV as a proxy of neural regulation of the heart (rate) depends on two factors referring to the importance of the underlying function (i.e., physiology) and ease of use (i.e., bioengineering).

It is important to point out that these factors, although related to the same aim (detect information on neural regulation), belong to different logical classes. Hence what pertains to physiology should be treated separately from what belongs to methodology. Said otherwise, HRV is *not* a measure of neural activity, although it might provide *information* [9] about neural regulation based on standard experiments employing classical stimulation and ablation protocols [10, 24]. In humans, stimulation of the system can easily be obtained by having the person stand up, thereby inducing a compensatory sympathetic increase and parasympathetic withdrawal (shorthand: "shift of the autonomic balance") [12].

The (metonymic) risk of equating RR V indices to "activity" of the nervous system (*ceci n'est pas un chapeau* [Magritte]) was probably

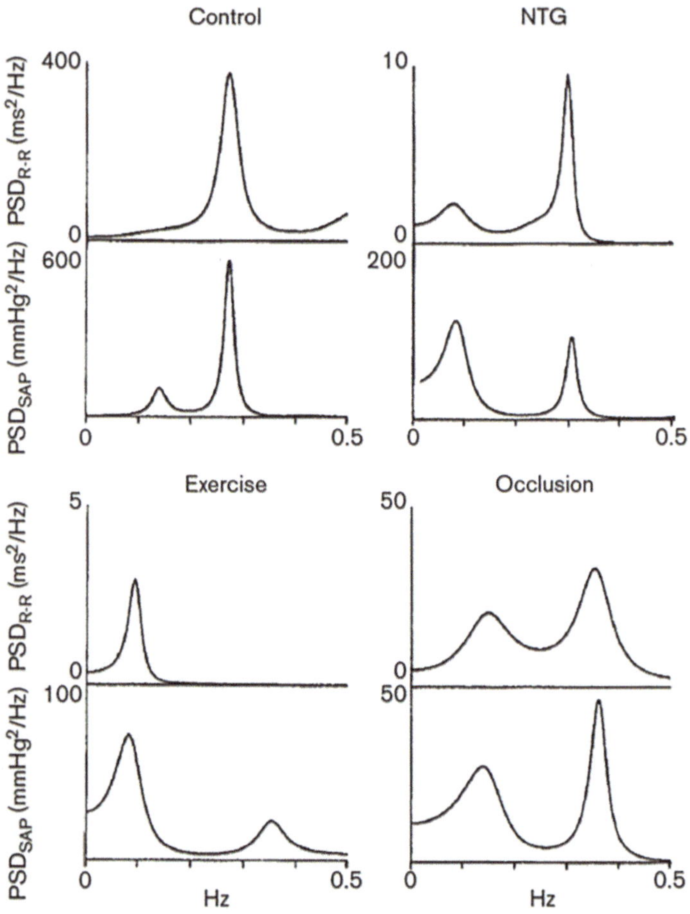

Fig. 13.5 Spectral analysis of RR interval (upper tracings in each panel) and systolic arterial pressure (SAP) (lower tracings in each panel) variabilities in conscious dogs at rest (CONTROL) and during experimental maneuvers leading to a sympathetic predominance (i.e., nitroglycerin infusion [NTG], treadmill exercise [Exercise], and transient acute coronary artery occlusion [Occlusion]). Note at control the presence of a single major HF component in the RR interval auto-spectrum; in SAP, a smaller LF component is also evident. During sympathetic activation, spectral distribution is altered in favor of low frequency; simultaneously, a drastic decrease in RR variance occurs (notice different scales on ordinates). PSD, power spectral density. (Reproduced from Ref. [6] with permission)

overlooked by a few investigators, and still now (over)interpretations might hover over ANS studies, for example, the interpretation of findings may be that at-rest LF oscillations (raw value) are mediated entirely by the vagus and that on standing by both the vagus and sympathetic arms; the respiratory (HF, raw values) oscillations are solely mediated by the vagus [24].

Based on a different model, acknowledging the obligatory nature of dual autonomic innervation—also derived from direct electrophysiological experiments in cats [11]—we looked at the effects of stimulation and ablation in both animals and humans, focusing not only on raw values of LF and HF but also on the relative power [12]. We saw, for example, that the shift in balance was particularly evident with nomalized-unit evaluation in both animals (using various stimuli to increase the excitatory set of the system: nitroglycerine, mild exercise, or coronary occlusion) (Fig. 13.5) and humans [17]. In animals, by surgery we could selectively abolish cardiac sympathetic pathways (afferent and efferent). We also showed that other conditions (transient myocardial ischemia, moderate exercise) could increase normalized LF while decreasing RR variance and thus raw values of spectral components [6]. It was apparent that the relative power of spectral components had the capacity to follow more closely the changes in the sympathovagal (or sympathetic–parasympathetic) balance. Importantly the presence of oscillations at LF and HF frequency is found simultaneously in MSNA and vagal efferent activity. This suggests that the relative power of RR V oscillations could be used to seek information about the performance of neurovisceral regulation [23]. A new unitary index, ANSI, is therefore proposed as a proxy of the performance of the entire system as a self-organizing [9] complex structure aiming at unitary goals [3].

Oscillations in HRV: Spectral Analysis

We will not delve into the technical aspects of methodologies that have been abundantly treated by several excellent reviews (starting with, e.g., [16]). We will only recall that the majority of studies employed parametric fast Fourier transformation algorithms to extract hidden oscillations; only a smaller fraction of studies employed non-parametric auto-regressive algorithms, which are less sensitive to the intrinsic nonlinearities and noise of the RR V signal [12]. The simplicity of obtaining the necessary signal with miniaturized instrumentation puts the use of HRV within everybody's reach. Hence, the importance of focusing on ECG rather than heart beats to obtain the tachogram for analysis, limiting analysis to sinus beat series, and avoiding slow breathing (i.e., entrainment) [25]. In brief, HRV measurement is easy to perform, non-invasive, and cost effective; however, it has several limitations, both methodological and practical.

First, according to specific algorithms, the number of extracted variables might vary. Furthermore, the rich data set might contain redundant variables [16], thus contributing to confound meaning and impairing usability. Finally, the interpretation of HRV indices varies according to the specific context (rest, standing, stress, drugs, etc.) and individual characteristics, such as age and sex or the presence of disease, such as diabetes.

In this context, the results of classical [3] and more recent studies [26] focusing on the hierarchical design of neural visceral regulation and providing evidence for common central mechanisms governing sympathetic and parasympathetic rhythmic activity [17] suggest the clinical usefulness of a unitary view of autonomic information, focusing on overall performance of regulation. We recently investigated whether a unitary Autonomic Nervous System Index of cardiac regulation (ANSI), as furnished by a radar plot [23] considering simultaneously the most informative spectral variables, could provide an easier appreciation of overall autonomic performance. We sought to verify whether a percentile rank transformation could allow the use of results from a large population as a benchmark of the information about autonomic regulation, against which individual autonomic performance could be tested. Hence, swapping physiology (and raw physical values) with information (and performance of process). A brief description of the methodology [12, 27] as in use in our laboratory follows.

Autonomic Evaluation

The day of autonomic evaluation, all subjects arrive in the clinic having avoided caffeinated beverages since waking as well as heavy physical exercise in the preceding 24 h. Recordings are performed between 09:00 am and 12:00 am in an air-conditioned, low-noise room. After a preliminary 10-min rest period in a supine position, ECG and respiratory activity are continuously recorded over a minimum 5-min period with the subject at rest and a 5-min period with the subject standing. Data are acquired with a PC, and a series of proxies of autonomic cardiac modulation are derived using an autoregressive spectral analysis tool [25]. In addition to RR interval (in msec) and RR-interval variability (assessed as total power [TP] in msec^2), the program automatically provides spectral components in both the LF (0.03–0.14 Hz) and HF (0.15–0.35 Hz) regions. The power of spectral components is assessed in both msec^2 and nu [12]. To include a simple evaluation of the effects of sympathetic activation as produced by active standing and the stand–rest difference, (Δ) in LFnu is computed.

Unbroken Nature of Neural Visceral Regulation: ANSI

Considering the fundamentally "unbroken" unitary nature of neural visceral regulation [16], we introduce a composite unitary autonomic nervous system index of cardiac regulation (ANSI) as a possible way to integrate the partial information spread across multiple autonomic variables (RR interval, TP, LF, and HF components [in both absolute and normalized units], LF/HF, and the stand–rest difference [in LFnu]) into a single comprehensive, heuristic parameter. ANSI is formally given by the areas of the octagon in the individual radar plots [28] that are built for each subject using eight HRV proxies, which are preliminarily scaled from 0 to 100 by the percentile rank transformation to share a range of variation and unit of measurement. To account for age and sex effects, percentile rank transformation is computed within the groups defined by the combinations of sex and age classes (with thresholds at 30 and 49 years) using a simple routine. ANSI, expressed in percentiles instead of raw, physical values, allows to rank individuals' overall autonomic condition against the reference population.

To minimize redundancy, a second more parsimonious, clinically manageable, version of ANSI is constructed by employing a decreased number of proxies (Fig. 13.6). The minimum number of proxies is selected from among the HRV variables recognized as being substantial by the combination of factor analysis [29],

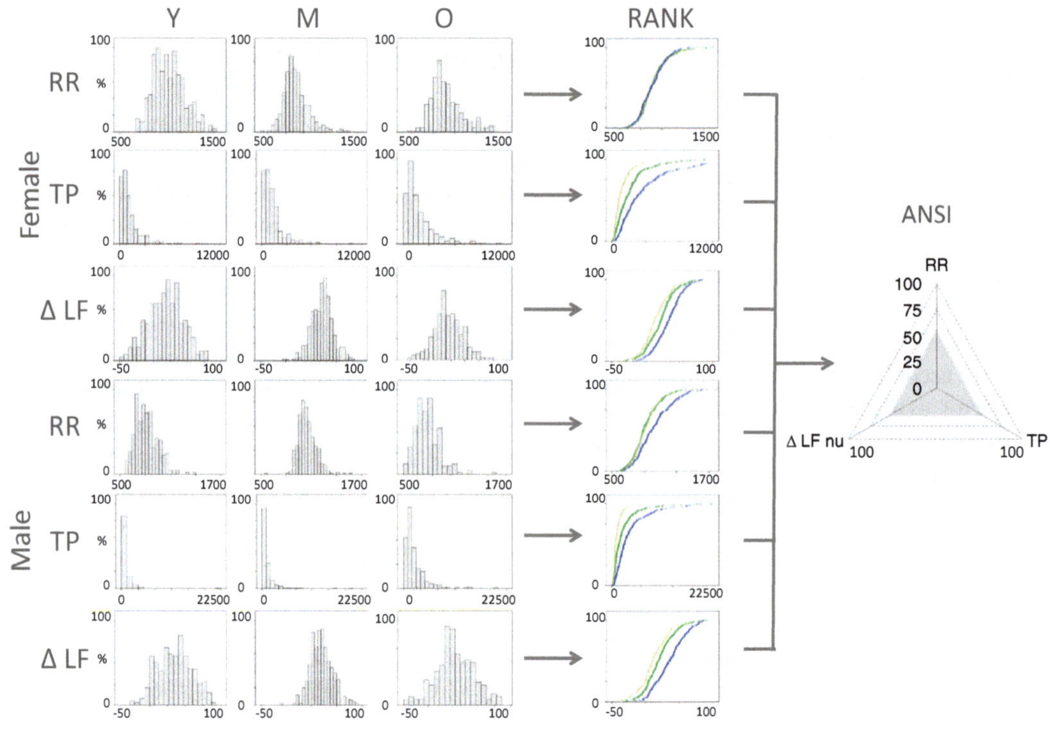

Fig. 13.6 The construction of ANSI$_3$ in its main steps. First, the procedure starts from the three selected HRV proxies (RR, RR power and ΔLFnu) and their distribution within sex and age classes jointly considered (left panels); then it proceeds through their within-group transformation according to the percentile rank (middle panels); and it ends by computing the indicator as the area of the triangle composing each individual radar plot, which is expressed in percentile (right). Notice that in this procedure the sex (and age) bias is eliminated. Abbreviations: Age subgroups with thresholds at 30 and 49 years, Y = young, M = middle, O = old; RR = RR interval, TP = RR interval power; Δ LF = stand–rest difference in LF nu. X values for RR are in msec, TP are msec^2; Δ LF are in nu. (Taken from Ref. [23] with permission)

which is carried out with the principal factor extraction method and varimax rotation (considering meaningful only loadings >0.4), and physiological underpinnings. A reduced ANSI score is regarded as a good synthesis of autonomic information comparable with ANSI8 if the linear correlation coefficient between the two indices is significant and high (i.e., >0.8). Subsequently, groups approximating the clinical status of individuals were formed by combining together the categories of systolic arterial pressure (with thresholds at 120 and 140 mmHg), body mass index (with thresholds at 25 and 30 kg/m^2), and smoking (no/yes). In this way, the benchmark population is composed of only ambulant, not hospitalized, patients who are devoid of symptoms or acute conditions.

Results

We report herein a summary of a recent study [23] on a benchmark population ($n = 1593$, age 39 ± 13 years), resolving the implications of sex and age differences. Descriptive anthropometric and autonomic data for the reference population, subdivided into age and sex classes, are listed in Table 13.1. Significant within-sex differences between age classes are evident. Table 13.2 further focuses on the clear sex-related difference in the subgroup of normal healthy subjects (free from risk factors, such as hypertension, obesity, or stress). Notice that females have lower BMI, arterial pressure, and mean RR as well as RR TP and RR LF in raw values, whereas HF absolute (a) is similar in both sex groups. Conversely, normalized-unit RR HF is greater in females (RR LF is lower in both absolute and normalized values). The stand–rest difference in normalized-unit LF is similar in both sexes.

Factor analysis applied to the eight HRV proxies (Table 13.3) to extract the potential tendency of spectral indices to form clusters of homogeneous meaning demonstrates that the first three factors reproduce a high percentage (variance accounted for [VAF] = 82.7%) of the total information spread across variables. Analysis further shows that taken individually, the first factor accounts for 44.0%, the second for 24.2%, and the third for 14.5% of the total variance. Moreover, factor loadings indicate aggregation of the HRV proxies into the following three clusters: normalized autonomic indices (LF nu, HF nu, LF/HF, and ΔLFnu [factor 1]), absolute indices (TP, LFa, and HFa [factor 2]), and RR interval (HR and RR [factor 3]). This suggests constructing the reduced ANSI with three selected proxies ($ANSI_3$), one for each factor.

In addition, binary logistic regression indicates that only the following variables carry significant information about sex prediction (71% correct prediction): RR ($p < 0.001$), RR LFnu ($p < 0.001$), RR HF nu ($p = 0.009$), and Δ stand–rest LFnu ($p < 0.001$). Interestingly, variance of RR—a major time domain index of HRV—does not add significantly to sex information. ANSI (Fig. 13.7), by design, is free from sex (and age) bias, and there is no difference between the sexes, for example, females 56.16 ± 27.79 versus males 54.43 ± 29.15 (not significant).

Discussion

Within the constraints of a complex system, neural visceral regulation can be depicted as the unitary result of a continuous dynamic balance between two opposing neural domains: an excitatory sympathetic one and an inhibitory parasympathetic one [6]. This is, on purpose, a simplified model because, for example, large nonlinearities are not addressed, and the opposing (excitatory–inhibitory) balance is not always appropriate (e.g., it does not consider possible parallel changes in either direction of the two autonomic arms).

However, this simplified model, as a first approximation, permits to argue about the (hypothetical) significance of the complex dynamics of HRV, for example, we can test experimentally whether increases in excitatory sympathetic drive are reflected in a change in spectral pattern: specifically a leftward shift of the spectral profile (Fig. 13.2). This implies that indices (or rather patterns [9]) from HRV can provide information about the underlying setting (prevailingly excitatory or inhibitory) of the entire system. The

Table 13.1 Descriptive anthropometric and autonomic data subdivided into sex and age sub-groups of the study population

		Young	Middle age	Old	
		Mean ± SD	Mean ± SD	Mean ± SD	Significance trend
Females					
		$n = 225$	$n = 341$	$n = 205$	
Age	Years	24.4 ± 3.4	40.2 ± 5.6	56.1 ± 5.5	$p < .001$
Weight	Kg	61.6 ± 13.6	69.3 ± 23.1	68.2 ± 17.0	$p < .001$
Height	Cm	165.0 ± 6.6	163.3 ± 6.5	160.7 ± 6.3	$p < .001$
BMI	Kg/m^2	22.6 ± 4.9	26.0 ± 8.3	26.4 ± 6.5	$p < .001$
HR	bpm	69.7 ± 12.2	69.5 ± 12.7	69.5 ± 13.3	$p = .840$
SAP	mmHg	109.1 ± 12.7	119.2 ± 18.8	134.5 ± 21.6	$p < .001$
DAP	mmHg	69.3 ± 9.6	76.3 ± 12.0	83.1 ± 12.4	$p < .001$
RR	ms	887.2 ± 155.0	893.7 ± 172.0	896.5 ± 178.1	$p = .566$
TP	ms^2	4073.7 ± 4195.7	2186.1 ± 2097.8	1250.4 ± 1180.9	$p < .001$
LFa	ms^2	1090.7 ± 1143.9	598.1 ± 732.2	365.7 ± 759.8	$p < .001$
HFa	ms^2	1586.8 ± 2157.5	558.4 ± 762.8	284.7 ± 510.0	$p < .001$
LF nu	nu	45.1 ± 20.6	50.8 ± 19.3	50.3 ± 20.5	$p = .006$
HF nu	nu	47.4 ± 21.2	41.0 ± 19.4	38.6 ± 20.2	$p < .001$
LF/HF	–	1.9 ± 3.5	2.4 ± 3.8	2.8 ± 5.3	$p = .020$
ΔLF nu	nu	34.0 ± 20.7	22.2 ± 21.7	15.7 ± 20.9	$p < .001$
Males					
		$n = 265$	$n = 361$	$n = 196$	
Age	Years	24.4 ± 3.6	40.1 ± 5.7	58.1 ± 7.4	$p < .001$
Weight	Kg	77.2 ± 13.3	81.0 ± 13.8	80.7 ± 17.2	$p = .011$
Height	Cm	179.8 ± 8.6	176.2 ± 7.7	171.7 ± 8.4	$p < .001$
BMI	Kg/m^2	23.8 ± 3.4	26.1 ± 4.1	27.4 ± 5.0	$p < .001$
HR	bpm	61.1 ± 13.0	67.2 ± 12.1	66.4 ± 11.3	$p < .001$
SAP	mmHg	122.2 ± 13.7	130.1 ± 18.3	138.7 ± 19.6	$p < .001$
DAP	mmHg	71.7 ± 9.4	82.6 ± 13.0	85.8 ± 11.9	$p < .001$
RR	ms	1028.9 ± 227.5	921.1 ± 181.8	931.8 ± 172.6	$p < .001$
TP	ms^2	4955.6 ± 8268.4	2959.0 ± 4690.2	1328.9 ± 1357.1	$p < .001$
LFa	ms^2	1332.4 ± 1830.5	893.2 ± 1302.3	406.3 ± 554.8	$p < .001$
HFa	ms^2	2003.6 ± 5403.5	727.8 ± 2239.4	177.2 ± 258.6	$p < .001$
LF nu	nu	49.0 ± 23.1	61.1 ± 22.0	58.1 ± 22.1	$p < .001$
HF nu	nu	46.1 ± 23.5	32.4 ± 21.4	32.4 ± 20.5	$p < .001$
LF/HF	–	2.1 ± 2.9	4.7 ± 7.4	4.2 ± 6.5	$p < .001$
ΔLF nu	nu	35.4 ± 23.7	20.1 ± 22.3	11.8 ± 22.6	$p < .001$

Abbreviations: *n* number of cases, *BMI* body mass index, *HR* heart rate, *SAP* systolic arterial pressure, *DAP* diastolic arterial pressure, *n* number of cases, *RR* RR interval, *TP* total power of RR variability, *LF* low-frequency component of RR variability, *a* absolute value, *HF* high-frequency component of RR variability, *nu* normalized unit, *Δ* stand–rest difference, *sig.* significance by trend test

autonomic nervous system becomes (or rather returns to being) connected to the central nervous system and to the sensory periphery in order to govern the various "bits" of unitary behavioral goals [30] that accompany everyday life from moment to moment. Early alterations in health status—such as hypertension, obesity, or stress—can thus be estimated from alterations in autonomic proxies. Obviously, the sex (and age) scalar difference in HRV indices might represent an important obstacle to simple clinical use of the methodology.

Table 13.2 Mean data for the study population of normal subjects ($n = 863$)

		Female		Male		
		Mean	SD	Mean	SD	*p*
Age	Years	34.70	11.67	33.39	11.88	
WEIGHT	kg	59.96	9.29	77.20	11.37	<0.001
HEIGHT	cm	163.97	6.61	178.00	8.85	<0.001
BMI	kg/m^2	22.30	3.14	24.32	2.67	<0.001
SAP	mmHg	110.55	12.74	121.06	12.70	<0.001
DAP	mmHg	70.47	8.62	73.12	8.91	<0.001
HR	b/min	68.43	11.26	62.81	12.61	<0.001
RRMean	msec	900.56	150.07	996.05	211.85	<0.001
RRTP	msec^2	3172.29	3453.82	4291.80	7326.63	0.006
RRLFa	msec^2	865.67	1031.88	1206.49	1752.07	0.001
RRHFa	msec^2	1136.65	1761.24	1593.64	4618.52	
RRLFnu	nu	46.16	19.36	51.96	22.78	<0.001
RRHFnu	nu	45.83	19.94	42.55	23.03	0.027
RRLFHF	–	1.88	3.31	2.53	3.54	0.006
RRLFHz	Hz	0.102	0.027	0.098	0.022	0.030
RRHFHz	Hz	0.27	0.06	0.26	0.06	0.044
ΔLFrrSTAND-REST	nu	30.14	20.00	29.82	24.48	

Abbreviations: *BMI* body mass index, *HR* heart rate, *SAP* systolic arterial pressure, *DAP* diastolic arterial pressure, *RR* RR interval, *TP* total power of RR variability, *LF* low-frequency component of RR variability, *a* absolute value, *HF* high-frequency component of RR variability, *nu* normalized unit, Hz Hertz, *Δ* stand–rest difference

Table 13.3 Factor analysis

Variance Accounted For (VAF) 82.7%			
Factor	1	2	3
VAF per factor	44.0%	24.2%	14.5%
	Loading		
HR			−.950
RR			.947
TP		.946	
LFa		.804	
HFa		.884	
LF nu	−.938		
HF nu	.923		
LF/HF	−.719		
ΔLF nu	.775		

Abbreviations: *RR* RR interval, *TP* total power of RR variability, *LF* low-frequency component of RR variability, *a* absolute value, *HF* high-frequency component of RR Variability, *nu* normalized unit, *Δ* stand–rest difference

Factor analysis is performed with principal factor extraction method and varimax rotation

With the emergence of personalized medicine, and the increasing availability of information technology for everybody, the value of indices capable of indicating the quality of control systems in maintaining health may grow, thus increasing their appeal to the market. In addition, it is conceivable that HRV will enjoy ever-increasing success. However, the methodological complexity involved is likely to act as a barrier to the acceptance of the necessary change of paradigm. We argue that a major facilitatory role might conversely be provided by using a single normalized (by rank) index (ANSI) [23], which can facilitate the practical use of the working of the autonomic nervous system. This approach permits to transcend the uncertainties of the physiological information carried by single indices. Thus, the focus might change from (scalar) measures of amplitude of sympathetic or parasympathetic activity, which are implicitly seen as separate (i.e., "autonomic"), to measures of performance of integrated neurovisceral regulation based on the dynamic interaction (pattern or "balance") of sympathetic and parasympathetic (oscillatory) control and the humoral milieu.

Irrespective of the specific methodology, age and sex are potent modifiers of (the amplitude) of

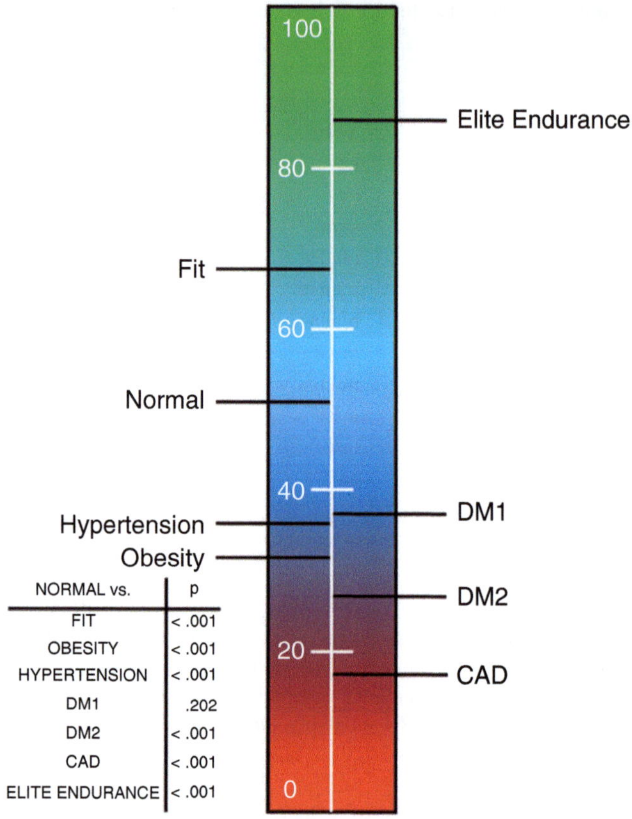

Fig. 13.7 Schematic representation of the position of the mean value of percentile rank for the various groups (left) and projected mean values of tests groups (elite endurance, diabetes type 1, diabetes type 2, coronary artery disease [CAD]). Significance of differences against normal (reference population of controls) is indicated in the bottom left panel (Dunnett's test). A 0–100 reference scale (white) is provided in the bar. (Taken from Ref. [23] with permission)

autonomic indices. The use of a unitary proxy (ANSI) permits to address directly the performance (with a 0–100 rank corresponding to the shift from poor to optimal) of overall autonomic regulation as defined against a benchmark of healthy, ambulatory population. Sex differences are recognized at the (scalar) level of raw individual indices, thus reinforcing the hypothesis that patterns and amplitudes provide a different kind of information [9].

We may regret the loss of the direct physiological information carried by the raw values of individual indices [31]. However, this may not be a limitation because mathematical manipulations might generate novel information [32] on the overall performance of the neural visceral regulation as the result of a unitary organization sub-serving complex behavioral dynamics. We might even argue that ANSI might supersede the previous view of the LF/HF ratio [12] as a comprehensive index of the balance between excitatory and inhibitory peripheral (sympathetic and parasympathetic) nerve activities. ANSI [23] could thus furnish a proxy of the overall setting of the visceral regulation independent of age and sex. However, only large-scale applications will tell if this approach truly represents an advancement in the clinical use of autonomic (or visceral) nervous system assessment.

References

1. Sacks O. The man who mistook his wife for a hat. New York: Summit Books; 1985.
2. Editorial. The soft science of medicine. Lancet. 2004;363 (9417):1247.
3. Hess WR. Nobel Lecture: The Central Control of the Activity of "Internal Organs". Nobelprize.org.Nobel Media AB 2014. Web. 2016. http://www.nobelprize.org/nobel_prizes/medicine/laureates/1949/hess-lecture.html
4. Langley JN. The autonomic nervous system (Pt. I). Cambridge: W. Heffer & Sons; 1921.
5. Burke RE. Sir Charles Sherrington's the integrative action of the nervous system: a centenary appreciation. Brain. 2007;130 (4):887–94.

6. Malliani A, Pagani M, Lombardi F, Cerutti S. Cardiovascular neural regulation explored in the frequency domain. Circulation. 1991;84(2):482–92.
7. Warner HR, Russell RO Jr. Effect of combined sympathetic and vagal stimulation on heart rate in the dog. Circ Res. 1969;24:567–73.
8. Sayers BM. Analysis of heart rate variability. Ergonomics. 1973;16(1):17–32.
9. Haken H. Synergetics: an introduction. Berlin: Springer Verlag; 1983.
10. Akselrod S, Gordon D, Ubel FA, Shannon DC, Berger AC, Cohen RJ. Power spectrum analysis of heart rate fluctuation: a quantitative probe of beat-to-beat cardiovascular control. Science. 1981;213(4504):220–2.
11. Schwartz PJ, Pagani M, Lombardi F, Malliani A, Brown AM. A cardiocardiac sympathovagal reflex in the cat. Circ Res. 1973;32(2):215–20.
12. Pagani M, Lombardi F, Guzzetti S, et al. Power spectral analysis of heart rate and arterial pressure variabilities as a marker of sympatho-vagal interaction in man and conscious dog. Circ Res. 1986;59(2):178–93.
13. Pagani M, Malliani A. Interpreting oscillations of muscle sympathetic nerve activity and heart rate variability. J Hypertens. 2000;18(s):1709–19.
14. Mochizuki Y, Shinomoto S. Analog and digital codes in the brain. Phys Rev E. 2014;89(2):022705.
15. Porta A, Guzzetti S, Montano N, et al. Entropy, entropy rate, and pattern classification as tools to typify complexity in short heart period variability series. IEEE Trans Biomed Eng. 2001;48(11):1282–91.
16. Task Force of the European Society of Cardiology and the North American Society of Pacing and Electrophysiology. Heart-rate variability: standards of measurements, physiological interpretation and clinical use. Circulation. 1996;93 (5):1043–65.
17. Pagani M, Montano N, Porta A, et al. Relationship between spectral components of cardiovascular variabilities and direct measures of muscle sympathetic nerve activity in humans. Circulation. 1997;95(6):1441–8.
18. Kandel ER, Schwartz JH, Jessell TM, editors. Principles of neural science. 4th ed. New York: McGraw Hill; 2000.
19. Massimini M, Porta A, Mariotti M, Malliani A, Montano N. Heart rate variability is encoded in the spontaneous discharge of thalamic somatosensory neurones in cat. J Physiol. 2000;526:387–96.
20. Voss A, Schroeder R, Heitmann A, Peters A, Perz S. Short-term heart rate variability—influence of gender and age in healthy subjects. PLoS One. 2015;10(3):e0118308.
21. Benjamin EJ, Blaha MJ, Chiuve SE, et al. Heart disease and stroke statistics-2017 update: a report from the American Heart Association. Circulation. 2017;135(10):e146–603.
22. Pelliccia F, Kaski JC, Crea F, Camici PG. Pathophysiology of Takotsubo syndrome. Circulation. 2017;135(24):2426–41.
23. Sala R, Malacarne M, Solaro N, Pagani M, Lucini D. A composite autonomic index as unitary metric for heart rate variability: a proof of concept. Eur J Clin Investig. 2017;47 (3):241–9.
24. Pomeranz B, Macaulay RJ, Caudill MA, et al. Assessment of autonomic function in humans by heart rate spectral analysis. Am J Phys. 1985;248(1 Pt 2):H151–3.
25. Lucini D, Marchetti I, Spataro A, et al. Heart rate variability to monitor performance in elite athletes: criticalities and avoidable pitfalls. Int J Cardiol. 2017;240:307–12.
26. Beissner F, Meissner K, Bar KJ, Napadow V. The autonomic brain: an activation likelihood estimation meta-analysis for central processing of autonomic function. J Neurosci. 2013;33(25):10503–11.
27. Badilini F, Pagani M, Porta A. Heartscope: a software tool addressing autonomic nervous system regulation. Comput Cardiol. 2005;32:259–62.
28. Saary MJ. Radar plots: a useful way for presenting multivariate health care data. J Clin Epidemiol. 2008;61(4):311–7.
29. Sala R, Spataro A, Malacarne M, et al. Discriminating between two autonomic profiles related to posture in Olympic athletes. Eur J Appl Physiol. 2016;116(4):815–22.
30. Engel BT. Psychosomatic medicine, behavioral medicine, just plain medicine. Psychosom Med. 1986;48(7):466–79.
31. Eckberg DL. Sympathovagal balance: a critical appraisal. Circulation. 1997;96(9):3224–32.
32. Lucini D, Solaro N, Pagani M. Autonomic differentiation map: a novel statistical tool for interpretation of Heart Rate Variability. Frontiers Physiol. 2018;9:401

Cardiovascular Allometry: Analysis, Methodology, and Clinical Applications 14

John K.-J. Li

Allometry. Artwork by Piet Michiels, Leuven, Belgium

J. K.-J. Li (✉)
Department of Biomedical Engineering,
Rutgers University, Piscataway, NJ, USA
e-mail: johnkjli@soe.rutgers.edu

Abstract

The classic works of "On Growth and Form" and "The Problem of Relative Growth" that began a century ago have so fittingly, albeit unintentionally, become pertinent to the

P. L. M. Kerkhof, V. M. Miller (eds.), *Sex-Specific Analysis of Cardiovascular Function*,
Advances in Experimental Medicine and Biology 1065, https://doi.org/10.1007/978-3-319-77932-4_14

modern-day clinical treatment strategy of the many patients with cardiovascular disease. This chapter uses allometry, which was established for comparative biology, to explore physiological and pathological differences due to differential growth, which may lead to differing diagnostic and treatment approaches for male versus female patients. Men and women have obvious differences in body and heart weights, as well as different geometries and structures of their blood vessels; the analysis in this chapter extends to their hemodynamic functional differences. This includes dimensional analysis to establish criteria for characterizing functions based on allometric formulations. The clinical applications of sex differences are analyzed for arterial stenosis, aneurysm, atherosclerosis, hypertension, and coronary revascularization. Allometric approaches are applied specifically to isolated cases of systolic hypertension to delineate the intermingled relations of aging and sex differences. This chapter aims to provide some preliminary insights into the usefulness of cardiovascular allometry. Its future impact on clinical diagnosis remains largely unexplored.

Keywords
Comparative hemodynamics · Allometry · Cardiovascular function · Sex differences

Allometry for Comparative Analysis of Cardiovascular Structure and Function

Simple Allometry

Modern comparative biological studies are credited to Huxley's [27] work on the "Problems of Relative Growth," in which he based many of his biological interpretations on allometric relations. His work must have been greatly influenced by D'Arcy Thompson's [84] monumental treatise, "On Growth and Form," which is based on the extensive analysis of numerous experimental observations. We shall first look at the role allometry plays among the species of *Homo sapiens* and then focus on the within-species aspects before applying it to examine differences in the sexes.

Generally speaking, the term "allometry" is defined as the change of proportions with an increase of size both within a single species and between adults of related groups. The commonly used allometric formula relates any measured physical quantity Y to body weight W, and a and b are derived or measured empirical constants. This results in the familiar power law,

$$Y = aW^b \tag{14.1}$$

This formula expresses simple allometry. Special cases are associated with this formula that have identifiable physiological meanings. For instance, when the exponent is 0, Y is proportional to a and is independent of body weight W. In other words, the physical variable is invariant of body weight. When the exponent b is 1/3, the variable is dependent on body length; when b is 2/3, Y is dependent on body surface area (BSA); and when $b = 1$, Y is simply proportional to body weight. This provides what is known as the basis of the "one-third power law" or geometric scaling as proposed by Lambert and Teissier [35]. It has recently been challenged by the "one-fourth power law," or the exponent in multiples of 1/4, as the basis of biological allometric formulation [90] and by compound allometry. This latter is seen in the clinical formulation of BSA estimation where simple allometry is substituted with compound allometry with BSA being dependent on both body weight and height of the male or the female.

The inverse proportionality applies when b is negative. On a log–log plot of the physical quantity with body weight, one can see that a straight line will result in a slope b. The slope can be positive or negative depending on the value of b. For comparison among species, a physical variable of interest is first measured for each of the species or sexes. The values are then plotted on a log–log graph. Because

$$\log Y = \log a + b \log W \tag{14.2}$$

plotting log Y on the ordinate and log W on the abscissa will result in a straight line with slope b and intercept a.

The allometric equation has proven to be powerful for the characterization of similarities among species. It is effective in relating a physiological phenomenon, either structural or functional, among mammals of grossly different body weight. However, allometry has been little studied in terms of its clinical applications and for identifying differences between the sexes.

Consideration of Similarity Criteria as Scaling Laws for the Cardiovascular System

The classic definition of a similarity criterion is when Y, formulated in terms of either product (s) or ratio(s) of physically measurable variables, remains constant despite changes in body weight and is dimensionless. Thus, the exponent b must necessarily be zero. In other words, similarity is present whenever any two, dimensionally identical, measurements occur in a constant ratio to each other. If such a ratio exists among different species or sexes, then a similarity criterion is established as the scaling law. This approach of establishing biological similarity criteria has been found to be very useful [4, 10, 21–24, 41, 45, 47, 61, 62, 80–82].

It has been recognized that there are gross similarities in the anatomical structure and physiological function of the mammalian cardiovascular system and between the sexes. To draw such conclusions based on observations alone is inadequate in that the physical principles governing such similarities are not present. To extrapolate one's observation of similarities from one species of a given size to another, one exercises visual "scaling." Thus, scaling can be used to form criteria that allow extrapolation or interpolation of observations.

Let us consider anatomical similarities through scaling by multiplication or division of body length dimensions based on simple allometry. Thus, the length of the aorta of a woman L_w is scaled to that of a man L_m by a scaling factor a_1,

$$L_w = a_1 L_m \tag{14.3}$$

Similarly, their heart weights can be scaled by a constant of proportionality, a_2,

$$W_{hw} = a_2 W_{hm} \tag{14.4}$$

With simple allometry, the length of the aorta is obviously proportional to the respective body weight's one-third power. Modeling is an important aspect of interpreting observed similarities and predicting experimental observations a priori. To be useful, an experimental test on a model should provide data that can be easily scaled to obtain either static or dynamic information that exists on the "prototype." This means the identifiable pertinent features of the prototype must be of sufficient accuracy. It is important that the two systems under observation have the same dimensional homogeneity within the same regime. Simply said, one would compare aortic blood flow and aortic diameter of a man to those, respectively, of a woman or a dog. One would not compare blood flow in the aorta of man to the blood flow of the femoral artery of a dog or a woman, nor compare blood pressure to flow because they have different dimensional variables, i.e., mmHg versus mL/s.

Observations of the anatomic cardiovascular structure among mammalian species have shown considerable geometric similarities. Geometric similarity implies the same shape or size and that corresponding linear dimensions of the cardiovascular systems are related by a constant scale factor. For instance, the length of the aorta from its root to its aorto-iliac bifurcation is about 65 cm in a 70-kg man. This length is approximately 45 cm in a 20 kg dog. The ratio of the two is 65/45 = 1.44. If the diameter of the dog aorta is 1.6 cm, this ratio can be used to calculate the aortic diameter of the man since

$$65/45 = D/1.6 \tag{14.5}$$

or D = 2.3 cm, which should be close to the actual size of the human aorta had we measured it. In terms of the sexes, both the aortic length and diameter differ among men and women; it is greater in men than women. Notice here that the aortic length and diameter both have the physical unit of cm. In addition, both length and diameter have the body weight exponent of 1/3 or $W^{1/3}$. It should be noted here that the growth charts used clinically for assessing the body heights and weights for boys and girls are significantly different.

One can also scale the total cross-sectional area of the vascular beds, the number of capillaries, or the size of the heart from one mammal to another. Numerous experimental studies have shown that heart weights in mammals are in a constant proportional to their body weights. Here again, we can see that the difference in body weights between man and woman leads to difference in their heart weights. This constant of proportionality gives a scaling factor such that we can obtain the heart weight of any mammal by simply knowing the body weight. As an example, if the body weights of man and a rabbit are 70 kg and 3 kg, respectively, and the man's heart weight–to–body weight ratio is 6/1000 or 0.6% (or 420 g), then the heart weight of the rabbit is:

$$W_h = 3 \times 0.6\% = 0.018\,\mathrm{kg} \text{ or } 18\,\mathrm{g} \quad (14.6)$$

If we extrapolate this to within-species comparison between a 70-kg man and a 60-kg woman, we can calculate the heart weight of the woman as:

$$W_h = 60 \times 0.6\% = 0.36\,\mathrm{kg} \text{ or } 360\,\mathrm{g} \quad (14.7)$$

which is quite a reasonable estimate. In patients with increased heart weight associated with obesity or left ventricular hypertrophy, such simple allometric estimates still provide a first estimate of the energetic demand and workload of the heart.

Comparative cardiovascular physiologists have provided many such criteria based on the above-mentioned approach, e.g., Stahl [80, 81] and Gunther [21]. For example, if we plot the aortic diameter to the length of the aorta for several mammals, the ratio of aortic length to its diameter is a constant. This ratio is easily calculated to be about 36.5 from the allometric equation of the aortic diameter,

$$D = 0.48\, W^{0.34} \quad (14.8)$$

and that for the length of the aorta [41, 45],

$$L_a = 17.5\, W^{0.31} \quad (14.9)$$

It should be noted here that the length of the aorta is seen to differ in man and woman and therefore so do the sizes of their respective aorta and the cross-sectional area. These differences undoubtedly lead to differences in aortic flow and stroke volume (SV).

Variations in the empirical values of a and b have been found to be dependent on sample size, range of body weight differences, and methods of measurement of the physical parameter, Y. The use of allometry by itself for comparative physiological interpretation has been subjected to criticism. For instance, one can easily obtain dimensionless numbers from allometric ratios with a zero exponent without any physiological significance. The ratio of a man's aortic length to the length of a woman's femoral artery can be a constant with zero exponent and dimensionless, but it does not bear any physiological relevance. Clearly, allometry must be used in conjunction with both physical and physiological relevance.

Despite the shortcomings, the use of allometric equations has its fundamental place in comparative physiology. Below we discuss the extent of the usefulness of simple allometry and its limitations when applied to the function of the cardiovascular system.

The Problem of Relative Growth in Relation to the Cardiovascular System

Proportional Growth

In his work, "On Growth and Form," D'Arcy Thompson [84] stressed that all organic forms are the result of *differential growth*. This includes general growth, which may be quantitatively different in space, or localized growth at any particular site. For instance, the total number of newly generated sarcomeres and the angiogenesis of new blood vessels occur at different locations and at different rates. The subject of differential growth received little consideration until the work of Huxley, who studied certain phases of "The Problems of Relative Growth" ([26] also the title of his book) and demonstrated the existence of empirical laws that appear to govern differential growth.

Huxley first studied known cases of differential growth and established a general law for such a growth. We now know that the establishment of one quantitative rule leads to the discovery of others. He postulated a hypothesis that the ratio between the intensity (or relative rate) of growth of the organ and that of the body remains constant over long periods of the animal's life. In terms of the cardiovascular system, heart weight in relation to body weight quickly comes to mind:

$$W_{\mathrm{h}} = aW^{\mathrm{b}} \tag{14.10}$$

because if $b = 1$, then there is proportional growth, and if b differs greatly from 1, then there is obviously a differential growth. Experimentally, the exponent b is found to be very close to 1, suggesting that the heart grows in proportion with increasing body weight. In pathological conditions, we see that both a and b can deviate substantially from their "normal" values. These a and b coefficients can also differ in man and woman and between infant, adult, and elderly.

Differential Growth and Growth in Time

The usefulness and obvious limitation of the simple allometric formula is that it does not explicitly address time dependence or effect of age. Thus, one can ignore the time relations of growth by relating the sizes of the blood vessel or heart size to body size. Thus, time or, specifically, aging dependence is neglected. Many equations exist for describing growth in time, t, e.g., exponential, linear, or constant rate, respectively,

$$Y = e^{at} \tag{14.11}$$

$$Y = at \tag{14.12}$$

or

$$\mathrm{d}Y/\mathrm{d}t = a \tag{14.13}$$

where a is a constant.

Deviations from Simple Allometry

One of the most serious difficulties, and often a neglected one, in studies of differential growth is how to decide whether the growth trend of the data is adequately represented by a straight line on a log–log or a double-logarithmic graph. Extrapolation of data points to extremes of body weight differences would also, in general, introduce large errors [92]. As an example, when the ratio of heart weight as a function of body weight measured in large mammals weighing more than 100 kg is used to estimate the similar ratio for a very small mammal weighing less than 10 g, such as a shrew, error can be considerable. Thus, extrapolating a male adult of 100 kg to his childhood body size as an infant can produce considerable error.

If different parts of the organ—such as the left anterior descending coronary artery or the smooth muscle cell in the tunica media of an artery—can show unequal constant differential growth ratios against body size, then the whole organ cannot exactly obey the allometry law against the same standard and vice versa. Thus, one can re-calibrate and re-formulate alternative allometric equations that can afford a better quantitative description than simple allometry. For instance, the DuBois formula is widely accepted for the calculation of BSA:

$$\mathrm{BSA} = a_1W^{\mathrm{b}_1} + a_2H^{\mathrm{b}_2} \tag{14.14}$$

Specifically, this is explicitly given as:

$$\mathrm{BSA} = W^{0.425}H^{0.725} \cdot 71.84 \tag{14.15}$$

which is dependent on both body weight W (kg) and height H (cm). This seems to suggest the relevance of cylindricality [88] for the form of *Homo sapiens*. The geometry of cylindricality is well represented in the cardiovascular system, e.g., arterial segment. It should be noted here that the BSA calculation previously given differs greatly from the simple allometry or the two-thirds power law,

$$\text{BSA} = aW^{2/3} \tag{14.16}$$

Hemodynamic Analysis Through Simple Allometry

Blood Pressure and Flow

Geometric similarities can describe many common anatomic structures in which the results of kinematic similarities are related to those of dynamic similitudes. For instance, the heart generates blood pressure and flow waveforms, which are transmitted to the arterial system. These waveforms should possess a certain similarity, and their amplitudes should be proportionally scaled. Indeed, aortic blood pressure waveforms in mammals exhibit striking similarity, e.g., they are of the same magnitude (*a* and *b* are constant for different mammals). However, we know that children have lower blood pressure and flow than adults and that women, in general, have lower blood pressure from childhood until adolescence compared with men, but the elderly have greater incidence of systolic hypertension. Blood flow waveforms are of the same shape across mammalian species, but they differ between the sexes as we shall discuss later in the text.

For two fluid mechanical systems to function similarly, both geometric and kinematic similarity must be possessed for the systems to be dynamically similar. This is where dimensional analysis and the Pi theorem come into play for the establishment of similarity criteria as we shall discuss now.

The close resemblance of fluid motion and blood flow has allowed hydrodynamic principles to be extrapolated to hemodynamics. Similarity criteria that were well established in hydrodynamics have been applied to arterial blood flow. For instance, the Reynolds number is essential for identifying viscous similitude and laminar-to-turbulent flow transitions.

Employing Huxley's allometric equation and Buckingham's Pi theorem, Gunther and DeLa Barra [22] deduced a dimensionless number for the mammalian circulation relating mean arterial pressure, mean blood velocity in the aorta, duration of cardiac cycle, basal oxygen consumption, total peripheral resistance, and total blood volume. Although the number is relatively constant, irrespective of mammalian body size, the parameters are often interdependent.

We have seen that heart weight in an adult mammal bears a constant proportion to its body weight. This constant fraction seems to hold for a number of mammals and for man and woman alike. The amount of blood ejected during each contraction of the heart, or the SV, also bears a constant proportion to body weight. This is because, grossly speaking, mammalian hearts are similar in general anatomical structure and in overall physiological function. Thus, for a given size of the heart, its SV is dictated by its heart size as can be seen from the simple allometric equations:

$$W_{\text{h}}(\text{LV}) = 2.61\ W^{1.10}\text{g} \tag{14.17}$$

$$\text{SV} = 0.74\ W^{1.03}\ \text{mL} \tag{14.18}$$

SV is the integral of flow during the ejection period,

$$\text{SV} = \int Q(t)dt \tag{14.19}$$

where $Q(t)$ is the instantaneous blood flow measured in the ascending aorta. Equivalently, this flow is the rate of change of the left ventricular volume,

$$Q(t) = \text{d}V(\text{t})/\text{d}t \tag{14.20}$$

SV has been considered one of the most important hemodynamic quantities in assessing ventricular function. Together with blood pressure, its magnitude bears a direct relation to the energy expenditure of the heart (the external mechanical work [EW] or the work performed by the heart to overcome the arterial load during ejection), which is given by:

$$\mathrm{EW} \sim \mathrm{p} \cdot \mathrm{SV} \tag{14.21}$$

This is seen as being equivalent to the area under the left-ventricular pressure-volume loop.

Aortic blood pressures in mammals are invariant of their body weights [25]. In allometric form, the mean arterial pressure can be expressed as:

$$p = 1.17\ 10^5\ W^{.033}\,\mathrm{dynes/cm^2} \tag{14.22}$$

or

$$p = 88\ W^{.033}\,\mathrm{mmHg}$$

The left ventricular EW can then be easily computed from the product of the mean arterial pressure and SV,

$$\begin{aligned}\mathrm{EW} &= 0.87 \times 10^5\ W^{1.063}\ \mathrm{ergs}\\ &= 0.0087\ W^{1.03}\ \mathrm{J}\end{aligned} \tag{14.23}$$

Therefore, in healthy mammals, a larger ventricle will generate a greater amount of EW simply because of its larger heart size [36, 37, 57]. This allometric relation also clearly demonstrates the difference in SV and EW between men and women merely because of their respective body weight and heart size.

Cardiac output, a familiar parameter used in physiology and medicine to assess cardiac function, is defined as the amount of blood ejected per minute and equivalently,

$$\mathrm{CO} = \mathrm{SV} \cdot \mathrm{f_h} \tag{14.24}$$

or

$$\begin{aligned}\mathrm{CO} &= \left(0.74\ W^{1.03}\right)\left(4.02\ W^{-0.25}\right) 60/1000\\ &= .178\ W^{0.78}\ \mathrm{L/min}\end{aligned} \tag{14.25}$$

because heart rate is given by

$$\mathrm{f_h} = 4.02\ W^{-0.25}\ \mathrm{s^{-1}} \tag{14.26}$$

These formulae can be easily applied to the estimation of heart rate and cardiac output in man and woman and from infancy to adulthood.

Peripheral Resistance to Flow

The total peripheral resistance for a mammalian systemic arterial tree can be calculated from the following relation:

$$\begin{aligned}R_s &= p/\mathrm{CO}\\ &= 2.8 \times 10^6\ W^{-.747}\mathrm{dyn.s.cm^{-5}}\end{aligned} \tag{14.27}$$

Thus, peripheral resistance follows the −3/4 power law and, incidentally, is inversely proportional to the metabolic rate (+3/4).

One can see that with simple manipulations of allometric formulae, a new and useful set of allometric equations can be produced. In the present example, we could easily obtain a good estimate of the peripheral resistance of the systemic circulation if the body weight is known. This is by far easier than carrying out experimental measurements to obtain this value, which is a difficult task in some situations. The derived allometric equation compares favorably with those reported by, e.g., Gunther and Guerra [24], who formulated the following:

$$R_s = 3.35\ 10^6\ W^{-.68}\ \mathrm{dynes\ s\ cm^{-5}} \tag{14.28}$$

Vascular input impedance, which has been used to represent the afterload to the heart, can also be readily obtained once frequency domain harmonics for pulsatile pressure and flow are secured [45, 47, 48].

Dimensional Analysis and the Pi Theorem

Mass, Length, and Time System and Buckingham's Pi Theorem

Of allometric calculations, the mass (M), length (L), and time (T) representation of a physical variable or parameter—or the so-called MLT system—is the most common. The physical quantities differ from physical constants. The former always possess units, whereas the latter are not always dimensionless (e.g., Planck's constant). The use of Buckingham's Pi theorem for dimensional analysis requires all physical quantities be expressed in MLT. This theorem has wide applications as shown below.

Dimensional homogeneity, another requirement to use the Pi theorem, was first proposed by Fourier in 1882, who stated that any equation applied to physical phenomena or involving physical measurements must be dimensionally homogeneous. Its usefulness can be found in the Navier-Stokes equations describing incompressible fluid flow in the longitudinal direction, e.g., in a blood vessel (in cylindrical coordinates). Every term in the equation has the dimension of a pressure gradient for flow in the z direction: $\mathrm{d}p/\mathrm{d}z$, or $M^1L^{-2}T^{-2}$.

Numerous dimensionless numbers have found their way using the dimensional matrix. The matrix comprises columns representing physical quantities, and rows are filled with basic units (M, L, T). To form a dimensional matrix, a priori knowledge of pertinent physical parameters is necessary. For instance, if 8 physical quantities are important for the description of blood flow in arteries, and there are 3 basic units (MLT) to represent them, then we are able to obtain 8–3 = 5 dimensionless Pi numbers. In general, the number of dimensionless Pi numbers is determined by the number of physical quantities minus the rank of the dimensional matrix.

To use the MLT system, one must first express explicitly any variable in its physical units using either the CGS (cm, g, s) or MKS (m, Kg, s) system or using SI units of representation. For instance, blood pressure is commonly measured in mmHg and must be converted to g/cm s^2. Thus, pressure (p) is given as force (F) per unit area (A),

$$p = \frac{F}{A} = [M][L]^{-1}[T]^{-2} \tag{14.29}$$

where A has the dimension of cm^2, or $[L]^2$, and force is mass times acceleration, Newton's second law,

$$F = m \cdot a = [M][L]/[T]^2 \tag{14.30}$$

where a is the acceleration in cm/s^2 ($[L]/[T]^2$). The left ventricular volume V has the unit of mL or cm^3 and a dimension of

$$V = [L]^3 \tag{14.31}$$

The aortic flow, Q, represented by the rate of change of ventricular volume, has the unit of mL/s or

$$Q = \frac{\mathrm{d}V}{\mathrm{d}t} = [L]^3/[T] \tag{14.32}$$

Linear flow velocity in a parabolic velocity profile in a femoral artery, or a blunt velocity profile in the ascending aorta, has the dimension of

$$v = \frac{\mathrm{d}z}{\mathrm{d}t} = [L]/[T] \tag{14.33}$$

or with a physical unit of cm/sec; z lies along the axis of blood flow direction.

Heart rate shown previously, in beats per second, has the dimension of

$$f_h = [T]^{-1} \tag{14.34}$$

which presents slight differences between the sexes and at different ages.

Dimensional Matrix for Hemodynamic Analysis

When formulating a dimensional matrix, it is necessary to identify the parameters considered pertinent to the problem at hand. These parameters must be expressed in terms of [M], [L], and [T]. For example, given arterial blood pressure (p), flow (Q), and heart rate (f_h), a dimensional matrix can be formed, thus:

$$\begin{array}{cccc} & p & Q & f_h \\ M & 1 & 0 & 0 \\ L & -1 & 3 & 0 \\ T & -2 & -1 & -1 \end{array} \tag{14.35}$$

This, therefore, is a 3 × 3 matrix or a square matrix.

As another example, the relationship between left ventricular wall tension (T) and left ventricular radius (r) and left ventricular pressure (Laplace law [38, 58, 71]), one can form a dimensional matrix in terms of the three parameters before the application of Buckingham's Pi theorem. This dimensional matrix is as follows:

$$\begin{array}{cccc} & T & p & r \\ M & 1 & 1 & 0 \\ L & 0 & -1 & 1 \\ T & -2 & -2 & 0 \end{array} \qquad (14.36)$$

Again, this is a 3 × 3 square matrix.

Allometry in Rheology and Vascular Mechanobiology

Dimensionless numbers provide useful scaling laws, particularly in multi-scale modeling and similarity transformation. Dimensional analysis is a powerful tool, but it is not limited to just mathematics, physics, and modeling; it has immense applicability to many biological phenomena [47].

Similarity criteria can be obtained in the analysis of blood flow. We can first form a dimensional matrix by incorporating parameters that are pertinent to the analysis. These are the fluid density (ρ) and viscosity (η), the diameter (D) of the blood vessel, and the velocities of the flowing blood (v) and the pulse wave (c). In terms of the dimensioning mass (M), length (L), and time (T) system, we can write down the following dimensional matrix:

$$\begin{array}{cccccc} & \rho & c & D & \eta & v \\ & (\mathrm{g/cm^3}) & (\mathrm{cm/s}) & (\mathrm{cm}) & (\mathrm{poise}) & (\mathrm{cm/s}) \\ M & 1 & 0 & 0 & 1 & 0 \\ L & -3 & 1 & 1 & -1 & 1 \\ T & 0 & -1 & 0 & -1 & -1 \\ & k_1 & k_2 & k_3 & k_4 & k_5 \end{array} \qquad (14.37)$$

where k_ns are Rayleigh indices referring to the exponents of the parameters. According to Buckingham's Pi theorem [36, 37, 40], two dimensionless Pi numbers (5–3 = 2) can be deduced.

Mathematically, we have the following,

$$\pi_i = \rho^{k1}\, c^{k2}\, D^{k3} \eta^{k4} v^{k5} \qquad (14.38)$$

or, in terms of MLT, then

$$\pi_i = \left(M^{k1}L^{-3k1}T^0\right)\left(M^0L^{k2}T^{-k2}\right)\left(M^0L^{k3}T^0\right) \left(M^0M^{k4}L^{-k4}T^{-k4}\right)\left(L^{k5}T^{-k5}\right) \qquad (14.39)$$

Because Pi numbers are dimensionless, the exponent must be zero. Equating the exponents of MLT to zero and solving, we obtain two Pi numbers or similarity criteria [36, 37]:

$$\pi_1 = \frac{\rho v D}{\eta} = \mathrm{Re} \text{ and } \pi_2 = \frac{c}{v} = \frac{1}{Ma} \qquad (14.40)$$

The first Pi number, or $\pi 1$, is clearly identifiable as the Reynolds number, Re. The second π_2 is the inverse of the Mach number, *Ma*. The *Ma* is equivalently seen as the ratio of flow velocity to pulse wave velocity (PWV) in terms of blood pulse wave propagation [52]. It is also termed the "velocity fluctuation ratio" (VFR). Recalling that to assume linearity of the arterial system, the flow velocity should be small compared with the PWV ($v << c$ [or PWV]) or the VFR should be small.

The requirements for dynamic similarity [72] are that two flows must possess both geometric and kinematic similarity. Thus, the effects of, for instance, viscous forces, pressure forces, and surface tension [45], must be considered. Here we have only the ratio of inertial forces to viscous forces, i.e., the Reynolds number, and the ratio of inertial forces to compressibility forces, i.e., the Mach number or the VFR. For a truly incompressible fluid, $c >> v$ such that $Ma = 0$. For the analysis of blood flow in arteries, both blood and arterial walls are normally assumed to be incompressible. The Poisson ratio (σ_p) for the aorta, which is the ratio of radial strain to longitudinal strain, is approximately 0.48, close to that of an incompressible material ($\sigma_p = 0.5$). As mentioned previously, the assumptions of linearity and linear-system analysis applied to hemodynamic studies often require the ratio $v/c << 1$ (flow velocity to PWV) or that the diameter of the blood vessel is small compared with the pulse propagation wavelength, $D/\lambda << 1$. This is justified during the large part of the cardiac cycle. At peak flow rates in early systole, however, the ratio of v/c is large (but not exceeding 1), so turbulence may ensue to produce nonlinear effects.

The Reynolds number varies with body length dimensions or the diameter of the aorta. One question that immediately arises concerns the resulting Reynolds numbers calculated for large mammals, such as the horse, showing that turbulence may occur for a large portion of the systole in the aorta. However, this may not necessarily be the case. It has been well documented that turbulence may not exist even for a Reynolds number greatly exceeding the critical value of 2000. It is only established that for a Reynolds number below 2000, turbulence does not normally occur. Again, this value was established under steady-flow conditions in rigid tubes. This differs from pulsatile flow in elastic arteries [42].

Arterial blood flow exhibits pulsatile characteristics, and peripheral outflow occurs mostly in diastole. In systole during ventricular ejection, the aorta distends as a reservoir to accommodate the flow as described by the classic Windkessel model of the arterial system [50]. In concert with the pulsation, this compliance of the aorta acts to protect the peripheral vascular beds from sudden surges in pressure and flow. The compliance, defined as the ratio of change in volume due to a change in pressure,

$$C = \mathrm{d}V/\mathrm{d}P \tag{14.41}$$

is proportional to body weight because pressure is practically an invariant of body weight. A larger volume change occurs in the aorta of a larger mammal, and the longer effective length of the aorta and a much slower heart rate both help to decrease the tendency of turbulence to reside in too large a portion of systole. Because the aorta contributes to a main portion of the total arterial compliance, C has been approximated in the clinical setting by the ratio of SV to pulse pressure (PP):

$$C = \mathrm{SV}/\mathrm{PP} \tag{14.42}$$

This simple relation has considerable clinical implications. Women with a smaller SV and a greater PP at age more than 55 years will likely incur a smaller total arterial compliance or an increased overall vascular stiffness than the corresponding male counterparts. The following section examines the sex differences more closely.

Allometric Analysis of Sex Differences in Cardiovascular Structure and Function Regarding Aging and Hypertension

Aging-induced alterations in the mechanical properties of blood vessel walls result in significantly increased vascular stiffness [47, 65]. These changes contribute in part to the prevalence of isolated systolic hypertension (ISH) in 8%–15% of people above 60 years old. Findings from several major clinical trials—namely, the Systolic Hypertension in the Elderly Program (SHEP) in 1991, the Systolic Hypertension in Europe (Syst-Eur) trial in 1997, and the Systolic Hypertension in China (Syst-China) trial in 1998—have shown that antihypertensive drug treatment must be prescribed for elderly patients with repeated systolic blood pressure measurement of 160 mmHg or higher [16, 34, 79], in other words, the elderly ISH group. The younger systolic hypertensive patients tend to be more related to sympathetic over-stimulation [2]. The underlying improvement in hemodynamics and mechanical properties with antihypertensive therapy still needs further understanding and quantification to improve the patient survival rate.

With our newly developed allometric hemodynamic model, we have shown that a greater increment in systolic pressure, along with the progression of aging, is a consequence of a comparatively much greater percentage reduction in compliance than an increase in peripheral resistance. This model provides an insight into the hemodynamic prevalence of the occurrence of ISH in the elderly. It should be made clear that natural aging does not necessarily lead to ISH, but the tendency of a proportionately larger increase in systolic pressure and a slight decrease in diastolic pressure inadvertently also increases the PP, thus making the elderly more susceptible to develop ISH. This aspect is well illustrated in Fig. 14.1. This model can be further expanded to include pressure-dependent properties of arterial compliance [46, 51, 54] and to account for male and female differences.

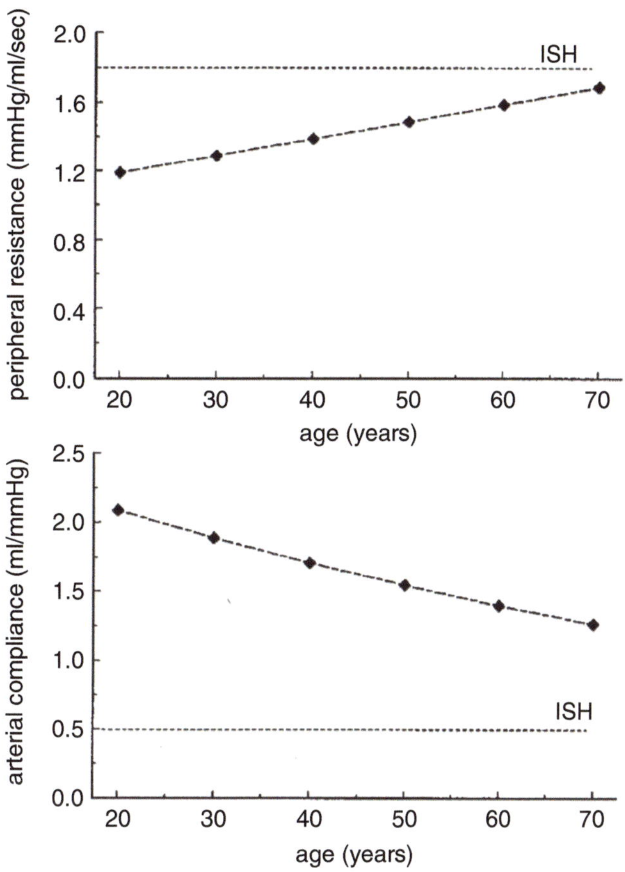

Fig. 14.1 A concurrent increase in peripheral resistance and decrease in compliance are necessary to develop ISH as shown. With advancing age, progressively increased peripheral resistance and decreased compliance make the elderly more prone to ISH.

The study of pulse waveform contours is important because of their relevance to the differential diagnosis of many forms of cardiovascular diseases. Their alterations are closely related to the mechanical properties of the vessel wall and vascular states and are linked to hypertension and atherosclerosis. In hypertension, for instance, increased vascular stiffness is always associated with an increased PP and systolic pressure. Increased wave reflections impede cardiac ejection and are detrimental to normal left ventricular function. This increased wave reflection can occur because of changing vascular bed characteristics or a modification of conduit arterial wall properties [3, 43]. The allometric hemodynamic model presented below would demonstrate that pulse wave reflection is significantly greater and arrives earlier in systole in an ISH subject compared with age-matched normal controls (Fig. 14.2). Increased wave reflection, particularly in systole, is a direct consequence of mismatching in proximal and distal vascular properties, which is attributable to large changes in vascular elastic stiffness or arterial compliance.

Parallel changes in increased stiffness of the heart and arteries have been recognized in several studies in both men and women [7, 68]. Systolic hypertension associated systolic and diastolic ventricular dysfunction are also common in men and women [93]. Understanding the dynamic interaction of the heart and the arterial system in patients with systolic hypertension is therefore important. Our previous investigations for analyzing such an interaction in men and women [31, 44, 49, 55] point to the importance of this direction.

We have established an allometric hemodynamic model for studying and comparing aging-related hemodynamic changes and systolic hypertension [56]. The model can explain and

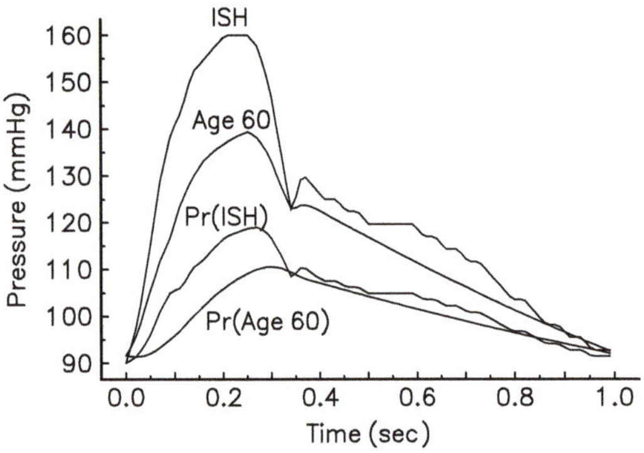

Fig. 14.2 Blood pressure waveforms in a 60-year-old healthy subject and an ISH patient are shown. The reflected wave is much more pronounced in the ISH patient (Pr(ISH)) than in the 60-year-old healthy subject (Pr (Age60)). The reflected wave arrives earlier in systole and with much greater amplitude in the ISH patient.

differentiate aging-induced vascular changes and ISH in terms of hemodynamics. Beyond normal aging, changes in compliance and peripheral resistance and significantly increased wave reflections account for the development of ISH. Therefore, therapeutic treatment must be aimed at decreasing wave reflections as well as improving vascular compliance.

The allometric hemodynamic model is based on a modified Windkessel model, which is popularly used in the clinical setting. The model consists of a characteristic impedance of the proximal aorta (Zo, [39]), arterial compliance (C), and peripheral resistance (Rs). The model is used to predict blood pressure waveform, and arterial compliance and peripheral resistance values, at the early adult ages of 20 years old and above. Blood pressure waveform is predicted with the measured normal aortic blood flow as input because it is known that SV and ejection fraction is preserved in systolic hypertension patients compared with normal subjects [2]. Arterial compliance and resistance were calculated for each decade beginning at adulthood at 20 years through 70 years based on the following allometric relations:

$$R_s = R_{so} + a \cdot (age - 20)\text{mmHg/mL/s} \quad (14.43)$$

$$C = C_o \exp(-b \cdot (age - 20))\text{mL/mmHg} \quad (14.44)$$

where Rso = 1.18 mmHg/mL/s and Co = 1.80 mL/mmHg are peripheral resistance and arterial compliance, respectively, for early adulthood at 20 years old. The negative exponent ($-b = -0.01$) in the compliance expression corresponds to a decrease in compliance with increasing age. Each change of decade in peripheral resistance follows a linear trend with $a = 0.01$ (Fig. 14.1).

Sex Differences in Vascular Geometry and Associated Risk Factors

Geometric Taper

Arterial diameters and lumen areas of the vascular tree can be determined from different imaging modalities, such as angiography, CT scan, ultrasound imaging, magnetic resonance imaging, or implanted sonomicrometers. For experimental studies, arteries in man and in dog retract some 25%–40% when removed. It is therefore necessary that in vivo lengths are restored and corresponding pressures set for mechanical measurements. Under normal conditions, greater distending pressure leads to greater lumen diameter seen on pressure-diameter recordings. Arterial vessel dimensions have been reported for dog [60] and for man [91]. The latter were used for constructing the analog model of the human systemic arterial tree.

"Geometrical taper" (Fig. 14.3) refers to a single continuous conduit, such as the aorta. This is a modification of the cylindrical tubular biological structure proposed by Wainwright [88]. The area change of the aortic cross section

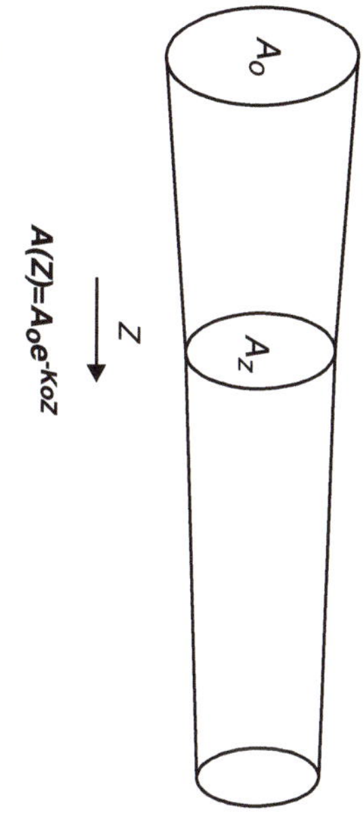

Fig. 14.3 Schematic diagram illustrating a blood vessel with geometric taper. The vessel diameter narrows with increasing distance (z) away from the origin. Geometric taper, an exponential function of distance, is normally calculated from the change in cross-sectional areas (A) as shown.

is close to an exponential form and can be expressed as:

$$A(z) = A(0)e^{-kz/r} \tag{14.45}$$

where

z = distance in the longitudinal axial direction along the vessel

r = vessel lumen radius (in cm)

k = taper factor, dimensionless

A(0) = cross-sectional area at the entrance of the vessel (in cm^2)

A(z) = cross-sectional area at distance z along the vessel (in cm^2)

The vessel area is calculated, assuming a circular cross section, as:

$$A = \pi r^2 \tag{14.46}$$

The taper factor, k, can be readily obtained as follows:

$$k = \frac{r}{z} \ln \frac{A(o)}{A(z)} \tag{14.47}$$

The taper factor, *k*, for the aorta has been reported to be in the range of 0.0314–0.0367 for 20–30-kg dogs [47]. This can be extrapolated for the human. The geometric taper factor can change substantially under varied vasoactive conditions and in disease conditions, such as atherosclerosis, stenosis, or aneurysm. When vasoactive drugs are administered that have differential effects on large and small arteries, changes in taper factors from normal can be rather pronounced.

Table 14.1 Measured external diameters and calculated taper factors in different arteries. A lower taper factor indicates more uniform longitudinal geometry

	d (cm)	k_o (cm^{-1})
Abdominal aorta	0.777	0.027 ± 0.007
Iliac artery	0.413	0.021 ± 0.005
Femoral artery	0.342	0.018 ± 0.007
Carotid artery	0.378	0.008 ± 0.004

An alternative formula to calculate taper factor per unit length, or k_o, is expressed as follows:

$$A(z) = A(0)e^{-k_o z} \tag{14.48}$$

The reported values of k_o obtained for the abdominal aorta, the iliac, and the femoral and carotid arteries are listed in Table 14.1. These are experimentally measured in vivo at a mean arterial pressure of approximately 90 mmHg in accordance with the allometric equation for mean blood pressure in mammals and in humans. The average body weight of dogs used is approximately 20 kg. It is obvious from these data that the taper factor is smaller for smaller vessels. Carotid arteries exhibit the least geometric taper. They are thus the best approximation to a geometrically uniform cylindrical vessel.

It is well known that women generally have smaller stature and thus shorter arterial length and diameter. The coronary arteries also exhibit branching and geometric tapering phenomena. For instance, in radiographic studies, Dodge et al. [13] showed in adults with normal heart size a global ejection fraction (>0.52) and end-diastolic volume (50–120 mL/m^2), and women had a significantly smaller epicardial artery diameter. Diameters in men increased and were significantly affected by left ventricular hypertrophy or dilated cardiomyopathy.

More recent studies with intravascular ultrasound also revealed sex differences [75] and the fact that women have smaller left main and left anterior descending coronary artery diameters. This finding was independent of body size. Kim et al. [32] showed, also with intravascular ultrasound, that the left main coronary arteries and lumen areas were smaller in women than in men. Luminal and external elastic membrane cross-sectional areas were also significantly smaller in women than in men. Both BSA and sex are independent predictors of the outcome. Allometrically speaking, they also concluded that body size thus has a greater influence than sex. Others have found that sex is an independent predictor of clinical outcomes after coronary revascularization (e.g., [67, 89]). The detriments seemed to be biased toward women because this revascularization patients population is generally smaller and older, and their situation is compounded with hypertension and diabetes.

Vascular Branching Effects

A pulse originating at the root of the aorta encounters several branching junctions before reaching different vascular beds. In relation to this, the number of generations of blood vessels is of important consideration in terms of blood flow. This can be found in the studies by Green [20] and Iberall [28]. Fractal analysis of vascular tree structures uses much of this information.

It has been found that alteration of pressure and flow through vascular junctions is more significantly affected by geometry than by elastic factors [40]. To this end, area ratios of daughter vessels to a mother vessel have been considered pertinent in governing blood pressure pulse transmission through vascular branching. Area ratios calculated for vascular branching junctions were approximately 1.08 at the aortic arch and 1.05 at the aorto-iliac junction [53]. These values are slightly larger than 1.0.

Karreman [30] used area ratio in his mathematical formulation of wave reflection at an arterial junction. By assuming that both the wall and fluid are non-viscous, and that wall thickness remains the same for an infinitely long tube, he arrived at a value of area ratio of 1.15 when the wave reflection is minimal for a bifurcation, such as at the bifurcation of abdominal aorta and left and right iliac arteries.

Geometric factors and plaque formation at vascular bifurcation depend significantly on local vessel properties as shown by Li [40]. Schulz and Rothwell [74] studied carotid artery bifurcations in terms of distribution of atherosclerotic plaque and stroke and found that the ratio of the internal to the common or external carotid artery and that of the outflow to the inflow area were greater in women than men. They concluded that women are more likely to have external common carotid artery stenosis. The area ratio concept has been studied extensively in the past (e.g., [19, 30, 40, 53]) because the branching daughter vessel's geometry and elastic properties can have a significant impact on the mother vessel in terms of outflow and wave reflections [53]. Fluid dynamic aspects (e.g., [14, 94]) also have indicated impact of regions of high shear stress and flow velocity ratio.

Hemodynamics at vascular branching junction has intrigued many researchers and clinicians. We have shown that blood pressure pulse wave reflection (Γv) at vascular branching junction is normally minimal for the forward wave [53]. When Γv is plotted against changes in elastic modulus (ΔE), as shown in Fig. 14.4, changes in vessel geometry produce a greater change in wave reflection at the vascular branching than corresponding percentage changes in elastic

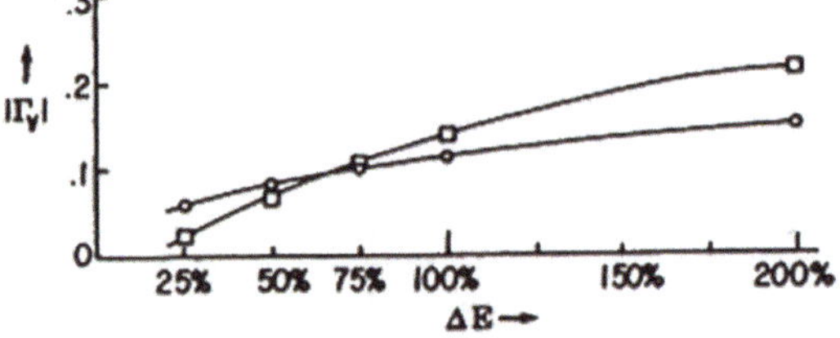

Fig. 14.4 Blood pressure pulse wave reflection (Γ_v) at vascular branching junction is plotted against changes in elastic modulus (ΔE), showing that geometric factors (square) dominate over elastic factors (circle)

modulus. In other words, reflection is increased with increasing branching vessel stiffness, but this increase is less pronounced when compared with the corresponding percentage reduction in branching vessel lumen radius (Fig. 14.4).

Hemodynamic Differences Between the Sexes in Stenosis and Atherosclerosis

Numerous modeling and experimental studies have been proposed to investigate the fluid mechanical factors contributing to atherosclerosis. Caro et al. [6] were earlier investigators to identify sites of atherogenesis as regions of decreased wall shear stress [5] and suggested that the transport of lipoprotein within the arterial wall and across the endothelium is a major factor in atherosclerosis. These common atherosclerotic sites have been illustrated by DeBakey et al. [11]. Thurbrikar and Robicsec [85] suggested the importance of pressure-induced arterial wall stress as an important factor in atherosclerosis.

Arterial stenosis, or the narrowing of the blood vessel, is associated with a serious hemodynamic consequence of pressure loss that develops across the stenosis. The pressure loss is primarily dependent on the flow rate and the geometry of the stenosis because the fluid properties of density and apparent viscosity are relatively constant. The fluid mechanics aspect of stenosis has been well studied because of its importance in the coronary arteries (e.g., [59]).

The close relation of atherosclerosis and stenosis and their morphological resemblance have indeed generated much interest in analyzing common fluid mechanical factors and consequences. In many instances, particularly in modeling studies, these two pathological conditions cannot be separated. Their importance in the manifestation of eventual diseases of the vascular system, however, is well recognized. Tarbell et al. [83] provided a comprehensive review of the relationships of fluid mechanics to arterial diseases and the role of gene expression.

In relation to allometry, several epidemiology and clinical studies have linked body height and mass to cardiovascular disease in both sexes (e.g., [1, 15, 18, 66]) based on the Framingham Study (e.g., [29]) and in different countries and in women (e.g., [69, 76, 87]). Based on the results of these previous investigations and those of comparative cardiovascular studies [45, 64, 90], Smulyan et al. [77] looked at body height effects on arterial hemodynamics in normal and end-stage renal disease (ESRD) patient groups and found that ESRD patients had shorter stature and greater body weight, faster carotid-to-femoral PWV, and augmentation index for both sexes. In general, body height tends to play an important role in terms of cardiovascular risk, particularly with advancing age. This included decreased pulmonary function, stroke, coronary artery diameters, and increased vascular stiffness in terms of shorter diastole and arrival time of reflected waves.

Vouyouka and Kent [86] reviewed sex-specific risk factors for arterial vascular disease in women and looked at three most prominent occurrences: carotid atherosclerosis, abdominal aortic aneurysm, and lower extremity occlusive disease.

Sonesson et al. [78] examined sex and age differences in compliance and diameter of the human aorta. They defined compliance, given previously in Eq. (14.41) as C = dV/dP, as the inverse of the pressure-strain elastic modulus expressed as PP obtained by auscultatory method to the corresponding fractional change in pulsatile diameter with respect to diastolic diameter (strictly speaking, this should have been the mean diameter):

$$Ep = \frac{\Delta P}{\left(\Delta D / D\right)} \qquad (14.49)$$

Pulsatile diameter is measured using an ultrasound echo tracking device. They found that the pressure-strain elastic modulus is greater in males than females for the abdominal aorta and that the differences increase significantly with age exponentially in males and linearly in females. This may be attributed to a comparatively lower blood pressure and PP in women.

Women have smaller diameters of the carotid and vertebral arteries [26, 73, 74], which are principal arteries perfusing the brain. In terms of augmentation index and wave reflections, women tend to have less compliance of these arteries [17]. The anatomic geometric differences contribute to somewhat greater outflow than inflow in women for the carotid arteries [8], as would be expected for carotid artery stenosis. This could be seen from Bernoulli's equation that there is faster velocity within the stenosis due to the greater pressure gradient.

Smoking, high cholesterol level, and age are common factors associated with carotid stenosis. However, in women [63], menopause, with the plaque being more stable as opposed to vulnerable, is associated with greater smooth muscle, which contributes to a more compliant tunica media and collagen, which in turn contributes to stiffer adventitia content than cholesterol content.

As discussed previously, women who undergo revascularization have more comorbidity, such as hypertension and diabetes, perhaps contributing to poorer outcome in coronary bypass surgery. This aspect is also noted in a more recent volumetric analysis with computed tomography by Dickerson et al. [12], but it was limited to only proximal larger coronary arteries. This indicates that factors other than diameter, e.g., tapering factor, arterial wall properties (see Chap. 19), underlying hemodynamics [2, 9, 47], and extravascular compression and collateral circulation, may play a role in the outcome, whereas sex differences have been little studied in terms of revascularization.

References

1. Allebeck P, Bergh C. Height, body mass index and mortality: do social factors explain the association? Public Health. 1992;106:375–82.
2. Berger DS, Li JK-J. Concurrent compliance reduction and increased peripheral resistance in the manifestation of isolated systolic hypertension. Am J Cardiol. 1990;65:67–71.
3. Berger D, Li JK-J, Laskey WK, Noordergraaf A. Repeated reflection of waves in the systemic arterial system. Am J Physiol (Heart & Circ Physiol). 1993;33(264):H269–81.
4. Calder WA III. Size, function and life history. New York: Dover; 1996.
5. Caro CG. Discovery of the role of wall shear in atherosclerosis. Arterioscler Thromb Vasc Biol. 2009;29:158–61.
6. Caro CG, Fitzgerald JM, Schroter RC. Atheroma and arterial wall shear observation, correlation and proposal of a shear dependent mass transfer mechanism for altherogenesis. Proc R Soc Lond B. 1971;177:109–33.
7. Chen C-H, Ting C-T, Lin S-J, Hsu T-L, S-J Ho PC, Chang M-S, O'Connor F, Spurgeon H, Lakatta E, Yin FCP. Which arterial and cardiac parameters best predict left ventricular mass? Circulation. 1998;98:422–8.
8. Comerota AJ, Salles-Cunha SX, Daoud Y, Jones L, Beebe HG. Gender differences in blood velocities across carotid stenosis. J Vasc Surg. 2004;40:939–44.
9. Coutinho T, Borlaug BA, Pellikka PA, Turner ST, Kullo IJ. Sex differences in arterial stiffness and ventricular-arterial interactions. J Am Coll Cardiol. 2013;61:96–103.
10. Dawson TH. Engineering design of the cardiovascular system of mammals. Englewood Cliffs: Prentice-Hall; 1991.
11. DeBakey ME, Lawrie GM, Glaeser DH. Patterns of atherosclerosis and their surgical significance. Ann Surg. 1985;201:115–31.
12. Dickerson JA, Nagaraja HN, Raman SV. Gender-related differences in coronary artery dimensions: a volumetric analysis. Clin Cardiol. 2010;33:E44–9.
13. Dodge JT Jr, Brown BG, Bolson EL, Dodge HT. Lumen diameter of normal human coronary arteries. Influence of age, sex, anatomic variation, and left ventricular hypertrophy or dilation. Circulation. 1992;86:232–46.
14. Fisher M, Fieman S. Geometric factors of the bifurcation in carotid atherogenesis. Stroke. 1990;21:267–71.
15. Fisher LD, Kennedy JW, Davis KB, et al. Association of sex, physical size, and operative mortality after coronary artery bypass in the Coronary Artery Surgery Study (CASS). J Thorac Cardiovasc Surg. 1982;84:334–41.
16. Gasowski J, Wang JG, Staessen JA. Clinical trials in isolated systolic hypertension. Curr Hypertens Rep. 10 1999; 1:387–393.
17. Gatzka CD, Cameron JD, Kingwell BA, Dart AM. Relation between coronary artery disease, aortic stiffness, and left ventricular structure in a population sample. Hypertension. 1998;32:575–8.
18. Gertler MM, Garn SM, White PD. Young candidates for coronary heart disease. JAMA. 1951;147:621–5.
19. Gosling RG, Dunbar G, King DH, Newman DL, Side CD, Woodcock JP, Fitzgerald DE, Keates JS, MacMillan D. The quantitative analysis of occlusive peripheral arterial disease by a non-intrusive ultrasonic technique. Angiology. 1971;22:52–5.
20. Green HD. Circulatory system: physical principles. In: Glasser O, editor. Medical physics 2. New York: Year Book Publishers; 1950.
21. Gunther B. Allometric ratios, invariant numbers and the theory of biological similarity. Physiol Rev. 1975;55:659.
22. Gunther B, DeLa Barra L. Physiometry of the mammalian circulatory system. Acta Physiol Lat-Am. 1966a;16:32.
23. Gunther B, DeLa Barra L. Theories of biological similarities, non-dimensional parameters and invariant numbers. Bull Math Biophys. 1966b;28:9–102.
24. Gunther B, Guerra B. Biological similarities. Acta Physiol Lat-Am. 1955;5:169.
25. Hales S. Statical essays containing haemostaticks. London: Innys and Manby; 1733.
26. Hansen F, Mangell P, Sonesson B, Länne T. Diameter and compliance in the human common carotid artery–Variations with age and sex. Ultrasound Med Biol. 1995;21(1):1–9.
27. Huxley, J.S. Problems of relative growth. Methuen, London, 1932.

28. Iberall AS. Anatomy and steady flow characteristics of the arterial system with an introduction of its pulsatile characteristics. Math Biosci. 1967;1:375–95.
29. Kannam JP, Levy D, Larson M, Wilson PWF. Short stature and risk for mortality and cardiovascular events: the Framingham Heart Study. Circulation. 1994;90:2241–7.
30. Karreman G. Some contributions to the mathematical biology of blood circulation. Reflections of pressure wave in the arterial system. Bull Math Biophys. 1952;14:327–50.
31. Kerkhof PLM, Heyndrickx GR, Li JK. Hemodynamic determinants and ventriculo-arterial coupling are sex-associated in heart failure patients. Conf Proc IEEE Eng Med Biol Soc. 2016 Aug; 2016:3286–3289. https://doi.org/10.1109/EMBC.2016.7591430. PMID: 28227219
32. Kim S, Apple S, GS Mintz TMM, DA Canos AM, Weissman NJ. The importance of gender on coronary artery size: in-vivo assessment by intravascular ultrasound. Clin Cardiol. 2004;27:291–4.
33. Kim H, Kim M, W Shim SO, Kim M, Park SM, Kim YH, Na JO, Shin MS, Yoon HJ, Shin GJ, Cho Y, Kim S, Hong K, Cho KI. Sex difference in the association between brachial pulse pressure and coronary artery disease: the Korean women's chest pain registry (KoROSE). J Clin Hypertens (Greenwich). 2017;19:38–44.
34. Kostis JB, Cabrera J, Cheng JQ, Cosgrove NM, Deng Y, Pressel SL, Davis BR. Association between chlorthalidone treatment of systolic hypertension and long-term survival. JAMA. 2011;306:2588–93.
35. Lambert R, Teissier G. Theorie de la similitude biologique. Ann Physiol Physiocochem Biol. 1927;3:212.
36. Li JK-JA. New similarity principle for cardiac energetics. Bull Math Biol. 1983a;45:1005–11.
37. Li JK-J. Hemodynamic significance of metabolic turnover rate. J Theor Biol. 1983b;103:333–8.
38. Li JK-J. Comparative cardiac mechanics: Laplace's law. J Theor Biol. 1986a;118:339–43.
39. Li JK-J. Time domain resolution of forward and reflected waves in the aorta. IEEE Trans Biomed Eng. 1986b;BME-33:783–5.
40. Li JK-J. Dominance of geometric over elastic factors in pulse transmission through arterial branching. Bull Math Biol. 1986c;48:97–103.
41. Li JK-J. Arterial system dynamics. New York: New York University Press; 1987.
42. Li JK-J. Laminar and turbulent flow in the mammalian aorta: Reynolds number. J Theor Biol. 1988;135:409–14.
43. Li JK-J. Increased arterial pulse wave reflections and pulsatile energy loss in acute hypertension. Angiol, J Vasc Dis. 1989;40:730–5.
44. Li JK-J. Feedback effects in heart-arterial system interaction. In: Sideman S, Beyar R, editors. Interactive phenomenon in the cardiac system. New York: Plenum; 1993. p. 325–33.
45. Li JK-J. Comparative cardiovascular dynamics of mammals. Boca Raton: CRC Press; 1996.
46. Li JK-J. A new description of arterial function: the compliance-pressure loop. Angiol, J Vasc Dis. 1998;49:543–8.
47. Li JK-J. The arterial circulation: physical principles and clinical application. Totowa: Humana Press; 2000.
48. Li JK-J. Dynamics of the vascular system. Singapore: World Scientific; 2004.
49. Li JK-J, Atlas G. Left ventricle–arterial system interaction in heart failure. Clin Med Insights: Cardiol. 2015;(Suppl):93–9. https://doi.org/10.4137/CMC.S18742.
50. Li JK-J, Noordergraaf A. Similar pressure pulse propagation and reflection characteristics in aortas of mammals. Am J Phys. 1991;261:R519–21.
51. Li JK-J, Zhu Y. Arterial compliance and its pressure-dependence in hypertension and vasodilation. Angiol J Vasc Dis. 1994;45:113–7.
52. Li JK-J, Melbin J, Riffle RA, Noordergraaf A. Pulse wave propagation. Circulation Res. 1981;49:442–52.
53. Li JK-J, Melbin J, Noordergraaf A. Directional disparity of pulse wave reflections in dog arteries. Am J Phys. 1984;247: H95–9.
54. Li JK-J, Cui T, Drzewiecki G. A nonlinear model of the arterial system incorporating a pressure-dependent compliance. IEEE Trans Biomed Eng. 1990;BME-37:673–8.
55. Li JK-J, Zhu Y, Nanna M. Computer modeling of the effects of aortic valve stenosis and arterial system afterload on left ventricular hypertrophy. Comput Biol Med. 1997;27:477–85.
56. Li JK-J, Zhu Y, O'Hara D, Khaw K. Allometric hemodynamic analysis of isolated systolic hypertension and aging. Cardiovasc Eng. 2007;7:135–9.
57. Loiselle DS, Gibbs CL. Species differences in cardiac energies. Am J Phys. 1979:490–8.
58. Martin RR, Haines H. Application of Laplace's law to mammalian hearts. Comp Biochem Physiol. 1970;34:959.
59. Mates RE, Gupta RL, Bell AC, Klocke FJ. Fluid dynamics of coronary artery stenosis. Circ Res. 1978;42:152–62.
60. McDonald DA. Blood flow in arteries, vol. 1960. London: Arnold; 1974.
61. McMahon TA. Size and shape in biology. Science. 1973;179:1201–4.
62. McMahon TA, Bonner JT. On size and life. New York: Scientific American Library; 1983.
63. Mehta LS, Beckie TM, DeVon HA, et al. Acute myocardial infarction in women: a scientific statement from the american heart association. Circulation. 2016;133:916–47.
64. Noordergraaf A, Li JK-J, Campbell KB. Mammalian hemodynamics: a new similarity principle. J Theor Biol. 1979;79:485.
65. O'Rourke MF, Hashimoto J. Mechanical factors in arterial aging: a clinical perspective. J Am Coll Cardiol. 2007;50:1–13.
66. Palmer JR, Rosenberg L, Shapiro S. Stature and the risk of myocardial infarction in women. Am J Epidemiol. 1990;132:27–32.
67. Peterson ED, Lansky AJ, Kramer J, Anstrom K, Lanzilotta MJ. Effect of gender on the outcomes of contemporary percutaneous coronary intervention. Am J Cardiol. 2001;88:359–64.
68. Redfield MM, Jacobsen SJ, Borlaug BA, Rodeheffer RJ, Kass DA. Age- and gender-related ventricular-vascular stiffening: a community-based study. Circulation. 2005;112:2254–62.
69. Rexrode KM, Hennekens CH, Willett WC, et al. A prospective study of body mass index, weight change, and risk of stroke in women. JAMA. 1997;277:1539–45.
70. Rich-Edwards JW, Manson JE, Stampfer MJ, et al. Height and the risk of cardiovascular disease in women. Am J Epidemiol. 1995;142:909–17.
71. Robard S, Williams F, Williams C. The spherical dynamics of the heart. Am Heart J. 1959;57:348–60.
72. Rosen R. Optimality principles in biology. London: Butterworth; 1967.
73. Scheel P, Ruge C, Schoning M. Flow velocity and flow volume measurements in the extracranial carotid and vertebral arteries in healthy adults: reference data and the effects of age. Ultrasound Med Biol. 2000;26:1261–6.
74. Schulz UG, Rothwell PM. Major variation in carotid bifurcation anatomy: a possible risk factor for plaque development? Stroke. 2001;32:2522–9.

75. Sheifer SE, Canes MR, Weinfurt KP, Arora UK, Mendekohn FO, Gersh BJ, Weissman NJ. Sex differences in coronary artery size assessed by intravascular ultrasound. Am Heart J. 2000;139:649–53.
76. Skurnick JH, Aladjem M, Aviv A. Sex differences in pulse pressure trends with age are cross-cultural. Hypertension. 2010;55:40–7.
77. Smulyan H, Marchais SJ, Pannier B, Guerin AP, Safar ME, London GM. Influence of body height on pulsatile arterial hemodynamic data. J Am Coll Cardiol. 1998;31:1103–9.
78. Sonesson B, Hansen F, Stale H, Lanne T. Compliance and diameter in the human abdominal aorta.- the influence of age and sex. Eur J Vasc Surg. 1993;7:690–7.
79. Staessen JA, Thijs L, Fagard R, O'Brien ET, Clement D, de Leeuw PW, Mancia G, Nachev C, Palatini P, Parati G, Tuomilehto J, Webster J. Predicting cardiovascular risk using conventional vs ambulatory blood pressure in older patients with systolic hypertension. Systolic hypertension in Europe Trial Investigators. JAMA. 1999;282:539–46.
80. Stahl WR. Similarity analysis of biological systems. Persp Biol Med. 1963a;6:291.
81. Stahl WR. The analysis of biological similarity. Adv Biol Med Phys. 1963b;9:356.
82. Stahl WR. Organ weights in primates and other mammals. Science. 1965;150:1039–42.
83. Tarbell JM, Shi Z-D, Dunn J, Hanjoong J. Fluid mechanics, arterial disease, and gene expression. Annu Rev Fluid Mech. 2014;46:591–614.
84. Thompson DW. On growth and form. Cambridge: Cambridge University Press; 1917.
85. Thubrikar MJ, Robicsec F. Pressure-induced arterial wall stress and atherosclerosis. Ann Thorac Surg. 1995;59:1594–603.
86. Vouyouka AG, Kent KC. Arterial vascular disease in women. J Vasc Surg. 2007;46:1295–302.
87. Waaler HT. Height, weight and mortality: the Norwegian experience. Acta Med Scand Suppl. 1984;679:1–56.
88. Wainwright SA. Axis and circumference. The cylindrical shape of plants and animals. Cambridge, MA: Harvard Univrsity Press; 1988.
89. Weintraube WS, Wenger NK, Kosinski AS, Douglas Jr JS, Liberman HA, Morris DC, King SB III. Percutaneous transluminal angioplasty in women compared with men. J Am Coll Cardiol. 1994;24:81–90.
90. West GB, Brown JH, Enquist BJ. A general model for the origin of allometric scaling laws in biology. Science. 1997;276:122–6.
91. Westerhof N, Bosman F, DeVries CJ, Noordergraaf A. Analog studies of the human systemic arterial tree. J Biomech. 1969;2:121–43.
92. White L, Haines H, Adams T. Cardiac output related to body weights in small mammals. Comp Biochem Physiol. 1968;27:559–65.
93. Zanchetti A, Grassi G, Mancia G. When should antihypertensive drug treatment be initiated and to what levels should systolic blood pressure be lowered? A critical reappraisal. J Hypertens. 2009;27:923–34.
94. Zarins CK, Giddens DP, Bharadvaj BK, Sottiurai VS, Mabon RF, Glagov S. Carotid bifurcation atherosclerosis. Quantitative correlation of plaque localization with flow velocity profiles and wall shear stress. Circ Res. 1983;53:502–14.

The Heart as a Psychoneuroendocrine and Immunoregulatory Organ

15

Carlo Dal Lin, Francesco Tona, and Elena Osto

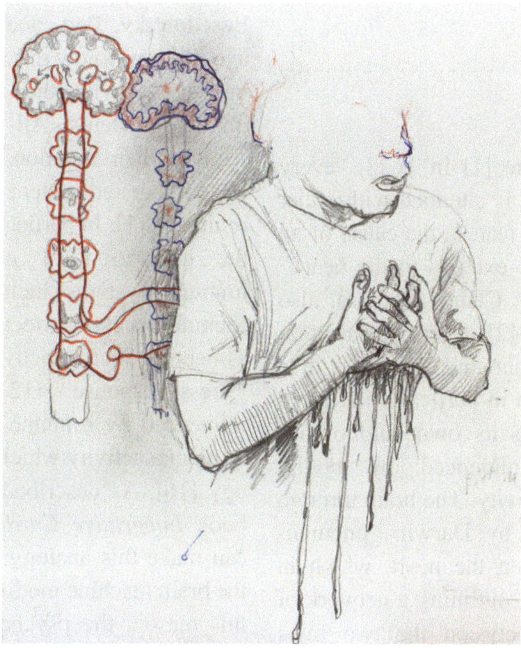

The heart as a multi-regulatory system. Artwork by Piet Michiels, Leuven, Belgium

C. Dal Lin · F. Tona
Department of Cardiac, Thoracic and Vascular Sciences, University of Padova, Padova, Italy

E. Osto (✉)
Laboratory of Translational Nutrition Biology, Federal Institute of Technology Zurich ETHZ, Zurich, Switzerland

Center for Molecular Cardiology, University of Zurich and University Heart Center, Cardiology, University Hospital Zurich, Zurich, Switzerland
e-mail: elena.osto@uzh.ch

Abstract

The heart can be viewed not just as muscle pump but also as an important checkpoint for a complex network of nervous, endocrine, and immune signals. The heart is able to process neurological signals independently from the brain and to crosstalk with the endocrine and immune systems. The heart communicates with the psyche through the neuro-endocrine-immune system in a highly integrated way, in

P. L. M. Kerkhof, V. M. Miller (eds.), *Sex-Specific Analysis of Cardiovascular Function*,
Advances in Experimental Medicine and Biology 1065, https://doi.org/10.1007/978-3-319-77932-4_15

order to maintain the homeostasis of the whole body with peculiarities specific to males and females.

Keywords
Alarmin · Beta-blocker · Brain-heart axis · Cardiokine · Emotion · Heart-brain interaction · Mental stress · Psychic factors in heart disease · Psychosocial stress · Neuroendocrine regulation · Immunoregulatory function · Pattern recognition receptor · Toll-like receptor · Sex-specific analysis · Review

Introduction

As William Harvey wrote [1] in 1628, "every affection of the mind that is attended with either pain or pleasure, hope or fear, is the cause of an agitation whose influence extends to the heart." Following the research of Claude Bernard, also Charles Darwin in 1872 [2] described the relationship between the brain and the heart in a simple and clear way and in particular indicated that the heart, which has its own autonomous rhythm, is strongly influenced by bodily sensations and mental activity. The brain through the vagus nerve (called by Darwin "pneumogastric nerve" [3]) acts on the heart, which in turn retraces on the brain, outlining a network of reciprocal connections between the two most important organs of the human body. In addition to the nervous system, we will see that the heart is also at the center of a dense endocrine and immune network.

Chronic psychological stress leads to continuous mechanical stress and chronic overstraining of the heart muscle machinery [4–6] resulting in immune activation and, if repair mechanisms are not sufficient, in negative cardiac remodeling and heart failure [7].

Psycho-Neuro-Endocrine-Immunology (PNEI): The Crosstalk Between Psyche and Biological Systems

Life is based on the proper integration between the major biological systems: a physiological truth that is often obscured by the current scientific view that seeks to study, understand, and cure the human being with a purely mechanistic approach: as if man was only the sum of genetically determined components performing single independent functions. Already Aristotle warned us: "the whole is more than the sum of its parts."

Thanks to the research [8–10] of Ader, Cohen, Besedowsky, Pert, and Felten, started 40 years ago, we now know that the psyche communicates incessantly with the nervous, endocrine, and immune systems. All components reciprocally regulate their function, thus leading to a single large integrated system of adaptation to the environment [11]. In particular, in human physiology, the integration of nervous, endocrine, and immune systems includes psychic regulation. Mental processes affect immune activity (as well as neuroendocrine activity – as in the case of the "stress response" [12, 13]) and are in turn influenced by immune activity [14, 15] as well as mental activity which modifies brain morphology [16]. As described by F. Bottaccioli in the book *Integrative Cardiology,* p. 143 [17], "we can make this analogy: the software running on the brain machine modifies the machine itself. For this reason, the psyche-brain system cannot be compared with a computer system. In this latter case, if one changes the software, the hardware does not follow suit, whereas in the first case (the psyche-brain system) the software modifies the hardware."

Neurotransmitters, hormones, and cytokines are the coded words spoken by the nervous, endocrine, and immune systems. Indeed, these terms are somehow reductive nowadays, since the distinction between neurotransmitters and cytokines has become less clear, because nerves can synthesize and release inflammatory substances such as histamine and cytokines such as interleukins (IL), i.e., IL-1 and IL-6 [9, 18]. On the other hand, immune cells can synthesize and release neurotransmitters and hormones, such as corticotropin-releasing hormone (CRH), adrenocorticotropic hormone (ACTH), endorphins, vasoactive intestinal peptides (VIPs), and others [9, 18]. In short, our body uses an "esperanto" type of "language" recognized by all its compartments [9, 18].

Sex differences in stress responses could be epigenetically inherited [19, 20] and can be found at all stages of life. These differences may relate to both the organizational and activational effects of gonadal hormones [21, 22] on brain structure, function [23], and neurochemistry (mainly serotonin (5-HT), corticotropin (CRF), and gamma-aminobutyric acid (GABA) signaling) and to genes on the sex chromosomes [24].

From puberty to menopause, adult women usually show lower hypothalamic-pituitary-adrenal axis (HPAA) and autonomic responses than men of the same age [25]. However, the HPAA response is higher in the luteal phase, when, for example, poststress-free cortisol levels approach those of men [26]. After menopause, there is an increase in sympathoadrenal responsiveness, which attenuates during oral hormone replacement therapy. Indeed the HPAA activity shows the same trends [25]. Interestingly, pregnancy is associated with an attenuated response of the sympathoadrenal and HPAA systems at least as assessed by biochemical stimulation [27]. Likely these sex differences in autonomic function are the result of estrogen exposure, which decreases sympathoadrenal responsiveness [28].

Moreover, immunocompetent cells in the brain are responsive to steroid hormones, and their role in sex-specific brain development is an emerging field of interest. In fact, men and women seem to possess different number, morphology, and signaling profile of immune cells in their brain, playing a crucial role in early-life programming of sex differences in the brain and behavior [29].

The presence of an integrated network limits the concept of "hierarchy" between organs and suggests a revision of the past mechanistic and reductionist assumptions oversimplifying the human being. In the PNEI network, all organs work at the same level exchanging signals to maintain the integrity of the whole system in relation to the environment [30–32]. The heart is one of these knots. Herein we will discuss how the cardiovascular system communicates in the PNEI network (Fig. 15.1).

The Heart in the (Regulatory) Network

As described by Francisco Torrent-Guasp [34–39], and illustrated in Fig. 15.2, the heart can be viewed as a helical muscle tube that produces two simple loops that start at the pulmonary artery and end in the aorta: a spiral horizontal basal loop that surrounds the right and left ventricular cavities, and changes direction to cause a second spiral, produced by almost vertically oriented fibers, giving rise to the helical configuration of the ventricular myocardial band. These anatomic structures are subsequently activated by a sort of peristaltic wave, starting at the right ventricle (just below the pulmonary artery) and progressing toward the aorta to produce a sequence of:

1. Narrowing, caused by the basal loop contraction
2. Shortening (related predominantly to the descendant segment contraction)
3. Lengthening (produced by the ascendant segment contraction)
4. Widening, as a consequence of several factors that act during ventricular myocardial relaxation [34–39]

This sequential activation, which is still under investigation, leads to the mechanical events responsible for ejection to empty and subsequent suction to fill [34] the ventricles.

The heart, if adequately nourished, continues to beat alone [43], regardless of the brain. Moreover, from embryology we know that the heart begins to beat before the brain is formed. A transplanted heart is not connected to the host nervous system, but can immediately satisfy the

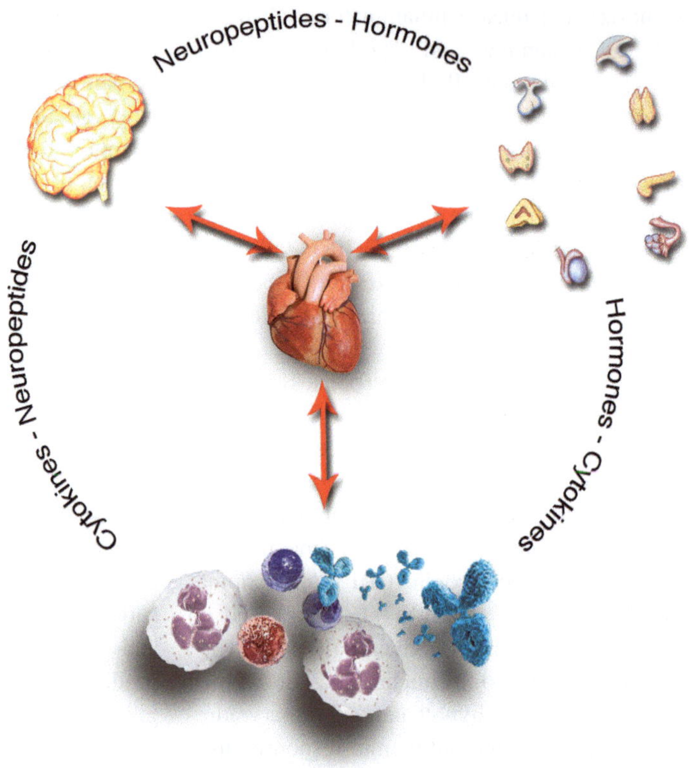

Fig. 15.1 The cardiovascular system in the psychoneurological, hormonal, and immune network: it receives and sends signals to the brain, to the immune, and to the endocrine system. (Modified from Dal Lin et al. [33])

physiological demands of its new host [44–46]. The heart rate is predominantly triggered by the rate of discharge of its dominating pacemaker (mainly the sinus atrial (SA) node and the atrium-ventricular (AV) node) whose action is fine-tuned in vivo from the balance between the sympathetic and parasympathetic nervous systems. A preserved heart rate variability is considered a sign of heart health [47–50], which highlights an important role of emotions in the cardiovascular equilibrium [51]. Indeed, the psyche-brain-heart connection is an important element that explains many otherwise unexplainable phenomena such as sudden deaths, coronary heart disease with normal cholesterol [52–54], and why women although more protected in terms of events than men have a higher mortality risk [55]. The brain-heart axis also explains why heart disorders affect brain functions (leading to stroke or cognitive dysfunction), and, on the contrary, psychoneurological pathologies (stroke, epilepsy, Parkinson's, Alzheimer's, depression, anxiety, psychosis) are accompanied by heart abnormalities [52–54, 56, 57].

Anatomical and Physiological Bases of the Brain-Heart Integration: Neurocardiology

Like the gastrointestinal system [58], the heart possesses a sufficiently extensive neuronal network to be characterized as its own "little brain" [59, 60]. The so-called intrinsic cardiac nervous system is connected to intrathoracic nervous ganglia, extrathoracic ganglions, spinal cord, and cortical nerve centers. Cardiac activity is not regulated only at the central level, but is predominantly established by beat-to-beat neurohormonal loops within autonomous intrathoracic nerve centers [61]. The anatomical and physiological details of the intrinsic cardiovascular system can be found in the book *Basic and Clinical Neurocardiology* by J.A. Armour and J.L. Ardell [61]. Below, a brief synthesis is presented to understand the deep connection between the heart and the brain.

In the heart there are three morphologic types of neurons (unipolar, bipolar, and multipolar, either intra- or inter-gangliar) with sensory and

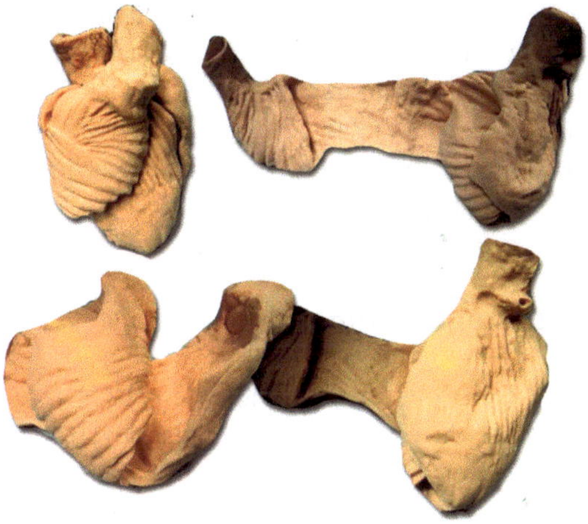

Fig. 15.2 The architectural arrangement of the normal ventricle as described by Buckberg et al. [40, 41]: "the helical ventricular myocardial band model displays the fiber orientation of the detached circumferential fibers or the basal loop (upper right), whereby its predominantly horizontal fibers differ from the conical apical loop. The helical component which contains oblique right- and left-handed fibers. The descending and ascending segments are superimposed (top right image). The lower left and right images display finer orientation when these segments are separated. The lower left image displays the right-handed helix or descending segment. It is connected to the myocardial fold, as its oblique fibers course downward toward the apex. The lower right image shows the overlying left-handed helix or ascending segment. This segment is longer, and its oblique fibers course upward from the apex toward the aorta." Modified from Buckberg [42] reproduced with permission from John Wiley and Sons

afferent or efferent activity (sympathetic and parasympathetic) and interneurons of connection. These neurons can communicate through various molecules: acetylcholine, catecholamines and neuropeptides (somatostatin, neuropeptide Y, VIP, substance P, opioids such as dimorphins A and B, Leu-enkephalin, calcitonin gene-related peptide (CGRP)), serotonin, histamine, ATP, nitric oxide (NO), or amino acids (glutamate, aspartate). These molecules transduce the signal directly or bind to specific receptors (nicotinic, muscarinic type 1, 2, 3, 4, alpha- and beta-adrenergic (mainly expressed at the apex and in the mid-ventricular regions [62]), P2Y, P2X, H1, Ang II, AT1, mu1), influencing membrane ionic voltage channels, the action potential transmission, and ultimately the excitability of the heart cells. In vitro studies show that a functional unit exists between cardiac neurons and cardiomyocytes, which lose contractile activity if separated from the neurons to which they are intimately linked [63].

As described by Shaffer et al. [64], "while efferent (descending) regulation of the heart by the autonomic nervous system is well known (with parasympathetic nerves that exert their effects more rapidly (<1 s) than sympathetic nerves (>5 s) [65]), newer data have suggested a more complex modulation of heart function by the intrinsic cardiac nervous system [66]." These intracardiac neurons (sensory, interconnecting, afferent, and motor neurons) [67] integrate the sympathetic and parasympathetic impulses with the afferent signals occurring from the mechanosensory and chemosensory neurons within the heart. Interestingly, approximately 85–90% of fibers in the vagus nerve carry cardiovascular afferent signals to the brain, to a greater extent than by any other major organ [68].

Cardiovascular afferent activity manifests in complex patterns occurring across time scales from milliseconds to minutes [61] with both short-term and long-term memory functions that can influence heart rhythm, blood pressure, and hormonal release [64, 69].

Moreover, the cardiac afferent activity from pressure-sensitive neurons in the heart, carotid arteries, and the aortic arch [70] seems to primarily modulate cognitive functions (such as sensory-motor and perceptual performance) as revealed by the heart-brain interaction studies by John and Beatrice Lacey [71]. Their research focused on activity occurring within a single cardiac cycle, and they confirmed that the cardiovascular activity influences perception and cognitive performance [72].

Velden and Wölk described that cognitive performance fluctuates at a rhythm around 10 Hz, demonstrating that afferent inputs from the heart synchronize cortical activity projecting on the neurons in the thalamus [73]. An important aspect of their work was the finding that "it is the 'pattern and stability' (the rhythm) of the heart's afferent inputs, rather than the number of neural bursts within the cardiac cycle, that are important in modulating thalamic activity, which in turn has global effects on brain function" [73].

Starting from this evidence, a growing body of research indicates that the afferent information processed by the intrinsic cardiac nervous system [60] can influence activity in the frontocortical areas [74–76] and motor cortex [77–79], affecting psychological factors, such as attention level, motivation [80], perceptual sensitivity [81], and emotional processing [82, 83]. Figure 15.3 describes the connections between the intrinsic cardiac neurons, the brainstem, the hypothalamus, the thalamus, the amygdala, and the cerebral cortex [66, 90].

Finally, according to the model of neurovisceral integration [74, 90–96], the information shared by the heart with the brain may also be coded by rhythmic [97] and electromagnetic patterns [98–100] which may represent the basis of intuitive-emotional processes [76, 84, 101, 102], awareness and feelings [103–105], and a rational, detached and "less egocentric" reasoning [106].

A disease arises when a disorganization occurs in this network of connections (intracardiac nervous system-intrathoracic nervous system-central nervous system): in fact, peculiar remodeling patterns accompany and often precede ischemic pathology, arrhythmias [107], heart failure, and arterial hypertension [61].

The Heart as an Endocrine Station

The heart serves and operates as a (loop) center to process and encode information [61] as it is also an endocrine gland producing its own hormones and neurotransmitters.

Since the early 1980s [108, 109], studies on the "endocrine heart" appeared. In addition to the well-known atrial natriuretic peptides (ANPs), cerebral natriuretic peptide (BNP) and C-type natriuretic peptide (CNP), many substances are produced and secreted into the blood by myocardial cells, fibroblasts, and endothelial and heart immune cells. All these molecules (called "cardiokines") maintain heart homeostasis and interact in cardiac remodeling in the context of acute or chronic cardiovascular affections [110], among others, the tumor necrosis factor alpha (TNF-alpha), beta-transforming growth factor beta (TGF-beta), GDF-15 (growth differentiation factor-15), activin-A, myostatin, adrenomedullin, and endocannabinoids [52].

Natriuretic Peptides

ANP (produced at the atrial level), BNP (produced at the atrial and ventricular level), and CNP (secreted by neurons and by the heart endothelium) have systemic action as they present brain, kidney, adrenal, adipose tissue, muscles, bones, pancreas, liver, immunitary cells, and platelets' receptors [111]. In addition to their antihypertensive activity, natriuretic peptides play an important metabolic role to ensure adequate energy reserves to the heart, affecting lipolysis processes and improving the glucose cellular uptake [112]. They also intervene in thermogenesis processes, converting white fat into brown fat [113].

The Heart as an Immune Organ

Inside the heart are usually present macrophages, dendritic cells, mast cells, and a small number of B and T lymphocytes [114]. Although essential for protecting the heart tissue from bacterial infections, they are activated in the event of tissue damage (such as myocardial infarction or myocarditis) and produce inflammation, whose excessive intensity and duration can lead to ventricular dilatation and heart failure [115].

Resident cardiac immune cells are triggered by two distinct orders of "alarmins" [116]: pathogen-derived (PAMPs) and damage-derived (DAMPs) molecules, which are released by dying, injured, or dysfunctional cells (with mitochondrial impairment [117, 118]) and recognized by specific pattern recognition receptors (PRRs) such as Toll-like receptors (TLRs).

It is interesting to note that physical and mental stress, through sympathetic mediated

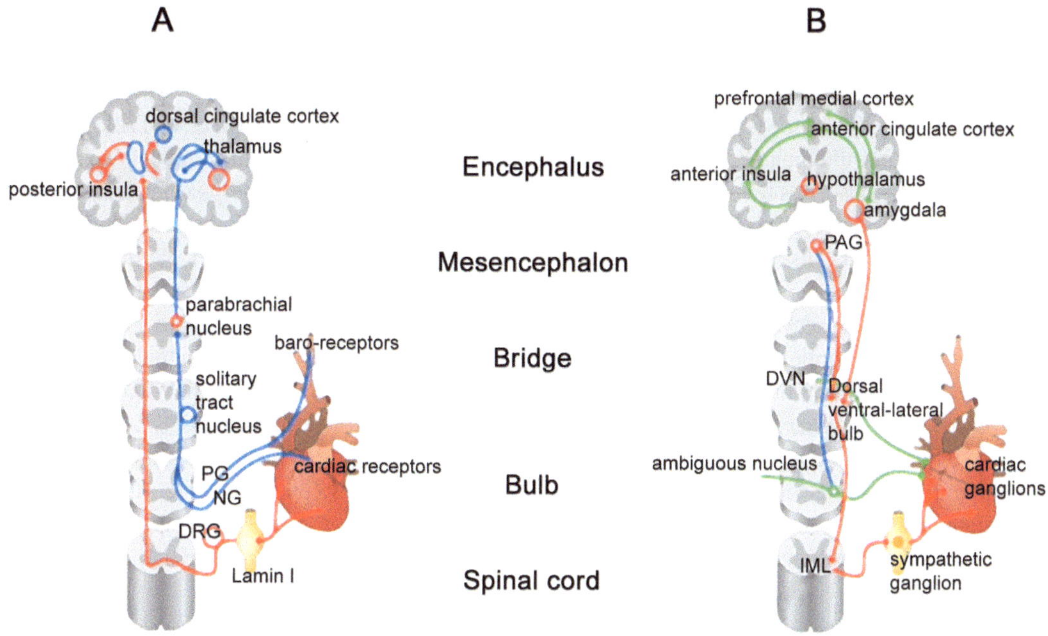

Fig. 15.3 (**a**) Schematic neural communication pathways interacting between the heart and the brain. The intrinsic cardiac nervous system integrates information from the extrinsic nervous system and from the sensory neurites within the heart. The heart sends afferents to the brain through the glossopharyngeal nerve (IX) and vagus nerve (X) (which connect to the petrosal (PG) and nodose (NG) ganglia) and through the nerves associated with the dorsal root ganglia (DRG) and lamin I of the spinal cord sensitive roots. Signals arrive to the brain at the back of the insula, in the middle cingulate cortex, in the hypothalamus, and in the locus coeruleus. Thus, cardiovascular afferents have connections to a number of brain centers involved in emotion and stress perception including thalamus, hypothalamus, and amygdala. Broad evidence implicates anger along with other emotions and mental stress in playing a significant role in myocardial ischemia, arrhythmias, and sudden death [52]. Indeed, these brain areas are connected with visual and acoustic cerebral areas, and this explains how visual [84–87] or acoustic stressors [88] may alter heart function for bad and for good [89]. (**b**) The pathways from the central nervous system (prefrontal medial cortex, anterior cingulate cortex, anterior insula, amygdala, and hypothalamus) reach the heart: the sympathetic way (red) descends into the spinal cord's intermediate mid-column to connect to the "star-shaped" ganglion, which connects to the intrinsic cardiovascular system. An increase in sympathetic activity is the principal method used to increase heart rate (HR) above the intrinsic level generated by the SA node. Following the onset of sympathetic stimulation, there is a delay of up to 5 s before the stimulation induces a progressive increase in HR, which reaches a steady level in 20–30 s if the stimulus is continuous [64]. The slowness of the response to sympathetic stimulation is in direct contrast to vagal stimulation, which is almost instantaneous. However, the effect on HR lasts longer, and even a short stimulus can affect HR for 5–10 s. Efferent sympathetic nerves target the SA node and AV node via the intrinsic cardiac nervous system and the bulk of the myocardium (heart muscle). Action potentials conducted by these motor neurons trigger norepinephrine and epinephrine release and binding to beta-adrenergic (β1) receptors located on cardiac muscle fibers. This speeds up spontaneous depolarization in the SA and AV nodes, increases HR, and strengthens the contractility of the atria and ventricles. The parasympathetic path (in green and blue) descends from the dorsal motor nucleus of the vagus nerve, from the solitary tract nucleus and the ambiguous nucleus and connects to the heart ganglia. The most obvious effect of vagal activity is to slow or even stop the heart. The vagus nerves are the primary nerves for the parasympathetic system and innervate the intrinsic cardiac nervous system and project to the SA node, AV node, and atrial cardiac muscle. Increased efferent activity in these nerves triggers acetylcholine release and binding to muscarinic (mainly M2) receptors. This decreases the rate of spontaneous depolarization in the SA and AV nodes, slowing HR. Because there is sparse vagal innervation of the ventricles, vagal activity minimally affects ventricular contractility. The response time of the sinus node is very short, and the effect of a single efferent vagal impulse depends on the phase of the cardiac cycle at which it is received. Thus, vagal stimulation results in an immediate response that typically occurs within the cardiac cycle in which it occurs and affects only one or two heartbeats after its onset. After cessation of vagal stimulation, HR rapidly returns to its previous level. An increase in HR can also be achieved by reduced vagal activity or vagal block. Thus, sudden changes in HR (up or down) between one beat and the next are parasympathetically mediated [64]. Modified from 81 reproduced with permission from Wolters Kluwer Health Inc. Abbreviations: DVN, dorsal motor nucleus of the vagus nerve; PAG, periaqueductal gray substance; IML spinal cord intermediate lateral column

neurohormonal mechanisms [52–54], can trigger an acute coronary syndrome (ACS) by promoting the secretion of inflammatory substances and inducing cardiac mast cells to release degradative and procoagulant enzymes [119].

In particular, mast cells in the heart have a high inflammatory potential represented by the mixture of substances present in their cytoplasmatic vesicles [120]. Autoptic studies revealed the presence of an important number of cardiac mast cells within the coronary arteries in subjects affected by hypertension, dilated cardiomyopathy, and mitral valve defects [121]. These highly - pro-inflammatory cells are susceptible to external environmental stimuli (e.g., PM 10 and PM 2.5 thin dust pollutants) [18, 122] but also to the "internal environment," such as to stress mediators (in particular CRH) and other urocortins, cardiac neurons neuropetides such as neurotensin and substance P, and IL-1 and IL-6 [123]. The inflammatory cascade triggered by cardiac mast cells during mental stress may produce an ACS with coronary spasm and thrombosis [119]. Finally, there is strong evidence that premenopausal female cardioprotection may at least partly be due to gender differences in cardiac mast cells [123]. A possible explanation for the differences between male and female cardiac mast cells may be that estrogen prevents the release of mast cell proteases or other products such as TNF-alpha [124].

A Neuro-Endocrine-Immune Symphony Plays on Coronary Endothelium

We recently reviewed [33, 52–54] the neuroendocrine and immune influences that act on the coronary endothelium, influencing its function and that can be studied through the evaluation of the coronary microvascular function [125]. In summary, as reported [52], "on endothelial cells acts a real neuro-endocrine-immune symphony in which the melody is played by the vitamin D [126], parathyroid hormone (PTH) [127], renin-angiotensin-aldosterone system (RAAS) axis [56, 128] in concert with thyroid and thyroid stimulating (TSH) hormones [129], growth hormone (GH) and insulin-like growth factor 1 (IGF-1) [130, 131], cortisol and ACTH [132], sex hormones [133, 134], insulin and glucagon-like protein-1 (GLP-1) [135], adipokines [56], oxytocin, vasopressin, and prolactin [136], melatonin [137, 138], bilirubin, heme catabolic pathway, and gamma-glutamyltransferase [139, 140]. Rounding out the orchestra is the immune system [141], with the familiar example of inflammation as a key process involved in the pathogenesis of atherosclerosis [47], the already discussed action of platelets and the autonomic balance, where the predominance of the sympathetic system on the parasympathetic, is a determining factor for endothelial dysfunction [56]."

In Addition to Cholesterol, Nutrition, and Sedentary Lifestyle, There Is a Psychological Risk at the Heart of Many Cardiac Disorders

Every year in the United States, about 450,000 people die of sudden cardiac death [142]. The causes can be multiple: congenital valve or coronary anomalies, hypertrophic cardiomyopathy, and, of course, coronary atherosclerosis. But despite the underlying coronary atherosclerosis, the main cause of sudden death is the alteration of heart rhythm due to a massive sympathetic stimulation of the heart [143]. In fact, at least 12% of myocardial infarctions and sudden deaths can also arise with healthy coronary vessels and even with an "antiatherosclerotic" lipid profile [143]. They happen mainly in females, with psychological disorders and high levels of emotional stress [143].

This evidence recalls what occurs in Takotsubo syndrome or stress-related heart disease. In this scenario, patients as a consequence of either positive or negative emotional events exhibit very high levels of catecholamines that cause left ventricular akinesia and its typical ballooning [144], as classic example, a woman who manifests a Takotsubo within a few hours of her husband's sudden death [145] or after joyful events such as becoming a grandmother, grandchildren visiting from abroad, etc. [146, 147].

Emotional stress and myocardial ischemia seem to be more of a female feature [148].

Mental Stress Myocardial Ischemia

There is sufficient evidence that psychosocial stress plays a paramount role in the onset of a cardiovascular disease [149], especially coronary artery disease and all its risk factors [150, 151]. Even adverse early life events, in particular during childhood and adolescence, predispose individuals to a greater rate of inflammatory-based diseases including cardiovascular ones through epigenetic signatures [152]. Nowadays it's well known that depression, anxiety, and post-traumatic stress disorder lead to cardiovascular disorders [153] and myocardial ischemia through neuroendocrine and immune mediators [154].

In order to assess the effects of stress on cardiac function, it is possible to use the mental stress-induced myocardial ischemia (MSIMI) test. This is a provocative test alternative to exercise and pharmacological stress-induced myocardial ischemia that uses psychological stimulations (mental arithmetic, simulated public speaking, problem-solving tasks, cognitive and psychomotor challenges and tasks involving the recall of negative emotions) rather than physical exercise [155]. Interestingly, stress-related ischemic alterations seen after the MSIMI have not been described during exercise/pharmacological stress [156]. MSIMI is frequent among patients with coronary arteriosclerosis. This ischemia is often asymptomatic, occurs at lower workload and oxygen demand than exercise-induced ischemia [157], has a negative prognostic impact (being not directly related to the severity of coronary stenosis) [158], and may not be affected by the action of beta-blockers [159]. MSIMI leads to coronary microvascular constriction [160, 161] and cardiac electrical instability [158].

Finally, mental stress, activating the sympathetic-adrenal-medullary axis, eliciting the release of catecholamines, determines the release of DAMPs and the activation of cardiac mast cells [119]. In turn, DAMPs can activate the innate immune response leading to sterile inflammation [162, 163], which can result in myocarditis and cardiomyopathy [115], as well as atherosclerosis [150], even in animal models [164].

The Woman's Heart

Traditionally, heart attacks have always been considered a male issue. This view is no longer the case today: in the United States and in many European countries, female mortality (even before menopause) for cardiovascular disease is even higher than male mortality [165]. This is very intriguing because in female, cholesterol and arteriosclerosis do not fully explain this evidence and contrast the established idea of a cardioprotective role of estrogens (just thinking to the increased risk of heart attack when taking birth control pills [166, 167]). Up to one third of women with cardiac ischemia have no coronary occlusion [168]. In particular, coronary spasms, plaque erosion (not angiographically critical), and arrhythmias play a greater role in cardiac events affecting women under the age of 60 [168]. From a clinical point of view, women have a milder symptomatology than men, which can easily lead to a delay in diagnosis and therapy [169]. In the medical history of these patients, often there are socioeconomic problems [18], strong conjugal stress, or episodes of violence and sexual abuses [170].

In Fig. 15.4 are depicted the traditional and nontraditional cardiovascular risk factors along with those specific for females.

As recently described [171], a woman's heart looks stiffer than the male's one. When heart failure occurs in women, left ventricular remodeling is oriented toward concentric hypertrophy. As a result, heart failure occurs in most cases with preserved ejection fraction (HFpEF). This is in contrast with heart failure in men, where the prevailing phenotype is heart failure with reduced ejection fraction (HFrEF). The renin-angiotensin system would appear to be less activated. Consequently, fibrosis is less important

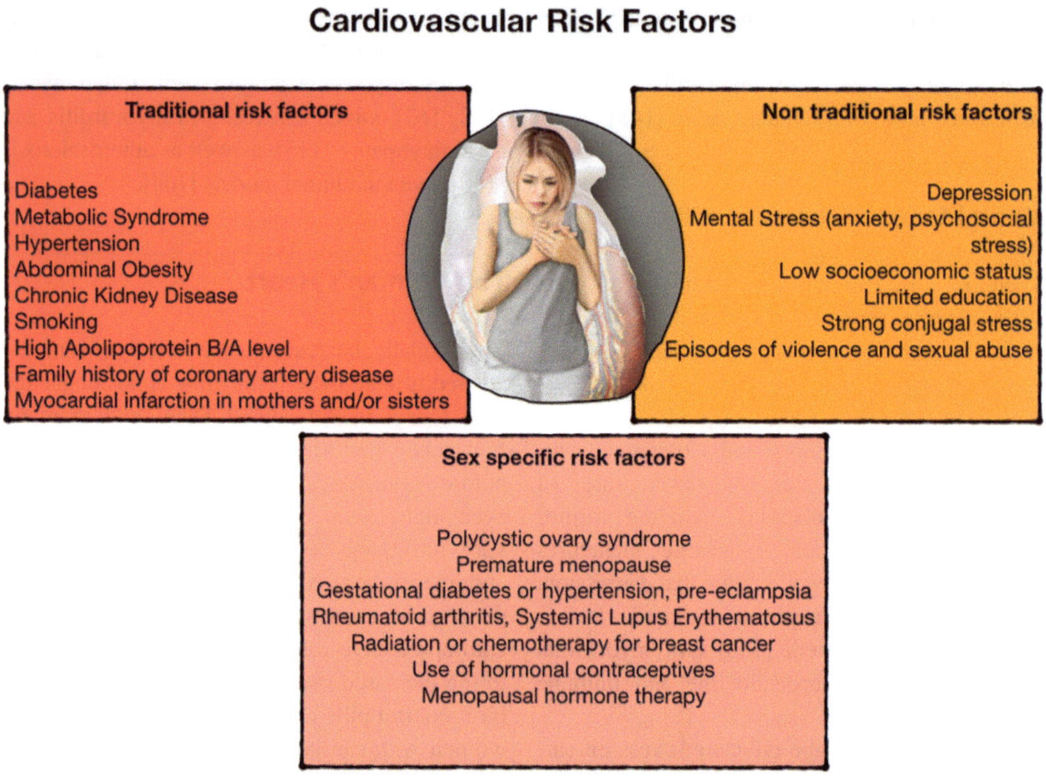

Fig. 15.4 Traditional and nontraditional cardiovascular risk factors and more women-specific risk factors. (Image modified from Gebhard C. Eur Heart J. 2017;38:1066–1068. Reproduced with permission from MediDesign Frank Geisler)

at a myocardial level, although there is a general stiffening of the circulatory system with an increase in the effective afterload. The reason for the preferential concentric hypertrophy and HFpEF instead of left ventricular dilation and HFrEF is still unknown. Some authors suggest that in normal conditions, in women the myocyte diameter is reduced with respect to men, and it may therefore be possible to increase the amount of contractile proteins without stretching the sarcomere [171].

Finally, the female heart seems to be stiffer: left ventricular diastolic elastance is higher for women than for men at comparable levels of filling pressure as shown in Fig. 15.5 [172].

Conclusions

The study of the heart and its connections with the psycho-neuro-endocrine-immune system leads at least to two types of conclusions, one theoretical and one practical.

As George Engel said about 40 years ago in *Science* [173], the reduction of complex phenomena in simple determinations (reductionism), the separation between "biological" and "psychological" phenomena (mind and body dichotomy), and the interpretation of life exclusively in physical or chemical terms (physicalism) are obstacles to the study of the human being and its physiopathology: obstacles that are causing very heavy consequences in terms of the effectiveness, costs [174], and credibility [175–180] of our care systems. Thanks to the PNEI, we have the tools for a scientific investigation of complex phenomena which, on different scales, determine the balance between health and illness and to study and rediscover therapeutic solutions that go beyond the current pharmacological vision. For example, as we have extensively documented in a recent book [17], the treatment of dyslipidemia can be

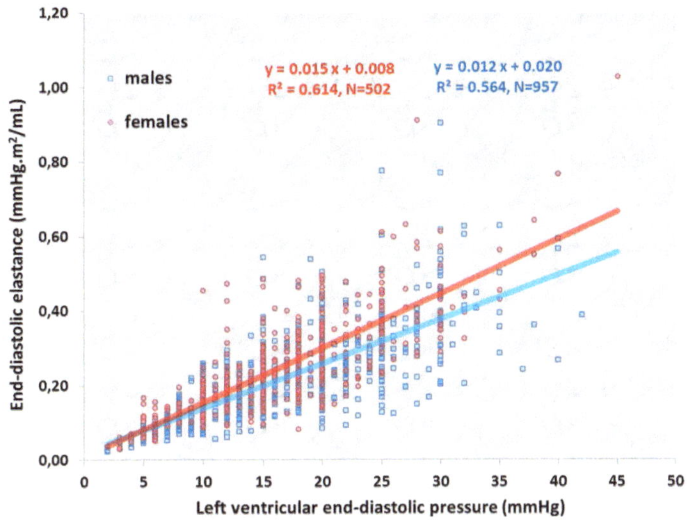

Fig. 15.5 Left ventricular diastolic elastance is higher in women compared to men. For women (N = 502) mean end-diastolic elastance is 0.2459 mmHg. m^2/mL (95% CI 0.2349–0.2570) and higher ($P < .0001$) compared to men ($N = 957$), showing an average of 0.2114 (95% CI 0.2046–0.2182). (From [172], with permission). *LV diastolic stiffness is higher for women than for men at comparable levels of filling pressure (EDP)*

effectively achieved by an integrated approach that includes, in addition to nutrition and physical activity, a range of phytotherapic products and proper management of mental stress, given that excessive cholesterol lowering increases the risk of heart attacks and general mortality [181, 182].

From a practical point of view, we count on very high standard of care in acute setting, while in the chronic and preventive context, we need to reconsider the management of patients with heart disease by investing more time and resources in terms of proper nutrition [183], physical activity, and stress management [54, 184, 185].

References

1. Harvey W. On the motion of the heart and blood in animals. Vol. 38, Part 3. The Harvard classics. New York: P.F. Collier & Son, 1909–14; Bartleby.com, 2001. 1628.
2. Darwin C. The expression of the emotions in man and animals. California Medicine. 1956;85.
3. Colzato LS, Sellaro R, Beste C. Darwin revisited: the vagus nerve is a causal element in controlling recognition of other's emotions. Cortex. 2017;92:95–102.
4. Cohen S, Janicki-Deverts D, Miller GE. Psychological stress and disease. JAMA. 2007;298:1685–7.
5. Dimsdale J, Psychological E. Stress and cardiovascular disease. J Am Col Cardiol. 2008;51:1237–46.
6. Carroll D, Phillips AC, Balanos GM. Metabolically exaggerated cardiac reactions to acute psychological stress revisited. Psychophysiology. 2009;46:270–5.
7. Zhang Y, Bauersachs J, Langer HF. Immune mechanisms in heart failure. Eur J Heart Fail. 2017. https://doi.org/10.1002/ejhf.942.
8. Ader R, Cohen N, Felten D. Psychoneuroimmunology: interactions between the nervous system and the immune system. Lancet. 1995;345:99–103.
9. Ader R. Psychoneuroimmunology, two-volume set. Elsevier; 2011.
10. Bottaccioli F. Epigenetica e psiconeuroendocrinoimmunologia. Edra S.p.A; 2014.
11. Verburg-van Kemenade BML, Cohen N, Chadzinska M. Neuroendocrine-immune interaction: evolutionarily conserved mechanisms that maintain allostasis in an ever-changing environment. Dev Comp Immunol. 2017;66:2–23.
12. McEwen BS. Protective and damaging effects of stress mediators: the good and bad sides of the response to stress. Metab Clin Exp. 2002;51:2–4.
13. Chrousos GP. Stress and disorders of the stress system. Nat Rev Endocrinol. 2009;5:374–81.
14. Dantzer R, Kelley KW. Twenty years of research on cytokine-induced sickness behavior. Brain Behav Immun. 2007;21:153–60.
15. Eisenberger NI, Moieni M, Inagaki TK, Muscatell KA, Irwin MR. In sickness and in health: the co-regulation of inflammation and social behavior. Neuropsychopharmacol Rev. 2017;42(1):242–53.
16. Draganski B, et al. Temporal and spatial dynamics of brain structure changes during extensive learning. J Neurosci. 2006;26:6314–7.
17. Fioranelli M. Integrative cardiology. Springer; 2017. https://doi.org/10.1007/978-3-319-40010-5
18. Bottaccioli F, Bottaccioli AG. Psiconeuroendocrinoimmunologia e scienza della cura integrata. Il manuale. Edra S.p.A; 2017.
19. Rodgers AB, Morgan CP, Bronson SL, Revello S, Bale TL. Paternal stress exposure alters sperm microRNA content and reprograms offspring HPA stress axis regulation. J Neurosci. 2013;33:9003–12.
20. Reynolds RM, Labad J, Buss C, Ghaemmaghami P, Räikkönen K. Transmitting biological effects of stress in utero: implications for mother and offspring. Psychoneuroendocrinology. 2013;38:1843–9.

21. Patchev VK, Almeida OF. Gender specificity in the neural regulation of the response to stress: new leads from classical paradigms. Mol Neurobiol. 1998;16:63–77.
22. Patchev VK, Hayashi S, Orikasa C, Almeida OF. Implications of estrogen-dependent brain organization for gender differences in hypothalamo-pituitary-adrenal regulation. FASEB J. 1995;9:419–23.
23. Bangasser DA, Wicks B. Sex-specific mechanisms for responding to stress. J Neurosci Res. 2017;95:75–82.
24. Bale TL, Epperson CN. Sex differences and stress across the lifespan. Nat Neurosci. 2015;18:1413–20.
25. Kajantie E, Phillips DIW. The effects of sex and hormonal status on the physiological response to acute psychosocial stress. Psychoneuroendocrinology. 2006;31:151–78.
26. Kudielka BM, Hellhammer DH, Wüst S, Kudielka BM. Why do we respond so differently? reviewing determinants of human salivary cortisol responses to challenge. Psychoneuroendocrinology. 2009;34:2–18.
27. Mastorakos G, Ilias I. Maternal and fetal hypothalamic-pituitary-adrenal axes during pregnancy and postpartum. Ann N Y Acad Sci. 2003;997:136–49.
28. Saleh TM, Connell BJ. Estrogen-induced autonomic effects are mediated by NMDA and GABAA receptors in the parabrachial nucleus. Brain Res. 2003;973:161–70.
29. Nelson LH, Lenz KM. The immune system as a novel regulator of sex differences in brain and behavioral development. J Neurosci Res. 2017;95:447–61.
30. Barabási A-L, Oltvai ZN. Network biology: understanding the cell's functional organization. Nat Rev Genet. 2004;5:101–13.
31. Ideker T, Krogan NJ. Differential network biology. Mol Syst Biol. 2012;8:1–9.
32. Lindfors E. Network biology. VTT Publication; 2011. p. 1–81. https://doi.org/10.1007/978-1-61779-276-2
33. Dal Lin C, Tona F, Osto E. Coronary microvascular function and beyond: the crosstalk between hormones, cytokines, and neurotransmitters. Int J Endocrinol. 2015;2015:1–17.
34. Torrent-Guasp F, et al. The structure and function of the helical heart and its buttress wrapping. I. The normal macroscopic structure of the heart. Semin Thorac Cardiovasc Surg. 2001;13:301–19.
35. Buckberg GD, et al. The structure and function of the helical heart and its buttress wrapping. IV. Concepts of dynamic function from the normal macroscopic helical structure. Semin Thorac Cardiovasc Surg. 2001;13:342–57.
36. Buckberg GD, Coghlan HC, Torrent-Guasp F. The structure and function of the helical heart and its buttress wrapping. V. Anatomic and physiologic considerations in the healthy and failing heart. Semin Thorac Cardiovasc Surg. 2001;13:358–85.
37. Buckberg GD, Coghlan HC, Torrent-Guasp F. The structure and function of the helical heart and its buttress wrapping. VI. Geometric concepts of heart failure and use for structural correction. Semin Thorac Cardiovasc Surg. 2001;13:386–401.
38. Buckberg GD, Coghlan HC, Hoffman JI, Torrent-Guasp F. The structure and function of the helical heart and its buttress wrapping. VII. Critical importance of septum for right ventricular function. Semin Thorac Cardiovasc Surg. 2001;13:402–16.
39. Kocica MJ, et al. The helical ventricular myocardial band: global, three-dimensional, functional architecture of the ventricular myocardium. Eur J Cardiothorac Surg. 2006;29 (Suppl 1):S21–40.
40. Buckberg GD, Hoffman JIE, Coghlan HC, Nanda NC. Ventricular structure–function relations in health and disease: Part II. Clinical considerations. Eur J Cardio-Thoracic Surg. 2015;47:778–87.
41. Buckberg GD, Hoffman JIE, Coghlan HC, Nanda NC. Ventricular structure–function relations in health and disease: Part I. The normal heart. Eur J Cardio-Thoracic Surg. 2015;47:587–601.
42. Buckberg G. The helical ventricular myocardial band during standard echocardiography: a structure-function relationship. Echocardiography. 2015;32:199–204.
43. Guyton AC. Textbook of medical physiology. Philadelphia: Elsevier Inc.; 2006.
44. Bernardi L, et al. Respiratory sinus arrhythmia in the denervated human heart. J Appl Physiol. 1989;67:1447–55.
45. Ordway GA, Charles JB, Randall DC, Billman GE, Wekstein DR. Heart rate adaptation to exercise training in cardiac-denervated dogs. J Appl Physiol. 1982;52:1586–90.
46. Mettauer B, et al. Exercising with a denervated heart after cardiac transplantation. Ann Transplant, Q Pol Transplant Soc. 2005;10:35–42.
47. Camm JA, Luscher TF, Serruys PW. ESC textbook of cardiovascular medicine. 2nd ed. Oxford: Oxford Medicine; 2009. p. 1–1398.
48. Pagani M, et al. Low and high frequency components of blood pressure variability. Ann N Y Acad Sci. 1996;783:10–23.
49. Pagani M, Lucini D, Montano N, Porta A, Malliani A. Physiological background of heart rate variability: do we understand it better? Card Electrophysiol Rev. 1999;3:274–8.
50. Lucini D, Pagani M. Exercise: should it matter to internal medicine? Eur J Intern Med. 2011;22:363–70.
51. Lane RD, et al. Neural correlates of heart rate variability during emotion. NeuroImage. 2009;44:213–22.
52. Dal Lin C, et al. Coronary microvascular and endothelial function regulation: crossroads of psychoneuroendocrine immunitary signals and quantum physics [Part A]. J Integr Cardiol. 2015;1:132–63.
53. Dal Lin C, et al. Coronary microvascular and endothelial function regulation: crossroads of psychoneuroendocrine immunitary signals and quantum physics [Part B]. J Integr Cardiol. 2015;1:164–88.
54. Dal Lin C, et al. Coronary microvascular and endothelial function regulation: crossroads of psychoneuroendocrine immunitary signals and quantum physics [Part C]. J Integr Cardiol. 2015;1:189–209.
55. Stock EO, Redberg R. Cardiovascular disease in women. Curr Probl Cardiol. 2012;37:450–526.
56. Pereira HV, José J, Almeida J, Sousa N. Stressed brain, diseased heart: a review on the pathophysiologic mechanisms of neurocardiology. Int J Cardiol. 2013;166:30–7.
57. van der Wall EE, van Gilst WH. Neurocardiology: close interaction between heart and brain. Neth Heart J. 2013;21:51–2.
58. Furness JB. The enteric nervous system and neurogastroenterology. Nat Rev Gastroenterol Hepatol. 2012;9:286–94.
59. Armour JA. Potential clinical relevance of the 'little brain' on the mammalian heart. Exp Physiol. 2008;93:165–76.
60. Armour JA. The little brain on the heart. Cleve Clin J Med. 2007;74(Suppl 1):S48–51.
61. Armour JA, Ardell JL. Basic and clinical neurocardiology. Oxford: Oxford University Press; 2004.
62. Lyon AR, Rees PSC, Prasad S, Poole-Wilson PA, Harding SE. Stress (Takotsubo) cardiomyopathy--a novel pathophysiological hypothesis to explain catecholamine-induced acute myocardial stunning. Nat Clin Pract Cardiovasc Med. 2008;5:22–9.

63. Murphy DA, O'Blenes S, Hanna BD, Armour JA. Functional capacity of nicotine-sensitive canine intrinsic cardiac neurons to modify the heart. Am J Phys. 1994;266:R1127–35.
64. Shaffer F, McCraty R, Zerr CL. A healthy heart is not a metronome: an integrative review of the heart's anatomy and heart rate variability. Front Psychol. 2014;5:1040.
65. Nunan D, Sandercock GRH, Brodie DA. A quantitative systematic review of normal values for short-term heart rate variability in healthy adults. PACE – Pacing Clin Electrophysiol. 2010;33:1407–17.
66. Kukanova B, Mravec B. Complex intracardiac nervous system. Bratisl Lek Listy. 2006;107:45–51.
67. Verkerk AO, et al. Functional NaV1.8 channels in intracardiac neurons: the link between SCN10A and cardiac electrophysiology. Circ Res. 2012;111:333–43.
68. Cameron OG. Visceral sensory neuroscience interoception. New York: Oxford University Press; 2002.
69. Randall DC, Evans JM, Billman GE, Ordway GA, Knapp CF. Neural, hormonal and intrinsic mechanisms of cardiac control during acute coronary occlusion in the intact dog. J Auton Nerv Syst. 1981;3:87–99.
70. Lacey BC, Lacey JI. Two-way communication between the heart and the brain. Significance of time within the cardiac cycle. Am Psychol. 1978;33:99–113.
71. McCraty R, Shaffer F. Heart rate variability: new perspectives on physiological mechanisms, assessment of self-regulatory capacity, and health risk. Glob Adv Heal Med. 2015;4:46–61.
72. Somsen RJM, Jennings JR, Van Der Molen MW. The cardiac cycle time effect revisited: temporal dynamics of the central-vagal modulation of heart rate in human reaction time tasks. Psychophysiology. 2004;41:941–53.
73. Velden M, Wolk C. Depicting cardiac activity over real time: a proposal for standardization. J Psychophysiol. 1987;1:173–5.
74. Lane RD, Reiman EM, Ahern GL, Thayer JF. Activity in medial prefrontal cortex correlates with vagal component of heart rate variability during emotion. Brain Cogn. 1982;47:97–100. (Academic Press)
75. Jennings JR, Sheu LK, Kuan DC-H, Manuck SB, Gianaros PJ. Resting state connectivity of the medial prefrontal cortex covaries with individual differences in high-frequency heart rate variability. Psychophysiology. 2016;53:444–54.
76. McCraty R, Atkinson M, Bradley RT. Electrophysiological evidence of intuition: part 2. A system-wide process? J Altern Complement Med. 2004;10:325–36.
77. Elam M, Yoa T, Svensson TH, Thoren P. Regulation of locus coeruleus neurons and splanchnic, sympathetic nerves by cardiovascular afferents. Brain Res. 1984;290:281–7.
78. Elam M, Svensson TH, Thoren P. Differentiated cardiovascular afferent regulation of locus coeruleus neurons and sympathetic nerves. Brain Res. 1985;358:77–84.
79. Elam M, Thorén P, Svensson TH. Locus coeruleus neurons and sympathetic nerves: activation by visceral afferents. Brain Res. 1986;375:117–25.
80. Schandry R, Montoya P. Event-related brain potentials and the processing of cardiac activity. Biol Psychol. 1996;42:75–85.
81. Montoya P, Schandry R, Müller A. Heartbeat evoked potentials (HEP): topography and influence of cardiac awareness and focus of attention. Electroencephalogr Clin Neurophysiol. 88:163–72.
82. Zhang JX, Harper RM, Frysinger RC. Respiratory modulation of neuronal discharge in the central nucleus of the amygdala during sleep and waking states. Exp Neurol. 1986;91:193–207.
83. Frysinger RC, Zhang JX, Harper RM. Cardiovascular and respiratory relationships with neuronal discharge in the central nucleus of the amygdala during sleep-waking states. Sleep. 1988;11:317–32.
84. Garfinkel SN, et al. Fear from the heart: sensitivity to fear stimuli depends on individual heartbeats. J Neurosci. 2014;34:6573–82.
85. Child N, et al. Effect of mental challenge induced by movie clips on action potential duration in normal human subjects independent of heart rate. Circ Arrhythm Electrophysiol. 2014;7:518–23.
86. Chen K-H, Aksan N, Anderson SW, Grafft A, Chapleau MW. Habituation of parasympathetic-mediated heart rate responses to recurring acoustic startle. Front Psychol. 2014;5:1288.
87. Cabrerizo M, et al. Induced effects of transcranial magnetic stimulation on the autonomic nervous system and the cardiac rhythm. Sci World J. 2014;2014(349):718.
88. Peters A, et al. Exposure to traffic and the onset of myocardial infarction. N Engl J Med. 2004;351:1721–30.
89. Trappe H-J. The effect of music on human physiology and pathophysiology. Music Med. 2012;4:100–5.
90. Palma J-A, Benarroch EE. Neural control of the heart: recent concepts and clinical correlations. Neurology. 2014;83:261–71.
91. Thayer JF, Lane RD. A model of neurovisceral integration in emotion regulation and dysregulation. J Affect Disord. 2000;61:201–16.
92. Park G, Thayer JF. From the heart to the mind: cardiac vagal tone modulates top-down and bottom-up visual perception and attention to emotional stimuli. Front Psychol. 2014;5:278.
93. Thayer JF, Brosschot JF. Psychosomatics and psychopathology: looking up and down from the brain. Psychoneuroendocrinology. 2005;30:1050–8.
94. Thayer JF, Ahs F, Fredrikson M, Sollers JJ, Wager TD. A meta-analysis of heart rate variability and neuroimaging studies: implications for heart rate variability as a marker of stress and health. Neurosci Biobehav Rev. 2012;36:747–56.
95. Thayer JF, Lane RD. Claude Bernard and the heart-brain connection: Further elaboration of a model of neurovisceral integration. Neurosci Biobehav Rev. 2009;33:81–8.
96. Thayer JF, Sternberg E. Beyond heart rate variability: vagal regulation of allostatic systems. Ann N Y Acad Sci. 2006;1088:361–72.
97. Lin P-F, et al. Correlations between the signal complexity of cerebral and cardiac electrical activity: a multiscale entropy analysis. PLoS One. 2014;9:e87798.
98. Aftanas LI, Brak IV, Reva NV, Pavlova SV. Brain oscillations and individual variability of cardiac defense in human. Ross Fiziol Zh Im I M Sechenova. 2013;99:1342–56.
99. Aftanas LI, Brak IV, Gilinskaia OM, Pavlov SV, Reva NV. Features of brain oscillatory activity and cardiac defense in treatment arterial hypertensives. Ross Fiziol Zh Im I M Sechenova. 2014;100:112–27.
100. Rahman SU, Hassan M. Heart's role in the human body: a literature review. ICCSS. 2013;2:1–6.
101. Garfinkel SN, et al. What the heart forgets: cardiac timing influences memory for words and is modulated by metacognition and interoceptive sensitivity. Psychophysiology. 2013;50:505–12.
102. Gray MA, et al. Emotional appraisal is influenced by cardiac afferent information. Emotion. 2012;12:180–91.
103. Craig AD. How do you feel — now? the anterior insula and human awareness. Nat Rev Neurosci. 2009;10:59–70.

104. Craig AD. How do you feel? interoception: the sense of the physiological condition of the body. Nat Rev Neurosci. 2002;3:655–66.
105. Craig AD. How do you feel?: an interoceptive moment with your neurobiological self. How do you feel?: an interoceptive moment with your neurobiological self. Princeton: Princeton University Press; 2014.
106. Grossmann I, Sahdra BK, Ciarrochi J, Haller J, Glück J. A heart and a mind: self-distancing facilitates the association between heart rate variability, and wise reasoning. Front Behav Neurosci. 2016. https://doi.org/10.3389/fnbeh.2016.00068.
107. Taggart P, Boyett MR, Logantha S, Lambiase PD. Anger, emotion, and arrhythmias: from brain to heart. Front Physiol. 2011;2:67.
108. Atlas SA, et al. Purification, sequencing and synthesis of natriuretic and vasoactive rat atrial peptide. Nature. 1984;309:717–9.
109. Cantin M, Genest J. The heart is an endocrine gland. Kardiol Pol. 1986;29:169–73.
110. Shimano M, Ouchi N, Walsh K. Cardiokines: recent progress in elucidating the cardiac secretome. Circulation. 2012;126: e327–32.
111. Gupta DK, Wang TJ. Natriuretic peptides and cardiometabolic health. Circ J. 2015;79:1647–55.
112. Zois NE, et al. Natriuretic peptides in cardiometabolic regulation and disease. Nat Rev Cardiol. 2014;11:403–12.
113. Bordicchia M, et al. Cardiac natriuretic peptides act via p38 MAPK to induce the brown fat thermogenic program in mouse and human adipocytes. J Clin Invest. 2012;122:1022–36.
114. Frangogiannis NG. The immune system and the remodeling infarcted heart. J Cardiovasc Pharmacol. 2014;63:185–95.
115. Epelman S, Liu PP, Mann DL. Role of innate and adaptive immune mechanisms in cardiac injury and repair. Nat Rev Immunol. 2015;15:117–29.
116. Bianchi ME. DAMPs, PAMPs and alarmins: all we need to know about danger. J Leukoc Biol. 2006;81:1–5.
117. Picard M, et al. Mitochondrial functions modulate neuroendocrine, metabolic, inflammatory, and transcriptional responses to acute psychological stress. Proc Natl Acad Sci. 2015;112:E6614–23.
118. Oka T, et al. Mitochondrial DNA that escapes from autophagy causes inflammation and heart failure. Nature. 2012;485:251–5.
119. Alevizos M, Karagkouni A, Panagiotidou S, Vasiadi M, Theoharides TC. Stress triggers coronary mast cells leading to cardiac events. Ann Allergy Asthma Immunol. 2014;112:309–16.
120. Gilfillan AM, Tkaczyk C. Integrated signalling pathways for mast-cell activation. Nat Rev Immunol. 2006;6:218–30.
121. Janicki JS, Brower GL, Levick SP. Mast cells: methods and protocols. New York: Springer; 2014. p. 121–39. https://doi.org/10.1007/978-1-4939-1568-2_8.
122. Metcalfe DD, Baram D, Mekori Y a. Mast cells. Physiol Rev. 1997;77:1033–79.
123. Levick SP, et al. Cardiac mast cells: the centrepiece in adverse myocardial remodelling. Cardiovasc Res. 2011;89:12–9.
124. Kim MS, Chae HJ, Shin TY, Kim HM, Kim HR. Estrogen regulates cytokine release in human mast cells. Immunopharmacol Immunotoxicol. 2001;23:495–504.
125. Flammer AJ, et al. The assessment of endothelial function: from research into clinical practice. Circulation. 2012;126:753–67.
126. Anderson JL, et al. Parathyroid hormone, vitamin D, renal dysfunction, and cardiovascular disease: dependent or independent risk factors? Am Heart J. 2011;162:331–339.e2.
127. Osto E, et al. Coronary microvascular dysfunction induced by primary hyperparathyroidism is restored after parathyroidectomy. Circulation. 2012;126:1031–9.
128. Davel AP, et al. Endothelial dysfunction in cardiovascular and endocrine-metabolic diseases: an update. Braz J Med Biol Res. 2011;44:920–32.
129. Abdu TA, Elhadd T, Pfeifer M, Clayton RN. Endothelial dysfunction in endocrine disease. Trends Endocrinol Metab. 2001;12:257–65.
130. Colao A, et al. The heart: an end-organ of GH action. Eur J Endocrinol. 2004;151:S93–S101.
131. Colao A. The GH-IGF-I axis and the cardiovascular system: clinical implications. Clin Endocrinol. 2008;69:347–58.
132. Fallo F, et al. Coronary microvascular function in patients with Cushing's syndrome. Endocrine. 2013;43:206–13.
133. Caretta N, et al. Low serum testosterone as a new risk factor for chronic rejection in heart transplanted men. Transplantation. 2013;96:501–5.
134. Caretta N, et al. Erectile dysfunction, penile atherosclerosis, and coronary artery vasculopathy in heart transplant recipients. J Sex Med. 2013;10:2295–302.
135. Sundell J, Knuuti J. Insulin and myocardial blood flow. Cardiovasc Res. 2003;57:312–9.
136. Japundžić-Žigon N. Vasopressin and oxytocin in control of the cardiovascular system. Curr Neuropharmacol. 2013;11:218–30.
137. Cos S, Alvarez-García V, González A, Alonso-González C, Martínez-Campa C. Melatonin modulation of crosstalk among malignant epithelial, endothelial and adipose cells in breast cancer (Review). Oncol Lett. 2014;8:487–92.
138. Tare M, et al. Maternal melatonin administration mitigates coronary stiffness and endothelial dysfunction, and improves heart resilience to insult in growth restricted lambs. J Physiol. 2014;592:2695–709.
139. Dutra FF, Bozza MT. Heme on innate immunity and inflammation. Front Pharmacol. 2014, May;5:115.
140. Çiftçi O, et al. Association between serum γ-glutamyltransferase levels and coronary microvascular function. Coron Artery Dis. 2013;24:201–8.
141. Pittman QJ. A neuro-endocrine-immune symphony. J Neuroendocrinol. 2011;23:1296–7.
142. Tomaselli GF. Introduction to a compendium on sudden cardiac death: epidemiology, mechanisms, and management. Circ Res. 2015;116:1883–6.
143. Daniel M, et al. Risk factors and markers for acute myocardial infarction with angiographically normal coronary arteries. Am J Cardiol. 2015;116:838–44.
144. Nef HM, Möllmann H, Akashi YJ, Hamm CW. Mechanisms of stress (Takotsubo) cardiomyopathy. Nat Rev Cardiol. 2010;7:187–93.
145. Y-Hassan S, Feldt K, Stålberg M. A missed penalty kick triggered coronary death in the husband and broken heart syndrome in the wife. Am J Cardiol. 2015;116:1639–42.
146. Katsanos S, Filippatou A, Ruschitzka F, Filippatos G. Positive emotions and Takotsubo syndrome: 'happy heart' or 'Diagoras' syndrome? Eur Heart J. 2016;37:2821–2.
147. Ghadri JR, et al. Happy heart syndrome: role of positive emotional stress in takotsubo syndrome. Eur Heart J. 2016;37:2823–9.
148. Murakami T, et al. Gender differences in patients with takotsubo cardiomyopathy: multi-center registry from Tokyo CCU Network. PLoS One. 2015;10:e0136655.

149. Tawakol A, et al. Relation between resting amygdalar activity and cardiovascular events: a longitudinal and cohort study. Lancet. 2017;389:834–45.
150. Inoue N. Stress and atherosclerotic cardiovascular disease. J Atheroscler Thromb. 2014:1–11. https://doi.org/10.5551/jat.21709.
151. Steptoe A, Kivimäki M. Stress and cardiovascular disease. Nat Rev Cardiol. 2012;9:360–70.
152. Saban KL, Mathews HL, DeVon HA, Janusek LW. Epigenetics and social context: implications for disparity in cardiovascular disease. Aging Dis. 2014;5:346–55.
153. Cohen BE, Edmondson D, Kronish IM. State of the art review: depression, stress, anxiety, and cardiovascular disease. Am J Hypertens. 2015;28:1295–302.
154. Legault SE, Freeman MR, Langer A, Armstrong PW. Pathophysiology and time course of silent myocardial ischaemia during mental stress: clinical, anatomical, and physiological correlates. Br Heart J. 1995;73:242–9.
155. Strike PC, Steptoe A. Systematic review of mental stress-induced myocardial ischaemia. Eur Heart J. 2003;24:690–703.
156. Pimple P, et al. Association between anger and mental stress-induced myocardial ischemia. Am Heart J. 2015;169:115–21.e2.
157. Rozanski A, et al. Mental stress and the induction of silent myocardial ischemia in patients with coronary artery disease. N Engl J Med. 1988;318:1005–12.
158. Wei J, et al. Meta-analysis of mental stress-induced myocardial ischemia and subsequent cardiac events in patients with coronary artery disease. Am J Cardiol. 2014;114:187–92.
159. Bairey CN, Krantz DS, DeQuattro V, Berman DS, Rozanski A. Effect of beta-blockade on low heart rate-related ischemia during mental stress. J Am Coll Cardiol. 1991;17:1388–95.
160. Ramadan R, et al. Myocardial ischemia during mental stress: role of coronary artery disease burden and vasomotion. J Am Heart Assoc. 2013;2:e000321.
161. Spieker LE, et al. Mental stress induces prolonged endothelial dysfunction via endothelin-A receptors. Circulation. 2002;105:2817–20.
162. Fleshner M. Stress-evoked sterile inflammation, danger associated molecular patterns (DAMPs), microbial associated molecular patterns (MAMPs) and the inflammasome. Brain Behav Immun. 2013;27:1–7.
163. Lugrin J, et al. Cutting edge: IL-1α is a crucial danger signal triggering acute myocardial inflammation during myocardial infarction. J Immunol. 2015;194:499–503.
164. Roth L, et al. Chronic intermittent mental stress promotes atherosclerotic plaque vulnerability, myocardial infarction and sudden death in mice. Atherosclerosis. 2015;242:288–94.
165. Nichols M, Townsend N, Scarborough P, Rayner M. Trends in age-specific coronary heart disease mortality in the European Union over three decades: 1980–2009. Eur Heart J. 2013;34:3017–27.
166. Mehta PK, Wei J, Wenger NK. Ischemic heart disease in women: a focus on risk factors. Trends in Cardiovasc Med. 2015;25:140–51.
167. Rossouw JE, Manson JE, Kaunitz AM, Anderson GL. Lessons learned from the Women's Health Initiative trials of menopausal hormone therapy. Obstet Gynecol. 2013;121:172–6.
168. Mehta LS, et al. Acute myocardial infarction in women: a scientific statement from the American Heart Association. Circulation. 2016;133:916–47.
169. D'Onofrio G, et al. Sex differences in reperfusion in young patients with ST-segment-elevation myocardial infarction: results from the VIRGO study. Circulation. 2015;131:1324–32.
170. Thurston RC, et al. Abuse and subclinical cardiovascular disease among midlife women: the study of women's health across the nation. Stroke. 2014;45:2246–51.
171. Razzolini R, Dal Lin C. Gender differences in heart failure. Ital J Gender-Specific Med. 2015;1:15–20.
172. Kerkhof PLM, Li JK-J, Kresh JY, Heyndrickx GR. Left ventricular diastolic elastance is higher in women compared to men and elevated with betablockade. FASEB J. 2015;29:799.5.
173. Engel G. The need for a new medical model: a challenge for biomedicine. Science (80-). 1977;196:129–36.
174. Fani Marvasti F, Stafford RS. From sick care to health care — reengineering prevention into the U.S. system. N Engl J Med. 2012;367:889–91.
175. Neuman J, Korenstein D, Ross JS, Keyhani S. Prevalence of financial conflicts of interest among panel members producing clinical practice guidelines in Canada and United States: cross sectional study. BMJ. 2011;343:d5621.
176. Giannoni A, et al. Do optimal prognostic thresholds in continuous physiological variables really exist? analysis of origin of apparent thresholds, with systematic review for peak oxygen consumption, ejection fraction and BNP. PLoS One. 2014;9:e105175.
177. Smith R. Medical journals are an extension of the marketing arm of pharmaceutical companies. PLoS Med. 2005;2:e138.
178. Benjamin DK, et al. Safety and transparency of pediatric drug trials. Arch Pediatr Adolesc Med. 2009;163:1080–6.
179. Ryan TJ. Dr Jerome Kassirer's book on the take: how medicine's complicity with big business can endanger your health: worthy of comment. Circulation. 2005;111:2552–4.
180. Rockey SJ, Collins FS. Managing financial conflict of interest in biomedical research. JAMA. 2010;303:2400–2.
181. Reddy VS, et al. Relationship between serum low-density lipoprotein cholesterol and in-hospital mortality following acute myocardial infarction (the lipid paradox). Am J Cardiol. 2015;115:557–62.
182. Cheng K-H, et al. Lipid paradox in acute myocardial infarction—the association with 30-day in-hospital mortality. Crit Care Med. 2015;43:1255–64.
183. Estruch R, Ros E, Salas-Salvado J, Covas M-I, Coreila D, Ards F, Gómez-Gracia E, Ruiz-Gutierrez V, Fiol M, Lapetra J, Lamuela-Raventos RM, Serra-Majem L, Pinto X, Basora J, Munoz MA, Sorli JV, Martinez JA, Martinez-Gonzalez MA, for the P. S. I.-. Primary prevention of cardiovascular disease with a Mediterranean diet. NEJM. 2013;368:1279.
184. Dal Lin C, et al. Thoughts modulate the expression of inflammatory genes and may improve the coronary blood flow in patients after a myocardial infarction. J Tradit Complement Med. 2017. https://doi.org/10.1016/j.jtcme.2017.04.011.
185. Anderson RH, et al. Assessment of the helical ventricular myocardial band using standard echocardiography. Echocardiography. 2015;32:1601–2.

Mitochondria and Sex-Specific Cardiac Function

16

Rosa Vona, Barbara Ascione, Walter Malorni, and Elisabetta Straface

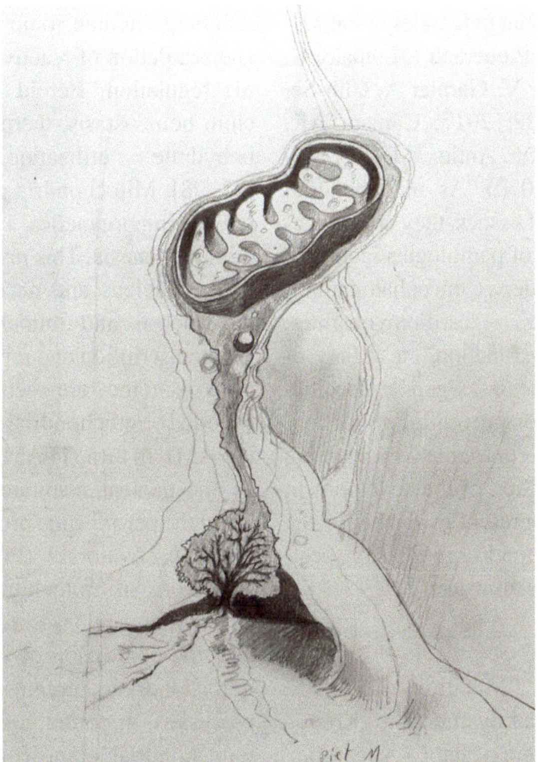

Cardiac mitochondria. Art work by Piet Michiels, Leuven, Belgium

R. Vona · B. Ascione · W. Malorni · E. Straface (✉)
Biomarkers Unit, Center for Gender-Specific Medicine, Istituto Superiore di Sanità, Rome, Italy
e-mail: rosa.vona@iss.it; barbara.ascione@iss.it; malorni@iss.it; elisabetta.straface@iss.it

Abstract

The focus of this chapter is the gender differences in mitochondria in cardiovascular disease. There is broad evidence suggesting that some of the gender differences in

P. L. M. Kerkhof, V. M. Miller (eds.), *Sex-Specific Analysis of Cardiovascular Function*, Advances in Experimental Medicine and Biology 1065, https://doi.org/10.1007/978-3-319-77932-4_16

cardiovascular outcome may be partially related to differences in mitochondrial biology (Ventura–Clapier R, Moulin M, Piquereau J, Lemaire C, Mericskay M, Veksler V, Garnier A, Clin Sci (Lond) 131(9):803–822, 2017)). Mitochondrial disorders are causally affected by mutations in either nuclear or mitochondrial genes involved in the synthesis of respiratory chain subunits or in their posttranslational control. This can be due to mutations of the mtDNA which are transmitted by the mother or mutations in the nuclear DNA. Because natural selection on mitochondria operates only in females, mutations may have had more deleterious effects in males than in females (Ventura–Clapier R, Moulin M, Piquereau J, Lemaire C, Mericskay M, Veksler V, Garnier A, Clin Sci (Lond) 131(9):803–822, 2017; Camara AK, Lesnefsky EJ, Stowe DF. Antioxid Redox Signal 13(3):279–347, 2010). As mitochondrial mutations can affect all tissues, they are responsible for a large panel of pathologies including neuromuscular disorders, encephalopathies, metabolic disorders, cardiomyopathies, neuropathies, renal dysfunction, etc. Many of these pathologies present sex/gender specificity. Thus, alleviating or preventing mitochondrial dysfunction will contribute to mitigating the severity or progression of the development of diseases. Here, we present evidence for the involvement of mitochondria in the sex specificity of cardiovascular disorders.

Keywords

Mitochondrial biology · Mutations of mtDNA · Mitochondrial dysfunction · Krebs cycle · Nuclear respiratory factor · Apoptotic bodies · Redox messenger · Reactive oxygen species · Calcium overload · Heart failure · Ischemic preconditioning · Autophagy · Mitophagy · Dynamin · Aging heart

Mitochondrial Functions

Mitochondria are crucial, multifunctional organelles, which actively regulate cellular homeostasis. They have a double membrane. The composition of the outer membrane is similar to that of other eukaryotic membranes. The inner membrane resembles prokaryotic membranes in composition and physical properties [102]. In the matrix, mitochondria contain a circular small genome (mtDNA) encoding for 13 proteins of the respiratory chain and for both ribosomal ribonucleic acid (rRNA) and transfer RNA (tRNA). Mutations in mtDNA result in severe neuromuscular diseases, mostly due to impaired energy production. The main function of mitochondria is the energy production as adenosine triphosphate (ATP) via citric cycle (tricarboxylic acid cycle, Krebs cycle). Other cell functions include ionic homeostasis, production and regulation of reactive oxygen species (ROS), pH regulation, steroid hormone synthesis, calcium homeostasis, thermogenesis, lipid and carbohydrate utilization, and cell death [82, 98]. Mitochondria proliferate by division of preexisting organelles, a process called mitochondrial biogenesis. This *process* is under the control of the nucleus and necessitates coordination of the nuclear and mitochondrial ones genomes. The transcription and replication of the mitochondrial genome are activated by the nuclear-encoded mitochondrial transcription factor A (TFAM). In turn, TFAM transcription is activated by the nuclear respiratory factors (NRF) 1 and 2 and the peroxisome proliferator-activated receptor γ coactivators 1 (PGC-1α or β), the master regulators of mitochondrial biogenesis. These coactivators and transcription factors also coordinate the expression of multiple nuclear-encoded mitochondrial proteins, which have to be processed, imported, and localized in the proper mitochondrial compartment with mitochondria-encoded proteins [122].

Mitochondria are essential for cardiac function. They play a major role in ATP supply, needed to support continuous cycles of contraction and relaxation, carry out synthesis of essential cellular components, calcium buffering [28], and trigger cell death signals [27]. Because the heart is highly dependent on mitochondrial oxidative energy, it is understandable that defects in mitochondrial structure and function can be found in association with different cardiovascular diseases. Indeed, abnormalities in the mitochondrial function cause cardiomyopathies, arrhythmias, and abnormalities of the conduction system. Mitochondrial dysfunction has been linked to several cardiovascular disorders, including hypertension, cardiac hypertrophy, ischemia/reperfusion, and heart failure [61, 117]. Importantly, mitochondria are directly involved in triggering of different and complexly interconnected programs controlling cell "fate" such as apoptosis, autophagy (mitophagy), and senescence, all involved in cardiovascular disorders [2]. Below will be expanded the individual programs in detail and their implications in cardiovascular diseases.

Mitochondria and Apoptosis

Mitochondria are instrumental for cell life and death. They play a central role in various forms of cell death, which are characterized by differential biochemical features, with predominant forms including apoptosis (caspase-dependent and caspase-independent), or necrosis. Besides amplifying and mediating extrinsic apoptotic pathways, mitochondria also play a central role in the integration and propagation of death signals originating from inside the cell such as DNA damage, oxidative stress, starvation, as well as those induced by radiation or chemotherapy [68]. Apoptosis, also known as programmed cell death, describes a particular mode of cell death that is characterized by a series of biochemical events that lead to a variety of morphological changes, including cell shrinkage, membrane blebbing or budding, nuclear fragmentation, chromatin condensation, and chromosomal DNA fragmentation. Ultimately, the cell is fragmented into compact membrane-enclosed structures, called "apoptotic bodies," which contain cytosol, condensed chromatin, and organelles. Apoptotic bodies are recognized by phagocytes and quickly removed preventing inflammation and tissue damage that might ensue upon cell lysis [1, 39]. Apoptosis is induced via two main ways involving either the mitochondria (the intrinsic pathway) or the activation of death receptors (the extrinsic pathway). Both pathways converge to induce the activation of caspases, executioners of cell death [129]. The link between the caspase signaling cascade and the mitochondria is provided by the Bcl-2 family member Bid. This protein is cleaved by caspase-8 and in its truncated form (tBID) translocates to the mitochondria where it induces the translocation, oligomerization, and insertion of other Bcl-2 family members, which control the integrity of the outer membranes of mitochondria. On the surface of the mitochondrion, Bcl-2 antiapoptotic proteins detect mitochondrial damage and activate two proapoptotic Bcl-2 effector proteins, Bax and Bak (Fig. 16.1). Interactions among these proteins break up the outer mitochondrial membrane and shape channels allowing the release of proteins present in the mitochondrial intermembrane space (such as the cytochrome-c, Smac/Diablo, and apoptosis-inducing factor (AIF)). Once cytochrome c is released, it binds to Apaf-1 to assemble the apoptosome, a complex that triggers the activation of the initiator procaspase-9 [64]. Activation of caspases induces a biochemical cascade, which leads to characteristic changes of cell morphology such as blebbing, shrinkage, nuclear and chromosomal DNA fragmentation, and chromatin condensation [1, 34, 111]. Permeabilization of the outer mitochondrial membrane is antagonized by antiapoptotic proteins such as Bcl-2, Bcl-W, Bcl-xL, A1/Bfl1, and MCL-1, which inhibit Bax and Bak functions. In addition to cytochrome c, other proteins are released to assist or potentiate apoptosis. For instance, Smac/Diablo antagonizes inhibitors of caspases, thereby enhancing caspase activation and apoptosis, and AIF that insures caspase-independent mitochondria-mediated apoptosis inducing chromatin condensation and DNA fragmentation [68].

The mitochondrial permeability transition is a permeability increase of the inner mitochondrial membrane mediated by a channel, the permeability transition pore (PTP) [7, 8]. PTP opening is affected by inducers like calcium and ROS or inhibitors like acidic matrix pH. While short-term opening may participate in physiological regulation of Ca^{2+} and ROS homeostasis, long-lasting opening of the PTP triggers mitochondrial swelling, rupture of the outer membrane, collapse of membrane potential, cessation of ATP synthesis, release of cytochrome c, and other proapoptotic factors which initiate the mitochondrial pathway of apoptosis [8].

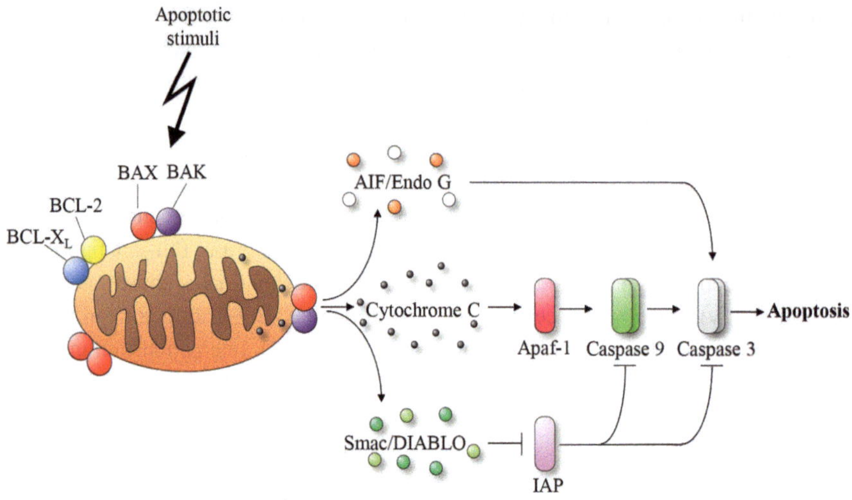

Fig. 16.1 Scheme of the apoptosis. The proteins of the BCl-2 family control the external mitochondrial membrane integrity. After apoptotic stimuli, BAX and BAK heterodimer break up the outer mitochondrial membrane and opening a channel, which allows the release of proteins present in the mitochondrial intermembrane space such as the cytochrome-c, Smac/DIABLO, and Endo G/AIF. The latter, in turn, led to caspase cascade activation inducing apoptotic modifications. Mitochondrial outer membrane permeabilization is antagonized by the antiapoptotic BCL-2 proteins, such as BCL-2, BCL-W, and BCL-Xl

As mentioned above, mitochondria are the primary intracellular site of oxygen consumption and the major source of ROS, most of them originating from the mitochondrial respiratory chain. In cells with high oxidative capacity as cardiomyocytes, mitochondria are an essential source of ROS and the most direct target for their damaging effects. Cardiomyocyte apoptosis, induced by the overproduction of ROS under ischemia or ischemia/reperfusion (I/R) or ischemic preconditioning (IPC), is an important pathological phenomenon in heart failure (HF). Therefore, a fine equilibrium between ROS production and removal determines the physiological versus pathological function of ROS. In fact, an excessive amount of ROS induces oxidative stress and promotes cell death under hypoxic conditions. Conversely, at physiological levels, ROS function as "redox messengers" in intracellular signaling [22]. For this reason mitochondria contain an arsenal of antioxidant systems with target specificity [68]. ROS can be removed by antioxidant enzymes such as superoxide dismutase, catalase, and glutathione peroxidase. The first line of defense against ROS is guaranteed by the presence of Mn^{2+}-SOD (SOD2) in the mitochondrial matrix, which results in superoxide anion dismutation and the subsequent generation of hydrogen peroxide. Although hydrogen peroxide is not a free radical, it is an oxidant and an intermediate in the chain of reactions that generate reactive free radicals, such as hydroxyl radical, which can oxidize mitochondrial components (proteins, lipids, DNA). Since most mitochondria lack catalase, the metabolism of hydrogen peroxide is mainly accomplished by mitochondrial glutathione, mGSH, with the participation of either GSH peroxidase or peroxiredoxins [22]. It has been demonstrated that a chronic exposure to ROS in the heart will induce apoptosis and fibrosis leading to heart dysfunction and remodeling [85].

Mitochondria play also an essential role in cell calcium homeostasis. They accumulate calcium along its electrochemical gradient, whereas calcium extrusion is an active process involving Na-dependent and Na-independent pathways. Calcium excess leads to uncoupling of oxidation from phosphorylation and mitochondrial membrane depolarization resulting in the opening of the PTP and finally in cell death [123]. The capacity of mitochondria to control excess calcium depends on the balance between calcium entry along the electrochemical gradient and calcium extrusion mechanisms. This plays a crucial role in cell homeostasis and cell death. Apoptosis is a physiological process, which is necessary for differentiation during embryogenesis. However, apoptosis is also triggered in pathological conditions. In cardiovascular diseases apoptosis may be induced also by mitochondrion-toxic agents. These agents are numerous and include pharmaceuticals, illicit drugs, exotoxins, and food ingredients [15, 63, 75, 103]. Several mitochondrial functions are prone to be affected by toxic agents. The most important among these is the respiratory chain. Dysfunction of the respiratory chain may result in decreased ATP production, increased ROS production, reduced antioxidative capacity, and reduced mitochondrial membrane potential or apoptosis [69]. A number of chemotherapeutic agents are cardiotoxic due to mitochondrial dysfunction [107]. Cardiotoxicity includes acute or chronic cardiovascular complications, which impair quality of life, even years after treatment. Chemotherapeutic agents associated with cardiotoxicity due to mitochondrial dysfunction include anthracyclines (e.g., doxorubicin), anthracenediones (e.g., mitoxantrone), alkylating agents (e.g., cyclophosphamide, cisplatin, ifosfamide, busulfan, mitomycin), vinca alkaloids (e.g., 5-fluorouracil, amsacrine, asparaginase), tyrosine-kinase inhibitors (e.g., imatinib, dasatinib, sunitinib, sorafenib), and other agents (e.g., trastuzumab, paclitaxel, etoposide, teniposide) (Simbre et al. 2005; [34]).

Mitochondria and Autophagy

Macroautophagy, hereafter referred to as autophagy, is a conserved process aimed at maintaining of cellular and tissue homeostasis under normal as well as stress conditions, including nutrient starvation, changes in metabolism, energy, and oxygen status. Autophagy is a degradation mechanism of cytoplasmic components, including damaged organelles, toxic protein aggregates, and intracellular pathogens [72]. Basal autophagy plays a key role in eukaryotic cells, degrading long half-life macromolecules and large supramolecular structures, including organelles such as mitochondria, peroxisomes, and endoplasmic reticulum [60]. This process is characterized by several steps. Initiation of autophagy describes the formation of the isolation membrane and phagophore, which then expands to engulf the cargo (protein aggregates or damaged organelles), thus forming the autophagosome, a double-membrane intracellular structure of reticular origin. This process is completed by autophagosome clearance, which occurs after fusion with lysosomes, enabling degradation of cargo by lysosomal enzymes [72] (Fig. 16.2). Degradation by-products, such as amino acids, can then be reused for the building of new

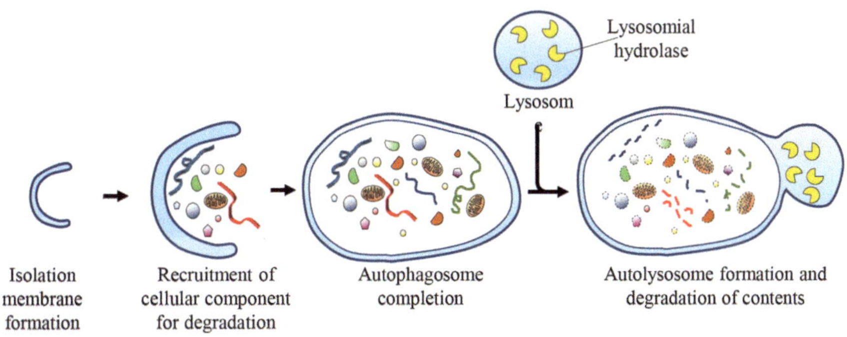

Fig. 16.2 Scheme of the autophagy process. Phagophore determines the onset of the autophagy. The progressive elongation of the phagophore is accompanied by the recruitment (specific or not) and degradation (autophagosome completion) of cellular components. The outer membrane of the autophagosome may fuse with a lysosome to form autophagolysosome. Finally, the engulfed material is degraded inside the autophagolysosome and recycled

macromolecules or for meeting metabolic demands [54, 127]. Initial phagophore formation requires the assembly of a complex consisting of BECLIN1, vacuolar protein sorting (VPS) 34, and VPS15 [49]. Next, the expansion of the membrane is mediated by two ubiquitin-like conjugation systems, microtubule-associated protein 1 light chain 3 (LC3) and autophagy protein (ATG) 12-ATG5, that promote assembly of the ATG16L complex and the conjugation of LC3 with phosphatidylethanolamine [48, 73]. Autophagy contributes to the maintenance of intracellular homeostasis in most cells of cardiovascular origin, including cardiomyocytes, endothelial cells, and arterial smooth muscle cells.

Mitophagy is an autophagic response that allows elimination of defective mitochondria and accelerates the mitochondrial turnover, thus preserving the pool of healthy organelles [95]. As the mitochondria occupy a critical position in the bioenergetics of the cardiovascular system, mitophagy is particularly important for cardiovascular homeostasis in health and disease [11]. Upstream, among others, two main actors in the regulation of mammalian mitophagy are the serine/threonine kinase PTEN-induced putative kinase 1 (PINK1) and the E3 ubiquitin ligase Parkin, which selectively promote the degradation of impaired mitochondria [78]. Under normal conditions, PINK1 is imported into the mitochondria, where it undergoes rapid cleavage by protease PARL, and maintained at low levels on the inner membrane. When the mitochondria present a decrease of membrane potential, PINK1 accumulates on the outer membrane. In healthy mitochondria, Parkin translocates from cytosol to damaged or uncoupled mitochondria, promoting ubiquitination of several mitochondrial outer membrane proteins, such as MFN1, MFN2, and VDAC. On damaged mitochondria recruitment of Parkin is selective and requires the participation of PINK1 [77]. Ubiquitination of proteins allows recruitment of p62/SQSTM1, an adaptor that interacts with ubiquitinated proteins and LC3, recruiting a phagophore to engulf the ubiquitinated mitochondrion [99]. Although Parkin-induced mitophagy has been shown to be dependent on the activity of PINK1, this process depends on DRP-1-mediated mitochondrial fission [35, 74, 114, 120] (Fig. 16.3). A recent study showed that Parkin mRNA and protein are present at low levels in normal mouse hearts but are upregulated after cardiac myocyte-specific deletion of the *Drp1* gene in adult mice [14]. Thus, *Drp1* deficiency appears to trigger Parkin-dependent over-activation of mitophagy leading to a severe myopathic phenotype. The authors propose that DRP1 helps in the maintaining mitochondrial quality control by promoting mitochondrial fission to segregate dysfunctional mitochondria that can then be targeted by mitophagy [109]. These data highlight the central role of mitochondrial dynamics in cardiac mitochondrial quality control, ensuring proper elimination of damaged and dysfunctional organelles [14].

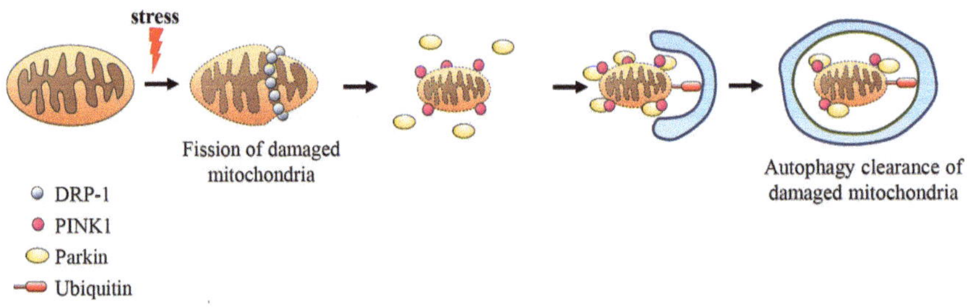

Fig. 16.3 Scheme of autophagy clearance of damaged mitochondria. Damaged mitochondria (usually with low membrane potential) undergo DRP1-mediated fission to initiate the process of mitophagy. Under reduced mitochondrial membrane potential, PINK1 accumulates on the outer mitochondrial membrane and allows the recruitment of the Parkin E3 ubiquitin ligase. Parkin induces ubiquitination of several surface proteins, which, in turn, trigger the damaged mitochondrion removal by an autophagosome

Mitophagy is also induced by Parkin-independent mechanisms that involve proteins or lipids already present on the outer membrane of the mitochondria. BNIP3 and BNIP3L/Nix, known for their role in apoptosis, interact directly with LC3 to initiate mitochondrial clearance [43, 91]. Fundc1 is a mitochondrial protein located on the outer membrane. It contains a binding domain for LC3 participating in mitochondrial engulfment. These pathways are triggered by different stimuli. As already mentioned, Parkin-PINK1 pathway is induced by a decrease in membrane potential, while Fundc1 has been reported to play an important role in the response to hypoxic stress [62]. All of them may be involved in cardiac stress [38]. Indeed, Parkin translocation to the mitochondria takes place during I/R [56]. It has been suggested that infarction-induced mitophagy is a beneficial homeostatic response to protect the heart [29]. Bnip3 is also activated during I/R and triggers mitophagy [57]

Mitochondria and Senescence

Cellular senescence is an irreversible growth arrest accompanied by inability to repair tissue damages occurring in response to various cellular stimuli, such as telomere erosion, DNA damage, and oxidative stress. Senescence is associated with the appearance of several biomarkers such as β-galactosidase activity and p53 activation [5, 12]. In model organisms, senescence is accompanied by mitochondrial malfunctions, which subtend the observed age-dependent decline in organ function [96]. Similarly, defects in mitochondrial function have also been observed in human. Moreover, some mitochondrial damages may predispose humans to certain age-related diseases. As a result, chronic diseases, including cardiovascular diseases, increase their prevalence with aging. Cardiac aging is characterized by the presence of hypertrophy, fibrosis, and defect in contractility, calcium handling, cell metabolism, and mitochondrial function [55]. The aging heart is generally associated with decreased protein quality control and dysfunctional mitochondria [106, 116].

Increased production of ROS emerged as the main proximate cause of aging. Mitochondria have been involved as key players, first because they are an important source of free radicals, especially in highly oxidative tissues, and second because they are a direct target of oxidative damage in aging cells [123]. As age increase, ROS tend to raise their contents as a result of the accumulation of damages to mitochondrial proteins, an imbalance between oxidative stress and antioxidant mechanisms, and declines in activity of mitochondrial respiratory chain complexes. Considering this, ROS regulate senescence diseases [86, 116]. There is increasing evidence that mitochondria can regulate cellular aging through the modulation of the metabolic profile of the cell [106, 113]. The key role of ROS generation and mitochondrial dysfunction in cardiac aging is supported by experiments that target mitochondrial ROS. For example, overexpression of the ROS scavenger enzyme catalase attenuates the development of hypertrophy, fibrosis, and diastolic dysfunction in the aging mouse heart [25, 51]. By contrast, prematurely aging mice with a mutation in mitochondrial DNA polymerase exhibit marked cardiac hypertrophy and fibrosis as well as systolic and diastolic dysfunction [26, 51]. Together, these observations suggest that ROS generation and mitochondrial damage contribute to cardiac aging.

ROS production by mitochondria may also be a significant mechanism by which cardiovascular risk factors lead to the formation of vascular lesions in a sex-specific manner [111]. Vascular diseases are instrumental in aging, as well as cardiac and neurological disorders.

Fusion and Fission of Mitochondria

Mitochondria are present in all cells except for erythrocytes. Depending on a tissue's function, mitochondria may be more or less prominent within a cell type. In the heart about 45% of

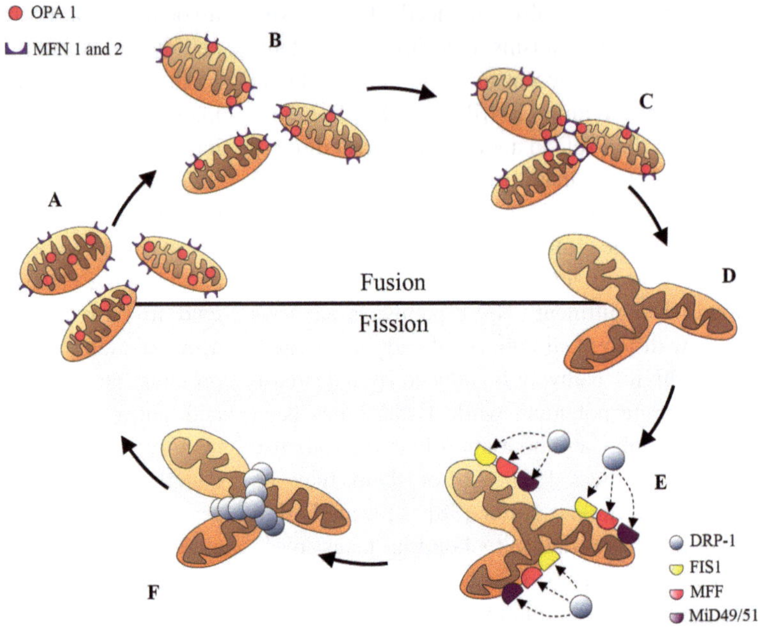

Fig. 16.4 Mitochondrial fusion and fission processes subtending the cell homeostasis (*A–D*) *Mitochondrial fusion*. The fusion process is driven by mitofusins (MFN1 and 2) and OPA1 usually located on the outer and inner mitochondrial membranes, respectively. Fused mitochondria both share materials (matrix components, damaged mitochondrial DNA) and bioenergetic properties (e.g., mitochondrial membrane potential) (*D–F*) *Mitochondrial fission*. DRP1 bind to FIS1, MFF, and MiD49/51 adapters starting to the moiety separation (fission)

the myocardial volume is taken up by mitochondria [31].

Interestingly, morphology of mitochondria differs among different cell types: for example, solitary organelles are retrieved in hepatocytes, whereas in many epithelial cells, they form an intricate network [23]. In adult cardiac myocytes, mitochondria localize within three subcellular distributions: interfibrillar, subsarcolemmal, and perinuclear. The mitochondrial morphology is mostly globular in the perinuclear region and predominantly rod shaped in the subsarcolemmal space, while inter-myofibrillar mitochondria have the same size of a sarcomere [31].

Ultrastructure and morphology of mitochondria are continuously regulated by fusion and fission balance events [6] (Fig. 16.4). Mitochondrial fusion is a complex sequential process, which involves integration of the outer mitochondrial membrane, inner mitochondrial membrane, and matrix content. The main regulators of these processes are the GTPase dynamin-related proteins: mitofusin 1 (MFN1) and mitofusin 2 (MFN 2), located in the outer mitochondrial membrane, and optical atrophy 1 (OPA1), located in the inner mitochondrial membrane [120]. MFN1 and MFN 2 are both needed for mitochondrial *fusion. In fact, their* loss results in a reduction of this process. Moreover, these proteins are partially redundant in their function; thus when MFN1 expression is decreased, mitochondrial fusion can be supported by MFN2 overexpression and vice versa [17]. Additionally, MFN2 has been implicated in several other physiological functions, such as modulation of energetic processes, endoplasmic reticulum-mitochondria coupling, and regulation of mitophagy, a process through which mitochondria are engulfed by autophagosomes and delivered to lysosomes for degradation [18, 120]. OPA1 is a transmembrane protein tightly associated with the mitochondrial inner membrane. The transcript coding for OPA1 is subjected to alternative splicing, giving rise to eight variants that are expressed in a variety of patterns across different tissues. In particular, both OPA1 short (88 kDa) and long (112 kDa) isoforms are necessary for mitochondrial fusion [108].

The opposite process (mitochondrial fission) results from mitochondria fragmentation and is regulated by the large GTPase dynamin-related protein (DRP-1), mitochondrial fission 1 (FIS1), mitochondrial fission factor (MIFF), and mitochondrial dynamic proteins of 49 and 51 kDa

(MiD49/51) [42, 97]. DRP-1 is a cytosolic protein, which lacks a mitochondrial destination sequence. It requires localization of FIS1 in the mitochondrial outer membrane to form the fission complex [10]. DRP1, like dynamins, functions as a mechanoenzyme, serving to constrict the mitochondrion physically, an early step in fission. However, in mammalian cells silencing FIS1 has little effect on DRP1 translocation to mitochondria [58]. MFF appears to be the protein acting as the DRP1 mitochondrial receptor. Reduction of MFF levels induces mitochondrial elongation a decrease of DRP1 translocation to mitochondria [36]. Likewise, MiD49 and MiD51 are involved in the mammalian fission machinery [81].

These processes play a central role in mitochondria quality control and are important for maintaining various cellular functions and viability. Fusion and fission balance is very important for mitochondrial participation in crucial cellular processes. The first is a necessary adaptation to nutrient starvation and increased metabolic demand [92]. The latter is required for mitochondrial proliferation following mitosis [50], apoptosis [128], and removing damaged mitochondria from the cells through mitophagy [118]. Mitochondrial fission is essential for maintenance and repair. It facilitates the removal of damaged components by partitioning them, so they can be targeted for removal and degradation by mitophagy. However, excessive mitochondrial fission and mitophagy can compromise the metabolic capacity of a cell [119]. Therefore, it is necessary to maintain an adequate equilibrium in the fission/fusion cycle. The connection between these processes and the onset of heart disease is not yet entirely clear. Although mitochondrial fusion and fission are most evident in neonatal cardiac myocytes, there is strong evidence that these processes are important in the adult cardiac tissue. In fact, the proteins involved in fusion and fission are highly expressed in the adult heart [30, 101]. More evidences for the importance of mitochondrial dynamics in the heart result from studies on the loss of functions of proteins involved in this process. For example, in cardiac myocytes isolated from murine models of an inducible ablation of MFN-1/2, fragmented mitochondria with abnormal cristae have been detected [83]. These morphological changes were accompanied by alterations in mitochondrial respiration, mitophagy, and mitochondrial biogenesis, which culminated in cardiomyopathy [21, 84]. Abnormalities in mitochondrial morphology were also observed in mice heterozygous for OPA1. These mitochondria, large and with abnormal cristae, manifest increased ability to accumulate Ca^{2+} and are marked by a delayed opening of the mitochondrial PTP coupled with increased sensitivity to prolonged mechanical stress. The decreased expression of OPA1 did not alter mitochondrial respiration [88]. Chen et al. described small and fragmented mitochondria in both human and rat models of heart failure, which were associated with decreased OPA1 levels [19, 20]. Decreased levels of Mfn2 mRNA were detected in neonatal rat cardiac myocytes exposed to phenylephrine to induce hypertrophy and in vivo models of cardiac hypertrophy [32]. Similar results were reported by Papanicolaou et al. They found that MFN2-deficient mice display modest cardiac hypertrophy accompanied by a slight functional deterioration [83]. Ong and coworkers found that inhibition of DRP1 increases the proportion of adult cardiac myocytes with elongated mitochondria and protects them against simulated I/R [79]. Moreover, Sharp et al. reported that DRP1 inhibition has therapeutic benefits even when administered after ischemia [104].

There are many evidences of the role played by the mitochondrial dynamics on vascular smooth muscle cells (VSMCs) and its effect on vascular diseases, which are instrumental for cardiac disorders. The principal function of VSMCs is the regulation of vascular tone and, consequently, blood pressure and blood flow. Unlike cardiac myocytes, VSMCs are highly plastic and undergo reversible changes in phenotype in response to environmental stimuli [46, 67]. When a vessel is damaged, these cells, which differentiated have a contractile phenotype characterized by little proliferation, proliferate and migrate toward the injury site [47]. Recently, a relationship has been proposed between the

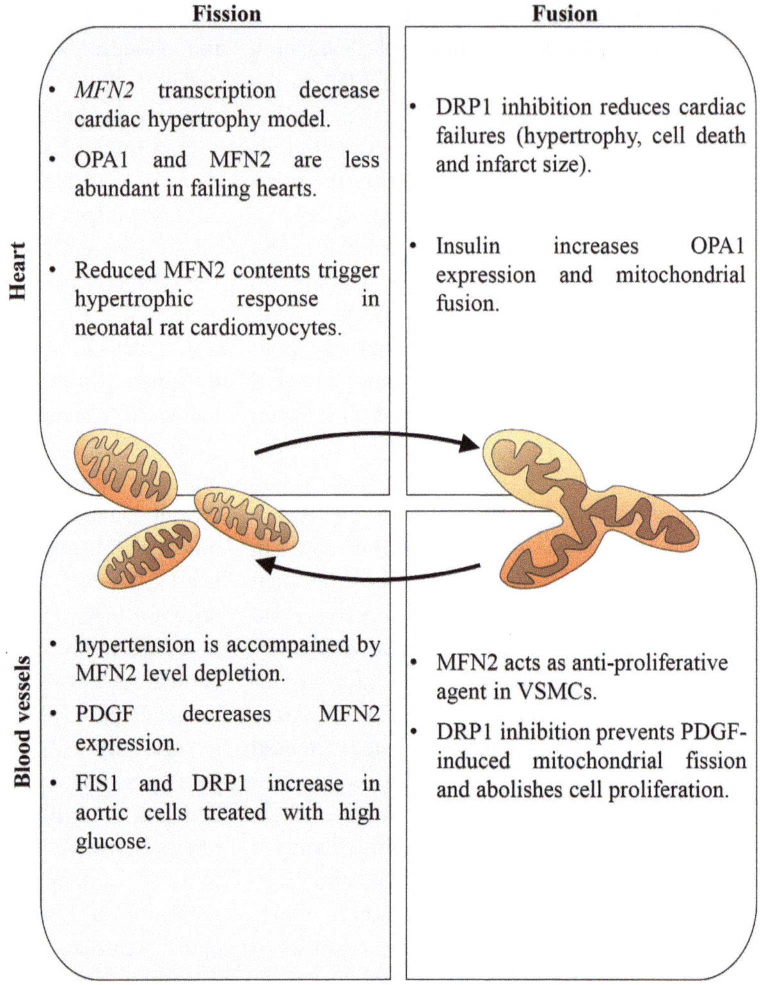

Fig. 16.5 Mitochondrial dynamics and cardiovascular diseases Summary of major molecular events related to the mitochondrial fission and fusion in cardiovascular diseases

VSMC proliferative phenotype and mitochondrial dynamics [16, 65]. VSMC proliferation can be induced by several factors, including platelet-derived growth factor (PDGF), insulin-like growth factor 1 (IGF-1), angiotensin II, and endothelin-1 [45, 90, 125]. Lately, it has been shown that PDGF induces mitochondrial fragmentation reducing MFN2 levels [100]. Shenouda et al. reported mitochondrial fragmentation and increased levels of FIS1 and DRP1 proteins in cultured human aortic endothelial cells incubated with high glucose medium [105]. Figure 16.5 summarizes the effects of proteins involved in fusion-fission mechanism on cardiovascular disease [120].

Mitochondria and Gender Difference

It is well known that gender affects several health issues. Women are more susceptible than men to depression, osteoporosis, asthma, smoke-induced lung cancer, and autoimmune diseases. However, not all medical problems show gender dimorphism. For example, males do not differ from females in infection responses [59]. Both clinical and experimental observations show that also in cardiovascular diseases, gender differences play a key role. The incidence of cardiovascular diseases increases with age in both sexes, although men and women are predisposed toward different cardiovascular diseases with old age [33, 40, 51]. Male/female differences in coronary artery disease, including a higher risk of obstructive disease in men and more microvascular disease in women [41], clearly contribute to sex differences in cardiovascular disease expression.

Men undergo higher health risks (ischemic heart diseases, hypertension, arrhythmias, and heart failure) than age-matched premenopausal women. This data, together with the observation that incidence of cardiovascular diseases tends to increases in women after menopause, support the notion that the incidence of cardiovascular diseases is associated with the decreasing levels of estrogens during menopause [93].

It has been shown that cardiomyocytes from female rats exhibit lower mitochondrial content, but female mitochondria are more efficient, more differentiated, and generate less ROS than the male ones [123].

As said above, mitochondria are particularly abundant in the heart, and their content and function are sex-dependent. It has been reported that the female cardiac muscle generates less mitochondrial ROS and exhibits lower oxidative damage than those of males. Several studies demonstrate that cardiovascular diseases such as myocardial infarction and atherosclerosis are underrepresented in premenopausal women compared to their male counterparts [4]. Although the mechanisms underlying sexual dimorphism in disorder development are uncertain, sex differences in the levels of biomarkers responding to oxidative stress have been observed in clinical and experimental studies. It has been shown that oxidative stress biomarkers are lower in healthy young women than in age-matched men [44, 89]. Moreover, studies on rat model systems shown that male produced more ROS than age-matched females [24].

Significant gender differences were also found in the uptake of Ca^{2+} by cardiac mitochondria [3]. Mitochondria from female rat heart showed lower Ca^{2+} uptake rates in physiological substrate solutions and maintained mitochondrial membrane potential in presence of Ca^{2+} at high dosage. Since mitochondrial calcium overload is an important factor in defining cardiac I/R injury, these differences may be explain why female myocardium suffers less injury with I/R [80, 123].

It is widely accepted that mitochondrial dysfunction, and particularly mitochondrial PTP opening, plays a major role in determining the extent of cardiac I/R injury. Recently, it has been proposed that the increased resistance of female heart mitochondria reflects regulation of mitochondrial PTP function rather than changes in the putative components [71]. In this context, sex hormones could be responsible for these effects because they have been reported to directly regulate several mediators of the mitochondrial biogenesis program. Regulation of mitochondrial function and biogenesis by estrogens/estrogen receptors has been extensively reviewed [19, 20, 53, 87]. Estrogen receptor may bind to mtDNA and is involved in the E2-induced expression of mtDNA and respiratory chain proteins. Several studies demonstrated that sex differences in oxidative stress markers are estrogen-dependent and that estrogen may exert protection in females by increasing antioxidant defenses and by diminishing ROS production [124]. Interestingly, a cross talk between mitochondrial function and sex-steroidal hormones during aging is reported [121]. Finally, estrogen may be considered a key factor subtending sex differences in mitochondrial function and morphology, ROS production, and antioxidant activity. In fact, estrogen by its receptor can upregulate the expression of NRF1, a key transcription factor regulating transcription of majority of mitochondrial respiratory chain complexes. Moreover, estrogen can also modulate the transcription of NRF1 indirectly through interaction with another transcription factor, PGC-1 [52].

Literature data confirm that premenopausal women experience less cardiovascular diseases compared with age-matched men. However, in postmenopausal women the rate of cardiovascular disease development and mortality from cardiac disease exceeds those of men. This corroborates the key role played by estrogens. Considering sex differences in the development of HF, it was shown that in hearts of female mice subjected to pressure overload, downregulation of metabolic and mitochondrial biogenesis transcription cascade genes are less important than in male hearts [126]. These differences could contribute to the protection of females against HF.

Sex-related specificity has also been found in HF induced by toxic agents like anthracyclines. Anticancer therapies involving anthracyclines are limited by their cardiotoxicity. His cardiotoxicity is considered as a complex multifactorial process involving oxidative stress and mitochondrial damage [112, 115]. It has been demonstrated that female rats seem much less sensitive to the cardiotoxic effects of doxorubicin (the main anthracycline) than male rats [37, 76].

Also in ischemic heart disease, the leading cause of morbidity and mortality in both men and women, women have lower risk before menopause [80, 94]. Ischemia and postischemic reperfusion cause a wide array of functional and structural injuries to mitochondria, due in part to excess production of ROS and calcium overload. This triggers PTP opening, decrease in ATP supply, and ultimately cell death [8]. Heart from female rat is more resistant to oxygen deprivation [9], and ischemic reperfusion injury induces lower infarct size in female than male rats [70]. A similar observation was made in patients with acute myocardial infarction. Primary percutaneous coronary intervention results in better myocardial salvage in women than men [70].

Sex-dependent ROS production by mitochondria may also be a significant mechanism by which cardiovascular risk factors lead to the formation of vascular lesions in a sex-specific manner [111].

Other sex differences may be revealed in response to cell stress. Exposure of rat VSMCs to ultraviolet radiation induces an upregulation of survival proteins in cells from females and an increased proapoptotic proteins and loss of mitochondrial membrane potential in cells from males [66]. Furthermore, cells from female rats show adhesion-associated resistance to apoptosis, which is apparently due to a more adhering phenotype, characterized by a higher propensity to undergo survival by autophagy [110].

Summing clearly, part of the protective effect of female sex in CVDs is linked to (i) better protection of mitochondrial function and content in female heart and vessels and (ii) better ability to handle calcium and to decrease ROS production.

Conclusions

Mitochondria are pivotal organelles for cell fate. Their involvement in cardiovascular diseases is increasingly acknowledged. Alterations of mitochondrial mass and function play an important role in CVDs. Particularly, can be hypothesized that a different mitochondrial function could be responsible for gender differences in CVDs. Future research should consider both sexes, not only to better understand the pathophysiology of the diseases but also to suggest more appropriate therapeutic interventions.

References

1. Alberts B, Johnson A, Lewis J, Raff M, Roberts K, Walter P. Apoptosis: programmed cell death eliminates unwanted cells. In: Molecular biology of the cell. Garland science, 5th ed. 2008. p. 1115. ISBN:978-0-8153-4105-5.5b.
2. Apostolova N, Blas-Garcia A, Esplugues JV. Mitochondria sentencing about cellular life and death: a matter of oxidative stress. Curr Pharm Des. 2011;17:4047–60.
3. Arieli Y, Gursahani H, Eaton MM, Hernandez LA, Schaefer S. Gender modulation of Ca^{2+} uptake in cardiac mitochondria. J Mol Cell Cardiol. 2004;37:507–13.
4. Barrett-Connor E. Sex differences in coronary heart disease. Why are women so superior? The 1995 Ancel Keys Lecture. Circulation. 1997;95:252–64.
5. Ben-Porath I, Weinberg RA. The signals and pathways activating cellular senescence. Int J Biochem Cell Biol. 2005;37:961–76.
6. Bereiter-Hahn J, Voth M. Dynamics of mitochondria in living cells: shape changes, dislocations, fusion, and fission of mitochondria. Microsc Res Tech. 1994;27:198–219.
7. Bernardi P, Di Lisa F. The mitochondrial permeability transition pore: molecular nature and role as a target in cardioprotection. J Mol Cell Cardiol. 2015;78:100–6. a
8. Bernardi P, Di Lisa F, Fogolari F, Lippe G. From ATP to PTP and back: a dual function for the mitochondrial ATP synthase. Circ Res. 2015;116:1850–62. b
9. Besik J, Szarszoi O, Kunes J, Netuka I, Maly J, Kolar F, Pirk J, Ostádal B. Tolerance to acute ischemia in adult male and female spontaneouslyhypertensive rats. Physiol. Res. 2007; 56:267–74.
10. Bossy-Wetzel E, Barsoum MJ, Godzik A, Schwarzenbacher R, Lipton SA. Mitochondrial fission in apoptosis, neurodegeneration and aging. Curr Opin Cell Biol. 2003;15:706–16.

11. Bravo-San Pedro JM, Kroemer G, Galluzzi L. Autophagy and mitophagy in cardiovascular disease. Circ Res. 2017;120:1812–24.
12. Burova E, Borodkina A, Shatrova A, Nikolsky N. Sublethal oxidative stress induces the premature senescence of human mesenchymal stem cells derived from endometrium. Oxidative Med Cell Longev. 2013:474931.
13. Camara AK, Lesnefsky EJ, Stowe DF. Potential therapeutic benefits of strategies directed to mitochondria. Antioxid Redox Signal. 2010; 13(3):279–347.
14. Cao DJ, Lavandero S, Hill JA. Parkin gone wild: Unbridled ubiquitination. Circ Res. 2015;117:31–313.
15. Carvalho FS, Burgeiro A, Garcia R, Moreno AJ, Carvalho RA, Oliveira PJ. Doxorubicin-induced cardiotoxicity: from bioenergetic failure and cell death to cardiomyopathy. Med Res Rev. 2014; 34(1):106–35.
16. Chalmers S, Saunter C, Wilson C, Coats P, Girkin JM, McCarron JG. Mitochondrial motility and vascular smooth muscle proliferation. Arterioscler Thromb Vasc Biol. 2012;32:3000–11.
17. Chan DC. Mitochondrial fusion and fission in mammals. Annu Rev Cell Dev Biol. 2006;22:79–99.
18. Chen Y, Dorn GW. PINK1-phosphorylated mitofusin2 is a Parkin receptor for culling damaged mitochondria. Science. 2013;340:471–5.
19. Chen L, Gong Q, Stice JP, Knowlton AA. Mitochondrial OPA1, apoptosis, and heart failure. Cardiovasc Res. 2009b;84:91–9.
20. Chen JQ, Cammarata PR, Baines CP, Yager JD. Regulation of mitochondrial respiratory chain biogenesis by estrogens/estrogen receptors and physiological, pathological and pharmacological implications. Biochim Biophys Acta. 2009a;1793:1540–70.
21. Chen Y, Liu Y, Dorn GW. Mitochondrial fusion is essential for organelle function and cardiac homeostasis. Circ Res. 2011;109:132–1331.
22. Circu ML, Aw TY. Reactive oxygen species, cellular redox systems, and apoptosis. Free Radic Biol Med. 2010;48:749–62.
23. Collins TJ, Berridge MJ, Lipp P, Bootman MD. Mitochondria are morphologically and functionally heterogeneous within cells. EMBO J. 2002;21:1616–27.
24. Colom B, Oliver J, Roca P, Garcia-Palmer FJ. Caloric restriction and gender modulate cardiac muscle mitochondrial H2O2 production and oxidative damage. Cardiovasc Res. 2007;74:456–65.
25. Dai DF, Santana LF, Vermulst M, Tomazela DM, Emond MJ, MacCoss MJ, Gollahon K, Martin GM, Loeb LA, Ladiges WC, Rabinovitch PS. Overexpression of catalase targeted to mitochondria attenuates murine cardiac aging. Circulation. 2009;119(21):2789–97.
26. Dai DF, Chen T, Wanagat J, Laflamme M, Marcinek DJ, Emond MJ, Ngo CP, Prolla TA, Rabinovitch PS. Age-dependent cardiomyopathy in mitochondrial mutator mice is attenuated by overexpression of catalase targeted to mitochondria. Aging Cell. 2010;9(4):536–44.
27. Danial NN, Korsmeyer SJ. Cell death: critical control points. Cell. 2004;116:205–19.
28. de Brito OM, Scorrano L. An intimate liaison: spatial organization of the endoplasmic reticulum-mitochondria relationship. EMBO J. 2010;29:2715–23.
29. Delbridge LM, Mellor KM, Taylor DJ, Gottlieb RA. Myocardial autophagic energy stress responses–macroautophagy, mitophagy, and glycophagy. Am J Physiol Heart Circ Physiol. 2015;308(10):H1194–204.
30. Delettre C, Griffoin J-M, Kaplan J, Dollfus H, Lorenz B, Faivre L, Lenaers G, Belenguer P, Hamel C. Mutation spectrum and splicing variants in the OPA1 gene. Hum Genet. 2001;109:584–91.
31. Dorn GW, Scorrano L. Two close, too close: sarcoplasmic reticulum-mitochondrial crosstalk and cardiomyocyte fate. Circ Res. 2010;107:689–99.
32. Fang L, Moore X-L, Gao X-M, Dart AM, Lim YL, Du X-J. Down-regulation of mitofusin-2 expression in cardiac hypertrophy in vitro and in vivo. Life Sci. 2007;80:2154–60.
33. Fang L, Moore XL, Chan W, White DA, Chin-Dusting J, Dart AM. Decreased fibrocyte number is associated with atherosclerotic plaque instability in man. Cardiovasc Res. 2012; 95(1):124–33.
34. Finsterer J, Ohnsorge P. Influence of mitochondrion-toxic agents on the cardiovascular system. Regul Toxicol Pharmacol. 2013;67:434–45.
35. Fischer F, Hamann A, Osiewacz HD. Mitochondrial quality control: an integrated network of pathways. Trends Biochem Sci. 2012;37:284–92.
36. Gandre-Babbe S, van der Bliek AM. The novel tail-anchored membrane protein Mff controls mitochondrial and peroxisomal fission in mammalian cells. Mol Biol Cell. 2008;19:2402–12.
37. Gonzalez Y, Pokrzywinski KL, Rosen ET, Mog S, Aryal B, Chehab LM, Vijay V, Moland CL, Desai VG, Dickey JS, Rao VA. Reproductive hormone levels and differential mitochondria-related oxidative gene expression as potential mechanisms for gender differences in cardiosensitivity to Doxorubicin in tumor-bearing spontaneously hypertensive rats. Cancer Chemother Pharmacol. 2015 Sep;76(3):447–59.
38. Gottlieb RA, Thomas A. Mitophagy and mitochondrial quality control mechanisms in the heart. Curr Pathobiol Rep. 2017;5(2):161–9.
39. Green D. Means to an end: apoptosis and other cell death mechanisms. Cold Spring Harbor: Cold Spring Harbor Laboratory Press. 2011. ISBN:978-0-87969-888-1.
40. Greiten LE, Holditch SJ, Arunachalam SP, Miller VM. Should there be sex-specific criteria for the diagnosis and treatment of heart failure? J Cardiovasc Transl Res. 2014;7:139–55.
41. Gulati M, Shaw LJ, Bairey Merz CN. Myocardial ischemia in women: lessons from the NHLBI WISE study. Clin Cardiol. 2012;35:141–8.
42. Hall AR, Hausenloy DJ. The shape of things to come: mitochondrial fusion and fission in the adult heart. Cardiovasc Res. 2012;94:391–2.
43. Hanna RA, Quinsay MN, Orogo AM, Giang K, Rikka S, Gustafsson ÅB. Microtubule-associated protein 1 light chain 3 (LC3) interacts with Bnip3 protein to selectively remove endoplasmic reticulum and mitochondria via autophagy. J Biol Chem. 2012;287(23):19094–104.
44. Ide T, Tsutsui H, Ohashi N, Hayashidani S, Suematsu N, Tsuchihashi M, et al. Greater oxidative stress in healthy young men compared with premenopausal women. Arterioscler Thromb Vasc Biol. 2002;22:438–42.
45. Janakidevi K, Fisher MA, Del Vecchio PJ, Tiruppathi C, Figge J, Malik AB. Endothelin-1 stimulates DNA synthesis and proliferation of pulmonary artery smooth muscle cells. Am J Phys Cell Physiol. 1992;263:C1295–301.
46. Jay DB, Papaharalambus CA, Seidel-Rogol B, Dikalova AE. Lass'egue B, Griendling KK. Nox5 mediates PDGF-induced proliferation in human aortic smooth muscle cells. Free Radic Biol Med. 2008;45:329–35.

47. Johnson JL, van Eys GJ, Angelini GD, George SJ. Injury induces dedifferentiation of smooth muscle cells and increased matrix-degrading metalloproteinase activity in human saphenous vein. Arterioscler Thromb Vasc Biol. 2001;21:114–1151.
48. Kabeya Y, Mizushima N, Yamamoto A, Oshitani-Okamoto-S, Ohsumi Y, Yoshimori T. LC3, GABARAP and GATE16 localize to autophagosomal membrane depending on form-II formation. J Cell Sci. 2004;117:2805–12.
49. Kang R, Zeh HJ, Lotze MT, Tang D. The Beclin 1 network regulates autophagy and apoptosis. Cell Death Differ. 2011;118:571–80.
50. Kashatus DF, Lim KH, Brady DC, et al. RALA and RALBP1 regulate mitochondrial fission at mitosis. Nat Cell Biol. 2011;13:1108–15.
51. Keller KM, Howlett SE. Sex differences in the biology and pathology of the aging heart. Can J Cardiol. 2016;32 (9):1065–73.
52. Khalifa AR, Abdel-Rahman EA, Mahmoud AM, Ali MH, Noureldin M, Saber SH, Mohsen M, Ali SS. Sex-specific differences in mitochondria biogenesis, morphology, respiratory function, and ROS homeostasis in young mouse heart and brain. Physiol Rep. 2017;5(6):pii: e13125.
53. Klinge CM. Estrogenic control of mitochondrial function and biogenesis. J Cell Biochem. 2008;105:1342–51.
54. Koleini N, Kardami E. Autophagy and mitophagy in the context of doxorubicin-induced cardiotoxicity. Oncotarget. 2017;8(28):46663–80.
55. Kovacic JC, Moreno P, Hachinski V, Nabel EG, Fuster V. Cellular senescence, vascular disease, and aging: Part 1 of a 2-part review. Circulation. 2011;123(15):1650–60.
56. Kubli DA, Zhang X, Lee Y, Hanna RA, Quinsay MN, Nguyen CK, Jimenez R, Petrosyan S, Murphy AN, Gustafsson AB. Parkin protein deficiency exacerbates cardiac injury and reduces survival following myocardial infarction. J Biol Chem. 2013;288(2):915–26.
57. Lee Y, Lee HY, Hanna RA, Gustafsson ÅB. Mitochondrial autophagy by Bnip3 involves Drp1-mediated mitochondrial fission and recruitment of Parkin in cardiac myocytes. Am J Physiol Heart Circ Physiol. 2011;301(5):H1924–31.
58. Lee Y, Jeong S-Y, Karbowski M, Smith CL, Youle RJ. Roles of the mammalian mitochondrial fission and fusion mediators Fis1, Drp1, and Opa1 in apoptosis. Mol Biol Cell. 2004;15:5001–11.
59. Leinwand LA. Sex is a potent modifier of the cardiovascular system. J Clin Invest. 2003;112:302–4.
60. Levine B, Kroemer G. Autophagy in aging, disease and death: the true identity of a cell death impostor. Cell Death Differ. 2012;16:1–2.
61. Limongelli G, Masarone D, Pacileo G. Mitochondrial disease and the heart. Heart. 2017;103:390–8.
62. Liu L, Feng D, Chen G, Chen M, Zheng Q, Song P, Ma Q, Zhu C, Wang R, Qi W, Huang L, Xue P, Li B, Wang X, Jin H, Wang J, Yang F, Liu P, Zhu Y, Sui S, Chen Q. Mitochondrial outer-membrane protein FUNDC1 mediates hypoxia-induced mitophagy in mammalian cells. Nat Cell Biol. 2012;14(2):177–85.
63. Liu Y, Zhou L, Xu HF, Yan L, Ding F, Hao W, Cao JM, Gao X. A preliminary experimental study on the cardiac toxicity of glutamate and the role of □-amino-3-hydroxy-5-methyl-4-isoxazolepropionic acid receptor in rats. Chin Med J (Engl). 2013; 126(7):1323–32
64. Llambi F, Green DR. Apoptosis and oncogenesis: give and take in the BCL-2 family. Curr Opin Genet Dev. 2011;21:12–20.
65. Marsboom G, Toth PT, Ryan JJ, Hong Z, Wu X, Fang Y-H, Thenappan T, Piao L, Zhang HJ, Pogoriler J, Chen Y, Morrow E, Weir EK, Rehman J, Archer SL. Dynamin-related protein 1-mediated mitochondrial mitotic fission permits hyperproliferation of vascular smooth muscle cells and offers a novel therapeutic target in pulmonary hypertension. Circ Res. 2012;11:1484–97.
66. Malorni W, Straface E, Matarrese P, Ascione B, Coinu R, Canu S, Galluzzo P, Marino M, Franconi F. Redox state and gender differences in vascular smooth muscle cells. FEBS Lett. 2008; 582(5):635–42.
67. Martin-Garrido A, Williams HC, Lee M, Seidel-Rogol B, Ci X, Dong J-T, Lassègue B, Martín AS, Griendling KK. Transforming growth factor β inhibits platelet derived growth factor-induced vascular smooth muscle cell proliferation via Akt-independent, Smad-mediated Cyclin D1 downregulation. PLoS One. 2013;e79657:8.
68. Marí M, Morales A, Colell A, García-Ruiz C, Kaplowitz N, Fernández-Checa JC. Mitochondrial glutathione: features, regulation and role in disease. Biochim Biophys Acta. 2013;1830:3317–28.
69. Meyer JN, Leung MC, Rooney JP, Sendoel A, Hengartner MO, Kisby GE, Bess AS. Mitochondria as a target of environmental toxicants. Toxicol Sci. 2013; 134(1):1–17.
70. Mehilli J, Ndrepepa G, Kastrati A, Nekolla SG, Markwardt C, Bollwein H, Pache J, Martinoff S, Dirschinger J, Schwaiger M, Schömig A. Gender and myocardial salvage after reperfusion treatment in acute myocardial infarction. J Am Coll Cardiol. 2005; 45(6):828–31.
71. Milerová M, Drahota Z, Chytilová A, Tauchmannová K, Houštěk J, Ošťádal B. Sex difference in the sensitivity of cardiac mitochondrial permeability transition pore to calcium load. Mol Cell Biochem. 2016;412:147–54.
72. Mizushima N, Levine B, Cuervo AM, Klionsky DJ. Autophagy fights disease through cellular self-digestion. Nature. 2008;451:1069–75.
73. Mizushima N, Noda T, Ohsumi Y. Apg16p is required for the function of the Apg12p–Apg5p conjugate in the yeast autophagy pathway. EMBO J. 1999;18:3888–96.
74. Miyamoto S, Brown JH. Drp1 and mitochondrial autophagy lend a helping hand in adaptation to pressure overload. Circulation. 2016;133:1225–7.
75. Morris PD, Robinson T, Channer KS. Reversible heart failure: toxins, tachycardiomyopathy and mitochondrial abnormalities. Postgrad. Med. J. 2012; 88: 706–12.
76. Moulin M, Piquereau J, Mateo P, Fortin D, Rucker-Martin C, Gressette M, Lefebvre F, Gresikova M, Solgadi A, Veksler V, Garnier A, Ventura-Clapier R. Sexual dimorphism of doxorubicin-mediated cardiotoxicity: potential role of energy metabolism remodeling. Circ Heart Fail. 2015; 8(1):98–108.
77. Narendra DP, Jin SM, Tanaka A, Suen DF, Gautier CA, Shen J, Cookson MR, Youle RJ. PINK1 is selectively stabilized on impaired mitochondria to activate Parkin. PLoS Biol. 2010;8(1):e1000298.
78. Narendra D, Tanaka A, Suen DF, Youle RJ. Parkin is recruited selectively to impaired mitochondria and promotes their autophagy. J Cell Biol. 2008;183:795–803.
79. Ong S-B, Subrayan S, Lim SY, Yellon DM, Davidson SM, Hausenloy DJ. Inhibiting mitochondrial fission protects the heart against ischemia/reperfusion injury. Circulation. 2010;121:2012–22.
80. Ostadal B, Netuka I, Maly J, Besik J, Ostadalova I. Gender differences in cardiac ischemic injury and protection–experimental aspects. Exp Biol Med (Maywood). 2009;234:1011–9.

81. Otera H, Wang C, Cleland MM, Setoguchi K, Yokota S, Youle RJ, Mihara K. Mff is an essential factor for mitochondrial recruitment of Drp1 during mitochondrial fission in mammalian cells. J Cell Biol. 2010;191:1141–58.
82. Palmeira CM, Moreno AJ. Mitochondrial bioenergetics. Methods and protocols. Springer Methods in Molecular Biology. 2012, 810. ISBN:978-1-61779-381-3.
83. Papanicolaou KN, Khairallah RJ, Ngoh GA, Chikando A, Luptak I, O'Shea KM, Riley DD, Lugus JJ, Colucci WS, Ledere WJ, Stanley WC, Walsh K. Mitofusin-2 maintains mitochondrial structure and contributes to stress-induced permeability transition in cardiac myocytes. Mol Cell Biol. 2011;31:1309–28.
84. Papanicolaou KN, Kikuchi R, Ngoh GA, Coughlan KA, Dominguez I, Stanley WC, Walsh K. Mitofusins 1 and 2 are essential for postnatal metabolic remodeling in heart. Circ Res. 2012; 111(8):1012–26.
85. Pei H, Yang Y, Zhao H, Li X, Yang D, Li D, Yang Y. The role of mitochondrial functional proteins in ROS production in ischemic heart diseases. Oxidative Med Cell Longev. 2016;2016:5470457.
86. Petersen KF, Befroy D, Dufour S, Dziura J, Ariyan C, Rothman DL, DiPietro L, Cline GW, Shulman GI. Mitochondrial dysfunction in the elderly: possible role in insulin resistance. Science. 2003;300:1140–2.
87. Pierdominici M, Ortona E, Franconi F, Caprio M, Straface E, Malorni W. Gender specific aspects of cell death in the cardiovascular system. Curr Pharm Des. 2011;17:1046–55.
88. Piquereau J, Caffin F, Novotova M, Prola A, Garnier A, Mateo P, Fortin D, Huynh LH, Nicolas V, Alavi MV, Brenner C, Ventura-Clapier R, Veksler V, Joubert F. Down-regulation of OPA1 alters mouse mitochondrial morphology, PTP function, and cardiac adaptation to pressure overload. Cardiovasc Res. 2012;94:408–17.
89. Powers RW, Majors AK, Lykins DL, Sims CJ, Lain KY, Roberts JM. Plasma homocysteine and malondialdehyde are correlated in an age- and genderspecific manner. Metabolism. 2002;51:1433–8.
90. Pukac L, Huangpu J, Karnovsky MJ. Platelet-derived growth factor-BB, insulin-like growth factor-I, and phorbol ester activate different signaling pathways for stimulation of vascular smooth muscle cell migration. Exp Cell Res. 1998;242:548–60.
91. Quinsay MN, Thomas RL, Lee Y, Gustafsson AB. Bnip3-mediated mitochondrial autophagy is independent of the mitochondrial permeability transition pore. Autophagy. 2010;6(7):855–62.
92. Rambold AS, Kostelecky B, Elia N, et al. Tubular network formation protects mitochondria from autophagosomal degradation during nutrient starvation. Proc Natl Acad Sci U S A. 2011;108:10190–5.
93. Regitz-Zagrosek V. Therapeutic implications of the gender-specific aspects of cardiovascular disease. Nat Rev Drug Discov. 2006;5:425–39.
94. Regitz-Zagrosek V. Sex and gender differences in symptoms of myocardial ischaemia. Eur. Heart J. 2011; 32:3064–066.
95. Ren J, Taegtmeyer H. Too much or not enough of a good thing – the Janus faces of autophagy in cardiac fuel and protein homeostasis. J Mol Cell Cardiol. 2015;84:223–6.
96. Rockstein M, Brandt KF. Enzyme changes in flight muscle correlated with aging and flight ability in the male housefly. Science. 1963;139:1049–51.
97. Rosdah AA, Holien J K, Delbridge LM, Dusting GJ, Lim SY. Mitochondrial fission- a drug target for cytoprotection or cytodestruction? Pharmacol Res Perspect. 2016;4:e00235.
98. Rossier MF. T channels and steroid biosynthesis: in search of a link with mitochondria. Cell Calcium. 2006;40:155–64.
99. Saito T, Sadoshima J. Molecular mechanisms of mitochondrial autophagy/mitophagy in the heart. Circ Res. 2015;116 (8):1477–90.
100. Salabei JK, Hill BG. Mitochondrial fission induced by platelet-derived growth factor regulates vascular smooth muscle cell bioenergetics and cell proliferation. Redox Biol. 2013;1:542–51.
101. Santel A, Frank S, Gaume B, Herrler M, Youle RJ, Fuller MT. Mitofusin-1 protein is a generally expressed mediator of mitochondrial fusion in mammalian cells. J Cell Sci. 2003;116:2763–74.
102. Scorrano L. Keeping mitochondria in shape: a matter of life and death. Eur J Clin Investig. 2013;43:886–93.
103. Shafiee H, Mohammadi H, Rezayat SM, Hosseini A, Baeeri M, Hassani S, Mohammadirad A, Bayrami Z, Abdollahi M. Prevention of malathion-induced depletion of cardiac cells mitochondrial energy and free radical damage by a magnetic magnesium-carrying nanoparticle. Toxicol Mech Methods. 2010; 20(9):538–43.
104. Sharp WW, Fang YH, Han M, Zhang HJ, Hong Z, Banathy A, Morrow E, Ryan JJ, Archer SL. Dynamin-related protein 1 (Drp1)-mediated diastolic dysfunction in myocardial ischemia-reperfusion injury: therapeutic benefits of Drp1 inhibition to reduce mitochondrial fission. FASEB J. 2014;28:316–26.
105. Shenouda SM, Widlansky ME, Chen K, Xu G, Holbrook M, Tabit CE, Hamburg NM, Frame AA, Caiano TL, Kluge MA, Duess M-A, Levit A, Kim B, Hartman M-L, Joseph L, Shirihai OS, Vita JA. Altered mitochondrial dynamics contributes to endothelial dysfunction in diabetes mellitus. Circulation. 2011;124:444–53.
106. Shirakabe A, Ikeda Y, Sciarretta S, Zablocki DK, Sadoshima J. Aging and autophagy in the heart. Circ Res. 2016;118:1563–76.
107. Simbre VC, Duffy SA, Dadlani GH, Miller TL, Lipshultz SE. Cardiotoxicity of cancer chemotherapy: implications for children. Paediatr Drugs. 2005; 7(3):187–202.
108. Song Z, Chen H, Fiket M, Alexander C, Chan DC. OPA1 processing controls mitochondrial fusion and is regulated by mRNA splicing, membrane potential, and Yme1L. J Cell Biol. 2007;178:749–55.
109. Song M, Gong G, Burelle Y, Gustafsson AB, Kitsis RN, Matkovich SJ, Dorn GW. Interdependence of Parkin-mediated mitophagy and mitochondrial fission in adult mouse hearts. Circ Res. 2015;117:346–51.
110. Straface E, Vona R, Gambardella L, Ascione B, Marino M, Bulzomi P, Canu S, Coinu R, Rosano G, Malorni W, Franconi F. Cell sex determines anoikis resistance in vascular smooth muscle cells. FEBS Lett. 2009; 583(21):3448–54.
111. Straface E, Vona R, Campesi I, Franconi F. Mitochondria can orchestrate sex differences in cell fate of vascular smooth muscle cells from rats. Biol Sex Differ. 2015;16:6–34.
112. Stěrba M, Popelová O, Lenčo J, Fučíková A, Brčáková E, Mazurová Y, Jirkovský E, Simůnek T, Adamcová M, Mičuda S, Stulík J, Geršl V. Proteomic insights into chronic anthracycline cardiotoxicity. J Mol Cell Cardiol. 2011;50 (5):849–62.
113. Sun N, Youle RJ, Finkel T. The mitochondrial basis of aging. Mol Cell. 2016;61:654–66.
114. Tanaka A, Cleland MM, Xu S, Narendra DP, Suen DF, Karbowski M, Youle RJ. Proteasome and p97 mediate mitophagy and degradation of mitofusins induced by Parkin. J Cell Biol. 2010;191:1367–80.

115. Tokarska-Schlattner M, Zaugg M, Zuppinger C, Wallimann T, Schlattner U. New insights into doxorubicin-induced cardiotoxicity: the critical role of cellular energetics. J Mol Cell Cardiol. 2006;41:389–405.
116. Tocchi A, Quarles EK, Basisty N, Gitari L, Rabinovitch PS. Mitochondrial dysfunction in cardiac aging. Biochim Biophys Acta. 2015;1847(11):1424–33.
117. Torrealba N, Aranguiz P, Alonso C, Rothermel BA, Lavandero S. Mitochondria in structural and functional cardiac remodeling. Adv Exp Med Biol. 2017;982:277–306.
118. Twig G, Elorza A, Molina AJ, et al. Fission and selective fusion govern mitochondrial segregation and elimination by autophagy. EMBO J. 2008;27:433–46.
119. Twig G, Shirihai OS. The interplay between mitochondrial dynamics and mitophagy. Antioxid Redox Signal. 2011;14:1939–51.
120. Vásquez-Trincado C, García-Carvajal I, Pennanen C, Parra V, Hill JA, Rothermel BA, Lavandero S. Mitochondrial dynamics, mitophagy and cardiovascular disease. J Physiol. 2016;594(3):509–25.
121. Velarde MC. Mitochondrial and sex steroid hormone crosstalk during aging. Longev Healthspan. 2014;3:2.
122. Ventura-Clapier R, Garnier A, Veksler V. Transcriptional control of mitochondrial biogenesis. The central role of PGC-1α. Cardiovasc Res. 2008;79:208–17.
123. Ventura-Clapier R, Moulin M, Piquereau J, Lemaire C, Mericskay M, Veksler V, Garnier A. Mitochondria: a central target for sex differences in pathologies. Clin Sci (Lond). 2017;131(9):803–22.
124. Viña J, Borrás C, Gambini J, Sastre J, Pallardó FV. Why females live longer than males: control of longevity by sex hormones. Sci Aging Knowl Environ. 2005;2005:pe17.
125. Watanabe T, Saotome M, Nobuhara M, Sakamoto A, Urushida T, Katoh H, Satoh H, Funaki M, Hayashi H. Roles of mitochondrial fragmentation and reactive oxygen species in mitochondrial dysfunction and myocardial insulin resistance. Exp Cell Res. 2014;323:314–25.
126. Witt H, Schubert C, Jaekel J, Fliegner D, Penkalla A, Tiemann K, Stypmann J, Roepcke S, Brokat S, Mahmoodzadeh S, Brozova E, Davidson MM, Ruiz Noppinger P, Grohé C, Regitz-Zagrosek V. Sex-specific pathways in early cardiac response to pressure overload in mice. J Mol Med. 2008;86:1013–24.
127. Yin Z, Pascual C, Klionsky DJ. Autophagy: machinery and regulation. Microb Cell. 2016;3:588–96.
128. Youle RJ, Karbowski M. Mitochondrial fission in apoptosis. Nat Rev Mol Cell Biol. 2005;6:657–63.
129. Zhivotovsky B. Caspases: the enzymes of death. Essays Biochem. 2003;39:25–40.

Sex Differences in the Coronary System 17

Viviany R. Taqueti

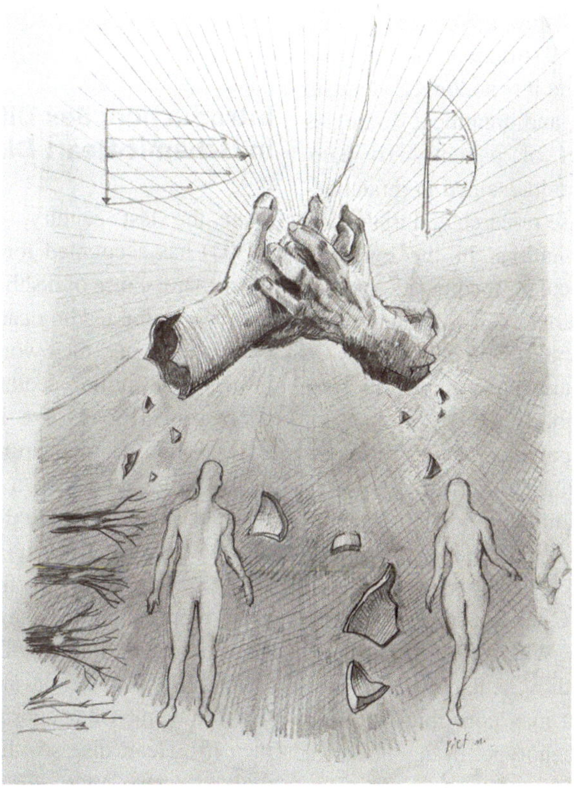

Artwork by Piet Michiels, Leuven, Belgium

V. R. Taqueti (✉)
Heart and Vascular Center; Noninvasive Cardiovascular Imaging Program, Departments of Medicine (Cardiology) and Radiology (Nuclear Medicine and Molecular Imaging), Brigham and Women's Hospital, Harvard Medical School, Boston, MA, USA
e-mail: vtaqueti@bwh.harvard.edu

Abstract

Cardiovascular disease remains the leading cause of morbidity and mortality for both women and men. Emerging evidence supports that ischemic heart disease (IHD) may manifest differently in women and men, in ways

P. L. M. Kerkhof, V. M. Miller (eds.), *Sex-Specific Analysis of Cardiovascular Function*,
Advances in Experimental Medicine and Biology 1065, https://doi.org/10.1007/978-3-319-77932-4_17

ranging from the clinical presentation, diagnosis, and management of disease to the basic biology and biomechanics of cardiomyocyte function and the coronary circulation. Women consistently present with a higher burden of symptoms and comorbidities as compared with men and experience worse outcomes. These data have proved perplexing given the decreased likelihood of women to demonstrate obstructive coronary artery disease (CAD) on coronary angiography. Reported sex differences have long been influenced by the practice of defining heart disease primarily as obstructive CAD, but obstructive plaque is now recognized as neither necessary nor sufficient to explain symptoms of IHD, and it is no longer adequate to tailor diagnostic and treatment strategies only to this subset of patients. To date, women remain underrepresented in guideline-changing heart disease research and trials, creating important limitations in the evidence base for cardiovascular medicine. Smaller epicardial coronary arteries in women as compared to men, coupled with differences in shear stress and inflammatory mediators over the life span, may modify the development of CAD in susceptible patients into a diffuse pattern with more contribution from coronary vasomotor dysfunction than focal obstruction. Newer studies corroborate that symptomatic women are more likely than men to present with nonobstructive CAD and coronary microvascular dysfunction. When present, these processes increase cardiovascular risk in both women and men but may constitute an especially malignant phenotype in a subset of severely affected women, with implications for the management of not only CAD but also heart failure with preserved ejection fraction. This represents a state-of-the-art review of sex differences in the coronary system, with an eye toward how diverse pathophysiological processes may contribute to IHD phenotypes prevalent in women and men. Beyond providing women and men with equitable optimal care according to current paradigms, understanding the pathophysiology of IHD beyond a conventional focus on obstructive CAD is needed to address what is likely a combination of biological as well as environmental determinants of their prognosis.

Keywords

Atherosclerosis · Cardiovascular disease · Coronary flow reserve · Coronary microvascular dysfunction · Heart failure with preserved ejection fraction · Ischemic heart disease · Nonobstructive coronary artery disease

Introduction: Sex Differences in Ischemic Heart Disease

Over the last century, cardiovascular disease (CVD) has accounted for more deaths than any other major cause of death in the United States [1] and is now the leading cause of mortality among both women and men worldwide [1–4], resulting in nearly 18 million deaths in 2015. Deaths from CVD primarily involve ischemic heart disease (IHD), such as myocardial infarction (MI) and heart failure, and also include those associated with stroke and peripheral arterial disease. Focusing within the cardiac system, approximately 630,000 Americans die from heart disease each year, representing 1 in every 4 deaths, and the numbers are similar for women and men (Fig. 17.1) [4]. Yet awareness of heart disease risk for women has substantially lagged that for men [5]. Heart disease—historically synonymous with coronary artery disease (CAD)—has been traditionally defined anatomically as obstructive atherosclerosis involving the epicardial coronary arteries. There is now greater understanding that IHD occurs in the presence of an inadequate blood supply to the myocardium, which may or may not result from obstructive atherosclerotic narrowing in the epicardial coronary arteries [6, 7].

Indeed, as demonstrated by recent national [1, 4] and global [2, 3] statistics, IHD poses a

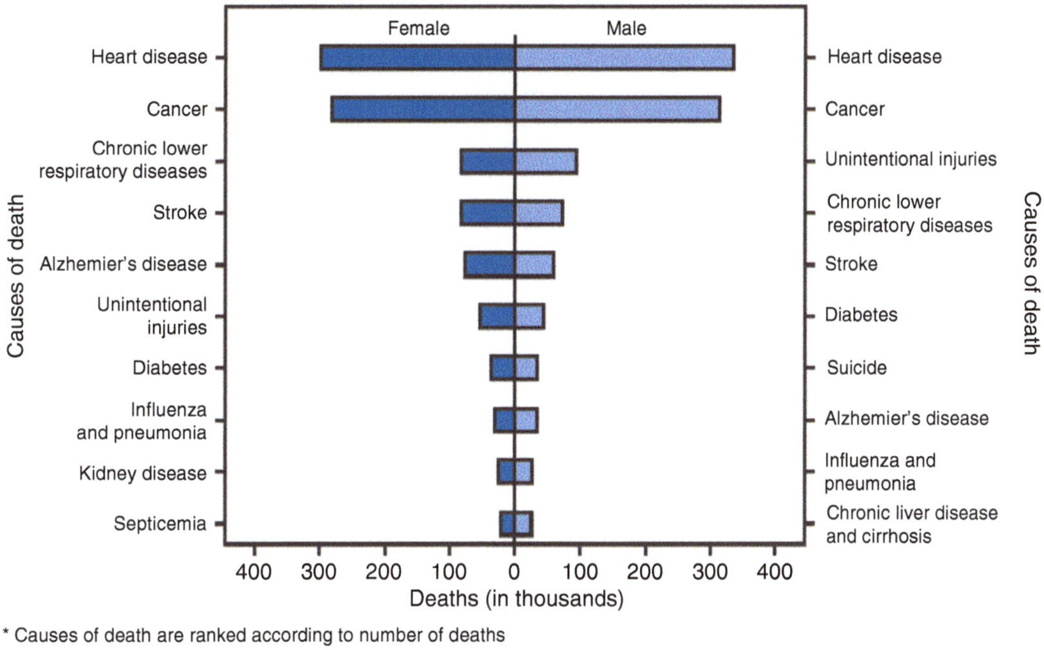

Fig. 17.1 Number of deaths from ten leading causes, by sex (United States 2015). (Source: National Vital Statistics System, Centers for Disease Control and Prevention, 2015 [4])

major threat to both women and men across their life spans. Emerging evidence supports that IHD may manifest differently in women and men, in ways ranging from the clinical presentation, diagnosis, and management of disease to the basic biology and biomechanics of cardiomyocyte function and the coronary circulation. This finding has led to calls to expand conventional tools developed more than a half-century ago for the diagnosis and management of (primarily obstructive) CAD to address the full spectrum of IHD impacting women as well as men. The following represents a state-of-the-art review of sex differences in the coronary system, with an eye toward how diverse pathophysiological factors and processes may contribute to IHD phenotypes prevalent in women and men.

Sex Differences in the Epidemiology of IHD: Reframing the "Gender Gap"

Over the last three decades, case fatality rates for heart disease in the United States have been similar or higher for women as compared to men (Fig. 17.2) [1]. Although this finding partly reflects that women outnumber men in older populations at greatest risk for IHD, women often present with a higher burden of comorbidities and experience worse IHD outcomes as compared to men. While trends in the United States suggest dramatic declines in cardiac deaths for both women and men over the last two decades, this decrease has not been uniform for all individuals, especially young women [8]. At the same time, important sex differences in the rates of IHD diagnosis, utilization of care, response to therapy, and clinical outcomes have been described [9–12]. Compared with men, women have a higher prevalence of persistent angina, nonobstructive CAD, coronary microvascular dysfunction (CMD), spontaneous coronary artery dissection, stress-induced cardiomyopathy, and heart failure with preserved ejection fraction (HFpEF) [13–21]. IHD risk factors including diabetes mellitus [22] and atrial fibrillation [23] are associated with higher rates of vascular complications in women versus men. Women presenting with acute coronary syndromes experience higher mortality as compared with men [24–27] and are

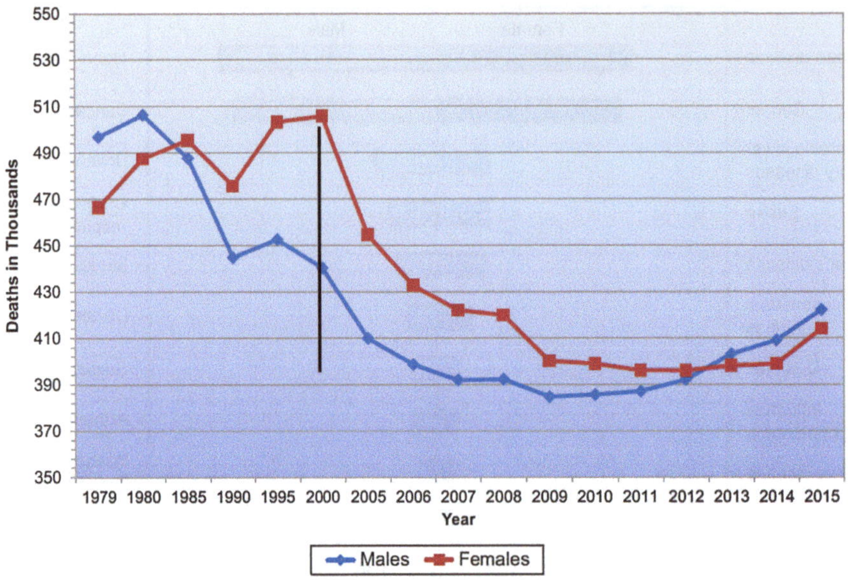

Fig. 17.2 Cardiovascular disease mortality trends for males and females (United States 1979–2015). (Source: National Center for Health Statistics and National Heart, Lung, and Blood Institute [1])

referred for cardiac transplantation at later stages of heart failure [28]. There is also underutilization in women of cardiac devices [29], including implantable cardiac defibrillators [30] and cardiac resynchronization therapy (CRT) [31], despite subgroup analyses of randomized controlled trial data showing that female sex is associated with improved responsiveness to CRT [32].

Thus, the previous assumption that heart disease in women is the same as that in men, only occurring about a decade later, represents an oversimplification and underscores the importance of sex-specific research in cardiovascular care [33]. There is urgent need for randomized and comparative trial data in female as well as male patients [34]. To date, women have been underrepresented in guideline-changing CVD research and trials (Fig. 17.3), often comprising less than a third of enrolled patients [35]. This phenomenon has obscured important sex-specific differences in the pathobiology of disease, which are now becoming more apparent. Central among these has been the recognition that a variety of disorders, not simply obstructive CAD, may result in ischemic symptoms and worse cardiac outcomes in patients, especially women [6, 10, 36]. As such, the heart disease "gender gap" likely reflects not only inconsistencies in awareness and application of guideline-directed management of CAD in women but also fundamental limitations in how heart disease is defined and managed in the population. Beyond providing women and men with equitable optimal cardiovascular care according to current paradigms, understanding the underlying biology of IHD beyond a conventional focus on obstructive CAD is needed to address what is likely a combination of biological as well as environmental determinants of their prognosis.

Sex Differences in Risk Factors of IHD

Traditional IHD disease risk factors include older age, smoking, hypertension, hyperlipidemia, obesity, insulin resistance, and a family history of premature atherosclerosis. These factors play important roles in the development of IHD in both women and men, although the prevalence of certain risk factors differ between the sexes, and some are stronger predictors of IHD in women. The overall incidence of IHD in women lags that in men by about a decade, suggesting loss of a cardioprotective effect in postmenopausal

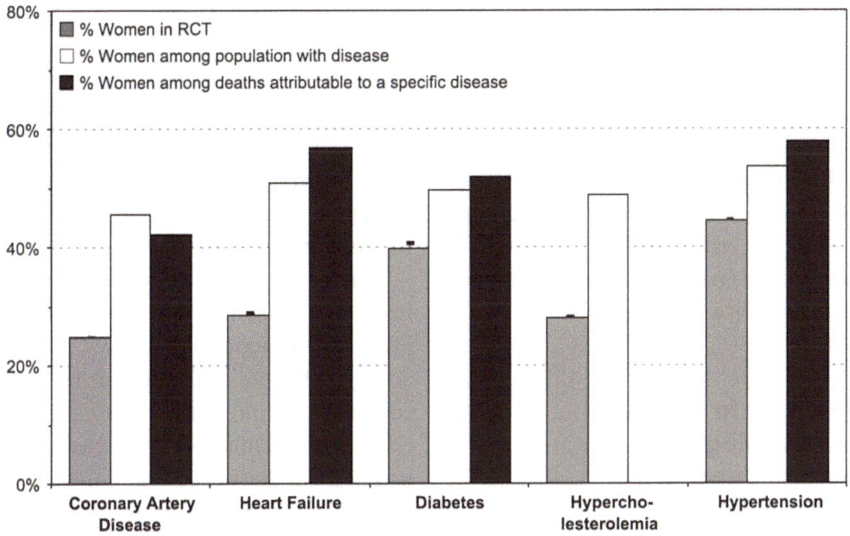

Fig. 17.3 Percentage of women enrolled in randomized clinical trials (RCTs) of cardiovascular disease, compared with relevant disease and death statistics. (Reproduced with permission [35])

women [37] that remains incompletely understood. Differences in the pattern of smoking between women and men have decreased over time [1], and smoking appears to confer a 25% increased adjusted relative risk of major adverse cardiovascular events (MACE) in women versus men, as reported in a meta-analysis of 86 prospective trials encompassing 3.9 million participants [38].

Systolic blood pressure rises steeply in older women, and mild-moderate hypertension has been associated with more cardiovascular complications in women than men (risk-factor-adjusted hazard ratio of 2.5 versus 1.6 in women and men, respectively) [39]. Nonetheless, no apparent sex difference was observed in the relationship of systolic blood pressure and MACE in a contemporary meta-analysis of 124 cohort studies of 1.2 million individuals, 44% of whom were women [40]. Although female lipid profiles worsen with menopause, elevated total cholesterol seems to confer a clinically similar (albeit statistically lower) risk of MACE in women as in men, according to a recent meta-analysis of 97 cohort studies including over 1 million individuals [41]. Despite a higher prevalence of obesity in nonwhite women as compared to nonwhite men [1], no significant sex differences have been observed between increasing body mass index (BMI) and adjusted risk of MACE [42].

In contrast, although the prevalence of diabetes mellitus is now similar in women and men, robust evidence exists for a greater excess risk of IHD among diabetic women than men. The relative risk for fatal IHD associated with diabetes was 40–50% higher in women than men, as reported in meta-analyses of up to 64 prospective cohort studies of nearly 860,000 diabetic patients [22, 43]. Diabetes appears to nullify any cardioprotective effects associated with younger age in women. The transition from normoglycemia to overt diabetes in women appears to accompany a greater decline in health than in men, such that women who develop diabetes experience a heavier burden of risk factors, including a higher BMI, than diabetic men.

In summary, while the effects of hypertension, hyperlipidemia, and obesity appear mostly similar between the sexes, prolonged smoking and diabetes seem significantly more hazardous for women than men. The mechanisms underlying these observed sex differences in risk factor effects are not well understood. Besides menopause, additional female-specific or female-predominant risk factors for IHD include: (1) pregnancy-related complications [44], such as gestational diabetes, pregnancy-induced hypertension, and preeclampsia; (2) emotional stress [45], such as that linked to stress cardiomyopathy [46]; and (3) autoimmune disorders characterized by chronic inflammation, such as systemic lupus erythematosus and rheumatoid arthritis [47], which underscore the pathological role of inflammation in atherosclerosis [48, 49].

Sex Differences in Clinical Presentation, Diagnosis and Management of IHD

Early reported sex differences [50] in the clinical presentation, diagnosis and management of IHD have been influenced by the long-standing practice of defining heart disease as obstructive CAD and tailoring diagnostic and treatment strategies to this subset of patients. We now recognize that obstructive CAD, as defined anatomically on coronary angiography, is neither necessary nor sufficient to explain symptoms of IHD, which commonly include angina and dyspnea in both women and men [51, 52]. Indeed, women present more frequently than men with symptoms of angina [14] but are less likely to manifest anatomic obstructive CAD. In a contemporary cohort of 11,223 symptomatic patients (42% women) referred for non-urgent coronary angiography, one-third of men, but two-thirds of women, had no obstructive CAD (Fig. 17.4), and these patients still experienced elevated risk of MACE (Fig. 17.5) [15]. Among patients with stable angina who are found to have obstructive CAD, sex differences also exist in the extent and severity of disease, with women less likely than men to have obstructive multivessel disease [36, 53]. Consistent sex differences in angiographic findings have been demonstrated not only in stable IHD but also in patients presenting with acute coronary syndromes, including unstable angina and MI [54–56]. In autopsy evaluations of patients who died of IHD, women also demonstrated less extensive and less obstructive CAD than men, despite pathologic evidence of MI [57], with more evidence of plaque erosion than plaque rupture [58]. In addition, in patients with obstructive CAD, women are less likely than men to demonstrate coronary collateralization [59]. Despite consistently documented lower angiographic disease burden and more often preserved left ventricular (LV) function as compared with men, women with IHD have similar adverse outcomes [1, 60].

Due to a greater symptom burden and rate of functional disability in women, coupled with a lower prevalence of obstructive CAD by coronary angiography, the evaluation of IHD in women as compared with men can present unique challenges to clinicians. The accuracy of standard noninvasive diagnostic testing for ischemia, such as stress testing with exercise electrocardiography, echocardiography, or nuclear myocardial perfusion imaging (MPI), can vary significantly when evaluated against a gold standard of finding anatomic obstructive CAD. This has been particularly relevant for women, whose symptoms are less likely to be explained by findings on coronary angiography and whose abnormal stress tests in the absence of obstructive CAD are more likely to be interpreted as "false positives" [61, 62]. Yet, women with angina and confirmed ischemia have increased mortality from IHD [63]. As discussed later on, the recent application of modern clinical diagnostic tools, such as cardiac computed tomography angiography (CCTA) and positron emission tomography (PET), is changing the paradigm of how disease is diagnosed [64, 65], broadening definitions of CAD and ischemia, respectively, to better reflect pathological phenotypes more prevalent in certain patients, especially women.

Management strategies for IHD, with the goal of ameliorating symptoms and improving survival, have also largely been the same for women and men while focused predominantly on obstructive CAD. As a function of this, symptomatic women are less likely than men to be candidates for invasive approaches, such as percutaneous coronary intervention (PCI), which targets flow-limiting obstructive coronary artery stenoses, and particularly coronary artery bypass grafting (CABG), which is most often reserved for multivessel obstructive disease. These

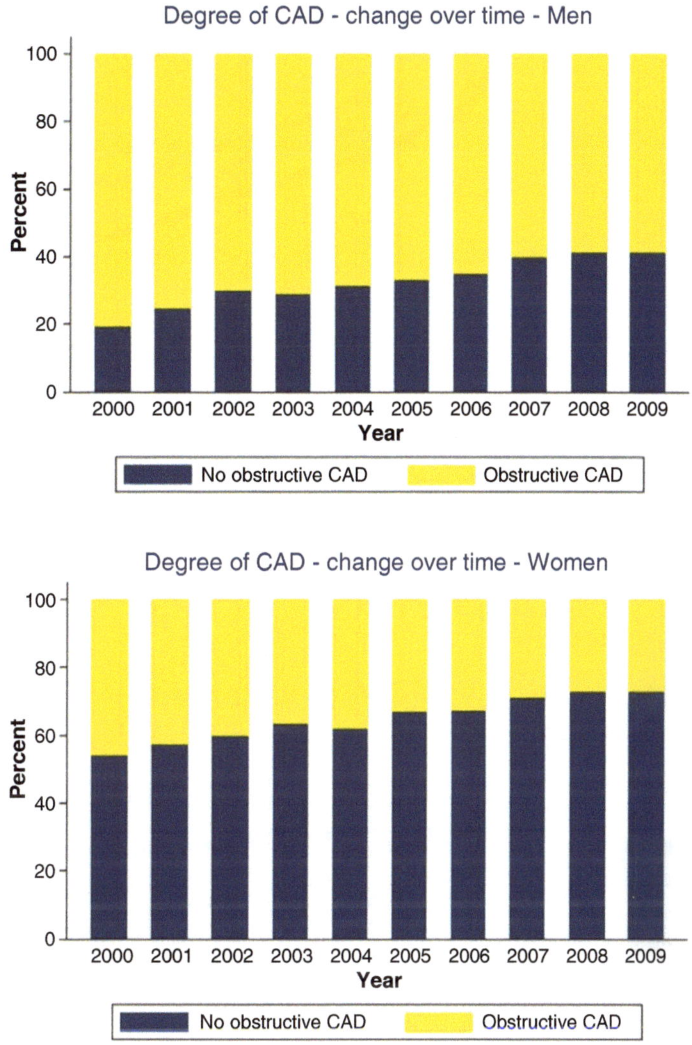

Fig. 17.4 Degree of coronary artery disease (CAD) by sex and year of invasive coronary angiography examination. (Reproduced with permission [15])

individuals may present with severe ischemia or acute MI prior to being diagnosed with nonobstructive coronary arteries (phenomena recently described as INOCA [66] and MINOCA [67, 68], respectively), comprising an even higher-risk subset of patients. Furthermore, women undergoing PCI or standard treatment for acute MI demonstrate increased risk of bleeding and vascular complications as compared to men [69, 70]. This finding suggests that the beneficial impact of invasive interventions, particularly in stable IHD, may be lower in women if not tailored to key biological differences involving not only their burden of obstructive atherosclerosis, but also the size of their vessels, effective circulating volumes, and metabolism of drugs.

Separate from differences in biology, there have been multiple reports of sex differences in the use of evidence-based interventions, including contemporary guideline-directed medical therapies (GDMT) [10]. In a 2002 survey of 3779 patients (42% female) with stable angina, women were less likely than men to receive optimal secondary prevention with antiplatelet and lipid-lowering therapies, even after angiographic documentation of disease [53]. At 1-year follow-up, women with confirmed CAD were less likely than men to report complete resolution of angina and more than twice as likely to suffer death or

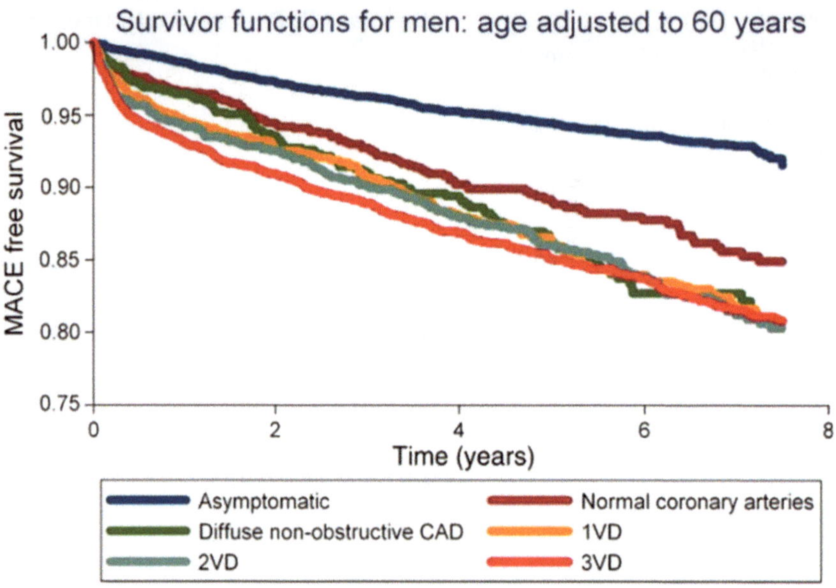

Numbers at risk	0	2	4	6
Asymptomatic	2359	2231	2101	1738
Normal CA	1214	854	597	367
Dif. non-obstr. CAD	869	557	362	174
1VD	1475	1072	783	474
2VD	1105	806	583	342
3VD	1783	1312	984	632

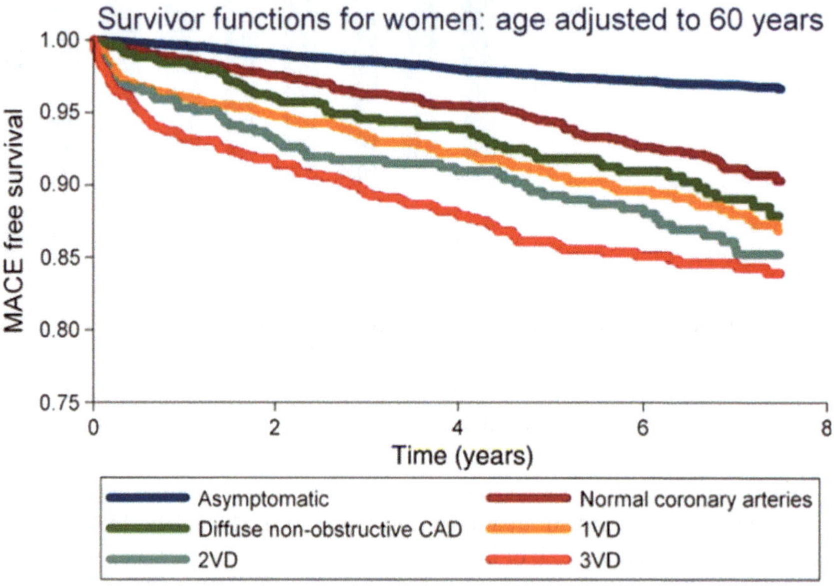

Numbers at risk	0	2	4	6
Asymptomatic	3346	3213	3044	2600
Normal CA	2237	1597	1155	721
Dif. non-obstr. CAD	809	527	336	187
1VD	777	567	411	252
2VD	377	274	209	143
3VD	471	333	256	161

Fig. 17.5 Major adverse cardiovascular event (MACE)-free survival by sex and vessel disease (VD) involvement on invasive coronary angiography. (Reproduced with permission [15])

nonfatal MI, even after multivariable adjustment for baseline risk factors including age, the presence of diabetes, abnormal LV function and severity of CAD, or interim revascularization. In 49,358 older patients hospitalized for CAD from 2003 to 2009, women (47%) were less likely than men to receive optimal GDMT at discharge and more likely to have higher mortality if they received suboptimal care [71]. Of note, authors found that the sex disparity in mortality disappeared with optimal care, and up to 69% of the excess mortality observed in older women could potentially be reduced by providing equitable optimal care, including timely initiation of aspirin, lipid-lowering medications and smoking cessation counseling. This finding is important because sex differences in response to medical therapies have been described, for example, greater coronary atheroma regression in response to high-intensity statins [72, 73] and greater benefit in cardiovascular outcomes with lesser degrees of LDL reduction [74] in women compared with men. These data underscore that women (and men) may especially benefit from a tailored approach that (1) achieves not only equitable standards of quality care as currently defined, (2) but also addresses fundamental differences in their biological and pathophysiologic manifestations of IHD.

Sex Differences in Coronary Anatomy and Function

Sex Differences in Coronary Artery Size and Blood Flow

The coronary arterial system represents a continuous network of functionally distinct vessel segments of decreasing size (Fig. 17.6) [75, 76]. The proximal large epicardial coronary arteries (400 μm to 2–5 mm in diameter) give way to small prearterioles (100–400 μm) and smaller intramural arterioles (<100 μm), which interface directly with the coronary capillary bed (<10 μm). The epicardial arteries have a primary capacitance function and exhibit minimal resistance to coronary flow under normal conditions, with their diameter primarily regulated by shear stress. In contrast, the prearterioles and arterioles make up most of the resistance circuit of the heart. Specifically, the prearterioles serve a critical role in the autoregulation of coronary blood flow via changes in vascular tone in response to local

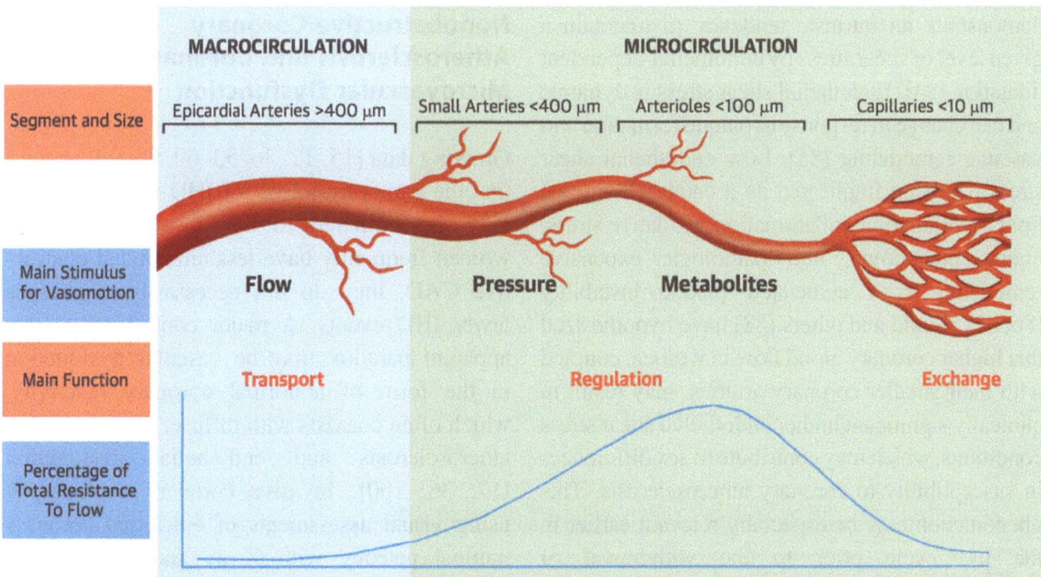

Fig. 17.6 Schematic of the anatomy and function of the coronary circulation. (Reproduced with permission [76])

shear stress and luminal pressure changes, and the arterioles are instrumental in the matching of myocardial oxygen demand with blood supply via changes in perfusion of the low-resistance coronary capillary bed in response to local tissue metabolism [77, 78].

Women have smaller epicardial coronary arteries than men, even after accounting for body habitus and LV mass [79–81]. In 710 patients (46% female) being evaluated for suspected CAD with CCTA and found to have limited coronary artery calcium scores (CAC, <100), women demonstrated significantly smaller diameters of all major epicardial coronary arteries, and these differences persisted after multivariable adjustment for age, BMI, body surface area, and LV mass [80]. Myocardial perfusion, or coronary blood flow, is typically also higher in women than in men at both rest and under hyperemic conditions [16, 82, 83]. In 1218 patients (67% female) being evaluated for suspected CAD with stress testing using PET and found to have no evidence of myocardial perfusion defects, women demonstrated higher coronary blood flows at both rest and peak stress when averaged over the entire left ventricle [16], resulting in a similar global coronary flow reserve (CFR) when compared to men.

With changes in blood flow, coronary arteries demonstrate an intrinsic tendency to maintain a given level of shear stress by endothelial-dependent dilatation [84]. Endothelial shear stress is dynamic and can change in response to plaque formation and vascular remodeling [85]. Low endothelial shear stress has been implicated as a catalyst for focal lipid accumulation, inflammation, oxidative stress, matrix breakdown, and pathologic expansive remodeling with associated plaque instability [86, 87]. Gould and others [88] have hypothesized that higher coronary blood flow in women, coupled with their smaller coronary arteries, may result in clinically significant higher endothelial shear stress conditions, which may contribute to sex differences in susceptibility to coronary atherosclerosis. This phenomenon may be especially relevant earlier in the life cycle prior to the withdrawal of estrogens, which may interact with pressure- and shear stress-dependent mechanisms of arteriolar vasomotor function to impact upon release of endothelial mediators including nitric oxide, prostaglandins, and endothelium-derived hyperpolarizing factor [89]. The burden of coronary atherosclerosis is indeed lower in women than in men, particularly at younger ages. The distribution of CAC stratified by age in 9341 asymptomatic study participants (40% female, median age 54 ± 10 years) is illustrated in Fig. 17.7, with women demonstrating a slower rate of rise in CAC over the life span [90]. A similar pattern has been shown in asymptomatic cohorts from the Framingham Heart Study [91] and the Multi-Ethnic Study of Atherosclerosis [92].

Differences in shear stress and their associated effects on mechanoreceptor-induced intracellular cascades may affect not only the susceptibility to, but also the anatomical pattern of CAD in women and men. While the mitigating effects of high shear stress on atheroma progression may not entirely prevent the development of CAD in susceptible patients, they may modify it into a diffuse pattern of disease with decreased contributions from focal obstruction. Such an effect would be consistent with previously described clinical and pathological observations of sex differences in the presentation of IHD.

Nonobstructive Coronary Atherosclerosis and Coronary Microvascular Dysfunction

Growing data [15, 17, 36, 53–60, 93–99] support that the pathophysiology of IHD in women may vary as compared to that in men. Although women frequently have less anatomical obstructive CAD, they do not necessarily experience fewer IHD events. A major contributor to this apparent paradox may be vascular dysfunction in the form of abnormal coronary reactivity, which often coexists with diffuse, nonobstructive atherosclerosis and endothelial dysfunction [17, 36, 100]. Invasive coronary angiography using visual assessments of epicardial coronary luminal patency with X-ray and contrast dye remains a cornerstone of modern cardiovascular care for the diagnosis of obstructive CAD.

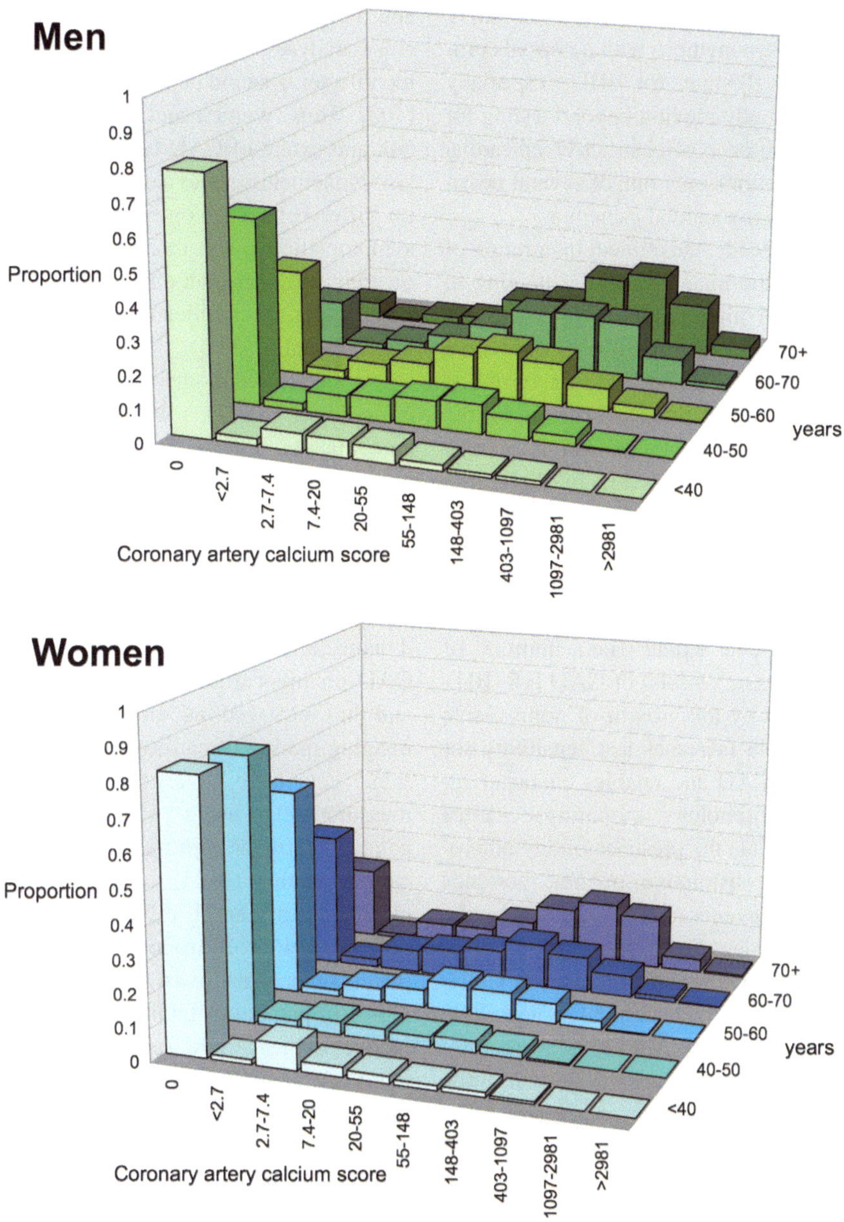

Fig. 17.7 Distribution of coronary artery calcium scores (logarithmic scale) by sex and age in asymptomatic individuals. (Reproduced with permission [90])

However, this technique has limited ability to identify diffuse atherosclerosis and small-vessel dysfunction, which may contribute to MACE including acute coronary syndromes, heart failure and death from plaque erosion, impaired vasoreactivity, and CMD with resultant myocardial ischemia [54, 56–58]. The addition of invasive fractional flow reserve, an assessment of the pressure drop across a focal epicardial stenosis, to coronary angiography has proven beneficial to identify lesion-specific ischemia and guide revascularization, [101] but may miss the integrated contribution of diffuse atherosclerosis and small-vessel disease to myocardial ischemia [102, 103].

It is now well recognized that a normal coronary angiogram is not synonymous with a normal coronary circulation. Testing for IHD, especially in women, is currently moving beyond testing for the presence or absence of obstructive epicardial CAD, which represents only one of several possible contributors to myocardial ischemia.

Over the last decade, the clinical integration of advanced diagnostic imaging tools is helping to redefine IHD and highlight the importance of nonobstructive CAD and CMD. Noninvasive approaches using CCTA and PET, for example, have provided very sensitive assessments for the evaluation of anatomic atherosclerotic plaque and functional ischemia, respectively. Regarding atherosclerotic plaque, recent observational studies have revealed stepwise incremental risk for future adverse events in patients along a continuum of both severity (i.e., mild, moderate, or severe stenosis) and extent (i.e., number of involved segments or vessels) of CAD [15, 104–106]. Accelerated by the growth of noninvasive CCTA, which both increases test sensitivity for the diagnosis of CAD and enables characterization of plaque morphology, mounting evidence now supports that (1) the presence of any atherosclerotic plaque, obstructive or not, portends increased risk of events and (2) the higher the overall plaque burden that is present, the higher the risk. From the international multicenter CONFIRM registry, in which 23,854 consecutive patients without known CAD (33% asymptomatic) underwent CCTA between 2005 and 2009, per-patient nonobstructive and obstructive CAD conferred increased risk of mortality in a stepwise manner as compared with absence of CAD, which was associated with very low risk [106]. In a subsequent sex-specific subgroup analysis of patients with no or nonobstructive CAD ($n = 18{,}158$), nonobstructive CAD was associated with a modest adjusted risk of MACE (composite of death and nonfatal MI) that was similarly increased in both women and men [107]. Further, in a smaller subset of patients with longer follow-up ($n = 5632$ followed for 5 years, 30% with nonobstructive CAD), there was no interaction of sex on the association between per-vessel extent of obstructive CAD and MACE, and authors concluded that exploratory analyses of atherosclerotic burden did not identify sex-specific patterns predictive of MACE [108]. Thus, women and men with comparable risk and extent of CAD had comparable prognosis. Nonetheless, prevalent patterns of disease do differ between women and men, with more symptomatic women than men manifesting nonobstructive rather than obstructive CAD [15, 60], with important implications for diagnosis and management.

Neither conventional angiography nor CCTA can detect CMD, which is defined not anatomically, but functionally as a reduced CFR in the absence of flow-limiting CAD. Global CFR, calculated as the ratio of hyperemic to rest absolute myocardial blood flow averaged over the left ventricle, is an integrated marker of coronary vasomotor dysfunction that measures the hemodynamic effects of focal, diffuse, and small-vessel CAD on myocardial tissue perfusion [82, 109] and has emerged as an important prognostic imaging marker of cardiovascular risk. Observational data have consistently shown that CFR measurements using PET MPI distinguish patients at low or high risk for MACE, including cardiac death [110–113], beyond comprehensive clinical assessment, LVEF, myocardial perfusion defects, low-level troponin elevation [96], or plaque severity on invasive coronary angiography [114]. In parallel to findings with atherosclerotic plaque, evidence now supports that (1) the existence of impaired CFR, whether in the presence or absence of obstructive CAD (e.g., CMD), portends increased risk of cardiovascular events and (2) the more severely impaired the overall global CFR, the higher the risk. Whereas a CFR ≥ 2 effectively excluded high-risk angiographic CAD and was associated with low rates of annualized cardiac death [115], event rates for patients with CFR <2 increased exponentially as CFR decreased [7, 112, 116]. Because CFR is a measure of not only the effects of epicardial CAD but also of diffuse atherosclerosis and CMD on myocardial tissue perfusion, worse prognosis in patients with CFR <2 may be related to coronary vasomotor dysfunction arising from a mix of pathophysiologic CAD phenotypes.

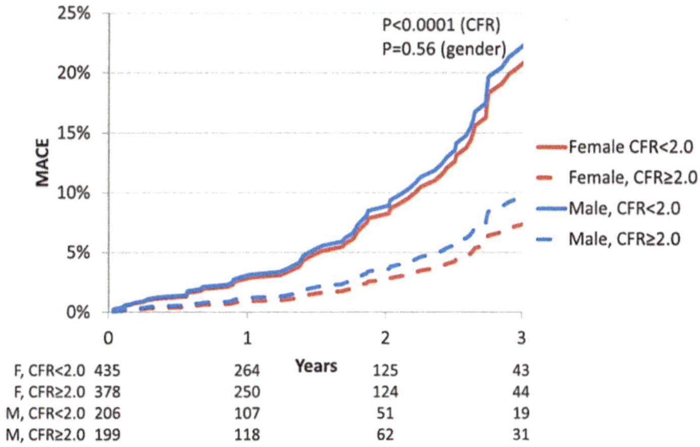

Fig. 17.8 Adjusted cumulative rate of major adverse cardiac events (MACE) by sex and coronary flow reserve (CFR) in patients with normal myocardial perfusion. (Reproduced with permission [16])

These findings highlight the morbidity associated with diffuse atherosclerosis and CMD. Similar to CTA findings for nonobstructive CAD, cardiac PET MPI studies support that CMD is common in symptomatic patients, affecting approximately 50% of patients with normal MPI and LVEF referred for testing [16]. Furthermore when present, CMD was associated with MACE independently of sex (Fig. 17.8) [16]. That is to say, both women and men with CMD (CFR <2 in the presence of normal cardiac perfusion) experienced worse outcomes, although this phenotype was twice as prevalent in women as in men. This result was consistent even in patients with a CAC score of 0 [16], and global CFR, but not CAC, provided significant incremental risk stratification over clinical risk score for prediction of MACE [117]. Thus, symptomatic patients who do not demonstrate regional ischemia associated with flow-limiting CAD may have diffuse atherosclerosis and CMD for which a more sensitive, quantitative assessment of ischemia may better diagnose abnormalities and identify novel targets for systemic therapies. Although not a uniquely female disorder, this pattern of abnormalities may be more prognostically useful in women because they often occur absent obstructive CAD, and may be especially relevant in those with diabetes and/or metabolic syndrome, MINOCA, INOCA, and HFpEF.

There is also emerging evidence that women with very low CFR may be at an especially elevated risk of cardiovascular events. In symptomatic patients referred for invasive coronary angiography after PET MPI, women had a lower pretest probability of CAD and a lower burden of obstructive CAD relative to men, but were not protected from MACE (Fig. 17.9a, b) [36], consistent with previously described epidemiologic trends. Whereas outcomes for men were more closely associated with presence or absence of severely obstructive CAD (Fig. 17.9c, d), in women, only those with impaired CFR demonstrated a significantly increased adjusted risk of MACE (Fig. 17.9e, f). The excess cardiovascular risk in women relative to men referred for coronary angiography was independently associated with and mediated by impaired CFR, not obstructive CAD (p for interaction = 0.04, Fig. 17.10a). Whereas most men with severely impaired CFR were found to have ≥1 vessel CAD on coronary angiography, most women with similarly impaired CFR demonstrated ≤1 vessel CAD (Fig. 17.10b). In adjusted analysis, approximately 40% of this observed differential effect of sex on outcomes was mediated by CFR [36].

A very low CFR may represent a crucial link to understanding the hidden biological risk of IHD among women. Previous data [114] support that revascularization, especially by CABG, in certain individuals with severely impaired CFR may be beneficial. That sex differences on outcomes of cardiovascular events are amplified in those with severely impaired CFR further suggests that certain patients (i.e., with very low

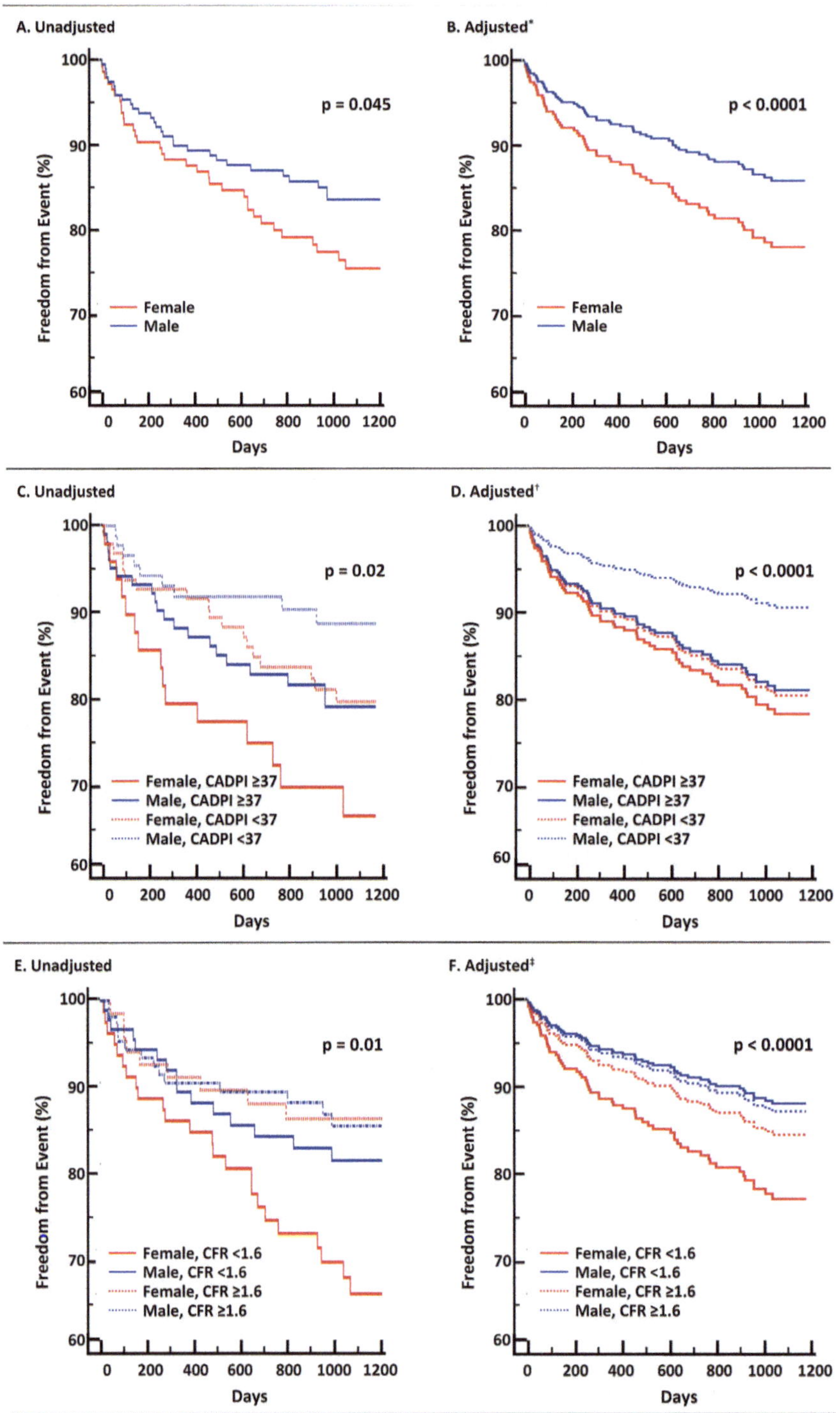

*Adjusted for nonwhite race; pretest clinical score; history of PCI, hypertension, and insulin use; BMI >27 kg/m^2; LVEF <50%; and LV ischemia >10%

†Adjusted for nonwhite race; pretest clinical score; history of PCI, hypertension, and insulin use; BMI >27 kg/m^2; LVEF <50%; LV ischemia >10%; and time-dependent revascularization with PCI or CABG within 90d of noninvasive imaging

‡Adjusted for nonwhite race; pretest clinical score; history of PCI, hypertension, and insulin use; BMI >27 kg/m^2; LVEF <50%; LV ischemia >10%; CAD prognostic index; time-dependent revascularization with PCI or CABG within 90d of noninvasive imaging

Fig. 17.9 Freedom from major adverse cardiovascular events (MACE) according to sex (**a** and **b**), sex and angiographic disease (**c** and **d**), or sex and coronary flow reserve (**e** and **f**) in patients referred for myocardial perfusion imaging and invasive coronary angiography. (Reproduced with permission [36])

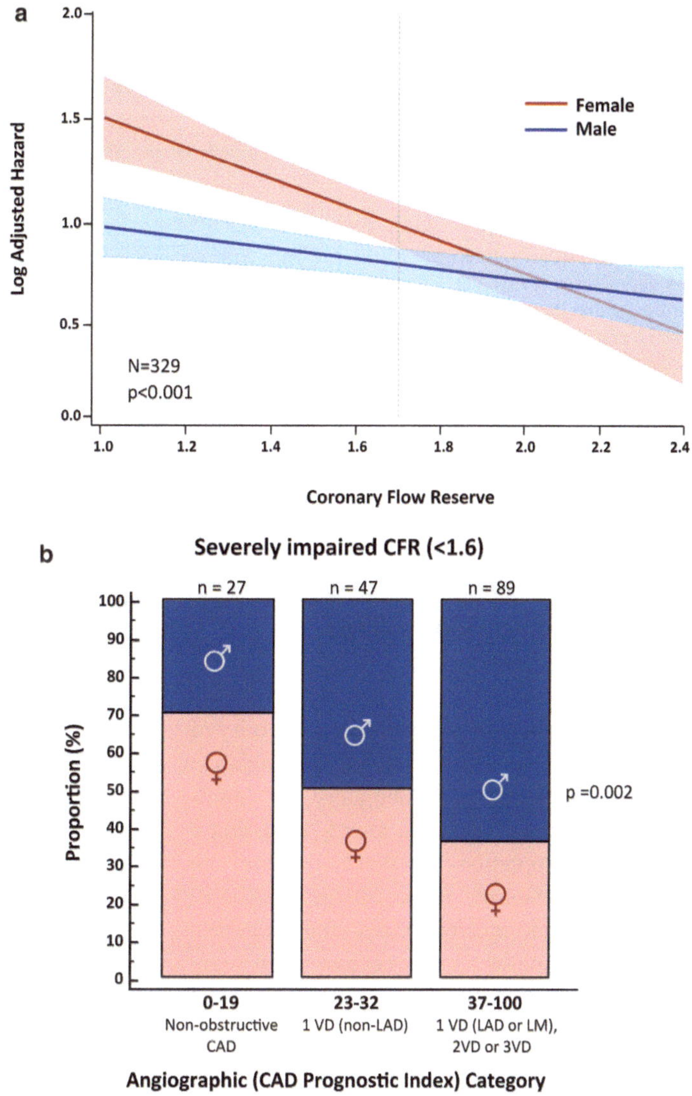

Fig. 17.10 Log adjusted hazard for major adverse cardiovascular events (MACE) by sex and coronary flow reserve (CFR), (**a**) Patients with severely impaired CFR (<1.6) by angiographic disease and sex categories, (**b**) *CAD* indicates coronary artery disease, *VD* vessel disease, *LAD* left anterior descending artery, *LM* left main artery. (Reproduced with permission [36])

CFR and less obstructive CAD, a phenotype more prevalent in women and less amenable to focal revascularization) may be at especially high risk (Fig. 17.11) [36]. Instead of being interpreted as demonstrating a "false positive" (or in some cases negative) traditional ischemic evaluation, patients with impaired CFR and less obstructive CAD may be at significantly increased CVD risk despite having access to revascularization, a tool fundamentally targeted to the management of obstructive CAD. Thus, while providing optimal, equitable guideline-directed care remains a critical goal for managing women with IHD, doing so according to current paradigms may be insufficient to address what is likely a combination of biological as well as environmental determinants of their prognosis. Additional research is needed to determine precisely what is optimal care for this subset of vulnerable patients with a predominance of women.

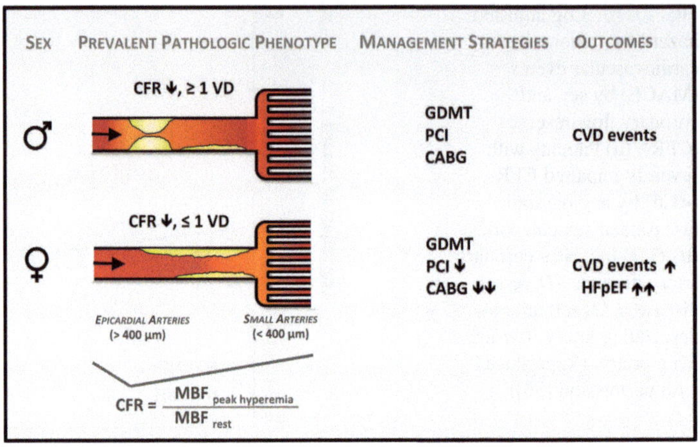

Fig. 17.11 Conceptual model of prevalent pathological phenotypes in women and men with ischemic heart disease and possible impact on cardiovascular management strategies and outcomes. *CABG* indicates coronary artery bypass surgery, *CFR* coronary flow reserve, *CVD* cardiovascular disease, *GDMT* guideline-directed medical therapy, *MBF* myocardial blood flow, *PCI* percutaneous coronary intervention, and *VD* vessel disease. (Reproduced with permission [36])

Remaining Knowledge Gaps and Future Directions

We have come to recognize that for a majority of patients seeking an explanation for their angina in the cardiac catheterization laboratory, no significantly obstructing lesion are found [118], and a substantial proportion of these patients are women. Nonetheless, these patients may still have significant atherosclerosis and ischemia, and their prognosis is not necessarily benign. Women as compared to men commonly present with higher rates of baseline comorbidities, including not only older age but also hypertension, diabetes mellitus, obesity, chronic renal failure, peripheral vascular disease, HFpEF, and inflammatory diseases such as rheumatoid arthritis [10, 25–27, 54–56]. Many of these disease processes are associated with diffuse atherosclerosis and microvascular ischemia, and parallel themes are emerging between nonobstructive CAD and CMD. In particular, two key findings regarding these entities in symptomatic patients have emerged: (1) they are common in both women and men but more prevalent in women, and (2) they are associated with similarly increased risk of IHD events in both women and men, except for in patients with severely impaired CFR, where women demonstrate even greater risk [36].

This latter subgroup is particularly interesting and represents a clinically unmet need that is ripe for future investigation. Specifically, in cases where impaired CFR stems not from obstructive CAD (with no opportunity for revascularization to mitigate CVD risk), a novel therapeutic strategy to systemically target IHD may be warranted. These cases of severe CMD, which often coexist with nonobstructive CAD, may provide a clue as to a common mechanism underlying IHD risk in both women and men. Such a mechanism may involve inflammation [119], endothelial dysfunction [120], and increased cardiomyocyte oxygen demand with ensuing microvascular ischemia, myocardial injury [96], and impaired cardiac mechanics [97, 121]. Thus, clearer understanding of the relationship between coronary vasomotor dysfunction and CAD comorbid conditions, including insulin resistance and heart failure, may guide development of novel systemic therapies to harness the benefit of more “complete revascularization” [114, 122–125] in a manner not defined by anatomy alone. As such, new imaging tools may represent important biomarkers not only for prospective studies evaluating the role of ischemia and revascularization, but also of novel anti-inflammatory

[126, 127], lipid-lowering [128], glucose-lowering [129], and neurohormonal-modulating [130] agents on cardiovascular outcomes.

Conclusion

While prominent sex differences are apparent in certain anatomical measures of CAD, IHD represents a continuum of disease in women and men, with both anatomical and functional manifestations. The integration of functional and anatomical parameters associated with CAD (i.e., quantification of coronary flow reserve with visualization of anatomic atherosclerotic plaque) may enhance our understanding of sex differences in cardiac risk and lead to improved algorithms for diagnosing and treating women and men with IHD. Many of the hypothesis-generating studies reviewed here illustrate the importance of performing sex-specific analyses to advance beyond current limitations in cardiovascular care. Insights like these may inform future trials and, possibly, the implementation of sex-specific thresholds within clinical guidelines for more optimal patient management. Doing so may lead us to a more complete understanding of the pathobiology of IHD and its manifestations in large subsets of patients, reframing sex-specific medicine as a form of precision medicine with the goal of improving outcomes for all.

Funding This work was supported by NIH K23HL135438 to Dr. Taqueti.

Disclosures None.

References

1. Benjamin EJ, Virani SS, Callaway CW, Chang AR, Cheng S, Chiuve SE, Cushman M, Delling FN, Deo R, de Ferranti SD, Ferguson JF, Fornage M, Gillespie C, Isasi CR, Jimenez MC, Jordan LC, Judd SE, Lackland D, Lichtman JH, Lisabeth L, Liu S, Longenecker CT, Lutsey PL, Matchar DB, Matsushita K, Mussolino ME, Nasir K, O'Flaherty M, Palaniappan LP, Pandey DK, Reeves MJ, Ritchey MD, Rodriguez CJ, Roth GA, Rosamond WD, Sampson UKA, Satou GM, Shah SH, Spartano NL, Tirschwell DL, Tsao CW, Voeks JH, Willey JZ, Wilkins JT, Wu JH, Alger HM, Wong SS, Muntner P. Heart disease and stroke statistics-2018 update: a report from the American Heart Association. Circulation. 2018;137:e67-492.
2. Atlas Writing Group, Timmis A, Townsend N, Gale C, Grobbee R, Maniadakis N, Flather M, Wilkins E, Wright L, Vos R, Bax J, Blum M, Pinto F, Vardas P. European society of cardiology: cardiovascular disease statistics 2017. Eur Heart J. 2018;39:508–79.
3. The top 10 causes of death – World Health Organization, 2015. Fact sheet, updated Jan 2017. http://www.who.int/mediacentre/factsheets/fs310/en/
4. Number of deaths from 10 leading causes, by sex — National Vital Statistics System, United States, 2015. Morb Mortal Wkly Rep. 2017;66:413. https://doi.org/10.15585/mmwr.mm6615a8.
5. Mosca L, Hammond G, Mochari-Greenberger H, Towfighi A, Albert MA, American Heart Association Cardiovascular D, Stroke in W, Special Populations Committee of the Council on Clinical Cardiology CoE, Prevention CoCNCoHB. Fifteen-year trends in awareness of heart disease in women: results of a 2012 American heart association national survey. Circulation. 2013;127:1254–63. e1251-1229
6. Marzilli M, Merz CN, Boden WE, Bonow RO, Capozza PG, Chilian WM, DeMaria AN, Guarini G, Huqi A, Morrone D, Patel MR, Weintraub WS. Obstructive coronary atherosclerosis and ischemic heart disease: an elusive link! J Am Coll Cardiol. 2012;60:951–6.
7. Taqueti VR, Di Carli MF. Clinical significance of noninvasive coronary flow reserve assessment in patients with ischemic heart disease. Curr Opin Cardiol. 2016;31:662–9.
8. Wilmot KA, O'Flaherty M, Capewell S, Ford ES, Vaccarino V. Coronary heart disease mortality declines in the United States from 1979 through 2011: evidence for stagnation in young adults, especially women. Circulation. 2015;132:997–1002.
9. Wenger NK. Clinical characteristics of coronary heart disease in women: emphasis on gender differences. Cardiovasc Res. 2002;53:558–67.
10. Shaw LJ, Bugiardini R, Merz CN. Women and ischemic heart disease: evolving knowledge. J Am Coll Cardiol. 2009;54:1561–75.
11. Bairey Merz CN. Sex, death, and the diagnosis gap. Circulation. 2014;130:740–2.
12. Jacobs AK. Coronary intervention in 2009: are women no different than men? Circ Cardiovasc Interv. 2009;2:69–78.
13. Gulati M, Cooper-DeHoff RM, McClure C, Johnson BD, Shaw LJ, Handberg EM, Zineh I, Kelsey SF, Arnsdorf MF, Black HR, Pepine CJ, Merz CN. Adverse cardiovascular outcomes in women with nonobstructive coronary artery disease: a report from the women's ischemia syndrome evaluation study and the St James women take heart project. Arch Intern Med. 2009;169:843–50.
14. Hemingway H, Langenberg C, Damant J, Frost C, Pyorala K, Barrett-Connor E. Prevalence of angina in women versus men: a systematic review and meta-analysis of international variations across 31 countries. Circulation. 2008;117:1526–36.
15. Jespersen L, Hvelplund A, Abildstrom SZ, Pedersen F, Galatius S, Madsen JK, Jorgensen E, Kelbaek H, Prescott E. Stable angina pectoris with no obstructive coronary artery disease is associated with increased risks of major adverse cardiovascular events. Eur Heart J. 2012;33:734–44.
16. Murthy VL, Naya M, Taqueti VR, Foster CR, Gaber M, Hainer J, Dorbala S, Blankstein R, Rimoldi O, Camici PG, Di Carli MF. Effects of sex on coronary microvascular dysfunction and cardiac outcomes. Circulation. 2014;129:2518–27.

17. Pepine CJ, Anderson RD, Sharaf BL, Reis SE, Smith KM, Handberg EM, Johnson BD, Sopko G, Bairey Merz CN. Coronary microvascular reactivity to adenosine predicts adverse outcome in women evaluated for suspected ischemia results from the national heart, lung and blood institute WISE (Women's Ischemia Syndrome Evaluation) study. J Am Coll Cardiol. 2010;55:2825–32.
18. Reis SE, Holubkov R, Conrad Smith AJ, Kelsey SF, Sharaf BL, Reichek N, Rogers WJ, Merz CN, Sopko G, Pepine CJ, Investigators W. Coronary microvascular dysfunction is highly prevalent in women with chest pain in the absence of coronary artery disease: results from the NHLBI WISE study. Am Heart J. 2001;141:735–41.
19. Scantlebury DC, Borlaug BA. Why are women more likely than men to develop heart failure with preserved ejection fraction? Curr Opin Cardiol. 2011;26:562–8.
20. Templin C, Ghadri JR, Diekmann J, Napp LC, Bataiosu DR, Jaguszewski M, Cammann VL, Sarcon A, Geyer V, Neumann CA, Seifert B, Hellermann J, Schwyzer M, Eisenhardt K, Jenewein J, Franke J, Katus HA, Burgdorf C, Schunkert H, Moeller C, Thiele H, Bauersachs J, Tschope C, Schultheiss HP, Laney CA, Rajan L, Michels G, Pfister R, Ukena C, Bohm M, Erbel R, Cuneo A, Kuck KH, Jacobshagen C, Hasenfuss G, Karakas M, Koenig W, Rottbauer W, Said SM, Braun-Dullaeus RC, Cuculi F, Banning A, Fischer TA, Vasankari T, Airaksinen KE, Fijalkowski M, Rynkiewicz A, Pawlak M, Opolski G, Dworakowski R, MacCarthy P, Kaiser C, Osswald S, Galiuto L, Crea F, Dichtl W, Franz WM, Empen K, Felix SB, Delmas C, Lairez O, Erne P, Bax JJ, Ford I, Ruschitzka F, Prasad A, Luscher TF. Clinical features and outcomes of takotsubo (stress) cardiomyopathy. N Engl J Med. 2015;373:929–38.
21. Tweet MS, Hayes SN, Pitta SR, Simari RD, Lerman A, Lennon RJ, Gersh BJ, Khambatta S, Best PJ, Rihal CS, Gulati R. Clinical features, management, and prognosis of spontaneous coronary artery dissection. Circulation. 2012;126:579–88.
22. Huxley R, Barzi F, Woodward M. Excess risk of fatal coronary heart disease associated with diabetes in men and women: meta-analysis of 37 prospective cohort studies. BMJ. 2006;332:73–8.
23. Emdin CA, Wong CX, Hsiao AJ, Altman DG, Peters SA, Woodward M, Odutayo AA. Atrial fibrillation as risk factor for cardiovascular disease and death in women compared with men: systematic review and meta-analysis of cohort studies. BMJ. 2016;532:h7013.
24. Champney KP, Frederick PD, Bueno H, Parashar S, Foody J, Merz CN, Canto JG, Lichtman JH, Vaccarino V, Investigators N. The joint contribution of sex, age and type of myocardial infarction on hospital mortality following acute myocardial infarction. Heart. 2009;95:895–9.
25. Bucholz EM, Butala NM, Rathore SS, Dreyer RP, Lansky AJ, Krumholz HM. Sex differences in long-term mortality after myocardial infarction: a systematic review. Circulation. 2014;130:757–67.
26. Gupta T, Kolte D, Khera S, Agarwal N, Villablanca PA, Goel K, Patel K, Aronow WS, Wiley J, Bortnick AE, Aronow HD, Abbott JD, Pyo RT, Panza JA, Menegus MA, Rihal CS, Fonarow GC, Garcia MJ, Bhatt DL. Contemporary sex-based differences by age in presenting characteristics, use of an early invasive strategy, and inhospital mortality in patients with non-ST-segment-elevation myocardial infarction in the United States. Circ Cardiovasc Interv. 2018;11: e005735.
27. Khera S, Kolte D, Gupta T, Subramanian KS, Khanna N, Aronow WS, Ahn C, Timmermans RJ, Cooper HA, Fonarow GC, Frishman WH, Panza JA, Bhatt DL. Temporal trends and sex differences in revascularization and outcomes of ST-segment elevation myocardial infarction in younger adults in the United States. J Am Coll Cardiol. 2015;66:1961–72.
28. Regitz-Zagrosek V, Petrov G, Lehmkuhl E, Smits JM, Babitsch B, Brunhuber C, Jurmann B, Stein J, Schubert C, Merz NB, Lehmkuhl HB, Hetzer R. Heart transplantation in women with dilated cardiomyopathy. Transplantation. 2010;89:236–44.
29. Taqueti VR, Bairey Merz CN. Sex-specific precision medicine: targeting CRT-D and other cardiovascular interventions to those most likely to benefit. Eur Heart J. 2017;38:1495–7.
30. Hernandez AF, Fonarow GC, Liang L, Al-Khatib SM, Curtis LH, LaBresh KA, Yancy CW, Albert NM, Peterson ED. Sex and racial differences in the use of implantable cardioverter-defibrillators among patients hospitalized with heart failure. JAMA. 2007;298:1525–32.
31. Zusterzeel R, Selzman KA, Sanders WE, Canos DA, O'Callaghan KM, Carpenter JL, Pina IL, Strauss DG. Cardiac resynchronization therapy in women: US food and drug administration meta-analysis of patient-level data. JAMA Intern Med. 2014;174:1340–8.
32. Moss AJ, Hall WJ, Cannom DS, Klein H, Brown MW, Daubert JP, Estes NA 3rd, Foster E, Greenberg H, Higgins SL, Pfeffer MA, Solomon SD, Wilber D, Zareba W, M-CT I. Cardiac-resynchronization therapy for the prevention of heart-failure events. N Engl J Med. 2009;361:1329–38.
33. Wenger NK, Ouyang P, Miller VM, Bairey Merz CN. Strategies and methods for clinical scientists to study sex-specific cardiovascular health and disease in women. J Am Coll Cardiol. 2016;67:2186–8.
34. Maas AH, van der Schouw YT, Regitz-Zagrosek V, Swahn E, Appelman YE, Pasterkamp G, Ten Cate H, Nilsson PM, Huisman MV, Stam HC, Eizema K, Stramba-Badiale M. Red alert for women's heart: the urgent need for more research and knowledge on cardiovascular disease in women: proceedings of the workshop held in Brussels on gender differences in cardiovascular disease, 29 September 2010. Eur Heart J. 2011;32:1362–8.
35. Melloni C, Berger JS, Wang TY, Gunes F, Stebbins A, Pieper KS, Dolor RJ, Douglas PS, Mark DB, Newby LK. Representation of women in randomized clinical trials of cardiovascular disease prevention. Circ Cardiovasc Qual Outcomes. 2010;3:135–42.
36. Taqueti VR, Shaw LJ, Cook NR, Murthy VL, Shah NR, Foster CR, Hainer J, Blankstein R, Dorbala S, Di Carli MF. Excess cardiovascular risk in women relative to men referred for coronary angiography is associated with severely impaired coronary flow reserve, not obstructive disease. Circulation. 2017;135:566–77.
37. Mendelsohn ME, Karas RH. The protective effects of estrogen on the cardiovascular system. N Engl J Med. 1999;340:1801–11.
38. Huxley RR, Woodward M. Full hazards of smoking and benefits of stopping for women. Lancet. 2013;381:96–8.
39. Vasan RS, Larson MG, Leip EP, Evans JC, O'Donnell CJ, Kannel WB, Levy D. Impact of high-normal blood pressure on the risk of cardiovascular disease. N Engl J Med. 2001;345:1291–7.
40. Peters SA, Huxley RR, Woodward M. Comparison of the sex-specific associations between systolic blood pressure and the risk of cardiovascular disease: a systematic review and

meta-analysis of 124 cohort studies, including 1.2 million individuals. Stroke. 2013;44:2394–401.
41. Peters SA, Singhateh Y, Mackay D, Huxley RR, Woodward M. Total cholesterol as a risk factor for coronary heart disease and stroke in women compared with men: a systematic review and meta-analysis. Atherosclerosis. 2016;248:123–31.
42. Emerging Risk Factors C, Wormser D, Kaptoge S, Di Angelantonio E, Wood AM, Pennells L, Thompson A, Sarwar N, Kizer JR, Lawlor DA, Nordestgaard BG, Ridker P, Salomaa V, Stevens J, Woodward M, Sattar N, Collins R, Thompson SG, Whitlock G, Danesh J. Separate and combined associations of body-mass index and abdominal adiposity with cardiovascular disease: collaborative analysis of 58 prospective studies. Lancet. 2011;377:1085–95.
43. Peters SA, Huxley RR, Woodward M. Diabetes as risk factor for incident coronary heart disease in women compared with men: a systematic review and meta-analysis of 64 cohorts including 858,507 individuals and 28,203 coronary events. Diabetologia. 2014;57:1542–51.
44. Ray JG, Vermeulen MJ, Schull MJ, Redelmeier DA. Cardiovascular health after maternal placental syndromes (champs): population-based retrospective cohort study. Lancet. 2005;366:1797–803.
45. Samad Z, Boyle S, Ersboll M, Vora AN, Zhang Y, Becker RC, Williams R, Kuhn C, Ortel TL, Rogers JG, O'Connor CM, Velazquez EJ, Jiang W, Investigators R. Sex differences in platelet reactivity and cardiovascular and psychological response to mental stress in patients with stable ischemic heart disease: insights from the remit study. J Am Coll Cardiol. 2014;64:1669–78.
46. Bybee KA, Prasad A. Stress-related cardiomyopathy syndromes. Circulation. 2008;118:397–409.
47. Symmons DP, Gabriel SE. Epidemiology of CVD in rheumatic disease, with a focus on RA and SLE. Nat Rev Rheumatol. 2011;7:399–408.
48. Taqueti VR, Ridker PM. Inflammation, coronary flow reserve, and microvascular dysfunction: moving beyond cardiac syndrome X. JACC Cardiovasc Imaging. 2013;6:668–71.
49. Taqueti VR, Ridker PM. Lipid-lowering and anti-inflammatory benefits of statin therapy: more than meets the plaque. Circ Cardiovasc Imaging. 2017;10:e006676.
50. Diamond GA, Forrester JS. Analysis of probability as an aid in the clinical diagnosis of coronary-artery disease. N Engl J Med. 1979;300:1350–8.
51. Kreatsoulas C, Shannon HS, Giacomini M, Velianou JL, Anand SS. Reconstructing angina: cardiac symptoms are the same in women and men. JAMA Intern Med. 2013;173:829–31.
52. Canto JG, Goldberg RJ, Hand MM, Bonow RO, Sopko G, Pepine CJ, Long T. Symptom presentation of women with acute coronary syndromes: myth vs reality. Arch Intern Med. 2007;167:2405–13.
53. Daly C, Clemens F, Lopez Sendon JL, Tavazzi L, Boersma E, Danchin N, Delahaye F, Gitt A, Julian D, Mulcahy D, Ruzyllo W, Thygesen K, Verheugt F, Fox KM, Euro Heart Survey I. Gender differences in the management and clinical outcome of stable angina. Circulation. 2006;113:490–8.
54. Hochman JS, Tamis JE, Thompson TD, Weaver WD, White HD, Van de Werf F, Aylward P, Topol EJ, Califf RM. Sex, clinical presentation, and outcome in patients with acute coronary syndromes. Global use of strategies to open occluded coronary arteries in acute coronary syndromes IIB investigators. N Engl J Med. 1999;341:226–32.
55. Hochman JS, McCabe CH, Stone PH, Becker RC, Cannon CP, DeFeo-Fraulini T, Thompson B, Steingart R, Knatterud G, Braunwald E. Outcome and profile of women and men presenting with acute coronary syndromes: a report from TIMI IIIB. TIMI investigators. Thrombolysis in myocardial infarction. J Am Coll Cardiol. 1997;30:141–8.
56. Berger JS, Elliott L, Gallup D, Roe M, Granger CB, Armstrong PW, Simes RJ, White HD, Van de Werf F, Topol EJ, Hochman JS, Newby LK, Harrington RA, Califf RM, Becker RC, Douglas PS. Sex differences in mortality following acute coronary syndromes. JAMA. 2009;302:874–82.
57. Smilowitz NR, Sampson BA, Abrecht CR, Siegfried JS, Hochman JS, Reynolds HR. Women have less severe and extensive coronary atherosclerosis in fatal cases of ischemic heart disease: an autopsy study. Am Heart J. 2011;161:681–8.
58. Arbustini E, Dal Bello B, Morbini P, Burke AP, Bocciarelli M, Specchia G, Virmani R. Plaque erosion is a major substrate for coronary thrombosis in acute myocardial infarction. Heart. 1999;82:269–72.
59. Liu Z, Pericak-Vance MA, Goldschmidt-Clermont P, Seo D, Wang L, Rundek T, Beecham GW. Coronary collateralization shows sex and racial-ethnic differences in obstructive artery disease patients. PLoS One. 2017;12:e0183836.
60. Shaw LJ, Shaw RE, Merz CN, Brindis RG, Klein LW, Nallamothu B, Douglas PS, Krone RJ, McKay CR, Block PC, Hewitt K, Weintraub WS, Peterson ED, American College of Cardiology-National Cardiovascular Data Registry I. Impact of ethnicity and gender differences on angiographic coronary artery disease prevalence and in-hospital mortality in the American College of Cardiology-National Cardiovascular Data Registry. Circulation. 2008;117:1787–801.
61. Taqueti VR, Dorbala S, Wolinsky D, Abbott B, Heller GV, Bateman TM, Mieres JH, Phillips LM, Wenger NK, Shaw LJ. Myocardial perfusion imaging in women for the evaluation of stable ischemic heart disease: state-of-the-evidence and clinical recommendations. J Nucl Cardiol. 2017;24:1402–26.
62. Mieres JH, Gulati M, Bairey Merz N, Berman DS, Gerber TC, Hayes SN, Kramer CM, Min JK, Newby LK, Nixon JV, Srichai MB, Pellikka PA, Redberg RF, Wenger NK, Shaw LJ, American Heart Association Cardiac Imaging Committee of the Council on Clinical C, Cardiovascular I, Intervention Committee of the Council on Cardiovascular R, Intervention. Role of noninvasive testing in the clinical evaluation of women with suspected ischemic heart disease: a consensus statement from the American Heart Association. Circulation. 2014;130:350–79.
63. Hemingway H, McCallum A, Shipley M, Manderbacka K, Martikainen P, Keskimaki I. Incidence and prognostic implications of stable angina pectoris among women and men. JAMA. 2006;295:1404–11.
64. Taqueti VR, Dorbala S. The role of positron emission tomography in the evaluation of myocardial ischemia in women. J Nucl Cardiol. 2016;23:1008–15.
65. Paul TK, Sivanesan K, Schulman-Marcus J. Sex differences in nonobstructive coronary artery disease: recent insights and substantial knowledge gaps. Trends Cardiovasc Med. 2017;27:173–9.
66. Bairey Merz CN, Pepine CJ, Walsh MN, Fleg JL. Ischemia and no obstructive coronary artery disease (INOCA): developing evidence-based therapies and research agenda for the next decade. Circulation. 2017;135:1075–92.

67. Beltrame JF. Assessing patients with myocardial infarction and nonobstructed coronary arteries (MINOCA). J Intern Med. 2013;273:182–5.
68. Pasupathy S, Tavella R, Beltrame JF. Myocardial infarction with nonobstructive coronary arteries (MINOCA): the past, present, and future management. Circulation. 2017;135:1490–3.
69. Lansky AJ, Hochman JS, Ward PA, Mintz GS, Fabunmi R, Berger PB, New G, Grines CL, Pietras CG, Kern MJ, Ferrell M, Leon MB, Mehran R, White C, Mieres JH, Moses JW, Stone GW, Jacobs AK, American College of Cardiology F, American Heart A. Percutaneous coronary intervention and adjunctive pharmacotherapy in women: a statement for healthcare professionals from the American Heart Association. Circulation. 2005;111:940–53.
70. Mehran R, Pocock SJ, Nikolsky E, Clayton T, Dangas GD, Kirtane AJ, Parise H, Fahy M, Manoukian SV, Feit F, Ohman ME, Witzenbichler B, Guagliumi G, Lansky AJ, Stone GW. A risk score to predict bleeding in patients with acute coronary syndromes. J Am Coll Cardiol. 2010;55:2556–66.
71. Li S, Fonarow GC, Mukamal KJ, Liang L, Schulte PJ, Smith EE, DeVore A, Hernandez AF, Peterson ED, Bhatt DL. Sex and race/ethnicity-related disparities in care and outcomes after hospitalization for coronary artery disease among older adults. Circ Cardiovasc Qual Outcomes. 2016;9: S36–44.
72. Puri R, Nissen SE, Nicholls SJ. Statin-induced coronary artery disease regression rates differ in men and women. Curr Opin Lipidol. 2015;26:276–81.
73. Puri R, Nissen SE, Shao M, Ballantyne CM, Barter PJ, Chapman MJ, Erbel R, Libby P, Raichlen JS, Uno K, Kataoka Y, Nicholls SJ. Sex-related differences of coronary atherosclerosis regression following maximally intensive statin therapy: insights from saturn. JACC Cardiovasc Imaging. 2014;7:1013–22.
74. Truong QA, Murphy SA, McCabe CH, Armani A, Cannon CP, Group TS. Benefit of intensive statin therapy in women: results from prove IT-TIMI 22. Circ Cardiovasc Qual Outcomes. 2011;4:328–36.
75. Camici PG, Crea F. Coronary microvascular dysfunction. N Engl J Med. 2007;356:830–40.
76. De Bruyne B, Oldroyd KG, Pijls NHJ. Microvascular (dys) function and clinical outcome in stable coronary disease. J Am Coll Cardiol. 2016;67:1170–2.
77. Chilian WM. Coronary microcirculation in health and disease. Summary of an NHLBI workshop. Circulation. 1997;95:522–8.
78. Crea F, Camici PG, Bairey Merz CN. Coronary microvascular dysfunction: an update. Eur Heart J. 2013.
79. Dickerson JA, Nagaraja HN, Raman SV. Gender-related differences in coronary artery dimensions: a volumetric analysis. Clin Cardiol. 2010;33:E44–9.
80. Hiteshi AK, Li D, Gao Y, Chen A, Flores F, Mao SS, Budoff MJ. Gender differences in coronary artery diameter are not related to body habitus or left ventricular mass. Clin Cardiol. 2014;37:605–9.
81. Kucher N, Lipp E, Schwerzmann M, Zimmerli M, Allemann Y, Seiler C. Gender differences in coronary artery size per 100 g of left ventricular mass in a population without cardiac disease. Swiss Med Wkly. 2001;131:610–5.
82. Gould KL, Johnson NP, Bateman TM, Beanlands RS, Bengel FM, Bober R, Camici PG, Cerqueira MD, Chow BJ, Di Carli MF, Dorbala S, Gewirtz H, Gropler RJ, Kaufmann PA, Knaapen P, Knuuti J, Merhige ME, Rentrop KP, Ruddy TD, Schelbert HR, Schindler TH, Schwaiger M, Sdringola S, Vitarello J, Williams KA Sr, Gordon D, Dilsizian V, Narula J. Anatomic versus physiologic assessment of coronary artery disease. Role of coronary flow reserve, fractional flow reserve, and positron emission tomography imaging in revascularization decision-making. J Am Coll Cardiol. 2013;62:1639–53.
83. Johnson NP, Gould KL. Integrating noninvasive absolute flow, coronary flow reserve, and ischemic thresholds into a comprehensive map of physiological severity. JACC Cardiovasc Imaging. 2012;5:430–40.
84. Lupi A, Buffon A, Finocchiaro ML, Conti E, Maseri A, Crea F. Mechanisms of adenosine-induced epicardial coronary artery dilatation. Eur Heart J. 1997;18:614–7.
85. Koskinas KC, Feldman CL, Chatzizisis YS, Coskun AU, Jonas M, Maynard C, Baker AB, Papafaklis MI, Edelman ER, Stone PH. Natural history of experimental coronary atherosclerosis and vascular remodeling in relation to endothelial shear stress: a serial, in vivo intravascular ultrasound study. Circulation. 2010;121:2092–101.
86. Chatzizisis YS, Coskun AU, Jonas M, Edelman ER, Feldman CL, Stone PH. Role of endothelial shear stress in the natural history of coronary atherosclerosis and vascular remodeling: molecular, cellular, and vascular behavior. J Am Coll Cardiol. 2007;49:2379–93.
87. Koskinas KC, Sukhova GK, Baker AB, Papafaklis MI, Chatzizisis YS, Coskun AU, Quillard T, Jonas M, Maynard C, Antoniadis AP, Shi GP, Libby P, Edelman ER, Feldman CL, Stone PH. Thin-capped atheromata with reduced collagen content in pigs develop in coronary arterial regions exposed to persistently low endothelial shear stress. Arterioscler Thromb Vasc Biol. 2013;33:1494.
88. Patel MB, Bui LP, Kirkeeide RL, Gould KL. Imaging microvascular dysfunction and mechanisms for female-male differences in cad. JACC Cardiovasc Imaging. 2016;9:465–82.
89. Huang A, Kaley G. Gender-specific regulation of cardiovascular function: estrogen as key player. Microcirculation. 2004;11:9–38.
90. Pletcher MJ, Tice JA, Pignone M, McCulloch C, Callister TQ, Browner WS. What does my patient's coronary artery calcium score mean? Combining information from the coronary artery calcium score with information from conventional risk factors to estimate coronary heart disease risk. BMC Med. 2004;2:31.
91. Hoffmann U, Massaro JM, Fox CS, Manders E, O'Donnell CJ. Defining normal distributions of coronary artery calcium in women and men (from the Framingham heart study). *Am J Cardiol.* 2008;102:1136–41. 1141.e1131
92. McClelland RL, Chung H, Detrano R, Post W, Kronmal RA. Distribution of coronary artery calcium by race, gender, and age: results from the multi-ethnic study of atherosclerosis (MESA). Circulation. 2006;113:30–7.
93. Bairey Merz CN, Shaw LJ, Reis SE, Bittner V, Kelsey SF, Olson M, Johnson BD, Pepine CJ, Mankad S, Sharaf BL, Rogers WJ, Pohost GM, Lerman A, Quyyumi AA, Sopko G, Investigators W. Insights from the NHLBI-sponsored women's ischemia syndrome evaluation (WISE) study: Part II: Gender differences in presentation, diagnosis, and outcome with regard to gender-based pathophysiology of atherosclerosis and macrovascular and microvascular coronary disease. J Am Coll Cardiol. 2006;47:S21–9.
94. Shaw LJ, Bairey Merz CN, Pepine CJ, Reis SE, Bittner V, Kelsey SF, Olson M, Johnson BD, Mankad S, Sharaf BL, Rogers WJ, Wessel TR, Arant CB, Pohost GM, Lerman A, Quyyumi AA, Sopko G, Investigators W. Insights from the NHLBI-sponsored women's ischemia syndrome evaluation

(WISE) study: Part I: Gender differences in traditional and novel risk factors, symptom evaluation, and gender-optimized diagnostic strategies. J Am Coll Cardiol. 2006;47:S4–S20.
95. von Mering GO, Arant CB, Wessel TR, McGorray SP, Bairey Merz CN, Sharaf BL, Smith KM, Olson MB, Johnson BD, Sopko G, Handberg E, Pepine CJ, Kerensky RA, National Heart L, Blood I. Abnormal coronary vasomotion as a prognostic indicator of cardiovascular events in women: results from the national heart, lung, and blood institute-sponsored women's ischemia syndrome evaluation (WISE). Circulation. 2004;109:722–5.
96. Taqueti VR, Everett BM, Murthy VL, Gaber M, Foster CR, Hainer J, Blankstein R, Dorbala S, Di Carli MF. Interaction of impaired coronary flow reserve and cardiomyocyte injury on adverse cardiovascular outcomes in patients without overt coronary artery disease. Circulation. 2015;131:528–35.
97. Taqueti VR, Solomon SD, Shah AM, Desai AS, Groarke JD, Osborne MT, Hainer J, Bibbo CF, Dorbala S, Blankstein R, Di Carli MF. Coronary microvascular dysfunction and future risk of heart failure with preserved ejection fraction. Eur Heart J. 2018; 39:840-9.
98. Arbustini E, Dal Bello B, Morbini P, Grasso M, Diegoli M, Fasani R, Pilotto A, Bellini O, Pellegrini C, Martinelli L, Campagna C, Gavazzi A, Specchia G, Vigano M, Roberts WC. Frequency and characteristics of coronary thrombosis in the epicardial coronary arteries after cardiac transplantation. Am J Cardiol. 1996;78:795–800.
99. Shaw LJ, Berman DS, Maron DJ, Mancini GB, Hayes SW, Hartigan PM, Weintraub WS, O'Rourke RA, Dada M, Spertus JA, Chaitman BR, Friedman J, Slomka P, Heller GV, Germano G, Gosselin G, Berger P, Kostuk WJ, Schwartz RG, Knudtson M, Veledar E, Bates ER, McCallister B, Teo KK, Boden WE, Investigators C. Optimal medical therapy with or without percutaneous coronary intervention to reduce ischemic burden: results from the clinical outcomes utilizing revascularization and aggressive drug evaluation (Courage) trial nuclear substudy. Circulation. 2008;117:1283–91.
100. Naya M, Murthy VL, Blankstein R, Sitek A, Hainer J, Foster C, Gaber M, Fantony JM, Dorbala S, Di Carli MF. Quantitative relationship between the extent and morphology of coronary atherosclerotic plaque and downstream myocardial perfusion. J Am Coll Cardiol. 2011;58:1807–16.
101. Tonino PA, De Bruyne B, Pijls NH, Siebert U, Ikeno F, van' t Veer M, Klauss V, Manoharan G, Engstrom T, Oldroyd KG, Ver Lee PN, MacCarthy PA, Fearon WF, Investigators FS. Fractional flow reserve versus angiography for guiding percutaneous coronary intervention. N Engl J Med. 2009;360:213–24.
102. Johnson NP, Gould KL, Di Carli MF, Taqueti VR. Invasive FFR and noninvasive CFR in the evaluation of ischemia: what is the future? J Am Coll Cardiol. 2016;67:2772–88.
103. Motwani M, Motlagh M, Gupta A, Berman DS, Slomka PJ. Reasons and implications of agreements and disagreements between coronary flow reserve, fractional flow reserve, and myocardial perfusion imaging. J Nucl Cardiol. 2018;25:104–19.
104. Bittencourt MS, Hulten E, Ghoshhajra B, O'Leary D, Christman MP, Montana P, Truong QA, Steigner M, Murthy VL, Rybicki FJ, Nasir K, Gowdak LH, Hainer J, Brady TJ, Di Carli MF, Hoffmann U, Abbara S, Blankstein R. Prognostic value of nonobstructive and obstructive coronary artery disease detected by coronary computed tomography angiography to identify cardiovascular events. Circ Cardiovasc Imaging. 2014;7:282–91.
105. Maddox TM, Stanislawski MA, Grunwald GK, Bradley SM, Ho PM, Tsai TT, Patel MR, Sandhu A, Valle J, Magid DJ, Leon B, Bhatt DL, Fihn SD, Rumsfeld JS. Nonobstructive coronary artery disease and risk of myocardial infarction. JAMA. 2014;312:1754–63.
106. Min JK, Dunning A, Lin FY, Achenbach S, Al-Mallah M, Budoff MJ, Cademartiri F, Callister TQ, Chang HJ, Cheng V, Chinnaiyan K, Chow BJ, Delago A, Hadamitzky M, Hausleiter J, Kaufmann P, Maffei E, Raff G, Shaw LJ, Villines T, Berman DS, Investigators C. Age- and sex-related differences in all-cause mortality risk based on coronary computed tomography angiography findings results from the international multicenter CONFIRM (coronary CT angiography evaluation for clinical outcomes: an international multicenter registry) of 23,854 patients without known coronary artery disease. J Am Coll Cardiol. 2011;58:849–60.
107. Leipsic J, Taylor CM, Gransar H, Shaw LJ, Ahmadi A, Thompson A, Humphries K, Berman DS, Hausleiter J, Achenbach S, Al-Mallah M, Budoff MJ, Cademartiri F, Callister TQ, Chang HJ, Chow BJ, Cury RC, Delago AJ, Dunning AL, Feuchtner GM, Hadamitzky M, Kaufmann PA, Lin FY, Chinnaiyan KM, Maffei E, Raff GL, Villines TC, Gomez MJ, Min JK. Sex-based prognostic implications of nonobstructive coronary artery disease: results from the international multicenter CONFIRM study. Radiology. 2014;273:393–400.
108. Schulman-Marcus J, o Hartaigh B, Gransar H, Lin F, Valenti V, Cho I, Berman D, Callister T, DeLago A, Hadamitzky M, Hausleiter J, Al-Mallah M, Budoff M, Kaufmann P, Achenbach S, Raff G, Chinnaiyan K, Cademartiri F, Maffei E, Villines T, Kim YJ, Leipsic J, Feuchtner G, Rubinshtein R, Pontone G, Andreini D, Marques H, Shaw L, Min JK. Sex-specific associations between coronary artery plaque extent and risk of major adverse cardiovascular events: the CONFIRM long-term registry. JACC Cardiovasc Imaging. 2016;9:364–72.
109. Gould KL. Does coronary flow trump coronary anatomy? JACC Cardiovasc Imaging. 2009;2:1009–23.
110. Fukushima K, Javadi MS, Higuchi T, Lautamaki R, Merrill J, Nekolla SG, Bengel FM. Prediction of short-term cardiovascular events using quantification of global myocardial flow reserve in patients referred for clinical 82Rb PET perfusion imaging. J Nucl Med. 2011;52:726–32.
111. Herzog BA, Husmann L, Valenta I, Gaemperli O, Siegrist PT, Tay FM, Burkhard N, Wyss CA, Kaufmann PA. Long-term prognostic value of 13N-ammonia myocardial perfusion positron emission tomography added value of coronary flow reserve. J Am Coll Cardiol. 2009;54:150–6.
112. Murthy VL, Naya M, Foster CR, Hainer J, Gaber M, Di Carli G, Blankstein R, Dorbala S, Sitek A, Pencina MJ, Di Carli MF. Improved cardiac risk assessment with noninvasive measures of coronary flow reserve. Circulation. 2011;124:2215–24.
113. Ziadi MC, Dekemp RA, Williams KA, Guo A, Chow BJ, Renaud JM, Ruddy TD, Sarveswaran N, Tee RE, Beanlands RS. Impaired myocardial flow reserve on rubidium-82 positron emission tomography imaging predicts adverse outcomes in patients assessed for myocardial ischemia. J Am Coll Cardiol. 2011;58:740–8.
114. Taqueti VR, Hachamovitch R, Murthy VL, Naya M, Foster CR, Hainer J, Dorbala S, Blankstein R, Di Carli MF. Global coronary flow reserve is associated with adverse cardiovascular events independently of luminal angiographic severity and modifies the effect of early revascularization. Circulation. 2015;131:19–27.

115. Naya M, Murthy VL, Taqueti VR, Foster CR, Klein J, Garber M, Dorbala S, Hainer J, Blankstein R, Resnic F, Di Carli MF. Preserved coronary flow reserve effectively excludes high-risk coronary artery disease on angiography. J Nucl Med. 2014;55:248–55.
116. Murthy VL, Lee BC, Sitek A, Naya M, Moody J, Polavarapu V, Ficaro EP, Di Carli MF. Comparison and prognostic validation of multiple methods of quantification of myocardial blood flow with 82Rb PET. J Nucl Med. 2014;55:1952–8.
117. Naya M, Murthy VL, Foster CR, Gaber M, Klein J, Hainer J, Dorbala S, Blankstein R, Di Carli MF. Prognostic interplay of coronary artery calcification and underlying vascular dysfunction in patients with suspected coronary artery disease. J Am Coll Cardiol. 2013;61:2098–106.
118. Patel MR, Peterson ED, Dai D, Brennan JM, Redberg RF, Anderson HV, Brindis RG, Douglas PS. Low diagnostic yield of elective coronary angiography. N Engl J Med. 2010;362:886–95.
119. Recio-Mayoral A, Rimoldi OE, Camici PG, Kaski JC. Inflammation and microvascular dysfunction in cardiac syndrome X patients without conventional risk factors for coronary artery disease. JACC Cardiovasc Imaging. 2013;6:660–7.
120. Di Carli MF, Bianco-Batlles D, Landa ME, Kazmers A, Groehn H, Muzik O, Grunberger G. Effects of autonomic neuropathy on coronary blood flow in patients with diabetes mellitus. Circulation. 1999;100:813–9.
121. Paulus WJ, Tschope C. A novel paradigm for heart failure with preserved ejection fraction: comorbidities drive myocardial dysfunction and remodeling through coronary microvascular endothelial inflammation. J Am Coll Cardiol. 2013;62:263–71.
122. Comparison of coronary bypass surgery with angioplasty in patients with multivessel disease. The bypass angioplasty revascularization investigation (BARI) investigators. N Engl J Med. 1996;335:217–25.
123. Farkouh ME, Domanski M, Sleeper LA, Siami FS, Dangas G, Mack M, Yang M, Cohen DJ, Rosenberg Y, Solomon SD, Desai AS, Gersh BJ, Magnuson EA, Lansky A, Boineau R, Weinberger J, Ramanathan K, Sousa JE, Rankin J, Bhargava B, Buse J, Hueb W, Smith CR, Muratov V, Bansilal S, King S 3rd, Bertrand M, Fuster V, Investigators FT. Strategies for multivessel revascularization in patients with diabetes. N Engl J Med. 2012;367:2375–84.
124. Frye RL, August P, Brooks MM, Hardison RM, Kelsey SF, MacGregor JM, Orchard TJ, Chaitman BR, Genuth SM, Goldberg SH, Hlatky MA, Jones TL, Molitch ME, Nesto RW, Sako EY, Sobel BE. A randomized trial of therapies for type 2 diabetes and coronary artery disease. N Engl J Med. 2009;360:2503–15.
125. Serruys PW, Morice MC, Kappetein AP, Colombo A, Holmes DR, Mack MJ, Stahle E, Feldman TE, van den Brand M, Bass EJ, Van Dyck N, Leadley K, Dawkins KD, Mohr FW, Investigators S. Percutaneous coronary intervention versus coronary-artery bypass grafting for severe coronary artery disease. N Engl J Med. 2009;360:961–72.
126. Ridker PM, Everett BM, Thuren T, MacFadyen JG, Chang WH, Ballantyne C, Fonseca F, Nicolau J, Koenig W, Anker SD, Kastelein JJP, Cornel JH, Pais P, Pella D, Genest J, Cifkova R, Lorenzatti A, Forster T, Kobalava Z, Vida-Simiti-L, Flather M, Shimokawa H, Ogawa H, Dellborg M, Rossi PRF, Troquay RPT, Libby P, Glynn RJ, Group CT. Antiinflammatory therapy with canakinumab for atherosclerotic disease. N Engl J Med. 2017;377:1119–31.
127. Everett BM, Pradhan AD, Solomon DH, Paynter N, Macfadyen J, Zaharris E, Gupta M, Clearfield M, Libby P, Hasan AA, Glynn RJ, Ridker PM. Rationale and design of the cardiovascular inflammation reduction trial: a test of the inflammatory hypothesis of atherothrombosis. Am Heart J. 2013;166:199–207. e115
128. Sabatine MS, Giugliano RP, Keech AC, Honarpour N, Wiviott SD, Murphy SA, Kuder JF, Wang H, Liu T, Wasserman SM, Sever PS, Pedersen TR, Committee FS, Investigators. Evolocumab and clinical outcomes in patients with cardiovascular disease. N Engl J Med. 2017;376:1713–22.
129. Zinman B, Wanner C, Lachin JM, Fitchett D, Bluhmki E, Hantel S, Mattheus M, Devins T, Johansen OE, Woerle HJ, Broedl UC, Inzucchi SE, Investigators E-RO. Empagliflozin, cardiovascular outcomes, and mortality in type 2 diabetes. N Engl J Med. 2015;373:2117–28.
130. McMurray JJ, Packer M, Desai AS, Gong J, Lefkowitz MP, Rizkala AR, Rouleau JL, Shi VC, Solomon SD, Swedberg K, Zile MR, Investigators P-H, Committees. Angiotensin-neprilysin inhibition versus enalapril in heart failure. N Engl J Med. 2014;371:993–1004.

Cerebral Circulation in Men and Women

18

Aurangzeb Memon and Louise D. McCullough

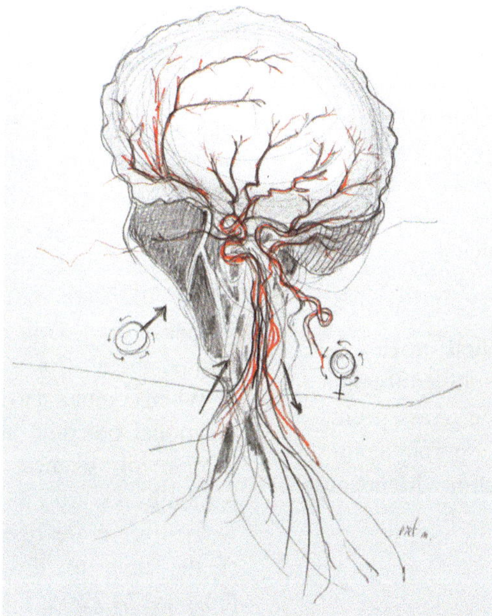

Cerebral circulation. Art work by Piet Michiels, Leuven, Belgium.

A. Memon
Department of Neurology, Providence Hospital, Mobile, AL, USA
e-mail: Aurangzeb.Memon@ascension.org

L. D. McCullough (✉)
Huffington Distinguished Chair of Neurology, Houston, TX, USA

Memorial Herman Hospital – TMC, Houston, Houston, TX, USA

The University of Texas Health Science Center, Houston, TX, USA
e-mail: Louise.D.McCullough@uth.tmc.edu

Abstract

Despite several advancements in stroke care, disparities continue to exist with regard to sex differences in cerebrovascular disease. These sex differences are due to a combination of several factors, many of which are unique to the female sex. Some of these unique factors, such as pregnancy and menopause, are related to hormonal changes seen throughout the female life cycle. Hormonal fluctuations,

P. L. M. Kerkhof, V. M. Miller (eds.), *Sex-Specific Analysis of Cardiovascular Function*,
Advances in Experimental Medicine and Biology 1065, https://doi.org/10.1007/978-3-319-77932-4_18

which impact the protective effects of the female sex hormones, can be induced by the use of hormonal contraception. Other risk factors, although present in both sexes, have a higher prevalence in elderly females, such as atrial fibrillation leading to cardioembolic strokes. Similarly, differences in premorbid modified Rankin Scale have an impact on the differences in stroke outcome between the two sexes. Clinical research aimed toward highlighting potential causes of these disparities has shown important differences in the calibers of blood vessels in the cerebral circulation between the two sexes, whereas basic science research has shown differences in circulating endothelial progenitor cell pools between males and females, with higher levels being more protective. With the increasing awareness of these sex differences, future research is being geared toward gender-specific modes of therapy, focusing on the molecular level, as well as the individual patient.

Keywords

Migrainous infarct · Cerebral stroke · Cardioembolic stroke · Modified Rankin Scale · Cerebral circulation · Sex-specific analysis · Barthel index · Neuroimaging · Pregnancy · Atrial fibrillation · Menopause · Preeclampsia

Introduction

Each year, 795,000 Americans suffer a stroke, while 130,000 die from their stroke [1]. Stroke is the leading cause of long-term disability and costs the United States approximately $33 billion each year [1]. While the overall incidence of stroke is declining in developed countries, the absolute number of strokes is increasing secondary to the aging of populations [2].

Investigations have shown that the underlying etiology and burden of stroke may differ between women and men. Differences in physiology between females and males may underlie some of the differences between the sexes seen in stroke. This chapter highlights emerging data on sex differences in stroke, focusing on clinical differences in pathophysiology and outcomes.

Sex Differences in Stroke Epidemiology

Men have a higher incidence of stroke until advanced age (85), after which the incidence is higher in women [2]. Women are more likely to die from their stroke than men [2]. This difference may be a result from a combination of a longer life expectancy of women, increased pre-stroke disability and the increasing incidence of stroke with advancing age. Women also have increased stroke severity and greater mortality after stroke [2], as well as a higher incidence of recurrent strokes (22% of women vs. 13% of men 40–69 years old; 28% of women vs. 23% of men $\geq$70 years old) [2]. Some of these differences are due to the higher rates of cardioembolic strokes in women [3].

When compared to males, females have poorer functional outcome after acute ischemic stroke (22.7% of women are fully recovered by 6 months vs. 26.7% of men) [4]. As a result, women are more likely to be severely disabled at the time of discharge compared to men (36.1 vs. 24.2%) [5]. Women are less likely than men to be discharged home after a stroke admission (40.9 vs. 50.6%) [6] and are more likely to be discharged to a long-term care facility (10 vs. 5%) [7].

Since women are older at the time of stroke, age is an important confounder in the analysis and understanding of sex differences. However, some of the disparities persist even after controlling for these factors. After controlling for confounding variables such as age, comorbidity, pre-stroke functional status, and stroke severity, women are less likely than men to receive a Barthel index of 95 or greater (OR 0.75; 95% CI 0.61–0.94) in some studies [8] . Even controlling for pre-stroke living conditions, more women than men still require institutional care at 3 months after stroke [9] and are less likely to achieve independence in activities of daily living [10]. However several new large epidemiological studies have suggested that these sex disparities are primarily due to poorer pre-stroke function in women. The Oxford Vascular Study found that when accounting for age and premorbid modified Rankin Scale (mRS) score, women did not have a higher risk of worse outcome after stroke when compared with men. Stroke-induced changes in mRS scores were not significantly different between the two sexes [11].

Conversely, while the crude survival rate was higher in men in the INSTRUCT trial, women were found to have a lower 1 year mortality rate after adjusting for confounding factors (advanced age, greater stroke severity, greater pre-stroke functional limitations, and presence of atrial fibrillation) [12]. This trial highlights the importance of considering the biological, clinical, and premorbid differences between the two sexes when studying differences in stroke outcome.

Figure 18.1 depicts sex differences in pre-stroke modified Rankin Scale, with 0 representing no symptoms, 5 representing severe disability, and 6 representing death. Notice the larger number of pre-stroke males in mRS categories 0 and 1 when compared to pre-stroke females.

Sex Differences in Acute Stroke Care

Sex disparities exist in the access of patients to acute stroke care, but the gap between the sexes appears to be closing with time and increasing awareness. Older studies had shown that women were less likely to be admitted to the hospital within the first 3 h of stroke onset [13], and even when presenting within an appropriate time window, a meta-analysis of several European and US studies showed that women were less likely to receive IV-tPA for acute stroke treatment [14]. Although good evidence is lacking, reasons for this disparity were thought to be due to delayed presentation, as more elderly women live alone and have an unknown time of symptom onset. Women are also more likely to present with more severe strokes, raising the possibility that IV tPA was considered to be contraindicated due to

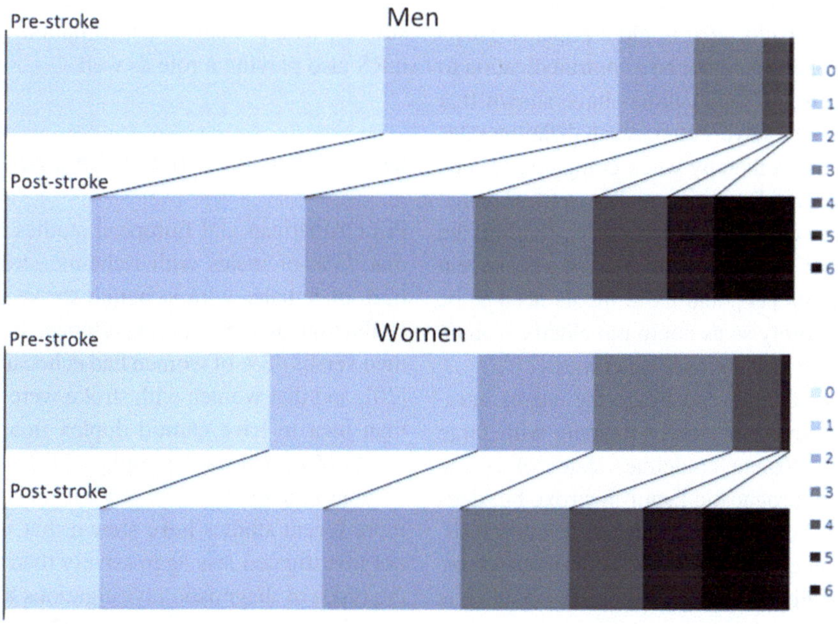

Fig. 18.1 Depicts sex differences in pre-stroke modified Rankin Scale, with 0 representing no symptoms, 5 representing severe disability, and 6 representing death. Notice the larger number of pre-stroke males in mRS categories 0 and 1 when compared to pre-stroke females

the risk of hemorrhage. A population-based cohort study from 2010 found that although men and women present with similar symptoms, women were more likely to be older at the time of stroke onset [14]. Given the advanced age, women were more likely to suffer from pre-stroke dementia, be in institutionalized living environments, and have greater stroke severity and higher rates of atrial fibrillation. This study found a higher 28-day mortality in women than in men, which is believed to be multifactorial, given the above confounding factors.

Importantly, emerging data is changing our view of how sex influences stroke outcome. In a recent study from 2016 examining a large stroke population, exclusion criteria for r-tPA in women and men were similar with regard to race, initial National Institutes of Health Stroke Scale score, warning signs, and contraindications, but that there still existed a disparity in treatment within the older age groups. More women were excluded from receiving IV tPA as they became older (>80 years) [15]. The lower rates of thrombolytic treatment likely reflects the fact that women are older at the time of stroke and fear on the physician's part of intracerebral hemorrhage due to frail vasculature and increased prevalence of amyloid angiopathy seen in older patients. However, age alone should not be a contraindication to tPA treatment, as some studies have shown that women tend to have better recanalization rates after reperfusion therapy when compared to men [16]. However, disparities in "defect-free care" and the use of early thrombolysis (within the "golden hour") remain even in some very recent registry reports [17], and these trends need to be monitored closely so as not to put elderly women at even higher risk for poor outcomes.

Recent guidelines encourage the use of intravenous tPA even in elderly patients with large strokes [18]. Recent guidelines state "older age is an adverse prognostic factor in stroke but does not modify the treatment effect of thrombolysis. Although older patients have poorer outcomes, higher mortality, and higher rates of sICH than those <80 years of age, compared with control subjects, intravenous alteplase provides a better chance of being independent at 3 months across all age groups" (Class I; Level of Evidence A). This is an important point as women are more likely to be older at the time of their stroke, and the potential for under treatment exists.

Sex Differences in Stroke Symptom Presentation

Whereas it is well recognized that the female sex can have atypical presentations for myocardial infarctions, differences in symptom presentation have been less well established in acute stroke. Previous observational studies had reported that women were more likely to present with atypical stroke symptoms such as pain, change in level of consciousness, or nonspecific or unclassifiable neurological symptoms, whereas men were more likely to report traditional stroke symptoms such as hemiparesis or imbalance [19]. However, more recent studies have shown that the most common presenting symptoms for stroke in the acute setting are similar between the two sexes [13, 20]. Again, given their advanced age, women were more likely to present with more severe strokes, with pre-stroke dementia and premorbid mRS also playing a role as well.

Sex Differences in Stroke Evaluation

Prior American and European studies had found that 71% of males with ischemic stroke versus 62% of females with ischemic stroke underwent evaluation of their carotid arteries, that 57% of men versus 48% of women had echocardiography [20], and that women with stroke were less likely than men to have carotid duplex imaging (32.8 vs. 44.0%), echocardiography (22.8 vs. 30.5%), or angiography (5.5 vs. 9.5%) [21]. However, more recent studies have shown that women are not investigated less aggressively than men, with the types of diagnostic investigations being comparable between the two sexes [22].

Evaluation/ treatment	Time frame assessed	Sex difference
Advanced neuroimaging (CTA/MRI/ MRA)	2005, 2010	No significant association was found between sex and advanced imaging [23]
ECHO Carotid duplex	2002–2007	ECHO 52.4% vs. 46.5%, $P < 0.05$ (M:F) Carotid 77.2% vs. 68.7%, $P < 0.05$ (M:F) Age, NIHSS, and prior stroke were independent predictors of diagnostic work-up; sex was not [24]
Emergency department triage	2010–2012	No sex difference in triage times [25]
Acute stroke unit care	2005–2012	No sex differences in stroke care and thrombolysis rates. Men received more MRI scans. Women had worse functional outcome at 3 months [22].
Emergency stroke care (imaging and tPA use)	2008	No significant difference in hospital mortality between the two sexes

Sex Differences in Stroke Prevention

Atrial Fibrillation

Patients with atrial fibrillation are up to five times more likely to suffer from an ischemic stroke when compared to patients who do not suffer from atrial fibrillation [26, 27]. There is a greater incidence of AF in men in all age groups but a higher prevalence of AF in women aged ≥75 years due to greater longevity in women [28]. Female sex is an established independent risk factor for embolic events in atrial fibrillation, as is evidenced by the additional point given to female gender in the CHA2DS2VASc score. It has also been observed that women with atrial fibrillation are more likely to suffer from a stroke when compared to men [29], especially in patients aged 75 or older [30]. . Whereas previously it was believed this was due to under-treatment of AF in women, more recent studies have shown that treatment is comparable between the two sexes [31]. Some mechanisms that have been suggested state hormonal differences and increasing levels of prothrombotic factors may be the underlying cause of the higher incidence of thromboembolic events seen in women suffering from AF [32, 33], with high circulating levels of Von Willebrand factor playing a key role [34].

Despite being on warfarin, women carried a significantly higher residual risk of stroke and systemic thromboembolism when compared to men, with the Atrial Fibrillation Follow-up Investigation of Rhythm Management (AFFIRM) study concluding that this was due to lower time in the therapeutic range of warfarin [35]. This conclusion, however, was not seen in the GARFIELD-AF study, which demonstrated similar times in the therapeutic range between the two genders [31]. Randomized controlled trials involving the use of non-vitamin K-dependent anticoagulants failed to demonstrate these differences between the two genders [36], but a recent study has shown a reduced risk of ischemic strokes in male patients taking rivaroxaban, with a higher bleeding risk seen in women taking rivaroxaban [37]. However, it has been suggested that women may have a higher net clinical benefit from NOAC therapy due to the high residual stroke risk in women treated with warfarin but lower stroke rates with NOAC therapy [38].

The following table highlights important sex differences in outcomes in recent clinical trials of atrial fibrillation.

Time period	Treatment	Sex difference
2010–2014	Majority on vitamin K antagonists	Increased stroke risk and less reduction in stroke risk with AC in women, no sex difference in bleeding risk [31]

(continued)

Time period	Treatment	Sex difference
2011–2013	Rivaroxaban and dabigatran	No overall sex difference in stroke risk, but increased bleeding risk in women taking rivaroxaban, with greater stroke reduction in men using rivaroxaban compared to women [37]
2012–2014	Majority on vitamin K antagonists	No sex difference in stroke risk or bleeding risk, reduction in stroke risk with AC similar between sexes but with possible greater reduction in women taking NOACs compared to women taking warfarin [38]

Vascular Risk Factors

Major vascular risk factors differ in prevalence between the sexes. Hypertension is an important risk factor for stroke because it is one of the most modifiable risk factors, with hypertension being more prevalent in women than in men [39]. Another modifiable risk factor, diabetes, may also impair sex-specific risk. It has been found that diabetic women have a 30% greater relative risk for stroke than diabetic men in a recent meta-analysis [40], with women in the age group of 55–74 being at highest risk, demonstrated in a recent study [41]. Women experience increased stroke risk in middle age [42]. The menopausal transition is associated with an increase in body weight and abdominal fat [43]. In many parts of the world, abdominal obesity is 2–10 times more common in women [44, 45] and may be a greater risk for ischemic stroke in women than in men [42, 46].

Pregnancy and Pregnancy-Related Complications

Pregnancy is a condition which is unique to females. Although the risk of stroke in pregnant women is fairly low, the risk appears to be highest in the postpartum period [47].

Preeclampsia, defined as hypertension with proteinuria, organ dysfunction, or uteroplacental dysfunction, is seen after 20 weeks of gestation. This condition is believed to be due to remodeling failure of the spiral arteries in the placenta, leading to placental hypoperfusion and hypoxia. This ultimately leads to excessive oxidative stress, which triggers a systemic inflammatory response causing endothelial dysfunction and vasoconstriction, which in turn causes systemic hypertension and end-organ hypoperfusion. An analysis of the Nationwide Inpatient Sample database revealed that women who suffered from preeclampsia had a fourfold increase in stroke during pregnancy [48]. Even in the absence of preeclampsia, women suffering from hypertension during pregnancy are 6–9 times more likely to have a stroke when compared to pregnant women not suffering from hypertension [49, 50].

Future risk factors for stroke are also seen increasingly in women having previously been diagnosed with preeclampsia, with an increased risk for future hypertension and diabetes mellitus [51], as well as increased cardiovascular mortality [52].

Currently there is a lack of risk factor scores that taking gestational complications in to account when determining a woman's risk for future cardiovascular disease. However, the American Heart Association has classified preeclampsia and lone gestation hypertension as risk factors for coronary artery disease and recommends active follow-up of risk factors [53].

Parity and Experimental Stroke

Interestingly, parity has paradoxical effects in health and disease in animal models, mirroring the clinical literature [54]. Multiparous mice exhibit increased body weights, elevated triglyceride and cholesterol levels, significant immune suppression, greater sedentary behavior, and muscle fatigue. While these attributes are generally associated with higher metabovascular risk, multiparous females demonstrated a surprising resistance to ischemic brain injury, higher VEGF levels, enhanced angiogenesis, and decreased and improved behavioral recovery at chromic time points after stroke. Pregnancy-associated changes in immunity and events at the fetal-placental interface have long-term consequences in the healthy and injured maternal brain. Inflammation and microglia/macrophage activity are chronically suppressed in the brain of multiparous mice, consistent with the systemic immune suppression induced by pregnancy seen by others [55].

Parity-associated neuroprotection may have evolved out of necessity, ensuring the survival of young offspring. Recent speculation submits that the shared blood system of the mother and fetus act as a biological anecdote to experimental plasmapheresis studies which have shown rejuvenating effects of youthful plasma factors on brain function in old mice [56]. We have found that critical window events at the fetal-placental interface during gestation include the bidirectional trafficking of cells between the mother and fetus, known as fetal "microchimerism." The increased immunosuppressive state during pregnancy facilitates the transfer and survival of microchimeric cells in the mother [57, 58] which have stem cell-like properties and multipotent potential and incorporate into the maternal bone marrow niche where they can persist for decades. These rare cells respond to sterile injuries and participate in regenerative recovery processes [59]. We have found that stroke mobilizes these cells and drives their migration to the ischemic brain. These fetal cells persist in the maternal population for months (or decades in women) after birth, escape immune surveillance, and contribute to poststroke angiogenesis. This is an important area for future exploration as these studies could help us understand the etiology and pathology of stroke risk and recovery in females.

Hormonal Contraception

Oral contraceptives, another unique risk factor for the female gender, increase the risk of stroke by twofold and the risk of venous thromboembolism by threefold when compared to women who do not use these drugs [60]. It is generally accepted that the lower the dose of estrogen, the lower the risk.

Migraine Headache with Aura

Migraine headache with aura is seen more commonly in women than in men, with a strong association between migraine headache with aura and stroke in women younger than 55 years of age [61]. This association is not seen in patients suffering migraines without auras. The mechanism behind this phenomenon is a topic of debate. Initially, the idea of migrainous infarct was entertained, in which ischemic infarct occurred during a migraine attack. However, most infarcts are seen to occur between migraine attacks, leading to much skepticism behind this theory.

Menopause and Hormone Replacement Therapy

Whereas it was previously believed that hormone replacement therapy after menopause increased the risk of stroke, this area was the topic of debate for several years. With continued debate, clinical trials began to investigate not only hormone replacement therapy after menopause but also the timing of hormone replacement therapy in relation to menopause, in order to assess the impact of hormone replacement therapy on cardiovascular disease and risk factors. In the ELITE trial (Vascular Effects of Early versus Late Postmenopausal Treatment with Estradiol) [62], the rate of carotid artery intima-media thickness progression was significantly lower in the estradiol group than in the placebo group among women who were less than 6 years past menopause but not among women who were 10 or more years past menopause, but there was no effect observed of timing of estradiol treatment relative to menopause and cardiac computed tomography measures of coronary atherosclerosis.

Although there were no significant differences in posttreatment coronary artery calcium and cardiac CT angiographic outcomes between the estradiol and placebo groups in this trial after 6 years of follow-up, results from the Women's Health Initiative study, with a mean of 7.4 years of follow-up, showed that coronary artery calcium scores were lower among women who were randomly assigned to receive conjugated estrogens compared to placebo [63].

According to the KEEPS study [64], there was significantly decreased risk of death, heart failure, or myocardial infarction when hormone replacement therapy was started early in post menopause, and hormone replacement therapy was not associated with an increased risk of cancer, stroke, deep vein thrombosis, or pulmonary embolism.

It has been suggested that the timing of hormone replacement therapy plays a vital role because the number or functional level of estrogen receptors and the health of the vasculature at the time of exposure to estrogen are important factors in determining whether the arterial wall responds positively to hormone therapy.

Basic Science Investigations into Sex Differences in Endothelial Cells

Endothelial progenitor cells (EPCs), which are cells derived from bone marrow, play a role in cardiovascular homeostasis [65]. Under normal conditions, EPCs in the peripheral blood provide a pool of cells that repair ongoing endothelial damage, but with ischemia, EPCs are mobilized from the marrow to peripheral blood leading to stimulation of compensatory angiogenesis [66], with reduced levels of circulating EPCs being demonstrated with risk factors of cardiovascular disease [67]. An Italian study showed fertile women had higher levels of EPC as compared to men, believed to be linked to higher levels of estradiol seen in this population. After menopause, EPC reduction, attributable to aging, to the change in the reproductive state, and to the worsened risk profile, may cause endothelial dysfunction and predispose to atherosclerosis [68].

Estrogen also causes an upregulation of endothelial nitric oxide synthase (eNOS) activity [69], with nitric oxide being an important vasodilator. In addition to this, estrogen also causes an upregulation of prostacyclin synthase and has beneficial effects in response to vascular injury and atherosclerosis by upregulating the expression of vascular endothelial growth factor (VEGF) and inhibiting endothelial cell apoptosis, as well as smooth muscle cell migration and proliferation [70].

There is a growing body of evidence indicating that excessive generation of reactive oxygen species by NADPH-oxidase contributes to oxidative stress and vascular dysfunction associated with cardiovascular and cerebrovascular diseases [71]. Animal models have shown that the activity and function of NADPH-oxidase are lower in the cerebral circulation of female versus male rats, with these differences dependent on estrogen such that estrogen deficiency results in higher NADPH-oxidase activity. The data from these animal models has suggested that estrogen suppresses the activity and function of NADPH-oxidase in the cerebral circulation.

In contrast to the above, it has been suggested that low levels of reactive oxygen species generated by NADPH-oxidase may serve to induce vasodilation in the cerebral circulation, indicating that there may potentially be a beneficial role of low levels of reactive oxygen species [72]. As more research and evidence emerges, this field of research will likely play a vital role in future treatment options, with the potential for treatment being targeted at the molecular level.

Figure 18.2 depicts endothelial progenitor cells (EPC) in the bone marrow being mobilized to the central nervous system for compensatory angiogenesis after suffering cerebral ischemia.

Sex Differences in Vessel Size

As mentioned previously, estrogen causes increased production of nitric oxide, an important vasodilator, from the endothelium [69]. In addition to nitric oxide, several other important vasodilators have been shown to be dependent on estrogen levels, including prostacyclin and endothelium-derived hyperpolarizing factor

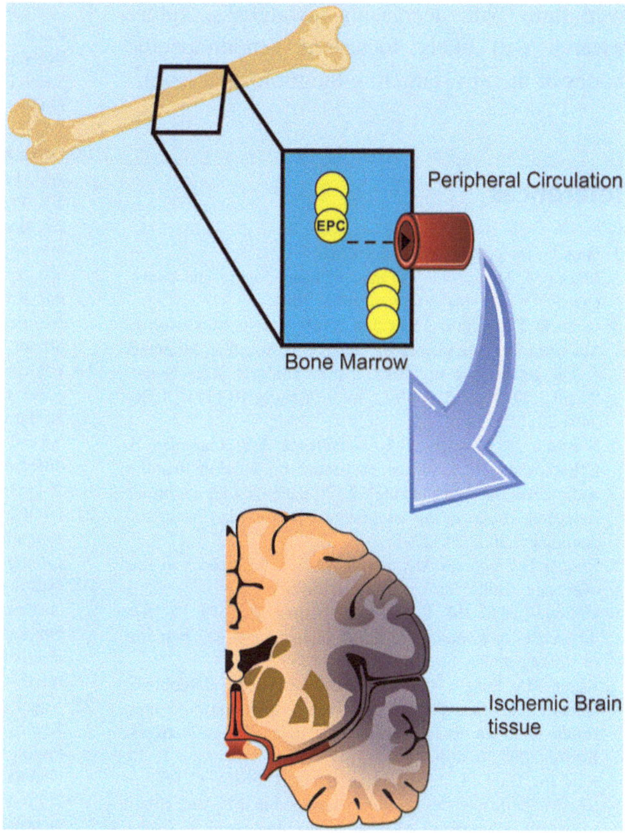

Fig. 18.2 Depicts endothelial progenitor cells (EPC) in the bone marrow being mobilized to the central nervous system for compensatory angiogenesis after suffering cerebral ischemia

[73]. In the setting of ischemia, increased activity of these vasodilators at different points in the vascular bed helps to maintain adequate cerebral blood flow, which explains the protective effects of estrogen [73].

In addition to these differences at the cellular level, an observation study from Brazil also demonstrated sex differences in caliber in the larger vessels of the posterior circulation. This study found that the calibers of the basilar and posterior cerebral arteries in males were larger than those of females [74].

Although speculative, the influences of estrogen on the above vasodilators may in part be a compensatory mechanism for smaller vessel size, though these gender differences in vessel calibers were only noted to be significant in the larger vessels of the posterior circulation.

Conclusions

Sex differences in the cerebrovascular system is a topic that is being increasingly recognized among researchers in the hope of improving individualized care and for better understanding of the etiology and pathophysiology of stroke in women. Whereas previous clinical trials showed women to have worse outcomes after stroke, more recent clinical trials have highlighted the multifactorial nature of these outcomes and the importance of considering pre-stroke functional status and living conditions in stroke outcomes. As women can suffer a stroke at any age, hormonal status is a key consideration to keep in mind when determining etiology and treatment for a stroke. In addition, while the gap in treatment disparities between the sexes has narrowed over time with increasing awareness, future research will likely focus on gender-specific modes of therapy, targeting the molecular level.

References

1. https://www.cdc.gov/stroke/facts.htm
2. Mackay J, Mensah G. The atlas of heart disease and stroke. Geneva: World Health Organization; 2004.
3. Giralt D, Domingues-Montanari S, Mendioroz M, Ortega L, Maisterra O, Perea-Gainza M, Delgado P, Rosell A, Montaner J. The gender gap in stroke: a meta-analysis. Acta Neurol Scand. 2012;125(2):83–90. https://doi.org/10.1111/j.1600-0404.2011.01514.x.
4. Niewada M, Kobayashi A, Sandercock PA, Kaminski B, Czlonkowska A. Influence of gender on baseline features and clinical outcomes among 17,370 patients with confirmed ischaemic stroke in the international stroke trial. Neuroepidemiology. 2005;24:123–8.
5. Roquer J, Campello AR, Gomis M. Sex differences in first ever acute stroke. Stroke. 2003;34:1581–5.
6. Holroyd-Leduc JM, Kapral MK, Austin PC, Tu JV. Sex differences and similarities in the management and outcome of stroke patients. Stroke. 2000;31:1833–7.
7. Kapral MK, Fang J, Hill MD, Silver F, Richards J, Jaigobin C, Cheung AM. Sex differences in stroke care and outcomes: results from the Registry of the Canadian Stroke Network. Stroke. 2005;36:809–14.
8. Reid JM, Dai D, Gubitz GJ, Kapral MK, Christian C, Phillips SJ. Gender differences in stroke examined in a 10-year cohort of patients admitted to a Canadian teaching hospital. Stroke. 2008;39:1090–5.
9. Glader EL, Stegmayr B, Norrving B, Terent A, Hulter-Asberg-K, Wester PO, Asplund K. Sex differences in management and outcome after stroke: a Swedish national perspective. Stroke. 2003;34:1970–5.
10. Gargano JW, Reeves MJ. Sex differences in stroke recovery and stroke-specific quality of life: results from a statewide stroke registry. Stroke. 2007;38:2541–8.
11. Renoux C, Rothwell PM, et al. Confounding by pre-morbid functional status in studies of apparent sex differences in severity and outcome of stroke. Stroke. 2017;48:2731–8. https://doi.org/10.1161/STROKEAHA.117.018187.
12. Phan HT, Gall S, et al. Sex differences in long-term mortality after stroke in the INSTRUCT (INternational STRoke oUtComes sTudy): a meta-analysis of individual participant data. Circ Cardiovasc Qual Outcomes. 2017;10:e003436.
13. Gall SL, Thrift AG, et al. Sex differences in presentation, severity, and management of stroke in a population based study. Neurology. 2010;74:974–81.
14. Reeves M, Bhatt A, Jajou P, Brown M, Lisabeth L. Sex differences in the use of intravenous rt-PA thrombolysis treatment for acute ischemic stroke, a meta-analysis. Stroke. 2009;40:1743–9.
15. Fredwall M, Sternberg S, Blackhurst D, Lee A, Leacock R, Nathaniel TI. Gender differences in exclusion criteria for recombinant tissue-type plasminogen activator. J Stroke Cerebrovasc Dis. 2016;25:2569–74.
16. Savitz SI, Schlaug G, Caplan L, Selim M. Arterial occlusive lesions recanalize more frequently in women than in men after intravenous tissue plasminogen activator administration for acute stroke. Stroke. 2005;36:1447–51.
17. Asdaghi N, Romano JG, Wang K, Ciliberti-Vargas MA, Koch S, Gardener H, Dong C, Rose DZ, Waddy SP, Robichaux M, Garcia EJ, Gonzalez-Sanchez JA, Burgin WS, Sacco RL, Rundek T. Sex disparities in ischemic stroke care: -FL-PR CReSD study (Florida-Puerto Rico Collaboration to Reduce Stroke Disparities). Stroke. 2016;47(10):2618–26.
18. Demaerschalk BM, Kleindorfer DO, Adeoye OM, Demchuk AM, Fugate JE, Grotta JC, Khalessi AA, Levy EI, Palesch YY, Prabhakaran S, Saposnik G, Saver JL, Smith EE, American Heart Association Stroke Council and Council on Epidemiology and Prevention. Scientific rationale for the inclusion and exclusion criteria for intravenous alteplase in acute ischemic stroke: a statement for healthcare professionals from the American Heart Association/American Stroke Association. Stroke. 2016;47(2):581–641.
19. Labiche LA, Chan W, Saldin KR, Morgenstern LB. Sex and acute stroke presentation. Ann Emerg Med. 2002;40:453–60.
20. Acciarresi M, De Luca P, Caso V, Agnelli G, D'Amore C, Alberti A, Venti M, Paciaroni M. Acute stroke symptoms: do differences exist between sexes? J Stroke Cerebrovasc Dis. 2014;23(10):2928–33.
21. Smith MA, Lisabeth LD, Brown DL, Morgenstern LB. Gender comparisons of diagnostic evaluation for ischemic stroke patients. Neurology. 2005;65:855–8.
22. Gattringer T, Ferrari J, Knoflach M, Seyfang L, Horner S, Niederkorn K, Culea V, Beitzke M, Lang W, Enzinger C, Fazekas F. Sex-related differences of acute stroke unit care: results from the Austrian stroke unit registry. Stroke. 2014;45 (6):1632–8.
23. Vagal A, Sanelli P, Sucharew H, Alwell KA, Khoury JC, Khatri P, Woo D, Flaherty M, Kissela BM, Adeoye O, Ferioli S, De Los Rios La Rosa F, Martini S, Mackey J, Kleindorfer D. Age, sex, and racial differences in neuroimaging use in acute stroke: a population based study. AJNR Am J Neuroradiol. 2017;38:1905–10. https://doi.org/10.3174/ajnr.A5340.
24. Turaj W, Slowik A, Wnuk M, Szczudlik A. Gender-related differences in diagnostic evaluation and outcome of ischemic stroke in Poland. Stroke. 2009;40(3):980–2.
25. Madisen TE, Choo EK, Seigel TA, Palms D, Silver B. Lack of gender disparities in emergency department triage of acute stroke patients. West J Emerg Med. 2015;16(1):203–9.
26. Go AS, Hylek EM, Phillips KA, et al. Prevalence of diagnosed atrial fibrillation in adults: national implications for rhythm management and stroke prevention: the Anticoagulation and Risk Factors in Atrial Fibrillation (ATRIA) Study. JAMA. 2001;285:2370–5.
27. Camm AJ, Lip GY, De Caterina R, et al. Focused update of the ESC Guidelines for the management of atrial fibrillation: an update of the 2010 ESC Guidelines for the management of atrial fibrillation. Developed with the special contribution of the European Heart Rhythm Association. Eur Heart J. 2012;33:2719–47.
28. Feinberg WM, Blackshear JL, Laupacis A, Kronmal R, Hart RG. Prevalence, age distribution, and gender of patients with atrial fibrillation. Analysis and implications. Arch Intern Med. 1995;155:469–73.
29. Madias C, Trohman RG. The link between atrial fibrillation and stroke in women. Women's Health. 2011;7:375–82.
30. Wagstaff AJ, Overvad TF, Lip GYH, Lane DA. Is female sex a risk factor for stroke and thromboembolism in patients with atrial fibrillation? A systematic review and meta-analysis. QJM, Int J Med. 2014;107(12):955–67.
31. Camm AJ, Accetta G, Al Mahmeed W, for the GARFIELD-AF Investigators, et al. Impact of gender on event rates at 1 year in patients with newly diagnosed non-valvular atrial

fibrillation: contemporary perspective from the GARFIELD-AF registry. BMJ Open. 2017;7:e014579. https://doi.org/10.1136/bmjopen-2016-014579.
32. Cove CL, Albert CM, Andreotti F, Badimon L, van Gelder IC, Hylek EM. Female sex as an independent risk factor for stroke in atrial fibrillation: possible mechanisms. Thromb Haemost. 2014;111:385–91.
33. Conway DS, Heeringa J, van Der Kuip DA, Chin BS, Hofman A, Witteman JC, et al. Atrial fibrillation and the prothrombotic state in the elderly: the Rotterdam Study. Stroke. 2003;34:413–7.
34. Wieberdink RG, van Schie MC, Koudstaal PJ, Hofman A, Witteman JC, de Maat MP, et al. High von Willebrand factor levels increase the risk of stroke: the Rotterdam Study. Stroke. 2010;41:2151–6.
35. Sullivan R, Zhang J, Zamba G, Lip GYH, Olshansky B. Relation of gender-specific risk of ischemic stroke in patients with atrial fibrillation to differences in warfarin anticoagulation control (from AFFIRM). Am J Cardiol. 2012;110:1799–802.
36. Pancholy SB, Sharma PS, Pancholy DS, et al. Meta-analysis of gender differences in residual stroke risk and major bleeding in patients with nonvalvular atrial fibrillation treated with oral anticoagulants. Am J Cardiol. 2014;113:485–90.
37. Shantha GPS, Bhave PD, Girotra S, Hodgson-Zingman D, Mazur A, Giudici M, Chrischilles E, Sarrazin MSV. Sex-specific comparative effectiveness of oral anticoagulants in elderly patients with newly diagnosed atrial fibrillation. Circ Cardiovasc Qual Outcomes. 2017;10:e003418.
38. Schnabel RB, Pecen L, Ojeda FM, et al. Heart. 2017. https://doi.org/10.1136/heartjnl-2016-310406.
39. Gall SL, Donnan G, Dewey HM, Macdonell R, Sturm J, Gilligan A, Srikanth V, Thrift AG. Sex differences in presentation, severity, and management of stroke in a population based study. Neurology. 2010;74:975–81.
40. Peters SA, Huxley RR, Woodward M. Diabetes as a risk factor for stroke in women compared with men: a systematic review and meta-analysis of 64 cohorts, including 775 385 individuals and 12 539 strokes. Lancet. 2014;383 (9933):1973–80.
41. Policardo L, Seghieri G, Francesconi P, Anichini R, Franconi F. Gender difference in diabetes-associated risk of first-ever and recurrent ischemic stroke. J Diabetes Complicat. 2015;29:713–7.
42. Towfighi A, Saver JL, Engelhardt R, Ovbiagele. A midlife stroke surge among women in the United States. Neurology. 2007;69(20):1898–904.
43. Davis SR, Castelo-Branco C, Chedraui P, Lumsden MA, Nappi RE, Shah D, Villaseca P. Writing Group of the International Menopause Society for World Menopause Day 2012: understanding weight gain at menopause. Climacteric. 2012;15(5):419–29.
44. Mauvais-Jarvis F. Sex differences in metabolic homeostasis, diabetes, and obesity. Biol Sex Differ. 2015;6:14.
45. Kelly T, Yang W, Chen CS, Reynolds K, He J. Global burden of obesity in 2005 and projections to 2030. Int J Obes. 2008;32 (9):1431–7.
46. Rodríguez-Campello A, Jiménez-Conde J, Ois Á, Cuadrado-Godia E, Giralt-Steinhauer E, Vivanco RM, Soriano-Tárraga C, Subirana I, Muñoz D, Gómez-González A, Puig-Pijoan A, Roquer J. Sex-related differences in abdominal obesity impact on ischemic stroke risk. Eur J Neurol. 2017;24(2):397–403.
47. Kuklina EV, Tong X, Bansil P, George MG, Callaghan WM. Trends in pregnancy hospitalizations that included a stroke in the United States from 1994 to 2007: reasons for concern? Stroke. 2011;42:2564–70.
48. James AH, Bushnell CD, Jamison MG, et al. Incidence and risk factors for stroke in pregnancy and the puerperium. Obstet Gynecol. 2005;106:509–16.
49. Lanska DJ, Kryscio RJ. Stroke and intracranial venous thrombosis during pregnancy and puerperium. Neurology. 1998;51:1622–8.
50. Lanska DJ, Kryscio RJ. Risk factors for peripartum and postpartum stroke and intracranial venous thrombosis. Stroke. 2000;31:1274–82.
51. Lykke JA, Langhoff-Roos J, Sibai BM, Funai EF, Triche EW, Paidas MJ. Hypertensive pregnancy disorders and subsequent cardiovascular morbidity and type 2 diabetes mellitus in the mother. Hypertension. 2009;53:944–51.
52. McDonald SD, Malinowski A, Zhou Q, Yusuf S, Devereaux PJ. Cardiovascular sequelae of preeclampsia/eclampsia: a systematic review and meta-analyses. Am Heart J. 2008;156:918–30.
53. Mosca L, Benjamin EJ, Berra K, et al. Effectiveness-based guidelines for the prevention of cardiovascular disease in women—2011 update: a guideline from the American Heart Association. Circulation. 2011;123:1243–62.
54. Ritzel RM, Patel AR, Spychala M, Verma R, Crapser J, Koellhoffer EC, Schrecengost A, Jellison ER, Zhu L, Venna VR, McCullough LD. Multiparity improves outcomes after cerebral ischemia in female mice despite features of increased metabovascular risk. Proc Natl Acad Sci U S A. 2017;114(28): E5673–82.
55. Racicot K, Kwon JY, Aldo P, Silasi M, Mor G. Understanding the complexity of the immune system during pregnancy. Am J Reprod Immunol. 2014;72(2):107–16.
56. Falick Michaeli T, Bergman Y, Gielchinsky Y. Rejuvenating effect of pregnancy on the mother. Fertil Steril. 2015;103 (5):1125–8.
57. Jeanty C, Derderian SC, Mackenzie TC. Maternal-fetal cellular trafficking: clinical implications and consequences. Curr Opin Pediatr. 2014;26(3):377–82.
58. Dutta P, Burlingham WJ. Microchimerism: tolerance vs. sensitization. Curr Opin Organ Transplant. 2011;16(4):359–65.
59. Mahmood U, O'Donoghue K. Microchimeric fetal cells play a role in maternal wound healing after pregnancy. Chimerism. 2014;5(2):40 52.
60. Urrutia P, Rachel R. Risk of acute thromboembolic events with oral contraceptive use: a systematic review and meta-analysis. Obstet Gynecol. 2013;122:380–9.
61. Kurth T, Slomke MA, Kase CS, et al. Migraine, headache, and the risk of stroke in women: a prospective study. Neurology. 2005;64:1020–6.
62. Hodis HN, Mack WJ, Henderson VW, Shoupe D, Budoff MJ, Hwang-Levine J, Li Y, Feng M, Dustin L, Kono N, Stanczyk FZ, Selzer RH, Azen SP, for the ELITE Research Group. Vascular effects of early versus late postmenopausal treatment with estradiol. N Engl J Med. 2016;374:1221–31.
63. Manson JE, Allison MA, Rossouw JE, et al. Estrogen therapy and coronary-artery calcification. N Engl J Med. 2007;356:2591–602.
64. Schierbeck LL, Rejnmark L, Tofteng CL, Stilgren L, Eiken P, Mosekilde L, Køber L, Jensen JB. Effect of hormone replacement therapy on cardiovascular events in recently postmenopausal women: randomised trial. BMJ. 2012;345:e6409. https://doi.org/10.1136/bmj.e6409.
65. O'Leary DH, Polak JF, Kronmal RA, Manolio TA, Burke GL, Wolfson SK, Group. ftCHSCR. Carotid-artery intima and

media thickness as a risk factor for myocardial infarction and stroke in older adults. N Engl J Med. 1999;340:14–22.
66. Corretti MC, Anderson TJ, Benjamin EJ. Guidelines for the ultrasound assessment of endothelial-dependent flow-mediated vasodilation of the brachial artery. J Am Coll Cardiol. 2002;39:257–65.
67. Fadini GP, de Kreutzenberg S, Albiero M, Coracina A, Pagnin E, Baesso E, Cignarella A, Bolego C, Plebani M, Nardelli GB, Sartore S, Agostini C, Avogaro A. Gender differences in endothelial progenitor cells and cardiovascular risk profile: the role of female estrogens. Arterioscler Thromb Vasc Biol. 2008;28:997–1004.
68. Fadini GP, de Kreutzenberg SV, Coracina A, Baesso I, Agostini C, Tiengo A, Avogaro A. Circulating CD34+ cells, metabolic syndrome, and cardiovascular risk. Eur Heart J. 2006;27:2247–55.
69. Hayashi S, Yamada K, Esaki T, et al. Estrogen increases endothelial nitric oxide by a receptor-mediated system. Biochem Biophys Res Commun. 1995;214:847–55.
70. Weiner CP, Lizasoain I, Baylis SA, et al. Induction of calcium-dependent nitric oxide synthases by sex hormones. Proc Natl Acad Sci U S A. 1994;91:5212–6.
71. Miller AA, Drummond GR, Sobey CG. Novel isoforms of NADPH oxidase in cerebral vascular control. Pharmacol Ther. 2006;111:928–48.
72. Miller AA, Drummond GR, Mast AE, Schmidt HHHW, Sobey CG. Effect of gender on NADPH-oxidase activity, expression, and function in the cerebral circulation: role of estrogen. Stroke. 2007;38:2142–9.
73. Davis CM, Siler DA, Alkayed NJEDHF. In the brain: influence of sex, vessel size and disease state. Women's Health (Lond Engl). 2011;7(3):293–303.
74. Stefani MA, Schneider FL, Marrone ACH, Severino AG. Influence of the gender on cerebral vascular diameters observed during the magnetic resonance angiographic examination of Willis circle. Braz Arch Biol Technol. 2013;56:1.

19 Arterial Wall Properties in Men and Women: Hemodynamic Analysis and Clinical Implications

John K.-J. Li

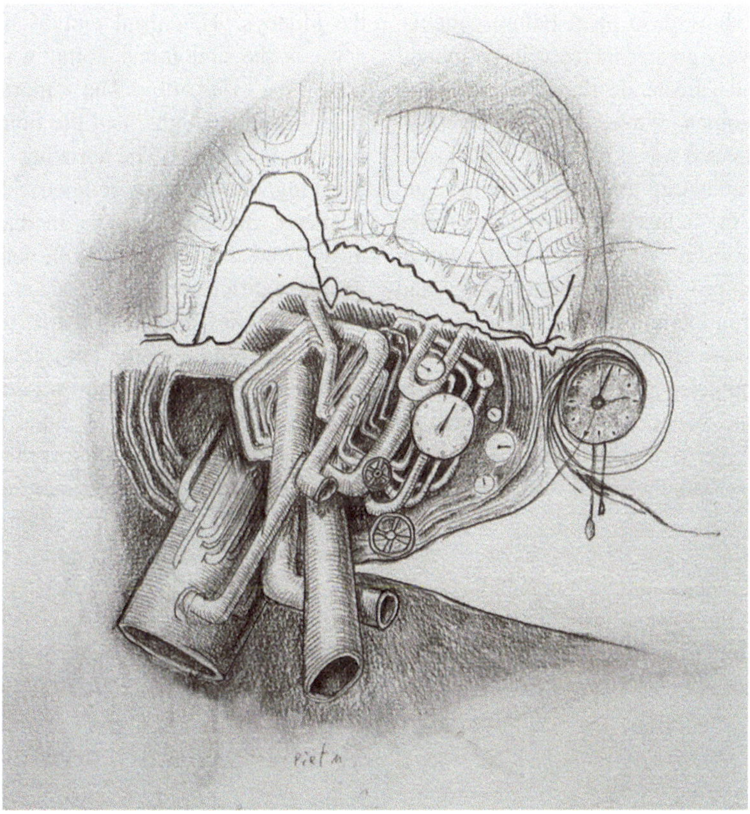

Arterial system with waveforms. Artwork by Piet Michiels, Leuven, Belgium.

J. K.-J. Li (✉)
Department of Biomedical Engineering,
Rutgers University, Piscataway, NJ, USA
e-mail: johnkjli@soe.rutgers.edu

Abstract

The properties of arterial walls are dictated by their underlying structure, which is responsible for the adequate perfusion of conduit

P. L. M. Kerkhof, V. M. Miller (eds.), *Sex-Specific Analysis of Cardiovascular Function*,
Advances in Experimental Medicine and Biology 1065, https://doi.org/10.1007/978-3-319-77932-4_19

branching arteries and their vascular beds. Beginning with the mechanobiology of arteries in terms of their composition and individual contributions to overall viscoelastic behavior in men and women, pressure–flow relations are analyzed and noted in terms of sex differences. Hemodynamic function in terms of indices of vascular stiffness—such as pressure–strain elastic modulus, pulse wave velocity, augmentation index, and cardio–ankle vascular index—are evaluated. They all showed differences between the sexes, and these differences also were shown among people of different cultures. Recent studies also showed, in heart failure patients, a comparatively greater increase in peripheral resistance and a greater decreased arterial compliance in women. Wave separation into forward and reflected waves allows elucidation of mechanical and drug-treated similarities and differences in induced hypertension. This may provide insight into treatment strategy in terms of improving mechanobiology and designing drug therapy for the sexes. Finally, modeling studies are useful in identifying how arterial compliance and its pressure dependence can be better used in differentiating aging- and hypertension-induced changes that differentially affect the sexes.

Keywords
Vascular mechanics · Hemodynamic monitoring · Pulse waveform analysis · Arterial compliance · Mechanobiology · Sex differences

Introduction

Arteries and the Arterial System

Men are generally taller, heavier, and have a larger supporting frame structure than women. Their respective arterial systems function similarly in terms of the circulation of blood, but they are somewhat dissimilar in terms of geometric and viscoelastic properties. These dissimilarities contribute to differences in the pulsatile pressure and flow waveforms between the sexes. The root of the aorta begins at the aortic valve, the outlet of which is the ascending aorta, which has the largest diameter and overall compliance. The first arteries branching off the aorta are the left and right main coronary arteries. The aortic arch junction is formed by the ascending aorta, the brachiocephalic artery, the left subclavian artery and the descending thoracic aorta.

The descending thoracic aorta has numerous arterial branches coming off at almost right angles: For example, the segmental arteries perfuse the spinal cord, and the renal arteries perfuse the kidneys. The distal end of the descending aorta is the abdominal aorta, which forms the aorto-iliac bifurcation. The femoral arteries perfuse the upper thighs, and the tibial arteries perfuse the lower legs. The aorta has, comparatively speaking, the greatest geometric taper, with its diameter decreasing with increasing distance away from the ventricle. Both aortic length and diameter differ between the sexes. The common carotid arteries are the longest, most relatively uniform vessels with the least geometrical tapering [28]. The brachial arteries perfuse the upper arms leading to the distal radial arteries at the wrists. Carotid, femoral, brachial and radial arteries are the most common and accessible sites for noninvasive clinical blood pressure monitoring and, consequently, carotid-to-femoral and brachial-to-radial pulse wave velocity (PWV) measurements for assessing vascular stiffness, which will be discussed later in the text.

Structure of the Arterial Wall

Grossly speaking, the arterial wall consists of elastin, collagen and smooth muscle cells embedded in a mucopolysaccharide ground substance. A cross-sectional view reveals the tunica intima, which is the innermost layer consisting of a thin layer (0.5–1 μm) of endothelial cells, connective tissue and basement membrane. The middle layer is the thick tunica media, which is separated from the intima by a prominent layer of elastic tissue, in other words, the internal lamina. The media

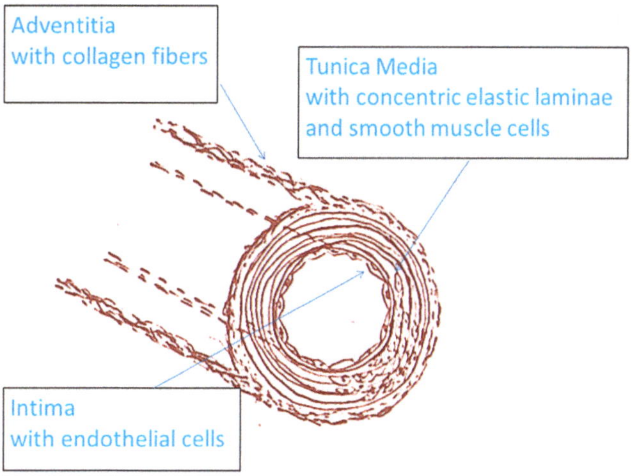

Fig. 19.1 Cross sections of the artery reveal three distinctive layers: the thin innermost tunica intima, the thick distensible tunica media, and the stiff outermost adventitia

contains elastin, smooth muscle cells and collagen fibers. The outermost layer is the adventitia, which is made up mostly of stiff, fibrous collagen. The difference in the wall composition separates arteries into large elastic arteries, such as the aorta, and smaller muscular arteries, such as the femoral and radial arteries.

Histological studies show that the elastic laminae are concentrically distributed and attached by smooth muscle cells and connective tissue (Fig. 19.1). In the longitudinal view, one can observe that the number of elastic laminae decreases with increasing distance from the aorta; however, the amount of smooth muscle increases, and the relative wall thickness increases. The wall thickness–to–radius ratio (h/r) is increased. The net stiffness is also increased, accounting for the increase in PWV toward the periphery as can be seen from the Moens-Korteweg formula relating propagating speed of the arterial pulse to conduit arterial elasticity and geometry [29]. The mechanical behavior of small peripheral arteries can be largely influenced by the behavior of the smooth muscle, particularly by the degree of its activation.

Gender-specific differences in aortic stiffness have been found between the sexes (e.g., [52]) and have been shown in the study of monkeys and rats [44]. In the latter study, elastin density decreased in male monkeys, and collagen increased significantly with aging in male and female rats. However, monkeys do not exhibit age-dependent increases in vascular diseases, such as hypertension, atherosclerosis and diabetes, observed in humans. Perhaps the relatively greater degree of acquired exercise and the low-fat diet (potassium-rich energy-generating banana is a favorite) indeed make a difference in monkeys. Generally, increased collagen and decreased elastin are associated with media thickening. Sex differences in the morphological structure of the tunica media of the human aorta have long been noted.

The extracellular matrix components, or those of elastin and collagen, govern the passive mechanical properties of the large and small arteries. Numerous studies exist on experimental animals and on humans [4, 51, 52]. Sex differences have also been reported. The contribution of smooth muscle to overall arterial wall properties and hemodynamic function has also been examined to a large extent. Smooth muscle activation becomes more important at greater blood pressures [3].

Mechanical Properties of Arteries

Vascular Stiffness and Arterial Wall Stress–Strain Relations

Arterial stiffness is traditionally expressed in terms of Young's modulus of elasticity, which gives a simple description of the elastic properties of the arterial wall. Young's modulus of elasticity (E) is defined by the ratio of tensile stress (σ_t) to tensile strain (ε_t). When the relationship between stress and strain is a linear one, then the material is said to be "Hookean," or simply, it obeys Hooke's law of elasticity. This normally applies to a purely elastic material. It is only valid for application to a cylindrical blood vessel when the radial and longitudinal deformations are small compared with the respective lumen diameter or length of the arterial segment.

The arterial wall is anisotropic, consisting of various components exhibiting an intertwined elastic behavior. In diseased conditions, such as vascular hypertrophy and hypertension, selective thickening in the tunica media is often accompanied by an increase in collagen and a decrease in elastin and/or a change in the level of smooth muscle activation. These observed changes are not uniform throughout the arterial wall, in other words, they are anisotropic. Nevertheless, the assumption of isotropy allows simpler clinical quantitative descriptions of the mechanical behavior of the arterial wall properties and simplifies mathematical computation.

For a cylindrical isotropic arterial segment with radius r, wall thickness h and segment length l, Young's modulus of elasticity in terms of tensile stress and tensile strain is:

$$E = \frac{\sigma_t}{\varepsilon_t} \tag{19.1}$$

Stress has the dimension of pressure or force (F) per unit area (A),

$$\sigma_t = \frac{F}{A} = P \tag{19.2}$$

where P is pressure.

Strain in the longitudinal direction, or along the length of the blood vessel, is expressed as the ratio of extension per unit length or the ratio of the amount stretched longitudinally to the length of the original vessel segment,

$$\varepsilon_t = \frac{\Delta l}{l} \tag{19.3}$$

Strain in the radial direction, or perpendicular to the vessel segment length, is the fraction of distention of the vessel lumen radius or diameter. It is given by:

$$\varepsilon_r = \frac{\Delta r}{r} \tag{19.4}$$

Changes in arterial wall thickness, h, often accompany radial changes. Detailed analysis of this aspect can be found elsewhere [29, 40]. When the arterial wall thickness, h, is considered, the relative volume (V) distensibility of the artery is given by:

$$\frac{1}{V}\frac{dV}{dP} = \frac{3rV}{2hE} \tag{19.5}$$

The magnitude of the h/r ratio separates arteries into "thin-walled" and "thick-walled" vessels. Thus, an increase in wall thickness or lumen radius alone can impact the overall distensibility of the artery. The consequence of this has been found to be particularly important in vascular hypertrophy. Thickening of the arterial wall is largely due to the remodeling in the tunica media.

Pressure–Strain Elastic Modulus

From the fractional change in pulsatile diameter, $\Delta D/D$, and the pulsatile change in pressure from systole to diastole or pulse pressure (PP), the difference between systolic (Ps) and diastolic (Pd) pressures, i.e., $\Delta P = Ps - Pd$, the pressure–strain elastic modulus (Ep) can be obtained, viz.:

$$Ep = \frac{\Delta P}{\left(\Delta D/D\right)} \tag{19.6}$$

Thus, simultaneous recordings of pressure–diameter relations allow Ep to be readily computed. Figure 19.2 illustrates such recordings using ultrasonic dimension gages for recording aortic diameter and catheterization for recording aortic pressure.

In general, aortic diameters are larger in men than in women and have a greater PP through early adulthood. Ep is more commonly obtained

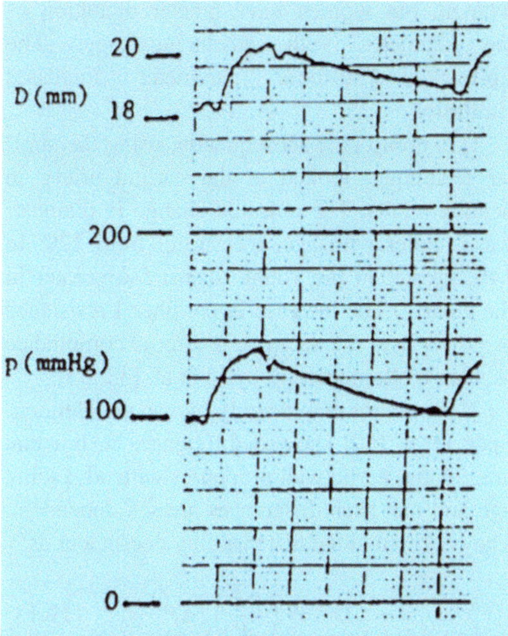

Fig. 19.2 Ultrasonic dimension gages were used to record the diameter of a dog aorta, together with aortic blood pressure, showing how Ep can be computed from the pressure–diameter relation $Ep = \mathrm{PP}/(\Delta D/D)$

noninvasively in the clinical setting with M-mode echocardiography for diameter measurement and brachial artery cuff for pulse pressure estimation.

To quantify the tension (T) exerted on the arterial wall due to intraluminal blood pressure distention, Laplace's law is useful. In the case of a blood vessel, there are two radii of curvature, one that is infinite in the longitudinal direction along the blood vessel axis and the other in the radial direction. Laplace's law is given as:

$$T = p \cdot r \tag{19.7}$$

This assumes the vessel has a thin wall or that the ratio of vessel wall thickness (h) to vessel lumen radius (r) is small, or $h/r \leq 1/10$. Here, p is the intramural–extramural pressure difference or the transmural pressure. When the vessel wall thickness is considered, the Lame equation is more relevant:

$$\sigma_t = \frac{pr}{h} \tag{19.8}$$

This relation is of particular importance in the analysis of an aneurysm, such as in abdominal aortic aneurysm (AAA) , where increased lumen radius is accompanied by decreased wall thickness such that a further increase in distending pressure can cause a rupture. In hypertension, however, T can be normalized by increasing the arterial wall thickness. Such chronical increase often leads to observed vascular hypertrophy.

To a good approximation, arteries are considered incompressible, with a Poisson ration of approximately 0.48. The Poisson ratio describes compressibility and is defined as the ratio of radial strain to longitudinal strain:

$$\sigma_n = \frac{\varepsilon_r}{\varepsilon_t} = \frac{\Delta r/r}{\Delta l/l} \tag{19.9}$$

When $\sigma_n = 0.5$, the material is said to be incompressible. This means that when a cylindrical artery is stretched, its volume remains unchanged.

Collagen is the stiffest wall component, with an Ep of 10^8–10^9 dynes/cm^2. This is some two orders of magnitude greater than those of elastin (1–6 $\times$ 10^6 dynes/cm^2) and smooth muscle (0.1–2.5 $\times$ 10^6 dynes/cm^2).

Elastin is relatively extensible, but it is not a purely Hookean material. In contrast, collagen is relatively inextensible because of its high elastic stiffness. Vascular smooth muscle can appreciably alter its elastic stiffness on activation. In addition, the mechanical properties of arterial vessel walls can be altered by neural mechanisms and by circulating catecholamines, such as norepinephrine. The overall composite of the arterial wall components operates in such a manner that elastin dominates the composite behavior at low pressures. At high pressures, collagen becomes more dominant. It has been found that Ep is a nonlinear function of pressure. The pressure dependence of the mechanical properties of arteries has been reported by several investigators (e.g., [3, 7, 36, 54]). With increasing positive transmural pressure, arterial vessel diameter is distended [53], while the corresponding compliance declines.

Longitudinally, along the arterial tree, we find that the number of elastic laminae decreases with increasing distance from the aorta, but the amount of smooth muscle increases and the wall thickness–to–radius ratio (*h*/*r*) increases. Overall vascular stiffness is thus increased away from the heart [28]. This latter phenomenon accounts for the observed large increase in pulse wave velocity. A longitudinal section also reveals a helical organization of the collagen fiber network. This network contributes significantly to the anisotropic properties of the arterial wall.

Sex Differences in Arterial Wall Properties: Arterial Compliance and Distensibility

Sonesson et al. [47] examined sex and age differences in compliance and diameter of the human aorta. They defined compliance, given by

$$C = dV/dP, \qquad (19.10)$$

as the inverse of the *Ep* expressed as the PP obtained by auscultatory method to the corresponding fractional change in pulsatile diameter with respect to diastolic diameter, *Dd*.

$$Ep = \frac{\Delta P}{\left({}^{\Delta D}/_{Dd}\right)} \qquad (19.11)$$

This definition differs from the conventional one when the mean diameter for the cardiac cycle is used for the denominator.

This group measured pulsatile diameter with an ultrasound echo-tracking device. They also calculated the arterial stiffness parameter based on a logarithmic relation between pressure and diameter for calculating a less pressure-dependent arterial compliance in vitro by Hayashi et al. [14] and modified for in vivo calculation by Kawasaki et al. [18], viz.:

$$\beta = \frac{\ln\left(\frac{Ps}{Pd}\right)}{\Delta D/Dd} \qquad (19.12)$$

where Ps and Pd are systolic and diastolic pressures, respectively. Results showed that *Ep* and β are both greater in males than females for the abdominal aorta and that the differences increase significantly with age, exponentially in males and linearly in females. This may attribute to a comparatively lower blood pressure in women, but women have greater dilatation of the diameter with advancing age. The Ep-tracked stiffness parameter increases similarly.

This group [13] subsequently, using the similar technique, looked at the carotid artery in healthy adults. The pulsatile change in diameter was smaller with age, measuring from 12% to 14% greater in the young to just 5% greater in the elderly. The increase in peripheral resistance is slower than the decrease in arterial compliance as shown allometrically by Li et al. [32, 35].

A population-based study by van der Heijden-Spek et al. [50] showed differences in brachial arterial and aortic wall properties with advancing age and that such differences are sex dependent. They computed the distensibility coefficient as:

$$DC = \left(\frac{\Delta A}{A}\right)/\Delta P \qquad (19.13)$$

and compliance coefficient

$$CC = \frac{\Delta A}{\Delta P} = \left(\frac{\Delta V}{L}\right)/\Delta P$$
$$= \pi\left(2D \cdot \Delta D + \Delta D^2\right)/4\Delta P \qquad (19.14)$$

Using the ultrasound echo-tracking device developed by themselves, Reneman's group [23] calculated DC and CC in the compliant aorta and muscular brachial artery and showed that, despite a larger brachial artery diameter and associated compliance in men, the distensibility was lower. With age, brachial artery dilatation is proportionally greater in women than in men, and brachial artery compliance did not decrease in men and increased in women. The opposite finding was found in the large compliant aorta, i.e., compliance decreased with increasing age in both sexes.

Hemodynamic Measurements and Model-Based Analysis of Arterial Wall Properties

The Windkessel Model

The idea of a lumped model of the arterial circulation was first described by Hales in 1733. Albeit largely qualitative, he did emphasize the storage properties of large arteries and the dissipative nature of small peripheral resistance arteries. In his description, the blood ejected by the heart during systole into the arterial system distends the large arteries, primarily the aorta. During diastole, the elastic recoil of these same arteries propels blood to perfuse the smaller peripheral resistance vessels. This initiated the earlier conceptual understanding that the distensibility of large arteries is important in allowing the transformation of intermittent outflow of the heart to steady outflow throughout the peripheral arteries. In other words, the large overall "compliance" of the large arteries protects the stiff peripheral vessels of organ vascular beds from the large swing of blood pressure due to pulsations. The significance of arterial pulsations remains a topic of debate.

The Windkessel, or air-bellow, is now credited to Frank [11], the original intent of whom was to obtain stroke volume (SV) from the aortic pressure pulse (PP) contour or the so-called pressure-derived flow [24].

Relating Arterial Wall Properties to Pulsatile Pressure and Flow

The usefulness of the Windkessel model, shown in Fig. 19.3, is in its ability to quantify arterial compliance, C, and peripheral resistance, Rs, from hemodynamic pressure (P) and flow (Q) measurements.

The amount of blood flow, Q_s, stored in the elastic aorta during each contraction is the difference between ventricular ejection or inflow, Q_i, to the large distensible aorta and the outflow, Q_o, to the small peripheral muscular arteries,

$$Q_o = (P - P_v)/R_s \quad (19.15)$$

The amount of outflow is equivalent to the pressure decrease from the arterial side (P) to the venous side (P_v) due to the peripheral resistance, R_s:

$$Q_o = (P - P_v)/R_s \quad (19.16)$$

At steady flow, assuming that P_v is small, or $Pv = 0$, we obtain a familiar expression for estimating the total peripheral resistance, and with the total inflow, $Q = Q_i$,

$$R_s = \bar{P}/\bar{Q} \quad (19.17)$$

or mean arterial pressure to mean arterial flow.

The storage property is described by arterial compliance, which expresses the amount of change in blood volume (dV) due to a change in distending pressure (dP) in the arterial lumen. In this case, we have

$$C = dV/dP \quad (19.18)$$

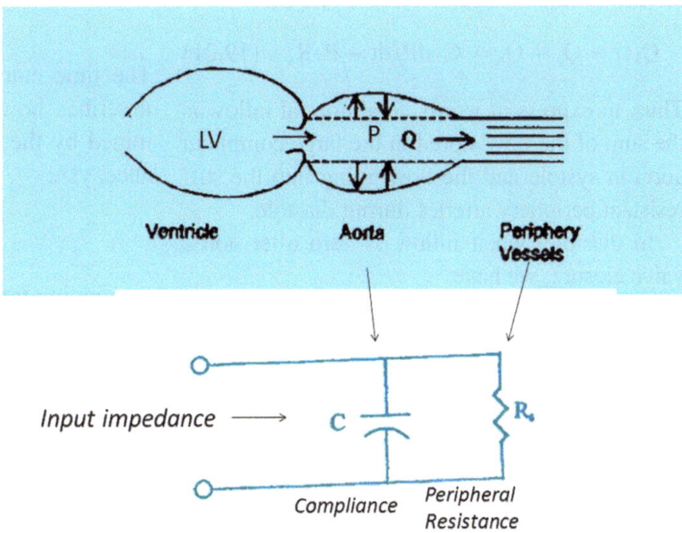

Fig. 19.3 Illustration of the left ventricle (LV) and the arterial circulation based on the two-element Windkessel model. The ventricle ejects into a compliant chamber representing the aorta; blood flow is stored in systole (solid line) and on diastolic elastic recoil (dotted line) stiff peripheral vessels are perfused. Compliance is represented by an energy-storing capacitor and the peripheral resistance by an energy-dissipating resistor

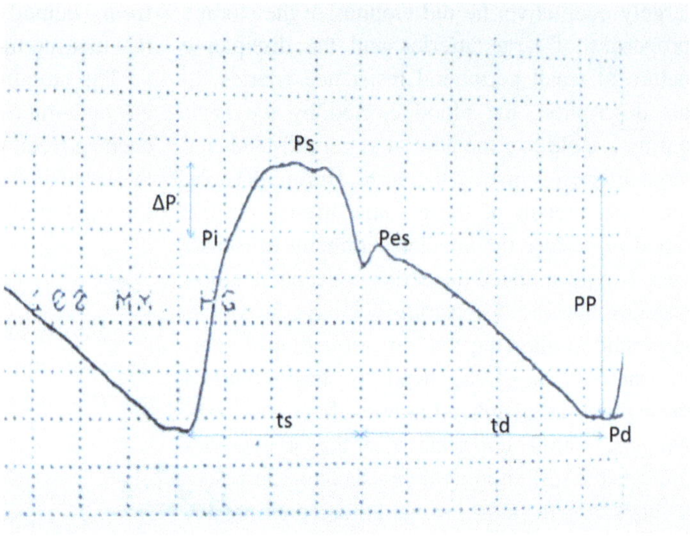

Fig. 19.4 Illustration of the measured aortic pressure pulse waveform for defining Ps, Pd, end-systolic pressure (Pes), PP (PP = Ps – Pd), inflection pressure (Pi) and augmented systolic pressure (ΔP = Ps − Pi)

For this reason, arterial compliance in the clinical setting has commonly been calculated from SV and PP, giving

$$C_v = SV/PP \qquad (19.19)$$

The amount of blood flow stored, or Q_s, due to arterial compliance is related to the rate of change in pressure, which is pulsatile, in distending the artery,

$$Q_s = C\ dP/dt \qquad (19.20)$$

Substituting for Q_s and Q_o, we obtain an expression relating the arterial pressure to flow incorporating the two Windkessel parameters, C and R_s:

$$Q(t) = Q_s + Q_o = C\ dP/dt + P/R_s \qquad (19.21)$$

Thus, to express in words, total arterial inflow is the sum of the flow stored in the large compliant aorta in systole and the flow going into the stiff resistant periphery arteries during diastole.

In diastole, when inflow is zero after aortic valve closure, we have

$$0 = C\ dP/dt + P/R_s \qquad (19.22)$$

or

$$dP/P = -dt/R_sC \qquad (19.23)$$

Thus, the rate of diastolic aortic pressure decrease after aortic valve closure is dependent on both the compliance of the arterial system and its total peripheral resistance. If we integrate both sides, we have

$$\ln P = t/R_sC \qquad (19.24)$$

or

$$P = P_o\ e^{-t/R_sC} \qquad (19.25)$$

which is valid for the diastolic period or $t = t_d$.

The diastolic aortic pressure decay from Pes to Pd follows a mono-exponential manner with a time constant τ (Fig. 19.4).

$$P_d = P_{es}\ e^{-td/\tau} \qquad (19.26)$$

The time constant of pressure decay, τ, which describes how fast the pressure falls, is determined by the product of resistance and compliance, viz.:

$$\tau = R_sC \qquad (19.27)$$

thus giving total arterial compliance

$$C = \frac{t_d}{R_s \ln \frac{P_{es}}{P_d}} \qquad (19.28)$$

An improved three-element Windkessel model (Fig. 19.5) of the arterial system, incorporating a characteristic impedance of the proximal aorta or Z_o, is now more widely used.

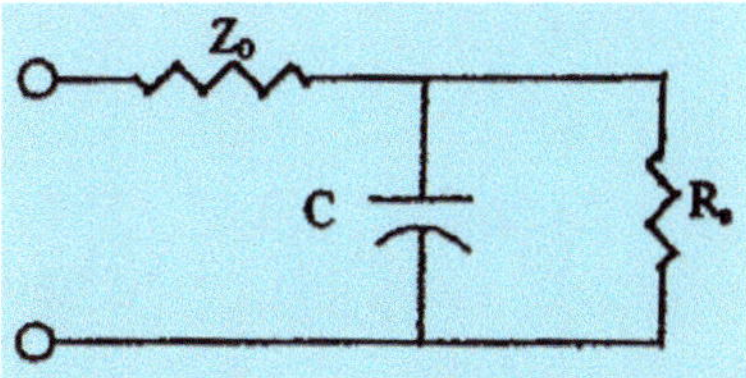

Fig. 19.5 An improved three-element Windkessel with Z_o representing the characteristic impedance of the proximal aorta

The Windkessel model compliance is determined in diastole only. Some investigators, such as Liu et al. [38], have used both the systolic and diastolic information contained in the pressure waveform and use the area method of calculating total arterial compliance:

$$C = \frac{\text{DPTI/SV}}{(\text{SPTI} + \text{DPTI}) \cdot \text{PP}} \quad (19.29)$$

where SPTI and DPTI are the systolic pressure–time integral and diastolic pressure–time integral, respectively.

Arterial Wall Properties Determined from Pulse Waveform Analysis

Pulse Wave Velocity

$$c_f = \frac{\Delta z}{\Delta t} \quad (19.30)$$

Pulse wave velocity has been popularly approximated by the so-called "foot-to-foot" velocity. Here, one simply estimates the pulse wave velocity (PVW) from the pulse transit time delay (PTT or Δt) of the "onset" or the "foot" between two pressure pulses measured at two different sites along an artery or the pulse propagation path. This requires, again, the simultaneous measurements of two pressures separated by a finite distance, Δz, normally 4–6 cm apart. A double-lumen catheter with two pressure ports connected to two pressure transducers or a Millar catheter with dual pressure sensors suffice for such measurement. The Moens-Korteweg formula relating PWV to stiffness is given by:

$$c_o = \sqrt{\frac{Eh}{2\rho r}} \quad (19.31)$$

where E is the Ep of the artery; h and r are the wall thickness and radius of the artery, respectively; and ρ is the density of blood. This formula is applicable to a single vessel, while foot-to-foot velocity has been obtained either for a single artery or over the pulse propagation path, e.g., over several arteries. Popular sites for noninvasive pulse wave velocity are the brachial, radial, carotid and femoral arteries. For instance, carotid-to-femoral pulse wave velocity has been used as an index of vascular stiffness change in the aorta as has carotid-to-radial pulse wave velocity (Fig. 19.6).

Wave Separation into Forward and Reflected Components

The amplification of pressure pulses has been attributed to the in-phase summation of reflected waves arising from structural and geometric nonuniformities. The microvascular beds have been recognized as the principal reflection sites. Thus, pulsatile pressure and flow waveforms contain information about the heart as well as the vascular system. Reflection in the vascular system has been suggested as a closed-end type because pressure is amplified and flow decreases, with arterioles being the major reflection site. By definition, reflected pressure and flow waves are 180° out of phase. This means an increase in reflection increases PP amplitude but decreases pulsatile flow amplitude.

Pressure (P) and flow (Q) waveforms measured at any site in the vascular system can be considered as the summation of a forward, or antegrade, traveling wave and a reflected, or retrograde, traveling wave:

$$P = P_f + P_r \quad (19.32)$$

$$Q = Q_f + Q_r \quad (19.33)$$

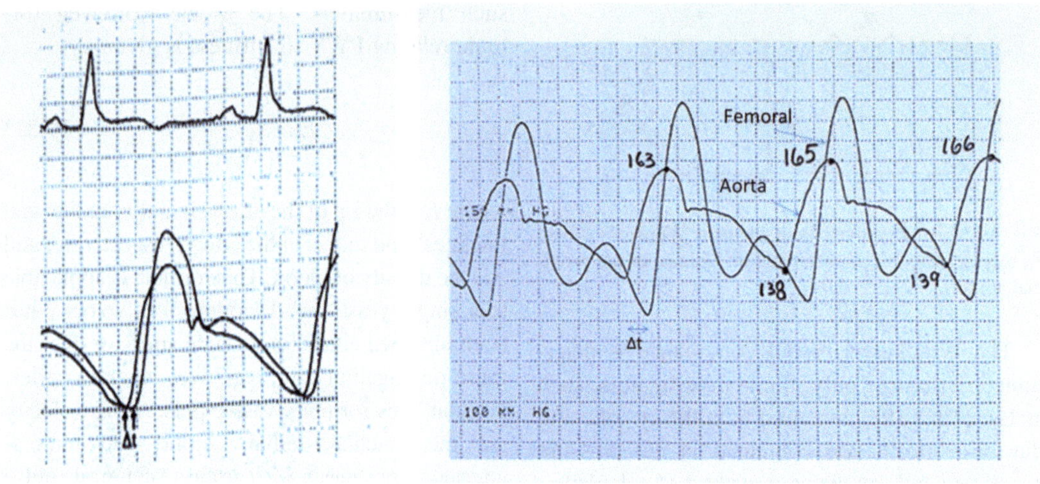

Fig. 19.6 PWV based on PTT (or Δt) of the onset of two pressures measured at a finite distance apart, in this case, those of the ascending and descending aortas (left panel). Pulse transit time delay is much more obvious when the two pressures are simultaneously measured at the ascending aorta and the femoral artery (right panel). Ps and Pd are displayed in this hypertensive subject. Femoral artery PP is significantly greater.

The forward and reflected pressure components can be resolved by means of the following set of equations:

$$P_f = (P + Q \cdot Zo)/2 \tag{19.34}$$

$$P_r = (P - Q \cdot Zo)/2 \tag{19.35}$$

where Z_o is the characteristic impedance, defined as the ratio of forward pressure to forward flow or, in other words, independent of wave reflections:

$$Z_o = \frac{P_f}{Q_f} = -\frac{P_r}{Q_r} \tag{19.36}$$

Z_o can be obtained from the average of the ratios of aortic pressure and flow during early systole when peripheral wave reflections exert minimal effect,

$$Zo = \frac{\Delta P}{\Delta Q} \tag{19.37}$$

or from the water-hammer formula,

$$Z_o = \frac{\rho c}{\pi r^2} \tag{19.38}$$

where ρ is the density of blood (1.06 g/cm^3), c is pulse wave velocity, and πr^2 is the cross-sectional area of the artery. The fast time-domain wave separation method by Li [28] is widely used.

Similarly, resolution of flow into its forward and reflected components (Fig. 19.7) can be obtained from a set of two equations:

$$Q_f = (Q + P/Z_o)/2 \tag{19.39}$$

$$Q_r = (Q - P/Z_o)/2 \tag{19.40}$$

Pulse pressure amplitude alone often does not indicate the underlying factors governing the morphology of blood pressure pulse waveforms. Thus, the routine clinically used cuff method for measuring Ps and Pd cannot be used to infer vascular stiffness or properties. One example is shown here in Fig. 19.8. Despite similar Ps and Pd induced either by a change in mechanical properties of the blood vessel wall or by a change in vasoactive states, the forward and reflected waveform morphologies and contributing factors differ [33]. A wave reflection-based distributed model [55] can be used to resolve the differences in vasoactive from mechanical factors.

The Augmentation Index and Transfer Function for Aortic Pressure Synthesis from Peripheral Pulse Tonometry

Because large-vessel properties can appreciably affect ventricular ejection, aortic compliance and

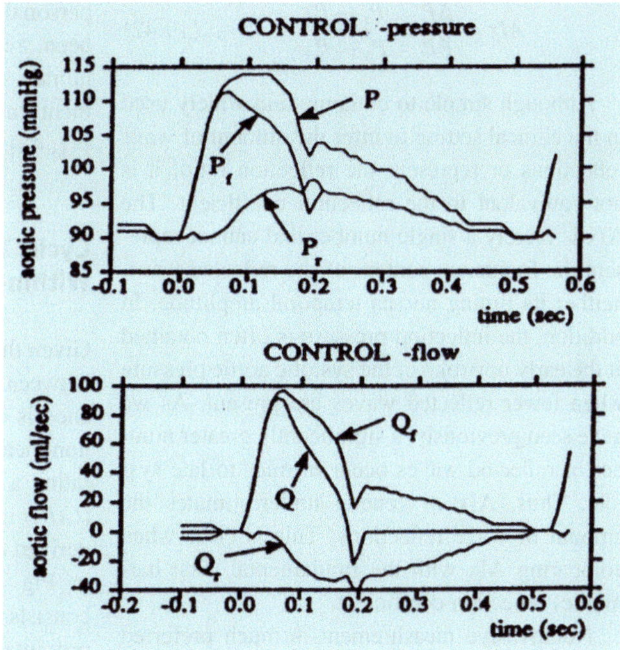

Fig. 19.7 Ascending aortic pressure, P, and flow, Q, separated into their respective forward (Pf, Qf) and reflected (Pr, Qr) components. Notice that wave reflection (+Pr and -Qr) has opposite effects on measured pressure and flow

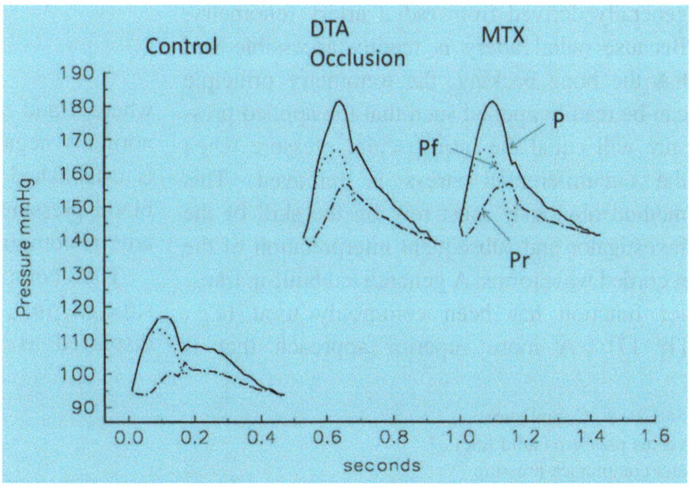

Fig. 19.8 Ascending aortic pressure and flow waveforms resolved into their respective forward (dotted line) and reflected (dash-dotted line) components during control (left), descending thoracic aorta (DTA) occlusion (center), and methoxamine-induced hypertension (MTX [right]). Although PPs are approximately the same due to mechanical (DTA) or vasoactive (MTX) interventions, the underlying morphology, mechanism, and forward and reflected waves differ

the wave reflections appearing in the aorta have become important markers for assessing arterial system behavior in the clinical setting. In addition, because frequency domain interpretation of wave reflections is complex and cumbersome to obtain, there exists a simplified index to interpret wave reflection in the aorta in the time domain. Ascending aortic pressure waveforms are defined in terms of their morphological differences and separated into different types. Peak Ps, Pd, PP, pressure at inflection point (P_i) where peak flow occurs, and augmented pressure (ΔP) are defined (see Fig. 19.4). Systolic pressure augmentation is given by:

$$\Delta P = P_s - P_i \tag{19.41}$$

and the corresponding augmentation index (AIx) for the aorta is given by

$$AIx = \frac{\Delta P}{PP} = \frac{P_s - P_i}{P_s - P_d} \quad (19.42)$$

Although simple to compute and widely used in the clinical setting to infer the amount of wave reflections or represent the reflection ratio, it is not equivalent to the reflection coefficient. The AIx is merely a single number and cannot represent the frequency content of the reflected wave, neither its timing nor its temporal amplitude. In addition, the inflection pressure is often obtained at the early upstroke of the systolic aortic pressure when fewer reflected waves are present. As we have seen previously, a significantly greater number of reflected waves occur in mid- to late systole. Thus, AIx in general underestimates the amount of wave reflections. This is clear when comparing AIx with the fundamental (first harmonic) reflection coefficient.

Noninvasive measurement is much preferred in the clinical setting. Since the central aortic pressure cannot be obtained noninvasively, it is generally derived from radial artery tonometry. Because radial artery is readily accessible and has the bone backing, the tonometry principle can be readily applied such that the applied pressure will equal the intravascular pressure when the circumferential stress is removed. This method obviously must rely on the skill of the investigator and subsequent interpretation of the recorded waveforms. A generalized built-in transfer function has been commonly used (e.g., [1, 17]). A more superior approach that is personalized, based on the Wiener system, has been recently proposed, and the results in human studies have shown significant improvement in estimating the central aortic pressure [41, 42].

Cyclic Variation of Arterial Compliance Within a Single Heart Beat

Given the differences in pulse pressure variations between the sexes, and because arterial compliance is pressure dependent, a more appropriate nonlinear model of the arterial system incorporating a pressure-dependent compliance element ($C(P)$) must be employed. This nonlinear model derived compliance-pressure loops are illustrated in Fig. 19.9 below [27, 30, 36]. The model consists of the characteristic impedance of the proximal aorta (Z_o), the peripheral resistance (Rs), and $C(P)$. The compliance is exponentially related to pressure and is expressed as

$$C(P) = a \cdot e^{b(P(t))} \quad (19.43)$$

where a and b are constants. The exponent b is normally negative. Thus, an inverse relationship is established between arterial compliance and blood pressure: with increasing blood pressure, arterial compliance decreases.

The compliance-pressure loop [27] constructed for each heart beat can be easily displayed as shown in Fig. 19.9 for varying

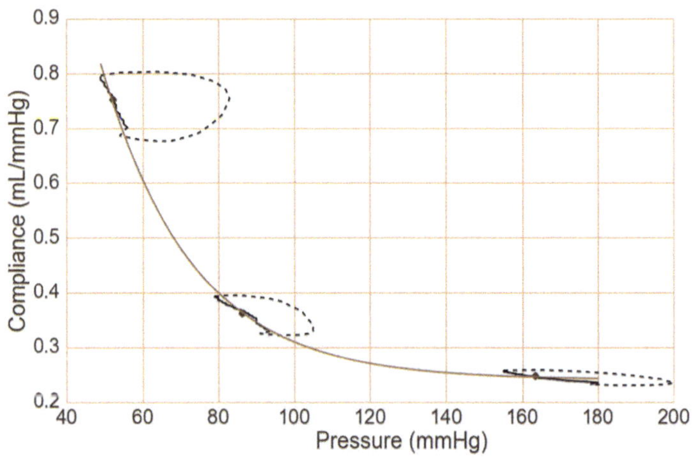

Fig. 19.9 Compliance versus pressure (solid line) and compliance-pressure loops (dotted lines) plotted for the control (middle), MTX-induced hypertension (right), and nitroprusside-induced vasodilation (left) cases (dog aorta). Note that overall compliance decreases with increasing pressure and that the loop area is compressed in hypertension and enlarged with a vasodilator

pressures. Very little work has been done in analyzing the beat-to-beat compliance-pressure relation for the sexes and the aging effects. It can be speculated, however, based on the $C(P)$ equation that the rate of change or the decrease in arterial compliance versus pressure (exponent b) is greater in elderly women. In addition, the beneficial effect of maximal arterial vasodilation-induced arterial compliance increment would be more limited than in their elderly male counterparts.

Sex Differences in Vascular Stiffness and Augmentation Index in Diseased Conditions

Abdominal Aortic Aneurym or AAA is among the more prominent cardiovascular disease and has been found to differentially affect the sexes, with lower prevalence but higher rupture rate in women than in men. Dobrin and Mrkvicka [6] have shown that depletion of elastin and collagen alters the tensile strength of the aorta and underlies the dilation of aneurysm. Analyzing wall properties by looking at the elastin synthesis and destruction in a small sample of men and women, Villard et al. [51] found that protein expression of elastin was less in women than men in cases of non-thrombolytic aneurysm.

It is well recognized that sex differences in vascular dysfunction exist, including differences in cardiovascular outcomes, particularly in relation to vascular stiffness (e.g., [9]). Women with end-stage renal disease (ESRD) have been suspected to have a higher mortality rate than men. Using the AIx for inferring systolic wave reflection and carotid-to-femoral artery pulse wave velocity as an index of vascular stiffness plus flow-mediated dilation (FMD) and velocity time integral (VTI) as a measure of microvascular endothelial function [8]. Guajardo et al. [12] investigated whether differences in vascular dysfunction lead to the different mortality differences in ESRD men and women because few studies have focused on sex-specific risk factors. Although they found little difference in pulse wave velocity, the AIx was significantly much greater in women than men. Mortality outcome was more associated with greater AIx and lower VTI and much less affected by PWV or FMD. Another interesting aspect is related to the Windkessel model parameters. In general, women have significantly decreased overall arterial compliance, C, and increased total peripheral resistance, Rs, than men. As such, the Windkessel time constant [i.e., $\tau = RsC = (1.32$ sec) vs. (0.97 sec)], as well as the τ/T ratio, is not significantly different because heart rates are similar.

Increased wave reflection and increased vascular stiffness both increase afterload to the heart, and are energeticlly wasterful [26, 28]. Hughes et al. [16] looked at whether their disproportional increase contributes to greater myocardial work in women. A generalized transfer function was used to derive aortic pressure from noninvasively measured radial arterial pressure.

Women in general have greater age-related increases in left ventricular mass (LVM) compared with men. This greater increased LVM leads to a greater oxygen demand and external work [25]. Our previous study also showed that women tend to experience what is known as “heart failure” with preserved ejection fraction, or HFpEF, which is associated with a greater incidence of diastolic dysfunction. The parallel or concurrent occurrence of LVM and hypertrophy with vascular hypertrophy has been reported [5]. Increased Ps in elderly women condition can be exacerbated by a ventricle with a double load [31].

Increased arterial stiffness associated with increased wave reflection, as evaluated by the AIx, has been reported in HFpEF patients. In a large community-based cohort study by Russo et al. [45], using a tonometer and 2D echocardiography, arterial stiffness and wave reflection were indirectly calculated based on PP/SV and total arterial compliance from the area method [38], AIx and PP amplification. AIx was based on the aortic pressure derived from a generalized transfer function. They showed that PP/SV stiffness and AIx were both greater and had a much lower total arterial compliance in women and that these were related to their associated LV diastolic dysfunction. It should be noted here that this study was centered on the elderly with average age being more than 70 years, and the majority of the subjects were Hispanic and black, both of which tend to have greater incidences of hypertension. In addition, there were more female subjects with greater number exhibiting hypertension and obesity. The difference in the central aortic pressure AIx appears to differ between the sexes even in pre-pubescent children [48].

Consideration of the vascular load in terms of the LV–arterial system interaction has also been studied [2, 19, 20, 26, 34]. Because the ventricle is intimately connected to the ascending aorta, the central aortic pressure is of primary importance. In terms of the interaction, left ventricular and arterial elastances are generally computed to interpret the coupling and the extent of interaction of the LV and the arterial system. Our recent study [20] showed that in heart failure patients, the left ventricular end-systolic and end-diastolic volumes are significantly ($P < 0.034$ and $P < 0.016$, respectively) smaller in women ($N = 67$), while ejection fraction, LV maximum elastance, LV and arterial system coupling index, and mean arterial pressure and PP are greater ($P < 0.008$). These data showed that these female patients exhibit a greater increase in their peripheral resistances and a greater decrease in their arterial compliances compared with male patients.

Hayward and Kelly [15] showed sex differences in central arterial pressure waveforms, carotid to be specific, with more striking differences as age increases. Augmentation obtained was given by AIx (%) $= -132$ *Height (m) + 243, i.e., AIx is inversely proportional to height. Peaking in late Ps in elderly females led to a greater AIx. Females were found to have a shorter diastolic pressure–time integral. The subendocardial viability ratio in women, given as the ratio of DPTI/TTI. where TTI is tension–time index, a well-known index of myocardial oxygen consumption,

$$\mathrm{SEVR} = \mathrm{DPTI}/\mathrm{TTI} \tag{19.44}$$

is significantly lower compared with male subjects despite similar TTI. However, sex differences in pulse wave reflection coefficients have been little studied because AIx cannot be fully related to the reflection coefficient. The latter is related to the whole cardiac cycle, while the former is only reflective of early systole [43, 55].

Hemodynamic Monitoring of Sex Differences in Different Cultures

In recent Korean studies [21, 22] looking at sex differences in arterial stiffness in middle-aged men and women, it was shown that aortic pulse wave velocity and brachial arterial pulse pressure amplification were both greater in men compared with women. Aortic PWV was assessed by carotid-to-femoral PWV as the common choice in the clinical setting. Radial artery tonometry–derived aortic AIx was obtained by way of transfer function. Pulse pressure amplification was obtained as the ratio of brachial arterial pressure to derived aortic pressure. Women have a greater AIx as shown by others. AIx was significantly correlated with age, Pd, and serum cholesterol levels.

Carotid-to-femoral PWV has been widely used as supposedly providing an indication of large-artery stiffness, i.e., of the aorta. It should be noted here that controversies exist because of the differing pulse transmission paths taken by the carotid and femoral arteries, in other words, opposite directions from the aortic arch junction.

A Chinese study by Liu et al. [37], based on radial artery monitoring, showed no difference between middle-aged men and women groups despite greater body weight and height in men. A Japanese study by Tomiyama et al. [49], using brachial–ankle pulse wave velocity (baPWV) as an index of vascular stiffness, showed that baPWV is greater in men than women. The difference, however, disappears when women reach menopause. This is attributed to the vasodilator effect of estrogen [10]. A study by Nishiwaki et al. [39], using the cardio–ankle vascular index (CAVI) initially proposed by Shirai et al. [46], showed that CAVI increases with age, as expected, and is greater in men than in women.

References

1. Chen CH, Nevo E, Fetics B, Pak PH, Yin FC, Maughan WL, et al. Estimation of central aortic pressure waveform by mathematical transformation of radial tonometry pressure. Circulation. 1977;95:1827–36.
2. Coutinho T, Borlaug BA, Pellikka PA, Turner ST, Kullo IJ. Sex differences in arterial stiffness and ventricular-arterial interactions. J Am Coll Cardiol. 2013;61:96–103.
3. Cox RH. Arterial wall mechanics and composition and the effects of smooth muscle activation. Am J Physiol. 1975;229:807–12.
4. D'Armiento J. Decreased elastin in vessel walls puts the pressure on. J Clin Invest. 2003;112(9):1308–10.
5. Devereux RB, de Simone G, Ganau A, Koren MJ, Mensah GA, Roman MJ. Left ventricular hypertrophy and hypertension. Clin Exp Hypertens. 1993;15:1025–32.
6. Dobrin PB, Mrkvicka R. Failure of elastin or collagen as possible critical connective tissue alterations underlying aneurysmal dilatation. Cardiovasc Surg. 1994;2:484–8.
7. Drzewiecki G, Field S, Mubarak I, Li JK-J. Effect of vascular growth pattern on lumen area and compliance using a novel pressure-area model for collapsible vessels. Am J Physiol (Heart & Circ Physiol). 1997;273:H2030–43.
8. Dubin RF, Guajardo I, Ayer A, Mills C, Donovan C, Beussink L, et al. Associations of macro- and microvascular endothelial dysfunction with subclinical ventricular dysfunction in end-stage renal disease. Hypertension. 2016;68:913–20.
9. Duprez D, Jacobs DR Jr. Arterial stiffness and left ventricular diastolic function. Does sex matter? Hypertension. 2012;60:283–4.
10. Farhat MY, Lavigne MC, Ramwell PW. The vascular protective effects of estrogen. FASEB J. 1996;10:615–24.
11. Frank O. Die Grundform des arteriellen pulses. Z Biol. 1899;37:483–526.
12. Guajardo I, Ayer A, AD Johnson PG, Mills C, Donovan C, Scherzer R, et al. Sex differences in vascular dysfunction and cardiovascular outcomes: the cardiac, endothelial function, and arterial stiffness in ESRD (CERES) study. Hemodial Int. 2017;22:93. https://doi.org/10.1111/hdi.12544.
13. Hansen F, P Mangell B, Sonesson TL. Diameter and compliance in the human common carotid artery – variations with age and sex. Ultrasound Med Biol. 1995;21:l–9.
14. Hayashi K, Handa H, Nagasawa S, Okumura A, Moritake K. Stiffness and elastic behavior of human intracranial and extracranial arteries. J Biomech. 1980;13:175–84.
15. Hayward CS, Kelly RP. Gender-related differences in the central arterial pressure waveform. J Am Coll Cardiol. 1997;30:1863–71.
16. Hughes WE, Spartano NL, Lefferts WK, Augustine JA, Heffernan KS. Sex differences in noninvasive estimates of left ventricular pressure energetics but not myocardial oxygen demand in young adults. Artery Research. 2014;8:197–204.
17. Karamanoglu M, O'Rourke MF, Avolio AP, Kelly RP. An analysis of the relationship between central aortic and peripheral upper limb pressure waves in man. Euro Heart J. 1993;14:160–7.
18. Kawasaki T, Sasayama S, Yagi S, Asakawa T, Hirai T. Non-invasive assessment of the age related changes in stiffness of major branches of the human arteries. Cardiovasc Res. 1987;21:678–87.
19. Kerkhof PLM, Kresh JY, Li JK-J, Heyndrickx GR. Left ventricular volume regulation in heart failure with preserved ejection fraction. Physiol Report. 2013;1(2):1–10.
20. Kerkhof PLM, GR Heyndrickx, JK-J Li. Hemodynamic determinants and ventriculo-arterial coupling are sex-associated in heart failure patients. In: Proceeding of 38th annual international conference of the IEEE engineering in medicine and biology society, IEEE X-plore; 2016.
21. Kim H, Kim M, W Shim SO, Kim M, Park SM, Kim YH, et al. Sex difference in the association between brachial pulse pressure and coronary artery disease: the Korean women's chest pain registry (KoROSE). J Clin Hypertens (Greenwich). 2017;19:38–44.
22. Kim J, Park DS Kim JB, Kim KS, Jeong JW, Park JC, Oh BH, et al. On behalf of the KAAS investigators. Gender difference in arterial stiffness in a multicenter cross-sectional study: the Korean Arterial Aging Study (KAAS). Pulse. 2014;2:11–7.
23. Kool MJF, van Merode T, Reneman RS, Hoeks APG, Struijker Boudier HA. Evaluation of reproducibility of a vessel wall movement detector system for assessment of large artery properties. Cardiovasc Res. 1994;28:610–4.
24. Li JK-J. Pressure-derived flow: a new method. IEEE Trans Biomed Eng BME. 1983;30:244–6.
25. Li JK-J. Regional left ventricular mechanics during myocardial ischemia. In: Sideman S, editor. Simulation and modeling of the cardiac system. Boston: Martinus Nijhoff Publishers; 1987. p. 451–62.
26. Li JK-J. Feedback effects in heart-arterial system interaction. In: Sideman S, Beyar R, editors. Interactive phenomenon in the cardiac system. New York: Plenum; 1993. p. 325–33.
27. Li JK-J. A new description of arterial function: the compliance-pressure loop. Angiology. J Vascular Diseas. 1998;49:543–8.
28. Li JK-J. The arterial circulation: physical principles and clinical applications. New York: Springer; 2000.
29. Li JK-J. Dynamics of the vascular system. Singapore: World Scientific Publishers; 2004.
30. Li JK-J, Cui T, Drzewiecki G. A nonlinear model of the arterial system incorporating a pressure-dependent compliance. IEEE Trans Biomed Eng BME. 1990;37:673–8.
31. Li JK-J, Zhu JY, Nanna M. Computer modeling of the effects of aortic valve stenosis and arterial system afterload on left ventricular hypertrophy. Comput Biol Med. 1997;27:477–85.
32. Li JK-J, Zhu Y, O'Hara D, Khaw K. Allometric hemodynamic analysis of isolated systolic hypertension and aging. Cardiovasc Eng. 2007;7:135–9.
33. Li JK-J, Zhu Y, Geipel PS. Pulse pressure, arterial compliance and wave reflection under differential vasoactive and mechanical loading. Cardiovasc Eng. 2010;10:170–5.
34. Li JK-J, Atlas G. Left ventricle–arterial system interaction in heart failure. Clin Med Insight: Cardiol. 2015;9(Suppl 1):93–9. https://doi.org/10.4137/CMC.S18742.
35. Li JK-J, Zhu Y, Noordergraaf A. A comparative approach to analysis and modeling of cardiovascular function. Molecular, cellular, and tissue engineering. In: Bronzino JD, Peterson DR, editors. Biomedical engineering handbook, 4th ed. Chapter 26, pp26. London/New York: CRC Press/Boca Raton; 2015. p. 1–12.

36. Li JK-J, Zhu Y. Arterial compliance and its pressure dependence in hypertension and vasodilation. Angiology, J Vasc Diseas. 1994;45:113–7.
37. Liu C, Zhao L, Liu C. Effects of blood pressure and sex on the change of wave reflection: evidence from Gaussian fitting method for radial artery pressure waveform. PLoS One. 2014;9(11):e112895.
38. Liu Z, Brin KP, Yin FC. Estimation of total arterial compliance: an improved method and evaluation of current methods. Am J Phys. 1986;251:H588–600.
39. Nishiwaki M, Kurobe K, Kiuchi A, Nakamura T. N Matsumoto. Sex differences in flexibility-arterial stiffness relationship and its application for diagnosis of arterial stiffening: a cross-sectional observational study. PLoS One. 2014; https://doi.org/10.1371/journal.pone.0113646.
40. Noordergraaf A. Circulatory system dynamics. New York: Academic; 1978.
41. Patel A, Li JK-J. Aortic pressure estimation using blind identification approach on single input multiple output non-linear Wiener systems. IEEE Trans Biomed Eng. 2017a. https://doi.org/10.1109/TBME.2017.2688425
42. Patel A, Li JK-J. Validation of a novel nonlinear black box Wiener System model for arterial pulse transmission. Comput Biol Med. 2017b;88:11. https://doi.org/10.1016/j.compbiomed.2017.06.020.
43. Phan TS, Li JK-J, Segers P, Reddy-Koppula M, Akers SR, Kuna ST, et al. Aging is associated with an earlier arrival of reflected waves without a distal shift in reflection sites. J Am Heart Assoc. 2016;5:e003733. https://doi.org/10.1161/JAHA.116.003733.
44. Qiu H, Depre C, Ghosh K, Resuello RG, Natividad F, Rossi F, et al. Mechanism of gender-specific differences in aortic stiffness with aging in nonhuman primates. Circulation. 2007;116:669–76.
45. Russo C, Jin Z, Palmieri V, Homma S, Rundek T, Elkind MSV, et al. Arterial stiffness and wave reflection sex differences and relationship with left ventricular diastolic function. Hypertension. 2012;60:362–8.
46. Shirai K, Utino J, Otsuka K, Takata M. A novel blood pressure-independent arterial wall stiffness parameter; Cardio-Ankle Vascular Index (CAVI). J Atheroscler Thromb. 2006;13:101–7.
47. Sonesson B, Hansen F, Stale H, Lanne T. Compliance and diameter in the human abdominal aorta- the influence of age and sex. Eur J Vasc Surg. 1993;7:690–7.
48. Stoner L, Faulknera J, Westruppb N, Lambrick D. Sexual differences in central arterial wave reflection are evident in prepubescent children. J Hypertens. 2015;33:304–7.
49. Tomiyama H, Yamashina A, Arai T, Hirose K, Koji Y, et al. Influences of age and gender on results of noninvasive brachial-ankle pulse wave velocity measurement–a survey of 12517 subjects. Atherosclerosis. 2003;166:303–9.
50. Van der Heijden-Spek JJ, Staessen JA, Fagard RH, Hoeks AP, Struijker Boudier HA, Van Bortel LM. Effect of age on brachial Artery Wall properties differs from the aorta and is gender dependent. A population study. Hypertension. 2000;35:637–42.
51. Villard C, Eriksson P, Swedenborg J, Hultgren R. Differences in elastin and elastolytic enzymes between men and women with abdominal aortic aneurysm. Aorta. 2014;2:179–85.
52. Wagenseil JE, Mecham RP. Elastin in large artery stiffness and hypertension. J Cardiovas Translat Res. 2012;5:264–73.
53. Weizsacker H.W. and K. Pascal. Anisotropic passive properties of blood vessel walls. In: Cardiovascular system dynamics: models and measurements, pp. 347–362. Eds. T. Kenner, R. Busse, H. Hinghofer-Szalkay, Plenum, 1982.
54. Wells SM, Langeille BL, Adamson SL. In vivo and in vitro mechanical properties of the sheep in thoracic aorta in the perinatal period and adulthood. Am J Phys. 1998;274: H1749–60.
55. Zhang H, Li JK-J. A novel wave reflection model of the human arterial system. Cardiovasc Eng. 2009;9:39–48.

Sex-Specific Characteristics of the Microcirculation

20

Virginia H. Huxley and Scott S. Kemp

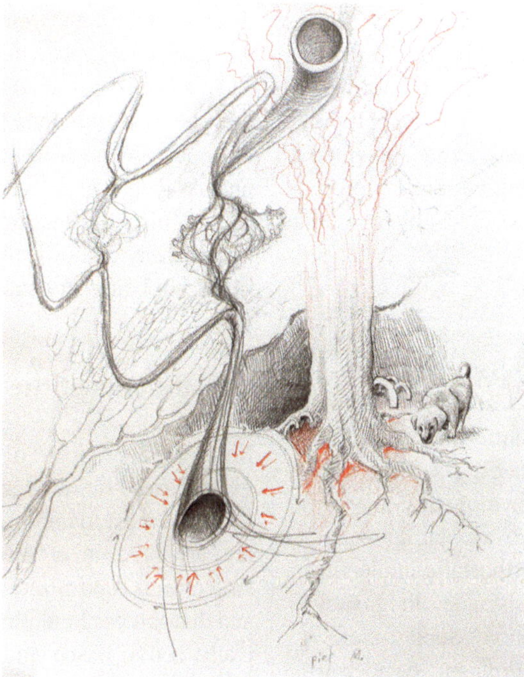

Art work by Piet Michiels, Leuven, Belgium

Abstract

The requirements of metabolizing tissue are both continuous and variable; accordingly, the microvasculature serving that tissue must be similarly dynamic. Just as it is recognized that males and females of the same species have differing metabolic requirements, is it not likely that the microvasculature serving these tissues will differ by sex? This section focusing on the constituents of the microcirculation identifies what is known presently about

V. H. Huxley (✉) · S. S. Kemp
Center for Gender Physiology, Department of Medical Pharmacology and Physiology, School of Medicine, University of Missouri, Columbia, MO, USA
e-mail: huxleyv@health.missouri.edu

P. L. M. Kerkhof, V. M. Miller (eds.), *Sex-Specific Analysis of Cardiovascular Function*,
Advances in Experimental Medicine and Biology 1065, https://doi.org/10.1007/978-3-319-77932-4_20

the role sex plays in matching metabolic demand with microvascular function and areas requiring additional study. Many of the identified sex differences are subtle and easily ignored. In the aggregate, though, they can profoundly alter phenotype, especially under stressful conditions including pregnancy, exercise, and disease states ranging from diabetes to heart failure. Although the features presently identified to "have sex" range from differences in growth, morphology, protein expression, and intracellular signaling, males and females alike achieve homeostasis, likely by different means. Studies of microvascular sexual dimorphism are also identifying age as an independent but interacting factor requiring additional attention. Overall, attempting to ignore either sex and/or age is inappropriate and will prevent the design and implementation of appropriate interventions to present, ameliorate, or correct microvascular dysfunction.

Keywords

Aquaporin · Arterioles · Barrier function · Blood flow regulation · Capillaries · Coronary microvascular dysfunction · Endothelium · Fibrinolysis · Fluid homeostasis · Glycocalyx · Hydrostatic pressure · Lymphatics · Venules · Microcirculation · Microvascular network · Myogenic response · Orthostatic intolerance · Pericyte · Peripheral resistance · Pregnancy · Rarefaction · Hypertension · Sepsis · Sex difference · Sexual dimorphism · Syndecan-1 · Vascular homeostasis

Introduction

The microvasculature constitutes the largest number of vessel segments and surface area of the blood vasculature. Despite its anatomical prominence, it is the least studied portion of the vasculature circulating blood reflecting the difficulties of access and visualization. Its tubular structures, the arterioles, capillaries, and venules, perform the majority of the "work." They control how much blood moves into and out of the tissues they penetrate and the movement of everything from respiratory gases, fluid, solutes, drugs, hormones, proteins, and cells between circulating blood and metabolizing tissue. In addition, they govern the production and modification of a growing list of circulating vasoactive mediators, peptides, growth factors, and hormones. The microvessels are also the sites most likely to experience changes in function early in the development of numerous acute life-threatening conditions such as sepsis and disseminated intravascular coagulation (DIC) [28, 36, 68] to chronic diseases spanning from diabetes, obesity, and Alzheimer's to heart failure [22, 41, 52, 67, 82, 113]. Even less studied is the influence of sex on these myriad functions in health or dysfunctions in disease. What work has been done and is now underway makes it clear that ignoring sex will lead to misunderstanding of the basis of vascular homeostasis or how it is modified in whichever sex is not being studied.

Constituents of the Microvasculature

Multiple definitions of the "microvasculature" exist based on size and anatomical location [105]. In most mammalian systems, microvessels are arranged in a network whose anatomical structure is optimized to the organ's function and distributes circulating blood close to metabolically active tissue and then removes it to the venous system. The three primary microvascular constituents are arterioles, capillaries, and venules.

Arterioles

Arterioles branch off of the feed arteries connecting an organ to the conduit vessels carrying blood from the heart at high velocity with high pulsatile pressure. Arterioles in mammals are <100 μm in diameter (in situ resting tone results in diameters ranging between 15 and 40 μm) [97]. Their walls consist of vascular smooth muscle (VSM) wrapping a smooth, continuous layer of endothelial cells (EC). The smallest of these vessel segments, the terminal arterioles, possess a single layer of VSM and resting diameter between 5 and 8 μm. Arteriolar constriction and dilation lead to fine, variable control of blood flow to and within the microvascular network. As capillaries lie downstream of the arterioles, changes in arteriolar tone also lead to control of the amount of materials delivered to the capillaries and of capillary hydrostatic pressure which controls transmural fluid movement. These vessels are dynamic as their structure can be changed both acutely and chronically in the face of changing hemodynamic conditions [76, 105].

Capillaries

Capillaries branch off of the terminal arterioles and are distinguished by the lack of VSM but presence of pericytes on the abluminal surface in contact with the EC lining [105]. The degree of pericyte coverage is organ dependent with coverage being highest in the capillaries of the retina and brain. It is not unusual for a single pericyte to have processes contacting more than one EC. The contribution of pericytes to the regulation of microvascular function, especially at the level of the capillaries, is just beginning to come into its own. These cells are referred to in the early literature as Rouget cells and discussed heavily by Krogh [65], with respect to their roll in participating in the regulation of capillary blood flow. Most recently the notion that pericyte contraction could influence blood flow within the capillary bed has re-emerged in discussion of the capillary no-reflow following ischemic insult in both the brain [37] and the heart [85]. Pericytes are also thought to play a role in regulation of barrier properties of intact exchange microvessels. Cross sections of fixed capillaries and venules often show pericyte foot processes in close proximity to the endothelial junctions of vessels that display a transient loss of barrier function in response to agents such as histamine and display massive endothelial gaps at times when permeability has returned back to normal levels [125]. The implication being that acute loss of barrier function is limited by the presence of the pericytes. To date, though, no studies have focused on sex differences in pericyte function or dysfunction.

Within the microvascular network are several orders of capillaries. Capillaries branching off of the terminal arterioles, the arterial capillaries, have blood flow diverging at the segment entrance and exit. True capillaries come off of the arterial capillaries, with flow diverging at the entrance and converging with flow in the next segment, the venular capillaries [98]. Mammalian capillary diameter across the capillary network ranges from 3 to 5 μm, and the EC thickness in areas away from the nucleus ranges from 0.1 to 0.3 μm facilitating the exchange of fluid and solutes. Exchange is driven by gradients in pressure driving fluid and gradients in concentration driving solute across the wall [105]. Additional factors favoring exchange are the high density/large number of capillary segments providing small distances between exchange elements and a huge surface area, the thin attenuated EC structure that provides a large area of lipid facilitating exchange of hydrophobic materials, and thin junctions favoring diffusion of hydrophilic solutes. The membranes of capillary EC also contain water channels, the aquaporins, through which water can flow exclusively, and vesicular structures, caveoli, that contain receptor and enzyme complexes that facilitate localization of interactions in the face of a flowing medium. In addition, the surface of the EC also possesses an extensive, charged carbohydrate structure, the glycocalyx, that also serves as a relatively unstirred region near the surface that also facilitates ligand interaction in a changing environment. The glycocalyx can act as a sensor of fluid flow and offers significant resistance to solute and fluid transport [44–46, 48, 55], and its structure/composition can be altered by disease [79, 102].

Venules

Flow, out of the capillary network through venular capillaries, converges into the venules [98]. These vessels, of larger diameter than their terminal arteriolar cousins, are generally not round but elliptical in cross section reflecting the sparse coverage by VSM. Pericytes are still found in the venular capillary segments that flow into the venules, but their coverage falls off with successive branches. These vessels coalesce into collecting venules (30–50 μm) and then into the muscular venules (50–100 μm). The venular portion of the microvascular network is involved with control of volume and modulation of inflammatory and immune responses [105]. The glycocalyx is also present on the venular endothelial cells; in addition to participating as a resistor to solute and fluid flux, the structure limits access to endothelial ligands mediating not only receptor access but also white cell interactions [70].

Lymphatics

The lymphatics, the "forgotten" portion of the vasculature, were first described by Hippocrates (460–377 B.C.) and then illustrated by the Greek physicians Herophilus (335–280 B.C.) and Erasistratus (304–250 B.C.) [12]. These vessels and the fluid they produced were again forgotten until the Italian anatomist Gasparo Asellius published his work (Gaspare [27]) almost a millennium after the systemic vasculature was described. While there is a more recent study of collecting and capillary lymphatic (the lymphatic vessels of sizes and distributions similar to the systemic microvessels) function, consideration of sex has received only passing interest. Relative to what has been shown with respect to sexual dimorphism in capillary function, it would be surprising if there were no sex-related differences in lymphatic function that are central to maintenance of fluid and solute homeostasis in healthy males and females.

The Common Constituent Separating "Inside from Out": The Endothelium

The entire network of microvessels is covered by a monolayer of EC. EC are involved in not only control of barrier function but also participate significantly in immune responses, inflammation, thrombosis, vascular remodeling, and regulation of arteriolar vessel tone. The EC are incredibly heterogeneous, likely reflecting these spectra of functions. The degree of heterogeneity reflects not only where in the vascular system (both organ and anatomical location) [96] the EC are located but also where within the microvascular network the EC reside [2, 117] and the sex of host [5, 54, 123]. One reason considering EC heterogeneity is important is that sampling a limited number of EC in vivo or in culture may reflect the behaviors of selected populations giving rise to a false or skewed understanding of the etiology of disease or approaches for ameliorating the disease state. In turn, EC dysfunction is a feature common to multiple chronic diseases including diabetes mellitus, heart failure, and stroke. This same constellation of diseases displays male/female differences with respect to incidence, onset, severity, morbidity, and mortality [72, 100, 120].

Sexual Dimorphisms in EC Phenotype that Influences Function

EC in culture are commonly referred to as taking on a "cobblestone" morphology. In fact, EC shape is a function of the composition of the substrata on or in which they are grown [73, 101]. On Matrigel™, EC will form tubelike structures; this characteristic is used as one verification of EC [6] to distinguish them from VSM, pericytes, white blood cells, or fibroblasts. EC plated on gelatin or collagen spread and, depending on vascular bed of origin, can actually take on multiple shapes. Recently it has been found that microvascular EC from skeletal muscle display sexual dimorphism with respect to shape in vitro when plated on gelatin [50]. Prior to plating, no sex difference existed in size – for both sexes the microvascular SKM EC were 16 μm in diameter (~2000 μm^3 volume); on plating EC from females spread further than SKM EC from males. While apparently not a large difference (>10%), the predictions are that the area of water-filled perimeter per capillary segment will be greater in females suggesting that hydrophilic solute flux would be greater in females than males. In another example, we find, on a per cell basis, the amount of lactate generated by microvascular EC in culture does not differ by sex [50]. The "but" is that if the differences in EC size are taken into account, a greater number of EC are required to cover a skeletal muscle capillary for a male than a female. The back-of-the-envelope calculation predicts a 50% greater lactate production per unit volume in equivalent-sized vessel segments of males relative to females. This prediction needs to be tested in vivo for both lactate and other EC-generated hormones and mediators.

Sexual Dimorphism in Arterioles

For a variety of historical reasons, most studies of microvascular arteriolar function have been conducted on animals of one sex or the other. Many studies of in vivo blood flow control, for example, have used the cremaster muscle in males because the tissue is thin facilitating study in the living tissue by light microscopy. Many small animal studies of exercise and low gravity are performed on females because females tend to use the exercise wheels spontaneously, and the hind limb-unweighted female does not suffer involution of the testis on unloading. In human studies of exercise, the subjects were often young males in a medical school or the military. Similarly, males were used in many studies because of the fear that reproductive hormone cycling would influence outcomes. The concern is that our perceptions of how the systems work are erroneous because a male is not a female and vice versa.

Architecture and Morphology (Rarefaction)

The microvasculature, particularly the arteriolar elements, changes with respect to structure and reactivity with sustained hypertension. The conclusion from the majority of studies is that the adaptation is a successive process of first increasing arteriolar tone to protect the downstream vessels from increases in capillary pressure, followed by structural remodeling of the arteriolar wall replacing VSM with matrix to maintain a fixed reduction in diameter, and finally vessel pruning (rarefaction). While female humans and mammals can develop hypertension, it occurs later in life and to a lesser degree [29, 95, 127]. In many animal models, it is difficult to make females hypertensive [26, 40, 119]. In one study, male and female rats were rendered hypertensive by removing a kidney and placing them on a high-salt diet [90]. The males developed hypertension well in advance of the females – by day 3 on the high-salt diet, mean arterial pressure (MAP) in the 85-day-old males was 135 ± 6 mmHg, whereas it was only 103 ± 3 mmHg in the females. At that point the skeletal muscle microvasculature was examined for changes in morphology. In the white *gastrocnemius* muscle, there was evidence of microvessel loss in the males but not the females. In fact, even following 4 weeks of the high-salt diet, the females remained free of evidence of microvascular remodeling and normotensive at 105 ± 5 mmHg, whereas MAP had risen to 160 ± 8 in their male counterparts where rarefaction was now discernable in both the white and red *gastrocnemius* as well as the *soleus* and *plantaris* muscles [90].

Vascular Reactivity (VSM Dilation/Constriction)

Arteriolar function is of particular importance because the state of vascular tone (constriction relative to dilatation) regulates blood flow distribution as well as capillary hydrostatic pressure. Sex plays a role in pressure-induced myogenic constriction of arterioles. The myogenic response in arterioles from female rats, for example, is smaller than that of males. It was suggested an increased release of NO and/or elevated eNOS activity related to the higher levels of estrogen in the female animals could account for the reduction in myogenic response [42, 126].

Adaptation to an environmental change also has been shown to differ by sex. In response to chronic hypoxia (CH) at the whole animal level equivalent, sex-independent changes in total peripheral resistance were observed, and in both sexes pressor responses were similarly reduced following CH. At the level of the mesenteric microvasculature, though, arteriolar reactivity, while reduced in males, remained unchanged by

CH in females; eNOS protein expression in these vessels was unchanged although EC calcium was elevated in CH females compared to controls [34]. This example illustrates that because one parameter, in this case total peripheral resistance, is sex-independent, it is incorrect to assume that all parameters are likewise independent of sex.

Only recently it has become accepted that coronary heart disease (CHD) and its manifestations differ by sex. Heart disease generally shows up at an earlier age in men than women [35]. In human females, CHD is a microvascular disease which manifests by increased arteriolar constriction and vasospasm not generally found in males [81, 91, 108]. In contrast, CHD in males is a macrovascular disease characterized by the presence of coronary occlusion and deposition of plaque [23]. Males adapt to CHD by growing coronary collateral vessels that bypass areas of occlusion. In contrast, collateral vessel formation in females is rare [61, 81]. The small size and large numbers of microvessels make them difficult to image and confirm coronary microvascular dysfunction in women especially using approaches developed for detecting CHD in males [91, 92]. Sex-sensitive arteriolar vascular reactivity is not limited to the heart. In the gut, responses to ischemia/reperfusion injury differ by sex. In male mice, 30 min of ischemia followed by 90 min of reperfusion was characterized by a loss of intestinal epithelial barrier integrity that paralleled increased endothelial/leukocyte interactions and reduction in blood flow resulting from a reduction in flow rate and the number of capillaries perfused. In females subjected to the same treatment, while loss of epithelial barrier function occurred, it was later in time and accompanied by similar changes in inflammatory response and degree of organ perfusion [116].

Onset, frequency, and pathophysiology of cardiovascular disease (CVD) outside of the heart involving the microcirculatory system are also influenced by sex [14, 25]. In the cerebrovasculature, sex differences have been described with respect to vascular anatomy [21] and also pharmacology [60]. In the brain, arteriolar responses to vasoactive compounds, including angiotensin [19, 114] and endothelin-1 [29, 71], and nitric oxide synthase activity [88, 115], are sexually dimorphic.

The incidence of stroke is higher in women than men, especially later in life [14, 21, 61, 99]. As with coronary microvessels, age and sex are significant, independent variables influencing cerebrovascular function. Recent work [1, 69] has shown cerebral blood flow to be higher in young women than men, a difference not found in older adults. Similarly, in younger women, the response to hypercapnia was greater than in age-matched males; following menopause this sex difference was no longer present [59]. An important consideration pointed out by Barnes [7] is that when sex differences in MAP and vascular architecture are considered along with the differences in responses to hypercapnia, cerebral blood flow in young women is actually lower than age-matched men. This consideration of multiple variables controlled by microvascular cerebral vessel function is especially important as a reduced response to hypercapnia is associated with elevated risk of stroke and increased cognitive decline, two identified sexually dimorphic risk factors [78].

In addition to the anatomical differences, sexual dimorphism exists with respect to the mediators of microvascular tone and pharmacological responses to vasoactive drugs verified in clinical studies using male subjects [86]. The potent vasoactive peptide endothelin (ET-1) is interesting as plasma levels of ET-1 differ by sex [3, 32, 33], as do the distribution, expression [33, 66], and activation of ET receptors and as do the mediators of the ET system. In females the ratio of ET-1 to ETB receptor activation is primary; in contrast, in males, ET-1/ETA receptor activation is of greatest importance [29]. The implication is that these differences contribute to the overall observation of lower blood pressure of females relative to males reflecting in part the predominance of ETA receptors on vascular smooth muscle that enhances constrictor action of ET-1 relative to ETB receptors predominant on EC-mediating vasodilator actions of ET-1. In rat coronary arterioles, age, as well as sex, influences the functional response. As females age, coronary arteriolar constriction to ET-1 increases, while in males with aging reduction in the constrictor response to ET-1 is observed [66]. In retinal arterioles, ET-1 sensitivity declines with age, especially in females, while expression levels of the two receptors, ETA and ETB, displayed no differences with either age or sex [71]. It was determined using a pharmacological approach that the age-related reduction in ET-1 response in males was mediated by ETA signaling pathways, while in females it was the ETB signaling pathways that mediated the attenuated ET-1 contractility with age.

A component that has been studied more intensely with respect to sex differences in blood pressure control is the autonomic nervous system (review, [57]). These studies are germane to microvascular, particularly arteriolar, function, given that sympathetic tone (including muscle sympathetic nerve activity) influences VSM contractile state via alpha-adrenergic receptor activation to increase peripheral vascular resistance. Elevated resistance, of course, leads to increased systemic blood pressure upstream of the arterioles and reduction of blood flow and hydrostatic pressure in the exchange microvessels downstream from the arterioles. A recent review by Barnes covers this subject in greater detail [7]. The highlights of the review are that alpha-adrenergic vasoconstriction is lower in young females than males [24], there are differences in beta-adrenergic receptors [39, 63], and while female sex hormones contribute to a reduction in tonic autonomic nervous support, the autonomic nervous system becomes a greater controller of blood pressure in females postmenopause.

Microvascular exchange tends to be equated with the capillary and venular elements of the network; in certain organs, like the heart, flux across arterioles appears to also contribute to net whole organ clearance. In the heart, arterioles are positioned anatomically in areas removed from the capillaries [58] and to be relatively leaky (protein reflection coefficient, σ, circa ~0.6 [16] compared to $\sigma > 0.8$ in skeletal muscle [124]). In porcine coronary arterioles, while basal arteriolar permeability to albumin did not differ by sex, the permeability response to adenosine with endurance exercise was greater in females than males [53, 123]. While skeletal muscle arterioles are tighter than capillaries and venules (e.g., have lower permeability), their EC layer requires a finite permeability to solutes to support VSM function. Arterioles from rat skeletal muscle demonstrated sex differences in the magnitude and direction of permeability responses to adenosine [121]. Differences in adenosine receptor isoform distribution, while demonstrating that A2b mediates EC permeability, were not found. Only when adenosine receptor signaling pathways were examined did it become evident that expression levels of the phosphodiesterase isoform 3 (PDE3) differed by sex [122].

Sexual Dimorphism in Capillary Function

Sex differences with respect to the primary function of capillary endothelial cells, fluid and solute exchange, have been demonstrated at multiple levels in a limited number of studies [15, 53, 54, 74, 89, 104, 112, 116]. In each it becomes evident that the differences can reflect the driving forces for exchange, EC morphology, and cell signaling, as well as EC responses to a myriad of mediators. While we often focus on the pathophysiology of loss of barrier function (e.g., increases in permeability with inflammation or trauma), interventions that "tighten" the barrier can also create pathology by limiting the passage of necessary materials (as appears to happen with hyperinsulinemia [104]) either out of or into the circulation. A well-functioning barrier exists in a state analogous to arteriolar tone – wherein permeability can increase or decrease in response to changes in tissue demand and environment [43, 47].

The amount of fluid or solute moved across the endothelial barrier is a function of the architecture of the pathways through and between the endothelial cells, the surface area for exchange, and the physicochemical properties of the barrier media (including the glycocalyx at the blood/EC interface and the extracellular matrix at the luminal and abluminal surfaces, respectively) and of the moving material (where size, shape, and charge can be influential). In a study of acute coronary syndrome, males were found to shed more syndecan-1 (a marker of glycocalyx damage) than females [79]. These data imply either a greater amount of sydecan-1, a denser glycocalyx, or higher protease activity in EC of males.

In the presence of natriuretic peptides (ANP, BNP, or CNP), components of the glycocalyx, including syndican-1, have been shown to be released into coronary effluent of isolated, perfused guinea pig hearts [56]. This result is interesting as (a) the atrial peptides can reversibly increase microvascular permeability [45, 46, 77] and hydraulic conductivity [47, 49, 51] and (b) the "normal" levels of ANP and BNP are higher in healthy women than men (Table 20.1). Further, preliminary studies from the same group find that syndican-1 and hyaluronan levels in the blood of healthy premenopausal women vary with the menstrual cycle, that syndican-1 levels of males are higher than cycling and postmenopausal females, and that the levels of syndican-1 have no temporal component in either males or postmenopausal females (personal communication). These results are intriguing because the changes in glycocalyx thickness with shedding could account for the well-known increases in premenstrual edema in females and be of importance in the changes in fluid balance observed during pregnancy.

In addition, especially for water and hydrophilic solutes the size of peptides, hormones, and proteins, the net transmural pressure gradient is of importance. Two of these factors appear to differ in males and females: the concentration of plasma protein (higher in males than females [52]) which sets the oncotic pressure gradient (1–3 mmHg greater in males than females [52]) drawing fluid out of the tissue toward the lumen and the hydrostatic pressure gradient largely a function of capillary blood pressure. Not only is arterial blood pressure 6–10 mmHg lower in females than males prior to menopause [95], but, from a more limited study, capillary pressures are 2 mmHg lower in females than males [111]. The net driving force for fluid and solute, the difference in these two values, appears to favor filtration in females over males; this conclusion is another thing that needs to be evaluated in vivo. A summary of published hemodynamic data for healthy males and females is presented in Table 20.2.

In a study of seizure activity in rats, it was found that nitric oxide synthase blockade with L-NAME induced an increase in blood-brain barrier (BBB) permeability (Evans blue dye extravasation), in females, not in males [15]. Surgical cessation of ovarian function in adult female swine also results in loss of meningeal microvascular barrier to fluorescently labeled protein [30]; replacement of estrogen in a pulsed dose (estrogen patches) restored barrier function (and changes in vessel architecture), whereas flat dose estrogen replacement was not effective [31]. Thus, not only can a hormone or vasoactive compounds be important in regulating microvascular function, but the rate of change (a spike) can be of importance. This appears to be also the case for vascular endothelial growth factor (VEGF) [8, 10], a potent mediator of permeability. Plasma levels of VEGF are higher in men than women and are correlated with the development of atherosclerosis [62], a disease process shown to occur more frequently in males.

At the level of the whole human, fluid balance can differ by sex. It is well known that the state of pregnancy results in profound changes in fluid distributions with resulting changes in hemodynamics [18, 80]. One supposition is that because females must possess the mechanisms to reversibly alter fluid distributions to accommodate pregnancy, the multiple factors regulating fluid and solute flux will differ between males and females.

Venules

The majority of vascular volume resides in the venous component of the microvasculature, and fluid exchange occurs predominantly across the capillaries. In the environment of microgravity, sex differences in volume regulation become apparent. Female astronauts during space flight experience a greater shift of fluid out of the plasma space than men and following return to

Table 20.1 Clinical chemistry/hematology reference values for adult healthy humans

Hormone	Units	Change with age	Male		Female						
				Peak a.m.		Follicular	Ovulatory/ mid-cycle	Luteal	Postmenopause	Statistics	Ref
Testosterone *T*	ng/dL	M ↓	350–1000		9–55				5–32	Range	1
		F ↓	618(488–786)*		24.6 (17.7–34.3)					M(75–25)	10
					18.3 ± 1.2	15.4 ± 1.6	22.7 ± 1.7 +	18.7 ± 1.5	10.5 ± 1.3 +	X ± SE	16
			750 ± 380*		10 ± 4					X ± SD	9
				571 ± 29						X ± SE	11
Free-T	pg/mL	M ↓	32–168*		<3					Range	1
		F ↓	119 (95–152*)		2.2 (1.4–3.2)	2.2				M(75–25)	10
					1.8 ± 0.15	1.5 ± 0.13	2.0 ± 0.17 +	1.6 ± 0.21	1.2 ± 0.1	X ± SD	16
				78 ± 8						X ± SE	11
Sex hormone binding globulin	nmol/L	M↑	10–80*		20–130					Range	1
		F ↓	37.0 (0.78–1.66) *		89.7 (58.7–132.7)					M(75–25)	10
					85.4 ± 6.0	85.3 ± 9.2	99.0 ± 8.8	103 ± 10.7+	67.7 ± 8.4+	X ± SD	16
				23.1 ± 2.3						X ± SE	11
Estrogen E2	pg/mL	M NC	11–43			12–233	41–398	22–341	<5–138	Range	1
		F ↓				31.8 ± 19.8			22.0 ± 14.0	X ± SE	21
						48 (37)	96 (111.0)+	122 (91.5)+		M(75–25)	12
					55.4 ± 10.3	68.1 ± 18.6	98.1 ± 19.0 +	106.8 ± 8.3+	1.3 ± 0.3+	X ± SE	16
						56.1 ± 6.4		132.2 ± 13.0		X ± SE	4
		Day				51 ± 10		86 ± 15+		X ± SE	5
		Night				50 ± 16		60 ± 8 ^			
			16 ± 14.4*		76.9 ± 71.0					X ± SD	9
				16 ± 1.7						X ± SE	11
Progesterone P4	ng/mL	M NC	<0.2–0.8			0.2–1.5	0.8–3.0	1.7–27.0	<0.2–0.8	Range	1
		F ↓				0.76 ± 0.59			0.67 ± 0.57	X ± SE	21

(continued)

Table 20.1 (continued)

Hormone	Units	Change with age	Male		Female					Statistics	Ref
				Peak a.m.		Follicular	Ovulatory/ mid-cycle	Luteal	Postmenopause		
						0.4 (0.3)	1.7 (3.0)	8.9 (7.4)[+]		M(75–25)	12
						0.22 ± 0.02		8.13 ± 1.55		X ± SE	4
		Day				0.3 ± 0.0		8.8 ± 1.9		X ± SE	5
		Night				0.2 ± 0.0		7.7 ± 1.5			
Follicle Stimulating Hormone FSH	mIU/mL	M↑	1.5–12.4			3.5–12.5	4.7–21.5	1.7–7.7	25.8–134.8	Range	1
		F↑				6.4 (2.5)	6.3 (5.3)	3.1 (2.5)		M(75–25)	12
					7.6 ± 0.88	6.5 ± 0.58	6.9 ± 0.7	3.0 ± 0.45	81.0 ± 6.9	X ± SE	16
				5.6 ± 0.9						X ± SE	11
Luteinizing Hormone LH	mIU/mL	M↑	1.7–8.6			2.4–12.6	14–95.6	1–11.4	7.7–58.5	Range	1
		F↑				4.7 (3.0)	8.5 (10.9)	4.3 (4.6)		M(75–25)	12
	IU/L				5.4 ± 0.35	5.1 ± 0.5	11.2 ± 2.8	3.8 ± 0.56	37.2 ± 3.7	X ± SE	16
				11.7 ± 0.9						X ± SE	11
Aldosterone	ng/dL					3.1 ± 0.3		8.5 ± 1.0		X ± SE	4
		Day	9.4 ± 2.3			8.1 ± 1.5		18.9 ± 2.8		X ± SE	5
		Night	7.8 ± 1.0			6.1 ± 1.5		11.2 ± 2.6			
			9 (7–13)		10 (7–15)					M(75–25)	23
Cortisol	ug/dL	M NC F?		16.6 ± 1.6						X ± SE	11
		Day	7.7 ± 1.2			8.4 ± 1.4		9.9 ± 0.9		X ± SE	5
		Night	3.5 ± 0.7			5.1 ± 0.9		4.1 ± 0.7			
Renin Activity	ng/(mL hr)					0.36 ± 0.05		1.21 ± 0.18		X ± SE	4
		Day	1.2 ± 0.2			1.1 ± 0.3		1.4 ± 0.2+		X ± SE	5
		Night	1.7 ± 0.4+			0.8 ± 0.1		2.0 ± 0.4+^			
NT-proBNP	ng/L	M↑	0–95		0–178					range	1
		F↑	16.2 (8.1–28.8)		42.9 (25.7–72.2)					M(75–25)	10

Brain Natriuretic Peptide, BNP	pg/mL	M↑	8.0 ± 12.8		13.9 ± 18.9					X ± SE	20
		F↑	5.4 (4.0, 13.4)		9.7 (4.0, 18.9)					M(75–25)	23
			7.7 ± 7.0		12.2 ± 10.2					X ± SE	6
NT-proANP	pmol/L	M↑	284 ± 157		379 ± 177					X ± SE	20
		F↑	273 (188–392)		343 (248–476)					M(75–25)	23
ANP	pg/ml	M↑				53.0 ± 5.8		35.2 ± 4.6		X ± SE	4
		F↑	16.7 ± 10.0		18.8 ± 11.7					X ± SE	6
Insulin	pmol/L	M ↓	59.3 ± 58.3		53.4 ± 24.9					X ± SD	9
		F↑	60.0 ± 53.5		69.8 ± 54.4					X ± SD	15
	mU/L		12.5		10.1				13.9	Median	22
HOMA-IR			1.01 (0.78–1.66)		0.89 (0.65–1.31)					M(75–25)	10
			1.8 ± 2.0		1.5 ± 0.7					X ± SD	9
			1.35 ± 1.31		1.49 ± 1.27					X ± SD	15
Fasting Glucose	mg/dL	M NC				80 ± 2			89 ± 3	X ± SE	21
		F↑	100 ± 8.2		95.5 ± 7.5					X ± SD	14
			105.6 ± 25.1		99.6 ± 26.0					X ± SD	23
	mmol/L		4.6 ± 0.4		4.4 ± 0.5					X ± SD	9
			5.98 ± 0.46		4.76 ± 0.45					X ± SD	15
			5.42		5.12				5.33	Median	22
Endothelin-1 ET-1	pg/mL	M↑				1.39 ± 0.41			1.74 ± 0.42	X ± SE	21
		F↑	1.4 ± 0.07*		1.0 ± 0.08					X ± SE	19
Leptin	ng/mL		4.5 ± 4.0		15.3 ± 8.2					X ± SD	9
Total Cholesterol	mg/dL	M NC				163 ± 6			216 ± 7	X ± SE	21
		F↑				166 (39.0)	162.0 (64.0)	161.0 (36.0)		M(75–25)	12
			200 ± 35		211 ± 38					X ± SD	23
	mmol/L		5.42 ± 1.13		5.43 ± 0.92					X ± SD	15
			5.54		5.38				6.12	Median	22
HDL	mg/dL	M ↓				67 ± 3			85 ± 4	X ± SE	22
		F↑				50.0 (17.0)	52.0 (16.0)	51.0 (16.0)		M(75–25)	12
			56.8 ± 12.6		72.4 ± 16.6					X ± SD	14

(continued)

Table 20.1 (continued)

Hormone	Units	Change with age	Male		Female					Statistics	Ref
				Peak a.m.		Follicular	Ovulatory/ mid-cycle	Luteal	Postmenopause		
			44 ± 12		58 ± 16					X ± SD	23
	mmol/L		1.22		1.5				1.44	Median	22
LDL	mg/dL					82 ± 6			117 ± 6	X ± SE	21
		F NC				102.0 (33.0)	98.5 (32.0)	97.0 (34)		M(75–25)	12
Triglycerides	mmol/L	M ↓				79 ± 9	53.0		73 ± 7	X ± SE	21
		F NC, ↑				56.0 (33)	(27.0)	51.0 (28.0)		M(75–25)	12
			1.31 ± 0.74		1.01 ± 0.53					X ± SD	15
			1.58		1.06				1.45	Median	22
	mg/dL		0.9 ± 0.13		0.70 ± 0.11					M(75–25)	10
Creatinine			1.25 ± 17		1.07 ± 0.15					X ± SE	20
ICAM-1	ng/mL	M↑				214.8 ± 9.2		215.7 ± 6.0		X ± SE	8
		F NC	264 ± 63		249 ± 60					X ± SD	15
			285 ± 92		309 ± 111					X ± SD	2
E-selectin	ng/mL					44.6 ± 5.1		41.4 ± 4.8		X ± SE	8
			56.8 ± 28.2		46.2 ± 23.7					X ± SD	15
			56 ± 20		50 ± 20					X ± SD	2
L-selectin	ng/mL		1107 ± 471		1140 ± 485					X ± SD	15
P-selectin	ng/mL		143 ± 44		123 ± 34					X ± SD	15
			205 (40–2550)		206 (22–5700)					M(75–25)	2
Thrombomodulin	ng/mL		48 ± 16		40 ± 12					X ± SD	2
Total Homocysteine	μmol/L		11.3(11.2–11.5)*		9.6(9.5–9.69)					X ± 95% CI	13
			9.72 (8.24–11.7) *		8.29 (6.96–10.1)					M(75–25)	23
CRP	mg/L	M↑	1 ± 1.5*		0.7 ± 1.2					X ± SD	14
		F↑	1.79 (0.84–3.50) *		2.25 (0.93–5.39)					M(75–25)	23

HCT	%		40–52*	35–47			Range	1
			40.3 ± 1.0*	39.5 ± 0.7			X ± SE	4
		Day	43 ± 1*		37 ± 1	39 ± 0	X ± SE	5
		Night	42 ± 1*		36 ± 0	37 ± 1		
Hemoglobin	g/dL		13.2–17.7	11.9–15.5			Range	1
			15.2 ± 1.0*	12.8 ± 0.6			X ± SD	3
RBC count	10^{12}/L		5.01 ± 0.31*	4.44 ± 0.33			X ± SD	15
Platelet Count	10^9/L	M ↓	239 ± 51.0*	249.3 ± 54.3			X ± SD	15
		F ↓	249 (244–254)*	272 (268–276)			X ± 95% CI	18
			235 ± 59*	261 ± 64			X ± SD	17
Leukocyte count	10^9/L		6.69 ± 1.66*	6.91 ± 1.48			X ± SD	15
Osmolarity	mOsm/ kgH_2O				287.4 ± 0.9	282.3 ± 0.9 +	X ± SE	4
			288 ± 1		288	284 ± 1 +	X ± SD	3

*M≠F; + differs from follicular; ^day≠night; # ≠postmenopause; $P < 0.05$ or better (refer to specific papers); *NC* no change, *HOMA* homeostasis model assessment of insulin resistance

Mean ± standard deviation (X ± SD); mean ± standard error of the mean X ± SE; mean ± 95% confidence interval X ± 95%CI; median (75%, 25%) M (75–25)

References

1. University of Iowa
2. Blann AD, Daly RJ, Amiral J. The influence of age, gender and ABO blood group on soluble endothelial cell markers and adhesion molecules. Br J Haematol. 1996;92:498–500
3. Bouwsema MM, Tedjasaputra V, Stickland MK. Are there sex differences in the capillary blood volume and diffusing capacity response to exercise ? J Appl Physiol. 2017;122:460–9
4. Chapman AB, Zamudio S, Woodmansee W, Merouani A, Osorio F, Johnson A, Moore LG, Dahms T, Coffin C, Abraham WT, Schrier RW. Systemic and renal hemodynamic changes in the luteal phase of the menstrual cycle mimic early pregnancy. Am J Physiol Ren Physiol. 1997;273(30):F777–82
5. Claybaugh JR, Sato AK, Crosswhite LK, Hassell LH. Effects of time of day, gender, and menstrual cycle phase on the human response to a water load. Am J Phys Regul Integr Comp Phys. 2000;279:966–73
6. Clerico A, Del Ry S, Maffei S, Prontera C, Emdin M, Giannessi D, Del Ry S, Maffei S, Prontera C, Emdin M, Giannessi D. The circulating levels of cardiac natriuretic hormones in healthy adults: effects of age and sex. Clin Chem Lab Med Walter de Gruyter. 2002;40(4):371–7
Conlan MG, Folsom AR, Finch A, Davis CE, Sorlie P, Marcucci G, Wu KK. Associations of factor VIII and von Willebrand factor with age, race, sex, and risk factors for atherosclerosis. The Atherosclerosis Risk in Communities (ARIC) Study. Thromb Haemost. 1993;70(3):380–5
8. Elhadd TA, Neary R, Abdu TAM, Kennedy G, Hill A, McLaren M, Akber M, Belch JJ, Clayton RN. Influence of the hormonal changes during the normal menstrual cycle in healthy young women on soluble adhesion molecules, plasma homocysteine, free radical markers and lipoprotein fractions. Int Angiol. 2003;22:222–8
9. Guerra B, Fuentes T, Delgado-Guerra S, Guadalupe-Grau A, Olmedillas H, Santana A, Ponce-Gonzalez JG, Dorado C, Calbet JAL. Gender Dimorphism in Skeletal Muscle Leptin Receptors, Serum Leptin and Insulin Sensitivity. PLoS ONE. Edited by A. Lucia. Public Library of Science, 2008;3(10):e3466
10. Lam CSP, Cheng S, Choong K, Larson MG, Murabito JM, Newton-Cheh C, Bhasin S, McCabe EL, Miller KK, Redfield MM, Vasan RS, Coviello AD, Wang TJ. Influence of Sex and Hormone Status on Circulating Natriuretic Peptides. J Am Coll Cardiol. 2011;58(6):618–26

(continued)

Table 20.1 (continued)

11. Montanini V, Simoni M, Chiossi G, Baraghini GF, Velardo A, Baraldi E, Marrama P. Age-related changes in plasma dehydroepiandrosterone sulphate, cortisol, testosterone and free testosterone circadian rhythms in adult men. Horm Res. 1988;29:1–6.
12. Mumford SL, Schisterman EF, Siega-Riz AM, Browne RW, Gaskins AJ, Trevisan M, Steiner AZ, Daniels JL, Zhang C, Perkins NJ, Wactawski-Wende J. A longitudinal study of serum lipoproteins in relation to endogenous reproductive hormones during the menstrual cycle: Findings from the BioCycle study. J Clin Endocrinol Metab. 2010;95(9):80–5
13. Nygard O, Vollset SE, Refsum H, Stensvold I, Nordrehaug JE, Ueland M, Kvale G. Total plasma homocysteine and cardiovascular risk profile. The Hordland homocysteine study. JAMA. 1995;274(April):1526–33
14. Oda E, Abe M, Kato K, Watanabe K, Veeraveedu PT, Aizawa Y. Gender differences in correlations among cardiovascular risk factors. Gend Med 2006;3(3):196–205.
15. Ponthieux, A., Herbeth, B., Droesch, S., Lambert, D., Visvikis, S., Molecule-, I. A., Ponthieux, A., Herbeth, B., Droe-, S., Lambert, D., Visvikis, S. and National, I. Age- and sex-related reference values for serum adhesion molecule concentrations in healthy individuals: intercellular adhesion molecule-1 and E-, P-, and L- Selectin Clin Chem. 2003;49 (9):1996–8
16. Rothman MS, Carlson NE, Xu M, Wang C, Swerdloff R, Lee P, Goh VHH, Ridgway EC, Wierman ME. Reexamination of Testosterone, Dihydrotestosterone, Estradiol and Estrone Levels across the Menstrual Cycle and in Postmenopausal Women Measured by Liquid Chromatography Tandem Mass Spectrometry. Steroids. 2012;76(720):177–82
17. Santimone I, Di Castelnuovo A, De Curtis A, Spinelli M, Cugino D, Zito F, Donati MB, Cerletti C, De Gaetano G, Iacoviello L, Cuore-sano A, Paolo G. White blood cell count, sex and age are major determinants of heterogeneity of platelet indices in an adult general population: results from the MOLI-SANI project. Haematologica. 2011;96(8):1180–8
18. Segal JB, Moliterno AR. Platelet counts differ by sex, ethnicity, and age in the United States. Ann Epidemiol. 2006;16(2):123–30
19. Takashi M, Yanagisawa M, Iida K, Ajisaka R, Suzuki N, Fujino M, Goto K, Masaki T, Sugishita Y. Age- and sex-related variation of plasma endothelin-1 concentration in normal and hypertensive subjects. Am Heart J. 1992;123(4):1092–3
20. Wang TJ, Larson MG, Levy D, Leip EP, Benjamin EJ, Wilson PW, Sutherland P, Omland T, Vasan RS. Impact of age and sex on plasma natriuretic peptide levels in healthy adults. Am J Cardiol. 2002;90(3):254–8
21. Wenner MM, Sebzda KN, Kuczmarski AV, Pohlig RT, Edwards DG. ETB receptor contribution to vascular dysfunction in postmenopausal women. Am J Phys Regul Integr Comp Phys. 2017;313(1):R51–7
22. Willeit J. The role of insulin in age-related sex differences of cardiovascular risk profile and morbidity. Atherosclerosis. 1997;130(1–2):183–9
23. Xanthakis V, Enserro DM, Murabito JM, Polak JF, Wollert KC, Januzzi JL, Wang TJ, Tofler G, Vasan RS. Ideal cardiovascular health associations with biomarkers and subclinical disease and impact on incidence of cardiovascular disease in the framingham offspring study. Circulation. 2014;130(19):1676–83

Table 20.2 Hemodynamic reference values for adult healthy humans at rest

Parameter	Units	Change with age	Male	Female				Statistics	Ref
					Follicular	Luteal	Menopause		
HR	bpm	M↓			66 ± 9	65 ± 9		X ± SE	7
		F↓			60 ± 2		57 ± 2	X ± SE	7
					62 ± 2	65 ± 2		X ± SE	2
			67.2 ± 16.6	71.0 ± 16.6				X ± SD	9
SV	mL		94 ± 17	97 ± 16				X ± SD	1
CO	L/min		6.9 ± 0.8	7.2 ± 1.1				X ± SD	1
					5.6 ± 0.4	6.6 ± 0.4+		X ± SE	2
Systolic BP	mmHg	M↑	121 ± 13	113 ± 14				X ± SD	3
		F↑	118 ± 10	115 ± 12				X ± SD	6
				109 ± 3			123 ± 3#	X ± SE	7
			121 ± 8.7*+		113.5 ± 8.7			X ± SE	5
			136.5 ± 12.5[1]				125 ± 12.3 [1#]		
			130 ± 17	126 ± 20				X ± SD	8
Diastolic BP	mmHg	M↑	78 ± 9	73 ± 9				X ± SD	3
		F↑	73 ± 7	70 ± 7				X ± SD	6
					67 ± 2		80 ± 2+	X ± SE	7
			70.6 ± 10.0*	66.3 ± 7.6				X ± SE	5
			81.1 ± 5.9[1]*				73.0 ± 5.9[1]		
			78 ± 9	74 ± 9				X ± SD	8
MAP	mmHg	M↑	92.5 ± 6.7	88.6 ± 7.2	81 ± 2		94 ± 2+	X ± SE	7,10
		F↑			81.7 ± 2.0	75.4 ± 2.3		X ± SE	2
Capillary Pressure	mmHg	M NSD	18.0 ± 2.5*	15 ± 2.4				X ± SE	5
		F↑	18.4 ± 2.0[1]				17.6 ± 3.4[1]		

*M≠F, 20–50 years; +differs from follicular; # ≠ postmenopause (>55 years); $P < 0.05$ or better (refer to specific papers);[1] >50 years

Mean ± standard deviation (X ± SD); mean ± standard error of the mean X ± SE; *NSD* not significantly different, *HR* heart rate, *BP* (arterial) blood pressure, *MAP* mean arterial pressure, *bpm* beats per minute, *SV* stroke volume, *CO* cardiac output

References

1. Bouwsema MM, Tedjasaputra V, Stickland MK. Are there sex differences in the capillary blood volume and diffusing capacity response to exercise ? J Appl Physiol. 2017;122:460–9

2. Chapman AB, Zamudio S, Woodmansee W, Merouani A, Osorio F, Johnson A, Moore LG, Dahms T, Coffin C, Abraham WT, Schrier RW. Systemic and renal hemodynamic changes in the luteal phase of the menstrual cycle mimic early pregnancy. Am J Physiol Ren Physiol. 1997;273(30):F777–82

3. Lam CSP, Cheng S, Choong K, Larson MG, Murabito JM, Newton-Cheh C, Bhasin S, McCabe EL, Miller KK, Redfield MM, Vasan RS, Coviello AD, Wang TJ. Influence of sex and hormone status on circulating natriuretic peptides. J Am Coll Cardiol. 2011;58(6):618–26

4. Montanini V, Simoni M, Chiossi G, Baraghini GF, Velardo A, Baraldi E, Marrama P. Age-related changes in plasma dehydroepiandrosterone sulphate, cortisol, testosterone and free testosterone circadian rhythms in adult men. Horm Res. 1988;29:1–6

5. Shore AC, Sandeman DD, Tooke JE. Capillary pressure, pulse pressure amplitude, and pressure waveform in healthy volunteers. Am J Physiol Heart Circ Physiol. 1995;268(1):H147–54

6. Wang TJ, Larson MG, Levy D, Leip EP, Benjamin EJ, Wilson PW, Sutherland P, Omland T, Vasan RS. Impact of age and sex on plasma natriuretic peptide levels in healthy adults. Am J Cardiol. 2002;90(3):254–8

7. Wenner MM, Sebzda KN, Kuczmarski AV, Pohlig RT, Edwards DG. ETB receptor contribution to vascular dysfunction in postmenopausal women. Am J Phys Regul Integr Comp Phys. 2017;313(1):R51–7

(continued)

Table 20.2 (continued)

8. Xanthakis V, Enserro DM, Murabito JM, Polak JF, Wollert KC, Januzzi JL, Wang TJ, Tofler G, Vasan RS. Ideal cardiovascular health associations with biomarkers and subclinical disease and impact on incidence of cardiovascular disease in the Framingham offspring study. Circulation. 2014;130(19):1676–83
9. Bjerregaard P. Mean 24 hour heart rate, minimal heart rate and pauses in healthy subjects 40–79 years of age. Eur Heart J. 1983;4(1):44–51
10. Kerkhof PLM, Kuznetsova T, Kresh JY, Handly N. Cardiophysiology illustrated by comparing ventricular volumes in healthy adult males and females. Adv Exp Med Biol. 2018;1065:123–38

gravity experience greater orthostatic intolerance [38, 75]. In response to orthostatic stress, males compensate by changing vascular resistance (microvascular), while females change heart rate. In a follow-up study to match sex-specific countermeasures (lower body negative pressure, LBNP) for changes in orthostatic stress (head-down tilt), it was found that compared to men tibial microvascular blood flow and oxygenation were both lower in women [38]. The conclusion of the study was that whereas LBNP is a suitable intervention to combat orthostatic changes in men, another approach needs to be designed for women.

It is in the venules draining the capillary tree that the majority of white blood cell/endothelial cell (WBC/EC) interactions are observed, and it is in these vessels, with their lower hydrostatic pressure and relatively low flow rates, that increases in vascular leakage to proteins as well as white cell diapedesis are observed [105]. Venules are also the site of integrin expression by EC in response to cytokines that promote WBC/EC interactions [20, 87]. In turn, inflammatory cytokine action is sensitive to the presence of estrogen and testosterone, respectively [64, 118].

Under basal conditions observations of in situ venular perfusion demonstrate no discernable difference between males and females with respect to WBC/EC interactions in the hamster cheek pouch [89]. While basal white cell/EC interactions did not differ in this preparation, permeability responses to ischemia/reperfusion (I/R) injury in the venous vasculature did. Following the I/R insult, fluorescently labeled dextrans were observed to egress more from the venules of males relative to the females clearly demonstrating that the loss of barrier function was apparently independent of WBC/EC interactions [89].

That said, in porcine coronary venules and rat skeletal muscle venules, differences in both basal venular permeability to protein and permeability responses with sex have been documented [53, 54, 123]. A study of protein clearance from mesenteric venular vessels in sexually mature rats [103] illustrates the complexity of protein and fluid movement under basal and stimulated conditions. While basal leak index (LI) of fluorescently labeled albumin did not differ by sex, acute treatment with insulin increased LI in males and was without effect in females. Analysis of protein clearing into the suffusate demonstrated lower total protein clearance and no difference albumin clearance, for males and females, respectively. In males, insulin reduced both total protein and albumin clearance rates; insulin reduced total protein clearance and was without effect on protein clearance in the females. Further, in response to acute insulin, the reduction in protein clearance was greater in the males than in females [103]. This study, given the mismatches between venular albumin LI, albumin clearance, and total protein clearance, illustrates that protein flux occurs across microvascular elements other than the venules and that male/female sex (and insulin) interacts with different elements governing protein and consequent fluid flux.

Sepsis results in loss of barrier integrity and, like inflammation, is accompanied by increases in WBC/EC interactions. Female animals and women appear less prone to develop sepsis than males [68, 94, 106, 110]. Among the factors implicated in this difference are estrogen and its interactions with EC and WBC [83], estrogen receptor distribution [106], sex differences in glucocorticoid action [17], and inflammatory status [9], to name but a few identified factors. A final component of venous function that has received attention with respect to sexual dimorphism is clot formation [94, 99] as well as fibrinolysis [99]. The complex interactions of the coagulation cascade and their associations with respect to genomic sex and sex hormones exist and are discussed in the review by [99].

Conclusions and Missing Information (e.g., Future Studies Are Needed)

There are multiple components displaying sex differences with respect to the myriad of microvascular functions. Figure 20.1 illustrates a selection of sexually dimorphic features across the microvasculature from the arterioles regulating blood flow into the exchange vessels; the capillaries where gas, fluid, and solute move

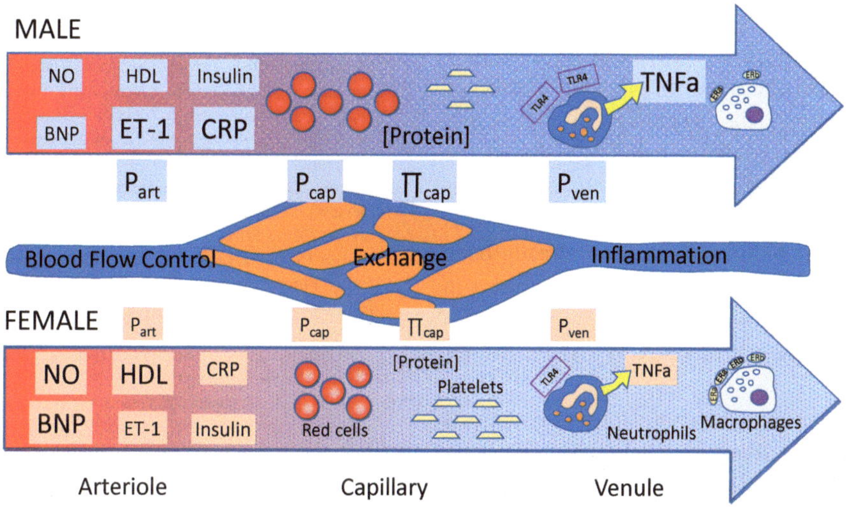

Fig. 20.1 Diagram of sexually dimorphic constituents identified in male and female humans influencing the three primary functions of the microvasculature:

1. Control of blood flow via changes in VSM tone of the arterioles
2. Control of gas, fluid, and solute exchange via changes in capillary barrier function
3. Control of inflammation and immune function in the venules

Overall, hydrostatic pressures, Part, Pcap and Pven, and VSM tone are higher in males than females reflecting in part the relatively higher levels of vasoconstrictors (ET-1) to vasodilators (BNP, NO) in the blood of males. Blood from males carries a greater number of red blood cells; while the amount of hemoglobin per cell does not vary greatly by sex, the larger number of cells means a higher net Hgb and thus a higher O_2 carrying capacity favoring a higher O_2 delivery in males. The higher protein content of plasma in the males results in a higher oncotic protein force that will offset the higher fluid movement out of the capillaries by Pcap in males. The number of platelets tends to be higher in females than males suggesting a higher tendency for formation of clots in females than males. While white cell numbers tend to be higher or not different between males and females, it is the sex differences in receptor density that appears to play a role in sex differences in WBC responses. On exposure to LPS, neutrophils via TLR4 receptor activation release greater amounts of the cytokine TNFa in males than females. Macrophages of females express higher ERa and ERb receptors; the response to LPS is mediated by ERa in both sexes but to a greater extent in females. Abbreviations: *art* arterioler, *BNP* B-type natriuretic peptide, *cap* capillary, *ERa* estrogen receptor alpha, *ERb* estrogen receptor beta, *Hgb* hemoglobin, *LPS* lipopolysaccharide, *NO* nitric oxide, *TLR* Toll-like receptor, *TNFa* tumor necrosis factor alpha, *ven* venule, *VSM* vascular smooth muscle, *WBC* white blood cell (count)

Table 20.1 has most of these data (not the macrophage (Campesi et al., 2017) or neutrophil (Aomatsu et al., 2013) data, though)

Aomatsu M, Kato T, Kasahara E, Kitagawa S. Gender difference in tumor necrosis factor-α production in human neutrophils stimulated by lipopolysaccharide and interferon-γ. Biochem Biophys Res Commun. 2013;441:220–5.

Campesi J, Marino M, Montella A, Pais S, Franconii F. Sex differences in estrogen receptor α and β levels and activation status in LPS-stimulated human macrophages. J Cell Physiol. 2017;232:340–5.

between tissue and the circulating blood; and the venules where inflammatory processes occur. At present, several studies appear to have disparate, contradictory outcomes. This apparent "noise" likely reflects the early point in time of this research as all the systems described actually are multifactorial and most of the factors in health, no less disease, have yet to be studied equally in males and females. Further, as stated earlier, much of the microvasculature is heterogeneous with the form and function of the microvascular elements being married to organ function. Obviously healthy males and females are in fluid homeostasis but how they achieve this state differs by sex. The major implication is that for a given stressor, the responses in males and females, and the resulting pathology should the stressor not be removed, are likely to differ. Multiple comorbidities such as diabetes, hypertension, and heart failure, which have already been shown to display sex differences with respect to onset, frequency, morbidity, and mortality, also impact microvascular function [4, 11, 13, 84, 93, 107].

The present state of confusion likely reflects the lack of data. Resolution requires more comprehensive research to assemble the comprehensive anatomical, functional, and biochemical pictures of the female and the male microvasculatures. Once developed, sex needs to be considered when diagnosing, implementing preventive strategies, and treating diseases involving microvascular functions. The review of the literature further illustrated the importance of age (in Tables 20.1 and 20.2, several reported age-dependent changes are summarized), an independent but interacting variable. Consideration of sex and age as interdependent variables is just beginning, and we would be remiss to not note that the contribution of ethnicity has been omitted in this review as there are several differences between Caucasian, Asian, African, and Hispanics that have been observed. A good example are the differences to platelet concentration, noted in Table 20.2, that while significantly higher in women than men also vary significantly by ethnicity [109].

References

1. Aanerud J, Borghammer P, Rodell A, Jónsdottir KY, Gjedde A. Sex differences of human cortical blood flow and energy metabolism. J Cereb Blood Flow Metab. 2017;37(7):2433–40. https://doi.org/10.1177/0271678X16668536.
2. Aird WC. Phenotypic heterogeneity of the endothelium: I. Structure, function, and mechanisms. Circ Res. 2007;100(2):158–73. https://doi.org/10.1161/01.RES.0000255691.76142.4a.
3. Ak G, Buyukberber S, Sevinc A, Turk HM, Ates M, Sari R, Savli H, Cigli A. The relation between plasma endothelin-1 levels and metabolic control, risk factors, treatment modalities, and diabetic microangiopathy in patients with Type 2 diabetes mellitus. J Diabetes Complicat. 2001;15(3):150–7. https://doi.org/10.1016/S1056-8727(01)00137-4.
4. Arnetz L, Ekberg N, Alvarsson M. Sex differences in type 2 diabetes: focus on disease course and outcomes. Diabetes Metab Syndr Obes Targets Ther. 2014;7:409–20.
5. Augustin HG, Koh GY. Organotypic vasculature: from descriptive heterogeneity to functional pathophysiology. Science. 2017;2379(August):1–22. https://doi.org/10.1126/science.aal2379.
6. Baatout S. Endothelial differentiation using Matrigel (review). Anticancer Res. 1997;17(1A):451–5. https://doi.org/10.2307/1276261.
7. Barnes JN. Sex specific factors regulating pressure and flow. Exp Physiol. 2017;102:1385. https://doi.org/10.1113/EP086531.
8. Bates DO. Vascular endothelial growth factors and vascular permeability. Cardiovasc Res. 2010;87(2):262–71. https://doi.org/10.1093/cvr/cvq105.
9. Casimir GJA, Duchateau J. Gender differences in inflammatory processes could explain poorer prognosis for males. J Clin Microbiol. 2011;49(1):478–9. https://doi.org/10.1128/JCM.02096-10.
10. Castle-Miller J, Bates DO, Tortonese DJ. Mechanisms regulating angiogenesis underlie seasonal control of pituitary function. Proc Natl Acad Sci U S A. 2017;114(12):E2514–23. https://doi.org/10.1073/pnas.1618917114.
11. Chen Q, Williams R, Healy CL, Wright CD, Wu SC, O'Connell TD. An association between gene expression and better survival in female mice following myocardial infarction. J Mol Cell Cardiol. 2010;49(5):801–11. https://doi.org/10.1016/j.yjmcc.2010.08.002.
12. Choi I, Lee S, Hong YK. The new era of the lymphatic system: no longer secondary to the blood vascular system. Cold Spring Harb Perspect Med. 2012;2(4):1–23. https://doi.org/10.1101/cshperspect.a006445.
13. Dantas APV, Franco M d CP, Silva-Antonialli MM, Tostes RCA, Fortes ZB, Nigro D, Carvalho MHC. Gender differences in superoxide generation in microvessels of hypertensive rats: role of NAD(P)H-oxidase. Cardiovasc Res. 2004;61(1):22–9. https://doi.org/10.1016/j.cardiores.2003.10.010.
14. Davis CM, Fairbanks SL, Alkayed NJ. Mechanism of the sex difference in endothelial dysfunction after stroke. Transl Stroke Res. 2013;4(4):381–9. https://doi.org/10.1007/s12975-012-0227-0.
15. Diler AS, Üzüm G, Akgün Dar K, Aksu U, Atukeren P, Ziylan YZ. Sex differences in modulating blood brain barrier permeability by NO in pentylenetetrazol-induced epileptic seizures. Life Sci. 2007;80(14):1274–81. https://doi.org/10.1016/j.lfs.2006.12.039.

16. Dongaonkar RM, Stewart RH, Geissler HJ, Laine GA. Myocardial microvascular permeability, interstitial oedema, and compromised cardiac function. Cardiovasc Res. 2010;87(2):331–9. https://doi.org/10.1093/cvr/cvq145.
17. Duma D, Collins JB, Chou JW, Cidlowski JA. Sexually dimorphic actions of glucocorticoids provide a link to inflammatory diseases with gender differences in prevalence. Sci Signal. 2010;3(143):ra74. https://doi.org/10.1126/scisignal.2001077.
18. Duvekot JJ, Cheriex EC, Pieters FAA, Peeters PLH, Menheere LPCA. Early pregnancy changes in hemodynamics and volume homeostasis are consecutive adjustments triggered by a primary fall in systemic vascular tone. Am J Obstet Gynecol. 1993;169(6):1382–92. https://doi.org/10.1016/0002-9378(93)90405-8.
19. Ely D, Boehme S, Dunphy G, Hart M, Chiarappa F, Miller B, Martins AS, Turner M, Milsted A. The Sry3 Y chromosome locus elevates blood pressure and renin-angiotensin system indexes. Gend Med. 2011;8(2):126–38. https://doi.org/10.1016/j.genm.2010.11.014.
20. Eriksson M, Sartono E, Martins CL, Balé C, Garly ML, Whittle H, Aaby P, Pedersen BK, Yazdanbakhsh M, Erikstrup C, Benn CS. A comparison of ex vivo cytokine production in venous and capillary blood. Clin Exp Immunol. 2007;150:469–76. https://doi.org/10.1111/j.1365-2249.2007.03515.x.
21. Faber JE, Moore SM, Lucitti JL, Aghajanian A, Zhang H. Sex differences in the cerebral collateral circulation. Transl Stroke Res. 2017;8(3):273–83. https://doi.org/10.1007/s12975-016-0508-0.
22. Fall T, Hågg S, Ploner A, Mågi R, Fischer K, Draisma HHM, Sarin AP, Benyamin B, Ladenvall C, Åkerlund M, Kals M, Esko T, Nelson CP, Kaakinen M, Huikari V, Mangino M, Meirhaeghe A, Kristiansson K, Nuotio ML, Kobl M, Grallert H, Dehghan A, Kuningas M, De Vries PS, De Bruijn RFAG, Willems SM, Heikkila K, Silventoinen K, Pietilainen KH, Legry V, Giedraitis V, Goumidi L, Syvanen AC, Strauch K, Koenig W, Lichtner P, Herder C, Palotie A, Menni C, Uitterlinden AG, Kuulasmaa K, Havulinna AS, Moreno LA, Gonzalez-Gross M, Evans A, Tregouet DA, Yarnell JWG, Virtamo J, Ferrieres J, Veronesi G, Perola M, Arveiler D, Brambilla P, Lind L, Kaprio J, Hofman A, Stricker BH, Van Duijn CM, Ikram MA, Franco OH, Cottel D, Dallongeville J, Hall AS, Jula A, Tobin MD, Penninx BW, Peters A, Gieger C, Samani NJ, Montgomery GW, Whitfield JB, Martin NG, Groop L, Spector TD, Magnusson PK, Amouyel P, Boomsma DI, Nilsson PM, Jarvelin MR, Lyssenko V, Metspalu A, Strachan DP, Salomaa V, Ripatti S, Pedersen NL, Prokopenko I, Mccarthy MI, Ingelsson E. Age- and sex-specific causal effects of adiposity on cardiovascular risk factors. Diabetes. 2015;64(5):1841–52. https://doi.org/10.2337/db14-0988.
23. Franconi F, Rosano G, Basili S, Montella A, Campesi I. Human cells involved in atherosclerosis have a sex. Int J Cardiol. 2017;228(February):983–1001. https://doi.org/10.1016/j.ijcard.2016.11.118.
24. Freedman RR, Sabharwal SC, Desai N. Sex differences in peripheral vascular adrenergic receptors. Circ Res. 1987;61:581–7.
25. Friedrich V, Bi W, Sehba FA. Sexual dimorphism in gene expression after aneurysmal subarachnoid hemorrhage. Neurol Res. 2015;37(12):1054–9. https://doi.org/10.1080/01616412.2015.1115211.
26. Galipeau D, Verma S, McNeill JH. Female rats are protected against fructose-induced changes in metabolism and blood pressure. Am J Physiol Heart Circ Physiol. 2002;283(6):H2478–84. https://doi.org/10.1152/ajpheart.00243.2002.
27. Asellius G. De lactibus sive lacteis venis. Milan: Mediolani; 1627.
28. Gill SE, Taneja R, Rohan M, Wang L, Mehta S. Pulmonary microvascular albumin leak is associated with endothelial cell death in murine sepsis-induced lung injury in vivo. PLoS One. 2014;9(2):1–12. https://doi.org/10.1371/journal.pone.0088501.
29. Gillis EE, Sasser JM, Sullivan JC. Endothelin, sex, and pregnancy: unique considerations for blood pressure control in females. Am J Phys Regul Integr Comp Phys. 2016;310(8):691–6. https://doi.org/10.1152/ajpregu.00427.2015.
30. Glinskii OV, Abraha TW, Turk JR, Rubin LJ, Huxley VH, Glinsky VV. Microvascular network remodeling in dura mater of ovariectomized pigs: role for angiopoietin-1 in estrogen-dependent control of vascular stability. Am J Physiol Heart Circ Physiol. 2007;293(2):H1131–7. https://doi.org/10.1152/ajpheart.01156.2006.
31. Glinskii OV, Huxley VH, Glinskii VV, Rubin LJ, Glinsky VV. Pulsed estrogen therapy prevents post-OVX porcine dura mater microvascular network weakening via a PDGF-BB-dependent mechanism. PLoS One. 2013;8(12):1–9. https://doi.org/10.1371/journal.pone.0082900.
32. Gohar EY, Giachini FR, Pollock DM, Tostes RCA. Role of the endothelin system in sexual dimorphism in cardiovascular and renal diseases. Life Sci. 2016a;159:20–9. https://doi.org/10.1016/j.lfs.2016.02.093. Elsevier Inc
33. Gohar EY, Yusuf C, Pollock DM. Ovarian hormones modulate endothelin A and B receptor expression. Life Sci. 2016b;159:148–52. https://doi.org/10.1016/j.lfs.2016.01.010. Elsevier Inc
34. Gonzales RJ, Bryant JM, Naik JS, Resta TC, Walker BR. Gender differences in mesenteric vasoconstrictor reactivity following chronic hypoxia. Microcirculation. 2008;15(6):473–84. https://doi.org/10.1080/10739680801891348.
35. Grines CL. Acute coronary syndromes: acute MI in women – the fountain of youth has run dry. Nat Rev Cardiol. 2015;12(6):322–3. https://doi.org/10.1038/nrcardio.2015.66.. Nature Publishing Group
36. Hack CE, Zeerleder S. The endothelium in sepsis: source of and a target for inflammation. Crit Care Med. 2001;29(7 Suppl):S21–7. https://doi.org/10.1097/00003246-200107001-00011.
37. Hall CN, Reynell C, Gesslein B, Hamilton NB, Mishra A, Sutherland BA, O'Farrell FM, Buchan AM, Lauritzen M, Attwell D. Capillary pericytes regulate cerebral blood flow in health and disease. Nature. 2014;508(7494):55–60. https://doi.org/10.1038/nature13165.
38. Harm DL, Jennings RT, Meck JV, Powell MR, Putcha L, Sams CP, Schneider SM, Shackelford LC, Smith SM, Whitson PA. Invited review: gender issues related to spaceflight: a NASA perspective. J Appl Physiol. 2001;91:2374–83.
39. Hart EC, Charkoudian N,Wallin BG, Curry TB, Eisenach J, Joyner MJ. Sex and ageing differences in resting arterial pressure regulation: the role of the beta-adrenergic receptors. J Physiol. 2011, 589:5285–97.
40. He J, Gu D, Chen J, Jaquish CE, Rao DC, Hixson JE, Chen J, Duan X, Huang J, Chen C-S, Kelly TN, Bazzano LA, Whelton PK. Gender difference in blood pressure responses to dietary sodium intervention in the GenSalt study. J Hypertens. 2009;27(1):48–54. https://doi.org/10.1097/HJH.0b013e328316bb87.

41. He P. Leucocyte/endothelium interactions and microvessel permeability: coupled or uncoupled? Cardiovasc Res. 2010;87(2):281–90. https://doi.org/10.1093/cvr/cvq140.
42. Huang A, Sun D, Koller A, Kaley G. Gender difference in myogenic tone of rat arterioles is due to estrogen-induced, enhanced release of NO. Am J Physiol Heart Circ Physiol. 1997;272(4 Pt 2):H1804–9.
43. Huxley VH. Physiologic regulation of capillary permeability. J Reconstr Microsurg. 1988a;4(4):341–6.
44. Huxley VH, Curry F-RE. Albumin modulation of capillary permeability: test of an adsorption mechanism. Am J Physiol Heart Circ Physiol. 1985;248(2 Pt 2):H264–73.
45. Huxley VH, Curry FE, Powers MR, Thipakorn B. Differential action of plasma and albumin on transcapillary exchange of anionic solute. Am J Physiol Heart Circ Physiol. 1993a;264:H1428–37.
46. Huxley, V. H., McKay, M. K., Meyer Jr., D. J., Williams, D. A. and Zhang, R. S. (1993b) 'Vasoactive hormones and autocrine activation of capillary exchange barrier function.', *Blood Cells*, 19(2), pp. 309–20–4.
47. Huxley VH, Meyer DJ Jr. Atrial natriuretic peptide (ANP)-induced increase in capillary albumin and water flux. Adv Exp Med Biol. 1988;242:23–31.
48. Huxley VH, Meyer DJ Jr. Capillary permeability: an albumin component attenuates active changes in Lp. Am J Physiol Heart Circ Physiol. 1990a;259(5 Pt 2):H1357–64.
49. Huxley VH, Meyer DJ Jr. Capillary permeability: atrial peptide action is independent of "protein effect". Am J Physiol Heart Circ Physiol. 1990b;259(5 Pt 2):H1351–6.
50. Huxley VH, Sieveking S, Kemp S, Schramm C. Phenotypic sexual dimorphism of micro- and macrovascular endothelium with respect to size, growth, and wound repair. FASEB J. 2017;31(Suppl 1.) p. 678.14
51. Huxley VH, Tucker VL, Verburg KM, Freeman RH. Increased capillary hydraulic conductivity induced by atrial natriuretic peptide. Circ Res. 1987;60(2):304–7. https://doi.org/10.1161/01.RES.60.2.304.
52. Huxley VH, Wang J. Cardiovascular sex differences influencing microvascular exchange. Cardiovasc Res. 2010;87(2):230–42. https://doi.org/10.1093/cvr/cvq142.
53. Huxley VH, Wang JJ, Sarelius IH. Adaptation of coronary microvascular exchange in arterioles and venules to exercise training and a role for sex in determining permeability responses. Am J Physiol Heart Circ Physiol. 2007;293(2): H1196–205. https://doi.org/10.1152/ajpheart.00069.2007.
54. Huxley VH, Wang J, Whitt SP. Sexual dimorphism in the permeability response of coronary microvessels to adenosine. Am J Physiol Heart Circ Physiol. 2004;288(4): H2006–13. https://doi.org/10.1152/ajpheart.01007.2004.
55. Huxley VH, Williams DA. Role of a glycocalyx on coronary arteriole permeability to proteins: evidence from enzyme treatments. Am J Physiol Heart Circ Physiol. 2000;278(4): H1177–85.
56. Jacob M, Saller T, Chappell D, Rehm M, Welsch U, Becker BF. Physiological levels of A-, B- and C-type natriuretic peptide shed the endothelial glycocalyx and enhance vascular permeability. Basic Res Cardiol. 2013;108(347):1–9. https://doi.org/10.1007/s00395-013-0347-z.
57. Joyner MJ, Casey DP. Regulation of increased blood flow (hyperemia) to muscles during exercise: a hierarchy of competing physiological needs. Physiol Rev. 2015;95(2):549–601.
58. Kassab GS, Rider CA, Tang NJ, Fung YC. Morphometry of pig coronary arterial trees. Am J Physiol Heart Circ Physiol. 1993;265(1):H3350–H365.
59. Kastrup A, Dichgans J, Niemeier M, Schabet M. Changes of cerebrovascular CO_2 reactivity during normal aging. Stroke. 1998;29:1311–4. https://doi.org/10.1161/01.STR.29.7.1311.
60. Kaufman MJ, Levin JM, Maas LC, Kukes TJ, Villafuerte RA, Dostal K, Lukas SE, Mendelson JH, Cohen BM, Renshaw PF. Cocaine-induced cerebral vasoconstriction differs as a function of sex and menstrual cycle phase. Biol Psychiatry. 2001;49(9):774–81. https://doi.org/10.1016/S0006-3223(00)01091-X.
61. Kautzky-Willer A, Harreiter JJ, Pacini G. Sex and gender differences in risk, pathophysiology and complications of type 2 diabetes mellitus. Endocr Rev. 2016;37(3):278–316. https://doi.org/10.1210/er.2015-1137.
62. Kimura K, Hashiguchi T, Deguchi T, Horinouchi S, Uto T, Oku H, Setoyama S, Maruyama I, Osame M, Arimura K. Serum VEGF—as a prognostic factor of atherosclerosis. Atherosclerosis. 2007;194:182–8. https://doi.org/10.1016/j.atherosclerosis.2006.07.025.
63. Kneale BJ, Chowienczyk PJ, Brett SE, Coltart DJ, Ritter JM. Gender differences in sensitivity to adrenergic agonists of forearm resistance vasculature. J Am Coll Cardiol. 2000; 36:1233–8.
64. Kovats S. Estrogen receptors regulate innate immune cells and signaling pathways. Cell Immunol. 2015;294:63–9. https://doi.org/10.1016/j.cellimm.2015.01.018.
65. Krogh A. The anatomy and physiology of capillaries. New Haven: Yale University Press; 1922.
66. LeBlanc AJ, Chen B, Dougherty PJ, Reyes RA, Shipley RD, Korzick DH, Muller-Delp JM. Divergent effects of aging and sex on vasoconstriction to endothelin in coronary arterioles. Microcirculation. 2013;20(5):365–76. https://doi.org/10.1111/micc.12028.
67. LeBlanc AJ, Nevitt CD. Targeting the vessel underdogs: therapeutic approaches for microvessel dysfunction in the heart. Crit Rev Biomed Eng. 2015;43(5–6):473–89. https://doi.org/10.1615/CritRevBiomedEng.2016016083.
68. Levi M, Van Der Poll T. Endothelial injury in sepsis. Intensive Care Med. 2013;39:1839–42. https://doi.org/10.1007/s00134-013-3054-1.
69. Liu W, Lou X, Ma L. Use of 3D pseudo-continuous arterial spin labeling to characterize sex and age differences in cerebral blood flow. Neuroradiology. 2016;58(9):943–8. https://doi.org/10.1007/s00234-016-1713-y.
70. Lukasz A, Hillgruber C, Oberleithner H, Kusche-Vihrog K, Pavenstädt H, Rovas A, Hesse B, Goerge T, Kümpers P. Endothelial glycocalyx breakdown is mediated by angiopoietin-2. Cardiovasc Res. 2017;113(6):671–80. https://doi.org/10.1093/cvr/cvx023.
71. MacIntyre JN, Slusar JE, Zhu J, Dong AX, Howlett SE, Kelly ME. Age-associated alterations in retinal arteriole reactivity to endothelin-1 differ between the sexes. Mech Ageing Dev. 2012;133(9–10):611–9. https://doi.org/10.1016/j.mad.2012.08.001.
72. Macintyre S, Hunt K, Sweeting H. Gender differences in health: are things really as simple as they seem? Soc Sci Med. 1996;42(4):617–24. https://doi.org/10.1016/0277-9536(95)00335-5.
73. Madri JA, Williams SK, Wyatt T, Mezzlo C. Capillary endothelial cell by matrix components cultures: phenotypic modulation by matrix components. J Cell Biol. 1983;97:153–65.
74. Maggioli E, McArthur S, Mauro C, Kieswich J, Kusters DHM, Reutelingsperger CPM, Yaqoob M, Solito E. Estrogen protects the blood-brain barrier from inflammation-induced disruption and increased lymphocyte trafficking. Brain Behav Immun. 2015;51:212–22. https://doi.org/10.1016/j.bbi.2015.08.020.

75. Mark S, Scott GBI, Donoviel DB, Leveton LB, Mahoney E, Charles JB, Siegel B. The impact of sex and gender on adaptation to space: executive summary. J Women's Health (Larchmt). 2014;23(11):941–7. https://doi.org/10.1089/jwh.2014.4914.
76. Martinez-Lemus, L. A. (2011) 'MiniReview The dynamic structure of arterioles', *Basic & Clinical Pharmacology & Toxicology*, 110, pp. 5–11. doi: https://doi.org/10.1111/j.1742-7843.2011.00813.x.
77. McKay MK, Huxley VH. ANP increases capillary permeability to protein independent of perfusate protein composition. Am J Physiol Heart Circ Physiol. 1995;268(3 Pt 2): H1139–48.
78. Miller VM, Garovic VD, Kantarci K, Barnes JN, Jayachandran M, Mielke MM, Joyner MJ, Shuster LT, Rocca WA. Sex-specific risk of cardiovascular disease and cognitive decline: pregnancy and menopause. Biol Sex Differ. 2013;4(1):6. https://doi.org/10.1186/2042-6410-4-6.
79. Miranda CH, de Carvalho Borges M, Schmidt A, Marin-Neto JA, Pazin-Filho A. Evaluation of the endothelial glycocalyx damage in patients with acute coronary syndrome. Atherosclerosis. 2016;247:184–8. https://doi.org/10.1016/j.atherosclerosis.2016.02.023.
80. Mudrovcic, N., Arefin, S., Van Craenenbroeck, A. H. and Kublickiene, K. (2017) Endothelial maintenance in health and disease: importance of sex differences Pharmacol Res 119, pp. 48–60. doi: https://doi.org/10.1016/j.phrs.2017.01.011.
81. Muller-Delp JM. The coronary microcirculation in health and disease. Physiology. 2013;2013:1–24. https://doi.org/10.1155/2013/238979.
82. Murthy VL, Naya M, Taqueti VR, Foster CR, Gaber M, Hainer J, Dorbala S, Blankstein R, Rimoldi O, Camici PG, Di Carli MF. Effects of sex on coronary microvascular dysfunction and cardiac outcomes. Circulation. 2014;129 (24):2518–27. https://doi.org/10.1161/CIRCULATIONAHA.113.008507.
83. Nathan L, Pervin S, Singh R, Rosenfeld M, Chaudhuri G. Estradiol inhibits leukocyte adhesion and transendothelial migration in rabbits in vivo: possible mechanisms for gender differences in atherosclerosis. Circ Res. 1999;85(4):377–85. https://doi.org/10.1161/01.RES.85.4.377.
84. Nussbaum C, Cavalcanti Fernandes Heringa A, Mormanova Z, Puchwein Schwepcke AF, Bechtold-Dalla Pozza S, Genzel-Boroviczény O. Early microvascular changes with loss of the glycocalyx in children with type 1 diabetes. J Pediatr. 2014;164(3):584–9. https://doi.org/10.1016/j.jpeds.2013.11.016.
85. O'Farrell FM, Attwell D. A role for pericytes in coronary no-reflow. Nat Rev Cardiol. 2014;11(7):427–32. https://doi.org/10.1038/nrcardio.2014.58.
86. Oertelt-Prigione S, Regitz-Zagrosek V. Gender aspects in cardiovascular pharmacology. J Cardiovasc Transl Res. 2009;2(3):258–66. https://doi.org/10.1007/s12265-009-9114-9.
87. Olyslaegers DA, Desmarets LM, Dedeurwaerder A, Dewerchin HL, Nauwynck HJ. Generation and characterization of feline arterial and venous endothelial cell lines for the study of the vascular endothelium. BMC Vet Res. 2013;9 (1):170. https://doi.org/10.1186/1746-6148-9-170.
88. Pak KJ, Geary GG, Duckles SP, Krause DN. Male–female differences in the relative contribution of endothelial vasodilators released by rat tail artery. Life Sci. 2002;71:1633–42.
89. Panazzolo DG, Da Silva LHA, Cyrino FZGDA, Sicuro FL, Kraemer-Aguiar LG, Bouskela E. Gender differences in microcirculation: observation using the hamster cheek pouch. Clinics (Sao Paulo). 2013;68(12):1537–42. https://doi.org/10.6061/clinics/2013(12)10.
90. Papanek PE, Rieder MJ, Lombard JH, Greene AS. Gender-specific protection from microvessel rarefaction in female hypertensive rats. Am J Hypertension 1998;11 (8 I):998–1005. https://doi.org/10.1016/S0895-7061(98)00114-9.
91. Patel H, Rosengren A, Ekman I. Symptoms in acute coronary syndromes: does sex make a difference? Am Heart J. 2004;148:27–33. https://doi.org/10.1016/j.ahj.2004.03.005.
92. Patel MB, Bui LP, Kirkeeide RL, Gould KL. Imaging microvascular dysfunction and mechanisms for female-male differences in CAD. J Am Coll Cardiol Img. 2016;9 (4):465–82. https://doi.org/10.1016/j.jcmg.2016.02.003.
93. Peters SAE, Huxley RR, Woodward M. Diabetes as a risk factor for stroke in women compared with men: a systematic review and meta-analysis of 64 cohorts, including 775,385 individuals and 12,539 strokes. Lancet. 2014;383 (9933):1973–80. https://doi.org/10.1016/S0140-6736(14)60040-4.
94. Pietropaoli AP, Glance LG, Oakes D, Fisher SG. Gender differences in mortality in patients with severe sepsis or septic shock. Gend Med. 2010;7(5):422–37. https://doi.org/10.1016/j.genm.2010.09.005.
95. Reckelhoff JF. Gender differences in the regulation of blood pressure. J Am Heart Assoc. 2001;37:1199–208. https://doi.org/10.1161/01.HYP.37.5.1199.
96. Regan, E. R. and Aird, W. C. (2012) 'Dynamical systems approach to endothelial heterogeneity.', Circ Res, 111(1), pp. 110–130. doi: https://doi.org/10.1161/CIRCRESAHA.111.261701.
97. Rhodin JAG. The ultrastructure of mammalian arterioles and precapillary sphincters. J Ultrastruct Res. 1967;18 (1–2):181–223. https://doi.org/10.1016/S0022-5320(67)80239-9.
98. Rhodin JAG. Ultrastructure of mammalian venous capillaries, venules, and small collecting veins. J Ultrastruct Res. 1968;25(5):452–500.
99. Roy-O'Reilly M, McCullough LD. Sex differences in stroke: the contribution of coagulation. Exp Neurol. 2014;259:16–27. https://doi.org/10.1016/j.expneurol.2014.02.011.
100. Sadrzadeh Rafie AH, Stefanick ML, Sims ST, Phan T, Higgins M, Gabriel A, Assimes T, Narasimhan B, Nead KT, Myers J, Olin J, Cooke JP. Sex differences in the prevalence of peripheral artery disease in patients undergoing coronary catheterization. Vasc Med. 2010;15(6):443–50. https://doi.org/10.1177/1358863X10388345.
101. Sage H, Pritzl P, Bornstein P. Secretory phenotypes of endothelial cells in culture: comparison of aortic, venous, capillary, and corneal endothelium. Arterioscler Thromb Vasc Biol. 1981;1(6):427–42. https://doi.org/10.1161/01.ATV.1.6.427.
102. Salmon AH, Satchell SC. Endothelial glycocalyx dysfunction in disease: albuminuria and increased microvascular permeability. J Pathol. 2012;226(4):562–74. https://doi.org/10.1002/path.3964.
103. Sasaki R. Roles of sex and insulin on microvascular exchange function. Columbia: University of Missouri; 2007.
104. Sasaki R, Whitt SP, Huxley VH. Permeability response of the rat mesenteric microvasculature to insulin. Clin Hemorheol Microcirc. 2006;34(1–2):259–63.

105. Scallan JP, Huxley VH, Korthuis RJ. Capillary fluid exchange regulation, functions, and pathology, colloquium series on integrated systems physiology: from molecule to function. San Rafael: Morgan & Claypool Life Sciences; 2010.
106. Schröder J, Kahlke V, Staubach KH, Zabel P, Stüber F. Gender differences in human sepsis. Arch Surg. 1998;133(11):1200–5. https://doi.org/10.1001/archsurg.133.11.1200.
107. Schuetz P, Yano K, Sorasaki M, Ngo L, St Hilaire M, Lucas JM, Aird WC, Shapiro NI. Influence of diabetes on endothelial cell response during sepsis. Diabetologia. 2011;54(5):996–1003. https://doi.org/10.1007/s00125-011-2059-y.
108. Schwertz D, Penckofer S. Sex differences and the effects of sex hormones on hemostasis and vascular reactivity. Heart Lung J Crit Care. 2001;30(6):401–26. https://doi.org/10.1067/mhl.2001.118764.
109. Segal JB, Moliterno AR. Platelet counts differ by sex, ethnicity, and age in the United States. Ann Epidemiol. 2006;16(2):123–30. https://doi.org/10.1016/j.annepidem.2005.06.052.
110. Sharawy N, Ribback S, Al-Banna N, Lehmann C, Kern H, Wendt M, Cerny V, Dombrowski F, Pavlovic D. Estradiol receptors agonists induced effects in rat intestinal microcirculation during sepsis. Microvasc Res. 2013;85:118–27. https://doi.org/10.1016/j.mvr.2012.10.002.
111. Shore AC, Sandeman DD, Tooke JE. Capillary pressure, pulse pressure amplitude, and pressure waveform in healthy volunteers. Am J Physiol Heart Circ Physiol. 1995;268(1):H147–54.
112. Siamwala JH, Macias BR, Lee PC, Hargens AR. Gender differences in tibial microvascular flow responses to head down tilt and lower body negative pressure. Phys Rep. 2017;5(4):e13143. https://doi.org/10.14814/phy2.13143.
113. Sorop O, Olver TD, Van DeWouw J, Heinonen I, Van Duin RW, Duncker DJ, Merkus D. The microcirculation: a key player in obesity-associated cardiovascular disease. Cardiovasc Res. 2017;113(9):1035–45. https://doi.org/10.1093/cvr/cvx093.
114. Sullivan JC, Rodriguez-Miguelez P, Zimmerman MA, Harris RA. Differences in angiotensin (1–7) between men and women. Am J Physiol Heart Circ Physiol. 2015;308(9):H1171–6. https://doi.org/10.1152/ajpheart.00897.2014.
115. Sun H, Mayhan WG. Sex difference in nitric oxide synthase-dependent dilatation of cerebral arterioles during long-term alcohol consumption. Alcohol Clin Exp Res. 2005;29(3):430–6. https://doi.org/10.1097/01.ALC.0000156117.87892.22.
116. Szabó A, Vollmar B, Boros M, Menger MD. Gender differences in ischemia-reperfusion-induced microcirculatory and epithelial dysfunctions in the small intestine. Life Sci. 2006;78(26):3058–65. https://doi.org/10.1016/j.lfs.2005.12.012.
117. Thurston G, Baluk P, McDonald D. Determinants of endothelial cell phenotype in venules. Microcirculation. 2000;7(1):67–80. https://doi.org/10.1111/j.1549-8719.2000.tb00743.x.
118. Tipton AJ, Sullivan JC. Sex differences in T cells in hypertension. Clin Ther. 2014;36(12):1822–900. https://doi.org/10.1016/j.clinthera.2014.07.011. Elsevier
119. Tostes RCA, David FL, Carvalho MHC, Nigro D, Scivoletto R, Fortes ZB. Gender differences in vascular reactivity to endothelin-1 in deoxycorticosterone-salt hypertensive rats. J Cardiovasc Pharmacol. 2000;36(5 Suppl 1):S99–101.
120. Tostes RCA, Fortes ZB, Callera GE, Montezano AC, Touyz RM, Webb RC, Carvalho MHC. Endothelin, sex and hypertension. Clin Sci. 2008;114(2):85–97. https://doi.org/10.1042/CS20070169.
121. Wang J. Modulation of coronary and skeletal muscle exchange by Adensoine: role of adenosine receptors. Columbia: University of Missouri; 2005.
122. Wang J, Bingaman S, Huxley VH. Intrinsic sex-specific differences in microvascular endothelial cell phosphodiesterases. Am J Physiol Heart Circ Physiol. 2010;298(4):H1146–54. https://doi.org/10.1152/ajpheart.00252.2009.
123. Wang J, Whitt SP, Rubin LJ, Huxley VH. Differential coronary microvascular exchange responses to adenosine: roles of receptor and microvessel subtypes. Microcirculation (New York, NY: 1994). 2005;12(4):313–26. https://doi.org/10.1080/10739680590934736.
124. Wolf MB. A three-pathway pore model describes extensive transport data from mammalian microvascular beds and frog microvessels. Microcirculation. 2002;9:497–511. https://doi.org/10.1038/sj.mn.7800163.
125. Wu NZ, Baldwin AL. Transient venular permeability increase and endothelial gap formation induced by histamine. Am J Physiol Heart Circ Physiol. 1992;262(4 Pt 2):H1238–47.
126. Wu Y, Huang A, Sun D, Falck JR, Koller A, Kaley G. Gender-specific compensation for the lack of NO in the mediation of flow-induced arteriolar dilation. Am J Physiol Heart Circ Physiol. 2001;280:H2456–61.
127. Zimmerman MA, Sullivan JC. Hypertension: What's sex got to do with it? Physiology. 2013;28:234–44. https://doi.org/10.1152/physiol.00013.2013.

Sex-Specific Ventricular and Vascular Adaptations to Exercise

21

Jill N. Barnes and Qi Fu

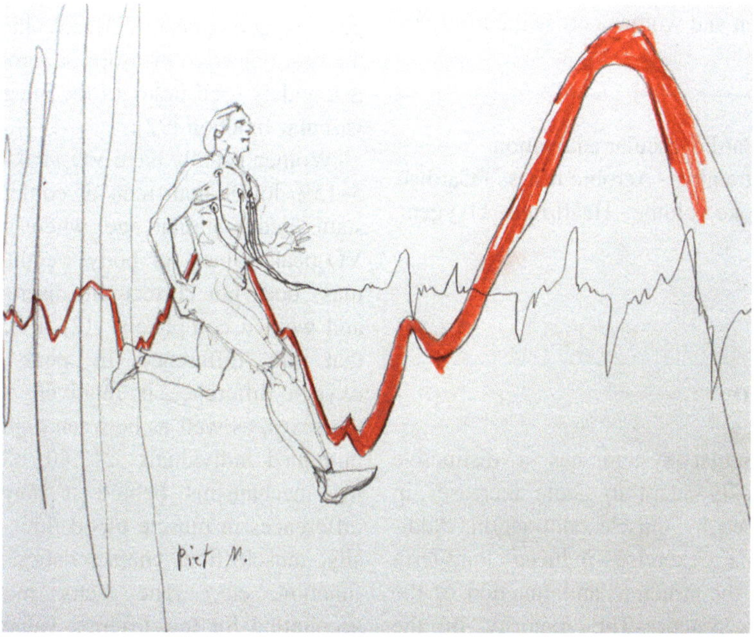

Cardiovascular adaptation during exercise. Art work by Piet Michiels, Leuven, Belgium

J. N. Barnes (✉)
University of Wisconsin-Madison, Madison, WI, USA
e-mail: jnbarnes@wisc.edu

Q. Fu
Institute for Exercise and Environmental Medicine, Dallas, TX, USA

University of Texas Southwestern Medical Center, Dallas, TX, USA
e-mail: QiFu@texashealth.org

Abstract

Increasing data suggest that there are sex differences in ventricular and vascular adaptations to aerobic (endurance) exercise, which may be attributed to different physical and physiological features in men and women. Despite that cardiovascular control during acute exercise at the same relative work rate

P. L. M. Kerkhof, V. M. Miller (eds.), *Sex-Specific Analysis of Cardiovascular Function*, Advances in Experimental Medicine and Biology 1065, https://doi.org/10.1007/978-3-319-77932-4_21

(e.g., the percentage of peak oxygen uptake) appears to be similar between the sexes, women have blunted responses or adaptations to prolonged (e.g., ≥1 year) exercise training compared with men. Currently, there is little evidence to suggest that exercise-induced vascular adaptations are different between men and women. Furthermore, sex differences in skeletal muscle adaptations to exercise, and how this influences cardiovascular function, remain unclear. Identifying potential differences and the mechanisms behind such exercise-induced adaptations is important for the optimization of exercise interventions between men and women across the life span.

Keywords
Exercise · Cardiovascular adaptation · Endurance training · Aerobic fitness · Cardiac output · Stroke volume · Heart rate · Oxygen uptake

Introduction

The cardiovascular system has a distinctive ability to quickly adapt to acute increases in workload caused by muscle contraction. Additionally, regular exercise induces long-term adaptations in the structure and function of the cardiovascular system. The majority of the research in this area has focused on adaptations in men, but emerging evidence on the adaptations in women now exists. Men and women have different physical and physiological features (e.g., body size, muscle mass, sex hormones, etc.), which may impact their responses to exercise. This chapter will focus on reviewing the cardiovascular adaptations to exercise and similarities and differences between men and women. Understanding sex-specific differences in exercise adaptations is important not only to physiologists but also to clinicians, as these differences could impact exercise rehabilitation for patient populations.

Cardiac Contribution to Aerobic Fitness

During aerobic (endurance) exercise, the uptake and transport of oxygen is required for oxidative phosphorylation and the efficient production of adenosine triphosphate to support the metabolic demands of the body. Oxygen uptake is a function of cardiac output (e.g., central factor) and total system arteriovenous oxygen difference (e.g., peripheral factor) [84]. The ability to increase one or both of these variables determines a person's aerobic capacity [42], and the upper limit for whole-body oxygen uptake is called the peak oxygen uptake (VO_2peak) [53]. VO_2peak is the best objective measure of aerobic fitness and is a widely used index of the integrity of cardiovascular function [92].

Women usually have VO_2peak values that are 5–15% lower than men of comparable training status and a similar age, whereas correction of VO_2peak values for body weight or lean body mass does not correct the discrepancy in men and women completely [12]. It has been found that the difference in peak arteriovenous oxygen difference is relatively small between the sexes, as well as between highly trained and untrained individuals [21, 40, 68, 94], though the mechanisms behind it may differ (e.g., differences in muscle blood flow, capillary density, muscle fiber characteristics, mitochondrial function, etc.). The factor most commonly accounting for the different values of VO_2peak in men and women is cardiac output. Indeed, lower cardiac output was found to contribute to the lower absolute and relative VO_2peak in women [115].

Cardiac output is the product of heart rate and stroke volume. Previous studies have demonstrated that peak heart rate during exercise is similar between men and women regardless of fitness levels [21, 30, 40, 68], while advancing age is associated with a decrease in peak heart rate in both sexes [68]. Indeed, the most consistent sex difference in cardiovascular responses during submaximal and maximal exercise is a lower stroke volume of women and a smaller increase in stroke volume from rest to exercise [21, 38, 39, 68, 69]. It has been proposed that the smaller stroke volume and its attenuated response during exercise in women are mainly attributed to a smaller heart size, particularly left ventricular volume and mass [7, 32, 41]. The difference in heart size has also been assumed to account for a majority of the difference in VO_2peak in both sexes [41, 61]. Additionally, sex-related differences in myocardial remodeling could influence left ventricular contractility, which may also account for the smaller stroke volume and its attenuated response during exercise in women [76, 111]. Indeed, Aksut et al. found that during submaximal exercise, the decrease in left ventricular end-systolic volume was less in young women than in young men [1]. However, left ventricular contractility was reported to be greater in women than men after the age of 65 years [19]. Thus, older women have a greater ability to eject stroke volume (i.e., hypercontractile), and left ventricular end-systolic volume could be lower in older women than in older men [26]. Conversely, women, especially older women, have reduced blood volume compared with age-matched men, which also contributes to the smaller stroke volume and cardiac output at rest and during exercise [7, 61, 115]. These results suggest that sex may be an important factor determining the central hemodynamic response to exercise in humans.

In contrast, some investigators reported no differences in stroke volume index, cardiac index, left ventricular end-diastolic and end-systolic volume indexes in the time course, or magnitude of changes with respect to oxygen uptake, expressed as the percentage of VO_2peak in men and women aged <40 years (Fig. 21.1a) [26, 101]. Whereas in people over the age of 60 years, all cardiac volume indexes are consistently larger in men across relative work rates, indicating greater reliance on the Frank-Starling mechanism to augment cardiac index (Fig. 21.1b) [26].

A potential important aspect when comparing people of different body sizes is the methods for scaling or normalization. Ratiometrically it can be used when there is a linear relationship between body size and cardiovascular parameters. But often this is not the case and allometric scaling is needed. Indeed, when stroke volume is scaled to lean body mass, sex differences become less

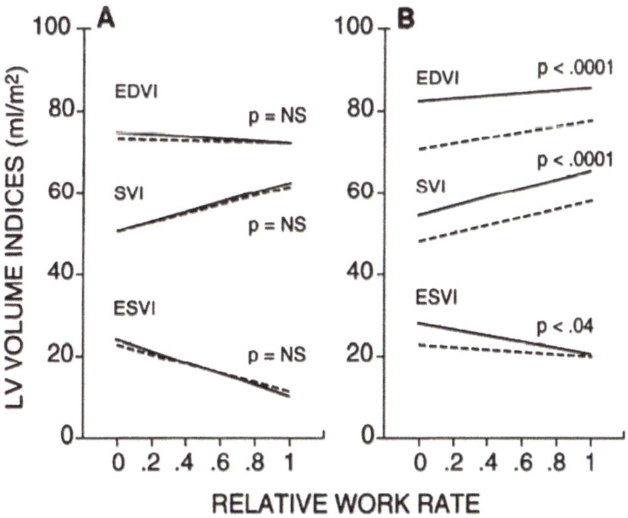

Fig. 21.1 Left ventricular (LV) volume indexes across relative work rates from rest to peak effort in young (aged <40 years; **a**) and older (aged >60 years; **b**) males (solid lines) and females (dashed lines). (**a**) Analysis of covariance shows no significant interaction or overall difference in cardiac volumes between young men and women. (**b**) In older group, all cardiac volume indexes are consistently larger in men across relative work rates compared with women. (Adapted with permission from Fleg et al. [26])

noticeable [12]. Lean body mass has been suggested as a possible factor causing differences in exercise cardiac output [89]. An earlier study by Freedson et al. [29] showed that in young men and young women matched on VO_2peak (L/min and mL/kg.min), men had a lower cardiac output and stroke volume and a higher arteriovenous oxygen difference during submaximal exercise (bicycle ergometer test at 35% of VO_2peak). When these variables were expressed independent of lean body mass, the above differences became nonsignificant [29]. These results support the notion that cardiac output differences between the sexes during submaximal exercise are due, in part, to differences in lean body mass. On the other hand, blood volume and hemoglobin mass remain lower in women compared with men even when scaled for body mass [7, 12, 61]. Blood volume, hemoglobin, and hematocrit values are 15–20% lower in women, including elite athletes; these factors probably contribute to the lower VO_2peak in women and lower average values for endurance exercise performance [12, 42].

Previous studies have demonstrated that there are rather substantial physiological and morphological sex differences between untrained or sedentary men and women, but these differences become less noticeable in highly trained male and female athletes who are competing in the same event or sport [116, 77, 119, 112]. For example, Zwiren et al. found that in men and women with similar training backgrounds and nonsignificantly different mean VO_2peak in mL/kg in fat-free mass.min, the differences in cardiovascular responses between the sexes at various percentages of VO_2peak were smaller than previously thought [119]. It seems that the sex difference in certain cardiovascular responses during exercise is a consequence of different levels of physical condition of the men and women. In accordance with this assumption, it was found that the increase in heart rate during submaximal and maximal exercise was greatest in untrained women and smallest in highly trained men at the same absolute work rate, which was expressed as VO_2 (mL/kg.min); however, when the heart rate responses at the same relative work rate, namely, the percentage of VO_2peak were compared, there were no differences among groups (Fig. 21.2) [30]. These results suggest that cardiovascular control during acute exercise is similar in men and women no matter whether they are trained or untrained. Although there are some physiological differences that may affect the mechanism of the changes, the overall response of the cardiovascular system to acute exercise is similar in men and women [61, 67].

One of the key principles in exercise physiology in humans is the remarkably constant relationship between the increase in oxygen uptake

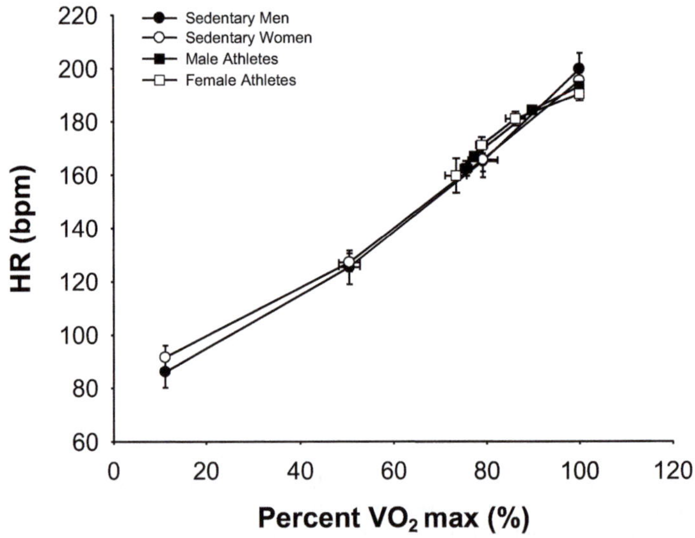

Fig. 21.2 Comparisons of heart rate (HR) responses during submaximal and maximal exercise in sedentary men and women and during submaximal exercise in highly trained male and female athletes at the same relative work rate, expressed as the percentage of maximal oxygen uptake (VO_2max), which represents a matter of cardiovascular control during exercise in healthy humans. (Adapted with permission from Fu and Levine [30])

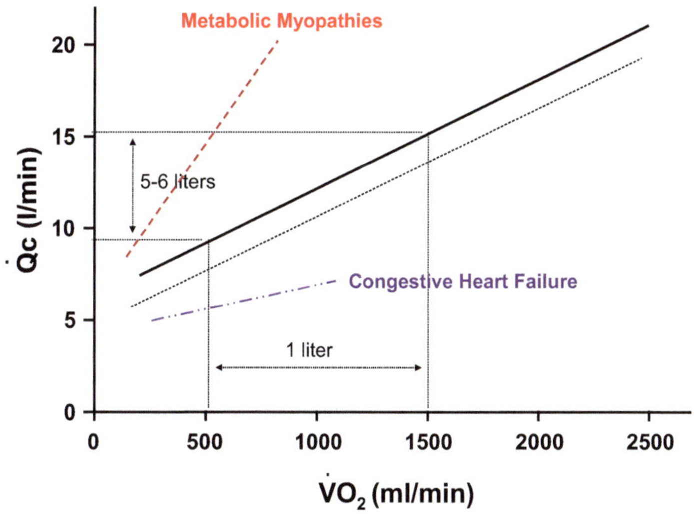

Fig. 21.3 Relationship between increases in oxygen uptake (VO_2) and the corresponding increases in cardiac output (Qc) during exercise in humans, which in most cases is 5–6/1. When this relationship is depressed, like in patients with congestive heart failure, it may be a sign of severe underlying disease with impending decompensation. Conversely, when it is exaggerated, like in patients with metabolic myopathies, it gives strong clues to the processes regulating Qc. (Adapted with permission from Fu and Levine [31])

(work rate) and the corresponding increase in cardiac output. Regardless of sex, age, or aerobic fitness, in general, about 5–6 liters of cardiac output are required for every liter of oxygen uptake above rest (Fig. 21.3) [30, 31, 33, 52, 79, 84]. When this relationship is depressed, it may be a sign of severe underlying disease with impending decompensation. Conversely, when it is exaggerated, like in patients with metabolic myopathies, it gives strong clues to the processes regulating cardiac output.

VO_2peak decreases with advancing age, and a lower stroke volume, heart rate, and arteriovenous oxygen difference at maximal exercise all contribute to the age-related decline in VO_2peak in both untrained and highly trained individuals [68, 92]. Nearly half of the age-related difference in VO_2peak is explained by a smaller stroke volume, and the remainder by a lower heart rate and arteriovenous oxygen difference at maximal exercise. It was found that age and sex each had a significant impact on aerobic fitness; VO_2peak declined with age approximately 40% from 20s to 80s years old in both sexes, but at any age it was greater in men than in women even after normalization for body weight [26]. Longitudinal studies showed that the rate of decline in VO_2peak in healthy adults was not constant across the age span, as assumed by cross-sectional studies, but accelerated markedly with each successive age decade, especially in men, regardless of physical activity habits [26]. Hossack and Bruce also found that the normal range of peak values for VO_2, heart rate, cardiac index, and stroke index during treadmill exercise testing decreased with age in both sexes, but men had a significantly greater reduction than women [38]. Fitzgerald et al. reported that the absolute (mL/kg.min per year) rate of decline in VO_2peak with age was greatest in highly trained women, next greatest in active women, and lowest in sedentary women; however, when expressed as percentage or relative decrease from mean levels at approximately 25 years of age, the rate of decline in VO_2peak was similar in the three groups [24]. Ogawa et al. also reported similar rates of decline in VO_2peak with age in untrained and highly trained men and women [68]. Similar to the previous findings in untrained women [15, 48, 49], it was found that menopausal status did not affect cardiovascular fitness in masters women runners [112].

Collectively, it seems that sex does not significantly affect cardiovascular control during endurance exercise in both untrained and highly trained individuals. For any task requiring a given absolute oxygen uptake, women are working at a higher percentage of their exercise capacity than men. This would result in a higher heart rate, greater stress, and a quicker onset of fatigue during exercise. If allowed to work at a similar percentage of their peak exercise capacity, men and women would have similar cardiovascular responses. Lastly, aerobic fitness declines with advancing age but accelerates markedly with each successive age decade, especially in men.

Left Ventricular and Myocardial Adaptations to Exercise Training

Regular exercise or exercise training promotes beneficial adaptations in the cardiovascular system which have cardioprotective effects [78]. Moderate to vigorous endurance training increases VO_2peak and exercise capacity through an increase in stroke volume or an increase in arteriovenous oxygen difference or both in humans. Depending on age and training duration, men and women could have different cardiovascular adaptations to exercise training even if they achieve similar increases in VO_2peak.

Previous studies have demonstrated that in young women, increases in stroke volume, arteriovenous oxygen difference, and VO_2peak with endurance training are similar to those in young men [12, 58, 59, 85, 95]. The increase in stroke volume after endurance training could be the result of left ventricular volume-overload hypertrophy leading to enhanced cardiac filling, increased myocardial contractile function (e.g., enhanced contractile response to β-adrenergic stimulation), reduced afterload, increased blood volume, or a combination of these factors [37, 40, 44, 81, 82]. Although the absolute increase in blood volume after exercise training was reported to be greater in young men compared with young women, the relative (percent) change did not differ between the sexes [58]. In addition, the increase in stroke volume in young men and young women appears to be independent of afterload because it was found that diastolic blood pressure remained unchanged in men but increased in women after training [58]. Recent research has shown that endurance training *up to 3 months* elicits similar increases in left ventricular mass and end-diastolic volume in initially sedentary young men and young women [40].

Despite similar adaptations in young men and young women reported in previous investigations, cross-sectional studies in athletes suggest that endurance training augments cardiovascular structure and function with apparently different phenotypes in males and females. Specifically, female athletes have smaller left ventricular dimensions and wall thickness compared with male athletes [17, 102, 114]. One possible explanation for this discrepancy may be the duration of training. In almost all previous training studies comparing young men versus young women, the intervention period is relatively short (e.g., $\leq$3 months) which could have missed or underestimated sex differences in cardiovascular adaptations. To support this notion, it was found that 3 months of intense training in male and female rowers caused similar cardiovascular adaptations between the sexes [4]. In addition, the alterations in cardiovascular function resulting from 20 weeks of endurance training were very similar across sex, age, and race with few exceptions [117]. However after prolonged (e.g., 1 year) intensive endurance training, initially sedentary young women were found to have a blunted cardiovascular response compared with young men [40].

Howden et al. [40] observed that previously sedentary young men progressively increased VO_2peak, left ventricular mass, and mean wall thickness before reaching a plateau from month 9 to 12 of training. In contrast, despite exactly the same training, the response in young women was markedly blunted, with VO_2peak, left ventricular mass, and mean wall thickness plateauing after only 3 months of training (Fig. 21.4) [40]. These results suggest that the development of left ventricular hypertrophy, which is in parallel with the increase in VO_2peak, is significantly blunted in young women. Interestingly, the

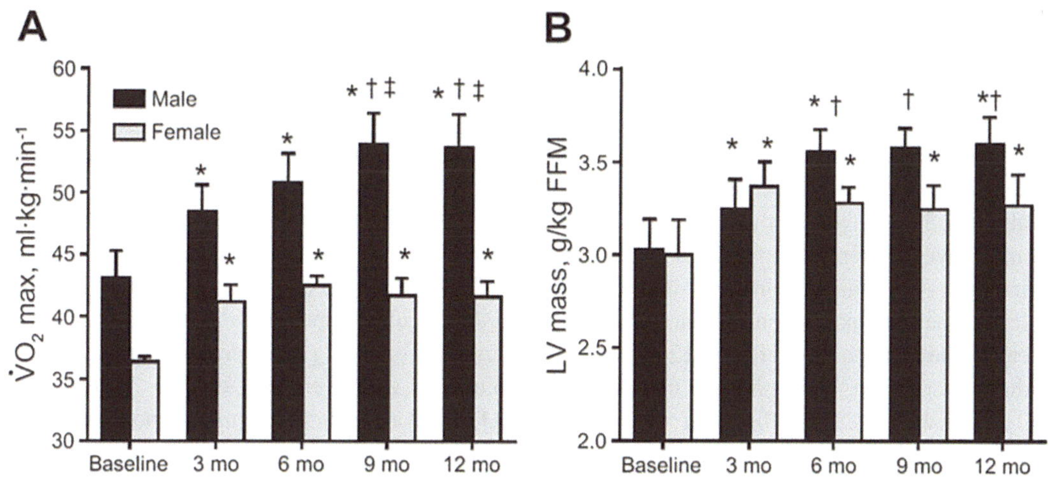

Fig. 21.4 Sex differences in cardiovascular adaptations to prolonged endurance training in initially sedentary young individuals. ***A*** and ***B***: effect of 1 year of endurance training on maximal oxygen uptake (VO_2max) indexed to baseline body mass in males and females (left), significant sex × time interaction $P = 0.084$ and changes in left ventricular (LV) mass measured by magnetic resonance imaging (MRI) scaled to baseline fat-free mass, significant sex × time interaction $P = 0.031$ (right). Both measured every 3 months during the training program. FFM, fat-free mass. Post hoc comparison with baseline (*), with *month* 3 (†), and with *month* 6 (‡) for $P < 0.05$ from liner mixed model. (Adapted with permission from Howden et al. [40])

response of left ventricular end-diastolic volume was not influenced by sex (e.g., increased by 20% in young men and 18% in young women after 1 year of training). However, young men had greater augmentation of the Frank-Starling mechanisms (the relationship between left ventricular filling pressure and stroke volume) after 1 year of training compared with young women, indicating a greater diastolic reserve in men [40]. This adaptation to training is extremely beneficial for performing exercise, allowing individuals to generate a higher stroke volume and cardiac output in response to an increase in left ventricular filling pressure. Indeed, an enhanced Frank-Starling mechanism has been shown to be a hallmark characteristic of endurance athletes [54].

The mechanisms underlying sex differences in left ventricular hypertrophy following prolonged endurance training in young individuals are unclear, but endogenous sex hormones (e.g., androgen and estrogen) may be involved. Regular exercise can increase circulating levels of testosterone, which promotes muscle growth through protein synthesis and inhibits protein degradation [108]. On the other hand, estrogen may attenuate or prevent the development of left ventricular hypertrophy with exercise training in premenopausal women via estrogen receptor β [27, 35, 74, 91]. In addition, estrogen-mediated atrial natriuretic factor (ANF) may also prevent left ventricular hypertrophy in young women [76]. Animal studies showed that estrogen exerted profound antihypertrophic effects on ventricular myocytes, by transactivation of the ANF gene and activation of the ANF receptor in an autocrine/paracrine manner, which in turn evokes cytoplasmic cyclic guanosine monophosphate signaling downstream of the guanylyl cyclase-A receptor [3]. On the other hand, male rats were found to exhibit higher cardiac angiotensin I-converting enzyme (ACE) and angiotensin-converting enzyme 2 (ACE2) activity, as well as left ventricular hypertrophy compared with female rats, while orchiectomy decreased the activity of these enzymes and hypertrophy, and ovariectomy increased hypertrophy and ACE2 but did not change ACE activity [18]. These results suggest that the interaction between sex hormones and the cardiac renin-angiotensin system components is also essential for left ventricular remodeling.

Healthy older individuals can respond to prolonged endurance exercise training with adaptations similar to those of younger individuals [34, 87]. However, older men and older women have different cardiovascular adaptations to training. For instance, Spina et al. found that in older women, stroke volume during maximal exercise did not increase after 9–12 months of endurance training even though increases in VO_2peak were similar between the sexes (increased by 22% in older women and 19% in older men) [94]. The entire increase in VO_2peak after training in older women was accounted for by a greater arteriovenous oxygen difference during maximal exercise [94]. Conversely, older men were found to increase stroke volume during maximal exercise and improve left ventricular systolic function with training [20, 88]. These results indicate a difference in the mechanisms for cardiovascular adaptations to endurance training in older men and older women.

A possible explanation for the lack of a training-induced increase in stroke volume during acute exercise in older women is that they do not develop left ventricular hypertrophy, and moreover, their myocardial function remains unchanged. Using radionuclide ventriculography, Spina et al. were unable to detect any left ventricular enlargement or changes in left ventricular ejection fraction in older women after training [96]. It was also found that endurance training had no effect on left ventricular function (β-adrenergic-mediated contractile function), left ventricular end-diastolic dimension, wall thickness, or mass in older women [97]. In addition, older women do not have any improvement in left ventricular filling dynamics after training, whereas older men do [93]. Furthermore, previous studies showed that older women were unable to expand blood volume with endurance training [98]. Estrogen deficiency might also explain why stroke volume does not increase in older women with exercise training [94]. In contrast, it was found that in older men, left ventricular end-diastolic dimension increased by 13% and wall thickness increased by 14%, while the thickness-to-radius ratio remained unchanged after training, indicating physiological eccentric hypertrophy [20]. Different from older women, blood volume increased after training in older men [69].

Collectively, there are sex differences in the cardiovascular adaptations to endurance exercise training. For young women, increases in stroke volume, arteriovenous oxygen difference, VO_2peak, and left ventricular remodeling with short-term (e.g., $\leq$3 months) endurance training are similar to those in young men. However, young women have blunted increases in VO_2peak and left ventricular hypertrophy after prolonged (e.g., $\geq$1 year) training. For older women, stroke volume during acute exercise does not increase after endurance training even though increases in VO_2peak are similar between the sexes. Lack of left ventricular remodeling and blood volume expansion may be responsible for a blunted cardiovascular response to exercise training in older women.

Ventricular-Vascular Coupling and the Role of Exercise

Structural and functional adaptations to exercise training involve the left ventricle, as discussed above, and the vasculature. With each myocardial contraction and subsequent ejection of blood from the ventricle, blood enters into the large central arteries in a pulsatile (noncontinuous) manner. Due to this high pulsatile flow, blood pressure in the large central arteries fluctuates with each myocardial contraction, with a typical systolic blood pressure of 120 mmHg and diastolic blood pressure of 80 mmHg in healthy humans. Blood pressure falls dramatically between the large central arteries and smaller arterioles where blood flow becomes more continuous. As the left ventricle undergoes remodeling to adapt to the stress of exercise training, so do the large central arteries.

The large central arteries must accommodate the pulsatile nature of blood flow from the left ventricle and the variable stroke volume when transitioning from rest to exercise. The large central arteries (including the aorta and carotid arteries) are distensible and compliant, and as the arteries narrow throughout the arterial tree, there are a gradual decrease in elastic compliance and increase in stiffness of these blood vessels.

The compliance of the large proximal arteries is reduced with advancing age, and similarly, arterial stiffness increases with age [60]. Women have a greater age-related decline in arterial compliance and distensibility compared with men [14, 110] and lower arterial compliance during the middle-aged and older years compared with men [107]. Importantly, the sex-specific differences appear to be dependent on the method of measurement and the sample size population [73, 107]. Large-scale studies like Framingham Heart Study indicate a statistically significant effect of sex on the age-related increase in aortic pulse wave velocity (PWV), a measure of arterial stiffness, in healthy adults although not all studies show sex differences in aortic PWV [107]. The reported sex differences in arterial compliance or distensibility may be due to the issue of scaling between men and women. Because men have a larger stature and greater diameters of the large central arteries, for a given change in pressure, there is a greater volume of flow through the vessel. Therefore, although there are sex differences in compliance or distensibility measurements, it is unclear if these differences have physiological or clinical implications.

Other factors may also influence sex differences in central arterial structure and function. Differences in sex hormones, primarily estrogen, may affect arterial structure and function, as menopause is associated with higher arterial stiffness (measured by PWV) [99]. Along these lines, PWV was reduced following 12 months of menopausal hormone therapy in postmenopausal women, suggesting a protective effect of estrogen [46]. Testosterone has sex-specific effects on PWV, with higher circulating androgen levels associated with greater PWV in postmenopausal women [16], yet higher total testosterone is correlated with lower PWV in men [109]. Another factor that may affect men and women differently is distribution of adipose tissue. More abdominal adiposity is associated with greater arterial stiffness, but this relationship is steeper in women, such that small increases in abdominal adiposity have more unfavorable effects in women compared to men [86]. Abdominal adiposity often increases during the postmenopausal years and is associated with greater cardiometabolic risk [71].

Regular aerobic (endurance) exercise is a common lifestyle intervention to prevent or reverse obesity and affects central arterial structure and function. In men, endurance training improved carotid arterial compliance, and endurance-trained men demonstrate more compliant arteries compared with their sedentary middle-aged and older counterparts [105]. Similarly, physically active women did not experience the age-related increases in aortic PWV when compared with sedentary women [104]. Thus, these studies suggest that habitual physical activity and/or endurance training may prevent or slow the age-associated increase in arterial stiffness.

Although there may be sex differences in baseline arterial stiffness measurements, currently there is little evidence to suggest that exercise-induced arterial adaptations are different between men and women. It should be noted that differences between men and women in response to exercise training have not been systematically tested or evaluated with different modes of exercise.

Any increases in the rigidity or stiffness of the central arteries will affect the central blood pressure. The idea that elevated arterial stiffness affects central and peripheral blood pressure has been reviewed extensively elsewhere [64, 65]. Briefly, an increase in arterial stiffness results in a faster velocity of wave travel (PWV) and an earlier return of the reflected pulse wave. If the reflected wave appears during the systolic phase of myocardial contraction, it distorts the aortic pressure waveform, which results in augmented aortic systolic pressure. Thus, higher arterial stiffness is associated with shorter wave reflection time and augmented aortic blood pressure.

Throughout the life span, there are sex differences in central hemodynamics such that women have shorter wave reflection time and greater wave augmentation [14, 57, 90]. The augmented pressure elevates aortic blood pressures to a greater extent than brachial blood pressures and can affect ventricular-vascular coupling [60, 106]. Often the sex differences in central hemodynamics are attributed to the shorter height in women (which means a shorter distance between the ventricle and the reflection sites in the periphery). However, even after correction for height, sex differences in central hemodynamic variables remain [57], suggesting that in women the central arteries may be more susceptible to age-related changes in arterial structure compared with men.

Importantly, the amount of pressure that the left ventricle needs to produce is dependent on the pressure in the proximal aorta. Ventricular pressure needs to increase sufficiently to open aortic valves allowing ejection of blood into the systemic circulation. Thus, aortic stiffness and elevated aortic blood pressure can limit left ventricular ejection of blood or places additional work on the left ventricle [65].

Throughout the systolic period of the cardiac cycle, ventricular pressure and aortic pressure values are similar. Yaginuma et al. [118] noted that aortic pressure is augmented in late systole due to wave reflection [118], which would require a similar augmentation in ventricular pressure. Aortic systolic pressure is related to left ventricular myocardial wall stress [55]. Indeed, there is an association between augmented systolic pressure due to wave reflection and impaired ventricular diastolic relaxation. There is also a relationship between the increase in aortic systolic pressure and myocardial ischemia [47]. Thus, ventricular-vascular coupling is a normal phenomenon where the structure and the function of the arteries determine arterial load, and the overall arterial load, in turn, affects the structure and function of the left ventricle. Ventricular-vascular coupling is a determinant of cardiac energetics, and pathophysiological changes in the left ventricular workload contribute to some cardiovascular disease states [9, 100].

The association between central hemodynamic variables and left ventricular function appears to be sex-dependent. Women with higher wave reflection have worsening left ventricular diastolic function, whereas, in men, there is no such association [90]. Similarly, there are distinct differences between ventricular-vascular coupling characteristics in men and women and how it changes with aging. Left ventricular performance (E_{LV}, left ventricular end-systolic elastance) is higher in men than women, and the increase with aging is more dramatic in women when compared with men (Fig. 21.5) [80]. The ratio of arterial load (E_A) compared with E_{LV} (E_A/E_{LV}) indicates ventricular-vascular coupling, which also shows differences between men and women. Across the life span, E_A/E_{LV} ratio is constant in men but reduced in women [11, 80]. This difference is primarily driven by the increases in E_{LV} compared with E_A with advancing age. Collectively, sex differences in the interaction between the left ventricle and the systemic arteries may explain the elevated risk for heart failure with preserved ejection fraction in women [11, 13, 45, 90].

The majority of studies investigating the ventricular or vascular response to exercise have been conducted in young healthy men. As more studies included women, it became apparent that sex differences exist during exercise [22, 63]. At maximal exercise, young men had higher E_{LV} and E_A values compared with young women [63]. These sex differences may be amplified by age and/or the presence of cardiovascular risk factors. For example, older women who were hypertensive have greater increase in E_A during exercise compared with older men with hypertension [72].

Interestingly, cardiorespiratory fitness (VO_2peak) is associated with ventricular-vascular coupling such that those with higher VO_2peak have lower E_A/E_{LV} ratio [22, 28]. The slope of the relationship between VO_2peak and ventricular-vascular coupling is different between men and women, with women have a greater/steeper slope compared with men. This suggests

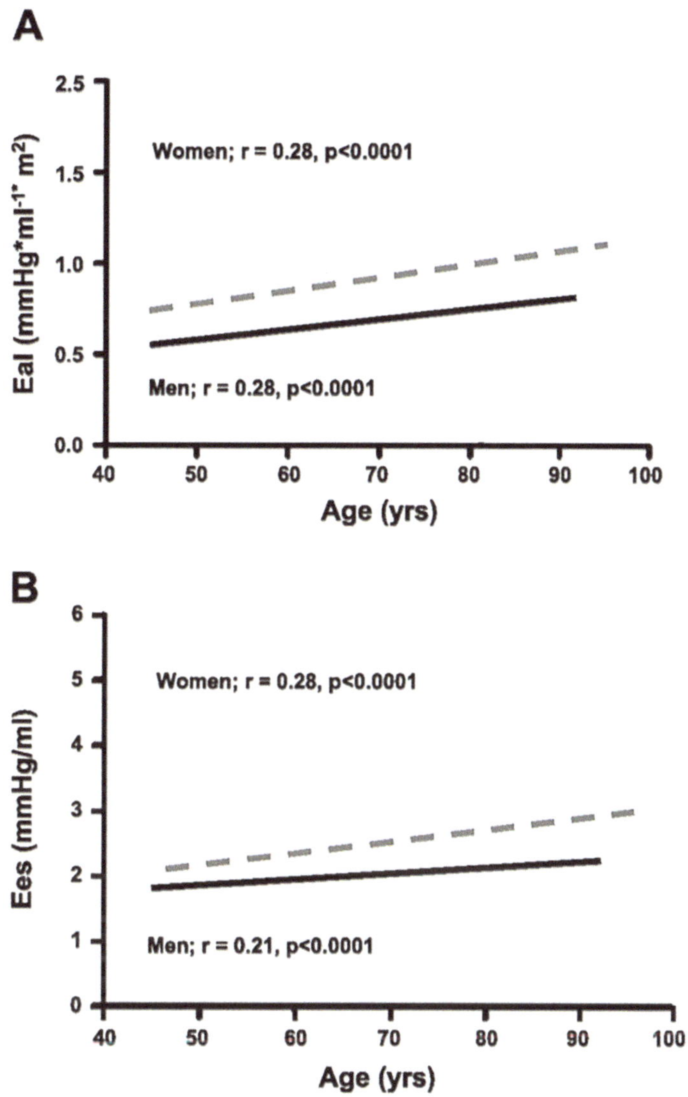

Fig. 21.5 Sex differences in the association between age and ventricular-vascular coupling variables. Arterial load (E_A) was indexed to body surface area (EaI) in panel A. Left ventricular performance or elastance (E_{LV}) was indexed to body surface area (shown as Ees) in panel B. Women (dashed line) had higher indexed arterial load and indexed left ventricular performance compared with men (solid line) at all ages. Additionally, the increase in left ventricular performance was significantly greater in women. Data is adapted from Redfield et al. [80], and figure is used with permission from Chantler et al. [10]

that higher cardiorespiratory fitness may have a disproportionate effect on E_A compared with E_{LV} [22]. This is consistent with the idea that endurance training reduces central arterial stiffness and wave reflection, which will lower the arterial load that the ventricle is pumping against. It is unclear what effect exercise training has on ventricular-vascular coupling during exercise, although a recent pilot study suggests that short-term exercise training can improve E_A/E_{LV} ratio in patients with metabolic syndrome [28]. Whether there are additional sex differences in such training effects during exercise is unknown [9].

Cardiorespiratory Fitness As a Vital Sign

Regular physical activity has been recommended as a primary means to improve overall health. In the landmark study by Blair et al., the association between VO_2peak and all-cause mortality was established [8]. Although physical activity is important in determining cardiorespiratory fitness, it is VO_2peak that is preferred over leisure-time physical activity for its ability to predict cardiovascular events [103]. High VO_2peak is

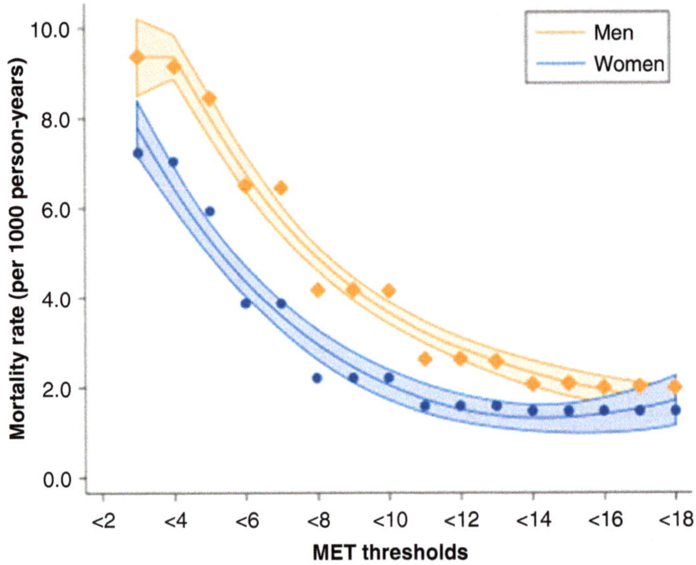

Fig. 21.6 Metabolic equivalent task (MET) thresholds achieved during maximal exercise test is associated with mortality rate (expressed as per 1000 person years). Women (shown in blue) achieved significantly lower MET thresholds compared with men (shown in orange). Shading represents 95% confidence interval. Used with permission from Al-Mallah et al. [2]

associated with reduced risk of cardiovascular events and all-cause mortality [5, 50].

The American Heart Association 2016 Scientific Statement suggests that cardiorespiratory fitness should be considered a vital sign [83]. The majority of studies investigating VO_2peak and mortality have focused on men, but a few studies have included women [36]. Sex-specific analysis suggests that the relationship between higher cardiorespiratory fitness and lower mortality is similar between men and women [50]. A recent meta-analysis quantitatively determined that increasing cardiorespiratory fitness by 1 MET translated to a 13% decline in all-cause mortality and 15% reduction in cardiovascular disease. When men and women were independently examined, the minimum METs associated with a reduction in risk at age 60 was 7 METs in men and 5 METs in women [50]. In one study, women who achieved 2.6 METs less than men during a stress test had a comparable prognosis of all-cause mortality (Fig. 21.6) [2]. Therefore in these few studies, cardiorespiratory fitness appears equally protective for men and women, but the threshold values for risk are different. Future studies should further evaluate cardiorespiratory fitness in both men and women to provide normative data to guide clinical diagnosis and treatment.

Exercise Is Medicine

Despite the benefits of exercise, over the last century, our technological improvements have greatly reduced the required physical activity in all facets of daily life. The World Health Organization estimates that globally, 25% of adults are not meeting the minimum physical activity recommendations (http://www.who.int/mediacentre/factsheets/fs385/en/). Thus, physical inactivity has emerged as a significant risk factor for chronic disease and morbidity.

In medicine and health care, exercise is an important tool in the prevention and treatment of many diseases, specifically cardiovascular diseases. Engaging in habitual exercise, particularly regular aerobic exercise, is associated with many favorable physiological adaptations. Habitual exercise is associated with lower risk of cardiovascular disease, hypertension, stroke, and reduced all-cause mortality [51]. In effort to increase physical activity, interventions have been designed for children, adults, and patient populations. The efficacy of the exercise intervention is somewhat dependent on sex differences in cardiovascular adaptations. For example, Martinez-Vizcaino et al. reported that school-based physical activity intervention

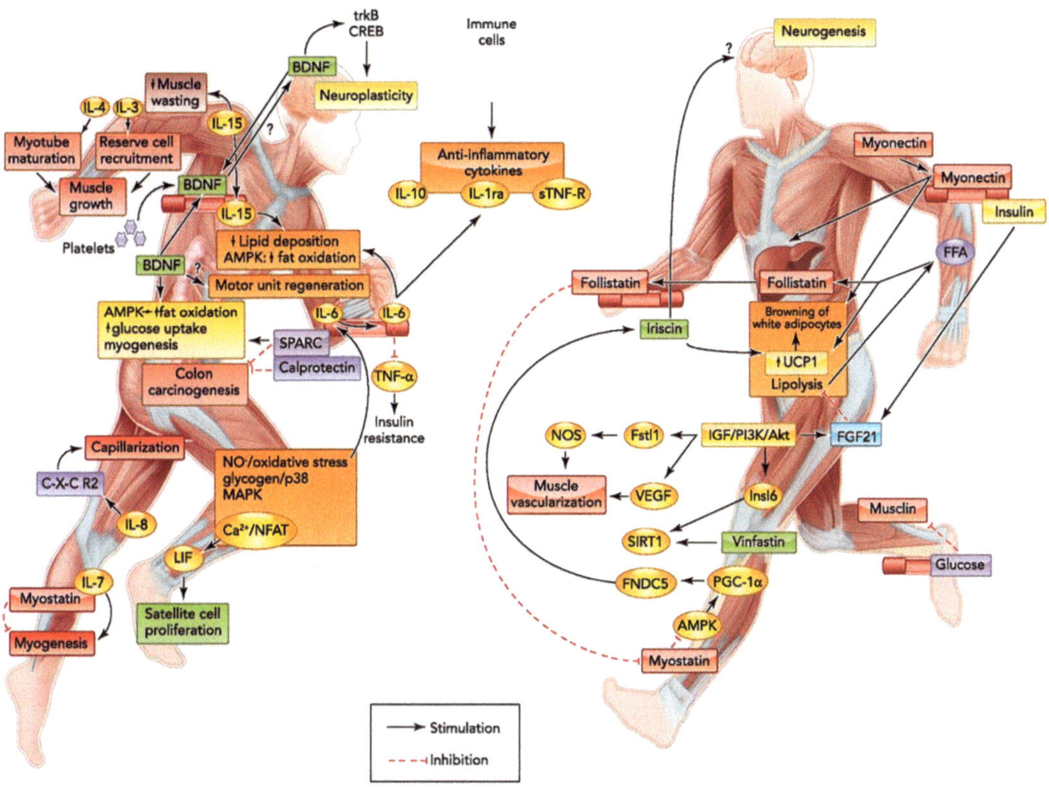

Fig. 21.7 Summary of the primary myokines, the signaling pathways involved, and the assumed physiological effects. AMPK, AMP-activated protein kinase; BDNF, brain-derived neurotrophic factor; CREB, cAMP response-element-binding protein; C-X-C R2, C-X-C receptor 2; FFA, free-fatty acid; FGF21, fibroblast growth factor 21; Fndc5, fibronectin type III domain-containing 5 protein; Fstl1, follistatin-like 1; IGF, insulin-like growth factor; IL-1ra, IL-1 receptor antagonist; Insl6, insulin-like 6; LIF, leukemia inhibitory factor; NO˙, nitric oxide; NOS, nitric oxide synthase; PGC-1α, peroxisome proliferator-activated receptor-γ coactivator 1α; PI3K, phosphatidylinositol 3-kinase; SIRT1, sirtuin 1; SPARC, secreted protein acidic and rich in cysteine; sTNF-R, soluble TNF receptors; trkB, tropomyosin receptor kinase; UCP1, uncoupling protein 1. Used with permission from Fiuza-Luces et al. [25]

reduced cardiometabolic risk factors in boys and girls, but the magnitude of improvement was greater in girls [56]. In adults over 60, an exercise intervention lowered cardiovascular risk factors (i.e., blood pressure) to a greater extent in women compared with men [62]. In contrast, a multicomponent intervention that included exercise training in patients after open-heart surgery reported a lower risk of future cardiovascular event in men, but this protection was not seen in women [43].

In addition, there may be sex differences in the response to an acute bout of exercise. During exercise, skeletal muscle manufactures numerous products, and these biologically active factors are secreted into the bloodstream during muscle contraction. Specifically, a myokine is defined as "a cytokine or peptide which is produced in the skeletal muscle cells and subsequently released into the circulation to exert endocrine or paracrine effects in other cells, tissues, or organs" [23]. These myokines are likely the underlying mechanisms behind the widespread, systemic effects of regular exercise (Fig. 21.7) [25, 113]. Interleukin-6 (IL-6), a prototypical myokine, is released during exercise, but the response is sex-specific [70, 75]. Treadmill running induces inflammatory cytokines (i.e., IL-6), and the cytokine expression was dependent on sex and the phase of the menstrual cycle in women. [66]. In another example, men demonstrated an increase in IL-6 after acute resistance exercise; however there was no change in IL-6 in women [6].

While IL-6 and other myokines are important in the systemic benefits of exercise, they are numerous. It takes decades of research to identify all of the myokines and factors released by contracting skeletal muscle. This is further complicated by the technical difficulty in tracing the physiological effects and determining the mechanism of action of such myokines [113]. The knowledge of these factors may have important therapeutic implications, but it will likely be impossible to synthesize a "polypill" product that could replicate the multitude of physiological effects of repeated muscle contraction.

The discovery of myokines has led to the pursuit of possible therapeutic implications of these myokines for disease and obesity prevention. Still others have suggested combining pharmacological approaches, creating a polypill in attempt to mimic the physiological effects of exercise. Fiuza-Luces et al. [25] compared this polypill approach with different modes of acute exercise on traditional cardiovascular risk factors [25]. While a combination of antihypertensive and statin medications did result in modest reductions in cardiovascular disease risk factors, this polypill combination was not superior to an endurance exercise intervention. Furthermore, there were no other beneficial changes that are typically seen with exercise interventions such as increases in cardiorespiratory fitness and decreases in adiposity.

A limited number of the beneficial effects of regular exercise can be replicated by acute pharmacological manipulation; yet, there is not yet a single low-cost regimen that mimics all of the physiological effects of repeated skeletal muscle contraction. Further, these pharmacological approaches are unable to mimic the favorable effects on psychological function, neuromuscular coordination, and well-being [25].

Summary

The cardiovascular response and adaptation to endurance exercise is well-characterized; however, much of this work has been conducted on only male subjects or a small sample of men and women combined. Thus, the sex differences during exercise, after exercise, and in response to exercise training are not completely known. Furthermore, sex differences in skeletal muscle adaptations to exercise and how these differences influence cardiovascular function are less known. Identifying potential differences and the mechanisms behind such exercise-induced adaptations is important for the optimization of exercise interventions between men and women across the life span.

References

1. Aksut SV, Pancholy S, Johnson J, Walter JD, DiMarzio D, Cave V, Cassel D, Heo J, Iskandrian AS. Comparison of left ventricular performance in healthy young women and men during exercise. J Nucl Cardiol. 1996;3(5):415–21.
2. Al-Mallah MH, Juraschek SP, Whelton S, Dardari ZA, Ehrman JK, Michos ED, Blumenthal RS, Nasir K, Qureshi WT, Brawner CA, Keteyian SJ, Blaha MJ. Sex differences in cardiorespiratory fitness and all-cause mortality: the Henry Ford Exercise testing (FIT) project. Mayo Clin Proc. 2016;91(6):755–62. https://doi.org/10.1016/j.mayocp.2016.04.002.
3. Babiker FA, De Windt LJ, van Eickels M, Thijssen V, Bronsaer RJ, Grohe C, van Bilsen M, Doevendans PA. 17beta-estradiol antagonizes cardiomyocyte hypertrophy by autocrine/paracrine stimulation of a guanylyl cyclase A receptor-cyclic guanosine monophosphate-dependent protein kinase pathway. Circulation. 2004;109(2):269–76. https://doi.org/10.1161/01.CIR.0000105682.85732.BD.
4. Baggish AL, Wang F, Weiner RB, Elinoff JM, Tournoux F, Boland A, Picard MH, Hutter AM Jr, Wood MJ. Training-specific changes in cardiac structure and function: a prospective and longitudinal assessment of competitive athletes. J Appl Physiol (1985). 2008;104(4):1121–8. https://doi.org/10.1152/japplphysiol.01170.2007.
5. Barnes JN, Joyner MJ. Physical activity and cardiovascular risk: 10 metabolic equivalents or bust. Mayo Clin Proc. 2013;88(12):1353–5. https://doi.org/10.1016/j.mayocp.2013.10.015.
6. Benini R, Prado Nunes PR, Orsatti CL, Barcelos LC, Orsatti FL. Effects of acute total body resistance exercise on hormonal and cytokines changes in men and women. J Sports Med Phys Fitness. 2015;55(4):337–44.
7. Best SA, Okada Y, Galbreath MM, Jarvis SS, Bivens TB, Adams-Huet B, Fu Q. Age and sex differences in muscle sympathetic nerve activity in relation to haemodynamics, blood volume and left ventricular size. Exp Physiol. 2014;99(6):839–48. https://doi.org/10.1113/expphysiol.2013.077248.
8. Blair SN, Kohl HW 3rd, Paffenbarger RS Jr, Clark DG, Cooper KH, Gibbons LW. Physical fitness and all-cause mortality. A prospective study of healthy men and women. JAMA. 1989;262(17):2395–401.
9. Chantler PD. Arterial ventricular uncoupling with age and disease and recoupling with exercise. Exerc Sport Sci Rev. 2017;45(2):70–9. https://doi.org/10.1249/JES.0000000000000100.

10. Chantler PD, Lakatta EG, Najjar SS. Arterial-ventricular coupling: mechanistic insights into cardiovascular performance at rest and during exercise. J Appl Physiol (1985). 2008a;105(4):1342–51. https://doi.org/10.1152/japplphysiol.90600.2008.
11. Chantler PD, Melenovsky V, Schulman SP, Gerstenblith G, Becker LC, Ferrucci L, Fleg JL, Lakatta EG, Najjar SS. The sex-specific impact of systolic hypertension and systolic blood pressure on arterial-ventricular coupling at rest and during exercise. Am J Physiol Heart Circ Physiol. 2008b;295(1):H145–53. https://doi.org/10.1152/ajpheart.01179.2007.
12. Charkoudian N, Joyner MJ. Physiologic considerations for exercise performance in women. Clin Chest Med. 2004;25(2):247–55. https://doi.org/10.1016/j.ccm.2004.01.001.
13. Chung AK, Das SR, Leonard D, Peshock RM, Kazi F, Abdullah SM, Canham RM, Levine BD, Drazner MH. Women have higher left ventricular ejection fractions than men independent of differences in left ventricular volume: the Dallas Heart Study. Circulation. 2006;113(12):1597–604. https://doi.org/10.1161/CIRCULATIONAHA.105.574400.
14. Coutinho T, Borlaug BA, Pellikka PA, Turner ST, Kullo IJ. Sex differences in arterial stiffness and ventricular-arterial interactions. J Am Coll Cardiol. 2013;61(1):96–103. https://doi.org/10.1016/j.jacc.2012.08.997.
15. Cowan MM, Gregory LW. Responses of pre- and postmenopausal females to aerobic conditioning. Med Sci Sports Exerc. 1985;17(1):138–43.
16. Creatsa M, Armeni E, Stamatelopoulos K, Rizos D, Georgiopoulos G, Kazani M, Alexandrou A, Dendrinos S, Augoulea A, Papamichael C, Lambrinoudaki I. Circulating androgen levels are associated with subclinical atherosclerosis and arterial stiffness in healthy recently menopausal women. Metabolism. 2012;61(2):193–201. https://doi.org/10.1016/j.metabol.2011.06.005.
17. D'Andrea A, Riegler L, Cocchia R, Scarafile R, Salerno G, Gravino R, Golia E, Vriz O, Citro R, Limongelli G, Calabro P, Di Salvo G, Caso P, Russo MG, Bossone E, Calabro R. Left atrial volume index in highly trained athletes. Am Heart J. 2010;159(6):1155–61. https://doi.org/10.1016/j.ahj.2010.03.036.
18. Dalpiaz PL, Lamas AZ, Caliman IF, Ribeiro RF Jr, Abreu GR, Moyses MR, Andrade TU, Gouvea SA, Alves MF, Carmona AK, Bissoli NS. Sex hormones promote opposite effects on ACE and ACE2 activity, hypertrophy and cardiac contractility in spontaneously hypertensive rats. PLoS One. 2015;10(5):e0127515. https://doi.org/10.1371/journal.pone.0127515.
19. De Bondt P, Van de Wiele C, De Sutter J, De Winter F, De Backer G, Dierckx RA. Age- and gender-specific differences in left ventricular cardiac function and volumes determined by gated SPET. Eur J Nucl Med. 2001;28(5):620–4.
20. Ehsani AA, Ogawa T, Miller TR, Spina RJ, Jilka SM. Exercise training improves left ventricular systolic function in older men. Circulation. 1991;83(1):96–103.
21. Fagard R, Lijnen P, Staessen J, Thijs L, Amery A. Effect of age on the hemodynamic response to posture in nonelderly hypertensive patients. Am J Hypertens. 1994;7(1):30–5.
22. Fahs CA, Rossow LM, Yan H, Ranadive SM, Agiovlasitis S, Wilund KR, Baynard T, Fernhall B. Resting and post exercise arterial-ventricular coupling in endurance-trained men and women. J Hum Hypertens. 2013;27(9):552–6. https://doi.org/10.1038/jhh.2013.7.
23. Febbraio MA, Hiscock N, Sacchetti M, Fischer CP, Pedersen BK. Interleukin-6 is a novel factor mediating glucose homeostasis during skeletal muscle contraction. Diabetes. 2004;53(7):1643–8.
24. Fitzgerald MD, Tanaka H, Tran ZV, Seals DR. Age-related declines in maximal aerobic capacity in regularly exercising vs. sedentary women: a meta-analysis. J Appl Physiol (1985). 1997;83(1):160–5.
25. Fiuza-Luces C, Garatachea N, Berger NA, Lucia A. Exercise is the real polypill. Physiology (Bethesda). 2013;28(5):330–58. https://doi.org/10.1152/physiol.00019.2013.
26. Fleg JL, O'Connor F, Gerstenblith G, Becker LC, Clulow J, Schulman SP, Lakatta EG. Impact of age on the cardiovascular response to dynamic upright exercise in healthy men and women. J Appl Physiol (1985). 1995;78(3):890–900.
27. Fliegner D, Schubert C, Penkalla A, Witt H, Kararigas G, Dworatzek E, Staub E, Martus P, Ruiz Noppinger P, Kintscher U, Gustafsson JA, Regitz-Zagrosek V. Female sex and estrogen receptor-beta attenuate cardiac remodeling and apoptosis in pressure overload. Am J Physiol Regul Integr Comp Physiol. 2010;298(6):R1597–606. https://doi.org/10.1152/ajpregu.00825.2009.
28. Fournier SB, Donley DA, Bonner DE, Devallance E, Olfert IM, Chantler PD. Improved arterial-ventricular coupling in metabolic syndrome after exercise training: a pilot study. Med Sci Sports Exerc. 2015;47(1):2–11. https://doi.org/10.1249/MSS.0000000000000388.
29. Freedson P, Katch VL, Sady S, Weltman A. Cardiac output differences in males and females during mild cycle ergometer exercise. Med Sci Sports. 1979;11(1):16–9.
30. Fu Q, Levine BD. Cardiovascular response to exercise in women. Med Sci Sports Exerc. 2005;37(8):1433–5.
31. Fu Q, Levine BD. Exercise and the autonomic nervous system. Handb Clin Neurol. 2013;117:147–60. https://doi.org/10.1016/B978-0-444-53491-0.00013-4.
32. Gardin JM, Savage DD, Ware JH, Henry WL. Effect of age, sex, and body surface area on echocardiographic left ventricular wall mass in normal subjects. Hypertension. 1987;9(2 Pt 2):II36–9.
33. Hagberg JM, Allen WK, Seals DR, Hurley BF, Ehsani AA, Holloszy JO. A hemodynamic comparison of young and older endurance athletes during exercise. J Appl Physiol (1985). 1985;58(6):2041–6.
34. Hagberg JM, Graves JE, Limacher M, Woods DR, Leggett SH, Cononie C, Gruber JJ, Pollock ML. Cardiovascular responses of 70- to 79-yr-old men and women to exercise training. J Appl Physiol (1985). 1989;66(6):2589–94.
35. Haines CD, Harvey PA, Leinwand LA. Estrogens mediate cardiac hypertrophy in a stimulus-dependent manner. Endocrinology. 2012;153(9):4480–90. https://doi.org/10.1210/en.2012-1353.
36. Harber MP, Kaminsky LA, Arena R, Blair SN, Franklin BA, Myers J, Ross R. Impact of cardiorespiratory fitness on all-cause and disease-specific mortality: advances since 2009. Prog Cardiovasc Dis. 2017;60:11. https://doi.org/10.1016/j.pcad.2017.03.001.
37. Higginbotham MB, Morris KG, Williams RS, McHale PA, Coleman RE, Cobb FR. Regulation of stroke volume during submaximal and maximal upright exercise in normal man. Circ Res. 1986;58(2):281–91.
38. Hossack KF, Bruce RA. Maximal cardiac function in sedentary normal men and women: comparison of age-related changes. J Appl Physiol Respir Environ Exerc Physiol. 1982;53(4):799–804.

39. Hossack KF, Kusumi F, Bruce RA. Approximate normal standards of maximal cardiac output during upright exercise in women. Am J Cardiol. 1981;47(5):1080–6.
40. Howden EJ, Perhonen M, Peshock RM, Zhang R, Arbab-Zadeh A, Adams-Huet B, Levine BD. Females have a blunted cardiovascular response to one year of intensive supervised endurance training. J Appl Physiol (1985). 2015;119(1):37–46. https://doi.org/10.1152/japplphysiol.00092.2015.
41. Hutchinson PL, Cureton KJ, Outz H, Wilson G. Relationship of cardiac size to maximal oxygen uptake and body size in men and women. Int J Sports Med. 1991;12(4):369–73. https://doi.org/10.1055/s-2007-1024696.
42. Joyner MJ. Physiological limiting factors and distance running: influence of gender and age on record performances. Exerc Sport Sci Rev. 1993;21:103–33.
43. Kadda O, Manginas A, Stavridis G, Balanos D, Kotiou M, Panagiotakos DB. Gender analysis in the outcomes of a lifestyle intervention among patients who had an open heart surgery. Angiology. 2016;67(1):66–74. https://doi.org/10.1177/0003319715577293.
44. Kanstrup IL, Marving J, Hoilund-Carlsen PF. Acute plasma expansion: left ventricular hemodynamics and endocrine function during exercise. J Appl Physiol (1985). 1992;73(5):1791–6.
45. Kawaguchi M, Hay I, Fetics B, Kass DA. Combined ventricular systolic and arterial stiffening in patients with heart failure and preserved ejection fraction: implications for systolic and diastolic reserve limitations. Circulation. 2003;107(5):714–20.
46. Kawecka-Jaszcz K, Czarnecka D, Olszanecka A, Rajzer M, Jankowski P. The effect of hormone replacement therapy on arterial blood pressure and vascular compliance in postmenopausal women with arterial hypertension. J Hum Hypertens. 2002;16(7):509–16. https://doi.org/10.1038/sj.jhh.1001431.
47. Kelly RP, Ting CT, Yang TM, Liu CP, Maughan WL, Chang MS, Kass DA. Effective arterial elastance as index of arterial vascular load in humans. Circulation. 1992;86(2):513–21.
48. Kilbom A. Physical training with submaximal intensities in women. I Reaction to exercise and orthostasis Scand J Clin Lab Invest. 1971;28(2):141–61.
49. Kilbom A, Astrand I. Physical training with submaximal intensities in women. II Effect on cardiac output Scand J Clin Lab Invest. 1971;28(2):163–75.
50. Kodama S, Saito K, Tanaka S, Maki M, Yachi Y, Asumi M, Sugawara A, Totsuka K, Shimano H, Ohashi Y, Yamada N, Sone H. Cardiorespiratory fitness as a quantitative predictor of all-cause mortality and cardiovascular events in healthy men and women: a meta-analysis. JAMA. 2009;301(19):2024–35. https://doi.org/10.1001/jama.2009.681.
51. Lee IM, Shiroma EJ, Lobelo F, Puska P, Blair SN, Katzmarzyk PT, Lancet Physical Activity Series Working G. Effect of physical inactivity on major non-communicable diseases worldwide: an analysis of burden of disease and life expectancy. Lancet. 2012;380(9838):219–29. https://doi.org/10.1016/S0140-6736(12)61031-9.
52. Levine BD. Exercise physiology for the clinician. In: Exercise and sports cardiology. New York: McGraw-Hill, Medical Publishing Division; 2001. p. 3–29.
53. Levine BD. VO_2max: what do we know, and what do we still need to know? J Physiol. 2008;586(1):25–34. https://doi.org/10.1113/jphysiol.2007.147629.
54. Levine BD, Lane LD, Buckey JC, Friedman DB, Blomqvist CG. Left ventricular pressure-volume and Frank-Starling relations in endurance athletes. Implications for orthostatic tolerance and exercise performance. Circulation. 1991;84(3):1016–23.
55. Lorell BH, Carabello BA. Left ventricular hypertrophy: pathogenesis, detection, and prognosis. Circulation. 2000;102(4):470–9.
56. Martinez-Vizcaino V, Sanchez-Lopez M, Notario-Pacheco B, Salcedo-Aguilar F, Solera-Martinez M, Franquelo-Morales P, Lopez-Martinez S, Garcia-Prieto JC, Arias-Palencia N, Torrijos-Nino C, Mora-Rodriguez R, Rodriguez-Artalejo F. Gender differences on effectiveness of a school-based physical activity intervention for reducing cardiometabolic risk: a cluster randomized trial. Int J Behav Nutr Phys Act. 2014;11:154. https://doi.org/10.1186/s12966-014-0154-4.
57. McEniery CM, Yasmin HIR, Qasem A, Wilkinson IB, Cockcroft JR, Investigators A. Normal vascular aging: differential effects on wave reflection and aortic pulse wave velocity: the Anglo-Cardiff collaborative trial (ACCT). J Am Coll Cardiol. 2005;46(9):1753–60. https://doi.org/10.1016/j.jacc.2005.07.037.
58. Mier CM, Domenick MA, Turner NS, Wilmore JH. Changes in stroke volume and maximal aerobic capacity with increased blood volume in men women. J Appl Physiol (1985). 1996;80(4):1180–6.
59. Mier CM, Domenick MA, Wilmore JH. Changes in stroke volume with beta-blockade before and after 10 days of exercise training in men and women. J Appl Physiol (1985). 1997;83(5):1660–5.
60. Mitchell GF, Parise H, Benjamin EJ, Larson MG, Keyes MJ, Vita JA, Vasan RS, Levy D. Changes in arterial stiffness and wave reflection with advancing age in healthy men and women: the Framingham Heart Study. Hypertension. 2004;43(6):1239–45. https://doi.org/10.1161/01.HYP.0000128420.01881.aa.
61. Mitchell JH, Tate C, Raven P, Cobb F, Kraus W, Moreadith R, O'Toole M, Saltin B, Wenger N. Acute response and chronic adaptation to exercise in women. Med Sci Sports Exerc. 1992;24(6 Suppl):S258–65.
62. Morita N, Okita K. Is gender a factor in the reduction of cardiovascular risks with exercise training? Circ J. 2013;77(3):646–51.
63. Najjar SS, Schulman SP, Gerstenblith G, Fleg JL, Kass DA, O'Connor F, Becker LC, Lakatta EG. Age and gender affect ventricular-vascular coupling during aerobic exercise. J Am Coll Cardiol. 2004;44(3):611–7. https://doi.org/10.1016/j.jacc.2004.04.041.
64. Nichols WW, Denardo SJ, Wilkinson IB, McEniery CM, Cockcroft J, O'Rourke MF. Effects of arterial stiffness, pulse wave velocity, and wave reflections on the central aortic pressure waveform. J Clin Hypertens (Greenwich). 2008;10(4):295–303.
65. Nichols WW, Nichols WW, DA MD. McDonald's blood flow in arteries: theoretic, experimental, and clinical principles. 6th ed. London: Hodder Arnold; 2011.
66. Northoff H, Symons S, Zieker D, Schaible EV, Schafer K, Thoma S, Loffler M, Abbasi A, Simon P, Niess AM, Fehrenbach E. Gender- and menstrual phase dependent regulation of inflammatory gene expression in response to aerobic exercise. Exerc Immunol Rev. 2008;14:86–103.
67. O'Toole ML. Gender differences in the cardiovascular response to exercise. Cardiovasc Clin. 1989;19(3):17–33.
68. Ogawa T, Spina RJ, Martin WH 3rd, Kohrt WM, Schechtman KB, Holloszy JO, Ehsani AA. Effects of aging, sex, and physical training on cardiovascular responses to exercise. Circulation. 1992;86(2):494–503.
69. Okazaki K, Ichinose T, Mitono H, Chen M, Masuki S, Endoh H, Hayase H, Doi T, Nose H. Impact of protein and

carbohydrate supplementation on plasma volume expansion and thermoregulatory adaptation by aerobic training in older men. J Appl Physiol (1985). 2009;107(3):725–33. https://doi.org/10.1152/japplphysiol.91265.2008.
70. Ostrowski K, Rohde T, Zacho M, Asp S, Pedersen BK. Evidence that interleukin-6 is produced in human skeletal muscle during prolonged running. J Physiol. 1998;508 (Pt 3):949–53.
71. Palmer BF, Clegg DJ. The sexual dimorphism of obesity. Mol Cell Endocrinol. 2015;402:113–9. https://doi.org/10.1016/j.mce.2014.11.029.
72. Park S, Ha JW, Shim CY, Choi EY, Kim JM, Ahn JA, Lee SW, Rim SJ, Chung N. Gender-related difference in arterial elastance during exercise in patients with hypertension. Hypertension. 2008;51(4):1163–9. https://doi.org/10.1161/HYPERTENSIONAHA.107.106690.
73. Pearson AC, Guo R, Orsinelli DA, Binkley PF, Pasierski TJ. Transesophageal echocardiographic assessment of the effects of age, gender, and hypertension on thoracic aortic wall size, thickness, and stiffness. Am Heart J. 1994;128 (2):344–51.
74. Pedram A, Razandi M, Lubahn D, Liu J, Vannan M, Levin ER. Estrogen inhibits cardiac hypertrophy: role of estrogen receptor-beta to inhibit calcineurin. Endocrinology. 2008;149(7):3361–9. https://doi.org/10.1210/en.2008-0133.
75. Phillips T, Childs AC, Dreon DM, Phinney S, Leeuwenburgh C. A dietary supplement attenuates IL-6 and CRP after eccentric exercise in untrained males. Med Sci Sports Exerc. 2003;35(12):2032–7. https://doi.org/10.1249/01.MSS.0000099112.32342.10.
76. Piro M, Della Bona R, Abbate A, Biasucci LM, Crea F. Sex-related differences in myocardial remodeling. J Am Coll Cardiol. 2010;55(11):1057–65. https://doi.org/10.1016/j.jacc.2009.09.065.
77. Pivarnik JM, Sherman NW. Responses of aerobically fit men and women to uphill/downhill walking and slow jogging. Med Sci Sports Exerc. 1990;22(1):127–30.
78. Powers SK, Smuder AJ, Kavazis AN, Quindry JC. Mechanisms of exercise-induced cardioprotection. Physiology (Bethesda). 2014;29(1):27–38. https://doi.org/10.1152/physiol.00030.2013.
79. Proctor DN, Beck KC, Shen PH, Eickhoff TJ, Halliwill JR, Joyner MJ. Influence of age and gender on cardiac output-VO_2 relationships during submaximal cycle ergometry. J Appl Physiol (1985). 1998;84(2):599–605.
80. Redfield MM, Jacobsen SJ, Borlaug BA, Rodeheffer RJ, Kass DA. Age- and gender-related ventricular-vascular stiffening: a community-based study. Circulation. 2005;112 (15):2254–62. https://doi.org/10.1161/CIRCULATIONAHA.105.541078.
81. Rerych SK, Scholz PM, Sabiston DC Jr, Jones RH. Effects of exercise training on left ventricular function in normal subjects: a longitudinal study by radionuclide angiography. Am J Cardiol. 1980;45(2):244–52.
82. Ross J Jr. Afterload mismatch and preload reserve: a conceptual framework for the analysis of ventricular function. Prog Cardiovasc Dis. 1976;18(4):255–64.
83. Ross R, Blair SN, Arena R, Church TS, Despres JP, Franklin BA, Haskell WL, Kaminsky LA, Levine BD, Lavie CJ, Myers J, Niebauer J, Sallis R, Sawada SS, Sui X, Wisloff U, American Heart Association Physical Activity Committee of the Council on L, Cardiometabolic H, Council on Clinical C, Council on E, Prevention, Council on C, Stroke N, Council on Functional G, Translational B, Stroke C. Importance of assessing cardiorespiratory fitness in clinical practice: a case for fitness as a clinical vital sign: a scientific statement from the American Heart Association. Circulation. 2016;134(24):e653–99. https://doi.org/10.1161/CIR.0000000000000461.
84. Rowell LB. Central circulatory adjustments to dynamic exercise. In: Rowell LB, editor. Human Cardiovascular Control. New York: Oxford University Press; 1993. p. 162–203.
85. Saltin B, Blomqvist G, Mitchell JH, Johnson RL Jr, Wildenthal K, Chapman CB. Response to exercise after bed rest and after training. Circulation. 1968;38(5 Suppl): VII1–78.
86. Scuteri A, Orru M, Morrell CH, Tarasov K, Schlessinger D, Uda M, Lakatta EG. Associations of large artery structure and function with adiposity: effects of age, gender, and hypertension. The SardiNIA study. Atherosclerosis. 2012;221(1):189–97. https://doi.org/10.1016/j.atherosclerosis.2011.11.045.
87. Seals DR, Hagberg JM, Hurley BF, Ehsani AA, Holloszy JO. Endurance training in older men and women. I Cardiovascular responses to exercise J Appl Physiol Respir Environ Exerc Physiol. 1984;57(4):1024–9.
88. Seals DR, Hagberg JM, Spina RJ, Rogers MA, Schechtman KB, Ehsani AA. Enhanced left ventricular performance in endurance trained older men. Circulation. 1994;89 (1):198–205.
89. Shepherd AP, Granger HJ, Smith EE, Guyton AC. Local control of tissue oxygen delivery and its contribution to the regulation of cardiac output. Am J Phys. 1973;225 (3):747–55.
90. Shim CY, Park S, Choi D, Yang WI, Cho IJ, Choi EY, Chung N, Ha JW. Sex differences in central hemodynamics and their relationship to left ventricular diastolic function. J Am Coll Cardiol. 2011;57(10):1226–33. https://doi.org/10.1016/j.jacc.2010.09.067.
91. Skavdahl M, Steenbergen C, Clark J, Myers P, Demianenko T, Mao L, Rockman HA, Korach KS, Murphy E. Estrogen receptor-beta mediates male-female differences in the development of pressure overload hypertrophy. Am J Physiol Heart Circ Physiol. 2005;288(2):H469–76. https://doi.org/10.1152/ajpheart.00723.2004.
92. Snell PG, Mitchell JH. The role of maximal oxygen uptake in exercise performance. Clin Chest Med. 1984;5(1):51–62.
93. Spina RJ, Miller TR, Bogenhagen WH, Schechtman KB, Ehsani AA. Gender-related differences in left ventricular filling dynamics in older subjects after endurance exercise training. J Gerontol A Biol Sci Med Sci. 1996;51 (3):B232–7.
94. Spina RJ, Ogawa T, Kohrt WM, Martin WH 3rd, Holloszy JO, Ehsani AA. Differences in cardiovascular adaptations to endurance exercise training between older men and women. J Appl Physiol (1985). 1993a;75(2):849–55.
95. Spina RJ, Ogawa T, Martin WH 3rd, Coggan AR, Holloszy JO, Ehsani AA. Exercise training prevents decline in stroke volume during exercise in young healthy subjects. J Appl Physiol (1985). 1992;72(6):2458–62.
96. Spina RJ, Ogawa T, Miller TR, Kohrt WM, Ehsani AA. Effect of exercise training on left ventricular performance in older women free of cardiopulmonary disease. Am J Cardiol. 1993b;71(1):99–104.
97. Spina RJ, Rashid S, Davila-Roman VG, Ehsani AA. Adaptations in beta-adrenergic cardiovascular responses to training in older women. J Appl Physiol (1985). 2000;89 (6):2300–5.
98. Stachenfeld NS, Mack GW, DiPietro L, Morocco TS, Jozsi AC, Nadel ER. Regulation of blood volume during training in post-menopausal women. Med Sci Sports Exerc. 1998;30 (1):92–8.

99. Staessen JA, van der Heijden-Spek JJ, Safar ME, Den Hond E, Gasowski J, Fagard RH, Wang JG, Boudier HA, Van Bortel LM. Menopause and the characteristics of the large arteries in a population study. J Hum Hypertens. 2001;15(8):511–8. https://doi.org/10.1038/sj.jhh.1001226.
100. Starling MR. Left ventricular-arterial coupling relations in the normal human heart. Am Heart J. 1993;125(6):1659–66.
101. Sullivan MJ, Cobb FR, Higginbotham MB. Stroke volume increases by similar mechanisms during upright exercise in normal men and women. Am J Cardiol. 1991;67 (16):1405–12.
102. Sun B, Ma JZ, Yong YH, Lv YY. The upper limit of physiological cardiac hypertrophy in elite male and female athletes in China. Eur J Appl Physiol. 2007;101(4):457–63. https://doi.org/10.1007/s00421-007-0517-5.
103. Talbot LA, Morrell CH, Metter EJ, Fleg JL. Comparison of cardiorespiratory fitness versus leisure time physical activity as predictors of coronary events in men aged < or = 65 years and > 65 years. Am J Cardiol. 2002;89(10):1187–92.
104. Tanaka H, CA DS, Seals DR. Absence of age-related increase in central arterial stiffness in physically active women. Arterioscler Thromb Vasc Biol. 1998;18(1):127–32.
105. Tanaka H, Dinenno FA, Monahan KD, Clevenger CM, CA DS, Seals DR. Aging, habitual exercise, and dynamic arterial compliance. Circulation. 2000;102(11):1270–5.
106. Tarumi T, Ayaz Khan M, Liu J, Tseng BY, Parker R, Riley J, Tinajero C, Zhang R. Cerebral hemodynamics in normal aging: central artery stiffness, wave reflection, and pressure pulsatility. J Cereb Blood Flow Metab. 2014;34(6):971–8. https://doi.org/10.1038/jcbfm.2014.44.
107. Vermeersch SJ, Rietzschel ER, De Buyzere ML, De Bacquer D, De Backer G, Van Bortel LM, Gillebert TC, Verdonck PR, Segers P. Age and gender related patterns in carotid-femoral PWV and carotid and femoral stiffness in a large healthy, middle-aged population. J Hypertens. 2008;26 (7):1411–9. https://doi.org/10.1097/HJH.0b013e3282ffac00.
108. Vingren JL, Kraemer WJ, Ratamess NA, Anderson JM, Volek JS, Maresh CM. Testosterone physiology in resistance exercise and training: the up-stream regulatory elements. Sports Med. 2010;40(12):1037–53. https://doi.org/10.2165/11536910-000000000-00000.
109. Vlachopoulos C, Ioakeimidis N, Miner M, Aggelis A, Pietri P, Terentes-Printzios D, Tsekoura D, Stefanadis C. Testosterone deficiency: a determinant of aortic stiffness in men. Atherosclerosis. 2014;233(1):278–83. https://doi.org/10.1016/j.atherosclerosis.2013.12.010.
110. Waddell TK, Dart AM, Gatzka CD, Cameron JD, Kingwell BA. Women exhibit a greater age-related increase in proximal aortic stiffness than men. J Hypertens. 2001;19 (12):2205–12.
111. Wainstein RV, Sasson Z, Mak S. Frequency-dependent left ventricular performance in women and men. Am J Physiol Heart Circ Physiol. 2012;302(11):H2363–71. https://doi.org/10.1152/ajpheart.01125.2011.
112. Wells CL, Boorman MA, Riggs DM. Effect of age and menopausal status on cardiorespiratory fitness in masters women runners. Med Sci Sports Exerc. 1992;24 (10):1147–54.
113. Whitham M, Febbraio MA. The ever-expanding myokinome: discovery challenges and therapeutic implications. Nat Rev Drug Discov. 2016;15(10):719–29. https://doi.org/10.1038/nrd.2016.153.
114. Whyte GP, George K, Sharma S, Firoozi S, Stephens N, Senior R, WJ MK. The upper limit of physiological cardiac hypertrophy in elite male and female athletes: the British experience. Eur J Appl Physiol. 2004;92(4–5):592–7. https://doi.org/10.1007/s00421-004-1052-2.
115. Wiebe CG, Gledhill N, Warburton DE, Jamnik VK, Ferguson S. Exercise cardiac function in endurance-trained males versus females. Clin J Sport Med. 1998;8(4):272–9.
116. Wilmore JH. The application of science to sport: physiological profiles of male and female athletes. Can J Appl Sport Sci. 1979;4(2):103–15.
117. Wilmore JH, Stanforth PR, Gagnon J, Rice T, Mandel S, Leon AS, Rao DC, Skinner JS, Bouchard C. Cardiac output and stroke volume changes with endurance training: the HERITAGE family study. Med Sci Sports Exerc. 2001;33 (1):99–106.
118. Yaginuma T, Noda T, Tsuchiya M, Takazawa K, Tanaka H, Kotoda K, Hosoda S. Interaction of left ventricular contraction and aortic input impedance in experimental and clinical studies. Jpn Circ J. 1985;49(2):206–14.
119. Zwiren LD, Cureton KJ, Hutchinson P. Comparison of circulatory responses to submaximal exercise in equally trained men and women. Int J Sports Med. 1983;4(4):255–9. https://doi.org/10.1055/s-2008-1026045.

Growth of Cardiovascular Structures from the Fetus to the Young Adult

22

Frederic Dallaire and Taisto Sarkola

Art work by Piet Michiels, Leuven, Belgium

F. Dallaire (✉)
Université de Sherbrooke, Sherbrooke, Canada
e-mail: frederic.a.dallaire@usherbrooke.ca

T. Sarkola
University of Helsinki, the Helsinki University Central Hospital/Children's Hospital, Helsinki, Finland

Abstract

The size, hemodynamics, and function of cardiovascular structures change dramatically from the early fetal life to late adolescence. The principal determinants of cardiovascular

P. L. M. Kerkhof, V. M. Miller (eds.), *Sex-Specific Analysis of Cardiovascular Function*,
Advances in Experimental Medicine and Biology 1065, https://doi.org/10.1007/978-3-319-77932-4_22

dimensions are related to the blood flow needed to meet metabolic demands. This demand is in turn tightly related to body size and body composition, keeping in mind that various tissues may have different metabolic rates. There is no simple model that links cardiac dimensions with a single body size measurement. Consequently, despite abundant scientific literature, few studies have proposed pediatric reference values that efficiently and completely account for the effect of body size. Other factors influence cardiovascular size and function in children, including sex. The influence of sex is multifactorial and not fully understood, but differences in body size and body composition play an important role. We will first review the determinants of cardiovascular size and function in children. We then explore the evaluation and normalization of cardiovascular size and function in pediatric cardiology in relation to the growth of cardiovascular structures during childhood, with a particular focus on sex differences.

Keywords
Fetus · Child · Adolescent · Growth · Puberty · Pediatric cardiology · Sex · Reference value · Nomogram · Indexing · Blood pressure · Echocardiography · ECG · Body composition · Lean body mass · Exercise testing · Artery size · Allometry · Z score

Introduction

It is common knowledge that cardiovascular dimensions are related to body size throughout the life span. There are numerous early observations that heart mass is proportional to body mass in several animal species [1]. In humans, the early work of Graham and collaborators put forward the notion that heart size was directly related to body surface area (BSA) in children [2, 3]. This notion was reinforced by several other studies showing a linear relationship between basal metabolic rate, cardiac output, and BSA (reviewed by Nevill et al. [4]). Despite the advantageous simplicity of a somewhat linear relationship between cardiovascular dimensions and body size, we now know that there is no simple scaling method that can explain the complex interaction of body size and cardiovascular size, especially in children (reviewed by Nevill et al. [4] and Batterham et al. [5]). Furthermore, it has been shown that the relation between body size and cardiovascular size may be different in healthy boys and girls [6–9]. Here, we review the determinants of cardiovascular size and function in children. We then discuss the implications of the growth of cardiovascular structures in the evaluation of cardiovascular size and function in pediatric cardiology, with a particular focus on sex differences.

Determinants of Cardiovascular Dimensions in the Healthy Child

Under the principle of minimal work, the optimal diameter of a vessel will be one that minimizes the energy needed to propel a given amount of blood flow (reviewed by Sluysmans and Colan [10]). Nature tends to find a compromise between the energy needed to overcome shear stress (inversely proportional to vessel diameter) and the energy needed to surmount blood inertia (proportional to vessel diameter). The blood flow required to support life varies greatly according to metabolic demands. It has been shown that under the principle of minimum work, the cardiovascular size in children is optimized to accommodate a flow that is two to four times greater than the resting cardiac output [10].

One can hypothesize that cardiovascular dimensions are likely related to the area under the curve of cardiac output over time, although this would be difficult to prove. On any given day, the total cardiac output will change according to many hemodynamic factors, but, in general, it should be closely related to the total amount of body mass that has been metabolically active during that day. Indeed, it has been observed many times that both cardiovascular dimensions and basal metabolic rates are closely related to body dimensions, with an even tighter correlation with surrogate metrics of metabolically active mass, such as lean body mass (reviewed by Batterham et al. [5]).

Many biological and morphological factors are associated with heart size, but their contributions are usually greatly reduced after adequate adjustment for body size and body composition. An interesting example of this is the demonstration by Blimkie and collaborators that the tight relation between peak oxygen consumption and heart size is mainly due to the close relationship between both these parameters and body size [11]. For the same height and weight, athletes have a different body composition compared with nonathletes, and this variation in muscle and fat mass explains a great proportion of the difference in peak VO_2 [12, 13]. There is also evidence that BSA – but not age – is a strong determinant of both arterial lumen and wall thickness during rapid growth in early childhood [14], although arterial wall thickening is also influenced by gender differences in blood pressure (and wall stress) later on in adolescence [15].

Between sexes, there are clear differences in heart size and aerobic fitness, despite adjustment for body size [6–9, 16]. This may in part be due to differences in body composition. Nevill and collaborators have shown that during puberty, at the peak of growth spurt, the proportion of muscle mass over total body mass increased in boys but not in girls [12]. This increase in muscle mass raises the resting basal metabolic rate which, in turn, impacts on cardiovascular growth. Interestingly, arterial lumen and wall thickness are larger in males compared with females in the upper body, including conduit arteries of the arms and head, but not in arteries of the legs [14], which could be explained by sex differences in upper versus lower body proportions or composition. Whether a precise determination of body composition could explain the entire variation in cardiovascular dimension between males and females is not known. Nevertheless, because measuring body composition precisely is not convenient, many studies reporting relation between body size and cardiovascular dimensions have rightfully stratified their results by sex.

Prenatal Growth of Cardiovascular Dimensions

It comes as no surprise that the main determinants of cardiovascular dimensions prior to birth are related to gestational age and fetal body size. It has also been known for decades that maternal height and weight, parity, ethnicity, and fetal sex influence fetal body growth overall [17–19]. However, the magnitude of the absolute sex difference in fetal body size seems relatively small and is not equally distributed in the fetal body [20]. Currently, prediction models to assess fetal cardiovascular dimensions in clinical settings are adjusted for gestational age or fetal biometry only [21–23].

It has also become apparent that disturbed fetal growth is inevitably linked with disturbed cardiovascular growth as well. Both the constitutional small fetal body size (traditional symmetric small for gestational age infant) and progressive intrauterine growth restriction (asymmetric and symmetric intrauterine growth restriction or fetal growth restriction) are linked with smaller cardiovascular dimensions. Although cardiovascular dimensions tend to follow body size in fetuses with abnormal growth, differences in fetal right and left ventricular shapes have been reported in intrauterine growth restriction [24, 25]. At the other end of the disturbed fetal growth spectrum, excessive fetal body growth may also present as constitutional or be progressive (maternal diabetes), which is reflected as a higher proportion of fat mass and cardiac hypertrophy in the latter form. In newborns and infants, cardiovascular dimensions have been adjusted for BSA, with little influence from sex and prematurity [26]. In premature infants, weight has also been used due to the relatively small variability in the weight range and the technical variability in the body length measurement [27]. Body surface area and weight do not, however, fully account for variability in body composition in the setting of disturbed fetal growth.

Much less is known about peripheral arterial morphology during the prenatal stage, which may be due to the technical challenges related to limited image resolution. So far, most studies have focused on the great arteries. More lately, predictors for peripheral conduit arteries have been reported for premature infants (surrogate for fetus) following abnormal intrauterine growth. In these studies, the dimensions of the arteries were strongly predicted by infant BSA and/or arterial end-organ size [28]. In asymmetric intrauterine growth restriction with brain sparing, neck arterial dimensions seem best predicted by head circumference.

Body Size Adjustment and Reference Values

Challenges in Cardiovascular Adjustment for Body Size in Pediatric Cardiology

The identification of an abnormal size in any cardiovascular structure is essential in pediatric and fetal cardiac imaging. In the growing child, clinicians must disentangle the physiologic effect of body size from the pathological effect of disease. Reference values and body size-adjusted measurements are thus a direct and important clinical application of allometry. The use of adjusted measurements, often in the form of Z scores, is now the norm in cardiovascular imaging [29, 30].

As mentioned above, there is no simple scaling method that allows easy adjustment of cardiovascular size with a single body size variable in children. There are several methods to determine reference values for cardiovascular size and function in children (review by Blais et al. [16] and Mawad et al. [31]). Briefly, among important aspects to be considered, one can name population selection, choice of regression models, parametric versus nonparametric approaches, choice of scaling variable, evaluation of residual values, and conformation to adequate distribution of residuals. Although several prediction equations and methods of adjustment have been proposed and published in pediatric cardiology, they are of variable quality and often prone to bias [32, 33, 31]. The method of indexing (dividing a parameter by a measure of body size) represents an interesting example of a clearly biased method of adjustment that is still consistently used despite decades of warning in the scientific literature [34, 35, 5, 4, 36, 37, 6, 38–43].

The choice of a body size variable to adjust measurements in children is a critically important aspect – and one that is much debated – in the determination of reference values. Estimation of BSA has often been viewed as the "best" variable to normalize cardiovascular structure dimensions, and most recent Z score equations are based on BSA [31, 32, 44]. However, variation in body composition that is not accounted for by estimation of BSA may represent a significant limitation of BSA-adjusted Z scores. In predicting cardiovascular structure size, it has been shown that multivariable models are often superior to models that incorporate only one body size variable such as BSA [45–47]. It is logical that a parameter adjusted for body size should become completely independent of body size. Ideally, this should also be true in populations with abnormal body habitus. The approach used to derive predicting equations in pediatric cardiology has improved, and the importance of careful validation to ensure absence of residual associations of Z scores with body size has gained attention [31, 45, 32]. However, with the growing population of overweight and obese children, Z score equations accounting for variability in body composition are still needed.

Cardiac and Vascular Normal Dimensions and Reference Values

The vast majority of recent reference values for cardiovascular dimensions in children are for echocardiographic measurements. In their recent review, Cantinotti and collaborators cite more than 50 articles reporting reference values in pediatric echocardiography [32]. The table lists studies reporting reference values for cardiovascular dimensions in pediatric echocardiography since the year 2000. Only studies including 400 subjects or more are listed. As previously reported, overall, there is a lack of validation of most of these prediction models to ensure that the adjusted measurement is completely independent of body size. Some authors analyzed residual associations to ensure that the adjusted measurement was independent of the body size parameter for which it was adjusted [26, 48, 45, 49, 37, 9, 6, 50, 51]. However, except for a few exceptions [45, 49, 37], residual associations with other body size parameters such as height, age, or body mass index are rarely reported.

As stated in the table, most authors proposing reference values for cardiovascular dimensions have assessed potential sex differences for the body size-adjusted parameter. About half of the studies present values stratified by sex, while others report values for both sexes combined due to nonsignificant sex difference or to sex difference considered negligible and clinically irrelevant.

In fetal echocardiography, Z score equations for different cardiovascular dimensions have mainly been adjusted for either gestational age or a parameter of fetal body size [52–54, 21, 22]. To our knowledge, other factors known to influence fetal body growth (maternal body size, parity, ethnicity, and fetal sex) have not been included in prediction models. However, using a parameter of fetal body size rather than gestational age in the adjustment should at least partly account for the sex difference. In the setting of congenital heart disease, obstruction of right or left heart flow may significantly impact on fetal cardiovascular growth with hypoplasia of cardiac structures and arteries. Detecting fetal cardiovascular dimensions that fall outside the normal range and a deterioration in the longitudinal growth of these structures is key in the attempt to predict the early postnatal course, treatments, and management needed. Lately, Z scores adjusted for fetal size have been used in defining thresholds scores for fetal interventions [55, 56] with the intention to improve the intrauterine growth of underdeveloped cardiovascular structures and, eventually, postnatal outcomes.

Recent and well-executed studies proposing pediatric reference values for other cardiac diagnostic modalities are scarce. For cardiac magnetic resonance imaging, there are few small single-center studies, and none have proposed Z score equations based on sound and complete parametric modelization [38, 57–59]. The largest was done on 114 prospectively recruited children and young adults (8–20 years of age) and used the very robust lambda mu sigma (LMS) modeling technique but on volumes already indexed for BSA or weight [60]. The study provided percentile limits but did not allow for the calculation of Z scores. However, it did show that indexed right ventricular volumes vary with age, thus casting doubts on their use to monitor ventricular dilatation over time. The only pediatric study on cardiac magnetic resonance imaging that did not rely on indexing was that of Buechel et al., which included only 52 patients [38].

Hemodynamics Parameters

Cardiovascular dimensions, cardiac output, compliance of heart chambers, preload, afterload, conduction properties, and heart rate change tremendously with growth from the fetal stage to young adulthood. All of these impact on cardiovascular hemodynamics such as filling properties, blood velocities, stroke volumes, myocardial performance, and time intervals of the cardiac cycle. Evaluating the independent contribution of each factor is complex as they are usually interrelated.

In the evaluation of diastolic function in children, it has been observed that body size, age, and heart rate influence filling properties [61–67]. Because body size and heart rate change concomitantly with age, evaluating their relative contribution has been challenging. Two recent reviews evaluated the variation and reference values of diastolic indices in children [32, 68]. The somewhat weak and nonlinear association of most diastolic function parameters with body size has impelled many authors to report mean values or centiles stratified for age [69, 63, 62], although some have attempted parametric modeling to produce Z score equations [70, 65]. To our knowledge, sex differences in diastolic function are negligible [70, 69, 62], and there are currently no sex-specific reference values available.

Myocardial strain and strain rate have gained popularity to assist in the assessment of myocardial function. The influence of body size and age on strain has been inconsistently reported. Some have suggested that longitudinal and circumferential strains vary with age [71–75], but this has not been confirmed by others [76–79]. Studies proposing reference values for strain have been recently reviewed [80, 81]. Most studies report mean and standard deviations stratified by age groups, except one that proposes Z scores adjusted for BSA [82]. No sex differences have been reported after adjustments for age or body size [72, 79, 82].

Electrocardiography

It is common knowledge that most ECG parameters change during infancy and childhood. Reference values for infant and children are however surprisingly scarce. The initial work of Davignon et al. [83] has been and is still used today although two more contemporary studies specifically proposing reference value in children are available [84, 85]. Eyeballing scatter plots reported for age in the Rijnbeek et al. study [85], one can appreciate that ECG parameters do not seem to follow longitudinal changes in body size and, consequently, cardiac size does not seem to fully explain ECG variations with age. Determinants of ECG changes are multifactorial and complex. A study of close to 1.5 million Brazilian pediatric and adult subjects showed that some ECG parameters remain stable throughout life (P-wave frontal axis), some change during childhood and stabilize in adolescence and adulthood (heart rate, QRS duration, and QT interval), and some display a constant change throughout life (QRS frontal axis, P-wave duration, and QTc interval) [86].

Sex differences in ECG parameters have been documented by several authors, especially for the QTc interval [85, 84, 87, 88, 86, 89, 90]. Interestingly, in Brazilian subjects [86], curves of ECG parameter mean values stratified by sex overlap in young children but then diverge during early adolescence. However, the study by Rijnbeek et al. [85] reports sex differences for several parameters prior to adolescence. Fortunately, they do report sex-specific reference values.

Cardiopulmonary Exercise Testing

Cardiopulmonary exercise testing is an important noninvasive method to assess cardiopulmonary fitness, which has diagnostic and prognostic value in the assessment of chronic disease in children. Peak aerobic performance and the way the body responds to exercise dramatically change during childhood [91–95].

In addition to the influence of body size on cardiovascular fitness, sex differences have been repeatedly observed in exercise parameters [16]. The fact that muscle mass is an important determinant of aerobic fitness may in part explain the sex differences found [12, 13, 96]. It is likely that other factors related to sex also play a role in the higher peak oxygen consumption found in males. As an example, Fig. 22.1 shows the relation between fat-free mass and peak VO_2 in 233 healthy children aged 12–17 years stratified by sex (unpublished results from our institution). For the same fat-free mass, boys consistently display higher VO_2 compared with girls. Similarly, as shown in Fig. 22.2, when peak VO_2 is adjusted for height, fat-free mass, and habitual physical activity, males still show higher mean values compared with females.

These differences between sexes are important enough to consider males and females separately when prediction models are constructed. Blais and collaborators recently reviewed 34 published studies proposing pediatric reference values for cardiopulmonary exercise testing [16]. They report that most studies show results stratified by sex and propose sex-specific sets of reference values. They also conclude that, unfortunately, high-quality non-biased reference values for children are limited [16]. Consequently, the data from the important but dated work of Cooper in the 1980s is often still used to estimate predicted values in children [97, 93, 94].

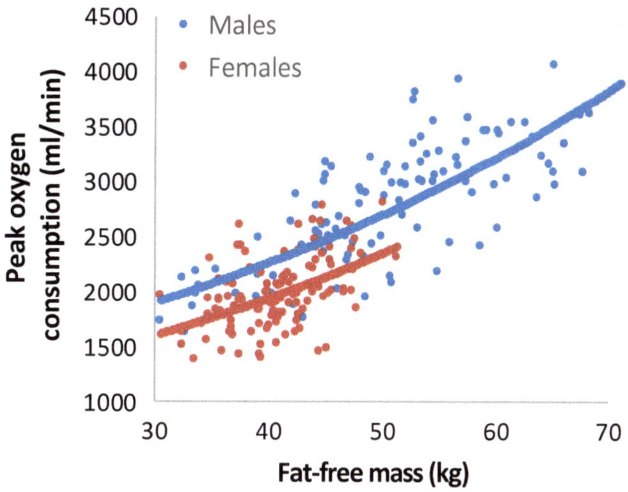

Fig. 22.1 Peak oxygen consumption according to fat-free mass stratified for sex

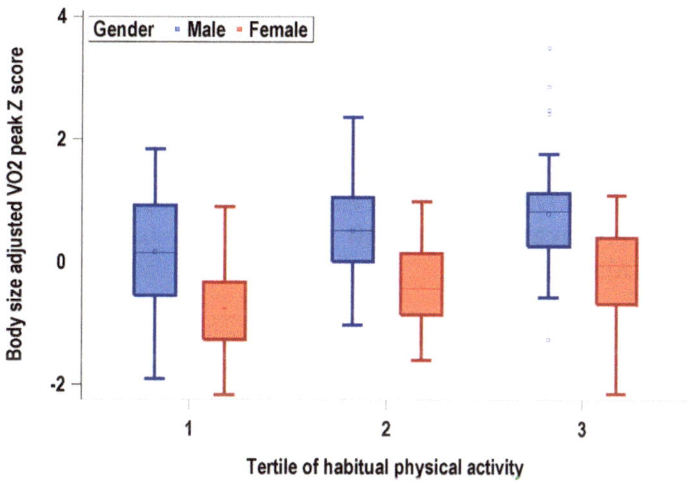

Fig. 22.2 Mean peak oxygen consumption after adjustment for height and fat-free mass and stratified by habitual physical activity and sex

Blood Pressure

Blood pressure is an easily available and commonly used parameter to assess hemodynamics in children, both in the setting of disease and for the assessment of cardiovascular risk. Resting blood pressure increases with age and body size to reach adult levels in adolescence. The heredity of blood pressure is known to be moderate with genetics explaining a substantial part of the variability (reviewed by Wang et al. [98]).

The most commonly used reference for outpatient blood pressure measurements is the American fourth report on the diagnosis, evaluation, and treatment of high blood pressure in children and adolescents [99]. It provides the normal range stratified for gender, age, and height. More recently, international blood pressure references for non-overweight children aged 6–17 years have been proposed with pooled data from seven nations [100]. Compared with the US Fourth Report, systolic pressures were lower but diastolic pressures higher. Higher pressure for males vs females were observed from >14 years of age, which is likely multifactorial but related to pubertal maturation. Currently, the definition of hypertension is based on the 95th percentile limit of the distribution of blood pressure in the population. A more appropriate approach would be to define pediatric hypertension based on the occurrence of target organ damage in the pediatric age or cardiovascular disease events in adult age. Reference values for 24-hour ambulatory blood pressure have been determined from different European populations for children aged 5–20 years [101]. They are reported separately for day and night periods and stratified for sex, age, and height.

Blood pressure impacts on arterial wall thickness and left ventricular mass. In later adolescence, differences in systolic blood pressure (and pulse pressure) explain gender differences in wall thickness [15]. Ethnic differences in carotid arterial layer thickness have also been reported with slightly thicker intima-media thickness in the healthy black pediatric population compared with the white [102]. This difference has at least partly been attributed to differences in lean body mass [103].

Impact of Disease and Treatment on Cardiovascular Growth and Maturation

An abnormal cardiovascular dimension is a major component in many forms of congenital heart malformations, and it is beyond the scope of this chapter to discuss in detail. The epidemiology of congenital heart diseases in relation to gender is addressed elsewhere [104].

Noncardiac disease, treatments, and interventions leading to long-term alterations in body growth and blood pressure and flow during childhood impact on cardiovascular growth as well. Examples include children with chronic kidney disease [105], obesity and type II diabetes [106], and childhood cancer survivors treated with radiotherapy [107]. The common feature in these populations is the negative impact of adverse body growth and composition, blood pressure, and disturbed flow on vessel growth including lumen dimension and wall thickness. Attenuation in the slope of wall thickening with age has also been reported when blood pressures and lipids have been lowered with an intervention [108, 109].

Summary

Body size changes tremendously from the early fetal life until the end of adolescence. There is a concomitant change in metabolic demands, cardiovascular size, and hemodynamics. The strong relation between body size and cardiovascular dimensions is however not linear, and predicting heart and vessel dimensions according to simple parameters of body size (height, weight, or BSA) has been challenging. This complexity may not have been sufficiently recognized in the scientific literature. Consequently, despite the large number of published reference values in pediatric cardiology, many are fraught with potential biases due to incomplete adjustment for body size. The effect of sex on cardiovascular size is well recognized, but there are areas of uncertainties regarding the contribution of body composition in explaining sex differences. In future studies, a careful examination of residual associations with potentially confounding variables (obesity, body composition, sex, and race) will greatly contribute to our understanding of cardiovascular growth.

Appendix

List of studies reporting echocardiography reference values for cardiovascular dimensions on ≥400 pediatric subjects published after 2000

Reference	Parameter(s) and population	Model and normalizing variable(s)	Information on gender
Cantinotti et al. (2017) [110]	17 measurements by 2D echo on 1151 children 0–17 years	Log-log model with BSA	Small sex difference for some parameter but deemed nonsignificant and no sex-specific values presented
Choudhry et al. (2017) [50]	LV dimension by M-mode in 553 preterm infants	Centile curves for weight using LMS method	Sex-specific curves provided
Chinali et al. (2016) [48]	LV mass by echocardiography (Devereux formula) in 400 children 0–18 years	LV mass indexed for height raised to the power of 2.16	No independent sex difference found
Hashimoto et al. (2016) [111]	Short-axis M-mode right to left ventricular end-diastolic ratio in 2769 Asian subjects 0–23 years	Only upper cutoff stratified by age is provided	No sex differences found
Foster et al. (2016) [49]	LV mass by Devereux formula (M-mode) on 1710 children 5–18 years	Centile curves with LMS method adjusted for estimated lean body mass	Sex-stratified data reported
Kobayashi et al. (2016) [51]	Coronary artery dimension of 3851 children 0–18 years	Centile curves with LMS adjusted for BSA	Sex-stratified data provided
Koestenberger et al. (2014 and 2016) [112, 113]	Right atrial and ventricular dimensions and area in 579 children 0–19 years	Several multivariable additive models presented for each measurements. Z score limits reported for age and height	Sex-specific prediction models for some parameters (when a significant difference was detected)
Dallaire et al. (2015) [45]	LV outflow track by 2D echocardiography on 1422 children 0–18 years	Multivariable additive model including weight and height	Sex-specific equations provided
Zhang et al. (2015) [114]	Coronary artery dimension on 506 Asian children 1–18 years	Polynomial third order adjusted for BSA	No sex difference found
Laser et al., (2015) [115]	LV mass by 3D echocardiography in 434 children 0–18 years	Left ventricular mass indexed for BSA, height, or weight and then expressed as percentile for age using the LMS method	Sex-stratified values from 7 to 18 years provided, no sex difference < 7 years of age
Cantinotti et al. (2014) [116]	Chamber diameters and areas in 1091 children 0–17 years	Z scores from log-log models normalized for BSA	No sex differences found
Cantinotti et al. (2014) [26]	Ventricular, valvular, and arterial dimensions in 445 Caucasian children 0–36 months	Z scores from log-log square root or linear models for BSA, depending on the parameter	No sex differences found
Dallaire et al. (2011) [6]	Coronary arteries in 1033 children 0.2–18 years	Linear model with square root of BSA	Small but clinically nonsignificant sex differences reported

(continued)

Reference	Parameter(s) and population	Model and normalizing variable(s)	Information on gender
Olivieri et al. (2009) [117]	Coronary arteries in 432 subjects 0–20 years	Log-log model using BSA	Sex differences not addressed
Pettersen et al. (2008) [118]	21 parameter by 2D and M-mode on 782 children 0–18 years	Third-order polynomial model for BSA after log transformation of the echocardiography parameter	Sex differences not addressed
Foster et al. (2008) [37]	LV mass by Devereux formula (M-mode) on 440 children 0–21 years	Centile curves with LMS method adjusted for height.	Sex-stratified curves not provided
Kaldararova et al. (2007) [119]	LVOT by 2D on 702 children 0–20 years	Median, 5th and 95th percentile according to strata of BSA	Sex differences not found
Bonatto et al. (2006) [120]	9 parameters by M-mode on 595 children 0–12 years	Centile curves for BSA, nonparametric analysis	Sex differences reported for 6 out of 9 parameters
Overbeek et al. (2006) [9]	M-mode LV dimensions in 747 children 0–18 years	Log-log model for weight	Sex differences not found

BSA body surface area, *LMS* lambda mu sigma, *LV* left ventricle

References

1. Prothero J. Heart weight as a function of body weight in mammals. Growth. 1979;43(3):139–50.
2. Graham TP Jr, Jarmakani JM, Canent RV Jr, Morrow MN. Left heart volume estimation in infancy and childhood. Reevaluation of methodology and normal values. Circulation. 1971;43(6):895–904.
3. Graham TP Jr., Jarmakani MM, Canent RV Jr., Capp MP, Spach MS. Characterization of left heart volumes and mass in normal children and in infants with intrinsic myocardial disease. Circulation. 1968;38(5):826–37.
4. Nevill AM, Bate S, Holder RL. Modeling physiological and anthropometric variables known to vary with body size and other confounding variables. Am J Phys Anthropol Suppl. 2005;41:141–53.
5. Batterham AM, George KP, Whyte G, Sharma S, McKenna W. Scaling cardiac structural data by body dimensions: A review of theory, practice, and problems. Int J Sports Med. 1999;20(8):495–502.
6. Dallaire F, Dahdah N. New equations and a critical appraisal of coronary artery z scores in healthy children. J Am Soc Echocardiogr. 2011;24(1):60–74.
7. Gautier M, Detaint D, Fermanian C, Aegerter P, Delorme G, Arnoult F, Milleron O, Raoux F, Stheneur C, Boileau C, Vahanian A, Jondeau G. Nomograms for aortic root diameters in children using two-dimensional echocardiography. Am J Cardiol. 2010;105(6):888–94.
8. Zilberman MV, Khoury PR, Kimball RT. Two-dimensional echocardiographic valve measurements in healthy children: Gender-specific differences. Pediatr Cardiol. 2005;26 (4):356–60.
9. Overbeek LI, Kapusta L, Peer PG, de Korte CL, Thijssen JM, Daniels O. New reference values for echocardiographic dimensions of healthy dutch children. Eur J Echocardiogr. 2006;7(2):113–21.
10. Sluysmans T, Colan SD. Theoretical and empirical derivation of cardiovascular allometric relationships in children. J Appl Physiol. 2005;99(2):445–57.
11. Blimkie CJ, Cunningham DA, Nichol PM. Gas transport capacity and echocardiographically determined cardiac size in children. J Appl Physiol Respir Environ Exerc Physiol. 1980;49(6):994–9.
12. Nevill AM, Holder RL, Baxter-Jones A, Round JM, Jones DA. Modeling developmental changes in strength and aerobic power in children. J Appl Physiol. 1998;84(3):963–70.
13. Nevill AM, Winter EM, Ingham S, Watts A, Metsios GS, Stewart AD. Adjusting athletes' body mass index to better reflect adiposity in epidemiological research. J Sports Sci. 2010;28(9):1009–16.
14. Sarkola T, Manlhiot C, Slorach C, Bradley TJ, Hui W, Mertens L, Redington A, Jaeggi E. Evolution of the arterial structure and function from infancy to adolescence is related to anthropometric and blood pressure changes. Arterioscler Thromb Vasc Biol. 2012;32(10):2516–24.
15. Doyon A, Kracht D, Bayazit AK, Deveci M, Duzova A, Krmar RT, Litwin M, Niemirska A, Oguz B, Schmidt BM, Sozeri B, Querfeld U, Melk A, Schaefer F, Wuhl E. Carotid artery intima-media thickness and distensibility in children and adolescents: reference values and role of body dimensions. Hypertension. 2013;62(3):550–6.
16. Blais S, Berbari J, Counil FP, Dallaire F. A systematic review of reference values in pediatric cardiopulmonary exercise testing. Pediatr Cardiol. 2015;36(6):1316.
17. Cawley RH, Mc KT, Record RG. Parental stature and birth weight. Am J Hum Genet. 1954;6(4):448–56.
18. Ounsted M, Ounsted C. Maternal regulation of intra-uterine growth. Nature. 1966;212(5066):995–7.
19. Lubchenco LO, Hansman C, Boyd E. Intrauterine growth in length and head circumference as estimated from live births at gestational ages from 26 to 42 weeks. Pediatrics. 1966;37 (3):403–8.
20. Schwarzler P, Bland JM, Holden D, Campbell S, Ville Y. Sex-specific antenatal reference growth charts for uncomplicated singleton pregnancies at 15–40 weeks of gestation. Ultrasound Obstet Gynecol. 2004;23(1):23–9.
21. Schneider C, McCrindle BW, Carvalho JS, Hornberger LK, McCarthy KP, Daubeney PE. Development of z-scores for fetal cardiac dimensions from echocardiography. Ultrasound Obstet Gynecol. 2005;26(6):599–605.
22. Lee W, Riggs T, Amula V, Tsimis M, Cutler N, Bronsteen R, Comstock CH. Fetal echocardiography: Z-score reference ranges for a large patient population. Ultrasound Obstet Gynecol. 2010;35(1):28–34.
23. Gagnon C, Bigras JL, Fouron JC, Dallaire F. Reference values and z scores for pulsed-wave doppler and m-mode measurements in fetal echocardiography. J Am Soc Echocardiogr. 2016;29:448–60.
24. Perez-Cruz M, Cruz-Lemini M, Fernandez MT, Parra JA, Bartrons J, Gomez-Roig MD, Crispi F, Gratacos E. Fetal cardiac function in late-onset intrauterine growth restriction vs small-for-gestational age, as defined by estimated fetal weight, cerebroplacental ratio and uterine artery doppler. Ultrasound Obstet Gynecol. 2015;46(4):465–71.
25. DeVore GR, Zaretsky M, Gumina DL, Hobbins JC. Abnormal Cardiovascular 24-Segment Sphericity Index of the Right and Left Ventricles in Fetuses with Growth Restriction. Ultrasound Obstet Gynecol. 2017 Jul 26. https://doi.org/10.1002/uog.18820. [Epub ahead of print] PubMed PMID: 28745414.
26. Cantinotti M, Scalese M, Murzi B, Assanta N, Spadoni I, Festa P, De Lucia V, Crocetti M, Marotta M, Molinaro S, Lopez L, Iervasi G. Echocardiographic nomograms for ventricular, valvular and arterial dimensions in caucasian children with a special focus on neonates, infants and toddlers. J Am Soc Echocardiogr. 2014;27(2):179–91. e172
27. Kampmann C, Wiethoff CM, Wenzel A, Stolz G, Betancor M, Wippermann CF, Huth RG, Habermehl P, Knuf M, Emschermann T, Stopfkuchen H. Normal values of m-mode echocardiographic measurements of more than 2000 healthy infants and children in central europe. Heart. 2000;83(6):667–72.
28. Olander RF, Sundholm JK, Ojala TH, Andersson S, Sarkola T. Neonatal Arterial Morphology Is Related to Body Size in Abnormal Human Fetal Growth. Circ Cardiovasc Imaging. 2016;Sep;9(9). pii: e004657. https://doi.org/10.1161/CIRCIMAGING.116.004657. PubMed PMID: 27601367.
29. Lopez L, Colan SD, Frommelt PC, Ensing GJ, Kendall K, Younoszai AK, Lai WW, Geva T. Recommendations for quantification methods during the performance of a pediatric echocardiogram: a report from the pediatric measurements writing group of the american society of echocardiography pediatric and congenital heart disease council. J Am Soc Echocardiogr. 2010;23(5):465–95.
30. Colan SD. The why and how of z scores. J Am Soc Echocardiogr. 2013;26(1):38–40.
31. Mawad W, Drolet C, Dahdah N, Dallaire F. A review and critique of the statistical methods used to generate reference values in pediatric echocardiography. J Am Soc Echocardiogr. 2013;26(1):29–37.
32. Cantinotti M, Kutty S, Franchi E, Paterni M, Scalese M, Iervasi G, Koestenberger M. Pediatric echocardiographic

nomograms: what has been done and what still needs to be done. Trends Cardiovasc Med. 2017;27(5):336–49.
33. Cantinotti M, Scalese M, Molinaro S, Murzi B, Passino C. Limitations of current echocardiographic nomograms for left ventricular, valvular and arterial dimensions in children: a critical review. J Am Soc Echocardiogr. 2012;25 (2):142–52.
34. Tanner JM. Fallacy of per-weight and per-surface area standards, and their relation to spurious correlation. J Appl Physiol. 1949;2(1):1–15.
35. Gutgesell HP, Rembold CM. Growth of the human heart relative to body surface area. Am J Cardiol. 1990;65 (9):662–8.
36. Neilan TG, Pradhan AD, King ME, Weyman AE. Derivation of a size-independent variable for scaling of cardiac dimensions in a normal paediatric population. Eur J Echocardiogr. 2009;10(1):50–5.
37. Foster BJ, Mackie AS, Mitsnefes M, Ali H, Mamber S, Colan SD. A novel method of expressing left ventricular mass relative to body size in children. Circulation. 2008;117 (21):2769–75.
38. Buechel EV, Kaiser T, Jackson C, Schmitz A, Kellenberger CJ. Normal right- and left ventricular volumes and myocardial mass in children measured by steady state free precession cardiovascular magnetic resonance. J Cardiovasc Magn Reson. 2009;11:19.
39. Jones NL, Summers E, Killian KJ. Influence of age and stature on exercise capacity during incremental cycle ergometry in men and women. Am Rev Respir Dis. 1989;140(5):1373–80.
40. Hansen JE, Sue DY, Wasserman K. Predicted values for clinical exercise testing. Am Rev Respir Dis. 1984;129(2 Pt 2):S49–55.
41. Sue DY, Hansen JE. Normal values in adults during exercise testing. Clin Chest Med. 1984;5(1):89–98.
42. Nes BM, Osthus IB, Welde B, Aspenes ST, Wisloff U. Peak oxygen uptake and physical activity in 13- to 18-year-olds: the young-hunt study. Med Sci Sports Exerc. 2013;45 (2):304–13.
43. Welsman JR, Armstrong N, Nevill AM, Winter EM, Kirby BJ. Scaling peak vo2 for differences in body size. Med Sci Sports Exerc. 1996;28(2):259–65.
44. Sluysmans T, Colan SD. Structural measurements and adjustment for growth. In: Lai WW, Mertens LL, Cohen SC, Geva T, editors. Echocardiography in pediatric and congenital heart disease. West Sussex: Wiley-Blackwell; 2009. p. 52–62.
45. Dallaire F, Bigras JL, Prsa M, Dahdah N. Bias related to body mass index in pediatric echocardiographic z scores. Pediatr Cardiol. 2015;36(3):667–76.
46. Bluemke DA, Kronmal RA, Lima JA, Liu K, Olson J, Burke GL, Folsom AR. The relationship of left ventricular mass and geometry to incident cardiovascular events: the mesa (multi-ethnic study of atherosclerosis) study. J Am Coll Cardiol. 2008;52(25):2148–55.
47. Brumback LC, Kronmal R, Heckbert SR, Ni H, Hundley WG, Lima JA, Bluemke DA. Body size adjustments for left ventricular mass by cardiovascular magnetic resonance and their impact on left ventricular hypertrophy classification. Int J Cardiovasc Imaging. 2010;26(4):459–68.
48. Chinali M, Emma F, Esposito C, Rinelli G, Franceschini A, Doyon A, Raimondi F, Pongiglione G, Schaefer F, Matteucci MC. Left ventricular mass indexing in infants, children, and adolescents: a simplified approach for the identification of left ventricular hypertrophy in clinical practice. J Pediatr. 2016;170:193–8.
49. Foster BJ, Khoury PR, Kimball TR, Mackie AS, Mitsnefes M. New reference centiles for left ventricular mass relative to lean body mass in children. J Am Soc Echocardiogr. 2016;29 (5):441–7. e442
50. Choudhry S, Salter A, Cunningham TW, Levy PT, Nguyen HH, Wallendorf M, Singh GK, Johnson MC. Normative left ventricular m-mode echocardiographic values in preterm infants up to 2 kg. J Am Soc Echocardiogr. 2017;30 (8):781–9. e784
51. Kobayashi T, Fuse S, Sakamoto N, Mikami M, Ogawa S, Hamaoka K, Arakaki Y, Nakamura T, Nagasawa H, Kato T, Jibiki T, Iwashima S, Yamakawa M, Ohkubo T, Shimoyama S, Aso K, Sato S, Saji T. A new z score curve of the coronary arterial internal diameter using the lambda-mu-sigma method in a pediatric population. J Am Soc Echocardiogr. 2016;29(8):794–801. e729
52. Sharland GK, Allan LD. Normal fetal cardiac measurements derived by cross-sectional echocardiography. Ultrasound Obstet Gynecol. 1992;2(3):175–81.
53. Tan J, Silverman NH, Hoffman JI, Villegas M, Schmidt KG. Cardiac dimensions determined by cross-sectional echocardiography in the normal human fetus from 18 weeks to term. Am J Cardiol. 1992;70(18):1459–67.
54. Shapiro I, Degani S, Leibovitz Z, Ohel G, Tal Y, Abinader EG. Fetal cardiac measurements derived by transvaginal and transabdominal cross-sectional echocardiography from 14 weeks of gestation to term. Ultrasound Obstet Gynecol. 1998;12(6):404–18.
55. McElhinney DB, Marshall AC, Wilkins-Haug LE, Brown DW, Benson CB, Silva V, Marx GR, Mizrahi-Arnaud A, Lock JE, Tworetzky W. Predictors of technical success and postnatal biventricular outcome after in utero aortic valvuloplasty for aortic stenosis with evolving hypoplastic left heart syndrome. Circulation. 2009;120(15):1482–90.
56. Lara DA, Morris SA, Maskatia SA, Challman M, Nguyen M, Feagin DK, Schoppe L, Zhang J, Bhatt A, Sexson-Tejtel SK, Lopez KN, Lawrence EJ, Andreas S, Wang Y, Belfort MA, Ruano R, Ayres NA, Altman CA, Aagaard KM, Becker J. Pilot study of chronic maternal hyperoxygenation and effect on aortic and mitral valve annular dimensions in fetuses with left heart hypoplasia. Ultrasound Obstet Gynecol. 2016;48(3):365–72.
57. Lorenz CH. The range of normal values of cardiovascular structures in infants, children, and adolescents measured by magnetic resonance imaging. Pediatr Cardiol. 2000;21 (1):37–46.
58. Lorenz CH, Walker ES, Morgan VL, Klein SS, Graham TP Jr. Normal human right and left ventricular mass, systolic function, and gender differences by cine magnetic resonance imaging. J Cardiovasc Magn Reson. 1999;1(1):7–21.
59. Robbers-Visser D, Boersma E, Helbing WA. Normal biventricular function, volumes, and mass in children aged 8 to 17 years. J Magn Reson Imaging. 2009;29(3):552–9.
60. Sarikouch S, Peters B, Gutberlet M, Leismann B, Kelter-Kloepping A, Koerperich H, Kuehne T, Beerbaum P. Sex-specific pediatric percentiles for ventricular size and mass as reference values for cardiac MRI: assessment by steady-state free-precession and phase-contrast mri flow. Circ Cardiovasc Imaging. 2010;3(1):65–76.
61. Schmitz L, Koch H, Bein G, Brockmeier K. Left ventricular diastolic function in infants, children, and adolescents. Reference values and analysis of morphologic and physiologic determinants of echocardiographic doppler flow signals during growth and maturation. J Am Coll Cardiol. 1998;32 (5):1441–8.

62. Eidem BW, McMahon CJ, Cohen RR, Wu J, Finkelshteyn I, Kovalchin JP, Ayres NA, Bezold LI, O'Brian Smith E, Pignatelli RH. Impact of cardiac growth on doppler tissue imaging velocities: a study in healthy children. J Am Soc Echocardiogr. 2004;17(3):212–21.
63. Cui W, Roberson DA, Chen Z, Madronero LF, Cuneo BF. Systolic and diastolic time intervals measured from doppler tissue imaging: normal values and z-score tables, and effects of age, heart rate, and body surface area. J Am Soc Echocardiogr. 2008;21(4):361–70.
64. Roberson DA, Cui W. Right ventricular tei index in children: effect of method, age, body surface area, and heart rate. J Am Soc Echocardiogr. 2007;20(6):764–70.
65. Roberson DA, Cui W, Chen Z, Madronero LF, Cuneo BF. Annular and septal doppler tissue imaging in children: normal z-score tables and effects of age, heart rate, and body surface area. J Am Soc Echocardiogr. 2007;20(11):1276–84.
66. Harada K, Rice MJ, Shiota T, Ishii M, McDonald RW, Reller MD, Sahn DJ. Gestational age- and growth-related alterations in fetal right and left ventricular diastolic filling patterns. Am J Cardiol. 1997;79(2):173–7.
67. Harada K, Suzuki T, Tamura M, Ito T, Shimada K, Takada G. Effect of aging from infancy to childhood on flow velocity patterns of pulmonary vein by doppler echocardiography. Am J Cardiol. 1996;77(2):221–4.
68. Cantinotti M, Lopez L. Nomograms for blood flow and tissue doppler velocities to evaluate diastolic function in children: a critical review. J Am Soc Echocardiogr. 2013;26(2):126–41.
69. Cantinotti M, Giordano R, Scalese M, Murzi B, Assanta N, Spadoni I, Crocetti M, Marotta M, Molinaro S, Kutty S, Iervasi G. Nomograms for mitral inflow doppler and tissue doppler velocities in caucasian children. J Cardiol. 2016;68 (4):288–99.
70. Dallaire F, Slorach C, Hui W, Sarkola T, Friedberg MK, Bradley T, Jaeggi E, Dragulescu A, Har RLH, Cherney DZI, Mertens L (2015) Reference values for pulse wave doppler and tissue doppler imaging in pediatric echocardiography. Circ Cardiovasc Imaging 8 (2):e002167. doi:10.002110.001161/CIRCIMAGING.002114.002167
71. Weidemann F, Eyskens B, Jamal F, Mertens L, Kowalski M, D'Hooge J, Bijnens B, Gewillig M, Rademakers F, Hatle L, Sutherland GR. Quantification of regional left and right ventricular radial and longitudinal function in healthy children using ultrasound-based strain rate and strain imaging. J Am Soc Echocardiogr. 2002;15(1):20–8.
72. Marcus KA, Mavinkurve-Groothuis AM, Barends M, van Dijk A, Feuth T, de Korte C, Kapusta L. Reference values for myocardial two-dimensional strain echocardiography in a healthy pediatric and young adult cohort. J Am Soc Echocardiogr. 2011;24(6):625–36.
73. Kaku K, Takeuchi M, Tsang W, Takigiku K, Yasukochi S, Patel AR, Mor-Avi V, Lang RM, Otsuji Y. Age-related normal range of left ventricular strain and torsion using three-dimensional speckle-tracking echocardiography. J Am Soc Echocardiogr. 2014;27(1):55–64.
74. Al-Naami GH. Torsion of young hearts: a speckle tracking study of normal infants, children, and adolescents. Eur J Echocardiogr. 2010;11(10):853–62.
75. Takahashi K, Al Naami G, Thompson R, Inage A, Mackie AS, Smallhorn JF. Normal rotational, torsion and untwisting data in children, adolescents and young adults. J Am Soc Echocardiogr. 2010;23(3):286–93.
76. Sato Y, Maruyama A, Ichihashi K. Myocardial strain of the left ventricle in normal children. J Cardiol. 2012;60 (2):145–9.
77. Kutty S, Deatsman SL, Nugent ML, Russell D, Frommelt PC. Assessment of regional right ventricular velocities, strain, and displacement in normal children using velocity vector imaging. Echocardiography. 2008;25(3):294–307.
78. Lorch SM, Ludomirsky A, Singh GK. Maturational and growth-related changes in left ventricular longitudinal strain and strain rate measured by two-dimensional speckle tracking echocardiography in healthy pediatric population. J Am Soc Echocardiogr. 2008;21(11):1207–15.
79. Zhang L, Gao J, Xie M, Yin P, Liu W, Li Y, Klas B, Sun J, Balluz R, Ge S. Left ventricular three-dimensional global systolic strain by real-time three-dimensional speckle-tracking in children: feasibility, reproducibility, maturational changes, and normal ranges. J Am Soc Echocardiogr. 2013;26(8):853–9.
80. Forsey J, Friedberg MK, Mertens L. Speckle tracking echocardiography in pediatric and congenital heart disease. Echocardiography. 2013;30(4):447–59.
81. Cantinotti M, Kutty S, Giordano R, Assanta N, Murzi B, Crocetti M, Marotta M, Iervasi G. Review and status report of pediatric left ventricular systolic strain and strain rate nomograms. Heart Fail Rev. 2015;20(5):601–12.
82. Dallaire F, Slorach C, Bradley T, Hui W, Sarkola T, Friedberg MK, Jaeggi E, Dragulescu A, Mahmud FH, Daneman D, Mertens L. Pediatric reference values and z score equations for left ventricular systolic strain measured by two-dimensional speckle-tracking echocardiography. J Am Soc Echocardiogr. 2016;29(8):786–93. e788
83. Davignon A, Rautaharju P, Boisselle E, Soumis F, Megelas M, Choguette A. Normal ECG standards for infants and children. Pediatr Cardiol. 1979;1:123–31.
84. Macfarlane PW, McLaughlin SC, Devine B, Yang TF. Effects of age, sex, and race on ecg interval measurements. J Electrocardiol. 1994;27(Suppl):14–9.
85. Rijnbeek PR, Witsenburg M, Schrama E, Hess J, Kors JA. New normal limits for the paediatric electrocardiogram. Eur Heart J. 2001;22(8):702–11.
86. Palhares DMF, Marcolino MS, Santos TMM, da Silva JLP, Gomes PR, Ribeiro LB, Macfarlane PW, Ribeiro ALP. Normal limits of the electrocardiogram derived from a large database of Brazilian primary care patients. BMC Cardiovasc Disord. 2017;17(1):152.
87. Pearl W. Effects of gender, age, and heart rate on qt intervals in children. Pediatr Cardiol. 1996;17(3):135–6.
88. Mason JW, Ramseth DJ, Chanter DO, Moon TE, Goodman DB, Mendzelevski B. Electrocardiographic reference ranges derived from 79,743 ambulatory subjects. J Electrocardiol. 2007;40(3):228–34.
89. Semizel E, Ozturk B, Bostan OM, Cil E, Ediz B. The effect of age and gender on the electrocardiogram in children. Cardiol Young. 2008;18(1):26–40.
90. Silvetti MS, Drago F, Ragonese P. Heart rate variability in healthy children and adolescents is partially related to age and gender. Int J Cardiol. 2001;81(2-3):169–74.
91. Rogowski MP, Guilkey JP, Stephens BR, Cole AS, Mahon AD. The influence of maturation on the oxygen uptake efficiency slope. Pediatr Exerc Sci. 2012;24(3):347–56.
92. Wasserman K, Hensen JE, Sue DY, Stringer WW, Sietsema KE, Sun X-G, Whipp BJ. Normal values. In: Principles of exercise testing and interpretation : Including pathophysiology and clinical applications. 5th ed. Philadelphia: Wolters Kluwer Health/Lippincott Williams & Wilkins; 2012. p. 154–80.
93. Cooper DM, Weiler-Ravell D, Whipp BJ, Wasserman K. Aerobic parameters of exercise as a function of body size during growth in children. J Appl Physiol. 1984;56 (3):628–34.
94. Cooper DM, Weiler-Ravell D, Whipp BJ, Wasserman K. Growth-related changes in oxygen uptake and heart rate

during progressive exercise in children. Pediatr Res. 1984;18(9):845–51.
95. Ten Harkel AD, Takken T, Van Osch-Gevers M, Helbing WA. Normal values for cardiopulmonary exercise testing in children. Eur J Cardiovasc Prev Rehabil. 2011;18(1):48–54.
96. Loftin M, Sothern M, Abe T, Bonis M. Expression of VO2 peak in children and youth, with special reference to allometric scaling. Sports Med. 2016;46(10):1451–60.
97. Cooper DM, Weiler-Ravell D. Gas exchange response to exercise in children. Am Rev Respir Dis. 1984;129(2 Pt 2): S47–8.
98. Wang X, Xu X, Su S, Snieder H. Familial aggregation and childhood blood pressure. Curr Hypertens Rep. 2015;17(1):509.
99. The fourth report on the diagnosis, evaluation, and treatment of high blood pressure in children and adolescents. Pediatrics. 2004;114(2 Suppl 4th Report):555–76.
100. Xi B, Zong X, Kelishadi R, Hong YM, Khadilkar A, Steffen LM, Nawarycz T, Krzywinska-Wiewiorowska M, Aounallah-Skhiri H, Bovet P, Chiolero A, Pan H, Litwin M, Poh BK, Sung RY, So HK, Schwandt P, Haas GM, Neuhauser HK, Marinov L, Galcheva SV, Motlagh ME, Kim HS, Khadilkar V, Krzyzaniak A, Romdhane HB, Heshmat R, Chiplonkar S, Stawinska-Witoszynska B, El Ati J, Qorbani M, Kajale N, Traissac P, Ostrowska-Nawarycz L, Ardalan G, Parthasarathy L, Zhao M, Zhang T. Establishing international blood pressure references among nonoverweight children and adolescents aged 6 to 17 years. Circulation. 2016;133(4):398–408.
101. Wuhl E, Witte K, Soergel M, Mehls O, Schaefer F. Distribution of 24-h ambulatory blood pressure in children: normalized reference values and role of body dimensions. J Hypertens. 2002;20(10):1995–2007.
102. Whincup PH, Nightingale CM, Owen CG, Rapala A, Bhowruth DJ, Prescott MH, Ellins EA, Donin AS, Masi S, Rudnicka AR, Sattar N, Cook DG, Deanfield JE. Ethnic differences in carotid intima-media thickness between UK children of black african-caribbean and white European origin. Stroke. 2012;43(7):1747–54.
103. Chowdhury SM, Henshaw MH, Friedman B, Saul JP, Shirali GS, Carter J, Levitan BM, Hulsey T. Lean body mass may explain apparent racial differences in carotid intima-media thickness in obese children. J Am Soc Echocardiogr. 2014;27(5):561–7.
104. Yoo BW. Epidemiology of congenital heart disease in boys and girls. Adv Exp Med Biol. 2018;162:191–3.
105. Schaefer F, Doyon A, Azukaitis K, Bayazit A, Canpolat N, Duzova A, Niemirska A, Sozeri B, Thurn D, Anarat A, Ranchin B, Litwin M, Caliskan S, Candan C, Baskin E, Yilmaz E, Mir S, Kirchner M, Sander A, Haffner D, Melk A, Wuhl E, Shroff R, Querfeld U. Cardiovascular phenotypes in children with CKD: The 4c study. Clin J Am Soc Nephrol. 2017;12(1):19–28.
106. Gao Z, Khoury PR, McCoy CE, Shah AS, Kimball TR, Dolan LM, Urbina EM. Adiposity has no direct effect on carotid intima-media thickness in adolescents and young adults: use of structural equation modeling to elucidate indirect & direct pathways. Atherosclerosis. 2016;246:29–35.
107. Vatanen A, Sarkola T, Ojala TH, Turanlahti M, Jahnukainen T, Saarinen-Pihkala UM, Jahnukainen K. Radiotherapy-related arterial intima thickening and plaque formation in childhood cancer survivors detected with very-high resolution ultrasound during young adulthood. Pediatr Blood Cancer. 2015;62(11):2000–6.
108. Litwin M, Niemirska A, Sladowska-Kozlowska J, Wierzbicka A, Janas R, Wawer ZT, Wisniewski A, Feber J. Regression of target organ damage in children and adolescents with primary hypertension. Pediatr Nephrol. 2010;25(12):2489–99.
109. Braamskamp M, Langslet G, McCrindle BW, Cassiman D, Francis GA, Gagne C, Gaudet D, Morrison KM, Wiegman A, Turner T, Miller E, Kusters DM, Raichlen JS, Martin PD, Stein EA, Kastelein JJP, Hutten BA. Effect of rosuvastatin on carotid intima-media thickness in children with heterozygous familial hypercholesterolemia: the Charon study (hypercholesterolemia in children and adolescents taking rosuvastatin open label). Circulation. 2017;136(4):359–66.
110. Cantinotti M, Giordano R, Scalese M, Murzi B, Assanta N, Spadoni I, Maura C, Marco M, Molinaro S, Kutty S, Iervasi G. Nomograms for two-dimensional echocardiography derived valvular and arterial dimensions in caucasian children. J Cardiol. 2017;69(1):208–15.
111. Hashimoto I, Watanabe K, Ichida F. Right to left ventricular diameter ratio >/=0.42 is the warning flag for suspecting atrial septal defect in preschool children: age- and body surface area-related reference values determined by m-mode echocardiography. Pediatr Cardiol. 2016;37(4):704–13.
112. Koestenberger M, Burmas A, Ravekes W, Avian A, Gamillscheg A, Grangl G, Grillitsch M, Hansmann G. Echocardiographic reference values for right atrial size in children with and without atrial septal defects or pulmonary hypertension. Pediatr Cardiol. 2016;37(4):686–95.
113. Koestenberger M, Nagel B, Ravekes W, Avian A, Burmas A, Grangl G, Cvirn G, Gamillscheg A. Reference values and calculation of z-scores of echocardiographic measurements of the normal pediatric right ventricle. Am J Cardiol. 2014;114(10):1590–8.
114. Zhang YQ, Chen SB, Huang GY, Zhang HY, Huang MR, Wang SS, Wu LP, Hong WJ, Shen R, Liu YQ, Zhu JX, Lu ZH. Coronary artery indexed diameter and z score regression equations in healthy Chinese Han children. J Clin Ultrasound. 2015;43(1):39–46.
115. Laser KT, Houben BA, Korperich H, Haas NA, Kelter-Klopping A, Barth P, Burchert W, DallaPozza R, Kececioglu D, Herberg U. Calculation of pediatric left ventricular mass: validation and reference values using real-time three-dimensional echocardiography. J Am Soc Echocardiogr. 2015;28(3):275–83.
116. Cantinotti M, Scalese M, Murzi B, Assanta N, Spadoni I, De Lucia V, Crocetti M, Cresti A, Gallotta M, Marotta M, Tyack K, Molinaro S, Iervasi G. Echocardiographic nomograms for chamber diameters and areas in Caucasian children. J Am Soc Echocardiogr. 2014;27(12):1279–92.
117. Olivieri L, Arling B, Friberg M, Sable C. Coronary artery z score regression equations and calculators derived from a large heterogeneous population of children undergoing echocardiography. J Am Soc Echocardiogr. 2009;22(2):159–64.
118. Pettersen MD, Du W, Skeens ME, Humes RA. Regression equations for calculation of z scores of cardiac structures in a large cohort of healthy infants, children, and adolescents: an echocardiographic study. J Am Soc Echocardiogr. 2008;21(8):922–34.
119. Kaldararova M, Balazova E, Tittel P, Stankovicova I, Brucknerova I, Masura J. Echocardiographic measurements of the aorta in normal children and young adults. Bratisl Lek Listy. 2007;108(10-11):437–41.
120. Bonatto RC, Fioretto JR, Okoshi K, Matsubara BB, Padovani CR, Manfrin TC, Gobbi Mde F, Martino RS, Bregagnollo EA. Percentile curves of normal values of echocardiographic measurements in normal children from the central-southern region of the state of Sao Paulo, Brazil. Arq Bras Cardiol. 2006;87(6):711–21.

Heart Function Analysis in Cardiac Patients with Focus on Sex-Specific Aspects

23

Peter L. M. Kerkhof, Richard A. Peace, Guy R. Heyndrickx, Lilian J. Meijboom, Ralf W. Sprengers, and Neal Handly

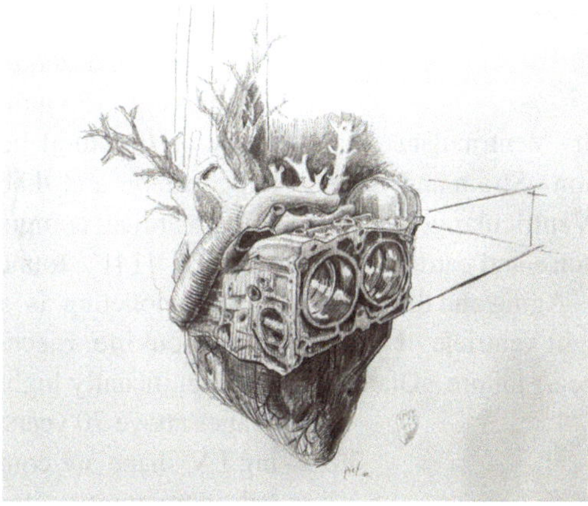

Repairing a diseased heart. Art work by Piet Michiels, Leuven, Belgium

P. L. M. Kerkhof (✉)
Department of Radiology and Nuclear Medicine, Amsterdam Cardiovascular Sciences, VU University Medical Center, Amsterdam, The Netherlands
e-mail: plm.kerkhof@vumc.nl

R. A. Peace
Department of Nuclear Medicine, Royal Victoria Infirmary, The Newcastle upon Tyne Hospitals NHS Foundation Trust, Newcastle upon Tyne, UK

G. R. Heyndrickx
Cardiovascular Center, Onze-Lieve-Vrouw Clinic, Aalst, Belgium

L. J. Meijboom · R. W. Sprengers
Department of Radiology & Nuclear Medicine, VU University Medical Center, Amsterdam, The Netherlands

N. Handly
Department of Emergency Medicine, Drexel University College of Medicine, Philadelphia, PA, USA

Abstract

Cardiac function is best described by investigating the pressure-volume relationships. This information permits description in terms of the ventricular volume regulation graph (VRG), estimation of systolic elastance, evaluation of lusitropic properties, and assessment of ventriculo-arterial coupling. Current techniques yield noninvasive determination of cardiac compartmental volumes, along with systolic/diastolic arterial pressure, while ventricular end-diastolic pressure can be inferred from an echocardiography-based surrogate measure. Ventricular volume is known to vary with age, as well as to be affected by intrinsic cardiac disease and abnormalities of the vascular

P. L. M. Kerkhof, V. M. Miller (eds.), *Sex-Specific Analysis of Cardiovascular Function*,
Advances in Experimental Medicine and Biology 1065, https://doi.org/10.1007/978-3-319-77932-4_23

system. Moreover, 35 years ago it has been shown in healthy adults that left ventricular volume is significantly smaller in women compared to men. This important observation has serious implications for several metrics which are routinely used in clinical practice, e.g., ejection fraction. The remarkable difference between ventricular size in men and women is also a powerful starting point for the study of aging and the investigation of interventions such as exercise. In this review we evaluate sex-specific characteristics of the VRG and the implications for various cardiac patient populations, during basal conditions and intervention such as exercise.

Keywords

Volume regulation graph · Ventricular function · Ejection fraction · Strain analysis · Torsion · Remodeling · Ventricular geometry · Cardiophysiology · Exercise and cardiac pump · Ventricular mass · Aging and the heart · Cardiac dimensions · Right ventricle · LAVI · Sex-specific analysis · Heart failure · Diabetes mellitus · Review

Die Figur ist etwas geändert worden,
da die in der früheren Arbeit S.9 gegebene Berechnung
eine kleine Berichtiging erfahren musste.

Otto Frank, in: *Die Grundform des arteriellen Pulses.* Z. Biol. 37 (1898) 483–526

Introduction

Thousands of publications are devoted to the measurement of the size of the heart which determination is found to be clinically important [12]. More specifically the volume of each of the four compartments has been estimated. Measurement techniques have evolved from the use of beeswax [19] to X-ray, MRI, and 3D echocardiography. The rationale for such efforts is clear: the heart is a pump, and to characterize the performance of any pump, both the pressure levels and volume displacements need to be known. While a technical pump is defined by certain "specifications", the biological pump displays an enormous range of plasticity, mainly depending on age and history (in terms of training state and pathologic events). Interestingly, significant differences in size have been observed in adults, both for healthy hearts [5, 54] and in patients with some type of cardiac (co)morbidity, as in heart failure (HF) [28]. Data collected by Kuznetsova et al. [38] are summarized in Table 23.1, illustrating typical differences between healthy men and women, notably regarding average ventricular volume at end-systole, indexed to body surface area (BSA).

Studying 1155 individuals without known coronary or structural heart disease, and free from hypertension and diabetes mellitus (DM), while using cardiac computed tomography, Gebhard et al. [14] found that left ventricular (LV) remodeling is a process which continues throughout life. Ejection fraction (EF) in females was significantly higher than in males but mostly at ages above 70 years. The findings about changing LV shape are confirmed, e.g., in a longitudinal community study which suggests that dynamic changes in LV geometric pattern over time are common. Higher blood pressure and greater body mass index (BMI) were found to be modifiable factors associated with the development of abnormal LV geometry, and such progression portends an adverse prognosis [39].

Cardiac dimension differences vary with aging, while among sexes these disparities are larger than those observed for arterial pressure and heart rate differences, both in healthy individuals (Table 23.1 and Fig. 23.1) and cardiac patients. Therefore, a dimensional comparison is attractive to pursue, and we primarily concentrate on the comparative study of ventricular size and related metrics in men and women.

LV remodeling may be intrinsically related to the senescence process, and in part driven by age-related comorbidities, such as long-standing hypertension and increased vascular stiffness. Initially they lead to asymptomatic alterations in diastolic function (i.e., LV stiffness) and

Table 23.1 Hemodynamic variables in healthy adults (N = 410)

Variable	Women	% M vs F	Men	*P*-value	Unit
Persons involved	215		195		Number
Age	43.4	−1.6%	42.7	0.62	Years
Systolic blood pressure	116.5	+4.5%	121.8	**<0.0001**	mmHg
Diastolic blood pressure	74.6	+4.3%	77.8	**<0.0001**	mmHg
Mean arterial pressure	88.6	+4.4%	92.5	**<0.0001**	mmHg
End-systolic volume index	17.7	+27.1%	22.5	**<0.0001**	mL/m^2
End-diastolic volume index	48.7	+21.7%	59.3	**<0.0001**	mL/m^2
Stroke volume index	31.0	+18.7%	36.8	**<0.0001**	mL/m^2
Cardiac output index	1.93	+10.4%	2.13	**<0.0001**	$L/min.m^2$
Ejection fraction	63.7	−2.8%	61.9	**0.0026**	%
Heart rate	62.5	−6.2%	58.6	**<0.0001**	Beats/min

Volume data obtained by 2D echocardiography and brachial pressure by cuff and Riva-Rocci method. Note that average values for all variables (with the exception of age) are significantly different in females (F) when compared to males (M). Based on data by Kuznetsova [38] and reported in Kerkhof et al. [33]

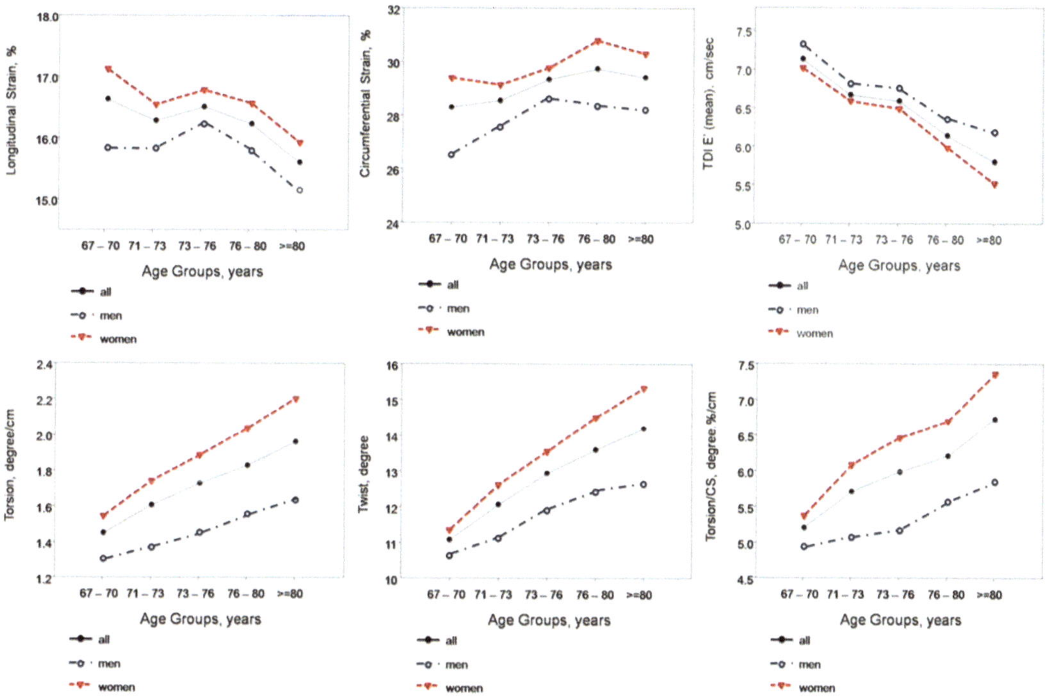

Fig. 23.1 Associations among age, measurement of left ventricular (LV) mechanics, or torsion in healthy elderly of both sexes. Graded reduction of longitudinal systolic shortening, including lower longitudinal strain (LS), lower mitral annulus relaxation velocity E′, and modest increase of circumferential fiber shortening (GCS) were shown by age quintiles, which were accompanied by larger LV torsion indices. All strain data and torsion to circumferential ratio are negative and presented as absolute values. Linear trend of *P* for all <0.05. *TDI* tissue Doppler imaging. (From Hung et al. [21]. Reproduced with permission)

longitudinal systolic function (as reflected by strain). It has been postulated that myocardial twist may increase as a response to the loss of longitudinal function that occurs with aging and that this compensation may help preserve EF [21]. Furthermore, these alterations have been noted to differ between men and women because a greater degree of cardiac torsion has been observed in women [22]. Alterations in LV structure also may occur with cardiac disease. In particular the development of HF is accompanied by LV remodeling.

Three-dimensional echocardiography (3DE) currently allows a comprehensive characterization of cardiac kinematics, enabling simultaneous assessment of various myocardial systolic components and torsion mechanics using speckle tracking. Using these methods, Hung et al. [21] investigated sex-related differences in cardiac mechanics in an elderly population without known HF, including 1105 participants. For both the left atrium (LA) and the LV, end-diastolic volume (EDV), end-systolic volume (ESV), LV stroke volume (SV), and EF were measured by the biplane modified Simpson method using apical 4- and 2-chamber views. The LV mass was assessed from LV linear dimensions and indexed to BSA. Right ventricular (RV) function was measured by fractional area change expressed as percentage of changes in cavity area from apical 4-chamber view. LV twist and torsion (twist per length, degree/cm or °/cm) were also estimated. Less negative global longitudinal strain (GLS) and global circumferential strain (GCS) indicate a functional decline in global myocardial deformation assessment, with more positive twist/torsion values representing better torque mechanics. The torsion to circumferential ratio (TCR) was also calculated. Figure 23.1 shows the results for five age groups, stratified for males and females, as well as for combined data. Women have smaller LV cavity size, greater concentricity, though demonstrated greater global longitudinal strain, global circumferential strain, and torsion than men (all $P < 0.05$). Overall, older age was associated with smaller LV EDV and ESV, lower SV, greater wall thicknesses, and larger LV mass index, resulting in greater LV concentricity in terms of higher mass to volume ratio. Increasing age was also associated with larger BSA-indexed LA volume, though unchanged RV area, RV fractional area change, or global LV EF. In general, women showed significantly smaller LV EDV and SV and higher EF than men (all $P < 0.001$). Conversely, men had greater LV wall thickness and larger LV mass index than women (both $P < 0.001$). Thus, in this asymptomatic, senescent population, a distinct, sex-specific pattern of cardiac remodeling was observed, while a significant increase of torsion was more pronounced in women [21].

Ventricular Volume Regulation in Children

During childhood and adolescence, the size of the heart in healthy individuals mainly relates to body mass (BM) or BSA. After indexation for BM or BSA, there is no real difference between boys and girls when LV dimensions are compared. Absence of a distinction is also observed for systolic and diastolic blood pressure in children up to 16 years [7]. Figure 23.2 illustrates the volume regulation graph (VRG) for boys and

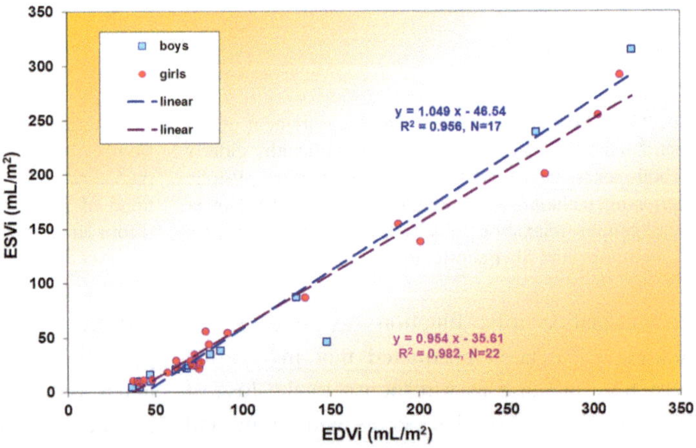

Fig. 23.2 Left ventricular end-systolic volume index (ESVi) and end-diastolic volume index (EDVi) in children (26 healthy and 13 with "intrinsic" myocardial disease). Age ranges from 1 month to 16 years. Average ESVi and EF are not significantly ($P > 0.15$) different for boys and girls (in contrast to findings in adults). (Data from Graham et al. [17])

girls (from neonates to 16 years), both healthy and those with cardiac disease. In these children the regression lines are not different for the two sexes. The usefulness of the VRG and the implications have been explained elsewhere [34].

Comparing Volumetric Characteristics of Adult Males and Females

Volume-related data are best represented in a diagram which relates ESV(i) to EDV(i) [4]. This choice applies to all four cardiac compartments [27], both in healthy individuals and patients with cardiac disease. An example regarding the LV in patients with anginal complaints is shown in Fig. 23.3. Data were obtained by gated SPECT. The linear regression lines for males and females are almost superimposed. However, average values for EF are significantly ($P < 0.0001$) higher in women.

The Heart and Other Pumps

Ziman [56] posed the question: *Which came first: science or technology*? The latter term implies "knowing how to do." We address this question in relation to the development of a pumping system, including the biological muscular pump known as the heart. It must be added that other biological pumps do exist: spiders, for example, use no muscles to extend their legs, sticking them out hydraulically instead, with pressures which may exceed 450 mmHg [52]. The physician Willem Kolff devoted most of his life to the development of an artificial kidney, extending his work to the development of an artificial heart [15]. An alternative route is based on assist devices [37]. Interestingly, Moscato et al. [43] were able to construct the ESV vs EDV relationship (i.e., the VRG representation), yielding high correlations in acute animal experiments using cardiac assist devices in sheep. Going back to

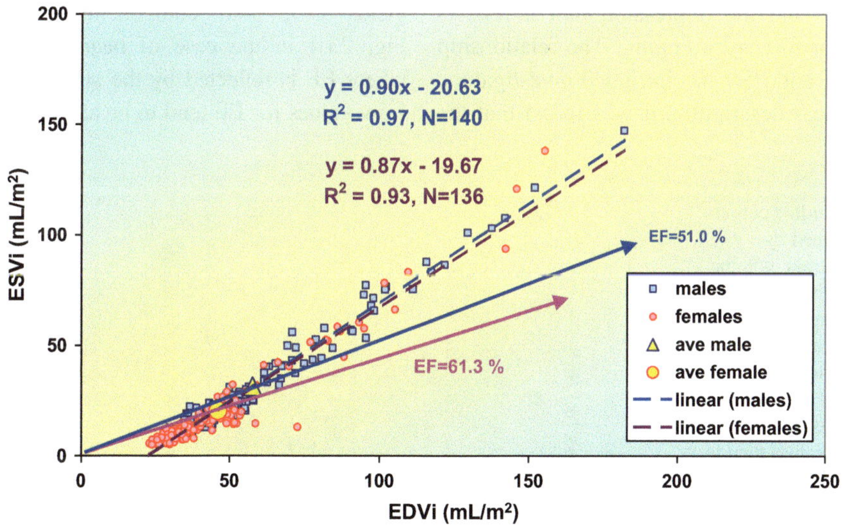

Fig. 23.3 *Left ventricular (LV) volume regulation in males and females, showing the higher average ejection fraction in women.* Distribution of end-systolic (ESVi) and end-diastolic (EDVi) volumes for 276 patients evaluated for chest pain and dyspnea, as determined by gated SPECT, presented in the LV volume domain. Regression lines for males and females are similar, but their average (ave) values (yellow triangle and circle, respectively) are significantly different, resulting in a substantially higher ($P < 0.0001$) average value for ejection fraction (EF) in females (see solid lines with arrowhead). (Data described by Peace et al. [44])

Ziman's question, we may ask if the concept of a cardiac pump promoted similar technical inventions, if mechanical insights helped to understand the biological pump, or rather if both ideas matured in parallel. Further details on technical and biological pumps are summarized elsewhere [35].

Ejection Fraction and Ventricular Volume

An earlier variant of the metric EF was introduced around 1955 to quantify indicator dilution curves. During steady-state conditions, it was observed that the washout fraction of tracer material (injected during diastole) which remained in the ventricle was rather constant [20]. This information yields residual volume (as it was called at that time), which is the same as our present end-systolic volume (ESV). The fraction was referred to as residual fraction and equals (1-EF). Subsequently, the EF gained popularity because of its ease of use and the impression that it reflects efficiency of forward pumping. The relationship between EF and ESV has intrigued investigators. Using a linear description, it was found that the correlation between EF and ESV is significantly less in patients using beta-blockers when compared to those not using this pharmaceutical [23]. Next, the relationship was described more accurately by using a universal nonlinear analytical description [23]. Effects of age and sex on the relationship between EF and ESV were studied applying a logarithmic approximation [44], which has been shown to work remarkably well [32]. The initial linear EF vs ESV approach [23] was further analyzed [51] to derive a performance index, while a few investigators have employed the slope as an indicator of survival by comparing various groups of cardiac patients [26, 31]. It must be emphasized that EF also depends on the imaging method employed to measure volumes [53].

Stroke Volume (SV), Heart Rate (HR), and Ejection Fraction (EF)

SV equals EDV minus ESV. HR multiplied by SV yields CO; both components are shown in Fig. 23.4 in the case of heart failure patients, where EF is reflected by the size of the bubbles. Low values for EF tend to be associated with low

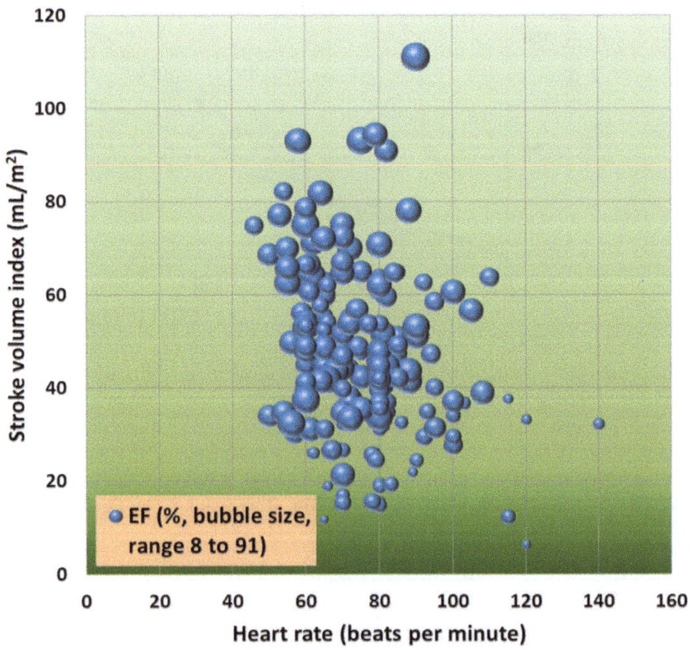

Fig. 23.4 Cardiac output index is obtained by multiplying stroke volume index (SVi) and heart rate (HR). This diagram also shows ejection fraction (EF) as reflected by the bubble size in 179 heart failure patients. Patients with low EF values tend to have low SVi and high HR. In this study group, the female and male patients have values for SVi and HR which are not significantly different, as described elsewhere [28]

levels of SV, but a wide variation can be seen. Although SV is (mathematically seen) an important component of EF, there is no clear correlation between SV and EF. Interestingly, in general there is no significant correlation between EF and EDV either [32]. However, EF is strongly associated with ESV, as we will see later (Fig. 23.7).

Left Ventricular Dimensions During Singleton Pregnancy

Pregnancy offers a physiological model to evaluate the adaptation of the healthy heart of the mother to transient preload and afterload changes [47]. Ideally, the performance and remodeling of the LV should be described during the uncomplicated singleton pregnancy while also explicitly considering shape changes. We used published data [16] and constructed a diagram showing average LV volume for five points in time during gestation in 34 women and one additional measurement at 48 weeks (Fig. 23.5). Remarkably, EF remains rather constant, while changes of EDV were most pronounced. Further details on hemodynamic adaptation can be found in Fu [13].

Ventricular-Arterial Coupling and Ejection Fraction

In an attempt to describe the efficiency of energy transfer from the ventricular pump to the adjacent vascular bed, a formalism has been introduced based on the elastance concept [26]. Ventricular end-systolic elastance (Emax) is derived from the pressure-volume loop [34], while effective arterial elastance (Ea) is defined by the ratio of end-systolic pressure and SV. Since Emax is the tangent to successive loops under constant conditions of contractility, this tangent has a slope and an intercept, assuming linearity which is reasonable when considering a sufficiently small range [34]. If this intercept is chosen along the abscissa, it is termed Vo (i.e., the volume where the Emax line intersects with the theoretical point where pressure would be zero). Hemodynamic coupling (k) between ventricle and afterload is then defined by the ratio Emax/Ea. Since pressure cancels out, k = SV/(ESV−Vo). Figure 23.6 shows k versus EF based on the classical study by Grossman et al. [18], covering the complete spectrum of EF values. Two scenarios are depicted: one with the

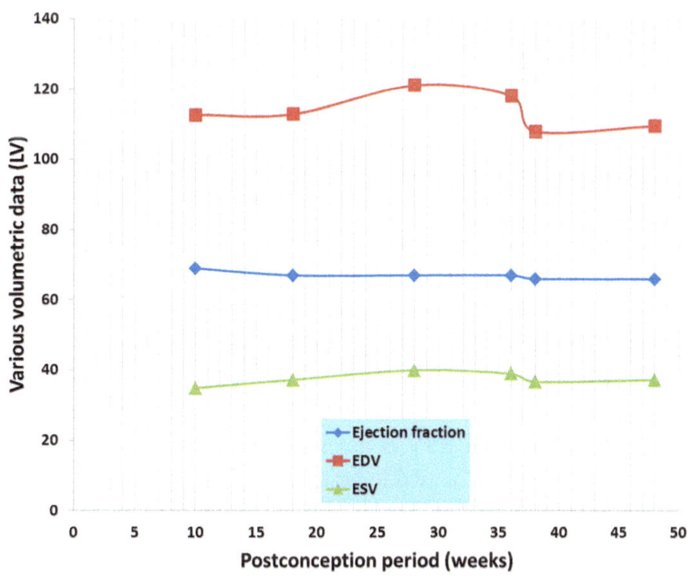

Fig. 23.5 Average values (N = 34) for end-diastolic volume (EDV) and end-systolic volume (ESV) for the left ventricle (LV) in healthy women with uncomplicated singleton pregnancy (Based on data published by Geva et al. [16])

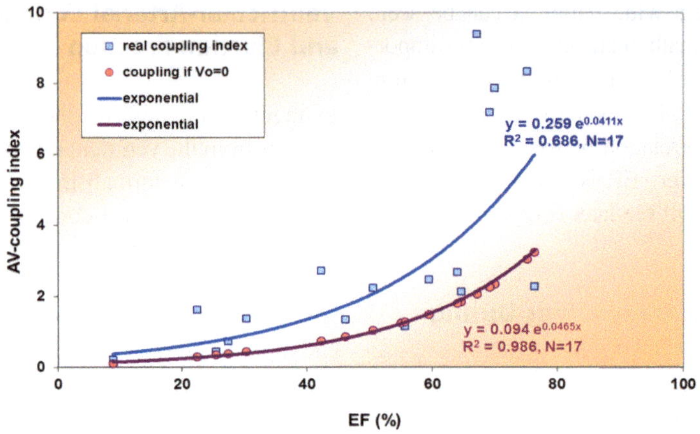

Fig. 23.6 Arterial-ventricular (AV) coupling is strongly related to ejection fraction (EF), simply based on their definitions. Data from patients described by Grossman et al. [18] are shown, to illustrate the effect of neglecting the volume intercept (Vo) which refers to the end-systolic pressure-volume relationship. Clearly, the coupling index is tightly connected to EF if the real value of Vo is omitted (circles) and even if this simplification is not considered (squares)

"real" values for Vo and the other assuming that for simplicity Vo can be neglected. In the latter case, the connection between k and EF is nearly perfect, meaning that both metrics are telling the same story. Surprisingly, the outcome is not much different if the actual value of Vo is considered, yielding a high correlation even for a not very sophisticated exponential approximation. In conclusion, the coupling index has not much more informational value than whatever may be revealed in the EF.

The Right Ventricle

Holt [20] was the first investigator to report values for ESV and EDV as measured in the RV, using a dilution method in dogs. Using data on both LV and RV as collected by Rominger et al. [46], we created a diagram showing EF versus ESV for both compartments (Fig. 23.7). Note that the range for ESV is much smaller for the RV, as expected. The two logarithmic regression lines almost coincide for the groups considered which span the spectrum from healthy volunteers to severe heart disease. In contrast, we found two distinct regression curves for LV and RV in post-Fallot repair patients [30], likely expressing consequences of the transformation following surgical intervention. Using data from Maffessanti et al. [41], we found for the (age group-based) average data available that the regression lines in the VRG representation were very close for men and women, with women having a significantly ($P = 0.01$) higher EF, similar as we have observed elsewhere for the LV [32]. This data set will be discussed later in more detail when describing age dependency.

In 18 patients with sepsis, while employing thermodilution to estimate RV size, Dhainaut et al. [9] found that SV is poorly correlated with EDV ($R = 0.32$) during military antishock trouser (MAST) inflation to manipulate preload. However, they found a high correlation ($R = 0.93$) for ESV vs EDV, which is in support of our VRG approach.

Left Atrium

The left atrium (LA) has an important reservoir function and may during late (ventricular) diastole actively assist LV filling (by the so-called atrial kick). The same study by Graham et al. [17] showing LV volumes in children (Fig. 23.2) also estimated LA size. Figure 23.8

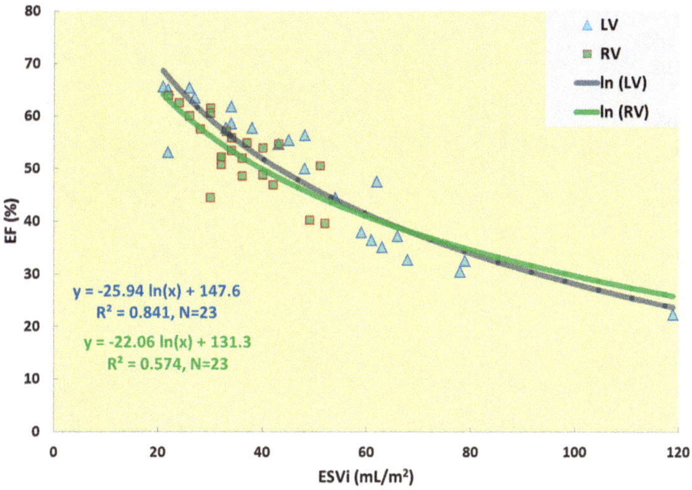

Fig. 23.7 Ejection fraction (EF) versus end-systolic volume index (ESVi) for left ventricle (LV) and right ventricle (RV), measured by fast cine MRI. The study included 52 healthy volunteers and 325 cardiac patients subdivided into 22 diagnostic groups without attention to possible sex differences (Based on diagnostic-group averaged data (with specified sex ratios of individual groups), as published by Rominger et al. [46]. Note that group sizes are unequal, ranging from 4 to 52 individuals)

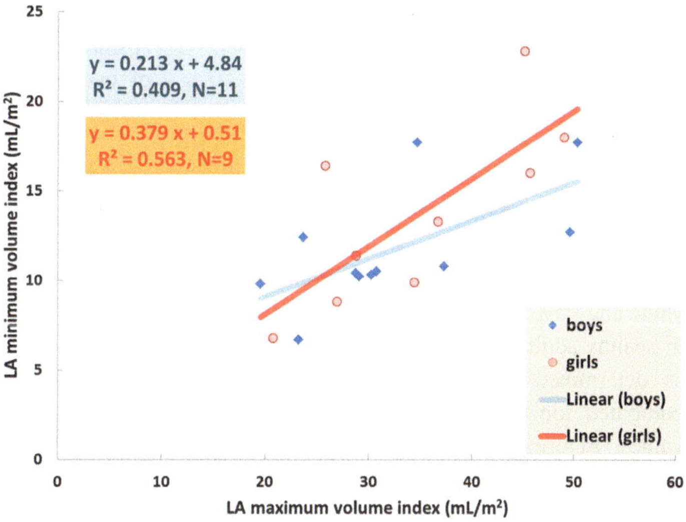

Fig. 23.8 Left atrium (LA) minimum volume index versus maximum volume index in 20 healthy children (9 girls). Age ranges from 3 months to 16 years. Average minimum and maximum (BSA indexed) volumes are not significantly different for boys and girls. (Data from Graham et al. [17])

illustrates the findings for the healthy subgroup ($N = 20$). Although the regression lines seem to differ, the numbers of participants involved are too limited for a meaningful statistical comparison of boys and girls. However, there is no significant sex-based difference between average minimum and maximum volumes, and this observation was also the case when the cardiac patients were included (not shown). These findings in children contrast to findings often seen in healthy adults. For example, Yu et al. [55] found in women a greater ($P < 0.001$) LA volume index (LAVI) than in their male counterparts in the risk factor (according to their CHA2DS2-VASc scores) ≥ 2 group.

In an editorial Di Tullio and Homma [10] note that LA minimum volume is strongly associated with LA EF in the entire population studied (1142

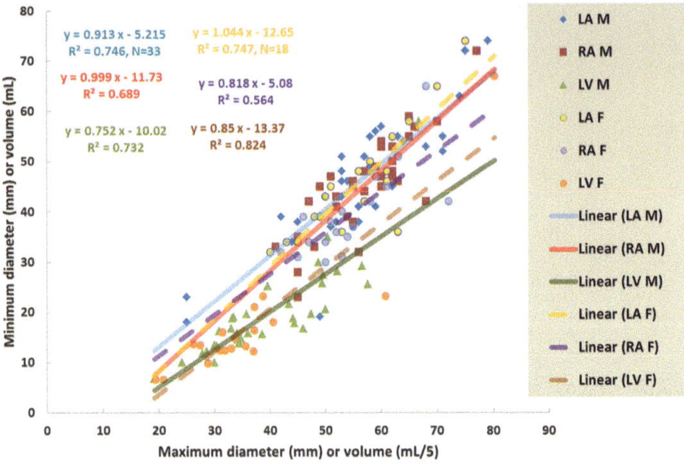

Fig. 23.9 Right atrial (RA), left atrial (LA), and left ventricular (LV) data shown in a combined volume regulation graph, where LV volumes are scaled so as to visually display a similar space as the atrial diameters (Data refer to 51 adult cardiac patients, including 33 males (M) and 18 females (F) investigated by cardiac MRI at VU University Medical Center, Amsterdam, the Netherlands)

healthy elderly). Furthermore, LA minimum volume is the strongest correlate of LA EF in every diagnostic subgroup they analyzed. This finding is to be expected, as it is essentially based on a mathematical triviality, which applies to all four cardiac compartments. A proof based on Monte Carlo simulation (i.e., employing random numbers) has been published elsewhere [29].

Right Atrium

Relatively few studies concern the right atrium (RA), and among those the majority concentrates on diameter or area. Actual RA volume (mean value and 95% confidence interval) was reported in healthy adults free of hypertension and obesity as determined by using 64-multidetector row computed tomography [40], yielding reference values. Figure 23.9 shows in a heterogeneous group of adult cardiac patients the MRI-based RA diameters, along with LA diameters and LV volumes (scaled so as to visually cover the diameter ranges for the atria) and stratified for sex.

Ventricular Dimensions Change with Aging

There has been some debate about the variation of LV EF in men and women during the aging process. An inevitable problem resides in the fact that physiological aging is difficult to delineate and separate from subclinical deterioration. In a gated SPECT study, Peace et al. [44] found in 127 patients with suspected coronary artery disease (CAD) for the LV that ESVi slightly decreases with age, while EF remained unchanged. A study in 442 dogs (280 male) with various types of cardiac disease [25] revealed that fractional shortening (FS), based on echocardiographically determined LV diameters, did not change with age (range a few months up to 15 years), while the slope in the diameter domain equivalent of the VRG tends to increase with age, when three groups are considered (with cutoff at 2 and 6 years).

Relatively few investigators have extended their studies to include the RV. Age-related findings obtained for RV ESV and EDV estimated by 3D echo in healthy adult men and women are illustrated in Fig. 23.10, based on data collected by Maffessanti et al. [41].

Another investigation compared LV volumes in both sexes for three age groups (Fig. 23.11) and confirmed larger ESVi and EDVi in males (N = 43), compared to females (N = 59) based on a gated SPECT study concerning individuals with a low likelihood of CAD [8]. The different volumes as observed in both sexes have implications for the values of EF, as illustrated in the VRG, where trajectories for constant EF

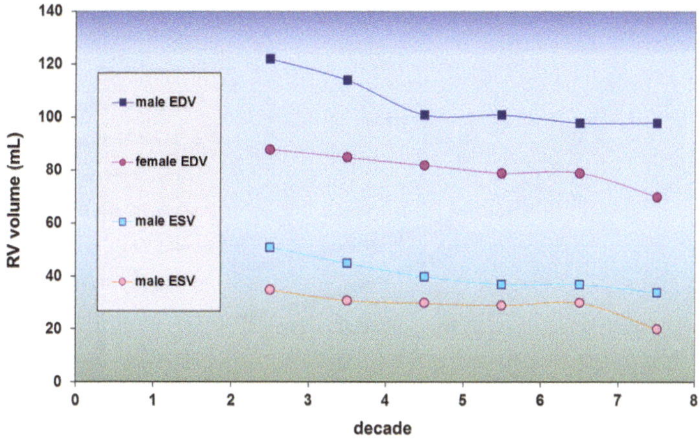

Fig. 23.10 Right ventricular (RV) end-systolic volume (ESV) and end-diastolic volume (EDV) as a function of age groups in healthy adults, estimated by 3D echo. Note that volumes are not indexed for body surface area. Ejection fraction was rather independent of age (not shown). Average values for six age groups with range 18–90 years (N = 507); first group <30 and last group >70 years. Thus, also in the RV, a larger EF (64.0 vs 61.7 %) is found in healthy women. (Based on data from Maffessanti et al. [41])

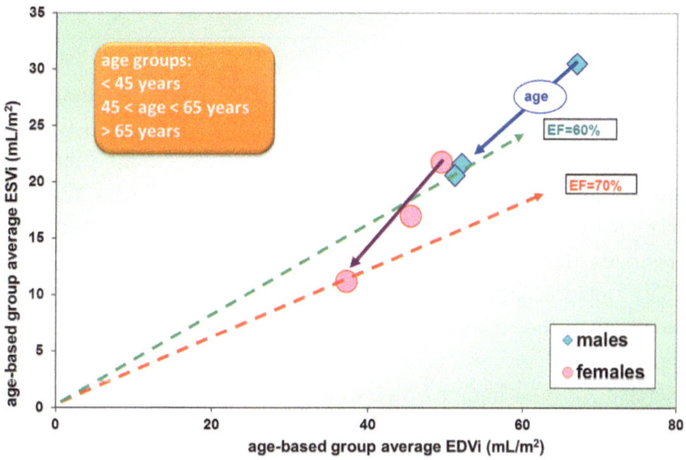

Fig. 23.11 The volume regulation graph for the left ventricle (healthy individuals N = 102), with group averages referring to ages < 45 and > 65 years and intermediate ages, stratified for sex. The lines with arrowhead indicate direction of increasing age for each sex. Note that the highest ejection fraction (EF) occurs in the youngest group of women. This value is considerably higher compared to the equivalent point in males. (Data from De Bondt et al. [8])

values of 60% and 70% are shown. In this study average EF increased with aging, notably in women.

The VRG During Exercise in Men and Women with Heart Disease

Apart from polypill considerations [36], it is now generally accepted that age-dependent appropriate levels and types of exercise constitute a natural and powerful means to maintain health. The impact of physical activity on the functioning and adaptability of the heart has been studied in humans and animal models. Data published by Rerych et al. [45] permit the analysis of effects of exercise in healthy volunteers (N = 30) and individuals with ischemic heart disease (N = 30), 10 with single coronary artery disease and 20 patients with multiple-vessel coronary disease. Radionuclide angiocardiograms were performed at rest and during exercise, with all individuals studied in the erect posture on a bicycle ergometer. In the controls, the mean HR

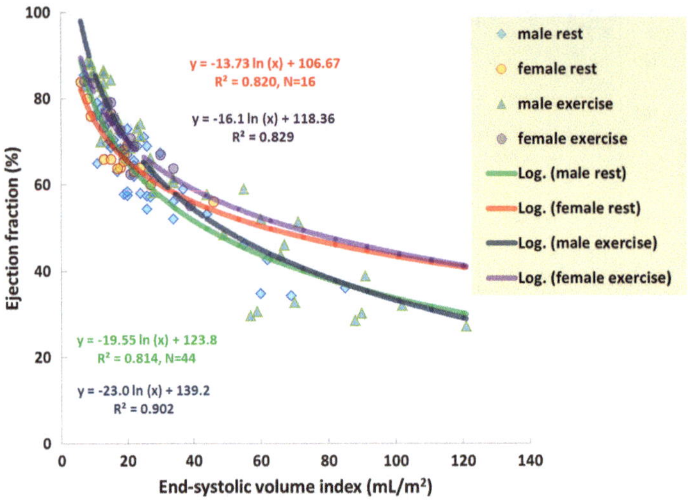

Fig. 23.12 Ejection fraction (EF) as determined by end-systolic volume index (ESVi) during rest and exercise in patients with coronary artery disease (N = 30) and healthy controls (N = 30). Regression curves show an excellent correlation between EF and ESVi and appear in the higher ESVi region to be elevated for women (N = 16), both at rest and during exercise, but this finding is not supported by presence of actual data points. Instead, the regression curves for ESVi < 50 mL/m^2 suggest a similar pattern for men and women, indicating on average comparable EF values at identical ESVi. (Data from Rerych et al. [45])

doubled and the cardiac output tripled during exercise. Their findings indicate that patients with ischemic myocardial disease respond to the stress of exercise by cardiac dilatation to maintain or increase SV at increased HR. Figure 23.12 illustrates the findings in terms of the EF vs ESVi relationship, subdivided for sex, and shows that the inverse nonlinear relationship applies to all groups, both at rest and during exercise. Only at higher ESVi values the curves seem to diverge for men and women. However, in that region (where ESVi > 50 mL/m^2), there are no data points for women, suggesting that the larger ESVi values often seen in men are responsible for this difference.

Cardiac Function in Patients with Diabetes Mellitus

The heart progressively remodels over the life course, and the impact of DM has been investigated. Over a 16-year period in 4062 Framingham Heart Study participants (mean age 45 years, 54% women; 11,485 person-observations), LV wall thickness, LV systolic and diastolic dimensions, and FS were studied in relation to age, sex, body mass index, blood pressure, smoking, and DM. With advancing age, LV dimensions decreased, whereas FS and LV wall thickness increased concomitantly. Female sex, greater blood pressure load, and presence of DM were found to attenuate the remodeling pattern, suggesting a mechanism for the preponderance of women with hypertension and individuals with DM among patients with diastolic HF [6]. We investigated LV volume in men and women with DM and found that average EF is significantly higher in women (Fig. 23.13), as also observed in healthy individuals [5] and in patients [44] with suspected CAD.

The VRG and Heart Failure

Heart failure (HF) can be manifest as different phenotypes [26], which are readily discerned in the VRG representation based on the selected cutoff value for EF. Figure 23.14 illustrates the three most common types with reduced (r),

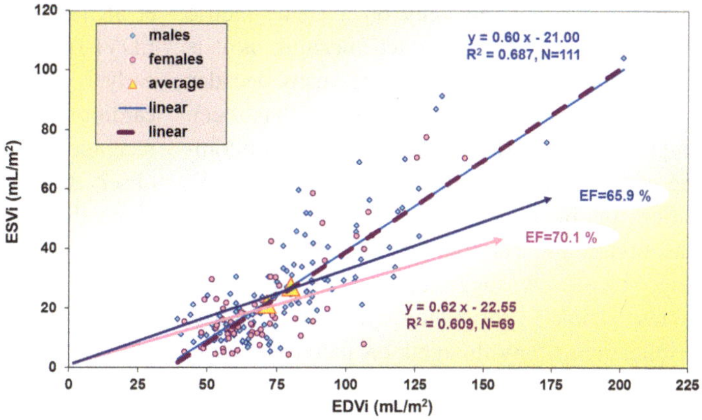

Fig. 23.13 Volume regulation graph for cardiac patients (N = 180) with diabetes mellitus (DM), stratified for males (M) and females (F). The regression lines for M and F (N = 69) almost coincide, but average ESVi and EDVi are significantly ($P = 0.032$) smaller in females. Although higher in women, no significant difference for ejection fraction (EF) was found. (Data from patients as described elsewhere [35])

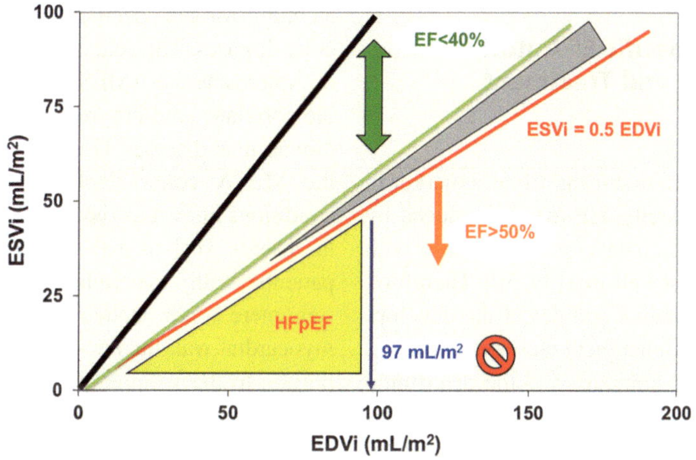

Fig. 23.14 In the volume domain, the three proposed phenotypes of heart failure (HF) can be easily recognized. The HFrEF type is confined to the upper area between the black identity line and the green line indicating that EF < 40%, whereas the HFpEF syndrome is located in the lower triangular region (yellow area). The gray wedge-shaped area is reserved for midrange HF. As a consequence of this separation, we find that the resulting regression lines are clearly different for the three groups, and thus their volume regulation differs, i.e., in HFpEF it is more "Starling-like," implying that SVi clearly increases as filling volume (EDVi) becomes larger. The fact that average EF for women is higher for most patient groups may imply that the cutoff levels for women are higher

midrange (m), and preserved (p) values for EF, respectively. They correspond to precisely defined areas which can clearly be identified in the LV volume domain [27]. In the past a borderline zone was introduced for EF values between 40% and 50%, where frequent transitions between HFpEF and HFrEF (in both directions) might occur. Currently, this zone is recognized as a separate phenotype, called midrange. However, since average EF is significantly higher in healthy adult women [5] and in cardiac patients [44] and is likely also the case in HF [28], establishing the same cutoff values for men and women is inappropriate, and thus the HF phenotypes should require separate analysis. Several studies have proposed to define a cutoff level at 55% for women with HFpEF, as reviewed elsewhere [27] and as also proposed by the *Cardiac Review and Evaluation Committee reviewing trastuzumab* (see below). Moreover, there are strong indications to not rely only on a vague metric such as EF, which has limited meaning as it is just a dimensionless ratio. In fact, many decades ago the potential weakness of EF has been emphasized [2, 24]. Another obvious shortcoming of the current guidelines resides in the application of plain *linear borders* between HF phenotypes as seen on the VRG presentation (Fig. 23.14), as suggested by the committee (s) responsible for formulating such rules, without justification or any thoughts on reconsideration. The application of machine learning could be of substantial importance to resolve the distinctions between HF groups, as shown elsewhere when addressing these issues [1].

Need for Sex-Specific EF Cutoff in Phenotyping and Treatment Considerations

There are concerns about using EF as a marker in studies of cardiotoxicity. HF may be induced by chemotherapy or treatments for breast cancer with radiation to the chest wall area [49, 50]. Therefore, it is important to predict and detect the development of cardiac dysfunction in these patients, e.g., by using imaging techniques and the determination of biomarkers. Cardiotoxicity, as defined by the guidelines of the *Cardiac Review and Evaluation Committee reviewing trastuzumab*, refers to a reduction of the EF of $\geq$5% to <55% with symptoms of HF or an asymptomatic reduction of the EF of $\geq$10% to <55% [48]. Other EF-based criteria have been applied, as reviewed elsewhere [42]. In one report, assessment of the EF fails to detect subtle alterations in LV function due to cardiotoxicity associated with chemotherapeutic regimens (such as anthracyclines and trastuzumab) administered to women with breast cancer [48]. Also, in that study parameters of diastolic function (i.e., mitral valve A and E wave filling velocities) and N-terminal pro-B-type natriuretic peptide did not predict cardiotoxicity. However, cardiac troponin plasma concentrations and longitudinal strain were found to predict the development of cardiotoxicity in those patients [48]. Furthermore, Bergamini et al. [3] found that a cutoff of 58 mL/m^2 for EDVi (rather than EF changes) may act as an early indicator of trastuzumab-related cardiotoxicity in human epidermal growth factor receptor II (HER2+) breast cancer patients. These observations highlight the current confusion about the metric EF. While traditional HF (i.e., with reduced EF) is often defined as EF < 50%, it appears that the threshold for women is higher [27] and that EF is not always a sensitive predictor [48]. The metric EF apparently cannot be reliably used as a single marker of cardiotoxicity, as cardiac damage seems to be more complicated than what the EF ratio can capture.

Apart from thc discussion about sex-specific cutoff values for EF, there is also increasing attention for sex-specific treatment of cardiovascular diseases. The goal of the Multi-Ethnic Study of Atherosclerosis (MESA) cohort was to examine correlates and progression of subclinical cardiovascular disease. This means that by design the MESA cohort included participants with conditions known as risk factors for cardiovascular disease, such as diabetes and hypertension. In patients with hypertension, cardiac myocyte sarcomeres are added in parallel, causing myocardial wall thickness to increase out of proportion to the volume of the LV cavity. Thus, a relative decrease in LV volume might be explained by an increase in LV mass. However, it was observed that in women, LV volume decreased with age even though there was no corresponding increase in LV mass. This finding suggests that some other mechanism may be diminishing LV volume in women or that the causal relationship between mass and volume in women may be different than that in men. One possible hypothesis is that these results reflect sex-specific variations in diffuse myocardial fibrosis, which may inhibit LV volume changes during remodeling. In the MESA cohort, there is evidence that diffuse myocardial fibrosis is greater in women. However, the fibrosis is associated with a greater age-related increase in men. The sex-specific LV mass and volume differences observed in the MESA study may also signal the need for sex-specific treatment approaches for the adverse effects of myocardial remodeling, such as in HF. In summary, in a cohort of individuals who were free of clinical cardiovascular disease at baseline, a longitudinal LV mass increase in men and a slight decrease in women are observed, while EDV decreased and mass-to-volume ratio increases with aging for both men and women. The longitudinal trends are congruent with previous cross-sectional observations in this cohort. These results suggest sex-specific differences in age-related cardiac remodeling [11].

Conclusions

Cardiac function can in detail be described by pressure-volume relationships. A reduced approach, not explicitly including information on pressure, employs the VRG description, which has been shown to be applicable to all four cardiac compartments. Examples have been presented regarding the LV in healthy children and those with myocardial disease (Fig. 23.2), with for the linear regression coefficients no difference between boys and girls. Similar regression results were found for the LV in adults with anginal complaints, indicating that average EF in women is higher ($P < 0.0001$) than in men (Fig. 23.3), and also for cardiac patients with DM (Fig. 23.13), but now without a sex-related difference in EF. The VRG concept was extended to LA and RA diameters (Fig. 23.9), yielding similar linear relationships for a heterogeneous group of cardiac patients. Interestingly, a high correlation was found for the inverse relationship between EF and ESVi in a combination of healthy individuals and cardiac patients (Fig. 23.7) for both LV and RV. This finding indicates that EF is closely connected to the actual value of ESVi, which observation is further supported by analysis of CAD patients at rest and during exercise (Fig. 23.12). The traditional subdivision of HF in three phenotypes (Fig. 23.14) becomes readily recognizable in the VRG representation and illustrates that the current HF paradigm may be based on an oversimplification, especially if EF cutoff values are considered identical for both sexes. ESVi seems to play a key role in various types of analysis, and, interestingly, this metric also refers to the largest percentwise difference between (healthy) men and women (Table 23.1). Not only is ESVi the primary determinant of EF but also of the ventriculo-arterial coupling index (k), which fact readily explains the almost trivial connection between k and EF (Fig. 23.7).

In retrospect, it is amazing to learn about the seemingly parallel developmental patterns regarding insight and design of mechanical and biological pump systems. Originating in Italy, the epicenter moved to England while also attracting engineers from France and Germany. During the past decades, both lines came to a synthesis with the introduction of artificial hearts and cardiac assist devices. The technical product is not yet as reliable and energy-efficient as the average biological pump and also requires maintenance. Despite the sophisticated measurement of pressure, volume, Doppler-derived velocities, strain (rate), and torsion, it is evident that our current understanding of the natural pump is far behind the technical insight concerning the simple technical imitations created thus far. Failure of a mechanical system is relatively easy to detect, and spare parts are usually available. In sharp contrast, in cases of a failing heart, we often must admit that defining the problem is not always precise and occasionally still a riddle.

References

1. Alonso-Betanzos A, Bol6n-Canedo V, Heyndrickx GR, Kerkhof PLM. Exploring guidelines for classification of major heart failure subtypes by using machine learning. Clin MedInsights Cardiol. 2015;9(Suppl 1):57–71.
2. Altschule MD. Limited usefulness of the so-called ejection fraction measurement in clinical practice. Chest. 1986;90:134–5.
3. Bergamini C, Torelli F, Ghiselli L, Rossi A, Trevisani L, Vinco G, et al. Left ventricular end-diastolic volume as early indicator of trastuzumab-related cardiotoxicity in HER2+ breast cancer patients: results from a single-center retrospective study. Minerva Cardioangiol. 2017;65:278–87.
4. Beringer JY, Kerkhof PLM. A unifying representation of ventricular volumetric indexes. IEEE Trans BME. 1998;45:365–71.
5. Buonanno C, Arbustini E, Rossi L, Dander B, Vassanelli C, Paris B, Poppi A. Left ventricular function in men and women. Another difference between sexes. Eur Heart J. 1982;3:525–8.
6. Cheng S, Xanthakis V, Sullivan LM, Lieb W, Massaro J, Aragam J, Benjamin EJ, Vasan RS. Correlates of

echocardiographic indices of cardiac remodeling over the adult life course: longitudinal observations from the Framingham Heart Study. Circulation. 2010;122:570–8. https://doi.org/10.1161/CIRCULATIONAHA.110.937821.
7. Dallaire F, Sarkola T. Growth of cardiovascular structures from the fetus to the young adult. In: Kerkhof PLM, Miller VM, editors. Sex-specific analysis of cardiovascular function. Cham: Springer; 2018; p. 347–60.
8. De Bondt P, Van de Wiele C, De Sutter J, De Winter F, De Backer G, Dierckx RA. Age- and gender-specific differences in left ventricular cardiac function and volumes determined by gated SPET. Eur J Nucl Med. 2001;28:620–4.
9. Dhainaut JF, Pinsky MR, Nouria S, Slomka F, Brunet F. Right ventricular function in human sepsis. Chest. 1997;112:1043–9.
10. Di Tullio MR, Homma S. Left atrial morphology and function: The other side of cardiovascular risk. Circ Cardiovasc Imaging. 2016;9:e004494. https://doi.org/10.1161/CIRCIMAGING.116.004494.
11. Eng J, McClelland RL, Gomes AS, Hundley WG, Cheng S, Wu CO, Carr JJ, Shea S, Bluemke DA, Lima JA. Adverse left ventricular remodeling and age assessed with cardiac MR imaging: The multi-ethnic study of atherosclerosis. Radiology. 2016;278:714–22. https://doi.org/10.1148/radiol.2015150982.
12. Ford LE. Heart size. Circ Res. 1976;39:297–303. https://doi.org/10.1161/01.RES.39.3.297.
13. Fu Q. Hemodynamic and electrocardiographic aspects of uncomplicated singleton pregnancy. In: Kerkhof PLM, Miller VM, editors. Sex-specific analysis of cardiovascular function. Cham: Springer; 2018; p. 413–31.
14. Gebhard C, Buechel RR, Stähli BE, Gransar H, Achenbach S, Berman DS, et al. Impact of age and sex on left ventricular function determined by coronary computed tomographic angiography: results from the prospective multicentre CONFIRM study. Eur Heart J Imag. 2017;18:990–1000.
15. Gedeon A. Science and technology in medicine. New York: Springer; 2006.
16. Geva T, Mauer MB, Striker L, Kirshon B, Pivarnik JM. Effects of physiologic load of pregnancy on left ventricular contractility and remodeling. Am Heart J. 1997;133:53–9.
17. Graham TP Jr, Jarmakani MM, Canent RV Jr, Capp MP, Spach MS. Characterization of left heart volumes and mass in normal children and in infants with intrinsic myocardial disease. Circulation. 1968;38:826–37.
18. Grossman W, Braunwald E, Mann T, McLaurin LP, Green LH. Contractile state of the left ventricle in man as evaluated from end-systolic pressure-volume relations. Circulation. 1977;56:845–52.
19. Hales S. Statical essays. London: Innys & Co; 1733.
20. Holt JP, Allensworth J. Estimation of the residual volume of the right ventricle of the dog's heart. Circ Res.1957;5:323–6.
21. Hung CL, Gonçalves A, Shah AM, Cheng S, Kitzman D, Solomon SD. Age- and sex-related influences on left ventricular mechanics in elderly individuals free of prevalent heart failure: The ARIC Study. Circ Cardiovasc Imaging. 2017;10. pii: e004510. doi:https://doi.org/10.1161/CIRCIMAGING.116.004510.
22. Kaku K, Takeuchi M, Tsang W, Takigiku K, Yasukochi S, Patel AR, Mor-Avi V, Lang RM, Otsuji Y. Age-related normal range of left ventricular strain and torsion using three-dimensional speckle-tracking echocardiography. J Am Soc Echocardiogr. 2014;27(1):55–64.
23. Kerkhof PLM, Baan J, Buis B, Arntzenius AC. Relations between ejection fraction and ventricular volume, and their alteration by chronic beta-blockade. Br Heart J. 1981;46:17–22.
24. Kerkhof PLM. Importance of end-systolic volume for the evaluation of cardiac pump performance. In: Chazov EI, Smirnov VN, Oganov RG, editors. Cardiology, an international perspective. New York: Plenum Press; 1984. p. 1339–52.
25. Kerkhof PL, Roos A, ter Haar G, Kocsis S, Pijnenburg HL, Stokhof AA. Age variance of left ventricular diameters in dogs with cardiac disease. J Gerontol A Biol Sci Med Sci. 1998;53: B25–31.
26. Kerkhof PLM, Kresh JY, Li JK-J, Heyndrickx GR. Left ventricular volume regulation in heart failure with preserved ejection fraction. Physiol Rep. 2013;1:e00007.
27. Kerkhof PLM. Characterizing heart failure in the ventricular volume domain. Clin Med Insights Cardiol. 2015;9 (S1):11–31.
28. Kerkhof PLM, Heyndrickx GR, Li JJ-K. Hemodynamic determinants and ventriculo-arterial coupling are sex-associated in heart failure patients. Conf Proc IEEE Eng Med Biol Soc. 2016;2016:3286–9. https://doi.org/10.1109/EMBC.2016.7591430.
29. Kerkhof PLM, Yoo BW, Merillon JP, Peace RA, Handly N. Monte Carlo method applied to the evaluation of the relationship between ejection fraction and its constituent components. Conf Proc IEEE Eng Med Biol Soc. 2017;2017:1295–8. https://doi.org/10.1109/EMBC.2017.8037069.
30. Kerkhof PLM, Yoo BW, van de Ven PM, Handly N. Sex-specific aspects of left and right ventricular volume regulation in patients following tetralogy of Fallot repair. Conf Proc IEEE Eng Med Biol Soc. 2017;2017:1303–6. https://doi.org/10.1109/EMBC.2017.8037071.
31. Kerkhof PL, Peace RA, Handly P. Importance of (measuring) the end-systolic volume index in predicting survival. Heart. 2018;104:180. https://doi.org/10.1136/heartjnl-2017-312051.
32. Kerkhof PL, van de Ven PM, Yoo BW, Peace RA, Heyndrickx GR, Handly N. Ejection fraction as related to basic components in the left and right ventricular volume domains. Int J Cardiol. 2018;255:105–10.
33. Kerkhof PLM, Kuznetsova T, Kresh JY, Handly N. Cardiophysiology illustrated by comparing ventricular volumes in healthy adult males and females. In: Kerkhof PLM, Miller VM, editors. Sex-specific analysis of cardiovascular function. Cham: Springer; 2018; p. 123–38.
34. Kerkhof PLM, Kuznetsova T, Ali R, Handly N. Left ventricular volume analysis as a basic tool to describe cardiac function. Adv Physiol Educ. 2018;42(1):130–9. https://doi.org/10.1152/advan.00140.2017.
35. Kerkhof PLM, Peace RA, Macfarlane PW. Sex- and age-related reference values in cardiology, with annotations and guidelines for interpretation. In: Kerkhof PLM, Miller VM, editors. Sex-specific analysis of cardiovascular function. Cham: Springer; 2018; p. 677–706.
36. Kerkhof PLM, Osto E. Women and men in the history of Western cardiology: some notes on their position as patients, role as investigational subjects, and impact as professionals. In: Kerkhof PLM, Miller VM, editors. Sex-specific analysis of cardiovascular function. Cham: Springer; 2018; p. 1–30.
37. Kresh JY, Kerkhof PL, Goldman SM, Brockman SK. Heart-mechanical assist device interaction. ASAIO Trans. 1986;32:437–43.
38. Kuznetsova T, Haddad F, Tikhonoff V, et al. Impact and pitfalls of scaling of left ventricular and atrial structure in population-based studies. J Hypertens. 2016;34:1186–94.

39. Lieb W, Gona P, Larson MG, Aragam J, Zile MR, Cheng S, Benjamin EJ, Vasan RS. The natural history of left ventricular geometry in the community: clinical correlates and prognostic significance of change in LV geometric pattern. JACC Cardiovasc Imaging. 2014;7:870–8. https://doi.org/10.1016/j.jcmg.2014.05.008.
40. Lin FY, Devereux RB, Roman MJ, et al. Cardiac chamber volumes, function, and mass as determined by 64-multidetector row computed tomography: mean values among healthy adults free of hypertension and obesity. JACC Cardiovasc Imaging. 2008;1:782–6. https://doi.org/10.1016/j.jcmg.2008.04.015.
41. Maffessanti F, Muraru D, Esposito R, Gripari P, Ermacora D, Santoro C, Tamborini G, Galderisi M, Pepi M, Badano LP. Age-, body size-, and sex-specific reference values for right ventricular volumes and ejection fraction by three-dimensional echocardiography: a multicenter echocardiographic study in 507 healthy volunteers. Circ Cardiovasc Imaging. 2013;6:700–10.
42. Martín M, Esteva FJ, Alba E, Khandheria B, Pérez-Isla L, García-Sáenz JA, Márquez A, Sengupta P, Zamorano J. Minimizing cardiotoxicity while optimizing treatment efficacy with trastuzumab: review and expert recommendations. Oncologist. 2009;14:1–11. https://doi.org/10.1634/theoncologist.2008-0137.
43. Moscato F, Vollkron M, Bergmeister H, Wieselthaler G, Leonard E, Schima H. Left ventricular pressure-volume loop analysis during continuous cardiac assist in acute animal trials. Artif Organs. 2007;31:369–76.
44. Peace RA, Adams PC, Lloyd JJ. Effect of sex, age, and weight on ejection fraction and end-systolic volume reference limits in gated myocardial perfusion SPECT. J Nucl Cardiol. 2008;15:86–93.
45. Rerych SK, Scholz PM, Newman GE, Sabiston DC Jr, Jones RH. Cardiac function at rest and during exercise in normals and in patients with coronary heart disease: evaluation by radionuclide angiocardiography. Ann Surg. 1978;187:449–64.
46. Rominger MB, Bachmann GF, Pabst W, Rau WS. Right ventricular volumes and ejection fraction with fast cine MR imaging in breath-hold technique: applicability, normal values from 52 volunteers, and evaluation of 325 adult cardiac patients. J Magn Reson Imaging. 1999;10:908–18.
47. Savu O, Jurcuţ R, Giuşcă S, van Mieghem T, Gussi I, Popescu BA, Ginghină C, Rademakers F, Deprest J, Voigt JU. Morphological and functional adaptation of the maternal heart during pregnancy. Circ Cardiovasc Imaging. 2012;5:289–97.
48. Sawaya H, Sebag IA, Plana JC, et al. Early detection and prediction of cardiotoxicity in chemotherapy-treated patients. Am J Cardiol. 2011;107:1375–80. https://doi.org/10.1016/j.amjcard.2011.01.006.
49. Shufelt C, Pacheco C, Tweet MS, Miller VM. Sex-specific physiology and cardiovascular disease. In: Kerkhof PLM, Miller VM, editors. Sex-specific analysis of cardiovascular function. Cham: Springer; 2018; p. 433–54.
50. Tadic M, Cuspidi C, Vasic D, Kerkhof PLM. Cardiovascular implications of diabetes, metabolic syndrome, thyroid disease and cardio-oncology in women. In: Kerkhof PLM, Miller VM, editors. Sex-specific analysis of cardiovascular function. Cham: Springer; 2018; p. 471–88.
51. Thormann J, Kramer W, Kindler M, Neuss H, Bahawar H, Schlepper M. Analysis of the efficacy of the new cardiotonic agent TA-064. Am Heart J. 1985;110:426–38.
52. Vogel S. Vital circuits – on pumps, pipes, and the workings of circulatory systems. New York: Oxford University Press; 1992.
53. Wood PW, Choy JB, Nanda NC, Becher H. Left ventricular ejection fraction and volumes: it depends on the imaging method. Echocardiography. 2014;31:87–100.
54. Yamada AT, Campos Neto Gde C, Soares J Jr, Giorgi MC, Araújo F, Meneghetti JC, Mansur AJ. Gender differences in ventricular volumes and left ventricle ejection fraction estimated by myocardial perfusion imaging: comparison of Quantitative Gated SPECT (QGS) and Segami software programs. Arq Bras Cardiol. 2007;88:285–90.
55. Yu HT, Lee JS, Kim TH, Uhm JS, Joung B, Hong GR, Lee MH, Shim CY, Pak HN. Advanced left atrial remodeling and appendage contractile dysfunction in women than in men among the patients with atrial fibrillation: Potential mechanism for stroke. J Am Heart Assoc. 2016;5(7). pii: e003361. doi:https://doi.org/10.1161/JAHA.116.003361.
56. Ziman J. The force of knowledge; the scientific dimension of society. Cambridge: Cambridge University Press; 1976.

Heart Transplantation Survival and Sex-Related Differences

24

Martina Previato, Elena Osto, Peter L. M. Kerkhof, Gareth Parry, and Francesco Tona

Heart transplant recipient George DeBord holds his own heart aloft (1968). Art work by Piet Michiels, Leuven, Belgium.

M. Previato · F. Tona (✉)
Department of Cardiac, Thoracic and Vascular Sciences, University of Padova, Padova, Italy
e-mail: francesco.tona@unipd.it

E. Osto (✉)
Laboratory of Translational Nutrition Biology, Federal Institute of Technology Zurich ETHZ, Zurich, Switzerland

Center for Molecular Cardiology, University of Zurich, Zurich, Switzerland

University Heart Center, Cardiology, University Hospital Zurich, Zurich, Switzerland
e-mail: elena.osto@uzh.ch

P. L. M. Kerkhof
Department of Radiology and Nuclear Medicine, Amsterdam Cardiovascular Sciences, VU University Medical Center, Amsterdam, The Netherlands

G. Parry
Department of Cardiopulmonary Transplantation, Freeman Hospital, Newcastle upon Tyne, UK

P. L. M. Kerkhof, V. M. Miller (eds.), *Sex-Specific Analysis of Cardiovascular Function*,
Advances in Experimental Medicine and Biology 1065, https://doi.org/10.1007/978-3-319-77932-4_24

Abstract

Orthotopic heart transplantation (OHT) is the "gold standard" treatment for patients with end-stage heart failure, with approximately 5000 transplants performed each year worldwide. Heart transplantation survival rates have progressively improved at all time points, despite an increase in donor and recipient age and comorbidity and greater recipient urgency; according to the registry of the International Society of Heart and Lung Transplantation (ISHLT), the median survival of patients posttransplantation is currently 12.2 years.

Long-term survival is sub-optimal, and outcomes after OHT remain constrained by the development of acute rejection and cardiac allograft vasculopathy (CAV). Moreover, donor organs are in short supply, making optimal organ utilization an ongoing priority. For these reasons, substantial interest continues to exist in identifying factors portending increased survival and improved organ utilization.

Keywords

Cardiac transplantation · Allograft vasculopathy · Sex mismatch · Donor organ · Microvasculopathy · Premenopausal female donor · Posttransplant mortality · Immunological difference · Transfusion history · United Network for Organ Sharing · Review

Historical Introduction

Christiaan Neethling Barnard (Beauford West 1922–2001 Coral Bay, Cyprus) performed the first human heart transplant at the Groote Schuur Hospital, Cape Town, on 3 December 1967. Louis Washkansky, a 53-year-old man, received the heart of a 25-year-old woman (Denise Darvall who was involved in a traffic accident), and he lived for 18 days after the operation. At the same time, this was the first case of sex mismatch between donor and recipient. Surgeons worldwide followed Barnard's lead and, by October 1971, recorded heart transplants numbered 178. Twenty-seven patients had survived, but only five of these had lived longer than 3 years. Forty-year-old Mrs. Dorothy Fischer was given a new heart in 1969 and became by then the longest surviving patient. Shortly after performing his historic operations, Barnard made it known that he was suffering from arthritis; his hands were becoming crippled at the joints. By 1997, there had been about 25,000 heart transplants and a survival rate, usually with good quality of life, of 65% at 5 years and 50% at 10 years [5, 7] (Fig. 24.1).

Fig. 24.1 Postal stamp issued in 1969 by the Republic of South Africa (RSA), honoring Dr. Barnard

Sex Mismatch Regarding Donor and Recipient

For many years, investigators have remained unsure to what extent donor and recipient sex influence outcomes in OHT [26]. Several early studies identified female donor sex to be an independent predictor of recipient mortality after OHT (e.g., Solomon et al. [33], Taylor et al. [36], Tsai et al. [37], Prendergast et al. [30]). The major limitation of these studies was that they failed to stratify patients by both donor and recipient sexes [6]. Given the fact that >75% of heart transplant recipients have traditionally been male, the investigation of the effect of donor sex on survival should have included separate examinations for male and female recipients.

This issue was addressed by further investigations which stratified both male and female recipients by the sex of the donor heart and highlighted that the sex matching or mismatching between donor and recipient (rather than the sex of the donor or recipient individually) is what influences outcomes. Multiple small, single-center studies [1, 18] and three large registry studies [19, 38, 39] all showed that adult male recipients (MR) receiving hearts from female donors (FD) had reduced survival when compared to male recipients receiving hearts from male donors.

Review of the Literature on Sex Mismatch in Cardiac Transplants

The study conducted by Al-Khaldi et al. [1] further stratified their cohort of 869 consecutive OHTs by recipient age. Sex mismatching (FD/MR), in agreement with other studies mentioned above, was associated with a reduced survival rate at 1, 5, and 10 years compared with the group of male donors/male recipients (MD/MR) in the >45 years of age sub-cohort (69%, 49%, and 27% vs 87%, 72%, and 45%, respectively). Instead, donor sex showed no effect on the long-term survival of male recipients <45 years of age. The absence of a survival difference in the group of MD/MR based on recipient age cutoff of 45 years suggests that the difference seen in the FD/MR group is likely due to factors associated with female donor gender rather than a direct effect of the recipient age. Moreover, female donor age did not appear to modulate survival of older male recipients, who had an equally bad prognosis whether the female donor age was above or below 45 years. Figure 24.2a shows short-term survival in males and females depending on sex (mis)match, while Fig. 24.2b illustrates long-term survival.

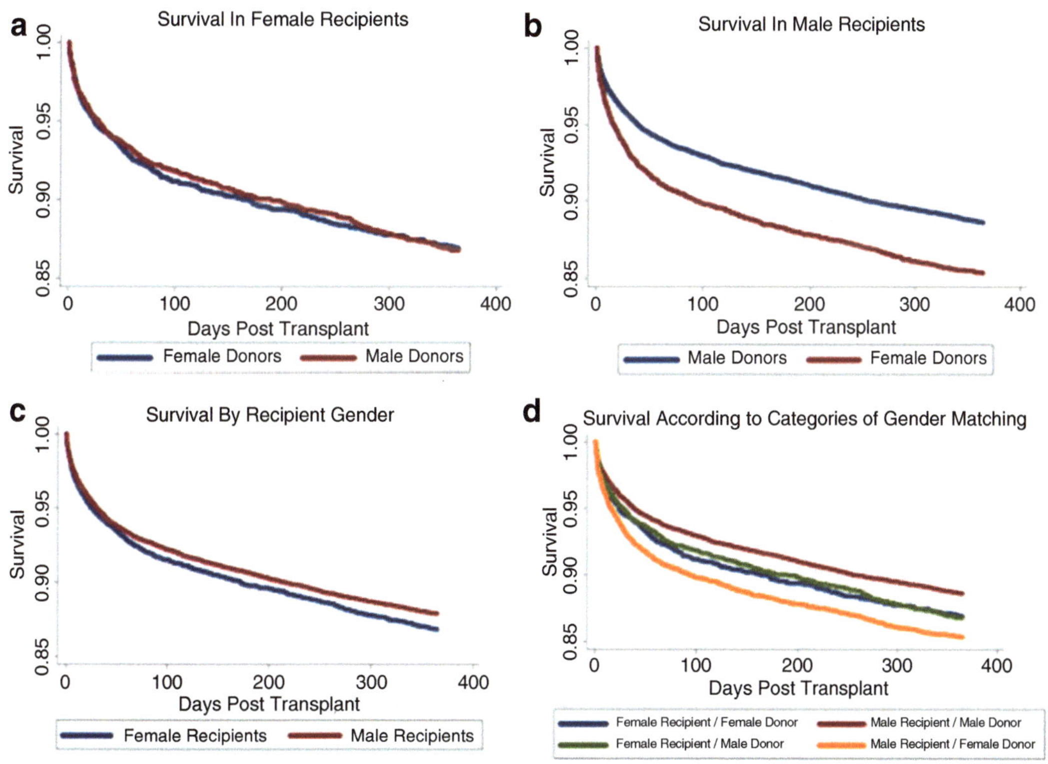

Fig. 24.2a Unadjusted Kaplan-Meier graphs of survival in the first year. (**a**) Survival according to donor sex in female recipients (hazard ratio [HR], 1.00; 95% confidence interval [CI], 0.88–1.14; p = 1.0). (**b**) Survival according to donor sex in male recipients (HR, 1.32; 95% CI, 1.22–1.43; p = 0.001). (**c**) Survival according to recipient sex (HR, 1.09; 95% CI, 1.01–1.18; p = 0.02). (**d**) Survival according to categories of sex matching (Reproduced from Reed et al. [], with permission)

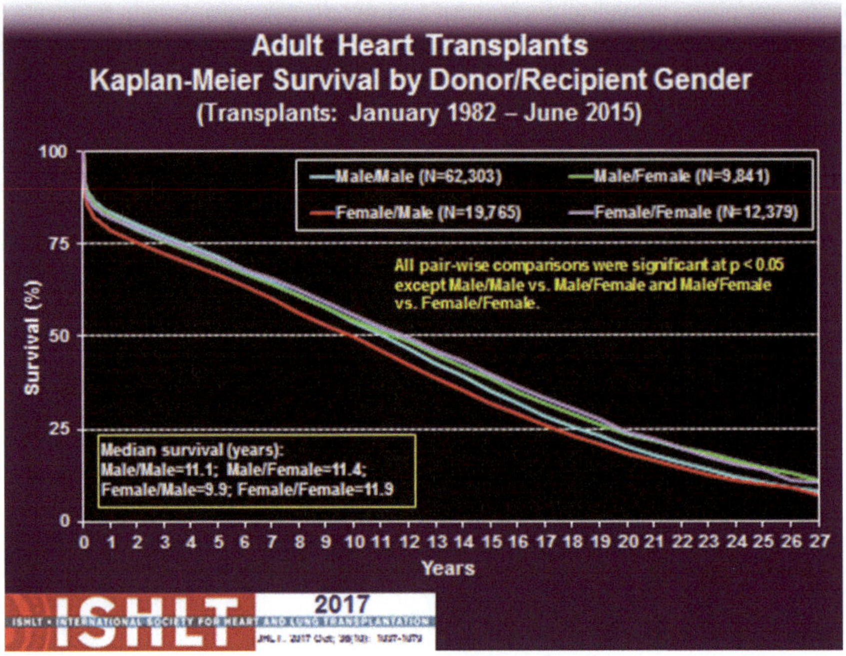

Fig. 24.2b Adult heart transplants Kaplan-Meier survival by donor/recipient gender. *Survival was calculated using the Kaplan-Meier method, which incorporates information from all transplants for whom any follow-up has been provided. Since many patients are still alive and some patients have been lost to follow-up, the survival rates are estimates rather than exact rates because the time of death is not known for all patients. The median survival is the estimated time point at which 50% of all of the recipients have died. Survival rates were compared using the log-rank test statistic. Adjustments for multiple comparisons were done using Scheffe's method* (From Lund et al. [] and http://www.ishlt.org/registries/slides.asp?slides=heartLungRegistry, reproduced with permission)

The analysis of the United Network for Organ Sharing (UNOS) database by Weiss et al. [38] further stratified recipients by UNOS status (that served as a surrogate for clinical acuity) and demonstrated that men listed for transplant as UNOS status 1a or 1b (or older UNOS status 1) receiving organs from female donors continued to show decreased survival, but interestingly this association did not persist when examining those male recipients listed as UNOS status 2.

While these studies seem to unanimously agree that female donor sex is a significant risk factor for posttransplant mortality in male recipients, data for female recipients are conflicting. The study conducted by Al-Khaldi et al. [1] found no association between donor sex and outcomes for the 213 female recipients in the series, independent of their age. The analysis of the UNOS database by Weiss et al. [38] showed that female recipients had a 3.6% lower overall survival at 5 years posttransplant, but no differences in mortality were observed based on donor sex; further stratification by UNOS status did not alter the primary findings.

However more recent studies with longer follow up found that adult female recipients had better survival when they received a heart from a female donor. Khush et al. [18] studied the outcomes of 60,584 adult recipients of OHT from the ISHLT Registry and demonstrated that female recipients of female allografts had a 10% decrease in adjusted mortality relative to female recipients of male allografts; these findings were similar for death-censored allograft survival.

The study by Kittleson et al. [19] showed that the10-year actuarial survival was significantly lower not only in the FD/MR group but also in the MD/FR group, compared to the MR/MD group (59% and 58% respectively, compared to

69%). Interestingly, this study also found that the survival curve of the FD/MR group separated from the other three groups (MD/MR, FD/FR, and MD/FR) within the first year posttransplant and continued with a lower survival over the 10-year follow-up period. For the MD/FR group, however, the survival curve did not diverge from that of the sex-matched groups until 4 years posttransplant, at which point it dropped and then paralleled the survival curve of the FD/MR group. This could explain why previous studies with a shorter follow-up (up to 5 years) found no differences in mortality based on donor sex for female recipients.

Although at this point it is the general view that male recipients receiving hearts from female donors have reduced survival and there is evidence suggesting that sex mismatch is undesirable in female recipients as well, none of the previously mentioned studies could definitively address the question of why sex mismatch leads to decreased survival.

Indeed, there are several potential mechanisms by which donor/recipient sex mismatch might affect outcomes in OHT patients [23]. Proposed hypotheses include:

- *Size mismatch.* Unique to the heart is the importance of physical size both for provision of adequate perfusion and to ensure adequate functional reserve [8]. It is known that female hearts are generally smaller relative to males [3]; the smaller mass of the female heart may not have the functional reserve required to supply the male body. It is common practice in numerous institutions to have size matching strategies that take into account donor-to-recipient body surface area (BSA) and/or body mass index (BMI) ratios in the organ allocating process; as a consequence many series report similar donor-to-recipient BSA and BMI ratios in all four of their donor/recipient sex strata (MD/MR, FD/FR, MD/FR, FD/MR). Some studies though hypothesized that the relatively smaller organ mass in the female donor compared with the male donor, even when corrected for BSA, might result in fewer functioning units and increased workload per individual unit; this could be one of the causes of reduced survival rates in male recipients of female donors.
- *Immunologic factors.* Immunological differences between men and women are obvious candidates [21]. It has been reported that female heart transplant recipients have higher risk of rejection and of hemodynamically significant issues [25] and require greater immunosuppression; this could be explained by an increased pretransplant immunoreactivity in females (as manifested by higher prevalence of HLA-B8 and DR3 haplotypes and antinuclear antibodies) and an earlier production of anti-HLA antibodies posttransplant [24]. Another possible explanation is that female T cells are more reactive than male T cells in their response to self-antigens and mismatched antigens, as demonstrated by a study performed with the Australian Bone Marrow Donor Registry [4].
- *Genetic and hormonal factors.* The influence of the Y chromosome on posttransplant outcome is currently under investigation. Studies in renal transplantation repeatedly showed worse outcomes in sex-mismatched transplants; a study by Tan et al. [35] demonstrated a strong association between the development of H-Y antibodies (46%) and the development of acute rejection (AR) in female recipients of male kidneys. Thus, it is possible that the development of H-Y antibodies could partially explain the poorer outcomes in female cardiac transplant recipients of male donor organs. Further study on the association between donor-recipient sex and cellular and antibody-mediated rejection is necessary to explore this hypothesis.

The relationship between sex hormones and immunological processes has been extensively documented but is not well understood; for example, estrogen levels in women receiving oral contraceptives are linked to a higher immunoglobulin M level, suggesting a role for estrogen in immunoglobulin production. Data from the Berlin Heart Center highlighted that endothelial allograft dysfunction and stenotic microvasculopathy occurred more frequently in allografts from premenopausal female to male transplants vs premenopausal female donor to female receiver, raising the question of hormonal influences on transplant immunoreactivity [12].

Olivetti et al. [27], in a study on the effect of aging in human hearts in males and females, showed that the male heart has a progressive myocardium loss of nearly 1 g/year, resulting in the loss of approximately 64 million cells, with increased myocyte cell volume in the remaining cells. In contrast, aging in the female heart was associated with preservation of ventricular myocardial mass and cell volume. If these effects are hormone-dependent, it is possible to hypothesize that, after transplanting a female heart in a male recipient, there will be a progressive loss of myocardial mass at a rate similar to that of the male heart (or even at a higher rate due to withdrawal of the female protective hormones and introduction of the male hormones) in the female donor heart that already has a smaller baseline ventricular mass. Sex differences in susceptibility to ischemia-reperfusion injury have also been proposed.

All of these theories, however, remain more or less speculative and require further investigation.

Evaluation of Cardiac Transplant Patients Using Imaging of Cardiac Chamber Dimensions

Ultimately, all transplantation-related effects due to the immunological, genetic, and hormonal factors involved, as well as aspects of size-associated (mis)match concerning cardiac dimension and coronary artery caliber, will likely translate into modification of ventricular size. Cardiac pump performance ("systolic function") is often evaluated by considering ejection fraction (EF). While this metric is relatively easy to determine, the interpretation has recently again been questioned [14, 20]. Therefore, we propose the combined study of the constituent components regarding EF [16], namely, end-systolic volume (ESV) and end-diastolic volume (EDV). An example is presented involving 101 cardiac transplant patients (age 4–67 years, 33 female recipients) carried out at the Freeman Hospital, Newcastle upon Tyne, UK. Left ventricular (LV) volumes were obtained by CT and images processed on a Siemens Syngo Via workstation. While at the time of Dr. Barnard and his colleagues names of donors and recipients were often disclosed for inclusion on the front page in newspapers, we nowadays have committee members who decide on how to carefully deal with these matters.

The key variables ESV and EDV are determinants of ventricular volume regulation as demonstrated in various diagnostic patient groups [14, 15]. Here we demonstrate that in cardiac transplant patients, these variables are clearly correlated (Fig. 24.3), with no significant correlation for FD to either FR or MR due to the limited number of participants in these subgroups and the relatively small volume ranges. For the whole group (Fig. 24.3 without stratification), we found ESV $=$ 0.379 EDV - 3.25, $R^2 = 0.466$, N $=$ 94. This finding is similar to what we found by calculating ESV vs EDV for data available in the literature. One longitudinal study concentrated on EF and EDV in 71 patients (87% males) during on average 5 years of follow-up and found no difference compared to a control group, unless rejection became manifest [34]. For their published data, we found ESV $=$ 0.384 EDV - 11.38, with $R^2 = 0.347$ for N $=$ 11 (time-related group averages).

Furthermore, EF is in general strongly associated with ESV in a (slightly) nonlinear and inverse manner, for both LV and right ventricle (RV). This notion has been documented using a logarithmic approach [28], as well as a robust description based on theoretical grounds [14]. The findings have been confirmed by studying various patient groups, including the LV and RV in post-Fallot repair patients [15]. Figure 24.4 shows the relationship between EF and ESV in the 101 heart transplant patients studied, yielding a high correlation for both the linear regression line and the logarithmic fit. In contrast, there is no significant association between EF and EDV in this group.

When stratified for the sex of the recipient, we see (Fig. 24.5) a similar pattern for the

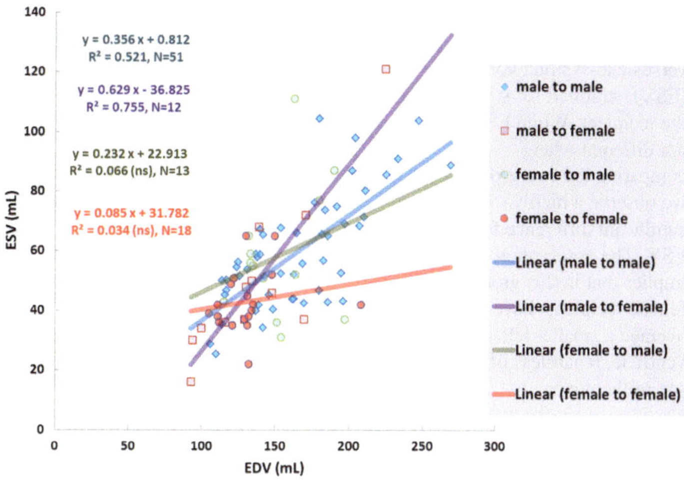

Fig. 24.3 Left ventricular volume regulation graph, showing end-systolic volume (ESV) versus end-diastolic volume (EDV) for cardiac transplant patients, stratified for donor and recipients sex. Participants shown (N = 94) are those for whom sex of donor and recipient was retrieved

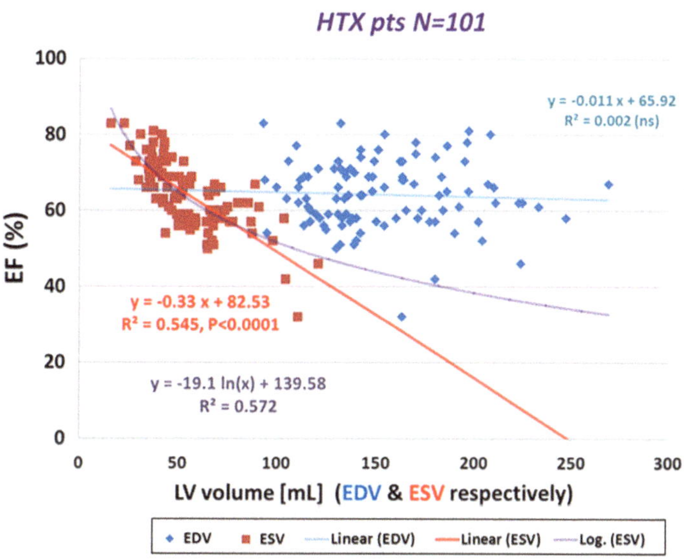

Fig. 24.4 Left ventricular ejection fraction (EF) versus either end-systolic volume (ESV) or end-diastolic volume (EDV) for cardiac transplant (HTX) patients. The relationship for EF vs ESV is intrinsically nonlinear, although in this case logarithmic fit shows only a small improvement compared to the linear approach. There is no significant correlation between EF and EDV

two groups. Remarkably, ESV is smaller ($P = 0.0005$) for the female recipients, regardless of the sex of the donor. This finding resembles the observations in numerous non-transplant studies [3, 16] and possibly suggests a sex-specific adaptation process regarding ventricular volume [17]. Future longitudinal studies are required to elucidate if this type of adaptation to the recipient can be confirmed. Interestingly, coronary artery size is also smaller in women compared to their matched male counterparts, independent of body size [13, 32], while this characteristic also seems to adjust following cardiac transplantation [11].

Traditionally, survival diagrams have been constructed to evaluate the outcome of sex-specific aspects of donor-recipient interaction. Prognosis has typically been associated with the use of the EF. However, values for EF may theoretically remain the same, while ESV and EDV change in unison and yield the same ratio. In contrast, analysis of the combination of ESV and EDV is more specific (Fig. 24.3). The observation that hearts in females are significantly smaller compared to males offers a fascinating starting point for further studies. Furthermore, in women there is more remodeling of coronary

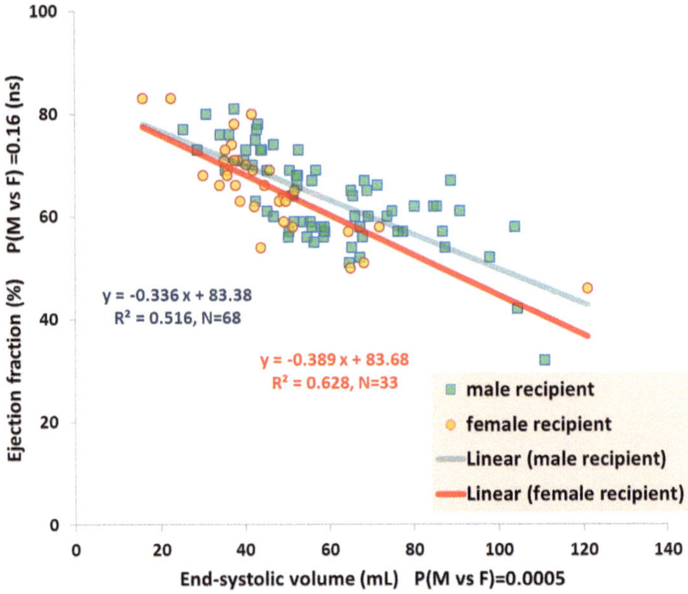

Fig. 24.5 Left ventricular ejection fraction (EF) versus end-systolic volume (ESV), stratified for sex of the recipient. While EF is not different when comparing the two groups, we observe a highly significant difference for ESV. The observation implies that in this group female recipients have on average a smaller left ventricle, regardless of the sex of the donor

arteries [32] which has been confirmed after cardiac transplant [11].

In summary, cardiac transplantation offers a unique window to the study of age- and sex-dependent adaptation in terms of LV remodeling and interaction with vascular properties. Therefore, available longitudinal data [10, 29] require detailed analysis, not only as a means of posttransplant follow-up for individual patients but also to enhance insight into sex-specific factors affecting remodeling and ventriculo-arterial adaptation as a function of time [9]. Furthermore, the impact of reinnervation [2] requires detailed study, particularly with attention to sex-specific aspects.

Conclusions

With the premises surveyed before, shall we redirect organ allocation avoiding sex mismatches? The available evidence supports the conclusion that donor-recipient sex matching should be performed whenever feasible, bearing in mind the complex nature of the matching process that must account for recipient acuity (waiting list status), donor and recipient blood type, and the presence of preformed antihuman leukocyte antigen antibodies in the transplant recipient. It is also important to remember that the amount of increased risk from sex mismatching has to be weighed against the underlying mortality risk of withholding transplant to find a more suitable donor.

It is important to keep in mind that sex mismatching is only one among many risk factors (e.g., ischemic heart failure etiology, prior transfusion history, receipt of a compatible but non-identical ABO group heart, etc.) associated with reduced survival after heart transplantation, many of which exert effects of a magnitude as large or larger.

Therefore, given the multiple donor and recipient factors that influence posttransplant outcomes, a rational response to a single donor or recipient factor should be made in the context for each specific case to provide the recipient with the optimum chance of successful long-standing heart transplant.

References

1. Al-Khaldi A, Oyer P, Robbins R. Outcome analysis of donor gender in heart transplantation. J Heart Lung Transplant. 2006;25:461–8.
2. Awad M, Czer LS, Hou M, Golshani SS, Goltche M, De Robertis M, Kittleson M, Patel J, Azarbal B, Kransdorf E,

Esmailian F, Trento A, Kobashigawa JA. Early denervation and later reinnervation of the heart following cardiac transplantation: a review. J Am Heart Assoc. 2016;5(11):e004070.
3. Buonanno C, Arbustini E, Rossi L, Dander B, Vassanelli C, Paris B, Poppi A. Left ventricular function in men and women. Another difference between sexes. Eur Heart J. 1982;3:525–8.
4. Butterworth M, McClellan B, Allansmith M. Influence of sex in immunoglobulin levels. Nature. 1967;214:1224–5.
5. Cooper DK. Christiaan Barnard and his contributions to heart transplantation. J Heart Lung Transplant. 2001;20:599–610.
6. Crandall BG, Renlund DG, Oconnell JB, Burton NA, Jones KW, Gay WA Jr, Doty DB, Karwande SV, Lee HR, Holland C. Increased cardiac allograft rejection in female heart transplant recipients. J Heart Lung Transplant. 1988;7:419–23.
7. Davies MK, Hollman A. Christiaan Neethling Barnard (1922–). Heart. 1997;77:406. PMCID: PMC484759
8. Eberlein M, Reed RM. Donor to recipient sizing in thoracic organ transplantation. World J Transplant. 2016;24 (6):155–64. https://doi.org/10.5500/wjt.v6.i1.155.
9. Eleid MF, Caracciolo G, Cho EJ, Scott RL, Steidley DE, Wilansky S, Arabia FA, Khandheria BK, Sengupta PP. Natural history of left ventricular mechanics in transplanted hearts: relationships with clinical variables and genetic expression profiles of allograft rejection. JACC Cardiovasc Imaging. 2010;3:989–1000.
10. Goland S, Siegel RJ, Burton K, De Robertis MA, Rafique A, Schwarz E, Zivari K, Mirocha J, Trento A, Czer LS. Changes in left and right ventricular function of donor hearts during the first year after heart transplantation. Heart. 2011;97 (20):1681–6.
11. Herity NA, Lo S, Lee DP, Ward MR, Filardo SD, Yock PG, Fitzgerald PJ, Hunt SA, Yeung AC. Effect of a change in gender on coronary arterial size: a longitudinal intravascular ultrasound study in transplanted hearts. J Am Coll Cardiol. 2003;41:1539–46.
12. Hiemann N, Knosalla E, Wellnhofer E, Lehmkuhl H, Hetzer R, Meyer R. Beneficial effect of female gender on long-term survival after heart transplantation. Transplantation. 2008;86:348–56.
13. Hiteshi AK, Li D, Gao Y, Chen A, Flores F, Mao SS, Budoff MJ. Gender differences in coronary artery diameter are not related to body habitus or left ventricular mass. Clin Cardiol. 2014;37:605–9.
14. Kerkhof PLM. Characterizing heart failure in the ventricular volume domain. Clin Med Insights Cardiol. 2015;9 (S1):11–31.
15. Kerkhof PLM, van de Ven PM, Yoo BW, Peace RA, Heyndrickx GR, Handly N. Ejection fraction as related to basic components in the left and right ventricular volume domains. Int J Cardiol. 2018a;255:105–10.
16. Kerkhof PLM, Kuznetsova T, Ali R, Handly N. Left ventricular volume analysis as a basic tool to describe cardiac function. Adv Physiol Educ. 2018b;42:130–9. https://doi.org/10.1152/advan.00140.2017.
17. Kerkhof PLM, Parry G. Sex-specific aspects in cardiac transplantation evaluated by left ventricular size in male and female recipients. FASEB J 2018;32 Suppl. 1, https://www.fasebj.org/doi/10.1096/fasebj.2018.32.1_supplement.901.1.
18. Khush K, Kubo J, Desai M. Influence of donor and recipient sex mismatch on heart transplant outcomes: Analysis of the International Society for Heart and Lung Transplantation Registry. J Heart Lung Transplant. 2012;31:459–66.
19. Kittleson MM, Shemin R, Patel JK, Ardehali A, Kawano M, Davis S, Moriguchi JD, Kobashigawa JA. Donor–recipient sex mismatch portends poor 10-year outcomes in a single-center experience. J Heart Lung Transplant. 2011;30:1018–22. https://doi.org/10.1016/j.healun.2011.03.018.
20. Konstam MA, Abboud FM. Ejection fraction: Misunderstood and overrated (Changing the paradigm in categorizing heart failure). Circulation. 2017;135:717–9.
21. Lietz K, John R, Kocher A, Schuster M, Mancini DM, Edwards NM, Itescu S. Increased prevalence of autoimmune phenomena and greater risk for alloreactivity in female heart transplant recipients. Circulation. 2001;104(suppl I):I-177–83.
22. Lund LH, et al. The Registry of the International Society for Heart and Lung Transplantation: Thirty-fourth Adult Heart Transplantation Report—2017; Focus Theme: Allograft ischemic time. J Heart Lung Transplantation. 2017;36:1037–46.
23. Matthews JC, Aaronson KD. Sex matters, but to what clinical avail? Circ Heart Fail. 2009;2(5):389–92. https://doi.org/10.1161/CIRCHEARTFAILURE.109.902460.
24. Mifsud NA, Purcell AW, Chen W, Holdworth R, Tait BD, McCluskey J. Immunodominance hierarchies and gender bias in direct TCD8-cell alloreactivity. Am J Transplant. 2008;8:121–32.
25. Mills RM, Naftel DC, Kirklin JK, Van Bakel AB, Jaski BE, Massin EK, Eisen HJ, Lee FA, Fishbein DP, Bourge RC. Heart transplant rejection with hemodynamic compromise: a multiinstitutional study of the role of endomyocardial cellular infiltrate. Cardiac Transplant Research Database. J Heart Lung Transplant. 1997;16:813–21.
26. McCarthy JF, McCarthy PM, Massad MG, Cook DJ, Smedira NG, Kasirajan V, Goormastic M, Hoercher K, Young JB. Risk factors for death after heart transplantation: does a single-center experience correlate with multicenter registries? Ann Thorac Surg. 1998;65:1574–8.
27. Olivetti G, Giordano G, Corradi D, et al. Gender differences and aging: effects on the human heart. JAMA. 1995;26:1068–79.
28. Peace RA, Adams PC, Lloyd JJ. Effect of sex, age, and weight on ejection fraction and end-systolic volume reference limits in gated myocardial perfusion SPECT. J. Nucl. Cardiol. 2008;15:86–93.
29. Peled Y, Lavee J, Arad M, Shemesh Y, Katz M, Kassif Y, et al. The impact of gender mismatching on early and late outcomes following heart transplantation. ESC. Heart Fail. 2017;4(1):31–9. https://doi.org/10.1002/ehf2.12107. Epub 2016 Aug 30.
30. Prendergast T, Furukawa S, Beyer AJ, Browne B, Eisen H, Jeevanandam V. The role of gender in heart transplantation. Ann Thorac Surg. 1998;65:88–94.
31. Reed RM, Netzer G, Hunsicker L, Mitchell BD, Rajagopal K, Scharf S, Eberlein M. Cardiac size and sex-matching in heart transplantation: size matters in matters of sex and the heart. JACC Heart Fail. 2014;2:73–83.
32. Sheifer SE, Canos MR, Weinfurt KP, et al. Sex differences in coronary artery size assessed by intravascular ultrasound. Am Heart J. 2000;139:649–53.
33. Solomon NA, McGiven JR, Alison PM, Ruygrok PN, Haydock DA, Coverdale HA, West TM. Changing donor and recipient demographics in a heart transplantation program: influence on early outcome. Ann Thorac Surg. 2004;77:2096–102.
34. Streeter RP, Nichols K, Bergmann SR. Stability of right and left ventricular ejection fractions and volumes after heart transplantation. J Heart Lung Transplant. 2005;24:815–8.
35. Tan JC, Wadia PP, Coram M, Grumet FC, Kambham N, Miller K, Pereira S, Vayntrub T, Miklos DB. H-Y antibody development associates with acute rejection in female patients with male kidney transplants. Transplantation. 2008;86:75–81.

36. Taylor DO, Edwards LB, Boucek MM, Trulock EP, Waltz DA, Keck BM, Hertz MI. Registry of the International Society for Heart and Lung Transplantation: twenty-third official adult heart transplantation report—2006. J Heart Lung Transplant. 2006;25:869–79.
37. Tsai FC, Marelli D, Bresson J, Gjertson D, Kermani R, Ardehali A, Esmailian F, Hamilton M, Fonarow GC, Moriguchi J, Plunkett M, Hage A, Tran J, Kobashigawa JA, Laks H. Recent trends in early outcome of adult patients after heart transplantation: a single-institution review of 251 transplants using standard donor organs. Am J Transplant. 2002;2:539–45.
38. Weiss E, Allen J, Patel N, et al. The impact of donor- recipient sex matching on survival after orthotopic heart transplantation. Circ Heart Fail. 2009;2:401–8.
39. Zeier M, Dohler B, Opelz G, Ritz E. The effect of donor gender on graft survival. J Am Soc Nephrol. 2002;13:2570–6.

Differences in Cardiovascular Aging in Men and Women

25

Alice E. Kane and Susan E. Howlett

Art work by Piet Michiels, Leuven, Belgium

A. E. Kane
Department of Pharmacology, Dalhousie University, Halifax, NS, Canada
e-mail: Alice.kane@dal.ca

S. E. Howlett (✉)
Department of Pharmacology, Dalhousie University, Halifax, NS, Canada

Department of Medicine (Geriatric Medicine), Dalhousie University, Halifax, NS, Canada
e-mail: Susan.Howlett@dal.ca

Abstract

Cardiovascular diseases increase dramatically with age in both men and women. While it is clear that advanced age allows more time for individuals to be exposed to risk factors in general, there is strong evidence that age itself is a major independent risk factor for cardiovascular disease. Indeed, there are distinct age-dependent cellular, structural, and functional changes in both the heart and blood

P. L. M. Kerkhof, V. M. Miller (eds.), *Sex-Specific Analysis of Cardiovascular Function*,
Advances in Experimental Medicine and Biology 1065, https://doi.org/10.1007/978-3-319-77932-4_25

vessels, even in individuals with no clinical evidence of cardiovascular disease. Studies in older humans and in animal models of aging indicate that this age-related remodeling is maladaptive. An emerging view is that the heart and blood vessels accumulate cellular and subcellular deficits with age and these deficits increase susceptibility to disease in older individuals. Aspects of this age-dependent remodeling of the heart and blood vessels differ between the sexes. There is also new evidence that these maladaptive changes are more prominent in older animals and humans with a high degree of frailty. These observations may help explain why men and women are susceptible to different cardiovascular diseases as they age and why frail older adults are most often affected by these diseases.

Keywords
Age · Aging · Animal models · Cardiovascular disease · Female · Fibrosis · Frailty · Frailty index · Gender · Heart · Heart failure · Hypertrophy · Male · Myocytes · Pathology · Physiology · Sex differences

Introduction

The risk of cardiovascular disease (CVD) development grows with old age in both sexes. There is evidence that this increase in risk is due to age-related changes to the structure and function of the cardiovascular system. There are clear changes that occur with advancing age at both the microscopic and macroscopic levels in the vasculature and the heart, and there is growing evidence that this maladaptive remodeling of the cardiovascular system with age is responsible for the higher prevalence of CVD in the older population. Additionally, there are clear male-female differences in the changes to the cardiovascular system with old age that may influence sex differences in CVD progression and outcomes. This review summarizes the main age-related structural and functional changes of the heart and vasculature. The implications of these changes in the development of CVD in aging are also discussed. Sex differences in cardiovascular changes with aging are highlighted, and the potential effect of these changes on CVD presentation is outlined. Finally, the association of frailty and CVD is discussed, and recent evidence for the association of frailty with age-related structural and functional cardiac changes is presented.

Impact of Age on the Gross Morphological Structure of the Heart and Vasculature

Advanced age is associated with distinct changes in the gross morphological structure of the heart and vasculature (Figs. 25.1 and 25.2). Atrial dilation and hypertrophy are common in older adults of both sexes even in the absence of CVD [16, 137]. This is likely as a result of adaptation to increased diastolic filling pressure and left ventricular (LV) diastolic dysfunction in old age. A recent observational study in humans found that age was associated with a greater risk of bigger atrial volume, but that male sex was protective [178]. However, another recent study using 3D echocardiography found an association between increased left atrial volume and age, but no effect of sex [6]. Interestingly, there appears to be a male-female difference in left atrial function with increasing age [121]. Studies in aged mice and rats have also shown evidence for increased atrial dilatation in old age, although sex differences have not yet been investigated [31, 98, 117]. Atrial fibrosis and ventricular fibrosis have a higher prevalence in older humans [40, 95, 105, 123], rodents [76, 98, 104], and sheep [72] compared to young. Most studies have not shown any evidence for differences in the prevalence of fibrosis between males and females, although a recent study of cardiac extracellular matrix proteins found that older women had a greater amount of collagen and other extracellular matrix proteins in the LV

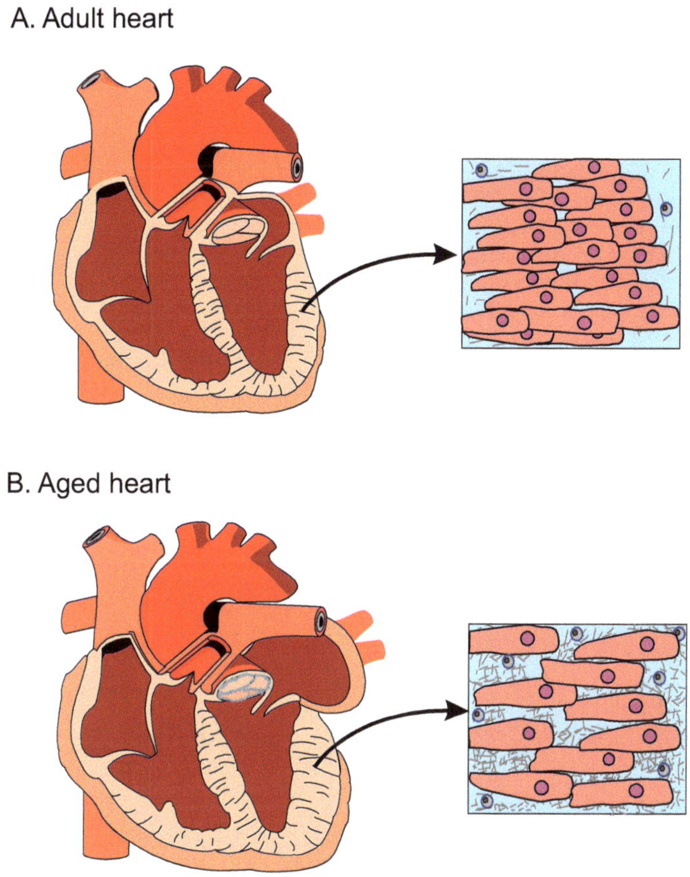

Fig. 25.1 **Age-related changes to the macroscopic and microscopic structure of the heart.** (**a**) A healthy young adult heart. (**b**) An aged heart, demonstrating changes at the macroscopic level from the young adult heart, including atrial dilatation, aortic valve calcification, and increased left ventricular (LV) wall thickness. At the microscopic level, the aged heart, compared to the young heart, shows increased fibrosis, an increase in the number of fibroblasts, a decrease in myocyte number, and myocyte hypertrophy

than older men, which may contribute to fibrosis development [39].

Recent studies suggest that LV hypertrophy, as evidenced by both thicker LV walls and greater LV mass, is more common with old age in humans [21, 42, 57]. The mechanism for this change is likely multifactorial and due to cellular changes in both the heart and vasculature, as well as the increase in fibrosis with old age. Several observational studies have shown evidence that greater LV wall thickness is more prominent in older women than men [21, 70]. Sex differences in LV mass changes with age are less clear, with some studies suggesting a greater increase in LV mass in men [42, 75] and some suggesting a greater increase in women [21, 57]. Some studies have actually found a decrease in LV mass in aged men without CVD, as a result of increased LV wall thickness but a decline in LV volume in both systole and diastole [70, 159]. This suggests a rise in LV concentric hypertrophy in old age, but perhaps no overall effect on LV mass [158]. Normalization of heart mass to body surface area (BSA) may provide important clarification of age-related changes in LV mass. Kaess et al. [79] in a large longitudinal study found an increase in unadjusted LV mass with age for men, but adjusting for BSA resulted in no age-related change. Interestingly this study saw a decline in LV mass with age for women, even when adjusting for BSA [79]. Animal studies have found thicker LV walls and greater LV mass with old age in mice and rats [31, 44, 48, 104, 117, 155], although sex differences were not investigated. More studies in this area would clarify age- and sex-related changes in LV hypertrophy. Studies of the right ventricle (RV) have found a reduction in RV mass and volume with age [84, 111]. Males appear to have greater RV mass and volume than females at all ages

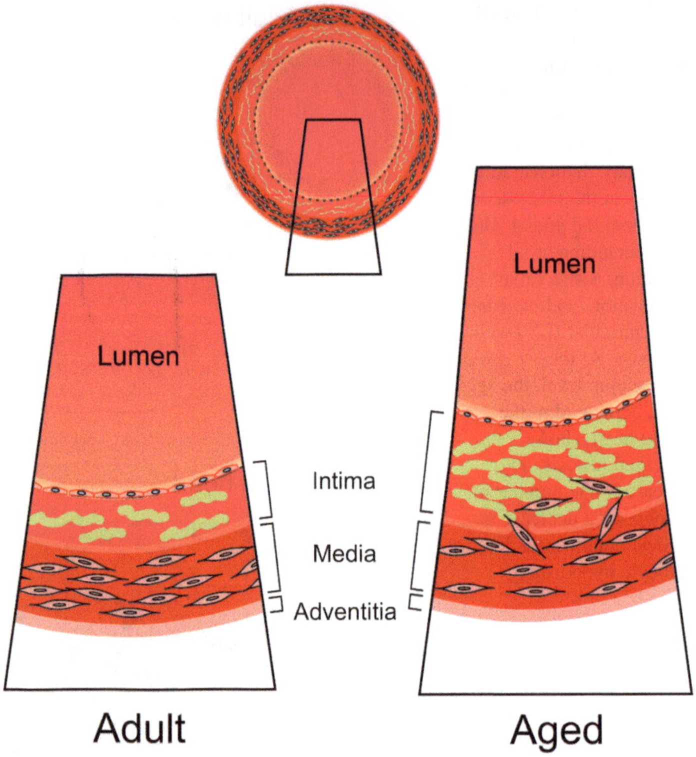

Fig. 25.2 **The central elastic arteries exhibit remodeling with age**. The three layers of the arterial wall are labeled as indicated. The lumen diameter of the large central elastic arteries such as the aorta increases with age. There is a marked increase in the intima plus media (IM) thickness, primarily as a result of an increase in the thickness of the intima. Enhanced collagen deposition and elastin degradation and fragmentation contribute to intimal remodeling in aging arteries. The number of vascular smooth muscle cells in the media declines with age, while the remaining cells hypertrophy and some cells migrate to the intima. Endothelial cell hypertrophy also increase with age

[111, 131], although males may have a greater loss of mass with age than females [84].

The aortic valve, located at the junction between the LV and the aorta, exhibits marked, macroscopic changes with advancing age, including thickening of the leaflets, lipid accumulation, and calcification [26, 135, 148]. There is also significant age-dependent remodeling of the arteries, even in individuals without overt signs of CVD. Some of these vascular changes are evident at the macroscopic level. Indeed, advanced age is associated with marked dilation of centrally located large elastic arteries, such as the aorta and carotid arteries [26, 53, 94]. Arterial wall thickness and lumen size also increase with age in both men and women [94]. This can be seen with the naked eye in postmortem specimens, as well as in radiography and echocardiography studies in older humans and in various animal models of aging [26, 53, 94]. Pronounced age-related changes in the heart and vasculature can also be seen at the microscopic level, as discussed below.

Age-Dependent Remodeling of the Heart and Vasculature at the Cellular Level

Microscopic Changes in the Heart

In combination with the macroscopic changes seen in the aging heart, there are corresponding changes in the ventricular myocytes and the ventricular and atrial fibroblasts (Fig. 25.1). Both human and animal studies have shown a reduction in the number of ventricular myocytes with old age [138, 139, 179], likely as a result of decreased regenerative ability of stem cells and increased apoptosis, necrosis, and autophagy [32, 68, 101, 154]. This loss of myocytes in the heart with old age results in both ventricular myocyte hypertrophy of the surviving myocytes and an elevated number of fibroblasts in older hearts from humans and animals. Olivetti and colleagues found an increase in ventricular myocyte volume with morphometric studies of human hearts [138, 139]. Animal studies have also shown greater myocyte size in aged rodents, rabbits, and monkeys (reviewed by [46, 85]). Proliferation of fibroblasts results in interstitial fibrosis and collagen accumulation in old age [40, 123]. Evidence of fibroblast proliferation and collagen accumulation is seen in animal studies using aged rodents and sheep [71, 72].

Interestingly, although human studies suggest more ventricular hypertrophy in whole female hearts than male hearts [21, 57, 70], the loss of ventricular myocytes, and subsequent myocyte hypertrophy and fibroblast proliferation, appears to be more prominent in males than females [106, 138, 139, 179]. This cellular-level sex difference was also seen in a study of nonhuman primates, where aged male monkeys had more myocyte loss and more myocyte hypertrophy than aged female monkeys [179]. Whether there is a sex difference in myocyte hypertrophy in other animal models is less clear, as few studies have included females. In those that did, some studies saw hypertrophy in female myocytes [35, 73], while others did not [65, 118]. Animal studies have not yet investigated whether there is a sex difference in fibroblast proliferation with aging.

Old age is also associated with a reduction in the number of pacemaker cells in the sinoatrial node (SAN) of the heart in humans [92, 123]. Animal studies in aged rats have also shown an age-related decline in the expression of genes involved in sinoatrial function [162]. Potential sex differences in pacemaker cell number and function with old age have not yet been explored.

Overall, increasing age is associated with changes to the myocytes, fibroblasts, and pacemaker cells as outlined in Fig. 25.1, and there may be some important sex differences in these microscopic changes, although more research is required.

Microscopic Remodeling in the Aging Vasculature

In addition to the readily visible modifications in arterial structure apparent at the macroscopic level, age-related remodeling can be seen at the microscopic level [26, 53, 94, 159, 171]. Age-associated changes occur in all three layers that make up the walls of the large elastic arteries. The innermost layer or *tunica intima* consists of connective tissue and an inner endothelial monolayer on a basement membrane [26, 52]. The *tunica media*, or middle layer, is composed of smooth muscle, separated by elastic lamellae, as well as collagen; it functions to allow the aorta to expand and contract with the flow of blood [26, 52]. The outermost *tunica adventitia* contains collagen fibers plus fibroblasts to provide the vessel with tensile strength [26, 52]. Age-dependent changes in structure and composition of these different layers can have marked effects on vascular function.

The most prominent changes in vascular structure in aging men and women are dilation of large elastic arteries and thickening of the vessel walls [53, 93, 159]. These alterations appear to be closely linked to the aging process, as similar changes occur in large elastic arteries from aged rodents, rabbits, and nonhuman primates [94]. Thickening of the arterial wall in humans has been quantified as an increase in the carotid wall intima plus media thickness (IMT). Studies have shown that carotid IMT increases by between two- and threefold by 90 years of age in men and women with no difference between the sexes [53, 69, 159]. One study did report that carotid artery IMT thickens with age more in men than in women, although this sex difference was eliminated after adjustment for cardiovascular risk factors [78]. A similar increase in IMT is seen in rodent models of aging and in older nonhuman primates [94]. The greater IMT with age is potentially important in terms of susceptibility to CVD in older adults, as an increase in IMT is a key risk factor for the development of atherosclerosis at any age [7].

Thickening of the arterial wall with age appears to result primarily from an increase in the thickness of the intima [26]. The levels of collagen Types I and III in the intima increase with age in humans and animal models, while the elastin content declines, and elastin fraying and fragmentation occur [63, 94, 132, 165,

171]. In addition, vascular smooth muscle cells migrate from the media to the intima [26, 69, 94]. Whether thickening of the media occurs with age is controversial, but the number of vascular smooth muscle cells in the media declines, and collagen production increases [26, 94, 132]. Interestingly, the elastic lamellae in the media become thicker and fracture with age [26, 140] and the media becomes calcified [69]. Collagen deposition in the adventitia also occurs with advancing age [52]. Repeated cycles of distention followed by elastic recoil over a lifetime have been proposed to promote collagen deposition and disrupt elastic fibers in aging arteries [132, 141]. Changes in collagen and elastin content, along with vascular calcification, are believed to have important effects on the distensibility (stiffness) of aging arteries.

In contrast to the intima and media, there are fewer studies of effects of age on the outer, tunica adventitia. As the function of adventitial collagen is to provide tensile strength to the aorta, deficits in adventitial collagen structure may promote aortic diseases in older adults. However, there is evidence that aging actually increases collagen I expression in the adventitial layer of the aorta in humans and animals [52, 165]. One quantitative assessment of collagen structure in human abdominal aortas showed that adventitial collagen structure was maintained in aging men, although whether adventitial collagen is also maintained in older women was not investigated [166]. Additional studies in this area would be of interest.

In addition to smooth muscle infiltration and an increase in connective tissues in the aging intima, there is evidence that the aging process modifies the structure of the endothelial cells themselves. Studies in human arteries have shown that endothelial cells hypertrophy with age and their shape becomes irregular [26, 94]. There is also a decline in the number of overlapping connections between endothelial cells with age, which may compromise the barrier function of this cell layer and render the endothelium more permeable [94, 124]. Vascular smooth muscle cells may actually infiltrate the subendothelial space [94]. There also is strong evidence that substances released by the endothelium can be modified in aging in both humans and in animal models [94, 163]. These age-associated alterations in endothelial cell structure affect vascular function, and this may differ between the sexes, as discussed in more detail in the next section.

In summary, there is growing evidence for age-dependent changes in the vasculature at both the macroscopic and microscopic levels. Key changes are illustrated in the diagram in Fig. 25.2. This age-related arterial remodeling is important. For example, the structural changes characteristic of arteries from normotensive older humans described here also occur in hypertensive patients at much younger ages [94, 171]. Thus, the aging heart and blood vessels may provide an ideal setting in which CVD can flourish, an idea that will be explored further below.

Functional Consequences of Age-Dependent Cardiovascular Remodeling in Healthy Males and Females

Changes to Myocardial Function in Aging

The microscopic and macroscopic changes in the heart and vasculature with aging, described in the previous sections, can lead to changes in both electrical and contractile cardiac function with old age. These include slowing of electrical conduction in the heart, a reduction in maximal heart rate, plus systolic and diastolic dysfunction. There is also evidence for some important sex differences in age-related functional changes as discussed further below.

Effects of Aging on Pacemaker Function, Cardiac Conduction, and Heart Rate

Old age is associated with changes in the electrical function of the heart. Studies in both humans and animal models have shown a slowing of conduction through the ventricles in old age [15, 123]. Recent studies have also found that aging is associated with increased P wave duration, and longer PR interval, indicating slowed conduction through the atria and AV node [76]. Bonda et al. [15] suggest that this slowing in conduction may be due to reduced expression of connexion-43 which is involved in myocyte cell connections. Dun and Boyden [37] suggest that there are changes to the electrophysiological properties of the atrial myocytes, which could also contribute to this change in function. Several animal studies have shown that aging is associated with longer action potentials in atrial myocytes, as a result of changes in the underlying ion currents [37]. A recent mouse study, however, found that old age was associated with reduced atrial action potential duration [76], which is consistent with studies finding reduced calcium currents in old atrial myocytes from dogs [36]. Sex differences in changes to electrical conduction in the heart have not yet been investigated, and more studies are needed.

Although aging does not appear to effect resting heart rate [44, 89, 93, 117], old age is associated with a reduction in maximal heart rate in response to stimulation such as exercise [93, 134]. The mechanisms for this age-related change appear to be both reduced pacemaker activity of the SAN [97] and a reduced responsiveness of the heart to ß-adrenergic stimulation [49]. SAN cells are reduced in number in old age [123], have lower ion channel expression [162], and have reduced spontaneous activity [97]. These mechanisms contribute to a reduction in the maximal heart rate that can bc initiated by the SAN. Human and animal studies have also demonstrated a reduction in the responsiveness of the heart to β-adrenergic receptor activation when the sympathetic nervous system is stimulated [17, 45, 49, 172]. In healthy younger adults, stimulation of the sympathetic nervous system by exercise or stress, for example, results in an increase in heart rate via norepinephrine activation of β-adrenergic receptors. In old age, this reduced responsiveness of the β-adrenergic receptors may contribute to the reduction in maximal heart rate. Few studies have investigated sex differences in changes to maximal heart rate with age or the associated mechanisms. One animal study found that there was an age-related decrease in β-adrenergic receptor sensitivity in old male monkeys, but not females [161], but more research is needed.

Effects of Aging on Systolic Function

Systolic function appears to decline in old age in in vivo studies in humans and animals [31, 33, 42, 44, 89, 155]. In support of this, there is evidence that the ability of ventricular myocytes to contract declines with age in rodent models [65, 73]. On the other hand, systolic function has been commonly measured using ejection fraction (EF), but the validity of EF as a measure of systolic function has recently been challenged in the literature [4, 43, 86, 90]. Although systolic function is generally thought to decline with age, not all studies have found this, especially when using EF as their primary outcome. Indeed, some studies observed improved LV EF with old age, associated with a decrease in end-systolic and end-diastolic volumes with age [58, 59]. Interestingly these studies found a greater increase in systolic function with age in women compared to men [58, 59]. In an observational MRI study of patients without CVD, De Bondt and colleagues found an increase in EF in females aged 65 and over, but not in males [34]. This effect was largely due to a decrease in end-systolic volume for older females that was not seen for older males [34], as shown in Fig. 25.3. Another study in a much older cohort, however, found a similar decline in stroke volume in both older males and females [42], suggesting that perhaps the decline in systolic function occurs at a later age in women than men.

Cardiac torsion, or twisting of the heart, increases with aging in response to reduced longitudinal systolic function [9, 24], and this adaptation may help to maintain overall systolic function and EF [75]. Interestingly, cardiac torsion is greater in women than men at all ages [174], which may contribute to the potential protection against declining systolic function seen in women compared to men [75]. Studies of RV systolic function have found an increase in RV EF with age, but no clear effect of sex [51, 84, 111]. One study investigating RV longitudinal strain as a potential measure of RV dysfunction

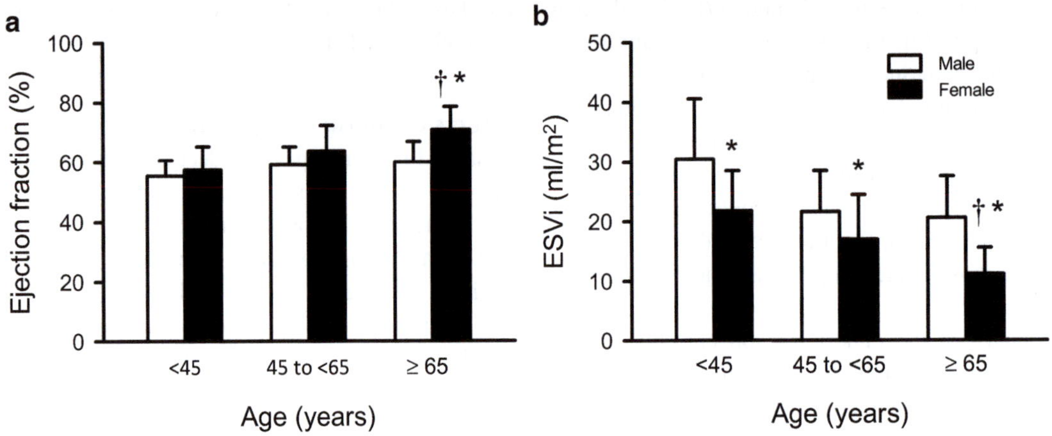

Fig. 25.3 Age- and sex-dependent changes in ejection fraction and end-systolic volume index. (**a**) Ejection fraction (EF) increases with age in women more than men. (**b**) End-systolic volume index (ESVi) declines with age in both sexes, and this decrease is particularly marked in women ≥65 years. Data have been replotted from Table 2 in [34]. Values represent the men ± SD. † denotes the effect of age is significant. * denotes the effect of sex is significant

found women had higher measures of RV longitudinal strain than men [131].

Animal studies in old rats and mice have shown reduced EF and stroke volume with aging in males but not females [31, 33, 44, 89, 155]. Studies of isolated myocytes from rodents have also found that myocytes from older animals have both smaller and slower contractions, compared to those from young animals, and that this decline in contractility is more prominent in cells from aging males than females [65, 73]. Cells isolated from aged male mice also display lower peak calcium transients, and peak calcium currents, which may provide a mechanism for this reduced contractility in aged male myocytes [46, 65, 73, 85]. Aged female myocytes from mice showed no change in these measures [46, 85], and studies in myocytes isolated from aged female sheep showed an increase in peak calcium currents or transients, compared to younger female sheep [35]. Animal studies that further explain the potential sex difference in declining systolic function with old age would be of interest. More studies in aging animals and humans, using additional methods of assessing systolic function rather than EF, will clarify the effects of age and sex on this outcome.

Age-Associated Diastolic Dysfunction and Atrial Remodeling

Diastolic dysfunction is commonly seen in the aging hearts of both men and women [16, 53, 93, 159], as well as in both male and female animal models of aging [31, 33, 44, 89, 117, 155]. Diastolic dysfunction is characterized by the impairment of relaxation of the LV, which results in slower and delayed LV filling [32, 93, 109]. Mechanisms for this impaired ventricular relaxation in aging include enhanced LV fibrosis [71], myocyte stiffness [67], and slower calcium transient decay in myocytes [46, 47]. Impairment of the relaxation of the LV can also result in maladaptive changes to the atria. Normal diastolic filling of the LV involves the early phase of passive filling during LV relaxation, followed by the late phase of active filling during atrial contraction. Delayed early filling of the LV, due to impaired ability of the LV to relax, results in a greater contribution of the atria to late diastolic filling [159]. This increased atrial contribution is accompanied by greater force of atrial contraction, as well as atrial dilation and hypertrophy as a result of increased diastolic filling pressure [53, 159]. Another study in an aged population investigated left atrial emptying fraction, as a measure of left atrial function, and found no effect of sex on this outcome [60].

Although diastolic dysfunction appears to occur with aging in both men and women, it is a major factor in the development of heart failure with preserved EF, which is more prevalent in women than in men, as discussed further below.

Vascular Function in Aging

It is clear that changes in arterial structure occur at both the macroscopic and microscopic levels with advancing age. This age-dependent remodeling of the large, central elastic arteries has a profound impact on cardiovascular function [29]. In some cases, these functional changes in the vasculature differ between the sexes. This may predispose men and women to different CVD as they age.

The Influence of Age on Blood Vessel Stiffness

One of the hallmarks of arterial aging is arterial "stiffening," which is described as an increase in the resistance of the arterial wall to expansion [53, 56]. The structural changes in the arterial wall described above are implicated in the age-related rise in arterial stiffness. Higher collagen content and calcification of the arteries are believed to increase arterial stiffness in older adults [20, 94, 132]. Other factors, including elastin degradation and fragmentation, also are thought to be key determinants of age-dependent atrial stiffening [20, 94, 132]. There is emerging evidence for differences in arterial stiffness between the sexes [27, 120]. Arterial stiffness is lower in premenopausal women than in age-matched men but it increases markedly after the onset of menopause [27, 99, 120]. Indeed, postmenopausal women may have stiffer central elastic arteries than men [27, 120], which suggests that female sex hormones such as estrogen may modulate arterial stiffness. In support of this, estrogen has been shown to increase the production of elastin and reduce collagen deposition in human arteries [27]. Changes in endothelial regulation of vascular tone and changes in other aspects of the vascular function also may contribute to the age-associated increase in arterial stiffness, although whether this differs between the sexes is unclear [132].

Arterial stiffness is thought to be responsible for the characteristic changes in blood pressure seen in older adults [3, 56]. In young individuals, the elasticity of the central arteries allows the aorta to store a large fraction of the ejected blood volume following each heartbeat, and aortic recoil during diastole then pushes the stored blood into circulation [56, 152]. By contrast in older individuals with stiff central arteries, the aorta and its major branches do not distend in response to the ejection of blood [56, 152]. This loss of elasticity with age means that blood flow is transmitted during systole, which leads to high systolic blood pressure [56]. As blood flow is transmitted in systole, the elastic recoil does not dissipate in diastole so that diastolic blood pressure decreases with age [56]. Thus, arterial stiffness causes an increase in systolic blood pressure, a decrease in diastolic blood pressure, and marked widening of the pulse pressure, as shown diagrammatically in Fig. 25.4. The lack of deflection of blood flow by the large central arteries augments the velocity at which the pulse wave travels in older adults [103, 141, 152]. There is strong evidence that a greater pulse wave velocity in older adults is an important risk factor for future adverse cardiovascular events [103, 141, 152].

In summary, the age-related increase in the stiffness of large elastic arteries helps explain characteristic changes in blood pressure with age and may contribute to sex differences in CVD expression in older adults.

Endothelial Dysfunction in Aging

The vascular endothelium is a monolayer of squamous epithelial cells that forms a semipermeable barrier between the blood and the vascular smooth muscle and plays a major role in the maintenance and regulation of blood flow [151]. The endothelium synthesizes and releases a variety of biologically active molecules, including nitric oxide, prostacyclin, endothelins, interleukins, endothelial growth factors, adhesion molecules, plasminogen inhibitors, and von Willebrand factor, in response to chemical and

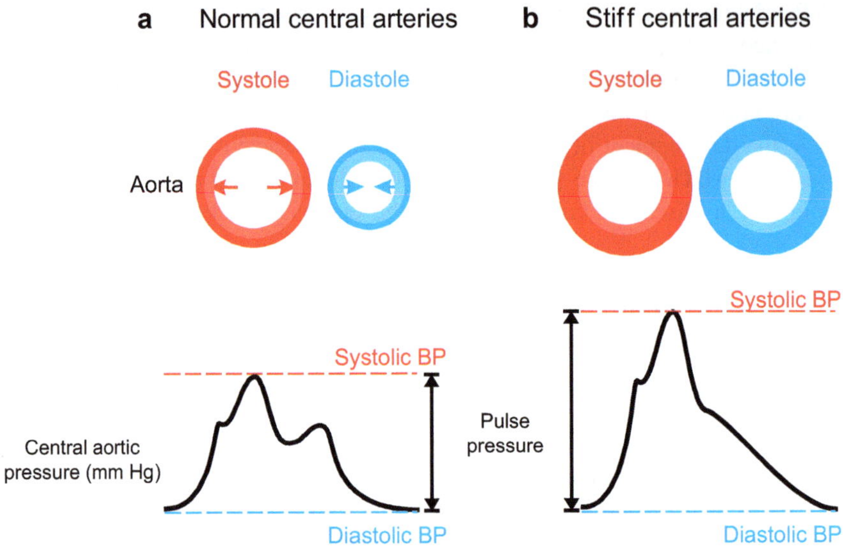

Fig. 25.4 **The age-related increase in central artery stiffness affects peripheral pressure.** (**a**) In younger adults, the central elastic arteries expand with each heartbeat, so that a portion of the stroke volume is transmitted in systole and the remainder is transmitted in diastole. (**b**) In older adults, the central elastic arteries stiffen and therefore do not expand with each contraction, so the majority of the stroke volume is transmitted in systole. This increases systolic blood pressure, decreases diastolic blood pressure, and widens pulse pressure in older adults

mechanical stimuli [150, 151]. These substances regulate many diverse functions such as arterial tone, angiogenesis, coagulation, and fibrinolysis [150]. Endothelial dysfunction is an early sign of vascular aging and it can appear in advance of obvious signs of CVD [55]. There is also emerging evidence for sex differences in endothelial dysfunction with age, as discussed in detail below.

Endothelial dysfunction in aging can be measured as a disruption in endothelium-dependent relaxation, which is mediated by nitric oxide. Nitric oxide is released from the endothelium by mechanical stimuli, including increased blood flow (shear stress), and by chemical stimuli, such as acetylcholine, bradykinin, or ATP [151]. When nitric oxide is released, it promotes vascular smooth muscle relaxation via an increase in intracellular cGMP, thereby preventing the interaction of the contractile filaments actin and myosin [83]. The aging endothelium is characterized by progressively impaired vasodilation in response to stimuli such as increased blood flow or vasodilatory compounds [55]. This diminished vasodilatory response is primarily attributable to reduced bioavailability of nitric oxide [55]. This leads to impairment in blood vessel relaxation with age, as well as a decrease in the vasoprotective, cardioprotective, and antiatherogenic effects of nitric oxide in older adults [55].

Nitric oxide is synthesized in endothelial cells by a constitutive enzyme called endothelial nitric oxide synthase (eNOS or NOS III; [83]). There is some evidence that the levels of eNOS are reduced in aging, which could help reduce nitric oxide bioavailability [53, 94]. Still, the primary mechanism responsible for the reduced bioavailability of nitric oxide in the aging vasculature is thought to involve the interaction of nitric oxide with reactive oxygen species to form peroxynitrate [69]. Peroxynitrate, in turn, oxidizes tetrahydrobiopterin (BH4), which is an essential cofactor for the synthesis of nitric oxide by eNOS [55, 70]. There is also evidence that age-dependent endothelial dysfunction occurs earlier in men than in women, although after menopause no sex difference is observed [169]. This suggests that estrogen exerts a protective role on endothelial function, potentially by preserving nitric oxide availability through activating its synthesis pathway and/or reduced production of reactive oxygen species [170]. Whether testosterone exerts beneficial or detrimental effects on endothelial function is not yet clear [170], and additional studies of the effects of age on endothelial function in both sexes are warranted.

There is strong evidence that endothelial dysfunction is an important cause of CVD, independent of age [53, 83]. Therefore age-dependent rise in endothelial dysfunction is likely to have a major impact on the risk of CVD in older adults.

Age-Dependent Alterations in Vascular Contractility and Reactivity

Age-dependent modifications in blood vessels may vary between vascular beds. The structural changes that lead to increased arterial stiffness are noticeable in large, elastic arteries, such as the aorta and carotid arteries, rather than in smaller, muscular arteries such as the brachial artery [132]. Still, stiffening of the central elastic arteries in aging can lead to high pulsations in the microvasculature and cause damage in vital organs such as the brain and kidney [141]. In addition, there is emerging evidence, in particular from animal studies, that the microvasculature undergoes functional and structural changes with age, although little is known about underlying mechanisms [102]. Whether this differs between the sexes is also unclear, although evidence suggests that microvascular dysfunction in the coronary arteries is more common in women than in men [120, 143]. Indeed, coronary microvascular dysfunction is thought to play a major role in the pathogenesis of ischemic heart disease in older women, whereas obstructive coronary artery disease is implicated in aging men [143]. Additional studies of the mechanisms responsible for male-female differences in the microvasculature would be of interest.

As discussed above, reduced bioavailability of nitric oxide released from the endothelial cells impairs blood vessel relaxation in aging [55]. However, the ability of blood vessels to dilate is mediated not only by the endothelium but also by the function of the vascular smooth muscle cells themselves. Whether age affects vascular smooth muscle function, however, is unclear. Some, but not all, studies in animal models suggest that the vasodilatory responses of vascular smooth muscle cells decline with age [129, 149]. This may depend on the vascular bed evaluated [149]. There is also variability in the results of studies in humans, potentially as a result of different vascular beds and small sample sizes in most studies [129, 149]. A recent systematic review and meta-analysis of human studies investigated the impact of age on vascular smooth muscle function measured as the endothelium-independent vasodilator responses to exogenous nitric oxide donors [129]. Results showed clearly that vascular smooth muscle function declined with age in two different types of vasculature, the central elastic arteries and the resistance arteries [129]. Interestingly, this age-associated decline in vascular smooth muscle function was more pronounced in women than in men [129]. Taken together, these observations suggest that endothelial-dependent and endothelial-independent vasodilatory responses are compromised in older adults, in particular females.

There is some evidence that vascular responsiveness to vasoconstrictors and vasodilators changes with age, although more work in this area is needed. Vasodilatory responsiveness to prostaglandins declines with age in human studies [149]. Vasoconstrictor responses to noradrenaline, endothelin, and potassium chloride have been reported to increase with age in some studies and not change in others, at least in animal models [149]. These variable results may be due to factors such as differences in the strain of animal and vascular bed investigated [149]. Whether arterial responsiveness to angiotensin receptor agonists is modified by age is not yet clear, although this has not been extensively investigated [130, 149]. Few studies have investigated the impact of age on vascular responsiveness in veins, but most studies report that age has little impact on the responsiveness of veins to a variety of pharmacologic agents [130]. Investigation of age-dependent alterations in vascular reactivity is an important area of inquiry that could reveal differences in responsiveness of the aging vasculature to drugs that target blood vessels in humans. Investigations of potential sex differences in vascular reactivity in aging are also required.

Age-Related Changes Promote Cardiovascular Diseases: Differences Between the Sexes

The age-related cardiovascular structural and functional changes outlined above promote CVD in aging men and women. Advancing age is associated with an increased risk of bradycardia and pacemaker implantation, atrial fibrillation (AF), both heart failure with reduced ejection fraction (HFrEF) and heart failure with preserved ejection fraction (HFpEF), aortic stenosis, aneurysms, isolated systolic hypertension, atherosclerosis, and ischemic heart disease (Table 25.1). There are some clear sex differences in the risk of developing these diseases as well as the risk of poor outcomes, and these will be discussed further below.

Bradycardia, Pacemaker Implantation, and Left Bundle Branch Block

There is higher risk of bradycardia and requirement for the implantation of pacemakers with old age in both men and women [2, 123, 167]. This is associated with age-related changes to the SAN including reduced numbers of SAN cells, reduced expression of ion channels in SAN cells, and the accompanying reduction in the speed of electrical conductivity in the heart. Several studies have shown higher rates of pacemaker implantation in men compared to women [119, 157, 167]. Two of these studies explored outcomes, including mortality, post-pacemaker implantation and did not see any clear sex difference [119, 157]. One study of patients with syncope (transient loss of consciousness often associated with bradycardia) showed that female sex was a significant risk factor for pacemaker implantation in this population [2]. Age-related sex differences in the risk of development of bradycardia, or of pacemaker implantation, as well as potential sex differences in outcomes post-pacemaker implantation require more investigation.

Aging is also associated with an increased prevalence of left bundle branch block (LBBB), as classified by a slowed QRS interval ($\geq$ 120 ms) [66, 77, 107, 180]. LBBB is associated with poor outcomes and other CVD, including heart failure [77, 177]. Risk factors for LBBB may include hypertension and reduced heart volume [77, 180]. There is no clear sex difference in the incidence of LBBB, with some studies showing greater prevalence in women [77, 107] and some showing greater prevalence in men [66]. Interestingly, women appear to benefit more than men from the main intervention to treat LBBB, cardiac resynchronization therapy, especially among those with shorter QRS intervals [180].

Atrial Fibrillation

AF is the most common sustained arrhythmia and is characterized by structural and electrical remodeling of the atria [176]. AF is a common risk factor for thromboembolic stroke and heart failure [115]. The risk of developing AF grows with advancing age [115]. Maladaptive remodeling of the atria including hypertrophy and fibrosis, as well as changes to the atrial myocytes, is associated with the greater risk of atrial fibrillation in old age. This increased risk of AF with old age appears to be particularly apparent in men [10, 153], although not all studies have seen a sex difference [127, 176]. Interestingly, Magnani and colleagues found that lower testosterone was associated with increased risk of AF in men aged over 80 years, suggesting a potential role of sex hormones and AF risk [112]. Although older men may be at a greater risk of developing atrial fibrillation, it is associated with worse outcomes in women including stroke [5, 30, 176], development of heart failure [108, 115], and increased mortality [11]. Yoshida and colleagues suggest that the higher risk of stroke in women with AF compared to men may be as a result of more left atrial remodeling resulting in

Table 25.1 Clinical consequences of cardiac and vascular aging in men and women

Age-related change[a]	Clinical outcome	Sex difference	Reference
↓ Expression of SAN node channels, SAN dysfunction	↑ Risk of bradycardia and pacemaker implantation	Not clear	Mizra et al. (2012), Ahmed et al. [2], and Uslan et al. [167]
Atrial hypertrophy and fibrosis Remodeling of ion channels in atrial myocytes	↑ Risk of atrial fibrillation	May be more prevalent in men but associated with worse outcomes in women	McManus et al. [115], Sherman et al. [153], Yoshida et al. [176], and Benjamin et al. [11]
↓ Electrical conduction Hypertension ↓ ESV	↑ Risk of left bundle branch block	Women benefit from interventions (CRT) more than men	Zusterzeel et al. [180], Haataja et al. [66], Liu et al. [107], and Jeong et al. [77]
↓ Numbers of ventricular myocytes and myocyte hypertrophy ↓ Myocyte contraction and ↓ Ca^{2+} transients ↓ Systolic function (↓ESV, ↓EDV)	↑ Risk of HFrEF	Risk is higher and outcomes are worse in men than in women	Dunlay and Roger [38], Nakada et al. [133], and Martínez-Sellés et al. [114]
Slow passive LV filling, diastolic dysfunction ↑ Left ventricular fibrosis, stiffness, and wall thickness ↓ Ca^{2+} removal, ↑ diastolic Ca^{2+}, and ↑ myocyte stiffness (titin changes) Atrial remodeling ↑ Vascular stiffness ↑ or ↔ EF	↑ Risk of HFpEF	More prevalent in women but associated with worse outcomes in men	Dunlay and Roger [38], Loffredo et al. [109], Greiten et al. [64], Kosmidou et al. [91], Nakada et al. [133], Lam et al. [95], and Martínez-Sellés et al. [114]
Aortic valve calcification, LV outflow obstruction ↓ aortic diameter	↑ Risk of aortic stenosis	May be more prevalent in men and associated with worse outcomes in men	Milin et al. [122], Cramariuc et al. [28], and Carroll et al. [19]
Elastin degradation and fragmentation plus changes in collagen in the aorta	↑ Abdominal aortic aneurysms	More common in men, ↑ rupture plus worse outcomes in women	Boese et al. [14] and Garcia et al. [54]
↑ Vascular stiffness	↑ Isolated systolic hypertension	↑ In postmenopausal women more than men	Lee and Oh [98], Coutinho [27], and Merz and Cheng [120]
↑ Intima-medial thickness	↑ Risk of atherosclerosis	Occurs in men and women	Bauer et al. [7]
Coronary microvascular dysfunction Endothelial dysfunction Hypertension	↑ Ischemic heart disease	More important contribution to ischemia in women	Merz and Cheng [120], Elias-Smale et al. [41], Park and Merz [143], and Virdis and Taddei [170]

[a]Abbreviations: *HFrEF* heart failure with reduced ejection fraction, *HFpEF* heart failure with preserved ejection fraction, *LV* left ventricle, *SAN* sinoatrial node, *ESVi* end-systolic volume, *EDVi* end-diastolic function, *CRT* cardiac resynchronization therapy, *EF* ejection fraction

reduced atrial function in women [176]. Pancholy and colleagues found that in a study of AF patients, women taking warfarin were at a greater risk of stroke than men taking warfarin, or women taking novel oral anticoagulants, suggesting that perhaps the higher risk of stroke in women may be a result of traditional anticoagulants being less effective in women [142]. Further studies are needed to understand the mechanisms behind the increased risk of poor outcomes, including stroke, in women with AF [5].

Heart Failure with Reduced Ejection Fraction (HFrEF)

There is an elevated risk of HFrEF, or systolic heart failure, with old age [38]. This is likely a result of the decline in systolic function at both the cellular and whole heart level that occurs in old age. The risk of developing HFrEF in old age appears to be higher for men than women [38, 133]. Heart failure commonly develops as a result of ischemic heart disease and the related cardiac remodeling. The severity of cardiac remodeling associated with ischemic heart disease appears to be higher in men compared to women [38, 41], and this may contribute to the higher risk of HFrEF in men. Evidence from animal studies supports this, with studies showing a greater prevalence in males of structural and functional changes in the heart such as reduced cellular contractility and decreased stroke volume and EF [31, 33, 44, 46, 73, 155, 181]. Outcomes after the development of HFrEF also appear to be worse in men than in women [38]. A meta-analysis of 31 studies found that overall there was a 23% increased risk of mortality in men with HFrEF compared to women [114]. Interestingly, other studies have found that although women with HFrEF have reduced mortality risk, they have a poorer quality of life [62, 145].

Interestingly, some studies have suggested that, as women have higher EF than men at all ages, there should be different cutoff values used in the definition of reduced EF for men and women [23, 80]. It is possible, then, that the higher prevalence of HFrEF in men with old age compared to women may be because of under-classification of reduced EF in women. In addition, whether EF is a good measure of myocardial contractility has recently been questioned [4, 12, 43, 86, 87], so whether distinctions between heart failure subtypes should be made based on EF is controversial [90]. More work in this area will clarify whether classification of heart failure by EF will continue to be the standard [64] and help clarify differences in risk for aging men and women.

Heart Failure with Preserved Ejection Fraction (HFpEF)

Old age is also associated with a greater risk of HFpEF, or diastolic heart failure [80, 109]. The increased risk of HFpEF in old age is associated with the age-related functional and structural cardiac changes of slow LV filling, diastolic dysfunction, atrial remodeling, thicker LV wall and fibrosis, and the corresponding cellular changes. Furthermore, age-related changes to the vasculature including increased aortic stiffness may also contribute to the development of HFpEF. Unlike HFrEF, the risk of developing HFpEF appears to be higher in women [13, 38, 64, 91, 133]. One contribution to this may be sex differences in risk factors, with a greater prevalence of older women than men having hypertension, which is a significant risk factor for HFpEF [64]. Additionally, sex differences in the underlying age-related structural and functional changes of the heart may contribute to this sex difference in HFpEF risk. Females are at a lower risk of developing systolic dysfunction in old age [34] due to increased prevalence of changes such as more LV wall thickness [21] and more cardiac torsion [75] which appear to maintain systolic function in aging women. This may then predispose women to developing HFpEF, rather than HFrEF [96]. As discussed above, the change in prevalence of HFpEF between males and females may also be due to the need for different definitions of reduced EF in women compared to men [23, 80]. Despite having less risk of developing HFpEF, many studies have shown that men have higher risk of poor outcomes as a result of HFpEF, compared to women [95, 114, 133]. Not all studies have found this, however, with several studies finding no sex difference in mortality risk for HFpEF patients [96, 164, 181]. Lam and colleagues suggest that one reason for this finding may be that in their study women had lower use of potentially beneficial medications such as angiotensin-converting enzyme inhibitors compared to men, which reduced the female survival advantage [96]. On the other hand, Zsilinszka and colleagues found no sex differences in management of heart failure in hospital, including medication use [181]. Further research into the sex differences in mechanisms leading to the development of heart failure and the risk of poor outcomes with heart failure in old age could be informative.

Aortic Stenosis

Age-related calcification of the aortic valve can lead to the development of aortic stenosis. This results in narrowing of the aortic valve, which obstructs LV outflow and increases the required effort of the heart [122]. Ultimately this can result in remodeling of the LV and the development of heart failure [135]. Cross-sectional studies have shown that the risk of aortic stenosis appears to be higher in men than in women [19, 122]. A recent longitudinal study of aortic stenosis patients found that although the progression of aortic stenosis was similar in men and women, women had lower risk of reduced EF, ischemic CV events, and all-cause mortality [28].

Aneurysms

Abdominal aortic aneurysms are balloon-like formations in the aorta that arise due to local weaknesses in the arterial wall. The occurrence of aortic diseases including aneurysms increases with age, and there is evidence for male-female differences [14, 54]. Although aneurysms are more common in men than in women, women are more likely to suffer from rupture, and they have worse outcomes after aneurysms are repaired [14, 54]. Abdominal aortic aneurysms are thought to develop, in part, as a result of changes in the structure and/or content of connective tissues (e.g., elastin and collagen) in the vessel wall [165]. Thus, the age-dependent changes in the structure of the aorta described earlier could contribute to the development of aneurysms in older adults.

There is evidence that that age-related disruption of elastin, including elastin degradation and fracture of the elastic lamellae in the media [26, 140], facilitates the development of abdominal aortic aneurysms [165]. Relatively little is known about sex differences in the effects of age on elastin, although the elasticity of central elastic arteries, including the carotid and the aorta, is similar in both sexes in humans and in animal models [50, 78]. Thus, differences in elastin are unlikely to explain male-female differences in aneurysms in aging. In theory, a decrease in adventitial collagen would compromise vessel integrity and could contribute to sex differences in aneurysms. However, collagen deposition actually increases with age in the adventitia of male rodents [52], although others report that adventitial collagen is unchanged, at least in older men [166]. Whether adventitial collagen changes with age in females is unclear. This could be important to investigate as women are at increased risk of rupture and experience poor outcomes after aneurysms are repaired [14, 54].

Isolated Systolic Hypertension

Isolated systolic hypertension can be defined as systolic blood pressure greater than 140 mm Hg with diastolic blood pressure less than 90 mm Hg [8]. This form of hypertension rises markedly with age and is the most common form of hypertension in older adults [3]. It is thought to be due, in part, to the age-associated increase in collagen, enhanced arterial calcification, and elastin degradation plus fragmentation that increase the stiffness of central elastic arteries [20, 94, 132]. Long-term exposure to modifiable risk factors for CVD also may contribute to isolated systolic hypertension in older adults [8].

Premenopausal women are less likely to have all forms of hypertension than men of a similar age, but rates rise in women after menopause, and by the age of 65 years, women are more likely to have hypertension than men [27, 99, 120]. Isolated systolic hypertension also is more common in older women than in older men, likely because of increased aortic stiffness in women [27, 99]. This is important, as isolated systolic hypertension is independently associated with all-cause mortality as well as with fatal and nonfatal cardiovascular events [8, 113]. As current antihypertensive therapies have limited effects on arterial stiffness [113], new therapies that target the age-related increase in central elastic artery stiffness could be beneficial for the treatment of this isolated systolic hypertension in older adults, especially women.

Ischemic Heart Disease

The incidence of ischemic heart disease rises with age in both sexes [143]. This may occur, at least in part, because the age-related increase in IMT itself increases the risk of developing atherosclerosis [7]. Still, there are important differences in the clinical presentation of ischemic heart disease in older men and women [41]. While older men typically develop obstructive, calcified atherosclerotic plaques, women initially present with more diffuse, less obstructive plaques [41]. Nonetheless, women do develop ischemic heart disease as they age and their prognosis is similar to that of men at the same age [143]. One explanation is that women develop more microvascular dysfunction in the coronary arteries than men [120, 143]. Indeed, there is evidence that coronary microvascular dysfunction makes a major contribution to ischemic heart disease in older women, while atherosclerosis plays a major role in aging men [143]. A better understanding of the mechanisms responsible for sex differences in the microvasculature would be revealing.

Vascular endothelial dysfunction is an early manifestation of vascular aging [41]. It impairs vasodilatory mechanisms and therefore can promote signs of ischemic heart disease in older adults, including dyspnea and chest pain [41]. The structural and functional changes in the aging endothelium are thought to contribute to endothelial dysfunction in both sexes. Men show signs of age-associated endothelial dysfunction earlier than women, so this may contribute to ischemic heart disease in men at all ages. As endothelial dysfunction rises markedly after menopause, it is likely to help promote ischemia in older women [170]. High blood pressure, which increases shear stress on the arterial wall and exacerbates endothelial dysfunction, also may contribute to ischemic heart disease in older individuals. This is especially true in women, who are more likely to develop hypertension as they age than men [27, 41, 120].

Frailty Influences Cardiovascular Aging and Disease Expression

Evidence that Frailty Affects Cardiovascular Disease Risk and Expression from Clinical Studies

Frailty is an age-related condition of increased risk of adverse health outcomes compared to others of the same chronological age [25, 74]. Frailty gives an indication of the overall health status of a person and can capture some of the unexplained heterogeneity in risk for people of the same age [168]. Clinically, frailty is commonly assessed using a frailty index (FI). The FI measures the number of health-related deficits a person has accumulated over their lifetime [125, 126, 146]. The number of deficits a person has are counted and divided by the number of deficits assessed to give an FI score between 0 and 1. The closer the score to 1, the more frail a person is considered to be [146]. Clinical studies have shown clear sex differences in the development of frailty, with women having higher FI scores than males at all ages [61]. Paradoxically, men also have a higher risk of mortality than women at all ages, a phenomenon that is not fully understood, but hypotheses include higher rates of self-reported deficits, more diseases that affect quality of life rather than mortality, and higher physiological reserve in women compared to men [61].

Frailty is related to a greater risk of CVD and an increased risk of cardiovascular mortality [1, 22, 156]. In particular, frailty appears to predispose patients to heart failure [87, 116]. Veronese and colleagues completed a recent meta-analysis of CVD morbidity and mortality in frailty. They analyzed results from 18 cohorts of older adults and found that frailty is associated with more risk of developing any-type CVD in both cross-sectional and longitudinal studies [169]. Furthermore frailty was associated with a clear increase in risk of cardiovascular mortality, as well as the specific risk of developing coronary heart disease or heart failure [169]. Interestingly, studies have also found associations between frailty and subclinical cardiovascular changes including LV hypertrophy, prehypertension, and undiagnosed injury to the myocardium [136]. Additionally, frail patients who have CVD have worse outcomes than non-frail patients, including hospitalization, institutionalization, and mortality [1, 156]. In particular, frail heart failure patients have much worse outcomes, including greater mortality risk, than non-frail heart failure patients [18, 110]. Frail patients are also at an increased risk of rehabilitation and further care requirements following cardiac surgery, as well as having greater postoperative mortality and morbidity risks [100, 156, 160]. More research is needed to understand the effect of sex differences in frailty on risks associated with CVD in frail men and women.

Although there is clinical evidence that frailty effects CVD risk and expression, the mechanisms of this effect have not been explored in humans. The development of tools to assess frailty in animal models of aging has resulted in vital work exploring the effect of frailty on the structure and function of the heart.

Emerging Work on Frailty and Cardiac Aging

FI tools for the assessment of frailty in aging animal models have been recently developed and validated [144, 173, 175]. These tools allow the quantification of health-related deficits in aging rodents and the calculation of FI scores for each animal. The mouse clinical FI contains 31 noninvasively assessed deficits [173] and has been shown to have good inter-rater reliability [46, 47, 81] and to be sensitive to interventions [82]. Importantly, the mouse clinical FI was recently shown to predict mortality risk [147]. This tool has been used in several recent studies to show that frailty is closely associated with many of the age-related changes in cardiac structure and function discussed in this review.

The first of these studies showed that cardiomyocyte hypertrophy in aged male and female mice was closely graded by FI score [144]. This was confirmed in a larger study of young and old male mice, which found that both cardiomyocyte hypertrophy and cardiac hypertrophy were correlated with and graded by FI score [48]. Additionally, measures of reduced ventricular contractility including lower LV developed pressure, and reduced rates of pressure development and decay were also correlated with and graded by FI score. Impaired contractility was similarly correlated with frailty at the cellular level, with a greater reduction in contraction velocity and size in cells from mice with higher FI scores. Interestingly, frailty was even related to changes at the subcellular level, with reductions in calcium currents, reduced peak calcium transients, and reduced protein expression of L-type calcium channels (CaV1.2) all correlated with and graded by FI score [48]. Each of these changes with frailty was also found to be greater in old mice compared to the young but with a large degree of heterogeneity and overlap, so that, for example, some of the older mice had myocyte contractions that were larger than many of the younger mice. Frailty was able to grade and explain some of this heterogeneity [48]. The close association of mouse cardiac hypertrophy and LV contractile dysfunction with frailty suggests that these maladaptive changes may also be associated with frailty in humans and may contribute to the higher susceptibility of frail older adults to diseases such as heart failure.

The association of frailty with the structure and function of the SAN and the atria has also been explored in young and old mice using the mouse clinical FI [76, 128]. Aged mice display, on average, reductions in electrical conduction time through the atria as shown by increased P wave duration, longer PR intervals, reduced atrial conduction velocity, and reduced atrial action potential durations. Each of these factors shows considerable variability within mice of the same age group but is correlated with and graded by FI score [76]. Similar findings were also seen in studies of the SAN from young and aged mice. Increasing frailty is correlated with reduced heart rate, decreased SAN recovery time and conduction velocity, and reduced SAN action potential duration [128]. Frailty is also associated with increasing atrial and SAN fibrosis and changes to extracellular matrix proteins in the atria and SAN [76, 128]. These changes to the structure and function of the atria and SAN can increase the likelihood of developing atrial fibrillation [76], as well as SAN dysfunction, which is the most common reason for pacemaker implantation [128]. Thus, there seems to be an association of frailty with maladaptive changes in the atria and SAN as well as in the ventricle. This has important implications for considering frailty in the assessment of CVD risk and in treatment strategies for the older frail population. The effect of frailty on the vasculature and on sex differences in age-related changes to the heart remains to be clarified. Initial animal studies have not seen any clear evidence of sex differences in frailty in mice [144, 173], although the larger studies have so far only used male mice [48, 76, 128]. More research is needed in this area.

With advanced age there is increased heterogeneity in health outcomes in both humans and animal models. As discussed in the sections above, there are changes to the cardiovascular system that have an increased risk of occurrence in the older population, on average, such as diastolic dysfunction. However, in the population as a whole, there is extensive heterogeneity in these outcomes. In fact, someone who is 70 years old may well have better diastolic function than someone who is 30 years old, or they may have worse diastolic function than someone who is 90 years old. Investigation of changes in the cardiovascular system within the context of the overall health of the patient is important in order to understand this heterogeneity in outcomes with aging. Frailty allows us to do this in both humans and now in animal models of aging [146, 173]. The use of animal FI tools to explore cardiac changes with aging has led to growing evidence that frailty is more closely associated with many of the structural and functional cardiac changes traditionally associated with aging, than chronological age itself. This has important implications for estimating risk of CVD and optimizing interventions for these diseases in the older and frail population.

Summary

There are major structural and functional changes to the cardiovascular system that occur in old age. These changes result in increased risk of CVD, as well as poor cardiovascular outcomes with increasing age. There is growing evidence for important sex differences in both the underlying age-related changes and the development of CVD in the older population. Additionally, frail older adults are at even greater risk of developing CVD, and there is emerging evidence that frailty is closely related to the structural and functional changes of the cardiovascular system that are commonly seen with aging. This has important implications for the assessment, prevention, and treatment of CVD in older, frail women and men.

References

1. Afilalo J, Alexander KP, Mack MJ, et al. Frailty assessment in the cardiovascular care of older adults. J Am Coll Cardiol. 2014;63:747–62.
2. Ahmed A, Frontera A, Carpenter A, et al. Clinical predictors of pacemaker implantation in patients with syncope receiving implantable loop recorder with or without ECG conduction abnormalities. PACE. 2015;38:934–41.
3. Al Ghatrif M, Lakatta EG. The conundrum of arterial stiffness, elevated blood pressure, and aging. Curr Hypertens Rep. 2015;17:12.
4. Alonso-Betanzos A, Bolón-Canedo V, Heyndrickx GR, et al. Exploring guidelines for classification of major heart failure subtypes by using machine learning. Clin Med Insights Cardiol. 2015;9:57–71.
5. Avgil Tsadok M, Jackevicius CA, Rahme E, et al. Sex differences in stroke risk among older patients with recently diagnosed atrial fibrillation. JAMA. 2012;307:1952–8.
6. Badano LP, Miglioranza MH, Mihaila S, et al. Left atrial volumes and function by three-dimensional echocardiography: reference values, accuracy, reproducibility, and comparison with two-dimensional echocardiographic measurements. Circ Cardiovasc Imaging. 2016;9:e004229.
7. Bauer M, Caviezel S, Teynor A, et al. Carotid intima-media thickness as a biomarker of subclinical atherosclerosis. Swiss Med Wkly. 2012;w13705:142.
8. Bavishi C, Goel S, Messerli FH. Isolated systolic hypertension: an update after. SPRINT. Am J Med. 2016;129:1251–8.
9. Beladan CC, Calin A, Rosca M, et al. Left ventricular twist dynamics: principles and applications. Heart. 2014;100 (9):731–40.
10. Benjamin EJ, Levy D, Vaziri SM, et al. Independent risk factors for atrial fibrillation in a population-based cohort: the Framingham heart study. JAMA. 1994;271:840–4.
11. Benjamin EJ, Wolf PA, D'Agostino RB. Impact of atrial fibrillation on the risk of death: the Framingham heart study. Circulation. 1998;98:946–52.
12. Beringer JY, Kerkhof PL. A unifying representation of ventricular volumetric indexes. IEEE Trans Biomed Eng. 1998;45:365–71.
13. Bhatia RS, Tu JV, Lee DS, et al. Outcome of heart failure with preserved ejection fraction in a population-based study. N Engl J Med. 2006;355:260–9.
14. Boese AC, Kim SC, Yin KJ, et al. Sex differences in vascular physiology and pathophysiology: estrogen and androgen signaling in health and disease. Am J Physiol Heart Circ Physiol. 2017. https://doi.org/10.1152/ajpheart.00217.2016.
15. Bonda TA, Szynaka B, Sokołowska M, et al. Remodeling of the intercalated disc related to aging in the mouse heart. J Cardiol. 2016;68:261–8.
16. Boyd AC, Schiller NB, Leung D, Ross DL, Thomas L. Atrial dilation and altered function are mediated by age and diastolic function but not before the eighth decade. JACC Cardiovasc Imaging. 2011;4:234–42.
17. Brodde OE, Leineweber K. Autonomic receptor systems in the failing and aging human heart: similarities and differences. Eur J Pharmacol. 2004;500:167–76.
18. Cacciatore F, Abete P, Mazzella F, et al. Frailty predicts long-term mortality in elderly subjects with chronic heart failure. Eur J Clin Investig. 2005;35:723–30.
19. Carroll JD, Carroll EP, Feldman T, et al. Sex-associated differences in left ventricular function in aortic stenosis of the elderly. Circulation. 1992;86:1099–107.
20. Cecelja M, Chowienczyk P. Molecular mechanisms of arterial stiffening. Pulse. 2016;4:43–8.
21. Chen C, Sung KT, Shih SC. Age, gender and load-related influences on left ventricular geometric remodelling, systolic mid-wall function, and NT-ProBNP in asymptomatic Asian population. PLoS One. 2016;11:e0156467.
22. Chen M. Frailty and cardiovascular disease: potential role of gait speed in surgical risk stratification in older adults. J Geriatr Cardiol. 2015;12:44–56.
23. Chung AK, Das SR, Leonard D, et al. Women have higher left ventricular ejection fractions than men independent of differences in left ventricular volume. Circulation. 2006;113:1597–604.
24. Codreanu I, Pegg TJ, Selvanayagam JB, et al. Normal values of regional and global myocardial wall motion in young and elderly individuals using navigator gated tissue phase mapping. Age. 2014;36:231–41.
25. Collard RM, Boter H, Schoevers RA, Oude Voshaar RC. Prevalence of frailty in community-dwelling older persons: a systematic review. J Am Geriatr Soc. 2012;60:1487–92.
26. Collins JA, Munoz JV, Patel TR, et al. The anatomy of the aging aorta. Clin Anat. 2014;27:463–6.
27. Coutinho T. Arterial stiffness and its clinical implications in women. Can J Cardiol. 2014;30:756–64.
28. Cramariuc D, Rogge BP, Lønnebakken MT, et al. Sex differences in cardiovascular outcome during progression of aortic valve stenosis. Heart. 2015;101:209–14.
29. Currie G, Delles C. Healthy vascular aging. Hypertension. 2017;70:229–31.
30. Dagres N, Nieuwlaat R, Vardas PE. Gender-related differences in presentation, treatment, and outcome of patients with atrial fibrillation in Europe. J Am Coll Cardiol. 2007;49:572–7.
31. Dai DF, Santana LF, Vermulst M, et al. Overexpression of catalase targeted to mitochondria attenuated murine cardiac aging. Circulation. 2009;199:2789–97.
32. Dai DF, Chen T, Johnson SC, et al. Cardiac aging: from molecular mechanisms to significance in human health and disease. Antioxid Redox Signal. 2012;16:1492–526.
33. Dai DF, Karunadharma PP, Chiao YA, et al. Altered proteome turnover and remodeling by short-term caloric restriction or rapamycin rejuvenate the aging heart. Aging Cell. 2014;13:529–39.
34. De Bondt P, Van de Wiele C, De Sutter J, De Winter F, De Backer G, Dierckx RA. Age- and gender-specific differences in left ventricular cardiac function and volumes determined by gated SPET. Eur J Nucl Med. 2001;28:620–4.
35. Dibb KM, Rueckschloss U, Eisner DA, et al. Mechanisms underlying enhanced cardiac excitation contraction coupling observed in the senescent sheep myocardium. J Mol Cell Cardiol. 2004;37:1171–81.
36. Dun W, Yagi T, Rosen MR, Boyden PA. Calcium and potassium currents in cells from adult and aged canine right atria. Cardiovasc Res. 2003;58:526–34.
37. Dun W, Boyden PA. Aged atria: electrical remodeling conducive to atrial fibrillation. J Interv Card Electrophysiol. 2009;25:9–18.
38. Dunlay SM, Roger VL. Gender differences in the pathophysiology, clinical presentation, and outcomes of ischemic heart failure. Curr Heart Fail Rep. 2012;9:267–76.
39. Dworatzek E, Baczko I, Kararigas G. Effects of aging on cardiac extracellular matrix in men and women. Proteomics Clin Appl. 2016;10:84–91.

40. Dzeshka MS, Lip GY, Snezhitskiy V, Shantsila E. Cardiac fibrosis in patients with atrial fibrillation. mechanisms and clinical implications J Am Coll Cardiol. 2015;66:943–59.
41. Elias-Smale SE, Günal A, Maas AH. Gynecardiology: distinct patterns of ischemic heart disease in middle-aged women. Maturitas. 2015;81:348–52.
42. Eng J, McClelland RL, Gomes AS, et al. Adverse left ventricular remodeling and age assessed with cardiac MR imaging: the multi-ethnic study of atherosclerosis. Radiology. 2015;278:714–22.
43. Faes TJ, Kerkhof PL. The volume regulation graph versus the ejection fraction as metrics of left ventricular performance in heart failure with and without a preserved ejection fraction: a mathematical model study. Clin Med Insights Cardiol. 2015;21:73–91.
44. Fannin J, Rice KM, Thulluri S, et al. Age-associated alterations of cardiac structure and function in the female F344xBN rat heart. Age. 2014;36:9684.
45. Farrell SR, Howlett SE. The age-related decrease in catecholamine sensitivity is mediated by ß1-adrenergic receptors linked to a decrease in adenylate cyclase activity in ventricular myocytes from male Fischer 344 rats. Mech Aging Dev. 2008;129:735–44.
46. Feridooni HA, Dibb KM, Howlett SE. How cardiomyocyte excitation, calcium release and contraction become altered with age. J Mol Cell Cardiol. 2015;83:62–72.
47. Feridooni HA, Sun MH, Rockwood K, Howlett SE. Reliability of a frailty index based on the clinical assessment of health deficits in male C57BL/6J mice. J Gerontol Ser A Biol Sci Med Sci. 2015b;70:686–93.
48. Feridooni HA, Kane AE, Ayaz O, et al. The impact of age and frailty on ventricular structure and function in C57BL/6J mice. J Physiol. 2017;595:3721–42.
49. Ferrara N, Komici K, Corbi G, et al. β-adrenergic receptor responsiveness in aging heart and clinical implications. Front Physiol. 2014;4:396.
50. Ferruzzi J, Bersi MR, Uman S, et al. Decreased elastic energy storage, not increased material stiffness, characterizes central artery dysfunction in fibulin-5 deficiency independent of sex. J Biomech Eng. 2015;137(3):0310071–03100714.
51. Fiechter M, Fuchs TA, Gebhard C, et al. Age-related normal structural and functional ventricular values in cardiac function assessed by magnetic resonance. BMC Med Imaging. 2013;13:6.
52. Fleenor BS. Large elastic artery stiffness with aging: novel translational mechanisms and interventions. Aging Dis. 2012;4:76–83.
53. Fleg JL, Strait J. Age-associated changes in cardiovascular structure and function: a fertile milieu for future disease. Heart Fail Rev. 2012;17:545–54.
54. Garcia M, Mulvagh SL, Merz CN, et al. Cardiovascular disease in women: clinical perspectives. Circ Res. 2016;118:1273–93.
55. Gargiulo S, Gamba P, Testa G, et al. The role of oxysterols in vascular aging. J Physiol. 2016;594:2095–113.
56. Gavish B, Izzo JL Jr. Arterial stiffness: going a step beyond. Am J Hypertens 2016;26:1223–1233.
57. Gebhard C, Stahli BE, Gebhard CE, et al. Age- and gender-dependent left ventricular remodeling. Echocardiography. 2013;30:1143–50.
58. Gebhard C, Stahli BE, Gebhard CE, et al. Gender- and age-related differences in rest and post-stress left ventricular cardiac function determined by gated SPECT. Int J Cardiovasc Imaging. 2014;30:1191–9.
59. Gebhard C, Buechel RR, Stahli BE, et al. Impact of age and sex on left ventricular function determined by coronary computed tomographic angiography: results from the prospective multicentre CONFIRM study. Eur Heart J Cardiovasc Imaging. 2016;18:990–1000.
60. Gonçalves A, Hung CL, Claggett B, et al. Left atrial structure and function across the spectrum of cardiovascular risk in the elderly: the atherosclerosis risk in communities study. Circ Cardiovasc Imaging. 2016;9:e004010.
61. Gordon EH, Peel NM, Samanta M. Sex differences in frailty: a systematic review and meta-analysis. Exp Gerontol. 2017;89:30–40.
62. Gottlieb SS, Khatta M, Friedmann E, et al. The influence of age, gender, and race on the prevalence of depression in heart failure patients. J Am Coll Cardiol. 2004;43(9):1542.
63. Greenwald SE. Aging of the conduit arteries. J Pathol. 2007;211:157–72.
64. Greiten LE, Holditch SJ, Arunachalam SP, Miller VM. Should there be sex-specific criteria for the diagnosis and treatment of heart failure? J Cardiovasc Transl Res. 2014;7:139–55.
65. Grandy SA, Howlett SE. Cardiac excitation-contraction coupling is altered in myocytes from aged male mice but not in cells from aged female mice. Am J Physiol Heart Circ Physiol. 2006;291:H2362–70.
66. Haataja P, Nikus K, Kahonen M, et al. Prevalence of ventricular conduction blocks in the resting electrocardiogram in a general population: the health 2000 survey. Int J Cardiol. 2013;176:1953–60.
67. Hamdani N, Bishu KG, von Frieling-Salewsky M, et al. Deranged myofilament phosphorylation and function in experimental heart failure with preserved ejection fraction. Cardiovasc Res. 2013;97:464–71.
68. Hariharan N, Sussman MA. Cardiac aging-getting to the stem of the problem. J Mol Cell Cardiol. 2015;83:32–6.
69. Harvey A, Montezano AC, Touyz RM. Vascular biology of aging-implications in hypertension. J Mol Cell Cardiol. 2015;83:112–21.
70. Hees PS, Fleg JL, Lakatta EG, Shapiro EP. Left ventricular remodeling with age in normal men versus women. novel insights using three-dimensional magnetic resonance imaging Am J Cardiol. 2002;90:1231–6.
71. Horn MA, Trafford AW. Aging and the cardiac collagen matrix: novel mediators of fibrotic remodeling. J Mol Cell Cardiol. 2015;93:175–85.
72. Horn MA, Graham HK, Richards MA, et al. Age-related divergent remodeling of the cardiac extracellular matrix in heart failure: collagen accumulation in the young and loss in the aged. J Mol Cell Cardiol. 2012;53:82–90.
73. Howlett SE. Age-associated changes in excitation-contraction coupling are more prominent in ventricular myocytes from male rats than in myocytes from female rats. Am J Physiol Heart Circ Physiol. 2010;298: H659–70.
74. Hubbard RE, Theou O. Frailty: enhancing the known knowns. Age Aging. 2012;41:574–5.
75. Hung C, Gonçalves A, Shah AM, et al. Age- and sex-related influences on left ventricular mechanics in elderly individuals free of prevalent heart failure. Circ Cardiovasc Imaging. 2017;10:e004510.
76. Jansen HJ, Moghtadaei M, Mackasey M, et al. Atrial structure, function and arrhythmogenesis in aged and frail mice. Sci Rep. 2017;7:44336.
77. Jeong JH, Kim JH, Park YH, et al. Incidence of and risk factors for bundle branch block in adults older than 40 years. Korean J Intern Med. 2004;19:171–8.
78. Juonala M, Kähönen M, Laitinen T, et al. Effect of age and sex on carotid intima-media thickness, elasticity and brachial

endothelial function in healthy adults: the cardiovascular risk in young Finns study. Eur Heart J. 2008;29:1198–206.
79. Kaess BM, Gona P, Larson MG, et al. Secular trends in echocardiographic left ventricular mass in the community: the Framingham heart study. Heart. 2013;99:1693–8.
80. Kaila K, Haykowsky MJ, Thompson RB, et al. Heart failure with preserved ejection fraction in the elderly: scope of the problem. Heart Fail Rev. 2012;17:555–62.
81. Kane AE, Hilmer SN, Huizer-Pajkos A, et al. Factors that impact on interrater reliability of the mouse clinical frailty index. Gerontol Ser A Biol Sci Med Sci. 2016;70:694–5.
82. Kane A, Hilmer S, Boyer D, et al. Impact of longevity interventions on a validated mouse clinical frailty index. Gerontol Ser A Biol Sci Med Sci. 2016b;71:333–9.
83. Kang KT. Endothelium-derived relaxing factors of small resistance arteries in hypertension. Toxicol Res. 2014;30:141–8.
84. Kawut SM, Lima JA, Barr G, et al. Sex and race differences in right ventricular structure and function: The MESA-right ventricle study. Circulation. 2011;123:2542–51.
85. Keller KM, Howlett SE. Sex differences in the biology and pathology of the aging heart. Can J Cardiol 2016; 32 (9):1065–73.
86. Kerkhof PL. Characterizing heart failure in the ventricular volume domain. Clin Med Insights Cardiol. 2015;25:11–31.
87. Kerkhof PLM, Kresh JY, Li JK, et al. Left ventricular volume regulation in heart failure with preserved ejection fraction. Physiol Rep. 2013;1:e0007.
88. Khan H, Kalogeropoulos AP, Georgiopoulou VV, et al. Frailty and risk for heart failure in older adults: the health, aging, and body composition study. Am Heart J. 2013;166:887–94.
89. Koch SE, Haworth KJ, Robbins N, et al. Age- and gender-related changes in ventricular performance in wild-type FVB/N mice as evaluated by conventional and vector velocity echocardiography imaging: a retrospective study. Ultrasound Med Biol. 2013;39:2034–43.
90. Konstam MA, Abboud FM. Ejection fraction: misunderstood and overrated (changing the paradigm in categorizing heart failure). Circulation. 2017;135:717–9.
91. Kosmidou I, Redfors B, Selker HP, et al. Infarct size, left ventricular function, and prognosis in women compared to men after primary percutaneous coronary intervention in ST-segment elevation myocardial infarction: results from an individual patient-level pooled analysis of 10 randomized trials. Eur Heart J. 2017;38:1656–63.
92. Lakatta EG. Cardiovascular regulatory mechanisms in advanced age. Physiol Rev. 1993;73:413–67.
93. Lakatta EG, Levy D. Arterial and cardiac aging: major shareholders in cardiovascular disease enterprises: part II: the aging heart in health: links to heart disease. Circulation. 2003;107:346–54.
94. Lakatta EG, Wang M, Najjar SS. Arterial aging and subclinical arterial disease are fundamentally intertwined at macroscopic and molecular levels. Med Clin North Am. 2009;93:583–604.
95. Lam CS, Carson PE, Anand IS, et al. Sex differences in clinical characteristics and outcomes in elderly patients with heart failure and preserved ejection fraction: the Irbesartan in heart failure with preserved ejection fraction (I-PRESERVE) trial. Circ Heart Fail. 2012;5:571–8.
96. Lam CSP, McEntegart M, Claggett B, et al. Sex differences in clinical characteristics and outcomes after myocardial infarction: insights from the valsartan in acute myocardial infarction trial (VALIANT). Eur J Heart Fail. 2015;17:301–12.
97. Larson ED, St. Clair JR, Sumner WA, et al. Depressed pacemaker activity of sinoatrial node myocytes contributes to the age-dependent decline in maximum heart rate. PNAS. 2013;110:18011–6.
98. Lau DH, Shipp NJ, Kelly DJ, et al. Atrial arrhythmia in aging spontaneously hypertensive rats: unraveling the substrate in hypertension and aging. PLoS One. 2013;8:e72416.
99. Lee HY, Oh BH. Aging and arterial stiffness. Circ J. 2010;74:2257–62.
100. Lee DH, Buth KJ, Martin BJ, et al. Frail patients are at increased risk for mortality and prolonged institutional care after cardiac surgery. Circulation. 2010;121(8):973.
101. Leon LJ, Gustafsson ÅB. Staying young at heart: autophagy and adaptation to cardiac aging. J Mol Cell Cardiol. 2015;95:78–85.
102. Li Y, Pagano PJ. Microvascular NADPH oxidase in health and disease. Free Radic Biol Med. 2017;109:33–47.
103. Lim MA, Townsend RR. Arterial compliance in the elderly: its effect on blood pressure measurement and cardiovascular outcomes. Clin Geriatr Med. 2009;25:191–205.
104. Lindsey ML, Goshorn DK, Squires CE, et al. Age-dependent changes in myocardial matrix metalloproteinase/tissue inhibitor of metalloproteinase profiles and fibroblast function. Cardiovasc Res. 2005;66:410–9.
105. Ling LH, Kistler PM, Ellims AH, et al. Diffuse ventricular fibrosis in atrial fibrillation: noninvasive evaluation and relationships with aging and systolic dysfunction. J Am Coll Cardiol. 2012;60:2402–8.
106. Liu CY, Liu YC, Wu C, et al. Evaluation of age-related interstitial myocardial fibrosis with cardiac magnetic resonance contrast-enhanced T_1 mapping. J Am Coll Cardiol. 2013;62:1280–7.
107. Liu L, Okamura T, Kadowaki T, et al. Bundle branch block and other cardiovascular disease risk factors: US–Japan comparison. Int J Cardiol. 2010;143:432–40.
108. Lloyd-Jones DM, Wang TJ, Leip EP, et al. Lifetime risk for development of atrial fibrillation: the Framingham heart study. Circulation. 2004;110:1042–6.
109. Loffredo FS, Nikolova AP, Pancoast JR, et al. Heart failure with preserved ejection fraction: molecular pathways of the aging myocardium. Circ Res. 2014;115:97–10.
110. Lupón J, González B, Santaeugenia S, et al. Prognostic implication of frailty and depressive symptoms in an outpatient population with heart failure. Rev Esp Cardiol. 2008;61:835–42.
111. Maceira AM, Prasad SK, Khan M, et al. Reference right ventricular systolic and diastolic function normalized to age, gender and body surface area from steady-state free precession cardiovascular magnetic resonance. Eur Heart J. 2006;27:2879–88.
112. Magnani JW, Moser CB, Murabito JM, et al. Association of sex hormones, aging, and atrial fibrillation in men: the Framingham heart study. Circ Arrhythm Electrophysiol. 2014;7:307–12.
113. Mancia G, Giannattasio C. Diagnostic and therapeutic problems of isolated systolic hypertension. J Hypertens. 2015;33:33–43.
114. Martínez-Sellés M, Doughty RN, Poppe K, et al. Gender and survival in patients with heart failure: interactions with diabetes and aetiology. Results from the MAGGIC individual patient meta-analysis. Eur J Heart Fail. 2012;14:473–9.
115. McManus DD, Rienstra M, Benjamin EJ. An update on the prognosis of patients with atrial fibrillation. Circulation. 2012;126:e143–6.

116. McNallan SM, Singh M, Chamberlain AM, et al. Frailty and healthcare utilization among patients with heart failure in the community. J Am Coll Cardiol. 2013;1:135–41.
117. Medrano G, Hermosillo-Rodriguez J, Pham T, et al. Left atrial volume and pulmonary artery diameter are noninvasive measures of age-related diastolic dysfunction in mice. J Gerontol A Biol Sci Med Sci. 2015;71:1141–50.
118. Mellor KM, Curl CL, Chandramouli C, et al. Aging-related cardiomyocyte functional decline is sex and angiotensin II dependent. Age (Dordr). 2014;36:9630.
119. Méndez-Bailón M, Muñoz-Rivas N, Jiménez-García R, et al. Women with atrial fibrillation and type 2 diabetes have a higher incidence of hospitalization and undergo ablation or pacemaker implantation less frequently than men. Eur J Intern Med. 2017;42:67–73.
120. Merz AA, Cheng S. Sex differences in cardiovascular aging. Heart. 2016;102:825–31.
121. Miglioranza MH, Badano LP, Mihăilă S, et al. Physiologic determinants of left atrial longitudinal strain: a two-dimensional speckle-tracking and three-dimensional echocardiographic study in healthy volunteers. J Am Soc Echocardiogr. 2016;29:1023–34.
122. Milin AC, Vorobiof G, Aksoy O, Ardehali R. Insights into aortic sclerosis and its relationship with coronary artery disease. J Am Heart Assoc. 2014;3:e001111.
123. Mirza M, Strunets A, Shen WK, et al. Mechanisms of arrhythmias and conduction disorders in older adults. Clin Geriatr Med. 2012;28:555–73.
124. Mistriotis P, Andreadis ST. Vascular aging: molecular mechanisms and potential treatments for vascular rejuvenation. Aging Res Rev. 2017;1:94–116.
125. Mitnitski AB, Mogilner AJ, MacKnight C, Rockwood K. The mortality rate as a function of accumulated deficits in a frailty index. Mech Aging Dev. 2002;123:1457–60.
126. Mitnitski A, Song X, Rockwood K. Assessing biological aging: the origin of deficit accumulation. Biogerontology. 2013;14:709–17.
127. Miyasaka Y, Barnes ME, Gersh BJ, et al. Secular trends in incidence of atrial fibrillation in olmsted county, Minnesota, 1980 to 2000, and implications on the projections for future prevalence. Circulation. 2006;114:119–25.
128. Moghtadaei M, Jansen HJ, Mackasey M, et al. The impacts of age and frailty on heart rate and sinoatrial node function. J Physiol. 2016;594:7105–26.
129. Montero D, Pierce GL, Stehouwer CD, et al. The impact of age on vascular smooth muscle function in humans. J Hypertens. 2015;33:445–53.
130. Moore A, Mangoni AA, Lyons D, Jackson SH. The cardiovascular system. Br J Clin Pharmacol. 2003;56:254–60.
131. Muraru D, Onciul S, Peluso D, et al. Sex- and method-specific reference values for right ventricular strain by 2-dimensional speckle-tracking echocardiography. Circ Cardiovasc Imaging. 2016;9:e003866.
132. Najjar SS, Scuteri A, Lakatta EG. Arterial aging: is it an immutable cardiovascular risk factor? Hypertension. 2005;46:454–62.
133. Nakada Y, Kawakami R, Nakano T, et al. Sex differences in clinical characteristics and long-term outcome in acute decompensated heart failure patients with preserved and reduced ejection fraction. Am J Physiol Heart Circ. 2016;310:H813–20.
134. Nes BM, Janszky I, Wisløff U, et al. Age-predicted maximal heart rate in healthy subjects: the HUNT fitness study. Scand J Med Sci Sports. 2013;23:697–704.
135. New SE, Aikawa E. Cardiovascular calcification: an inflammatory disease. Circ J. 2011;75:1305–13.
136. Newman AB, Gottdiener JS, McBurnie MA, et al. Associations of subclinical cardiovascular disease with frailty. J Gerontol A Biol Sci Med Sci. 2001;56:M158–66.
137. Nikitin NP, Witte KKA, Thackray SDR, et al. Effect of age and sex on left atrial morphology and function. Eur J Echocardiogr. 2003;4:36–42.
138. Olivetti G, Melissari M, Capasso JM, Anversa P. Cardiomyopathy of the aging human heart. Myocyte loss and reactive cellular hypertrophy. Circ Res. 1991;68:1560–8.
139. Olivetti G, Giordano G, Corradi D, et al. Gender differences and aging: effects on the human heart. J Am Coll Cardiol. 1995;26:1068–79.
140. O'Rourke MF, Hashimoto J. Mechanical factors in arterial aging: a clinical perspective. J Am Coll Cardiol. 2007;50:1–13.
141. O'Rourke MF, Adji A, Namasivayam M, et al. Arterial aging: a review of the pathophysiology and potential for pharmacological intervention. Drugs Aging. 2011;28:779–95.
142. Pancholy SB, Sharma PS, Pancholy DS, et al. Meta-analysis of gender differences in residual stroke risk and major bleeding in patients with nonvalvular atrial fibrillation treated with oral anticoagulants. Am J Cardiol. 2014;113:485–90.
143. Park SM, Merz CN. Women and ischemic heart disease: recognition, diagnosis and management. Korean Circ J. 2016;46:433–42.
144. Parks RJ, Fares E, Macdonald JK, et al. A procedure for creating a frailty index based on deficit accumulation in aging mice. Gerontol Ser A Biol Sci Med Sci. 2012;67:217–27.
145. Riedinger MS, Dracup KA, Brecht M, et al. Quality of life in women with heart failure, normative groups, and patients with other chronic conditions. AJCC. 2002;11:211–9.
146. Rockwood K, Mitnitski A. Frailty in relation to the accumulation of deficits. Journals Gerontol Biol Sci Med Sci. 2007;62:722–7.
147. Rockwood K, Blodgett JM, Theou O, et al. A frailty index based on deficit accumulation quantifies mortality risk in humans and in mice. Sci Rep. 2017;7:43068.
148. Roosens B, Bala G, Droogmans S, et al. Occurrence of cardiovascular calcifications in normal, aging rats. Exp Gerontol. 2012;47:614–9.
149. Rubio-Ruiz ME, Pérez-Torres I, Soto ME, et al. Aging in blood vessels. Medicinal agents FOR systemic arterial hypertension in the elderly. Aging Res Rev. 2014;18:132–47.
150. Sader MA, Celermajer DS. Endothelial function, vascular reactivity and gender differences in the cardiovascular system. Cardiovasc Res. 2002;53:597–604.
151. Sandow SL, Senadheera S, Grayson TH, et al. Calcium and endothelium-mediated vasodilator signaling. Adv Exp Med Biol. 2012;740:811–31.
152. Sethi S, Rivera O, Oliveros R, et al. Aortic stiffness: pathophysiology, clinical implications, and approach to treatment. Integr Blood Press Control. 2014;7:29–34.
153. Sherman DG, Kim SG, Boop BS, et al. Occurrence and characteristics of stroke events in the atrial fibrillation follow-up investigation of sinus rhythm management (AFFIRM) study. JAMA. 2005;165:1185–91.
154. Sheydina A, Riordon DR, Boheler KR. Molecular mechanisms of cardiomyocyte aging. Clin Sci. 2011;121:315–29.
155. Shinmura K, Tamaki K, Sano M, et al. Impact of long-term caloric restriction on cardiac senescence: caloric restriction ameliorates cardiac diastolic dysfunction associated with aging. J Mol Cell Cardiol. 2011;50:117–27.

156. Singh M, Stewart R, White H. Importance of frailty in patients with cardiovascular disease. Eur Heart J. 2014;35:1726–31.
157. Speedie A, Mathew C, Shahi RS, Calton R. Long-term mortality in patients with permanent pacemaker implantation. J Assoc Physicians India. 2016;64:18–22.
158. Steenman M, Lande G. Cardiac aging and heart disease in humans. Biophys Rev. 2017;9:131–7.
159. Strait JB, Lakatta EG. Aging-associated cardiovascular changes and their relationship to heart failure. Heart Fail Clin. 2012;8:143–64.
160. Sündermann S, Dademasch A, Rastan A, et al. One-year follow-up of patients undergoing elective cardiac surgery assessed with the comprehensive assessment of frailty test and its simplified form. Interact Cardiovasc Thorac Surg. 2011;13:119–23.
161. Takagi G, Asai K, Vatner SF, et al. Gender differences on the effects of aging on cardiac and peripheral adrenergic stimulation in old conscious monkeys. Am J Physiol Heart Circ Physiol. 2003;285:H527–34.
162. Tellez JO, Mczewski M, Yanni J, et al. Aging-dependent remodeling of ion channel and Ca2+ clock genes underlying sino-atrial node pacemaking. Exp Physiol. 2011;96:1163–78.
163. Thorin E, Thorin-Trescases N. Vascular endothelial aging, heartbeat after heartbeat. Cardiovasc Res. 2009;84:24–32.
164. Tribouilloy C, Rusinaru D, Mahjoub H, et al. Prognosis of heart failure with preserved ejection fraction: a 5 year prospective population-based study. Eur Heart J. 2008;29:339–47.
165. Tsamis A, Krawiec JT, Vorp DA. Elastin and collagen fibre microstructure of the human aorta in aging and disease: a review. J R Soc Interface. 2013;10:20121004.
166. Urabe G, Hoshina K, Shimanuki T, et al. Structural analysis of adventitial collagen to feature aging and aneurysm formation in human aorta. J Vasc Surg. 2016;63:1341–50.
167. Uslan DZ, Tleyjeh IM, Baddour LM, et al. Temporal trends in permanent pacemaker implantation: a population-based study. Am Heart J. 2008;155:896–903.
168. Vaupel JW, Manton KG, Stallard E. The impact of heterogeneity in individual frailty on the dynamics of mortality. Demography. 1979;16:439–54.
169. Veronese N, Ceredac E, Stubbs B, et al. Risk of cardiovascular disease morbidity and mortality in frail and pre-frail older adults: results from a meta-analysis and exploratory meta-regression analysis. Aging Res Rev. 2017;35:63–73.
170. Virdis A, Taddei S. Endothelial aging and gender. Maturitas. 2012;71:326–30.
171. Wang M, Monticone RE, Lakatta EG. Arterial aging: a journey into subclinical arterial disease. Curr Opin Nephrol Hypertens. 2010;19:201–7.
172. White M, Roden R, Minobe W, et al. Age-related changes in beta-adrenergic neuroeffector systems in the human heart. Circulation. 1994;90:1225–38.
173. Whitehead JC, Hildebrand B, Sun M, et al. A clinical frailty index in aging mice: comparisons with frailty index data in humans. Gerontol Biol Sci Med Sci. 2014;69:621–32.
174. Yoneyama K, Gjesdak O, Choi E, et al. Age, gender and hypertension-related remodeling influences left ventricular torsion assessed by tagged cardiac magnetic resonance in asymptomatic individuals: the multi-ethnic study of atherosclerosis. Circulation. 2012;126:2481–90.
175. Yorke A, Kane AE, Hancock Friesen CL, et al. Development of a rat clinical frailty index. Gerontol Biol Sci Med Sci. 2017;72:897–903.
176. Yoshida K, Obokata M, Kurosawa K, et al. Effect of sex differences on the association between stroke risk and left atrial anatomy or mechanics in patients with atrial fibrillation. Circ Cardiovasc Imaging. 2016;9:e004999.
177. Zannad F, Huvelle E, Dickstein K, et al. Left bundle branch block as a risk factor for progression to heart failure. Eur J Heart Fail. 2007;9:7–14.
178. Zemrak F, Ambale-Venkatesh B, Captur G, et al. Left atrial structure in relationship to age, sex, ethnicity, and cardiovascular risk factors: MESA (Multi-Ethnic Study of Atherosclerosis). Circ Cardiovasc Imaging. 2017;10:e005379.
179. Zhang XP, Vatner SF, Shen YT, et al. Increased apoptosis and myocyte enlargement with decreased cardiac mass; distinctive features of the aging male, but not female, monkey heart. J Mol Cell Cardiol. 2007;43:487–91.
180. Zusterzeel R, Selzman KA, Sanders WE, et al. US food and drug administration meta-analysis of patient-level data. JAMA Intern Med. 2014;174:1340–8.
181. Zsilinszka R, Shrader P, DeVore AD, et al. Sex differences in the management and outcomes of heart failure with preserved ejection fraction in patients presenting to the emergency department. J Card Fail. 2016;22:781–8.

26 Hemodynamic and Electrocardiographic Aspects of Uncomplicated Singleton Pregnancy

Qi Fu

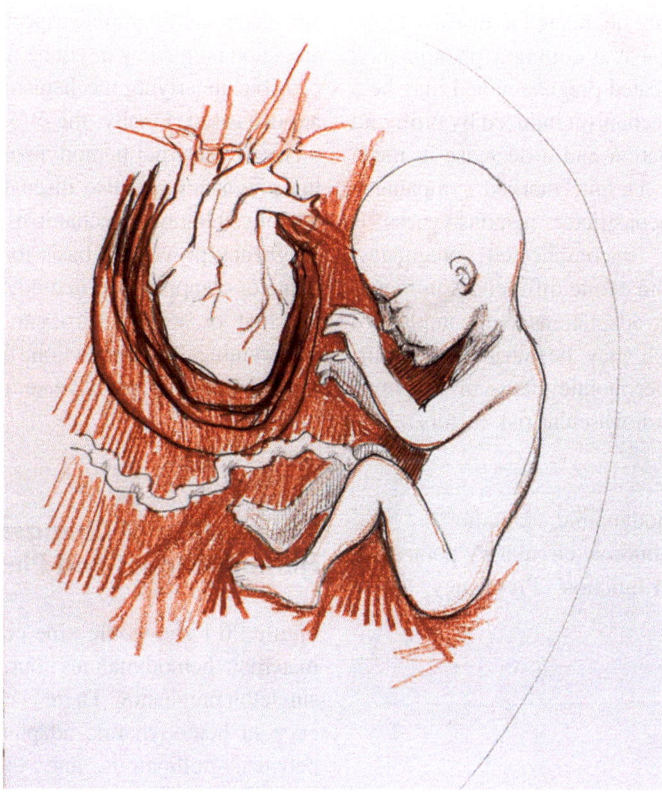

Art work by Piet Michiels, Leuven, Belgium

Q. Fu (✉)
Internal Medicine, Cardiology Division, The University of Texas Southwestern Medical Center, Institute for Exercise and Environmental Medicine, Texas Health Presbyterian Hospital Dallas, Dallas, TX, USA
e-mail: QiFu@TexasHealth.org

Abstract

Pregnancy is associated with significant changes in maternal hemodynamics, which are triggered by profound systemic vasodilation and mediated through the autonomic

P. L. M. Kerkhof, V. M. Miller (eds.), *Sex-Specific Analysis of Cardiovascular Function*, Advances in Experimental Medicine and Biology 1065, https://doi.org/10.1007/978-3-319-77932-4_26

nervous system as well as the renin-angiotensin-aldosterone system. Vascular function changes to help accommodate an increase in intravascular volume due to blood volume expansion associated with pregnancy while maintaining the efficiency of ventricular-arterial coupling and diastolic perfusion pressure. The heart undergoes physiological (*eccentric*) hypertrophy due to increased volume load and cardiac stroke work, whereas the functional change of the left ventricle remains controversial. There are changes in cardiac electrical activity during pregnancy which can be detected in the electrocardiogram that are not related to disease. Sympathetic activation is a common phenomenon during uncomplicated pregnancy and may be a compensatory mechanism induced by profound systemic vasodilation and a decrease in mean arterial pressure. Despite marked sympathetic activation, vasoconstrictor responsiveness is blunted during uncomplicated pregnancy. There are race and ethnic differences in maternal hemodynamic adaptations to uncomplicated pregnancy, which may be attributed to differences in socioeconomic status or in prevalence rates of cardiovascular risk factors.

Keywords
Gestation · Hemodynamic adaptations · Hormones · Autonomic circulatory control · Cardiac-vascular function · Pregnancy

Introduction

In humans, pregnancy elicits dramatic changes in maternal hemodynamics, which begin as early as 4–5 weeks of gestation [34, 77, 112] and reach a plateau in the second trimester [29, 38, 47, 120, 121]. These changes are required to meet the increased metabolic demands of the mother, as well as to ensure adequate uteroplacental circulation for fetal development and growth [100, 125]. Inadequate hemodynamic adaptations during pregnancy can lead to the development of maternal cardiovascular complications (e.g., gestational hypertension, preeclampsia, eclampsia, etc.) and/or poor fetal outcomes (e.g., intrauterine growth restriction, preterm birth, low birth weight, etc.). Maternal hemodynamic adaptations during pregnancy have been proposed to occur through the autonomic nervous system, in particular, the sympathetic division [49, 58]. Additionally, pregnancy-induced blood volume expansion, alterations in hormonal factors (e.g., estrogen, progesterone, relaxin, etc.), and cardiac-vascular remodeling also contribute importantly to the adequate adaptations of maternal hemodynamics.

In this chapter, we review the hemodynamic and electrocardiographic aspects of uncomplicated singleton pregnancy in young healthy women. The possible underlying mechanisms for these changes are discussed. Finally, the effects of race and ethnicity on maternal hemodynamic adaptations during pregnancy are also discussed. Understanding the physiological mechanisms of uncomplicated pregnancy provides a basis for identifying inadequate or compromised hemodynamic responses in the first or second trimester, which may offer opportunities of prediction, prevention, and/or early treatment for adverse maternal and fetal outcomes.

Time Course of Changes in Maternal Hemodynamics and Plasma Volume

Figure 26.1 shows the time course of changes in maternal hemodynamics during uncomplicated singleton pregnancy. There is no significant difference in hemodynamic adaptations to pregnancy between nulliparous and multiparous women [99]. However, maternal hemodynamics can be affected by age and physical characteristics, such as height, weight, and fitness [156]. Specifically, cardiac output and stroke volume are smaller, while total peripheral resistance and blood pressure are greater in older women compared with younger women during pregnancy. Moreover, total peripheral resistance is smaller in taller than shorter women, resulting in an increase in stroke volume.

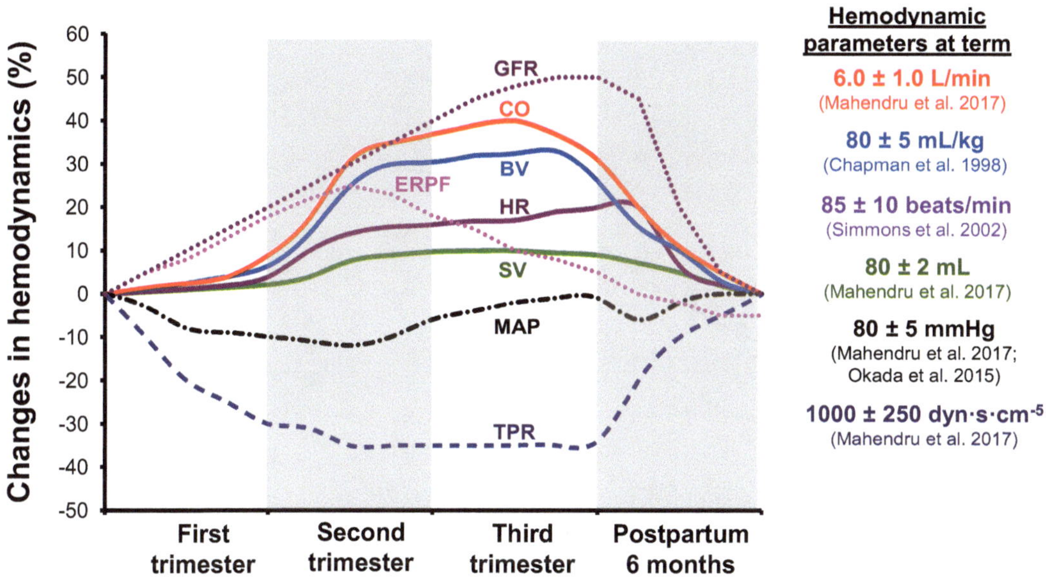

Fig. 26.1 Time course of changes in maternal hemodynamics, blood volume, and renal function during uncomplicated singleton pregnancy in healthy young women. *BV* blood volume, *CO* cardiac output, *ERPF* estimated renal plasma flow, *GFR* glomerular filtration rate, *HR* heart rate, *MAP* mean arterial pressure, *SV* stroke volume, *TPR* total peripheral resistance. (Adapted and reproduced with permission from [92])

There is a rapid reduction in mean arterial pressure of about 6–8 mmHg from preconception to early pregnancy (e.g., ≤8 weeks of gestation), driven predominantly by a decrease in diastolic pressure with minor to no change in systolic pressure [34, 77, 99, 112]. The blood pressure reduction reaches a nadir (by about 10%) in the second trimester [99] and, thereafter, gradually returns to the pre-pregnancy level or even increases in the third trimester due to an increase in cardiac output [34, 69, 99, 109, 112]. Mean arterial pressure decreases by about 5 mmHg shortly after delivery [112] and returns to the pre-pregnancy level by 6 months of postpartum [99].

Heart rate progressively increases during pregnancy and reaches the peak in the third trimester [92, 99, 112]. Compared to pre-pregnancy, the percent increase in heart rate at term is 15–20% [38, 69, 99, 112]. Heart rate decreases after delivery and is about 5–10 beats per minute lower during postpartum compared with the pre-pregnancy level [112].

Cardiac output starts to increase from the fifth week of gestation and reaches the peak by the second to third trimester of pregnancy. The increase in cardiac output is driven by a decrease in total peripheral resistance [34] and an increase in stroke volume in the first trimester, and by both increases in heart rate and stroke volume in the second and third trimesters [29, 44, 64, 74, 107]. Cardiac output increases by about 40% during uncomplicated singleton pregnancy [29, 64, 74, 96, 107, 126, 155] and returns to the pre-pregnancy level as early as 2 weeks of postpartum [74].

Stroke volume gradually increases during pregnancy, reaches the peak in the second trimester by about 8–10%, and remains constant in the third trimester. The increase in stroke volume during pregnancy is a combination of increased preload (due to blood volume expansion and increased cardiac filling) and decreased afterload (due to systemic vasodilation), while myocardial contractility does not seem to change during uncomplicated pregnancy [117] or even decreases progressively and modestly throughout gestation [107]. Stroke volume decreases quickly by about 10% after delivery due to a decrease in preload (blood loss during labor) and an increase in

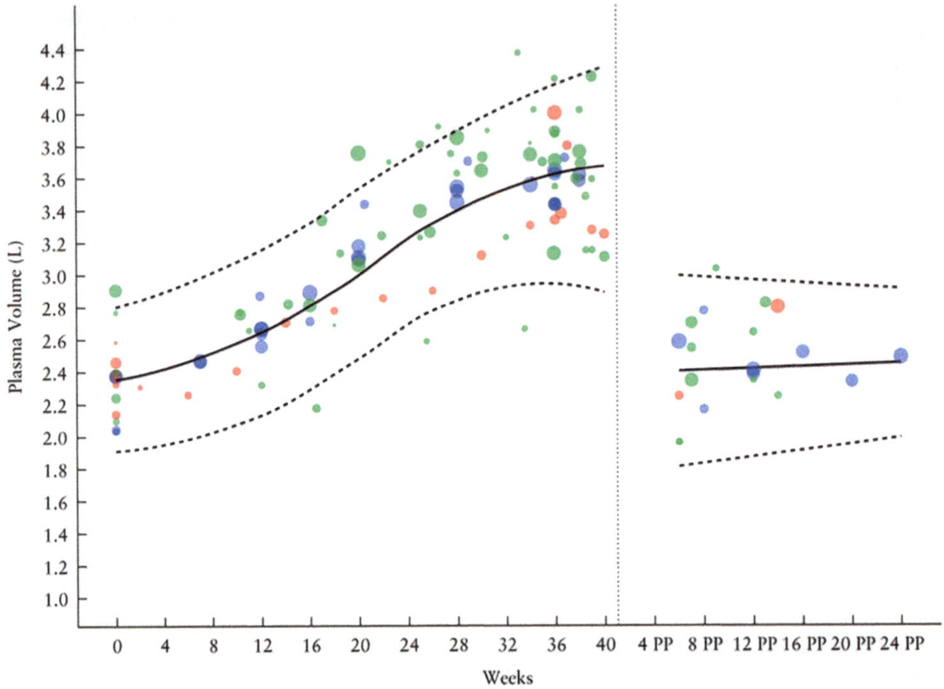

Fig. 26.2 Absolute plasma volume increase during pregnancy (left) and postpartum (PP; right). Size of individual plots indicates sample size of point estimate and their color indicates quality assigned to the study: red, low quality; green, moderate quality; blue, high quality. Curve fit is weighted by inverse variance and plotted against weighted 5th and 95th percentiles. (Reproduced with permission from [43])

afterload, and it returns to the pre-pregnancy level after 2 weeks of postpartum [74].

Total peripheral resistance decreases rapidly by 30–40% during the first trimester, remains low throughout gestation [29, 64, 126], and returns to the pre-pregnancy level 3–6 months of postpartum (by −2%) [64, 126].

Plasma volume starts to increase as early as the first week of gestation even before placental implantation, with the steepest increase occurring during the second trimester (Fig. 26.2) [43, 75]. Plasma volume continues to increase in the third trimester with a pooled maximum increase of 1.13 l (95% CI, 1.07–1.19), an increase of about 45% in uncomplicated singleton pregnancy compared with nonpregnant conditions [43]. Blood volume also increases by approximately 30% during pregnancy (Fig. 26.1), driven primarily by the increase in plasma volume rather than an increase in red blood cell volume, and it returns to the nonpregnant level after 6 weeks of postpartum [43]. Red blood cell mass increases up to 40% via erythropoiesis during uncomplicated pregnancy, which is less compared with the increase in plasma volume, leading to a "physiological anemia" from hemodilution [125]. It has been proposed that pregnancy "resets" both volume and osmoreceptors to a new steady state [46, 147], facilitating salt and water retention to expand blood volume.

Mechanisms for Hemodynamic Adaptations During Pregnancy

Figure 26.3 depicts a schematic overview of the possible mechanisms underlying maternal hemodynamic adaptations during uncomplicated pregnancy. A decrease in total peripheral resistance has been considered as a trigger for changes in maternal hemodynamics during pregnancy [29, 47]. Mean arterial pressure falls due to the

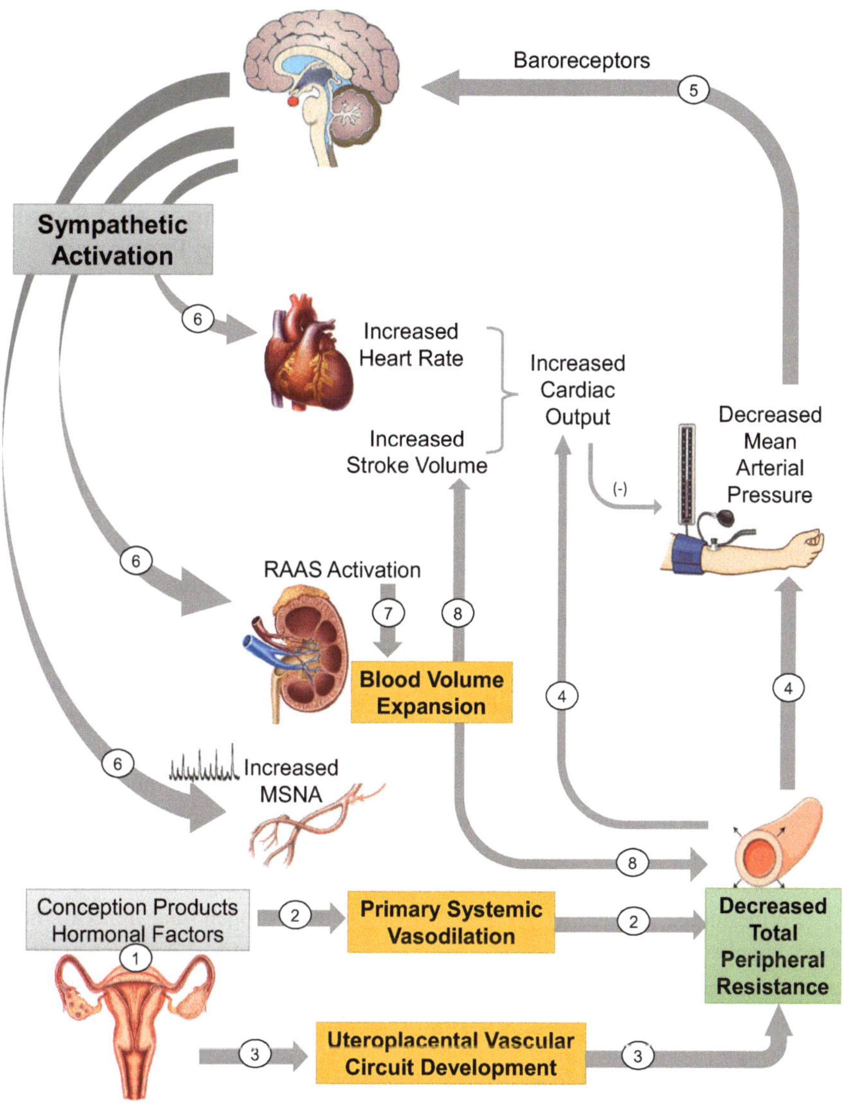

Fig. 26.3 A schematic overview of the possible mechanisms underlying maternal hemodynamic adaptations during uncomplicated pregnancy in humans. *RAAS* the renin-angiotensin-aldosterone system, *MSNA* muscle sympathetic nerve activity

decrease in total peripheral resistance, which activates the sympathetic nervous system and the renin-angiotensin-aldosterone system (RAAS) through unloading of the arterial baroreceptors [32, 33, 147]. There are several explanations for the decrease in total peripheral resistance during pregnancy, which include (1) primary systemic vasodilation, (2) development of a new uteroplacental vascular circuit likened to an arteriovenous fistula, and (3) retention of fluids leading to expansion of blood volume and loading of the cardiopulmonary baroreceptors [58].

Primary systemic vasodilation is supported by the early onset of the decrease in total peripheral resistance, beginning in the first 2 weeks after conception, and therefore, mean arterial pressure decreases early during pregnancy [34]. These changes precede activation of the RAAS and blood volume expansion [116]. The mechanisms responsible for primary system vasodilation in early pregnancy are unclear, but the endothelial release of nitric oxide has been proposed as one major determining factor. Estrogen upregulates nitric oxide synthesis [163], and thus, an increase in circulating estradiol concentration during early pregnancy [77, 112, 148] could stimulate an increase in nitric oxide synthesis [164]. Both estrogen and nitric oxide have profound vasodilator effects, which can decrease total peripheral resistance [5, 71, 132]. Besides estrogen, relaxin also increases during pregnancy [145] (Fig. 26.4 *left panel*). Relaxin is a 6 kDa peptide hormone produced by the corpus luteum and secreted into maternal blood during pregnancy, and its circulating level is higher in the first than in either succeeding trimester [145]. Evidence indicates that relaxin has a vasodilatory effect, which is mediated by its major receptor, the relaxin/insulin-like family peptide 1 receptor in the arteries [39, 111]. In addition, activation of endothelial nitric oxide synthase is also involved in the rapid vasodilatory responses of relaxin [39].

Development of the new uteroplacental vascular circuit is not sufficient to explain the decrease in total peripheral resistance, as much of the decrease occurs outside the uteroplacental circulation [42]. Furthermore, women demonstrate hemodynamic changes qualitatively similar to early pregnancy in the mid-luteal phase of the menstrual cycle when both estrogen and progesterone levels are high [35] and in an exaggerated form during ovarian hyperstimulation [11]. These results suggest that the initial hemodynamic changes of pregnancy could occur early before the development of an intact maternal-fetal-placental unit [58].

Blood volume expansion is attributed to activation of the RAAS [8, 34, 43, 130, 166] via the baroreflex mechanism, which increases the circulatory volume and restores relative vascular under-filling due to profound systemic vasodilation [43]. Additionally, products of conception are thought to be a prerequisite for the expansion of blood volume [34]. Previous studies using oral estradiol and progesterone supplement in postmenopausal women found that both hormones activated the RAAS [70, 78]. In contrast, it was demonstrated that progesterone activated while estrogen inhibited the RAAS [131]. Blood volume expansion causes systemic vasodilation via loading of cardiopulmonary baroreceptors [28].

Hormonal Changes and Renal Function During Pregnancy

Plasma renin activity and aldosterone concentration increase early on during pregnancy [34, 77, 112] (Fig. 26.4 *right panel*), which contribute importantly to salt and water retention and blood volume expansion. Though estrogen increases angiotensinogen or renin substrate, which also activates the RAAS through non-hemodynamic mechanisms [166], plasma renin activity and aldosterone concentration are increased during pregnancy out of proportion to the enhanced substrate availability [8, 34, 130, 166]. Aldosterone concentration continues to rise during the third trimester independent of a change in plasma renin activity [34, 112], suggesting that factors other than hemodynamic activation of the RAAS, such as markedly increased progesterone levels, may be important in the later stages of pregnancy [34].

Hormonal changes associated with pregnancy, especially estrogen and relaxin, have specific renal vasodilating effects presumably mediated by nitric oxide, overriding secondary activation of other renal vasoconstricting systems such as the RAAS [34, 35, 40]. Renal vasodilation results in increases in renal plasma flow and glomerular filtration and a decrease in renal vascular resistance during pregnancy [34, 39, 137]. Serial renal function studies by Sims and Krantz [137] showed that the estimated renal plasma flow was approximately 25% higher than control (nonpregnant) values throughout the first and second trimesters, declined to control values during the third trimester, and then decreased to values significantly below the control values for many months during the postpartum (Fig. 26.1). The glomerular filtration rate was increased progressively by about 50% throughout gestation, returning to the range of control values early in the postpartum (Fig. 26.1) [137]. As a result of the increase in filtration rate, the concentration of

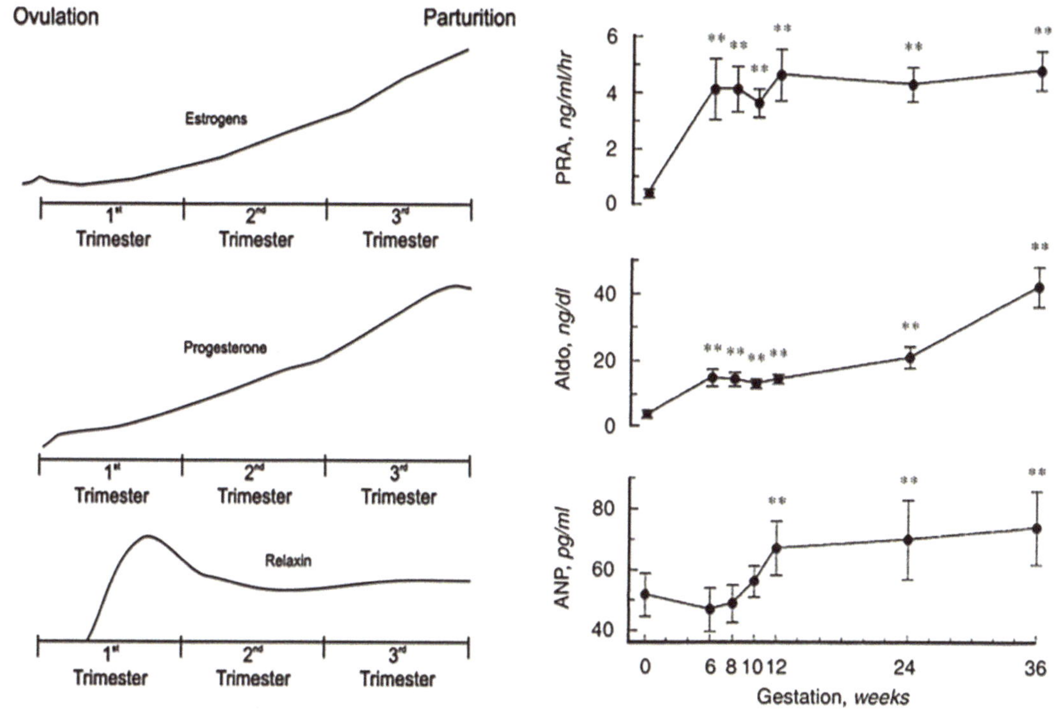

Fig. 26.4 Changes in sex hormones (left) and renal-adrenal hormones (right) during uncomplicated singleton pregnancy. *PRA* plasma renin activity, *Aldo* aldosterone, *ANP* atrial natriuretic peptide. (Adapted and reproduced with permissions from [17, 34])

urea and creatinine in the plasma was reduced to approximately one-half and two-thirds, respectively, of the concentrations in the control values [137].

Alterations in Vascular Function

Arterial Compliance

Global (systemic) arterial compliance was reported to increase by about 30% in the first trimester and remain high thereafter throughout gestation in previous studies [117]. Both conduit and resistant vessels contribute to the increase in global arterial compliance during uncomplicated pregnancy [117]. These changes are adaptive in that they help accommodate the increased intravascular volume while maintaining the efficiency of ventricular-arterial coupling and diastolic perfusion pressure. Reduced smooth muscle tone may be responsible for the increased arterial compliance in pregnancy. It is also possible that vascular remodeling associated with pregnancy accounts for the increase in arterial compliance. Hormonal effects, in particular, estrogen, may be the cause of vascular remodeling. Estrogen receptors have been found in the vascular smooth muscle and endothelium. Increased circulating levels of estrogen during pregnancy causes vasodilation through endothelium-dependent (related to acetylcholine) and endothelium-independent (nitride-responsive) pathways [63, 165]. In addition, increased aortic blood velocities noted early in pregnancy could enhance shear stress, which induces release of endothelial relaxing factors such as nitric oxide and prostaglandin I_2 [117]. However, pulse wave analysis in normal pregnancy showed that augmentation pressure and index, a measure of systemic arterial stiffness, significantly decreased from the first to the second trimester and then rose again in the third trimester [60, 85]. These findings are inconsistent with those reported by Poppas et al.

[117]. Interestingly, all carotid artery elastic parameters indicate significant stiffening from the first trimester to the third trimester, which is reversed after delivery [103]. The carotid artery exhibits regionally specific stiffening during uncomplicated pregnancy, which appears to represent a qualitatively different change in arterial elastic behavior.

Endothelial Function

The peripheral arterial vasculature is maintained in a state of active vasodilation by continuous synthesis of endothelium-derived nitric oxide from L-arginine [153]. An increase in nitric oxide-mediated dilation in the cardiovascular adaptation to pregnancy was reported in rats [37, 41]. Although an enhanced nitric oxide-mediated vasodilation was observed in the hand during uncomplicated pregnancy in humans [164], whether other vascular beds respond in the same way as the hand remains unknown. Flow-mediated dilation, a noninvasive method using high-resolution Doppler ultrasound, is an established approach to assess endothelial function and is highly dependent on the capacity of endothelial cells to release nitric oxide [3]. Cross-sectional studies in healthy young women showed that flow-mediated dilation increased in the third trimester [45, 127, 128, 133]. Only one longitudinal study with a very small sample size ($n = 5$) has been conducted in healthy women, showing that there was a tendency toward a progressive increase in flow-mediated dilation from the first to the third trimester [45]. Thus, longitudinal studies with more women are needed to determine the time course of changes in endothelium-dependent vasodilation during uncomplicated pregnancy in humans.

Venous Capacitance and Compliance

In addition to profound arterial vasodilation, uncomplicated pregnancy is also characterized by venous dilation and increased intravascular volume [124]. Maternal venous system is important in the regulation of intravascular volume and blood pressure. Decreases in venous distensibility shift the intravascular volume from the peripheral to central circulation [124], which could impact maternal hemodynamics. Venous capacitance was found to increase in the second and third trimester of pregnancy [65]. Previous studies showed that venous capacitance in the forearm and leg was similar in pregnant women in the third trimester, while in nonpregnant women venous capacitance was greater in the leg than in the forearm [102]. Furthermore, it was found that venous distensibility or compliance increased, greater in the calf than in the forearm, as pregnancy progressed in healthy young women, and reached maximal values in the third trimester [14, 124, 139, 140]. It is suggested that the increase in limb venous distensibility during uncomplicated pregnancy reflects adaptation to the changes in circulation, such as an increase in plasma and blood volume [124]. In addition, hormonal changes associated with pregnancy may be also responsible for the increase in venous distensibility.

Structural and Functional Adaptations of the Heart

Changes in hemodynamics and hormones during pregnancy can influence the structure and function of the maternal heart [126, 135, 143, 152]. For instance, the increase in cardiac output starting from the first trimester places a volume load on the heart; in addition, hormonal changes, in particular, estrogen and relaxin, may directly or indirectly affect the heart [135]. Pregnancy-induced structural and functional changes of the heart are usually reversed by 3–6 months of postpartum [52, 126, 135, 143].

Physiological Remodeling of the Heart

There are important morphological adaptations of the heart during pregnancy. For example, atrial enlargement begins in the first trimester [122] and reaches the peak in the third trimester around 30 weeks of gestation [126, 152], which is

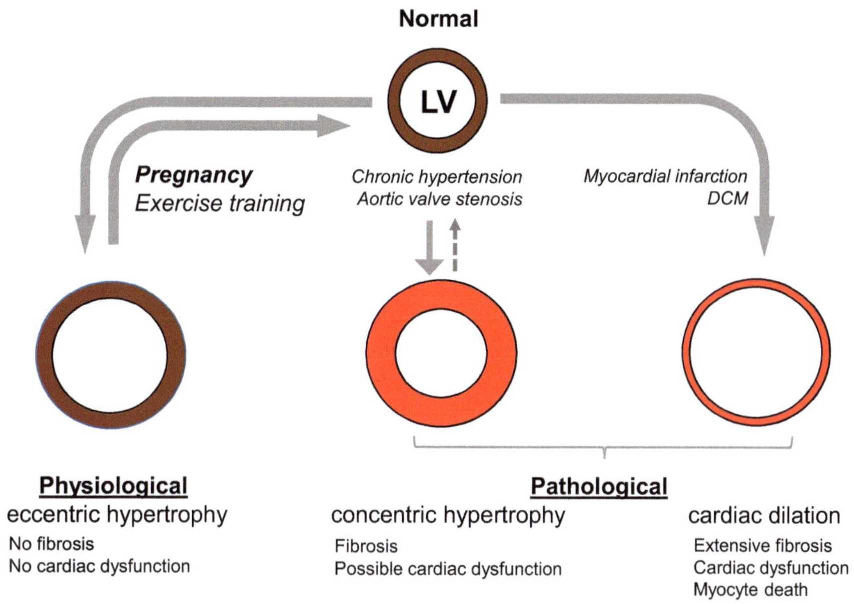

Fig. 26.5 Left ventricular remodeling in the physiological and pathophysiological conditions. *LV* left ventricle, *DCM* dilated cardiomyopathy

associated with blood volume expansion and an increase in preload. The maximum increase in left atrial dimension is about 14% during uncomplicated singleton pregnancy [84]. Left ventricular end-diastolic volume increases by 5–10% during the second trimester [126], while left ventricular mass increases by 30–50% and peaks around the third trimester [114, 117, 120, 121, 126, 169]. Left ventricular wall thickness also increases with a mean increase of 20–30% during pregnancy [62, 96, 120, 121, 135]. There is no significant change in the ratio of left ventricular wall thickness to chamber dimension during uncomplicated pregnancy, which is consistent with *eccentric hypertrophy* in response to increased preload and cardiac stroke work [107, 123, 135, 143] (Fig. 26.5). The heart undergoes remodeling similar to that observed after aerobic exercise training or in athletes [94], and it changes from an ellipsoid to a more spherical shape during uncomplicated pregnancy [10, 101]. Hypertrophy of the left ventricle may cause a decrease in diastolic compliance [104], whereas hormonal changes, such as estrogen- and relaxin-induced increases in nitric oxide bioavailability, may have an opposite effect [115]. These structural adaptations could be reversed by 3–6 months of postpartum [52, 126, 135, 143]. Animal studies demonstrated that pregnancy-induced left ventricular hypertrophy was associated with increased cardiac angiogenesis, lack of fibrosis, and decreased expression of remodeling enzymes that were reversed postpartum [150]. The myocardium can rapidly reserve hypertrophy and remodeling, and there does not appear to be any long-term consequence of uncomplicated pregnancy [135].

Changes in Left Ventricular Function

Changes in geometry and loading of the heart during pregnancy may affect functional evaluation of the left ventricle. Surprisingly, results regarding left ventricular diastolic function in uncomplicated pregnancy are controversial; some studies found worsened diastolic function in the third trimester, while others reported no change or even improved diastolic function [12, 52, 80, 104, 152, 169]. These discrepancies could be explained by the different methodology used, differences in experimental design (e.g., cross-sectional versus longitudinal), and/or heterogeneity of the study population. Prolongations of mitral E wave duration and deceleration time were found during uncomplicated pregnancy, which is consistent with increased passive filling of the left ventricle during early diastole [135].

Results regarding left ventricular systolic function during uncomplicated pregnancy are also controversial [12, 44, 52, 107, 117, 169]. Ejection-phase indices of left ventricular function, including systolic fractional shortening and mean velocity of circumferential fiber thickening, were reported to increase [89], remain unchanged [84], or decrease [120, 121] during pregnancy. However, stroke work – an indication of global ventricular performance [9] – was found to increase significantly during uncomplicated pregnancy [126].

Pregnancy and Cardiac Electrophysiology

The physiological left ventricular hypertrophy, changes in hormones associated with pregnancy, and a displacement of the heart due to the expanding uterus may lead to alterations in cardiac electrical activity, which could be detected in the electrocardiogram that are not related to disease. The most consistent electrocardiogram changes during uncomplicated pregnancy include a decrease in the R-R interval, an increase in heart rate, and a decrease in heart rate variability [15, 17, 30, 108, 141, 158, 161]. The underlying mechanisms may be pregnancy-induced sympathetic activation and vagal withdrawal, as well as intrinsic remodeling of the ion channels that control cardiac pacemaking [17, 50]. Changes in the electrocardiogram can be detected early in pregnancy.

Cross-sectional studies have demonstrated that in pregnant women the absolute value of the QT interval in the first trimester is shorter compared with nonpregnant women, and this observation sustains throughout gestation [15, 30, 90]. However, the QT interval is dependent on the heart rate or R-R interval; when the QT interval is corrected for the heart rate (QT_c), it is significantly longer (but still within the normal range) in pregnant women than in nonpregnant women [1, 90, 146]. The lengthened QT_c interval may be attributed to pregnancy-induced changes in sex hormones, in particular, the progesterone-to-estrogen ratio [2]. T wave dispersion (e.g., increases in width and amplitude) has also been found during uncomplicated pregnancy [7, 17] (Fig. 26.6), which may be due to differences in repolarization of the endocardial, mid-myocardial, and epicardial regions [7]. It is also possible that dispersion may reflect heterogeneity in repolarization from apex to base [36, 66, 83], which may be modulated by sympathetic activity [17].

There is an increased susceptibility to cardiac arrhythmias during pregnancy and postpartum [17, 66]. Sinus tachycardia is common, especially in the third trimester [4]. Ectopic beats and nonsustained arrhythmia are encountered in more than 50% of pregnant women investigated for palpitations [4]. Maternal premature ventricular contractions and maternal premature atrial contractions are also common in pregnancy [54, 66]. Changes in ion channel expression and behavior may be responsible for the changes in the electrocardiogram during pregnancy, but the actual role of ion channels in electrical remodeling remains unknown [17]. Sodium channels and changes in calcium homoeostasis may also contribute to cardiac electrical remodeling.

Autonomic Circulatory Control in Pregnancy

Maternal hemodynamic adaptations during pregnancy occur through the autonomic nervous system [49, 58], but the actual role of sympathetic neural control in uncomplicated pregnancy remains largely unclear.

Vasomotor Sympathetic Activity

Earlier human research on vasomotor sympathetic activity during pregnancy focused exclusively on plasma norepinephrine concentration, which ranged from increased to decreased compared

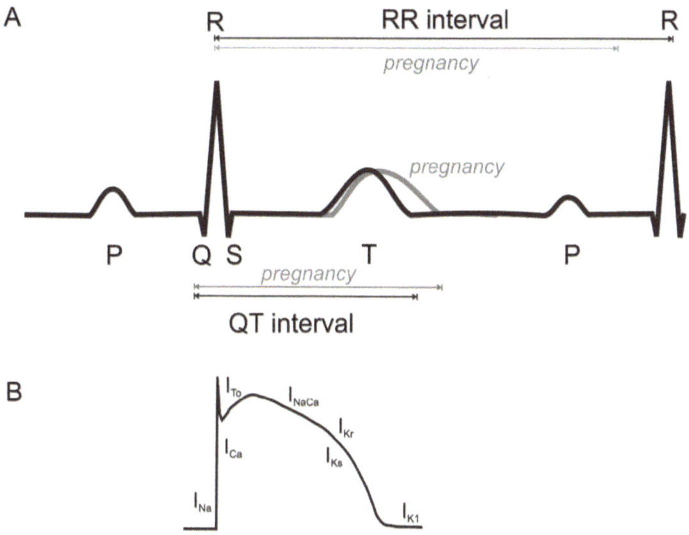

Fig. 26.6 Changes in the electrocardiogram during uncomplicated singleton pregnancy in humans. (**A**) The major deflections of the human electrocardiogram. The P wave is associated with atrial depolarization. The QRS complex is the result of ventricular depolarization, and the T-wave repolarization. The shape, duration and timing of the T wave are affected by heterogeneity of repolarization of the ventricle. In pregnancy, the RR duration is reduced, i.e., heart rate increases. In addition, the QT interval increases, as does T-wave dispersion. (**B**) Human ventricular cellular action potentials. The rapid upstroke of cellular action potentials is well co-ordinated across the ventricle, leading to the short QRS duration. Conversely, the repolarization of ventricular action potentials varies across the ventricle, leading to the broad T wave. The dominant ion channels during various parts of the action potential are indicated. (Reproduced with permission from [17])

with that of nonpregnant women [13, 34, 149, 170]. Plasma norepinephrine concentration is an insensitive measure of vasomotor sympathetic activity, since it can be affected by many factors, such as efferent nerve discharges, synaptic transmitter release, reuptake mechanisms, clearance, regional blood flow, and/or plasma volume [57]. In contrast, the microneurographic technique for obtaining direct intraneural recordings of postganglionic efferent nerve activity allows a precise, quantitative, and reproducible assessment of sympathetic neural vasoconstrictor activity [154, 160] (Fig. 26.7). It has been demonstrated that sympathetic outflow to the skeletal muscle is similar to that of the kidney and the heart under physiologic conditions [81, 82].

Microneurographic studies found that vasomotor sympathetic activity increased early on (even within the first week of conception) during uncomplicated pregnancy [77], continued to increase throughout gestation, and returned to the pre-pregnancy level within 10 weeks of postpartum [67, 112, 151]. These results suggest that sympathetic activation is a common phenomenon during pregnancy in humans. Marked sympathetic activation during uncomplicated pregnancy may be a compensatory mechanism induced by profound systemic vasodilation and the decrease in mean arterial pressure. Furthermore, animal studies have shown that estrogen receptors are located in the brain centers such as the nucleus tractus solitarius, the ventrolateral medulla, and the area postrema [136]. The elevated circulating estrogen concentration during pregnancy may increase sympathetic outflow through a central mechanism [72]. It is also possible that sympathetic activation is a reflex response to the presence of the uterine "shunt." There are other circumstances where arteriovenous anastomoses have been documented to cause sympathetic activation, such as cirrhosis [56]. However, the uterine "shunt" may not be obvious

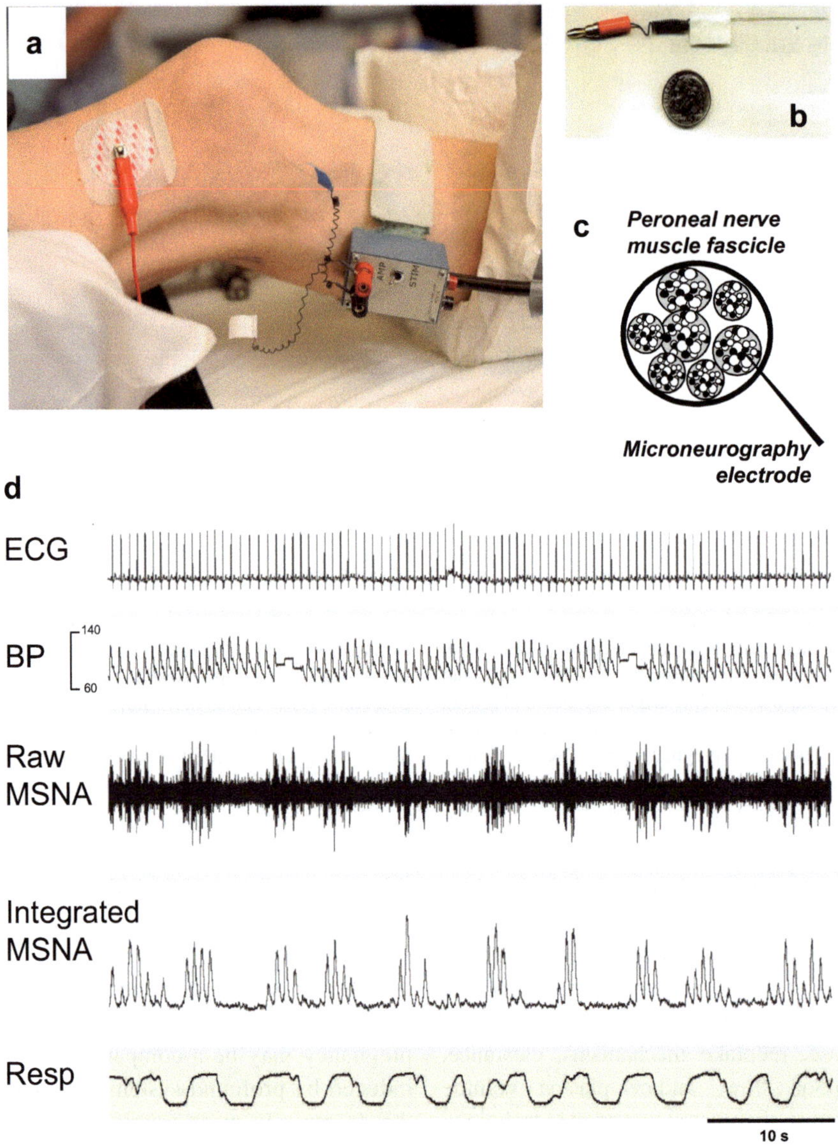

Fig. 26.7 The microneurographic technique for assessment of postganglionic efferent sympathetic nerve discharge (action potential) to the skeletal muscle vasculature in humans. (**a**) A picture showing the microneurographic recording in humans. The preamplifier is attached to the calf; the reference electrode (with a blue flag) is inserted into the skin a few centimeters away from the recording electrode (with a white flag) that is inserted into the peroneal nerve. (**b**) The size of the electrode. (**c**) Demonstration of a tungsten microelectrode inserted into a muscle fascicle of the peroneal nerve. (**d**) Representative tracings of electrocardiogram (ECG), blood pressure (BP), original signals filtered at 700–2000 Hz (Raw) and integrated multiunit muscle sympathetic nerve activity (MSNA), and respiration (Resp) from one human subject

during very early pregnancy in humans. Animal research suggests that angiotensin II may contribute to about 50% sympathetic activity in transgenic mice positive for human renin and angiotensinogen [95]. Indeed, it was found that pregnancy-induced increase in vasomotor sympathetic activity was positively correlated to direct renin [77]. Finally, it is possible that pregnancy-induced increase in aldosterone concentration also contributes to sympathetic activation. In support of this notion, a positive relationship between the change in muscle sympathetic nerve activity and the change in aldosterone from pre-pregnancy to early pregnancy was reported in healthy young women [77].

Previous investigations showed that vasomotor sympathetic activity increased in women with uncomplicated pregnancy and was even greater in hypertensive pregnant women during the third trimester [67, 68]. It has been proposed that the marked sympathetic activation during the latter months of normal pregnancy helps to return the arterial pressure to the pre-pregnancy level, but when the increase in vasomotor sympathetic activity is excessive, hypertension ensues. This notion seems to be supported by the findings of Schobel et al. [129] and Fischer et al. [55] showing that gestational hypertension and preeclampsia are in a state of sympathetic overactivity, which normalizes after delivery.

Vasoconstrictor Responsiveness

Despite marked sympathetic activation, total peripheral resistance remains low during uncomplicated pregnancy [29, 64, 126]. Animal studies demonstrated that normal pregnancy was associated with a reduced vasoconstrictor response to angiotensin II, noradrenaline, and vasopressin [86, 113, 142]. In humans, reduced vascular responsiveness to α-adrenergic stimulation [91, 110] and angiotensin II infusion [16, 97] has been reported during uncomplicated pregnancy. The reduced vasoconstrictor responsiveness during pregnancy may be explained by an increase in nitric oxide bioavailability. A previous study in rats showed that chronic administration of NG-nitro-L-arginine methyl ester, a nitric oxide synthase inhibitor, during the second to third trimester led to an increase in vascular reactivity [86]. Indeed, resistance to the pressor effects of angiotensin II and sympathetic adrenergic stimulation during normal pregnancy is compatible with an increase in endothelial nitric oxide synthase activity [147]. Additionally, angiotensin receptors may play a role in vascular reactivity during pregnancy [142]. Angiotensin II binds to both angiotensin type I receptors and angiotensin type II receptors, which are associated with vasoconstriction and vasodilatation, respectively. When an angiotensin type II receptor blockade was administered to rats, it produced vascular reactivity similar to the nonpregnant control group [142], indicating that receptor density and/or sensitivity of angiotensin type II receptors is enhanced during pregnancy, resulting in a net vasodilatory effect. Finally, prostacyclin I_2 attenuates the pressor response to angiotensin II in uncomplicated pregnancy [27]. Various pregnancy-associated tissues, such as human myometrium, trophoblast, and placenta, synthesize prostacyclin in vitro. Prostacyclin is an extremely potent vasodilator, possibly acting as a circulating and a local hormone, and in its opposing action to angiotensin II. Indeed, prostacyclin is a primary regulator of arteriolar tone [27].

Baroreflex Function

The baroreflexes are important mechanisms for the overall regulation of arterial blood pressure in humans [59, 79, 144]. Abrupt decreases and increases in systolic pressure produce shortening and lengthening of the R-R interval, which is defined as the cardio-vagal baroreflex [48]. Rapid reductions and elevations in diastolic pressure elicit excitation and inhibition of vasomotor sympathetic activity, which is known as the sympathetic baroreflex [144, 159].

Results regarding the cardio-vagal baroreflex during uncomplicated pregnancy are quite controversial; increased [88, 91, 168], unchanged [6, 51, 138], or decreased [18, 25, 26, 93, 105, 118, 134, 141, 157] sensitivity has been reported. The discrepancies among studies may be attributable to differences in methodologies implemented to evaluate the baroreflex and/or different stages of gestation evaluated. A recent case study in one woman before, during, and immediately after her uncomplicated pregnancy showed that cardio-vagal baroreflex sensitivity peaked in the first trimester when compared with the pre-pregnancy level and progressively decreased throughout the second and third trimesters [73]. In contrast, studies in women with gestational hypertension and preeclampsia consistently reported impaired cardio-vagal baroreflex sensitivity [31, 53, 106, 119, 134, 162].

Surprisingly, there is little information available about the sympathetic baroreflex during uncomplicated pregnancy in humans. One cross-sectional study found that sympathetic baroreflex sensitivity was reduced in pregnant women during the third trimester relative to nonpregnant controls despite that blood pressure was similar between the groups [151]. However, a longitudinal case study involving a series of repeated measures over time showed that although sympathetic baroreflex sensitivity was variable throughout pregnancy, the majority of the values indicated a greater sensitivity in pregnancy compared with the pre-pregnancy level [73]. These preliminary observations need to be confirmed in more women during pregnancy.

Race/Ethnic Differences in Maternal Hemodynamics

Maternal hemodynamic adaptations to pregnancy can be affected by race or ethnicity. For example, it was found that African American women exhibited statistically significant higher blood pressure readings (but within normal ranges) throughout pregnancy compared with other ethnic groups [98]. African American women have a higher rate of overweight and obesity, which may put additional strain on their cardiovascular system [98] and increase their risk factor for gestational hypertensive disorders [61]. In addition, it was found that serum and tissue creatine kinase activity was higher in African Americans than in other ethnic groups [20–22, 24], while creatine kinase is associated with blood pressure in the general population with a 14 mmHg increase in systolic pressure per log creatine kinase increase. High creatine kinase might enhance contractility through greater adenosine triphosphate-buffering capacity at myosin ATPase, but also lower nitric oxide bioavailability because of a greater creatine demand with high creatine kinase [76]. Compared with Caucasians, African American women display greater resistance artery contractility to norepinephrine during uncomplicated pregnancy and evidence of higher vascular creatine kinase activity with attenuated nitric oxide synthesis [23]. Conversely, Asian women were found to have lower mean arterial pressure, less sympathetic activation, but more upregulated renal-adrenal responses compared with Caucasian women during uncomplicated pregnancy [112]. These results provide insights into the pathophysiological mechanisms for racial differences in the prevalence of gestational hypertension and preeclampsia.

The Generation R Study in the Netherlands showed that Dutch pregnant women had higher systolic blood pressure levels as compared with women in other ethnic groups in each trimester of pregnancy [19]. Compared with Dutch women, Turkish and Moroccan women had lower diastolic blood pressure levels in each trimester. These differences remain after adjusting for education and lifestyle factors. Similarly, another Dutch study among 2413 pregnant women found Dutch women to have the highest systolic blood pressure levels in pregnancy and the lowest diastolic blood pressure levels for Mediterranean women [87]. One possible explanation might be that the maternal cardiovascular system of the various ethnic groups adapts differently during pregnancy because of genetic differences or different environmental factors [19]. Some of these ethnic differences may be attributed to differences in socioeconomic status or in prevalence rates of cardiovascular risk factors, such as obesity, smoking, or physical inactivity [87, 167].

Summary

There are dramatic alterations in maternal hemodynamics during uncomplicated pregnancy, which occur as early as 4 weeks of gestation. These alterations are triggered by profound systemic vasodilation and are mediated through the autonomic control mechanisms (though the actual role of the sympathetic nervous system in pregnancy remains to be elucidated). In addition, blood volume expansion and hormonal changes associated with pregnancy also contribute importantly to maternal hemodynamic adaptations. Pregnancy-induced hemodynamic and electrocardiographic changes are usually reversed by 3–6 months of postpartum. Understanding the physiology of uncomplicated pregnancy provides a basis for identifying inadequate or compromised hemodynamic responses in the first or second trimester, which may offer opportunities of prediction, prevention, and/or early treatment for adverse maternal and fetal outcomes.

Acknowledgments The author would like to thank Dr. Satyam Sarma for his constructive comments and suggestions on this chapter.

References

1. Task Force of the European Society of Cardiology and the North American Society of Pacing and Electrophysiology. Heart rate variability: standards of measurement, physiological interpretation and clinical use. Circulation. 1996;93:1043–65.
2. Abehsira G, Bachelot A, Badilini F, Koehl L, Lebot M, Favet C, Touraine P, Funck-Brentano C, Salem JE. Complex influence of gonadotropins and sex steroid hormones on QT interval duration. J Clin Endocrinol Metab. 2016;101:2776–84.
3. Accini JL, Sotomayor A, Trujillo F, Barrera JG, Bautista L, Lopez-Jaramillo P. Colombian study to assess the use of noninvasive determination of endothelium-mediated vasodilatation (CANDEV). Normal values and factors associated. Endothelium. 2001;8:157–66.
4. Adamson DL, Nelson-Piercy C. Managing palpitations and arrhythmias during pregnancy. Heart. 2007;93:1630–6.
5. Anderson RJ, Berl T, McDonald KM, Schrier RW. Prostaglandins: effects on blood pressure, renal blood flow, sodium and water excretion. Kidney Int. 1976;10:205–15.
6. Antonazzo P, Cetin I, Tarricone D, Lombardi F, Pardi G. Cardiac autonomic modulation in normal, high-risk, and in vitro fertilization pregnancies during the first trimester. Am J Obstet Gynecol. 2004;190:199–205.
7. Antzelevitch C, Shimizu W, Yan GX, Sicouri S. Cellular basis for QT dispersion. J Electrocardiol. 1998;30 (30 Suppl):168–75.
8. August P, Lenz T, Ales KL, Druzin ML, Edersheim TG, Hutson JM, Muller FB, Laragh JH, Sealey JE. Longitudinal study of the renin-angiotensin-aldosterone system in hypertensive pregnant women: deviations related to the development of superimposed preeclampsia. Am J Obstet Gynecol. 1990;163:1612–21.
9. Aurigemma GP, Zile MR, Gaasch WH. Contractile behavior of the left ventricle in diastolic heart failure: with emphasis on regional systolic function. Circulation. 2006;113:296–304.
10. Badke FR, Covell JW. Early changes in left ventricular regional dimensions and function during chronic volume overloading in the conscious dog. Circ Res. 1979;45:420–8.
11. Balasch J, Arroyo V, Carmona F, Llach J, Jimenez W, Pare JC, Vanrell JA. Severe ovarian hyperstimulation syndrome: role of peripheral vasodilation. Fertil Steril. 1991;56:1077–83.
12. Bamfo JE, Kametas NA, Nicolaides KH, Chambers JB. Maternal left ventricular diastolic and systolic long-axis function during normal pregnancy. Eur J Echocardiogr. 2007;8:360–8.
13. Barron WM, Mujais SK, Zinaman M, Bravo EL, Lindheimer MD. Plasma catecholamine responses to physiologic stimuli in normal human pregnancy. Am J Obstet Gynecol. 1986;154:80–4.
14. Barwin BN, Roddie IC. Venous distensibility during pregnancy determined by graded venous congestion. Am J Obstet Gynecol. 1976;125:921–3.
15. Baumert M, Javorka M, Seeck A, Faber R, Sanders P, Voss A. Multiscale entropy and detrended fluctuation analysis of QT interval and heart rate variability during normal pregnancy. Comput Biol Med. 2012;42:347–52.
16. Benjamin N, Rymer J, Todd SD, Thom M, Ritter JM. Sensitivity to angiotensin II of forearm resistance vessels in pregnancy. Br J Clin Pharmacol. 1991;32:523–5.
17. Bett GC. Hormones and sex differences: changes in cardiac electrophysiology with pregnancy. Clin Sci (Lond). 2016;130:747–59.
18. Blake MJ, Martin A, Manktelow BN, Armstrong C, Halligan AW, Panerai RB, Potter JF. Changes in baroreceptor sensitivity for heart rate during normotensive pregnancy and the puerperium. Clin Sci (Lond). 2000;98:259–68.
19. Bouthoorn SH, Gaillard R, Steegers EA, Hofman A, Jaddoe VW, van Lenthe FJ, Raat H. Ethnic differences in blood pressure and hypertensive complications during pregnancy: the generation R study. Hypertension. 2012;60:198–205.
20. Brewster LM, Clark JF, van Montfrans GA. Is greater tissue activity of creatine kinase the genetic factor increasing hypertension risk in black people of sub-Saharan African descent? J Hypertens. 2000;18:1537–44.
21. Brewster LM, Coronel CM, Sluiter W, Clark JF, van Montfrans GA. Ethnic differences in tissue creatine kinase activity: an observational study. PLoS One. 2012;7:e32471.
22. Brewster LM, Mairuhu G, Bindraban NR, Koopmans RP, Clark JF, van Montfrans GA. Creatine kinase activity is associated with blood pressure. Circulation. 2006;114:2034–9.
23. Brewster LM, Taherzadeh Z, Volger S, Clark JF, Rolf T, Wolf H, Vanbavel E, van Montfrans GA. Ethnic differences in resistance artery contractility of normotensive pregnant women. Am J Physiol Heart Circ Physiol. 2010;299:H431–6.
24. Brewster LM, van Bree S, Reijneveld JC, Notermans NC, Verschuren WM, Clark JF, van Montfrans GA, de Visser M. Hypertension risk in idiopathic hyperCKemia. J Neurol. 2008;255:11–5.
25. Brooks VL, Kane CM, Van Winkle DM. Altered heart rate baroreflex during pregnancy: role of sympathetic and parasympathetic nervous systems. Am J Phys. 1997;273:R960–6.
26. Brooks VL, Keil LC. Changes in the baroreflex during pregnancy in conscious dogs: heart rate and hormonal responses. Endocrinology. 1994;135:1894–901.
27. Broughton Pipkin F, Morrison R, O'Brien PM. Prostacyclin attenuates both the pressor and adrenocortical response to angiotensin II in human pregnancy. Clin Sci (Lond). 1989;76:529–34.
28. Bundgaard-Nielsen M, Wilson TE, Seifert T, Secher NH, Crandall CG. Effect of volume loading on the Frank-Starling

relation during reductions in central blood volume in heat-stressed humans. J Physiol. 2010;588:3333–9.
29. Capeless EL, Clapp JF. Cardiovascular changes in early phase of pregnancy. Am J Obstet Gynecol. 1989;161:1449–53.
30. Carpenter RE, D'Silva LA, Emery SJ, Uzun O, Rassi D, Lewis MJ. Changes in heart rate variability and QT variability during the first trimester of pregnancy. Physiol Meas. 2015;36:531–45.
31. Cat Genova G, Veglio F, Rabbia F, Milan A, Grosso T, Chiandussi L. Baroreflex sensitivity in secondary hypertension. Clin Exp Hypertens. 2001;23:89–99.
32. Chapleau MW, Hajduczok G, Sharma RV, Wachtel RE, Cunningham JT, Sullivan MJ, Abboud FM. Mechanisms of baroreceptor activation. Clin Exp Hypertens. 1995;17:1–13.
33. Chapleau MW, Heesch CM, Abboud FM. Prevention or attenuation of baroreceptor resetting by pulsatility during elevated pressure. Hypertension. 1987;9:III137–41.
34. Chapman AB, Abraham WT, Zamudio S, Coffin C, Merouani A, Young D, Johnson A, Osorio F, Goldberg C, Moore LG, Dahms T, Schrier RW. Temporal relationships between hormonal and hemodynamic changes in early human pregnancy. Kidney Int. 1998;54:2056–63.
35. Chapman AB, Zamudio S, Woodmansee W, Merouani A, Osorio F, Johnson A, Moore LG, Dahms T, Coffin C, Abraham WT, Schrier RW. Systemic and renal hemodynamic changes in the luteal phase of the menstrual cycle mimic early pregnancy. Am J Phys. 1997;273:F777–82.
36. Choi BR, Burton F, Salama G. Cytosolic Ca2+ triggers early afterdepolarizations and Torsade de Pointes in rabbit hearts with type 2 long QT syndrome. J Physiol. 2002;543:615–31.
37. Chu ZM, Beilin LJ. Mechanisms of vasodilatation in pregnancy: studies of the role of prostaglandins and nitric-oxide in changes of vascular reactivity in the in situ blood perfused mesentery of pregnant rats. Br J Pharmacol. 1993;109:322–9.
38. Clapp JF 3rd, Capeless E. Cardiovascular function before, during, and after the first and subsequent pregnancies. Am J Cardiol. 1997;80:1469–73.
39. Conrad KP. Emerging role of relaxin in the maternal adaptations to normal pregnancy: implications for preeclampsia. Semin Nephrol. 2011;31:15–32.
40. Conrad KP, Davison JM. The renal circulation in normal pregnancy and preeclampsia: is there a place for relaxin? Am J Physiol Renal Physiol. 2014;306:F1121–35.
41. Conrad KP, Joffe GM, Kruszyna H, Kruszyna R, Rochelle LG, Smith RP, Chavez JE, Mosher MD. Identification of increased nitric oxide biosynthesis during pregnancy in rats. FASEB J. 1993;7:566–71.
42. Curran-Everett D, Morris KG Jr, Moore LG. Regional circulatory contributions to increased systemic vascular conductance of pregnancy. Am J Phys. 1991;261:H1842–7.
43. de Haas S, Ghossein-Doha C, van Kuijk SM, van Drongelen J, Spaanderman ME. Physiological adaptation of maternal plasma volume during pregnancy: a systematic review and meta-analysis. Ultrasound Obstet Gynecol. 2017;49:177–87.
44. Desai DK, Moodley J, Naidoo DP. Echocardiographic assessment of cardiovascular hemodynamics in normal pregnancy. Obstet Gynecol. 2004;104:20–9.
45. Dorup I, Skajaa K, Sorensen KE. Normal pregnancy is associated with enhanced endothelium-dependent flow-mediated vasodilation. Am J Phys. 1999;276:H821–5.
46. Durr JA, Stamoutsos B, Lindheimer MD. Osmoregulation during pregnancy in the rat. Evidence for resetting of the threshold for vasopressin secretion during gestation J Clin Invest. 1981;68:337–46.
47. Duvekot JJ, Cheriex EC, Pieters FA, Menheere PP, Peeters LH. Early pregnancy changes in hemodynamics and volume homeostasis are consecutive adjustments triggered by a primary fall in systemic vascular tone. Am J Obstet Gynecol. 1993;169:1382–92.
48. Eckberg DL, Sleight P. Human baroreflex in health and disease. New York Oxford: Oxford University Press; 1992.
49. Ekholm EM, Piha SJ, Erkkola RU, Antila KJ. Autonomic cardiovascular reflexes in pregnancy. A longitudinal study. Clin Auton Res. 1994;4:161–5.
50. El Khoury N, Mathieu S, Marger L, Ross J, El Gebeily G, Ethier N, Fiset C. Upregulation of the hyperpolarization-activated current increases pacemaker activity of the sinoatrial node and heart rate during pregnancy in mice. Circulation. 2013;127(20):2009.
51. Eneroth-Grimfors E, Westgren M, Ericson M, Ihrman-Sandahl C, Lindblad LE. Autonomic cardiovascular control in normal and pre-eclamptic pregnancy. Acta Obstet Gynecol Scand. 1994;73:680–4.
52. Estensen ME, Beitnes JO, Grindheim G, Aaberge L, Smiseth OA, Henriksen T, Aakhus S. Altered maternal left ventricular contractility and function during normal pregnancy. Ultrasound Obstet Gynecol. 2013;41:659–66.
53. Faber R, Baumert M, Stepan H, Wessel N, Voss A, Walther T. Baroreflex sensitivity, heart rate, and blood pressure variability in hypertensive pregnancy disorders. J Hum Hypertens. 2004;18:707–12.
54. Ferrero S, Colombo BM, Ragni N. Maternal arrhythmias during pregnancy. Arch Gynecol Obstet. 2004;269:244–53.
55. Fischer T, Schobel HP, Frank H, Andreae M, Schneider KT, Heusser K. Pregnancy-induced sympathetic overactivity: a precursor of preeclampsia. Eur J Clin Investig. 2004;34:443–8.
56. Floras JS, Legault L, Morali GA, Hara K, Blendis LM. Increased sympathetic outflow in cirrhosis and ascites: direct evidence from intraneural recordings. Ann Intern Med. 1991;114:373–80.
57. Folkow B, Di Bona GF, Hjemdahl P, Toren PH, Wallin BG. Measurements of plasma norepinephrine concentrations in human primary hypertension. A word of caution on their applicability for assessing neurogenic contributions. Hypertension. 1983;5:399–403.
58. Fu Q, Levine BD. Autonomic circulatory control during pregnancy in humans. Semin Reprod Med. 2009;27:330–7.
59. Fu Q, Witkowski S, Levine BD. Vasoconstrictor reserve and sympathetic neural control of orthostasis. Circulation. 2004;110:2931–7.
60. Fujime M, Tomimatsu T, Okaue Y, Koyama S, Kanagawa T, Taniguchi T, Kimura T. Central aortic blood pressure and augmentation index during normal pregnancy. Hypertens Res. 2012;35:633–8.
61. Gaillard R, Steegers EA, Hofman A, Jaddoe VW. Associations of maternal obesity with blood pressure and the risks of gestational hypertensive disorders. The generation R study. J Hypertens. 2011;29:937–44.
62. Geva T, Mauer MB, Striker L, Kirshon B, Pivarnik JM. Effects of physiologic load of pregnancy on left ventricular contractility and remodeling. Am Heart J. 1997;133:53–9.
63. Gilligan DM, Badar DM, Panza JA, Quyyumi AA, Cannon RO 3rd. Effects of estrogen replacement therapy on peripheral vasomotor function in postmenopausal women. Am J Cardiol. 1995;75:264–8.
64. Gilson GJ, Samaan S, Crawford MH, Qualls CR, Curet LB. Changes in hemodynamics, ventricular remodeling,

and ventricular contractility during normal pregnancy: a longitudinal study. Obstet Gynecol. 1997;89:957–62.
65. Goodlin RC. Venous reactivity and pregnancy abnormalities. Acta Obstet Gynecol Scand. 1986;65:345–8.
66. Gowda RM, Khan IA, Mehta NJ, Vasavada BC, Sacchi TJ. Cardiac arrhythmias in pregnancy: clinical and therapeutic considerations. Int J Cardiol. 2003;88:129–33.
67. Greenwood JP, Scott EM, Stoker JB, Walker JJ, Mary DA. Sympathetic neural mechanisms in normal and hypertensive pregnancy in humans. Circulation. 2001;104:2200–4.
68. Greenwood JP, Stoker JB, Walker JJ, Mary DA. Sympathetic nerve discharge in normal pregnancy and pregnancy-induced hypertension. J Hypertens. 1998;16:617–24.
69. Grindheim G, Estensen ME, Langesaeter E, Rosseland LA, Toska K. Changes in blood pressure during healthy pregnancy: a longitudinal cohort study. J Hypertens. 2012;30:342–50.
70. Hassager C, Riis BJ, Strom V, Guyene TT, Christiansen C. The long-term effect of oral and percutaneous estradiol on plasma renin substrate and blood pressure. Circulation. 1987;76:753–8.
71. Haynes WG, Noon JP, Walker BR, Webb DJ. Inhibition of nitric oxide synthesis increases blood pressure in healthy humans. J Hypertens. 1993;11:1375–80.
72. He XR, Wang W, Crofton JT, Share L. Effects of 17beta-estradiol on the baroreflex control of sympathetic activity in conscious ovariectomized rats. Am J Phys. 1999;277: R493–8.
73. Hissen SL, El Sayed K, Macefield VG, Brown R, Taylor CE. Muscle sympathetic nerve activity peaks in the first trimester in healthy pregnancy: a longitudinal case study. Clin Auton Res. 2017;27:401–6.
74. Hunter S, Robson SC. Adaptation of the maternal heart in pregnancy. Br Heart J. 1992;68:540–3.
75. Hytten FE, Paintin DB. Increase in plasma volume during normal pregnancy. J Obstet Gynaecol Br Emp. 1963;70:402–7.
76. Iyengar MR. Creatine kinase as an intracellular regulator. J Muscle Res Cell Motil. 1984;5:527–34.
77. Jarvis SS, Shibata S, Bivens TB, Okada Y, Casey BM, Levine BD, Fu Q. Sympathetic activation during early pregnancy in humans. J Physiol. 2012;590:3535–43.
78. Jespersen CM, Arnung K, Hagen C, Hilden T, Nielsen F, Nielsen MD, Giese J. Effects of natural oestrogen therapy on blood pressure and renin-angiotensin system in normotensive and hypertensive menopausal women. J Hypertens. 1983;1 (4):361.
79. Johnson JM, Rowell LB, Niederberger M, Eisman MM. Human splanchnic and forearm vasoconstrictor responses to reductions of right atrial and aortic pressures. Circ Res. 1974;34:515–24.
80. Kametas NA, McAuliffe F, Hancock J, Chambers J, Nicolaides KH. Maternal left ventricular mass and diastolic function during pregnancy. Ultrasound Obstet Gynecol. 2001;18:460–6.
81. Kamiya A, Kawada T, Mizuno M, Miyamoto T, Uemura K, Seki K, Shimizu S, Sugimachi M. Baroreflex increases correlation and coherence of muscle sympathetic nerve activity (SNA) with renal and cardiac SNAs. J Physiol Sci. 2006;56:325–33.
82. Kamiya A, Kawada T, Yamamoto K, Michikami D, Ariumi H, Miyamoto T, Shimizu S, Uemura K, Aiba T, Sunagawa K, Sugimachi M. Dynamic and static baroreflex control of muscle sympathetic nerve activity (SNA) parallels that of renal and cardiac SNA during physiological change in pressure. Am J Physiol Heart Circ Physiol. 2005;289: H2641–8.
83. Kanai A, Salama G. Optical mapping reveals that repolarization spreads anisotropically and is guided by fiber orientation in Guinea pig hearts. Circ Res. 1995;77:784–802.
84. Katz R, Karliner JS, Resnik R. Effects of a natural volume overload state (pregnancy) on left ventricular performance in normal human subjects. Circulation. 1978;58:434–41.
85. Khalil A, Jauniaux E, Cooper D, Harrington K. Pulse wave analysis in normal pregnancy: a prospective longitudinal study. PLoS One. 2009;4:e6134.
86. Khalil RA, Crews JK, Novak J, Kassab S, Granger JP. Enhanced vascular reactivity during inhibition of nitric oxide synthesis in pregnant rats. Hypertension. 1998;31:1065–9.
87. Knuist M, Bonsel GJ, Zondervan HA, Treffers PE. Risk factors for preeclampsia in nulliparous women in distinct ethnic groups: a prospective cohort study. Obstet Gynecol. 1998;92:174–8.
88. Kuo CD, Chen GY, Yang MJ, Lo HM, Tsai YS. Biphasic changes in autonomic nervous activity during pregnancy. Br J Anaesth. 2000;84:323–9.
89. Lafleur L, Mitchell GF. Differences in the immunological activities of antibody-secreting cell precursors in mouse spleen selected on the basis of antigen-binding capacity. Eur J Immunol. 1975;5:648–52.
90. Lechmanova M, Kittnar O, Mlcek M, Slavicek J, Dohnalova A, Havranek S, Kolarik J, Parizek AQT. Dispersion and T-loop morphology in late pregnancy and after delivery. Physiol Res. 2002;51:121–9.
91. Leduc L, Wasserstrum N, Spillman T, Cotton DB. Baroreflex function in normal pregnancy. Am J Obstet Gynecol. 1991;165:886–90.
92. Liu LX, Arany Z. Maternal cardiac metabolism in pregnancy. Cardiovasc Res. 2014;101:545–53.
93. Lucini D, Strappazzon P, Vecchia LD, Maggioni C, Pagani M. Cardiac autonomic adjustments to normal human pregnancy: insight from spectral analysis of R-R interval and systolic arterial pressure variability. J Hypertens. 1999;17:1899–904.
94. Lusiani L, Ronsisvalle G, Bonanome A, Visona A, Castellani V, Macchia C, Pagnan A. Echocardiographic evaluation of the dimensions and systolic properties of the left ventricle in freshman athletes during physical training. Eur Heart J. 1986;7:196–203.
95. Ma X, Sigmund CD, Hingtgen SD, Tian X, Davisson RL, Abboud FM, Chapleau MW. Ganglionic action of angiotensin contributes to sympathetic activity in renin-angiotensinogen transgenic mice. Hypertension. 2004;43:312–6.
96. Mabie WC, DiSessa TG, Crocker LG, Sibai BM, Arheart KLA. A longitudinal study of cardiac output in normal human pregnancy. Am J Obstet Gynecol. 1994;170:849–56.
97. Magness RR, Cox K, Rosenfeld CR, Gant NF. Angiotensin II metabolic clearance rate and pressor responses in nonpregnant and pregnant women. Am J Obstet Gynecol. 1994;171:668–79.
98. Magriples U, Boynton MH, Kershaw TS, Duffany KO. Rising SS, Ickovics JR. Blood pressure changes during pregnancy: impact of race, body mass index, and weight gain. Am J Perinatol. 2013;30:415–24.
99. Mahendru AA, Everett TR, Wilkinson IB, Lees CC, McEniery CM. A longitudinal study of maternal cardiovascular function from preconception to the postpartum period. J Hypertens. 2014;32:849–56.
100. Mahendru AA, Foo FL, McEniery CM, Everett TR, Wilkinson IB, Lees CC. Change in maternal cardiac output from preconception to mid-pregnancy is associated with birth

weight in healthy pregnancies. Ultrasound Obstet Gynecol. 2017;49:78–84.
101. Marciniak A, Claus P, Sutherland GR, Marciniak M, Karu T, Baltabaeva A, Merli E, Bijnens B, Jahangiri M. Changes in systolic left ventricular function in isolated mitral regurgitation. A strain rate imaging study. Eur Heart J. 2007;28:2627–36.
102. Martin A, Brown MA, Bucci J, Whitworth JA. Measuring venous capacitance and blood flow in pregnancy. Aust N Z J Obstet Gynaecol. 1997;37:335–9.
103. Mersich B, Rigo J Jr, Besenyei C, Lenard Z, Studinger P, Kollai M. Opposite changes in carotid versus aortic stiffness during healthy human pregnancy. Clin Sci (Lond). 2005;109:103–7.
104. Mesa A, Jessurun C, Hernandez A, Adam K, Brown D, Vaughn WK, Wilansky S. Left ventricular diastolic function in normal human pregnancy. Circulation. 1999;99:511–7.
105. Moertl MG, Ulrich D, Pickel KI, Klaritsch P, Schaffer M, Flotzinger D, Alkan I, Lang U, Schlembach D. Changes in haemodynamic and autonomous nervous system parameters measured non-invasively throughout normal pregnancy. Eur J Obstet Gynecol Reprod Biol. 2009;144(Suppl 1):S179–83.
106. Molino P, Veglio F, Genova GC, Melchio R, Benedetto C, Chiarolini L, Rabbia F, Grosso T, Mulatero P, Chiandussi L. Baroreflex control of heart rate is impaired in pre-eclampsia. J Hum Hypertens. 1999;13:179–83.
107. Mone SM, Sanders SP, Colan SD. Control mechanisms for physiological hypertrophy of pregnancy. Circulation. 1996;94:667–72.
108. Nakagawa M, Katou S, Ichinose M, Nobe S, Yonemochi H, Miyakawa I, Saikawa T. Characteristics of new-onset ventricular arrhythmias in pregnancy. J Electrocardiol. 2004;37:47–53.
109. Nama V, Antonios TF, Onwude J, Manyonda IT. Mid-trimester blood pressure drop in normal pregnancy: myth or reality? J Hypertens. 2011;29:763–8.
110. Nisell H, Hjemdahl P, Linde B. Cardiovascular responses to circulating catecholamines in normal pregnancy and in pregnancy-induced hypertension. Clin Physiol. 1985;5:479–93.
111. Novak J, Parry LJ, Matthews JE, Kerchner LJ, Indovina K, Hanley-Yanez K, Doty KD, Debrah DO, Shroff SG, Conrad KP. Evidence for local relaxin ligand-receptor expression and function in arteries. FASEB J. 2006;20:2352–62.
112. Okada Y, Best SA, Jarvis SS, Shibata S, Parker RS, Casey BM, Levine BD, Fu Q. Asian women have attenuated sympathetic activation but enhanced renal-adrenal responses during pregnancy compared to Caucasian women. J Physiol. 2015;593:1159–68.
113. Paller MS. Mechanism of decreased pressor responsiveness to ANG II, NE, and vasopressin in pregnant rats. Am J Phys. 1984;247:H100–8.
114. Pandey AK, Banerjee AK, Das A, Bhawani G, Kumar A, Majumadar B, Bhattacharya AK. Evaluation of maternal myocardial performance during normal pregnancy and post partum. Indian Heart J. 2010;62:64–7.
115. Paulus WJ, Shah AMNO. cardiac diastolic function. Cardiovasc Res. 1999;43:595–606.
116. Phippard AF, Horvath JS, Glynn EM, Garner MG, Fletcher PJ, Duggin GG, Tiller DJ. Circulatory adaptation to pregnancy--serial studies of haemodynamics, blood volume, renin and aldosterone in the baboon (Papio hamadryas). J Hypertens. 1986;4:773–9.
117. Poppas A, Shroff SG, Korcarz CE, Hibbard JU, Berger DS, Lindheimer MD, Lang RM. Serial assessment of the cardiovascular system in normal pregnancy. Role of arterial compliance and pulsatile arterial load. Circulation. 1997;95:2407–15.
118. Quesnell RR, Brooks VL. Alterations in the baroreflex occur late in pregnancy in conscious rabbits. Am J Obstet Gynecol. 1997;176:692–4.
119. Rang S, Wolf H, van Montfrans GA, Karemaker JM. Serial assessment of cardiovascular control shows early signs of developing pre-eclampsia. J Hypertens. 2004;22:369–76.
120. Robson SC, Hunter S, Boys RJ, Dunlop W. Hemodynamic changes during twin pregnancy. A Doppler and M-mode echocardiographic study. Am J Obstet Gynecol. 1989a;161:1273–8.
121. Robson SC, Hunter S, Boys RJ, Dunlop W. Serial study of factors influencing changes in cardiac output during human pregnancy. Am J Phys. 1989b;256:H1060–5.
122. Robson SC, Hunter S, Dunlop W. Left atrial dimension during early puerperium. Lancet. 1987;2:111–2.
123. Roman MJ, Saba PS, Pini R, Spitzer M, Pickering TG, Rosen S, Alderman MH, Devereux RB. Parallel cardiac and vascular adaptation in hypertension. Circulation. 1992;86:1909–18.
124. Sakai K, Imaizumi T, Maeda H, Nagata H, Tsukimori K, Takeshita A, Nakano H. Venous distensibility during pregnancy. Comparisons between normal pregnancy and pre-eclampsia. Hypertension. 1994;24:461–6.
125. Sanghavi M, Rutherford JD. Cardiovascular physiology of pregnancy. Circulation. 2014;130:1003–8.
126. Savu O, Jurcut R, Giusca S, van Mieghem T, Gussi I, Popescu BA, Ginghina C, Rademakers F, Deprest J, Voigt JU. Morphological and functional adaptation of the maternal heart during pregnancy. Circ Cardiovasc Imaging. 2012;5:289–97.
127. Savvidou MD, Kametas NA, Donald AE, Nicolaides KH. Non-invasive assessment of endothelial function in normal pregnancy. Ultrasound Obstet Gynecol. 2000;15:502–7.
128. Savvidou MD, Vallance PJ, Nicolaides KH, Hingorani AD. Endothelial nitric oxide synthase gene polymorphism and maternal vascular adaptation to pregnancy. Hypertension. 2001;38:1289–93.
129. Schobel HP, Fischer T, Heuszer K, Geiger H, Schmieder RE. Preeclampsia -- a state of sympathetic overactivity. N Engl J Med. 1996;335:1480–5.
130. Sealey JE, Itskovitz-Eldor J, Rubattu S, James GD, August P, Thaler I, Levron J, Laragh JH. Estradiol- and progesterone-related increases in the renin-aldosterone system: studies during ovarian stimulation and early pregnancy. J Clin Endocrinol Metab. 1994;79:258–64.
131. Seely EW, Walsh BW, Gerhard MD, Williams GH. Estradiol with or without progesterone and ambulatory blood pressure in postmenopausal women. Hypertension. 1999;33:1190–4.
132. Shaamash AH, Elsnosy ED, Makhlouf AM, Zakhari MM, Ibrahim OA, EL-d HM. Maternal and fetal serum nitric oxide (NO) concentrations in normal pregnancy, pre-eclampsia and eclampsia. Int J Gynaecol Obstet. 2000;68:207–14.
133. Sierra-Laguado J, Garcia RG, Lopez-Jaramillo P. Flow-mediated dilatation of the brachial artery in pregnancy. Int J Gynaecol Obstet. 2006;93(1):60.
134. Silver HM, Tahvanainen KU, Kuusela TA, Eckberg DL. Comparison of vagal baroreflex function in nonpregnant women and in women with normal pregnancy, preeclampsia, or gestational hypertension. Am J Obstet Gynecol. 2001;184:1189–95.
135. Simmons LA, Gillin AG, Jeremy RW. Structural and functional changes in left ventricle during normotensive and preeclamptic pregnancy. Am J Physiol Heart Circ Physiol. 2002;283:H1627–33.

136. Simonian SX, Herbison AE. Differential expression of estrogen receptor and neuropeptide Y by brainstem A1 and A2 noradrenaline neurons. Neuroscience. 1997;76:517–29.
137. Sims EA, Krantz KE. Serial studies of renal function during pregnancy and the puerperium in normal women. J Clin Invest. 1958;37:1764–74.
138. Slangen BF, Out IC, Janssen BJ, Peeters LL. Blood pressure and heart rate variability in early pregnancy in rats. Am J Phys. 1997;273:H1794–9.
139. Spaanderman ME, Willekes C, Hoeks AP, Ekhart TH, Peeters LL. The effect of pregnancy on the compliance of large arteries and veins in healthy parous control subjects and women with a history of preeclampsia. Am J Obstet Gynecol. 2000;183:1278–86.
140. Stainer K, Morrison R, Pickles C, Cowley AJ. Abnormalities of peripheral venous tone in women with pregnancy-induced hypertension. Clin Sci (Lond). 1986;70:155–7.
141. Stein PK, Hagley MT, Cole PL, Domitrovich PP, Kleiger RE, Rottman JN. Changes in 24-hour heart rate variability during normal pregnancy. Am J Obstet Gynecol. 1999;180:978–85.
142. Stennett AK, Qiao X, Falone AE, Koledova VV, Khalil RA. Increased vascular angiotensin type 2 receptor expression and NOS-mediated mechanisms of vascular relaxation in pregnant rats. Am J Physiol Heart Circ Physiol. 2009;296: H745–55.
143. Stewart RD, Nelson DB, Matulevicius SA, Morgan JL, McIntire DD, Drazner MH, Cunningham FG. Cardiac magnetic resonance imaging to assess the impact of maternal habitus on cardiac remodeling during pregnancy. Am J Obstet Gynecol. 2016;214:640 e641-6.
144. Sundlof G, Wallin BG. Human muscle nerve sympathetic activity at rest. Relationship to blood pressure and age. J Physiol. 1978;274:621–37.
145. Szlachter BN, Quagliarello J, Jewelewicz R, Osathanondh R, Spellacy WN, Weiss G. Relaxin in normal and pathogenic pregnancies. Obstet Gynecol. 1982;59:167–70.
146. Tanindi A, Akgun N, Pabuccu EG, Gursoy AY, Yuce E, Tore HF, Duvan CI. Electrocardiographic P-wave duration, QT interval, T peak to end interval and Tp-e/QT ratio in pregnancy with respect to trimesters. Ann Noninvasive Electrocardiol. 2016;21:169–74.
147. Tkachenko O, Shchekochikhin D, Schrier RW. Hormones and hemodynamics in pregnancy. Int J Endocrinol Metab. 2014;12:e14098.
148. Tulchinsky D, Korenman SG. The plasma estradiol as an index of fetoplacental function. J Clin Invest. 1971;50 (7):1490.
149. Tunbridge RD, Donnai P. Plasma noradrenaline in normal pregnancy and in hypertension of late pregnancy. Br J Obstet Gynaecol. 1981;88:105–8.
150. Umar S, Nadadur R, Iorga A, Amjedi M, Matori H, Eghbali M. Cardiac structural and hemodynamic changes associated with physiological heart hypertrophy of pregnancy are reversed postpartum. J Appl Physiol (1985). 2012;113:1253–9.
151. Usselman CW, Skow RJ, Matenchuk BA, Chari RS, Julian CG, Stickland MK, Davenport MH, Steinback CD. Sympathetic baroreflex gain in normotensive pregnant women. J Appl Physiol (1985). 2015;119:468–74.
152. Valensise H, Novelli GP, Vasapollo B, Borzi M, Arduini D, Galante A, Romanini C. Maternal cardiac systolic and diastolic function: relationship with uteroplacental resistances. A Doppler and echocardiographic longitudinal study. Ultrasound Obstet Gynecol. 2000;15:487–97.
153. Vallance P, Collier J, Moncada S. Effects of endothelium-derived nitric oxide on peripheral arteriolar tone in man. Lancet. 1989;2:997–1000.
154. Vallbo AB, Hagbarth KE, Torebjork HE, Wallin BG. Somatosensory, proprioceptive, and sympathetic activity in human peripheral nerves. Physiol Rev. 1979;59:919–57.
155. van Oppen AC, Stigter RH, Bruinse HW. Cardiac output in normal pregnancy: a critical review. Obstet Gynecol. 1996;87:310–8.
156. Vinayagam D, Thilaganathan B, Stirrup O, Mantovani E, Khalil A. Maternal hemodynamics in normal pregnancy: reference ranges and role of maternal characteristics. Ultrasound Obstet Gynecol. 2018;51:665–71.
157. Visontai Z, Lenard Z, Studinger P, Rigo J Jr, Kollai M. Impaired baroreflex function during pregnancy is associated with stiffening of the carotid artery. Ultrasound Obstet Gynecol. 2002;20:364–9.
158. Voss A, Malberg H, Schumann A, Wessel N, Walther T, Stepan H, Faber R. Baroreflex sensitivity, heart rate, and blood pressure variability in normal pregnancy. Am J Hypertens. 2000;13:1218–25.
159. Wallin BG, Delius W, Sundlof G. Human muscle nerve sympathetic activity in cardiac arrhythmias. Scand J Clin Lab Invest. 1974;34:293–300.
160. Wallin BG, Fagius J. Peripheral sympathetic neural activity in conscious humans. Annu Rev Physiol. 1988;50:565–76.
161. Walther T, Wessel N, Baumert M, Stepan H, Voss A, Faber R. Longitudinal analysis of heart rate variability in chronic hypertensive pregnancy. Hypertens Res. 2005;28:113–8.
162. Wasserstrum N, Kirshon B, Rossavik IK, Willis RS, Moise KJ Jr, Cotton DB. Implications of sino-aortic baroreceptor reflex dysfunction in severe preeclampsia. Obstet Gynecol. 1989;74:34–9.
163. Weiner CP, Lizasoain I, Baylis SA, Knowles RG, Charles IG, Moncada S. Induction of calcium-dependent nitric oxide synthases by sex hormones. Proc Natl Acad Sci U S A. 1994;91:5212–6.
164. Williams DJ, Vallance PJ, Neild GH, Spencer JA, Imms FJ. Nitric oxide-mediated vasodilation in human pregnancy. Am J Phys. 1997;272:H748–52.
165. Williams JK, Adams MR, Herrington DM, Clarkson TB. Short-term administration of estrogen and vascular responses of atherosclerotic coronary arteries. J Am Coll Cardiol. 1992;20:452–7.
166. Wilson M, Morganti AA, Zervoudakis I, Letcher RL, Romney BM, Von Oeyon P, Papera S, Sealey JE, Laragh JH. Blood pressure, the renin-aldosterone system and sex steroids throughout normal pregnancy. Am J Med. 1980;68:97–104.
167. Winkleby MA, Kraemer HC, Ahn DK, Varady AN. Ethnic and socioeconomic differences in cardiovascular disease risk factors: findings for women from the Third National Health and Nutrition Examination Survey, 1988-1994. JAMA. 1998;280:356–62.
168. Yang CC, Chao TC, Kuo TB, Yin CS, Chen HI. Preeclamptic pregnancy is associated with increased sympathetic and decreased parasympathetic control of HR. Am J Physiol Heart Circ Physiol. 2000;278:H1269–73.
169. Zentner D, du Plessis M, Brennecke S, Wong J, Grigg L, Harrap SB. Deterioration in cardiac systolic and diastolic function late in normal human pregnancy. Clin Sci (Lond). 2009;116:599–606.
170. Zuspan FP. Catecholamines. Their role in pregnancy and the development of pregnancy-induced hypertension. J Reprod Med. 1979;23:143–50.

Sex-Specific Physiology and Cardiovascular Disease

27

Chrisandra L. Shufelt, Christine Pacheco, Marysia S. Tweet, and Virginia M. Miller

Cardiovascular risk factors and disease in women. Art work by Piet Michiels, Leuven, Belgium.

Abstract

Sex differences in cardiovascular diseases can be classified as those which are specific to one sex and those that differ in incidence, prevalence, etiology, symptomatology, response to treatment, morbidity, and mortality in one sex compared to the other. All sex differences in cardiovascular conditions have their basis in the combined expression of genetic and

C. L. Shufelt (✉) · C. Pacheco
Barbra Streisand Women's Heart Center, Cedars-Sinai Medical Center, Smidt Heart Insititute, Los Angeles, CA, USA
e-mail: Chrisandra.Shufelt@cshs.org; christine.pachecoclaudio@cshs.org

M. S. Tweet
Department of Cardiovascular Medicine, College of Medicine, Mayo Clinic, Rochester, MN, USA
e-mail: tweet.marysia@mayo.edu

V. M. Miller
Surgery and Physiology, Women's Health Research Center, College of Medicine, Mayo Clinic, Rochester, MN, USA
e-mail: miller.virginia@mayo.edu

P. L. M. Kerkhof, V. M. Miller (eds.), *Sex-Specific Analysis of Cardiovascular Function*, Advances in Experimental Medicine and Biology 1065, https://doi.org/10.1007/978-3-319-77932-4_27

hormonal differences between women and men. This chapter addresses how understanding basic mechanisms of hormone responses, imaging diagnostics, and integration of genomics and proteomics has advanced diagnosis and improved outcomes for cardiovascular conditions, apart from those related to pregnancy that are more prevalent in women. These conditions include obstructive coronary artery disease, coronary microvascular dysfunction, spontaneous coronary artery dissection, diseases of the cardiac muscle including heart failure and takotsubo cardiomyopathy, and conditions related to neurovascular dysregulation including hot flashes and night sweats associated with menopause and effects of exogenous hormones on vascular function. Improvement in technologies allowing for noninvasive assessment of neuronally mediated vascular reactivity will further improve our understanding of the basic etiology of the neurovascular disorders. Consideration of sex, hormonal status, and pregnancy history in diagnosis and treatment protocols will improve prevention and outcomes of cardiovascular disease in women as they age.

Keywords

Atrial fibrillation · Cardiomyopathy · Endothelial dysfunction · Coronary microvascular dysfunction · 17β-Estradiol · Heart failure · Hot flashes · Ischemic nonobstructive coronary artery disease · INOCA · Menopause · Migraine · SCAD · Spontaneous coronary artery dissection · Takotsubo · Vasomotor symptoms

Introduction

The fundamental origin of all sex differences in health and disease results from the presence of the sex chromosomes (XX for female and XY for males). Genes on these chromosomes influence the expression of genes on the autosomes, as well as direct the development of the reproductive organs and ultimately gonadal production of the sex steroids estrogen and testosterone [1, 2]. An important interaction between the sex chromosomes and sex hormones is through the gene for the androgen receptor located on the X chromosome. Variations in this gene will have greater effects in males who carry only one copy of the X chromosome. Females have two copies of the X chromosome but one is inactivated, reducing the gene dosage effect, which also results in mosaic distribution of the genetic variants on each X in various tissues of the body. These differences in genes and hormones provide the basis for all sex differences in cardiovascular regulatory mechanisms in health and disease. It is essential when considering the expression of various cardiovascular diseases to recall several points: (1) receptors for the sex steroids are present in most tissues of the body, but their relative expression may vary; (2) estrogen is a metabolic product of testosterone—thus both testosterone and estrogen are present in men and women but in different proportions; (3) concentrations of the sex steroids vary across the life span as evidenced by increased production at puberty, increases during pregnancy, and decreases in women during the transition to menopause. The dramatic changes in hormone-mediated physiological changes in the cardiovascular (and other) system of a woman's body with pregnancy demonstrate the activational effects of the sex steroids, that is, reversible phenotypes expressed in the presence of the hormones but diminished in their absence. Specific aspects of cardiovascular physiology associated with pregnancy are discussed in a chapter by Fu [3]. In this section, the focus will be on conditions of the coronary circulation and the heart and a group of conditions that require autonomic neurovascular regulation. The final section will discuss cardiovascular issues related to hormonal treatments in women.

The Coronary Circulation

Obstructive Coronary Artery Disease Coronary artery disease (CAD) is the leading cause of death in women in the United States, affecting 6.6 million women a year with higher mortality rates than their male counterparts, from 1985 to 2011 [4]. A recent reduction in these rates has been attributed to national and international efforts which focused on fostering understanding, awareness, and application of evidence-based therapy to women. Women who present with obstructive CAD are generally older than their male counterparts, have more cardiovascular comorbidities, and have a higher incidence of adverse cardiovascular outcomes including mortality following acute myocardial infarction (MI) [5]. Occlusion of the coronary arteries is caused by disruption of plaque resulting from accumulation of infiltrating macrophages and low-density lipoproteins within the vessel wall. Formation of thrombus at the site of the disrupted plaque occludes the artery limiting blood flow to the myocardium resulting in MI. The pathophysiology of CAD and MI is distinctive in women, who are less likely than men to present with plaque rupture [6]. The reasons for this difference remain unclear.

Chest pain is the most prevalent symptom of acute MI in both sexes. However, women are more likely to present with atypical symptoms, including pain in the upper back and neck, fatigue, nausea, and vomiting. These differences in symptom presentation, along with the long-standing notion that CAD and acute MI are uncommon in women, account, in part, for delays in seeking medical advice, poor recognition of symptoms, underdiagnosis, and suboptimal treatment in women [7–9]. Women with MI also are less likely to undergo coronary angiograms when compared to men, despite that this test is the diagnostic and therapeutic gold standard. When obstructive lesions exist, revascularization is most often the treatment of choice in the clinical setting of both ST-elevation MI (STEMI) and high-risk non-ST-elevation MI (NSTEMI) [10, 11]. However, revascularization of the occluded artery may be more challenging in women due to bleeding at the site of access and small and more tortuous coronary arteries. Future studies should aim at improving the treatment of obstructive CAD in women.

Coronary Microvascular Dysfunction Approximately 30–50% of women who undergo coronary angiography for evaluation of chest pain do not have obstructive CAD [12, 13]. Ischemia in the setting of no obstructive coronary artery disease (INOCA) can be due to coronary microvascular dysfunction (CMD). Clinical suspicion for CMD should be high in patients that present with persistent chest pain, ischemic changes on noninvasive stress testing, and no obstructive CAD by invasive or noninvasive diagnostic methods. Once thought to be a benign condition, CMD has been found to have a 2.5% annual rate of major adverse cardiovascular events (MACE) including stroke, heart failure, or MI [14, 15].

Risk factors for CMD are similar to traditional coronary risk factors for obstructive CAD such as smoking, hypertension, hyperlipidemia, and diabetes [16]. In addition, sex-specific risk factors have been identified such as autoimmune diseases [17], treatments for breast cancer with radiation to the chest wall area, as well as certain chemotherapy agents [18, 19]. The gold standard diagnostic test for CMD is an invasive coronary reactivity test (CRT) that uses vasoactive agents to functionally test the coronary microvasculature. During CRT, a Doppler guide wire is positioned in the proximal left anterior descending artery for sequential infusion of the vasoactive drugs adenosine, acetylcholine, and nitroglycerin to assess microvascular and macrovascular (epicardial) endothelial and non-endothelial function. Coronary flow reserve (CFR) is the calculated ratio of peak flow velocity to baseline flow after adenosine has been administered. Data from the National Heart, Lung, and Blood Institute-sponsored Women's Ischemia Syndrome Evaluation (WISE) study found that women with no obstructive CAD who have a low CFR to adenosine are at higher risk for major adverse cardiac events as compared to those with normal CFR [15].

Noninvasive imaging of those with nonobstructive coronaries with cardiac magnetic resonance imaging (CMRI) can be used to evaluate the coronary microvascular at both stress (adenosine) and rest. The characteristic finding to diagnoses of CMD by CMRI is a decreased perfusion pattern in the subendocardial region of the heart, which can also be seen in the affected coronary distribution of obstructed arteries. CMRI also has prognostic value in women with CMD [20, 21]. Cardiac positron emission tomography and transthoracic Doppler echocardiography are other noninvasive techniques that measure CFR, but they do not replace diagnosis by intracoronary CRT.

As with obstructive CAD, treatment of CMD is focused on lifestyle modification and aggressive risk factor control. In addition to its diagnostic utility, CRT allows to guide treatment based on which abnormal pathways (i.e., endothelium or smooth muscle) have been found. Medications to treat endothelial dysfunction include statins, angiotensin-converting enzyme inhibitors, and low-dose aspirin. Medications to treatment of non-endothelial dysfunction include beta-blockers, alpha-beta-blockers, and nitrates [22].

Spontaneous Coronary Artery Dissection Spontaneouscoronary artery dissection (SCAD) is a cause of acute coronary syndrome that affects women (81–92% of all cases) more than men. Women present with SCAD at an average age of between 42 and 53 years (range can be 17–71 years or older) and have minimal other cardiovascular risk factors [23–28].

SCAD is caused by formation of a hematoma within the wall of one or more coronary arteries, leading to obstruction of coronary artery blood flow to the myocardium. A tear in the artery is not always observed either because it may be too small to detect or perhaps hematoma occurs independent of a tear. Associated conditions for SCAD include extracoronary vascular abnormalities (i.e., fibromuscular dysplasia, coronary tortuosity), pregnancy and the postpartum state, extreme emotion/stress/exercise, and connective tissue disease.

SCAD is diagnosed on coronary angiography and previously considered rare with reported prevalence of 0.07–1.1% [29]. However, recent technological advancements in the catheterization laboratory with the ability to visualize intramural hematoma of the coronary wall using intravascular imaging techniques such as intravascular ultrasound or optical coherence tomography have improved both awareness and diagnosis of this condition. Recent studies indicate that SCAD is the etiology of as many as 35% of myocardial infarctions among women <50 years of age and that SCAD is the most common etiology of pregnancy-associated MI [27, 30].

Accurate diagnosis of SCAD is important as management strategies are different than those for coronary artery disease due to atherosclerosis. While atherosclerotic MI is best managed with revascularization, primarily with stenting, a similar approach for acute SCAD is associated with a high risk of complications [31]. SCAD can self-heal in the majority of conservatively managed patients. Therefore, as long as a patient has normal or near-normal distal coronary blood flow and is clinically stable, acute SCAD can be managed without revascularization [31].

While patients with history of SCAD fare better survival rates that matched controls with atherosclerosis [24], there is a significant burden of major adverse cardiac events including recurrent SCAD, recurrent myocardial infarction, recurrent chest pain, and heart failure [23–25]. Precipitators of recurrent SCAD and underlying mechanisms of SCAD itself remain poorly understood, requiring continued research into the causation, genetic predisposition, diagnostic and prognostic indicators, and treatment options for patients with recurrent SCAD with other comorbidities (e.g., migraine) [32].

The Heart

Atrial Fibrillation Atrial fibrillation (AF) is due to chaotic atrial electrical activity with dysfunctional or nonexistent contraction of the atria; this can be accompanied with irregular and/or rapid ventricular contraction predisposing to heart failure. However, the most serious and immediate risk from AF is the association of stroke; one proposed mechanism is due to formation and dislodgement of thrombus in the poorly contractile and often dilated left atrial appendage [33]. This risk can be reduced with chronic anticoagulation; appropriate patients to treat can be selected using risk stratification techniques including the CHA_2DS_2VASc and HAS-BLED scores to predict risk of stroke and bleeding, respectively [34]. Of note, the female sex is considered an important risk factor for stroke associated with atrial fibrillation and is accounted for on the CHA_2DS_2VASc tool [35, 36]. Despite many existing hypotheses, the exact pathophysiology as to why the female sex increases stroke risk in the setting of AF is not understood.

The proposed causes of AF are numerous. One hypothesis is that it may involve changes in function of the potassium channels on the atrial muscle leading to prolongation of the electrical refractory period. Variants in genes encoding subunits of the potassium channel, one of which is located on the X chromosome, may account for sex differences in etiology and prevalence of and mortality from AF [37, 38]. Antiarrhythmic drugs that prolong the QT interval increase risk for torsade de pointes in women, and this was the reason some such drugs were removed from the market [39]. Estrogenic compounds modulate potassium currents in atrial tissue [40, 41], but the relationship between endogenous and exogenous estrogen on incidence of AF in women is not established.

Radiofrequency catheter ablation is effective in reducing AF in women. However, women are often older at the time of presentation, experience more bleeding events during procedures, and have higher recurrence rates than men, which may be due to the older age and stage of disease when the procedures are performed [42–44].

Heart Failure HF, an important cause of morbidity and mortality [45], is caused by abnormal ventricular diastolic filling, reduced systolic ejection of blood, or a combination of both. It is a clinical diagnosis with a complex range of etiologies and mechanisms. Patients can present with symptoms such as dyspnea, fatigue, exercise intolerance, and fluid retention including edema and ascites [46]. Defining the prevalence of HF is challenging due to varying definitions and diagnosis methods. Regardless, over one million US hospitalizations in 2010 coded HF as the first discharge diagnosis, and one in eight deaths mentions HF on the death certificate [45].

Left ventricular (LV) HF can be subcategorized into HF with ejection fraction (EF) preserved (HFpEF) and HF where EF is reduced (HFrEF). The severity of HF is stratified based on structural changes to the heart and severity of symptoms and response to treatment (Table 27.1) [46].

In general, HFrEF refers to a patient whose LVEF is <40%. These patients may also have concurrent diastolic dysfunction with increased LV filling pressures. The diagnosis of HFpEF refers to a patient whose LVEF is >50%. In HFpEF, the diastolic relaxation of the LV is impaired [46]. On clinical evaluation, patients with HFpEF exhibit evidence of arterial and LV stiffness with chronically elevated LV filling pressures. At autopsy, patients with HFpEF have more cardiac hypertrophy, myocardial fibrosis, epicardial coronary atherosclerosis, and coronary microvascular rarefaction than matched controls possibly accounting for the observed clinical findings [47].

There is debate regarding the threshold for categorizing midrange EF and whether there is a need for sex-specific criterion for phenotyping HF [48]. For example, patients with HF and LVEF of 41–49% are considered borderline HFpEF, because the clinical characteristics, response to treatment, and outcomes are similar to those who have HFpEF [46]. The risk of HFpEF is similar between men and women after adjustment for age and risk factors [49]. However, HFrEF is less frequent among women than men [49]. A recent study suggests women have more reverse remodeling with standard medical therapy than men ($\geq$15% reduction in LV end-systolic volume index in those with HFrEF), resulting in fewer hospitalizations for HF and fewer cardiovascular and all-cause mortality [50].

Associated factors and predisposing conditions for HF include coronary artery disease

Table 27.1 Classification of heart failure by structure and function

ACCF/AHA stages of heart failure		NYHA functional classification		
A	At high risk for heart failure without structural heart disease or symptoms	**None**		
B	Structural heart disease without heart failure signs or symptoms	I	No limitation of physical activity	Ordinary physical activity does not cause heart failure symptoms
C	Structural heart disease with prior or current heart failure symptoms	I	No limitation of physical activity	Ordinary physical activity does not cause heart failure symptoms
		II	Slight limitation of physical activity	Comfortable at rest, but ordinary activity causes heart failure symptoms
		III	Marked limitation of physical activity	Comfortable at rest, but less than ordinary activity causes heart failure symptoms
		IV	Unable to carry on any physical activity without heart failure symptoms	Heart failure symptoms at rest
D	Refractory heart failure requiring specialized interventions	IV	Unable to carry on any physical activity without heart failure symptoms	Heart failure symptoms at rest

Table adapted from Yancy et al. [49]

ACCF American College of Cardiology Foundation, *AHA* American Heart Association, *NYHA* New York Heart Association

(major etiology of HFrEF), diabetes, cigarette smoking, hypertension, vascular heart disease, obesity, and age. There are also less common causes of HF such as SCAD, myocarditis, infiltrative disease, substance abuse (e.g., alcohol, methamphetamines), chemotherapy induced (e.g., doxorubicin, trastuzumab), and familial cardiomyopathy, which can be challenging to diagnose. Some patients have an unknown etiology of HF [45, 46].

Current therapies known to benefit those with HFrEF include beta-blockers and angiotensin-converting enzyme inhibitors (ACEI) and aldosterone antagonists. However, depending on specific patient needs and clinical characteristics, additional therapies such as hydralazine-nitrates, ivabradine, implantable cardiac defibrillators/chronic resynchronization therapy, angiotensin receptor-neprilysin inhibitors, and advanced therapies including mechanical support and cardiac transplantation can be considered for patient management [46, 49, 51].

Women have historically been underrepresented in clinical trials for the treatment of HFrEF, representing approximately a quarter of patients in studies examining the effects of beta-blockers on this condition [52]. Women appear to benefit from most recommended goal-directed medical therapies in HFrEF, although the degree of benefit may differ slightly from men. In a meta-analysis of studies reporting the effects of beta-blockers that includes 13,833 patients (24% women) from 11 trials, the benefits of beta-blockers remained significant in all age categories, with no interaction according to sex [52]. In a cohort of 27,837 patients with congestive heart failure (53% women) who filled prescriptions for ACEI, women had improved survival, although to a lesser degree than men [HR 0.80 (0.76–0.85) for women vs. HR 0.71 (0.67–0.75) for men] [53]. Although these findings remain to be confirmed in a randomized-controlled trial, a similar population study found that women diagnosed with heart failure appeared to have improved survival when they filled a prescription for angiotensin receptor blockers (ARB) as opposed to an ACEI, whereas men did not [HR 0.69 (0.59–0.80) for women vs. HR 1.10 (0.95–1.30) for men] [54].

The first left ventricular assist devices (LVAD) developed for HF could only be used in patients with a body surface area > 1.5 m^2, limiting their use in women with advanced HFrEF, who accounted for under a quarter of patients enrolled in early trials [55–57]. Despite current design-related limitations, women who received LVAD therapy while awaiting cardiac transplantation had greater risk of cerebrovascular complications but similar mortality rates to men (19% vs. 20%, $p = 0.89$) [58, 59]. Newer, smaller LVAD models have recently been developed and may allow more women with advanced HFrEF to benefit from this therapy.

Therapies for patients with HFpEF focus on reducing of volume overload and blood pressure, with a goal systolic blood pressure of less than 130 mmHg [46, 51]. HFpEF patients who have a lower heart rate at hospital dismissal are observed to have lower all-cause mortality [60].

Takotsubo Cardiomyopathy Also known as stress cardiomyopathy or broken heart syndrome, takotsubo cardiomyopathy is an acute-onset cardiomyopathy first described in 1990 [61]. Diagnostic criteria include characteristic wall motion abnormalities on left ventricular angiogram or cardiac echocardiogram extending beyond a single epicardial vascular distribution in the absence of atherosclerotic disease, most often akinesis of the apical and mid myocardial walls with hyperkinesis of the basal segments, resembling its namesake, a Japanese octopus or fish trap. Postmenopausal women account for 90% of diagnosed patients, and takotsubo cardiomyopathy is classically triggered by emotional stress or physical triggers, including anesthesia, infection, and respiratory distress [61]. Potential pathophysiologic mechanisms include high catecholamine serum levels, leading to an overstimulation of β_2-adrenoceptors and myocardial stunning, acute multivessel epicardial coronary artery spasm, and acute LV mid-chamber and/or outflow tract obstruction [62–65].

Clinical presentation for takotsubo cardiomyopathy is often indistinguishable from an acute MI, with over 75% of patients presenting with chest pain. Nearly all patients have ischemic ECG changes and may present with STEMI. Serum troponin levels are elevated in 90% of cases; however, peak levels are significantly lower than in patients with STEMI due to vessel occlusion. Brain-type natriuretic peptide (BNP) is elevated in the acute phase of the disease. Elderly patients and those with hypotension, ventricular arrhythmias, left ventricular outflow tract obstruction (LVOTO), mitral regurgitation, and thrombus formation are at higher risk for complications including cardiogenic shock, which can occur in up to 25% of cases [61]. Acute ischemic stroke, occurring in up to 14% of patients, can also complicate the clinical course [66].

Treatment of takotsubo cardiomyopathy includes cardiac monitoring, supportive measures (maintaining blood pressure, perfusion, hemodynamics, airway and fluid volume), as well as the initiation of goal-directed medical treatment for HF. It is important to identify whether there is concurrent LVOTO versus isolated ventricular systolic failure, which can be assessed by echocardiography, in order to delineate optimal treatment options and anticipated hemodynamic responses to therapies. Current expert consensus suggests 72 h observation in the intensive cardiac care unit for higher-risk patients [61]. In cases of cardiogenic shock specifically if LVOTO is present, the use of inotropic agents is discouraged to avoid adrenergic stimulation.

In conjunction with current HF treatment guidelines, beta-blockers and ACE inhibitors are recommended in the absence of contraindications such as acute cardiogenic shock. Beta-blockers have been considered to prevent detrimental myocardial response to sympathetic nervous system stimulation; however, in the largest reported takotsubo cohort which included 1750 patients, there was no association between the use of beta-blocking agents and improved survival [67]. In this same report, ACE inhibitors or angiotensin receptor antagonists were associated with significantly lower rates of death at 12-month follow-up (2% vs. 9%, $p < 0.001$) [67]. Cases of takotsubo cardiomyopathy demonstrate complete LV systolic function recovery at 3–6 months, and this characteristic is considered a diagnostic criterion by expert groups [61]. ECG changes and elevated BNP can persist up to a year following diagnosis. Takotsubo cardiomyopathy can recur in up to 22% of cases at 5 years, and one study reported that patients diagnosed with takotsubo are at similar 3-year mortality risk as patients who experience acute MI with evidence of occlusive atherosclerosis [61]. However, these findings remain to be confirmed, and future studies are needed to determine optimal medical therapy aimed at preventing complications, recurrence, and adverse cardiovascular events.

Peripartum Cardiomyopathy Peripartum cardiomyopathy is a form of systolic HF that presents in the last month of pregnancy or in the first 5 months postpartum in the absence of another underlying etiology (such as acute MI). In some cases, chronic HF persists and may lead to cardiac transplantation. It affects about 1:1000 births worldwide, but the prevalence varies by geographic locations [30, 68].

Although peripartum cardiomyopathy is the leading cause of maternal peripartum death, little is known regarding origin and underlying mechanisms of its progression. Preeclampsia and gestational diabetes may predispose women to peripartum cardiomyopathy, but the mediators connecting these conditions to the cardiomyopathy are incompletely defined. Some studies suggest that increased cleavage of prolactin by cathepsin D into the proapoptotic and antiangiogenic 16 kDa prolactin fragment may be involved [69]. Alternatively, other studies suggest an upregulation of soluble factors of placental origin (sFlt1, a soluble form of the vascular endothelial growth factor [VEGF] receptor) affects the coronary microcirculation [69–71]. Some women with peripartum cardiomyopathy bear mutations in titin, also known as connectin, which is a large sarcomeric protein involved in tethering the sarcomere to the cell wall [72]. Titin also binds a number of enzymes involved with maintaining high levels of ATP in regions of high energy demand in the sarcomere [73]. Whether any of these factors can be used as diagnostic, prognostic, or treatment targets for the disease remains to be determined.

Autonomic Neurovascular Dysregulation

Vascular Conditions Associated with Decreases in Ovarian Function With decreases or loss of ovarian function, about 75% of women experience hot flashes and night sweats [74]. Three contemporary longitudinal studies identified four patterns of the vasomotor instability relative to their onset prior to menopause, the intensity of events, and their duration relative to menopause (Fig. 27.1) [75–77]. These patterns are similar among women across the globe, suggesting

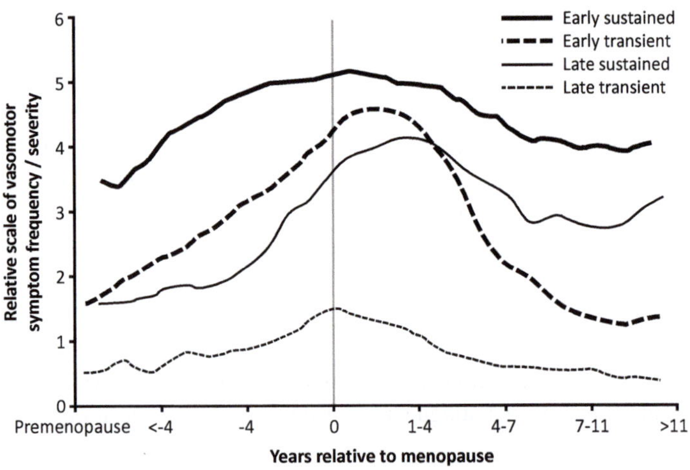

Fig. 27.1 Composite representation of four patterns of vasomotor instability relative to their onset prior to menopause, the intensity of events, and their duration relative to menopause reported in three contemporary studies [75–77]

physiological rather than cultural or environmental causes. Whether or not vasomotor instability places women at risk for various cardiovascular diseases is controversial. Part of the controversy is driven by the absence of reliable and systematic objective measurement of the hot flashes and night sweats. Most studies depend on self-report tools which may underestimate (or overestimate) their number, frequency, and severity. However, even with these unsophisticated reporting tools, the emerging data suggest that the severity of the hot flashes and night sweats may be associated with reduced endothelium-dependent vasodilatation, hypertension, increased carotid intima-media thickness, coronary artery calcification, and CMD [78–82]. The relationship between vasomotor irregularities and soluble markers associated with coagulation requires further study [83–85]. Resolving some of these controversies will require longitudinal studies that are designed to address the specific patterns of the hot flashes and night sweats with specific cardiovascular functions or events.

Hot flashes and night sweats involve the central thermoregulatory system in the hypothalamus and are mediated through autonomic control of the peripheral vasculature and sweat glands. Therefore, the onset, frequency, intensity, and duration of the symptoms may represent differences in underlying dysfunction in central and autonomic mechanisms and may involve other brain regions associated with sleep and mood [86]. Indeed, severe vasomotor symptoms may be related to obstructive sleep apnea [87], which is a risk factor for coronary artery calcification [88].

The characterization of autonomic function associated with expression of hot flashes and night sweats has not been fully characterized. However, diminished parasympathetic tone may be involved as heart rate variability decreases at menopause [89], and sleep-related decreases in blood pressure are not observed in women who experience hot flashes with insomnia [90].

Neurons of the hypothalamic-pituitary axis implicated in the development of hot flashes and night sweats are kisspeptin/neurokinin B/dynorphin (KNDy) neurons [91]. Although menopausal hormone therapy (MHT) consisting of 17β-estradiol, conjugated equine estrogen, or progesterone (alone or in combination with the estrogen) reduce both hot flashes and night sweats in women, antagonists of the KNDy pathway offer a potential alternative option to treat these symptoms [92].

In addition to menopausal hormone therapy [93], selective serotonin reuptake inhibitors also are effective in reducing hot flashes in women. Stimulation of serotonergic receptors in the brain alters both sympathetic and parasympathetic nerve activity (see [94] for review). Estrogen modulates gene transcription of serotonin in the neuronal nucleus, augments activity of the serotonin uptake transporter, and increases the sensitivity of 5-HT_{1A} receptors on postsynaptic neurons. Thus, declining levels of estrogen, as would occur at menopause, can decrease the activity of the serotonergic system. An interaction between the serotonergic and adrenergic systems [95] is suggested as both selective serotonin reuptake inhibitors (SSRIs) and clonidine, an alpha$_2$-adrenergic receptor agonist which activates presynaptic inhibition in the brain, are effective in relieving vasomotor symptoms in some women [96].

In addition to central autonomic control pathways mediating vasomotor symptoms, much remains to be learned about how peripheral pathways innervating the vascular smooth muscle (expression of adrenergic or cholinergic receptors), neurotransmitter synthesis, and receptor coupling to vasodilatory pathways such as those mediated by cyclic guanylate cyclase [97–99] might differ among women who experience various patterns of hot flashes.

Although hot flashes and night sweats are typically referred to as menopausal symptoms, the fact that they can occur in men with androgen depletion [91] and have different patterns of expression in women suggests that the changes in hormonal milieu may unmask underlying autonomic and vascular dysregulation.

Migraine is another condition that may be unmasked by changes in the hormonal milieu. The hypothesis that changes in sex hormones, in particular estrogen, contribute to the etiology of migraine is supported by the observations that, at puberty, migraines occur about three times in women than men and that migraine is associated with estrogen levels during the menstrual cycle, with pregnancy and postpartum, and during perimenopause. While hormonally related migraine is associated with abrupt declines in estrogen levels, as with menstrual-related migraine and postpartum and in perimenopause, it is unlikely that all migraine variants have a common hormonal etiology (see [100–102]). Genetic variants considered in the etiology of migraine include those for the estrogen receptor alpha; receptors and neurotransmitters associated with adrenergic, GABAergic, and nitroxidergic nerves; and enzymes associated with estrogen metabolism [100, 103–105]. However, systematic genomic studies of individuals within the various classifications of migraines may help to better understand comorbid conditions and provide insight into better treatment options [106]. Other emerging concepts for the etiology of migraine suggest an immunological component, mitochondrial dysregulation leading to oxidative stress, and iron deficiency [107–110].

Although migraines were once considered an "essentially benign condition" [111], associations with stroke and SCAD remain to be comprehensively examined with regard to migraine with and without aura or relative to hormone levels, hormonal responsiveness, and genomics [32, 112, 113].

Exogenous Hormones and Cardiovascular Disease

Exogenous hormones are used by women for contraception, replacement for primary and premature ovarian insufficiency and surgical oophorectomy, and treatment for vasomotor, mood, urogynecological symptoms, and osteoporosis following menopause. Metabolism of sex steroid hormones is complex, and genetic variants in any one of the numerous enzymes associated with synthesis, uptake, and catabolism [114] of the hormones, as well as variation in the hormone receptors [115], will affect responses to exogenous products. The pharmacogenomics of estrogen metabolism and response is an emerging area of investigation, and greater understanding of genetic variants associated with ovarian function and hormone response will allow for individualizing treatments to maximize benefits and reduce the risk of adverse events with the use of these products.

Hormonal Oral Contraceptives Oral contraceptives to suppress ovulation modify endogenous production of sex steroids from a functioning ovary and, thus, disrupt the hypothalamic-pituitary-ovary feedback regulatory pathway. Longitudinal randomized control trials evaluating adverse cardiovascular events and cardiovascular risk associated with oral hormone-based contraceptives are confounded by the fact that the composition of the products has changed over the years especially for the type of synthetic progestin allowing for longer cessation of ovulation and decreased frequency of menses (see Table 27.2 and Fig. 27.2) [116, 117]. Evaluation of adverse cardiovascular events in older women who may have used various formulations of oral contraceptives is also confounded by their pregnancy history and perhaps use of menopausal hormone therapy. In spite of these limitations, there are some data to support that women who use oral contraceptives have decreased incidence of heart disease [118, 119].

Similar to all oral medications, oral hormonal products will enter the enterohepatic circulation for first-pass metabolism in the liver. In addition to metabolism of the drugs in the liver, the hormones may stimulate production of proteins involved in coagulation and inflammation and alter production of triglycerides and lipoproteins. These effects of the hormones on liver metabolism may be important for treating women with polycystic ovarian syndrome [120]. There is an increased risk of venous and pulmonary thrombosis with some products (Table 27.2); therefore, the use of oral contraceptives is not recommended for women with known thrombophilia [116].

Table 27.2 Progestins in oral contraceptives ranked according to risk for venous thrombosis[a]

Oral contraceptive	Risk of thrombosis
First-generation <50 µg ethinyl estradiol with the progestins noretynodrel, norethisterone, and norethisterone acetate	6–12/10,000 women
Second-generation <50 µg ethinyl estradiol with the progestins norgestrel or levonorgestrel	5–7/10,000 women
Third-generation <50 µg ethinyl estradiol with the progestins desogestrel or gestodene or norgestimate	9–12/10,000 women
Fourth-generation <50 µg ethinyl estradiol with the progestins drospirenone, dienogest, or nomegestrol acetate	9–12/10,000 similar
Progestin-only (norethisterone, ethynodiol diacetate, levonorgestrel, desogestrel, lynestrenol)	2–3/10,000 women

[a]Derived from Table 27.1 of Gialeraki et al. [116] reprinted with permission

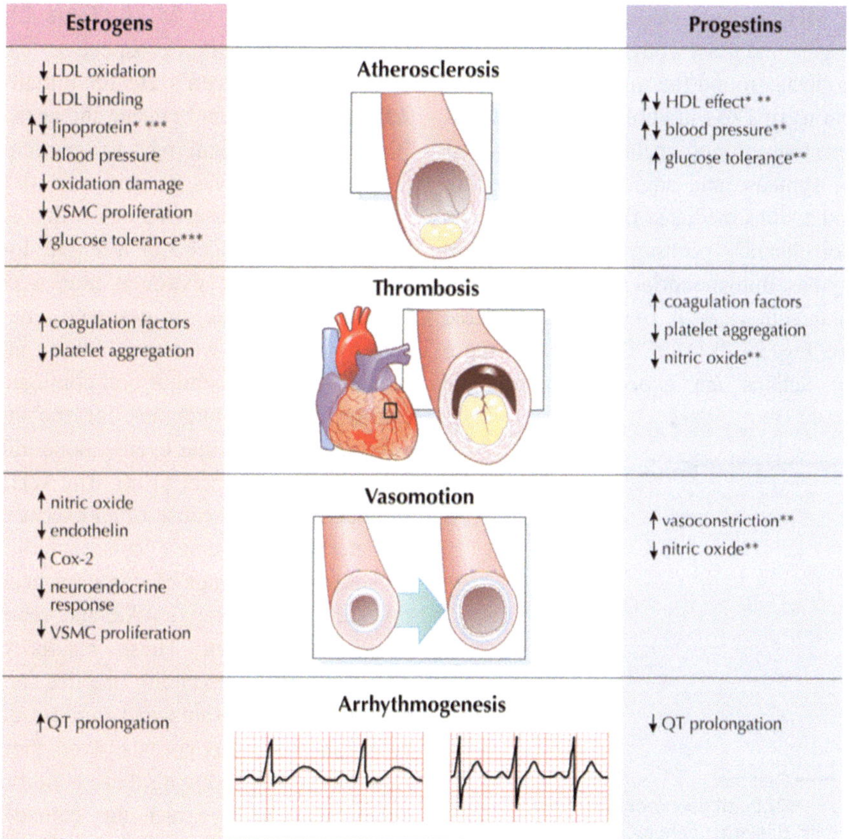

Fig. 27.2 Impact of hormonal contraception on mechanisms of cardiovascular disease. *Dependent on delivery route of estrogen, **dependent on type of progestin, ***dependent on the dose of estrogen. *Cox-2* cyclooxygenase-2, *HDL* high-density lipoprotein, *LDL* low-density lipoprotein, *VSMC* vascular smooth muscle cell. Reproduced with permission from reference [117]

Hormone Replacement for Ovarian Insufficiency and Early Oophorectomy Primary and premature ovarian insufficiencies are defined by loss of ovarian function before the age of 40 years, and early oophorectomy is defined as the surgical removal of the ovaries prior to 45 years of age.

These conditions are characterized by increases in 18 comorbid conditions of premature aging [121–123], including incidence of and mortality from cardiovascular disease and stroke [124–128]. Cardiovascular mortality is reduced by treatment with estrogen at least until the age of natural menopause (Fig. 27.3) [122–124].

Hormone Treatment at Menopause Until 2002, based on evidence from observational and epidemiological studies, it was accepted that the use of hormones (in particular, estrogen) during menopause reduced mortality due to cardiovascular disease [129–135]. The most frequently used formulation of MHT in those studies was conjugated equine estrogen, and most women initiated the use of these products around the time of menopause. Cellular actions of 17β-estradiol encompass regulation of gene transcription, including regulation of nitric oxide synthase and angiotensin-converting enzyme, and actions mediated through membrane receptors, ion channels, posttranslational modification of enzymes, mitochondrial function, and biochemical interactions with oxygen-derived free radicals (see Fig. 27.2) [100]. Collectively, these mechanistic actions are expected to promote vasodilatation, reduce adhesion of leukocytes to the vascular wall, reduce platelet activation, and thus slow development of atherosclerosis. In addition, oral estrogen products altered the serum lipid profile to reduce low-density lipoproteins and increase high-density lipoproteins, an effect that was not observed in women using statins [135, 136].

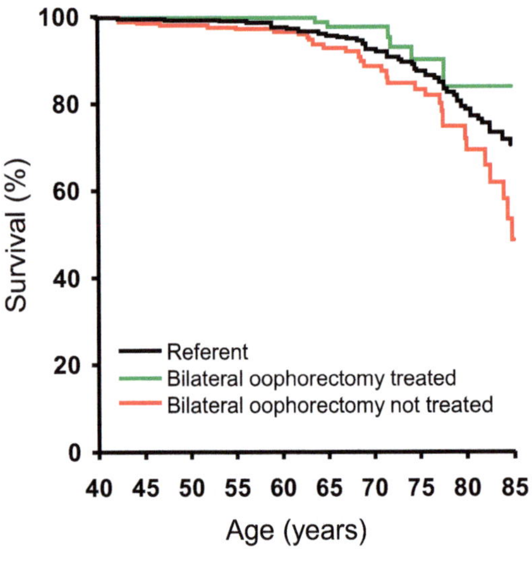

Fig. 27.3 A population study of cardiovascular mortality in women who underwent bilateral oophorectomy before age 45 years and were or were not treated with estrogen through age 45 years or longer. Reproduced with permission from reference [124]

However, observational and epidemiological data may be biased as women self-select to use MHT and may represent a healthy subset of women (healthy user bias). Several prospective trials have been conducted to try to address the preventive actions of MHT (Table 27.3) [137]. The first large-scale, prospective, randomized trial was the Women's Health Initiative (WHI) that enrolled women between the ages of 50 and 79 years. The mean age for participants was 63 years, which was as many as 10 years past the age of menopause. Women with a uterus were randomized to placebo or a single dose of oral conjugated equine estrogen plus a continuous combined synthetic progestogen, medroxyprogesterone acetate; women without a uterus were randomized to placebo or oral conjugated equine estrogen alone. Outcomes for the study were adverse cardiovascular events, bone mineral density, and breast cancer [138]. The WHI trial was stopped in 2002 because of an excess number of predetermined adverse events, including the cardiovascular outcomes of coronary heart disease, myocardial infarction, stroke, and venous thromboembolism [139]. These results drastically altered clinical prescribing practice for MHT for menopausal symptoms and chronic diseases of aging such as osteoporosis, even though MHT was effective in reducing these conditions. However, subsequently over the course of about 15 years, several criticisms of the WHI have emerged that need to be considered. First, women in the WHI were many years past menopause, and, thus, the results cannot be generalized to women who would use MHT around the peri- and immediate time past menopause [140, 141]. Evidence from experimental animals indicates that the timing of initiation of hormonal

Table 27.3 Summary of variables for open-label and prospective trials of the use of menopausal hormone therapy and cardiovascular outcomes since the WHI

Trial	Timing of treatment relative to the menopause	Formulation	Participant characterization	Primary cardiovascular outcomes	Duration
DOPS[a] [146]	3–24 months	Open label –	Women with and without a uterus 45–52 years of age with FSH >2 standard deviations over premenopausal values	Death; hospitalization for myocardial infarctions or heart failure	11 years of intervention; 16 years of follow-up
		Women with a uterus: triphasic estradiol and norethisterone acetate (2 mg synthetic 17β-estradiol for 12 days, 2 mg 17β-estradiol plus 1 mg norethisterone acetate for 10 days, and 1 mg 17β-estradiol for 6 days) Women without a uterus: oral 17β-estradiol (2 mg/day)	Exclusion: history of bone disease (including non-traumatic vertebral fractures on radiography), uncontrolled chronic disease, previous or current cancer or thromboembolic disease, current or past treatment with glucocorticoids for more than 6 months, current or previous use of hormone replacement therapy within the past 3 months, and alcohol or drug dependency		
EPAT [136]	Median 3.5 years	Oral micronized 17β-estradiol (1 mg/day)	Serum estradiol <20 pg/mL 45 years or older No preexisting cardiovascular disease; low-density lipoprotein cholesterol equal to or greater than 3.37 mmol/L (130 mg/dL)	Rate of change of carotid intima-medial thickness	2 years
KEEPS [153]	6–36 months	Oral CEE (0.45 mg/day) or transdermal 17β-estradiol (50 μg/day) each with oral progesterone (200 mg/day for 12 days in a 30-day cycle)	Women with a uterus between ages of 42 and 58 years Exclusion: history of cardiovascular disease or venous thrombosis; coronary calcification scores >50 Agatston units; body mass index >35 kg/m^2; triglyceride >400 mg/dL; LDL cholesterol <190 mg/dL; lipid-lowering medications; untreated hypertension (systolic >150 mmHg, diastolic >95 mmHg); diabetes; history of chronic diseases including cancer; smoking more than 10 cigarettes/day	Rate of change of carotid intima-medial thickness; coronary arterial calcification	4 years

(continued)

Table 27.3 (continued)

Trial	Timing of treatment relative to the menopause	Formulation	Participant characterization	Primary cardiovascular outcomes	Duration
ELITE [147]	<6 years or ≥ 10 years	Oral micronized 17β-estradiol (1 mg/day) plus vaginal progesterone (45 mg for 10 days in a 30-day cycle) for women with a uterus	Women with and without a uterus; median ages 55 and 65 years for the two groups, respectively	Rate of change of carotid intima-medial thickness; coronary arterial calcification	6–7 years (median 5 years)
			Exclusion: fasting plasma triglyceride level ≥ 500 mg/dL; diabetes mellitus or fasting serum blood glucose >140 mg/dL; serum creatinine >2.0 mg/dL; uncontrolled hypertension (systolic/diastolic blood pressure > 160/110 mmHg); untreated thyroid disease; life-threatening disease with prognosis <5 years; history of deep vein thrombosis, pulmonary embolism; breast cancer		

Reproduced with permission from Miller and Harman [137]

[a]Abbreviations: *DOPS* Danish Osteoporosis Prevention Study, *EPAT* Estrogen in the Prevention of Atherosclerosis Trial, *KEEPS* Kronos Early Estrogen Prevention Study, *ELITE* Early versus Late Intervention Trial with Estradiol, *CEE* conjugated equine estrogen, *LDL* low-density lipoprotein cholesterol

treatments close to the time of oophorectomy or menopause (the timing hypothesis) is critical to prevent the initiation and progression of cardiovascular disease [142]. Indeed, the subsequent sub-analysis of data from the WHI that stratified women by menopausal age at the time of treatment randomization suggests that women who were within 5 years of menopause had reduced coronary artery calcification, fewer myocardial infarctions, and lower all-cause mortality compared to those who were randomized after that time point [143–145]. In addition, the Danish Osteoporosis Prevention Study (DOPS), which began at the same time as the WHI, enrolled women between the ages of 45 and 58 years to estrogen or no treatment. Although these women were treated with different formulations of hormone products (triphasic estradiol and norethisterone acetate, a type of birth control pill with estrogen dose constant and variable progestin dose, or 2 mg estradiol a day if they had a hysterectomy), after 11 years of follow-up, the number of myocardial infarctions and incidence of HF and death were reduced in the treated women compared to controls, not receiving treatment [146].

The Early versus Late Intervention Trial with Estradiol (ELITE) directly tested the timing hypothesis. This study enrolled two age groups of women: an early group aged 55 years and a late group aged 65 years. After 5 years of randomization to oral 17β-estradiol, the progression of atherosclerotic plaque measured by intima-medial thickness of the carotid artery was lower in the early aged treated compared to early placebo group. However, the rate of progression of carotid plaque was similar between the treated and placebo in the late aged group [147].

Another criticism of the WHI) was that a conclusion based on single dose and formulation of conjugated equine estrogen was generalized to all hormonal products including those that contain the endogenous form of estrogen 17β-estradiol and those that might be delivered transdermally [141]. Conjugated equine estrogen is a mixture of steroid metabolites with a ratio of estrone to 17-β-estradiol that can range from 6-20 times more estrone than 17β-estradiol [148], sulfonated estrone and estradiol, equilin, and other methylated products. The binding affinity of these different ligands for estrogen receptors varies and cannot be compared by dose equivalency with 17β-estradiol. In addition, oral products, which undergo metabolism in the liver before reaching the systemic circulation, will affect production of procoagulant and inflammatory proteins perhaps increasing risk for thrombosis. Indeed, in the Estrogen and Thromboembolism Risk (ESTHER) multicenter case-controlled study, the odds ratio for developing an idiopathic venous thromboembolism was about four times higher in women using oral compared to those using transdermal products. In addition, the use of micronized progesterone also had a lower odds ratio for developing venous thromboembolism than the use of synthetic progestogens [149, 150]. The risk of venous thromboembolism with the use of MHT will also be affected by prothrombotic mutations [151, 152].

The Kronos Early Estrogen Prevention Study (KEEPS) was designed to provide a direct comparison of oral conjugated equine estrogen (0.45 mg/day) and transdermal 17β-estradiol (50 μg/day) both with pulsed micronized progesterone compared to placebo on slowing progression of atherosclerosis [153]. Unlike women enrolled in the WHI, women in KEEPS were within 3 years of menopause and were at low risk for cardiovascular disease based on conventional risk factors of body mass index, serum lipid profile, insulin sensitivity, and blood pressure [154]. This phenotypic profile was similar to women in early observational studies and reflects a "healthy user" profile that with subsequent analysis of the WHI was one that demonstrated reduced cardiovascular risk with the treatments [136, 155, 156]. In KEEPS after 4 years of treatment, the rate of change of intima-media thickness in the carotid artery did not differ among groups [157]. However, there were no reports of venous thromboembolisms, and both formulations reduced vasomotor symptoms [158]. However, the products differed in regard to effectiveness in improving mood (oral conjugated equine estrogen being better than the transdermal 17β-estradiol) and sexual function (17β-estradiol was better than the oral conjugated equine estrogen) [159, 160]. Variants in genes associated with innate immunity may have contributed to effects of the products on the cardiovascular outcomes measured in KEEPS [161, 162].

In summary, evaluation of the effects of sex hormones on cardiovascular function in women needs to consider the timing of the initiation of treatment; the type, dose, and formulation of the product; the existing cardiovascular risk profile of the woman; and the potential pharmacogenomic effects of the hormones on the outcome of interest.

Conclusion

Sex differences in cardiovascular health and disease have origin in the presence of sex chromosomes, and the production of sex steroid hormones that varies across the life span, is influenced by pregnancy, and by the use of exogenous hormone products. Diagnosis and treatment of some cardiovascular diseases in women have been hampered by the lack of appreciation of symptom presentation in women, poor understanding of cellular mechanisms causing disease, and not treating women according to standard guidelines of care. However, much remains to be learned regarding identification of unique factors which may place women at risk for early development of cardiovascular diseases such as those that might be related to neurovascular autonomic dysregulation. In addition, research is needed to optimize diagnosis and treatment for conditions such as microvascular disease, SCAD, and takotsubo cardiomyopathies and to develop prognostic indicators for women who might be at risk for recurrence of these conditions. Longitudinal studies are needed to determine effects of various types of hormonal treatments on cardiovascular function. Large-scale clinical trials of new treatment options for cardiovascular conditions need to enroll women, and the results of such studies need to be analyzed by sex in order to inform individualized care that will lead to improved outcomes.

References

1. Bellott DW, Hughes JF, Skaletsky H, Brown LG, Pyntikova T, Cho TJ, Koutseva N, Zaghlul S, Graves T, Rock S, et al. Mammalian Y chromosomes retain widely expressed dosage-sensitive regulators. Nature. 2014;508(7497):494–9.
2. Arnold AP, Lusis AJ. Understanding the sexome: measuring and reporting sex differences in gene systems. Endocrinology. 2012;153(6):2551–5.
3. Fu Q. Hemodynamic and electrocardiographic aspects of uncomplicated singleton pregnancy. In: Kerkhof PLM, Miller VM, editors. Sex-specific analysis of cardiovascular function. New York: Springer; 2018. page numbers.
4. Mozaffarian D, Benjamin EJ, Go AS, Arnett DK, Blaha MJ, Cushman M, de Ferranti S, Despres JP, Fullerton HJ, Howard VJ, et al. Heart disease and stroke statistics--2015 update: a report from the American Heart Association. Circulation. 2015;131(4):e29–322.
5. Pancholy SB, Shantha GP, Patel T, Cheskin LJ. Sex differences in short-term and long-term all-cause mortality among patients with ST-segment elevation myocardial infarction treated by primary percutaneous intervention: a meta-analysis. JAMA Intern Med. 2014;174(11):1822–30.
6. Falk E, Nakano M, Bentzon JF, Finn AV, Virmani R. Update on acute coronary syndromes: the pathologists' view. Eur Heart J. 2013;34(10):719–28.
7. Roswell RO, Kunkes J, Chen AY, Chiswell K, Iqbal S, Roe MT, Bangalore S. Impact of sex and contact-to-device time on clinical outcomes in acute ST-segment elevation myocardial infarction-findings from the National Cardiovascular Data Registry. J Am Heart Assoc. 2017;6(1).
8. Choi K, Shofer FS, Mills AM. Sex differences in STEMI activation for patients presenting to the ED 1939. Am J Emerg Med. 2016;34(10):1939–43.
9. Mehta LS, Beckie TM, DeVon HA, Grines CL, Krumholz HM, Johnson MN, Lindley KJ, Vaccarino V, Wang TY, Watson KE, et al. Acute myocardial infarction in women: a scientific statement from the American Heart Association. Circulation. 2016;133(9):916–47.
10. O'Gara PT, Kushner FG, Ascheim DD, Casey DE Jr, Chung MK, de Lemos JA, Ettinger SM, Fang JC, Fesmire FM, Franklin BA, et al. 2013 ACCF/AHA guideline for the management of ST-elevation myocardial infarction: executive summary: a report of the American College of Cardiology Foundation/American Heart Association Task Force on Practice Guidelines: developed in collaboration with the American College of Emergency Physicians and Society for Cardiovascular Angiography and Interventions. Catheter Cardiovasc Interv. 2013;82(1):E1–27.
11. Jneid H, Anderson JL, Wright RS, Adams CD, Bridges CR, Casey DE Jr, Ettinger SM, Fesmire FM, Ganiats TG, Lincoff AM, et al. 2012 ACCF/AHA focused update of the guideline for the management of patients with unstable angina/non-ST-elevation myocardial infarction (updating the 2007 guideline and replacing the 2011 focused update): a report of the American College of Cardiology Foundation/American Heart Association Task Force on Practice Guidelines. J Am Coll Cardiol. 2012;60(7):645–81.
12. Reis SE, Holubkov R, Conrad Smith AJ, Kelsey SF, Sharaf BL, Reichek N, Rogers WJ, Merz CN, Sopko G, Pepine CJ, et al. Coronary microvascular dysfunction is highly prevalent in women with chest pain in the absence of coronary artery disease: results from the NHLBI WISE study. Am Heart J. 2001;141(5):735–41.

13. Jespersen L, Hvelplund A, Abildstrom SZ, Pedersen F, Galatius S, Madsen JK, Jorgensen E, Kelbaek H, Prescott E. Stable angina pectoris with no obstructive coronary artery disease is associated with increased risks of major adverse cardiovascular events. Eur Heart J. 2012;33(6):734–44.
14. von Mering GO, Arant CB, Wessel TR, McGorray SP, Bairey Merz CN, Sharaf BL, Smith KM, Olson MB, Johnson BD, Sopko G, et al. Abnormal coronary vasomotion as a prognostic indicator of cardiovascular events in women: results from the National Heart, Lung, and Blood Institute-Sponsored Women's Ischemia Syndrome Evaluation (WISE). Circulation. 2004;109(6):722–5.
15. Pepine CJ, Anderson RD, Sharaf BL, Reis SE, Smith KM, Handberg EM, Johnson BD, Sopko G, Bairey Merz CN. Coronary microvascular reactivity to adenosine predicts adverse outcome in women evaluated for suspected ischemia results from the National Heart, Lung and Blood Institute WISE (Women's Ischemia Syndrome Evaluation) study. J Am Coll Cardiol. 2010;55(25):2825–32.
16. Camici PG, Crea F. Coronary microvascular dysfunction. N Engl J Med. 2007;356(8):830–40.
17. Ishimori ML, Martin R, Berman DS, Goykhman P, Shaw LJ, Shufelt C, Slomka PJ, Thomson LE, Schapira J, Yang Y, et al. Myocardial ischemia in the absence of obstructive coronary artery disease in systemic lupus erythematosus. JACC Cardiovasc Imaging. 2011;4(1):27–33.
18. Saiki H, Petersen IA, Scott CG, Bailey KR, Dunlay SM, Finley RR, Ruddy KJ, Yan E, Redfield MM. Risk of heart failure with preserved ejection fraction in older women after contemporary radiotherapy for breast cancer. Circulation. 2017;135(15):1388–96.
19. Sandoo A, Kitas GD, Carmichael AR. Breast cancer therapy and cardiovascular risk: focus on trastuzumab. Vasc Health Risk Manag. 2015;11:223–8.
20. Doyle M, Weinberg N, Pohost GM, Bairey Merz CN, Shaw LJ, Sopko G, Fuisz A, Rogers WJ, Walsh EG, Johnson BD, et al. Prognostic value of global MR myocardial perfusion imaging in women with suspected myocardial ischemia and no obstructive coronary disease: results from the NHLBI-sponsored WISE (Women's Ischemia Syndrome Evaluation) study. JACC Cardiovasc Imaging. 2010;3(10):1030–6.
21. Freed BH, Narang A, Bhave NM, Czobor P, Mor-Avi V, Zaran ER, Turner KM, Cavanaugh KP, Chandra S, Tanaka SM, et al. Prognostic value of normal regadenoson stress perfusion cardiovascular magnetic resonance. J Cardiovasc Magn Reson. 2013;15:108.
22. Mehta PK, Bairey Merz CN. Treatment of angina in subjects with evidence of myocardial ischemia and no obstructive coronary artery disease. In: Bonow RO, editor. Braunwald's heart disease. Philadelphia: Elsevier; 2011.
23. Tweet MS, Hayes SN, Pitta SR, Simari RD, Lerman A, Lennon RJ, Gersh BJ, Khambatta S, Best PJ, Rihal CS, et al. Clinical features, management, and prognosis of spontaneous coronary artery dissection. Circulation. 2012;126 (5):579–88.
24. Tweet MS, Hayes SN, Codsi E, Gulati R, Rose CH, Best PJM. Spontaneous coronary artery dissection associated with pregnancy. J Am Coll Cardiol. 2017;70(4):426–35.
25. Saw J, Humphries K, Aymong E, Sedlak T, Prakash R, Starovoytov A, Mancini GBJ. Spontaneous coronary artery dissection: clinical outcomes and risk of recurrence. J Am Coll Cardiol. 2017;70(9):1148–58.
26. Lettieri C, Zavalloni D, Rossini R, Morici N, Ettori F, Leonzi O, Latib A, Ferlini M, Trabattoni D, Colombo P, et al. Management and long-term prognosis of spontaneous coronary artery dissection. Am J Cardiol. 2015;116 (1):66–73.
27. Nakashima T, Noguchi T, Haruta S, Yamamoto Y, Oshima S, Nakao K, Taniguchi Y, Yamaguchi J, Tsuchihashi K, Seki A, et al. Prognostic impact of spontaneous coronary artery dissection in young female patients with acute myocardial infarction: a report from the Angina Pectoris-Myocardial Infarction Multicenter Investigators in Japan. Int J Cardiol. 2016;207:341–8.
28. Rogowski S, Maeder MT, Weilenmann D, Haager PK, Ammann P, Rohner F, Joerg L, Rickli H. Spontaneous coronary artery dissection: angiographic follow-up and long-term clinical outcome in a predominantly medically treated population. Catheter Cardiovasc Interv. 2017;89(1):59–68.
29. Tweet MS, Gulati R, Hayes SN. What clinicians should know about spontaneous coronary artery dissection. Mayo Clin Proc. 2015;90(8):1125–30.
30. Elkayam U. Clinical characteristics of peripartum cardiomyopathy in the United States: diagnosis, prognosis, and management. J Am Coll Cardiol. 2011;58(7):659–70.
31. Tweet MS, Eleid MF, Best PJ, Lennon RJ, Lerman A, Rihal CS, Holmes DR Jr, Hayes SN, Gulati R. Spontaneous coronary artery dissection: revascularization versus conservative therapy. Circ Cardiovasc Interv. 2014;7(6):777–86.
32. McGrath-Cadell L, McKenzie P, Emmanuel S, Muller DW, Graham RM, Holloway CJ. Outcomes of patients with spontaneous coronary artery dissection. Open Heart. 2016;3(2): e000491.
33. Ko D, Rahman F, Schnabel RB, Yin X, Benjamin EJ, Christophersen IE. Atrial fibrillation in women: epidemiology, pathophysiology, presentation, and prognosis. Nat Rev Cardiol. 2016;13(6):321–32.
34. January CT, Wann LS, Alpert JS, Calkins H, Cigarroa JE, Cleveland JC Jr, Conti JB, Ellinor PT, Ezekowitz MD, Field ME, et al. 2014 AHA/ACC/HRS guideline for the management of patients with atrial fibrillation: executive summary: a report of the American College of Cardiology/American Heart Association Task Force on practice guidelines and the Heart Rhythm Society. Circulation. 2014;130 (23):2071–104.
35. Camm AJ, Savelieva I. Female gender as a risk factor for stroke associated with atrial fibrillation. Eur Heart J. 2017;38 (19):1480–4.
36. Renoux C, Coulombe J, Suissa S. Revisiting sex differences in outcomes in non-valvular atrial fibrillation: a population-based cohort study. Eur Heart J. 2017;38(19):1473–9.
37. Abbott GW. KCNE4 and KCNE5: K(+) channel regulation and cardiac arrhythmogenesis. Gene. 2016;593(2):249–60.
38. Sabeh MK, MacRae CA. The genetics of atrial fibrillation. Curr Opin Cardiol. 2010;25(3):186–91.
39. Heinrich J. General Accounting Office, GAO-01-286R Drugs Withdrawn From Market. 2001.
40. Liu H, Jin MW, Xiang JZ, Huang Y, Sun HY, Chiu SW, Lau CP, Li GR. Raloxifene inhibits transient outward and ultra-rapid delayed rectifier potassium currents in human atrial myocytes. Eur J Pharmacol. 2007;563(1–3):61–8.
41. Kocic I, Gruchala M, Petrusewicz J. Selective inhibition of pinacidil effects by estrogen in guinea pig heart. Int J Cardiol. 2006;110(1):22–6.
42. Ko D, Rahman F, Martins MA, Hylek EM, Ellinor PT, Schnabel RB, Benjamin EJ, Christophersen IE. Atrial fibrillation in women: treatment. Nat Rev Cardiol. 2017;14 (2):113–24.
43. Shah SV, Kruse J, Andrei AC, Li Z, Malaisrie SC, Knight BP, Passman RS, McCarthy PM. Gender differences in

outcomes after surgical ablation of atrial fibrillation. J Thorac Cardiovasc Surg. 2016;151(2):391–8. e392
44. Zylla MM, Brachmann J, Lewalter T, Hoffmann E, Kuck KH, Andresen D, Willems S, Eckardt L, Tebbenjohanns J, Spitzer SG, et al. Sex-related outcome of atrial fibrillation ablation: Insights from the German Ablation Registry. Heart Rhythm. 2016;13(9):1837–44.
45. Benjamin EJ, Blaha MJ, Chiuve SE, Cushman M, Das SR, Deo R, de Ferranti SD, Floyd J, Fornage M, Gillespie C, et al. Heart disease and stroke statistics-2017 update: a report from the American Heart Association. Circulation. 2017;135(10): e146–603.
46. Yancy CW, Jessup M, Bozkurt B, Butler J, Casey DE Jr, Drazner MH, Fonarow GC, Geraci SA, Horwich T, Januzzi JL, et al. 2013 ACCF/AHA guideline for the management of heart failure: a report of the American College of Cardiology Foundation/American Heart Association Task Force on Practice Guidelines. J Am Coll Cardiol. 2013;62(16):e147–239.
47. Mohammed SF, Hussain S, Mirzoyev SA, Edwards WD, Maleszewski JJ, Redfield MM. Coronary microvascular rarefaction and myocardial fibrosis in heart failure with preserved ejection fraction. Circulation. 2015;131(6):550–9.
48. Kerkhof PL, Peace RA, Heyndrickx GR, Handly N. Heart function analysis in cardiac patients with focus on sex-specific aspects. In: Kerkhof PLM, Miller VM, editors. Sex-specific analysis of cardiovascular function. New York: Springer; 2018. page numbers.
49. Yancy CW, Jessup M, Bozkurt B, Butler J, Casey DE Jr, Colvin MM, Drazner MH, Filippatos GS, Fonarow GC, Givertz MM, et al. 2017 ACC/AHA/HFSA focused update of the 2013 ACCF/AHA guideline for the management of heart failure: a report of the American College of Cardiology/American Heart Association Task Force on clinical practice guidelines and the Heart Failure Society of America. J Card Fail. 2017;23(8):628–51.
50. Lam PH, Dooley DJ, Deedwania P, Singh SN, Bhatt DL, Morgan CJ, Butler J, Mohammed SF, Wu WC, Panjrath G, et al. Heart rate and outcomes in hospitalized patients with heart failure with preserved ejection fraction. J Am Coll Cardiol. 2017;70(15):1861–71.
51. Dunlay SM, Roger VL, Redfield MM. Epidemiology of heart failure with preserved ejection fraction. Nat Rev Cardiol. 2017;14(10):591–602.
52. Kotecha D, Manzano L, Krum H, Rosano G, Holmes J, Altman DG, Collins PD, Packer M, Wikstrand J, Coats AJ, et al. Effect of age and sex on efficacy and tolerability of beta blockers in patients with heart failure with reduced ejection fraction: individual patient data meta-analysis. BMJ. 2016;353:i1855.
53. Keyhan G, Chen SF, Pilote L. Angiotensin-converting enzyme inhibitors and survival in women and men with heart failure. Eur J Heart Fail. 2007;9(6–7):594–601.
54. Hudson M, Rahme E, Behlouli H, Sheppard R, Pilote L. Sex differences in the effectiveness of angiotensin receptor blockers and angiotensin converting enzyme inhibitors in patients with congestive heart failure--a population study. Eur J Heart Fail. 2007;9(6–7):602–9.
55. Cook JL, Grady KL, Colvin M, Joseph SM, Brisco MA, Walsh MN, gen VADWG. Sex differences in the care of patients with advanced heart failure. Circ Cardiovasc Qual Outcomes. 2015;8(2 Suppl 1):S56–9.
56. Rose EA, Gelijns AC, Moskowitz AJ, Heitjan DF, Stevenson LW, Dembitsky W, Long JW, Ascheim DD, Tierney AR, Levitan RG, et al. Long-term use of a left ventricular assist device for end-stage heart failure. N Engl J Med. 2001;345 (20):1435–43.
57. Stevenson LW, Miller LW, Desvigne-Nickens P, Ascheim DD, Parides MK, Renlund DG, Oren RM, Krueger SK, Costanzo MR, Wann LS, et al. Left ventricular assist device as destination for patients undergoing intravenous inotropic therapy: a subset analysis from REMATCH (Randomized Evaluation of Mechanical Assistance in Treatment of Chronic Heart Failure). Circulation. 2004;110(8):975–81.
58. Bogaev RC, Pamboukian SV, Moore SA, Chen L, John R, Boyle AJ, Sundareswaran KS, Farrar DJ, Frazier OH, HeartMate IICI. Comparison of outcomes in women versus men using a continuous-flow left ventricular assist device as a bridge to transplantation. J Heart Lung Transplant. 2011;30 (5):515–22.
59. Blumer V, Mendirichaga R, Hernandez GA, Zablah G, Chaparro SV. Sex-specific outcome disparities in patients receiving continuous-flow left ventricular assist devices: a systematic review and meta-analysis. ASAIO J. 2017. https://doi.org/10.1097/MAT.0000000000000695.
60. Aimo A, Vergaro G, Castiglione V, Barison A, Pasanisi E, Petersen C, Chubuchny V, Giannoni A, Poletti R, Maffei S, et al. Effect of sex on reverse remodeling in chronic systolic heart failure. JACC Heart Fail. 2017;5(10):735–42.
61. Lyon AR, Bossone E, Schneider B, Sechtem U, Citro R, Underwood SR, Sheppard MN, Figtree GA, Parodi G, Akashi YJ, et al. Current state of knowledge on Takotsubo syndrome: a Position Statement from the Taskforce on Takotsubo Syndrome of the Heart Failure Association of the European Society of Cardiology. Eur J Heart Fail. 2016;18(1):8–27.
62. Abraham J, Mudd JO, Kapur NK, Klein K, Champion HC, Wittstein IS. Stress cardiomyopathy after intravenous administration of catecholamines and beta-receptor agonists. J Am Coll Cardiol. 2009;53(15):1320–5.
63. Becker LC, Pepine CJ, Bonsall R, Cohen JD, Goldberg AD, Coghlan C, Stone PH, Forman S, Knatterud G, Sheps DS, et al. Left ventricular, peripheral vascular, and neurohumoral responses to mental stress in normal middle-aged men and women. Reference Group for the Psychophysiological Investigations of Myocardial Ischemia (PIMI) Study. Circulation. 1996;94(11):2768–77.
64. Wittstein IS, Thiemann DR, Lima JA, Baughman KL, Schulman SP, Gerstenblith G, Wu KC, Rade JJ, Bivalacqua TJ, Champion HC. Neurohumoral features of myocardial stunning due to sudden emotional stress. N Engl J Med. 2005;352(6):539–48.
65. Merli E, Sutcliffe S, Gori M, Sutherland GG. Tako-Tsubo cardiomyopathy: new insights into the possible underlying pathophysiology. Eur J Echocardiogr. 2006;7(1):53–61.
66. de Gregorio C, Grimaldi P, Lentini C. Left ventricular thrombus formation and cardioembolic complications in patients with Takotsubo-like syndrome: a systematic review. Int J Cardiol. 2008;131(1):18–24.
67. Templin C, Ghadri JR, Diekmann J, Napp LC, Bataiosu DR, Jaguszewski M, Cammann VL, Sarcon A, Geyer V, Neumann CA, et al. Clinical features and outcomes of takotsubo (stress) cardiomyopathy. N Engl J Med. 2015;373(10):929–38.
68. Arany Z, Elkayam U. Peripartum cardiomyopathy. Circulation. 2016;133(14):1397–409.
69. Hilfiker-Kleiner D, Kaminski K, Podewski E, Bonda T, Schaefer A, Sliwa K, Forster O, Quint A, Landmesser U, Doerries C, et al. A cathepsin D-cleaved 16 kDa form of prolactin mediates postpartum cardiomyopathy. Cell. 2007;128(3):589–600.
70. Patten IS, Rana S, Shahul S, Rowe GC, Jang C, Liu L, Hacker MR, Rhee JS, Mitchell J, Mahmood F, et al. Cardiac

angiogenic imbalance leads to peripartum cardiomyopathy. Nature. 2012;485(7398):333–8.
71. Damp J, Givertz MM, Semigran M, Alharethi R, Ewald G, Felker GM, Bozkurt B, Boehmer J, Haythe J, Skopicki H, et al. Relaxin-2 and soluble Flt1 levels in peripartum cardiomyopathy: results of the multicenter IPAC study. JACC Heart Fail. 2016;4(5):380–8.
72. Ware JS, Li J, Mazaika E, Yasso CM, DeSouza T, Cappola TP, Tsai EJ, Hilfiker-Kleiner D, Kamiya CA, Mazzarotto F, et al. Shared genetic predisposition in peripartum and dilated cardiomyopathies. N Engl J Med. 2016;374(3):233–41.
73. Lange S, Auerbach D, McLoughlin P, Perriard E, Schafer BW, Perriard JC, Ehler E. Subcellular targeting of metabolic enzymes to titin in heart muscle may be mediated by DRAL/FHL-2. J Cell Sci. 2002;115(Pt 24):4925–36.
74. Freeman EW, Sherif K. Prevalence of hot flushes and night sweats around the world: a systematic review. Climacteric. 2007;10(3):197–214.
75. Mishra GD, Kuh D. Health symptoms during midlife in relation to menopausal transition: British prospective cohort study. BMJ. 2012;344:e402.
76. Tepper PG, Brooks MM, Randolph JF Jr, Crawford SL, El Khoudary SR, Gold EB, Lasley BL, Jones B, Joffe H, Hess R, et al. Characterizing the trajectories of vasomotor symptoms across the menopausal transition. Menopause. 2016;23(10):1067–74.
77. Avis NE, Crawford SL, Greendale G, Bromberger JT, Everson-Rose SA, Gold EB, Hess R, Joffe H, Kravitz HM, Tepper PG, et al. Duration of menopausal vasomotor symptoms over the menopause transition. JAMA Intern Med. 2015;175(4):531–9.
78. Huang AJ, Sawaya GF, Vittinghoff E, Lin F, Grady D. Hot flushes, coronary heart disease, and hormone therapy in postmenopausal women. Menopause. 2009;16(4):639–43.
79. Jackson EA, El Khoudary SR, Crawford SL, Matthews K, Joffe H, Chae C, Thurston RC. Hot flash frequency and blood pressure: data from the study of women's health across the nation. J Womens Health (Larchmt). 2016;25(12):1204–9.
80. Thurston RC, Sutton-Tyrell K, Everson-Rose SA, Hess R, Matthews KA. Hot flashes and subclinical cardiovascular disease. Findings from the study of women's health across the nation heart study. Circulation. 2008;118:1234–40.
81. Muka T, Oliver-Williams C, Colpani V, Kunutsor S, Chowdhury S, Chowdhury R, Kavousi M, Franco OH. Association of vasomotor and other menopausal symptoms with risk of cardiovascular disease: a systematic review and meta-analysis. PLoS One. 2016;11(6):e0157417.
82. Stuenkel CA, Davis SR, Gompel A, Lumsden MA, Murad MH, Pinkerton JV, Santen RJ. Treatment of symptoms of the menopause: an endocrine society clinical practice guideline. J Clin Endocrinol Metab. 2015;100(11):3975–4011.
83. Svartberg J, von Muhlen D, Kritz-Silverstein D, Barrett-Connor E. Vasomotor symptoms and mortality: the Rancho Bernardo Study. Menopause. 2009;16(5):888–91.
84. Allison MA, Manson JE, Aragaki A, Langer RD, Rossouw J, Curb D, Martin LW, Phillips L, Stefanick ML, Cochrane BB, et al. Vasomotor symptoms and coronary artery calcium in postmenopausal women. Menopause. 2010;17(6):1136–45.
85. Szmuilowicz ED, Manson JE, Rossouw JE, Howard BV, Margolis KL, Greep NC, Brzyski RG, Stefanick ML, O'Sullivan MJ, Wu C, et al. Vasomotor symptoms and cardiovascular events in postmenopausal women. Menopause. 2011;18(6):603–10.
86. Cintron D, Lipford M, Larrea-Mantilla L, Spencer-Bonilla G, Lloyd R, Gionfriddo MR, Gunjal S, Farrell AM, Miller VM, Murad MH. Efficacy of menopausal hormone therapy on sleep quality: systematic review and meta-analysis. Endocrine. 2017;55(3):702–11.
87. Gao C, Kapoor E, Lipford M, Miller V, Schroeder D, Mara K, Faubion S. Association of vasomotor symptoms and sleep apnea risk in midlife women. Menopause. 2017;25(4). https://doi.org/10.1097/GME.0000000000001020. (in press)
88. Kwon Y, Duprez DA, Jacobs DR, Nagayoshi M, McClelland RL, Shahar E, Budoff M, Redline S, Shea S, Carr JJ, et al. Obstructive sleep apnea and progression of coronary artery calcium: the multi-ethnic study of atherosclerosis study. J Am Heart Assoc. 2014;3(5):e001241.
89. Mercuro G, Podda A, Pitzalis L, Zoncu S, Mascia M, Melis GB, Rosano GM. Evidence of a role of endogenous estrogen in the modulation of autonomic nervous system. Am J Cardiol. 2000;85(6):787–9. A789
90. de Zambotti M, Trinder J, Javitz H, Colrain IM, Baker FC. Altered nocturnal blood pressure profiles in women with insomnia disorder in the menopausal transition. Menopause. 2017;24(3):278–87.
91. Rance NE, Dacks PA, Mittelman-Smith MA, Romanovsky AA, Krajewski-Hall SJ. Modulation of body temperature and LH secretion by hypothalamic KNDy (kisspeptin, neurokinin B and dynorphin) neurons: a novel hypothesis on the mechanism of hot flushes. Front Neuroendocrinol. 2013;34(3):211–27.
92. Prague JK, Roberts RE, Comninos AN, Clarke S, Jayasena CN, Nash Z, Doyle C, Papadopoulou DA, Bloom SR, Mohideen P, et al. Neurokinin 3 receptor antagonism as a novel treatment for menopausal hot flushes: a phase 2, randomised, double-blind, placebo-controlled trial. Lancet. 2017;389(10081):1809–20.
93. Cintron D, Lahr BD, Bailey KR, Santoro N, Lloyd R, Manson JE, Neal-Perry G, Pal L, Taylor HS, Wharton W, et al. Effects of oral versus transdermal menopausal hormone treatments on self-reported sleep domains and their association with vasomotor symptoms in recently menopausal women enrolled in the Kronos Early Estrogen Prevention Study (KEEPS). Menopause. 2017.
94. McCall RB, Clement ME. Role of serotonin1A and serotonin2 receptors in the central regulation of the cardiovascular system. PharmacolRev. 1994;46:231–43.
95. Deecher D, Andree TH, Sloan D, Schechter LE. From menarche to menopause: Exploring the underlying biology of depression in women experiencing hormonal changes. Psychoneuroendocrinology. 2008;33:3–17.
96. Santoro N. Symptoms of menopause: hot flushes. Clin Obstet Gynecol. 2008;51(3):539–48.
97. Tuomikoski P, Ebert P, Groop PH, Haapalahti P, Hautamaki H, Ronnback M, Ylikorkala O, Mikkola TS. Evidence for a role of hot flushes in vascular function in recently postmenopausal women. Obstet Gynecol. 2009;113(4):902–8.
98. Tuomikoski P, Ebert P, Groop PH, Haapalahti P, Hautamaki H, Ronnback M, Ylikorkala O, Mikkola TS. Effect of hot flushes on vascular function: a randomized controlled trial. Obstet Gynecol. 2009;114(4):777–85.
99. Sassarini J, Fox H, Ferrell W, Sattar N, Lumsden MA. Vascular function and cardiovascular risk factors in women with severe flushing. Clin Endocrinol (Oxf). 2011;74(1):97–103.
100. Miller VM, Duckles SP. Vascular actions of estrogens: functional implications. PharmacolRev. 2008;60(2):210–41.
101. Martin VT, Pavlovic J, Fanning KM, Buse DC, Reed ML, Lipton RB. Perimenopause and menopause are associated

with high frequency headache in women with migraine: results of the American Migraine Prevalence and Prevention Study. Headache. 2016;56(2):292–305.
102. Pavlovic JM, Allshouse AA, Santoro NF, Crawford SL, Thurston RC, Neal-Perry GS, Lipton RB, Derby CA. Sex hormones in women with and without migraine: evidence of migraine-specific hormone profiles. Neurology. 2016;87 (1):49–56.
103. An X, Fang J, Lin Q, Lu C, Ma Q, Qu H. New evidence for involvement of ESR1 gene in susceptibility to Chinese migraine. J Neurol. 2017;264(1):81–7.
104. CoSkun S, Yucel Y, Cim A, Cengiz B, Oztuzcu S, Varol S, Ozdemir HH, Uzar E. Contribution of polymorphisms in ESR1, ESR2, FSHR, CYP19A1, SHBG, and NRIP1 genes to migraine susceptibility in Turkish population. J Genet. 2016;95(1):131–40.
105. Sutherland HG, Champion M, Plays A, Stuart S, Haupt LM, Frith A, MacGregor EA, Griffiths LR. Investigation of polymorphisms in genes involved in estrogen metabolism in menstrual migraine. Gene. 2017;607:36–40.
106. Hershey A, Horn P, Kabbouche M, O'Brien H, Powers S. Genomic expression patterns in menstrual-related migraine in adolescents. Headache. 2012;52(1):68–79.
107. Loewendorf AI, Matynia A, Saribekyan H, Gross N, Csete M, Harrington M. Roads less traveled: sexual dimorphism and mast cell contributions to migraine pathology. Front Immunol. 2016;7:140.
108. Borkum JM. Migraine triggers and oxidative stress: a narrative review and synthesis. Headache. 2016;56(1):12–35.
109. Calhoun AH, Gill N. Presenting a new, non-hormonally mediated cyclic headache in women: end-menstrual migraine. Headache. 2017;57(1):17–20.
110. Finsterer J, Zarrouk-Mahjoub S. Mitochondrial vasculopathy. World J Cardiol. 2016;8(5):333–9.
111. Bousser MG, Welch KMA. Relation between migraine and stroke. Lancet Neurol. 2005;4:533–42.
112. Bushnell CD, Hurn P, Colton C, Miller VM, del Zoppo G, Elkind MSV, Stern BJ, Herrington D, Ford-Lynch G, Gorelick P, et al. Advancing the study of stroke in women. Summary and recommendations for future research from an NINDS-sponsored multidisciplinary working group. Stroke. 2006. (in press). https://doi.org/10.1161/01.STR.0000236053.37695.15.
113. Androulakis XM, Kodumuri N, Giamberardino LD, Rosamond WD, Gottesman RF, Yim E, Sen S. Ischemic stroke subtypes and migraine with visual aura in the ARIC study. Neurology. 2016;87(24):2527–32.
114. Moyer AM, de Andrade M, Weinshilboum RM, Miller VM. Influence of SULT1A1 genetic variation on age at menopause, estrogen levels, and response to hormone therapy in recently postmenopausal white women. Menopause. 2016;23(8):863–9.
115. Sudhir K, Chou TM, Chatterjee K, Smith EP, Williams TC, Kane JP, Malloy MJ, Korach KS, Rubanyi GM. Premature coronary artery disease associated with a disruptive mutation in the estrogen receptor gene in a man. Circulation. 1997;96:3774–7.
116. Gialeraki A, Valsami S, Pittaras T, Panayiotakopoulos G, Politou M. Oral contraceptives and HRT risk of thrombosis. Clin Appl Thromb Hemost. 2016. https://doi.org/10.1177/1076029616683802.
117. Shufelt CL, Bairey Merz CN. Contraceptive hormone use and cardiovascular disease. J Am Coll Cardiol. 2009;53 (3):221–31.
118. Samson ME, Adams SA, Merchant AT, Maxwell WD, Zhang J, Bennett CL, Hebert JR. Cardiovascular disease incidence among females in South Carolina by type of oral contraceptives, 2000–2013: a retrospective cohort study. Arch Gynecol Obstet. 2016;294(5):991–7.
119. Dinger J, Do Minh T, Heinemann K. Impact of estrogen type on cardiovascular safety of combined oral contraceptives. Contraception. 2016;94(4):328–39.
120. Dokras A. Noncontraceptive use of oral combined hormonal contraceptives in polycystic ovary syndrome-risks versus benefits. Fertil Steril. 2016;106(7):1572–9.
121. Shuster LT, Gostout BS, Grossardt BR, Rocca WA. Prophylactic oophorectomy in premenopausal women and long-term health. Menopause Int. 2008;14(3):111–6.
122. Rocca WA, Gazzuola-Rocca L, Smith CY, Grossardt BR, Faubion SS, Shuster LT, Kirkland JL, Stewart EA, Miller VM. Accelerated accumulation of multimorbidity after bilateral oophorectomy: a population-based cohort study. Mayo Clin Proc. 2016;91(11):1577–89.
123. Rocca WA, Gazzuola Rocca L, Smith CY, Grossardt BR, Faubion SS, Shuster LT, Kirkland JL, Stewart EA, Miller VM. Bilateral oophorectomy and accelerated aging: cause or effect? J Gerontol A Biol Sci Med Sci. 2017;72(9):1213–7.
124. Rivera CM, Grossardt BR, Rhodes DJ, Brown RD, Roger VL, Melton LJ, Rocca WA. Increased cardiovascular mortality after early bilateral oophorectomy. Menopause. 2009;16 (1):15–23.
125. Daan NM, Muka T, Koster MP, Roeters van Lennep JE, Lambalk CB, Laven JS, Fauser CG, Meun C, de Rijke YB, Boersma E, et al. Cardiovascular risk in women with premature ovarian insufficiency compared to premenopausal women at middle age. J Clin Endocrinol Metab. 2016;101 (9):3306–15.
126. Roeters van Lennep JE, Heida KY, Bots ML, Hoek A. collaborators of the Dutch Multidisciplinary Guideline Development Group on Cardiovascular Risk Management after Reproductive D: Cardiovascular disease risk in women with premature ovarian insufficiency: a systematic review and meta-analysis. Eur J Prev Cardiol. 2016;23 (2):178–86.
127. Sullivan SD, Sarrel PM, Nelson LM. Hormone replacement therapy in young women with primary ovarian insufficiency and early menopause. Fertil Steril. 2016;106(7):1588–99.
128. Rocca WA, Grossardt BR, Miller VM, Shuster LT, Brown RD Jr. Premature menopause or early menopause and risk of ischemic stroke. Menopause. 2012;19(3):272–7.
129. Criqui MH, Suarez L, Barrett-Connor E, McPhillips J, Wingard DL, Garland C. Postmenopausal estrogen use and mortality. Results from a prospective study in a defined, homogeneous community. Am J Epidemiol. 1988;128:606–14.
130. Falkeborn M, Persson I, Adami HO, Bergstrom R, Eaker E, Lithell H, Mohsen R, Naessen T. The risk of acute myocardial infarction after oestrogen and oestrogen-progestogen replacement. Br J Obstet Gynaecol. 1992;99 (10):821–8.
131. Henderson BE, Paganini-Hill A, Ross RK. Estrogen replacement therapy and protection from acute myocardial infarction. Am J Obstet Gynecol. 1988;159(2):312–7.
132. Petitti DB, Perlman JA, Sidney S. Noncontraceptive estrogens and mortality: long-term follow-up of women in the Walnut Creek Study. Obstet Gynecol. 1987;70(3 Pt 1):289–93.

133. Stampfer MJ, Willett WC, Colditz GA, Rosner B, Speizer FE, Hennekens CH. A prospective study of postmenopausal estrogen therapy and coronary heart disease. N Engl J Med. 1985;313(17):1044–9.
134. Wolf PH, Madans JH, Finucane FF, Higgins M, Kleinman JC. Reduction of cardiovascular disease-related mortality among postmenopausal women who use hormones: evidence from a national cohort. Am J Obstet Gynecol. 1991;164:489–94.
135. Bush TL, Barrett-Connor E, Cowan LD, Criqui MH, Wallace RB, Suchindran CM, Tyroler HA, Rifkind BM. Cardiovascular mortality and noncontraceptive use of estrogen in women: results from the Lipid Research Clinics Program Follow-up Study. Circulation. 1987;75:1102–9.
136. Hodis H, Mack W, Lobo R, Shoupe D, Sevanian A, Mahrer P, Selzer R, Liu C, Liu C, Azen S. Estrogen in the prevention of atherosclerosis. a randomized, double-blind, placebo-controlled trial. Ann Intern Med. 2001;135(11):939–53.
137. Miller VM, Harman SM. An update on hormone therapy in postmenopausal women: mini-review for the basic scientist. AmJPhysiol (Heart CircPhysiol 42). 2017;313(5):H1013–21.
138. Anderson G, Cummings S, Freedman LS, Furberg C, Henderson M, Johnson SR, Kuller L, Manson J, Oberman A, Prentice RL, et al. Design of the women's health initiative clinical trial and observational study. Control Clin Trials. 1998;19(1):61–109.
139. Rossouw JE, Anderson GL, Prentice RL, LaCroix AZ, Kooperberg C, Stefanick ML, Jackson RD, Beresford SA, Howard BV, Johnson KC, et al. Risks and benefits of estrogen plus progestin in healthy postmenopausal women: principal results From the Women's Health Initiative randomized controlled trial. JAMA. 2002;288(3):321–33.
140. Langer RD, Simon JA, Pines A, Lobo RA, Hodis HN, Pickar JH, Archer DF, Sarrel PM, Utian WH. Menopausal hormone therapy for primary prevention: why the USPSTF is wrong. Menopause. 2017;24(10):1101–12.
141. Langer RD. The evidence base for HRT: what can we believe? Climacteric. 2017;20(2):91–6.
142. Clarkson TB, Melendez GC, Appt SE. Timing hypothesis for postmenopausal hormone therapy: its origin, current status, and future. Menopause. 2013;20(3):342–53.
143. Hsia J, Langer RD, Manson JE, Kuller L, Johnson KC, Hendrix SL, Pettinger M, Heckbert SR, Greep N, Crawford S, et al. Conjugated equine estrogens and coronary heart disease. Arch Intern Med. 2006;166:357–65.
144. Manson J, Allison M, Rossouw JE, Carr J, Langer R, Hsia J, Kuller L, Cochrane B, Hunt J, Ludlam S, et al. Estrogen therapy and coronary-artery calcification. N Engl J Med. 2007;356(25):2591–602.
145. Rossouw J, Prentice R, Manson J, Wu L, Barad D, Barnabei V, Ko M, LaCroix A, Margolis K, Stefanick M. Postmenopausal hormone therapy and risk of cardiovascular disease by age and years since menopause. JAMA. 2007;297(13):1465.
146. Schierbeck L, Rejnmark L, Tofteng C, Stilgren L, Eiken P, Mosekilde L, Kober L, Jensen J. Effect of hormone replacement therapy on cardiovascular events in recently postmenopausal women: randomized trial. BMJ. 2012;345:e6409.
147. Hodis HN, Mack WJ, Henderson VW, Shoupe D, Budoff MJ, Hwang-Levine J, Li Y, Feng M, Dustin L, Kono N, et al. Vascular effects of early versus late postmenopausal treatment with estradiol. N Engl J Med. 2016;374(13):1221–31.
148. Faigle JW, Schenkel L. Pharmacokinetics of estrogens and progestogens, vol. 1. London: Chuyrchill Livingston; 1998.
149. Scarabin PY, Oger E, Plu-Bureau G. Differential association of oral and transdermal oestrogen-replacement therapy with venous thromboembolism risk. Lancet. 2003;362:428–32.
150. Canonico M, Oger E, Plu-Bureau G, Conard J, Meyer G, Levesque H, Trillot N, Barrellier MT, Wahl D, Emmerich J, et al. Estrogen and Thromboembolism Risk (ESTHER) Study Group. Hormone therapy and venous thromboembolism among postmenopausal women: impact of the route of estrogen administration and progestogens: the ESTHER study. Circulation. 2007;115(7):840–5.
151. Straczek C, Oger E, de Jonage-Canonico BY, Plu-Bureau G, Conard J, Meyer G, Alhenc-Gelas M, Levesque H, Trillot N, Barrellier M-T, et al. Prothrombotic mutations, hormone therapy, and venous thromboembolism among postmenopausal women. Impact of the route of estrogen administration. Circulation. 2005;112:3495–500.
152. de Andrade M, Armasu SM, McCauley BM, Petterson TM, Heit JA. Identification of genetic interaction with risk factors using a time-to-event model. Int J Environ Res Public Health. 2017;14(10). https://doi.org/10.3390/ijerph14101228.
153. Harman SM, Brinton EA, Cedars M, Lobo R, Manson JE, Merriam GR, Miller VM, Naftolin F, Santoro N. KEEPS: the Kronos Early Estrogen Prevention Study. Climacteric. 2005;8:3–12.
154. Miller VM, Black DM, Brinton EA, Budoff MJ, Cedars MI, Hodis HN, Lobo RA, Manson JE, Merriam GR, Naftolin F, et al. Using basic science to design a clinical trial: baseline characteristics of women enrolled in the Kronos Early Estrogen Prevention Study (KEEPS). J Cardiovasc Transl Res. 2009;2(3):228–39.
155. Bray PF, Larson JC, Lacroix AZ, Manson J, Limacher MC, Rossouw JE, Lasser NL, Lawson WE, Stefanick ML, Langer RD, et al. Usefulness of baseline lipids and C-reactive protein in women receiving menopausal hormone therapy as predictors of treatment-related coronary events. Am J Cardiol. 2008;101(11):1599–605.
156. Wild RA, Wu C, Curb JD, Martin LW, Phillips L, Stefanick M, Trevisan M, Manson JE. Coronary heart disease events in the Women's Health Initiative hormone trials: effect modification by metabolic syndrome: a nested case-control study within the Women's Health Initiative randomized clinical trials. Menopause. 2013;20:254–60.
157. Harman SM, Black DM, Naftolin F, Brinton EA, Budoff MJ, Cedars MI, Hopkins PN, Lobo RA, Manson JEM, Merriam GR, Miller VM, et al. Arterial imaging outcomes and cardiovascular risk factors in recently menopausal women: a randomized trial. Ann Intern Med. 2014;161(4):249–60.
158. Santoro N, Allshouse A, Neal-Perry G, Pal L, Lobo RA, Naftolin F, Black DM, Brinton EA, Budoff MJ, Cedars MI, et al. Longitudinal changes in menopausal symptoms comparing women randomized to low-dose oral conjugated estrogens or transdermal estradiol plus micronized progesterone versus placebo: the Kronos Early Estrogen Prevention Study. Menopause. 2017;24(3):238–46.
159. Gleason CE, Dowling NM, Wharton W, Manson JE, Miller VM, Atwood CS, Brinton EA, Cedars MI, Lobo RA, Merriam GR, et al. Effects of hormone therapy on cognition and mood in recently postmenopausal women: findings from the randomized, controlled KEEPS-Cognitive and Affective Study. PLoS medicine. 2015;12(6):e1001833.
160. Taylor HS, Tal A, Pal L, Li F, Black DM, Brinton EA, Budoff MJ, Cedars MI, Du W, Hodis HN, et al. Effects of

oral vs transdermal estrogen therapy on sexual function in early postmenopause: ancillary study of the Kronos Early Estrogen Prevention Study (KEEPS). JAMA Intern Med. 2017;177(10):1471–9.
161. Miller VM, Petterson TM, Jeavons EN, Lnu AS, Rider DN, Heit JA, Cunningham JM, Huggins GS, Hodis HN, Budoff MJ, et al. Genetic polymorphisms associated carotid artery intima-media thickness and coronary artery calcification in women of the Kronos Early Estrogen Prevention Study. Physiol Genomics. 2013;45(2):79–88.
162. Miller VM, Jenkins GD, Biernacka JM, Heit JA, Huggins GS, Hodis HN, Budoff MJ, Lobo RA, Taylor HS, Manson JE, et al. Pharmacogenomics of estrogens on changes in carotid artery intima-medial thickness and coronary arterial calcification: Kronos Early Estrogen Prevention Study. Physiol Genomics. 2016;48(1):33–41.

Cardiovascular Sequels During and After Preeclampsia

28

Nina D. Paauw and A. Titia Lely

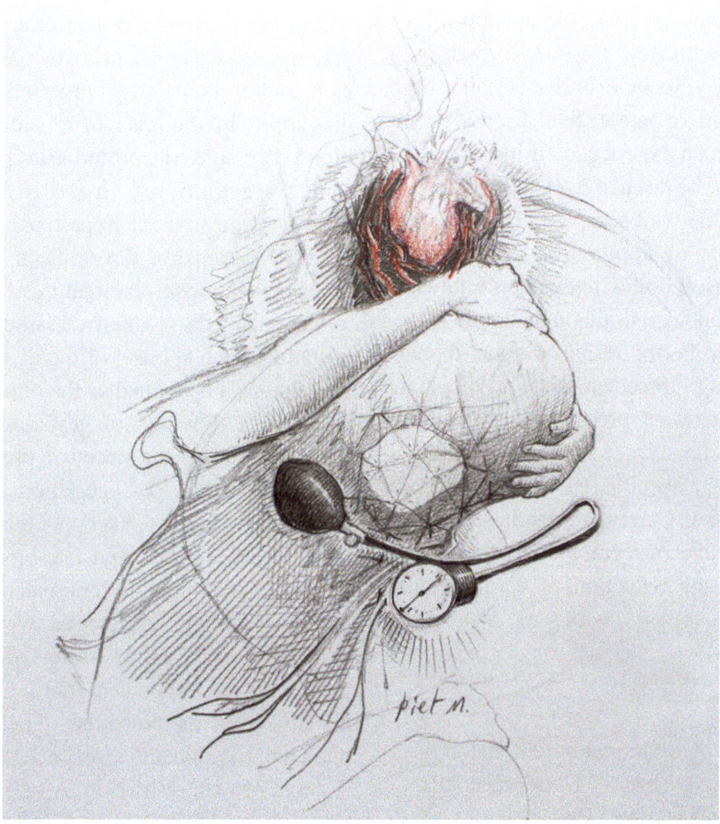

Pregnancy-specific disorders. Artwork by Piet Michiels, Leuven, Belgium.

N. D. Paauw · A. T. Lely (✉)
Department of Obstetrics, Wilhelmina Children's Hospital Birth Center, University Medical Center Utrecht, University of Utrecht, Utrecht, The Netherlands
e-mail: A.T.Lely@umcutrecht.nl

Abstract

Preeclampsia is a pregnancy-specific disorder complicating 2%–8% of pregnancies worldwide and characterized by de novo development

P. L. M. Kerkhof, V. M. Miller (eds.), *Sex-Specific Analysis of Cardiovascular Function*, Advances in Experimental Medicine and Biology 1065, https://doi.org/10.1007/978-3-319-77932-4_28

of hypertension and proteinuria. Current understanding of the pathophysiology of preeclampsia is limited. A main feature is disrupted spiral artery remodeling in the placenta, which restricts the blood flow to the placenta, which in turn leads to decreased uteroplacental perfusion. Impaired blood flow through the placenta might result in fetal growth restriction and secretion of several factors by the placenta—mainly pro-inflammatory cytokines and anti-angiogenic factors—which spread into the maternal circulation, leading to endothelial dysfunction, which subsequently results in disrupted maternal hemodynamics. To date, no treatment options are available apart from termination of pregnancy. Despite normalization of the maternal vascular disturbances after birth, it has become apparent that formerly preeclamptic women experience an increased risk to develop cardiovascular and kidney disease later in life. One well-accepted concept is that the development of preeclampsia is an indicator of maternal susceptibility to develop future cardiovascular conditions, although the increased risk might also be the result of organ damage caused during preeclampsia. Given the associations between preeclampsia and long-term complications, preeclampsia is acknowledged as woman-specific risk factor for cardiovascular disease. Current research focuses on finding effective screening and prevention strategies for the reduction of cardiovascular disease in women with a history of preeclampsia.

Keywords

Preeclampsia · Gestational hypertension · Maternal hemodynamics · Uteroplacental perfusion · Placentation · Placental dysfunction · Syncytiotrophoblast · Pro-inflammatory cytokines · Anti-angiogenic factor · s-Flt1 · Endoglin · Renin-angiotensin aldosterone system · Fetal growth restriction · Renal function

Preeclampsia

Definition and Clinical Features

Preeclampsia is a pregnancy-specific disorder that complicates 2%–8% of pregnancies worldwide [1, 2]. Traditionally the condition is diagnosed in the case of de novo development of hypertension ($\geq$ 140/90 mmHg) and proteinuria during the second half of pregnancy (>20 weeks of gestation) [3]. According to the latest definitions, however, new-onset thrombocytopenia, renal insufficiency, neurological complications, liver involvement and fetal growth restriction may substitute for new-onset proteinuria [4]. In addition, preeclampsia can occur with chronic hypertension and/or preexisting proteinuria and is then diagnosed in the case of a sudden increase in blood pressure or proteinuria [4]. Within the same spectrum of hypertensive pregnancy disorders, gestational hypertension is diagnosed when hypertension develops in the absence of other preeclampsia-like features.

Preeclampsia is often classified based on the gestational age at onset of the disease; early-onset preeclampsia is defined as the onset of preeclampsia before 34 weeks of gestation and late-onset preeclampsia, as the onset of preeclampsia at or after 34 weeks of gestation. Although the presenting features overlap, this distinction is made because early- and late-onset preeclampsia are associated with different maternal and neonatal outcomes. In the developed world, the burden associated with earlier onset of the disease is mostly related to neonatal morbidity and mortality because of the increased risk of perinatal death and neonatal complications associated with prematurity and growth restriction. Another reason to distinguish early- and late-onset preeclampsia is that they are thought to have a different etiologic background and therefore are often assessed separately in pathophysiologic studies. A less frequently used classification of preeclampsia is based on the severity of the disease, with severe preeclampsia defined by blood pressure > 160 mmHg systolic or > 110 mmHg diastolic [5].

From the diagnostic criteria, it becomes apparent that preeclampsia is a systemic disorder that may affect multiple maternal organs. Therefore, the syndrome of hemolysis, elevated liver enzymes and low platelets can occur in addition to preeclamptic features and is characterized by acute liver injury, hemolysis and thrombocytopenia. Other severe maternal complications that can occur as a result of exacerbation of organ damage include acute renal failure, eclamptic insults and pulmonary edema. It is important to note that these complications can occur $\leq$7 days after delivery. In the developing world, however, these complications contribute heavily to maternal morbidity and mortality; in developed countries these complications are uncommon [6, 7]. A recent comparison between two developed countries shows an incidence of eclampsia of 5.4 in 10,000 deliveries in the Netherlands versus 2.7 in 10,000 deliveries in the United Kingdom (2004/2005 annual reports LEMMON study and UK national wide report). Women with eclampsia in the Netherlands were not managed according to guidelines with respect to blood pressure control [6].

Pathophysiology

Current understanding of the pathophysiology of preeclampsia is limited. The syndrome is thought to be a result of a complex interplay between the placenta and maternal constitutional factors; the maternal immune response plays an important role. Recent literature describes preeclampsia as a three-stage model [8] (Fig. 28.1). According to this model, preeclampsia originates very early during pregnancy from incomplete toleration of the allogeneic fetus, and thereby disrupts the balanced intra-uterine environment (first stage). These disruptions impede normal placental development by interfering with adequate placentation by decreasing immune-regulated remodeling of the spiral arteries in the decidua (second stage). Because of incomplete spiral artery remodeling, the spiral arteries cannot transform into widely dilated vessels, which are required to create the high-caliber, low-resistance placental bed [9]. As a consequence, blood flow to the placenta is restricted, leading to decreased uteroplacental perfusion, which might lead to ischemia reperfusion injury and oxidative stress of the chorionic villi [10]. Several studies reported that within the impaired oxygenated placenta, the syncytiotrophoblast produces and secretes several factors, mainly pro-inflammatory cytokines [11], and anti-angiogenic factors, such as the anti-angiogenic factors soluble fms-like tyrosine kinase-1 (sFlt-1) and soluble endoglin (sEng). These factors then spread into the maternal circulation and lead to generalized activation of the inflammatory system and endothelial dysfunction, resulting in the clinical manifestations of the syndrome (third stage). As pointed out, a complete understanding of the complex is incomplete, alternated volume homeostasis and sympathetic system overactivity, which are thought to be important mediators of the disease [12–14] in addition to the immune system and angiogenic imbalance. Recent research suggests that corin, a transmembrane protease that converts pro-atrial natriuretic peptide to active atrial natriuretic peptide, is involved in the pathogenesis of preeclampsia [15, 16].

It is postulated that both placental factors and maternal susceptibility are required for the development of preeclampsia. For late-onset preeclampsia, it seems that disrupted placentation is less important in the development of the disease because evidence of placental dysfunction is seen only in the minority of cases [17]. It has been suggested that late-onset preeclampsia can be considered as a process of restricted intervillous perfusion near term. In addition to this three-stage model, recent theories propose that a small uterus and a decreased capacity of cytotrophoblast progenitors might also limit placental health and result in a clinical manifestation of preeclampsia in the presence of maternal susceptibility factors [18]. Conditions associated with increased placental mass, such as multi-fetal gestations and hydatidiform mole, also predispose women to preeclampsia.

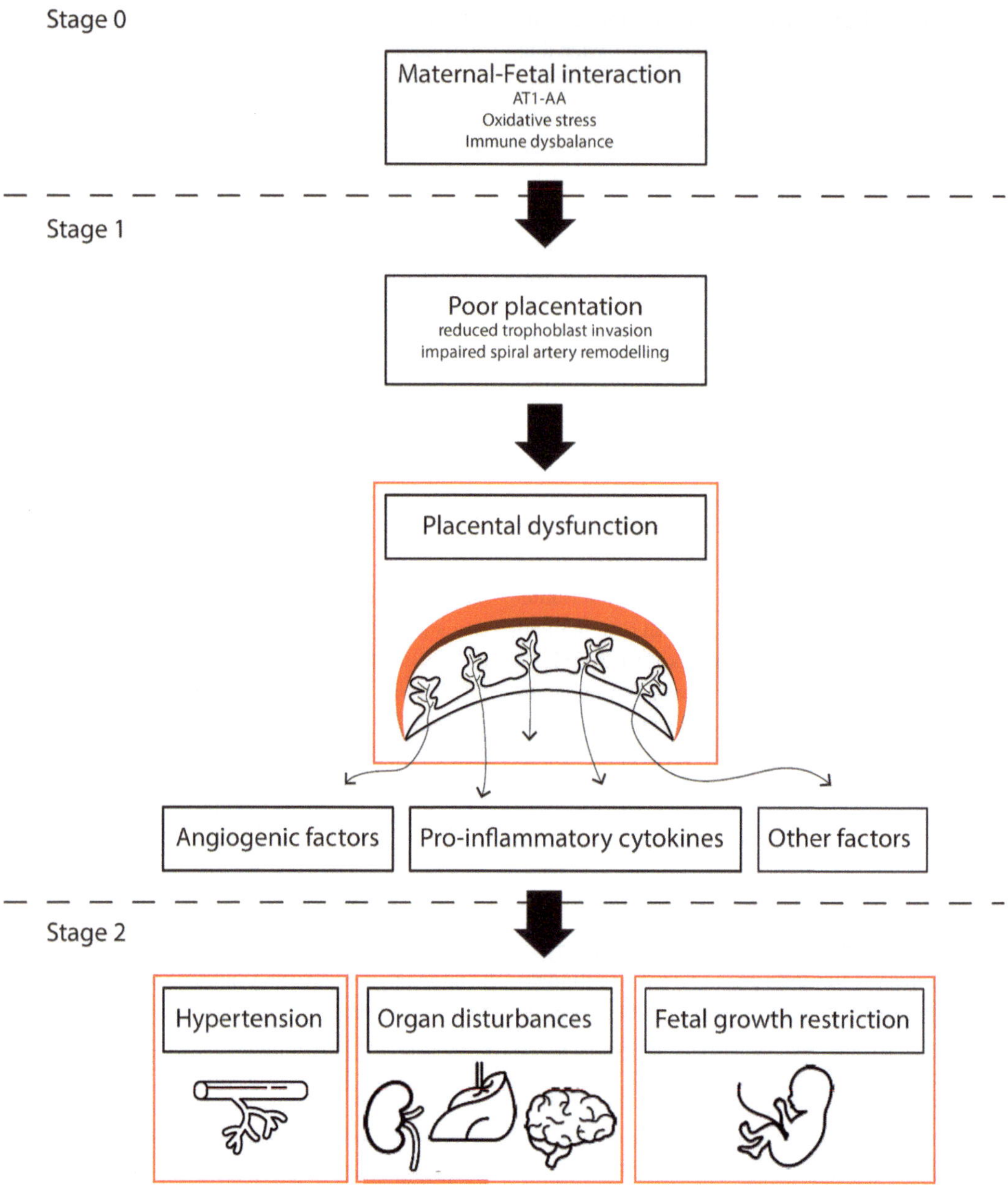

Fig. 28.1 Integrated model of the complex pathophysiology of pre-eclampsia. Three-stage model of the disease, including early abnormalities at the maternal—fetal interface (stage 0), placenta dysfunction (stage 1) and the maternal syndrome (stage 2)

The Role of the Placenta

Delivery of the placenta typically resolves within 48–72 h, implicating the placenta to play a central role in the pathogenesis of preeclampsia. In line with the three-stage model for preeclampsia, abnormal uterine artery Doppler ultrasound—which is consistent with decreased uteroplacental perfusion—is observed before the clinical onset of preeclampsia. The maternal syndrome of preeclampsia is often accompanied by intra uterine fetal growth restriction. Characteristics of placenta derived from pregnancies complicated by preeclampsia show evidence of placental hypoperfusion and ischemia. Findings include both villous and vascular histopathological lesions consisting of acute atherosis placental infarcts, intimal thickening, necrosis, atherosclerosis and endothelial damage [19]. Although the prevalence of such lesions is approximately four- to seven-fold greater in preeclampsia, placental lesions are not specific to the diagnosis of preeclampsia. The severity of the gross placental pathology appears to be correlated with the severity of the clinical disease.

The Role of Maternal Factors

Importantly, disrupted placental development is not the sole basis underlying the pathogenesis of preeclampsia because interaction with maternal constitutional factors is crucial to determine the clinical outcome. Several medical conditions—including chronic hypertension, diabetes mellitus, renal disease, obesity and antiphospholipid syndrome—are associated with increased risk for preeclampsia [20]. Advanced maternal age is also an independent risk factor for preeclampsia. These conditions are suggested to result in decreased reserve capacity of the cardiovascular system, which is required for the normal adaptations of pregnancy, thereby increasing maternal susceptibility to the development of preeclampsia.

Pregnancy is accompanied by large adaptations in the maternal hemodynamic systems, including an increase in plasma volume $\leq$40% without being accompanied by a decrease in systemic resistance to keep blood pressure in the normal range. To establish this, decreased sensitivity of the renin-angiotensin aldosterone system [21], increased compliance of the vascular wall [22] and increased production of nitric oxide (NO) by endothelial cells [23] are important mechanisms. Furthermore, these adaptations are accompanied by increased cardiac output requiring the reserve capacity of the heart and increase glomerular filtration requiring the use of reserve capacity of the kidney. Impairments in these adaptive mechanisms are proposed to be involved in the development of the maternal syndrome of preeclampsia, as outlined in the next paragraph. Therefore, it has been proposed that pregnancy performs as a "stress test" by temporarily unmasking these limited reserves. As mentioned previously, the maternal syndrome of preeclampsia is often accompanied by fetal growth restriction (caused by placenta dysfunction). Fetal growth restriction caused by placenta insufficiency also can occur without preeclampsia: This latter phenomenon gives support that maternal susceptibility factors (complex interactions) are present for the development of the maternal syndrome.

In addition to the limited hemodynamic reserves because of continual factors, the initial disturbances in renin-aldosterone-*angiotensin* II axis, the imbalance in angiogenic factors regulating vasculogenesis, the excessive oxidative stress and syncytiotrophoblast debris and the genetic susceptibility may also all have roles in the pathogenesis of preeclampsia by disrupting initial placentation [24].

Disrupted Maternal Hemodynamics

Increased Sensitivity of the Renin-Angiotensin Aldosterone System

During preeclampsia, plasma volume expansion and the decrease in systemic vascular resistance are less pronounced. In response, most circulating components of the renin-angiotensin aldosterone system (RAAS), that is, angiotensin II, are lower during preeclampsia compared with healthy pregnant women. However, importantly, angiotensin II sensitivity is increased in preeclamptic women compared with healthy pregnant women [21, 25]. Because angiotensin II is an important vasoconstrictor agent, non-adaptation of angiotensin II sensitivity might play an important role in the development of hypertension during pregnancy. The exact mechanisms behind the increased angiotensin II sensitivity are unknown, but possibly alternated placental and or vascular AT1-type 1 expression, or also heterodimerization of AT1 receptors with bradykinin receptors, are involved [26]. Other mechanisms, such as increased angiotensin 1–7 expression, AT1-type 1 receptor autoantibodies and hemopexin could also be involved [26].

Vascular Dysfunction

In general, preeclampsia is characterized by endothelial dysfunction, inducing impaired regulation and secretion of the endothelium-derived vasoactive factors and increasing systemic vascular resistance [27]. Nitric oxide bioavailability is decreased during preeclampsia [28] and the balance between the prostaglandins, prostacyclin and thromboxane, is disrupted [29].

Disruptions in Cardiac Function

In addition, changes in cardiac function are observed, which are thought to be a result of the increased afterload caused by higher total vascular resistance. Studies have demonstrated concentric hypertrophy during preeclampsia, suggesting significant pressure overload [30]. With regard to systolic and diastolic function, several studies have shown impairments during preeclampsia, sometimes already present before the onset of overt signs and symptoms [31]. In general, severe dysfunction is reported in early-onset preeclampsia compared with late-onset disease [32]. Some studies suggest that the disruptions in cardiac function might be caused by endothelial dysfunction, especially affecting the shortening of the longitudinal muscle fibers , which are vulnerable to ischemia or changes in wall stress because of their sub-endocardial location. Other studies have shown a correlation of cardiac dysfunction with sFtl-1 and sEng levels independent of blood pressure [33]. Theoretically, increased levels of sFlt-1 and sEng might be involved in coronary vasoconstriction, leading to myocardial ischemia [34–36], development of cardiac fibrosis and development of peripartum cardiomyopathy [37].

Disruptions in Kidney Function

During preeclamptic pregnancy, the kidneys are strongly affected by systemic endothelial dysfunction, which leads to proteinuria. A lower glomerular filtration rate (GFR) is observed, whereas in normal pregnancy renal hyperfiltration occurs with a 40%–60% increase in GFR as a result of the increase in plasma volume [38]. Because no differences were found in effective renal plasma flow between preeclamptic women and healthy pregnant women, the lower eGFR causes a relative state of hypofiltration [39]. The mechanism of hypofiltration during preeclampsia is not elucidated. Both (renal) hemodynamics and secondary changes to structural renal changes and podocyte alterations have been proposed [38, 40]. Histologically, glomerular endotheliosis is seen [40], which is characterized by fibrin deposition, endothelial swelling and detachment, decreased density and size of endothelial fenestrae, thickening of the glomerular basement membrane and loss of capillary space [39, 40].

Therapy

To date, there is no treatment option to cure preeclampsia apart from termination of pregnancy (delivery of the placenta), which can result in premature birth, often remote, before term. Several options—including several anti-oxidants, progesterone, bed rest and a low-sodium diet—have been studied as a potential treatment for preeclampsia. For all these options, no beneficial effect has been shown. Any intervention that could improve placental function and/or delay delivery and prolong fetal gestation could be in favor neonatal outcomes. Currently trials with aspirin and calcium supplementation (in populations where there is inadequate intake) have a small beneficial effect in preventing preeclampsia in high-risk population and are effective when started early in gestation [41, 42]. In addition, experimental therapies with decreasing anti-angiogenic factors, such as statins and NO donors, are under investigation [43, 44]. Another strategy to restore the angiogenic balance is the administration of VEGF-121 of placenta-like growth factor. VEGF-121 already has been shown to decrease hypertension and proteinuria in animal models of preeclampsia [45] and currently human trials testing adenovirus delivery of VEGF are underway. Removal of sFtl-1 by apheresis is another approach to restore the angiogenic imbalance during preeclampsia, and it has shown promising first results regarding the prolongation of pregnancy [46].

Preeclampsia as an Indicator for Future Maternal Cardiovascular Health

Long-Term Complications After Preeclampsia

During the last decades it has become more apparent that the implications of these disorders are not limited to the pregnancy itself but rather extend to later life. Despite normalization of the maternal vascular disturbance after birth, studies consistently show that formerly preeclamptic women experience an increased risk to develop cardiovascular and metabolic events later in life as outlined below.

Hypertension

Mildly increased systolic and diastolic blood pressure in women with a history of preeclampsia are often found in the first years postpartum [47, 48]. A recent meta-analysis showed that women with a history of preeclampsia have an approximately three-fold increased risk to develop hypertension later in life. There seems to be a dose-effect relationship, with increased severity (systolic blood pressure >160 mmHg or diastolic >110 mmHg) and multiple episodes of preeclampsia, thus increasing the risk even further [49, 50]. In addition, it was shown that early onset of preeclampsia (< 34 weeks) is accompanied by a higher risk compared with women with a history of late-onset preeclampsia [51]. It is estimated that formerly preeclamptic women develop chronic hypertension and CVD 6–8 years earlier compared with women having a history of normotensive pregnancy [52, 53]. Women with only growth restriction also are at increased risk, although to a lesser extent, and also here severity of the disease seems to contribute to the increased risk [54, 55].

Cardiovascular Disease

Studies have shown consistently that women with a history of preeclampsia have an approximately two-fold increased risk to develop future cardiovascular disease, including ischemic heart disease, stroke and death from cardiovascular [56, 57]. In a recent meta-analysis, this association remained significant after adjusting for age, body-mass index (BMI) and diabetes mellitus. For coronary heart disease, it was shown that the occurrence of major coronary events was greater in women with a history of preeclampsia combined with a child born small-for-gestational age (up to three-fold increased risk) and greatest in women with a history of preeclampsia combined with preterm delivery (up to five-fold increased risk) [58]. Growth restriction without signs of preeclampsia was also accompanied by a greater risk: A meta-analyses of four studies showed a relative risk of 1.71 [53]. Additional studies suggest that this risk is greater after more than one pregnancy complicated by intrauterine growth restriction (IUGR) [59] and increases with IUGR severity [54]. Stroke in women with a history of preeclampsia was found to occur approximately 8–10 years earlier than in women with normal pregnancy outcomes [60].

Renal Disease

Overall it has been reported that formerly preeclamptic women have a 5- to 12-fold increased risk to develop end-stage renal disease (ESRD) later in life. Vikse et al. were the first to demonstrate an increased risk, and they showed that the increased risk was present even after correction for traditional risk factors. They also showed that the risk triples when women had preeclampsia during more than one pregnancy or a pregnancy complicated by low birth weight or preterm delivery. Because ESRD has a low incidence in the general population, studies also tried to assess the risk on chronic kidney disease (CKD). Given the few studies available, CKD is estimated to be increased by 4–10 times [61, 62]. High BMI seems to be an important factor in the development of ESRD, with a recent report showing a significant association between preeclampsia and ESRD, which disappeared after adjustment for obesity [63]. At this moment, all of the abovementioned studies have the limitation that they are based on the use of diagnostic codes and not on eGFR, which might result in underestimation.

In contrast, overestimation might have occurred by misdiagnosis of preeclampsia in women with undiagnosed pre-existing kidney disease.

Apart from the risk of kidney disease, disturbances in albuminuria and eGFR are observed in women with a history preeclampsia, which may contribute to both cardiovascular and kidney disease in later life. At this moment, no longitudinal assessment of these parameters has been published to establish the link with clinical manifestations; however, reports at fixed time points postpartum have shown mild disturbances. Regarding albuminuria, it was shown by a meta-analysis that the occurrence of micro-albuminuria is as high as 31% in women with previous pre-eclampsia versus 7% in controls at a weighted mean of 7.1 years postpartum [64]. This study showed, again, that there is a relation between the severity of preeclampsia and the increase in risk, with a four-fold increased risk of albuminuria after mild preeclampsia and an eight-fold increased risk after severe preeclampsia. However, smaller studies included in the meta-analysis reported a lower incidences of proteinuria in formerly preeclamptic women [48, 65] and showed that albuminuria slowly resolves in the years postpartum [66]. This discrepancy between findings might be caused by the fact that the meta-analysis did not correct for confounders, such as BMI and comorbidity, which seem to play an important role. Regarding kidney function, most studies have reported no significant differences in formerly preeclamptic women compared with healthy parous controls [64]. Only recently have subtle changes in renal function and renal hemodynamics in formerly preeclamptic women been reported. A few studies found signs of high normal eGFR at approximately 10 years after early-onset preeclampsia [65, 67], which might indicate an early stage of hyperfiltration. Indeed, one study observed increased filtration fraction in women with formerly early-onset preeclampsia [68]. Hyperfiltration might be the first sign for future kidney problems in formerly preeclamptic women because it might contribute to progressive renal damage [69].

Heart Failure

The risk for heart failure, with a reported four-fold increased risk, is highest after preeclampsia [57]. The prevalence of asymptomatic heart disease (heart failure stage B) is already increased in the first year after index pregnancy, with 1 in the 4 women having a history of preeclampsia meeting the criteria. Regarding the disruption in cardiac function during preeclamptic pregnancy, it is thought that heart failure after preeclampsia is mainly the result of generalized endothelial dysfunction. A recent meta-analysis showed that endothelial dysfunction is widely present in the years after preeclampsia [70] and another showed that endothelial dysfunction correlates with subclinical and symptomatic heart failure [71]. In contrast, it has been suggested that heart failure shares a common pathogenic background with preeclampsia because there is a large overlap in biomarkers for both diseases [72]. In addition to heart failure, a clear association also has been shown between preeclampsia and late risk of cardiomyopathy [73].

Metabolic Disease

In the years after index pregnancy, formerly preeclamptic women present with unfavorable metabolic profiles, such as high BMI, insulin resistance and high cholesterol and triglycerides [49, 74]. In a recent meta-analysis, most of these metabolic parameters (including glucose, insulin, triglycerides, total cholesterol, high-density lipoprotein cholesterol, low-density lipoprotein cholesterol and homocysteine levels) were confirmed to be associated with previous preeclampsia compared with women having a history of healthy pregnancies [75]. Moreover, the increased insulin resistance is reflected in the greater rates of diabetes mellitus in formerly preeclamptic women [57]. Regarding hypertension and stroke, it has been shown that type 2 diabetes mellitus develops, on average, 8–10 years earlier than in women with normal pregnancy outcomes [76].

Disrupted Maternal Hemodynamics, Blood Pressure Regulation and Vascular Function

Not only during pregnancy, but also after pregnancy, is the RAAS and vascular function found to be disrupted. These pathways are highly suggested be involved in a mechanistic link with increased risk for future cardiovascular disease. Regarding the RAAS, it has been described that the sensitivity of the system remains increased postpartum [77–80]. In addition, a higher sodium sensitivity index and significant salt-induced increases in blood pressure were found in women with a history of preeclampsia [80]. Furthermore, formerly preeclamptic women have greater sympathetic activity in blood pressure regulation causing lower baroreflex sensitivity and alterations in extracellular volume [81]. Sustained sympathetic overactivity after PE may contribute to target organ damage and increases cardiovascular morbidity and mortality independent of an increase in blood pressure. Indeed, heighted sympathetic activation is a key indicator of poor prognosis and survival in many disease conditions.

Another contributing factor to the risk for cardiovascular disease seems to be the presence of endothelial dysfunction after preeclampsia. As already mentioned, endothelial dysfunction is thought to play a major in the development of heart failure after preeclampsia. A meta-analysis assessing endothelial function, as measured by flow-mediated dilation after preeclampsia, clearly showed that endothelial function develops at 3 years postpartum compared with controls having a history of healthy pregnancy [70]. Too few studies have been performed at later ages to draw conclusions on long-term differences. Regarding other measures of vascular function, such as carotid–femoral pulse-wave velocity and augmentation index follow-up, has been performed at older ages. A meta-analysis of these studies showed evidence of vascular dysfunction with adverse effects found to be more pronounced in women <40 years old [82].

Implications for the Child Born After Preeclampsia

During the last decades, it has become more apparent that long-term complications of preeclampsia are not limited to the mother but also affect the exposed child. Several studies have shown that children born after preeclampsia have higher blood pressure compared with children of mothers who did not have preeclampsia [83–87]. This was confirmed by a meta-analysis of 18 studies, which showed a small but significant increase in blood pressure in young childhood and adolescence after preeclamptic birth [88]. Few data are available of the long-term cardiovascular manifestations in those children, with only one study reporting that offspring born from pregnancies complicated by preeclampsia are at increased risk of stroke [89]. However, for children born with low birth weight, this association is clearly established, with large population-based observational studies reporting increased risk of developing cardiovascular events in later life [90–92]. In addition to cardiovascular events, an association with metabolic diseases has clearly been established. These observations are consistent with the concept of developmental origins of health and disease, commonly attributed to David Barker (the “Barker hypothesis”), which suggests that conditions affecting specific sensitive developmental periods program organ structure and function throughout life [93]. The concept of intergenerational passage of unfavorable cardiovascular risk profiles is supported by the results of work with animals [94, 95].

Mechanisms Involved in Long-Term Complications After Preeclampsia

The exact mechanisms behind the increased risk of long-term complications have not yet been elucidated. One well-accepted concept is that the development of pregnancy disorders is an indicator of maternal susceptibility to develop

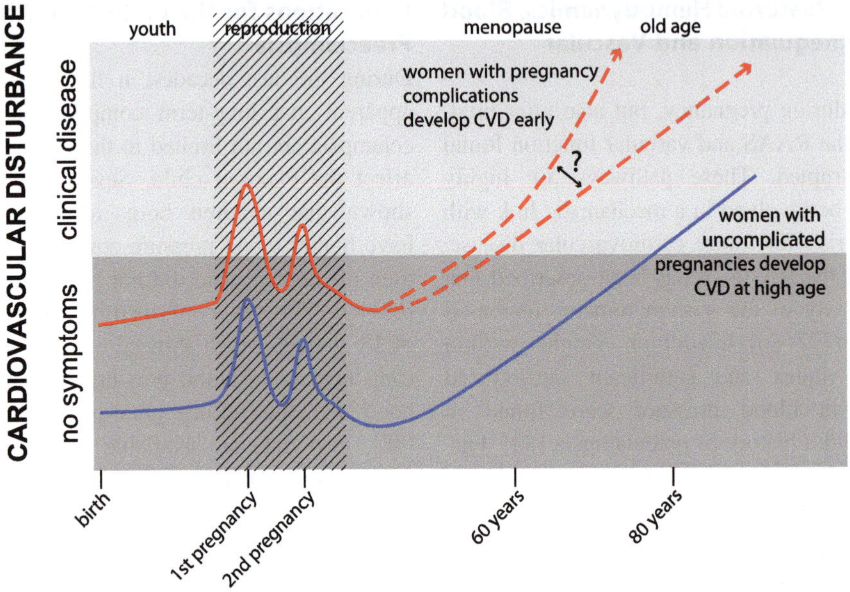

Fig. 28.2 Pregnancy as a critical window for cardiovascular health in the mother. Pregnancy is a window during which we can identify mothers at high risk to develop cardiovascular disease (CVD) later in life. During fetal life, both genetic factors and intra-uterine environment contributed to confer a higher risk of CVD development (i.e., setting the baseline). In addition, maternal exposure to a pregnancy with clinical disease might directly contribute to long-term abnormalities by prolonging disturbances and/or damage. Whether preeclampsia itself contributes to an increased risk of pre-existing factors remains to be established (dotted line) (theory based on Sattar and Greer [96])

future cardiovascular conditions ("pregnancy as a "stress test" for cardiovascular health") (Fig. 28.2). As mentioned earlier, adaptation to pregnancy demands increased cardiovascular and renal effort, and it has been proposed that pregnancy might temporally unmask limited cardiovascular reserves. This theory is consistent with the requirement of constitutional maternal factors for development of the disease and suggests that development of preeclampsia can help to identify, in advance, susceptible/high-risk subjects. Although the hierarchy in pathways is unclear, clearly women with a history of preeclampsia have a predisposition to develop long-term complications.

Alternatively, the increased cardiovascular and kidney risk may actually result from disturbances caused by organ damage during preeclampsia. This is supported by the finding that the increased long-term cardiovascular mortality holds even for previously healthy women, without any overt vascular risk factors, who develop preeclampsia [97]. Contributing factors are, for instance, persistent loss of podocytes [98] leading to proteinuria and renal disease over the long-term. Other proposed pathways include persistence of disrupted pathways as observed during preeclampsia, such as mildly increased inflammatory response, increased RAAS activity, imbalance of angiogenic factors [99] and increased arterial stiffness [100–102] (Fig. 28.3).

Some studies in humans tried to address the question of whether preeclampsia independently contributes to the development of long-term complications. Although some studies have suggested an independent contribution of preeclampsia to the increase in blood pressure in later life as well as risk of ESRD and endothelial dysfunction [103–105], it remains difficult to discern cause and effect in human settings due to the inability to correct for all preexisting factors. Therefore this question was addressed in various animal models of preeclampsia. Several strategies exist—including injection of an sFlt-1 expressing

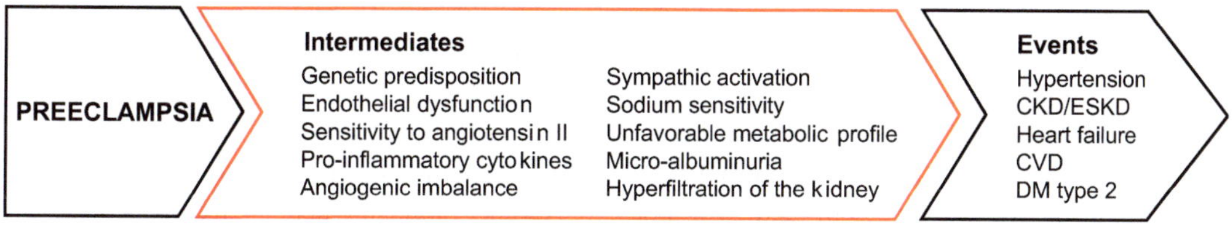

Fig. 28.3 The presence of classical CVD risk factors, mechanistic disturbances and clinical manifestations before the onset of cardiovascular events in women with a history of preeclampsia offers the opportunity for screening and prevention to decrease the risk of long-term events. The best intermediates to use for screening and the best time to screen remain to be established

adenovirus [106], injection of LPS [107] and decrease of uterine blood flow—to induce a preeclampsia-like phenotype with hypertension and proteinuria in pregnant animals [108]. In both models it was shown that blood pressure was neither increased nor disrupted during the postpartum period compared with animals that were not exposed during pregnancy [77, 109, 110]. Only slight decreases, within the normal range, of cardiac, renal and mesenteric-vessel function were found postpartum after decreased uterine perfusion compared with healthy pregnant animals [110, 111]. The degree of these changes was marginal and therefore unlikely to contribute to cardiovascular manifestations in the absence of other triggers. Together these studies indicate that in the absence of a pre-existing cardiovascular profile, exposure to preeclampsia has minimal adverse effects on the cardiovascular system. However, it seems that animals are hyper-responsive to a second "hit" after resolution of the preeclamptic features postpartum. For example, it was shown in the sFlt-1 model that carotid injury after preeclampsia elicited increased vascular remodeling response, increased smooth-muscle cell proliferation and increased fibrosis [112]. In addition, in the LPS model, increased sensitivity of blood pressure and increased proteinuria on angiotensin II infusion was observed compared with never-exposed rats [77]. These studies suggest that although preeclampsia does not spontaneously cause vascular abnormalities, preeclampsia might lead to increased long-term susceptibility to vascular disorders, which are unmasked on additional triggers. The findings do, however, emphasize that external influences, such as diet and other lifestyle factors, might amplify the long-term consequences of preeclampsia.

Toward Screening and Prevention

Given the above-mentioned associations between preeclampsia and long-term complications, several health organizations have acknowledged preeclampsia as a woman-specific risk factor for cardiovascular disease in their guidelines and have implemented recommendations for screening and prevention strategies. For example, the American Heart Association guideline, Prevention of CVD in Women, recommends careful monitoring of risk factors after preeclampsia and advises medical treatment with thiazide diuretics at a blood pressure $\geq$ 140/90 mm Hg [113]. The American Stroke Association (ASA) recommends to consider evaluation of all women with a history of preeclampsia and eclampsia 6 months to 1 year postpartum and to treat abnormal cardiovascular risk factors, including hypertension, obesity, smoking and dyslipidemia [114]. Other guidelines, such as the national Dutch guideline for cardiovascular risk management after reproductive disorders, recommends screening for risk factors at the age of 50 years after a pregnancy complicated by preeclampsia [53].

The development of such guidelines is a good first step toward decreasing CVD burden in a specific group of women at high risk for long-term complications. As becomes apparent from the above-mentioned criteria, recently established guidelines all propose different approaches for screening and prevention in women with a history of preeclampsia. This is due to low levels of for the optimal time points and parameters for screening. In addition, it seems that the widely used 10-year estimates for a cardiovascular event, such as the Framingham Risk Score, do not help to distinguish women who will and will not develop long-term complications because of young age and premenopausal state. Furthermore, no randomized controlled studies have been performed to test the effectiveness of various treatment regimens in the prevention of disease. Research is warranted to develop optimal evidence-based and cost-effective strategies.

To develop adequate screening strategies, much focus has been on mapping the time-course of cardiovascular risk factors in women with a history of preeclampsia. This, however, has not resulted in the correct prediction of women at highest risk yet because the widely used 10-year prediction scheme for a cardiovascular event—such as the Framingham Risk Score—cannot help to distinguish women who will and will not develop long-term complications because of young age and premenopausal state [115]. Therefore, new approaches for screening for cardiovascular disease are currently under investigation. One of these approaches is assessing biomarkers. Although various markers have been shown to be increase at ≤20 years after pregnancy [115], their implementation in risk assessment after preeclampsia is still much debated. Another promising approach is the study of surrogate endpoints, which can be performed using advanced imaging techniques [115].

One of the studied imaging techniques is the determination of carotid intima-media thickness (cIMT). This technique can detect offer an opportunity for the early identification of preclinical atherosclerotic load in formerly preeclamptic women [116]. Although the knowledge of cIMT adds little information to the 10-year prediction models, it was recently shown that this measurement is valuable to predict events in high-risk patients [116]. During preeclampsia and ≤10 years after preeclampsia, a great atherosclerotic burden, as measured by cIMT, was identified in the meta-analyses. This suggests that cIMT may offer an opportunity to detect women with an atherosclerotic burden after preeclampsia, even before menopause. Nevertheless, the role of cIMT for diagnosing vascular disease for screening for vascular disease after preeclampsia remains to be validated.

Another promising imaging technique to detect women at high risk is coronary artery calcium scoring (CACS). Compared with cIMT, the disadvantage might be that this measurement might only have value at later age (> 45 years) when calcification in atherosclerotic lesions becomes evident. To date, few studies have assessed CACS in formerly preeclamptic women. Two retrospective studies showed a positive association between CACS and self-reported hypertension in pregnancy [117, 118]. More recently, a prospective study investigating women at 30 years after preeclampsia, compared with women having a history of normal pregnancy, showed that women with a history of preeclampsia are at increased risk to have a CACS score > 50 [119]. As for cIMT, it must be validated whether this measure will be effective to distinguish between women who will develop cardiovascular events and those who will not.

Regarding optimal prevention strategies in formerly preeclamptic women, additional research is also required [120]. At this moment, strategies for the prevention of cardiovascular disease are mainly derived from studies performed in older men. To obtain more insights, studies on lifestyle interventions and effective medication are required. The first study on postpartum lifestyle intervention after preeclampsia showed promising effects on body constitution and fat intake in a non-randomized cohort [121]. Although it seems likely that women are highly motivated to optimize their lifestyle after pregnancy [122], lifestyle modification is known to be difficult to achieve. Drug therapy with RAAS inhibition and/or statins might be a more potent solution to decrease cardiovascular and renal risk. Randomized controlled trials are required to determine the most effective treatments for decreasing blood pressure and correcting metabolic disturbances in relatively young women.

Until we find effective therapies specific for post-preeclamptic women, these women should be screened and managed as well as possible with general CVD-prevention measures (blood pressure control and lifestyle advice). Ultimately, a multidisciplinary approach will lead to the establishment of evidence-based guidelines with a customized and structured approach of formerly preeclamptic patients that may help to decrease the burden of CVD and renal disease in this population.

References

1. Hernandez-Diaz S, Toh S, Cnattingius S. Risk of pre-eclampsia in first and subsequent pregnancies: prospective cohort study. BMJ. 2009;338:b2255.
2. Sibai B, Dekker G, Kupferminc M. Pre-eclampsia. Lancet. 2005;365(9461):785–99.
3. American College of Obstetricians and Gynecologists, Task force Hypertension in Pregnancy. Hypertension in pregnancy. Report of the American College of Obstetricians and Gynecologists' Task Force on Hypertension in Pregnancy. Obstet Gynecol. 2013;122(5):1122–31.
4. Tranquilli AL, Dekker G, Magee L, et al. The classification, diagnosis and management of the hypertensive disorders of pregnancy: a revised statement from the ISSHP. Pregnancy Hypertens. 2014;4(2):97–104.
5. Brown MA, Lindheimer MD, de Swiet M, Van Assche A, Moutquin JM. The classification and diagnosis of the hypertensive disorders of pregnancy: statement from the International Society for the Study of Hypertension in Pregnancy (ISSHP). Hypertens pregnancy. 2001;20(1):IX–XIV.
6. Schaap TP, Knight M, Zwart JJ, et al. Eclampsia, a comparison within the International Network of Obstetric Survey Systems. BJOG. 2014;121(12):1521–8.
7. A review of maternal deaths in South Africa during 1998. National Committee on Confidential Enquiries into Maternal Deaths. S Afr Med J. 2000;90(4):367–73.
8. Redman CW, Sargent IL. Immunology of pre-eclampsia. Am J Reprod Immunol (New York, NY 1989). 2010;63 (6):534–43.
9. Pijnenborg R, Vercruysse L, Hanssens M. The uterine spiral arteries in human pregnancy: facts and controversies. Placenta. 2006;27(9–10):939–58.
10. Burton GJ, Woods AW, Jauniaux E, Kingdom JC. Rheological and physiological consequences of conversion of the maternal spiral arteries for uteroplacental blood flow during human pregnancy. Placenta. 2009;30(6):473–82.
11. Lau SY, Barrett CJ, Guild S-J, Chamley LW. Necrotic trophoblast debris increases blood pressure during pregnancy. J Reprod Immunol. 2013;97(2):175–82.
12. Schobel HP, Fischer T, Heuszer K, Geiger H, Schmieder RE. Preeclampsia—a state of sympathetic overactivity. N Engl J Med. 1996;335(20):1480–5.
13. van Beck E, Peeters LL. Pathogenesis of preeclampsia: a comprehensive model. Obstet Gynecol Surv. 1998;53 (4):233–9.
14. Fischer T, Schobel HP, Frank H, Andreae M, Schneider KTM, Heusser K. Pregnancy-induced sympathetic overactivity: a precursor of preeclampsia. Eur J Clin Invest. 2004;34 (6):443–8.
15. Cui Y, Wang W, Dong N, et al. Role of corin in trophoblast invasion and uterine spiral artery remodelling in pregnancy. Nature. 2012;484(7393):246–50.
16. Zhou Y, Wu Q. Role of corin and atrial natriuretic peptide in preeclampsia. Placenta. 2013;34(2):89–94.
17. Raymond D, Peterson E. A critical review of early-onset and late-onset preeclampsia. Obstet Gynecol Surv. 2011;66 (8):497–506.
18. Redman CW, Sargent IL. Latest advances in understanding preeclampsia. Science. 2005;308(5728):1592–4.
19. Falco ML, Sivanathan J, Laoreti A, Thilaganathan B, Khalil A. Placental histopathology associated with pre-eclampsia: systematic review and meta-analysis. Ultrasound Obstet Gynecol. 2017;50(3):295–301.
20. Sibai BM, Lindheimer M, Hauth J, et al. Risk factors for preeclampsia, abruptio placentae, and adverse neonatal outcomes among women with chronic hypertension. National Institute of Child Health and Human Development Network of Maternal-Fetal Medicine Units. N Engl J Med. 1998;339(10):667–71.
21. Verdonk K, Visser W, Van Den Meiracker AH, Danser AHJ. The renin-angiotensin-aldosterone system in pre-eclampsia: the delicate balance between good and bad. Clin Sci (Lond). 2014;126(8):537–44.
22. Khalil A, Jauniaux E, Cooper D, Harrington K. Pulse wave analysis in normal pregnancy: a prospective longitudinal study. PLoS One. 2009;4(7):e6134.
23. Boeldt DS, Yi FX, Bird IM. eNOS activation and NO function: pregnancy adaptive programming of capacitative entry responses alters nitric oxide (NO) output in vascular endothelium—new insights into eNOS regulation through adaptive cell signaling. J Endocrinol. 2011;210(3):243–58.
24. Chaddha V, Viero S, Huppertz B, Kingdom J. Developmental biology of the placenta and the origins of placental insufficiency. Semin Fetal Neonatal Med. 2004;9 (5):357–69.
25. Gant NF, Chand S, Whalley PJ, MacDonald PC. The nature of pressor responsiveness to angiotensin II in human pregnancy. Obstet Gynecol. 1974;43(6):854.
26. van der Graaf AM, Toering TJ, Faas MM, Lely AT. From preeclampsia to renal disease: a role of angiogenic factors and the renin-angiotensin aldosterone system? Nephrol Dial Transplant [Internet]. 2012 [cited 2014 Oct 16];27(Suppl 3): iii51–7. Available from: http://www.ncbi.nlm.nih.gov/pubmed/23115142
27. Johal T, Lees CC, Everett TR, Wilkinson IB. The nitric oxide pathway and possible therapeutic options in pre-eclampsia. Br J Clin Pharmacol. 2014;78(2):244–57.
28. López-Jaramillo P, Arenas WD, García RG, Rincon MY, López M. The role of the L-arginine-nitric oxide pathway in preeclampsia. Ther Adv Cardiovasc Dis [Internet]. 2008;2 (4):261–75. Available from: http://www.embase.com/search/results?subaction=viewrecord&from=export&id=L352219675
29. Friedman SA. Preeclampsia: a review of the role of prostaglandins. Obstet Gynecol. 1988;71(1):122–37.
30. Valensise H, Novelli GP, Vasapollo B, et al. Maternal diastolic dysfunction and left ventricular geometry in gestational hypertension. Hypertens (Dallas, Tex 1979). 2001;37 (5):1209–15.
31. Melchiorre K, Sutherland GR, Baltabaeva A, Liberati M, Thilaganathan B. Maternal cardiac dysfunction and remodeling in women with preeclampsia at term. Hypertens (Dallas, Tex 1979). 2011;57(1):85–93.
32. Borges VTM, Zanati SG, PeraColi MTS, et al. Maternal hypertrophy and diastolic disfunction and brain natriuretic

peptide concentration in early and late pre-Eclampsia. Ultrasound Obstet Gynecol. 2017;51:591–23.
33. Shahul S, Medvedofsky D, Wenger JB, et al. Circulating antiangiogenic factors and myocardial dysfunction in hypertensive disorders of pregnancy. Hypertension. 2016;67(6):1273–80.
34. Hammadah M, Georgiopoulou VV, Kalogeropoulos AP, et al. Elevated soluble Fms-like tyrosine kinase-1 and placental-like growth factor levels are associated with development and mortality risk in heart failure. Circ Heart Fail. 2016;9(1):e002115.
35. Groarke JD, Choueiri TK, Slosky D, Cheng S, Moslehi J. Recognizing and managing left ventricular dysfunction associated with therapeutic inhibition of the vascular endothelial growth factor signaling pathway. Curr Treat Options Cardiovasc Med. 2014;16(9):335.
36. Kapur NK, Heffernan KS, Yunis AA, et al. Elevated soluble fms-like tyrosine kinase-1 levels in acute coronary occlusion. Arterioscler Thromb Vasc Biol. 2011;31(2):443–50.
37. Kapur NK, Wilson S, Yunis AA, et al. Reduced endoglin activity limits cardiac fibrosis and improves survival in heart failure. Circulation. 2012;125(22):2728–38.
38. Cornelis T, Odutayo A, Keunen J, Hladunewich M. The kidney in normal pregnancy and preeclampsia. Semin Nephrol. 2011;31(1):4–14.
39. Lafayette RA, Druzin M, Sibley R, et al. Nature of glomerular dysfunction in pre-eclampsia. Kidney Int. 1998;54(4):1240–9.
40. Stillman IE, Karumanchi SA. The glomerular injury of preeclampsia. J Am Soc Nephrol. 2007;18(8):2281–4.
41. Rolnik DL, Wright D, Poon LC, et al. Aspirin versus placebo in pregnancies at high risk for preterm preeclampsia. N Engl J Med. 2017;377(7):613–22.
42. López-Jaramillo P. Prevention of preeclampsia with calcium supplementation and its relation with the L-arginine:nitric oxide pathway. Braz J Med Biol Res [Internet]. 1996;29(6):731–41. Available from: http://www.embase.com/search/results?subaction=viewrecord&from=export&id=L26225692
43. Ahmed A. Molecular mechanisms and therapeutic implications of the carbon monoxide/hmox1 and the hydrogen sulfide/CSE pathways in the prevention of pre-eclampsia and fetal growth restriction. Pregnancy Hypertens. 2014;4(3):243–4.
44. Holwerda KM, Faas MM, van Goor H, Lely AT. Gasotransmitters: a solution for the therapeutic dilemma in preeclampsia? Hypertension [Internet] 2013 [cited 2014 Oct 16]; 62(4):653–9. Available from: http://www.ncbi.nlm.nih.gov/pubmed/23959552
45. Woods AK, Hoffmann DS, Weydert CJ, et al. Adenoviral delivery of VEGF121 early in pregnancy prevents spontaneous development of preeclampsia in BPH/5 mice. Hypertens (Dallas, Tex 1979). 2011;57(1):94–102.
46. Thadhani R, Hagmann H, Schaarschmidt W, et al. Removal of soluble Fms-like tyrosine kinase-1 by dextran sulfate apheresis in preeclampsia. J Am Soc Nephrol. 2016;27(3):903–13.
47. Drost JT, Arpaci G, Ottervanger JP, et al. Cardiovascular risk factors in women 10 years post early preeclampsia: the Preeclampsia Risk EValuation in FEMales study (PREVFEM). Eur J Prev Cardiol. 2012;19(5):1138–44.
48. Mangos GJ, Spaan JJ, Pirabhahar S, Brown MA. Markers of cardiovascular disease risk after hypertension in pregnancy. J Hypertens. 2012;30(2):351–8.
49. Magnussen EB, Vatten LJ, Smith GD, Romundstad PR. Hypertensive disorders in pregnancy and subsequently measured cardiovascular risk factors. Obstet Gynecol. 2009;114(5):961–70.
50. Lykke JA, Langhoff-Roos J, Sibai BM, Funai EF, Triche EW, Paidas MJ. Hypertensive pregnancy disorders and subsequent cardiovascular morbidity and type 2 diabetes mellitus in the mother. Hypertension. 2009;53(6):944–51.
51. Lazdam M, de la Horra A, Diesch J, et al. Unique blood pressure characteristics in mother and offspring after early onset preeclampsia. Hypertension. 2012;60(5):1338–45.
52. Suzuki H, Watanabe Y, Arima H, Kobayashi K, Ohno Y, Kanno Y. Short- and long-term prognosis of blood pressure and kidney disease in women with a past history of preeclampsia. Clin Exp Nephrol [Internet]. 2008 [cited 2014 Nov 30];12(2):102–9. Available from: http://www.ncbi.nlm.nih.gov/pubmed/18180874
53. Heida KY, Bots ML, de Groot CJ, et al. Cardiovascular risk management after reproductive and pregnancy-related disorders: a Dutch multidisciplinary evidence-based guideline. Eur J Prev Cardiol. 2016;23(17):1863–79.
54. Lykke JA, Paidas MJ, Triche EW, Langhoff-Roos J. Fetal growth and later maternal death, cardiovascular disease and diabetes. Acta Obstet Gynecol Scand. 2012;91(4):503–10.
55. Berends AL, de Groot CJ, Sijbrands EJ, et al. Shared constitutional risks for maternal vascular-related pregnancy complications and future cardiovascular disease. Hypertension. 2008;51(4):1034–41.
56. Brown MC, Best KE, Pearce MS, Waugh J, Robson SC, Bell R. Cardiovascular disease risk in women with pre-eclampsia: systematic review and meta-analysis. Eur J Epidemiol. 2013;28(1):1–19.
57. Wu P, Haththotuwa R, Kwok CS, et al. Preeclampsia and future cardiovascular health: a systematic review and meta-analysis. Circ Cardiovasc Qual Outcomes. 2017;10(2):e003497.
58. Riise HKR, Sulo G, Tell GS, et al. Incident coronary heart disease after preeclampsia: role of reduced fetal growth, preterm delivery, and parity. J Am Heart Assoc. 2017;6(3):e004158.
59. Nilsson PM, Li X, Sundquist J, Sundquist K. Maternal cardiovascular disease risk in relation to the number of offspring born small for gestational age: national, multi-generational study of 2.7 million births. Acta Paediatr. 2009;98(6):985–9.
60. Zoet GA, Linstra KM, Bernsen MLE, et al. Stroke after pregnancy disorders. Eur J Obstet Gynecol Reprod Biol. 2017;215:264–6.
61. Männistö T, Mendola P, Vääräsmäki M, et al. Elevated blood pressure in pregnancy and subsequent chronic disease risk. Circulation [Internet]. 2013 [cited 2014 Oct 16];127(6):681–90. Available from: http://www.pubmedcentral.nih.gov/articlerender.fcgi?artid=4151554&tool=pmcentrez&rendertype=abstract
62. Wu C-C, Chen S-H, Ho C-H, et al. End-stage renal disease after hypertensive disorders in pregnancy. Am J Obstet Gynecol [Internet]. 2014 [cited 2014 Oct 16];210(2):147.e1–8. Available from: http://www.ncbi.nlm.nih.gov/pubmed/24060448
63. Kattah AG, Scantlebury DC, Agarwal S, et al. Preeclampsia and ESRD: the role of shared risk factors. Am J Kidney Dis. 2016;69(4):498–505.
64. McDonald SD, Han Z, Walsh MW, Gerstein HC, Devereaux PJ. Kidney disease after preeclampsia: a systematic review and meta-analysis. Am J Kidney Dis [Internet]. 2010 [cited 2014 Oct 16];55(6):1026–39. Available from: http://www.ncbi.nlm.nih.gov/pubmed/20346562
65. Sandvik MK, Hallan S, Svarstad E, Vikse BE. Preeclampsia and prevalence of microalbuminuria 10 years later. Clin J Am

Soc Nephrol [Internet]. 2013 [cited 2014 Oct 16];8 (7):1126–34. Available from: http://www.pubmedcentral.nih.gov/articlerender.fcgi?artid=3700700&tool=pmcentrez&rendertype=abstract

66. Berks D, Steegers EA, Molas M, Visser W. Resolution of hypertension and proteinuria after preeclampsia. Obstet Gynecol. 2009;114(6):1307–14.
67. Paauw ND, Joles JA, Drost JT, et al. High-normal estimated glomerular filtration rate in early-onset preeclamptic women 10 years postpartum. Hypertension. 2016;68(6):1407–14.
68. Toering TJ, van der Graaf AM, Visser FW, et al. Higher filtration fraction in formerly early-onset preeclamptic women without comorbidity. Am J Physiol Physiol. 2015;308(8):F824–31.
69. Helal I, Fick-Brosnahan GM, Reed-Gitomer B, Schrier RW. Glomerular hyperfiltration: definitions, mechanisms and clinical implications. Nat Rev. 2012;8(5):293–300.
70. Weissgerber TL, Milic NM, Milin-Lazovic JS, Garovic VD. Impaired flow-mediated dilation before, during, and after preeclampsia: a systematic review and meta-analysis. Hypertension. 2016;67(2):415–23.
71. Breetveld NM, Ghossein-Doha C, van Neer J, et al. Decreased endothelial function and increased subclinical heart failure in women, many years after pre-eclampsia. Ultrasound Obstet Gynecol. 2017;
72. Alma LJ, Bokslag A, Maas AHEM, Franx A, Paulus WJ, de Groot CJM. Shared biomarkers between female diastolic heart failure and pre-eclampsia: a systematic review and meta-analysis. ESC Hear Fail. 2017;4(2):88–98.
73. Behrens I, Basit S, Lykke JA, et al. Association between hypertensive disorders of pregnancy and later risk of cardiomyopathy. JAMA. 2016;315(10):1026–33.
74. van Rijn BB, Nijdam M-E, Bruinse HW, et al. Cardiovascular disease risk factors in women with a history of early-onset preeclampsia. Obstet Gynecol [Internet] 2013 [cited 2014 Oct 16];121(5):1040–8. Available from: http://www.ncbi.nlm.nih.gov/pubmed/23635741
75. Hermes W, Franx A, van Pampus MG, et al. Cardiovascular risk factors in women who had hypertensive disorders late in pregnancy: a cohort study. Am J Obstet Gynecol [Internet]. 2013 [cited 2014 Oct 16];208(6):474.e1–8. Available from: http://www.ncbi.nlm.nih.gov/pubmed/23399350
76. Heida KY, Franx A, van Rijn BB, et al. Earlier Age of Onset of Chronic Hypertension and Type 2 Diabetes Mellitus After a Hypertensive Disorder of Pregnancy or Gestational Diabetes Mellitus. Hypertens (Dallas, Tex 1979). 2015;66 (6):1116–22.
77. van der Graaf AM, Toering TJ, van der Wiel MWK, et al. Angiotensin II responsiveness after preeclampsia: translational data from an experimental rat model and early-onset human preeclampsia. J Hypertens. 2017;35(12):2468–78.
78. Spaanderman ME, Ekhart TH, de Leeuw PW, Peeters LL. Angiotensin II sensitivity in nonpregnant formerly preeclamptic women and healthy parous controls. J Soc Gynecol Investig. 2004;11(6):416–22.
79. Hladunewich MA, Kingdom J, Odutayo A, et al. Postpartum assessment of the renin angiotensin system in women with previous severe, early-onset preeclampsia. J Clin Endocrinol Metab [Internet]. 2011 [cited 2014 Oct 16];96(11):3517–24. Available from: http://www.ncbi.nlm.nih.gov/pubmed/21880804
80. Saxena AR, Karumanchi SA, Brown NJ, Royle CM, McElrath TF, Seely EW. Increased sensitivity to angiotensin II is present postpartum in women with a history of hypertensive pregnancy. Hypertension [Internet]. 2010 [cited 2014 Oct 16];55(5):1239–45. Available from: http://www.pubmedcentral.nih.gov/articlerender.fcgi?artid=2880505&tool=pmcentrez&rendertype=abstract
81. Courtar DA, Spaanderman ME, Aardenburg R, Janssen BJ, Peeters LL. Low plasma volume coincides with sympathetic hyperactivity and reduced baroreflex sensitivity in formerly preeclamptic patients. J Soc Gynecol Investig. 2006;13 (1):48–52.
82. Grand'Maison S, Pilote L, Okano M, Landry T, Dayan N. Markers of vascular dysfunction after hypertensive disorders of pregnancy: a systematic review and meta-analysis. Hypertension. 2016;68(6):1447–58.
83. Palti H, Rothschild E. Blood pressure and growth at 6 years of age among offsprings of mothers with hypertension of pregnancy. Early Hum Dev. 1989;19(4):263–9.
84. Seidman DS, Laor A, Gale R, Stevenson DK, Mashiach S, Danon YL. Pre-eclampsia and offspring's blood pressure, cognitive ability and physical development at 17-years-of-age. Br J Obstet Gynaecol. 1991;98(10):1009–14.
85. Oglaend B, Forman MR, Romundstad PR, Nilsen ST, Vatten LJ. Blood pressure in early adolescence in the offspring of preeclamptic and normotensive pregnancies. J Hypertens. 2009;27(10):2051–4.
86. Tenhola S, Rahiala E, Halonen P, Vanninen E, Voutilainen R. Maternal preeclampsia predicts elevated blood pressure in 12-year-old children: evaluation by ambulatory blood pressure monitoring. Pediatr Res. 2006;59(2):320–4.
87. Fraser A, Nelson SM, Macdonald-Wallis C, Sattar N, Lawlor DA. Hypertensive disorders of pregnancy and cardiometabolic health in adolescent offspring. Hypertension. 2013;62(3):614–20.
88. Davis EF, Newton L, Lewandowski AJ, et al. Pre-eclampsia and offspring cardiovascular health: mechanistic insights from experimental studies. Clin Sci (Lond). 2012;123 (2):53–72.
89. Kajantie E, Eriksson JG, Osmond C, Thornburg K, Barker DJ. Pre-eclampsia is associated with increased risk of stroke in the adult offspring: the Helsinki birth cohort study. Stroke. 2009;40(4):1176–80.
90. Barker DJ, Osmond C, Golding J, Kuh D, Wadsworth ME. Growth in utero, blood pressure in childhood and adult life, and mortality from cardiovascular disease. BMJ. 1989;298(6673):564–7.
91. Barker DJ, Osmond C. Low birth weight and hypertension. BMJ. 1988;297(6641):134–5.
92. Law CM, Shiell AW. Is blood pressure inversely related to birth weight? The strength of evidence from a systematic review of the literature. J Hypertens. 1996;14(8):935–41.
93. Barker DJ. Adult consequences of fetal growth restriction. Clin Obstet Gynecol. 2006;49(2):270–83.
94. Gallo LA, Tran M, Moritz KM, Jefferies AJ, Wlodek ME. Pregnancy in aged rats that were born small: cardiorenal and metabolic adaptations and second-generation fetal growth. FASEB J. 2012;26(10):4337–47.
95. Gallo LA, Tran M, Cullen-McEwen LA, et al. Transgenerational programming of fetal nephron deficits and sex-specific adult hypertension in rats. Reprod Fertil Dev. 2014;26(7):1032–43.
96. Sattar N, Greer IA. Pregnancy complications and maternal cardiovascular risk: opportunities for intervention and screening? BMJ. 2002;325(7356):157–60.
97. Craici I, Wagner S, Garovic VD. Preeclampsia and future cardiovascular risk: formal risk factor or failed stress test? Ther Adv Cardiovasc Dis. 2008;2(4):249–59.
98. White WM, Garrett AT, Craici IM, et al. Persistent urinary podocyte loss following preeclampsia may reflect subclinical renal injury. PLoS One. 2014;9(3):e92693.

99. Rasmussen LG, Lykke JA, Staff AC. Angiogenic biomarkers in pregnancy: defining maternal and fetal health. Acta Obstet Gynecol Scand. 2015;94(8):820–32.
100. Elvan-Taspinar A, Bots ML, Franx A, Bruinse HW, Engelbert RH. Stiffness of the arterial wall, joints and skin in women with a history of pre-eclampsia. J Hypertens. 2005;23(1):147–51.
101. Yinon Y, Kingdom JCP, Odutayo A, et al. Vascular dysfunction in women with a history of preeclampsia and intrauterine growth restriction: insights into future vascular risk. Circulation [Internet]. 2010 [cited 2014 Oct 16];122(18):1846–53. Available from: http://www.ncbi.nlm.nih.gov/pubmed/20956209
102. Everett TR, Mahendru AA, McEniery CM, Wilkinson IB, Lees CC. Effect of S-nitrosoglutathione on arterial stiffness in women with severe preeclampsia and healthy controls. Arch Dis Child Fetal Neonatal Ed [Internet]. 2012;97:A30. Available from: http://www.embase.com/search/results?subaction=viewrecord&from=export&id=L71393968
103. Vikse BE, Irgens LM, Karumanchi SA, Thadhani R, Reisæter AV, Skjærven R. Familial factors in the association between preeclampsia and later ESRD. Clin J Am Soc Nephrol [Internet]. 2012 [cited 2014 Oct 16];7(11):1819–26. Available from: http://www.pubmedcentral.nih.gov/articlerender.fcgi?artid=3488941&tool=pmcentrez&rendertype=abstract
104. Romundstad PR, Magnussen EB, Smith GD, Vatten LJ. Hypertension in pregnancy and later cardiovascular risk: common antecedents? Circulation. 2010;122(6):579–84.
105. Chambers JC, Fusi L, Malik IS, Haskard DO, De Swiet M, Kooner JS. Association of maternal endothelial dysfunction with preeclampsia. JAMA. 2001;285(12):1607–12.
106. Maynard SE, Min J, Merchan J, et al. Excess placental soluble fms-like tyrosine kinase 1 (sFlt1) may contribute to endothelial dysfunction, hypertension, and proteinuria in preeclampsia. J Clin Invest. 2003;111(5):649–58.
107. Faas MM, Schuiling GA, Baller JF, Visscher CA, Bakker WW. A new animal model for human preeclampsia: ultra-low-dose endotoxin infusion in pregnant rats. Am J Obstet Gynecol. 1994;171(1):158–64.
108. Li J, LaMarca B, Reckelhoff JF. A model of preeclampsia in rats: the reduced uterine perfusion pressure (RUPP) model. AJP Hear Circ Physiol. 2012;303(1):H1–8.
109. Bytautiene E, Lu F, Tamayo EH, et al. Long-term maternal cardiovascular function in a mouse model of sFlt-1-induced preeclampsia. Am J Physiol Heart Circ Physiol [Internet]. 2010 [cited 2014 Oct 16];298(1):H189–93. Available from: http://www.ncbi.nlm.nih.gov/pubmed/19915174
110. Paauw ND, Joles JA, Spradley FT, et al. Exposure to placental ischemia impairs postpartum maternal renal and cardiac function in rats. Am J Physiol Regul Integr Comp Physiol. 2017;312(5):R664–70.
111. Brennan L, Morton JS, Quon A, Davidge ST. Postpartum vascular dysfunction in the reduced Uteroplacental perfusion model of preeclampsia. PLoS One. 2016;11(9):e0162487.
112. Pruthi D, Khankin EV, Blanton RM, et al. Exposure to experimental preeclampsia in mice enhances the vascular response to future injury. Hypertension. 2015;65(4):863–70.
113. Mosca L, Benjamin EJ, Berra K, et al. Effectiveness-based guidelines for the prevention of cardiovascular disease in women—2011 update: a guideline from the american heart association. Circulation [Internet]. 2011 [cited 2014 Oct 16];123(11):1243–62. Available from: http://www.pubmedcentral.nih.gov/articlerender.fcgi?artid=3182143&tool=pmcentrez&rendertype=abstract
114. Bushnell C, McCullough LD, Awad IA, et al. Guidelines for the prevention of stroke in women: a statement for healthcare professionals from the American Heart Association/American Stroke Association. Stroke. 2014;45(5):1545–88.
115. Zoet GA, Koster MPH, Velthuis BK, et al. Determinants of future cardiovascular health in women with a history of preeclampsia. Maturitas [Internet]. 2015;82(2):153–61. Available from: https://doi.org/10.1016/j.maturitas.2015.07.004
116. Milic NM, Milin-Lazovic J, Weissgerber TL, Trajkovic G, White WM, Garovic VD. Preclinical atherosclerosis at the time of pre-eclamptic pregnancy and up to 10 years postpartum: systematic review and meta-analysis. Ultrasound Obstet Gynecol. 2017;49(1):110–5.
117. Sabour S, Franx A, Rutten A, et al. High blood pressure in pregnancy and coronary calcification. Hypertens (Dallas, Tex 1979). 2007;49(4):813–7.
118. Cassidy-Bushrow AE, Bielak LF, Rule AD, et al. Hypertension during pregnancy is associated with coronary artery calcium independent of renal function. J Womens Health (Larchmt). 2009;18(10):1709–16.
119. White WM, Mielke MM, Araoz PA, et al. A history of preeclampsia is associated with a risk for coronary artery calcification 3 decades later. Am J Obstet Gynecol [Internet]. 2016;214(4):519.e1–8. Available from: https://doi.org/10.1016/j.ajog.2016.02.003
120. Paauw ND, Luijken K, Franx A, Verhaar MC, Lely AT. Long-term renal and cardiovascular risk after preeclampsia: towards screening and prevention. Clin Sci. 2016;130(4):239–46.
121. Berks D, Hoedjes M, Raat H, Duvekot JJ, Steegers EA, Habbema JD. Risk of cardiovascular disease after pre-eclampsia and the effect of lifestyle interventions: a literature-based study. BJOG [Internet]. 2013 [cited 2014 Oct 16];120(8):924–31. Available from: http://www.ncbi.nlm.nih.gov/pubmed/23530583
122. Cusimano MC, Pudwell J, Roddy M, Cho C-KJ, Smith GN. The maternal health clinic: an initiative for cardiovascular risk identification in women with pregnancy-related complications. Am J Obstet Gynecol. 2014;210(5):438.e1–9.

Cardiovascular Implications of Diabetes, Metabolic Syndrome, Thyroid Disease, and Cardio-Oncology in Women

29

Marijana Tadic, Cesare Cuspidi, Dragan Vasic, and Peter L. M. Kerkhof

Art work by Piet Michiels, Leuven, Belgium.

M. Tadic (✉)
Department of Cardiology, Charité-University-Medicine Berlin, Berlin, Germany

C. Cuspidi
University of Milan-Bicocca and Istituto Auxologico Italiano, Clinical Research Unit, Meda, Italy

D. Vasic
Clinic of Vascular and Endovascular Surgery, Clinical Centre of Serbia, Belgrade, Serbia

P. L. M. Kerkhof
Department of Radiology and Nuclear Medicine, Amsterdam Cardiovascular Sciences, VU University Medical Center, Amsterdam, The Netherlands

Abstract

Cardiovascular disease may be associated with several comorbidities, including diabetes mellitus, thyroid disorders, and the metabolic syndrome, which are predominantly observed in women and often starting at particular ages. In addition, common treatment options for carcinomas frequently seen in women may induce serious cardiotoxic effects. We review the scope of the problem, the pathophysiologic mechanisms involved, as well as the resulting abnormalities regarding cardiac structure and

P. L. M. Kerkhof, V. M. Miller (eds.), *Sex-Specific Analysis of Cardiovascular Function*,
Advances in Experimental Medicine and Biology 1065, https://doi.org/10.1007/978-3-319-77932-4_29

function as observed by using imaging techniques.

Keywords
Comorbidities in women with heart disease · Menopause · Estrogen · Diabetes mellitus · Thyroid disease · Metabolic syndrome · Cardiac remodeling · Cardio-oncology · Heart failure · Left ventricle · Right ventricle · Obesity · Review

Introduction

Due to more pronounced awareness, the number of studies that demonstrate a significant sex difference in terms of epidemiology, pathophysiology, and therapy for various diseases is steadily increasing. Sex-related differences can be ascribed to sex hormones, their expression, and their effects on organs. Additionally, women demonstrate more pronounced changes in hormone status and body composition due to reproductive factors varying during their lifetime. Also, distinct lifestyles observed between sexes should not be ignored. This review summarizes current insight regarding prevalent comorbidities observed in connection with cardiovascular disease as far as predominantly diagnosed in women.

Diabetes Mellitus as a Comorbidity

While diabetes mellitus (DM) is more prevalent in men, the impact on the cardiovascular system appears to be more pronounced in women. A large study conducted in Canada in the period from 1995 to 2005 shows that prevalence of DM among men was >19%, while it was <16% in women [1]. In 2013, there were 14 million more men affected by DM than women [2]. It was reported that >50% of DM patients are middle aged, while incidence increases with age in both sexes, reaching the highest rates in women older than 75 years [2].

The possible reasons for the reported sex-specific distinctions might be differences in obesity and insulin resistance. The prevalence of overweight females in 2013 raised to 38%, similar to the prevalence among men (37%) [2]. However, overweight or obesity is more prevalent among women older than 45 years, whereas males are overweight at younger age [2]. Nevertheless, recent studies showed that men develop DM with lower body mass index (BMI) [3, 4]. One investigation included 51,920 men and 43,137 women from Scotland and showed that average BMI closest to date of diagnosis of type 2 DM was 31.83 kg/m^2 in men, contrasting ($p = 0.001$) a value of 33.69 kg/m^2 in women [3].

It seems that obesity significantly increases total cardiovascular mortality in women with DM because the Emerging Risk Factors Collaboration study showed that the hazard ratio (HR) for cardiovascular disease in women with DM (compared to women without) was 2.59 (95%CI: 2.29–2.93), and significantly lower in men HR 1.89 (95%CI: 1.73–2.06) [5].

Insulin resistance is higher in men [6]. Middle-aged men have higher levels of fasting glucose and triglycerides and lower HDL cholesterol than women of a similar age even after adjustment for BMI [5]. Fat distribution also varies by gender, and men have greater visceral fat in comparison with women. Exposure to high circulating lipid levels induces an inflammatory response that is accompanied by microvascular dysfunction, correlating with the amount of visceral adipose tissue (Fig. 29.1).

Diabetes, Heart, and Gender

Diabetic cardiomyopathy is more commonly seen in women. Pathophysiologic factors that could explain the higher incidence in women are presented in Table 29.1.

Considering the fact that DM is one of the most important risk factors for development of heart failure with preserved ejection fraction (EF), it is of great importance to recognize early signs of diabetic cardiomyopathy. The major problem is that criteria for its diagnosis are not clear, and most of them are based on traditional echocardiographic parameters such as pulsed Doppler and

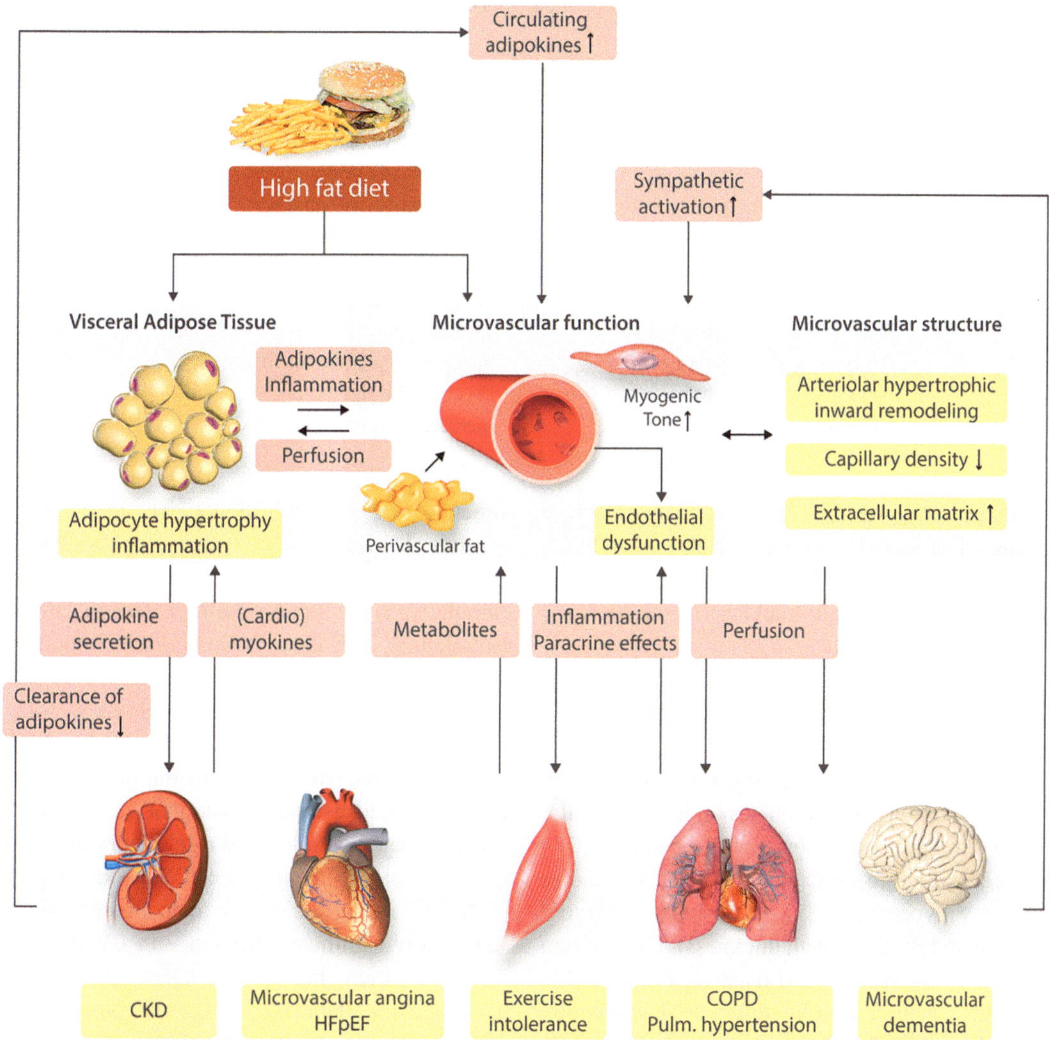

Fig. 29.1 Proposed mechanisms of obesity-related microvascular dysfunction predisposing to multi-organ disease. High fat diet on a regular basis changes the composition of visceral adipose tissue and induces a low-grade local inflammatory response, which together modify the secretion of adipokines. Simultaneously, high fat diet results in endothelial dysfunction throughout the body, which not only alters vascular tone, and contributes to development of microvascular insulin resistance, but also influences vascular structure and perivascular inflammation. In concert, these microvascular changes impair organ perfusion and organ function, thereby further contributing to altered release and clearance of metabolites and neurohumoral factors, like adipokines, inflammatory cytokines, as well as (cardio)myokines. Global microvascular dysfunction in obese subjects therefore is a common pathway that contributes to exercise intolerance and predisposes to development of chronic kidney disease, microvascular dementia, coronary microvascular angina, COPD, and pulmonary hypertension. *CKD* chronic kidney disease, *HFpEF* heart failure with preserved ejection fraction, *COPD* chronic obstructive pulmonary disease. (With permission reproduced from: Sorop O, Olver TD, van de Wouw J, Heinonen I, van Duin RW, Duncker DJ, Merkus D. The microcirculation: a key player in obesity-associated cardiovascular disease. Cardiovasc Res. 2017;113: 1035–1045. doi: https://doi.org/10.1093/cvr/cvx093)

not new parameters obtained by speckle-tracking imaging.

The introduction of new echocardiographic speckle-tracking technique enables detailed insight in assessment of cardiac mechanics. The latter is an angle-independent and significantly less load-dependent method. Tissue Doppler imaging provides information about cardiac

Table 29.1 Pathophysiological differences between genders that could favor the occurrence of diabetes

Males	Females
Increased fasting glucose	Increased insulin resistance
Diabetes develops at earlier age	Diabetes develops with higher BMI
–	Gestational diabetes
Increased overweight incidence	Increased obesity incidence
Abdominal obesity	Abdominal + peripheral obesity
Increased fatty liver	Unchanged incidence of fatty liver
Decreased androgens	Increased androgens
Increased erectile dysfunction	Increased polycystic ovary syndrome
Unchanged depression incidence	Increased depression incidence
–	Psychosocial risk factors
Increased neuropathy	Increased cardiovascular risk

mechanics in only small portion of myocardium included in sample volume. On the other hand, speckle-tracking imaging gives mechanical information in the total wall thickness of the heart.

Diabetes and Ventricular Function

A large number of studies show the detrimental influence of DM or prediabetes on left ventricular (LV) structure, function, and mechanics [7–9]. LV structure is usually defined in studies with LV hypertrophy (LV mass index, interventricular and posterior wall thickness). LV function is defined by systolic and diastolic function. LV systolic function usually refers to LV ejection fraction, while LV diastolic function is mainly defined by pulsed and tissue Doppler parameters (E/A, E/e′). LV mechanics refers to strain.

Studies reveal an increase in LV longitudinal strain after improvement of glycemic control [10] or deterioration in LV longitudinal and circumferential strain during the 2-year follow-up in diabetic patients with usual hypoglycemic therapy [11]. Our previous study also reported significant influence of DM and prediabetes on right ventricular (RV) function and mechanics [12]. A major problem is that the vast majority of these studies did not consider the difference between genders.

Novel investigations also revealed that DM impacts layer-specific longitudinal and circumferential LV strain [13, 14]. The authors found that diabetic-induced LV remodeling, as well as RV changes, occurs in endocardium-epicardium direction [14, 15]. Additional risk factors such as arterial hypertension, obesity, dyslipidemia, renal function impairment, or smoking have a cumulative and not only additive effect on cardiac remodeling. The latter is particularly important because diabetic patients rarely have isolated diabetes and mostly have several listed risk factors.

Interesting study that included 1266 individuals free from hypertension, diabetes, and cardiovascular disease showed that glucose level was associated with left and right ventricular diastolic function only in women, but not in men [16]. The reanalysis of our data did not show statistically significant difference in LV multidirectional strain between genders among the patients with diabetes [7]. On the other hand, our previous study in a population of subjects with the metabolic syndrome clearly showed that different risk factors were associated with LV hypertrophy [17]. Namely, increased blood pressure and fasting glucose were independently of other metabolic syndrome criteria related with LV hypertrophy only in women, whereas only increased blood pressure was associated with LV hypertrophy in men [17]. Similar results were also obtained for RV remodeling in the metabolic syndrome [18]. The multivariate analysis revealed that a combination of increased blood pressure, impaired fasting glucose, and dyslipidemia was related with RV hypertrophy solely in women [18]. Increased systolic blood pressure, impaired fasting glucose, and abdominal obesity were independently associated with RV diastolic dysfunction in women, whereas increased systolic blood pressure was the only independent predictor in men. The influence of gender on LV hypertrophy were reported earlier [19, 20]. It was found that glycemic control was related with greater wall thickness, LV mass, and diastolic dysfunction only in women [19]. Furthermore, investigators showed that female sex was the only independent predictor of LV hypertrophy in diabetic patients [20].

One should be aware of the fact that some authors already reported the gender-specific normal values for longitudinal LV strain in the large sample of global population with high percentage of cardiovascular risk factors (hypertension, smoking, drinking alcohol, diabetes) [21]. The authors claimed that normal LV longitudinal strain for women was $> -18.8\%$ and for men $>17.4\%$ [21]. This possibly could be the reason for the difference in LV longitudinal strain between women and men with DM.

There are several possible mechanisms that could explain the influence of diabetes on cardiac remodeling. First, diabetes is characterized by increased production of free oxygen radicals that induce accumulation of collagen in the interstitium and subsequent myocardial fibrosis [22], which stimulates loss of elasticity and increased ventricular stiffness, and ultimately leads to diastolic dysfunction, hypertrophy, and systolic dysfunction [23]. Second, the autonomic cardiac neuropathy in diabetes contributes to deteriorated systolic function and elevated diastolic filling pressure, which results in worsening of overall myocardial function. Third, the mutual relationship between insulin resistance and biohumoral systems such as renin-angiotensin-aldosterone (RAA) and sympathetic nervous system might also clarify cardiac remodeling in diabetic patients [24]. Fourth, diabetes is related to obesity that could both directly and indirectly induce cardiac damage. Obesity is associated with increased preload which further induces LV and RV dilatation, cardiac dysfunction, and hypertrophy.

Higher morbidity in females with diabetes potentially could be explained by sex hormones that significantly influence energy metabolism, body structure, vascular function, and inflammatory reactions. Furthermore, psychosocial stress seems to have higher impact on women than on men [22–24].

Metabolic Syndrome (MetSy) as a Comorbidity

The term metabolic syndrome (MetSy) was coined in 1977 and now represents a cluster of five cardiovascular risk factors, including arterial hypertension, abdominal obesity, high-density lipoprotein cholesterol, triglyceride level, and fasting blood glucose. At least three factors must be present to meet the requirements for diagnosing the syndrome. Prevalence is high (20–30% of the general population), and this figure increases with age in a gender-specific manner. In a population younger than 50 years, its prevalence is somewhat higher in men, whereas after the age of 50 years, its prevalence is significantly higher in women [25].

Sex-related factors are mainly associated with hyperandrogenism, insulin resistance, abdominal obesity, and HDL cholesterol reduction that occurred during and after menopause. However, one should not forget to consider psychosocial factors that are more pronounced in women than men [25].

The MetSy is defined differently in women and men. All existing definitions are using gender-specific cutoff values for dyslipidemia and obesity [26–31]. Only values for HDL, BMI, and waist circumference changed over time. However, the gender-specific cutoff criteria persist until today. Even in the last definition from the National Cholesterol Education Program's Adult Treatment Panel III (NCEP-ATP III), there are gender-specific cutoff values for HDL (<1.04 mmol/L for men and < 1.3 mmol/L for women) and waist circumference (>102 cm for men and > 88 cm for women) [31].

Probably the factor most responsible for a gender-specific pattern of the MetSy refers to hormonal changes that happen during menopause. During this period the balance between estrogen and testosterone modifies, and fat now accumulates in the abdomen and not at the gluteal and hip regions, as is the case before menopause

[32, 33]. This visceral fat is associated with increased insulin resistance, elevated free fatty acid level, and increased activity of hepatic lipase [33, 34]. This could also explain why perimenopausal and postmenopausal women are at the higher risk of the MetSy development.

Metabolic Syndrome and Mortality in Both Sexes

A large number of studies confirmed higher cardiovascular morbidity and mortality among men, and this is usually ascribed to the protective influence of estrogen. However, it is possible that some conditions such as insulin resistance could significantly change – increase cardiovascular morbidity and mortality in women.

The classical Rancho Bernardo Study reports that the relative hazard of ischemic heart disease mortality in diabetic patients compared to controls was 1.9 in men and 3.3 in women, after adjusting for age, systolic blood pressure, cholesterol, body mass index, and cigarette smoking [35].

A large meta-analysis that included 64 cohorts involving 858,507 subjects and 28,203 coronary heart disease events showed that risk ration for coronary heart disease in women was 2.82 (95% CI: 2.35–3.38) and 2.16 (95% CI: 1.82–2.56) in men [36]. The adjusted risk ratio for coronary heart disease was 44% higher in women with diabetes than in men with diabetes (RR 1.44 [95% CI: 1.27–1.63]).

The Atherosclerosis Risk in Communities reported that men and women with the MetSy over a period of 11 years had approximately 1.5 and 2 times higher risk to develop coronary heart disease than controls [37]. The risk for ischemic stroke occurrence in the subjects with the MetSy was similar for both genders.

The Hoorn Study that involved 615 men and 749 women aged 50–75 years without diabetes or cardiovascular disease at baseline used NCEP-ATP III definition and showed that the MetSy was associated with twofold increase in age-adjusted risk of fatal cardiovascular disease in men and nonfatal cardiovascular disease in women [38]. The risk for nonfatal cardiovascular disease was slightly higher in women (RR 2.05; 95% CI: 1.29–3.26) than in men (RR 1.88; 95% CI: 1.24–2.87).

The study that involved 11 prospective European cohort studies and included 6156 men and 5356 women with no diabetes, aged 30–89 years, with a median follow-up of 8.8 years, showed that the age-adjusted MetSy prevalence was somewhat higher in men (15.7%) than in women (14.2%) [39]. The risk ratios for all-cause mortality in subjects with the MetSy did not significantly differ between genders: 1.44 (95% CI: 1.17–1.84) in men and 1.38 (95% CI: 1.02–1.87) in women after adjustment for age, cholesterol levels, and smoking [39]. However, the risk for cardiovascular mortality was higher among women with the MetSy: 2.26 (95% CI: 1.61–3.17) in men and 2.78 (95% CI: 1.57–4.94) in women [39].

Metabolic Syndrome and Cardiac Remodeling in Both Sexes

The MetSy and its components have different influence on cardiac remodeling, both LV and RV. Our previous studies showed that the MetSy and its criteria are associated with LV and RV hypertrophy, as well as LV and RV diastolic dysfunction [40, 41].

Multiple regression analysis showed that systolic blood pressure and waist circumference were independently associated with LV mass index [40]. Furthermore, systolic blood pressure, waist circumference, and triglyceride level were independently associated with LV diastolic function (mitral E/e′) [40]. Interestingly, systolic blood pressure, waist circumference, glucose level, and left ventricular mass index were independently associated with the RV hypertrophy, whereas waist circumference, glucose level, and LV mass index were independently associated with RV diastolic function (tricuspid E/e′) [41]. However, gender-specific analysis showed the difference between men and women [17, 18].

Analysis of individual MetSy factors demonstrated that increased blood pressure was independently associated with LV hypertrophy in both genders, whereas impaired fasting glucose was independently related with LV hypertrophy only in women [17]. The same study showed that the combination of impaired glucose level, abdominal obesity, and dyslipidemia was related with LV hypertrophy only in women. Increased blood pressure and fasting glucose, as well as elevated triglyceride level, were independently associated with LV diastolic dysfunction in women, whereas the only independent predictor in men was elevated blood pressure [17]. The combination of hypertension, increased fasting glucose, and dyslipidemia was independent predictor of LV diastolic dysfunction merely in women. Aijaz et al. published similar findings and showed that LV mass index was greater and LV diastolic dysfunction more prevalent in women with isolated metabolic syndrome (without diabetes and hypertension) than in women without metabolic syndrome, whereas there was no difference in men [42]. However, after including the patients with hypertension or diabetes, the stepwise increase in LV mass index and prevalence of LV diastolic dysfunction from subjects without MetSy, across individuals with isolated MetSy to patients with MetSy was obtained in both genders [42]. Other studies showed that the prevalence of LV hypertrophy was approximately 2.3- and 1.9-fold higher in men and women with MetSy than in controls, but there was no difference between genders [43].

Our more recent investigation revealed the negative influence of the MetSy on LV mechanics [44]. 2D global longitudinal and circumferential strains were significantly reduced in the MetSy group, whereas global radial strain was not different between the MetSy and control group. The increasing number of the MetSy criteria progressively decreased LV longitudinal strain from the individuals with no MetSy criteria to the individuals with all five criteria [44]. However, this was not the case with LV circumferential and radial strains. The combination of increased blood pressure, abdominal obesity, and increased fasting glucose level was related with the higher level of LV strain impairment comparing with other clusters of the MetSy components. The multivariate analysis of the MetSy criteria showed that 24-h mean blood pressure, waist circumference, and fasting glucose level were independently associated with 2D global longitudinal and circumferential myocardial strain [44]. The interaction between gender and MetSy significantly influenced only LV longitudinal strain.

The previous investigation from our study group showed that increased blood pressure and abdominal obesity were independently associated with RV hypertrophy in women, whereas the only independent predictor in men was systolic blood pressure [18]. Furthermore, the analysis revealed that increased blood pressure, impaired fasting glucose, and dyslipidemia were a combination of the MetSy risk factors related with RV hypertrophy in women, but not in men [18]. Increased systolic blood pressure, impaired fasting glucose, and waist circumference were independently associated with RV diastolic function in women, whereas increased systolic blood pressure was the only independent predictor in men. Impaired fasting glucose, abdominal obesity, and dyslipidemia were a combination of MetSy criteria, which was independently associated with RV diastolic dysfunction only in women. Other research showed that prevalence of RV hypertrophy was significantly higher in both hypertensive men and women with MetSy than in their non-MetSy counterparts. However, the percentage was higher among women than in men (45% vs. 60%) [43].

Our investigation that studies the influence of the MetSy on RV strain reported that 2D RV global longitudinal strain was significantly decreased in the MetSy group compared with the control subjects [44]. Interestingly, 3D RV EF was also significantly reduced in the MetSy subjects ($55 \pm 4\%$ vs. $58 \pm 4\%$; $p < 0.001$). The multivariate analysis of MetSy criteria showed that systolic blood pressure, waist circumference, and fasting glucose were independently associated with RV systolic function (3D RV EF) and mechanics (RV longitudinal strain) [44].

The possible mechanisms underlying the interaction of gender on the association between the MetSy and LV remodeling remain hypothetical. Increased large artery stiffness characteristic for the MetSy [46] possibly could increase LV mass and impair LV diastolic function by elevating cardiac afterload. However, the increase in LV mass in women was independent of increased arterial stiffness in the Hoorn Study [38]. On the other hand, insulin resistance could reverse the positive cardiovascular effects of estrogen in women and therefore increase LV mass. Since insulin resistance incidence is higher in females, it is quite reasonable to accept the hypothesis that abdominal obesity, incubator of insulin resistance, has a more important role in cardiac remodeling in women than in men.

Thyroid Dysfunction and Outcome in Women

Thyroid disorders represent typical health problems in the female population. The occurrence of hyperthyroidism and hypothyroidism is 0.5–4% in a population without iodine deficit; however it is five to ten times higher in women than in men.

A recent study that involved 11,359 participants (57 ± 6 years, 58% women) detected mild hypothyroidism in 542 subjects (4.8%), moderate or severe hypothyroidism in 255 individuals (2.2%), mild hyperthyroidism in 378 individuals (3.3%), and moderate and severe hyperthyroidism in 206 individuals (1.8%) [47]. The percentage of females ranged from 68.6% to 82%. The risk of stroke or myocardial infarction during 22.5-year median follow-up was similar in the patients from both parts of the spectrum of thyroid disorders (i.e., hypo- and hyperthyroidism) [47]. Large meta-analysis that included 1,898,314 participants showed that patients with hypothyroidism had significantly higher risks of ischemic heart disease (RR 1.13; 95% CI: 1.01–1.26), myocardial infarction (RR: 1.15; 95% CI: 1.05–1.25), cardiac mortality (RR: 1.96; 95% CI: 1.38–2.80), and all-cause mortality (RR: 1.25; 95% CI: 1.13–1.39) than euthyroid subjects [48]. The same study demonstrated that subclinical hypothyroidism (especially with thyrotropin level $\geq$ 10 mIU/L) was also related with higher risks of ischemic heart disease and cardiac mortality [48].

The large recently published prospective study that involved 75,076 women aged 20–89 years followed participants for 28 years and reported that women with hyperthyroidism had an elevated risk of breast cancer mortality after 60 years of age (HR 2.04; 95% CI: 1.16–3.60) compared to euthyroid women [49]. Hypothyroid women had elevated mortality risks for diabetes mellitus (HR 1.58; 95% CI: 1.03–2.41), cardiovascular disease (HR 1.20; 95% CI: 1.01–1.42), and cerebrovascular disease (HR 1.45, 95% CI: 1.01–2.08) [49].

The other recent investigation involved 85,856 hyperthyroid patients and 847,057 matched population-based controls [50]. During the follow-up of 9.2 years, mortality was the highest in the first 3 months after diagnosis of hyperthyroidism (HR 4.62; 95% CI: 4.40–4.85) and remained increased during follow-up. The risk for all investigated cardiovascular events was increased, with the highest risk in the first 3 months after hyperthyroidism diagnosis was established [50]. The risk was highest for atrial fibrillation (HR 7.32; 95% CI: 6.58–8.14) and arterial embolism (HR 6.08; 95% CI: 4.30–8.61), but the risks of venous thromboembolism, acute myocardial infarction, and ischemic and nonischemic stroke were also increased two to three times [50].

Hypothyroidism and Cardiac Remodeling

The influence of thyroid hormones on cardiac function is not yet established because of conflicting results from existing studies.

There are several mechanisms that could explain the effect of hypothyroidism on the LV remodeling. First, hypothyroidism induces the activity reduction of some enzymes responsible for intracellular calcium handling that modifies the expression of contractile protein [51]. Second, tissue changes such as dehydration, collagen alteration, myocyte orientation, or capillary distribution might contribute to cardiac dysfunction [51]. Third, hypothyroidism is related with other cardiovascular risk factors (obesity, dyslipidemia, diabetes, hypertension, and metabolic syndrome) that also induce cardiac remodeling. Additionally, hypothyroidism is associated with hemodynamic changes (increase in peripheral vascular resistance and arterial stiffness, as well as increased cardiac afterload), which might provoke cardiac impairment. These parameters are associated with decreased LV volumes and reduced oxygen consumption, which further could decrease cardiac performance and induce heart failure. Studies showed that diastolic heart failure was strongly related with hypothyroidism and more prevalent in women [52].

The most of investigations agree that hypothyroidism, as well as subclinical hypothyroidism, induces an increase in LV mass and LV diastolic dysfunction [53–56]. However, these studies were generally based on pulsed or tissue Doppler indices which possibly could not detect subtle and subclinical changes in cardiac function and mechanics.

Our recent study included female with subclinical hypothyroidism, followed them for 1 year, and showed that 2D LV mechanics is significantly impaired in these patients during the whole cardiac cycle [57]. It was revealed that 2D global longitudinal and circumferential strains, as well as corresponding systolic and early diastolic strain rates, were significantly reduced in women with subclinical hypothyroidism in comparison with the healthy controls [57]. The 3D analyses demonstrated that 3D LV end-diastolic volume, stroke volume, cardiac output, and 3D LV EF were significantly reduced in women with subclinical hypothyroidism before therapy in comparison with the controls or the patients after 1 year of treatment. 3D LV mass index was lower in the controls and the treated subclinical hypothyroidism patients compared with the patients at baseline. 3D LV global longitudinal, circumferential, area, and radial strains were significantly decreased in the hypothyroid patients at the baseline in comparison with the healthy controls [57]. However, after a 1-year levothyroxine treatment, LV systolic and diastolic function significantly improved in women with subclinical hypothyroidism [57]. Interestingly, LV diastolic function in patients did not completely recover, and it was still deteriorated in comparison with the healthy controls. These findings show that normalization of TSH level does not automatically mean the normalization of LV mechanics and function, in particular for LV diastolic function that obviously requires more time for improvement. Previous studies demonstrated normalization of LV diastolic function after 1-year levothyroxine therapy [58, 59]. However, evaluation of LV diastolic function in their investigations was performed by pulsed and tissue Doppler, while we used speckle-tracking imaging which is a more precise imaging tool. These results concurred with previous investigations which used tissue Doppler-derived strain to evaluate LV longitudinal strain in hypothyroid patients [60]. The researchers reported that longitudinal strain and systolic strain rate of each of 16 LV segments were significantly reduced in hypothyroid patients, but they did not analyze LV circumferential and radial strains nor 3D parameters of LV function and deformation [60].

Similar findings were also reported for the RV, e.g., our investigation in female patients showed that RV and right atrial functions were significantly depressed with subclinical hypothyroidism [61].

The association between right heart and hypothyroidism is intriguing because it could not be easily explained from a pathophysiological standpoint. Such a relationship has been proposed by several investigators [62–66] and can be justified by several mechanisms: (1) changed intracellular calcium handling could be one of the most important mechanisms; (2) altered myocyte direction and capillary distribution; (3) reduced cardiac oxygen consumption related to elevated afterload, reduced contractility, and decreased efficiency; (4) reduced degradation of myocardial matrix and increased insulin growth factor 1 which might induce cardiac hypertrophy and further RV dysfunction; (5) other cardiovascular risk factors related to hypothyroidism (dyslipidemia, hypertension); (6) increased pulmonary vascular resistance and pulmonary hypertension in hypothyroid patients; (7) ventricular interaction through interventricular septum, which transduces pressure and volume overload from the LV to RV, as well as by transmission of increased LV filling pressure across the pulmonary circulation to the RV.

Our study revealed that RV function assessed by 3D RV EF and RV longitudinal strain was impaired in the females with subclinical hypothyroidism [61]. Furthermore, levothyroxine therapy significantly improved RV and right atrial longitudinal strain, even though right heart remodeling was not entirely reversible even 1 year after restoring euthyroid status. This investigation showed that RV diastolic function was deteriorated in the patients with subclinical hypothyroidism which concurs with earlier investigations [62–66]. The impact of hypothyroidism on RV systolic function is debatable. The findings from our study showed that RV systolic function, estimated by 3D echocardiography and 2D strain, was significantly impaired in the hypothyroid women which was not detected by conventional echocardiographic parameters – TAPSE or tricuspid systolic velocity obtained by tissue Doppler [62–66]. Previous investigation that used cardiac magnetic resonance showed that hypothyroid patients had significantly decreased cardiac preload and elevated afterload with a subsequent reduction in stroke volume and cardiac output [67], which was in agreement with our results [61]. This might a result of thickened RV wall in the female patients with subclinical hypothyroidism before levothyroxine therapy which was previously reported in overt hypothyroid patients [68].

Hyperthyroidism and Cardiac Remodeling

Although the mechanism of LV remodeling in hyperthyroidism has not established yet, there are several possible mechanisms that could explain cardiovascular changes: (1) reduced peripheral resistance and peripheral vasodilatation, (2) increased activity of RAA system, (3) decreased renal perfusion, (4) increased LV volumes and cardiac output. All these modifications might stimulate deterioration of LV systolic and diastolic mechanical function, as well as promote LV hypertrophy [69].

Investigations which address the association between hyperthyroidism and cardiac remodeling were mostly focused on LV mass and diastolic function. These studies showed significantly increased LV mass and impaired LV diastolic function among patients with clinical or subclinical hyperthyroidism [70, 71]. However, after normalization of thyroid function by radioactive iodine treatment, a significant reduction in both LV mass index and cardiac index was documented [72].

Our research comparing a treated and untreated patients with subclinical hyperthyroidism demonstrated impaired LV multidirectional myocardial deformation, as evaluated with 2D and 3D strain, in the untreated group. [73]. Both LV systolic and diastolic mechanical LV functions were deteriorated in subjects with subclinical hyperthyroidism comparing with healthy controls [73]. LV volumes and cardiac output, evaluated by 3D echocardiography, were increased in individuals with subclinical hyperthyroidism. Finally, TSH and FT4 levels correlated with LV structure and strain in the whole study population that included healthy controls and patients with subclinical hyperthyroidism [73]. These findings are in agreement with details reported elsewhere in the literature [74, 75].

The study by Abdulrahman et al. involved 25 patients (18 women) with a history of differentiated thyroid carcinoma being on long-term TSH-suppressive levothyroxine therapy and reported that subclinical hyperthyroidism affected LV longitudinal and circumferential strains [74]. The authors observed a U-shaped association between a range of thyroid hormone levels (from hyper- to hypothyroid concentrations) and myocardial longitudinal and circumferential strain [75]. Our study showed that LV systolic function, estimated by 2D and 3D LV ejection fraction, in patients with subclinical hyperthyroidism was in the normal range and did not differ from healthy controls. In contrast, LV deformation, assessed by 2D and 3D strain, was deteriorated in the subjects with subclinical hyperthyroidism [73]. The reason for preserved LV systolic function, despite reduced LV strain, potentially might lie in the paradoxically increased LV twist that we found in the subjects with subclinical hyperthyroidism. LV twist represents a complex cardiac motion that consists of circumferential motion of the apex with respect to the base of the heart, enabling a more effective contraction by the LV myocytes. Our findings suggest that 3D LV volumes and cardiac output are elevated in the individuals with subclinical hyperthyroidism, which further implies hemodynamic overload and possibly explains the observed increased LV twist in this group of patients. Considering the fact that LV ejection fraction was not different between healthy controls and subjects with subclinical hyperthyroidism, it appears that this traditional echocardiographic parameter is not enough for the estimation of LV systolic function in subjects with subclinical hypo- or hyperthyroidism.

There are several important clinical implications of these findings. First, these results regarding reduced LV strain might potentially clarify cardiovascular morbidity and mortality in the patients with subclinical thyroid disorders. It has already been known that LV deformation and particularly LV longitudinal strain represent good predictors of the overall and cardiovascular outcome [21]. Second, LV strain evaluation could provide the information about the necessity of therapy and timing of introduction. Earlier studies showed that LV remodeling in subclinical thyroid dysfunction was reversible, but strain evaluation would provide information about possible improvement much sooner, and therapy could be potentially stopped earlier. Third, assessment of LV longitudinal strain could be very useful and facilitate the follow-up of these patients.

There are very limited data about the influence of hyperthyroidism on RV remodeling. However, our study group provided data regarding RV remodeling in the patients with subclinical hyperthyroidism [76]. In particular our research revealed that RV diastolic function was deteriorated in these individuals comparing with the healthy controls [76]. Furthermore, findings showed that RV volumes evaluated by 3D echocardiography were larger, whereas 3D RV EF was reduced in the subjects with subclinical hyperthyroid participants. Interestingly, RV and right atrial longitudinal strains were significantly reduced in these subjects.

There are only a few investigations on the influence of hyperthyroidism on RV remodeling. One study reported that RV diastolic function was deteriorated in patients with overt hyperthyroidism, while RV systolic function assessed by the tissue Doppler imaging was better in this group of patients [77]. Nacar et al. showed that right atrial volume index was higher in the subjects with subclinical hyperthyroidism, but without statistical significance [78]. Our results were in line with these findings. Recently, Teasdale et al. demonstrated significant improvement in cardiac output, exercise capacity, and physical quality on resolution of hyperthyroidism [79]. Exercise capacity was assessed using the 6-minute-walk-distance (6MWT), and quality of life was assessed by Medical Outcome Study 36-item Short-Form Health Status Survey.

Several potential mechanisms might illuminate the association between RV remodeling and hyperthyroidism (either overt or subclinical). It is already known that subclinical hyperthyroidism is related to mild hyperkinetic cardiovascular conditions that could induce chronic hemodynamic overload, including RV dilatation and hypertrophy. Additionally, one should not forget that mild pulmonary hypertension is present in 35% and 36% of patients with Graves′ disease or toxic multinodular goiter, respectively [80]. Investigators showed that pulmonary vascular resistance was significantly higher in hyperthyroid patients. However, there was no association between pulmonary hypertension and TSH, fT3, and fT4 levels [80].

Cardio-Oncology

The prevalence of oncology patients is steadily rising, and it is expected that the incidence will increase up to 45% by 2030 [81]. Fortunately, despite this alarming trend, the mortality rate was reduced by 20–30% in the last 10 years [82]. Modern chemotherapy has significantly decreased mortality and improved survival of oncology patients. However, follow-up of patients treated with novel chemotherapeutic agents also showed the large spectrum of cardiotoxic manifestations [83].

Radiation Effects and Two Types of Chemotherapy-Related Cardiotoxicity

Several factors may influence the severity of cardiotoxicity: chemotherapy type, cumulative dosage of chemotherapy, age, concurrent or previous radiation therapy, and comorbidities such as hypertension, diabetes, renal failure, etc. The majority of chemotherapy-induced cardiotoxicity mechanisms can be divided in two groups [84]: (1) Type 1 is irreversible and induced by anthracyclines; (2) Type 2 is reversible and provoked by monoclonal antibodies such as trastuzumab.

The vast majority of available investigations concerns the influence of chemo- and radiotherapy on the LV. The development of imaging techniques, particularly echocardiography, enabled comprehensive assessment of LV structure, function, and mechanics in cancer patients.

However, data about the influence of chemotherapy and radiotherapy on RV remodeling are limited and conflicting [84]. Nevertheless, the number of studies that emphasize the predictive importance of RV remodeling in patients with various cardiovascular conditions is increasing, which is the reason why the assessment of RV function and structure also became important in patients who underwent chemo- and/or radiotherapy.

Detailed analysis of treatment-related side effects is particularly important in breast and genital cancers typical for women. Therefore, we describe mechanisms of radiation-induced changes of the heart, as well as the impact of chemotherapy used in the treatment of these diseases as far as based on anthracyclines, which have been shown to be cardiotoxic agents.

Potential Cellular Mechanisms During Radiation Therapy

The prognosis of patients with breast cancers has improved with a 5-year survival of approximately 90% in recent years [85, 86], partially because of the increased use of adjuvant therapies such as chemo- and radiotherapy. Recent advances in techniques substantially reduce the radiation exposure to hearts during radiotherapy of mamma carcinoma, thus lowering the incidence of cardiovascular disease. Yet, current low radiation levels for breast cancers still impact on the coronary microvasculature, with subsequent inflammatory and thrombotic changes, resulting in capillary loss, and reduced myocardial tissue perfusion [87]. These alterations cause high risk, particularly for heart failure with preserved ejection fraction (HFpEF) [88]. As a consequence the most important competing mortality risk for breast cancer survivors derives from radiation-induced HFpEF [89]. Previous studies have demonstrated a connection between exposure and subsequent microvascular rarefaction, but LV diastolic function has only recently been evaluated [87]. Importantly, the same underlying mechanism responsible for radiation exposure-related HFpEF appears to be manifest in metabolic risk-related [90] HFpEF in humans [91] and the disturbance in nitric oxide (NO)-cGMP-PKG signaling seen in the microvascular endothelium resulting from microvascular inflammation (Fig. 29.2).

Potential Cellular Mechanisms of Cardiac Remodeling During Chemotherapy

Anthracyclines induce the production of highly reactive oxygen species that interact with iron and generate very toxic and active hydroxyl radicals that provoke intracellular damage. Furthermore, anthracyclines might also directly impact cardiomyocytes by topoisomerase II blockage which is very active in cancer cells. It must be noted that there are two isoenzymes – alpha is active in cancer cells and beta in cardiomyocytes. However, they are structurally very similar, which is why anthracyclines affect both isoenzymes and induce anthracycline cardiomyopathy (Type 1 cardiotoxicity) that is dose-dependent and typically irreversible because of stimulation of myofibrillar disarray and myocardial cell death [92].

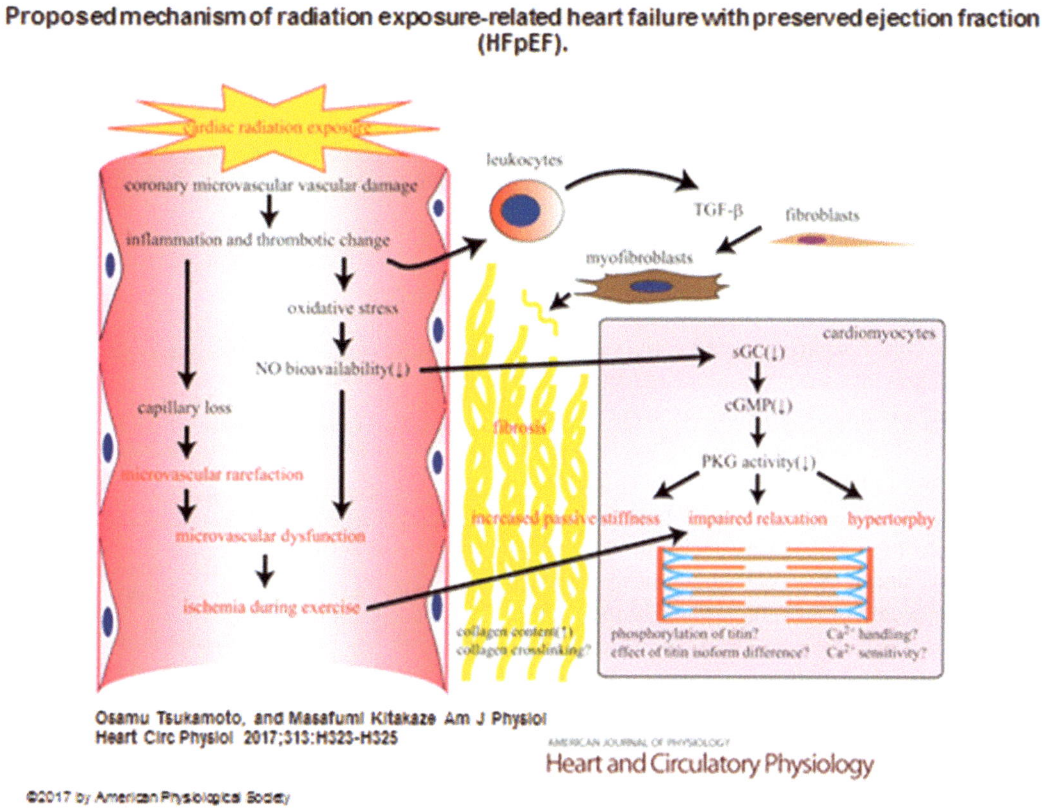

Fig. 29.2 Proposed mechanism of radiation exposure-related heart failure with preserved ejection fraction (HFpEF). Cardiac radiation exposure induces coronary vascular inflammation followed by microvascular rarefaction, cardiac fibrosis, and depressed nitric oxide (NO)-PKG signaling in cardiomyocytes, all of which contribute to the development of HFpEF. TGF, transforming growth factor; sGC, soluble guanylate cyclase. (From Tsukamoto O, Kitakaze M. Radiation-induced HFpEF model as a potential tool for the exploration of novel therapeutic targets. Am J Physiol Heart Circ Physiol. 2017 Aug 1;313(2):H323–H325. doi: https://doi.org/10.1152/ajpheart.00307.2017. Reproduced with permission from American Physiological Society)

Trastuzumab is a monoclonal antibody that inhibits receptor tyrosine-protein kinase expressed on cancer cells. The problem is that the same enzyme is also expressed on cardiomyocytes and its inhibition provokes dilated cardiomyopathy [93]. Trastuzumab-induced cardiotoxicity (Type 2 cardiotoxicity) is not dose-related, and it is generally reversible after withdrawal of treatment.

Cardiac Remodeling in Oncology Patients

Considering the fact that anthracyclines and trastuzumab are used in protocols for breast cancer treatment, it is of great importance to establish the effect of these agents on cardiac remodeling. Besides, attention should be given to adequate imaging protocols that would timely detect these cardiac changes. It has been shown that the incidence of cardiomyopathy was significantly lower in patients treated only with trastuzumab than treated with anthracyclines [94]. Slamon et al. reported that cardiac dysfunction of New York Heart Association classes III and IV was present in 27% of patients who received anthracycline, cyclophosphamide, and trastuzumab; in 8% of individuals treated with anthracycline or cyclophosphamide; in 13% of participants treated with combined paclitaxel and trastuzumab; and in 1% of patients who received only paclitaxel [94].

The majority of investigations agree about the devastating influence of chemotherapy on LV EF and LV longitudinal strain in women with breast cancer [95–100]. Earlier studies showed the impact of chemotherapy on LV systolic and diastolic function evaluated by conventional echocardiographic parameters. However, investigations published in the last decade that used speckle-tracking imaging revealed early signs of LV dysfunction that occur before the reduction of LV EF, traditionally considered the cornerstone metric of LV systolic function. These newer studies showed that particularly LV longitudinal strain was influenced by chemotherapy based on anthracycline and trastuzumab in women with breast cancer. A recent study showed that LV end-diastolic volume rather than EF may act as an early indicator of trastuzumab-related cardiotoxicity in HER2+ breast cancer patients [101].

Unfortunately, to our knowledge there is no data available regarding the influence of gender on LV remodeling in cancer patients in general as far as treated with chemotherapy.

Recent studies reported negative influence of chemotherapy on RV function and mechanics in women with breast cancer [97, 102–106]. Boczar et al. detected a significant worsening of RV longitudinal strain after only 3 months of anthracycline-based chemotherapy in females with breast cancer [105]. Chang et al. showed that RV longitudinal strain represents an independent predictor of dyspnea in chemotherapy treated patients with breast cancer independently of LV and RV systolic and diastolic function [107]. Bosignore et al. explained that chemotherapy in early-stage breast cancer females provokes a spectrum of alterations: angiogenesis inhibition, changes of hemodynamics, adverse inotropic response, vascular dysfunction, reduction of muscle contractile strength, and oxygen transport impairment [108].

Detailed evaluation of LV function in the cancer patients is crucial because it determines further treatment. The assessment of RV function should also be considered as an important component of echocardiographic evaluation. Patients suffering from chemotherapy-induced cardiac failure should be treated by both a cardiologist and oncologist who should make a joined decision about future therapy, and particularly regarding (dis)continuation of chemotherapy.

References

1. Lipscombe LL, Hux JE. Trends in diabetes prevalence, incidence, and mortality in Ontario, Canada 1995–2005: a population-based study. Lancet. 2007;369:750–6.
2. International Diabetes Federation. IDF diabetes atlas. 6th ed. Brussels: International Diabetes Federation; 2013.
3. Logue J, Walker JJ, Colhoun HM, et al. Scottish diabetes research network epidemiology group. Do men develop type 2 diabetes at lower body mass indices than women? Diabetologia. 2011;54:3003–6.
4. Paul S, Thomas G, Majeed A, et al. Women develop type 2 diabetes at a higher body mass index than men. Diabetologia. 2012;55:1556–7.
5. Emerging Risk Factors Collaboration, Sarwar N, Gao P, Seshasai SR, et al. Diabetes mellitus, fasting blood glucose concentration, and risk of vascular disease: a collaborative meta-analysis of 102 prospective studies. Lancet. 2010;375:2215–22.
6. Sattar N. Gender aspects in type 2 diabetes mellitus and cardiometabolic risk. Best Pract Res Clin Endocrinol Metab. 2013;27(4):501–7.
7. Tadic M, Ilic S, Cuspidi C, Stojcevski B, Ivanovic B, Bukarica L, Jozika L, Celic V. Left ventricular mechanics in untreated normotensive patients with type 2 diabetes mellitus: a two- and three-dimensional speckle tracking study. Echocardiography. 2015;32(6):947–55.
8. Tadic M, Ilic S, Cuspidi C, Kocijancic V, Celic V. Prediabetes, diabetes and left heart deformation. Rev Esp Cardiol (Engl Ed). 2014;67(12):1062–4.
9. Jørgensen PG, Jensen MT, Mogelvang R, Fritz-Hansen T, Galatius S, Biering-Sørensen T, Storgaard H, Vilsbøll T, Rossing P, Jensen JS. Impact of type 2 diabetes and duration of type 2 diabetes on cardiac structure and function. Int J Cardiol. 2016;221:114–21.
10. Leung M, Wong VW, Hudson M, Leung DY. Impact of improved glycemic control on cardiac function in type 2 - diabetes mellitus. Circ Cardiovasc Imaging. 2016;9(3): e003643.
11. Roos CJ, Scholte AJ, Kharagjitsingh AV, Bax JJ, Delgado V. Changes in multidirectional LV strain in asymptomatic patients with type 2 diabetes mellitus: a 2-year follow-up study. Eur Heart J Cardiovasc Imaging. 2014;15 (1):41–7.
12. Tadic M, Celic V, Cuspidi C, Ilic S, Pencic B, Radojkovic J, Ivanovic B, Stanisavljevic D, Kocabay G, Marjanovic T. Right heart mechanics in untreated normotensive patients with prediabetes and type 2 diabetes mellitus: a two- and three-dimensional echocardiographic study. J Am Soc Echocardiogr. 2015;28(3):317–27.
13. Enomoto M, Ishizu T, Seo Y, Yamamoto M, Suzuki H, Shimano H, Kawakami Y, Aonuma K. Subendocardial systolic dysfunction in asymptomatic normotensive diabetic patient. Circ J. 2015;79(8):1749–55.

14. Vukomanovic V, Tadic M, Suzic-Lazic J, Kocijancic V, Celic V. The relationship between heart rate variability and left ventricular layer-specific deformation in uncomplicated diabetic patients. Int J Cardiovasc Imaging. 2017;33(4):481–90.
15. Tadic M, Vukomanovic V, Cuspidi C, Suzic-Lazic J, Pencic-Popovic B, Radojkovic J, Babic R, Celic V. The relationship between right ventricular deformation and heart rate variability in asymptomatic diabetic patients. J Diabetes Complicat. 2017;31(7):1152–7.
16. Dalen H, Thorstensen A, Romundstad PR, Aase SA, Stoylen A, Vatten LJ. Cardiovascular risk factors and systolic and diastolic cardiac function: a tissue Doppler and speckle tracking echocardiographic study. J Am Soc Echocardiogr. 2011;24(3):322–32.
17. Tadic MV, Ivanovic BA, Petrovic M, Celic V, Neskovic A. Gender influence on left ventricular structure and function in metabolic syndrome. Are women at greater risk? J Clin Ultrasound. 2013;41(9):538–45.
18. Tadic MV, Ivanovic BA, Petrovic M. Is Gender responsible for everything? The relationship between sex and right ventricular remodeling in metabolic syndrome. Echocardiography. 2013;30(7):778–85.
19. Kannel WB, McGee DL. Diabetes and cardiovascular disease. The Framingham study. JAMA. 1979;241:2035–8.
20. Kiencke S, Handschin R, von Dahlen R, Muser J, Brunner-Larocca HP, Schumann J, Felix B, Berneis K, Rickenbacher P. Pre-clinical diabetic cardiomyopathy: prevalence clinical diabetic cardiomyopathy: prevalence, clinical diabetic cardiomyopathy: prevalence, screening, and outcome. Eur J Heart Fail. 2010;12(9):951–7.
21. Kuznetsova T, Cauwenberghs N, Knez J, Yang WY, Herbots L, D'hooge J, Haddad F, Thijs L, Voigt JU, Staessen JA. Additive prognostic value of left ventricular systolic dysfunction in a population-based cohort. Circ Cardiovasc Imaging. 2016;9(7):e004661.
22. Young ME, Mcnulty P, Taegtmeyer H. Adaptation and maladaptation of the heart in diabetes: part II potential mechanisms. Circulation. 2002;105:1861–70.
23. Factor SM, Minase T, Sonnenblick EH. Clinical and morphological features of human hypertensive-diabetic cardiomyopathy. Am Heart J. 1980;99:446–58.
24. Kalil GZ, Haynes WG. Sympathetic nervous system in obesity-related hypertension: mechanisms and clinical implications. Hypertens Res. 2012;35(1):4–16.
25. Santilli F, D'Ardes D, Guagnano MT, Davi G. Metabolic syndrome: sex-related cardiovascular risk and therapeutic approach. Curr Med Chem 2017 Jul 10. [Epub ahead of print].
26. Alberti KG, Zimmet PZ. Definition, Diagnosis and classification of diabetes mellitus and its complications. Part 1: diagnosis and classification of diabetes mellitus provisional report of a WHO consultation. Diabet Med. 1998;15:539–53.
27. Balkau B, Charles MA. Comment on the provisional report from the WHO consultation. European Group for the Study of insulin resistance (EGIR). Diabet Med. 1999;16:442–3.
28. Executive Summary of The Third Report of The National Cholesterol Education Program (NCEP). Expert panel on detection, evaluation, and treatment of high blood cholesterol in adults (adult treatment panel III). JAMA. 2001;285:2486–97.
29. Alberti KG, Zimmet P, Shaw J. The IDF epidemiology task force consensus group. The metabolic syndrome: a new worldwide definition. Lancet. 2005;366:1059–62.
30. Einhorn D, Reaven GM, Cobin RH, Ford E, Ganda OP, Handelsman Y, et al. American College of Endocrinology position statement on the insulin resistance syndrome. Endocr Pract. 2003;9:237–52.
31. Grundy SM, Cleeman JI, Daniels SR, Donato KA, Eckel RH, Franklin BA, Gordon DJ, Krauss RM, Savage PJ, Smith SC Jr, et al. Diagnosis and management of the metabolic syndrome: an American Heart Association/National Heart, lung and blood institute scientific statement. Circulation. 2005;112:2735–52.
32. Phillips GB, Jing T, Heymsfield SB. Does insulin resistance, visceral adiposity, or a sex hormone alteration underlie the metabolic syndrome? Studies in women. Metabolism. 2008;57(6):838–44.
33. Kozakowski J, Gietka-Czernel M, Leszczyńska D, Majos A. Obesity in menopause – our negligence or an unfortunate inevitability? Prz Menopauzalny. 2017;16(2):61–5.
34. Pradhan AD. Sex differences in the metabolic syndrome: implications for cardiovascular health in women. Clin Chem. 2014;60:44–52.
35. Barrett-Connor EL, Cohn BA, Wingard DL, Edelstein SL. Why is diabetes mellitus a stronger risk factor for fatal ischemic heart disease in women than in men? The Rancho Bernardo study. JAMA. 1991;265(5):627–31.
36. Peters SA, Huxley RR, Woodward M. Diabetes as risk factor for incident coronary heart disease in women compared with men: a systematic review and meta-analysis of 64 cohorts including 858,507 individuals and 28,203 coronary events. Diabetologia. 2014;57(8):1542–51.
37. McNeill AM, Rosamond WD, Girman CJ, Golden SH, Schmidt MI, East HE, Ballantyne CM, Heiss G. The metabolic syndrome and 11-year risk of incident cardiovascular disease in the atherosclerosis risk in communities study. Diabetes Care. 2005;28(2):385–90.
38. Dekker JM, Girman C, Rhodes T, Nijpels G, Stehouwer CD, Bouter LM, Heine RJ. Metabolic syndrome and 10-year cardiovascular disease risk in the Hoorn study. Circulation. 2005;112(5):666–73.
39. Hu G, Qiao Q, Tuomilehto J, Balkau B, Borch-Johnsen K, Pyorala K, DECODE Study Group. Prevalence of the metabolic syndrome and its relation to all-cause and cardiovascular mortality in nondiabetic European men and women. Arch Intern Med. 2004;164(10):1066–76.
40. Ivanovic BA, Tadic MV, Simic DV. Are all criteria of metabolic syndrome equally harmful? Acta Cardiol. 2011;66(2):189–96.
41. Tadic M, Ivanovic B, Grozdic I. Metabolic syndrome impacts the right ventricle: true or false? Echocardiography. 2011;28(5):530–8.
42. Aijaz B, Ammar KA, Lopez-Jimenez F, Redfield MM, Jacobsen SJ, Rodeheffer RJ. Abnormal cardiac structure and function in the metabolic syndrome: a population-based study. Mayo Clin Proc. 2008;83(12):1350–7.
43. Cuspidi C, Valerio C, Sala C, Negri F, Esposito A, Masaidi M, Giudici V, Zanchetti A, Mancia G. Metabolic syndrome and biventricular hypertrophy in essential hypertension. J Hum Hypertens. 2009;23(3):168–75.
44. Tadic M, Cuspidi C, Majstorovic A, Pencic B, Backovic S, Ivanovic B, Scepanovic R, Martinov J, Kocijancic V, Celic V. Does the metabolic syndrome impact left-ventricular mechanics? A two-dimensional speckle tracking study. J Hypertens. 2014;32(9):1870–8.
45. Tadic M, Cuspidi C, Sljivic A, Andric A, Ivanovic B, Scepanovic R, Ilic I, Jozika L, Marjanovic T, Celic V. Effects of the metabolic syndrome on right heart mechanics and function. Can J Cardiol. 2014;30(3):325–31.
46. Schillaci G, Pirro M, Vaudo G, Mannarino MR, Savarese G, Pucci G, Franklin SS, Mannarino E. Metabolic syndrome is

associated with aortic stiffness in untreated essential hypertension. Hypertension. 2005;45:1078–82.
47. Martin SS, Daya N, Lutsey PL, Matsushita K, Fretz A, McEvoy JW, Blumenthal RS, Coresh J, Greenland P, Kottgen A, Selvin E. Thyroid function, cardiovascular risk factors, and incident atherosclerotic cardiovascular disease: the Atherosclerosis Risk in Communities (ARIC) study. J Clin Endocrinol Metab 2017 Jun 12. [Epub ahead of print].
48. Ning Y, Cheng YJ, Liu LJ, Sara JD, Cao ZY, Zheng WP, Zhang TS, Han HJ, Yang ZY, Zhang Y, Wang FL, Pan RY, Huang JL, Wu LL, Zhang M, Wei YX. What is the association of hypothyroidism with risks of cardiovascular events and mortality? A meta-analysis of 55 cohort studies involving 1,898,314 participants. BMC Med. 2017;15(1):21.
49. Journy NMY, Bernier MO, Doody MM, Alexander BH, Linet MS, Kitahara CM. Hyperthyroidism, Hypothyroidism, and cause-specific mortality in a large cohort of women. Thyroid. 2017;27(8):1001–10.
50. Dekkers OM, Horváth-Puhó E, Cannegieter SC, Vandenbroucke JP, Sørensen HT, Jørgensen JO. Acute cardiovascular events and all-cause mortality in patients with hyperthyroidism: a population-based cohort study. Eur J Endocrinol. 2017;176(1):1–9.
51. Fazio S, Palmieri EA, Lombardi G, Biondi B. Effects of thyroid hormone on the cardiovascular system. Recent Prog Horm Res. 2004;59:31–50.
52. Vargas-Uricoechea H, Bonelo-Perdomo A. Thyroid dysfunction and heart failure: mechanisms and associations. Curr Heart Fail Rep. 2017;14(1):48–58.
53. Vitale G, Galderisi M, Lupoli GA, Celentano A, Pietropaolo I, Parenti N, De Divitiis O, Lupoli G. Left ventricular myocardial impairment in subclinical hypothyroidism assessed by a new ultrasound tool: pulsed tissue Doppler. J Clin Endocrinol Metab. 2002;87(9):4350–5.
54. Arinc H, Gunduz H, Tamer A, Seyfeli E, Kanat M, Ozhan H, Akdemir R, Uyan C. Tissue Doppler echocardiography in evaluation of cardiac effects of subclinical hypothyroidism. Int J Cardiovasc Imaging. 2006;22(2):177–86.
55. Martins RM, Fonseca RH, Duarte MM, Reuters VS, Ferreira MM, Almeida C, Buescu A, Teixeira Pde F, Vaisman M. Impact of subclinical hypothyroidism treatment in systolic and diastolic cardiac function. Arq Bras Endocrinol Metabol. 2011;55(7):460–7.
56. Meena CL, Meena RD, Nawal R, Meena VK, Bharti A, Meena LP. Assessment of left ventricular diastolic dysfunction in sub-clinical hypothyroidism. Acta Inform Med. 2012;20(4):218–20.
57. Tadic M, Ilic S, Kostic N, Caparevic Z, Celic V. Subclinical hypothyroidism and left ventricular mechanics: a three-dimensional speckle tracking study. J Clin Endocrinol Metab. 2014;99(1):307–14.
58. Monzani F, Di Bello V, Caraccio N, Bertini A, Giorgi D, Giusti C, Ferrannini E. Effect of levothyroxine on cardiac function and structure in subclinical hypothyroidism: a double blind, placebo-controlled study. J Clin Endocrinol Metab. 2001;86(3):1110–5.
59. Jagdish A, Singh H, Batra A, Siwach SB, Kumar R, Gupta R. An echocardiographic study on the effect of levothyroxine therapy on cardiac function and structure in hypothyroidism. JIACM. 2009;10(1–2):27–31.
60. Tiryakioglu SK, Tiryakioglu O, Ari H, Basel MC, Ozkan H, Bozat T. Left ventricular longitudinal myocardial function in overt hypothyroidism: a tissue Doppler echocardiographic study. Echocardiography. 2010 May;27(5):505–11.
61. Tadic M, Ilic S, Celic V. Right ventricular and right atrial function and deformation in patients with subclinical hypothyroidism: a two- and three-dimensional echocardiographic study. Eur J Endocrinol. 2013;170(1):77–85.
62. Turhan S, Tulunay C, Ozduman Cin M, Gursoy A, Kilickap M, Dincer I, Candemir B, Gullu S, Erol C. Effects of thyroxine therapy on right ventricular systolic and diastolic function in patients with subclinical hypothyroidism: a study by pulsed wave tissue Doppler imaging. J Clin Endocrinol Metab. 2006 Sep;91(9):3490–3.
63. Oner FA, Yurdakul S, Oner E, Uzum AK, Erguney M. Evaluation of the effect of L-thyroxin therapy on cardiac functions by using novel tissue Doppler-derived indices in patients with subclinical hypothyroidism. Acta Cardiol. 2011 Feb;66(1):47–55.
64. Oner FA, Yurdakul S, Oner E, Arslantaş MK, Usta M, Ergüney M. Evaluation of ventricular functions using tissue Doppler echocardiography in patients with subclinical hypothyroidism. Turk Kardiyol Dern Ars. 2011 Mar;39 (2):129–36.
65. Kosar F, Sahin I, Turan N, Topal E, Aksoy Y, Taskapan C. Evaluation of right and left ventricular function using pulsed-wave tissue Doppler echocardiography in patients with subclinical hypothyroidism. J Endocrinol Investig. 2005 Sep;28(8):704–10.
66. Arinc H, Gunduz H, Tamer A, Seyfeli E, Kanat M, Ozhan H, Akdemir R, Celebi H, Uyan C. Evaluation of right ventricular function in patients with thyroid dysfunction. Cardiology. 2006;105(2):89–94.
67. Ripoli A, Pingitore A, Favilli B, Bottoni A, Turchi S, Osman NF, De Marchi D, Lombardi M, L'Abbate A, Iervasi G. Does subclinical hypothyroidism affect cardiac pump performance? Evidence from a magnetic resonance imaging study. J Am Coll Cardiol. 2005;45(3):439–45.
68. Kosar F, Sahin I, Aksoy Y, Uzer E, Turan N. Usefulness of pulsed-wave tissue Doppler echocardiography for the assessment of the left and right ventricular function in patients with clinical hypothyroidism. Echocardiography. 2006;23 (6):471–7.
69. Biondi B, Palmieri EA, Klain M, Schlumberger M, Filetti S, Lombardi G. Subclinical hyperthyroidism: clinical features and treatment options. Eur J Endocrinol. 2005;152:1–9.
70. Marcisz C, Jonderko G, Wróblewski T, Kurzawska G, Mazur F. Left ventricular mass in patients with hyperthyroidism. Med Sci Monit. 2006;12(11):CR481–6.
71. Yue WS, Chong BH, Zhang XH, Liao SY, Jim MH, Kung AW, Tse HF, Siu CW. Hyperthyroidism-induced left ventricular diastolic dysfunction: implication in hyperthyroidism-related heart failure. Clin Endocrinol. 2011;74(5):636–43.
72. Mark PD, Andreassen M, Petersen CL, Kjaer A, Faber J. Treatment of subclinical hyperthyroidism: effect on left ventricular mass and function of the heart using magnetic resonance imaging technique. Endocr Connect. 2015;4 (1):37–42.
73. Tadic M, Ilic S, Cuspidi C, Marjanovic T, Celic V. Subclinical hyperthyroidism impacts left ventricular deformation: 2D and 3D echocardiographic study. Scand Cardiovasc J. 2015;49(2):74–81.
74. Abdulrahman RM, Delgado V, Ng AC, Ewe SH, Bertini M, Holman ER, et al. Abnormal cardiac contractility in long-term exogenous subclinical hyperthyroid patients as demonstrated by two-dimensional echocardiography speckle tracking imaging. Eur J Endocrinol. 2010;163:435–41.
75. Abdulrahman RM, Delgado V, Hoftijzer HC, Ng AC, Ewe SH, Marsan NA, et al. Both exogenous subclinical

hyperthyroidism and short-term overt hypothyroidism affect myocardial strain in patients with differentiated thyroid carcinoma. Thyroid. 2011;21:471–6.
76. Tadic M, Celic V, Cuspidi C, Ilic S, Zivanovic V, Marjanovic T. How does subclinical hyperthyroidism affect right heart function and mechanics? J Ultrasound Med. 2016;35(2):287–95.
77. Arinc H, Gunduz H, Tamer A, Seyfeli E, Kanat M, Ozhan H, et al. Evaluation of right ventricular function in patients with thyroid dysfunction. Cardiology. 2006;105(2):89–94.
78. Nacar AB, Acar G, Yorgun H, Akçay A, Özkaya M, Canpolat U, et al. The effect of antithyroid treatment on atrial conduction times in patients with subclinical hyperthyroidism. Echocardiography. 2012;29(8):950–5.
79. Teasdale SL, Inder WJ, Stowasser M, Stanton T. Hyperdynamic right heart function in graves' hyperthyroidism measured by echocardiography normalises on restoration of euthyroidism. Heart Lung Circ. 2017;26 (6):580–5.
80. Zuhur SS, Baykiz D, Kara SP, Sahin E, Kuzu I, Elbuken G. Relationship among pulmonary hypertension, autoimmunity, thyroid hormones and dyspnea in patients with hyperthyroidism. Am J Med Sci. 2017;353(4):374–80.
81. O'Hare M, Sharma A, Murphy K, Mookadam F, Lee H. Cardio-oncology Part I: chemotherapy and cardiovascular toxicity. Expert Rev Cardiovasc Ther. 2015;13(5):511–8.
82. Coleman MP, Forman D, Bryant H, et al. Cancer survival in Australia, Canada, Denmark, Norway, Sweden, and the UK, 1995-2007 (the international Cancer Benchmarking Partnership): an analysis of population-based cancer registry data. Lancet. 2011;377:127–38.
83. Smith BD, Smith GL, Hurria A, Hortobagyi GN, Buchholz TA. Future of cancer incidence in the United States: burdens upon an aging, changing nation. J Clin Oncol. 2009;27 (17):2758–65.
84. Moudgil R, Yeh ET. Mechanisms of cardiotoxicity of cancer chemotherapeutic agents: cardiomyopathy and beyond. Can J Cardiol. 2016;32(7):863–70.
85. Tsukamoto O, Kitakaze M. Radiation-induced HFpEF model as a potential tool for the exploration of novel therapeutic targets. Am J Physiol Heart Circ Physiol. 2017;313(2): H323–5.
86. Jemal A, Siegel R, Ward E, Hao Y, Xu J, Thun MJ. Cancer statistics. CA Cancer J Clin. 2009;59:225–49.
87. Seemann I, Gabriels K, Visser NL, Hoving S, te Poele JA, Pol JF, Gijbels MJ, Janssen BJ, van Leeuwen FW, Daemen MJ, Heeneman S, Stewart FA. Irradiation induced modest changes in murine cardiac function despite progressive structural damage to the myocardium and microvasculature. Radiother Oncol. 2012;103:143–50.
88. Saiki H, Petersen IA, Scott CG, Bailey KR, Dunlay SM, Finley RR, Ruddy KJ, Yan E, Redfield MM. Risk of heart failure with preserved ejection fraction in older women after contemporary radiotherapy for breast cancer. Circulation. 2017;135:1388–96.
89. Abdel-Qadir H, Austin PC, Lee DS, Amir E, Tu JV, Thavendiranathan P, Fung K, Anderson GM. A population-based study of cardiovascular mortality following early-stage breast cancer. JAMA Cardiol. 2017;2:88–93.
90. Paulus WJ, Tschöpe C. A Novel paradigm for heart failure with preserved ejection fraction: comorbidities drive myocardial dysfunction and remodeling through coronary microvascular endothelial inflammation. J Am Coll Cardiol. 2013;62:263–71.
91. Hamdani N, Franssen C, Lourenço A, Falcão-Pires I, Fontoura D, Leite S, Plettig L, López B, Ottenheijm CA, Becher PM, González A, Tschöpe C, Díez J, Linke WA, Leite-Moreira AF, Paulus WJ. Myocardial titin hypophosphorylation importantly contributes to heart failure with preserved ejection fraction in a rat metabolic risk model. Circ Heart Fail. 2013;6:1239–49.
92. Vejpongsa P, Yeh ET. Topoisomerase 2β: a promising molecular target for primary prevention of anthracycline-induced cardiotoxicity. Clin Pharmacol Ther. 2014;95 (1):45–52.
93. Crone SA, Zhao YY, Fan L, Gu Y, Minamisawa S, Liu Y, Peterson KL, Chen J, Kahn R, Condorelli G, Ross J Jr, Chien KR, Lee KF. ErbB2 is essential in the prevention of dilated cardiomyopathy. Nat Med. 2002;8(5):459–65.
94. Slamon DJ, Leyland-Jones B, Shak S, Fuchs H, Paton V, Bajamonde A, Fleming T, Eiermann W, Wolter J, Pegram M, Baselga J, Norton L. Use of chemotherapy plus a monoclonal antibody against HER2 for metastatic breast cancer that overexpresses HER2. N Engl J Med. 2001;344(11):783–92.
95. Piotrowski G, Gawor R, Stasiak A, Gawor Z, Potemski P, Banach M. Cardiac complications associated with trastuzumab in the setting of adjuvant chemotherapy for breast cancer overexpressing human epidermal growth factor receptor type 2 – a prospective study. Arch Med Sci. 2012;8 (2):227–35.
96. Nakano S, Takahashi M, Kimura F, Senoo T, Saeki T, Ueda S, Tanno J, Senbonmatsu T, Kasai T, Nishimura S. Cardiac magnetic resonance imaging-based myocardial strain study for evaluation of cardiotoxicity in breast cancer patients treated with trastuzumab: a pilot study to evaluate the feasibility of the method. Cardiol J. 2016;23(3):270–80.
97. Grover S, Leong DP, Chakrabarty A, Joerg L, Kotasek D, Cheong K, Joshi R, Joseph MX, DePasquale C, Koczwara B, Selvanayagam JB. Left and right ventricular effects of anthracycline and trastuzumab chemotherapy: a prospective study using novel cardiac imaging and biochemical markers. Int J Cardiol. 2013;168(6):5465–7.
98. Fei HW, Ali MT, Tan TC, Cheng KH, Salama L, Hua L, Zeng X, Halpern EF, Taghian A, MacDonald SM, Scherrer-Crosbie M. Left ventricular global longitudinal strain in HER-2+ breast cancer patients with anthracyclines and trastuzumab who develop cardiotoxicity is associated with subsequent recovery of left ventricular ejection fraction. Echocardiography. 2016;33(4):519–26.
99. Guerra F, Marchesini M, Contadini D, Menditto A, Morelli M, Piccolo E, Battelli N, Pistelli M, Berardi R, Cascinu S, Capucci A. Speckle-tracking global longitudinal strain as an early predictor of cardiotoxicity in breast carcinoma. Support Care Cancer. 2016;24(7):3139–45.
100. Tan TC, Bouras S, Sawaya H, Sebag IA, Cohen V, Picard MH, Passeri J, Kuter I, Scherrer-Crosbie M. Time trends of left ventricular ejection fraction and myocardial deformation indices in a cohort of women with breast cancer treated with anthracyclines, taxanes, and trastuzumab. J Am Soc Echocardiogr. 2015;28(5):509–14.
101. Bergamini C, Torelli F, Ghiselli L, Rossi A, Trevisani L, Vinco G, et al. Left ventricular end-diastolic volume as early indicator of trastuzumab-related cardiotoxicity in HER2+ breast cancer patients: results from a single-center retrospective study. Minerva Cardioangiol. 2017;65:278–87.
102. Dercle L, Giraudmaillet T, Pascal P, Lairez O, Chisin R, Marachet MA, Ouali M, Rousseau H, Bastié D, Berry I. Is TOMPOOL (gated blood-pool SPECT processing software) accurate to diagnose right and left ventricular dysfunction in a clinical setting? J Nucl Cardiol. 2014;21(5):1011–22.
103. Cottin Y, Touzery C, Coudert B, Richebourg S, Cohen M, Toubeau M, Louis P, Wolf JE, Brunotte F. Diastolic or

systolic left and right ventricular impairment at moderate doses of anthracycline? A 1-year follow-up study of women. Eur J Nucl Med. 1996;23(5):511–6.
104. Lange SA, Ebner B, Wess A, Kögel M, Gajda M, Hitschold T, Jung J. Echocardiography signs of early cardiac impairment in patients with breast cancer and trastuzumab therapy. Clin Res Cardiol. 2012;101(6):415–26.
105. Boczar KE, Aseyev O, Sulpher J, Johnson C, Burwash IG, Turek M, Dent S, Dwivedi G. Right heart function deteriorates in breast cancer patients undergoing anthracycline-based chemotherapy. Echo Res Pract. 2016;3(3):79–84.
106. Kılıçaslan B, Özdoğan Ö, Demir Pişkin G, Kahya Eren N, Dursun H. Echocardiographic signs of right ventricle changes after Trastuzumab treatment in breast cancer patients with erb-2 overexpression. Anatol J Cardiol. 2015;15(2):143–8.
107. Chang WT, Shih JY, Feng YH, Chiang CY, Kuo YH, Chen WY, Wu HC, Cheng JT, Wang JJ, Chen ZC. The early predictive value of right ventricular strain in epirubicin-induced cardiotoxicity in patients with breast cancer. Acta Cardiol Sin. 2016;32(5):550–9.
108. Bonsignore A, Warburton D. The mechanisms responsible for exercise intolerance in early-stage breast cancer: what role does chemotherapy play? Hong Kong Physiother J. 2013;31:2–11.

Sex-Specific Cardiovascular Comorbidities with Associations in Dermatologic and Rheumatic Disorders

30

Peter L. M. Kerkhof and Irina Khamaganova

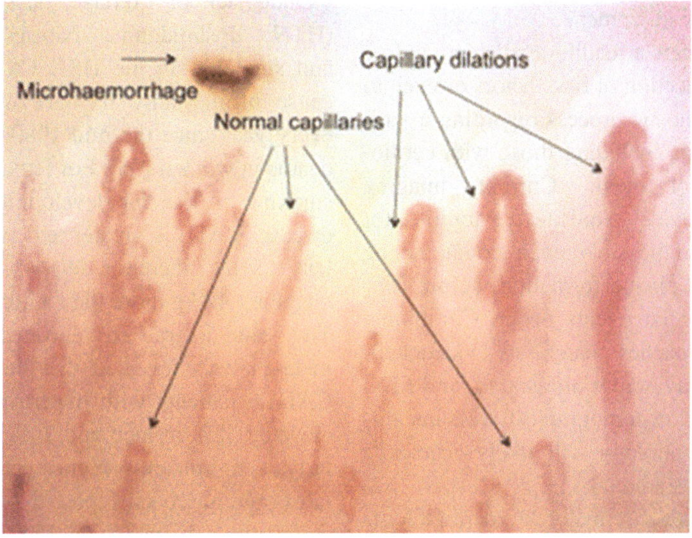

Primary Raynaud's phenomenon and nailfold videocapillaroscopy. *Example of irregular, apical, and venous branch dilations, as well as presence of microhemorrhages. Normal capillaries are also visible (magnification ×200). Reproduced from Pizzorni et al. Clin Rheumatol. 2017;36: 1637–1642, with permission.*

Abstract

Cardiology, dermatology, and rheumatology form a fascinating triad. Many skin and joint disorders are associated with cardiovascular comorbidities because they share etiologic elements. Female predominance is often remarkable and likely related to autoimmune pathology. Although studies have shown that X-encoded genes may be involved in the differences in immunity between males and females, other studies have also shown that sex chromosomes are irrelevant and that estrogens and androgens are responsible for the differences. The elevated immune activity

P. L. M. Kerkhof (✉)
Department of Radiology and Nuclear Medicine, Amsterdam Cardiovascular Sciences, VU University Medical Center, Amsterdam, The Netherlands
e-mail: plm.kerkhof@VUmc.nl

I. Khamaganova
Department of Skin Diseases & Cosmetology, Pirogov Russian National Research Medical University, Moscow, Russia

P. L. M. Kerkhof, V. M. Miller (eds.), *Sex-Specific Analysis of Cardiovascular Function*, Advances in Experimental Medicine and Biology 1065, https://doi.org/10.1007/978-3-319-77932-4_30

in females provides a beneficial position in coping with a pathogenic stimulus but may also enhance their susceptibility to autoimmunity. The complexity of the immune system and its role as a defensive force against infection requires an armamentarium to precisely identify and selectively control inflammatory processes or cells which promote atherosclerosis. On the other hand, the inflammation in skin diseases seems to be an active source of diverse proinflammatory cytokines and chemokines which can predispose to cardiovascular comorbidities. Also, it has been shown that comorbidity disproportionately accelerates risk in women.

The skin offers a readily available window to facilitate detection of risk factors or even to assist the diagnostic process regarding a variety of disorders, including those with cardiovascular involvement. Current imaging techniques provide exquisite capabilities for diagnosing and possibly even counteracting atherosclerotic plaque formation, before serious cardiovascular events occur. Combining imaging approaches (such as videocapillaroscopy, intravascular ultrasound, and FDG positron emission tomography) with insights based on immunology will likely accelerate advances in this area.

We review major dermatologic manifestations and rheumatologic disorders which are associated with cardiac and vascular abnormalities. In particular we discuss sex-specific aspects concerning incidence and severity of cardiovascular disease associated with systemic sclerosis, rheumatoid arthritis, systemic lupus erythematosus, psoriasis, atopic dermatitis, and hidradenitis suppurativa.

Keywords

Cardio-dermatology · Sex-specific risk factors · Heart failure · Stroke · Psoriasis · Atopic dermatitis · Hidradenitis suppurativa · Systemic sclerosis · Systemic lupus erythematosus · Rheumatoid arthritis · Ankylosing spondylitis · Inflammatory bowel disease · Rosacea · Immune-mediated inflammatory disease · Ischemic heart disease · Inflammasome · Neutrophil extracellular trap · Inflammation imaging · Sex hormones · Pulse wave velocity · Carotid intima-media thickness · Nailfold videocapillaroscopy · PET scan · Review

Introduction: Sex Disparities

Mortality due to cardiovascular disease (CVD) is high and mainly modulated by comorbidities and unhealthy lifestyle habits. Classical risk factors include lack of exercise, obesity, hypertension (HTN), dyslipidemia, diabetes mellitus (DM), and smoking. In the USA CVD is the leading cause of mortality in both women and men, every year since the mid-1980s leading to more deaths in women [58]. For females, the impact of smoking on risk for developing coronary artery disease (CAD) is greater and essentially equivalent to the increase in risk associated with weighing 42 kg more than a nonsmoking female. The same is true for elevated triglycerides; a meta-analysis revealed that the relative risk of CAD in patients with hypertriglyceridemia was elevated 32% in men and 76% in women. Similarly, a significantly higher cardiovascular mortality rate has been reported in diabetic women compared to diabetic men. Evidence suggests that the compilation of traditional risk factors currently used to assess risk of CAD is underestimated in women. Several novel risk factors for CAD in women have been suggested in recent years, such as anemia, metabolic syndrome (MetSy), and (elevated) high-sensitivity C-reactive protein (hs-CRP) levels. Incorporation of these elements into assessments of CAD risk seems warranted [58]. Besides, women may also be affected by specific risk factors, notably (pre) eclampsia and polycystic ovary syndrome (PCOS). Also, women suffer (more often than men) from chronic inflammatory autoimmune diseases. The latter category includes systemic lupus erythematosus (SLE), rheumatoid arthritis (RA), and systemic sclerosis (SySc). Psoriasis appears equally distributed among both sexes. In

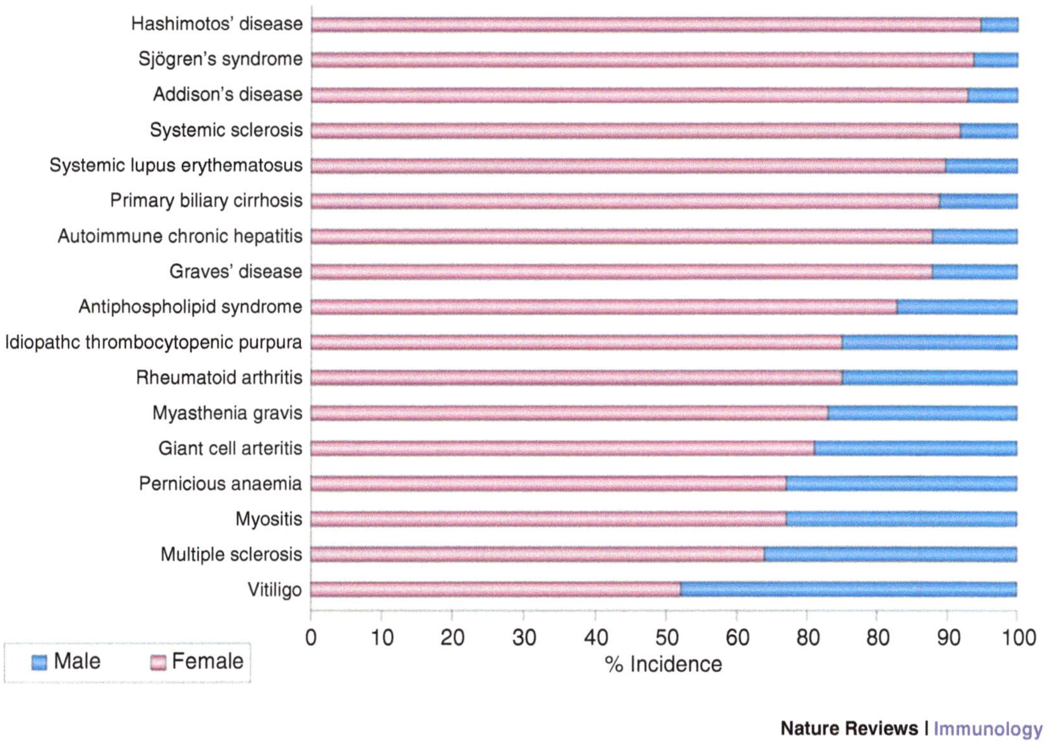

Fig. 30.1 Sex distribution of the most important autoimmune diseases. (From Libert et al. [49], reproduced with permission)

addition, Raynaud's phenomenon is often seen in SLE, RA, scleroderma, Sjögren's syndrome, and polymyositis. Indeed, the percentwise incidence of certain autoimmune diseases, with most showing possible involvement of the cardiovascular system, differs among men and women (Fig. 30.1), as summarized by Libert et al. [49]. Autoimmune diseases affect approximately 7–10% of the population of the United States, while an estimated 46.1 million American adults suffer from arthritis and rheumatologic conditions [50].

A distinction between sexes in comparative studies is most meaningful when specific age groups are considered. Influences of sex hormones vary with the circulating levels, which change over the life span (Fig. 30.2), having consequences for the incidence of various diseases. Also, the immune systems of men and women differ in significant ways, especially after puberty [30]. Other risk factors involved being equal, disease course, and survival depend on the relative strengths of the arsenal of protecting mechanisms.

Additional Risk Factors in Women

Traditional risk factors and the Framingham risk score (FRS) underestimate the risk of ischemic heart disease (IHD) in women. Mortality rate related to IHD is higher in women. The "typical" female presentation of signs and symptoms of IHD is more complex and multifactorial than that of men. Evidence suggests that additional risk assessment (blood inflammatory markers, evidence of plaque burden, and evidence of ischemia) may be of relatively greater importance in women due to this diagnostic uncertainty [6].

From the time of puberty, women have, on average, greater mean C-reactive protein (CRP) measures compared with men. The CRP difference is consistent with the higher prevalence in women of inflammatory-mediated autoimmune

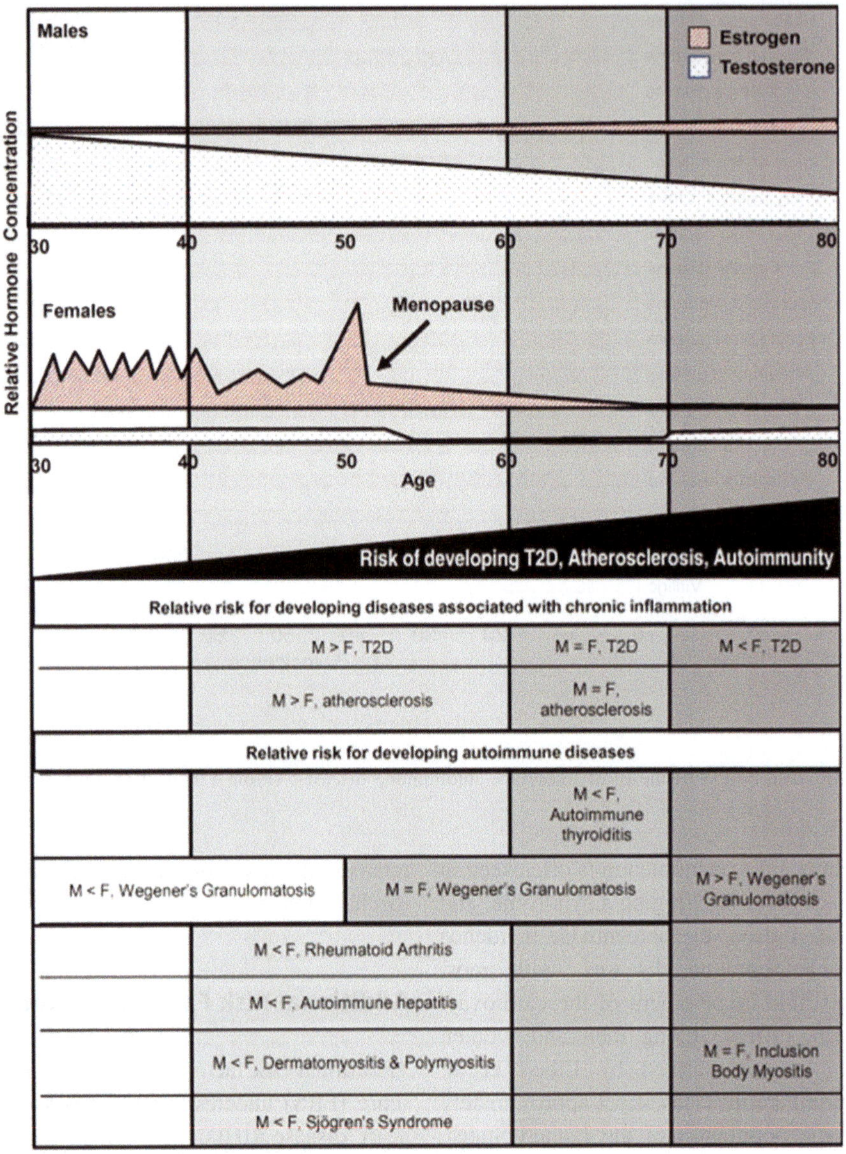

Fig. 30.2 Correlations between the relative levels of estrogen and testosterone and the incidence of type 2 diabetes (T2D), atherosclerosis, and autoimmunity in men and women between the ages 30 and 80 years. White and shaded columns demarcate periods in the life span during which chronic inflammatory and autoimmune diseases manifest clinically. A comparison of the incidence of each disease in men and women for each time period and disease is indicated. (From Gubbels Bupp [30], reproduced with permission)

diseases, such as RA or SLE, suggesting a prominent role for inflammation in the observed IHD sex differences. Indeed, the relative risk of imminent IHD events increases with levels of hs-CRP, acting synergistically with other risk factors to accelerate IHD risk in women. Furthermore, inflammatory measures are related to other IHD risk markers such as the cardiometabolic syndrome and type 2 diabetes (T2D). IHD risk assessment in women is likely improved by the use of multiple biomarkers [69].

In premenopausal women disruption of ovulatory cycling is associated with an increased risk of coronary atherosclerosis and CVD events. PCOS, prevalent in 10–13% of women, is linked with a clustering of risk factors, incident T2D, and IHD events. The cardiometabolic syndrome is a clustering of risk factors, including insulin resistance, dyslipidemia, HTN, or abdominal obesity. In women this syndrome is often associated with alterations in endogenous estrogens and androgens.

Disadvantages of Being Male

Males have only one X chromosome. From an evolutionary point of view, this is a tremendous disadvantage, as every newly arisen recessive mutation on the X chromosome will be phenotypically manifested. This survival disadvantage is shown by the existence of numerous X-linked primary immunodeficiencies [49]. These are diseases that typically do not affect females or cause less severe defects in females. To avoid double dosage of proteins in females, one of the X chromosomes is randomly silenced during X chromosome inactivation, which occurs in the early stages of female embryogenesis; however, the pseudo-autosomal regions of the X chromosome escape inactivation. The process of X chromosome inactivation results in female cellular mosaicism: in a female, approximately half of the cells express genes derived from the maternal X chromosome and the other half express genes derived from the paternal X chromosome. As a consequence, deleterious or disadvantageous mutations that occur in an X chromosome-linked gene will result in the functional loss of the protein in all cells in a male but in only half of cells in a female [49]. The X chromosome has numerous genes which, directly or indirectly, are involved in immunity and naturally occurring variations in one gene copy might result in two distinct alleles with different regulatory and response capacities. For females, this means additional physiological diversity: not only do heterozygous females avoid the effects of deleterious gene mutations, they also benefit from added diversity when facing new immune challenges, such as microbial infections [49].

The Skin as Mirror and Log-Book

Many skin and skin-related diseases affect the sexes unequally. One study in a mostly white population showed that females had a higher incidence of connective tissue diseases (scleroderma, morphea, dermatomyositis, primary Sjögren's syndrome, SLE, hidradenitis suppurativa, erythromelalgia, venous stasis syndrome, and venous ulcers). In that study males had a higher incidence of psoriasis and psoriatic arthritis and basal cell and squamous cell carcinoma [2]. The skin forms an easily accessible window to study associated diseases, such as those referring to the cardiovascular system. Similar as in dermatology or rheumatology, certain types of CVD can (notably already during an early phase) be studied with the aid of ophthalmologic investigations. Retinal funduscopy is an established noninvasive technique to inspect smaller blood vessels and is gaining renewed interest with the introduction of advanced digital analysis methods, also permitting evaluation support at remote sites using transmitted images [12]. Several inflammatory diseases with cardiovascular involvement have a prominent dermatologic component, and, here too, major progress can be made by applying technological advances in the field of optics. Close inspection of the skin may reveal important signals about (risk factors for) CVD. For example, following the detection of prominent splinter nail hemorrhages in a patient, a positron emission tomography (PET) scan revealed vascular inflammation in the aortic arch and its proximal branches, consistent with giant cell arteritis [70], and seborrheic keratosis was shown to be the strongest independent skin-related predictor of atrioventricular (AV) block in aging people [65].

Visual inspection of the skin can be enhanced by suitable optical techniques. In cases of psoriasis, it is clear that angiogenesis represents a key phenomenon and that insight in the microcirculation within psoriatic lesions would be informative. Recently, an optical perfusion camera was developed which corrects images for the heartbeat. The system permits local evaluation of the superficial cutaneous microcirculation before and during topical treatment in patients with psoriasis [34] and reveals heterogeneity in perfusion intensity. Furthermore, nailfold videocapillaroscopy (being a combination of an optical microscope and a digital video camera) has been applied to evaluate scleroderma, where decreased peripheral blood perfusion is usually associated with microvascular impairment [66].

Additional Types of Imaging

Imaging techniques in general hold a great future. Another example is the PET scan which studies glycolytic rates in macrophages within a vessel wall affected by inflammatory lesions [78]. Plaque composition (including fibrous tissue, fibro-fatty tissue, necrotic core, and dense calcium) is evaluated by ultrasound intravascular imaging, and in CAD patients plaque burden can be assessed by a so-called near-infrared spectroscopy-derived lipid core index, showing that women have more favorable plaque characteristics than men, despite their worse risk profile [79]. Carotid intima-media thickness (CIMT) is an imaging biomarker useful for the noninvasive evaluation of atherosclerosis, with a relatively greater risk predicted for women than men. Furthermore, coronary artery calcium (CAC) is highly correlated with traditional risk factors. Women with a high-risk CAC score combined with multiple risk factors show a 10% greater IHD event risk than men, which fact supports the notion that risk in women is disproportionately accelerated by comorbidities [69]. Imaging has also gained a prominent role in the study of vascular inflammation, as further discussed below.

Inflammatory Pathogenesis and Vascular Imaging

The seminal work by Libby [48] has emphasized the pivotal role of inflammation in the development of atherosclerosis. Inflammation is a fundamental biological process, directed to help maintain and restore homeostasis. In principle, inflammation is beneficial, but a continuous trigger may entail a chronic process, possibly implying functional decline. Any acute imbalance of homeostasis, e.g., by burn, bleeding, surgery, trauma, or infection, can lead to acute inflammation. This process potentially becomes uncontrolled and reaches systemic proportions, occasionally advancing to a life-threatening form of inflammation, such as during sepsis [49].

Inflammasomes are multi-protein complexes, and genetic mutations were first recognized to result in autoinflammatory diseases, which are characterized by the absence of both autoantibodies and autoreactive T/B cells. Multiple factors that contribute to lowering the threshold of immunity and inflammasomes play a key role in this threshold. For example, IL-1β and IL-18 further perpetuate Th17 responses and endothelial cell damage, which potentiate a number of autoimmune diseases, including CVD, SLE, and synovitis in RA [88]. Interleukin (IL)-32 not only modulates important inflammatory pathways including tumor necrosis factor (TNF)α, IL-6 or IL-1β but also involves the modulation of endothelial cell function and the serum concentration of high-density lipoprotein [18]. Many chronic inflammatory disorders such as chronic obstructive pulmonary disease (COPD), inflammatory bowel disease (IBD), and RA deserve attention in this survey because of their association with CVD.

Vascular Inflammation Imaging

CVD may be viewed as a systemic inflammatory disorder related to monocytes, macrophages, neutrophils, and T cells. Current research into the natural history of atheroma has implicated various immune cells, including phagocytes, lymphocytes, plasmacytoid dendritic cells, and neutrophils. The cellular processes involved are summarized in Fig. 30.3, which also illustrates the importance of PET to study the inflammatory process in the arterial wall by using the ^{18}F-fluorodeoxyglucose (FDG) tracer. This glucose analogue is used for imaging an increased metabolic rate in the presence of inflammation and

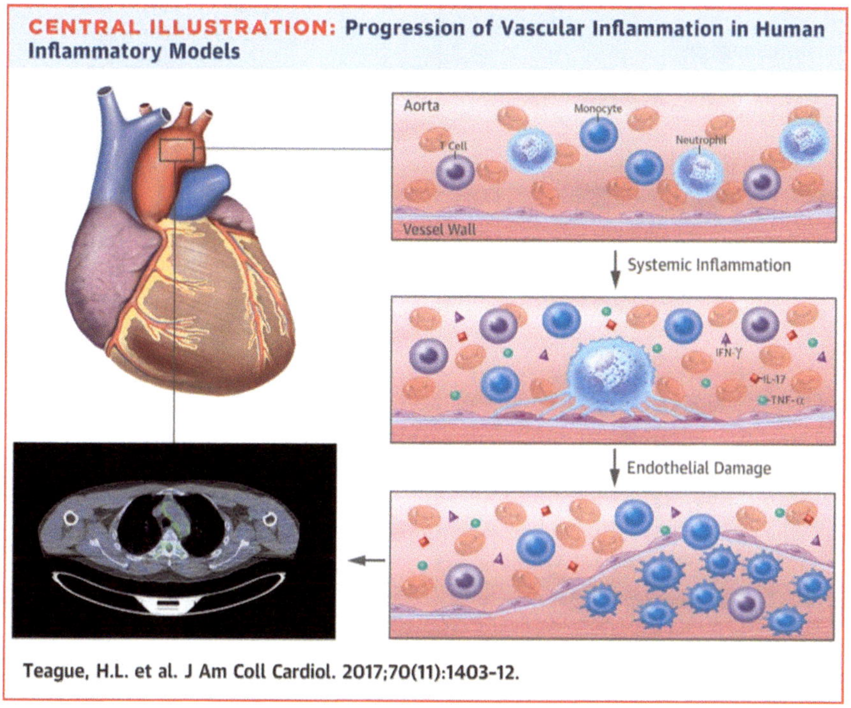

Fig. 30.3 A gross illustration of the aortic arch has been taken in cross section to magnify the vessel wall. Neutrophil activation due to systemic inflammation leads to the formation of neutrophil extracellular traps and may initiate damage to the endothelium. Monocytes and T cells then infiltrate the lesion. Monocytes differentiate into macrophages, where they proliferate to sustain their population. Macrophages within the vessel wall have high glycolytic rates and take up the ^{18}F-fluorodeoxyglucose (FDG) tracer, which is detectable by FDG positron emission tomography in human models of inflammation. *IFN* interferon, *IL* interleukin, *TNF* tumor necrosis factor. (Reproduced from Teague et al. [78], with permission)

hypoxia by detecting glycolytic rates of macrophages [78].

However, the type of changes may be variable in different diseases. For instance, the OLINK high-throughput proteomic assay was performed in atopic dermatitis (AD) in comparison with psoriasis. Ten proteins were increased in serum of both diseases, including Th1 (IFN-γ, CXCL9, TNF-β) and Th17 (CCL20) markers. The examined markers of atherosclerosis (fractalkine/CX3CL1, CCL8, M-CSF, HGF), T-cell development/activation (CD40L, IL-7, CCL25, IL-2RB, IL-15RA, CD6), and angiogenesis (VEGF-A) were significantly increased only in AD. In fact, AD blood signature was largely different compared to psoriasis and was characterized by dysregulation of inflammatory and cardiovascular risk markers, strongly supporting its systemic nature beyond atopic/allergic association [14].

Cellular and Molecular Inflammatory Pathways

In non-psoriatic individuals, the pathophysiology of obesity, aberrant adipocyte metabolism, diabetes, and CVDs involves immune-mediated or inflammatory pathways. An immune-mediated inflammatory disease (IMID) may impact these comorbid conditions through shared genetic risks, common environmental factors, or common inflammatory pathways that are co-expressed in IMIDs and target organs. Pathogenic immune pathways in psoriasis are now well explored, and a large number of inflammatory mediators have been identified in skin lesions, permitting consideration of possible mechanistic links between CVD and skin inflammation, increased risks of obesity or metabolic alterations. Well-

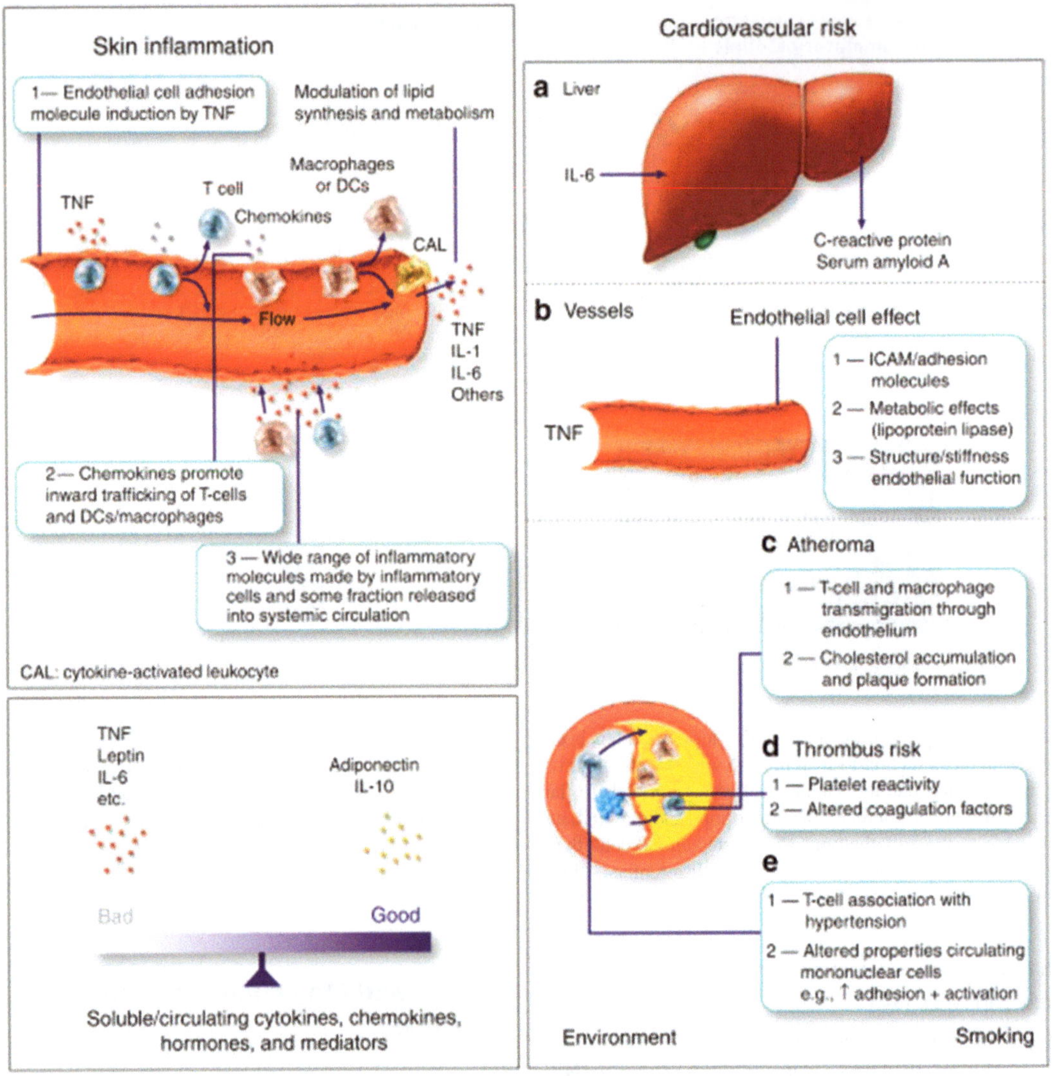

Fig. 30.4 "Cardiovascular risk factors": some potential cellular and molecular inflammatory pathways that could be triggered in the skin (left) and that would then reach target tissues for cardiovascular risk (right). Cytokine (chemokine)-activated leukocytes (CAL) in cutaneous sites could either enter the skin tissue or circulate after rolling on inflamed endothelial cells in psoriasis lesions. These cells and cytokines released into the systemic circulation, e.g., tumor necrosis factor (TNF), IL-1, or IL-6, may alter the function of hepatocytes (*a*), vascular cells (*b*), atheroma (*c*), thrombus risk (*d*), or leukocyte physiology (*e*) to increase cardiovascular risk factors or overt pathological pathways, as detailed in the right side of this figure. *DCs* dendritic cells. (From Davidovici et al. [19], reproduced with permission)

established risk factors for CVD can originate from inflammation in other tissues [19], as summarized in Fig. 30.4.

Rosacea is a chronic facial skin disorder with neurovascular, acneiform, and glandular components. The associated inflammatory response may be excessive and is triggered or exacerbated, e.g., by sunlight, menopausal flushing, spicy food, or alcohol consumption. An example of coincidence of skin inflammation and vascular changes is found for rosacea patients having significantly ($P < 0.001$) higher epicardial fat thickness and CIMT than healthy subjects [7]. Sex-specific differences have not been described for rosacea.

Immunology and Inflammation of the Vascular Wall

Inflammation is a hallmark of atherosclerosis, in addition to endothelial dysfunction. Monocytes are major cell components of the innate immune system that accumulate quickly in response to injury or infection. A circulating subset of Ly6C^{high} monocytes act as potent inflammatory mediators [60]. These cells exit the bone marrow, accumulate in the vessel wall, and differentiate into macrophages, which are sustained through self-renewal. In addition to the accumulation of monocytes in atherosclerotic lesions, these immune cells contribute to the biological response following a myocardial infarction (MI). Monocytes are both destructive and protective, leading to infarct rupture and healing, respectively. This results in heart failure (HF) when an overabundance of monocytes interferes with the healing process. Differentiated leukocytes, especially monocytes and neutrophils, take up residence in end-organ tissues, giving rise to plaques and inflammation. The sympathetic innervation of blood vessels plays a role in the increased emergency supply of leukocytes. In the periphery, increased adhesion molecule expression augments leukocyte recruitment. Stress elevates norepinephrine levels in the bone marrow and activates bone marrow stem cells. Collectively, these mechanisms suggest a multi-organ communication system that activates the bone marrow through the sympathetic innervation system, increasing hematopoietic stem and progenitor cell proliferation and thus enhancing leukocytosis [78]. Furthermore, the role of neutrophils is recognized in atherosclerotic plaque development, as they are the initial immune cells to infiltrate inflammatory sites. The antimicrobial action of neutrophils is indispensable to combat infection. However, their involvement also yields tissue damage and toxic debris. A newly identified defense mechanism is the ability to create a network, called neutrophil extracellular trap (NET), which formation process is coined "NETosis." Nowadays, it is evident that NETs themselves are proinflammatory, induce endothelial and tissue damage, and are highly prothrombotic. Additionally, NETs provide a communication platform between neutrophils and macrophages, culminating in atherosclerotic plaque destabilization [78].

Fibrotic Disorders

Fibrotic diseases represent a variety of disorders that are characterized by the development of severe organ fibrosis without an obvious cause. They are not well understood and include devastating diseases such as idiopathic pulmonary fibrosis (IPF) and scleroderma, often affecting the cardiovascular system and having a poor prognosis. End-stage fibrotic diseases, such as IPF, scleroderma, myelofibrosis, and kidney, pancreatic, and cardiac fibrosis, converge in the activation of a transcription factor in fibroblasts, suggesting that c-*jun* is a central mediator [84].

Arterial Stiffness and Chronic Inflammation

Arterial stiffness increases with inflammation of the vessel wall. Various metrics are available to noninvasively estimate arterial properties as reflected by arterial stiffness, among them the augmentation index (AIx) and pulse wave velocity (PWV). These measurement techniques are described elsewhere [4]. Alternatively, a photoplethysmograph transducer is placed on the index finger to estimate a systolic index (SI) which reportedly correlates with PWV [21]. A growing number of studies suggest that chronic inflammatory diseases are responsible for part of the excess cardiovascular risk. Compared with control subjects, carotid-femoral PWV and AIx are significantly higher in patients, including those with IBD, RA, psoriasis, SLE, and systemic sclerosis (SySc), as described by Dregan [21]. Moreover, a significant relationship between aortic stiffness and left ventricular systolic and diastolic dysfunction was reported in patients with IBD [89].

Inflammatory Joint Diseases

Patients with any type of inflammatory joint disease (IJD), including RA, have an increased risk of premature death compared with the general population, mainly because of the risk of CVD. While pathogenic mechanisms and clinical expression of cardiovascular comorbidities vary, it appears that atherosclerosis is associated with all IJDs. The combination of traditional risk factors and inflammation is mainly responsible for the increased cardiovascular risk in patients with IJDs [56].

Rheumatoid Arthritis

Rheumatic illnesses include diseases with evidence of autoimmunity as well as the common, near ubiquitous, osteoarthritis. These diseases are generally more common among women compared with men. In contrast to SLE, RA, and osteoarthritis, data concerning rheumatic diseases in relation to menopause are scarce, but it is known that menopause may affect the risk and course. Osteoporosis, an integral part of inflammatory rheumatic diseases, worsens by menopause. Hormone replacement therapy has been studied but reported effects vary [77]. Autoimmune rheumatic diseases can affect the coronary vessels, myocardium, pericardium, heart valves, and the conduction system. The diagnosis of these unique cardiac complications necessitates medical awareness and a high index of suspicion. Increased risk of advanced atherosclerosis plays a pivotal role in the development of cardiac comorbidities. Yet, other immune-mediated mechanisms may contribute to the pathogenesis. Cartilage degradation in the synovial joint constitutes the hallmark of RA, mediated by persistently activated fibroblast-like synoviocytes (FLS) that express matrix-degrading enzymes [44]. Patients' optimal care requires coordination between the primary caregiver, the rheumatologist, immunologist, and cardiologist [57]. Patients with RA have a twofold higher risk of sudden cardiac death (SCD), possibly resulting from conduction disorders. The increased risk might already be present at the clinical onset of arthritis. One study found that the prevalence of conduction disorders in patients with early arthritis was 12.5%, similar to the general population and not associated with observed changes in inflammation markers. However, a high cholesterol was associated with a prolonged QRS time, emphasizing the importance of traditional CV risk factor management in arthritis, apart from treatment of disease activity. Furthermore, a relationship ($P = 0.01$) between the improvement in disease activity and heart rate decrease was found. This remarkable finding possibly implies a 10-year CV mortality risk difference of 24% [81].

Incidence M/F Prevalence of 1%, with women three times more often affected, but both sexes have equal prevalence of erosive disease. Peak onset in fifth decade but on average somewhat earlier in females.

Symptoms, Signs and Laboratory Findings Insidious onset with morning stiffness, swelling of affected joints and arthralgia. Erythrocyte sedimentation rate (ESR) and CRP elevated, while various antibodies can be detected. Speckle tracking can detect subclinical myocardial dysfunction in RA ([8]).

Association with CVD Atherosclerosis with increased risk of MI, cerebrovascular accident (CVA), and pericarditis, while myocarditis is rare (often associated with active disease), HF common (mostly diastolic), valvular heart disease is rare (mostly clinically insignificant), and conduction abnormalities uncommon and mostly RBBB or AV-block [56].

Possible CV complications SCD, deep vein thrombosis, pulmonary embolus.

Ankylosing Spondylitis (AS, Morbus Bechterew)

Ankylosing spondylitis (AS), psoriatic arthritis, reactive arthritis (formerly termed Reiter syndrome), and arthritis associated with IBD belong to the seronegative spondyloarthropathies. AS is a chronic inflammatory disease of the joints of the axial skeleton with gradual onset.

Incidence M/F With onset usually before the age of 40 years, there is male predominance in AS, while symptoms are also more prominent in men. Genetic factors seem to be involved, with increased risk when associated with HLA-B27 positive testing (8% in whites and 4% in blacks). The course of AS appears to exhibit sex-specific differences, with a delayed confirmation of diagnosis and distinct adipose tissue associations in relation to disease activity in women.

Symptoms, Signs and Laboratory Findings Bouts of backache with progressive motion limitations, arthritis of peripheral joints, elevated ESR, but the severity of the disease varies greatly. Muscle wasting, expressed as fat-free mass index below the reference population, was detected in male TNF-α blocker-naïve AS patients, especially in those with active disease. Women had higher volumes of body fat than men but near the median of the reference population. The relationships between fat content and disease activity emphasize the complex association between adipose tissue and inflammation [38]. Disease activity can be assessed by CRP and ESR measurements.

Association with CVD Cardiac conduction abnormalities and aortic valve regurgitation are relatively often associated with AS, but arterial stiffness is found not to be increased [22]. One prospective study evaluated cardiac involvement in 15 AS patients using cardiac magnetic resonance (CMR), based on previous transthoracic echocardiography (TTE) screening. LV diastolic dysfunction was the most common finding on TTE (79%), followed by aortic root dilatation (14%), right ventricular (RV) dilatation (7%), and RV dysfunction (7%). LV ejection fraction (EF) below 55% was observed in five patients (33%). Myocardial extracellular volume (ECV) quantified by CMR T1 mapping correlated with the CRP concentration ($R = 0.78$, $P < 0.01$) and ESR level (Spearman's rank $R = 0.73$, $P < 0.01$). ECV has the potential to function as a marker for disease monitoring [11].

Dermatologic Abnormalities and Cardiovascular Disease

Systemic Sclerosis (Scleroderma)

Systemic sclerosis (SySc) is an autoimmune disease characterized by local (80% of cases) or diffuse (20%) skin fibrosis and vascular injury to major visceral organs. This disorder has been reviewed elsewhere [26]. Until recently, cardiac involvement was not a focus of SySc research because pulmonary and renal impairment are more prominent and predictive of the overall morbidity and mortality. However, arrhythmias, conduction disturbances, myocardial fibrosis, and MI have been shown to be associated with worse outcome in these patients [40, 75]. Speckle-tracking imaging and 3D echocardiography offer insight into cardiac function and mechanics. Multilayer strain assessment provides comprehensive information about deformation in each myocardial layer and was found to be significantly reduced in SySc patients [75].

Incidence M/F Female predominance (ratio 5:1) but increased mortality in males due to faster progression in males. Prevalence ranges from 8 to 489 per million, being higher in the USA and Japan.

Symptoms, Signs and Laboratory Findings Skin thickening (fingers), sclerodactyly, digital pitting scars, bibasilar pulmonary fibrosis.

Associated CVD Raynaud's phenomenon (mostly in females), arrhythmias, conduction disturbances, myocardial fibrosis, ischemia, cardiac remodeling, and deterioration of all cardiac chambers.

Possible CV Complications Postmenopause is the main risk factor for developing isolated pulmonary hypertension [68], and hormone replacement therapy may prevent the development of this complication in patients with limited cutaneous involvement [9].

Systemic Lupus Erythematosus (SLE)

SLE is a chronic multi-organ autoimmune inflammatory condition with fluctuating course, mostly (90%) seen in adult women (during childbearing years). The disorder has recently been reviewed [86]. Atherosclerosis is common in SLE. Actually, CVD which comprises coronary heart disease and stroke is one of the most important causes of death in these patients. The risks of developing clinical CVD and subclinical atherosclerosis are increased in SLE, which observation is not fully explained by traditional risk factors such as smoking, HTN, and elevated cholesterol. Therefore, it is believed that immune dysfunction also contributes to CVD risk in SLE. Abnormalities in serum lipid profile and the autoantibody and T-lymphocyte response to lipids may play a role in development of atherosclerosis. Future algorithms incorporating new biomarkers such as proinflammatory high-density lipoprotein and the use of imaging techniques (e.g., carotid ultrasound scanning) may become increasingly valuable [15]. Patients with SLE are at excess risk of cardiovascular events (CVEs). The relative importance of SLE disease activity, medications, or traditional risk factors in this increased risk has been investigated in cohort of 1874 patients. It was found that the event rate (14.1/1000 person-years) was 2.66 times what would be expected in the general population based on FRS and not associated with duration of SLE. Persons using 20 mg/day or more of corticosteroids had a substantial increase in risk even after adjustment for disease activity [52]. The range of cardiovascular involvement in SLE is broad including atherosclerosis and vascular inflammation/vasculitis, while Raynaud's phenomenon (20%) often antedates other features of SLE. The impact of CVEs on both mortality and morbidity is tremendous: the incidence of CAD both in its acute MI and chronic (with angina and chronic HF) forms is over seven times greater in patients with SLE than in healthy controls, even after accounting for traditional CVD risk factors, including sex, age, and lipid profile. In particular, younger female patients with SLE (aged 35–44 years) have an over 50 times greater risk of having an MI compared with the Framingham dataset. When considering the typical bimodal mortality pattern found in SLE, the real importance of CVD becomes apparent: a decade after the diagnosis of SLE, one of the leading causes of death is MI. Patients with SLE also have a greater risk for stroke, with an overall prevalence that can reach 20% and high recurrence rate and greater mortality than matched controls [16]. Furthermore, the prevalence of subclinical atherosclerosis is considerably higher in patients with SLE. Studies in multiple units, using different imaging techniques such as vascular ultrasound and electron beam tomography, have consistently shown that patients with SLE have significantly higher prevalence of atherosclerotic plaque compared to healthy controls. [16]. In another study a 44% prevalence of abnormal stress myocardial perfusion by CMR was observed in the absence of obstructive CAD in 18 female SLE patients with anginal chest pain. These findings are consistent with the hypothesis that anginal chest pain in SLE patients without obstructive CAD is due to myocardial ischemia potentially caused by microvascular coronary dysfunction [39]. The reason why SLE is such a dramatic risk factor for atherosclerosis and CVD is yet to be fully explained. Decisive evidence associating specific lupus-related factors to the development of atherosclerosis has been proved difficult to obtain, perhaps because even in very large cohort studies, the actual number of patients with CVEs is small. In multivariable analysis, only non-modifiable risk factors, namely, age and male gender, remained significant [16]. The presence of longstanding systemic inflammation associated with persistently active SLE could contribute to plaque formation and disruption. On the other hand, it has been found that patients with lupus have a high prevalence of traditional CVD risk factors such as HTN, altered lipid profile, and impaired glucose tolerance, which to some extent may be the result of chronic treatment with corticosteroids. However, to date, no undisputed correlation has been found between corticosteroid use and atherosclerosis in SLE [52]. However, there are also conflicting data suggesting that there is an increased CVD risk among patients who are undertreated with steroids, which could imply that poorer disease control is more relevant in determining a higher CVD risk than steroid treatment per se [16].

Etiology Idiopathic or drug-related (e.g., hydralazine, methyldopa, quinidine), the latter with equal sex ratio.

Significant gaps in knowledge remain about the effect of estrogen on cardiovascular risk factors during SLE in humans. Studies in women with SLE were not designed to determine the effect of estrogen or hormone therapy on blood pressure even though HTN is highly prevalent, and risk of premature ovarian failure could necessitate use of hormone therapy in women with SLE [28]. SLE is an autoimmune disease of unknown cause with multiple genes, environmental factors, and sex hormones likely playing roles in its pathogenesis. SLE is characterized by a loss of tolerance to self-antigens which leads to the production of autoantibodies to the nucleus, most commonly anti-double-stranded DNA (anti-dsDNA) antibodies. These autoantibodies contribute to immune complex formation and can deposit within virtually every tissue in the body, leading to inflammation and tissue injury, thus producing the clinical symptoms [28].

Incidence M/F The overall age-adjusted incidence (American College of Rheumatology (ACR) definition) per 100,000 persons is 5.5 and prevalence 72.8 in the USA. Among females, the incidence is 9.3 per 100,000, and the prevalence 128.7 per 100,000. The latter is 2.3-fold higher in black than in white persons and tenfold higher in females than in males. Among incident cases, the mean age at diagnosis is 39.3 years. Black SLE patients have a higher proportion of renal disease and end-stage renal disease (ESRD) (40.5% and 15.3%, respectively) as compared to white SLE patients (18.8% and 4.5%, respectively), as shown in the Michigan study [72]).

Symptoms, Signs and Laboratory Findings (Malar/discoid) Rash associated with photosensitivity, fever, malaise, arthritis (90%), (hemolytic) anemia (60%), leukopenia (45%), lymphopenia, thrombocytopenia (30%), antinuclear antibody (ANA) positive (98%), and anti-native DNA (60%).

Association with CVD Atherosclerosis increased risk (mainly in young women), pericarditis common (often asymptomatic), myocarditis often subclinical (but high prevalence in autopsy studies), valvular heart disease is rare, and conduction abnormalities are rare and mostly concern sinus tachycardia [56]. Among 115 patients with SLE compared to matched controls in one study, a longer disease duration was independently associated with higher systolic blood pressure, blood glucose levels, and accelerated rates of increased total cholesterol [41]. HF carries an increased risk in SLE patients. In one study among 45,284,540 individuals, 0.21% were identified with SLE and 0.22% with a new diagnosis of HF. The incidence of HF was markedly higher in the SLE group compared with controls, as were other cardiovascular risk factors. In regression analysis, SLE was an independent predictor of HF ($P < 0.0001$). RR of HF was highest in young males with SLE for age 20–24, with an overall trend of increasing absolute risk but decreasing RR with advancing age in both sexes. Renal involvement in SLE correlated with earlier and higher incidence of HF. The findings suggest that patients with SLE have significantly higher risk of developing HF (Fig. 30.5) and a worse cardiovascular risk profile compared with the general population [43].

The mechanisms underlying the elevated risk of HF, especially in younger patients, remain speculative. While direct myocardial involvement in SLE has been documented, lupus myocarditis

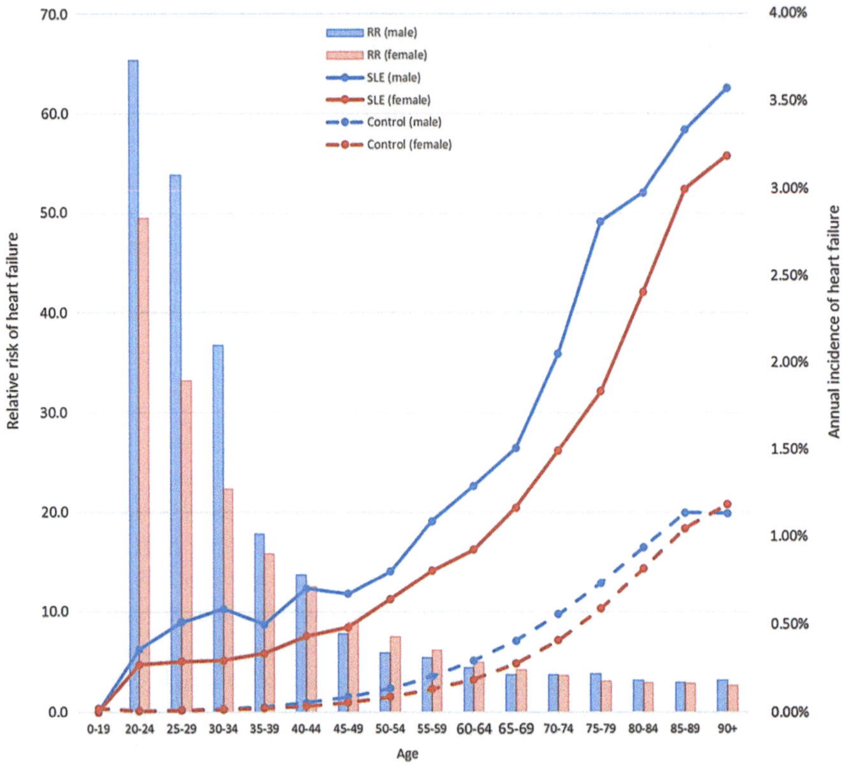

Fig. 30.5 Annual incidence and relative risk (RR) of heart failure over age, by sex, and disease status. SLE, systemic lupus erythematosus. (Reproduced from Kim et al. [43], with permission)

is rare and does not explain the high HF risk found (Fig. 30.5). However, the observed risks are likely due to a combination of microvascular disease and chronic inflammation. In that study, the most prevalent risk factors for HF in the SLE population were HTN, dyslipidemia, anemia, thyroid disease, and conduction system disease [43].

Possible CV Complications Vasculitis, endocarditis, pulmonary fibrosis, stroke syndromes, HF, venous thrombosis.

Giant Cell Arteritis (GCA, Temporal Arteritis, Arteriitis Temporalis, Horton-Brown Syndrome)

Giant cell arteritis (GCA) is the most common form of large vessel vasculitis and frequently associated with polymyalgia rheumatica. CVD and cerebrovascular disease are both increased in patients with GCA, with an HR of 2.06 (95% CI 1.72–2.46) for MI and HR 1.28 (95% CI 1.06–1.54) for CVA in patients versus controls. The risk of events is highest in the first year, potentially implicating high-dose glucocorticoid use or increased levels of inflammation, as seen in the general population and other rheumatic diseases [64].

Incidence M/F Incidence 25 per 100,000 in persons >50 years, mostly Caucasians. 80% in women with highest incidence of 7.4 per 10,000 person-years in women aged 70–79.

Symptoms, Signs and Laboratory Findings (chiefly temporal) Headache, amaurosis, diplopia, abnormal blood chemistry, and findings related to polymyalgia rheumatica.

Association with CVD One study evaluated the risk of CVA and CVD in a parallel cohort study of 5827 patients with GCA and 37,090 age-, sex-, and location-matched controls. Patients with GCA, compared with controls, had an increased risk of CVA (1.45, 1.31–1.60), for CVD (1.49, 1.37–1.62), or either (1.47, 1.37–1.57). In the GCA cohort, predictors of "CVA or CVD" included increasing age > 80 years versus <65 years (1.98, 1.62–2.42), male sex (1.20, 1.05–1.38), and socioeconomic status, most deprived quintile versus least deprived (1.34, 1.01–1.78). Thus, patients with GCA are more likely to develop CVA or CVD than controls [64].

Possible CV Complications MI and CVA.

Raynaud's Phenomenon (RP): Primary and Secondary Types

RP is paroxysmal bilateral digital ischemia, leading to pallor and cyanosis followed by rubor due to intense hyperemia. Various aspects have recently been reviewed [85]. It mainly affects young women. Primary type (50% of cases) is benign; secondary type can cause digital ischemia or gangrene. RP is often the earliest symptom of digital microcirculatory damage, associated with the development of connective tissue disease. Nailfold videocapillaroscopy detects and quantifies the microvascular changes that characterize secondary RP and can identify morphological patterns specific to various SySc microangiopathy stages [66].

Symptoms, Signs and Laboratory Findings Pallor of fingers (rarely toes).

Prevalence M/F The reported prevalence ranges from less than 1% (in men) and up to 20% (in women) depending on definitions and population selected [27]. Clear geographic variation, with prevalence being lowest in Japan and highest in France.

CVD Associations Occlusive arterial disease, coronary vasospasm, PAH.

Psoriasis and Psoriatic Arthritis

Psoriasis is a common benign, chronic but unpredictable autoimmune disease, now classified as an IMID of the skin. This inflammatory disease with both a genetic basis and known environmental triggers primarily affects the skin. Patients with various IMIDs, including psoriasis, are at higher risk of developing "systemic" comorbidities such as MetSy. For recent reviews, see Nestle et al. [54] and Lowes et al. [51]. Many patients with psoriasis also have serious other health conditions such as DM, heart disease, and depression. The most common type of psoriasis is called plaque psoriasis. *Psoriatic arthritis* (PsA) eventually occurs in 10–20% of these patients and is associated with obesity, T2D, HTN, the MetSy, fatty liver, and an increased risk of cardiovascular events, while the prevalence of Crohn's disease and subclinical colitis is also increased [63]. *Arthritis mutilans* may manifest telescoping of digits. PsA is different from more common types of arthritis (such as osteoarthritis or RA). Methotrexate is often used to treat PsA.

Incidence M/F Equal occurrence in both sexes, with prevalence around 1/500 in the USA, implying that an estimated 6.7 million adults have psoriasis, usually between ages 10 and 30 years.

Symptoms, Signs and Laboratory Findings Sharply defined plague formation, i.e., patches of thick red skin covered by silvery scales typically found on the elbows, knees, scalp, lower back, face, palms, and soles of feet. Pruritus may be severe. Parakeratosis. Nail pitting and onycholysis. Arthritis in certain types (20%). Leukocytosis and elevated ESR. The enzyme L-kynureninase has been identified as a novel inflammatory factor in psoriasis and other inflammatory diseases [33]. A laser Doppler imaging system has been developed to evaluate the superficial cutaneous microcirculation before and during topical treatment in psoriasis [34].

Association with CVD Microvascular dysfunction of coronary arteries has been found as evaluated by coronary flow reserve [59], CAD, MI. Notably, well-established risk factors for CVD can originate from inflammation in other tissues [19]. *Psoriasis vulgaris* is associated with systemic comorbid conditions, including DM, MetSy, besides CAD, and MI. This has led to the concept of the "psoriatic march," i.e., psoriasis as a chronic inflammatory condition driving CVD [13]. Furthermore, patients with moderate-to-severe psoriasis treated with tumor necrosis factor (TNF)-alpha inhibitors showed reductions in the proinflammatory marker CRP level and arterial intima-media thickness—both surrogate markers for CVD. Because vascular inflammation in patients with psoriasis correlates with disease severity, it stands to reason that it might also correlate with disease duration. In this context, a relationship between psoriasis vulgaris duration, vascular inflammation, and major adverse cardiovascular events (MACE) such as MI, ischemic stroke, or death due to CVD was hypothesized [24]. These authors found that in a cohort the risk for MACE increased by 1% per year of psoriasis duration, adding one more important concept, namely, that this risk is time dependent. Thus, the longer an individual has moderate-to-severe psoriasis, the higher the risk for MACEs.

Patients with moderate-to-severe psoriasis have an increased risk of CVD [67]. CVD risks arise from comorbidities such as obesity and DM that are associated with *psoriatic arthritis* [10]. CVD has been noted in children as well. The pediatric psoriasis was associated with obesity, HTN, diabetes, arrhythmia, and valvular heart disease. The highest odds of CVD risk factors in hospitalized children depended on race/ethnicity and age and were noted in blacks and Hispanics and children ages 0–9 years, but there were no sex differences [46]. In adults, women have less severe forms of psoriasis based on a scoring system analysis [32]. An increase in MI was noted in psoriasis [47]. Some methods of psoriasis systemic treatment, such as methotrexate and TNF inhibitors, reduce the CVD risk [1, 71]. On the other hand, the CVD may be under-recognized in psoriasis patients. For example, coronary artery calcification detected by computed tomography was higher than in previous investigations without tomography [37].

Atopic Dermatitis (AD, Atopic Eczema)

Atopic dermatitis (AD) is a chronic relapsing inflammatory pediatric allergic skin disease, which may persist into adulthood. A genetic predisposition is common. The barrier dysfunction, alterations in cell-mediated immune responses, IgE-mediated hypersensitivity, and environmental factors are crucial. Loss of function mutations in filaggrin have been implicated in severe AD because of a potential increase in transepidermal water loss, pH alterations, and dehydration [17].

Established adult AD comorbidities include psychiatric and autoimmune diseases. In a recent meta-analysis the risk of CVD and T2D was compared between adult patients with and without AD. Sixteen out of 2855 publications were included in the qualitative analysis, and no association was observed between AD and T2D, HTN, CVA, or MI, but a positive association was observed with angina pectoris. While adults with AD in some populations have increased prevalence of cardiovascular risk factors, such as obesity and smoking, the authors concluded that it is unlikely that AD represents an independent and clinically relevant risk factor for cardiometabolic disease [80]. AD and psoriasis are known to be comorbid with CVD. Such cases are known as "inflammatory skin march" [25].

Incidence M/F Prevalence of adult AD ranges from 2.1% to 4.9% across countries. The prevalence is generally lower for males vs females and decreases with age [5]. Incidence of AD has increased two- to threefold in industrialized countries, impacting approximately 15–20% of children and 1–3% of adults worldwide [3].

Symptoms, Signs and Laboratory Findings Dry skin, itching, and redness of the skin, as well as lichenificated lesions, are typical. White dermographism is a well-known sign of AD, but not obligatory. There is no definitive laboratory test. However, hypereosinophilia and increased IgE often accompany clinical features.

Association with CVD Recent investigations demonstrate that the association between AD and CVD was at most modest, but in more refined cohorts the cardiovascular risk profile and genetic architecture was comparable [55]. Patients with AD are at significantly higher risk of obesity and metabolic disorders, thus impacting on CVDs [31]. The AD blood signature was found to be largely different compared to psoriasis, with dysregulation of inflammatory and cardiovascular risk markers, strongly supporting its systemic nature beyond atopic/allergic association [14].

Hidradenitis Suppurativa (HS, Acne Inversa, Morbus Verneuil)

HS is a complex inflammatory suppurative cutaneous disorder, mainly with axillary and groin distribution, showing tendency to recurrence and possibly progressing to a more advanced stage. The lesions start during late puberty and may last through the age of 50 years. Unfortunately, there is a relative paucity of knowledge about HS [35, 36]. The primary defect in HS pathophysiology is proposed to stem from hair follicle obstruction at the site where the apocrine gland drains, leading to a cyst followed by rupture, with a robust immune response [36]. A specific genetic signature and environmental factors (including smoking), microbial colonization, and adiposity, all contribute to the HS phenotype [62]. A comparison of ultrasound imaging of carotid intima-media thickness in 60 HD patients to detect subclinical atherosclerosis found that the FRS underestimates risks involved [29]. In addition, a population-based cohort study showed that HS was associated with a significantly increased risk of adverse cardiovascular outcomes and all-cause mortality [23].

Incidence M/F 6/100,000, and more common in women (75%) between 20 and 50 years. Worldwide prevalence is estimated between 0.05 and 4%, depending on methodology and population studied. In the USA officially classified as an orphan disease.

Symptoms, Signs and Laboratory Findings The HS lesion consists of nodules and cysts, which are painful and cause suppuration with odiferous discharge with subsequent scarring. There is no diagnostic test, and the culture of exudate from a lesion shows a panoply of organisms, mostly common skin commensals.

Association with CVD and Other Diseases Atherosclerosis [83] and MetSy. Some patients with HS may be suffering from unrecognized cardiovascular risk factors, and therefore dermatologists have a key role in detecting these conditions. Besides, associated rheumatologic disorders such as spondyloarthropathy or inflammatory arthritis may occur [36]. Also, a connection between HS and IBD has been reported [20, 87].

Sarcoidosis

Sarcoidosis is a systemic connective tissue disease (CTD) of unknown etiology characterized by the presence of noncaseating granulomas in any organ, most commonly the lungs and intrathoracic lymph nodes. A diagnosis of sarcoidosis should be suspected in any young or middle-aged adult presenting with unexplained cough, shortness of breath, or constitutional symptoms, especially among blacks or Scandinavians. This multisystem disease may involve almost any organ system; therefore, it results in various clinical manifestations. Cutaneous sarcoidosis occurs in up to one third of patients with systemic sarcoidosis. Recognition of cutaneous lesions is important because they provide a visible clue to the diagnosis and are an easily accessible source of tissue for histologic examination. Because lesions can exhibit many different morphologies, cutaneous sarcoidosis is known as one of the "great imitators" in dermatology.

Sarcoidosis is the result of noncaseating granuloma formation due to ongoing inflammation that causes the accumulation of activated T cells and macrophages, which then secrete cytokines and TNF-α. The precise cause of sarcoidosis is unknown. However, most studies suggest that in a susceptible individual this follows from an exaggerated immune response to an antigen related to certain environmental factors, microbes, along with strong evidence of a genetic predisposition [73].

Symptoms and Findings Papules, plaques, lupus pernio, scar sarcoidosis, and rare morphologies such as alopecia, ulcers, hypopigmented patches, and ichthyosis [42].

CVD Associations Coronary flow reserve is reduced in sarcoidosis [45]. Systemic sclerosis, vasculitis, and sarcoidosis are among CTDs with serious cardiovascular involvement; this is due to multiple causative factors such as myopericarditis, micro-/macrovascular disease, coronary artery disease, myocardial fibrosis, pulmonary hypertension, and finally HF [53]. Despite aggressive treatment, some patients may develop life-threatening cardiac complications from severe, progressive disease, ultimately requiring heart transplantation [73]. Also, 18F-FDG PET scans have been employed to evaluate the prognostic impact of extent, severity, and heterogeneity of abnormalities in suspected cardiac sarcoidosis [74].

The Problem of Vulnerable Plaque

Once atheroma was regarded as a graveyard of acellular lipid debris enrobed by a capsule of proliferated smooth muscle cells [48]. Gradually a prominent role for inflammation in atherosclerosis and its complications was appreciated. Current notion that inflammation and immune response contribute to atherogenesis has garnered increased interest in underlying processes. Formerly focused on luminal narrowing due to the bulk of atheroma, modern concepts recognize the biological attributes as key determinants of its clinical significance [48]. Circulating monocytes infiltrate the plaque and differentiate into macrophages, contributing to an inflammatory environment associated with higher risk of cardiovascular events. While the pivotal role of circulating monocytes is firmly established, the identification of specific monocyte subsets that may be particularly atherogenic continues. For example, using multivariate linear regression, the relation between monocyte phenotype, notably surface receptor expression, and arterial wall inflammation in 79 patients (54% males) with peripheral artery disease, chronic kidney disease, or RA was analyzed, using 18F-FDG PET/CT scan assessment of arterial wall inflammation and extensive monocyte characterization. The monocyte chemokine receptor CCR2, essential for transmigration, was found to significantly ($P = 0.015$) correlate with PET/CT-based arterial wall inflammation, independent of traditional cardiovascular risk factors and statin use [82].

Conclusions

In general, females are healthier and live longer than males, and this is true for several mammalian species, including humans. Additionally, the course and survival rates from illnesses—including infectious diseases, sepsis, trauma, or injury—are better in women than in men. The immunological advantage of women has been long known. Several studies have shown the superior ability of women to produce more antibodies and serum IgM. In contrast, the greater susceptibility of males to infections is evident from birth. Boys are more prone to septicemia and meningitis, and the incidence of tuberculosis is higher in males from infancy to adulthood. Also, males experience more frequent and severe infections caused by bacteria or viruses. The X chromosome is partly responsible for the hyperresponsiveness of the female immune system. Indeed, it has been proposed that males suffer the equivalent of inbreeding depression involving 5% of their protein-coding genome (the X chromosome transcriptome), leading to a higher rate of mortality, from infancy to adulthood [49].

Imaging techniques, such as videocapillaroscopy, intravascular ultrasound, FDG PET scan, (tissue) Doppler imaging, and speckle tracking, have great potential for the study of inflammatory diseases, especially in combination with immunological investigations.

Substantial sex-related differences have been noted regarding symptoms, signs, disease course, and outcome in cardiovascular comorbidities, emphasizing the need for intensive coordination between the primary caregiver, the rheumatologist/dermatologist, immunologist, and cardiologist [57].

References

1. Ahlehoff O, Skov L, Gislason G, Gniadecki R, Iversen L, Bryld LE, Lasthein S, Lindhardsen J, Kristensen SL, Torp-Pedersen C, Hansen PR. Cardiovascular outcomes and systemic anti-inflammatory drugs in patients with severe psoriasis: 5-year follow-up of a Danish nationwide cohort. J Eur Acad Dermatol Venereol. 2015;29(6):1128–34. https://doi.org/10.1111/jdv.12768.
2. Andersen LK, Davis MDP. Sex differences in the incidence of skin and skin-related diseases in Olmsted County, Minnesota, United States, and a comparison with other rates published worldwide. Int J Dermatol. 2016;55(9):939–55. https://doi.org/10.1111/ijd.13285.
3. Avena-Woods C. Overview of atopic dermatitis. Am J Manag Care. 2017;23(8 Suppl):S115–23.
4. Avolio AP, Kuznetsova T, Heyndrickx GR, Kerkhof PLM, Li JK-J. Arterial flow, pulse pressure and pulse wave velocity in men and women at various ages. In: Kerkhof PLM, Miller VM, editors. Sex-specific analysis of cardiovascular function. New York: Springer; 2018. https://doi.org/10.1007/978-3-319-77932-4_10.
5. Barbarot S, Auziere S, Gadkari A, Girolomoni G, Puig L, Simpson EL, Margolis DJ, de Bruin-Weller M, Eckert L. Epidemiology of atopic dermatitis in adults: results from an international survey. Allergy. 2018. doi: https://doi.org/10.1111/all.13401.
6. Bairey Merz CN, Shaw LJ, Reis SE, et al. Insights from the NHLBI-Sponsored Women's Ischemia Syndrome Evaluation (WISE) Study: Part II: gender differences in presentation, diagnosis, and outcome with regard to gender-based pathophysiology of atherosclerosis and macrovascular and microvascular coronary disease. J Am Coll Cardiol. 2006;47 (3 Suppl):S21–9.
7. Belli AA, Altun I, Altun I. Rosacea patients had significantly higher epicardial fat thickness and carotid intima-media thickness volumes than controls. An Bras Dermatol. 2017;92 (6):820–5. https://doi.org/10.1590/abd1806-4841.20176832.
8. Benacka O, Benacka J, Blazicek P, Belansky M, Payer J, Killinger Z, Lietava J. Speckle tracking can detect subclinical myocardial dysfunction in rheumatoid arthritis patients. Bratisl Lek Listy. 2017;118(1):28–33. https://doi.org/10.4149/BLL_2017_006.
9. Beretta L, Caronni M, Origgi L, Ponti A, Santaniello A, Scorza R. Hormone replacement therapy may prevent the development of isolated pulmonary hypertension in patients with systemic sclerosis and limited cutaneous involvement. Scand J Rheumatol. 2006;35:468–71.
10. Bergman M, Lundholm A. Mitigation of disease- and treatment-related risks in patients with psoriatic arthritis. Arthritis Res Ther. 2017;19(1):63. https://doi.org/10.1186/s13075-017-1265-5.
11. Biesbroek PS, Heslinga SC, Konings TC, van der Horst-Bruinsma IE, Hofman MBM, van de Ven PM, Kamp O, van Halm VP, Peters MJL, Smulders YM, van Rossum AC, Nurmohamed MT, Nijveldt R. Insights into cardiac involvement in ankylosing spondylitis from cardiovascular magnetic resonance. Heart. 2017;103(10):745–52. https://doi.org/10.1136/heartjnl-2016-310667.
12. Biousse V, Bruce BB, Newman NJ. Ophthalmoscopy in the 21st century: the 2017 H. Houston Merritt Lecture. Neurology. 2018;90(4):167–75. https://doi.org/10.1212/WNL.0000000000004868.
13. Boehncke WH, Boehncke S, Tobin AM, Kirby B. The 'psoriatic march': a concept of how severe psoriasis may drive cardiovascular comorbidity. Exp Dermatol. 2011;20:303–7.
14. Brunner PM, Suárez-Fariñas M, He H, Malik K, et al. The atopic dermatitis blood signature is characterized by increases in inflammatory and cardiovascular risk proteins. Sci Rep. 2017;7(1):8707. https://doi.org/10.1038/s41598-017-09207-z.
15. Croca SC, Rahman A. Imaging assessment of cardiovascular disease in systemic lupus erythematosus. Clin Dev Immunol. 2012;2012:694143. https://doi.org/10.1155/2012/694143.
16. Croca S, Rahman A. Atherosclerosis in systemic lupus erythematosus. Best Pract Res Clin Rheumatol. 2017;31:364–72. https://doi.org/10.1016/j.berh.2017.09.012.
17. David Boothe W, Tarbox JA, Tarbox MB. Atopic dermatitis: pathophysiology. Adv Exp Med Biol. 2017;1027:21–37. https://doi.org/10.1007/978-3-319-64804-0_3.
18. Damen MSMA, Popa CD, Netea MG, Dinarello CA, Joosten LAB. Interleukin-32 in chronic inflammatory conditions is associated with a higher risk of cardiovascular diseases. Atherosclerosis. 2017;264:83–91. https://doi.org/10.1016/j.atherosclerosis.2017.07.005.
19. Davidovici BB, Sattar N, Prinz J, Puig L, Emery P, Barker JN, van de Kerkhof P, Ståhle M, Nestle FO, Girolomoni G, Krueger JG. Psoriasis and systemic inflammatory diseases: potential mechanistic links between skin disease and co-morbid conditions. J Invest Dermatol. 2010;130:1785–96.
20. Deckers IE, Benhadou F, Koldijk MJ, Del Marmol V, Horváth B, Boer J, van der Zee HH, Prens EP. Inflammatory bowel disease is associated with hidradenitis suppurativa: results from a multicenter cross-sectional study. J Am Acad Dermatol. 2017;76(1):49–53. https://doi.org/10.1016/j.jaad.2016.08.031.
21. Dregan A. Arterial stiffness association with chronic inflammatory disorders in the UK Biobank study. Heart. 2018. pii: heartjnl-2017-312610. doi:https://doi.org/10.1136/heartjnl-2017-312610.
22. Dzieża-Grudnik A, Sulicka J, Strach M, et al. Arterial stiffness is not increased in patients with short duration rheumatoid arthritis and ankylosing spondylitis. Blood Press. 2017;26:115–21.
23. Egeberg A, Gislason GH, Hansen PR. Risk of major adverse cardiovascular events and all-cause mortality in patients with hidradenitis suppurativa. JAMA Dermatol. 2016;152 (4):429–34. https://doi.org/10.1001/jamadermatol.2015.6264.

24. Egeberg A, Skov L, Joshi AA, Mallbris L, Gislason GH, Wu JJ, Rodante J, Lerman JB, Ahlman MA, Gelfand JM, Mehta NN. The relationship between duration of psoriasis, vascular inflammation, and cardiovascular events. J Am Acad Dermatol. 2017;77:650–656.e3. https://doi.org/10.1016/j.jaad.2017.06.028.
25. Furue M, Kadono T. "Inflammatory skin march" in atopic dermatitis and psoriasis. Inflamm Res. 2017;66(10):833–42. https://doi.org/10.1007/s00011-017-1065-z.
26. Gabrielli A, Avvedimento EV, Krieg T. Mechanisms of disease: scleroderma. N Engl J Med. 2009;360:1989–2003. REVIEW
27. Garner R, Kumari R, Lanyon P, Doherty M, Zhang W. Prevalence, risk factors and associations of primary Raynaud's phenomenon: systematic review and meta-analysis of observational studies. BMJ Open. 2015;5(3):e006389. https://doi.org/10.1136/bmjopen-2014-006389.
28. Gilbert EL, Ryan MJ. Estrogen in cardiovascular disease during systemic lupus erythematosus. Clin Ther. 2014;36:1901–12. https://doi.org/10.1016/j.clinthera.2014.07.021.
29. González-López MA, Lacalle M, Mata C, et al. Carotid ultrasound is useful for the cardiovascular risk stratification in patients with hidradenitis suppurativa. PLoS One. 2018;13 (1):e0190568. https://doi.org/10.1371/journal.pone.0190568.
30. Gubbels Bupp MR. Sex, the aging immune system, and chronic disease. Cell Immunol. 2015;294:102–10.
31. Guttman-Yassky E, Krueger JG, Lebwohl MG. Systemic immune mechanisms in atopic dermatitis and psoriasis with implications for treatment. Exp Dermatol. 2017. doi:https://doi.org/10.1111/exd.13336.
32. Hägg D, Sundström A, Eriksson M, Schmitt-Egenolf M. Severity of psoriasis differs between men and women: a study of the clinical outcome measure Psoriasis Area and Severity Index (PASI) in 5438 Swedish Register patients. Am J Clin Dermatol. 2017;18(4):583–90. https://doi.org/10.1007/s40257-017-0274-0.
33. Harden JL, Lewis SM, Lish SR, Suárez-Fariñas M, Gareau D, Lentini T, Johnson-Huang LM, Krueger JG, Lowes MA. The tryptophan metabolism enzyme L-kynureninase is a novel inflammatory factor in psoriasis and other inflammatory diseases. J Allergy Clin Immunol. 2016;137:1830–40. https://doi.org/10.1016/j.jaci.2015.09.055.
34. Hendriks AG, Steenbergen W, Hondebrink E, van Hespen JC, van de Kerkhof PC, Seyger MM. Whole field laser Doppler imaging of the microcirculation in psoriasis and clinically unaffected skin. J Dermatolog Treat. 2014;25:18–21. https://doi.org/10.3109/09546634.2013.777382.
35. Hoffman LK, Ghias MH, Garg A, Hamzavi IH, Alavi A, Lowes MA. Major gaps in understanding and treatment of hidradenitis suppurativa. Semin Cutan Med Surg. 2017;36 (2):86–92. https://doi.org/10.12788/j.sder.2017.024.
36. Hoffman LK, Ghias MH, Lowes MA. Pathophysiology of hidradenitis suppurativa. Semin Cutan Med Surg. 2017;36 (2):47–54. https://doi.org/10.12788/j.sder.2017.017.
37. Honma M, Shibuya T, Iwasaki T, Iinuma S, Takahashi N, Kishibe M, Minami-Hori M, Ishida-Yamamoto A. Prevalence of coronary artery calcification in Japanese patients with psoriasis: a close correlation with bilateral diagonal earlobe creases. J Dermatol. 2017;44(10):1122–8. https://doi.org/10.1111/1346-8138.13895.
38. Ibáñez Vodnizza S, Visman IM, van Denderen C, Lems WF, Jaime F, Nurmohamed M, van der Horst-Bruinsma IE. Muscle wasting in male TNF-α blocker naïve ankylosing spondylitis patients: a comparison of gender differences in body composition. Rheumatology (Oxford). 2017;56(9):1566–72. https://doi.org/10.1093/rheumatology/kex187.
39. Ishimori ML, Martin R, Berman DS, Goykhman P, Shaw LJ, Shufelt C, et al. Myocardial ischemia in the absence of obstructive coronary artery disease in systemic lupus erythematosus. JACC Cardiovasc Imaging. 2011;4 (1):27–33.
40. Ivanovic BA, Tadic MV, Zlatanovic MM, Damjanov NN, Ostojić PM, Bonaci-Nikolic BN. Which factors impact myocardial function in systemic sclerosis? Echocardiography. 2012;29:307–17. https://doi.org/10.1111/j.1540-8175.2011.01560.x.
41. Karp I, Abrahamowicz M, Fortin PR, Pilote L, Neville C, Pineau CA, Esdaile JM. Longitudinal evolution of risk of coronary heart disease in systemic lupus erythematosus. J Rheumatol. 2012;39(5):968–73. https://doi.org/10.3899/jrheum.111127.
42. Katta R. Cutaneous sarcoidosis: a dermatologic masquerader. Am Fam Physician. 2002;65(8):1581–4.
43. Kim CH, Al-Kindi SG, Jandali B, Askari AD, Zacharias M, Oliveira GH. Incidence and risk of heart failure in systemic lupus erythematosus. Heart. 2017;103:227–33. https://doi.org/10.1136/heartjnl-2016-309561.
44. Klingenberg R, Lüscher TF. Rheumatoid arthritis and coronary atherosclerosis: two cousins engaging in a dangerous liaison. Eur Heart J. 2015;36(48):3423–5.
45. Kul S, Kutlu GA, Guvenc TS, Kavas M, Demircioglu K, Yilmaz Y, Yakar HI, Kanbay A, Boga S, Caliskan M. Coronary flow reserve is reduced in sarcoidosis. Atherosclerosis. 2017;264:115–21. https://doi.org/10.1016/j.atherosclerosis.2017.05.005.
46. Kwa L, Kwa MC, Silverberg JI. Cardiovascular comorbidities of pediatric psoriasis among hospitalized children in the United States. J Am Acad Dermatol. 2017;77(6):1023–9. https://doi.org/10.1016/j.jaad.2017.08.034.
47. Lebwohl M. Does treatment of psoriasis reduce cardiovascular comorbidities? J Invest Dermatol. 2017;137(8):1612–3. https://doi.org/10.1016/j.jid.2017.06.001.
48. Libby P. Inflammation in atherosclerosis. Nature. 2002;420:868–74.
49. Libert C, Dejager L, Pinheiro I. The X chromosome in immune functions: when a chromosome makes the difference. Nat Rev Immunol. 2010;10:594–604. https://doi.org/10.1038/nri2815.
50. Lockwood MM, O'Malley PG. Admitting what we do not know in rheumatologic disease: The first step toward learning more. JAMA Intern Med. 2018. doi:https://doi.org/10.1001/jamainternmed.2017.8557.
51. Lowes MA, Suárez-Fariñas M, Krueger JG. Immunology of psoriasis. Annu Rev Immunol 2014;32:227–55. doi: https://doi.org/10.1146/annurev-immunol-032713-120225. Review. PMID: 24655295.
52. Magder LS, Petri M. Incidence of and risk factors for adverse cardiovascular events among patients with systemic lupus erythematosus. Am J Epidemiol. 2012;176:708–19.
53. Mavrogeni S, Markousis-Mavrogenis G, Koutsogeorgopoulou L, Kolovou G. Cardiovascular magnetic resonance imaging: clinical implications in the evaluation of connective tissue diseases. J Inflamm Res. 2017;10:55–61. https://doi.org/10.2147/JIR.S115508. eCollection 2017
54. Nestle FO, Kaplan DH, Barker J. Psoriasis. N Engl J Med. 2009;361:496–509. https://doi.org/10.1056/NEJMra0804595.
55. Nijsten T. Atopic dermatitis and comorbidities: added value of comprehensive dermatoepidemiology. J Invest Dermatol. 2017;137(5):1009–11. https://doi.org/10.1016/j.jid.2017.02.003.
56. Nurmohamed MT, Heslinga M, Kitas GD. Cardiovascular comorbidity in rheumatic diseases. Nat Rev Rheumatol. 2015;11:693–704. https://doi.org/10.1038/nrrheum.2015.112.

57. Nussinovitch, editor. The heart in rheumatic, autoimmune and inflammatory diseases. London: Academic Press; 2017.
58. O'Callaghan KM. Solutions for disparities for women with heart disease. J Cardiovasc Transl Res. 2009;2(4):518–25. https://doi.org/10.1007/s12265-009-9125-6.
59. Osto E, Piaserico S, Maddalozzo A, Forchetti G, Montisci R, Famoso G, Giovagnoni A, Peserico A, Iliceto S, Tona F. Impaired coronary flow reserve in young patients affected by severe psoriasis. Atherosclerosis. 2012;221(1):113–7. https://doi.org/10.1016/j.atherosclerosis.2011.12.015.
60. Phinikaridou A, Andia ME, Passacquale G, Ferro A, Botnar RM. Noninvasive MRI monitoring of the effect of interventions on endothelial permeability in murine atherosclerosis using an albumin-binding contrast agent. J Am Heart Assoc. 2013;2(5):e000402. https://doi.org/10.1161/JAHA.113.000402.
61. Pizzorni C, Sulli A, Smith V, Ruaro B, Trombetta AC, Cutolo M, Paolino S. Primary Raynaud's phenomenon and nailfold videocapillaroscopy: age-related changes in capillary morphology. Clin Rheumatol. 2017;36(7):1637–42. https://doi.org/10.1007/s10067-016-3442-3.
62. Prens E, Deckers I. Pathophysiology of hidradenitis suppurativa: an update. J Am Acad Dermatol 2015;73 (5 Suppl 1):S8–11. doi: https://doi.org/10.1016/j.jaad.2015.07.045. Review. PMID: 26470623.
63. Ritchlin CT, Colbert RA, Gladman DD. Psoriatic arthritis. N Engl J Med. 2017;376(21):2095–6. https://doi.org/10.1056/NEJMc1704342.
64. Robson JC, Kiran A, Maskell J, Hutchings A, Arden N, Dasgupta B, Hamilton W, Emin A, Culliford D, Luqmani R. Which patients with giant cell arteritis will develop cardiovascular or cerebrovascular disease? A Clinical Practice Research Datalink Study. J Rheumatol. 2016;43(6):1085–92. https://doi.org/10.3899/jrheum.151024.
65. Roshdy HS, Soliman MH, El-Dosouky II, Ghonemy S. Skin aging parameters: a window to heart block. Clin Cardiol. 2017. doi:https://doi.org/10.1002/clc.22848.
66. Ruaro B, Sulli A, Smith V, Pizzorni C, Paolino S, Alessandri E, Cutolo M. Microvascular damage evaluation in systemic sclerosis: the role of nailfold videocapillaroscopy and laser techniques. Reumatismo. 2017;69(4):147–55. https://doi.org/10.4081/reumatismo.2017.959.
67. Ryan C, Kirby B. Psoriasis is a systemic disease with multiple cardiovascular and metabolic comorbidities. Dermatol Clin. 2015;33(1):41–55. https://doi.org/10.1016/j.det.2014.09.004.
68. Scorza R, Caronni M, Bazzi S, Nador F, Beretta L, Antonioli R, Origgi L, Ponti A, Marchini M, Vanoli M. Post-menopause is the main risk factor for developing isolated pulmonary hypertension in systemic sclerosis. Ann N Y Acad Sci. 2002;966:238–46.
69. Shaw LJ, Bugiardini R, Merz CN. Women and ischemic heart disease: evolving knowledge. J Am Coll Cardiol. 2009;54 (17):1561–75. https://doi.org/10.1016/j.jacc.2009.04.098.
70. Singh S, Michelena HI, Warrington KJ. The nails give it away. J Am Coll Cardiol. 2011;57:996. https://doi.org/10.1016/j.jacc.2010.03.111.
71. Sobchak C, Eder L. Cardiometabolic disorders in psoriatic disease. Curr Rheumatol Rep. 2017;19(10):63. https://doi.org/10.1007/s11926-017-0692-2.
72. Somers EC, Marder W, Cagnoli P, et al. Population-based incidence and prevalence of systemic lupus erythematosus: the Michigan Lupus Epidemiology and Surveillance program. Arthritis Rheumatol. 2014;66:369–78. https://doi.org/10.1002/art.38238.
73. Soto-Gomez N, Peters JI, Nambiar AM. Diagnosis and management of sarcoidosis. Am Fam Physician. 2016;93 (10):840–8.
74. Sperry BW, Tamarappoo BK, Oldan JD, Javed O, Culver DA, Brunken R, Cerqueira MD, Hachamovitch R. Prognostic Impact of Extent, Severity, and Heterogeneity of Abnormalities on 18F-FDG PET Scans for Suspected Cardiac Sarcoidosis. JACC Cardiovasc Imaging. 2017. pii: S1936-878X(17)30612-5.
75. Tadic M, Zlatanovic M, Cuspidi C, Ivanovic B, Stevanovic A, Damjanov N, Kocijancic V, Celic V. The relationship between left ventricular deformation and heart rate variability in patients with systemic sclerosis: Two- and three-dimensional strain analysis. Int J Cardiol. 2017;236:145–50. https://doi.org/10.1016/j.ijcard.2017.02.043.
76. Tadic M, Cuspidi C, Vasic D, Kerkhof PLM. Cardiovascular implications of diabetes, metabolic syndrome, thyroid disease and cardio-oncology in women. In: Kerkhof PLM, Miller VM, editors. Sex-specific analysis of cardiovascular function. New York: Springer; 2018. https://doi.org/10.1007/978-3-319-77932-4_29.
77. Talsania M, Scofield RH. Menopause and rheumatic disease. Rheum Dis Clin N Am. 2017;43:287–302. https://doi.org/10.1016/j.rdc.2016.12.011.
78. Teague HL, Ahlman MA, Alavi A, et al. Unraveling vascular inflammation: from immunology to imaging. J Am Coll Cardiol. 2017;70:1403–12.
79. Ten Haaf ME, Rijndertse M, Cheng JM, et al. Sex differences in plaque characteristics by intravascular imaging in patients with coronary artery disease. EuroIntervention. 2017;13 (3):320–8. https://doi.org/10.4244/EIJ-D-16-00361.
80. Thyssen JP, Halling-Overgaard AS, Andersen YMF, Gislason G, Skov L, Egeberg A. The association between cardiovascular disease and type 2 diabetes in adults with atopic dermatitis: a systematic review and meta-analysis. Br J Dermatol. 2017. https://doi.org/10.1111/bjd.16215.
81. Turk SA, Heslinga SC, Dekker J, Britsemmer L, van der Lugt V, Lems WF, van Schaardenburg D, Nurmohamed MT. The relationship between cardiac conduction times, cardiovascular risk factors, and inflammation in patients with early arthritis. J Rheumatol. 2017;44:580–6.
82. Verweij SL, Duivenvoorden R, Stiekema LCA, Nurmohamed NS, van der Valk FM, Versloot M, Verberne HJ, Stroes ESG, Nahrendorf M, Bekkering S, Bernelot Moens SJ. CCR2 expression on circulating monocytes is associated with arterial wall inflammation assessed by 18F-FDG PET/CT in patients at risk for cardiovascular disease. Cardiovasc Res. 2018;114:468–75. https://doi.org/10.1093/cvr/cvx224.
83. Vinkel C, Thomsen SF. Risk factors for cardiovascular disease in patients with hidradenitis suppurativa. J Eur Acad Dermatol Venereol. 2017;31(9):e411–3. https://doi.org/10.1111/jdv.14225.
84. Wernig G, Chen SY, Cui L, et al. Unifying mechanism for different fibrotic diseases. Proc Natl Acad Sci U S A. 2017;114:4757–62.
85. Wigley FM, Flavahan NA. Raynaud's phenomenon. N Engl J Med. 2016;375:556–65.
86. Wu GC, Liu HR, Leng RX, et al. Subclinical atherosclerosis in patients with systemic lupus erythematosus: a systemic review and meta-analysis. Autoimmun Rev. 2016;15:22–37.
87. Wu GC, Leng RX, Lu Q, et al. Subclinical atherosclerosis in patients with inflammatory bowel diseases: a systematic review and meta-analysis. Angiology. 2017;68:447–61.
88. Yang CA, Chiang BL. Inflammasomes and human autoimmunity: a comprehensive review. J Autoimmun. 2015;61:1–8. https://doi.org/10.1016/j.jaut.2015.05.001.
89. Zanoli L, Boutouyrie P, Fatuzzo P, et al. Inflammation and aortic stiffness: an individual participant data meta-analysis in patients with inflammatory bowel disease. J Am Heart Assoc. 2017;6:e007003.

31 The Role of Sex in the Pathophysiology of Pulmonary Hypertension

Craig K. Docherty, Katie Yates Harvey, Kirsty M. Mair, Sinead Griffin, Nina Denver, and Margaret R. MacLean

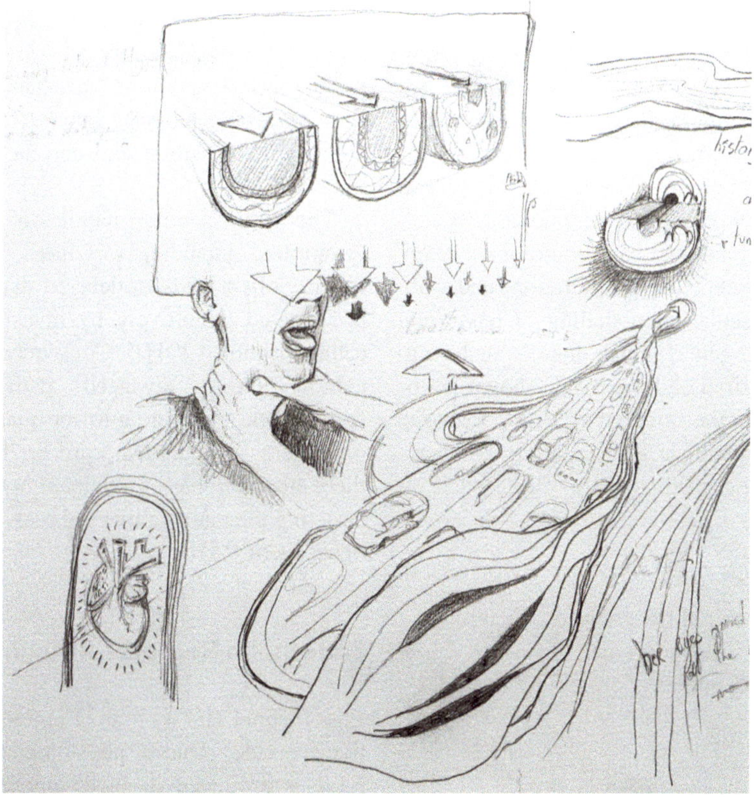

Right heart failure in a PAH patient. Artwork by Piet Michiels, Leuven, Belgium

C. K. Docherty · K. Y. Harvey · K. M. Mair
S. Griffin · N. Denver · M. R. MacLean (✉)
Research Institute of Cardiovascular and Medical Sciences, College of Medical, Veterinary and Life Sciences, University of Glasgow, Glasgow, UK
e-mail: mandy.maclean@glasgow.ac.uk

Abstract

Pulmonary arterial hypertension (PAH) is a progressive disease characterised by increased pulmonary vascular resistance and pulmonary artery remodelling as result of increased

P. L. M. Kerkhof, V. M. Miller (eds.), *Sex-Specific Analysis of Cardiovascular Function*, Advances in Experimental Medicine and Biology 1065, https://doi.org/10.1007/978-3-319-77932-4_31

vascular tone and vascular cell proliferation, respectively. Eventually, this leads to right heart failure. Heritable PAH is caused by a mutation in the bone morphogenetic protein receptor-II (BMPR-II). Female susceptibility to PAH has been known for some time, and most recent figures show a female-to-male ratio of 4:1. Variations in the female sex hormone estrogen and estrogen metabolism modify FPAH risk, and penetrance of the disease in BMPR-II mutation carriers is increased in females. Several lines of evidence point towards estrogen being pathogenic in the pulmonary circulation, and thus increasing the risk of females developing PAH. Recent studies have also suggested that estrogen metabolism may be crucial in the development and progression of PAH with studies indicating that downstream metabolites such as 16α-hydroxyestrone are upregulated in several forms of experimental pulmonary hypertension (PH) and can cause pulmonary artery smooth muscle cell proliferation and subsequent vascular remodelling. Conversely, other estrogen metabolites such as 2-methoxyestradiol have been shown to be protective in the context of PAH. Estrogen may also upregulate the signalling pathways of other key mediators of PAH such as serotonin.

Keywords
Pulmonary arterial hypertension · Vascular remodelling · Sex hormones · Estrogen · Estrogen metabolites · Serotonin

Introduction

Pulmonary arterial hypertension (PAH) is a debilitating disease characterised by an increase in pulmonary artery pressure and pulmonary vascular resistance. The World Health Organization (WHO) categorises PAH patients into five distinct functional classifications (Table 31.1) [116].

Occurrence of both idiopathic and familial PAH ranges from 6–10 cases per million in the population, whereas the number of those patients developing PAH as a result of other illness is estimated to be much higher, although not reported [8]. PAH is an incurable vasculopathy which affects the arteries of the pulmonary circulation and contributes to the morbidity and mortality of adult and paediatric patients with a wide range of lung and heart diseases. PAH is characterised by a progressive obstruction of distal pulmonary arteries and formation of plexiform lesions leading to elevated pulmonary vascular pressures and right heart failure. The pathogenesis of PAH is complex and involves numerous pathophysiological phenotypes including pulmonary artery endothelial cell (PAEC) dysfunction, pulmonary artery smooth muscle cell (PASMC) proliferation, apoptotic resistance, metabolic shift (Warburg effect), impaired angiogenesis, phenotypic transition and chronic inflammation [57, 105, 133].

The non-specific nature of early PAH symptoms, including dizziness, syncope and fatigue, often leads to delayed diagnosis, which is achieved definitively by invasive right heart catheterisation [91]. Consequently, patients exhibit more advanced pathophysiological features and therefore a lower quality of life. To date there is no known cure for PAH although there are a number of treatment options available targeting the main pathways involved in the pathogenesis of PAH.

Endothelin Receptor Antagonists

Endothelin-1 (ET-1) is a 21 amino acid, vasoactive peptide. Under physiological conditions, ET-1 is produced in small amounts, mainly in endothelial cells, acting as an autocrine/paracrine mediator. The biological effects of ET-1 are transduced by two receptor subtypes, the ET_A and ET_B receptors. In the vasculature, the ET_A receptor is mainly located on vascular smooth muscle cells, whilst ET_B receptors are localised on both endothelial and vascular smooth muscle cells. Activation of ET_A and ET_B receptors on vascular smooth muscle results in vasoconstriction [87, 112]. The activation of endothelial ET_B

Table 31.1 Clinical classification of pulmonary hypertension

Classification of pulmonary hypertension
1. Pulmonary arterial hypertension
1.1. Idiopathic PAH (IPAH)
1.2. Heritable PAH (HPAH)
1.4.1. BMPR-II
1.4.2. ALK-1, ENG, SMAD9, CAV1, KCNK3
1.4.3. Unknown
1.3. Drug and toxin induced
1.4. Associated with
1.4.1. Connective tissue disease
1.4.2. HIV infection
1.4.3. Portal hypertension (PPHTN)
1.4.4. Congenital heart disease
1.4.5. Schistosomiasis
1.4.6. Chronic haemolytic anaemia
1.5. Persistent pulmonary hypertension of the newborn (PPHN)
2. Pulmonary hypertension due to left heart disease
2.1 Left ventricular systolic dysfunction
2.2 Left ventricular diastolic dysfunction
2.3 Valvular disease
2.4 Congenital/acquired left ventricular inflow/outflow tract obstruction and congenital cardiomyopathies
3. Pulmonary hypertension due to lung diseases and/or hypoxia
3.1 Chronic obstructive pulmonary disease (COPD)
3.2 Interstitial lung disease
3.3 Other pulmonary diseases with mixed restrictive and obstructive patterns
3.4 Sleep-related breathing behaviour
3.5 Alveolar hypoventilation disorders
3.6 Chronic exposure to high altitudes
3.7 Developmental lung diseases
4. Chronic thromboembolic pulmonary hypertension (CTEPH)
5. Pulmonary hypertension with unclear multifactorial mechanisms
5.1 Hematologic disorders: chronic haemolytic anaemia, myeloproliferative disorders, splenectomy
5.2 Systemic disorders: sarcoidosis, pulmonary histiocytosis, lymphangioleiomyomatosis
5.3 Metabolic disorders: glycogen storage disease, Gaucher disease, thyroid disorders
5.4 Others: tumoral obstruction, fibrosing mediastinitis, chronic renal failure, segmental PH

receptors located on the endothelium stimulates the release of nitric oxide (NO), and prostacyclin [53] prevents apoptosis and plays a minor role in endothelial-dependent vasodilation [115]. The lungs are the major site for both clearance and production of circulating ET-1: up to 50% of circulating ET-1 is cleared through the ET_B receptors [29].

In PAH an upregulation of the ET-1 system occurs. ET-1 plasma levels are elevated and the increase is correlated with right atrial pressure, pulmonary artery oxygen saturation and pulmonary vascular resistance [120]. ET-1 precursor, prepro-ET-1, is also abundantly overexpressed in endothelial cells from patients with PAH [46] and stimulates proliferation of PASMCs and vasoconstriction [145]. ET-1 receptor antagonists, bosentan, macitentan and ambrisentan, are currently approved for the treatment of PAH, preventing the aberrant activity of ET-1 observed in patients. Bosentan and macitentan are dual ET_A and ET_B receptor antagonist, whereas ambrisentan is a selective ET_A receptor antagonist [21].

Prostacyclin Analogues

Prostacyclin is a member of the eicosanoid family of mediators, which include prostaglandins, thromboxanes and leukotrienes. It is a potent vasodilator and inhibitor of platelet aggregation. Prostacyclin is synthesised from arachidonic acid by the actions of cyclooxygenase and prostacyclin synthase, primarily in endothelial cells. In PAH dysregulation of prostacyclin metabolic pathways occurs. A reduction in prostacyclin levels are observed, characterised by a decrease in urinary metabolites [16] and a decrease in prostacyclin synthase expression in the lungs of PAH patients [134]. Several prostacyclin analogues including epoprostenol, iloprost and treprostinil are used clinically in the treatment of PAH [43]. In addition to their vasodilator properties, prostacyclin analogues can also have inhibitory effects on PASMC proliferation and migration [19]. Hence, prostacyclin agonists and analogues are used to treat severe PAH. The most recent member of this drug class, selexipag, a selective prostacyclin receptor agonist, was approved by the Food and Drug Administration (FDA) for PAH in 2015 [69, 114].

Phosphodiesterase-5 (PDE-5) Inhibitors

The cyclic nucleotides, cyclic adenosine monophosphate (cAMP) and cyclic guanosine monophosphates (cGMP) are important mediators of pulmonary vasodilation via activation of protein kinase A (PKA) and protein kinase G (PKG), respectively [90]. NO is a well-established mediator involved in activation of the cGMP pathway. Phosphodiesterases (PDEs) catalyse the hydrolysis of cAMP and cGMP and promote vasodilation by increasing the concentration of both cyclic nucleotides intracellularly. PDE-5 is cGMP selective and is highly expressed in the lung [80]. Inhibitors of PDE-5 promote vasodilation via cGMP and are effective in PAH management. PDE-5 inhibitor drugs approved for the treatment of PAH include sildenafil, tadalafil and vardenafil [44, 107].

Soluble Guanylate Cyclase Stimulator

Soluble guanylate cyclase (sGC) is the primary NO receptor. On binding to NO, sGC catalyses the synthesis of cGMP, which promotes vasodilation but also inhibits vascular smooth muscle cell proliferation, leukocyte recruitment, platelet aggregation and vascular remodelling. There is dysregulation of NO production, sGC activity and cGMP degradation in PAH [119]. This enzyme is a direct target to activate NO-sGC-cGMP pathway, and targeting of sGC stimulation by riociguat has consistently improved response to exercise following treatment in pulmonary hypertension patients [74]. This occurs via vasodilation within the pulmonary arterial bed, thus improving haemodynamics and vascular tone.

Sex As a Risk Factor in PAH

In the 1950s, Dresdale first documented an increased frequency of female PAH patients compared to men [26]. The increased prevalence of PAH in females remains, evidenced by recent epidemiological studies highlighting that 70–80% of PAH patients are female [6]. This is in contrast to other demographics, which have changed over time, i.e. age of diagnosis ranges from 50 to 65 years in contemporary registries compared to 36 years previously reported [54]. As the age of diagnosis has increased, patients are presenting with more co-morbidities and are therefore more difficult to manage clinically. However, reasons underlying the imbalanced female-to-male ratio regarding disease prevalence remain obscure. Conversely, male PAH patients exhibit poorer survival than female patients [59], leading to investigation of the phenomenon known as the 'sex paradox' in pulmonary hypertension. Recent studies suggest that males have poorer adaptive remodelling of the right ventricle (RV) in response to increased afterload – this may explain the sex differences in survival in PAH [59]. Such paradoxical observations suggest a role for complex sex hormone signalling and processing pathways in the development and progression of PAH [38]. This sex paradox also seems to be dependent on the age of PAH patients with the age of onset being earlier in women than men, perhaps due to altered estrogen production at menopause. The influence of sex hormones in women presenting with PAH is further suggested by the onset of PAH during pregnancy and/or post-partum period [122]. Whilst treatment guidelines now suggest that female PAH patients do not take estrogen-based contraceptives [50], current PAH therapies do not take sex bias into account despite recent studies having shown sex differences in treatment response [41, 113]. For example, Gabler and colleagues reported that ET-1 receptor antagonists had greater efficacy in females than males [41]. A more recent study found increased likelihood of response to PDE-5 inhibitor, tadalafil, in the treatment of PAH in men compared with women [86]. Sex-specific heterogeneity in treatment response may reflect differences in PAH pathobiology and afford the opportunity to inform individual treatment decisions and provide the basis for exploring potential differences in mechanisms of disease between sexes.

Sex Differences in the BMPR-II Signalling Pathway in Pulmonary Hypertension

Bone morphogenetic protein (BMP) receptor type II (BMPR-II) is a serine/threonine transmembrane type II receptor kinase, belonging to the transforming growth factor (TGF)-β super family. BMPR-II is involved in processes such as development, embryogenesis and adult tissue homeostasis. BMP ligands cause the formation of a heterodimer between BMPR-II and a type I receptor. Following heterodimer formation and the phosphorylation of the type I receptor, there is phosphorylation of Smad 1/5/9 proteins which in turn leads to the upregulation of the expression of nuclear inhibitors of differentiation 1 and 3 (Id1/Id3); both Id1 and Id3 act to repress proliferation. The proliferation, migration and apoptosis of both endothelial cells and smooth muscle cells within the vasculature can be influenced by BMP signalling [45]. Mutations in the gene encoding for BMPR-II is the most well-established risk factor for heritable PAH. Over 300 different mutations sites within the BMPR-II gene have been identified in PAH patients and are associated with PAH development and progression. Approximately 75% of familial PAH cases and 20% of idiopathic PAH cases are associated with a loss-of-function mutation in this gene [79]. The penetrance of BMPR-II mutations is particularly low, only approximately 20% of male carriers develop PAH, whilst approximately 45% of female carriers develop the disease [100]. Therefore, other genetic and environmental risk factors must exist. Alternative splicing of BMPR-II is thought to play a role in the penetrance of the disease. BMPR-II can be alternatively spliced producing two different isoforms – the full transcript and one lacking exon 12. Exon 12 is the largest exon within BMPR-II and encodes a cytoplasmic domain [78]. The ratio of the isoform-B to isoform-A is increased in PAH patients when compared to unaffected BMPR-II mutation carriers. Therefore it has been suggested that expression of isoform B is involved in the penetrance of the disease [20].

A reduction in BMPR-II signalling and therefore a reduction in the anti-proliferative Id1/3 lead to increased proliferation of the PASMCs [96] and subsequent pathological remodelling of the pulmonary vasculature. It has been demonstrated that PAH patients harbouring a BMPR-II mutation develop the disease at an earlier age and have worse haemodynamic profiles at the time of diagnosis than those patients without a BMPR-II mutation. As a result, BMPR-II mutation carriers have an increased mortality [35].

It is well established that a greater number of females develop PAH. There are sex-specific differences in the expression and activity of BMPR-II, even under basal conditions, which could contribute to the greater number of females within the PAH patient population. Female PASMCs obtained from non-PAH subjects have been shown to have reduced expression of BMPR-II at both gene and protein levels when compared to male PASMCs obtained from non-PAH subjects [83]. Furthermore, the downstream factors in the BMPR-II signalling pathway, Smad 1 and Id1/3, were also downregulated in female PASMCs. This suggests that females could be predisposed to PAH because they have basally lower levels of BMPR-II. Interestingly, when male non-PAH

PASMCs were stimulated with estrogen, no changes were observed in the expression of BMPR-II and Smad 1, yet there was a reduction in mRNA and protein levels of the Id genes [83]. Therefore, estrogen could be responsible for driving the suppression of the BMPR-II pathway and the predisposition of females to PAH. Other sex-specific differences in BMPR-II expression have been highlighted in PAH patients. BMPR-II gene expression in lymphocytes was approximately 20% lower in female patients than in male patients [4]. When comparing whole lung BMPR-II expression in male and female mice, BMPR-II gene expression was lower in the female compared to male mouse. The same study also highlighted an estrogen response element (ERE) in the BMPR-II promotor. This ERE was found to be a highly conserved functional binding site for estrogen receptor alpha (ERα) [4]. Therefore, it is suggested that estrogen is able to reduce the expression of BMPR-II through the direct binding of ERα to the BMPR-II promoter. In addition, we have demonstrated decreased BMPR-II signalling in lungs from the Sugen/hypoxic rat model of pulmonary hypertension [82]. Inhibition of estrogen synthesis with an aromatase inhibitor reversed the experimental pulmonary hypertension in the female rats and restored the deficient BMPR-II signalling [82], suggesting that endogenous estrogen can supress BMPR-II signalling in vivo. This strengthens the hypothesis that estrogen can cause sex-specific differences in BMPR-II expression, which ultimately results in the predisposition of females to PAH. Despite dysfunctional BMPR-II signalling being well characterised in the lung, the downstream effects in the RV are not well understood. A recent clinical study suggested BMPR-II mutation carriers have poorer RV function despite similar afterload and cardiac adaptation in mutation carrier and noncarrier groups [137]. This study suggests that BMPR-II signalling can influence RV dysfunction in PAH. Currently, little is known regarding sex-specific differences in BMPR-II expression in the RV and the potential effects this might have on RV function in PAH.

The Influence of Sex Hormones in Pulmonary Hypertension

A recent comprehensive study by Ventetuolo and colleagues characterised the relationship of sex and haemodynamics in PAH [140]. Ventetuolo et al. also recently demonstrated robust clinical data outlining significantly increased plasma estrogen levels in male idiopathic PAH patients vs control subjects [138]. Differential effects of estrogen on the pulmonary circulation and RV and/or differential metabolism of estrogen to mitogenic or anti-proliferative metabolites may well explain the differences observed in outcomes of animal models as well as the notable sex difference in the incidence of the disease.

One hypothesis that may explain the female susceptibility in PAH is that the female hormone estrogen is pathogenic in the pulmonary circulation. The estrogen pathway is known to be affected in diseases such as systemic lupus erythematosus and breast cancer, both of which have a higher female prevalence [101]. Progesterone, testosterone, DHEA, as well as estrogen and metabolites have been shown to affect pulmonary vascular tone and cell proliferation [68]. Male rodents with hypoxia- or monocrotaline-induced PH develop more severe experimental PH than females, whilst there is no sex bias in Sugen/hypoxic rodent models [81, 136]. In experimental models of PH (hypoxia-induced and monocrotaline-induced) where male animals are dosed with exogenous estrogen, this exerts protective, vasodilatory and anti-proliferative effects [67]. Exogenously administered estrogen can improve RV function in the Sugen/hypoxic model of PH by stimulating RV contractility and protecting against pulmonary vascular remodelling [75]. However, inhibiting endogenous estrogen with an aromatase inhibitor reverses Sugen/hypoxic-induced PH in females and in BMPR-II mutant mice without affecting RV function implying that local endogenous estrogens are protective [14, 23, 82]. However, eliminating endogenous circulating estrogen via ovariectomy also abolishes the PH phenotype in female-susceptible transgenic mice including serotonin transporter (SERT) overexpressing (SERT+) mice [142], Smad $1^{+/-}$ mice [83] and dexfenfluramine-treated mice [24].

The Role of Sex Hormones and Sex Hormone Metabolites in Pulmonary Arterial Hypertension

Estrogen Biosynthesis and Metabolism

The female sex hormones are steroid hormones that comprise both estrogens and progesterones. Steroid hormones have multifactorial functions throughout the body and are synthesised from cholesterol in the gonads, adrenal glands and placenta and to a lesser extent in the adipose tissue, liver and skin [65]. Estrogens are the primary female sex hormone involved in development and maintenance of the female reproductive system. Evidence also exists for a significant role for estrogens in other systems including the cardiovascular system [65]. Synthesis of estrogen can also take place in the liver and adipose tissue, and the reported expression of estrogen-synthesising enzymes in vascular smooth muscle cells and endothelial cells [125] implicates the importance of 'local' estrogen synthesis and autocrine/paracrine effects out with the reproductive organs.

Estrogens exist as three major naturally occurring isoforms: estrone, estrogen and estriol. Estrogen is the predominant circulating hormone in premenopausal women, whilst estrone is important during the menopause and estriol during pregnancy. In the follicular phase, circulating concentrations of estrogen are about ~0.4 nM, and during ovulation these levels rise to ~2.2 nM. After menopause, estrogen levels are reduced up to 20-fold [89]. The biosynthesis of estrogen is initiated by the synthesis of androstenedione from the precursor cholesterol. Androstenedione provides an intermediate stage in metabolism from which estrogen can be synthesised by cytochrome P450 enzyme (CYP) 19A1 (aromatase) which converts androstenedione to estrone, which in turn is converted to estrogen by 17β hydroxysteroid dehydrogenase 1 (17βHSD1). Alternatively, reduction of androstenedione to testosterone can occur via 17β hydroxysteroid dehydrogenase 2 (17βHSD2), followed by aromatisation of testosterone to estrogen.

As a predominantly female hormone, estrogen is assumed detrimental in generating the high prevalence of female PAH patients. For example, polymorphisms in aromatase, the estrogen synthesising enzyme, and the gene for ERα (*Esr1*) are associated with elevated estrogen levels in the lungs of female patients and predispose to porto-pulmonary hypertension [108, 143]. In agreement, heightened exposure to exogenous estrogen through both use of oral contraceptives and hormone replacement therapy is associated with PAH [85]. On the contrary, there is evidence to suggest hormone replacement therapy can prevent the development of PH in patients with systemic sclerosis [10].

Aromatase

Certain tissues can modulate their own estrogenic milieu by the local conversion of testosterone or androstenedione to estrogen and estrone, by the activity of aromatase which is expressed within the pulmonary arteries [49, 82]. Circulating aromatase precursors act as a reservoir for the synthesis of estrogens within extra-gonadal sites, namely, androstenedione, dehydroepiandrosterone (DHEA) and DHEA-sulphate (DHEA-S) [117]. Locally synthesised estrogens can act in a paracrine or autocrine fashion creating a potentially potent estrogenic milieu that may be superior to the effects of circulating levels. The aromatase gene is present in gonadal and extragonadal tissues with its site-specific transcription being driven by different signalling pathways at alternative sites in the promoter region.

Expression of aromatase has also been defined within the lung tissue giving rise to the presence of a local pulmonary vasculature estrogenic microenvironment. The aromatase inhibitor anastrozole has shown remarkable therapeutic potential in the Sugen/hypoxia rat model and hypoxic rat model where it successfully reversed the PH phenotype in female rats, and this correlated with reduced plasma estrogen levels and restored BMPR-II signalling [82, 126]. Inhibition of aromatase by anastrozole is now widely recognised for use in estrogen-sensitive breast cancer [39]. The therapeutic potential in PAH of anastrozole has very recently been shown to be safe and effective in a small proof-of-concept study [63], where anastrozole significantly reduced estrogen levels and improved 6-min walk distance (6MWD). These results are promising and suggest that longer and larger Phase II trials of anastrozole are warranted.

Oxidative Metabolism of Estrogens: Cytochrome P450 Enzymes

Estrogen can be rapidly metabolised by oxidation. Oxidative metabolism occurs primarily in the liver producing non-estrogenic metabolites ready for elimination from the body. Estrogen and estrone are in equilibrium with 17βHSD1, and oxidation/reduction occurs at the carbon (C)-17 position and favours formation of estrone. Further metabolism of estrogen occurs, for example, at the C16, C4 and C2 positions producing biologically active metabolites, including 16-α-hydroxyestrone (16αOHE1). CYPs contain an iron (Fe^{3+}) docking site for oxygen, which in the presence of NADPH is reduced (Fe^{2+}). CYP enzymes are critically involved in xenobiotic metabolism, as well as metabolism of eicosanoids and steroids and cholesterol synthesis, amongst others. Metabolism by the activity of CYP enzymes has a genotoxic capacity via the generation of reactive oxygen intermediates that can interact with nucleic acids and proteins and initiate tumorigenesis [99]. Specifically, CYP1A1, CYP1A2, CYP1B1 and CYP3A4 appear critical in the formation of 2- and 4-hydroxy derivatives of estrogen. The 2-hydroxylation pathway is the major metabolic pathway in the liver, whereas 4-hydroxylation constitutes a relatively minor pathway. The 2-and 4-hydroxy derivatives are then further converted/deactivated to 2- and 4-methoxy metabolites by the catechol-O-methyltransferase (COMT), reducing the capacity to redox cycle/generate quinones and semiquinones. In a regulatory feedback mechanism, estrogen can promote its own breakdown by driving the transcription of the human CYP1B1 gene via an ERE in the promoter region [132].

CYP1B1 in PH

There is evidence that CYP1B1 may play an important role in the pathogenesis of PAH. CYP1B1 is highly expressed in pulmonary arteries from patients with PAH, and the CYP1B1 inhibitor, 2,3′,4, 5′-tetramethoxystilbene (TMS), can reverse PH in Sugen/hypoxic and hypoxic models of PH ([143]). In addition it increases survival in the monocrotaline rat model of pulmonary hypertension [60]. TMS also abolished estrogen-induced proliferation in hPASMCs from controls and patients [143]. Genetic polymorphisms in CYP1B1 and the estrogen metabolites produced by this enzyme have also been identified as factors influencing the penetrance of the disease in individuals with BMPR-II mutations [3]. Indeed a CYP1B1 SNP in tight linkage disequilibrium with SNPs associated with pulmonary hypertension has been identified, suggesting these pathways may underpin sexual dimorphism in RV failure [139].

Serotonin and dexfenfluramine can also increase the expression of CYP1B1 in PASMCs [24, 60, 142]. Indeed, TMS and/or CYP1B1 deletion in mice reverses the development of PH in SERT+ mice or dexfenfluramine-treated mice [24, 60]. Dexfenfluramine upregulated expression of CYP1B1 in PASMCs from PAH patients and the associated dexfenfluramine-mediated proliferation was ablated by the CYP1B1 inhibitor, TMS. Estrogen also increased the expression of CYP1B1 in PAH-derived PASMCs [24].

The enzyme CYP1B1 is involved in both the metabolism of estrogen and arachidonic acid. In the lungs, epoxyeicosatrienoic acids (EETs) and hydroxyeicosatetraenoic acids (HETEs) have inflammatory and mitogenic effects. 20-HETEs increase NADPH oxidase (Nox)-derived reactive oxygen species (ROS) production in bovine PAECs [15, 88] and are implicated in the development of hypoxic PH [148]. Notably, EETs have opposite effects in the systemic circulation, which may explain the plausible divergent effects of estrogen in the systemic versus the pulmonary circulation [148].

16α-Hydroxyestrone in the Development of PH

16αOHE1 has been shown to be a biologically active and significantly more estrogenic than estrogen at the classical estrogen receptors ERα and ERβ [4, 97]. Urinary levels of 16αOHE1 were elevated in a hypoxia-induced PH model, and when administered to mice, 16αOHE1 caused a significant increase in RV hypertrophy (RVH), pulmonary artery remodelling and RV systolic pressure (RVSP) [143]. 16αOHE1 also induced proliferation in PASMCs confirming this pro-proliferative phenotype in the human setting [143]. Indeed, an increase in plasma 16αOHE1 levels have been demonstrated in 4,40-methylenedianiline (DAPM)-induced PH in rats providing further evidence for a role for 16αOHE1 in the development of experimental PH [13]. Austin et al. [3] established a link between altered estrogen metabolism and penetrance in heritable PAH. Here, a decreased ratio of 2-hydroxyestradiol/16αOHE1 was observed in affected BMPR-II mutation carriers vs unaffected BMPR-II mutation carriers, thus suggesting a shift to the pro-proliferative phenotype observed in heritable PAH [3]. The molecular mechanism for 16αOHE1-induced proliferation appears to be complex and is, at least in part, mediated by ROS. 16αOHE1 induces superoxide and hydrogen peroxide production and downregulates the protective effects of nuclear factor erythroid-derived like factor 2 (Nrf-2) through Nox1 and ERα [56]. Together, these studies further emphasise the potential role 16αOHE1 may play in the development of PAH.

2-Methoxyestradiol As an Anti-proliferative Metabolite

There is an increasing body of evidence suggesting that some estrogen metabolites can be protective in the setting of PAH. The 2-methoxyestrogens are perhaps the best example of the estrogen metabolites known to have beneficial effects in the pulmonary vasculature. Indeed, 2-methoxyestradiol (2ME2) has been studied extensively in the context of many cancers, and much research has centred around the metabolites' anti-proliferative and potentially pro-apoptotic nature within this setting. 2ME2 is metabolised from its precursor metabolite 2-hydroxyestradiol (2OHE2) by the enzyme, COMT. 2ME2 is also readily converted to its less biologically active form 2-methoxyestrone (2ME1) by 17βHSD2 [125].

It is only within the last decade or so that several studies have outlined the potential beneficial effects of 2ME2 in the context of pulmonary hypertension. Tofovic et al. have demonstrated that 2ME2 reversed indices such as RVH, RVSP and pulmonary artery remodelling and increased survival in the monocrotaline and bleomycin models of experimental pulmonary hypertension [127, 129]. 2ME2 has also been shown to reduce human PASMC and human lung fibroblast proliferation in a concentration-dependent manner [130], thus demonstrating that 2ME2 has beneficial effects in PASMC remodelling as well as pulmonary fibrosis and associated adventitial remodelling. Indeed, further preclinical studies have shown that, in the monocrotaline-induced PH model, combination therapy of 2ME2 and sildenafil significantly reduced pulmonary artery remodelling and inflammatory responses as well as increasing survival [128]. 2-Ethoxyestradiol, a synthetic analogue of 2ME2, also attenuates experimental PH and is ten times more potent than 2ME2 both in vitro and in vivo [131]. The exact mechanism to which 2ME2 exerts its potent anti-proliferative effect are unclear. However, there is evidence to suggest that 2ME2 reduces endothelial cell production of the potent mitogen ET-1 through alterations in mitogen-activated protein kinase (MAPK) activity; this reduction in ET-1 was also demonstrated to be ERα and ERβ independent [27]. Further work is required to elucidate the exact mechanism of action of 2ME2 and whether it may well be useful in a clinical setting.

Other Potential Protective Estrogen Metabolites

To date there is limited data demonstrating other beneficial estrogen metabolites; however, some studies have shown that the hydroxy metabolite 2OHE2 can reverse monocrotaline-induced PH in male rats through a reduction in RVH and pulmonary artery remodelling [129]. 2OHE2 has anti-proliferative effects in rat cardiac fibroblasts, and it has been demonstrated that the anti-proliferative effect of 2OHE2 is mediated through conversion to 2ME2 by COMT. Furthermore, 2ME2 and 2OHE2's anti-proliferative effects are independent of ERα and ERβ [27]. 4-Hydroxyestradiol (4OHE2) is anti-proliferative in male human PASMCs [83]. Interestingly, 4OHE2 is pro-proliferative in the breast adenocarcinoma cell line, MCF-7, suggesting possible tissue- and cell-specific effects [111].

Other Sex Hormones of Interest

Although estrogen and its metabolites have been the focus of research, other sex hormones are biologically active in the pulmonary vasculature. For example, the female steroid hormone progesterone has been shown to be vasodilatory in systemic and pulmonary arteries mediated through an endothelium-dependent NO-induced induction of cGMP as well as inhibition of voltage-gated and receptor-operated calcium channels [47, 73]. Furthermore, progesterone is anti-proliferative in aortic smooth muscle cells [72]; however, whether it is also anti-proliferative in the pulmonary vasculature is unknown. Preclinically, progesterone attenuates monocrotaline-induced pulmonary hypertension in rats by reducing RVSP, RVH and vascular remodelling whilst also significantly increasing survival [124]. However, the effects of progesterone on the pulmonary vasculature in the clinical setting are as yet unknown and an avenue open to future investigation.

The male sex hormone testosterone has also been postulated to be beneficial in the setting of pulmonary hypertension, not only because fewer males are afflicted with the disease but also because it has been shown to be a potent vasodilator in isolated pulmonary arteries of both rats [34] and humans [118]. However, in human pulmonary arteries, the vasodilatory property of testosterone is much more efficacious in males [110]. Indeed, Jones et al. have demonstrated that testosterone is likely to exert this vasodilatory effect via calcium antagonistic effects on voltage-gated calcium channels independent of the androgen receptor [62]. However, in the cardiac setting testosterone has been shown to promote maladaptive RV remodelling and fibrosis, suggesting a detrimental effect on overall RV function in the context of increased afterload [51]. This observation is in line with current clinical data where males with PAH have worse clinical outcomes than females [9, 58]. There is little clinical data regarding testosterone treatment in PAH; however, there were no differences observed in plasma testosterone levels between male healthy and PAH patients in a recent study by Ventetuolo and colleagues [138].

DHEA, the precursor steroid hormone to testosterone, can inhibit hypoxia-induced vasoconstriction of the pulmonary arteries via opening of Ca^{2+}-activated large-conductance potassium channels (BK_{Ca}) [36]. In vivo, DHEA also prevents and reverses hypoxia-induced PH in rats by increasing the expression and functionality of BK_{Ca} channels to increase vasodilation of the pulmonary arteries [11]. Indeed, other groups have demonstrated the beneficial effects of DHEAon the pulmonary vasculature [102] as well as its ability to improve cardiac endpoints such as increasing tricuspid annular plane systolic excursion (TAPSE), RV internal diameter during diastole and cardiac index (CI) ultimately increasing right ventricular function [2]. Interestingly, DHEA has also been shown to inhibit Src/STAT3 activation and upregulate BMPR-II expression in PASMCs isolated from human PAH patients, further outlining its protective effects in the pulmonary vasculature [104]. Preliminary clinical studies involving DHEA have been promising; for example, in a small clinical trial, oral DHEA treatment improved the 6MWD and pulmonary haemodynamics in COPD-induced PH [28]. Furthermore, lower plasma DHEA-S levels have been found in male PAH patients, further suggesting a potential protective role for the hormone in the pulmonary vasculature [138].

Estrogen Receptors and Sex-Specific Differences

It has become apparent that estrogen receptors impact a multitude of biological functions including inflammatory response, cardiovascular function [33] and bone homeostasis [7]. The majority of estrogenic activity is via canonical estrogenic signalling. Upon high-affinity binding with estrogen, these nuclear receptors act as transcription factors regulating gene activity [71]. Estrogen signalling targets three main classes of estrogen receptors: the nuclear ERα, ERβ and a transmembrane G protein-coupled estrogen receptor (GPER). All are located in various target organs mediating genomic and non-genomic estrogen signalling. Significantly, in PAH these receptors localise to the PASMCs that may account for estrogen-dependant proliferation and remodelling ([146]). Structurally, ERα and ERβ both display sequence homology comprising of five distinct domains; two activation domains within the n-terminal and ligand-binding domains regulate the transcriptional activity of ER [66]. Due to alternative splicing, several isoforms of each also exist with at least three ERα and five ERβ isoforms discovered to date.

ERα is highly expressed in female hPASMCs from PAH patients and mediates estrogen-induced proliferation via MAPK and Akt signalling [146]. High levels of ERα lead to human PASMC proliferation and remodelling, characteristic in PAH, through various canonical signalling pathways [5]. The second receptor ERβ may play a more beneficial role in protection against PAH modulating angiogenesis through the production of various angiogenic factors such as vascular endothelial growth factor (VEGF) and NO (de [22]) upon estrogen treatment. Interestingly use of an ERβ agonist induced reversal of PAH symptoms with a prevention of the savour by application of ERβ antagonists. Therefore, postulations suggest that targeting ERβ in future therapeutic strategies may lead to the development of PAH treatments [135].

The third class and more recently discovered transmembrane GPER binds estrogen, causing rapid non-genomic effects [146]. In vascular physiology, GPER initiates rapid non-genomic effects via opening of ion channels and activation of various signalling enzymes. Studies have shown that GPER may have a beneficial role in the pulmonary vascualture and RV [68]. The potential beneficial effects of GPER include anti-fibrotic effects, enhanced cardiac contractibility and most notably the ability to mediate the protective effects of estrogen. Whilst GPER agonists do not exert a proliferative effect on hPASMCs, they have been shown to reduce RV overload in PH rats, and so GPERs may play a role in the development of experimental PH [1, 146].

Serotonin and Sex Differences in Pulmonary Hypertension

5-Hydroxytryptamine (5-HT), or 'serotonin', was chemically identified by Rapport et al. [106] as one of the major vasoconstricting substances in defibrinated blood, originating from platelets [106]. Serotonin was subsequently found to also act as a major neurotransmitter, being involved in a variety of processes carried out by the central nervous system, as well as regulating several functions in the periphery. The biosynthesis pathway converts dietary tryptophan to 5-hydroxytryptophan by the action of the enzyme tryptophan hydroxylase (TPH). This process is the rate-limiting step in the generation of serotonin. Two isoforms of TPH exist; TPH1 is responsible for synthesis of serotonin in the periphery, whilst TPH2 is expressed abundantly in the brain [98]. 5-Hydroxytryptophan is then decarboxylated by a ubiquitous amino acid decarboxylase, resulting in the formation of serotonin. The physiological effects of serotonin are then mediated by 14 different 5-HT receptor subtypes. These receptors are divided into seven distinct classes (5-HT_1–5-HT_7), mainly on the basis of their structural and functional characteristics. Serotonin is also a substrate for SERT, which actively uptakes serotonin into cells. In the lungs, serotonin is locally released from pulmonary neuroendocrine cells and neuroepithelial cell bodies distributed throughout the airways. Synthesis of serotonin can also occur in PAECs. Furthermore, the lungs play an important role in the removal of serotonin from the circulation, with as much as 95% being taken up or inactivated. Under normal circumstances, pulmonary tissue is exposed to low levels of serotonin as most serotonin is stored in platelets, thus removing it from the circulation [42, 123, 144]. However, under hypoxic conditions [61], and during situations involving mechanical strain [103], large amounts of serotonin are secreted. In the pulmonary circulation, serotonin promotes PASMC proliferation, vasoconstriction and thrombosis, processes involved in the development of PAH.

An increased incidence of PH due to the use of the anorexigenic drugs aminorex and dexfenfluramine, SERT substrates and indirect serotonergic agonists stimulated interest in the role of serotonin in PAH, leading to the 'serotonin hypothesis of PAH'. Aminorex increases plasma levels of serotonin by inducing its release from platelets and attenuating its breakdown [37, 147]. Dexfenfluramine also increases levels of circulating serotonin and interacts with SERT, stimulating the release of serotonin from platelets and inhibiting its reuptake [12, 40].

A body of evidence now exists further implicating serotonin and the development of PAH. For instance, some cohorts of patients with PAH have elevated plasma levels of serotonin, in addition to a decrease in the levels of serotonin stored in platelets due to platelet pool storage disease [52]. Furthermore, in PAECs from PAH patients, increased TPH1 expression and serotonin synthesis has been observed, and this is thought to contribute to increased PASMC proliferation [31]. Knockdown of the TPH1 gene also has beneficial effects in animal models of PH [92, 95]. Indeed, synthesis of serotonin has also been associated with the development of the Sugen/hypoxia model of PH. Inhibition of the VEGF receptor by Sugen 5416 combined with hypoxia generates a preclinical model of PH that recapitulates human PAH more closely than previously characterised models [48]. This model is associated with increased expression of TPH1, the rate-limiting enzyme in serotonin synthesis [17, 18].

5-HT receptors, in particular the 5-HT1B receptor and SERT, contribute to the actions of serotonin in PH. Serotonin is a potent vasoconstrictor and in human pulmonary arteries mediates its effects via the 5-HT1B receptor [93], which is also overexpressed in PASMCs from female PAH patients [141] and in the pulmonary circulation of larger experimental models of PH such as pigs [109]. In addition to the role of 5-HT1B receptors in vasoconstriction, they also play a role in mediating cellular proliferation and vascular remodelling. Several cellular studies have demonstrated the mitogenic effects of serotonin on pulmonary artery endothelial, smooth muscle and fibroblast cells. Moreover, in experimental models of PH genetic ablation and pharmacological inhibition of the 5-HT1B receptor attenuate PH and inhibit pulmonary vascular remodelling [55, 64]. SERT is also critical in mediating the mitogenic effects of serotonin in the pulmonary circulation. Augmented proliferative capabilities of PASMCs from patients with PAH have been associated with increased expression levels of SERT [32, 84]. The enhanced proliferation of these cells to 5-HT or serum can be abolished by SERT inhibitors such as citalopram and fluoxetine [84]. In PAH patients, SERT expression is also located mainly in the medial layer of remodelled vessels [32]. Hypoxia causes increased SERT expression and activity, augmenting mitogenic responses in PASMCs [30]. Serotonin-induced proliferation requires SERT-dependent generation of ROS, activation of the extracellular signal-regulated kinase (ERK) pathway and the induction of the transcriptional factor GATA-4. Activation of the Rho kinase pathway by internalised serotonin may also play a role in SERT-mediated proliferation [70, 76, 121]. Both the serotonin receptors and SERT can also interact in a cooperative fashion to mediate serotonin-induced contraction and proliferation of PASMCs [70, 76, 94]. Through its actions via SERT and its receptors, serotonin plays a role in the development of PAH, contributing to both vasoconstriction and vascular remodelling. Serotonin can also influence other pathways associated with PH and may act as a 'second hit' as exogenous administration of serotonin uncovers a PH phenotype in BMPR-II $^{+/-}$mice [77].

Models of PH have now been characterised that mimic the female susceptibility to PH observed in humans. These models all involve serotonin-dependent mechanisms and include SERT+ mice, mice overexpressing the calcium-binding protein S100A4/mts1 (that function downstream of serotonin) and mice dosed with the indirect serotonergic agonist dexfenfluramine [24, 25, 142]. In each of these models, a PH phenotype is only observed in female animals and is estrogen-dependent. For instance, removal of female sex hormones by ovariectomy reverses the PH phenotype in SERT+ mice, and the phenotype can be reinstated by replacing estrogen in these animals. Serotonin mechanisms may also be important in regulating estrogen metabolism as SERT overexpression and dexfenfluramine can increase CYP1B1 expression [24, 60]. Furthermore, estrogen has been shown to increase the expression of TPH1, SERT and the 5-HT1B receptor in human PASMCs [24, 142]. Indeed, increased 5-HT1B receptor expression in human PASMCs for female PH patients may be the consequence of an estrogen-induced decrease in miRNA-96 expression and contribute to the development of PH in females [141].

Concluding Statement

Female sex is now recognised as a significant risk factor in the development of PAH. An increasing body of evidence now suggests that estrogen and some of its metabolites may play a role in the pathogenesis of PAH as well as experimental PH (Fig. 31.1). Interactions of estrogen with serotonin and other key pathways may contribute to the development of PH.

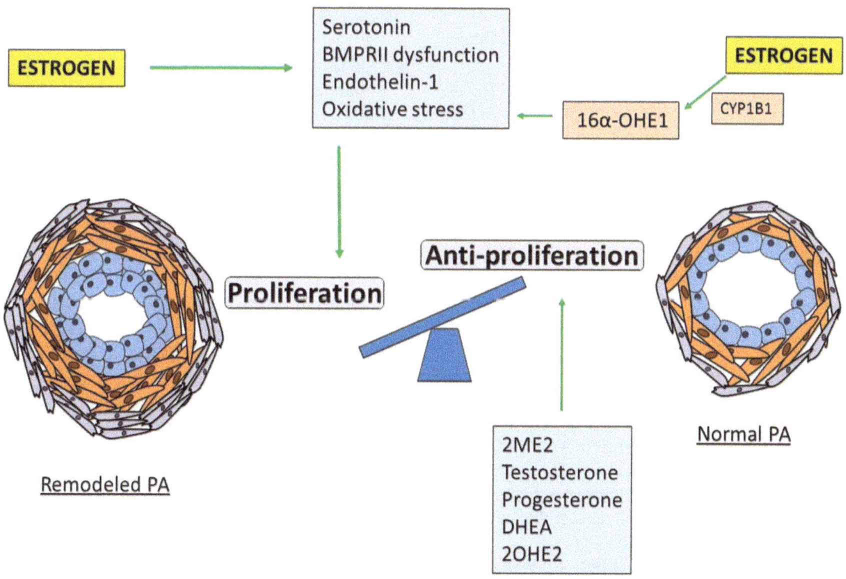

Fig. 31.1 The potential effects of sex hormones and estrogen metabolites on pulmonary artery (PA) remodelling. In remodelled PAs, excess proliferation is observed in endothelial cells (blue), smooth muscle cells (orange) and fibroblasts (purple) eventually leading to reduced luminal area. Estrogen itself can increase pulmonary artery smooth muscle cell proliferation; however, it has also been postulated to influence the serotonin and endothelin-1 pathways as well as reducing BMPR-II signalling, all of which are recognised to be pro-proliferative mechanisms. The cytochrome P450 1B1 isoform (CYP1B1) enzyme converts estrogen to 16-α-hydroxyestrone (16αOHE1), a metabolite identified to be pro-proliferative by increasing oxidative stress responses. Other sex hormones such as testosterone, dehydroepiandrosterone (DHEA) and progesterone are known to be anti-proliferative and have been shown to reverse experimental PH. Some estrogen metabolites are also known to be anti-proliferative and reverse experimental PH such as 2-hydroxyestradiol (2OHE2) and 2-methoxyestradiol (2ME2)

References

1. Alencar AK, Montes GC, Montagnoli T, Silva AM, Martinez ST, Fraga AG, Wang H, Groban L, Sudo RT, Zapata-Sudo G. Activation of GPER ameliorates experimental pulmonary hypertension in male rats. Eur J Pharm Sci. 2017;97:208–17.
2. Alzoubi A, Toba M, Abe K, O'Neill KD, Rocic P, Fagan KA, Mcmurtry IF, Oka M. Dehydroepiandrosterone restores right ventricular structure and function in rats with severe pulmonary arterial hypertension. Am J Physiol Heart Circ Physiol. 2013;304:H1708–18.
3. Austin ED, Cogan JD, West JD, Hedges LK, Hamid R, Dawson EP, Wheeler LA, Parl FF, Loyd JE, Phillips JA 3rd. Alterations in oestrogen metabolism: implications for higher penetrance of familial pulmonary arterial hypertension in females. Eur Respir J. 2009;34:1093–9.
4. Austin ED, Hamid R, Hemnes AR, Loyd JE, Blackwell T, Yu C, Phillips Iii JA, Gaddipati R, Gladson S, Gu E, West J, Lane KB. BMPR2 expression is suppressed by signaling through the estrogen receptor. Biol Sex Differ. 2012;3:6.
5. Awad KS, West JD, De Jesus Perez V, Maclean M. Novel signaling pathways in pulmonary arterial hypertension (2015 Grover conference series). Pulm Circ. 2016;6(3):285–94.
6. Badesch D, Raskob G, Elliott C, Krichman A, Farber H, Frost A, Barst R, Benza R, Liou T, Turner M, Giles S, Feldkircher K, Miller D, Mcgoon M. Pulmonary arterial hypertension baseline characteristics from the reveal registry. Chest. 2010;137:376–87.
7. Bado I, Gugala Z, Fuqua SAW, Zhang XH. Estrogen receptors in breast and bone: from virtue of remodeling to vileness of metastasis. Oncogene. 2017;36:4527–37.
8. Barst RJ. Pulmonary hypertension: past, present and future. Ann Thorac Med. 2008;3:1.
9. Benza RL, Miller DP, Gomberg-Maitland M, Frantz RP, Foreman AJ, Coffey CS, Frost A, Barst RJ, Badesch DB, Elliott CG, Liou TG, Mcgoon MD. Predicting survival in pulmonary arterial hypertension: insights from the registry to evaluate early and long-term pulmonary arterial hypertension disease management (REVEAL). Circulation. 2010;122:164–72.
10. Beretta L, Caronni M, Origgi L, Ponti A, Santaniello A, Scorza R. Hormone replacement therapy may prevent the development of isolated pulmonary hypertension in patients with systemic sclerosis and limited cutaneous involvement. Scand J Rheumatol. 2006;35:468–71.
11. Bonnet S, Dumas-De-La-Roque E, Begueret H, Marthan R, Fayon M, Dos Santos P, Savineau JP, Baulieu EE. Dehydroepiandrosterone (DHEA) prevents and reverses chronic hypoxic pulmonary hypertension. Proc Natl Acad Sci U S A. 2003;100:9488–93.
12. Buczko W, De Gaetano G, Garattini S. Effect of fenfluramine on 5-Hydroxytryptamine uptake and release by rat blood platelets. Br J Pharmacol. 1975;53:563–8.
13. Carroll-Turpin M, Hebert V, Chotibut T, Wensler H, Krentzel D, Varner KJ, Burn BR, Chen YF, Abreo F, Dugas TR. 4,4′-Methylenedianiline alters serotonergic transport in a novel, sex-specific model of pulmonary arterial hypertension in rats. Toxicol Sci. 2015;147:235–45.
14. Chen X, Austin ED, Talati M, Fessel JP, Farber-Eger EH, Brittain EL, Hemnes AR, Loyd JE, West J. Oestrogen inhibition reverses pulmonary arterial hypertension and associated metabolic defects. Eur Respir J. 2017;50:1602337.
15. Choudhary D, Jansson I, Stoilov I, Sarfarazi M, Schenkman J. Metabolism of Retinoids and arachidonic acid by human and mouse cytochrome P4501B1. Drug Metab Dispos. 2004;32:840–7.
16. Christman BW, Mcpherson CD, Newman JH, King GA, Bernard GR, Groves BM, Loyd JE. An imbalance between the excretion of thromboxane and prostacyclin metabolites in pulmonary hypertension. N Engl J Med. 1992;327:70–5.
17. Ciuclan L, Bonneau O, Hussey M, Duggan N, Holmes AM, Good R, Stringer R, Jones P, Morrell NW, Jarai G, Walker C, Westwick J, Thomas M. A novel murine model of severe pulmonary arterial hypertension. Am J Respir Crit Care Med. 2011;184:1171–82.
18. Ciuclan L, Hussey MJ, Burton V, Good R, Duggan N, Beach S, Jones P, Fox R, Clay I, Bonneau O, Konstantinova I, Pearce A, Rowlands DJ, Jarai G, Westwick J, Maclean MR, Thomas M. Imatinib attenuates hypoxia-induced pulmonary arterial hypertension pathology via reduction in 5-hydroxytryptamine through inhibition of tryptophan hydroxylase 1 expression. Am J Respir Crit Care Med. 2013;187:78–89.
19. Clapp LH, Finney P, Turcato S, Tran S, Rubin LJ, Tinker A. Differential effects of stable prostacyclin analogs on smooth muscle proliferation and cyclic AMP generation in human pulmonary artery. Am J Respir Cell Mol Biol. 2002;26:194–201.
20. Cogan J, Austin E, Hedges L, Womack B, West J, Loyd J, Hamid R. Role of BMPR2 alternative splicing in heritable pulmonary arterial hypertension penetrance. Circulation. 2012;126:1907–16.
21. Correale M, Totaro A, Lacedonia D, Montrone D, Di Biase M, Barbaro Foschino MP, Brunetti ND. Novelty in treatment of pulmonary fibrosis: pulmonary hypertension drugs and others. Cardiovasc Hematol Agents Med Chem. 2013;11:169–78.
22. De Jesus Perez VA. Making sense of the estrogen paradox in pulmonary arterial hypertension. Am J Respir Crit Care Med. 2011;184(6):629–30. United States
23. Dean A, Nilsen M, Loughlin L, Salt IP, Maclean MR. Metformin reverses development of pulmonary hypertension via aromatase inhibition. Hypertension. 2016;68:446–54.
24. Dempsie Y, Macritchie NA, White K, Morecroft I, Wright AF, Nilsen M, Loughlin L, Mair KM, Maclean MR. Dexfenfluramine and the Oestrogen-metabolizing enzyme CYP1B1 in the development of pulmonary arterial hypertension. Cardiovasc Res. 2013;99:24–34.
25. Dempsie Y, Nilsen M, White K, Mair KM, Loughlin L, Ambartsumian N, Rabinovitch M, Maclean MR. Development of pulmonary arterial hypertension in mice over-expressing S100A4/Mts1 is specific to females. Respir Res. 2011;12:159.
26. Dresdale DT, Schultz M, Michtom RJ. Primary pulmonary hypertension. I Clinical and hemodynamic study. Am J Med. 1951;11:686–705.
27. Dubey RK, Jackson EK, Keller PJ, Imthurn B, Rosselli M. Estradiol metabolites inhibit Endothelin synthesis by an estrogen receptor-independent mechanism. Hypertension. 2001;37:640–4.
28. Dumas De La Roque E, Savineau JP, Metivier AC, Billes MA, Kraemer JP, Doutreleau S, Jougon J, Marthan R, Moore N, Fayon M, Baulieu EE, Dromer C. Dehydroepiandrosterone (DHEA) improves pulmonary hypertension in chronic obstructive pulmonary disease (COPD): a pilot study. Ann Endocrinol (Paris). 2012;73:20–5.

29. Dupuis J, Stewart DJ, Cernacek P, Gosselin G. Human pulmonary circulation is an important site for both clearance and production of endothelin-1. Circulation. 1996;94:1578–84.
30. Eddahibi S, Fabre V, Boni C, Martres MP, Raffestin B, Hamon M, Adnot S. Induction of serotonin transporter by hypoxia in pulmonary vascular smooth muscle cells. Relationship with the Mitogenic action of serotonin. Circ Res. 1999;84:329–36.
31. Eddahibi S, Guignabert C, Barlier-Mur AM, Dewachter L, Fadel E, Dartevelle P, Humbert M, Simonneau G, Hanoun N, Saurini F, Hamon M, Adnot S. Cross talk between endothelial and smooth muscle cells in pulmonary hypertension: critical role for serotonin-induced smooth muscle hyperplasia. Circulation. 2006;113:1857–64.
32. Eddahibi S, Humbert M, Fadel E, Raffestin B, Darmon M, Capron F, Simonneau G, Dartevelle P, Hamon M, Adnot S. Serotonin transporter overexpression is responsible for pulmonary artery smooth muscle hyperplasia in primary pulmonary hypertension. J Clin Invest. 2001;108:1141–50.
33. El Khoudary SR. Gaps, limitations and new insights on endogenous estrogen and follicle stimulating hormone as related to risk of cardiovascular disease in women traversing the menopause: a narrative review. Maturitas. 2017;104:44–53.
34. English KM, Jones RD, Jones TH, Morice AH, Channer KS. Gender differences in the vasomotor effects of different steroid hormones in rat pulmonary and coronary arteries. Horm Metab Res. 2001;33:645–52.
35. Evans JD, Girerd B, Montani D, Wang XJ, Galie N, Austin ED, Elliott G, Asano K, Grunig E, Yan Y, Jing ZC, Manes A, Palazzini M, Wheeler LA, Nakayama I, Satoh T, Eichstaedt C, Hinderhofer K, Wolf M, Rosenzweig EB, Chung WK, Soubrier F, Simonneau G, Sitbon O, Graf S, Kaptoge S, Di Angelantonio E, Humbert M, Morrell NW. BMPR2 mutations and survival in pulmonary arterial hypertension: an individual participant data meta-analysis. Lancet Respir Med. 2016;4:129–37.
36. Farrukh IS, Peng W, Orlinska U, Hoidal JR. Effect of Dehydroepiandrosterone on hypoxic pulmonary vasoconstriction: a Ca(2+)-activated K(+)-channel opener. Am J Phys. 1998;274:L186–95.
37. Fishman AP. Aminorex to fen/Phen: an epidemic foretold. Circulation. 1999;99:156–61.
38. Foderaro A, Ventetuolo CE. Pulmonary arterial hypertension and the sex hormone paradox. Curr Hypertens Rep. 2016;18:84.
39. Forbes JF, Cuzick J, Buzdar A, Howell A, Tobias JS, Baum M. Effect of anastrozole and tamoxifen as adjuvant treatment for early-stage breast cancer: 100-month analysis of the ATAC trial. Lancet Oncol. 2008;9:45–53.
40. Fristrom S, Airaksinen MM, Halmekoski J. Release of platelet 5-hydroxytryptamine by some anorexic and other sympathomimetics and their acetyl derivatives. Acta Pharmacol Toxicol (Copenh). 1977;41:218–24.
41. Gabler NB, French B, Strom BL, Liu Z, Palevsky HI, Taichman DB, Kawut SM, Halpern SD. Race and sex differences in response to endothelin receptor antagonists for pulmonary arterial hypertension. Chest. 2012;141:20–6.
42. Gaddum JH, Hebb CO, Silver A, Swan AA. 5-Hydroxytryptamine; pharmacological action and destruction in perfused lungs. Q J Exp Physiol Cogn Med Sci. 1953;38:255–62.
43. Galie N, Hoeper MM, Humbert M, Torbicki A, Vachiery JL, Barbera JA, Beghetti M, Corris P, Gaine S, Gibbs JS, Gomez-Sanchez MA, Jondeau G, Klepetko W, Opitz C, Peacock A, Rubin L, Zellweger M, Simonneau G, Guidelines, E. S. C. C. F. P. Guidelines for the diagnosis and treatment of pulmonary hypertension: the Task Force for the Diagnosis and Treatment of Pulmonary Hypertension of the European Society of Cardiology (ESC) and the European Respiratory Society (ERS), endorsed by the International Society of Heart and Lung Transplantation (ISHLT). Eur Heart J. 2009;30:2493–537.
44. Gao XF, Zhang JJ, Jiang XM, Ge Z, Wang ZM, Li B, Mao WX, Chen SL. Targeted drugs for pulmonary arterial hypertension: a network meta-analysis of 32 randomized clinical trials. Patient Prefer Adherence. 2017;11:871–85.
45. Garcia De Vinuesa A, Abdelilah-Seyfried S, Knaus P, Zwijsen A, Bailly S. Bmp signaling in vascular biology and dysfunction. Cytokine Growth Factor Rev. 2016;27:65–79.
46. Giaid A, Yanagisawa M, Langleben D, Michel RP, Levy R, Shennib H, Kimura S, Masaki T, Duguid WP, Stewart DJ. Expression of endothelin-1 in the lungs of patients with pulmonary hypertension. N Engl J Med. 1993;328:1732–9.
47. Glusa E, Graser T, Wagner S, Oettel M. Mechanisms of relaxation of rat aorta in response to progesterone and synthetic progestins. Maturitas. 1997;28:181–91.
48. Gomez-Arroyo J, Saleem SJ, Mizuno S, Syed AA, Bogaard HJ, Abbate A, Taraseviciene-Stewart L, Sung Y, Kraskauskas D, Farkas D, Conrad DH, Nicolls MR, Voelkel NF. A brief overview of mouse models of pulmonary arterial hypertension: problems and prospects. Am J Physiol Lung Cell Mol Physiol. 2012;302:L977–91.
49. Harada N, Honda SI, Hatano O. Aromatase inhibitors and enzyme stability. Endocr Relat Cancer. 1999;6:211–8.
50. Hemnes AR, Kiely DG, Cockrill BA, Safdar Z, Wilson VJ, Al Hazmi M, Preston IR, Maclean MR, Lahm T. Statement on pregnancy in pulmonary hypertension from the Pulmonary Vascular Research Institute. Pulm Circ. 2015;5:435–65.
51. Hemnes AR, Maynard KB, Champion HC, Gleaves L, Penner N, West J, Newman JH. Testosterone negatively regulates right ventricular load stress responses in mice. Pulm Circ. 2012;2:352–8.
52. Herve P, Launay JM, Scrobohaci ML, Brenot F, Simonneau G, Petitpretz P, Poubeau P, Cerrina J, Duroux P, Drouet L. Increased plasma serotonin in primary pulmonary hypertension. Am J Med. 1995;99:249–54.
53. Hirata Y, Emori T, Eguchi S, Kanno K, Imai T, Ohta K, Marumo F. Endothelin receptor subtype B mediates synthesis of nitric oxide by cultured bovine endothelial cells. J Clin Invest. 1993;91:1367–73.
54. Hoeper MM, Gibbs JSR. The changing landscape of pulmonary arterial hypertension and implications for patient care. Eur Respir Rev. 2014;23(134):450–7.
55. Hood KY, Mair KM, Harvey AP, Montezano AC, Touyz RM, Maclean MR. Serotonin signaling through the 5-HT1B receptor and NADPH oxidase 1 in pulmonary arterial hypertension. Arterioscler Thromb Vasc Biol. 2017;37:1361–70.
56. Hood KY, Montezano AC, Harvey AP, Nilsen M, Maclean MR, Touyz RM. Nicotinamide adenine dinucleotide phosphate oxidase-mediated redox signaling and vascular remodeling by 16alpha-Hydroxyestrone in human pulmonary artery cells: implications in pulmonary arterial hypertension. Hypertension. 2016;68(3):796–808.
57. Humbert M, Montani D, Perros F, Dorfmüller P, Adnot S, Eddahibi S. Endothelial cell dysfunction and cross talk between endothelium and smooth muscle cells in pulmonary arterial hypertension. Vasc Pharmacol. 2008;49:113–8.
58. Humbert M, Sitbon O, Yaici A, Montani D, O'Callaghan DS, Jais X, Parent F, Savale L, Natali D, Gunther S, Chaouat A, Chabot F, Cordier JF, Habib G, Gressin V, Jing ZC, Souza R, Simonneau G. Survival in incident and prevalent cohorts of

patients with pulmonary arterial hypertension. Eur Respir J. 2010;36:549–55.
59. Jacobs W, Van De Veerdonk MC, Trip P, De Man F, Heymans MW, Marcus JT, Kawut SM, Bogaard HJ, Boonstra A, Vonk Noordegraaf A. The right ventricle explains sex differences in survival in idiopathic pulmonary arterial hypertension. Chest. 2014;145:1230–6.
60. Johansen AK, Dean A, Morecroft I, Hood K, Nilsen M, Loughlin L, Anagnostopoulou A, Touyz RM, White K, Maclean MR. The serotonin transporter promotes a pathological estrogen metabolic pathway in pulmonary hypertension via cytochrome P450 1B1. Pulm Circ. 2016;6:82–92.
61. Johnson DE, Georgieff MK. Pulmonary neuroendocrine cells. Their secretory products and their potential roles in health and chronic lung disease in infancy. Am Rev Respir Dis. 1989;140:1807–12.
62. Jones RD, English KM, Pugh PJ, Morice AH, Jones TH, Channer KS. Pulmonary vasodilatory action of testosterone: evidence of a calcium antagonistic action. J Cardiovasc Pharmacol. 2002;39:814–23.
63. Kawut SM, Archer-Chicko CL, Dimichele A, Fritz JS, Klinger JR, Ky B, Palevsky HI, Palmisciano AJ, Patel M, Pinder D, Propert KJ, Smith KA, Stanczyk F, Tracy R, Vaidya A, Whittenhall ME, Ventetuolo CE. Anastrozole in Pulmonary Arterial Hypertension (AIPH): a randomized, double-blind placebo-controlled trial. Am J Respir Crit Care Med. 2016;195(3):360–8.
64. Keegan A, Morecroft I, Smillie D, Hicks MN, Maclean MR. Contribution of the 5-HT(1B) receptor to hypoxia-induced pulmonary hypertension: converging evidence using 5-HT(1B)-receptor knockout mice and the 5-HT (1B/1D)-receptor antagonist GR127935. Circ Res. 2001;89:1231–9.
65. Knowlton AA. Estrogen and cardiovascular disease: aging and estrogen loss at the heart of the matter? Futur Cardiol. 2012;8:9–12.
66. Kumar R, Zakharov MN, Khan SH, Miki R, Jang H, Toraldo G, Singh R, Bhasin S, Jasuja R. The dynamic structure of the estrogen receptor. J Amino Acids. 2011;2011:1.
67. Lahm T, Albrecht M, Fisher A, Selej M, Patel N, Brown J, Justice M, Brown M, Van Demark M, Trulock K, Dieudonne D, Reddy J, Presson R, Petrache I. 17 beta-Estradiol attenuates hypoxic pulmonary hypertension via estrogen receptor-mediated effects. Am J Respir Crit Care Med. 2012;185:965–80.
68. Lahm T, Tuder R, Petrache I. Progress in solving the sex hormone paradox in pulmonary hypertension. Am J Phys Lung Cell Mol Phys. 2014;307:L7–L26.
69. Lau EMT, Giannoulatou E, Celermajer DS, Humbert M. Epidemiology and treatment of pulmonary arterial hypertension. Nat Rev Cardiol. 2017;14:603–14.
70. Lawrie A, Spiekerkoetter E, Martinez EC, Ambartsumian N, Sheward WJ, Maclean MR, Harmar AJ, Schmidt AM, Lukanidin E, Rabinovitch M. Interdependent serotonin transporter and receptor pathways regulate S100A4/Mts1, a gene associated with pulmonary vascular disease. Circ Res. 2005;97:227–35.
71. Lee HR, Kim TH, Choi KC. Functions and physiological roles of two types of estrogen receptors, Erα and Erβ, identified by estrogen receptor knockout mouse. Lab Anim Res. 2012;28:71–6.
72. Lee WS, Harder JA, Yoshizumi M, Lee ME, Haber E. Progesterone inhibits arterial smooth muscle cell proliferation. Nat Med. 1997;3:1005–8.
73. Li HF, Zheng TZ, Li W, Qu SY, Zhang CL. Effect of progesterone on the contractile response of isolated pulmonary artery in rabbits. Can J Physiol Pharmacol. 2001;79:545–50.
74. Lian TY, Jiang X, Jing ZC. Riociguat: a soluble guanylate cyclase stimulator for the treatment of pulmonary hypertension. Drug Des Devel Ther. 2017;11:1195–207.
75. Liu A, Schreier D, Tian L, Eickhoff JC, Wang Z, Hacker TA, Chesler NC. Direct and indirect protection of right ventricular function by estrogen in an experimental model of pulmonary arterial hypertension. Am J Physiol Heart Circ Physiol. 2014;307:H273–83.
76. Liu Y, Suzuki YJ, Day RM, Fanburg BL. Rho kinase-induced nuclear translocation of ERK1/ERK2 in smooth muscle cell mitogenesis caused by serotonin. Circ Res. 2004;95:579–86.
77. Long L, Maclean MR, Jeffery TK, Morecroft I, Yang X, Rudarakanchana N, Southwood M, James V, Trembath RC, Morrell NW. Serotonin increases susceptibility to pulmonary hypertension in BMPR2-deficient mice. Circ Res. 2006;98:818–27.
78. Machado RD, Aldred MA, James V, Harrison RE, Patel B, Schwalbe EC, Gruenig E, Janssen B, Koehler R, Seeger W, Eickelberg O, Olschewski H, Elliott CG, Glissmeyer E, Carlquist J, Kim M, Torbicki A, Fijalkowska A, Szewczyk G, Parma J, Abramowicz MJ, Galie N, Morisaki H, Kyotani S, Nakanishi N, Morisaki T, Humbert M, Simonneau G, Sitbon O, Soubrier F, Coulet F, Morrell NW, Trembath RC. Mutations of the TGF-beta type II receptor BMPR2 in pulmonary arterial hypertension. Hum Mutat. 2006;27:121–32.
79. Machado RD, Eickelberg O, Elliott CG, Geraci MW, Hanaoka M, Loyd JE, Newman JH, Phillips JA, Soubrier F 3rd, Trembath RC, Chung WK. Genetics and genomics of pulmonary arterial hypertension. J Am Coll Cardiol. 2009;54:S32–42.
80. Maclean M, Johnston E, Mcculloch K, Pooley L, Houslay M, Sweeney G. Phosphodiesterase isoforms in the pulmonary arterial circulation of the rat: changes in pulmonary hypertension. J Pharmacol Exp Ther. 1997;283:619–24.
81. Mair KM, Johansen AK, Wright AF, Wallace E, Maclean MR. Pulmonary arterial hypertension: basis of sex differences in incidence and treatment response. Br J Pharmacol. 2014a;171:567–79.
82. Mair KM, Wright AF, Duggan N, Rowlands DJ, Hussey MJ, Roberts S, Fullerton J, Nilsen M, Loughlin L, Thomas M, Maclean MR. Sex-dependent influence of endogenous estrogen in pulmonary hypertension. Am J Respir Crit Care Med. 2014b;190:456–67.
83. Mair KM, Yang XD, Long L, White K, Wallace E, Ewart MA, Docherty CK, Morrell NW, Maclean MR. Sex affects bone morphogenetic protein type II receptor signaling in pulmonary artery smooth muscle cells. Am J Respir Crit Care Med. 2015;191:693–703.
84. Marcos E, Fadel E, Sanchez O, Humbert M, Dartevelle P, Simonneau G, Hamon M, Adnot S, Eddahibi S. Serotonin-induced smooth muscle hyperplasia in various forms of human pulmonary hypertension. Circ Res. 2004;94:1263–70.
85. Masi A. Pulmonary-hypertension and oral-contraceptive usage. Chest. 1976;69:451–3.
86. Mathai SC, Hassoun PM, Puhan MA, Zhou Y, Wise RA. Sex differences in response to tadalafil in pulmonary arterial hypertension. Chest. 2015;147(1):188–97.
87. Mcculloch KM, Docherty CC, Morecroft I, Maclean MR. EndothelinB receptor-mediated contraction in human pulmonary resistance arteries. Br J Pharmacol. 1996;119:1125–30.

88. Medhora M, Chen Y, Gruenloh S, Harland D, Bodiga S, Zielonka J, Gebremedhin D, Gao Y, Falck JR, Anjaiah S, Jacobs ER. 20-HETE increases superoxide production and activates NAPDH oxidase in pulmonary artery endothelial cells. Am J Physiol Lung Cell Mol Physiol. 2008;294: L902–11.
89. Mendelsohn ME, Karas RH. The protective effects of estrogen on the cardiovascular system. N Engl J Med. 1999;340:1801–11.
90. Montani D, Chaumais MC, Guignabert C, Günther S, Girerd B, Jaïs X, Algalarrondo V, Price LC, Savale L, Sitbon O, Simonneau G, Humbert M. Targeted therapies in pulmonary arterial hypertension. Pharmacol Ther. 2014;141:172–91.
91. Montani D, Günther S, Dorfmüller P, Perros F, Girerd B, Garcia G, Jaïs X, Savale L, Artaud-Macari E, Price LC, Humbert M, Simonneau G, Sitbon O. Pulmonary arterial hypertension. Orphanet J Rare Dis. 2013;8:97.
92. Morecroft I, Dempsie Y, Bader M, Walther DJ, Kotnik K, Loughlin L, Nilsen M, Maclean MR. Effect of tryptophan hydroxylase 1 deficiency on the development of hypoxia-induced pulmonary hypertension. Hypertension. 2007;49:232–6.
93. Morecroft I, Heeley RP, Prentice HM, Kirk A, Maclean MR. 5-hydroxytryptamine receptors mediating contraction in human small muscular pulmonary arteries: importance of the 5-HT1B receptor. Br J Pharmacol. 1999;128:730–4.
94. Morecroft I, Loughlin L, Nilsen M, Colston J, Dempsie Y, Sheward J, Harmar A, Maclean MR. Functional interactions between 5-hydroxytryptamine receptors and the serotonin transporter in pulmonary arteries. J Pharmacol Exp Ther. 2005;313:539–48.
95. Morecroft I, White K, Caruso P, Nilsen M, Loughlin L, Alba R, Reynolds PN, Danilov SM, Baker AH, Maclean MR. Gene therapy by targeted adenovirus-mediated knockdown of pulmonary endothelial Tph1 attenuates hypoxia-induced pulmonary hypertension. Mol Ther. 2012;20:1516–28.
96. Morrell NW, Adnot S, Archer SL, Dupuis J, Jones PL, Maclean MR, Mcmurtry IF, Stenmark KR, Thistlethwaite PA, Weissmann N, Yuan JX, Weir EK. Cellular and molecular basis of pulmonary arterial hypertension. J Am Coll Cardiol. 2009;54:S20–31.
97. Mueck AO, Seeger H, Lippert TH. Estradiol metabolism and malignant disease. Maturitas. 2002;43:1–10.
98. Nakamura K, Hasegawa H. Developmental role of tryptophan hydroxylase in the nervous system. Mol Neurobiol. 2007;35:45–54.
99. Nebert DW, Dalton TP. The role of cytochrome P450 enzymes in endogenous Signalling pathways and environmental carcinogenesis. Nat Rev Cancer. 2006;6:947–60.
100. Newman JH, Trembath RC, Morse JA, Grunig E, Loyd JE, Adnot S, Coccolo F, Ventura C, Phillips JA, Knowles JA 3rd, Janssen B, Eickelberg O, Eddahibi S, Herve P, Nichols WC, Elliott G. Genetic basis of pulmonary arterial hypertension: current understanding and future directions. J Am Coll Cardiol. 2004;43:33S–9S.
101. Ohta A, Nagai M, Nishina M, Tomimitsu H, Kohsaka H. Age at onset and gender distribution of systemic lupus erythematosus, polymyositis/dermatomyositis, and systemic sclerosis in Japan. Mod Rheumatol. 2013;23:759–64.
102. Oka M, Karoor V, Homma N, Nagaoka T, Sakao E, Golembeski SM, Limbird J, Imamura M, Gebb SA, Fagan KA, Mcmurtry IF. Dehydroepiandrosterone upregulates soluble guanylate cyclase and inhibits hypoxic pulmonary hypertension. Cardiovasc Res. 2007;74:377–87.
103. Pan J, Copland I, Post M, Yeger H, Cutz E. Mechanical stretch-induced serotonin release from pulmonary neuroendocrine cells: implications for lung development. Am J Physiol Lung Cell Mol Physiol. 2006;290:L185–93.
104. Paulin R, Meloche J, Jacob MH, Bisserier M, Courboulin A, Bonnet S. Dehydroepiandrosterone inhibits the Src/STAT3 constitutive activation in pulmonary arterial hypertension. Am J Physiol Heart Circ Physiol. 2011;301:H1798–809.
105. Rabinovitch M. Molecular pathogenesis of pulmonary arterial hypertension. J Clin Invest. 2012;122:4306–13.
106. Rapport MM, Green AA, Page IH. Crystalline serotonin. Science. 1948;108:329–30.
107. Rashid J, Patel B, Nozik-Grayck E, Mcmurtry IF, Stenmark KR, Ahsan F. Inhaled sildenafil as an alternative to oral sildenafil in the treatment of pulmonary arterial hypertension (PAH). J Control Release. 2017;250:96–106.
108. Roberts K, Fallon M, Krowka M, Brown R, Trotter J, Peter I, Tighiouart H, Knowles J, Rabinowitz D, Benza R, Badesch D, Taichman D, Horn E, Zacks S, Kaplowitz N, Kawut S, Dis PVCL, Dis PVCL. Genetic risk factors for Portopulmonary hypertension in patients with advanced liver disease. Am J Respir Crit Care Med. 2009;179:835–42.
109. Rondelet B, Van Beneden R, Kerbaul F, Motte S, Fesler P, Mcentee K, Brimioulle S, Ketelslegers JM, Naeije R. Expression of the serotonin 1b receptor in experimental pulmonary hypertension. Eur Respir J. 2003;22:408–12.
110. Rowell KO, Hall J, Pugh PJ, Jones TH, Channer KS, Jones RD. Testosterone acts as an efficacious vasodilator in isolated human pulmonary arteries and veins: evidence for a biphasic effect at physiological and supra-physiological concentrations. J Endocrinol Investig. 2009;32:718–23.
111. Seeger H, Wallwiener D, Kraemer E, Mueck AO. Comparison of possible carcinogenic estradiol metabolites: effects on proliferation, apoptosis and metastasis of human breast Cancer cells. Maturitas. 2006;54:72–7.
112. Seo B, Oemar BS, Siebenmann R, Von Segesser L, Luscher TF. Both ETA and ETB receptors mediate contraction to endothelin-1 in human blood vessels. Circulation. 1994;89:1203–8.
113. Shapiro S, Pollock D, Gillies H, Henig N, Allard M, Blair C, Anglen C, Kohan D. Frequency of edema in patients with pulmonary arterial hypertension receiving Ambrisentan. Am J Cardiol. 2012;110:1373–7.
114. Sharma K. Selexipag for the treatment of pulmonary arterial hypertension. Expert Rev Respir Med. 2015;10(1):1–3.
115. Shichiri M, Kato H, Marumo F, Hirata Y. Endothelin-1 as an autocrine/paracrine apoptosis survival factor for endothelial cells. Hypertension. 1997;30:1198–203.
116. Simonneau G, Gatzoulis MA, Adatia I, Celermajer D, Denton C, Ghofrani A, Gomez Sanchez MA, Krishna Kumar R, Landzberg M, Machado RF, Olschewski H, Robbins IM, Souza R. Updated clinical classification of pulmonary hypertension. J Am Coll Cardiol. 2013;62: D34–41.
117. Simpson ER. Sources of estrogen and their importance. J Steroid Biochem Mol Biol. 2003;86:225–30.
118. Smith AM, Bennett RT, Jones TH, Cowen ME, Channer KS, Jones RD. Characterization of the Vasodilatory action of testosterone in the human pulmonary circulation. Vasc Health Risk Manag. 2008;4:1459–66.
119. Stasch JP, Evgenov OV. Soluble guanylate cyclase stimulators in pulmonary hypertension. Handb Exp Pharmacol. 2013;218:279–313.
120. Stewart DJ, Levy RD, Cernacek P, Langleben D. Increased plasma endothelin-1 in pulmonary hypertension: marker or mediator of disease? Ann Intern Med. 1991;114:464–9.

121. Suzuki YJ, Day RM, Tan CC, Sandven TH, Liang Q, Molkentin JD, Fanburg BL. Activation of GATA-4 by serotonin in pulmonary artery smooth muscle cells. J Biol Chem. 2003;278:17525–31.
122. Terek D, Kayikcioglu M, Kultursay H, Ergenoglu M, Yalaz M, Musayev O, Mogulkoc N, Gunusen I, Akisu M, Kultursay N. Pulmonary arterial hypertension and pregnancy. J Res Med Sci. 2013;18(1):73–6.
123. Thomas DP, Vane JR. 5-hydroxytryptamine in the circulation of the dog. Nature. 1967;216:335–8.
124. Tofovic PS, Zhang X, Petrusevska G. Progesterone inhibits vascular remodeling and attenuates Monocrotaline-induced pulmonary hypertension in estrogen-deficient rats. Prilozi. 2009a;30:25–44.
125. Tofovic SP. Estrogens and development of pulmonary hypertension: interaction of estradiol metabolism and pulmonary vascular disease. J Cardiovasc Pharmacol. 2010;56:696–708.
126. Tofovic SP, Jackson EK. Complexities of oestradiol pharmacology in pulmonary arterial hypertension. Eur Respir J. 2013;41(6):1465.
127. Tofovic SP, Jones T, Petrusevska G. Dose-dependent therapeutic effects of 2-Methoxyestradiol on Monocrotaline-Induced pulmonary hypertension and vascular Remodelling. Prilozi. 2010a;31:279–95.
128. Tofovic SP, Jones TJ, Bilan VP, Jackson EK, Petrusevska G. Synergistic therapeutic effects of 2-methoxyestradiol with either sildenafil or bosentan on amelioration of monocrotaline-induced pulmonary hypertension and vascular remodeling. J Cardiovasc Pharmacol. 2010b;56:475–83.
129. Tofovic SP, Salah EM, Mady HH, Jackson EK, Melhem MF. Estradiol metabolites attenuate monocrotaline-induced pulmonary hypertension in rats. J Cardiovasc Pharmacol. 2005;46:430–7.
130. Tofovic SP, Zhang X, Jackson EK, Zhu H, Petrusevska G. 2-methoxyestradiol attenuates bleomycin-induced pulmonary hypertension and fibrosis in estrogen-deficient rats. Vasc Pharmacol. 2009b;51:190–7.
131. Tofovic SP, Zhang X, Zhu H, Jackson EK, Rafikova O, Petrusevska G. 2-Ethoxyestradiol is antimitogenic and attenuates monocrotaline-induced pulmonary hypertension and vascular remodeling. Vasc Pharmacol. 2008;48:174–83.
132. Tsuchiya Y, Nakajima M, Yokoi T. Cytochrome P450-mediated metabolism of estrogens and its regulation in human. Cancer Lett. 2005;227:115–24.
133. Tuder RM, Archer SL, Dorfmüller P, Erzurum SC, Guignabert C, Michelakis E, Rabinovitch M, Schermuly R, Stenmark KR, Morrell NW. Relevant issues in the pathology and pathobiology of pulmonary hypertension. J Am Coll Cardiol. 2013;62:D4–12.
134. Tuder RM, Cool CD, Geraci MW, Wang J, Abman SH, Wright L, Badesch D, Voelkel NF. Prostacyclin synthase expression is decreased in lungs from patients with severe pulmonary hypertension. Am J Respir Crit Care Med. 1999;159:1925–32.
135. Umar S, Iorga A, Matori H, Nadadur RD, Li J, Maltese F, Van Der Laarse A, Eghbali M. Estrogen rescues preexisting severe pulmonary hypertension in rats. Am J Respir Crit Care Med. 2011;184(6):715–23.
136. Umar S, Rabinovitch M, Eghbali M. Estrogen paradox in pulmonary hypertension: current controversies and future perspectives. Am J Respir Crit Care Med. 2012;186:125–31.
137. Van Der Bruggen CE, Happe CM, Dorfmuller P, Trip P, Spruijt OA, Rol N, Hoevenaars FP, Houweling AC, Girerd B, Marcus JT, Mercier O, Humbert M, Handoko ML, Van Der Velden J, Vonk Noordegraaf A, Bogaard HJ, Goumans MJ, De Man FS. Bone morphogenetic protein receptor type 2 mutation in pulmonary arterial hypertension: a view on the right ventricle. Circulation. 2016;133:1747–60.
138. Ventetuolo CE, Baird GL, Barr RG, Bluemke DA, Fritz JS, Hill NS, Klinger JR, Lima JA, Ouyang P, Palevsky HI, Palmisciano AJ, Krishnan I, Pinder D, Preston IR, Roberts KE, Kawut SM. Higher estradiol and lower Dehydroepiandrosterone-sulfate levels are associated with pulmonary arterial hypertension in men. Am J Respir Crit Care Med. 2016a;193:1168–75.
139. Ventetuolo CE, Mitra N, Wan F, Manichaikul A, Barr RG, Johnson C, Bluemke DA, Lima JA, Tandri H, Ouyang P, Kawut SM. Oestradiol metabolism and androgen receptor genotypes are associated with right ventricular function. Eur Respir J. 2016b;47:553–63.
140. Ventetuolo CE, Praestgaard A, Palevsky HI, Klinger JR, Halpern SD, Kawut SM. Sex and Haemodynamics in pulmonary arterial hypertension. Eur Respir J. 2014;43:523–30.
141. Wallace E, Morrell NW, Yang XD, Long L, Stevens H, Nilsen M, Loughlin L, Mair KM, Baker AH, Maclean MR. A sex-specific microRNA-96/5-hydroxytryptamine 1B axis influences development of pulmonary hypertension. Am J Respir Crit Care Med. 2015;191:1432–42.
142. White K, Dempsie Y, Nilsen M, Wright AF, Loughlin L, Maclean MR. The serotonin transporter, gender, and 17beta oestradiol in the development of pulmonary arterial hypertension. Cardiovasc Res. 2011;90:373–82.
143. White K, Johansen A, Nilsen M, Ciuclan L, Wallace E, Paton L, Campbell A, Morecroft I, Loughlin L, Mcclure J, Thomas M, Mair K, Maclean M. Activity of the estrogen-metabolizing enzyme cytochrome P450 1B1 influences the development of pulmonary arterial hypertension. Circulation. 2012;126:1087–U202.
144. Wiersma DA, Roth RA. Clearance of 5-hydroxytryptamine by rat lung and liver: the importance of relative perfusion and intrinsic clearance. J Pharmacol Exp Ther. 1980;212:97–102.
145. Wort SJ, Woods M, Warner TD, Evans TW, Mitchell JA. Endogenously released Endothelin-1 from human pulmonary artery smooth muscle promotes cellular proliferation: relevance to pathogenesis of pulmonary hypertension and vascular remodeling. Am J Respir Cell Mol Biol. 2001;25:104–10.
146. Wright AF, Ewart M-A, Mair K, Nilsen M, Dempsie Y, Loughlin L, Maclean MR. Oestrogen receptor alpha in pulmonary hypertension. Cardiovasc Res. 2015;106:206–16.
147. Zheng Y, Russell B, Schmierer D, Laverty R. The effects of Aminorex and related compounds on brain monoamines and metabolites in CBA mice. J Pharm Pharmacol. 1997;49:89–96.
148. Zhu D, Ran Y. Role of 15-lipoxygenase/15-hydroxyeicosatetraenoic acid in hypoxia-induced pulmonary hypertension. J Physiol Sci. 2012;62:163–72.

Sex Differences in Heart Failure

32

Gianluigi Savarese and Domenico D'Amario

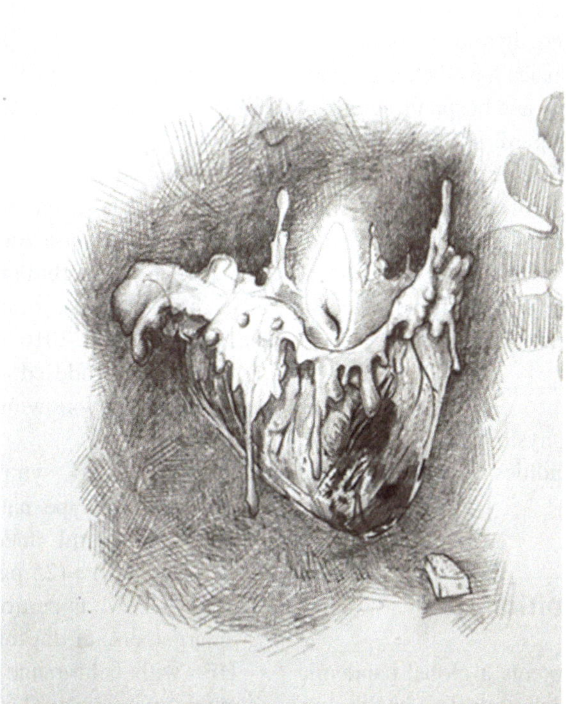

The failing heart. Art work by Piet Michiels, Leuven, Belgium

G. Savarese (✉)
Division of Cardiology, Department of Medicine, Karolinska Institutet, Stockholm, Sweden
e-mail: gianluigi.savarese@ki.se

D. D'Amario
Polo di Scienze Cardiovascolari e Toraciche, Università Cattolica del Sacro Cuore, Fondazione Policlinico A. Gemelli, Rome, Italy

Abstract

Heart failure (HF) represents a global pandemic health problem with a high impact on health-care costs, affecting about 26 million adults worldwide. The overall HF prevalence and incidence are ~2% and ~0.2% per year, respectively, in Western countries, with half of

P. L. M. Kerkhof, V. M. Miller (eds.), *Sex-Specific Analysis of Cardiovascular Function*, Advances in Experimental Medicine and Biology 1065, https://doi.org/10.1007/978-3-319-77932-4_32

the HF population with reduced ejection fraction (HFpEF) and half with preserved (HFpEF) or mid-range ejection fraction (HFmrEF). Sex differences may exist in HF. More males have HFrEF or HFmrEF and an ischemic etiology, whereas more females have HFpEF and hypertension, diastolic dysfunction, and valvular pathologies as HF etiologies. Females are generally older, have a higher EF, higher frequency of HF-related symptoms, and lower NYHA functional status. Generally, it is observed that female HF patients tend to have more comorbidities such as atrial fibrillation, diabetes, hypertension, anemia, iron deficiency, renal disease, arthritis, frailty, depression, and thyroid abnormalities. However, overall, females have better prognosis in terms of mortality and hospitalization risk compared with men, regardless of EF. Potential sex differences in HF characteristics may be underestimated because of the underrepresentation of females in cardiovascular research and, in particular, the sex imbalance in clinical trial enrollment may avoid to identify sex-specific differences in treatments' benefit.

Keywords
Heart failure · Pathophysiology · Sex differences · Comorbidities · Treatments

Heart Failure: Definition

Heart failure (HF) represents a global pandemic health problem with a high impact on health-care costs, affecting about 26 million adults worldwide [1, 2].

The current European Society of Cardiology (ESC) guidelines define HF as a clinical syndrome characterized by typical symptoms (e.g., breathlessness, ankle swelling, and fatigue) that may be accompanied by signs (e.g., elevated jugular venous pressure, pulmonary crackles, and peripheral edema) caused by a structural and/or functional cardiac abnormality, resulting in a reduced cardiac output and/or elevated intracardiac pressures at rest or during stress [3]. In the past, HF has been defined as systolic or diastolic HF [4, 5]. However, this definition did not consider that both systolic and diastolic dysfunction often coexist; that is the reason why it was abandoned [6]. Later, left ventricular (LV) ejection fraction (EF), mainly because of its prognostic role and its use as inclusion criteria in HF trials, has been considered suitable for HF patients' classification, with EF <35% defined as HF with reduced EF (HFrEF) and EF>50% as HF with preserved EF (HFpEF) [7]. EF <35–40% has been widely investigated by randomized clinical trials, and several effective treatments are available in this category of patients [8]. EF >45–50% has become more recently topic of investigation, with several trials unsuccessfully testing potential treatments in this field but many steps forward in patients' characterization [9–12]. EF 40–50% has represented a gray zone for a long time [13]. Patients with EF 40–50% have been inconsistently characterized, studied, included in trials, and treated in daily clinical practice. Finally, based on the differences in gender, age, and comorbidity distribution and to the differential response to therapies reported across the EF spectrum, 2016 ESC guidelines on HF [3, 14, 15] proposed three HF phenotypes according to the following criteria:

- HFpEF: (1) HF symptoms/signs; (1) LVEF ≥50%; (2) B-type natriuretic peptide (BNP) levels >35 pg/mL and/or N-terminal proBNP (NT-proBNP) >125 pg/mL; and (3) structural disease (LV hypertrophy and/or left atrial enlargement) or diastolic dysfunction
- HF with mid-range (HFmrEF): (1) HF symptoms/signs; (1) LVEF 40–49%; (2) B-type natriuretic peptide (BNP) levels >35 pg/mL and/or N-terminal proBNP (NT-proBNP) >125 pg/mL; and (3) structural disease (LV hypertrophy and/or left atrial enlargement) or diastolic dysfunction
- HFrEF: (1) HF symptoms/signs; (2) LVEF<40%

Thus, EF = 40–50% has been finally recognized as an independent entity and named HFmrEF, but studies to define its epidemiology, characteristics, and prognosis in order to develop treatments and care strategies are needed.

Prevalence and Incidence

The prevalence of HF is ~2% in Europe and North America, raising with age, exceeding 10% after the age of 65 years [16–20]. Epidemiological surveys do not report any significant difference in prevalence between males and females, with higher rates in younger males and older females [17, 18, 21–25]. Approximately half of the HF population has HFrEF and half HFpEF or HFmrEF [19, 26, 27], with more males with HFrEF or HFmrEF and more females with HFpEF [13, 27, 28] (Fig. 32.1).

The incidence of HF in Western countries has been estimated to be ~0.2% per year [16, 29], with rates approximately doubling with each advancing decade after the age of 45 years [30]. Although males have been shown to report higher HF incidence as compared with females at all the ages [23, 24, 29, 31–34], males and females report a similar 20% lifetime risk of developing HF because of the longer life span in females [35]. Additionally, a higher HFpEF incidence has been observed in females and a higher HFrEF incidence in males [24, 36] with females but not men developing HFpEF later in life than HFrEF [31].

The aging of the population, together with the improved survival in HF patients, is expected to determine a dramatic increase of HF prevalence, in particular of HFpEF, despite stable or even decreasing incidence rate in both males and females [12, 37].

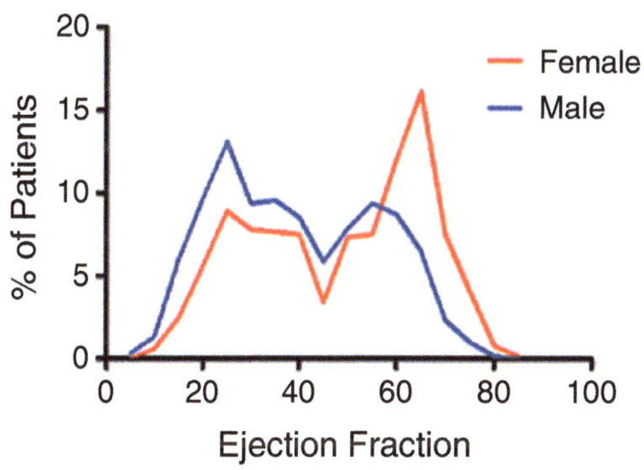

Fig. 32.1 The distribution of ejection fraction according to sex in patients with heart failure (defined by Framingham criteria) from Olmsted County, Minnesota, USA. (Reproduced with permission from Dunlay, S. M. et al. Longitudinal changes in ejection fraction in heart failure patients with preserved and reduced ejection fraction [14])

Prognosis

Despite the advancements in treatments and diagnostic technology, the outcome in HF is still poor [38] with high rates of mortality and rehospitalization in both HFrEF and HFpEF/HFmrEF patients [38].

Sex differences in outcomes have been investigated in different settings.

In chronic HF, females show lower short- and long-term mortality and hospitalization rates as compared with men regardless of EF, especially when HF etiology is not ischemic [28, 39–47]. Studies in advanced HF, enrolling mainly patients with HFrEF, confirm female gender to be associated with lower mortality/hospitalization rates [48–50], whereas in acute decompensated HF, males and females share the same in-hospital mortality risk regardless of EF [51] and the same long-term outcome [52–54] but with higher EF associated with improved prognosis in males [54].

Quality of Life

Patients with HF report significant impairment in quality of life [55]. Whether there is any difference in quality of life between men and females is debated and mainly depending on the method used to assess it [56–60]. Indeed, some studies reported females perceiving less quality of life and more impairment in physical functioning as compared with males [56, 60], others similar quality of life in males vs. females [59], and others that females are more often in ideal cardiovascular health and report less premature life years lost as compared with males [57, 61].

Etiology and Pathophysiology

Etiology of HF is predominantly ischemic (due to established coronary artery disease (CAD)), followed by hypertension and valvular heart disease [62–64]. However, differences in clinical, etiological, and laboratory characteristics exist between males and females [65–70]. Even though females make up nearly half of HF patients worldwide, inclusion and representation of females in HF-oriented randomized controlled trials remain low, resulting in modest robustness of gender-specific data regarding the prognosis, survival outcomes, and treatment modalities [66, 67]. The BEST study investigated sex differences in advanced HF with impaired EF and revealed that females are younger, more likely to be black, have higher EF and right ventricular EF values, lower norepinephrine plasma levels and lower prevalence of atrial fibrillation with higher prevalence of left bundle branch block, higher heart rate, and increased cardiothoracic ratio compared to males [50]. Additionally, females had heavier risk factor burden than males and were more likely to report macro- and microvasculature abnormalities with more diffuse atherosclerotic disease accompanied with endothelial and smooth muscle dysfunction [71, 72] and cardiac-specific sympathetic activation [73].

Ischemic etiology for HF is significantly more common among males (females are less likely to have an obstructive CAD prior to hospitalization and accompanying risk factors), while hypertension, diastolic dysfunction, and valvular pathologies are more common HF etiologies among females [44, 50, 64, 65, 70, 74–76] (Fig. 32.2). Most of HF clinical trials and registry analyses report that females are significantly older than males and are smoking less, have more diabetes mellitus (DM) and higher LVEF values (consequently having a higher prevalence of HFpEF), and exhibit more severe HF symptoms.

Peripheral blood analyses showed lower levels of inflammatory and extracellular matrix remodeling biomarkers in females vs. males with HF [69]. A primary female sex hormone, estrogen, especially 17β-estradiol, has a well-established role in protecting heart by blunting cytotoxic, apoptotic, ischemic, and hypertrophic stimuli [78, 79]. Furthermore, female-specific risk

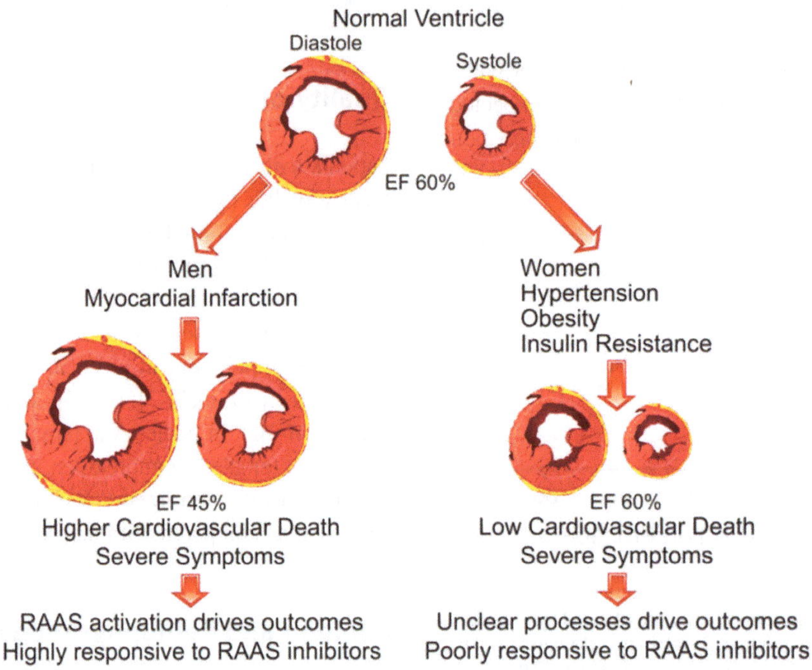

Fig. 32.2 Different pathophysiological characterizations of heart failure in males and females according to EF distribution. Reproduced with permission from Borlaug BA. Defining HFpEF: where do we draw the line? [77]

factors and conditions such as hypertensive disorders during pregnancy, peripartum cardiomyopathy, menopause, and the use of hormonal therapies for contraception can all have pathophysiologic implications in HF development [80–83].

From the pathophysiological standpoint, cardiomyocytes obtained from female animal models demonstrated significantly higher survival compared to males when exposed to oxidative stress-induced cell death. This was attributed to the activation of Akt signaling pathway and the inhibition of pro-apoptotic pathways mediated through estrogen receptor-alpha (ER-α), that is expressed to a significantly higher degree in females [84]. Recent preclinical findings obtained from HFpEF animal models suggest that myocardial microvascular inflammatory endothelial activation and oxidative stress reduced nitric oxide (NO)-dependent signaling from endothelial cells to cardiomyocytes, thus contributing to the increased cardiomyocyte stiffness and hypertrophy [85]. Clinically, sex differences in HF population are also observed in plasma B-type natriuretic peptide (BNP) levels, peak oxygen consumption (peak-VO2), LVEF, and LV mass index, suggesting that myocardial adaptations to hemodynamic overload are sex-specific and that induction of LV hypertrophy during myocardial dysfunction is more rapid and prominent in males [86–88].

In summary, these data show that data collected in females with HF cannot be assumed to similarly apply to males [89].

Cardiac Structure and Function

Females, on average, have smaller body sizes, and, therefore, their hearts and coronary vessels are smaller, resulting in LV volumetric measures being smaller in females than in males even after adjustment for body surface area [90, 91]. In contrast to males, females have lower prevalence of structural heart disease (LV dysfunction and/or CAD) and thus, less risk of sudden cardiac arrest [92, 93].

Among HFpEF patients, females have higher indexed LV wall thickness, higher LVEF, and significantly worse diastolic function, while global LV strain, LV mass, and LV volumes indexed for height are similar to males, as assessed by echocardiography [94]. Furthermore, LV stiffness is significantly higher in females than males, and there is a trend toward the association between female gender and higher prevalence of abnormal LV geometry (concentric or eccentric hypertrophy or concentric remodeling) [94]. Ventricular and arterial stiffness, as measures of central hemodynamic profile, physiologically increase with age in both sexes and are associated with worse LV diastolic function. These associations are found to be significantly more dominant in females, independently of body size and heart rate [95, 96].

However, it should be considered that among patients with HFpEF, resting echocardiographic parameters, including global longitudinal strain, poorly identify patients with HFpEF, since approximately 40% of them do not have LV hypertrophy or concentric remodeling, 30% do not have increased left atrial sizes, and 33% have normal echocardiographic indices of diastolic function at rest [97, 98].

Moreover, since indirect evidences suggest that on exercise, increased external constraint to LV filling might contribute to impaired use of the Starling mechanism in HFpEF patients, future studies should aim to identify useful exercise parameters, in order to achieve a simple, reproducible, and standardized exercise-based echocardiographic methodology able to characterize and to predict prognosis in HFpEF, even independently of gender [97].

Finally, the 2016 ESC guidelines on HF introduced HFmrEF as a potentially distinct entity in the HF spectrum. To date, the few observations available suggest that its characteristics are overall intermediate between those of HFpEF and HFrEF, with HFmrEF more similar to HFrEF for prevalence and incidence of ischemic heart disease [15, 20].

Taken together, these data suggest that diastolic dysfunction is more pronounced in females compared to males. Therefore, females are approximately two times more likely to develop HFpEF than males [99].

Females seem to be better protected in terms of cardiomyocyte apoptosis, exhibiting smaller infarct sizes and less adverse ventricular remodeling after myocardial infarction when compared to males [67, 100]. This is probably due to a significantly more enhanced myocyte protection from ischemic injury, oxidative stress, and pressure/volume overload [84, 101–104] in females vs. males. Moreover, female hormone-related cardioprotection, that blunts adverse cardiac remodeling due to volume-stressed myocardium, is linked with altered mast cell phenotype and/or prevention of mast cell activation, whereas this is not shown in male rats [105]. Myocardial contractile reserve, defined as sarcoplasmic reticulum calcium load, is significantly higher in the myocardium of male animals compared to females, suggesting fundamental sex differences in cellular calcium regulation under conditions of increased stress [106].

Clinical Characteristics and Presentation

Nearly three decades ago, the Studies of Left Ventricular Dysfunction (SOLVD) trials showed different clinical characteristics of HF patients according to sex and race [107]. Particularly in HFpEF, females are generally older, have higher LVEF and more HF-related symptoms, and display more severe HF (as assessed by NYHA functional status). Similar findings have been observed also in HFrEF, with females exhibiting higher LVEF but worse NYHA class [108]. Acute dyspnea is a hallmark symptom of HF, and increased BNP plasma levels significantly correlate with age, female sex, disease severity, and prognosis [109]. In the acute setting, increased BNP levels (>500 pg/mL) are significantly stronger predictors of death in females than males (5.1 vs. 1.8-fold increase in mortality odds, respectively), with higher mortality among females vs. males with BNP > 500 pg/mL (68% vs. 46%, respectively) [110]. However, During acute HF hospitalizations, females exhibit earlier dyspnea relief and improvement in general well-being compared to males [111]. Moreover, hypertension, diuretic and anti-depressant use are more prominent in females vs. males [44, 69]. In the I-PRESERVE trial enrolling HFpEF patients, females were more likely to be obese and have chronic kidney disease, hypertension and nonischemic HF etiology, whereas they were less likely to have chronic obstructive pulmonary disease [45].

The EuroHeart Failure survey II and studies on African and Asian populations confirmed that atrial fibrillation and preserved LVEF were more common in females [112]. Moreover, Asian females are observed to be more likely anemic and have lower mean hemoglobin, creatinine, and uric acid plasma levels [113, 114].

Comorbidities

Comorbidities are the rule rather than the exception in HF and are associated with disease severity (Fig. 32.3). The analysis of the European Heart Failure Pilot survey showed that the majority of HF patients (74%) had at least one comorbidity, while the presence of DM, chronic kidney disease (CKD) and anemia was independently related to increased mortality and/or HF hospitalization risk [115]. In general, the increased risk for morbidity and mortality associated with these comorbidities increases with age and is similar in patients with HFpEF and HFrEF [116, 117]. Results from an US community-based study, enrolling a cohort of patients with first diagnosis of HF, showed that DM, chronic obstructive pulmonary disease (COPD), anemia, and creatinine clearance <30 mL/min/1.73 m^2 were independent predictors of hospitalization together with male sex [118]. Generally, it is observed that female HF patients tend to have more comorbidities such as DM, hypertension, anemia, iron deficiency, renal disease, arthritis, frailty, depression, and thyroid abnormalities compared to males [116].

- *Diabetes mellitus*: DM conferred a five-fold increased risk of HF in females and a two-fold increased risk in men, as showed in the

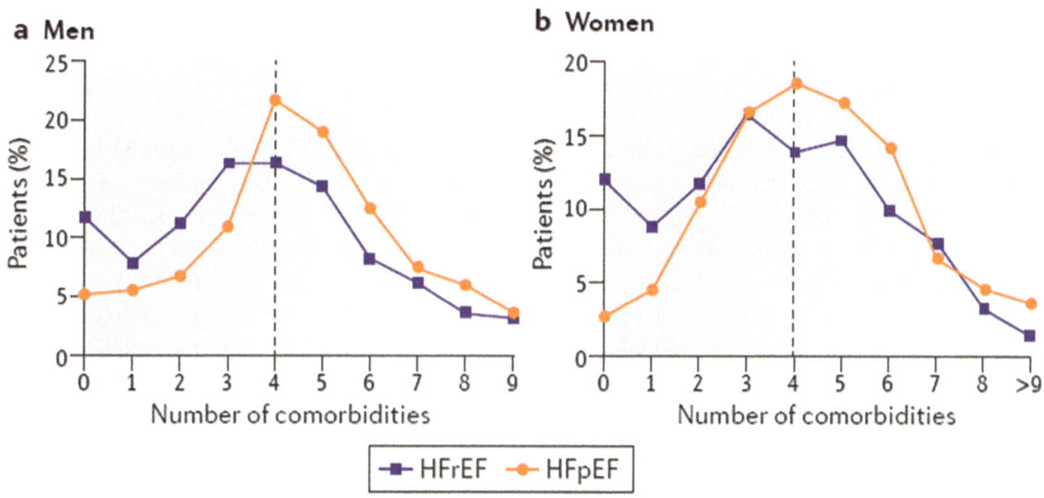

Fig. 32.3 Multimorbidity in heart failure. Frequency distribution of the number of comorbid conditions in men and females with heart failure with preserved (HFpEF) or reduced ejection fraction (HFrEF). Patients with HFpEF report higher number of comorbidities. (Modified with permission from [119])

Framingham Heart Study [120]. Among individuals with DM, regardless of CAD status, HF occurred approximately two-fold more commonly in females than in men. These findings imply that females are more susceptible to the deleterious impact of DM on the heart [120]. Women with DM had greater evidence of adverse LV remodeling. Insulin resistance was associated with increased LV mass in females alone, but this relation was attenuated by obesity [120].

- *Hypertension*: Kenchaiah and colleagues reported that, even after adjustments for comorbidities, the presence of hypertension was associated with a 3.4-fold increased risk of HF in females and a two-fold increase in males [120]. Therefore, in the Framingham Heart Study, although the prevalence of hypertension was similar in both sexes (about 60%), the population attributable fraction was much higher in females than in males [120].
- *Chronic kidney disease*: CKD is one of the most common comorbidities in HF population, especially among elderly, and is a well-established traditional risk factor for poor outcomes [121]. The cardiorenal syndrome [122] involves inflammatory, immunologic, neurohormonal, and stress-mediated mechanisms along with metabolic and nutritional changes and significant alterations in hemodynamic and acid-base/fluid status [123, 124]. According to data from Acute Decompensated Heart Failure National Registry (ADHERE), kidney disease was present in more than 60% of HF patients, and 52% of them were females [125]. Generally, CKD has higher prevalence among females than men in HF; however, there are also studies report higher prevalence among men or no significant sex differences [45, 116, 126].
- *Anemia and iron deficiency*: There is high prevalence of anemia in HF. Anemia is associated with worse symptoms and adverse clinical outcomes [127, 128]. Anemia and iron deficiency can coexist, but it should be stressed that anemia in HF may be not caused by iron deficiency and iron deficiency can be an isolated laboratory finding not causing anemia [129]. Importantly, anemia should always be approached in the context of cardiorenal-anemia syndrome or "continuum", and its severity is often related to renal function [116]. Recent data suggest that one-third of HF patients have anemia, most commonly

normocytic, while iron deficiency is present in 43.2–68.0% of patients with anemia and among 14.7–35.3% of those without anemia [130–132]. Iron deficiency seems to be much more common than anemia across the overall HF population, with nearly two-fold higher prevalence according to some reports [116, 131, 132]. Anemia has a different gender-specific impact on all-cause death. Indeed, it increases odds of death in men with HFrEF but not with HFpEF, while among females, it is the opposite [133]. Across various HF populations, prevalence of anemia is higher in women than men, but the putative pathophysiological mechanisms for that are still not clear [133–137].

- *Frailty assessment*: Frailty is an important prognostic factor that is observed in a substantial proportion of HF patients. It can be considered as a biological phenotype or an accumulation of deficits measured by the deficit index [138]. The biological phenotype of frailty has been associated with a two-fold increased odds of death, whereas 0.1 increase in the deficit index is associated with a 44% increased risk of death [138]. Frailty, detected by standardized geriatric scales and/or clinical parameters such as weak grip strength, physical exhaustion, slowness, low activity, and unintentional weight loss >10 lb over a period of 1 year [138], has not been widely analyzed according to gender in HF patients. Frailty is reported to occur in about 40–55% of HF patients, more commonly in those > or = 70 years and in females [138–141]. Moreover, clinical parameters are usually difficult to be evaluated in the elderly and reduce the certainty of diagnosis, especially in HFpEF that is the most frequent HF type in this subset of patients. The multifaceted scenario in the aging population is further jeopardized by the concomitant changes in the peripheral vasculature and pulmonary and skeletal muscle function, together with the comorbid conditions such as chronic obstructive pulmonary disease, that make difficult the evaluation of HF sign and symptoms, in particular in presence of HFpEF.
- *Chronic obstructive pulmonary disease*: COPD occurs in approximately a third of HF patients, with a slightly higher prevalence in HFpEF compared with HFrEF (5), despite the potential for bias related to the increased likelihood of COPD diagnosis in HFpEF patients. Pitt et al. showed that COPD is a critical independent predictor for hospitalization in male patients with HF [118].
- *Depression*: Prevalence of depression in HF is substantial, ranging from 30 to 48%, and is significantly associated with worse quality of life [139, 142, 143]. Gender-wise, HF females are more likely to suffer from depression than HF males (64% vs. 44%) [142]. Even more in females, the effect of depression on all-cause mortality in HF persists for many years after the initial diagnosis, accelerating physical decline [144, 145].

Treatments

Over the last 30 years, revolutionary improvements have characterized HF therapy, from pharmacological to device treatments. Gender differences in HF treatments have been mainly evaluated in subgroup and post hoc analyses of randomized clinical trials, but the limited number of females enrolled in these studies prevents guidelines to recommend differences in therapies by gender [146]. Furthermore, not only the mode of treatment but also the physiological response to HF-directed therapy might differ between females and men [65]. For a long time, there has been a notion that women are undertreated or treated less aggressively in the context of acute coronary syndromes (ACS) and HF. In the setting of acute myocardial infarction (AMI), females have been shown to receive less aggressive treatment in the early phases of AMI; however, early mortality, defined as 30-day mortality, was similar between genders [147, 148]. In the setting of HF, females seem to receive more diuretics and antidepressants but less angiotensin-converting enzyme inhibitor (ACEI) therapy compared to males [69].

Three comprehensive meta-analyses including trials on ACEI reported a significant reduction of overall mortality and overall mortality or HF hospitalization in men, but the similar reduction in risk observed in females was not statistically significant. However, this evidence may be more likely to be explained by the larger confidence intervals for the hazard ratios in the female analysis due to the limited number of females enrolled in trials, rather than by any gender difference in ACEI efficacy [149, 150]. Indeed, observational data report that ACEIs significantly reduce mortality rates in both males and females, even if the protective effect appears to be greater in men [151]. Trials on angiotensin receptor blockers (ARB) and sacubitril/valsartan have not shown any gender difference in terms of outcome [152–154]. When comparing ACEIs with ARBs, there are signals for ARBs reducing mortality more than ACEIs in females but no difference in men [155]. This may be explained by the modulating effect of estrogen on angiotensin II receptor expression, by the ACE2 gene on the X chromosomes or by the higher discontinuation rates for ACEIs due to cough in females vs. males [146]. Mineralocorticoid-receptor antagonist [156, 157] and beta-blocker [158–161] trials have reported no gender differences in outcomes [149]. Post hoc analyses of the DIG trial reported different prognosis in females vs. males receiving digoxin [162]. A serum digoxin concentration of 1.2–2.0 ng/mL was associated with higher risk of mortality in females but not in males, whereas a serum digoxin concentration of 0.5–0.9 ng/mL was associated with reduced mortality in men but neutral effect in women [163].

Regarding HF devices, females are less likely to receive an implantable cardioverter defibrillator (ICD) or a cardiac resynchronization therapy-defibrillator (CRT-D) as compared with men. One of the explanations for this phenomenon may be the observed higher risk of complications [164–166]. Both genders show an ICD/CRT-D-associated reduction of mortality, with some studies reporting major benefit in terms of mortality and hospitalization rates, less inappropriate shocks, higher CRT response, and ICD efficacy in females [167–171]. Cardiac resynchronization therapy-pacemaker (CRT-P) efficacy has shown not to be different by gender [172, 173], but again the limited number of females included in trials prevents any definitive conclusion. Regarding left ventricular assist devices, several studies report no gender differences in mortality, time to first infection, bleeding, or malfunction with either pulsatile- or continuous-flow devices, but females have shown an increased risk of a first neurological event [174–177]. Women and men show the same prognosis after heart transplant [178].

Final Considerations

HF represents a major global health issue, and over the last years, it has become more and more evident that it is not "just a man's problem". Understanding the underlying differences between males and females may represent a step forward to precision medicine and tailoring therapies and health care according to the individual characteristics of patients. Gender differences in HF epidemiology, clinical characteristics, and symptoms exist and have been previously reported. However, they may be underestimated because of the underrepresentation of females in cardiovascular research and, in particular, the sex imbalance in clinical trial enrollment may avoid to identify sex-specific differences in treatments' benefit. Even though women's health research continues to advance, there is still a long way to go.

References

1. Ponikowski P, Anker SD, AlHabib KF, Cowie MR, Force TL, Hu S, Jaarsma T, Krum H, Rastogi V, Rohde LE, Samal UC, Shimokawa H, Budi Siswanto B, Sliwa K, Filippatos G. Heart failure: preventing disease and death worldwide. ESC Heart Fail. 2014;1(1):4–25.
2. Benjamin EJ, Blaha MJ, Chiuve SE, Cushman M, Das SR, Deo R, de Ferranti SD, Floyd J, Fornage M, Gillespie C, Isasi

CR, Jimenez MC, Jordan LC, Judd SE, Lackland D, Lichtman JH, Lisabeth L, Liu S, Longenecker CT, Mackey RH, Matsushita K, Mozaffarian D, Mussolino ME, Nasir K, Neumar RW, Palaniappan L, Pandey DK, Thiagarajan RR, Reeves MJ, Ritchey M, Rodriguez CJ, Roth GA, Rosamond WD, Sasson C, Towfighi A, Tsao CW, Turner MB, Virani SS, Voeks JH, Willey JZ, Wilkins JT, Wu JH, Alger HM, Wong SS, Muntner P, American Heart Association Statistics C, Stroke Statistics S. Heart disease and stroke statistics-2017 update: a report from the American Heart Association. Circulation. 2017;135(10):e146–603.
3. Ponikowski P, Voors AA, Anker SD, Bueno H, Cleland JG, Coats AJ, Falk V, Gonzalez-Juanatey JR, Harjola VP, Jankowska EA, Jessup M, Linde C, Nihoyannopoulos P, Parissis JT, Pieske B, Riley JP, Rosano GM, Ruilope LM, Ruschitzka F, Rutten FH, van der Meer P, Authors/Task Force M. 2016 ESC Guidelines for the diagnosis and treatment of acute and chronic heart failure: the Task Force for the diagnosis and treatment of acute and chronic heart failure of the European Society of Cardiology (ESC) developed with the special contribution of the Heart Failure Association (HFA) of the ESC. Eur Heart J. 2016;37(27):2129–200.
4. Zile MR, Baicu CF, Gaasch WH. Diastolic heart failure--abnormalities in active relaxation and passive stiffness of the left ventricle. N Engl J Med. 2004;350(19):1953–9.
5. Baicu CF, Zile MR, Aurigemma GP, Gaasch WH. Left ventricular systolic performance, function, and contractility in patients with diastolic heart failure. Circulation. 2005;111 (18):2306–12.
6. Yip G, Wang M, Zhang Y, Fung JW, Ho PY, Sanderson JE. Left ventricular long axis function in diastolic heart failure is reduced in both diastole and systole: time for a redefinition? Heart. 2002;87(2):121–5.
7. McMurray JJ, Adamopoulos S, Anker SD, Auricchio A, Bohm M, Dickstein K, Falk V, Filippatos G, Fonseca C, Gomez-Sanchez MA, Jaarsma T, Kober L, Lip GY, Maggioni AP, Parkhomenko A, Pieske BM, Popescu BA, Ronnevik PK, Rutten FH, Schwitter J, Seferovic P, Stepinska J, Trindade PT, Voors AA, Zannad F, Zeiher A, Guidelines ESCCfP. ESC Guidelines for the diagnosis and treatment of acute and chronic heart failure 2012: the Task Force for the diagnosis and treatment of acute and chronic heart failure 2012 of the European Society of Cardiology. Developed in collaboration with the Heart Failure Association (HFA) of the ESC. Eur Heart J. 2012;33(14):1787–847.
8. Burnett H, Earley A, Voors AA, Senni M, McMurray JJ, Deschaseaux C, Cope S. Thirty years of evidence on the efficacy of drug treatments for chronic heart failure with reduced ejection fraction: a network meta-analysis. Circ Heart Fail. 2017;10(1). https://doi.org/10.1161/CIRCHEARTFAILURE.116.003529.
9. Massie BM, Carson PE, McMurray JJ, Komajda M, McKelvie R, Zile MR, Anderson S, Donovan M, Iverson E, Staiger C, Ptaszynska A, Investigators IP. Irbesartan in patients with heart failure and preserved ejection fraction. N Engl J Med. 2008;359(23):2456–67.
10. Yusuf S, Pfeffer MA, Swedberg K, Granger CB, Held P, McMurray JJ, Michelson EL, Olofsson B, Ostergren J. Effects of candesartan in patients with chronic heart failure and preserved left-ventricular ejection fraction: the CHARM-Preserved Trial. Lancet. 2003;362(9386):777–81.
11. Pitt B, Pfeffer MA, Assmann SF, Boineau R, Anand IS, Claggett B, Clausell N, Desai AS, Diaz R, Fleg JL, Gordeev I, Harty B, Heitner JF, Kenwood CT, Lewis EF, O'Meara E, Probstfield JL, Shaburishvili T, Shah SJ, Solomon SD, Sweitzer NK, Yang S, McKinlay SM, Investigators T. Spironolactone for heart failure with preserved ejection fraction. N Engl J Med. 2014;370(15):1383–92.
12. Lam CS, Donal E, Kraigher-Krainer E, Vasan RS. Epidemiology and clinical course of heart failure with preserved ejection fraction. Eur J Heart Fail. 2011;13 (1):18–28.
13. Lam CS, Solomon SD. The middle child in heart failure: heart failure with mid-range ejection fraction (40–50%). Eur J Heart Fail. 2014;16(10):1049–55.
14. Dunlay SM, Roger VL, Weston SA, Jiang R, Redfield MM. Longitudinal changes in ejection fraction in heart failure patients with preserved and reduced ejection fraction. Circ Heart Fail. 2012;5(6):720–6.
15. Koh AS, Tay WT, Teng THK, Vedin O, Benson L, Dahlstrom U, Savarese G, CSP L, Lund LH. A comprehensive population-based characterization of heart failure with mid-range ejection fraction. Eur J Heart Fail. 2017;19:1624–34.
16. Bleumink GS, Knetsch AM, Sturkenboom MC, Straus SM, Hofman A, Deckers JW, Witteman JC, Stricker BH. Quantifying the heart failure epidemic: prevalence, incidence rate, lifetime risk and prognosis of heart failure The Rotterdam Study. Eur Heart J. 2004;25(18):1614–9.
17. Zarrinkoub R, Wettermark B, Wandell P, Mejhert M, Szulkin R, Ljunggren G, Kahan T. The epidemiology of heart failure, based on data for 2.1 million inhabitants in Sweden. Eur J Heart Fail. 2013;15(9):995–1002.
18. Mosterd A, Hoes AW, de Bruyne MC, Deckers JW, Linker DT, Hofman A, Grobbee DE. Prevalence of heart failure and left ventricular dysfunction in the general population; The Rotterdam Study. Eur Heart J. 1999;20(6):447–55.
19. Hogg K, Swedberg K, McMurray J. Heart failure with preserved left ventricular systolic function; epidemiology, clinical characteristics, and prognosis. J Am Coll Cardiol. 2004;43(3):317–27.
20. Savarese G, Lund LH. Global public health burden of heart failure. Card Fail Rev. 2017;3(1):7–11.
21. Schocken DD, Arrieta MI, Leaverton PE, Ross EA. Prevalence and mortality rate of congestive heart failure in the United States. J Am Coll Cardiol. 1992;20 (2):301–6.
22. Ho KK, Pinsky JL, Kannel WB, Levy D. The epidemiology of heart failure: the Framingham Study. J Am Coll Cardiol. 1993;22(4 Suppl A):6A–13A.
23. Ohlmeier C, Mikolajczyk R, Frick J, Prutz F, Haverkamp W, Garbe E. Incidence, prevalence and 1-year all-cause mortality of heart failure in Germany: a study based on electronic healthcare data of more than six million persons. Clin Res Cardiol. 2015;104(8):688–96.
24. Gomez-Soto FM, Andrey JL, Garcia-Egido AA, Escobar MA, Romero SP, Garcia-Arjona R, Gutierrez J, Gomez F. Incidence and mortality of heart failure: a community-based study. Int J Cardiol. 2011;151(1):40–5.
25. Ni H. Prevalence of self-reported heart failure among US adults: results from the 1999 National Health Interview Survey. Am Heart J. 2003;146(1):121–8.
26. Bursi F, Weston SA, Redfield MM, Jacobsen SJ, Pakhomov S, Nkomo VT, Meverden RA, Roger VL. Systolic and diastolic heart failure in the community. JAMA. 2006;296(18):2209–16.
27. Goyal P, Almarzooq ZI, Horn EM, Karas MG, Sobol I, Swaminathan RV, Feldman DN, Minutello RM, Singh HS, Bergman GW, Wong SC, Kim LK. Characteristics of hospitalizations for heart failure with preserved ejection fraction. Am J Med. 2016;129(6):635 e15–26.

28. Owan TE, Hodge DO, Herges RM, Jacobsen SJ, Roger VL, Redfield MM. Trends in prevalence and outcome of heart failure with preserved ejection fraction. N Engl J Med. 2006;355(3):251–9.
29. Cowie MR, Wood DA, Coats AJ, Thompson SG, Poole-Wilson PA, Suresh V, Sutton GC. Incidence and aetiology of heart failure; a population-based study. Eur Heart J. 1999;20(6):421–8.
30. Kannel WB. Incidence and epidemiology of heart failure. Heart Fail Rev. 2000;5(2):167–73.
31. Meyer S, Brouwers FP, Voors AA, Hillege HL, de Boer RA, Gansevoort RT, van der Harst P, Rienstra M, van Gelder IC, van Veldhuisen DJ, van Gilst WH, van der Meer P. Sex differences in new-onset heart failure. Clin Res Cardiol. 2015;104(4):342–50.
32. Roger VL, Weston SA, Redfield MM, Hellermann-Homan JP, Killian J, Yawn BP, Jacobsen SJ. Trends in heart failure incidence and survival in a community-based population. JAMA. 2004;292(3):344–50.
33. Johansson S, Wallander MA, Ruigomez A, Garcia Rodriguez LA. Incidence of newly diagnosed heart failure in UK general practice. Eur J Heart Fail. 2001;3(2):225–31.
34. Ceia F, Fonseca C, Mota T, Morais H, Matias F, de Sousa A, Oliveira A, Investigators E. Prevalence of chronic heart failure in Southwestern Europe: the EPICA study. Eur J Heart Fail. 2002;4(4):531–9.
35. Lloyd-Jones DM, Larson MG, Leip EP, Beiser A, D'Agostino RB, Kannel WB, Murabito JM, Vasan RS, Benjamin EJ, Levy D. Lifetime risk for developing congestive heart failure: the Framingham Heart Study. Circulation. 2002;106(24):3068–72.
36. Brouwers FP, de Boer RA, van der Harst P, Voors AA, Gansevoort RT, Bakker SJ, Hillege HL, van Veldhuisen DJ, van Gilst WH. Incidence and epidemiology of new onset heart failure with preserved vs. reduced ejection fraction in a community-based cohort: 11-year follow-up of PREVEND. Eur Heart J. 2013;34(19):1424–31.
37. Gerber Y, Weston SA, Redfield MM, Chamberlain AM, Manemann SM, Jiang R, Killian JM, Roger VL. A contemporary appraisal of the heart failure epidemic in Olmsted County, Minnesota, 2000 to 2010. JAMA Intern Med. 2015;175(6):996–1004.
38. Fonarow GC, Stough WG, Abraham WT, Albert NM, Gheorghiade M, Greenberg BH, O'Connor CM, Sun JL, Yancy CW, Young JB, Investigators O-H, Hospitals. Characteristics, treatments, and outcomes of patients with preserved systolic function hospitalized for heart failure: a report from the OPTIMIZE-HF Registry. J Am Coll Cardiol. 2007;50(8):768–77.
39. Wenger NK, Speroff L, Packard B. Cardiovascular health and disease in women. N Engl J Med. 1993;329 (4):247–56.
40. Ho KK, Anderson KM, Kannel WB, Grossman W, Levy D. Survival after the onset of congestive heart failure in Framingham Heart Study subjects. Circulation. 1993;88 (1):107–15.
41. Vaartjes I, Hoes AW, Reitsma JB, de Bruin A, Grobbee DE, Mosterd A, Bots MI. Age- and gender-specific risk of death after first hospitalization for heart failure. BMC Public Health. 2010;10:637.
42. Martinez-Selles M, Doughty RN, Poppe K, Whalley GA, Earle N, Tribouilloy C, McMurray JJ, Swedberg K, Kober L, Berry C, Squire I, Meta-Analysis Global Group In Chronic Heart F. Gender and survival in patients with heart failure: interactions with diabetes and aetiology. Results from the MAGGIC individual patient meta-analysis. Eur J Heart Fail. 2012;14(5):473–9.
43. O'Meara E, Clayton T, McEntegart MB, McMurray JJ, Pina IL, Granger CB, Ostergren J, Michelson EL, Solomon SD, Pocock S, Yusuf S, Swedberg K, Pfeffer MA, Investigators C. Sex differences in clinical characteristics and prognosis in a broad spectrum of patients with heart failure: results of the Candesartan in Heart failure: assessment of reduction in mortality and morbidity (CHARM) program. Circulation. 2007;115(24):3111–20.
44. Deswal A, Bozkurt B. Comparison of morbidity in women versus men with heart failure and preserved ejection fraction. Am J Cardiol. 2006;97(8):1228–31.
45. Lam CS, Carson PE, Anand IS, Rector TS, Kuskowski M, Komajda M, McKelvie RS, McMurray JJ, Zile MR, Massie BM, Kitzman DW. Sex differences in clinical characteristics and outcomes in elderly patients with heart failure and preserved ejection fraction: the Irbesartan in Heart Failure with Preserved Ejection Fraction (I-PRESERVE) trial. Circ Heart Fail. 2012;5(5):571–8.
46. Rodriguez F, Wang Y, Johnson CE, Foody JM. National patterns of heart failure hospitalizations and mortality by sex and age. J Card Fail. 2013;19(8):542–9.
47. Parashar S, Katz R, Smith NL, Arnold AM, Vaccarino V, Wenger NK, Gottdiener JS. Race, gender, and mortality in adults > or =65 years of age with incident heart failure (from the Cardiovascular Health Study). Am J Cardiol. 2009;103 (8):1120–7.
48. Adams KF Jr, Sueta CA, Gheorghiade M, O'Connor CM, Schwartz TA, Koch GG, Uretsky B, Swedberg K, McKenna W, Soler-Soler J, Califf RM. Gender differences in survival in advanced heart failure. Insights from the FIRST study. Circulation. 1999;99(14):1816–21.
49. Martinez-Selles M, Dominguez M, Martinez E, Garcia Fernandez MA, Garcia E. Women with left ventricular ejection fraction < or = 20% have better prognosis than men. Int J Cardiol. 2007;120(2):276–8.
50. Ghali JK, Krause-Steinrauf HJ, Adams KF, Khan SS, Rosenberg YD, Yancy CW, Young JB, Goldman S, Peberdy MA, Lindenfeld J. Gender differences in advanced heart failure: insights from the BEST study. J Am Coll Cardiol. 2003;42(12):2128–34.
51. Hsich EM, Grau-Sepulveda MV, Hernandez AF, Peterson ED, Schwamm LH, Bhatt DL, Fonarow GC. Sex differences in in-hospital mortality in acute decompensated heart failure with reduced and preserved ejection fraction. Am Heart J. 2012;163(3):430-7–437 e1-3.
52. Nakada Y, Kawakami R, Nakano T, Takitsume A, Nakagawa H, Ueda T, Nishida T, Onoue K, Soeda T, Okayama S, Takeda Y, Watanabe M, Kawata H, Okura H, Saito Y. Sex differences in clinical characteristics and long-term outcome in acute decompensated heart failure patients with preserved and reduced ejection fraction. Am J Physiol Heart Circ Physiol. 2016;310(7):H813–20.
53. Zsilinszka R, Shrader P, DeVore AD, Hardy NC, Mentz RJ, Pang PS, Peacock WF, Fonarow GC, Hernandez AF. Sex differences in the management and outcomes of heart failure with preserved ejection fraction in patients presenting to the emergency department with acute heart failure. J Card Fail. 2016;22(10):781–8.
54. Kajimoto K, Minami Y, Sato N, Otsubo S, Kasanuki H, Investigators of the Acute Decompensated Heart Failure Syndromes Registry. Gender differences in left ventricular ejection fraction and outcomes among patients hospitalized for acute decompensated heart failure. Am J Cardiol. 2017;119:1623–30.
55. Hobbs FD, Kenkre JE, Roalfe AK, Davis RC, Hare R, Davies MK. Impact of heart failure and left ventricular systolic dysfunction on quality of life: a cross-sectional study

comparing common chronic cardiac and medical disorders and a representative adult population. Eur Heart J. 2002;23 (23):1867–76.
56. Chin MH, Goldman L. Gender differences in 1-year survival and quality of life among patients admitted with congestive heart failure. Med Care. 1998;36(7):1033–46.
57. Stewart S, Ekman I, Ekman T, Oden A, Rosengren A. Population impact of heart failure and the most common forms of cancer: a study of 1 162 309 hospital cases in Sweden (1988 to 2004). Circ Cardiovasc Qual Outcomes. 2010;3(6):573–80.
58. Hoekstra T, Jaarsma T, van Veldhuisen DJ, Hillege HL, Sanderman R, Lesman-Leegte I. Quality of life and survival in patients with heart failure. Eur J Heart Fail. 2013;15 (1):94–102.
59. Riegel B, Moser DK, Carlson B, Deaton C, Armola R, Sethares K, Shively M, Evangelista L, Albert N. Gender differences in quality of life are minimal in patients with heart failure. J Card Fail. 2003;9(1):42–8.
60. Friedman MM. Gender differences in the health related quality of life of older adults with heart failure. Heart Lung. 2003;32(5):320–7.
61. Simon M, Boutouyrie P, Narayanan K, Gaye B, Tafflet M, Thomas F, Guibout C, Perier MC, Pannier B, Jouven X, Empana JP. Sex disparities in ideal cardiovascular health. Heart (British Cardiac Society). 2017;103:1595–601.
62. He J, Ogden LG, Bazzano LA, Vupputuri S, Loria C, Whelton PK. Risk factors for congestive heart failure in US men and women: NHANES I epidemiologic follow-up study. Arch Intern Med. 2001;161(7):996–1002.
63. Lee DS, Gona P, Vasan RS, Larson MG, Benjamin EJ, Wang TJ, Tu JV, Levy D. Relation of disease pathogenesis and risk factors to heart failure with preserved or reduced ejection fraction: insights from the framingham heart study of the national heart, lung, and blood institute. Circulation. 2009;119(24):3070–7.
64. Dunlay SM, Weston SA, Jacobsen SJ, Roger VL. Risk factors for heart failure: a population-based case-control study. Am J Med. 2009;122(11):1023–8.
65. Hsich EM, Pina IL. Heart failure in women: a need for prospective data. J Am Coll Cardiol. 2009;54(6):491–8.
66. Taylor AL. Heart failure in women. Curr Heart Fail Rep. 2015;12(2):187–95.
67. Dunlay SM, Roger VL. Gender differences in the pathophysiology, clinical presentation, and outcomes of ischemic heart failure. Curr Heart Fail Rep. 2012;9(4):267–76.
68. Kishi S, Reis JP, Venkatesh BA, Gidding SS, Armstrong AC, Jacobs DR Jr, Sidney S, Wu CO, Cook NL, Lewis CE, Schreiner PJ, Isogawa A, Liu K, Lima JA. Race-ethnic and sex differences in left ventricular structure and function: the Coronary Artery Risk Development in Young Adults (CARDIA) Study. J Am Heart Assoc. 2015;4(3):e001264.
69. Meyer S, van der Meer P, van Deursen VM, Jaarsma T, van Veldhuisen DJ, van der Wal MH, Hillege HL, Voors AA. Neurohormonal and clinical sex differences in heart failure. Eur Heart J. 2013;34(32):2538–47.
70. Azad N, Kathiravelu A, Minoosepeher S, Hebert P, Fergusson D. Gender differences in the etiology of heart failure: a systematic review. J Geriatr Cardiol. 2011;8 (1):15–23.
71. Jacobs AK. Coronary intervention in 2009. Are women no different than men? Circ Cardiovasc Interv. 2009;2 (1):69–78.
72. Crea F, Battipaglia I, Andreotti F. Sex differences in mechanisms, presentation and management of ischaemic heart disease. Atherosclerosis. 2015;241(1):157–68.
73. Mitoff PR, Gam D, Ivanov J, Al-hesayen A, Azevedo ER, Newton GE, Parker JD, Mak S. Cardiac-specific sympathetic activation in men and women with and without heart failure. Heart. 2011;97(5):382–7.
74. Galvao M, Kalman J, DeMarco T, Fonarow GC, Galvin C, Ghali JK, Moskowitz RM. Gender differences in in-hospital management and outcomes in patients with decompensated heart failure: analysis from the Acute Decompensated Heart Failure National Registry (ADHERE). J Card Fail. 2006;12 (2):100–7.
75. Frazier CG, Alexander KP, Newby LK, Anderson S, Iverson E, Packer M, Cohn J, Goldstein S, Douglas PS. Associations of gender and etiology with outcomes in heart failure with systolic dysfunction: a pooled analysis of 5 randomized control trials. J Am Coll Cardiol. 2007;49 (13):1450–8.
76. Petrie MC, Dawson NF, Murdoch DR, Davie AP, McMurray JJ. Failure of women's hearts. Circulation. 1999;99 (17):2334–41.
77. Borlaug BA. Defining HFpEF: where do we draw the line? Eur Heart J. 2016;37(5):463–5.
78. Luo T, Kim JK. The role of estrogen and estrogen receptors on cardiomyocytes: an overview. Can J Cardiol. 2016;32 (8):1017–25.
79. Bouma W, Noma M, Kanemoto S, Matsubara M, Leshnower BG, Hinmon R, Gorman JH 3rd, Gorman RC. Sex-related resistance to myocardial ischemia-reperfusion injury is associated with high constitutive ARC expression. Am J Physiol Heart Circ Physiol. 2010;298(5):H1510–7.
80. Harvey RE, Coffman KE, Miller VM. Women-specific factors to consider in risk, diagnosis and treatment of cardiovascular disease. Womens Health (Lond). 2015;11 (2):239–57.
81. Ntusi NB, Mayosi BM. Aetiology and risk factors of peripartum cardiomyopathy: a systematic review. Int J Cardiol. 2009;131(2):168–79.
82. Arany Z, Elkayam U. Peripartum cardiomyopathy. Circulation. 2016;133(14):1397–409.
83. Hilfiker-Kleiner D, Sliwa K. Pathophysiology and epidemiology of peripartum cardiomyopathy. Nat Rev Cardiol. 2014;11(6):364–70.
84. Wang F, He Q, Sun Y, Dai X, Yang XP. Female adult mouse cardiomyocytes are protected against oxidative stress. Hypertension. 2010;55(5):1172–8.
85. Franssen C, Chen S, Unger A, Korkmaz HI, De Keulenaer GW, Tschope C, Leite-Moreira AF, Musters R, Niessen HW, Linke WA, Paulus WJ, Hamdani N. Myocardial microvascular inflammatory endothelial activation in heart failure with preserved ejection fraction. JACC Heart Fail. 2016;4 (4):312–24.
86. Tasevska-Dinevska G, Kennedy LM, Cline-Iwarson A, Cline C, Erhardt L, Willenheimer R. Gender differences in variables related to B-natriuretic peptide, left ventricular ejection fraction and mass, and peak oxygen consumption, in patients with heart failure. Int J Cardiol. 2011;149(3):364–71.
87. Elmariah S, Goldberg LR, Allen MT, Kao A. Effects of gender on peak oxygen consumption and the timing of cardiac transplantation. J Am Coll Cardiol. 2006;47 (11):2237–42.
88. Luchner A, Brockel U, Muscholl M, Hense HW, Doring A, Riegger GA, Schunkert H. Gender-specific differences of cardiac remodeling in subjects with left ventricular dysfunction: a population-based study. Cardiovasc Res. 2002;53 (3):720–7.
89. Greiten LE, Holditch SJ, Arunachalam SP, Miller VM. Should there be sex-specific criteria for the diagnosis

and treatment of heart failure? J Cardiovasc Transl Res. 2014;7(2):139–55.
90. Salton CJ, Chuang ML, O'Donnell CJ, Kupka MJ, Larson MG, Kissinger KV, Edelman RR, Levy D, Manning WJ. Gender differences and normal left ventricular anatomy in an adult population free of hypertension. A cardiovascular magnetic resonance study of the Framingham Heart Study Offspring Cohort. J Am Coll Cardiol. 2002;39 (6):1055–60.
91. Wingate S. Cardiovascular anatomy and physiology in the female. Crit Care Nurs Clin North Am. 1997;9(4):447–52.
92. Chugh SS, Uy-Evanado A, Teodorescu C, Reinier K, Mariani R, Gunson K, Jui J. Women have a lower prevalence of structural heart disease as a precursor to sudden cardiac arrest: The Ore-SUDS (Oregon Sudden Unexpected Death Study). J Am Coll Cardiol. 2009;54(22):2006–11.
93. Jessup M, Pina IL. Is it important to examine gender differences in the epidemiology and outcome of severe heart failure? J Thorac Cardiovasc Surg. 2004;127 (5):1247–52.
94. Gori M, Lam CS, Gupta DK, Santos AB, Cheng S, Shah AM, Claggett B, Zile MR, Kraigher-Krainer E, Pieske B, Voors AA, Packer M, Bransford T, Lefkowitz M, McMurray JJ, Solomon SD, Investigators P. Sex-specific cardiovascular structure and function in heart failure with preserved ejection fraction. Eur J Heart Fail. 2014;16(5):535–42.
95. Russo C, Jin Z, Palmieri V, Homma S, Rundek T, Elkind MS, Sacco RL, Di Tullio MR. Arterial stiffness and wave reflection: sex differences and relationship with left ventricular diastolic function. Hypertension. 2012;60(2):362–8.
96. Shim CY, Park S, Choi D, Yang WI, Cho IJ, Choi EY, Chung N, Ha JW. Sex differences in central hemodynamics and their relationship to left ventricular diastolic function. J Am Coll Cardiol. 2011;57(10):1226–33.
97. Vinereanu D, Margulescu AD. The fallacy of resting echocardiographic parameters of cardiac function in heart failure with preserved ejection fraction: add global longitudinal strain to the list. Eur J Heart Fail. 2017;19(7):901–3.
98. DeVore AD, McNulty S, Alenezi F, Ersboll M, Vader JM, Oh JK, Lin G, Redfield MM, Lewis G, Semigran MJ, Anstrom KJ, Hernandez AF, Velazquez EJ. Impaired left ventricular global longitudinal strain in patients with heart failure with preserved ejection fraction: insights from the RELAX trial. Eur J Heart Fail. 2017;19(7):893–900.
99. Scantlebury DC, Borlaug BA. Why are women more likely than men to develop heart failure with preserved ejection fraction? Curr Opin Cardiol. 2011;26(6):562–8.
100. Mehilli J, Ndrepepa G, Kastrati A, Nekolla SG, Markwardt C, Bollwein H, Pache J, Martinoff S, Dirschinger J, Schwaiger M, Schomig A. Gender and myocardial salvage after reperfusion treatment in acute myocardial infarction. J Am Coll Cardiol. 2005;45 (6):828–31.
101. Cavasin MA, Tao Z, Menon S, Yang XP. Gender differences in cardiac function during early remodeling after acute myocardial infarction in mice. Life Sci. 2004;75 (18):2181–92.
102. Fliegner D, Schubert C, Penkalla A, Witt H, Kararigas G, Dworatzek E, Staub E, Martus P, Ruiz Noppinger P, Kintscher U, Gustafsson JA, Regitz-Zagrosek V. Female sex and estrogen receptor-beta attenuate cardiac remodeling and apoptosis in pressure overload. Am J Physiol Regul Integr Comp Physiol. 2010;298(6):R1597–606.
103. Gardner JD, Brower GL, Janicki JS. Gender differences in cardiac remodeling secondary to chronic volume overload. J Card Fail. 2002;8(2):101–7.
104. Guerra S, Leri A, Wang X, Finato N, Di Loreto C, Beltrami CA, Kajstura J, Anversa P. Myocyte death in the failing human heart is gender dependent. Circ Res. 1999;85 (9):856–66.
105. Lu H, Melendez GC, Levick SP, Janicki JS. Prevention of adverse cardiac remodeling to volume overload in female rats is the result of an estrogen-altered mast cell phenotype. Am J Physiol Heart Circ Physiol. 2012;302(3):H811–7.
106. Petre RE, Quaile MP, Rossman EI, Ratcliffe SJ, Bailey BA, Houser SR, Margulies KB. Sex-based differences in myocardial contractile reserve. Am J Physiol Regul Integr Comp Physiol. 2007;292(2):R810–8.
107. Johnstone D, Limacher M, Rousseau M, Liang CS, Ekelund L, Herman M, Stewart D, Guillotte M, Bjerken G, Gaasch W, et al. Clinical characteristics of patients in studies of left ventricular dysfunction (SOLVD). Am J Cardiol. 1992;70(9):894–900.
108. Ng AC, Wong HS, Yong AS, Sindone AP. Impact of gender on outcomes in chronic systolic heart failure. Int J Cardiol. 2007;117(2):214–21.
109. Wang TJ, Larson MG, Levy D, Leip EP, Benjamin EJ, Wilson PW, Sutherland P, Omland T, Vasan RS. Impact of age and sex on plasma natriuretic peptide levels in healthy adults. Am J Cardiol. 2002;90(3):254–8.
110. Christ M, Laule-Kilian K, Hochholzer W, Klima T, Breidthardt T, Perruchoud AP, Mueller C. Gender-specific risk stratification with B-type natriuretic peptide levels in patients with acute dyspnea: insights from the B-type natriuretic peptide for acute shortness of breath evaluation study. J Am Coll Cardiol. 2006;48(9):1808–12.
111. Meyer S, Teerlink JR, Metra M, Ponikowski P, Cotter G, Davison BA, Felker GM, Filippatos G, Greenberg BH, Hua TA, Severin T, Qian M, Voors AA. Sex differences in early dyspnea relief between men and women hospitalized for acute heart failure: insights from the RELAX-AHF study. Clin Res Cardiol. 2017;106(4):280–92.
112. Nieminen MS, Harjola VP, Hochadel M, Drexler H, Komajda M, Brutsaert D, Dickstein K, Ponikowski P, Tavazzi L, Follath F, Lopez-Sendon JL. Gender related differences in patients presenting with acute heart failure. Results from EuroHeart Failure Survey II. Eur J Heart Fail. 2008;10(2):140–8.
113. Tsuchihashi-Makaya M, Hamaguchi S, Kinugawa S, Goto K, Goto D, Furumoto T, Yamada S, Yokoshiki H, Takeshita A, Tsutsui H. Sex differences with respect to clinical characteristics, treatment, and long-term outcomes in patients with heart failure. Int J Cardiol. 2011;150(3):338–9.
114. Andreotti F, Conti E, Lanza GA, Crea F. Sex, survival bias, and mortality following acute myocardial infarction. Ital Heart J. 2003;4(8):508–10.
115. van Deursen VM, Urso R, Laroche C, Damman K, Dahlstrom U, Tavazzi L, Maggioni AP, Voors AA. Co-morbidities in patients with heart failure: an analysis of the European Heart Failure Pilot Survey. Eur J Heart Fail. 2014;16(1):103–11.
116. Hopper I, Kotecha D, Chin KL, Mentz RJ, von Lueder TG. Comorbidities in heart failure: are there gender differences? Curr Heart Fail Rep. 2016;13(1):1–12.
117. Mentz RJ, Kelly JP, von Lueder TG, Voors AA, Lam CS, Cowie MR, Kjeldsen K, Jankowska EA, Atar D, Butler J, Fiuzat M, Zannad F, Pitt B, O'Connor CM. Noncardiac comorbidities in heart failure with reduced versus preserved ejection fraction. J Am Coll Cardiol. 2014;64(21):2281–93.
118. Pitt B. Male gender, diabetes, COPD, anemia, and creatinine clearance < 30 mL/min predicted hospitalization

after heart failure diagnosis. Ann Intern Med. 2010;152(8): JC4-2–3.
119. Dunlay SM, Roger VL, Redfield MM. Epidemiology of heart failure with preserved ejection fraction. Nat Rev Cardiol. 2017;14(10):591–602.
120. Kenchaiah S, Vasan RS. Heart failure in women – insights from the Framingham Heart Study. Cardiovasc Drugs Ther. 2015;29(4):377–90.
121. Silverberg D, Wexler D, Blum M, Schwartz D, Iaina A. The association between congestive heart failure and chronic renal disease. Curr Opin Nephrol Hypertens. 2004;13 (2):163–70.
122. Ronco C, Ronco F. Cardio-renal syndromes: a systematic approach for consensus definition and classification. Heart Fail Rev. 2012;17(2):151–60.
123. Schefold JC, Filippatos G, Hasenfuss G, Anker SD, von Haehling S. Heart failure and kidney dysfunction: epidemiology, mechanisms and management. Nat Rev Nephrol. 2016;12(10):610–23.
124. Vassalotti JA, Centor R, Turner BJ, Greer RC, Choi M, Sequist TD. Practical approach to detection and management of chronic kidney disease for the primary care clinician. Am J Med. 2016;129(2):153–162.e7.
125. Heywood JT, Fonarow GC, Costanzo MR, Mathur VS, Wigneswaran JR, Wynne J, Committee ASA, Investigators. High prevalence of renal dysfunction and its impact on outcome in 118,465 patients hospitalized with acute decompensated heart failure: a report from the ADHERE database. J Card Fail. 2007;13(6):422–30.
126. Conde-Martel A, Arkuch ME, Formiga F, Manzano-Espinosa L, Aramburu-Bodas O, Gonzalez-Franco A, Davila-Ramos MF, Suarez-Pedreira I, Herrero-Domingo A, Montero-Perez-Barquero M. Gender related differences in clinical profile and outcome of patients with heart failure. Results of the RICA Registry. Rev Clin Esp. 2015;215 (7):363–70.
127. Kosiborod M, Curtis JP, Wang Y, et al. Anemia and outcomes in patients with heart failure: a study from the national heart care project. Arch Intern Med. 2005;165 (19):2237–44.
128. Arora NP, Ghali JK. Iron deficiency anemia in heart failure. Heart Fail Rev. 2013;18(4):485–501.
129. van Veldhuisen DJ, Anker SD, Ponikowski P, Macdougall IC. Anemia and iron deficiency in heart failure: mechanisms and therapeutic approaches. Nat Rev Cardiol. 2011;8 (9):485–93.
130. Cleland JG, Zhang J, Pellicori P, Dicken B, Dierckx R, Shoaib A, Wong K, Rigby A, Goode K, Clark AL. Prevalence and outcomes of anemia and hematinic deficiencies in patients with chronic heart failure. JAMA Cardiol. 2016;1(5):539–47.
131. Ebner N, Jankowska EA, Ponikowski P, Lainscak M, Elsner S, Sliziuk V, Steinbeck L, Kube J, Bekfani T, Scherbakov N. The impact of iron deficiency and anaemia on exercise capacity and outcomes in patients with chronic heart failure. Results from the studies investigating co-morbidities aggravating heart failure. Int J Cardiol. 2016;205:6–12.
132. de Silva R, Rigby AS, Witte KK, Nikitin NP, Tin L, Goode K, Bhandari S, Clark AL, Cleland JG. Anemia, renal dysfunction, and their interaction in patients with chronic heart failure. Am J Cardiol. 2006;98(3):391–8.
133. Kajimoto K, Minami Y, Sato N, Otsubo S, Kasanuki H. Gender differences in anemia and survival in patients hospitalized for acute decompensated heart failure with preserved or reduced ejection fraction. Am J Cardiol. 2017;120(3):435–42.
134. Ogah OS, Davison BA, Sliwa K, Mayosi BM, Damasceno A, Sani MU, Mondo C, Dzudie A, Ojji DB, Kouam C, Suliman A, Schrueder N, Yonga G, Ba SA, Maru F, Alemayehu B, Edwards C, Cotter G. Gender differences in clinical characteristics and outcome of acute heart failure in sub-Saharan Africa: results of the THESUS-HF study. Clin Res Cardiol. 2015;104(6):481–90.
135. Cohen-Solal A, Damy T, Terbah M, Kerebel S, Baguet J-P, Hanon O, Zannad F, Laperche T, Leclercq C, Concas V, Duvillié L, Darné B, Anker S, Mebazaa A. High prevalence of iron deficiency in patients with acute decompensated heart failure. Eur J Heart Fail. 2014;16(9):984–91.
136. Kajimoto K, Sato N, Takano T. Association between anemia, clinical features and outcome in patients hospitalized for acute heart failure syndromes. Eur Heart J: Acute Card Care. 2014;4(6):568–76.
137. Yamauchi T, Sakata Y, Takada T, Nochioka K, Miura M, Tadaki S, Ushigome R, Sato K, Onose T, Tsuji K. Prognostic impact of anemia in patients with chronic heart failure–with special reference to clinical background: report from the CHART-2 Study. Circ J. 2015;79(9):1984–93.
138. McNallan SM, Chamberlain AM, Gerber Y, Singh M, Kane RL, Weston SA, Dunlay SM, Jiang R, Roger VL. Measuring frailty in heart failure: a community perspective. Am Heart J. 2013;166(4):768–74.
139. Altimir S, Lupon J, Gonzalez B, Prats M, Parajon T, Urrutia A, Coll R, Valle V. Sex and age differences in fragility in a heart failure population. Eur J Heart Fail. 2005;7(5):798–802.
140. Gonzalez-Moneo MJ, Sanchez-Benavides G, Verdu-Rotellar JM, Cladellas M, Bruguera J, Quinones-Ubeda S, Enjuanes C, Pena-Casanova J, Comin-Colet J. Ischemic aetiology, self-reported frailty, and gender with respect to cognitive impairment in chronic heart failure patients. BMC Cardiovasc Disord. 2016;16(1):163.
141. Kaila K, Haykowsky MJ, Thompson RB, Paterson DI. Heart failure with preserved ejection fraction in the elderly: scope of the problem. Heart Fail Rev. 2012;17(4-5):555–62.
142. Gottlieb SS, Khatta M, Friedmann E, Einbinder L, Katzen S, Baker B, Marshall J, Minshall S, Robinson S, Fisher ML, Potenza M, Sigler B, Baldwin C, Thomas SA. The influence of age, gender, and race on the prevalence of depression in heart failure patients. J Am Coll Cardiol. 2004;43(9):1542–9.
143. Jiang W, Alexander J, Christopher E, et al. Relationship of depression to increased risk of mortality and rehospitalization in patients with congestive heart failure. Arch Intern Med. 2001;161(15):1849–56.
144. Freedland KE, Hesseler MJ, Carney RM, Steinmeyer BC, Skala JA, Dávila-Román VG, Rich MW. Major depression and long-term survival of patients with heart failure. Psychosom Med. 2016;78(8):896–903.
145. Vaccarino V, Kasl SV, Abramson J, Krumholz HM. Depressive symptoms and risk of functional decline and death in patients with heart failure. J Am Coll Cardiol. 2001;38(1):199–205.
146. Ghali JK, Lindenfeld J. Sex differences in response to chronic heart failure therapies. Expert Rev Cardiovasc Ther. 2008;6(4):555–65.
147. Gan SC, Beaver SK, Houck PM, MacLehose RF, Lawson HW, Chan L. Treatment of acute myocardial infarction and 30-day mortality among women and men. N Engl J Med. 2000;343(1):8–15.

148. Berg J, Bjorck L, Nielsen S, Lappas G, Rosengren A. Sex differences in survival after myocardial infarction in Sweden, 1987–2010. Heart. 2017;103(20):1625–30.
149. Shekelle PG, Rich MW, Morton SC, Atkinson CS, Tu W, Maglione M, Rhodes S, Barrett M, Fonarow GC, Greenberg B, Heidenreich PA, Knabel T, Konstam MA, Steimle A, Warner Stevenson L. Efficacy of angiotensin-converting enzyme inhibitors and beta-blockers in the management of left ventricular systolic dysfunction according to race, gender, and diabetic status: a meta-analysis of major clinical trials. J Am Coll Cardiol. 2003;41(9):1529–38.
150. Garg R, Yusuf S. Overview of randomized trials of angiotensin-converting enzyme-inhibitors on mortality and morbidity in patients with heart-failure (Vol 273, Pg 1450, 1995). Jama-J Am Med Assoc. 1995;274(6):462–2.
151. Keyhan G, Chen SF, Pilote L. Angiotensin-converting enzyme inhibitors and survival in women and men with heart failure. Eur J Heart Fail. 2007;9(6-7):594–601.
152. Young JB, Dunlap ME, Pfeffer MA, Probstfield JL, Cohen-Solal A, Dietz R, Granger CB, Hradec J, Kuch J, McKelvie RS, McMurray JJ, Michelson EL, Olofsson B, Ostergren J, Held P, Solomon SD, Yusuf S, Swedberg K, Candesartan in Heart Failure Assessment of Reduction in M, Morbidity I, Committees. Mortality and morbidity reduction with Candesartan in patients with chronic heart failure and left ventricular systolic dysfunction: results of the CHARM low-left ventricular ejection fraction trials. Circulation. 2004;110(17):2618–26.
153. Cohn JN, Tognoni G, Valsartan Heart Failure Trial I. A randomized trial of the angiotensin-receptor blocker valsartan in chronic heart failure. N Engl J Med. 2001;345 (23):1667–75.
154. McMurray JJ, Packer M, Desai AS, Gong J, Lefkowitz MP, Rizkala AR, Rouleau JL, Shi VC, Solomon SD, Swedberg K, Zile MR, Investigators P-H, Committees. Angiotensin-neprilysin inhibition versus enalapril in heart failure. N Engl J Med. 2014;371(11):993–1004.
155. Hudson M, Rahme E, Behlouli H, Sheppard R, Pilote L. Sex differences in the effectiveness of angiotensin receptor blockers and angiotensin converting enzyme inhibitors in patients with congestive heart failure – a population study. Eur J Heart Fail. 2007;9(6-7):602–9.
156. Zannad F, McMurray JJ, Krum H, van Veldhuisen DJ, Swedberg K, Shi H, Vincent J, Pocock SJ, Pitt B, Group E-HS. Eplerenone in patients with systolic heart failure and mild symptoms. N Engl J Med. 2011;364(1):11–21.
157. Pitt B, Zannad F, Remme WJ, Cody R, Castaigne A, Perez A, Palensky J, Wittes J. The effect of spironolactone on morbidity and mortality in patients with severe heart failure. Randomized Aldactone Evaluation Study Investigators. N Engl J Med. 1999;341(10):709–17.
158. Packer M, Bristow MR, Cohn JN, Colucci WS, Fowler MB, Gilbert EM, Shusterman NH. The effect of carvedilol on morbidity and mortality in patients with chronic heart failure. U.S. Carvedilol Heart Failure Study Group. N Engl J Med. 1996;334(21):1349–55.
159. Packer M, Coats AJ, Fowler MB, Katus HA, Krum H, Mohacsi P, Rouleau JL, Tendera M, Castaigne A, Roecker EB, Schultz MK, DeMets DL, Carvedilol Prospective Randomized Cumulative Survival Study G. Effect of carvedilol on survival in severe chronic heart failure. N Engl J Med. 2001;344(22):1651–8.
160. Leizorovicz A, Lechat P, Cucherat M, Bugnard F. Bisoprolol for the treatment of chronic heart failure: a meta-analysis on individual data of two placebo-controlled studies – CIBIS and CIBIS II. Cardiac Insufficiency Bisoprolol Study. Am Heart J. 2002;143(2):301–7.
161. Ghali JK, Pina IL, Gottlieb SS, Deedwania PC, Wikstrand JC, Group M-HS. Metoprolol CR/XL in female patients with heart failure: analysis of the experience in Metoprolol Extended-Release Randomized Intervention Trial in Heart Failure (MERIT-HF). Circulation. 2002;105 (13):1585–91.
162. Rathore SS, Wang Y, Krumholz HM. Sex-based differences in the effect of digoxin for the treatment of heart failure. N Engl J Med. 2002;347(18):1403–11.
163. Adams KF Jr, Patterson JH, Gattis WA, O'Connor CM, Lee CR, Schwartz TA, Gheorghiade M. Relationship of serum digoxin concentration to mortality and morbidity in women in the digitalis investigation group trial: a retrospective analysis. J Am Coll Cardiol. 2005;46(3):497–504.
164. MacFadden DR, Crystal E, Krahn AD, Mangat I, Healey JS, Dorian P, Birnie D, Simpson CS, Khaykin Y, Pinter A, Nanthakumar K, Calzavara AJ, Austin PC, Tu JV, Lee DS. Sex differences in implantable cardioverter-defibrillator outcomes: findings from a prospective defibrillator database. Ann Intern Med. 2012;156(3):195–203.
165. Curtis LH, Al-Khatib SM, Shea AM, Hammill BG, Hernandez AF, Schulman KA. Sex differences in the use of implantable cardioverter-defibrillators for primary and secondary prevention of sudden cardiac death. JAMA. 2007;298 (13):1517–24.
166. Hernandez AF, Fonarow GC, Liang L, Al-Khatib SM, Curtis LH, LaBresh KA, Yancy CW, Albert NM, Peterson ED. Sex and racial differences in the use of implantable cardioverter-defibrillators among patients hospitalized with heart failure. JAMA. 2007;298(13):1525–32.
167. Masoudi FA, Go AS, Magid DJ, Cassidy-Bushrow AE, Gurwitz JH, Liu TI, Reynolds K, Smith DH, Reifler LM, Glenn KA, Fiocchi F, Goldberg R, Gupta N, Peterson PN, Schuger C, Vidaillet H, Hammill SC, Greenlee RT. Age and sex differences in long-term outcomes following implantable cardioverter-defibrillator placement in contemporary clinical practice: findings from the Cardiovascular Research Network. J Am Heart Assoc. 2015;4(6):e002005.
168. Tompkins CM, McNitt S, Polonsky B, Daubert JP, Wang PJ, Moss AJ, Zareba W, Kutyifa V. Sex differences in inappropriate ICD device therapies: MADIT-II and MADIT-CRT. J Cardiovasc Electrophysiol. 2017;28(1):94–102.
169. Moss AJ, Zareba W, Hall WJ, Klein H, Wilber DJ, Cannom DS, Daubert JP, Higgins SL, Brown MW, Andrews ML, Multicenter Automatic Defibrillator Implantation Trial III. Prophylactic implantation of a defibrillator in patients with myocardial infarction and reduced ejection fraction. N Engl J Med. 2002;346(12):877–83.
170. Albert CM, Quigg R, Saba S, Estes NA 3rd, Shaechter A, Subacius H, Howard A, Levine J, Kadish A, Investigators D. Sex differences in outcome after implantable cardioverter defibrillator implantation in nonischemic cardiomyopathy. Am Heart J. 2008;156(2):367–72.
171. Chatterjee NA, Borgquist R, Chang Y, Lewey J, Jackson VA, Singh JP, Metlay JP, Lindvall C. Increasing sex differences in the use of cardiac resynchronization therapy with or without implantable cardioverter-defibrillator. Eur Heart J. 2017;38(19):1485–94.
172. Cleland JG, Daubert JC, Erdmann E, Freemantle N, Gras D, Kappenberger L, Tavazzi L, Cardiac Resynchronization-Heart Failure Study I. The effect of cardiac resynchronization on morbidity and mortality in heart failure. N Engl J Med. 2005;352(15):1539–49.

173. Bristow MR, Saxon LA, Boehmer J, Krueger S, Kass DA, De Marco T, Carson P, DiCarlo L, DeMets D, White BG, DeVries DW, Feldman AM, Comparison of Medical Therapy P, Defibrillation in Heart Failure I. Cardiac-resynchronization therapy with or without an implantable defibrillator in advanced chronic heart failure. N Engl J Med. 2004;350(21):2140–50.
174. Hsich EM, Naftel DC, Myers SL, Gorodeski EZ, Grady KL, Schmuhl D, Ulisney KL, Young JB. Should women receive left ventricular assist device support?: findings from INTERMACS. Circ Heart Fail. 2012;5(2):234–40.
175. Sherazi S, Kutyifa V, McNitt S, Papernov A, Hallinan W, Chen L, Storozynsky E, Johnson BA, Strawderman RL, Massey HT, Zareba W, Alexis JD. Effect of gender on the risk of neurologic events and subsequent outcomes in patients with left ventricular assist devices. Am J Cardiol. 2017;119(2):297–301.
176. Meeteren JV, Maltais S, Dunlay SM, Haglund NA, Beth Davis M, Cowger J, Shah P, Aaronson KD, Pagani FD, Stulak JMA. multi-institutional outcome analysis of patients undergoing left ventricular assist device implantation stratified by sex and race. J Heart Lung Transplant. 2017;36(1):64–70.
177. Morris AA, Pekarek A, Wittersheim K, Cole RT, Gupta D, Nguyen D, Laskar SR, Butler J, Smith A, Vega JD. Gender differences in the risk of stroke during support with continuous-flow left ventricular assist device. J Heart Lung Transplant. 2015;34(12):1570–7.
178. Tsao CI, Chen RJ, Chou NK, Ko WJ, Chi NH, Yu HY, Chen YS, Chen SC, Wang SS. The influence of gender on survival after heart transplantation. Transplant Proc. 2008;40(8):2634–5.

Sex-Related Aspects of Biomarkers in Cardiac Disease

33

Alma M. A. Mingels and Dorien M. Kimenai

Biomarkers in Cardiac Disease
Art work by Piet Michiels, Leuven, Belgium

A. M. A. Mingels (✉) · D. M. Kimenai
Department of Clinical Chemistry, Central Diagnostic Laboratory, Maastricht University Medical Center, Maastricht, The Netherlands
e-mail: alma.mingels@mumc.nl

Abstract

Biomarkers play an important role in the clinical management of cardiac care. In particular, cardiac troponins (cTn) and natriuretic peptides

P. L. M. Kerkhof, V. M. Miller (eds.), *Sex-Specific Analysis of Cardiovascular Function*,
Advances in Experimental Medicine and Biology 1065, https://doi.org/10.1007/978-3-319-77932-4_33

are the cornerstones for the diagnosis of acute myocardial infarction (AMI) and for the diagnosis of heart failure (HF), respectively. Current guidelines do not make a distinction between women and men. However, the commonly used "one size fits all" algorithms are topic of debate to improve assessment of prognosis, particularly in women. Due to the high-sensitivity assays (hs-cTn), lower cTn levels (and 99th percentile upper reference limits) were observed in women as compared with men. Sex-specific diagnostic thresholds may improve the diagnosis of AMI in women, though clinical relevance remains controversial and more trials are needed. Also other diagnostic aspects are under investigation, like combined biomarkers approach and rapid measurement strategies. For the natriuretic peptides, previous studies observed higher concentrations in women than in men, especially in premenopausal women who might benefit from the cardioprotective actions. Contrary to hs-cTn, natriuretic peptides are particularly incorporated in the ruling-out algorithms for the diagnosis of HF and not ruling-in. Clinical relevance of sex differences here seems marginal, as clinical research has shown that negative predictive values for ruling-out HF were hardly effected when applying a universal diagnostic threshold that is independent from sex or other risk factors. Apart from the diagnostic issues of AMI in women, we believe that in the future most sex-specific benefits of cardiac biomarkers can be obtained in patient follow-up (guiding therapy) and prognostic applications, fitting modern ideas on preventive and personalized medicine.

Keywords

Acute myocardial infarction · Age dependence · Biomarkers · Cardiac troponin I · Cardiac troponin T · Estrogen · Heart failure · Natriuretic peptides · BNP · NT-proBNP · Pregnancy · Review

Introduction

Biomarkers play an important role in the clinical management of cardiac care and risk stratification for cardiovascular diseases (CVD) [72, 84] and heart failure [68, 94]. Although over the last decades the number of deaths attributable to CVD has declined globally [59], the reduction on CVD-related deaths is less pronounced for women than men [65]. With respect to sex-specific analysis, the common cardiac care guidelines do not make a distinction between women and men on the use of cardiac biomarkers [68, 72, 84, 94]. However, the awareness has been raised that sex inequalities, and hence, the use of the "one size fits all" algorithms that comprises cardiac biomarkers, may hamper the diagnosis and treatment in women [22, 72].

A number of biomarkers such as lactate dehydrogenase, creatine kinase (CK), or creatine kinase muscle-brain (CKMB) type were used in the past for the diagnosis of acute myocardial infarction (AMI) (Fig. 33.1). Though most of these biomarkers are run in every clinical laboratory today, also for other purposes, their main disadvantage is the lack in cardiac specificity and so none of them are preferred anymore in the diagnosis of cardiac diseases [2, 72, 84].

At this point, cardiac troponins (cTn) are, due to their outperformance on cardiac specificity, incorporated in common guidelines and clinical practice for diagnosis, risk stratification, and prognosis of acute coronary syndrome (ACS) (Fig. 33.1) [72, 84].

That no sex-based differences are made in cTn algorithms can be explained from a historical point of view, since the "older" assays were not able to detect any sex differences. Due to the analytical improvement of these laboratory assays over the last years, nowadays, even basal levels of cTn in healthy individuals can be measured [5]. This improvement resulted in the observation

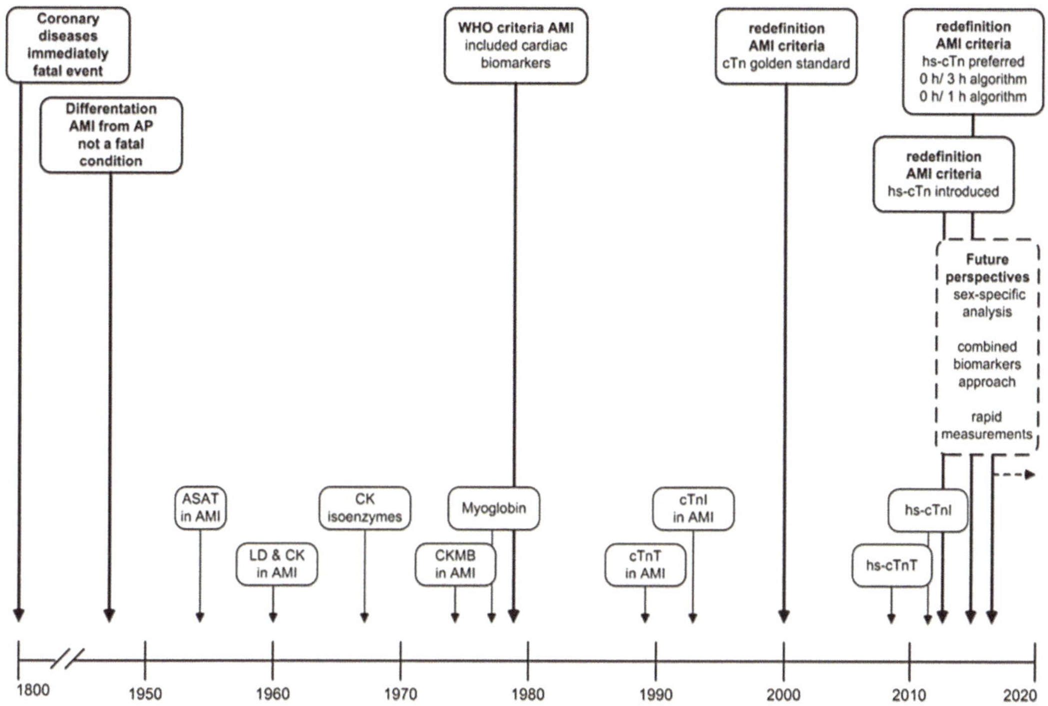

Fig. 33.1 History of cardiac biomarkers for the diagnosis of acute myocardial infarction (AMI). Abbreviations: *AMI* acute myocardial infarction, *AP* angina pectoris, *ASAT* aspartate aminotransferase, *CK* creatine kinase, *CKMB* creatine kinase muscle-brain type, *cTnI* cardiac troponin I, *cTnT* cardiac troponin T, *hs-cTnI* high-sensitivity cardiac troponin I, *hs-cTnT* high-sensitivity cardiac troponin T, *LD* lactate dehydrogenase, *WHO* World Health Organization

that disparities between women and men on cTn levels became apparent [58]. Hence, the technical progress has led to the ongoing debate whether the "one size fits all" strategy is sufficiently sensitive for women.

Also regarding the use of natriuretic peptides, no sex-based differences are incorporated in the guidelines and clinical practice [68, 94], despite the fact that it has clearly been shown that natriuretic peptide concentrations differ between men and women [70, 71].

In this chapter we will provide an overview of the current knowledge and existing gaps on sex differences of cardiac biomarkers. We focus on cTn as the gold standard for the diagnosis of AMI, and we focus on natriuretic peptides (B-type natriuretic peptic, BNP and N-terminal proBNP, NT-proBNP) in the area of diagnosis and prognosis of heart failure (HF).

Cardiac Troponins

The Basics of Cardiac Troponins

The contractile apparatus of cardiac and skeletal muscle cells consists of thick and thin filaments that slide along each other upon muscle contraction and relaxation. Thick filaments are built up from myosin; thin filaments are built up from actin with tropomyosin wrapped around and at regular sites a troponin complex of troponin T (TnT), troponin I (TnI), and troponin C (TnC). TnT, TnI, and TnC are tropomyosin binding, inhibition, and calcium binding components, respectively [4]. Upon stimulation of the myocytes, calcium ions are released and bound to TnC, and subsequently there is a conformational change of the troponin complex, resulting in accessible binding sites between actin and myosin and leading to contraction of the muscles. TnT and TnI isoforms are characterized by a cardiac-specific N-terminal extension, while TnC in cardiac and slow skeletal muscle is similar (Swiss-Prot, P19429, P45379, and P63316, respectively), which makes them highly valuable for detection of cardiac injury and thus for the management of cardiac diseases. It has been estimated that around 5–10% of the total cTnT and cTnI content is present free and unbound in the cytosol, but this statement has been questioned by others. TnC, on the other hand, is not a cardiac-specific biomarker and will be therefore left out of consideration in this chapter.

After cardiac injury, e.g., in acute myocardial infarction (AMI), cardiac troponin T (cTnT) and cardiac troponin I (cTnI) are released in the bloodstream and detectable for diagnostic and prognostic purposes [72, 84]. In AMI patients, a main peak is typically shown within 1 day after onset of symptoms of acute chest pain and remains elevated for approximately 1 week, as depicted in Fig. 33.2. It was hypothesized that the first peak reflects the relatively fast release of cytosolic cTn, while the second broader peak represents the slower dissociation of cTn from the sarcomeres. If this is true, it is quite remarkable that cTnT and cTnI show a slight difference in their return to baseline. We believe this is due to differences between cTnT and cTnI clearance, as there are differences observed between cTnT and cTnI in patients with reduced kidney function [26], but also both proteins differ in their biochemical composition and their molecular weights (cTnT and cTnI being 37 kDa and 26 kDa, respectively), plus they are both susceptible to degradation [16, 47].

Cardiac Troponins in Healthy Women and Men

In the past, due to the restriction of conventional cTn assays, clinicians and scientists could easily discriminate "patients with AMI" from "patients without AMI" by the ability to detect positive cTn levels in case of clinical cardiac injury. However, a higher prevalence of cTn levels, particularly cTnT, was also seen in patients with kidney diseases [1, 13, 35]. Furthermore, also in particular subsets of individuals with several cardiac phenotypes like left ventricular hypertrophy (LVH) and congestive heart failure (CHF), elevated troponins were present [87]. The prominent factor that shed new light on the thought that cTn are not only the result of clinical cardiac injury was the technical improvement of the cTn assays. At this point, using the so-called "high-sensitivity" assays, we now know that cTn levels are also present in apparently healthy individuals [5]. In particular subgroups, and now

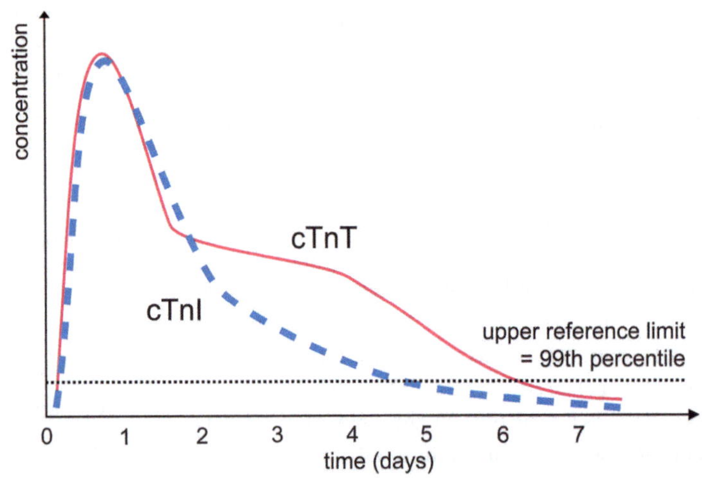

Fig. 33.2 Release patterns of cardiac troponin after AMI. Reprinted with permission from InTechOpen [39]. Abbreviations: *AMI* acute myocardial infarction, *cTnI* cardiac troponin I, *cTnT* cardiac troponin T

to an even larger and more obvious extent, a higher prevalence of cTn elevations is observed, mainly in individuals who are at risk for cardiovascular events, like in individuals with diabetes mellitus, hypertension, and reduced kidney function [35, 45, 54, 79, 87]. The current guidelines listed the conditions whereby elevated levels of cTn are observed other than AMI to consider for further diagnostic workup (Table 33.1) [72].

With the shift to more sensitive cTn assays, also sex inequalities of cTn levels in apparently healthy women and men became evident [58]. Although it is nowadays well established that basal levels of cTn are significantly lower in women than men, the underlying mechanisms of these divergence are still not completely unraveled and understood, but these are most likely multifactorial [30].

The pathogenesis of coronary artery disease (CAD) between women and men seems to differ. Women with CAD are more often presented with endothelial dysfunction than men, while men are more often present with more localized, i.e., less diffuse, coronary disease [40, 88]. The sex hormone estrogen seems to play a protective role in the pathogenesis of several cardiac pathologies [30]. It turned out that estrogen attenuates the processes of atherosclerosis, LVH, and cardiomyocyte apoptosis, possibly resulting into lower cTn levels in women than men [28, 67, 88].

Another important contributor to sex-based differences of cTn levels seems to be the left ventricular (LV) anatomy. The LV mass differs between women and men, whereby men have a higher LV mass as compared with women [24]. Previous studies showed a strong relationship between LV mass and circulating cTn levels [23, 61], and the higher cardiac mass of men probably leads to more release of cTn, which results into a higher basal circulating cTn levels in men than women.

Cardiac Troponins for the Diagnosis of Acute Myocardial Infarction

The cornerstones in the diagnosis of AMI are the clinical presentation, electrocardiography

Table 33.1 Conditions or procedures whereby elevated levels of cTn are observed other than AMI

Most frequent conditions
Tachyarrhythmias
Heart failure
Critical illness (e.g., shock/sepsis/burns)
Myocarditis (incl. myocardial extension of endocarditis or pericarditis)
Takotsubo cardiomyopathy
Structural heart disease (e.g., aortic stenosis)
Aortic dissection
Pulmonary embolism, pulmonary hypertension
Renal dysfunction and associated cardiac disease
Other conditions or procedures
Coronary spasm
Acute neurological event (e.g., stroke or subarachnoid hemorrhage)
Cardiac contusion or cardiac procedures (CABG, PCI, ablation, pacing, cardioversion, or endomyocardial biopsy)
Hypo- and hyperthyroidism
Infiltrative diseases (e.g., amyloidosis, hemochromatosis, sarcoidosis, scleroderma)
Myocardial drug toxicity or poisoning (e.g., doxorubicin, 5-fluorouracil, Herceptin, snake venoms)
Extreme endurance efforts
Rhabdomyolysis

Adapted from ESC guidelines 2016 [72]

Abbreviations: *AMI* acute myocardial infarction, *CABG* coronary artery bypass graft, *cTn* cardiac troponin, *PCI* percutaneous coronary intervention

(ECG) findings, and cardiac troponin measurements (Fig. 33.3). For patients with ST-elevation myocardial infarction (STEMI), the ECG findings are the gold standard to establish AMI diagnosis. However, patients with non-ST-elevation myocardial infarction (NSTEMI) show no diagnostic or inconclusive ECG elevations, whereby the clinical presentation of the patient, in conjunction with cTn measurements, establishes the diagnosis of AMI.

The current guidelines require a significant "rise and/or fall" of high-sensitivity cardiac troponin T (hs-cTnT) or I (hs-cTnI) between serial measurements with at least one value above the 99th percentile upper reference limit of hs-cTn from a healthy reference population [72, 84]. Although the ESC guidelines recommend this 0 h/3 h algorithm, they incorporated also an 0 h/1 h algorithm as alternative approach [72]. The 0 h/1 h algorithm handles other diagnostic thresholds for hs-cTnT and hs-cTnI, dependent of which hs-cTn assay is used [72].

The diagnosis of AMI in women is particularly challenging, as women with suspected ACS are more likely to present with an atypical clinical presentation and/or inconclusive ECG, which may hamper the diagnosis of AMI [15, 32, 43]. Besides, women have a higher prevalence of unrecognized (silent) AMIs as compared with men (women 54% vs men 33%) [25]. All in all this may lead to worse prognosis after AMI in women, and therefore an optimized diagnostic cTn algorithm is of great clinical importance.

Over the last years, we may already have booked some progress on closing the diagnostic gap between women and men. As stated previously, the common guidelines prefer hs-cTn assays above the conventional cTn assays, which have been introduced in Europe around 2012 [84]. Hs-cTn assays should meet two analytical criteria: (1) the 99th percentile upper reference limit of hs-cTn should not be lower than the 10% coefficient of variation (CV) cutoff, and (2) the proportion of individuals with measurable

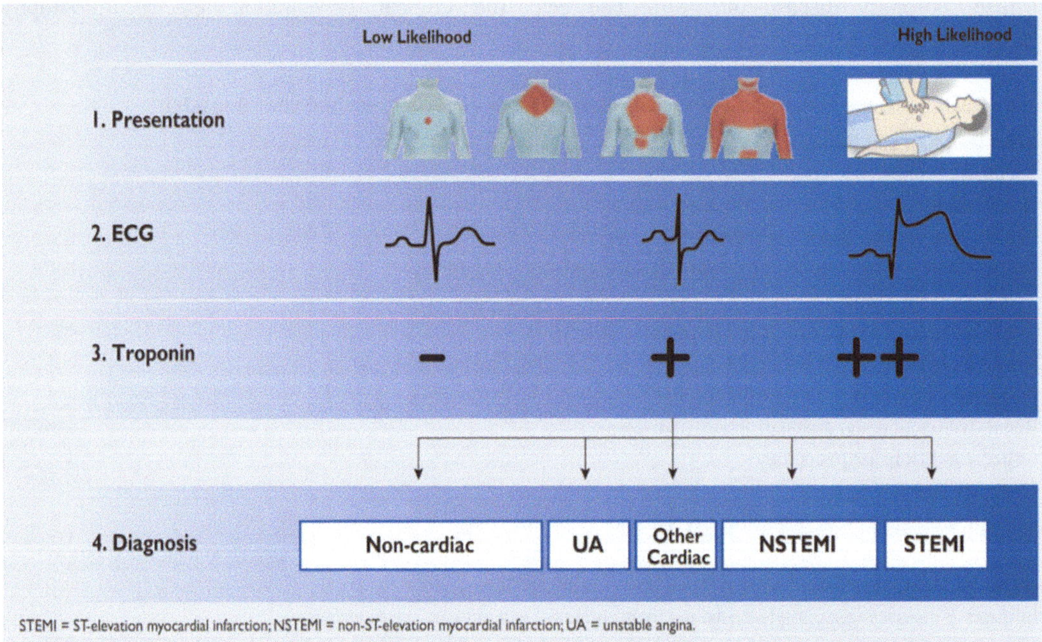

Fig. 33.3 Diagnostic assessment tools for AMI. Reprinted with permissions from Oxford University Press [72]. Abbreviations: *ECG* electrocardiography, *NSTEMI* non-ST-elevation myocardial infarction, *STEMI* ST-elevation myocardial infarction, *UA* unstable angina

cTn concentrations (above the assay's limit of detection) should be ≥50% [7]. Troponin assays that do not meet these criteria are defined as conventional cTn assays. Today there are two assays available which meet the high-sensitivity criteria: the hs-cTnT assay from Roche and Architect hs-cTnI assay from Abbott [7]. Whereas clinical cutoff values of conventional cTn assays were limited to 10% CV, the technical improvement of hs-cTn assays has led that the clinical cutoff values of hs-cTn assays are lowered corresponding to the 99th percentile upper reference limits from a reference population (hs-cTnT, 14 ng/L; hs-cTnI, 26 ng/L, package insert). This enabled that cTn concentrations could now be detected also in other high-risk CVD patients, as described above (Table 33.1). Somewhat misleading is that the higher prevalence of elevated cTn levels found by hs-cTn assays resulted into more "false-positive" results. Most of these unexpected hs-cTn elevations are "true positive" for (subclinical) myocardial injury (rather than AMI) and thus still require diagnostic workup [86]. Thus, the change from conventional cTn assays to hs-cTn assays, and thereby the lowering of the AMI cutoff values, has led to a higher proportion of individuals suitable for beneficial therapies [31]. Thus far, no studies investigated the clinical impact of change over from conventional to hs-cTn assays for diagnosis of AMI in women and men separately. However, hypothetically, the change over from cTn to hs-cTn assays may have led that unrecognized AMIs in women who were missed in the past could now be detected with hs-cTn assays. At this moment the available evidence supporting this hypothesis is scarce and this topic is currently under investigation.

Another important aspect is that women have lower troponin levels as compared with men [58]. A number of studies determined sex-specific 99th percentile upper reference limits of hs-cTnT and hs-cTnI and showed that women-specific thresholds are remarkably lower as compared with men-specific thresholds [5, 37, 44, 58]. As currently a universal diagnostic threshold of hs-cTn is recommended for diagnosing AMI, which is higher than women-specific thresholds, this may contribute to underdiagnosis of AMI in women. In addition, the recommended universal hs-cTnT and hs-cTnI thresholds (14 ng/L vs 26 ng/L, respectively) are not biologically equivalent [89]. A direct comparison of (sex-specific) 99th percentile upper reference limits of hs-cTnT and hs-cTnI from a single reference population revealed numerically similar 99th percentile upper reference limits (hs-cTnT, 15 ng/L; hs-cTnI, 13 ng/L) and enhanced further investigation into sex-specific analysis and downward adjustment of hs-cTnI threshold (Fig. 33.4) [44].

Whether these sex inequalities are of such clinical relevance that the use of universal hs-cTn thresholds hampers the diagnosis of AMI in women, and sex-specific thresholds should be incorporated in diagnostic cTn algorithms, remains controversial. The consideration of sex-specific analysis should be carefully weighted, as this implementation also will result into more complexity of acute cardiac care management. Thus far, the limited number of studies who investigated these issue showed contradictory findings [9, 73, 76, 80, 81, 85]. Two studies recommended to remain the universal thresholds, as they did not observe differences in sex-specific diagnostic and prognostic performance of hs-cTnT on clinical outcome [9], even when sex-specific thresholds were applied [73]. Contrary to these findings, Schofer et al. found sex disparities in the diagnostic performance of AMI using hs-cTnI, particularly into the rule-in performance [80]. The hypothesis that sex-specific thresholds reduce the underdiagnosis of AMI in women is reinforced by the findings of Shah et al., showing that the use of sex-specific hs-cTnI thresholds resulted into a doubling of AMI diagnosis in women but also that the prevalence of AMI diagnosis became similar for women and men [81]. They concluded that sex-specific clinical decision limits of hs-cTnI should be considered for further investigation to close the diagnostic gap between women and men. Two studies showed that applying sex-specific hs-cTnI thresholds did not lead to excessively false-positive AMI diagnosis [76, 85]. Altogether, this finding suggests that sex inequalities may be more relevant for the hs-cTnI assay than for the

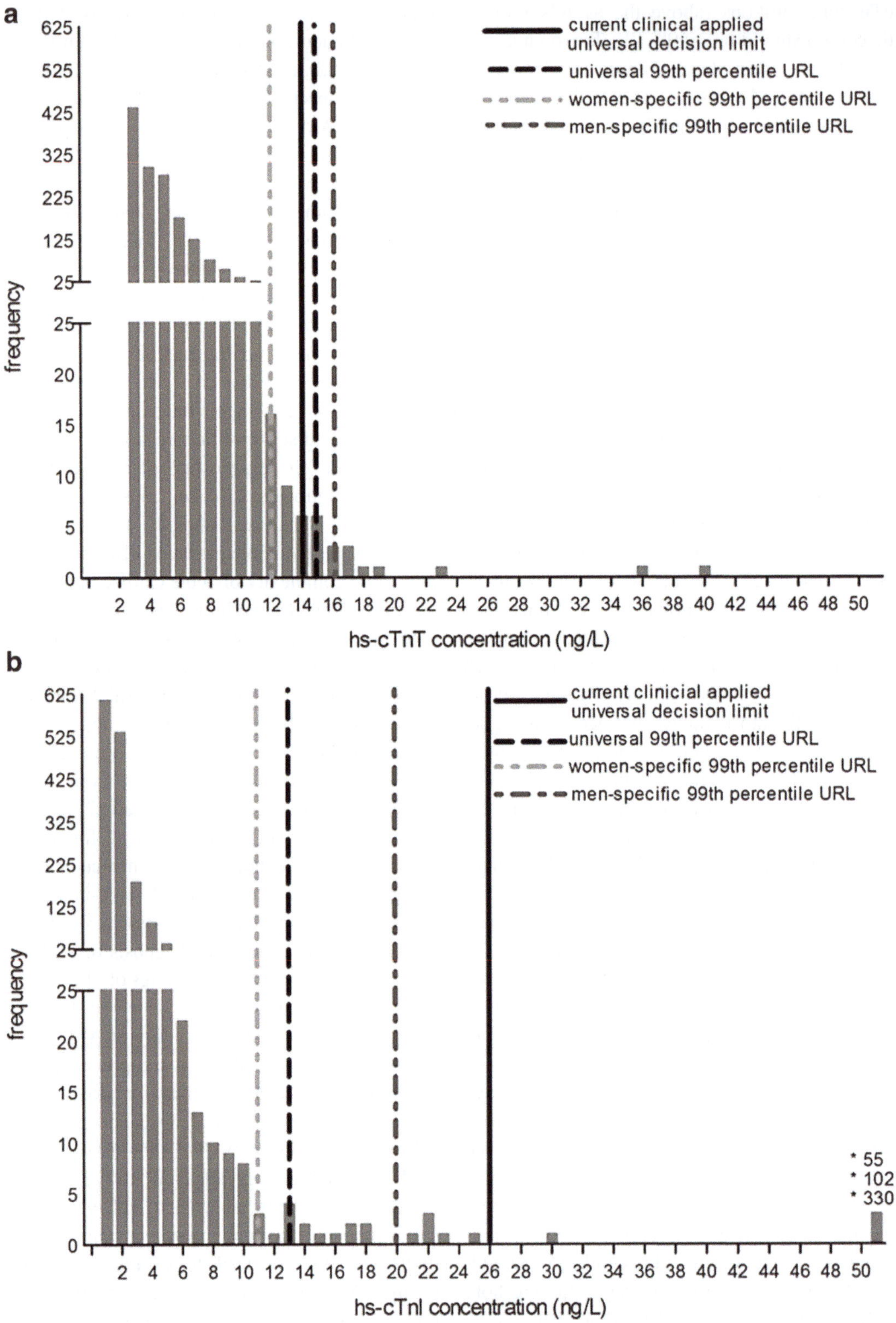

Fig. 33.4 Universal and sex-specific 99th percentile upper reference limits of hs-cTnT (**a**) and hs-cTnI (**b**). Adapted from Kimenai et al. with permission from BMJ Publishing Group Ltd. [44]. Abbreviations: *hs-cTnI* high-sensitivity cardiac troponin I, *hs-cTnT* high-sensitivity cardiac troponin T, *URL* upper reference limit

hs-cTnT assay. Large multicenter randomized controlled trials (e.g., clinical trials.gov NCT01852123) are currently ongoing that should clarify the clinical impact of incorporation of sex-specific cardiac troponin thresholds in the diagnostic algorithm of AMI.

Cardiac Troponins As Prognostic Biomarker

Despite conventional cTn assays were able to detect troponin levels in the clinical range seen in AMI patients, they were not suitable for further risk stratification or prognosis in lower-risk populations. With the shift to hs-cTn assays, cTn could also be detected in the general population, which makes cTn a new promising biomarker for risk stratification and prognosis of CVD morbidity and mortality in the general population. Thus far, a number of epidemiological studies have shown that basal hs-cTn levels are associated with cardiac morbidity and mortality in the general population [11, 23, 82, 90].

Due to fact that there are pathophysiological differences between sexes resulting into different cTn levels in women and men (Sect. "Cardiac troponins in healthy women and men"), the clinical impact of sex on the associations between hs-cTn and CVD mortality is a topic of debate [21, 29, 63, 77]. While some studies showed no interaction of sex on the relationships with troponin and CVD mortality [29, 77], other studies did [21, 63] (Table 33.2). The inconsistency between these studies is probably a result of the divergence of age. Besides the sex effect, also age modifies the association between hs-cTn and mortality. In individuals ≥70 years, the impact of sex seems not of clinical importance, as nonsignificant interaction terms of sex were found (all-cause mortality, $P_{\text{interaction}} = 0.74$; incident CVD, $P_{\text{interaction}} = 0.71$) [29]. However, in the lower age range, both for hs-cTnT and hs-cTnI, higher prognostic values on CVD mortality and all-cause mortality were observed in women than in men, whereby the risk on mortality of both sexes increased in older individuals ≥65 years [21, 51, 63, 96].

The prognostic impact of sex-specific hs-cTn thresholds in the subgroup of patients with suspected ACS on CVD mortality is also not clarified yet [20, 60, 73, 81]. Shah et al. observed that both women and men susceptive of AMI that were reclassified as having AMI after using sex-specific thresholds had the highest risk on death and recurrent AMI after 1 year as compared with subjects without having AMI [81]. In line with these results, Cullen et al. showed that particularly for women, sex-specific hs-cTnI thresholds improved the identification of women for CVD events [20, 81]. However, thus far, these observations are not confirmed for sex-specific hs-cTnT thresholds [60, 73]. Mueller-Hennessen stated that age might be the crucial factor that should be taken into account, instead of sex-specific analysis [60]. Further study is needed to draw conclusions about this statement, and whether or not sex-specific thresholds of hs-cTnT and/or hs-cTnI improve the long-term prognosis in individuals with suspected ACS.

Natriuretic Peptides

The Basics of Natriuretic Peptides

The heart muscle has an endocrine phenotype as it produces the cardiac hormones atrial natriuretic peptide (ANP) and B-type natriuretic peptide (BNP). ANP (28 amino acids (aa), Swiss-Prot P01160) is primarily derived from cardiomyocytes from the atria, and BNP (32 aa, Swiss-Prot P16860) is derived from cardiomyocytes from the ventricles, although it was first identified in porcine brain (brain-type). ANP and BNP are quite similar peptides that are released upon stimulation by stretching of the myocyte, as often seen with an overfilled heart, thereby causing excretion of sodium and water. Besides promoting natriuresis and diuresis functionalities, their counterregulatory function also comprises vasodilation and inhibition of the renin-angiotensin-aldosterone and sympathetic nervous system. The release of ANP is regulated by its release from storage granules into the

Table 33.2 Characteristics and main findings of studies investigating clinical impact of sex on the relationship with hs-cTn and CVD morbidity and mortality in the general population

Publication	Population	Median follow-up	Women/ men	Outcome (HR, 95% CI)
Zeller 2014	Scottish Heart Health Extended Cohort (30–74 years)	20 years	7742/7598	**CVD events per cubic root hs-cTnI increase**
				Women: 1.35 (1.18–1.53)
				Men: 1.26 (1.15–1.38)
				Coronary mortality per cubic root hs-cTnI increase
				Women: 1.53 (1.23–1.92)
				Men: 1.35 (1.21–1.50)
Dallmeier 2015	The ActiFE study (≥65 years)	4.0 years	618/804	**All-cause mortality per ln hs-cTnT increase**
				Women: 3.67 (2.31–5.81)
				Men: 2.15 (1.61–2.87)
				All-cause mortality per ln hs-cTnI increase
				Women: 3.33 (2.15–5.18)
				Men: 1.92 (1.55–2.38)
Eggers 2015	PIVUS study (70 years)	10 years	502/502	**All-cause mortality per ln hs-cTnI increase**
				Women: 1.59 (1.11–2.28)
				Men: 1.38 (1.12–1.70)
				Cardiac morbidity: incident CVD per ln hs-cTnI increase
				Women: 1.23 (0.82–1.86)
				Men: 1.14 (0.88–1.47)
Omland 2015	HUNT study (≥20 years)	13.9 years	5281/4431	**CVD mortality per 1-SD in ln hs-cTnI increase**
				Women: 1.44 (1.31–1.58)
				Men: 1.10 (1.00–1.20)
				All-cause mortality per 1-SD in ln hs-cTnI increase
				Women: 1.33 (1.24–1.42)
				Men: 1.08 (1.01–1.15)

Abbreviations: *AMI* acute myocardial infarction, *CVD* cardiovascular disease, *HF* heart failure, *hs-cTnI* high-sensitivity cardiac troponin I, *hs-cTnT* high-sensitivity cardiac troponin T, *ln* natural log-transformed

circulation, while on the other hand the release of BNP is regulated on the level of its gene expression. C-type natriuretic peptide (CNP) (Swiss-Prot P23582) and D-type natriuretic peptide (DNP) are also part of the natriuretic peptide family but should not be considered as cardiac peptides in humans.

The bioactive hormonal end products ANP and BNP are derived from their precursors, first in the form of preprohormones and next as prohormones. Human proBNP (108 aa) is cleaved at position 73–76, leading to the inactive side-product N-terminal fragment proBNP 1–76 (NT-proBNP) and the active C-terminal fragment proBNP 77–108 being known as BNP or BNP-32. Intact proBNP has an expected mass of 11 kDa, but in the mid-region (proBNP 36–71), glycosylation occurs to a fully or partially extent leading to unclarified masses that are estimated to be around 25 kDa. For the bioactivity of BNP, the ring structure is essential, formed by a disulfide bond, as is synthesized in the endoplasmic reticulum and as is necessary for receptor binding and biological activity.

Natriuretic peptide concentrations in the blood circulation are fully integrated into clinical practice as important diagnostic and prognostic biomarkers. A number of different commercial immunoassays are available, especially for BNP and NT-proBNP; for NT-proBNP there is just one source of antibodies and calibrators (Roche Diagnostics). In 2005 the IFCC initiated recommendations on analytical and pre-analytical quality specifications and to improve assay harmonization [6, 8, 75]. BNP is especially susceptible to degradation that may affect antibody affinity. Only EDTA-anticoagulated plastic tubes are acceptable and samples should be immediately put on ice and measured as soon as possible, since BNP is not stable at room temperature and at 2–8 °C it is stable for just a few hours [95]. For NT-proBNP the main issue is the potential of cross-reactivity with split products of NT-proBNP and proBNP [50]. The extent of glycosylation does not play a role since antibodies are selected outside the region with glycosylation sites. NT-proBNP is measurable in serum or plasma (10% lower in EDTA plasma) and is stable for at least 48 h at room temperature [95].

Upon stretching of cardiomyocytes, proBNP is split into equimolar amounts of BNP and NT-proBNP, and indeed, they are used for similar clinical purposes as is pointed out further in this chapter [68, 94]. However, their concentrations in the blood circulation are absolutely different and values are not interchangeable, also for other reasons than the previously discussed pre-analytics and stability. Namely, BNP and NT-proBNP are both cleared by the kidneys, but BNP is besides that also cleared by natriuretic peptide receptor type C (NPR-C) and neural endopeptidases. This results in a half-life for BNP of approximately 25 min, and for NT-proBNP, this is estimated to be twice as long with approximately 120 min. Finally, both peptides have different molecular weights, with the conversion factor for BNP from pmol/L to pg/mL being 3.460 and for NT-proBNP 8.457.

Natriuretic Peptides in Healthy Women and Men

Early studies already illustrated significantly higher natriuretic peptide concentrations in women than in men [19, 70, 71] suggesting a close relation between the cardiac endocrine function and the sex steroid hormones. Figure 33.5 illustrates a typical NT-proBNP distribution as found in our reference population [58]. The sex effect remained true even after correction for differences in body composition and LV mass [49].

Interestingly, sex-based differences in natriuretic peptide concentrations seem to be more pronounced for premenopausal women than for postmenopausal women. It was observed that BNP and NT-proBNP concentrations were lower in postmenopausal women as compared to premenopausal women [48, 49] though this difference could not be confirmed by others [17]. Also, BNP and NT-proBNP concentrations in postmenopausal women seem to become closer to the concentrations as found in men, as was especially true for older adults of 70–75 years and older [21, 56], while others reported higher NT-proBNP concentrations for women independent of their hormonal and menopausal status [19]. Table 33.3 illustrates that we also found higher NT-proBNP concentrations for women when compared to men, irrespective of their age, with the lowest NT-proBNP concentrations for women who almost reached or just reached their menopause (40–50 years). The latter was found by some [19] but not by all [21, 56] as discussed in more detail in Sect "Natriuretic peptides and age".

It has been hypothesized that the sex effect on natriuretic peptides is caused by a stimulatory effect of estrogens and an inhibitory effect of androgens, as schematically illustrated in Fig. 33.6. NT-proBNP levels indeed were inversely associated with androgen concentrations (total and free testosterone) [17, 48], though the association with estradiol was found to be weak or not significant at all [17, 74]. The latter is not completely surprising because of the following issues. The biological variation in

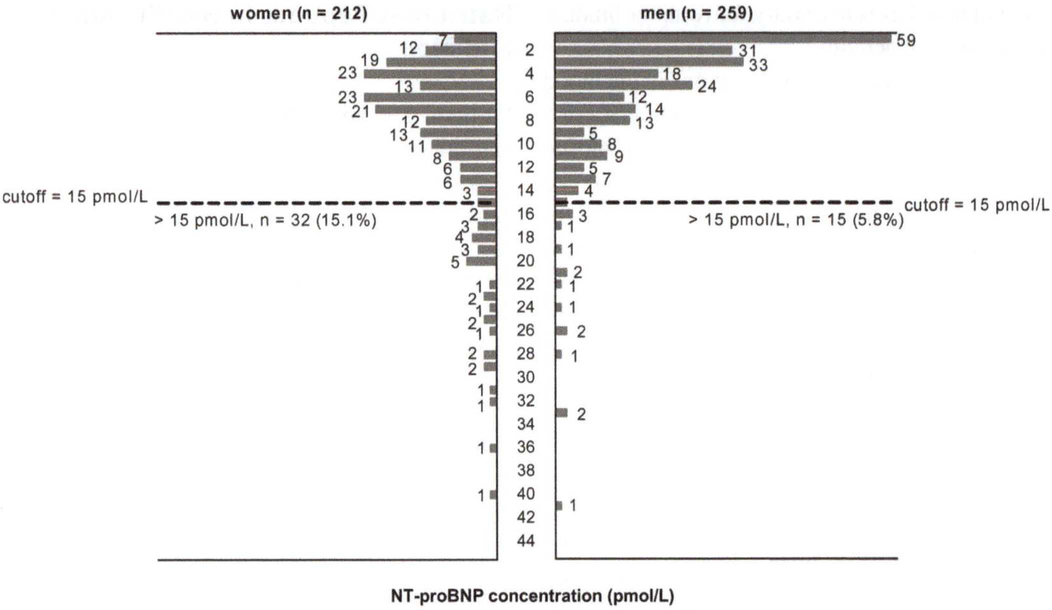

Fig. 33.5 Distribution of NT-proBNP concentrations (pmol/L) in healthy women (left, $n = 212$, mean age 50 ± 11 years) and men (right, $n = 259$, mean age 53 ± 10 years) [58]. Indicated are the number of individuals with elevated concentrations above the universal diagnostic threshold (125 pg/mL/15 pmol/L) used to rule out chronic HF

Table 33.3 NT-proBNP (pmol/L) concentrations in a healthy population [58], stratified by sex and age

Age	*n*	Total	Women	Men
NT-proBNP (pmol/L), median [IQR]				
<30 years	17	3.7 [2.0–6.3]	5.6 [3.1–6.6]	1.6 [0.9–3.5]
30–39 years	49	5.2 [2.4–9.7]	7.6 [3.2–10.7]	3.9 [1.4–6.7]
40–49 years	144	4.0 [1.9–6.5]	4.8 [3.7–8.0]	2.7 [1.1–5.1]
50–59 years	177	6.2 [2.8–9.5]	7.2 [5.0–12.1]	3.4 [1.7–8.5]
≥60 years	84	9.6 [4.9–13.5]	11.7 [6.6–17.3]	7.6 [4.2–12.4]
NT-proBNP ≥ 15 pmol/L, *n* (%)				
<30 years	17	1 (6)	1 (9)	0 (0)
30–39 years	49	5 (10)	4 (14)	1 (5)
40–49 years	144	5 (3)	4 (6)	1 (1)
50–59 years	177	22 (12)	15 (19)	7 (7)
≥60 years	84	17 (20)	10 (33)	7 (13)

NT-proBNP (pmol/L) concentrations are expressed as median (IQR) and as n (%) (>125 pg/mL or 15 pmol/L, universal diagnostic threshold to rule out chronic HF)
Abbreviation: *HF* heart failure

estradiol concentrations is great, up to fivefold during a menstrual cycle. Strikingly, NT-proBNP concentrations were lower in the midcycle phase than in the follicular or luteal phase and thus completely opposite to the fluctuations of estradiol [48]. Moreover, estradiol ranges in cycling women are ≥2 times greater than in men, while testosterone ranges in women are hardly to mention in comparison to the concentrations found in men (total testosterone, factor 10; free testosterone, factor 30). There might also be an alternative explanation via the indirect role of sex hormone-

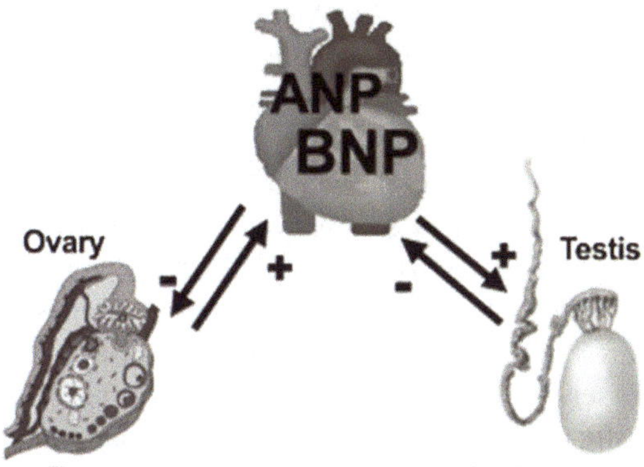

Fig. 33.6 Schematic representation of the relationship between cardiac endocrine and gonadal functions in women and men, showing a stimulatory effect of estrogens and an inhibitory effect of androgens. Reprinted with permission from Elsevier [18]. Abbreviations: *ANP* atrial natriuretic peptide, *BNP* brain-type natriuretic peptide, *SHBG* sex hormone-binding globulin

binding globulin (SHBG) which synthesis is stimulated by estrogens and inhibited by androgens [18] and SHBG was indeed positively associated with BNP and NT-proBNP concentrations [17]. Anyway, the explanation for this sex hormone phenomenon is not understood yet. The current hypothesis is that fertile women benefit from the cardioprotective actions of natriuretic peptides, including vasodilation and diuresis/natriuresis, as is also noticed by the lower risk for cardiovascular events for women during their cycling [59].

Clinical studies demonstrate that also interventions that affect sex hormones in women and men fit with previous observations. For instance, BNP and NT-proBNP concentrations were higher in premenopausal women who receive estrogens for contraception [48] or postmenopausal women who receive estrogens for hormone replacement therapy [52], as compared to premenopausal women. Moreover, androgen receptor blockage and, to a lesser extent, androgen suppression in men with prostate cancer also result in increased NT-proBNP concentrations [27].

Natriuretic Peptides During Pregnancy

Of special clinical interest is the relation of natriuretic peptides with pregnancy. Normal pregnancy goes along with median concentrations about twice that of nonpregnant controls, rising early in pregnancy and remaining high throughout gestation until ≈72 h after delivery [14]. Complications including acute HF may be triggered by eclampsia or preeclampsia and in such situations NT-proBNP might possibly help in the diagnosis [83] though this is not (yet) widely established in routine clinical care [38]. It has been proven though that in preeclampsia NT-proBNP may reflect ventricular stress and subclinical cardiac dysfunction worsening if fetal growth restriction is present [36].

Natriuretic Peptides and Age

Also the influence of age on baseline natriuretic peptide concentrations is very strong [8] and might even more be important in clinical patients than in healthy individuals [19]. Table 33.3 shows that we indeed found substantially higher NT-proBNP concentrations for older adults. Moreover, as mentioned before, NT-proBNP concentrations in our population were higher for women than for men and this was true for all age categories ($P < 0.05$), similar to what was found by some [19], while others found equal NT-proBNP concentrations for postmenopausal women and men [21, 56]. When focusing on the age effect, it is immediately clear that many of the older healthy individuals exceed the diagnostic cutoff that is used to rule out (chronic) HF. This is especially the case for adults with an age of 70–75 years and older, as also illustrated by Fig. 33.7. It is therefore crucial to remind that the current cutoffs are designed for ruling out HF but that other diagnostic tools are necessary for further diagnostic workup in the ruling in [68, 94].

Other clinical factors that influence natriuretic peptide concentrations are body mass index or obesity, ethnicity, and non-HF pathologies [68, 91, 94] and are not described here in great detail. Explicit attention should be given to renal impairment which substantially increases NT-proBNP concentrations and to a lesser extent also BNP concentrations [3, 55].

Natriuretic Peptides for the Diagnosis of Heart Failure

The definition of HF is mainly based on the measurement of the left ventricular ejection fraction (LVEF) and includes a wide range of patients, from those with normal to reduced LVEF. Table 33.4 shows that natriuretic peptides are also part of the criteria for HF, and with the recent ESC and ACC/AHA guideline updates [68, 95] their role became even more important for diagnostic purposes, establishing disease severity, and prognosis.

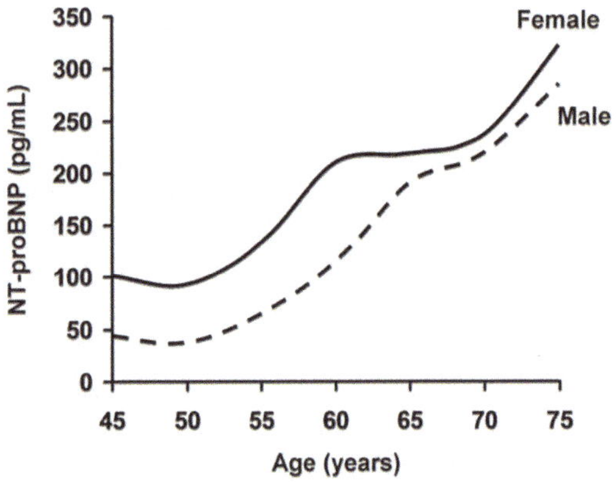

Fig. 33.7 Age- and sex-specific 80th percentile upper reference limits for NT-proBNP concentrations in healthy individuals and stage A/B HF subjects. The universal diagnostic threshold to rule out chronic HF is set at 125 pg/mL or 15 pmol/L. Reprinted with permission from Elsevier [56]

In the diagnostic workup, two algorithms are prescribed as summarized in the following part [68]. They use either BNP or NT-proBNP, though absolute values and diagnostic thresholds are not interchangeable [94], and to a lesser extent also mid-regional proANP (MR-proANP). There are also several analytical and clinical aspects like sex and age that should be considered when using natriuretic peptides for clinical use [8].

- In the non-acute setting [68], the diagnosis of HF is assessed by clinical history, physical examination, and ECG. If all elements are normal, HF is highly unlikely and other diagnoses need to be considered. If at least one element is abnormal, natriuretic peptides should be measured, if available, to identify those who need echocardiography. The upper limit of normal in the non-acute setting for BNP is 35 pg/mL (10 pmol/L), and for NT-proBNP, it is 125 pg/mL (15 pmol/L).
- In the acute setting [68], the diagnostic workup is somewhat different. Upon presentation with acute dyspnea and the suspicion of acute HF, several diagnostic tests are recommended to differentiate cardiac causes of acute dyspnea from noncardiac causes. Natriuretic peptides should be assessed in all patients, and here higher thresholds should be used, namely, for BNP 100 pg/mL (35 pmol/L) and for

Table 33.4 Definition of heart failure (HF)

Criteria	HFrEF	HFmrEF	HFpEF
1	Symptoms ± signs	Symptoms ± signs	Symptoms ± signs
2	LVEF <40%	LVEF 40–49%	LVEF ≥50%
3		1. Elevated levels of natriuretic peptides[a]	1. Elevated levels of natriuretic peptides[a]
		2. At least one additional criterion[b]	2. At least one additional criterion[b]

Adapted from ESC guidelines 2016; see also for further details [68]

Abbreviations: *HFrEF* heart failure with reduced ejection fraction, *HFmrEF* heart failure with midrange ejection fraction, *HFpEF* heart failure with preserved ejection fraction, *LVEF* left ventricular ejection fraction

[a]BNP >35 pg/mL (10 pmol/L); NT-proBNP >125 pg/mL (15 pmol/L)

[b]Relevant structural disease or diastolic dysfunction

NT-proBNP 300 pg/mL (35 pmol/L) and/or MR-proANP 120 pmol/L. Other diagnostic tests that are recommended include ECG; X-thorax; other laboratory assessments like cTn, urea, creatinine, electrolytes (sodium, potassium), glucose, complete blood count, liver function tests, and thyroid-stimulating hormone (TSH); and finally, echocardiography.

The use of natriuretic peptides is especially recommended for ruling out HF in patients who present with dyspnea, since their negative predictive values are very high (0.94–1.00), both in the non-acute and the acute settings [41, 46, 68]. This remains true even when using a universal diagnostic threshold that are regardless of age [41] or sex [46] or any of the other risk factors. It was first thought to use age-dependent cutoffs (US Food and Drug Administration, FDA-cleared cutoffs), 125 pg/mL (15 pmol/L) for patients younger than 75 years and 450 pg/mL (50 pmol/L) for patients of 75 years and older, and/or renal-dependent cutoffs for eGFR <60 ml/min/1.72m^2 [3, 55]. However, clinical research has inevitably shown that negative predictive values were hardly affected when applying a single and relatively low cutoff that is independent from risk factors and thus easier to use in daily practice and still safe for all types of patients.

In contrast, natriuretic peptides are less appropriate for establishing the diagnosis of HF with positive predictive values in the non-acute setting of only 0.44–0.57 and in the acute setting of 0.66–0.67 [68]. This could be mainly explained by the numerous cardiovascular and non-cardiovascular causes of elevated natriuretic peptides, as summarized in Table 33.5 [68, 91, 94] with atrial fibrillation, age, and renal failure being important, if not the most important factors [68].

The severity of CHF is significantly associated with natriuretic peptide concentrations as, for instance, determined by the New York Heart Association classification [41, 53]. Concentrations are lower for HF patients with preserved ejection fraction (HFpEF) than for patients with reduced ejection fraction (HFrEF) [12], but diagnostic values apply similarly to both groups [68].

Table 33.5 Causes of elevated natriuretic peptide concentrations

Cardiac
HF, including RV syndromes
Acute coronary syndromes
Heart muscle disease, including LVH
Valvular heart disease
Pericardial disease
Atrial fibrillation
Myocarditis
Cardiac surgery
Cardioversion
Toxic-metabolic myocardial insults, including cancer chemotherapy
Noncardiac
Advancing age
Anemia
Renal failure
Pulmonary: obstructive sleep apnea, severe pneumonia
Pulmonary hypertension
Critical illness
Bacterial sepsis
Severe burns

Adapted from ACC guidelines 2017 [94]
Abbreviations: *HF* heart failure, *LVH* left ventricular hypertrophy, and *RV* right ventricular

Natriuretic Peptides for the Management of Heart Failure

Over the last 15 years, it has been hypothesized that natriuretic peptides might be helpful as objective measures in the management of HF patients. Most of the prospective clinical trials investigated whether it is useful to titrate the pharmacologic therapy to a fixed target concentration. Results of single studies have been quite conflicting, unfortunately, whereas meta-analyses overall show promising effects. The most recent and largest meta-analysis by Savarese et al. included 2686 patients from 12 studies [78]. They reported benefits on both hospitalization and mortality of the HF patients. Significant results were obtained specifically for NT-proBNP-guided studies, in contrast to the BNP-guided studies. The final evidence was expected from the GUIDE-IT (Guiding Evidence Based Therapy Using Biomarker Intensified Treatment in Heart Failure) study by investigating high-risk patients with relatively advanced HFrEF [33], but NT-proBNP-guided therapy turned out not to be more effective in hospitalization or mortality than usual care strategy nor were doses of medical therapy different between the groups; this study was therefore stopped for futility when 894 of the 1100 patients were enrolled [34].

Sex-based differences for natriuretic peptide-guided therapy have not been reported, though this is different for the influence of age. The benefits for NT-proBNP guidance appear at least to be true for individuals of 75 years or younger [10, 12, 66] and for patients with HFrEF but not for HFpEF [12]. It is conceivable that the more frequent presence of comorbidities in elderly may prevent or even promote potentially harmful up-titration of drugs.

Unfortunately, the concept of serial sampling of natriuretic peptides is quite limited because of a great biological variation which is of course crucial for successful HF management. Changes in concentration of more than 50 to 100% are necessary to be sure that the changes observed are "real" (exceeding reference change values) and are indeed due to therapy interventions [92, 93]. It was recently validated and confirmed that the biological variation is independent from the clinical condition of the patient, thus being equal in healthy individuals and chronic HF patients [57].

Natriuretic Peptides for Prevention, Prognosis, and Risk Stratification

Numerous prognostic markers of death and hospitalization have been identified in HF patients. Extensive review and meta-analysis illustrate that especially age and to a lesser extent also sex are strong predictors for cardiac death and hospitalization in patients with HF, if not the most important ones [64, 69]. In the top ten of the most prominent predictors is also space for a couple of biomarkers that are related to renal function (sodium, creatinine, urea), oxygen supply (hemoglobin), and cardiac function (NT-proBNP) [64].

The use of natriuretic peptides as prognostic predictors remains however challenging for daily practice [68]. Nevertheless, the recent American guideline included new level 1 recommendations on the use of natriuretic peptides as powerful predictors with a specific role for (1) screening in patients at risk for developing HF (presence of hypertension, diabetes mellitus, or vascular disease), (2) baseline levels on admission to the hospital in case of acutely decompensated HF, (3) a predischarge level during hospitalization, and (4) in case of chronic HF to consider also biomarkers of myocardial injury (cTn) or fibrosis [94]. It seems almost logical that sex and probably also age are very important predictors that should always be considered when using these applications, but future evidence has to proof this first.

An interesting sex aspect is the benefit that fertile women gather from their higher baseline natriuretic peptide concentrations, as previously discussed in Sect. "Natriuretic peptides in healthy women and men" [59]. The survival advantage of women though was canceled out in case of congenital heart disease what might be related to the higher prevalence of severe pulmonary hypertension [62]. Also, despite that, marked sex differences were found in acute HF patients with preserved versus reduced ejection fractions with opposite associations of anemia and LVEF requiring further attention [42]. Even in community-dwelling older adults, sex-based differences remain an ongoing research topic where associations of NT-proBNP with all-cause mortality were substantially stronger among women [21].

Conclusions and Future Perspectives

Taken together, the cardiac biomarkers cTn, NT-proBNP, and BNP are the cornerstones in the diagnosis and clinical management of AMI and HF, respectively. Current cardiac care guidelines do not make a distinction between women and men and make use of "one size fits all" algorithms.

The debate on sex-specific analysis mainly concerns the underdiagnosis of AMI in women. Due to the improvement of the cTn assays, a new era arose where cTn levels became measurable in apparently healthy individuals, whereby lower cTn levels (and 99th percentile upper reference limits) were observed in women as compared with men. The evidence for sex-specific diagnostic thresholds remains though controversial, and ongoing and future trials will clarify whether or not sex-specific analysis should be incorporated in the diagnostic algorithm of AMI.

Also other aspects of cTn algorithms are under investigation that may improve the diagnosis of AMI. For instance, combined biomarker approaches are investigated, as promising new biomarkers such as copeptin, heart fatty acid-binding protein, ST2, and growth differentiation factor-15 may have additive diagnostic value over hs-cTn. Up till now, however, the results are quite controversial, probably due to the already high diagnostic performance of hs-cTn. Furthermore, the hs-cTn assays allow further investigation to rapid measurement strategies, which could lead to faster diagnosis of AMI and, eventually, resulting into better prognosis after AMI.

For the natriuretic peptides, previous studies observed higher natriuretic peptide concentrations in women, especially in premenopausal women, than in men. Natriuretic peptides are particularly incorporated in the ruling-out algorithms for the diagnosis of HF and, in contrast, to hs-cTn, where clinical research has shown that negative predictive values for ruling out HF were hardly affected when applying a universal diagnostic threshold of natriuretic peptides that is independent from sex or other risk factors.

Finally, we believe that in the future most sex-specific benefits of cardiac biomarkers can be obtained in the field of patient follow-up and in the field of cardiac risk prevention and risk stratification. Both are unfinished areas and fit the modern ideas of preventive and personalized medicine: from guiding therapy and prognostic applications for patients who are already diagnosed to patients at risk and in the future possibly to even healthy individuals.

References

1. Abbas NA, John RI, Webb MC, et al. Cardiac troponins and renal function in nondialysis patients with chronic kidney disease. Clin Chem. 2005;51:2059–66.
2. Alpert JS, Thygesen K, Antman E, Bassand JP. Myocardial infarction redefined--a consensus document of The Joint European Society of Cardiology/American College of Cardiology Committee for the redefinition of myocardial infarction. J Am Coll Cardiol. 2000;36:959–69.
3. Anwaruddin S, Lloyd-Jones DM, Baggish A, et al. Renal function, congestive heart failure, and amino-terminal pro-brain natriuretic peptide measurement: results from the ProBNP Investigation of Dyspnea in the Emergency Department (PRIDE) Study. J Am Coll Cardiol. 2006;47:91–7.
4. Apple FS, Collinson PO, Biomarkers ITFoCAoC. Analytical characteristics of high-sensitivity cardiac troponin assays. Clin Chem. 2012a;58:54–61.
5. Apple FS, Ler R, Murakami MM. Determination of 19 cardiac troponin I and T assay 99th percentile values from a common presumably healthy population. Clin Chem. 2012b;58:1574–81.
6. Apple FS, Panteghini M, Ravkilde J, et al. Quality specifications for B-type natriuretic peptide assays. Clin Chem. 2005;51:486–93.
7. Apple FS, Sandoval Y, Jaffe AS, Ordonez-Llanos J, Bio-Markers ITFoCAoC. Cardiac troponin assays: guide to understanding analytical characteristics and their impact on clinical care. Clin Chem. 2017;63:73–81.
8. Apple FS, Wu AH, Jaffe AS, et al. National Academy of Clinical Biochemistry and IFCC Committee for Standardization of Markers of Cardiac Damage Laboratory Medicine practice guidelines: analytical issues for biomarkers of heart failure. Circulation. 2007;116:e95–8.
9. Balmelli C, Meune C, Twerenbold R, et al. Comparison of the performances of cardiac troponins, including sensitive assays, and copeptin in the diagnostic of acute myocardial infarction

and long-term prognosis between women and men. Am Heart J. 2013;166:30–7.
10. Bayes-Genis A, Lupon J, Jaffe AS. Can natriuretic peptides be used to guide therapy? EJIFCC. 2016;27:208–16.
11. Blankenberg S, Salomaa V, Makarova N, et al. Troponin I and cardiovascular risk prediction in the general population: the BiomarCaRE consortium. Eur Heart J. 2016;37:2428–37.
12. Brunner-La Rocca HP, Eurlings L, Richards AM, et al. Which heart failure patients profit from natriuretic peptide guided therapy? a meta-analysis from individual patient data of randomized trials. Eur J Heart Fail. 2015;17:1252–61.
13. Buiten MS, de Bie MK, Rotmans JI, et al. Serum cardiac troponin-I is superior to troponin-T as a marker for left ventricular dysfunction in clinically stable patients with end-stage renal disease. PLoS One. 2015;10:e0134245.
14. Canobbio MM, Warnes CA, Aboulhosn J, et al. Management of pregnancy in patients with complex congenital heart disease: a scientific statement for healthcare professionals from the American Heart Association. Circulation. 2017;135: e50–87.
15. Canto JG, Goldberg RJ, Hand MM, et al. Symptom presentation of women with acute coronary syndromes: myth vs reality. Arch Intern Med. 2007;167:2405–13.
16. Cardinaels EP, Mingels AM, van Rooij T, et al. Time-dependent degradation pattern of cardiac troponin T following myocardial infarction. Clin Chem. 2013;59:1083–90.
17. Chang AY, Abdullah SM, Jain T, et al. Associations among androgens, estrogens, and natriuretic peptides in young women: observations from the Dallas Heart Study. J Am Coll Cardiol. 2007;49:109–16.
18. Clerico A, Fontana M, Vittorini S, Emdin M. The search for a pathophysiological link between gender, cardiac endocrine function, body mass regulation and cardiac mortality: proposal for a working hypothesis. Clin Chim Acta. 2009;405:1–7.
19. Costello-Boerrigter LC, Boerrigter G, Redfield MM, et al. Amino-terminal pro-B-type natriuretic peptide and B-type natriuretic peptide in the general community: determinants and detection of left ventricular dysfunction. J Am Coll Cardiol. 2006;47:345–53.
20. Cullen L, Greenslade JH, Carlton EW, et al. Sex-specific versus overall cut points for a high sensitivity troponin I assay in predicting 1-year outcomes in emergency patients presenting with chest pain. Heart. 2016;102:120–6.
21. Dallmeier D, Denkinger M, Peter R, et al. Sex-specific associations of established and emerging cardiac biomarkers with all-cause mortality in older adults: the ActiFE study. Clin Chem. 2015;61:389–99.
22. Daniels LB, Maisel AS. Cardiovascular biomarkers and sex: the case for women. Nat Rev Cardiol. 2015;12:588–96.
23. de Lemos JA, Drazner MH, Omland T, et al. Association of troponin T detected with a highly sensitive assay and cardiac structure and mortality risk in the general population. JAMA. 2010;304:2503–12.
24. de Simone G, Devereux RB, Daniels SR, Meyer RA. Gender differences in left ventricular growth. Hypertension. 1995;26:979–83.
25. de Torbal A, Boersma E, Kors JA, et al. Incidence of recognized and unrecognized myocardial infarction in men and women aged 55 and older: the Rotterdam Study. Eur Heart J. 2006;27:729–36.
26. deFilippi C, Seliger SL, Kelley W, et al. Interpreting cardiac troponin results from high-sensitivity assays in chronic kidney disease without acute coronary syndrome. Clin Chem. 2012;58:1342–51.
27. Dockery F, Bulpitt CJ, Agarwal S, et al. Anti-androgens increase N-terminal pro-BNP levels in men with prostate cancer. Clin Endocrinol (Oxf). 2008;68:59–65.
28. Donaldson C, Eder S, Baker C, et al. Estrogen attenuates left ventricular and cardiomyocyte hypertrophy by an estrogen receptor-dependent pathway that increases calcineurin degradation. Circ Res. 2009;104:265–75. 211p following 275
29. Eggers KM, Johnston N, Lind L, Venge P, Lindahl B. Cardiac troponin I levels in an elderly population from the community--The implications of sex. Clin Biochem. 2015;48:751–6.
30. Eggers KM, Lindahl B. Impact of sex on cardiac troponin concentrations-A critical appraisal. Clin Chem. 2017;63:1457–64.
31. Eggers KM, Lindahl B, Melki D, Jernberg T. Consequences of implementing a cardiac troponin assay with improved sensitivity at Swedish coronary care units: an analysis from the SWEDEHEART registry. Eur Heart J. 2016;37:2417–24.
32. Elsaesser A, Hamm CW. Acute coronary syndrome: the risk of being female. Circulation. 2004;109:565–7.
33. Felker GM, Ahmad T, Anstrom KJ, et al. Rationale and design of the GUIDE-IT study: guiding evidence based therapy using biomarker intensified treatment in heart failure. JACC Heart Fail. 2014;2:457–65.
34. Felker GM, Anstrom KJ, Adams KF, et al. Effect of natriuretic peptide-guided therapy on hospitalization or cardiovascular mortality in high-risk patients with heart failure and reduced ejection fraction: a randomized clinical trial. JAMA. 2017;318:713–20.
35. Freda BJ, Tang WH, Van Lente F, Peacock WF, Francis GS. Cardiac troponins in renal insufficiency: review and clinical implications. J Am Coll Cardiol. 2002;40:2065–71.
36. Giannubilo SR, Pasculli A, Tidu E, et al. Relationship between maternal hemodynamics and plasma natriuretic peptide concentrations during pregnancy complicated by preeclampsia and fetal growth restriction. J Perinatol. 2017;37:484–7.
37. Gore MO, Seliger SL, Defilippi CR, et al. Age- and sex-dependent upper reference limits for the high-sensitivity cardiac troponin T assay. J Am Coll Cardiol. 2014;63:1441–8.
38. Royal College of Obstetricians and Gynaecologists, National Collaborating Centre for Women's and Children's Health. Hypertension in pregnancy: the management of hypertensive disorders during pregnancy. London: National Institute for Health and Clinical Excellence: Guidance; 2010.
39. Jacobs EJH, Mingels AMA, Dieijen van-Visser MP. Cardiac biomarkers in end-stage renal disease. In: Sahay M, editor. Chronic kidney disease and renal transplantation. InTech Publication; 2012. p. 147–60.
40. Jaffe AS, Apple FS. High-sensitivity cardiac troponin assays: isn't it time for equality? Clin Chem. 2014;60:7–9.
41. Januzzi JL Jr, Camargo CA, Anwaruddin S, et al. The N-terminal Pro-BNP investigation of dyspnea in the emergency department (PRIDE) study. Am J Cardiol. 2005;95:948–54.
42. Kajimoto K, Minami Y, Sato N, et al. Gender differences in anemia and survival in patients hospitalized for acute decompensated heart failure with preserved or reduced ejection fraction. Am J Cardiol. 2017;120:435–42.
43. Khan NA, Daskalopoulou SS, Karp I, et al. Sex differences in acute coronary syndrome symptom presentation in young patients. JAMA Intern Med. 2013;173:1863–71.
44. Kimenai DM, Henry RM, van der Kallen CJ, et al. Direct comparison of clinical decision limits for cardiac troponin T and I. Heart. 2016;102:610–6.
45. Kimenai DM, Martens RJH, Kooman JP, et al. Troponin I and T in relation to cardiac injury detected with electrocardiography in a population-based cohort – The Maastricht Study. Sci Rep. 2017;7:6610.
46. Krauser DG, Chen AA, Tung R, et al. Neither race nor gender influences the usefulness of amino-terminal pro-brain natriuretic peptide testing in dyspneic subjects: a ProBNP

Investigation of Dyspnea in the Emergency Department (PRIDE) substudy. J Card Fail. 2006;12:452–7.
47. Labugger R, Organ L, Collier C, Atar D, Van Eyk JE. Extensive troponin I and T modification detected in serum from patients with acute myocardial infarction. Circulation. 2000;102:1221–6.
48. Lam CS, Cheng S, Choong K, et al. Influence of sex and hormone status on circulating natriuretic peptides. J Am Coll Cardiol. 2011;58:618–26.
49. Lew J, Sanghavi M, Ayers CR, et al. Sex-based differences in cardiometabolic biomarkers. Circulation. 2017;135:544–55.
50. Luckenbill KN, Christenson RH, Jaffe AS, et al. Cross-reactivity of BNP, NT-proBNP, and proBNP in commercial BNP and NT-proBNP assays: preliminary observations from the IFCC Committee for Standardization of Markers of Cardiac Damage. Clin Chem. 2008;54:619–21.
51. Lyngbakken MN, Rosjo H, Holmen OL, et al. Gender, high-sensitivity troponin I, and the risk of cardiovascular events (from the Nord-Trondelag Health Study). Am J Cardiol. 2016;118:816–21.
52. Maffei S, Del Ry S, Prontera C, Clerico A. Increase in circulating levels of cardiac natriuretic peptides after hormone replacement therapy in postmenopausal women. Clin Sci (Lond). 2001;101:447–53.
53. Maisel AS, Krishnaswamy P, Nowak RM, et al. Rapid measurement of B-type natriuretic peptide in the emergency diagnosis of heart failure. N Engl J Med. 2002;347:161–7.
54. Martens RJ, Kimenai DM, Kooman JP, et al. Estimated glomerular filtration rate and albuminuria are associated with biomarkers of cardiac injury in a population-based cohort study: The Maastricht study. Clin Chem. 2017;63:887–97.
55. McCullough PA, Duc P, Omland T, et al. B-type natriuretic peptide and renal function in the diagnosis of heart failure: an analysis from the Breathing Not Properly Multinational Study. Am J Kidney Dis. 2003;41:571–9.
56. McKie PM, Cataliotti A, Lahr BD, et al. The prognostic value of N-terminal pro-B-type natriuretic peptide for death and cardiovascular events in healthy normal and stage A/B heart failure subjects. J Am Coll Cardiol. 2010;55:2140–7.
57. Meijers WC, van der Velde AR, Muller Kobold AC, et al. Variability of biomarkers in patients with chronic heart failure and healthy controls. Eur J Heart Fail. 2017;19:357–65.
58. Mingels A, Jacobs L, Michielsen E, et al. Reference population and marathon runner sera assessed by highly sensitive cardiac troponin T and commercial cardiac troponin T and I assays. Clin Chem. 2009;55:101–8.
59. Mozaffarian D, Benjamin EJ, Go AS, et al. Heart disease and stroke statistics-2016 update: a report from the American Heart Association. Circulation. 2016;133:e38–360.
60. Mueller-Hennessen M, Lindahl B, Giannitsis E, et al. Diagnostic and prognostic implications using age- and gender-specific cut-offs for high-sensitivity cardiac troponin T – Sub-analysis from the TRAPID-AMI study. Int J Cardiol. 2016;209:26–33.
61. Neeland IJ, Drazner MH, Berry JD, et al. Biomarkers of chronic cardiac injury and hemodynamic stress identify a malignant phenotype of left ventricular hypertrophy in the general population. J Am Coll Cardiol. 2013;61:187–95.
62. Oliver JM, Gallego P, Gonzalez AE, et al. Impact of age and sex on survival and causes of death in adults with congenital heart disease. Int J Cardiol. 2017;245:119–24.
63. Omland T, de Lemos JA, Holmen OL, et al. Impact of sex on the prognostic value of high-sensitivity cardiac troponin I in the general population: the HUNT study. Clin Chem. 2015;61:646–56.
64. Ouwerkerk W, Voors AA, Zwinderman AH. Factors influencing the predictive power of models for predicting mortality and/or heart failure hospitalization in patients with heart failure. JACC Heart Fail. 2014;2:429–36.
65. Pagidipati NJ, Peterson ED. Acute coronary syndromes in women and men. Nat Rev Cardiol. 2016;13:471–80.
66. Pfisterer M, Buser P, Rickli H, et al. BNP-guided vs symptom-guided heart failure therapy: the Trial of Intensified vs Standard Medical Therapy in Elderly Patients With Congestive Heart Failure (TIME-CHF) randomized trial. JAMA. 2009;301:383–92.
67. Piro M, Della Bona R, Abbate A, Biasucci LM, Crea F. Sex-related differences in myocardial remodeling. J Am Coll Cardiol. 2010;55:1057–65.
68. Ponikowski P, Voors AA, Anker SD, et al. 2016 ESC Guidelines for the diagnosis and treatment of acute and chronic heart failure: The Task Force for the diagnosis and treatment of acute and chronic heart failure of the European Society of Cardiology (ESC) Developed with the special contribution of the Heart Failure Association (HFA) of the ESC. Eur Heart J. 2016;37:2129–200.
69. Rahimi K, Bennett D, Conrad N, et al. Risk prediction in patients with heart failure: a systematic review and analysis. JACC Heart Fail. 2014;2:440–6.
70. Raymond I, Groenning BA, Hildebrandt PR, et al. The influence of age, sex and other variables on the plasma level of N-terminal pro brain natriuretic peptide in a large sample of the general population. Heart. 2003;89:745–51.
71. Redfield MM, Rodeheffer RJ, Jacobsen SJ, et al. Plasma brain natriuretic peptide concentration: impact of age and gender. J Am Coll Cardiol. 2002;40:976–82.
72. Roffi M, Patrono C, Collet JP, et al. 2015 ESC Guidelines for the management of acute coronary syndromes in patients presenting without persistent ST-segment elevation: Task Force for the Management of Acute Coronary Syndromes in Patients Presenting without Persistent ST-Segment Elevation of the European Society of Cardiology (ESC). Eur Heart J. 2016;37:267–315.
73. Rubini Gimenez M, Twerenbold R, Boeddinghaus J, et al. Clinical effect of sex-specific cutoff values of high-sensitivity cardiac troponin T in suspected myocardial infarction. JAMA Cardiol. 2016;1:912–20.
74. Saenger AK, Dalenberg DA, Bryant SC, Grebe SK, Jaffe AS. Pediatric brain natriuretic peptide concentrations vary with age and sex and appear to be modulated by testosterone. Clin Chem. 2009;55:1869–75.
75. Saenger AK, Rodriguez-Fraga O, Ler R, et al. Specificity of B-type natriuretic peptide assays: cross-reactivity with different BNP, NT-proBNP, and proBNP peptides. Clin Chem. 2017;63:351–8.
76. Sandoval Y, Smith SW, Schulz KM, et al. Diagnosis of type 1 and type 2 myocardial infarction using a high-sensitivity cardiac troponin I assay with sex-specific 99th percentiles based on the third universal definition of myocardial infarction classification system. Clin Chem. 2015;61:657–63.
77. Saunders JT, Nambi V, de Lemos JA, et al. Cardiac troponin T measured by a highly sensitive assay predicts coronary heart disease, heart failure, and mortality in the Atherosclerosis Risk in Communities Study. Circulation. 2011;123:1367–76.
78. Savarese G, Trimarco B, Dellegrottaglie S, et al. Natriuretic peptide-guided therapy in chronic heart failure: a meta-analysis of 2686 patients in 12 randomized trials. PLoS One. 2013;8:e58287.
79. Scheven L, de Jong PE, Hillege HL, et al. High-sensitive troponin T and N-terminal pro-B type natriuretic peptide are associated with cardiovascular events despite the cross-

sectional association with albuminuria and glomerular filtration rate. Eur Heart J. 2012;33:2272–81.
80. Schofer N, Brunner FJ, Schluter M, et al. Gender-specific diagnostic performance of a new high-sensitivity cardiac troponin I assay for detection of acute myocardial infarction. Eur Heart J Acute Cardiovasc Care. 2017;6:60–8.
81. Shah AS, Griffiths M, Lee KK, et al. High sensitivity cardiac troponin and the under-diagnosis of myocardial infarction in women: prospective cohort study. BMJ. 2015;350:g7873.
82. Sze J, Mooney J, Barzi F, Hillis GS, Chow CK. Cardiac troponin and its relationship to cardiovascular outcomes in community populations – a systematic review and meta-analysis. Heart Lung Circ. 2016;25:217–28.
83. Tanous D, Siu SC, Mason J, et al. B-type natriuretic peptide in pregnant women with heart disease. J Am Coll Cardiol. 2010;56:1247–53.
84. Thygesen K, Alpert JS, Jaffe AS, et al. Third universal definition of myocardial infarction. Eur Heart J. 2012;33:2551–67.
85. Trambas C, Pickering JW, Than M, et al. Impact of high-sensitivity troponin I testing with sex-specific cutoffs on the diagnosis of acute myocardial infarction. Clin Chem. 2016;62:831–8.
86. Twerenbold R, Boeddinghaus J, Nestelberger T, et al. Clinical use of high-sensitivity cardiac troponin in patients with suspected myocardial infarction. J Am Coll Cardiol. 2017;70:996–1012.
87. Wallace TW, Abdullah SM, Drazner MH, et al. Prevalence and determinants of troponin T elevation in the general population. Circulation. 2006;113:1958–65.
88. Westerman S, Wenger NK. Women and heart disease, the underrecognized burden: sex differences, biases, and unmet clinical and research challenges. Clin Sci (Lond). 2016;130:551–63.
89. Wildi K, Gimenez MR, Twerenbold R, et al. Misdiagnosis of myocardial infarction related to limitations of the current regulatory approach to define clinical decision values for cardiac troponin. Circulation. 2015;131:2032–40.
90. Willeit P, Welsh P, Evans JDW, et al. High-sensitivity cardiac troponin concentration and risk of first-ever cardiovascular outcomes in 154,052 participants. J Am Coll Cardiol. 2017;70:558–68.
91. Wu AH, Jaffe AS, Apple FS, et al. National Academy of Clinical Biochemistry laboratory medicine practice guidelines: use of cardiac troponin and B-type natriuretic peptide or N-terminal proB-type natriuretic peptide for etiologies other than acute coronary syndromes and heart failure. Clin Chem. 2007;53:2086–96.
92. Wu AH, Smith A. Biological variation of the natriuretic peptides and their role in monitoring patients with heart failure. Eur J Heart Fail. 2004;6:355–8.
93. Wu AH, Smith A, Wieczorek S, et al. Biological variation for N-terminal pro- and B-type natriuretic peptides and implications for therapeutic monitoring of patients with congestive heart failure. Am J Cardiol. 2003;92:628–31.
94. Yancy CW, Jessup M, Bozkurt B, et al. 2017 ACC/AHA/HFSA Focused Update of the 2013 ACCF/AHA Guideline for the Management of Heart Failure: a report of the American College of Cardiology/American Heart Association Task Force on Clinical Practice Guidelines and the Heart Failure Society of America. J Am Coll Cardiol. 2017;23:628–51.
95. Yeo KT, Wu AH, Apple FS, et al. Multicenter evaluation of the Roche NT-proBNP assay and comparison to the Biosite Triage BNP assay. Clin Chim Acta. 2003;338:107–15.
96. Zeller T, Tunstall-Pedoe H, Saarela O, et al. High population prevalence of cardiac troponin I measured by a high-sensitivity assay and cardiovascular risk estimation: the MORGAM Biomarker Project Scottish Cohort. Eur Heart J. 2014;35:271–81.

Cardiac Rehabilitation for Women

34

Vera Bittner

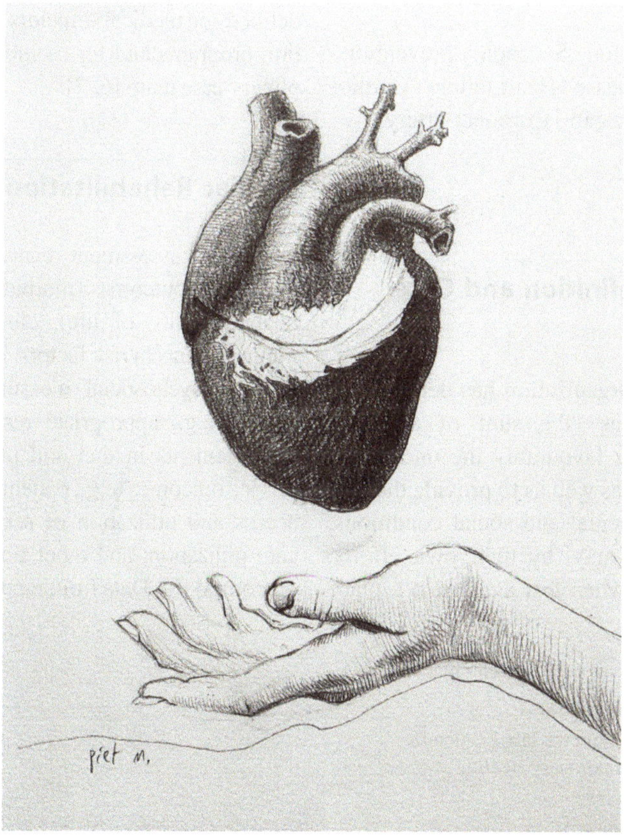

Art work by Piet Michiels, Leuven, Belgium

V. Bittner (✉)
Division of Cardiovascular Disease, University of Alabama at Birmingham, Birmingham, AL, USA
e-mail: vbittner@uab.edu

Abstract

Cardiac rehabilitation provides comprehensive secondary prevention services and improves outcomes among patients with coronary heart

P. L. M. Kerkhof, V. M. Miller (eds.), *Sex-Specific Analysis of Cardiovascular Function*, Advances in Experimental Medicine and Biology 1065, https://doi.org/10.1007/978-3-319-77932-4_34

disease, systolic heart failure, and other cardiac rehabilitation indications. Worldwide, cardiac rehabilitation services are underutilized. Underutilization is more common among women than in men. This chapter reviews barriers to referral, enrollment, and adherence among women and potential interventions to improve women's participation in cardiac rehabilitation. Gender similarities and differences in clinical, behavioral, and health outcomes of cardiac rehabilitation are addressed, and gaps in the evidence base are discussed, pointing to areas of future research.

Keywords
Cardiac rehabilitation · Secondary prevention · Coronary heart disease · Heart failure · Cardio-oncology · Gender gap · Coronary artery dissection

Introduction: Definition and Core Components

The World Health Organization has defined cardiac rehabilitation as "the sum of activities required to influence favourably the underlying cause of the disease, as well as to provide the best possible physical, mental and social conditions, so that the patients may, by their own efforts, preserve or resume when lost as normal a place as possible in the community" [1]. Cardiac rehabilitation has evolved from supervised exercise training after myocardial infarction to the provision of comprehensive secondary prevention services for coronary heart disease and multiple other disease states (Table 34.1).

Core components of cardiac rehabilitation (Table 34.2) have been codified by multiple professional organizations and form the basis for certification of cardiac rehabilitation programs in some countries [2–5]. Specifics of each core component should conform to current national guidelines. Regular program review and updates as appropriate are thus essential. Core competencies have been defined for medical directors of cardiac rehabilitation programs and for members of the multidisciplinary care team [6, 7].

Cardiac Rehabilitation Outcomes

Outcomes assessment can be broadly divided into health outcomes (morbidity, mortality, health-related quality of life), clinical outcomes (e.g., improvements in risk factors, in functional capacity, and in psychosocial measures), behavioral outcomes (e.g., appropriate response to symptoms, improvements in diet and physical activity), and service outcomes (e.g., patient and staff satisfaction, access and utilization of services, patient health-care utilization, and other financial and economic outcomes) [8]. Data from randomized clinical trials,

Table 34.1 Indications for cardiac rehabilitation

US Medicare-covered indications
Myocardial infarction within the last 12 months
Coronary angioplasty or coronary stenting
Stable angina pectoris
Coronary artery bypass surgery
Heart valve repair or replacement
Heart or heart-lung transplant
Stable chronic systolic heart failure
Peripheral arterial disease
Potential indications, not currently covered by most insurance providers
Diastolic heart failure
Metabolic syndrome
"Pre-habilitation" prior to cardiac surgery or percutaneous structural cardiac interventions
Cardio-oncology patients

clinical databases, claims data, and national registries document clinically and statistically significant improvements in these outcome measures across patient populations, disease states, different care settings, and across countries [4, 9].

A 2016 Cochrane systematic review and meta-analysis of randomized controlled trials of cardiac rehabilitation for coronary heart disease (including data from 63 studies through July 2014) demonstrated a 26% reduction in cardiovascular mortality (RR: 0.74; 95% CI 0.64–0.86) and an 18% reduction in the risk of hospital admission (RR 0.82; 95% CI 0.70–0.96). In contrast to several prior reviews summarizing earlier trials, this new analysis showed no effect on total mortality, and there was no impact on recurrent myocardial infarction or revascularization. Improvements in health-related quality of life were evident in the majority of studies [10]. Observational data in less selected contemporary patient populations tend to show significant reductions in total mortality estimated as high as 53% in a recent meta-analysis [11]. Studies that match for propensity to participate in cardiac rehabilitation to decrease the potential for confounding provide similar estimates [12–14]. Several analyses have suggested a "dose response" with greater mortality benefit as the number of attended cardiac rehabilitation sessions increases [13, 15, 16]. A similar "dose response" is also apparent in health services expenditures including hospitalizations, physician visits, diagnostic tests, and drugs among patients eligible for cardiac rehabilitation [15]. Compared to those who were not referred to cardiac rehabilitation, patients with high cardiac rehabilitation participation rates (≥67% of sessions) had 53% lower costs, and patients with lower participation rates had 38% lower costs, while those who were referred but did not attend had 23% higher costs over 3 years of follow-up (Fig. 34.1) [15]. It has been estimated that an

Table 34.2 Core components for cardiac rehabilitation

Patient assessment
Exercise training
Physical activity counseling
Nutritional counseling
Weight management
Blood pressure management
Lipid management
Diabetes management
Tobacco cessation
Psychosocial management

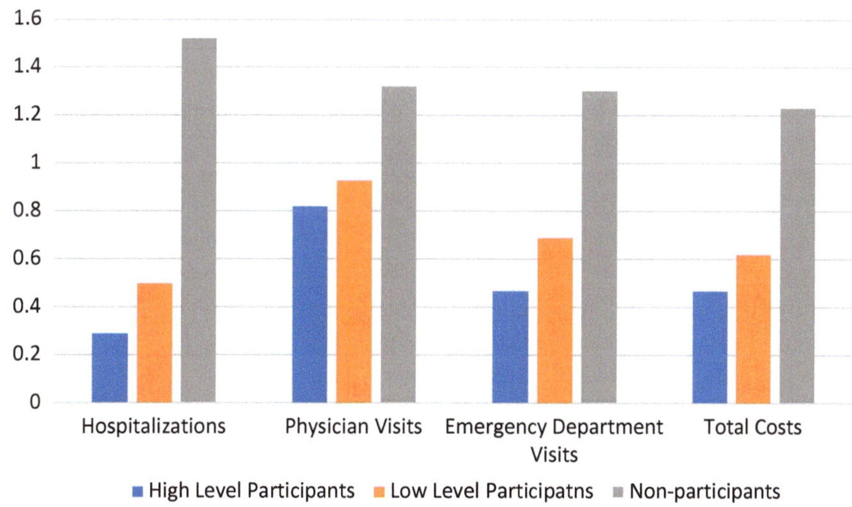

Fig. 34.1 Relative difference in daily costs as a function of participation in cardiac rehabilitation. Marked reductions in costs are evident for high adherers of cardiac rehabilitation, intermediate costs for those who are lower adherers, and increased costs for those who were referred, but did not attend cardiac rehabilitation. Non-referred patients are the referent group (relative difference = 1). (Based on data from Alter et al. [15])

increase in cardiac rehabilitation participation rates from 20 to 70% could save 25,000 lives and 180,000 hospitalizations annually among cardiac rehabilitation-eligible individuals in the United States [17].

Mortality benefits and cost savings are not apparent for cardiac rehabilitation in the setting of chronic systolic heart failure, but exercise training is associated with a reduction in hospital admissions up to 12 months of follow-up [18, 19].

Subgroup data consistently show that both women and men who participate in cardiac rehabilitation benefit in health and clinical outcomes and in behavioral and psychosocial outcomes. Utilization of cardiac rehabilitation services, however, is lower in women than in men, and the degree of improvement in the various outcome domains may vary by gender. The remainder of this chapter will focus on these differences.

Gender Gaps in Referral, Enrollment, and Adherence

Cardiac rehabilitation services are underutilized. Most of the available data focus on individuals after myocardial infarction or with other manifestations of coronary heart disease. Utilization rates for other eligible diagnoses such as systolic heart failure are even lower [20].

Most recent US data published by the Centers for Disease Control and Prevention indicate participation rates for cardiac rehabilitation services of 33.7% in 2013 and 35.5% in 2015 [21]. Comparable rates were 34% in Canada [22], 36.5% in Europe (with considerable between country variation) [23], and < 10% in Japan [24] suggesting that underutilization of cardiac rehabilitation is a global issue. Within countries, access to cardiac rehabilitation varies geographically depending on the number and distribution of cardiac rehabilitation programs, depends on insurance coverage if there is no universal coverage, and is associated with a multitude of provider and patient characteristics [25]. Among referred patients, many do not enroll and, among those who enroll, adherence is often low and dropout rates are substantial [25]. Few patients thus realize the full potential benefit of cardiac rehabilitation participation.

Women are less likely to be referred to cardiac rehabilitation, are less likely to enroll, and tend to have poorer adherence and higher dropout rates. For individuals eligible for cardiac rehabilitation (Table 34.1), Samayoa et al., aggregating data from studies considered to be of "good quality," reported enrollment rates for men from 22.1 to 67% and for women from 14.3 to 38.2% [26]. The summary odds ratio for female versus male enrollment in these studies was 0.56 (95% CI 0.45–0.70; $P < 0.00001$). Data were comparable when all studies were included in the meta-analysis. More recent multivariably adjusted data from the Centers for Disease Control and Prevention show cardiac rehabilitation utilization rates after myocardial infarction of 36.4% among men and 28.8% among women [21].

Supervia et al. recently published a systematic review of barriers to women's cardiac rehabilitation referral, enrollment, and completion, stratifying by patient-, provider-, and system-related barriers [27]. Many of these barriers are not unique to women but may be present disproportionately among female patients eligible for cardiac rehabilitations at least in part due to later age of onset of cardiac rehabilitation-eligible disease states among women compared to men. However, provider referral bias may play a significant role. When presented with 32 case scenarios of cardiac rehabilitation-eligible post-surgical patients, nearly one third of 36 physicians (42% of whom were women) showed evidence of gender bias, judging female patients less likely to benefit from cardiac rehabilitation compared to male patients with comparable characteristics. This judgment was not affected by the physician's gender, length in practice, or subspecialty [28].

Adherence to cardiac rehabilitation among enrolled patients shows significant heterogeneity. In an analysis of the Medicare database, the median number of sessions attended was 25 with an interquartile range of 13–34 [16]. More than 40% of patients attended 30 or more sessions, while 13% attended <6 sessions, and only 18% attended all 36 sessions.

Attendance varied by indication for cardiac rehabilitation and severity of illness as indicated by the importance of comorbidities and the pronounced negative impact of hospitalization during the course of cardiac rehabilitation. Advanced age, nonwhite race, and female gender also negatively predicted adherence. However, the R^2 for the overall model was only 0.05 suggesting that administrative databases provide only a limited understanding of adherence determinants. Marzolini and colleagues analyzed determinants of non-completion in a large clinical dataset from Canada [29]. Women were more likely to withdraw than men (35% vs. 29%). The relative importance of reasons for withdrawal differed by gender with lack of interest, a new cardiac diagnosis, musculoskeletal issues, and lack of transportation being most important among women. Smaller, often single center, studies among women have pointed toward psychological factors such as symptoms of depression and anxiety and lack of social support [30, 31]. Younger women have the lowest attendance rates. Clinical experience suggests that barriers often cluster in individual patients, making it less likely that single factor interventions will be successful.

Cardiac Rehabilitation Outcomes Stratified by Gender

Propensity matched analyses from Olmsted County in Minnesota showed a comparable post myocardial infarction survival benefit in women and men associated with participation in cardiac rehabilitation [14]. Goel et al. replicated these analyses for a population of individuals in Olmsted County who had undergone percutaneous coronary intervention, and Pack et al. did the same for those who had undergone coronary artery bypass surgery [32, 33]. After percutaneous intervention, the mortality reduction over 10 years ranged from 45 to 47% depending on statistical technique employed (matched pair analysis, propensity matching, or landmark analysis) [32]. After CABG, participation in cardiac rehabilitation was associated with a 46% reduction in mortality at 10 years (hazard ratio (HR) 0.54; 95% confidence interval, 0.40–0.74; $P < 0.001$) and a 10-year absolute risk reduction of 12.7% (number needed to treat of 8) [33]. Both analyses showed a significant mortality benefit for women and men associated with participation in cardiac rehabilitation and no gender disparity [32, 33]. Mortality benefit of a similar order of magnitude and without gender disparity was also evident among individuals who had undergone combined coronary artery bypass grafting and valvular surgery [34].

Suaya and colleagues using claims data from Medicare reported on 5-year mortality after hospitalization for coronary conditions or cardiac revascularization among users and nonusers of cardiac rehabilitation, stratified by age and gender [13]. Among men, the absolute differences in mortality in favor of cardiac rehabilitation users ranged from 5.7% among 65–74 year-old individuals, to 9.8% among those 75–84 years old, and 14.9% among those 85 years and older. For women, the respective mortality differences were 8.2%, 13.4%, and 19.5%. Overall, the cumulative 5-year mortality reduction was greater in women than men (10.4% vs. 7.1%). A large population-based study in the Netherlands reached the opposite conclusion: the mortality benefit among women was lower than among men in a mixed population of cardiac rehabilitation attendees including those after acute coronary syndrome and those who had undergone percutaneous or surgical revascularization [35]. The investigators reported hazard ratios for mortality for 6–12 months, 6–24 months, and 6–48 months. While the point estimates for the adjusted hazard ratios for women were consistently below 1 (ranging from 0.67 to 0.79), the mortality reduction reached statistical significance only for the 6–24 month period. Among men, point estimates of the hazard ratio ranged from 0.45 to 0.62 and were statistically significant for all time frames [35]. In a contemporary cohort of participants in a French acute coronary syndrome registry, the mortality reduction associated with cardiac rehabilitation was numerically greater among men than women (51% vs. 37%), but there was no statistically significant interaction by gender [36].

In clinical trials of cardiac rehabilitation, participants randomized to the intervention group have lower rehospitalization rates than their usual care counterparts, but gender-specific estimates are not available due to underrepresentation of women in these clinical trials (<15%). The analyses showed no decrease in subsequent myocardial infarction rates or need for revascularization following the index admission [10]. Analyses from observational datasets have reported lower rates of subsequent myocardial infarction among cardiac rehabilitation participants, but do not provide gender-specific estimates for this endpoint [14, 16].

In the Heart Failure–A Controlled Trial Investigating Outcomes of Exercise TraiNing (HF-ACTION), exercise training in patients with systolic heart failure was associated with a nonsignificant reduction in the primary endpoint of all-cause mortality or hospitalization (hazard ratio [HR], 0.93 [95% CI, 0.84–1.02]; $p = 0.13$) and nonsignificant reductions in the secondary endpoints of mortality, cardiovascular mortality or cardiovascular hospitalization, and cardiovascular mortality or heart failure hospitalization [37]. In an exploratory analysis of the impact of gender on outcomes, exercise training in women was associated with a significant reduction in the combined endpoint of all-cause mortality and hospitalization (HR 0.74; 95% CI, 0.59–0.92) compared with men (HR 0.99; 95% CI, 0.86–1.13) with a significant treatment by gender interaction ($p = 0.027$) [38]. Only one study to date has explored the link between survival after heart transplantation and participation in cardiac rehabilitation and showed that attendance at cardiac rehabilitation within the first 90 days after transplantation was a predictor of improved survival; gender-specific estimates are not available [39].

Exercise capacity is a powerful prognostic factor among healthy individuals and patients with coronary artery disease [40]. Functional capacity at entry and functional capacity at the end of cardiac rehabilitation are both associated with long-term prognosis (combined outcome of mortality, myocardial infarction, and heart failure hospitalization) [41]. Functional capacity is thus a key outcome in cardiac rehabilitation. The greatest predictor of exercise capacity at the end of cardiac rehabilitation is aerobic capacity at entry to the cardiac rehabilitation program. Aerobic capacity at entry is higher among men than women. Due to a greater age-related decline in aerobic capacity among men than women, the gender difference in aerobic capacity is most pronounced at younger ages and narrows progressively among those in their 70s and 80s [42]. While most studies show improvements in exercise capacity during cardiac rehabilitation for men and women, the degree of improvement may differ by gender. Gee et al. retrospectively analyzed data in 758 men and 346 women and showed that men had not only greater exercise capacity at entry to the cardiac rehabilitation program but increased their exercise capacity to a greater degree (2.16 METS) than women (1.65 METS), a result that held up after adjusting for age, body mass index, and indication for cardiac rehabilitation [43]. The authors acknowledge, however, that they were unable to assess the impact of important comorbidities which may have affected the degree of improvement and which may have varied in prevalence or severity between women and men.

Gender differences may be further modified by ethnicity. African-American participants tend to have worse risk factor profiles than their Caucasian counterparts at entry to cardiac rehabilitation [44, 45]. In a retrospective analysis of prospectively collected data from a single-center cardiac rehabilitation program, women had less improvement in 6 min walk distance than men over the course of cardiac rehabilitation. African-American women had the least improvement even after adjustment for age, number of sessions attended, and baseline exercise capacity [44]. Lower gains in functional capacity among women, and particularly among African-American women, were also reported by Johnson et al., using steady-state MET levels during exercise in cardiac rehabilitation as the outcome [45].

Cardiac events and hospitalization among the elderly are potential threats to their ability to live independently. A small study in 10 women and 12 men assessed their ability to do basic activities of daily living and household chores pre- and post-cardiac rehabilitation. At entry, women had lower scores than men, both women and men improved by their post-cardiac rehabilitation assessment, but women improved to a greater degree than men. As a result, the gender difference was no longer statistically significant after cardiac rehabilitation [46]. Patients undergoing transcatheter aortic valve replacement are often elderly and frail. Inpatient rehabilitation for 3 weeks improved submaximal and maximal exercise capacity in a cohort of 136 such patients, over half women, and also improved mental and physical function scores [47]. These results require confirmation in larger populations, but the data suggest that participation in cardiac rehabilitation may be particularly important among older women to preserve their ability to live independently and that assessment of frailty should become routine in cardiac rehabilitation programs [48].

Exercise training improves functional capacity among patients with systolic heart failure. In the HF-ACTION trial, peak oxygen consumption and 6 min walk distance were significantly lower among women than men at baseline despite similar health status and left ventricular ejection fraction [49]. However, the improvement in peak oxygen consumption (ml/kg/min) between baseline and 3 months was numerically similar among men (adjusted model estimate of least-squares mean 0.50, 95% CI 0.22–0.79) and women (adjusted model estimate of least-squares mean 0.73, 95% CI 0.27–1.19), and the gender-by-treatment interaction was not statistically significant ($p = 0.42$), despite lower exercise adherence among women [38]. Exercise training also improves exercise capacity among individuals with heart failure and preserved ejection fraction as shown in a recent meta-analysis of 6 trials (266 patients, 55% women) [50]. Gender-specific results were not reported. Exercise capacity is also favorably affected by cardiac rehabilitation participation among recipients of left ventricular assist devices and recipients of cardiac transplants, but the studies are small and include too few women to provide gender-specific estimates [39, 51, 52].

Physical activity levels tend to be lower in women than men at entry to cardiac rehabilitation. In the Wisconsin Outcomes Registry of 8929 individuals who completed the cardiac rehabilitation program, only 10.6% of women and 17.1% of men reported ≥150 min of physical activity per week at program entry [53]. Self-reported physical activity above this threshold improved significantly in both genders (to 57.7% among women and 69.7% in men), but the adjusted odds of achieving the milestone of ≥150 min/week at graduation from the cardiac rehabilitation program were significantly lower among women than men (adjusted odds ratio (AOR), 0.66; 95% CI, 0.59–0.74) [53].

Achievement of secondary prevention goals is similar among women and men, including guideline determined pharmacologic therapies (aspirin, renin-angiotensin system inhibition, beta-blockade and statins), numerical goals for blood pressure, lipoprotein measures, and blood glucose, indices of body weight, and smoking cessation [53]. Little is known about gender differences in long-term retention of benefits achieved in cardiac rehabilitation. In an analysis by Gupta et al., retention of gains achieved during cardiac rehabilitation at 1 year was generally similar among women and men, although women had better retention of self-reported dietary behavior than men [54]. As all studies of longer-term outcomes, this study was limited to participants who returned for their reassessment visit, and caution is advised in extrapolating results to all patients who complete a cardiac rehabilitation program.

Most, but not all, studies suggest that psychosocial distress at entry to cardiac rehabilitation tends to be higher among women than men. Depressive symptoms at entry to cardiac rehabilitation are a risk factor for non-completion of cardiac rehabilitation among women [30]. Whether early recognition of depression and treatment of depression could change adherence is unknown. Psychosocial distress levels improve during the course of cardiac rehabilitation, but with considerable variability between participants. Few studies have specifically examined gender differences. Hazelton et al. evaluated psychosocial characteristics pre- and post-cardiac rehabilitation among 380 women and men at a Southeastern US academic facility [55]. They evaluated self-reported depressive symptoms, anxiety, panic, anger, and relationship satisfaction. Psychosocial distress was generally higher among women than men. Cardiac rehabilitation did not affect relationship satisfaction in either gender (men remained generally satisfied, women generally dissatisfied). Depressive symptoms and anxiety improved in both genders, while women had a more favorable reduction in panic symptoms. Clinically relevant anger, however, improved only among men, not among women [55]. The reasons for these differences are unknown.

Cardiac Rehabilitation in Spontaneous Coronary Artery Dissection and Microvascular Angina

Spontaneous coronary artery dissection is a rare cause of myocardial infarction which tends to occur almost exclusively in younger women without major atherosclerotic risk factors. Data on safety and efficacy of cardiac rehabilitation are limited. Investigators from the Mayo Clinic published a case series of nine women in 2015 showing improvements in exercise capacity and psychosocial measures without adverse events [56]. Subjective benefits of participation were confirmed through surveys among patients participating in the Mayo Clinic Spontaneous Coronary Artery Dissection Registry, and adverse effects were not apparent. In this group, participation rates in cardiac rehabilitation were higher than usually seen among more typical patients with atherosclerotic coronary artery disease [57]. Whether additional benefits accrue with cardiac rehabilitation initiatives dedicated to individuals with spontaneous coronary artery dissection is unclear [58].

Data on safety and efficacy of cardiac rehabilitation in microvascular angina, a cause of myocardial ischemia that disproportionately affects women, are similarly limited. Szot et al. studied 55 women with microvascular angina and documented not only improvements in functional capacity with cardiac rehabilitation but also improvements in quality of life and myocardial perfusion by radioisotope imaging [59, 60]. Improvements in exercise capacity, metabolic parameters, and psychosocial variables were also seen in a small randomized trial by Asbury et al. [61].

Interventions to Close the Gender Gap

Interventions that have been shown to improve utilization of cardiac rehabilitation services for women and men are detailed in a Presidential Advisory from the American Heart Association and the excellent systematic review by Supervia [25, 27]. As is evident from the compilations, while many interventions have shown benefit, most interventions have been tested only in a single setting, few of the studies are randomized controlled trials, women tend to be underrepresented, and gender-specific data on intervention outcomes are often lacking.

Systematizing referral to cardiac rehabilitation is fairly easy to implement, eliminates referral bias and errors of omission, and increases referral rates [25, 27]. It is clear, however, that care pathways such as Get With The Guidelines which automate referral and increase referral rates are not sufficient to improve enrollment rates [62]. The Cardiac Rehabilitation Care Continuity through Automatic Referral Evaluation (CRCARE) study compared usual care with systematic referral, liaison referral, and combined systematic and liaison strategies [63]. While all strategies resulted in better referral and enrollment rates than usual care, the combined strategy had the best results with referral and enrollment rates of 84% and 70% compared to 29% and 26% with usual care, respectively. Importantly, the combined strategy eliminated the gender gap. Referral bias in favor of men was apparent in the liaison referral only arm, suggesting that automating the referral process is a critical step in eliminating the gender gap. Additional research is needed to determine whether the strength of referral differed by gender in the liaison-only arm or whether women and men respond differently to verbal descriptions of a cardiac rehabilitation program and encouragement to attend cardiac rehabilitation. A hospitalized woman's perception of how likely it is that she will attend cardiac rehabilitation after discharge is an important predictor of subsequent enrollment [64]. Whether this perception is amenable to change is unknown.

Serial program changes such as standardizing the number of prescribed cardiac rehabilitation sessions, showing patients a video outlining the benefits of cardiac rehabilitation, and implementing a dual incentive program for both patients and staff using small token gifts and recognizing adherence can lead to sustained improvements in cardiac rehabilitation adherence over time and appear to be effective for women and men to a similar degree [65]. Taking the concept of motivational programming a step further, a small pilot study in 13 patients (among them 4 women) suggested recently that financial incentives among patients with limited resources may significantly increase program adherence [66].

Several investigators have shown gender differences in illness perception, recovery goals after myocardial infarction, psychosocial characteristics, and comfort with the traditional in-center cardiac rehabilitation environment giving rise to the notion that women-specific programming may enhance utilization rates, adherence, and potentially outcomes among women [67]. A small pilot trial in Toronto tested whether individualized telephone coaching designed to support self-management prior to intake in the cardiac rehabilitation program could impact women's attendance at the program intake visit. Attendance at the intake visit was higher in the intervention than usual care group (57.6% vs. 33.3%, $p = 0.048$), but neither attendance beyond the intake visit nor program outcomes were reported [68]. Midence et al. compared outcomes of cardiac rehabilitation in a small group of women randomized to attend mixed-sex or women-only cardiac rehabilitation groups [69]. In adjusted analyses, outcomes such as changes in exercise capacity and risk factors were equivalent in the two program types. There was a suggestion in the data that women randomized to the women-only cardiac rehabilitation group had lower levels of anxiety and depression than their counterparts assigned to mixed-sex cardiac rehabilitation groups.

Beckie et al. randomized women who had been referred to cardiac rehabilitation to traditional mixed-gender cardiac rehabilitation versus a gender-tailored women-only intervention. The 12-week intervention consisted of 36 electrocardiogram-monitored exercise sessions and 10 psychoeducational sessions based on the transtheoretical stages of change and motivational interviewing strategies [70]. Mirroring the multitude of barriers to participation in traditional cardiac rehabilitation, recruitment into the trial proved challenging [71]. Ultimately, 252 women were randomized. Attendance at exercise (90% vs. 77%) and educational sessions (87% vs. 56%) were better in the intervention group than in the traditional cardiac rehabilitation group [31]. The two programs equally improved exercise capacity [72] and heart rate recovery after exercise [73]. The group randomized to the women-only intervention had greater improvements in inflammatory markers, depressive symptoms, quality of life, and perceptions of health [73–77]. To date, these results have not been reproduced in other cohorts.

Given women's higher burden of barriers to participation in traditional center-based cardiac rehabilitation programs, home-based programs have been increasingly advocated to broaden access to cardiac rehabilitation among women. In a recent Cochrane review, Anderson et al. included 23 trials among 2890 patients with history of acute myocardial infarction, coronary revascularization, or heart failure and found no differences in major outcomes such as mortality, exercise capacity, changes in cardiovascular risk factors, and health-related quality of life between home and in-center cardiac rehabilitation programs [78]. There was a slightly higher rate of program completion among the participants assigned to the home-based intervention (risk ratio 1.04; 95% confidence interval 1.0–1.08). The authors emphasized, however, that only 19% of the participants were women, and 4 of the 23 studies didn't include any women. In a randomized study assigning women to mixed-sex supervised cardiac rehabilitation, a women-only supervised program, or home-based cardiac rehabilitation, Grace et al. found no improvement in adherence with alternate programming [79]. It is thus unclear whether home-based programs could narrow or eliminate the gender gap in cardiac rehabilitation participation.

Gaps in Evidence and Future Directions

Much is to be learned about optimizing the cardiac rehabilitation experience and cardiac rehabilitation outcomes in general and for women in particular. While data suggest that more rehabilitation sessions correlate with better prognosis, we do not know the optimal duration of cardiac rehabilitation nor the optimal frequency of sessions for various rehabilitation outcomes in different disease states, and we don't know whether these differ by gender. While most in the field agree that a course of cardiac rehabilitation should include both aerobic training and resistance training, it is likely that the "optimal exercise prescription" will not only vary by specific rehabilitation outcomes but that it will vary by a multitude of patient characteristics with gender being just one of many variables to be considered. Behavior change and maintenance of behavior change are difficult to achieve. There are no large high quality head-to-head studies comparing different strategies to achieve behavior change in cardiac rehabilitation and their impact on outcomes stratified by gender. Data on long-term outcomes of cardiac rehabilitation are limited at best. Investigations of rehabilitation settings other than the classic in-center exercise, evaluation of women-specific programming, and investigations of the potential impact of "smart-technologies" are in their infancy, and it is too early to tell whether these will help to lessen gender disparities in enrollment and adherence and improve outcomes of cardiac rehabilitation. Cardiac rehabilitation is a multimodality intervention which, in aggregate, improves outcomes. However, we know little about the impact of individual components of cardiac rehabilitation on outcomes nor how impact of individual components may differ by gender and other patient characteristics. Hopefully, analyses of large registries and clinical trials (including pragmatic trials within large registries) will close some of the many evidence gaps. Once clinical trials have shown benefits for specific cardiac rehabilitation approaches, clinical trial results must be translated into clinical practice. Our ability to apply evidence-based approaches is contingent on appropriate representation of patients eligible for cardiac rehabilitation and clear descriptions of the interventions utilized. As shown throughout this chapter, many studies in cardiac rehabilitation do not provide gender-specific data, in part due to underrepresentation of women in published studies. Furthermore, a recent analysis of 63 clinical trials of exercise-based cardiac rehabilitation showed that over half the publications lacked a complete description of the exercise schedule and did not adequately describe its tailoring and progression through the course of the cardiac rehabilitation program [80]. Clinical trials of cardiac rehabilitation should aim for sample sizes that allow robust subgroup analyses by gender, and authors and editors must work together to improve the reporting of such trials.

Lack of definitive data should not dissuade us from fully implementing cardiac rehabilitation for women (and men) with qualifying diagnoses/disease states. Automated referral should be standard in all care settings to eliminate gender bias in referral, and additional strategies should be employed to maximize attendance in traditional cardiac rehabilitation programs or their community or home-based equivalents (Table 34.3). Given lower long-term healthcare expenditures among cardiac rehabilitation attendees, health systems and third party payors would be well advised to incentivize cardiac rehabilitation attendance instead of discouraging attendance by co-pays that are often unaffordable for patients.

Conclusion

Underutilization of cardiac rehabilitation services is widespread and is more pronounced among women than men. Automating referral to cardiac

Table 34.3 Selected strategies to improve referral, enrollment, and adherence among patients eligible for cardiac rehabilitation

Automated referral
Liaison referral
Peer navigation
Early access after the index event
Increasing awareness (providers and patients)
Strong encouragement by the provider
Gender-tailored programming
Community- or home-based programming
Telemedicine
Smartphone- or web-based interventions
Financial incentives for participation

rehabilitation eliminates referral bias and, combined with other interventions, can substantially improve participation rates. Participation in cardiac rehabilitation improves clinical, behavioral, and health outcomes among patients with a variety of cardiovascular diagnoses and decreases health service utilization and costs. Innovative delivery models need additional validation but may ultimately make "cardiac rehabilitation for all" feasible even in low resource settings.

Sources of Funding None.

Disclosures Dr. Bittner has research support from Amgen, Astra Zeneca, Bayer Healthcare, Esperion, and Sanofi.

References

1. WHO expert committee on Rehabilitation After Cardiovascular Diseases WSEODC. Rehabilitation after cardiovascular diseases, with special emphasis on developing countries. 1993.
2. AACVPR. AACVPR Program Certification.
3. Balady GJ, Williams MA, Ades PA, et al. Core components of cardiac rehabilitation/secondary prevention programs: 2007 update: a scientific statement from the American Heart Association exercise, cardiac rehabilitation, and prevention committee, the council on clinical cardiology; the councils on cardiovascular nursing, epidemiology and prevention, and nutrition, physical activity, and metabolism; and the American Association of Cardiovascular and Pulmonary Rehabilitation. Circulation. 2007;115:2675–82.
4. Piepoli MF, Corra U, Benzer W, et al. Secondary prevention through cardiac rehabilitation: from knowledge to implementation. A position paper from the cardiac rehabilitation section of the European Association of Cardiovascular Prevention and Rehabilitation. Eur J Cardiovasc Prev Rehabil. 2010;17:1–17.
5. Woodruffe S, Neubeck L, Clark RA, et al. Australian Cardiovascular Health and Rehabilitation Association (ACRA) core components of cardiovascular disease secondary prevention and cardiac rehabilitation 2014. Heart Lung Circ. 2015;24:430–41.
6. Hamm LF, Sanderson BK, Ades PA, et al. Core competencies for cardiac rehabilitation/secondary prevention professionals: 2010 update: position statement of the American Association of Cardiovascular and Pulmonary Rehabilitation. J Cardiopulm Rehabil Prev. 2011;31:2–10.
7. King M, Bittner V, Josephson R, Lui K, Thomas RJ, Williams MA. Medical director responsibilities for outpatient cardiac rehabilitation/secondary prevention programs: 2012 update: a statement for health care professionals from the American Association of Cardiovascular and Pulmonary Rehabilitation and the American Heart Association. Circulation. 2012;126:2535–43.
8. Sanderson BK, Southard D, Oldridge N. Writing G. AACVPR consensus statement. Outcomes evaluation in cardiac rehabilitation/secondary prevention programs: improving patient care and program effectiveness. J Cardpulm Rehabil. 2004;24:68–79.
9. Kachur S, Chongthammakun V, Lavie CJ, et al. Impact of cardiac rehabilitation and exercise training programs in coronary heart disease. Prog Cardiovasc Dis. 2017;60:103–14.
10. Anderson L, Oldridge N, Thompson DR, et al. Exercise-based cardiac rehabilitation for coronary heart disease: cochrane systematic review and meta-analysis. J Am Coll Cardiol. 2016;67:1–12.
11. Sumner J, Harrison A, Doherty P. The effectiveness of modern cardiac rehabilitation: a systematic review of recent observational studies in non-attenders versus attenders. PLoS One. 2017;12:e0177658.
12. Kureshi F, Kennedy KF, Jones PG, et al. Association between cardiac rehabilitation participation and health status outcomes after acute myocardial infarction. JAMA Cardiol. 2016;1:980–8.
13. Suaya JA, Stason WB, Ades PA, Normand SL, Shepard DS. Cardiac rehabilitation and survival in older coronary patients. J Am Coll Cardiol. 2009;54:25–33.
14. Witt BJ, Jacobsen SJ, Weston SA, et al. Cardiac rehabilitation after myocardial infarction in the community. J Am Coll Cardiol. 2004;44:988–96.
15. Alter DA, Yu B, Bajaj RR, Oh PI. Relationship between cardiac rehabilitation participation and health service expenditures within a universal health care system. Mayo Clin Proc. 2017;92:500.
16. Hammill BG, Curtis LH, Schulman KA, Whellan DJ. Relationship between cardiac rehabilitation and long-term risks of death and myocardial infarction among elderly Medicare beneficiaries. Circulation. 2010;121:63–70.
17. Ades PA, Keteyian SJ, Wright JS, et al. Increasing cardiac rehabilitation participation from 20% to 70%: a road map from the million hearts cardiac rehabilitation collaborative. Mayo Clin Proc. 2017;92:234–42.
18. Reed SD, Whellan DJ, Li Y, et al. Economic evaluation of the HF-ACTION (heart failure: a controlled trial investigating outcomes of exercise training) randomized controlled trial: an exercise training study of patients with chronic heart failure. Circ Cardiovasc Qual Outcomes. 2010;3:374–81.
19. Taylor RS, Sagar VA, Davies EJ, et al. Exercise-based rehabilitation for heart failure. Cochrane Database Syst Rev. 2014;27(4):CD003331.
20. Forman DE, Sanderson BK, Josephson RA, Raikhelkar J, Bittner V. American College of Cardiology's prevention of cardiovascular disease S. Heart failure as a newly approved

diagnosis for cardiac rehabilitation: challenges and opportunities. J Am Coll Cardiol. 2015;65:2652–9.
21. Fang J, Ayala C, Luncheon C, Ritchey M, Loustalot F. Use of outpatient cardiac rehabilitation among heart attack survivors – 20 states and the District of Columbia, 2013 and four states, 2015. MMWR Morb Mortal Wkly Rep. 2017;66:869–73.
22. Grace SL, Bennett S, Ardern CI, Clark AM. Cardiac rehabilitation series: Canada. Prog Cardiovasc Dis. 2014;56(5):530.
23. Humphrey R, Guazzi M, Niebauer J. Cardiac rehabilitation in Europe. Prog Cardiovasc Dis. 2014;56:551–6.
24. Goto Y. Current state of cardiac rehabilitation in Japan. Prog Cardiovasc Dis. 2014;56:557–62.
25. Balady GJ, Ades PA, Bittner VA, et al. Referral, enrollment, and delivery of cardiac rehabilitation/secondary prevention programs at clinical centers and beyond: a presidential advisory from the American Heart Association. Circulation. 2011;124:2951–60.
26. Samayoa L, Grace SL, Gravely S, Scott LB, Marzolini S, Colella TJ. Sex differences in cardiac rehabilitation enrollment: a meta-analysis. Can J Cardiol. 2014;30:793–800.
27. Supervia M, Medina-Inojosa JR, Yeung C, et al. Cardiac rehabilitation for women: a systematic review of barriers and solutions. Mayo Clin Proc. 2017;92:565.
28. Beckstead JW, Pezzo MV, Beckie TM, Shahraki F, Kentner AC, Grace SL. Physicians' tacit and stated policies for determining patient benefit and referral to cardiac rehabilitation. Med Decis Mak. 2014;34:63–74.
29. Marzolini S, Brooks D, Oh PI. Sex differences in completion of a 12-month cardiac rehabilitation programme: an analysis of 5922 women and men. Eur J Cardiovasc Prev Rehabil. 2008;15:698–703.
30. Sanderson BK, Bittner V. Women in cardiac rehabilitation: outcomes and identifying risk for dropout. Am Heart J. 2005;150:1052–8.
31. Beckie TM, Beckstead JW. Predicting cardiac rehabilitation attendance in a gender-tailored randomized clinical trial. J Cardiopulm Rehabil Prev. 2010;30:147–56.
32. Goel K, Lennon RJ, Tilbury RT, Squires RW, Thomas RJ. Impact of cardiac rehabilitation on mortality and cardiovascular events after percutaneous coronary intervention in the community. Circulation. 2011;123:2344–52.
33. Pack QR, Goel K, Lahr BD, et al. Participation in cardiac rehabilitation and survival after coronary artery bypass graft surgery: a community-based study. Circulation. 2013;128:590–7.
34. Goel K, Pack QR, Lahr B, et al. Cardiac rehabilitation is associated with reduced long-term mortality in patients undergoing combined heart valve and CABG surgery. Eur J Prev Cardiol. 2015;22:159–68.
35. de Vries H, Kemps HM, van Engen-Verheul MM, Kraaijenhagen RA, Peek N. Cardiac rehabilitation and survival in a large representative community cohort of Dutch patients. Eur Heart J. 2015;36:1519–28.
36. Pouche M, Ruidavets JB, Ferrieres J, et al. Cardiac rehabilitation and 5-year mortality after acute coronary syndromes: the 2005 French FAST-MI study. Arch Cardiovasc Dis. 2016;109:178–87.
37. O'Connor CM, Whellan DJ, Lee KL, et al. Efficacy and safety of exercise training in patients with chronic heart failure: HF-ACTION randomized controlled trial. JAMA. 2009;301:1439–50.
38. Pina IL, Bittner V, Clare RM, et al. Effects of exercise training on outcomes in women with heart failure: analysis of HF-ACTION (heart failure-a controlled trial investigating outcomes of exercise TraiNing) by sex. JACC Heart Fail. 2014;2:180–6.
39. Rosenbaum AN, Kremers WK, Schirger JA, et al. Association between early cardiac rehabilitation and long-term survival in cardiac transplant recipients. Mayo Clin Proc. 2016;91:149–56.
40. Blaha MJ, Hung RK, Dardari Z, et al. Age-dependent prognostic value of exercise capacity and derivation of fitness-associated biologic age. Heart. 2016;102:431–7.
41. Brawner CA, Abdul-Nour K, Lewis B, et al. Relationship between exercise workload during cardiac rehabilitation and outcomes in patients with coronary heart disease. Am J Cardiol. 2016;117:1236–41.
42. Ades PA, Savage PD, Brawner CA, et al. Aerobic capacity in patients entering cardiac rehabilitation. Circulation. 2006;113:2706–12.
43. Gee MA, Viera AJ, Miller PF, Tolleson-Rinehart S. Functional capacity in men and women following cardiac rehabilitation. J Cardiopulm Rehabil Prev. 2014;34:255–62.
44. Sanderson BK, Mirza S, Fry R, Allison JJ, Bittner V. Secondary prevention outcomes among black and white cardiac rehabilitation patients. Am Heart J. 2007;153(6):980.
45. Johnson D, Sacrinty M, Mehta H, Douglas C, Paladenech C, Robinson K. Cardiac rehabilitation in African Americans: evidence for poorer outcomes compared with whites, especially in women and diabetic participants. Am Heart J. 2015;169:102–7.
46. Johnston M, MacDonald K, Manns P, Senaratne M, Rodgers W, Haennel RG. Impact of cardiac rehabilitation on the ability of elderly cardiac patients to perform common household tasks. J Cardiopulm Rehabil Prev. 2011;31:100–4.
47. Eichler S, Salzwedel A, Reibis R, et al. Multicomponent cardiac rehabilitation in patients after transcatheter aortic valve implantation: predictors of functional and psychocognitive recovery. Eur J Prev Cardiol. 2017;24:257–64.
48. Vigorito C, Abreu A, Ambrosetti M, et al. Frailty and cardiac rehabilitation: a call to action from the EAPC cardiac rehabilitation section. Eur J Prev Cardiol. 2017;24:577–90.
49. Pina IL, Kokkinos P, Kao A, et al. Baseline differences in the HF-ACTION trial by sex. Am Heart J. 2009;158:S16–23.
50. Pandey A, Parashar A, Kumbhani D, et al. Exercise training in patients with heart failure and preserved ejection fraction: meta-analysis of randomized control trials. Circ Heart Fail. 2015;8:33–40.
51. Kerrigan DJ, Williams CT, Ehrman JK, et al. Cardiac rehabilitation improves functional capacity and patient-reported health status in patients with continuous-flow left ventricular assist devices: the Rehab-VAD randomized controlled trial. JACC Heart Fail. 2014;2:653–9.
52. Mahfood Haddad T, Saurav A, Smer A, et al. Cardiac rehabilitation in patients with left ventricular assist device: a SYSTEMATIC REVIEW AND META-ANALYSIS. J Cardiopulm Rehabil Prev. 2017;37:390.
53. Turk-Adawi KI, Oldridge NB, Vitcenda MJ, Tarima SS, Grace SL. Secondary prevention recommendation attainment with cardiac rehabilitation: is there a gender disparity? Womens Health Issues. 2016;26:278–87.
54. Gupta R, Sanderson BK, Bittner V. Outcomes at one-year follow-up of women and men with coronary artery disease discharged from cardiac rehabilitation: what benefits are maintained? J Cardiopulm Rehabil Prev. 2007;27:11–8. quiz 19-20
55. Hazelton G, Williams JW, Wakefield J, Perlman A, Kraus WE, Wolever RQ. Psychosocial benefits of cardiac rehabilitation among women compared with men. J Cardiopulm Rehabil Prev. 2014;34:21–8.

56. Silber TC, Tweet MS, Bowman MJ, Hayes SN, Squires RW. Cardiac rehabilitation after spontaneous coronary artery dissection. J Cardiopulm Rehabil Prev. 2015;35:328–33.
57. Krittanawong C, Tweet MS, Hayes SE, et al. Usefulness of cardiac rehabilitation after spontaneous coronary artery dissection. Am J Cardiol. 2016;117:1604–9.
58. Chou AY, Prakash R, Rajala J, et al. The first dedicated cardiac rehabilitation program for patients with spontaneous coronary artery dissection: description and initial results. Can J Cardiol. 2016;32:554–60.
59. Szot W, Zajac J, Kostkiewicz M, Owoc J, Bojar I. Cardiac rehabilitation: a good measure to improve quality of life in peri- and postmenopausal women with microvascular angina. Ann Agric Environ Med. 2015;22:390–5.
60. Szot W, Zajac J, Kubinyi A, Kostkiewicz M. The effects of cardiac rehabilitation on overall physical capacity and myocardial perfusion in women with microvascular angina. Kardiol Pol. 2016;74:431–8.
61. Asbury EA, Slattery C, Grant A, Evans L, Barbir M, Collins P. Cardiac rehabilitation for the treatment of women with chest pain and normal coronary arteries. Menopause. 2008;15:454–60.
62. Mazzini MJ, Stevens GR, Whalen D, Ozonoff A, Balady GJ. Effect of an American Heart Association get with the guidelines program-based clinical pathway on referral and enrollment into cardiac rehabilitation after acute myocardial infarction. Am J Cardiol. 2008;101:1084–7.
63. Gravely S, Anand SS, Stewart DE, Grace SL, Investigators C. Effect of referral strategies on access to cardiac rehabilitation among women. Eur J Prev Cardiol. 2014;21:1018–25.
64. Sanderson BK, Shewchuk RM, Bittner V. Cardiac rehabilitation and women: what keeps them away? J Cardiopulm Rehabil Prev. 2010;30:12–21.
65. Pack QR, Johnson LL, Barr LM, et al. Improving cardiac rehabilitation attendance and completion through quality improvement activities and a motivational program. J Cardiopulm Rehabil Prev. 2013;33:153–9.
66. Gaalema DE, Savage PD, Rengo JL, Cutler AY, Higgins ST, Ades PA. Financial incentives to promote cardiac rehabilitation participation and adherence among Medicaid patients. Prev Med. 2016;92:47–50.
67. Bjarnason-Wehrens B, Grande G, Loewel H, Voller H, Mittag O. Gender-specific issues in cardiac rehabilitation: do women with ischaemic heart disease need specially tailored programmes? Eur J Cardiovasc Prev Rehabil. 2007;14:163–71.
68. Price JAD. A pilot trial of a coaching intervention designed to increase women's attendance at cardiac rehabilitation intake. Toronto: Graduate Department of Nursing Science: University of Toronto; 2012.
69. Midence L, Arthur HM, Oh P, Stewart DE, Grace SL. Women's health Behaviours and psychosocial well-being by cardiac rehabilitation program model: a randomized controlled trial. Can J Cardiol. 2016;32:956–62.
70. Beckie TMA. Behavior change intervention for women in cardiac rehabilitation. J Cardiovasc Nurs. 2006;21:146–53.
71. Beckie TM, Mendonca MA, Fletcher GF, Schocken DD, Evans ME, Banks SM. Examining the challenges of recruiting women into a cardiac rehabilitation clinical trial. J Cardiopulm Rehabil Prev. 2009;29:13–21. quiz 22-3
72. Beckie TM, Beckstead JW, Kip K, Fletcher G. Physiological and exercise capacity improvements in women completing cardiac rehabilitation. J Cardiopulm Rehabil Prev. 2013;33:16–25.
73. Beckie TM, Beckstead JW, Kip KE, Fletcher G. Improvements in heart rate recovery among women after cardiac rehabilitation completion. J Cardiovasc Nurs. 2014;29:38–47.
74. Beckie TM, Beckstead JW, Groer MW. The influence of cardiac rehabilitation on inflammation and metabolic syndrome in women with coronary heart disease. J Cardiovasc Nurs. 2010;25:52–60.
75. Beckie TM, Beckstead JW. The effects of a cardiac rehabilitation program tailored for women on global quality of life: a randomized clinical trial. J Womens Health (Larchmt). 2010;19:1977–85.
76. Beckie TM, Beckstead JW, Schocken DD, Evans ME, Fletcher GF. The effects of a tailored cardiac rehabilitation program on depressive symptoms in women: a randomized clinical trial. Int J Nurs Stud. 2011;48:3–12.
77. Beckie TM, Beckstead JW. The effects of a cardiac rehabilitation program tailored for women on their perceptions of health: a randomized clinical trial. J Cardiopulm Rehabil Prev. 2011;31:25–34.
78. Anderson L, Sharp GA, Norton RJ, et al. Home-based versus centre-based cardiac rehabilitation. Cochrane Database Syst Rev. 2017;6:CD007130.
79. Grace SL, Midence L, Oh P, et al. Cardiac rehabilitation program adherence and functional capacity among women: a randomized controlled trial. Mayo Clin Proc. 2016;91:140–8.
80. Abell B, Glasziou P, Hoffmann T. Reporting and replicating trials of exercise-based cardiac rehabilitation: do we know what the researchers actually did? Circ Cardiovasc Qual Outcomes. 2015;8:187–94.

Stroke Rehabilitation: Therapy Robots and Assistive Devices

35

Verena Klamroth-Marganska

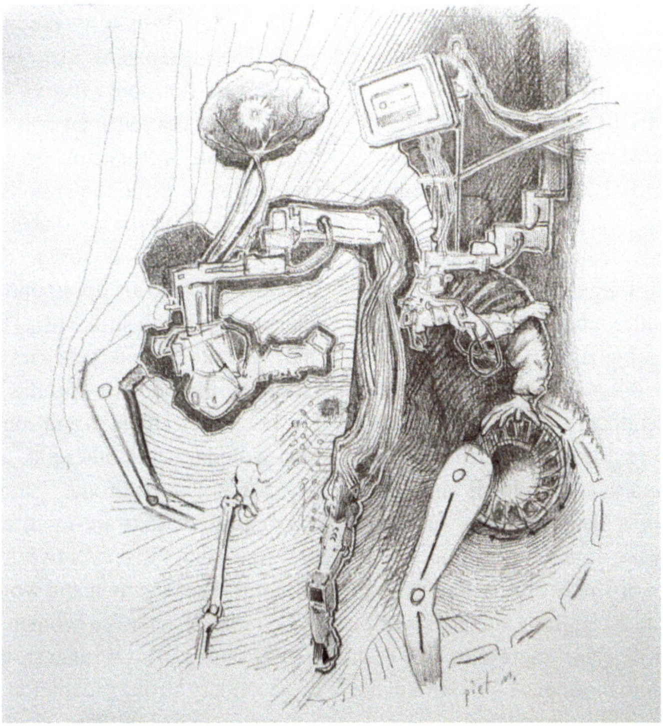

Therapy robots and assistive devices. Art work by Piet Michiels, Leuven, Belgium.

V. Klamroth-Marganska (✉)
Department of Health Sciences and Technology (D-HEST), Sensory-Motor Systems (SMS) Lab, Institute of Robotics and Intelligent Systems (IRIS), ETH Zurich, Zurich, Switzerland

Medical Faculty, Reharobotics Group, Spinal Cord Injury Center, Balgrist University Hospital, University of Zurich, Zurich, Switzerland
e-mail: verena.klamroth@hest.ethz.ch

Abstract

Motor impairments after stroke are often persistent and disabling, and women are less likely to recover and show poorer functional outcomes. To regain motor function after stroke, rehabilitation robots are increasingly integrated into clinics. The devices fall into

P. L. M. Kerkhof, V. M. Miller (eds.), *Sex-Specific Analysis of Cardiovascular Function*, Advances in Experimental Medicine and Biology 1065, https://doi.org/10.1007/978-3-319-77932-4_35

two main classes: robots developed to train lost motor function after stroke (therapy devices) and robots designed to compensate for lost skills (i.e., assistive devices). The article provides an overview of therapeutic options with robots for motor rehabilitation after stroke.

Keywords
Rehabilitation robots · Brain injury · Sex differences · Motor function · Neurorehabilitation · Therapy device · Telerehabilitation · Locomotor training · Multiplayer strategy

Sex Disparities in Stroke Epidemiology, Pathophysiology, Treatment, and Outcomes After Stroke

Stroke is a worldwide health concern and one of the main causes of disability. An estimated 17 million people worldwide experience a stroke each year [1]. In high-income countries, the incidence decreases, probably due to improved risk factor control and better healthcare interventions. However, the incidence in low-income and middle-income countries, and thus the global number, is increasing [2]. With a declining rate of stroke mortality and increasing aging populations worldwide, the prevalence of stroke survivors rises and is estimated to reach 70 million subjects worldwide by 2030 [1]. It is clear, therefore, that both the financial and human impacts of stroke are immense for patient, family, and society.

There are differences between women and men regarding stroke which are increasingly being recognized and have even led to specific guidelines for the prevention of stroke in women by the American Heart Association [3]. For instance, the incidence of stroke is different in men and women. During most of the lifetime, men are at a higher risk to experience a stroke compared to women. But in the young population, the risk for women is higher than in men of the same age, probably due to oral contraceptive use and pregnancy. Moreover, in postmenopausal women, a steep increase in the incidence of stroke occurs [4]. Due to the higher life expectancy of women and much higher incidence of stroke in older ages, women suffer more strokes than men. Furthermore, there are more overall stroke deaths among women. Stroke is the second leading cause of death in women and only the fifth leading cause in men in the USA [5]. But of note, women are on average 4.5 years older when they get their first stroke, and when adjusting for age, the death rates are similar between men and women [6, 7].

Sex disparities in the etiology of stroke are also well documented: cardioembolism is the main cause of stroke in women, whereas large and small vessel disease is the main cause among men. The differences may be due not only to genetic and hormonal factors but also to lifestyle differences between men and women. Many significant risk factors for stroke are modifiable: Hypertension, smoking, body fat distribution, diet, physical inactivity, diabetes mellitus, alcohol intake, psychosocial stress, depression, cardiac causes, and ratio of apolipoproteins B to A1 are all associated with stroke [8]. They are controllable and should be targets for population-based programs that are sex-specific. For instance, men are more often smokers and consume more alcohol than women, while other risk factors such as obesity/metabolic syndrome, atrial fibrillation, and migraine with aura are more common in women [9]. Accordingly, preventive measures may have different effects in men and women: Aspirin seems to be more effective in women at preventing first-ever strokes [10]. Of interest, the effect of aspirin on secondary stroke prevention seems not different between sexes [11, 12].

Several studies suggest that different cell pathways are activated in males and females after ischemic stroke [13]. This may also have an influence on the effect of treatment. For instance, in animal models, progesterone treatment after neonatal brain injury resulted in reduced tissue loss only in males but not in females [14]. Similarly, there is some evidence from animal studies that differences in the effectiveness of steroid hormones after stroke exist. While estrogens appeared to be effective in young adult rodent males and ovariectomized females, they might have been hazardous in intact females [15]. In humans, there exist differences between sexes in the acute care which is provided. A European study showed that investigations such as Doppler examination, echocardiogram, and angiography were less frequently performed in women than in men [16]. Women are also less likely to receive intravenous alteplase treatment for thrombolytic therapy while in hospital [12]. And these differences cannot be explained by age or severity of stroke alone. So, what is the reason for these disparities between men and women? It is assumed that a complex interplay of demographic factors, pre-existing health state, and psychosocial functioning, together with age and disease severity, may explain the described differences [12].

The outcome after stroke is mainly defined by motor impairment. It affects about 80% of stroke survivors and typically involves movements of the face, arm, and leg on one side of the body [17]. So far, there is hardly any complete cure for the damage to the central nervous system. Although the successes of acute treatment after neurological diseases such as stroke are considerable (e.g., by intravenous thrombolysis), the therapies are applicable only in narrow time windows and rarely lead to a complete decline in symptoms. Therefore, the motor consequences of stroke continue to be treated mainly with permanent forms of physical therapy.

Physiotherapy and occupational therapy aim primarily at restoring and maintaining the motor function to perform activities of daily living (ADLs). No major differences between men and women were observed regarding the amount of physical therapy received [16]. Nevertheless, sex seems to be a prognostic factor in stroke rehabilitation [18]. Several studies have shown that women have a worse functional outcome with more disabilities and handicaps within the first 3 months after stroke [12, 17]. Results among studies in this subacute phase are divergent and may be partly attributed to age at stroke onset or other demographic or clinical characteristics [17]. Few studies have assessed functional outcome beyond these 3 months. In a study where observation was prolonged to cover the whole first year after stroke, deterioration was more common in women [18]. Male patients have a higher probability to be independent in both ADLs and stair climbing [18]. This better outcome in men may be partly attributable to differences in muscular strength between men and women. Strength tends to decline more with age in women than men and plays an important role for an independent living [19, 20]. Muscle weakness may be directly responsible for compromised motor function after stroke, and several studies showed that strengthening interventions increase strength and improve activity [21]. But rather than just strength or fitness, motor control is essential for skilled use of the limbs.

Neurorehabilitation of Motor Function After Stroke

Various therapeutic approaches for stroke rehabilitation exist, and a large debate about the relative effectiveness of these approaches on motor function after stroke is ongoing [22]. Take, for example, constraint-induced movement therapy (CIMT). CIMT is a promising therapy form where the less affected limb is restricted. The affected arm must be used instead to perform tasks and exercises which are tailored to the person's capacity. Although the therapy is well known and is supposed to be a useful tool for recovering the ability to perform ADLs, a Cochrane review found that CIMT was associated with only limited improvements in motor impairment and motor function, and in the author's opinion, these benefits did "not convincingly reduce disability" [23]. Besides the limited outcome, the inclusion criteria for CIMT are strict, and the majority of patients (>75%) does not even meet the minimum motor ability required for CIMT [24]. CIMT is one example that highlights the generally modest therapeutic success of conventional therapy forms. As a consequence, neurological research is strongly focused on optimizing learning strategies in order to boost recovery.

There is strong evidence that a therapy that includes intensive, high-repetition, task-oriented training improves motor function after stroke [22]. A Cochrane review on such repetitive and task-oriented training after stroke showed improvements, although small, in arm and hand functions as well as walking distance and walking ability [25].

How is high repetition in neurorehabilitation defined? From animal studies, we can conclude that large doses of practice in the order of hundreds of daily repetitions of upper extremity practice and thousands of daily repetitions of gait may be required [26]. But this amount is seldom reached in neurorehabilitation therapy. A study investigated the numbers of repetitions performed in a typical training session of physical and occupational therapies post-stroke [26]. The average number of gait steps in a training session was 357. In upper limb rehabilitation, the average number of repetitions per training session reached a mere 32. How can we explain this discrepancy between the required and achieved numbers? Tasks performed with the upper limb are often complex. Most tasks are composed of several movement components, namely reaching for, grasping, moving/manipulating, and then releasing an object [27]. An intense training with a high number of repetitions of these coordinated movements requires a lot of physical and mental effort by the therapist who is assisting the patient.

Robots can relieve the therapist. Robots, as described here, are machines that are equipped with advanced control techniques to interact with the user. As they enable to offer highly repetitive therapy over a longer period, robotic devices for the treatment of the upper and lower extremities are increasingly being integrated into everyday clinical practice.

Robots for Motor Training of the Upper Extremity

A variety of therapy robots for neurorehabilitation are under development and clinically tested. More than 120 devices for therapy of the upper extremity only are described [28]. The diverse technical solutions differ in actuation, sensors, and control strategies. The aim of the devices is to provide motion assistance during physical therapy. Regarding the mechanical structure, there are two main categories: end-effector-based and exoskeleton-based robots. Both types allow for intense, high-repetition therapy. They differ in the interaction with the user. End-effectors are attached to the human limb at only one, usually distal, point (e.g., the lower arm). Movement of the end-effector changes the position of the arm at this connection point. It also indirectly changes the position of other limb segments (e.g., the upper arm) and creates interaction torques in each limb segment that are not fully determined by the device [28]. In patients, the movement may provoke pathologic compensatory movements that can be harmful. Modular systems overcome this limitation by combining several end-effector devices that are connected at the single limb segments [29].

Exoskeletons are orthoses that surround the arm like a scaffold. The device segments align to the anatomical segments of the human arm to which they are attached at multiple locations. The rotational axes of the device correspond largely to the rotational axes of the human arm. Exoskeletons require more complex compensation and control algorithms than end-effectors. But they allow more naturalistic movements of the arm. The arm positions can be precisely defined in each segment, and each joint can be controlled independently [30].

ARMin is an example of an exoskeleton robot for therapy of the upper extremity after brain injury. It allows for functional shoulder, elbow, and wrist movements and hand opening/closing (Fig. 35.1 [31, 32]). As the device is connected with a graphical display (i.e., a screen), the user can perform games, for example, air hockey, where highly repetitive movements are presented in a motivating scenario. Instant audio, visual, and haptic feedbacks are given. The level of game difficulty can be adapted to the abilities of the user. Additionally, the device provides assistance when the user is not able to execute a movement alone (assistance as needed) [33]. For example, when the user attempts to catch a ball in the game and does not succeed to finalize the movement, the robot guides the arm to the target point. This guidance is supposed to support motor learning by providing sensorimotor input. Besides games, ARMin allows to train activities that are relevant in daily life (ADL), for example,

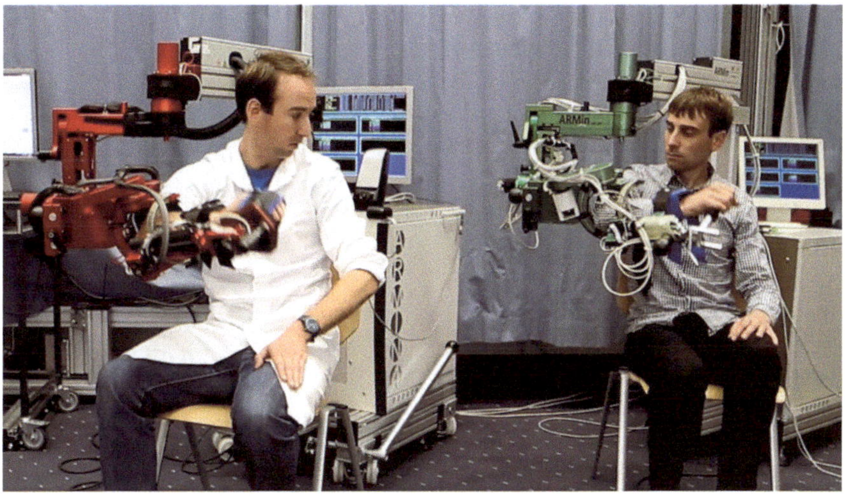

Fig. 35.1 Two ARMins with subjects for Beam Me In setup

cooking or buying a train ticket from an automatic ticket counter.

The ARMin robot was already successfully tested in subjects in the chronic phase (i.e., minimum of 6 months) post-stroke [34]. After 8 weeks of therapy (i.e., 24 h of therapy), robot-assisted therapy with ARMin led to significantly higher gains in motor function than did conventional (physical and occupational) therapy. Furthermore, the same gains were achieved faster with the robot. It seems that particularly patients with severe impairment benefit from therapy with ARMin. Looking at the other outcomes of the clinical study, strength gains were significantly higher with conventional therapy. Secondary analysis of the data implies that the gains in motor function in the conventional group were partly attributable to these strength gains [35]. In contrast, the ARMin group seemed to gain rather motor skills (i.e., the ability to reliably deliver accurate execution of motor tasks) [36]. No difference in outcome between men and women could be observed in this study.

Robots for Motor Training of the Lower Extremity

Three months after stroke, 20% of patients remain wheelchair bound, and most of those who are ambulatory present with a disrupted, asymmetric walking pattern [37]. Motor weakness and balance disturbances make locomotor training challenging as it may require two or more therapists to assist the legs and trunk during movement. Robotic devices are helpful tools in training of gait by providing body weight support and guidance of the legs. Some devices are equipped with sensors that provide information regarding forces and torques, angles, and movement trajectories and give feedback to the user.

As for the upper extremity, the devices can be differentiated into end-effectors and exoskeletons. An example of a stationary end-effector robot is the Gangtrainer® (Reha-Stim). The robot device guides the legs through two footplates which simulate the bilateral joint stance and swing phases of walking in a coordinated way, on fixed trajectories. An extension of the Gangtrainer® is the HapticWalker®, a robot with separately programmable footplates which allow also more sophisticated walking tasks, such as stair climbing, and the simulation of gait perturbations such as stumbling [38]. These functions come at the cost of bigger dimensions of the device. A smaller version of the HapticWalker® for clinical use is the Geo-System® that enables the practice of simulated floor walking and stair descending and ascending [39].

Regarding exoskeletons for the lower extremity, several solutions combine the orthosis with a harness for body weight support and a treadmill.

The devices can be adjusted in walking speed, the amount of body weight support, and the amount of assistance provided to the individual needs of the users. Examples of commercially available devices are the Lokomat® (Hocoma), the ReoAmbulator® (Motorika), and the LokoHelp® (Woodway).

Overground gait trainers are mobile exoskeleton solutions. In rehabilitation, they are mainly developed for training purposes. But the use is rapidly increasing toward assistive purposes for use at home and even in the community. The wearable robots surround the body like a scaffold. Using crutches, the user can stand, walk, get up from the chair, and sit down. Even sideways movements and climbing stairs are possible. The rigid yet movable metal structures can enable even completely paralyzed people to walk.

Some groups develop soft mobile exoskeletons. Such lightweight and flexible exoskeletons (i.e., exosuits) can be discretely worn under clothing. The straps on the legs mimic the muscles and ligaments of the body. The power of the motors is transmitted via cables to the joints. Subjects who can stand but are restricted in mobility, being due to age or a partial weakness, is given assistance in everyday life by these devices. An example is the Myosuit (Fig. 35.2) [40]. If the user needs help with physically demanding activities such as transfer from sitting to standing, ascending or descending stairs, or walking short distances, the soft exoskeleton can aid in the movement [40, 41].

An exosuit for unilateral use assists ankle plantarflexion and dorsiflexion and was successfully tested in hemiparetic subjects post-stroke where it led to a reduction in paretic hip hiking and circumduction during walking [42].

Summarizing the current status of clinical research on electromechanical-assisted gait training, a Cochrane review found moderate-quality evidence that such training combined with physiotherapy improves recovery of independent walking in people after stroke more than does physiotherapy alone [43]. The increase in walking ability was not dependent on the type of device used. Whether end-effector devices improve walking velocity more than exoskeleton devices needs studies that aim at comparing the different devices.

Fig. 35.2 Subject wearing the Myosuit for stair climbing

Robots for Basic Research

Robots bear more advantages than only intense, high-repetition training in a motivating environment. Robots can provide optimized learning strategies which are not achieved with conventional therapy forms. Error amplification is an example of a learning strategy that was successfully implemented with robots. The principle behind error amplification is to amplify the trajectory errors throughout a movement, usually haptically by the robot or visually by the display. This amplification may bring the error to the attention of the performer. The success of error amplification depends on the subjects' skill level [44]. It seems helpful particularly in those stroke patients who are only moderately affected and helps to learn the spatial components of trajectories [45]. For more severely affected patients, not the haptic amplification of an error but rather the haptic guidance through a movement is beneficial. This passive movement alone as a kind of "proprioceptive observation" can teach the movement and induces cortical changes [46, 47].

In addition to providing new treatment options, robots may deepen the understanding about the mechanisms and the time course of the recovery of motor control. The abnormalities in motor control after stroke are not entirely subject-specific. Typical patterns of impairments are observed across subjects and are part of the recovery process (e.g., pathological synergies) [48]. Robots that are equipped with sensors can reveal these patterns in much more depth and finer time resolution than clinical tests or motion-capturing systems. The latter provide kinematic information (e.g., position, velocity, and acceleration) of a movement. Robots with haptic sensors can additionally record kinetic information, i.e., forces or torques [49].

An example is the time course of recovery. It is generally accepted that true recovery of motor control, i.e., the ability to perform a movement in the same manner as before injury, takes place in the first 3–6 months after stroke [50]. Improvements beyond this sensitive period may be mediated almost entirely by compensation, meaning that alternative movement patterns which are different from those originally used before the injury are learned (for instance, using the trunk for reaching movements of the arm).

By analyzing movements that were performed in a robot, it could be recently shown that true recovery (i.e., motor control) and compensation are dissociated already early after stroke. The true recovery of arm motor control seems to plateau after only 5 weeks [51]. Beyond this time, compensation and strength gains dominate the functional improvement.

Analyzing data recorded during robotic therapy on large groups of patients and application of machine learning will change our understanding of the different variables affecting recovery after stroke. It may reveal how sex can impede or promote stroke recovery. This is important since little is known about differences among men and women regarding long-term recovery. Novel ways to improve neurorehabilitation therapies will emerge. The rehabilitation therapy will be individually adapted depending on the phase, the observed deficits, and the severity of impairment. The goal is to improve performance in the early as well as in later chronic stages.

Rehab Gyms

Another important advantage of robots is the possibility that a single therapist can effectively supervise multiple subjects. Rehab gyms allow such a training scenario. A therapist supervises a group of patients who train on different robotic devices. Subjects can even interact through the devices by playing games in virtual scenarios. Users can compete against each other or play with each other against the computer. The input device can be a robot or other devices such as Nintendo Wii® or a computer mouse. Although both users play the same game, the difficulty of the virtual game is customized for each individual. For instance, the resistance to the movement can be increased for the "better" player or the "worse" player receives more support during a movement. Even ADLs such as cooking can be practiced together in virtual reality. This so-called multiplayer strategy allows social interaction with others during therapy, being patients, friends, or family members. Whether this cooperation promotes motivation and gains in motor function is currently being investigated in a clinical study involving the partner. Preliminary data suggests that women and men differ in preferences how these games should be set up. Men seem to enjoy competitive gaming scenarios more than do women.

Robot-Assisted Telerehabilitation

Remote therapy (i.e., telerehabilitation) with robotic devices can improve the access to therapy and provides an opportunity to increase the duration and the frequency of training. Several telerehabilitation systems with remote auditory and/or visual interactions between therapist and patient already exist and were clinically tested in stroke patients [52, 53]. Remote connection with not only visual and auditory but also haptic devices is increasingly investigated. To do so, therapy robots are placed at the home of patients. They communicate with the therapist to exchange information regarding daily usage and functional performance (e.g., training angles achieved, difficulty levels across exercises) [54, 55].

We extended haptic telerehabilitation by providing both the therapist and the patient with a robot (Fig. 35.2). This system enables therapists to haptically interact with the patient, i.e., to guide the patient's arm through a movement, but also to assess the impairment of the arm remotely. As the therapist can feel the patient impairment via the robot in his own limb, we call it the "Beam Me In" setup.

The Role of the Therapist in Robot-Assisted Therapy

Robots do not replace the therapist but provide him with a tool that allows him to use his skills and knowledge. But while conventional therapy allows the therapist to interact directly with the patient and feel the constraints of a limb, in robot-assisted therapy, it is the therapist's primary responsibility to operate the device. Interactions of the therapist with the robot are not intended. They are interpreted by the robot as an external disturbance. We aim to integrate the therapist into the robot-assisted therapy. In the BridgeT project, the therapist can operate the ARMin robot (see above) via handgrips that are equipped with sensors. The therapist teaches the device by directly guiding it through movements that the patient should perform [56]. The device learns from the therapist, repeats the movement, and adjusts the difficulty to the patient's abilities, e.g., if the patient becomes tired or stronger during the therapy.

Conclusion

Although sex differences regarding stroke are recognized, little information on differences among men and women regarding motor recovery after stroke is available. Robots with sensors have the potential to monitor slight changes in motor function. Thus, they are suitable to deepen the understanding of the individual recovery process and to find patterns among different groups, e.g., among men and women. Eventually, robot-assisted therapy will optimize treatment strategies for patients of both sexes.

References

1. Feigin VL, et al. Global and regional burden of stroke during 1990–2010: findings from the Global Burden of Disease Study 2010. Lancet. 2014;383(9913):245–55.
2. Feigin VL, Lawes CM, Bennett DA, Barker-Collo SL, Parag V. Worldwide stroke incidence and early case fatality reported in 56 population-based studies: a systematic review. Lancet Neurol. 2009;8(4):355–69.
3. Bushnell C, et al. Guidelines for the prevention of stroke in women. Stroke. 2014;45(5):1545–88.
4. W. G. f. t. W. s. H. I. Investigators. Risks and benefits of estrogen plus progestin in healthy postmenopausal women: principal results From the Women's Health Initiative randomized controlled trial. JAMA. 2002;288(3):321–33.
5. Lackland DT, et al. Factors influencing the decline in stroke mortality. Stroke. 2014;45(1):315–53.
6. N. C. f. H. Statistics. Health, United States, 2011. Hyattsville: US Department of Health and Human Services, Centers for Disease Control and Prevention, National Center for Health Statistics; 2012.
7. Appelros P, Stegmayr B, Terént A. Sex differences in stroke epidemiology. Stroke. 2009;40(4):1082–90.
8. O'Donnell MJ, et al. Risk factors for ischaemic and intracerebral haemorrhagic stroke in 22 countries (the INTERSTROKE study): a case-control study. Lancet. 2010;376(9735):112–23.
9. Olsen TS, Dehlendorff C, Andersen KK. Sex-related time-dependent variations in post-stroke survival–evidence of a female stroke survival advantage. Neuroepidemiology. 2007;29(3–4):218–25.
10. Berger JS, Roncaglioni MC, Avanzini F, Pangrazzi I, Tognoni G, Brown DL. Aspirin for the primary prevention of cardiovascular events in women and men: a sex-specific meta-analysis of randomized controlled trials. JAMA. 2006;295(3):306–13.
11. A. Trialists'Collaboration. Collaborative meta-analysis of randomised trials of antiplatelet therapy for prevention of death, myocardial infarction, and stroke in high risk patients. BMJ. 2002;324(7329):71–86.
12. Reeves MJ, et al. Sex differences in stroke: epidemiology, clinical presentation, medical care, and outcomes. Lancet Neurol. 2008;7(10):915–26.
13. Gibson CL. Cerebral ischemic stroke: is gender important? J Cereb Blood Flow Metab. 2013;33(9):1355–61.
14. Peterson BL, Won S, Geddes RI, Sayeed I, Stein DG. Sex-related differences in effects of progesterone following neonatal hypoxic brain injury. Behav Brain Res. 2015;286:152–65.
15. Gibson CL, Gray LJ, Murphy SP, Bath PM. Estrogens and experimental ischemic stroke: a systematic review. J Cereb Blood Flow Metab. 2006;26(9):1103–13.
16. Di Carlo A, et al. Sex differences in the clinical presentation, resource use, and 3-month outcome of acute stroke in Europe. Stroke. 2003;34(5):1114–9.

17. Langhorne P, Coupar F, Pollock A. Motor recovery after stroke: a systematic review. Lancet Neurol. 2009;8(8):741–54.
18. Paolucci S, et al. Is sex a prognostic factor in stroke rehabilitation? Stroke. 2006;37(12):2989–94.
19. Bassey E, Harries U. Normal values for handgrip strength in 920 men and women aged over 65 years, and longitudinal changes over 4 years in 620 survivors. Clin Sci. 1993;84 (3):331–7.
20. Rantanen T, Era P, Heikkinen E. Physical activity and the changes in maximal isometric strength in men and women from the age of 75 to 80 years. J Am Geriatr Soc. 1997;45 (12):1439–45.
21. Ada L, Dorsch S, Canning CG. Strengthening interventions increase strength and improve activity after stroke: a systematic review. Aust J Physiother. 2006;52(4):241–8.
22. Veerbeek JM, et al. What is the evidence for physical therapy poststroke? A systematic review and meta-analysis. PloS One. 2014;9(2):e87987.
23. Corbetta D, Sirtori V, Castellini G, Moja L, Gatti R. Constraint-induced movement therapy for upper extremities in people with stroke. The Cochrane Library. 2015.
24. Bowden MG, Woodbury ML, Duncan PW. Promoting neuroplasticity and recovery after stroke: future directions for rehabilitation clinical trials. Curr Opin Neurol. 2013;26 (1):37–42.
25. French B, et al. Repetitive task training for improving functional ability after stroke. The Cochrane Library. 2016.
26. Lang CE, et al. Observation of amounts of movement practice provided during stroke rehabilitation. Arch Phys Med Rehabil. 2009;90(10):1692–8.
27. Lang CE, et al. Dose response of task-specific upper limb training in people at least 6 months poststroke: A phase II, single-blind, randomized, controlled trial. Ann Neurol. 2016;80(3):342–54.
28. Maciejasz P, Eschweiler J, Gerlach-Hahn K, Jansen-Troy A, Leonhardt S. A survey on robotic devices for upper limb rehabilitation. J Neuroeng Rehabil. 2014;11(3). https://doi.org/10.1186/1743-0003-11-3.
29. Lo AC, et al. Robot-assisted therapy for long-term upper-limb impairment after stroke. N Engl J Med. 2010;362 (19):1772–83.
30. Keller U, Klamroth V, van Hedel HJ, Riener R. ChARMin: a robot for pediatric arm rehabilitation. In: Robotics and Automation (ICRA), 2013 I.E. International Conference on. IEEE; 2013. p. 3908–13.
31. Guidali M, Erne R, Riener R, Lambercy O, Gassert R, "Instrumented handles for an arm rehabilitation robot," In: Automed Workshop; 2010.
32. Nef T, Guidali M, Riener R. ARMin III – arm therapy exoskeleton with an ergonomic shoulder actuation. Appl Bionics Biomech. 2009;6(2):127–42.
33. Guidali M, Schlink P, Duschau-Wicke A, Riener R. Online learning and adaptation of patient support during ADL training. In: Proceedings of IEEE International Rehabilitation Robotics (ICORR) conference; 2011. p. 1–6.
34. Klamroth-Marganska V, et al. Three-dimensional, task-specific robot therapy of the arm after stroke: a multicentre, parallel-group randomised trial. Lancet Neurol. 2014;13 (2):159–66.
35. Valero-Cuevas FJ, Klamroth-Marganska V, Winstein CJ, Riener R. Robot-assisted and conventional therapies produce distinct rehabilitative trends in stroke survivors. J NeuroEng Rehabil. 2016;13(1):92.
36. Kitago T, Krakauer JW. Motor learning principles for neurorehabilitation. Handb Clin Neurol. 2013;110:93–103.
37. Jørgensen HS, Nakayama H, Raaschou HO, Olsen TS. Recovery of walking function in stroke patients: the Copenhagen Stroke Study. Arch Phys Med Rehabil. 1995;76 (1):27–32.
38. Hesse S, Sarkodie-Gyan T, Uhlenbrock D. Development of an advanced mechanised gait trainer, controlling movement of the centre of mass, for restoring gait in non-ambulant subjects-Weiterentwicklung Eines Mechanisierten Gangtrainers mit Steuerung des Massenschwerpunktes zur Gangrehabilitation Rollstuhlpflichtiger Patienten. Biomed Tech/Biomed Eng. 1999;44(7–8):194–201.
39. Hesse S, Waldner A, Tomelleri C. Innovative gait robot for the repetitive practice of floor walking and stair climbing up and down in stroke patients. J Neuroeng Rehabil. 2010;7(1):30.
40. Schmidt K, et al. The Myosuit: bi-articular anti-gravity exosuit that reduces hip extensor activity in sitting transfers. Front Neurorobot. 2017;11:57.
41. Asbeck AT, De Rossi SM, Holt KG, Walsh CJ. A biologically inspired soft exosuit for walking assistance. Int J Robot Res. 2015;34(6):744–62.
42. Awad LN, et al. A soft robotic exosuit improves walking in patients after stroke. Sci Transl Med. 2017;9(400):eaai9084.
43. Mehrholz J, Thomas S, Werner C, Kugler J, Pohl M, Elsner B. Electromechanical-assisted training for walking after stroke. The Cochrane Library. 2017.
44. Marchal-Crespo L, Michels L, Jaeger L, López-Olóriz J, Riener R. Effect of error augmentation on brain activation and motor learning of a complex locomotor task. Front Neurosci. 2017;11:526.
45. Cesqui B, Aliboni S, Mazzoleni S, Carrozza M, Posteraro F, Micera S. On the use of divergent force fields in robot-mediated neurorehabilitation. In: Biomedical Robotics and Biomechatronics, 2008. BioRob 2008. 2nd IEEE RAS & EMBS International Conference on. IEEE; 2008. p. 854–861.
46. Carel C, et al. Neural substrate for the effects of passive training on sensorimotor cortical representation: a study with functional magnetic resonance imaging in healthy subjects. J Cereb Blood Flow Metab. 2000;20(3):478–84.
47. Heuer H, Lüttgen J. Robot assistance of motor learning: a neuro-cognitive perspective. Neurosci Biobehav Rev. 2015;56:222–40.
48. Brunnstrom S. Motor testing procedures in hemiplegia: based on sequential recovery stages. Phys Ther. 1966;46(4):357–75.
49. van Kordelaar J, van Wegen E, Kwakkel G. Impact of time on quality of motor control of the paretic upper limb after stroke. Arch Phys Med Rehabil. 2014;95(2):338–44.
50. Krakauer JW. Motor learning: its relevance to stroke recovery and neurorehabilitation. Curr Opin Neurol. 2006;19(1):84–90.
51. Cortes JC, et al. A short and distinct time window for recovery of arm motor control early after stroke revealed with a global measure of trajectory kinematics. Neurorehabil Neural Repair. 2017;31(6):552–60.
52. Agostini M, et al. Telerehabilitation and recovery of motor function: a systematic review and meta-analysis. J Telemed Telecare. 2015;21(4):202–13.
53. Chen J, Jin W, Zhang X-X, Xu W, Liu X-N, Ren C-C. Telerehabilitation approaches for stroke patients: systematic review and meta-analysis of randomized controlled trials. J Stroke Cerebrovasc Dis. 2015;24(12):2660–8.
54. Butler A, Bay C, Wu D, Richards K, Buchanan S. Expanding tele-rehabilitation of stroke through in-home robot. 2014.
55. Ivanova E, Minge M, Schmidt H, Thüring M, Krüger J. User-centered design of a patient's work station for haptic robot-based telerehabilitation after stroke. Curr Dir Biomed Eng. 2017;3(1):39–43.
56. Just F, Baur K, Riener R, Klamroth-Marganska V, Rauter G. Online adaptive compensation of the ARMin Rehabilitation Robot. In: Biomedical Robotics and Biomechatronics (BioRob), 2016 6th IEEE international conference on IEEE; 2016. p. 747–752.

Precision Medicine and Personalized Medicine in Cardiovascular Disease

36

Gemma Currie and Christian Delles

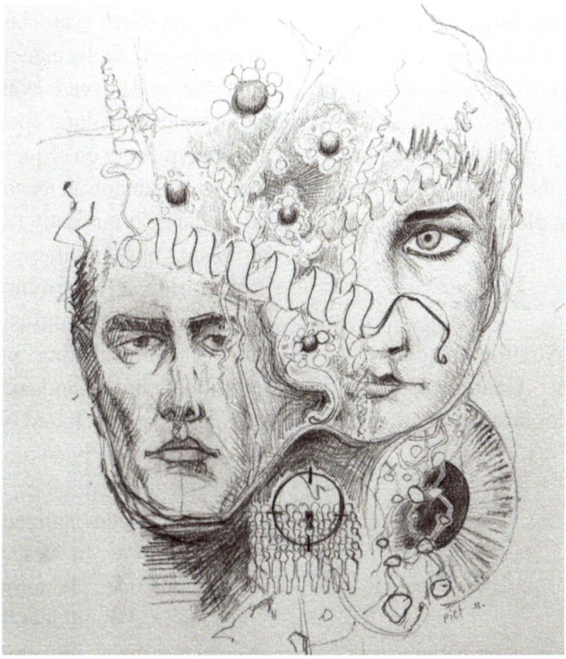

Precision medicine. Artwork by Piet Michiels, Leuven, Belgium

Abstract

Precision medicine aims to offer "the right treatment to the right patient at the right time." In cardiovascular medicine the potential of precision medicine applies to all stages of the disease development and includes risk prediction, preventative measures, and targeted therapeutic approaches. Precision medicine will benefit from new developments in the area of genomics and other omics but equally heavily depends on established biomarkers, functional tests, and imaging. Cardiovascular medicine often relies on noninvasive diagnostic

G. Currie · C. Delles (✉)
Institute of Cardiovascular & Medical Sciences, University of Glasgow, BHF Glasgow Cardiovascular Research Centre, Glasgow, Scotland, UK
e-mail: Christian.Delles@glasgow.ac.uk

P. L. M. Kerkhof, V. M. Miller (eds.), *Sex-Specific Analysis of Cardiovascular Function*,
Advances in Experimental Medicine and Biology 1065, https://doi.org/10.1007/978-3-319-77932-4_36

procedures and symptom-based disease management. In contrast, other clinical disciplines including oncology and immunology have already moved to molecular diagnostics that lend themselves to precision medicine approaches. There are opportunities to implement precision medicine approaches by focusing on common diseases such as hypertension, conditions with diagnostic and prognostic uncertainty such as angina, and conditions that are associated with high mortality and involve costly and potentially harmful interventions such as dilated cardiomyopathy and cardiac resynchronization therapy. Sex and gender issues have not yet been fully explored in precision medicine although the opportunity to use molecular data to more accurately manage men and women with cardiovascular disease has been acknowledged. A mindshift is required in order to fully exploit the potential of precision medicine to tackle the global burden of cardiovascular diseases.

Keywords
Precision medicine · Stratified medicine · Personalized medicine · Genomics · Proteomics · Metabolomics · Biomarker · Sex · Gender · Big data · Prevention

Introduction

The concept of personalized medicine is not new. Doctors have always tried to see features of health and disease in the context of an individual patient's medical, social, and personal circumstances. The notion of "treating the patient, not the disease" is one of the basic principles of clinical medicine that extends to sex and gender issues and features throughout this book. At the same time, it is evident that patients share factors that determine risk, natural progression of disease, and response to treatment. Identification of such shared factors has been one of the main achievements of recent years, facilitating transition from an experience-driven approach in individual patients to evidence-based medicine based on data derived from experimental work and large clinical trials.

In cardiovascular medicine we are used to biomarker-based diagnostic criteria, criteria-led therapeutic approaches, and monitoring of therapeutic approaches using imaging and other techniques. Prominent examples include the definitions of acute coronary syndromes based on cardiac troponins [1], hypertension based on blood pressure or thresholds [2], diabetes based on measures of glycemic control [3], and heart failure not only based on symptoms but also on ejection fraction [4]. In all these examples depending on precise characterization of disease, the appropriate therapeutic approach can be chosen. With recent advances in molecular medicine, in particular with the widespread availability of genetic and genomic data in large population and the ability to genotype or even sequence the whole genome of individual patients at reasonable cost, cardiovascular medicine is currently facing new challenges and promising opportunities.

Unfortunately, terminology in the area of precision medicine is confusing and reflects the conflict between treating patients based on their individual clinical and social features and treating groups of patients based on shared characteristics. The matter has been complicated by the firm belief of most doctors that they have always employed a personalized approach and do not need a new word describing their established clinical practice. The terms "stratified medicine," "personalized medicine," and "precision medicine" have been introduced but are often used interchangeably. In this chapter we will use the expression "precision medicine" but not without briefly acknowledging the development of terminology (Fig. 36.1).

The Medical Research Council in the UK states that "*Stratified medicine* is based on identifying subgroups of patients with distinct mechanisms of disease, or particular responses to treatments. This allows us to identify and develop treatments that are effective for particular groups of patients. Ultimately stratified medicine will ensure that the "right patient gets the right treatment at the right time" [5]. Where such stratification is increasingly precise and "subgroups of

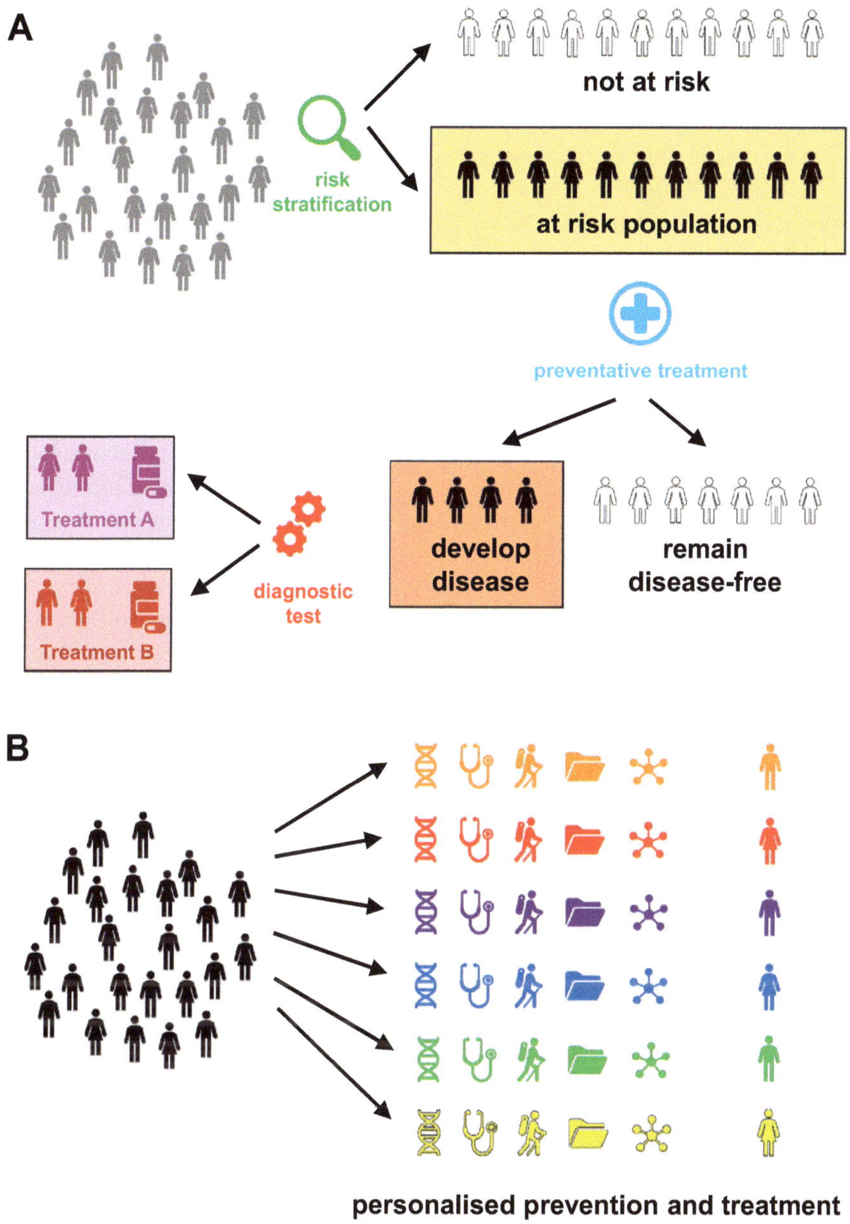

Fig. 36.1 Idealized principles of stratified and personalized medicine approaches. (**a**) Stratified approaches group individuals based on criteria such as sex, age, genetic background, ethnicity, biomarker studies, or imaging studies into groups that will then be managed uniformly. The panel shows a general population that undergoes testing for risk stratification. In this example we do not know how well the test performs as nothing is known about the group that is considered not to be at risk. We have chosen a condition that is associated with higher risk in women. Of the at-risk population, 40% develop disease despite preventative treatment. Here we do not know whether this is because of poor efficacy of the treatment or because the test wrongly classified subjects into the at-risk group despite being at no risk. Treatment of disease will then be based on further tests but also on other aspects including age and sex. In this example Treatment A is only appropriate for female patients. (**b**) Personalized approaches collect as much information as possible including genetic/genomic, imaging, and functional studies, individual lifestyle factors, and medical records and use integrative analysis methods to develop precise and individualized management strategies that will include all areas from disease prevention to therapeutic strategies. As all patients are seen individually, sex and gender aspects are only one of many factors and do not immediately feature in this figure

patients" are sufficiently small, this "stratified" approach evolves into a "personalized" approach. In the 2016 strategy paper "Improving Outcomes Through *Personalized Medicine*," the National Health Service (NHS) in England has described changes in the way how medicine will be delivered and how new targeted treatment approaches will be developed [6]. In the context of cardiovascular medicine, it is encouraging to read in this document that "Personalised medicine is important not only for the 1 in 17 people who have a rare disease, or for those living with cancer, but also for the many others who have or are at risk of developing other common diseases." Despite this, the document is mainly based on the opportunities provided by the era of genomic medicine, and we will explore in this chapter how other biomarkers and omics techniques can contribute to precision medicine. The term *precision medicine* has indeed been used by a US National Research Council publication *Toward Precision Medicine* [7], and particularly after the launch of President Obama's Precision Medicine Initiative in January 2015 [8], it has become the preferred term by many working in this area of medicine.

More recently the areas covered by precision medicine have been further expanded, recognizing the broad medical and societal aspects of this new approach. The abovementioned NHS document defines four areas covered by personalized medicine: *p*rediction and *p*revention of disease, more *p*recise diagnosis, and targeted and *p*ersonalized interventions [6]. These four "P"s can be further complemented by the *p*articipatory nature of modern medicine, involving patients and the wider public especially when it comes to radical changes in the way we practice medicine [9].

There are a number of excellent overview articles available that explore the development and opportunities of precision medicine in more detail, and we would like to refer the reader to these documents [10, 11]. Moreover, many chapters of this book already directly or indirectly address areas of precision medicine by defining differences, for example, between men and women in pathophysiology, risk, treatment goals, and response to treatment development. In this chapter we will therefore not provide a comprehensive review of precision medicine in cardiovascular diseases but rather guide the reader through this rapidly evolving area along selected examples.

Finally, it is important to see the development of precision medicine as part of the battle against chronic diseases worldwide. With global improvements in social circumstances and nutrition and the increasingly successful fight against infectious diseases and cancer, cardiovascular diseases are the major driver of morbidity and mortality worldwide [12]. From a health-economic perspective, further reductions in the prevalence of cardiovascular and cardiometabolic diseases such as hypertension and diabetes will have far greater impact than similar efforts in the area of cancer [13]. While it is not our intention to discourage the community from fighting against cancer and other conditions, we would like to emphasize that new approaches including precision medicine strategies are required to tackle the global cardiovascular disease challenge.

Is Current Cardiovascular Medicine Imprecise?

The present book covers a wide range of cardiovascular conditions, and from the information provided, one would not immediately assume that cardiovascular medicine is imprecise. It is indeed true that precision medicine will be less of a revolution of contemporary clinical practice than an evolution, building on currently employed predictive, diagnostic, and therapeutic strategies. The interest in precision medicine has, however, grown with the availability of mature molecular medicine methods, particularly in the area of genomics. There is a strong belief that the increasingly detailed molecular characterization of disease has the potential to translate into equally precise therapies. This belief has been fueled by successes in other areas of medicine which we will briefly mention below in Sect. "Precision Medicine in the Non-cardiovascular Field." Despite apparent successes in prevention and treatment of cardiovascular diseases, there is still room for improvement. Many of these potential improvements are related to recognition of sex differences in development and progression of cardiovascular diseases as highlighted throughout this book.

Definition of Disease The precise definition of a disease phenotype is a notoriously underexplored area in cardiovascular medicine. In contrast to cancer and immunological and neurological diseases, cardiovascular diseases are often defined by symptoms or relatively simple diagnostic tests. For example, the stratification of the majority of patients with *diabetes* into two major forms depending on insulin production and insulin resistance does not take individual differences in disease pathogenesis into account. We have only recently learned from clinical experience in patients with maturity onset diabetes of the young (MODY), patients with "double diabetes," other states of insulin resistance such as overweight and obesity, pancreatogenic diabetes, and differences in incidence of diabetes across ethnicities that the categorization into "Type 1" and "Type 2" does not always meet clinical needs [14]. Recent data from genome-wide association studies (GWAS) provide novel insights into the pathophysiology of diabetes and may translate into more precise preventative and therapeutic approaches [15].

Similarly, *hypertension* is exclusively defined on a readout (blood pressure) which does not take the pathophysiological origin of the condition into account. Constructs such as "prehypertension" or "high normal blood pressure" suggest that the disease process starts early but do not define its pathophysiology. Furthermore, the dichotomous classification into "primary" and "secondary" forms of hypertension is too simplistic from a mechanistic point of view. Firstly, we have, for example, learned that with more precise assessment of renin and aldosterone concentrations combined with specific sampling techniques, functional tests, and up-to-date imaging, the number of patients with primary aldosteronism is much higher than originally thought [16, 17]. There are similar examples including novel forms of renal hypertension based on newly detected mutations in electrolyte transporters and their regulation [18] and new insights in the relationship between salt and the pathogenesis of hypertension beyond osmotic effects [19]. Secondly, even in patients with "primary" hypertension, the notion that this condition is simply the result of many small changes in numerous contributing pathways may not be correct. People with primary hypertension are still different from each other; knowing more about the relative contribution of key pathways including the renin-angiotensin-aldosterone system, the sympathetic nervous system, and the immune system or the role of large and small vessels in the pathogenesis of hypertension can help in finding the right treatment. The National Institute for Health and Care Excellence (NICE) guidance on hypertension [20], for example, stratifies patients according to age as a surrogate of plasma renin activity [21] to decide on the most appropriate first-line treatment. This approach has recently been refined by the Prevention And Treatment of Hypertension With Algorithm-based therapy (PATHWAY)-2 study where patients with the lowest plasma renin concentrations were most likely to benefit from spironolactone as antihypertensive agent [22].

Stratified Treatment The above example on treatment of *hypertension* illustrates that more precise information on individual pathogenetic factors of disease can indeed translate into more precise treatments. The definition of *heart failure* remains based on clinical symptoms, but the recognition of differences in etiology, natural progression, and response to treatment of heart failure with reduced as opposed to preserved ejection fraction now explains the poor response rate to established treatment in some patients and has led to the design of specific studies addressing particularly those with heart failure with preserved ejection fraction who do not respond to the majority of conventional therapies [23, 24]. This exemplifies how a previously imprecise disease definition ("heart failure") has become more precise recently by taking ejection fraction into account, and there is room for further precision by recognizing sex differences in ejection fraction as outlined in more detail in Chapter 8 in this book. One can speculate that other features including molecular factors will also help to reach a more precise diagnosis and thereby inform more precise treatment.

Response to Treatment Despite the increasing knowledge of pathophysiology and attempts to provide evidence-based therapies, the overall response to treatment of cardiovascular diseases remains poor. Of those who are diagnosed with *hypertension* and started on drug therapy, only about a third achieve target blood pressure, with some variation across regions and countries [25]. Similar figures have been seen for *diabetes* and other cardiometabolic and cardiovascular conditions [12] and are a major challenge not only to individual patients but also to healthcare systems. For some commonly used drugs including *warfarin* and *clopidogrel*, the response to treatment and thereby the required doses depend on the genetic makeup, and genotype-led prescription has been implemented in some healthcare systems [26, 27].

Therefore, while contemporary cardiovascular medicine is not necessarily imprecise, it appears evident that more precise definition of disease, while considering differences between men and women, has the potential to translate into targeted treatments with improved response rates (Fig. 36.2). Molecular methods including genomics have the potential to add significantly to this precision, but as the above examples have shown, precision medicine also critically involves established biomarkers such as cardiac troponins and imaging techniques, particularly in the cardiovascular field.

Domains of Precision Medicine Throughout the Life Course

Cardiovascular but also other chronic diseases typically develop over a long period of time. Preventative measures as well as intervention strategies will therefore be required in a large number of people over many years or even lifelong. It has been argued that with the availability of efficient, low-risk, and low-cost medicines including aspirin and statins, pharmacological preventative strategies can be prescribed to everybody and only depending on age [28]. This polypill concept is indeed attractive and may well reduce incidence and prevalence of cardiovascular diseases substantially. The same is true for other preventative measures such as healthy diet and other lifestyle measures where there is unopposed agreement that these should apply to everybody and not require stratification. However, the "dosage" and types of lifestyle modification may well require some thought in individual patients and move even such basic measures into the area of precision medicine.

In contrast to the polypill concept, it has also been argued that preventative measures should focus on those at highest risk and therapeutic measures should be optimized in order to achieve the best possible effect. This would define groups or individual patients who are more or less likely to be at risk in the first place and subsequently benefit from preventative or therapeutic approaches. Health economic benefits of a precision medicine approach clearly depend on a number of factors including:

- *Costs of the stratification test.* Such tests can be inexpensive as in the case of testing for NT-proBNP to support the diagnosis of heart failure; more costly, for example, if echocardiography is involved for subtyping of heart failure and diagnosis of concomitant disease; and very expensive if it involves genome sequencing, MRI scanning, or proteomic analysis.
- *Precision of the stratification test.* False-positive and false-negative test results will add imprecision rather than precision, and especially where a single test result determines clinical management, precision of the test has to be high.
- *Population-specific risk.* As with any diagnostic testing, the results of a test have to be interpreted in the context of population risk or, more generally, pretest probability. Conditions with high prevalence (e.g., hypertension in the elderly) are less likely to benefit from sophisticated stratification approaches than rarer conditions (e.g., insulin resistance after gestational diabetes). Similarly, for drugs

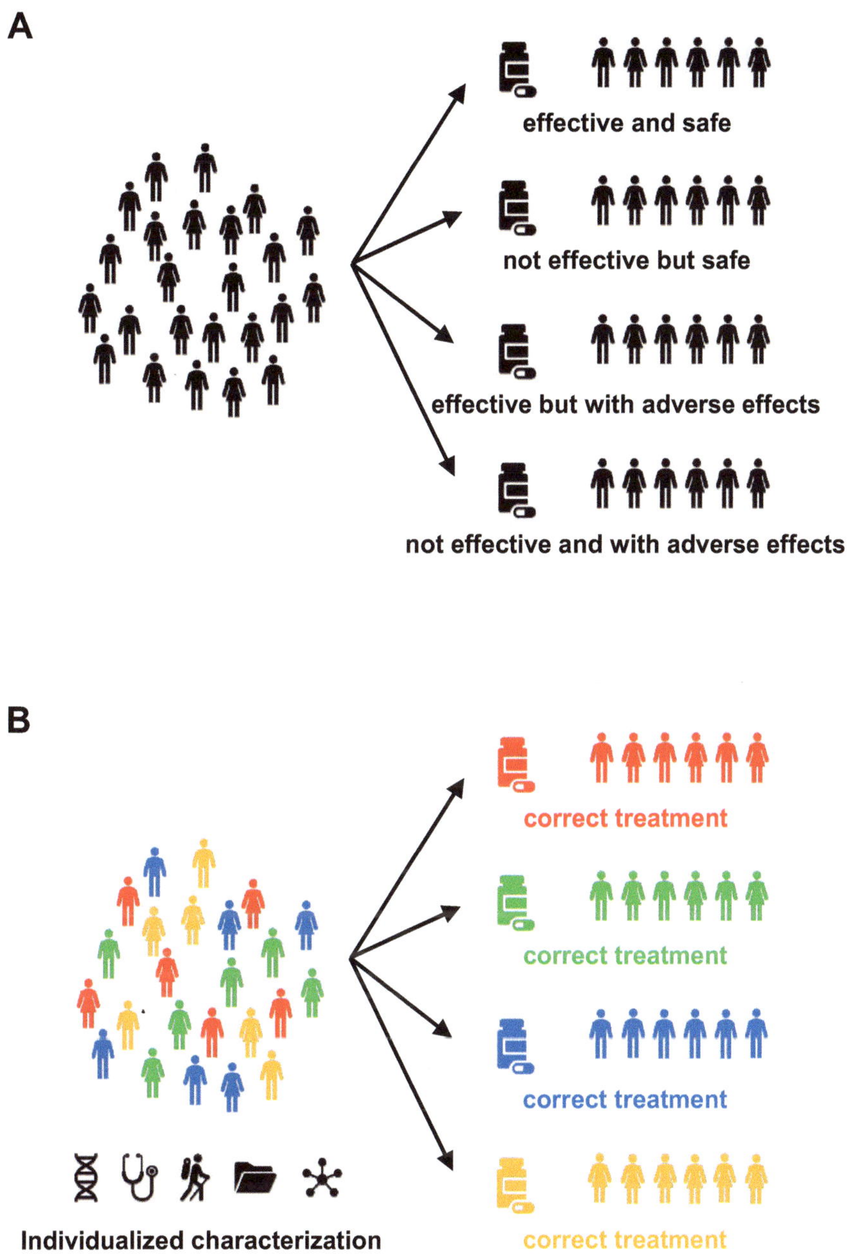

Fig. 36.2 Precision medicine and response to treatment. (**a**) Conventional strategies assign treatments to large groups of individuals but take the risk that treatment may not be effective and/or not safe. (**b**) In precision medicine individuals undergo comprehensive individualized characterization to decide about the right treatment for the right person. Depending on test performance, this precise treatment will be effective and free of adverse effects. In this example, sex plays a role in the allocation of treatment with the blue and orange treatments being the right choice only for men and women, respectively

that work well in the vast majority of patients, stratification for response to treatment may be less attractive than for drugs with less robust therapeutic effects.

- *Relationship between pathophysiology, test, and therapeutic approach.* The closer a test is related to the pathophysiology of disease and to the mechanism of action of the proposed treatment, the better will be its precision and thereby its usefulness as stratification tool or to inform personalized approaches.

- *Natural course of disease.* Diseases with greater risks of unfavorable outcomes including disability and death and diseases reaching these outcomes faster than others may benefit more from precision medicine approaches than those with more benign clinical course. This may be one of the reasons why cancer has attracted so much attention for implementation of precision medicine, but one should remember that conditions such as heart failure, pulmonary hypertension, and stroke can be associated with equally short or shorter lifespan compared to many types of cancer or lead to disability and associated personal and societal costs.
- *Onset and duration of treatment.* The costs of diagnostic testing have to be seen in the context of treatment resulting from such testing. In particular genetic testing can be performed early in life and may result in lifelong preventative or therapeutic approaches. The costs of such strategies have to be balanced against other factors in this list.
- *Cost of treatment.* As mentioned above, for low-cost treatments such as statins and aspirin, expensive stratification may be less attractive than for more expensive treatments including patent-protected drugs or interventional or surgical approaches. While cardiovascular drugs are not generally thought to be particularly expensive compared to, for example, drugs used in the treatment of cancers and rheumatological and immunological diseases, the long-term nature of treatment of cardiovascular condition, the advent of new and expensive drugs, and the increasing number of devices and interventions in cardiovascular medicine should be taken into account.
- *Efficacy and adverse effects of treatment.* These are again main factors that will decide if stratification and personalized approaches are indicated and cost-effective. Providing poorly efficient treatments to carefully selected patients may not necessarily be more cost-effective than prescribing such treatments to unselected populations.

As evident from the list above, the domains of precision medicine span across the whole life course and include risk prediction, preventative strategies, and therapeutic approaches. In a review article by Topol [29], this potential has been highlighted as "from prewomb to tomb" as precision medicine can indeed include conception planning, prenatal diagnostics, disease prevention, molecular diagnosis of disease, personalized treatment, prediction of lifespan, and molecular autopsy and genetic tests that can inform clinical management of family members and other patients.

Indeed, adoption of precision medicine principles by some of the leading stakeholders including the US Department of Veterans Affairs' VA Million Veteran Program [31], and the previously mentioned US Precision Medicine Initiative [8], the NHS in the UK [6], and large-scale biobanking initiatives including the UK Biobank [32] demonstrate the momentum in precision medicine.

Precision Medicine Within the Cardiovascular Continuum: Tools and Applications

Cardiovascular diseases have long been neglected in precision medicine initiatives. Compared to other conditions, cardiovascular diseases often develop slowly and over the course of many decades and are (wrongly) perceived to be more benign than other conditions including cancers and rheumatological diseases. However, due to the large amount of people who are potentially affected by cardiovascular diseases, these conditions are indeed an ideal case for precision medicine approaches. And in fact, the relatively slow development of many cardiovascular conditions provides ample opportunity to identify those at risk, offer preventative strategies, and start therapies early in the course of disease. Within the cardiovascular continuum [33], there are options to address risk factors, identify disease at subclinical stages, treat disease on the basis of diagnostic tests, and prevent further disease progression.

A number of cardiovascular disease areas are already subject to precision medicine approaches, and it appears that there is a particular focus on common conditions such as *hypertension* [34–36], conditions with diagnostic and therapeutic uncertainty such as *angina* and *coronary artery disease* [37, 38], and conditions that are associated with high mortality and/or require complex and expensive interventions such as *dilated cardiomyopathy* and *cardiac resynchronization therapy* [39–41]. It should be emphasized yet again that such approaches not only refer to novel molecular and genetic diagnostic approaches [34, 36, 38] but also to improved risk scores based on existing data [35, 37]. Here we will look into the tools that are available for precision medicine approaches.

Genetics and Genomics Recent advances in analysis of the human genome have led to the discovery of genetic variants that are robustly associated with cardiovascular conditions including hypertension, coronary artery disease, and chronic kidney disease [26]. Cardiovascular risk scores based on such genetic variants have been developed and shown to add significantly to the predictive power of traditional risk factors [42]. Genetic and genomic factors are indeed suitable precision medicine tools in areas of risk prediction and response to treatment as the genome is largely stable throughout the life course. Testing can therefore be performed at any point in life and inform clinical management. We have already referred to the potential of pharmacogenetics for treatment with warfarin and clopidogrel [26, 27], but this further extends to assessing the risk of adverse effects and response efficacy of statins, aspirin, β-blockers, dalcetrapib, and vitamin E [43] where some centers have started implementing such tests [27, 44].

"Higher" Omics The advantage of a stable genome can be a disadvantage in cardiovascular medicine where environmental factors including diet, exercise, smoking, and air pollution have a major impact on disease development and progression. Epigenetic tools and studies of gene expression (transcriptomics) and protein expression (proteomics) and the comprehensive assessment of the metabolome (metabolomics) of biofluids, tissue samples, and the gut microbiome [45] are better suited to reflect the interaction between the genetic background and the environment. It is a common feature of these "higher" omics that results are much more dependent on sample collection, analysis platforms, and data processing and interpretation as compared to genetic and genomic studies. Therefore the immediate use of such techniques in precision medicine approaches is currently limited for technical reasons, despite their enormous theoretical potential.

- *Epigenetics* generally refers to studies into gene function beyond changes in the DNA sequence. Epigenetic studies therefore include assessment of DNA damage, telomere length, DNA methylation, and studies into noncoding RNAs. Where such changes are assessed comprehensively, they are sometimes called epigenomic studies. Recent studies in humans have provided robust evidence for the critical role of epigenetic changes in the development of cardiovascular disease including hypertension [46], chronic kidney disease [47], acute coronary syndrome [48], and other complex diseases [49]. Indeed, the plasticity of the epigenome facilitates studies to examine the causal relationship between metabolic factors such as body mass index and the risk of cardiovascular diseases [50].
- *Proteomics* refers to studies into the expression of a large number of proteins in a biological sample. The majority of biomarkers in cardiovascular medicine such as NT-proBNP or cardiac troponins are indeed peptides or proteins, and the potential of proteomics to inform diagnosis and disease management is evident [51]. Proteomic signatures of cardiovascular conditions including

coronary artery disease [52], chronic kidney disease [53], and preeclampsia [54, 55] have been defined, and clinical proteomic studies have been conducted, for example, into the prediction of cardiovascular events [56] and adverse effects of treatment [57]. As with other "higher" omics studies, there is an ongoing effort to standardize and simplify proteomic analyses in order to translate these techniques into routine clinical care [51, 58].
- *Metabolomics* studies the expression of small molecules and metabolites in biological samples and is probably closest to the actual state of an organism compared to other omics techniques. This apparent advantage is, however, associated with disadvantages. Confounding factors such as diet or exercise, variation in sample handling, and performance of different metabolomic platforms are challenges related to current metabolomic approaches [59]. From the perspective of data analysis, the fact that a single metabolite can be substrate and product of multiple biochemical pathways is a particular problem of metabolomic studies. However, there is robust evidence of a relationship between the genome and the metabolome [60]. Metabolomic signatures of cardiovascular diseases including heart failure [61] and hypertension [62] have been described, and the potential of metabolomics in precision medicine has been recognized [63].

Biomarkers and Imaging We cannot emphasize enough that precision medicine is not necessarily dependent on large and complex omics-derived data but can also be informed by traditional biomarkers, imaging studies, and functional tests. Three points are worth mentioning in this context. First, in order to make medicine more precise, biomarkers have to add value to existing diagnostic tests and stratification approaches. This challenge is well known for traditional biomarkers but also affects other areas such as omics studies where strong association and low *P*-values do not necessarily translate into improved prediction on the individual patient level. Second, in order to implement biomarkers into clinical practice, their use has to be evaluated in robust trials supported by well-defined regulatory requirements. We will refer to this challenge that applies to precision medicine in general in more detail below. Third, in order to provide more precise information, biomarker analyses may have to be more precise as well. For example, studying lipid subfractions can explain residual risk of atherosclerotic events beyond standard lipid profiles [64] or dissect cardiovascular effects of drugs such as interleukin-6 blockers [65], and novel imaging modalities such as PET-CT have to be explored systematically [66] to define their role in clinical management and detailed characterization of disease [67]. This further extends to functional data such as assessment of blood pressure and fractional flow reserve not only as stratification tools but in the first instance to find out which patients will benefit from additional tests or more intensive use of existing tests [68, 69].

Big Data While considerable precision can be achieved by assessment of a limited number of risk factors and biomarkers (cardiovascular risk scores such as the Framingham score are prime examples), it is clear that residual risk and diagnostic accuracy will require larger amounts of data. Not all of these data have to be generated in new studies but are already available in routine practice. The challenges lie in the integration of health records, standardization of coding systems, and use of such data for precision medicine purposes. The introduction of electronic health records and systems to integrate data without breaching confidentiality in some healthcare systems has already paved the way to explore available information for more precise risk stratification and individual management of disease [70, 71]. Such efforts extend further from routine clinical data to data obtained from research, for example, by integrating genomic and multi-omic data with phenotypes and epidemiological data [72].

Bioinformatics and Modeling As mentioned above, the large amounts of data that are already available together with complex phenotypes and newly generated biomarker and omics data require new approaches to data analysis and integration. Traditional approaches such as regression analysis to test incremental benefits of new biomarkers may fail especially when large numbers of markers are explored; new analytical techniques have been proposed in this context and include artificial intelligence and machine-learning approaches [73, 74]. However, the need to improve analysis and modeling capacities also extends to functional and imaging data that are currently not exploited to their full potential and often not integrated across different imaging modalities. There are proposals to improve functional modeling especially in cardiac imaging by simplifying and standardizing imaging modalities, assessment of novel biomechanical parameters, and integration of different imaging modalities to inform precise patient management [75, 76].

Patient Factors The changes that precision medicine may bring to clinical practice have the potential to reshape the way we practice medicine. It is therefore important that all stakeholders including physicians, healthcare providers, and regulatory authorities work jointly toward this aim. Most importantly, however, patients and the wider public have to be an integral part of the move to precision medicine as areas such as confidentiality of data and adherence to therapy that are of key interest to patients are involved [9, 77]. This important aspect of implementation has, as yet, been underexplored although smaller studies indicate generally favorable attitudes toward implementation of precision medicine approaches in cardiovascular medicine and oncology [78, 79].

New Trial Designs Current clinical trials are in the first instance designed to test the effectiveness of preventative or therapeutic strategies. Inclusion and exclusion criteria are predefined and not subject to assessment within the trial. For example, the Justification for the Use of Statins in Primary Prevention: An Intervention Trial Evaluating Rosuvastatin (JUPITER) trial [80] has provided robust data on the effectiveness of statin therapy, where incidence of major cardiovascular events was reduced in people with CRP levels higher than 2.0 mg/dL. The study did, however, not aim to test whether people with lower CRP levels would also benefit from statin therapy or whether a different cutoff level would provide more precise stratification. Similarly, many studies into treatment of heart failure employ ejection fraction as one of the inclusion criteria but do not test whether the predefined ejection fraction cutoff provides the best possible stratification or whether, for example, echocardiography-based assessment of ejection fraction is as reliable as MRI-based assessment. Most recently, the Systolic Blood Pressure Intervention Trial (SPRINT) [81] provided evidence for reduced cardiovascular risk with lower blood pressure targets but has not assessed whether the specific method that was used to assess blood pressure in this trial is superior to assessment methods more commonly used in clinical practice. In order to robustly assess stratification tools that will inform precision medicine, these tools have to be tested as rigorously as treatments are currently being tested in clinical trials. Novel clinical trial designs that assess both the stratification test and efficacy of the intervention have been proposed in this context [24]. As an example, the ongoing Proteomic prediction and Renin angiotensin aldosterone system Inhibition prevention Of early diabetic nephRopathy In TYpe 2 diabetic patients with normoalbuminuria (PRIORITY) trial [82] employs a proteomics-based classifier to stratify patients with Type 2 diabetes and normoalbuminuria into those at low and high risk of progressing to microalbuminuria and tests the preventative effect of treatment with spironolactone in the high-risk arm. The study design focuses on assessment of the diagnostic test as primary outcome but uses the infrastructure of a stratified trial to study an interventional strategy at the same time.

Precision Medicine in the Non-cardiovascular Field

We have outlined the potential of stratified medicine in the cardiovascular field and the tools that are already available or have to be developed to support this task. However, even if the potential to tackle chronic diseases including cardiovascular diseases has been widely acknowledged, the majority of case studies and in fact mature clinical implementation are happening in non-cardiovascular areas of medicine, namely, oncology, immunology, infectious diseases, and rheumatology [83]. The reasons are manifold but provide opportunities for cardiovascular researchers to learn from successes but also from errors in other areas of medicine. We would like to briefly mention three notions that are being discussed in this context.

- Cancers are commonly believed to be associated with *high morbidity and mortality* and *high treatment costs*, which makes them ideally suited for any attempts to improve preventative and therapeutic strategies. It is an important task for the cardiovascular community to further raise awareness of the serious risks associated with conditions such as heart failure and pulmonary hypertension and the contribution of common and yet only suboptimally treated conditions such as hypertension, chronic kidney disease, and diabetes. This mindshift is required to engage stakeholders and the wider public in development and implementation of precision medicine approaches in cardiovascular diseases.
- A much more specific reason for the apparent success of precision medicine in non-cardiovascular areas is the fact that conditions such as cancer or organ transplantation are *limited to single or few organ systems* that can be subject to *representative and meaningful tissue-based molecular diagnosis*. Subtyping of, for example, pancreatic cancer can be performed on small biopsy specimens and inform risk stratification and treatment strategies, whereas such tissue-based molecular diagnosis is not possible or at least not generally implemented in cardiovascular conditions that develop over long periods of time, often involve multiple genetic variations, and affect multiple organ systems.
- Finally, while in vitro experiments and experimental models provide invaluable information in all areas of medicine, the philosophy of employing molecular pathology in *human samples* to understand disease mechanisms, secure the diagnosis, and inform and monitor treatment has always been present in areas such as oncology or rheumatology, although it is less established in cardiovascular medicine where noninvasive tests, biomarkers, and imaging are preferred [84]. In this context Feldman et al. [85] rightly state that "a principal difference between the translational sciences in HF [heart failure] and in cancer biology that might explain in part the relative lack of precision therapy in the treatment of HF is that HF research has been driven in large part by discoveries in investigational models of HF, whereas drug discovery in cancer has been driven by the elucidation of altered biology of human tumors."

In this context it is important to recognize that medical disciplines should not be seen in isolation and that patients can indeed be affected by more than one disease. The growing clinical need to address cardiovascular conditions in patients with cancer [86, 87] and patients with other, previously incurable, chronic diseases may help to build bridges between disciplines and to adopt existing stratification strategies into the wider management of cardiovascular diseases.

Sex Differences and Precision Medicine in Cardiovascular Diseases

Sex-specific aspects of cardiovascular medicine have been explored in detail throughout this book. Common themes across all cardiovascular disciplines include the underrepresentation of women in major cardiovascular trials [88], lack of sex-specific analysis of study data instead of simple statistical analysis for a binary "confounder," and lack of mechanistic studies into differences between men and women in the pathogenesis of cardiovascular diseases. It is therefore not surprising that these shortfalls also translate into the area of precision medicine. It is evident from other chapters in this book that men and women in many cases require different risk stratification, diagnostic tests, and therapeutic strategies, and we do not have to repeat these notions here. They are, however, a key principle of precision medicine in cardiovascular diseases with sex and gender being important stratification tools to facilitate personalized approaches. The key role of precision medicine to address sex and gender issues has been outlined recently by Reckelhoff and Cordozo [89]. The potential includes female-specific aspects of cardiovascular disease [90] but also men's health [91]. As with other areas in precision medicine, clinical decision will ultimately not only be determined by novel omics-based data but also by traditional but sex-specific risk assessment [37] and analysis of known pathways such as the renin-angiotensin-aldosterone system for sex-specific regulation [92].

As with many other clinical studies into diagnostic tests and therapeutic strategies, there are few omics studies available that specifically investigate sex and gender aspects beyond adjustment of data for biological sex. The potential especially of *epigenetic studies* to inform sex-specific precision medicine approaches has, however, been recognized [93]. Indeed, for example, *proteomic studies* can explore sex and gender differences throughout all stages of the cardiovascular continuum [94].

Among the most relevant and probably relatively easily addressable issues are different *responses to pharmacological treatment* between men and women [95]. Pharmacogenetic and pharmacogenomic studies could benefit enormously from deep integration of sex in the design and analysis of studies [96].

Challenges and Barriers

The implementation of precision medicine in the management of cardiovascular diseases faces challenges that are similar in other areas of medicine. We have already discussed technical challenges such as standardization of omics experiments and some of the requirements of tests and available data to serve as stratification tools.

The major challenge, however, lies in the acceptance of precision medicine by society and key stakeholders. In a recent blog article, Richard Smith mentions the example of mustard makers who make most of their money from the unused mustard leftover on the plate and draws parallels to the current prescription practice where the pharmaceutical industry earns money from drugs that are not effective and potentially even harmful in sizeable groups of patients [9]. It remains unproven, however, that more precise prescribing with better chances of success would necessarily reduce revenue for the pharmaceutical industry. In fact, more precise definition of groups of patients who will benefit from preventative and therapeutic approaches could even increase the target population for a specific drug.

Beyond the general societal challenges, we believe that one of the most important issues in precision medicine is the move from stratified and guideline-based approaches to increasingly personalized and still evidence-based strategies. The complexities of the technologies involved in this new area of medicine such as genomics, bioinformatics, and analysis of big data are by far beyond the current curriculum at medical schools and probably even further away from training programs for qualified and practicing doctors [97]. Precision medicine in cardiovascular diseases can only succeed if it is firmly embedded in training of future and present doctors, learning again from how colleagues in oncology practice medicine.

Summary and Conclusions

We appreciate that for a chapter on precision medicine we remain vague and imprecise throughout large parts of this article. This is mainly because of the lack of robust data that demonstrate specific successes of cardiovascular precision medicine and especially of the role of sex in precision medicine approaches – in contrast to the large amount of opinion papers, reviews, and policy documents. We would therefore like to propose a few specific steps that should help translating precision medicine into cardiology practice.

1. *Explore low-hanging fruit*. We suggest that areas with the greatest chances of success and impact should be used for well-designed studies into precision medicine approaches. These should include studies integrating sex in the design and analysis and probably focus on pharmacogenomics and imaging aspects.
2. *Make use of existing data*. While it is important to generate new and especially omics-based data, strategies to make use of existing data by forming international research consortia and using electronic health records should be supported.
3. *Improve diagnostic tests*. We have witnessed the potential of high-sensitivity biomarker assays for CRP and cardiac troponins and expect further potential for other biomarkers but also for imaging and functional studies where data individual integration and modeling are currently underused.
4. *Standardize phenotypes*. Data can only be integrated where the high precision of diagnostic tests and the quality of research data are matched by the quality of clinical phenotypes.
5. *Standardize and simplify omics techniques to support large-scale epidemiological studies*. The potential of omics studies has been extensively demonstrated in recent years. However, in order to move to clinical applicability, large-scale studies with reproducible techniques have to be designed.
6. *Change the mindset in cardiovascular medicine toward molecular diagnostics*. Unlike other disciplines, cardiovascular medicine relies largely on noninvasive tests rather than molecular diagnosis that may have the potential to more precisely describe the disease.
7. *Conduct stratified clinical trials*. Future clinical trials should move beyond testing of the intervention but include testing the stratification rules.

For all these developments, the involvement of patient representatives and the wider public is of key importance so that the transformative potential of precision medicine can be implemented in clinical practice.

Acknowledgments Our work is supported by grants from the European Commission (Cooperative Research Projects "sysVASC" (603288), "HOMAGE" (305507), and "PRIORITY" (101813)) and the British Heart Foundation (Centre of Research Excellence Award RE/13/5/30177).

References

1. Roffi M, Patrono C, Collet JP, Mueller C, Valgimigli M, Andreotti F, et al. 2015 ESC guidelines for the management of acute coronary syndromes in patients presenting without persistent ST-segment elevation: task force for the Management of Acute Coronary Syndromes in patients presenting without persistent ST-segment elevation of the European Society of Cardiology (ESC). Eur Heart J. 2016;37(3):267–315.
2. Mancia G, Fagard R, Narkiewicz K, Redon J, Zanchetti A, Bohm M, et al. 2013 ESH/ESC guidelines for the management of arterial hypertension: the task force for the management of arterial hypertension of the European Society of Hypertension (ESH) and of the European Society of Cardiology (ESC). J Hypertens. 2013;31(7):1281–357.
3. American Diabetes Association. 2. Classification and diagnosis of diabetes. Diabetes Care. 2017;40(Suppl 1):S11–24.
4. Ponikowski P, Voors AA, Anker SD, Bueno H, Cleland JG, Coats AJ, et al. 2016 ESC guidelines for the diagnosis and treatment of acute and chronic heart failure: the task force for the diagnosis and treatment of acute and chronic heart failure of the European Society of Cardiology (ESC)developed with the special contribution of the heart failure association (HFA) of the ESC. Eur Heart J. 2016;37(27):2129–200.
5. Medical Research Council. Stratified medicine. https://www.mrc.ac.uk/research/initiatives/stratified-medicine/. Accessed 7 Jan 2018.
6. National Health Service. Improving outcomes through personalised medicine. https://www.england.nhs.uk/wp-content/uploads/2016/09/improving-outcomes-personalised-medicine.pdf. Accessed 7 Jan 2018.
7. National Research Council (US) Committee on A framework for developing a new taxonomy of disease. Toward precision medicine: building a knowledge network for biomedical research and a new taxonomy of disease. Washington, DC: National Academic Press (US); 2011.
8. President Obama's Precision Medicine Initiative. https://obamawhitehouse.archives.gov/the-press-office/2015/01/30/fact-sheet-president-obama-s-precision-medicine-initiative. Accessed 7 Jan 2018.
9. Smith R. Stratified, personalised, or precision medicine. http://blogs.bmj.com/bmj/2012/10/15/richard-smith-stratified-personalised-or-precision-medicine/. Accessed 7 Jan 2018.
10. Dominiczak A, Delles C, Padmanabhan S. Genomics and precision medicine for clinicians and scientists in hypertension. Hypertension. 2017;69(4):e10–3.
11. Konig IR, Fuchs O, Hansen G, von Mutius E, Kopp MV. What is precision medicine? Eur Respir J. 2017;50 (4):1700391.
12. GBD 2013 Mortality and Causes of Death Collaborators. Global, regional, and national age-sex specific all-cause and cause-specific mortality for 240 causes of death, 1990-2013: a

systematic analysis for the global burden of disease study 2013. Lancet. 2015;385(9963):117–71.
13. Dzau VJ, Ginsburg GS, Van Nuys K, Agus D, Goldman D. Aligning incentives to fulfil the promise of personalised medicine. Lancet. 2015;385(9982):2118–9.
14. Chatterjee S, Khunti K, Davies MJ. Type 2 diabetes. Lancet. 2017;389(10085):2239–51.
15. Dorajoo R, Liu J, Boehm BO. Genetics of type 2 diabetes and clinical utility. Genes (Basel). 2015;6(2):372–84.
16. Rossi GP. Prevalence and diagnosis of primary aldosteronism. Curr Hypertens Rep. 2010;12(5):342–8.
17. Azizan EA, Brown MJ. Novel genetic determinants of adrenal aldosterone regulation. Curr Opin Endocrinol Diabetes Obes. 2016;23(3):209–17.
18. Ji W, Foo JN, O'Roak BJ, Zhao H, Larson MG, Simon DB, et al. Rare independent mutations in renal salt handling genes contribute to blood pressure variation. Nat Genet. 2008;40 (5):592–9.
19. Titze J. Sodium balance is not just a renal affair. Curr Opin Nephrol Hypertens. 2014;23(2):101–5.
20. National Institute for Health and Clinical Excellence. Hypertension in adults: diagnosis and management. https://www.nice.org.uk/guidance/cg127. Accessed 7 Jan 2018.
21. Buhler FR, Burkart F, Lutold BE, Kung M, Marbet G, Pfisterer M. Antihypertensive beta blocking action as related to renin and age: a pharmacologic tool to identify pathogenetic mechanisms in essential hypertension. Am J Cardiol. 1975;36 (5):653–69.
22. Williams B, MacDonald TM, Morant S, Webb DJ, Sever P, McInnes G, et al. Spironolactone versus placebo, bisoprolol, and doxazosin to determine the optimal treatment for drug-resistant hypertension (PATHWAY-2): a randomised, double-blind, crossover trial. Lancet. 2015;386(10008):2059–68.
23. Shah SJ. Precision medicine for heart failure with preserved ejection fraction: an overview. J Cardiovasc Transl Res. 2017;10(3):233–44.
24. Shah SJ. Innovative clinical trial designs for precision medicine in heart failure with preserved ejection fraction. J Cardiovasc Transl Res. 2017;10(3):322–36.
25. Olsen MH, Angell SY, Asma S, Boutouyrie P, Burger D, Chirinos JA, et al. A call to action and a lifecourse strategy to address the global burden of raised blood pressure on current and future generations: the lancet commission on hypertension. Lancet. 2016;388(10060):2665–712.
26. Musunuru K, Hickey KT, Al-Khatib SM, Delles C, Fornage M, Fox CS, et al. Basic concepts and potential applications of genetics and genomics for cardiovascular and stroke clinicians: a scientific statement from the American Heart Association. Circ Cardiovasc Genet. 2015;8(1):216–42.
27. Cavallari LH, Lee CR, Beitelshees AL, RM C-DH, Duarte JD, Voora D, et al. Multisite investigation of outcomes with implementation of CYP2C19 genotype-guided antiplatelet therapy after percutaneous coronary intervention. JACC Cardiovasc Interv. 2017;11(2):181–91.
28. Wald NJ, Law MR. A strategy to reduce cardiovascular disease by more than 80%. BMJ. 2003;326(7404):1419.
29. Topol EJ. Individualized medicine from prewomb to tomb. Cell. 2014;157(1):241–53.
30. Snyderman R. Personalized health care: from theory to practice. Biotechnol J. 2012;7(8):973–9.
31. US Department of Veterans Affairs. VA Million Veteran Program linked to precision medicine effort. https://www.research.va.gov/pubs/varqu/spring2016/13.cfm. Accessed 7 Jan 2018.
32. UK Biobank. http://www.ukbiobank.ac.uk/. Accessed 7 Jan 2018.
33. Dzau V, Braunwald E. Resolved and unresolved issues in the prevention and treatment of coronary artery disease: a workshop consensus statement. Am Heart J. 1991;121(4 Pt 1):1244–63.
34. Padmanabhan S, Joe B. Towards precision medicine for hypertension: a review of genomic, Epigenomic, and Microbiomic effects on blood pressure in experimental rat models and humans. Physiol Rev. 2017;97(4):1469–528.
35. Sun D, Liu J, Xiao L, Liu Y, Wang Z, Li C, et al. Recent development of risk-prediction models for incident hypertension: an updated systematic review. PLoS One. 2017;12(10): e0187240.
36. Savoia C, Volpe M, Grassi G, Borghi C, Agabiti Rosei E, Touyz RM. Personalized medicine-a modern approach for the diagnosis and management of hypertension. Clin Sci (Lond). 2017;131(22):2671–85.
37. Ladapo JA, Budoff M, Sharp D, Zapien M, Huang L, Maniet B, et al. Clinical utility of a precision medicine test evaluating outpatients with suspected obstructive coronary artery disease. Am J Med. 2017;130(4):482. e411–482.e417
38. Niccoli G, Montone RA, Lanza GA, Crea F. Angina after percutaneous coronary intervention: the need for precision medicine. Int J Cardiol. 2017;248:14–9.
39. Halliday BP, Cleland JGF, Goldberger JJ, Prasad SK. Personalizing risk stratification for sudden death in dilated cardiomyopathy: the past, present, and future. Circulation. 2017;136(2):215–31.
40. Kinnamon DD, Morales A, Bowen DJ, Burke W, Hershberger RE, DCM Consortium*. Toward genetics-driven early intervention in dilated cardiomyopathy: design and implementation of the DCM precision medicine study. Circ Cardiovasc Genet. 2017;10(6):e001826.
41. Nassif ME, Tang Y, Cleland JG, Abraham WT, Linde C, Gold MR, et al. Precision medicine for cardiac resynchronization: predicting quality of life benefits for individual patients-an analysis from 5 clinical trials. Circ Heart Fail. 2017;10(10): e004111.
42. Thanassoulis G, Peloso GM, Pencina MJ, Hoffmann U, Fox CS, Cupples LA, et al. A genetic risk score is associated with incident cardiovascular disease and coronary artery calcium: the Framingham heart study. Circ Cardiovasc Genet. 2012;5 (1):113–21.
43. Berinstein E, Levy A. Recent developments and future directions for the use of pharmacogenomics in cardiovascular disease treatments. Expert Opin Drug Metab Toxicol. 2017;13 (9):973–83.
44. Tuteja S, Limdi N. Pharmacogenetics in cardiovascular medicine. Curr Genet Med Rep. 2016;4(3):119–29.
45. Jie Z, Xia H, Zhong SL, Feng Q, Li S, Liang S, et al. The gut microbiome in atherosclerotic cardiovascular disease. Nat Commun. 2017;8(1):845.
46. Richard MA, Huan T, Ligthart S, Gondalia R, Jhun MA, Brody JA, et al. DNA methylation analysis identifies loci for blood pressure regulation. Am J Hum Genet. 2017;101 (6):888–902.
47. Chu AY, Tin A, Schlosser P, Ko YA, Qiu C, Yao C, et al. Epigenome-wide association studies identify DNA methylation associated with kidney function. Nat Commun. 2017;8 (1):1286.
48. Li J, Zhu X, Yu K, Jiang H, Zhang Y, Deng S, et al. Genome-wide analysis of DNA methylation and acute coronary syndrome. Circ Res. 2017;120(11):1754–67.
49. Ligthart S, Marzi C, Aslibekyan S, Mendelson MM, Conneely KN, Tanaka T, et al. DNA methylation signatures of chronic low-grade inflammation are associated with complex diseases. Genome Biol. 2016;17(1):255.

50. Mendelson MM, Marioni RE, Joehanes R, Liu C, Hedman AK, Aslibekyan S, et al. Association of Body Mass Index with DNA methylation and gene expression in blood cells and relations to Cardiometabolic disease: a Mendelian randomization approach. PLoS Med. 2017;14(1):e1002215.
51. Lindsey ML, Mayr M, Gomes AV, Delles C, Arrell DK, Murphy AM, et al. Transformative impact of proteomics on cardiovascular health and disease: a scientific statement from the American Heart Association. Circulation. 2015;132 (9):852–72.
52. Delles C, Schiffer E, von Zur Muhlen C, Peter K, Rossing P, Parving HH, et al. Urinary proteomic diagnosis of coronary artery disease: identification and clinical validation in 623 individuals. J Hypertens. 2010;28(11):2316–22.
53. Good DM, Zurbig P, Argiles A, Bauer HW, Behrens G, Coon JJ, et al. Naturally occurring human urinary peptides for use in diagnosis of chronic kidney disease. Mol Cell Proteomics. 2010;9(11):2424–37.
54. Carty DM, Siwy J, Brennand JE, Zurbig P, Mullen W, Franke J, et al. Urinary proteomics for prediction of preeclampsia. Hypertension. 2011;57(3):561–9.
55. Myers JE, Tuytten R, Thomas G, Laroy W, Kas K, Vanpoucke G, et al. Integrated proteomics pipeline yields novel biomarkers for predicting preeclampsia. Hypertension. 2013;61(6):1281–8.
56. Brown CE, McCarthy NS, Hughes AD, Sever P, Stalmach A, Mullen W, et al. Urinary proteomic biomarkers to predict cardiovascular events. Proteomics Clin Appl. 2015;9 (5–6):610–7.
57. Williams SA, Murthy AC, DeLisle RK, Hyde C, Malarstig A, Ostroff R, et al. Improving assessment of drug safety through proteomics: early detection and mechanistic characterization of the unforeseen harmful effects of Torcetrapib. Circulation. 2017;37(10):999–1010.
58. Yin X, Baig F, Haudebourg E, Blankley RT, Gandhi T, Muller S, et al. Plasma proteomics for epidemiology: increasing throughput with standard-flow rates. Circ Cardiovasc Genet. 2017;10(6):e001808.
59. Barba I, Andres M, Dorado DG. Metabolomics and heart diseases: from basic to clinical approach. Curr Med Chem. 2017. https://doi.org/10.2174/0929867324666171006151408.
60. Kettunen J, Tukiainen T, Sarin AP, Ortega-Alonso A, Tikkanen E, Lyytikainen LP, et al. Genome-wide association study identifies multiple loci influencing human serum metabolite levels. Nat Genet. 2012;44(3):269–76.
61. Delles C, Rankin NJ, Boachie C, McConnachie A, Ford I, Kangas A, et al. Nuclear magnetic resonance-based metabolomics identifies phenylalanine as a novel predictor of incident heart failure hospitalisation: results from PROSPER and FINRISK 1997. Eur J Heart Fail. 2017. https://doi.org/10.1002/ejhf.1076.
62. Menni C, Graham D, Kastenmuller G, Alharbi NH, Alsanosi SM, McBride M, et al. Metabolomic identification of a novel pathway of blood pressure regulation involving hexadecanedioate. Hypertension. 2015;66(2):422–9.
63. Beger RD, Dunn W, Schmidt MA, Gross SS, Kirwan JA, Cascante M, et al. Metabolomics enables precision medicine: "A white paper, community perspective". Metabolomics. 2016;12(10):149.
64. Lawler PR, Akinkuolie AO, Harada P, Glynn RJ, Chasman DI, Ridker PM, et al. Residual risk of atherosclerotic cardiovascular events in relation to reductions in very-low-density lipoproteins. J Am Heart Assoc. 2017;6(12):e007402.
65. McInnes IB, Thompson L, Giles JT, Bathon JM, Salmon JE, Beaulieu AD, et al. Effect of interleukin-6 receptor blockade on surrogates of vascular risk in rheumatoid arthritis: MEASURE, a randomised, placebo-controlled study. Ann Rheum Dis. 2015;74(4):694–702.
66. Dibble EH, Yoo DC. Precision medicine and PET/computed tomography in cardiovascular disorders. PET Clin. 2017;12 (4):459–73.
67. Teague HL, Ahlman MA, Alavi A, Wagner DD, Lichtman AH, Nahrendorf M, et al. Unraveling vascular inflammation: from immunology to imaging. J Am Coll Cardiol. 2017;70 (11):1403–12.
68. Futier E, Lefrant JY, Guinot PG, Godet T, Lorne E, Cuvillon P, et al. Effect of individualized vs standard blood pressure management strategies on postoperative organ dysfunction among high-risk patients undergoing major surgery: a randomized clinical trial. JAMA. 2017;318(14):1346–57.
69. Bundhun PK, Gupta C, Huang F. Should fraction flow reserve be considered an important decision-making tool to stratify patients with stable coronary artery disease for percutaneous coronary intervention?: a meta-analysis. Medicine (Baltimore). 2017;96(46):e8748.
70. Lucas JE, Bazemore TC, Alo C, Monahan PB, Voora D. An electronic health record based model predicts statin adherence, LDL cholesterol, and cardiovascular disease in the United States military health system. PLoS One. 2017;12(11): e0187809.
71. Wu PY, Cheng CW, Kaddi C, Venugopalan J, Hoffman R, Wang MD. Advanced big data analytics for -Omic data and electronic health records: toward precision medicine. IEEE Trans Biomed Eng. 2017;64(2):263–3.
72. Brody JA, Morrison AC, Bis JC, O'Connell JR, Brown MR, Huffman JE, et al. Analysis commons, a team approach to discovery in a big-data environment for genetic epidemiology. Nat Genet. 2017;49(11):1560–3.
73. Goldstein BA, Navar AM, Carter RE. Moving beyond regression techniques in cardiovascular risk prediction: applying machine learning to address analytic challenges. Eur Heart J. 2017;38(23):1805–14.
74. Krittanawong C, Zhang H, Wang Z, Aydar M, Kitai T. Artificial intelligence in precision cardiovascular medicine. J Am Coll Cardiol. 2017;69(21):2657–64.
75. Capelli C, Sauvage E, Giusti G, Bosi GM, Ntsinjana H, Carminati M, et al. Patient-specific simulations for planning treatment in congenital heart disease. Interface Focus. 2018;8 (1):20170021.
76. Mangion K, Gao H, Husmeier D, Luo X, Berry C. Advances in computational modelling for personalised medicine after myocardial infarction. Heart. 2017;104(7):550–7.
77. Ye S, Kronish IM. To persist or not to persist: learning from precision medicine to optimize statin adherence. Rev Esp Cardiol (Engl Ed). 2018;71(1):4–5.
78. Day S, Coombes RC, McGrath-Lone L, Schoenborn C, Ward H. Stratified, precision or personalised medicine? Cancer services in the 'real world' of a London hospital. Sociol Health Illn. 2017;39(1):143–58.
79. Mukherjee C, Sweet KM, Luzum JA, Abdel-Rasoul M, Christman MF, Kitzmiller JP. Clinical pharmacogenomics: patient perspectives of pharmacogenomic testing and the incidence of actionable test results in a chronic disease cohort. Per Med. 2017;14(5):383–8.
80. Ridker PM, Danielson E, Fonseca FA, Genest J, Gotto AM Jr, Kastelein JJ, et al. Rosuvastatin to prevent vascular events in men and women with elevated C-reactive protein. N Engl J Med. 2008;359(21):2195–207.
81. SPRINT Research Group, Wright JT Jr, Williamson JD, Whelton PK, Snyder JK, Sink KM, et al. A randomized trial of intensive versus standard blood-pressure control. N Engl J Med. 2015;373(22):2103–16.

82. Lindhardt M, Persson F, Currie G, Pontillo C, Beige J, Delles C, et al. Proteomic prediction and renin angiotensin aldosterone system inhibition prevention of early diabetic nephRopathy in TYpe 2 diabetic patients with normoalbuminuria (PRIORITY): essential study design and rationale of a randomised clinical multicentre trial. BMJ Open. 2016;6(3):e010310.
83. Academy of Medical Sciences. Realising the potential of stratified medicine. https://acmedsci.ac.uk/viewFile/51e915f9f09fb.pdf. Accessed 7 Jan 2018.
84. Latosinska A, Frantzi M, Vlahou A, Merseburger AS, Mischak H. Clinical proteomics for precision medicine: the bladder cancer case. Proteomics Clin Appl. 2017;12(2). https://doi.org/10.1002/prca.201700074.
85. Feldman AM, Kontos CD, McClung JM, Gerhard GS, Khalili K, Cheung JY. Precision medicine for heart failure: lessons from oncology. Circ Heart Fail. 2017;10(6):e004202.
86. Chang HM, Moudgil R, Scarabelli T, Okwuosa TM, Yeh ETH. Cardiovascular complications of Cancer therapy: best practices in diagnosis, prevention, and management: part 1. J Am Coll Cardiol. 2017;70(20):2536–51.
87. Chang HM, Okwuosa TM, Scarabelli T, Moudgil R, Yeh ETH. Cardiovascular complications of Cancer therapy: best practices in diagnosis, prevention, and management: part 2. J Am Coll Cardiol. 2017;70(20):2552–65.
88. Melloni C, Berger JS, Wang TY, Gunes F, Stebbins A, Pieper KS, et al. Representation of women in randomized clinical trials of cardiovascular disease prevention. Circ Cardiovasc Qual Outcomes. 2010;3(2):135–42.
89. Reckelhoff JF, Cordozo LLY. As precision medicine becomes more important, is it finally time for increased emphasis on gender medicine? Biochemist. 2017;39:4–5.
90. Miller VM, Rocca WA, Faubion SS. Sex differences research, precision medicine, and the future of Women's health. J Womens Health (Larchmt). 2015;24(12):969–71.
91. Mata DA, Katchi FM, Ramasamy R. Precision medicine and men's health. Am J Mens Health. 2017;11(4):1124–9.
92. Te Riet L, van Esch JH, Roks AJ, van den Meiracker AH, Danser AH. Hypertension: renin-angiotensin-aldosterone system alterations. Circ Res. 2015;116(6):960–75.
93. Giannakopoulou E, Konstantinou F, Ragia G, Tavridou A, Karaglani M, Chatzaki E, et al. Epigenetics-by-sex interaction for coronary artery disease risk conferred by the cystathionine gamma-Lyase gene promoter methylation. OMICS. 2017;21 (12):741–8.
94. Baetta R, Pontremoli M, Martinez Fernandez A, Spickett CM, Banfi C. Proteomics in cardiovascular diseases: unveiling sex and gender differences in the era of precision medicine. J Proteome. 2017;173:62–76.
95. Franconi F, Campesi I. Pharmacogenomics, pharmacokinetics and pharmacodynamics: interaction with biological differences between men and women. Br J Pharmacol. 2014;171(3):580–94.
96. Gaignebet L, Kararigas G. En route to precision medicine through the integration of biological sex into pharmacogenomics. Clin Sci (Lond). 2017;131(4):329–42.
97. Academy of Medical Sciences. Stratified, personalised or P4 medicine: a new direction for placing the patient at the centre of healthcare and health education. May 2015. https://acmedsci.ac.uk/viewFile/564091e072d41.pdf. Accessed 7 Jan 2018.

Big-Data Analysis, Cluster Analysis, and Machine-Learning Approaches

37

Amparo Alonso-Betanzos and Verónica Bolón-Canedo

Art work by Piet Michiels, Leuven, Belgium

Abstract

Medicine will experience many changes in the coming years because the so-called "medicine of the future" will be increasingly proactive, featuring four basic elements: predictive, personalized, preventive, and participatory. Drivers for these changes include the digitization of data in medicine and the availability of computational tools that deal with massive volumes of data. Thus, the need to apply machine-learning methods to medicine has increased dramatically in recent years while facing challenges related to an unprecedented large number of clinically relevant features and highly specific diagnostic tests. Advances regarding data-storage technology and the progress concerning genome studies have enabled collecting vast amounts of

A. Alonso-Betanzos (✉) · V. Bolón-Canedo
Department of Computer Science, University of A Coruña, A Coruña, Spain
e-mail: ciamparo@udc.es

P. L. M. Kerkhof, V. M. Miller (eds.), *Sex-Specific Analysis of Cardiovascular Function*,
Advances in Experimental Medicine and Biology 1065, https://doi.org/10.1007/978-3-319-77932-4_37

patient clinical details, thus permitting the extraction of valuable information. In consequence, big-data analytics is becoming a mandatory technology to be used in the clinical domain.

Machine learning and big-data analytics can be used in the field of cardiology, for example, for the prediction of individual risk factors for cardiovascular disease, for clinical decision support, and for practicing precision medicine using genomic information. Several projects employ machine-learning techniques to address the problem of classification and prediction of heart failure (HF) subtypes and unbiased clustering analysis using dense phenomapping to identify phenotypically distinct HF categories. In this chapter, these ideas are further presented, and a computerized model allowing the distinction between two major HF phenotypes on the basis of ventricular-volume data analysis is discussed in detail.

Keywords
Machine learning · Big-data analysis · Cluster analysis · Precision medicine · Heart failure phenotyping · Support vector machine

Introduction

In July 2016, the Food and Drug Administration approved the first monoclonal antibody that inactivates the components responsible for the degradation of low-density lipoprotein receptors in the liver, decreasing their blood levels to much lower numbers than those that can be achieved with statins. This development is relevant because it represents the first important step toward a new version, as an information science, of the field of medicine. Genomics pioneer L. Hood [1] calls this the "medicine of the future," which is changing its way of working from reactive to proactive toward "4P medicine" (powerfully predictive, personalized, preventive [a shift of focus from illness to wellness], and participatory). Several things will eventually drive this change, and among the most important will be the digitization of medicine and the development of computational tools—that is, machine-learning tools and methods, including preprocessing techniques needed to eliminate noise and irrelevant variables from the data—with the ability to deal with big data. Because data dimensionality is continuously increasing, it will be necessary to incorporate big-data analytics in order to be able to deal with the billions of data points that are expected for each individual patient in the next decade. Predictive medicine must correlate this high number of dimensional data sets with individual genotypes and phenotypes.

Thus, in this medicine of the future, machine learning [2] and big-data analytics [3–7] will be key disciplines. Data, often in large volumes, will be analyzed based on epidemiological variables, electronic health records (EHRs), genomic databases, and so on. These data will allow for the practice of preventive and precision medicine and the avoidance of medical errors that might occur because of medical doctors' distress due to their increasing and intensifying workloads, thereby increasing quality at affordable prices (see Fig. 37.1). For example, 10 years ago the expense of sequencing the genome of just one individual was approximately €200,000, whereas currently it is <€600.

Computerized and artificial intelligence (AI)–based methods for the analysis and interpretation of medical databases are not that new; these techniques found relatively early application in the medical sciences [8]. Machine learning and big-data analytics can be used in the field of cardiology in several ways, such as predicting an individual's risk for cardiovascular disease, clinical decision support, precision medicine using genomic information, and so on. Some works that can be found in the literature using machine-learning techniques investigate the problem of classification and prediction of heart failure (HF) subtypes [9] and unbiased clustering analysis using dense phenomapping to identify phenotypically distinct categories of preserved ejection fraction (HFpEF). Alonso-Betanzos et al. [10] propose a computerized model that allows a clearer distinction between the two major phenotypes of patients with HF based on

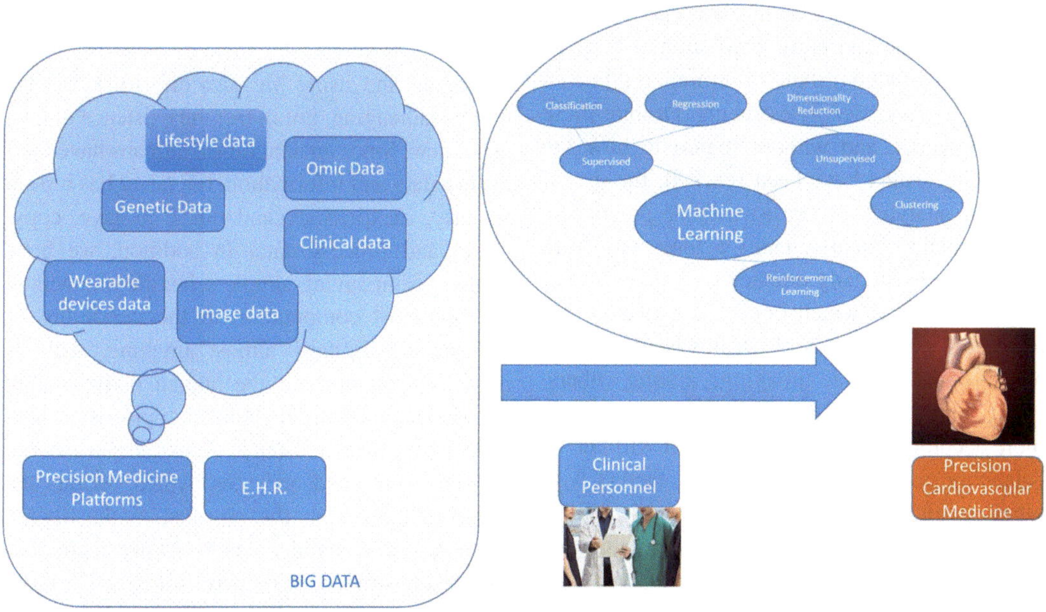

Fig. 37.1 The new scenario of big data and the application of artificial intelligence and machine-learning techniques to precision medicine. *EHR* electronic health record

ventricular volume data analysis. In that work, a gray zone—which might correspond to a third separate phenotype—was identified, thus corroborating the capacity of machine-learning techniques to discover knowledge in medicine. Narula et al. [11] used an ensemble machine-learning model to aid in cardiac phenotypic recognition, specifically for the automated discrimination of hypertrophic cardiomyopathy from the physiological hypertrophy seen in athletes. The model used the previous feature of selection preprocessing (using the information gain filter [12]) step-over features of cardiac tissue deformation. Three different models integrated the ensemble (support vector machine [SVM], random forest [RF], and artificial neural network), and majority voting was used to reach a final decision.

However, studies using big-data analytics in the field of cardiology are not yet frequent in the literature. Some investigators described a framework aiming at setting an "initial but timely step toward a more intelligent and learning health care system that will require innovative bonds among patients, clinicians, data scientists, and health care systems" [13]. Motwani et al. [14] used machine learning on a data set from patients (10,030 patients during a 5-year follow-up period) undergoing coronary computed tomographic angiography. The aim of the study was to compare the results of cardiovascular outcome prediction using machine learning (including automated feature selection and an ensemble algorithm for learning) with those of traditional prognosis, which was limited regarding the use of clinical and imaging findings. The results showed considerable improvement in predicting all-cause mortality of those patients. In another study, the investigators studied the use of a Bayesian statistical model to address the limited predictive capacity of existing risk scores derived from multi-variate analyses [15]. The prognostic model showed superior prediction of acute, early, and late right-ventricular failure after left-ventricular (LV) assist device (LVAD) therapy compared with the currently available risk-prediction model. In conclusion, these models might facilitate clinical decision making while screening candidates for LVAD therapy. The trade-offs between data requirements and model utility were analyzed by Ng et al. [16] and Spertus et al. [17], who concluded that machine-learning techniques should be more frequently used in health care—in particular in cardiovascular risk estimations and mortality predictions—because they can greatly contribute to minimizing bias in hospital performance assessments. Despite these seminal works, and without regard to how the promising results have been reached, the use of big-data analytics in cardiovascular practice is still incipient, and it remains a long way from being a reality in daily practice.

Some areas in which impact is expected in a few years are the fields of cardiovascular epidemiology and cardiac imaging, among others. Using data from EHRs will not only obtain more accurate predictions, because it will permit balancing primary well-known risk factors with other secondary less-investigated ones, it also crosses those risk factors with other illnesses, such as cancer or cerebrovascular diseases. At the same time, more general health models could be obtained in this way. Machine learning with big data will also permit the selection of sub-populations in specific geographic areas, thus opening a door to the design of local health policies and adequate resource planning, which are lacking in many countries. Image analysis also soon will probably see changes in patient classification, diagnosis, and visualization because multi-modal big data are increasingly being generated from echocardiography studies, computerized tomography studies, magnetic resonance studies, and so on. Patients will be empowered by the generalized use of mobile devices and apps, thus allowing for extra-hospital management of cardiac conditions, such as cardiac insufficiency, auricular fibrillation, etc., and helping decrease the incidence of costly patient re-hospitalization.

Some of the difficulties encountered are data standardization between and within hospitals; the need to render data anonymous; and data heterogeneity, complexity, and disorganization, which in turn leads to the need for preprocessing techniques aiming at removing noise, discretizing and filtering data, removing irrelevant variables, and so on. In addition to technical difficulties, some other aspects to be considered are the security and privacy of data, which is of special importance in a medical context.

Big Data

Managing and using big data effectively is currently challenging, but in fact the existence of data is not new. Since ancient times, humans have tried to save data and information, but never has it been so easy, inexpensive, and quick to save, copy, share, and process data. In addition, we have evolved from saving simple scientific numbers at the origin of computation to the possibility of representing digitally almost anything, such as music, travel, or even the human body, among others. This growing digitalization process is possible thanks to the existence of myriad sensors that register events and activities, which permits the transformation from the physical to the digital. Because digital entities can be easily replicated, saved, transmitted, modified, sold, and so on, health sectors, for example, are being transformed into information and knowledge services.

However, sensors are not the only difference. Until some years ago, we were happily living with our database relational model. However, some companies (such as Google and Yahoo) found out that the database model limited the type and quantity of data that could be saved and analyzed and that this fact was contrary to their business. Thus, they decided to confront certain problems in a non-traditional way, which included the fact that all data could be considered important in some way. All data have value and saving and analyzing large volumes of data was a key point in their new business proposal. The problem? The value of big data is really discovered only after analyzing large volumes of them. Since then, analyzing big-data analytics has become a major driver of the economy.

As mentioned, the use of data is not new in the field of health care, where researchers have always been involved in collecting and analyzing data. However, the new digitization context, toward which we are currently moving, implies a volume, variety, and velocity of data production that pose new opportunities and demands in terms of both scale and complexity. Those challenges are the main impetus behind the development of the National Institutes of Health Big Data to

Knowledge initiative [18] for addressing the opportunities and challenges presented by biomedical big data as well as the partnership between the National Cancer Institute and the United States Department of Energy to research years of cancer data and analyze them to develop new and more effective cancer treatments [19].

Several characteristics are required for data sets to be considered "big data" (see Fig. 37.2). Among the most important so-called five "Vs" of big data [20] are volume, velocity, variety, validity/veracity, and value.

- The *volume* of data that must be processed by algorithms is substantially large (on the order of petabytes [i.e. 1015 bytes] and zettabytes [i.e. 1021 bytes] and continuously growing. Large volumes of data demand different data-storage and -processing tools and new characteristics of the data-preparation and -preprocessing steps (see section "Preprocessing methods for big data").
- Data might appear in some context at high *velocities*, in other words, important volumes of data manifesting in short time periods. Being able to analyze this data in real time might be crucial for some applications, but this requires a specific infrastructure to manage data streams. There are use cases, however, in which velocity is not a problem. For example, many more "tweets" are generated per minute than magnetic resonance imaging (MRI) scan images, and while reacting to a negative tweet might be relevant for a company, a strict real-time response is rarely required for MRI diagnostics.
- Data currently come in different types (structured, semi-structured, and unstructured), and formats (text, images, audio, video, XML, etc.). This great *variety* increases complexity in data-saving and analytics solutions.
- The variety of data types and formats, with large volumes at being generated at high velocities, constitutes the ideal situation to raise doubts about their degree of date quality and/or *veracity* or *validity*. Are the data correct? Are they of good-enough quality? Can I simultaneously use data that have different degrees of precision or temporal or spatial scale? Are these data relevant for my problem? Can they lead me to "actionable" information? Data preprocessing techniques are mostly

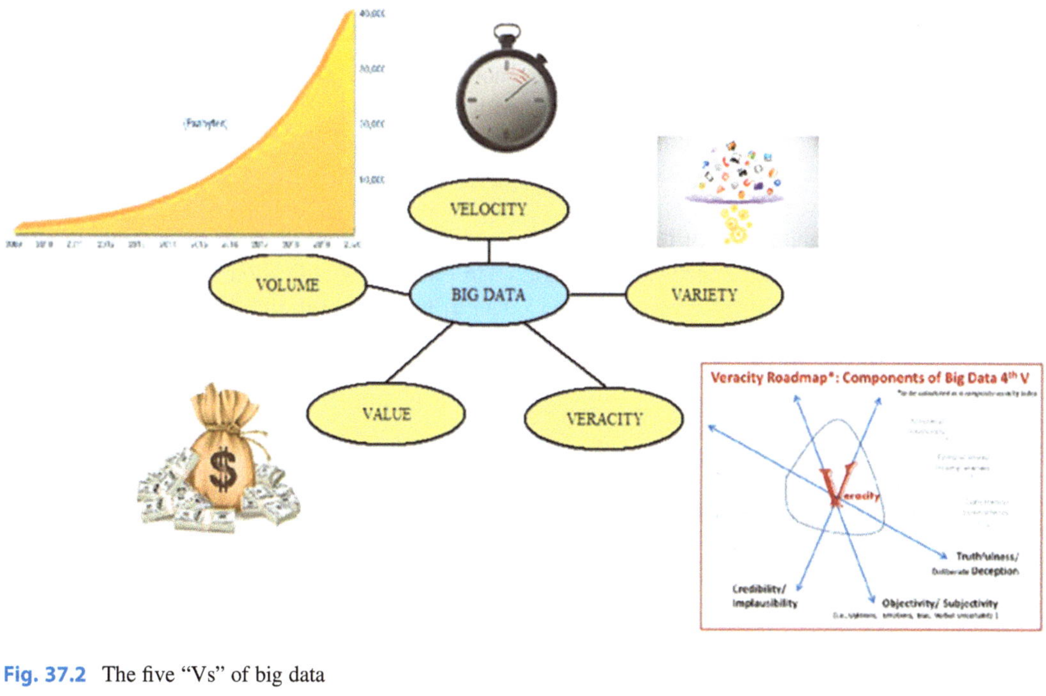

Fig. 37.2 The five "Vs" of big data

unavoidable because they remove noise, invalid data, or redundant features, for example. However, these processes also pose a problem and a challenge. Regarding the problem, on one hand their use implies an extra effort that perhaps is not justified if the results are not really affected; in contrast, if the information to be obtained is sensible, the preprocessing step should not be avoided. Thus, depending on the application, veracity could be mandatory or secondary. Regarding the challenge, most preprocessing techniques (such as discretization or feature selection) were not designed for the use of big data, and the data suffer when scalability is needed.

- Validity and veracity are critical determinants for users of big data because they affect the last "V": *value*. Value implies that the knowledge and information derived from the data must be useful for the company or entity. To derive that knowledge, big-data analytics must be used, which makes data scientists essential. Although some aspects could be automatized, management of the entire process (from designing the appropriate infrastructure needed to adequately visualize the results) is complex and requires high-level human expertise.

Many issues require planning and careful execution for the use of big data. Among the most important, especially in the field of health care, are security and privacy. The privacy of those individuals whose data are being managed is crucial, and addressing issues of this type requires deep understanding of the nature of the data, the relevant norms and regulations, and the techniques that should be used, such as, for example, anonymization. A careless analysis might reveal private information, thus opening a possibly unnoticed gap in privacy. Big-data security issues should be also considered, mostly because the types of databases that should be used do not provide as robust a built-in security mechanism as do traditional relational database-management systems. Similar situations might appear in the data-analysis phase, in which data might be distributed among several nodes, which is a common situation. If, for example, we are trying to derive prognostic models using patient data from different hospitals in a region or country, the machine-learning algorithms being used might need to interchange or combine intermediate results, again opening the possibility for inadvertent breaches of privacy. Thus, privacy preservation should be a requirement for the employed algorithms. Because the analysis of big data is different from traditional data analysis, a systematic methodology [20, 21] is needed to organize the diverse associated activities (see Fig. 37.3), and these must be managed by specialized data scientists.

Data Identification

In the first step, the business case to be confronted should be clearly identified, with an adequate motivation for the analysis, together with the types of data and the analytics needed. The projects goals also should be established. An important outcome of this stage is definition of the tools and other economic or personnel resources needed.

In the data-identification stage, the sources of data to be employed and their quantity, quality,

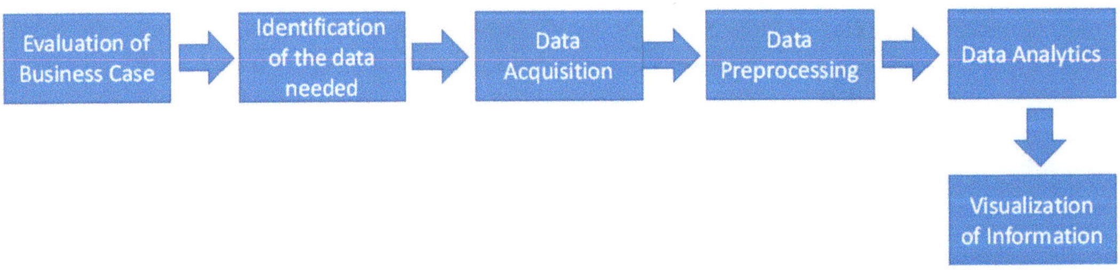

Fig. 37.3 Typical life cycle of big-data analytics

and format should be established. The data sources can be internal or external to the company. In the latter case, a list of data providers (this includes publicly available data sets) is needed. Finally, if personally identifiable information requires removal or masking, the involved requirements for anonymization and re-identification should be stated.

Data Collection

Data are collected from the data sources identified in the previous step during the acquisition phase. In addition, meta-data (such as data size, structure, date, time of creation, etc.) is added, if needed, to maintain the data during the rest of the methodology phases. Persistence of the data should be assured because fault-tolerance scenarios must be accounted for; therefore, data should be stored in a database.

Data Preprocessing

Preprocessing data implies several operations, such as removing noise, filtering data to remove some types possibly lacking value for posterior analysis, aggregating data that might be spread across several data sets, and removing irrelevant and redundant features in the data to simplify the data-analysis phase. Preprocessing is a fundamental step for assuring data quality and consistency, and it will be described in detail in section "Preprocessing methods for big data."

Data Analysis

The data-analysis stage is devoted to discovering patterns, correlations, predictions, or anomalies in the data to give answers to the questions raised in the first stage, thus making it possible to derive business value from the data. Because large volumes of data should be analyzed, specialized software tools and applications for machine learning and statistics are needed. The algorithms to be employed should obviously be scalable, and the type of architecture and tools to be employed depend on the restrictions of the specific use case. Stream-processing architecture produces real-time data insights because it computes one data element at a time. The data analysis is performed in almost real time, and thus immediate action can be taken in response. An example of this is acting in response to a patient's health status or the experience of a hospitalized patient. Tools are different for batch processing, which processes large volumes of data for which a quick response time is not critical; an example could be the elaboration of a monthly report on some activities of the hospital. Batch processing is more related to the volume and variety, whereas for stream processing velocity is most critical. Therefore, for some applications, batch-processed data might be outdated by the time it reaches health-care professionals. Of course, scenarios exist in which both types of data processing can be employed. For example, in marketing, batch data processing can be used to analyze the habits of consumers from historical data sets. Then health-care marketers can create tailored and targeted marketing campaigns that will ideally improve adherence and engagement from patients by establishing which communication channels will result in the best response rate from each group. From there onward, streaming processes can analyze which social media messages are most effective for each individual and take immediate action. Another example is the different use cases that can be derived from health services running on smartphones with sensors, which have become extremely popular, tracking regularly daily activities, such as sport, sleep, and diet habits across sport-oriented social networks. Because this is also an important phase, machine-learning methods will be described in more detail in section "Supervised and unsupervised machine-learning methods."

Data Visualization

Finally, data visualization consists of a presentation of the output (from the previous phase) in a format that allows business users to understand the results obtained and thus be able to make decisions based on them. This might comprise tables, graphs, information "blocks," and so on.

Choosing the appropriate infrastructure is crucial in big-data projects because companies build their competence around it. The selected infrastructure strongly affects big-data architecture for new products and services. Having the appropriate tools for storing, processing, and analyzing data is key. Among the most well-known tools are big-data databases (e.g., MongoDB, HBase, or Apache Cassandra); tools for transferring and aggregating data (e.g., Flume or Lucene); frameworks (e.g., Apache Hadoop [http://hadoop.apache.org/], Apache Spark [http://spark.apache.org/], or Apache Flink [https://flink.apache.org/]) with their corresponding machine-learning libraries (e.g., Mahout, MlLib, and Flink ML) or Spark components for graph analytics (e.g., GraphX and GraphFrames).

Preprocessing Methods for Big Data

As mentioned in the previous section, the advent of big data has brought an important number of challenges to the scientific community, which now must deal with unprecedented volumes in data and try to extract useful information from them. We continue to store data of all kinds, and usually a preprocessing step is necessary before applying an adequate machine-learning method (see section "Supervised and unsupervised machine-learning methods"). In fact, a typical scenario in health data is to be able to classify a patient as presenting with a particular condition or not (e.g., the risk of not catching a patient presenting with heart failure). From a machine-learning perspective, this is a classification task in which a learner or classifier must learn the characteristics of the data (i.e. variables about a given patient, such as age, sex, and the results of medical tests) and then provide a prediction. However, in many cases these variables are of different natures because attributes such as sex are discrete; other attributes, such as weight, have continuous values. Some classifiers can only deal with discrete data, so there exists an important preprocessing technique called "discretization."

Another problem typically encountered by machine-learning researchers facing medical data are that it is likely that some variables are not relevant for the prediction task. For example, sex can be an important factor for determining the risk of presenting with heart failure but completely irrelevant for other conditions. To solve this problem, feature selection is usually applied as a preprocessing step to remove those unimportant variables from the task at hand. In this section we will focus on two popular preprocessing techniques: discretization and feature selection.

Discretization

Discretization is a preprocessing technique that consists of transforming the continuous variables of a data set into discrete variables. By applying this technique, quantitative data are transformed into qualitative data, thus procuring a nonoverlapping division of a continuous domain. Discretization also can be considered as a data-reduction mechanism because it reduces data from a large domain of numeric values to a subset of discrete values [22].

In some cases, discretization is a mandatory step because some machine-learning algorithms used afterward are not able to handle continuous variables. For instance, this is the case with the popular classifier Naïve Bayes [23] and also with feature-selection methods, such as the information gain filter [24]. Apart from this, discretization can have a beneficial impact on the performance of learning algorithms, for example, in terms of speed (especially important in the context of big data and real-time learning) and accuracy. The basic discretization process is formed by four steps, which are detailed here and also can be seen in Fig. 37.4.

Step 1: Sort the continuous values for an attribute (either in ascending or descending order). It is crucial to choose an efficient sorting algorithm to perform this step.

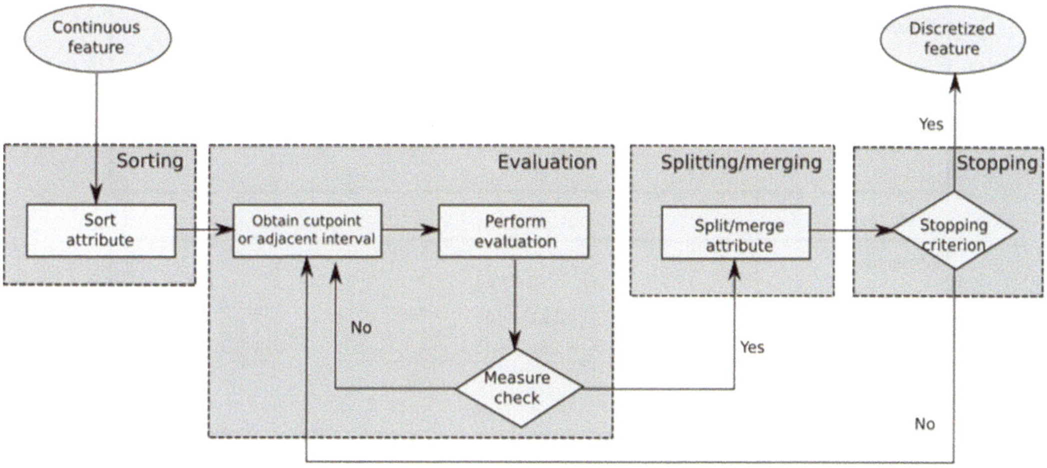

Fig. 37.4 Discretization process

Step 2: After sorting, select the best cut point of the best pair of adjacent intervals in the attribute range to split or merge them in the next step. It is necessary to define an evaluation measure of function, which can be correlation, gain, improvement in performance, or any other benefit according to the class label.

Step 3: Split or merge intervals according to the operation method of each discretization algorithm. To split, the possible cut points are the different continuous values in each attribute. To merge, the discretization algorithm tries to find the best adjacent intervals in each iteration.

Step 4: Stop according to some criterion or otherwise return to step 2. Usually a trade-off between a low number of intervals, good comprehension, and consistency is assumed.

A broad suite of discretization algorithms can be found in the literature, and the selection of one or another depends on the type of the data. For a complete taxonomy about discretization methods, see Ramírez et al. [22]. In the following text, some of the most popular methods will be described (including those in the popular Weka tool [25]).

- *Equal width*: This simple unsupervised discretization algorithm calculates the range of the variable and then divides it into equal parts. The resulting intervals will generally be unbalanced, with many items ending in a few of them and some much less populated. The split/merge step is disregarded in this simple method.
- *Equal frequency*: This algorithm obtains intervals that contain a constant number of items. The basic version of this method aims to obtain a fixed number of intervals, although this is suboptimal for some classification algorithms. Therefore, a variation called Proportional k-Interval Discretization (PKID) [26] can be used instead. This algorithm adjusts the number of intervals according to the number of samples.
- *Minimum Descriptive Length* (MDL): This popular method uses information-entropy minimization as a heuristic to calculate the most suitable cut points [27].

In a big-data scenario, the problem is that classical data-reduction methods were not designed to handle such a large amount of data, which makes their use difficult or even impossible in some cases. To solve this issue, in the past few years new implementations of the most popular methods have appeared that take advantage of distributed computational frameworks. For example, a distributed implementation of PKID is available in Spark MlLib. Moreover, a distributed implementation of MDL is available for Apache Spark [28], which leverages a

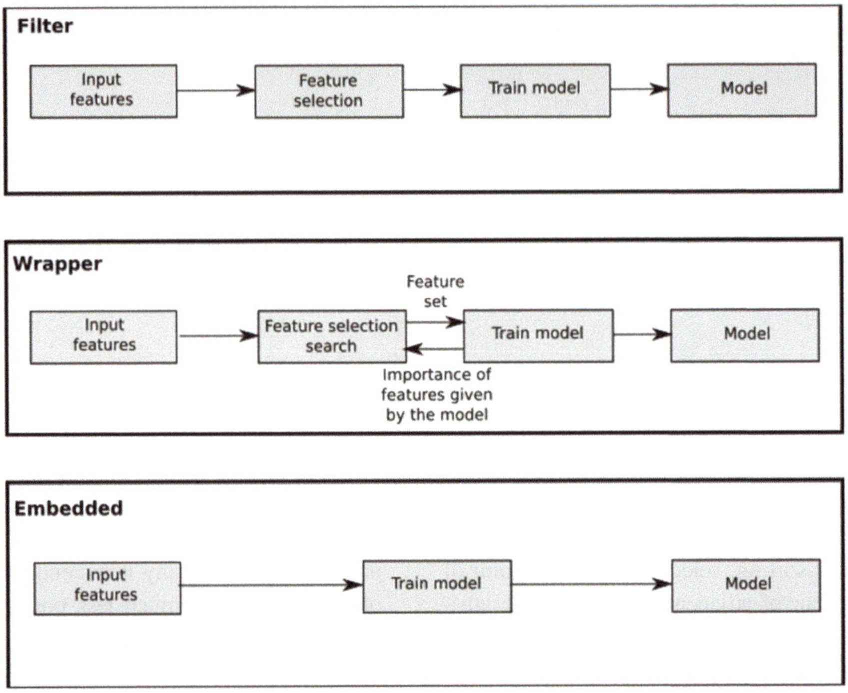

Fig. 37.5 Feature-selection approaches

computer cluster to speed up the sorting and cut point–evaluation steps involved in this method, thus enabling it to deal with large data sets.

Feature Selection

Analogous to the term "big data," the term "big dimensionality" has been coined to refer to the unprecedented number of features arriving at levels that render existing machine-learning methods inadequate [29]. Thus, dimensionality-reduction techniques, such as feature selection, have become almost essential. Feature selection is the process of selecting relevant variables (features), by removing the irrelevant and/or redundant ones, with the aim of obtaining better and simpler models. Because the process does not transform the original features (unlike feature-extraction methods), it obtains models that might be easier to interpret for researchers or medical practitioners. Moreover, it presents other benefits, such as enhancing generalization by decreasing overfitting and the confers the requirement of shorter training times (a crucial point in real-time application).

Typically, feature-selection methods are classified into filters, wrappers, and embedded methods based on their relationship with the learning algorithm (see Fig. 37.5). The simplest model is the filter, which relies on the general characteristics of training data and performing the feature-selection process as a preprocessing step with independence of the induction algorithm. Wrappers involve a learning algorithm as a black box, and they use their prediction performance to assess the relative usefulness of subsets of variables. Finally, embedded methods perform feature selection in the process of training and are usually specific to given learning machines. They learn which features best contribute to the accuracy of the model while the model is being created. Because of this interaction with the learning algorithm, wrappers and embedded methods tend to give more accurate subsets of features, but they are usually specific for a particular classifier and are computationally expensive (especially wrappers). In contrast, filters are advantageous for their low computational cost and good generalization abilities [30]. Each of the three approaches is extensively used, although the filter model is more adequate for big-data settings.

Considering that several algorithms exist for each one of the previously commented approaches, there is a vast body of feature-selection methods. We describe some of the most popular ones here:

- *Correlation-based feature selection*: This is a simple multi-variate filter algorithm that ranks feature subsets according to a correlation-based heuristic-evaluation function [31]. The bias of the evaluation function is toward subsets that contain features that are highly correlated with the class and not correlated with each other. Irrelevant features should be ignored because they will have low correlation with the class. Redundant features should be screened out because they will be highly correlated with one or more of the remaining features. The acceptance of a feature will depend on the extent to which it predicts classes in areas of the instance space not already predicted by other features.
- *Consistency-based*: This filter [32] evaluates the worth of a subset of features by the level of consistency in the class values when the training instances are projected onto the subset of attributes.
- *Information gain*: This filter [12] provides an ordered ranking of all features, and then a threshold is required.
- *ReliefF*: This filter [33] is an extension of the original Relief algorithm. The original Relief works by randomly sampling an instance from the data and then locating its nearest neighbor from the same class and from the opposite class. The values of the attributes of the nearest neighbors are compared with the sampled instance and used to update relevance scores for each attribute. The rationale is that a useful attribute should differentiate between instances from different classes and have the same value for instances from the same class. ReliefF adds the ability of dealing with multi-class problems and is also more robust and capable of dealing with incomplete and noisy data. ReliefF is applicable in all situations; it has low bias; it includes interaction among features; and it may capture local dependencies missed by other methods.
- *Minimum redundancy maximum relevance*: This filter [34] selects features that have the greatest relevance with the target class and that are also minimally redundant. In other words," it selects features that are maximally dissimilar to each other. Both optimization criteria (maximum relevance and minimum redundancy) are based on mutual information.
- *Recursive Feature Elimination for SVMs*: This embedded method [35] performs feature selection by iteratively training an SVM classifier with the current set of features and removing the least important feature indicated by the SVM.

Although feature selection is almost mandatory for machine-learning algorithms to be able to manage large dimensional data sets, most available methods were not developed considering this scenario, and their computational costs prevent their use in big-data settings. Recently, some approaches—such as employing graphical processing units (to implement faster versions [36] of well-known algorithms) or parallelization using MapReduce, Hadoop, or Apache Spark—have been developed to solve this problem.

Supervised and Unsupervised Machine-Learning Methods

As mentioned in the Introduction section, digitization seems to be an unstoppable process in the fields of medicine and health, among others. For the analysis of these ever-increasing amounts of data, thus being able to derive information and knowledge from them, it is unavoidable to use automated methods, which should be scalable to keep pace with the crescent input loads. Machine learning [37] is a sub-discipline of the field of AI that consists of a set of methods and algorithms that can learn from data and devise models for different processes, such as pattern recognition or prediction, for example. However, machine learning [38–41] is not a new field: It appeared due to the early interest of AI researchers in determining if computers could learn directly from data without being programmed to perform specific tasks. However, due to the appearance of big data, machine learning is currently a hot buzzword, and an area of very active research, which—together with other factors— have brought a "new spring to the step" of the field of AI. Although many machine-learning algorithms have been around for several years, the ability to automatically apply complex mathematical calculations to large quantities of data in competitive time is a recent development. Machine-learning algorithms learn a function f: X→Y. This function belongs to a certain "family" [42] and maps the input domain of data X to a certain output domain Y (a prediction, for example). The five main types of problems machine learning can solve [43] include:

1. Classification, where the algorithm must assign unseen inputs to a series of predefined classes
2. Regression, where the focus is predicting a continuous output
3. Clustering, where inputs must be labeled into unknown groups (unlike classification)
4. Density estimation, where the goal is finding the distribution of a set of inputs
5. Dimensionality reduction, where inputs are simplified by mapping them to lower dimensional spaces

These tasks can also be classified according to the nature of the available learning data, which is provided in the form of examples (xi, yi) € X Y. In this case, three basic forms of learning can be distinguished:

- *Supervised learning*, where a set of known patterns are used for training, that is, the training data set is labeled, and thus yi is the corresponding ground truth for xi, and the aim of the process is to classify data based on that *a priori* knowledge. This previous knowledge of the data set's instance classes (i.e. the value to be predicted) is used to learn predictive models from the data set of examples in order to classify unseen instances. One important aspect of supervised classification is the evaluation of algorithms by means of an evaluation function, which usually quantifies the generalization ability of the classifier. That is, the goal is to minimize the error or loss function, f^ = Y × Y→R, which quantifies the difference between the predicted output and the real ground-truth label for that sample. In real-world problems, the true classification error is unknown, and thus so is its underlying probability distribution. Therefore, it must be estimated from the data. Because the loss cannot be minimized directly on the test instances and their labels, because typically these are not available at training times, supervised algorithms aim to construct functions that generalize well to previously unseen data, not to those that perform optimally on the given training data set (thus overfitting the data). In training and evaluating the devised model, two sources of data are employed. The parameters of the model are set based on the train data only, and if the test data are generated from the same underlying process that has generated the training data, an unbiased estimate of the generalization performance can be obtained by measuring the test-data performance of the trained model. It is important to recall that the test performance should not be used to adjust the model parameters because in this case the measure of performance will lose its independence. In particular, the mean squared error (MSE) is the measure typically employed for evaluating estimations made by the algorithms. The MSE is the second-order moment of the error, and therefore it incorporates both the variance and the bias of the estimator. The most common supervised learning tasks are classification and regression.

As an example, we show the results of classification for the data set Heart Disease (Cleveland) from the University of California-Irvine learning repository (http://archive.ics.uci.edu/ml/). We used the well-known platform Weka [25] to

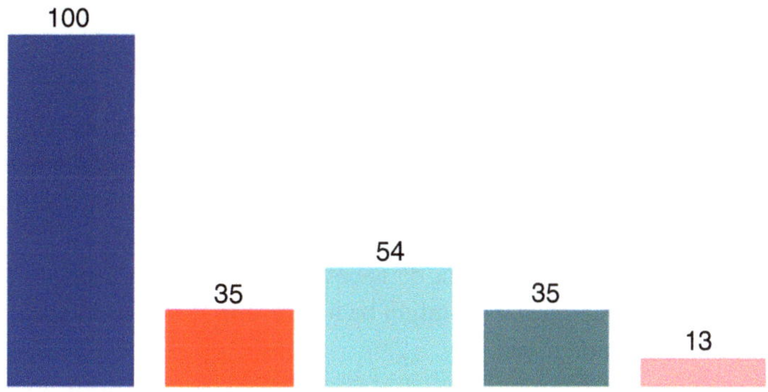

Fig. 37.6 Distribution of classes in the heart disease (Cleveland) data set

apply the selected algorithms. The Heart Disease data set has 13 attributes and 5 different classes as output (with 160, 35, 54, 35, and 13 samples respectively), with a total of 297 samples, because 6 of the 303 total have missing data and thus were eliminated from this study (see Fig. 37.6).

Several different classification algorithms can be used, and they show the results employing the RF classifier [44] because it is one of the state-of-the-art and more accurate classifiers. For this data set, the number of data available is not large, and thus if we must divide the data set into training (usually 80%) and test (20%) of the data, the estimation of the true error will be not very accurate. In these cases, a cross-validation procedure is often used. Cross-validation consists of making k partitions (folds) of the data, using k-1 for training and the remaining one for testing, and repeating the procedure k times until all folds have been used for testing the model. Thus, we are evaluating k models, and by averaging the results we have an idea of the variance of the learning algorithm with the variations in the training data and thus can obtain a more real approach to the error of the model. Cross-validation is conventionally applied with $k = 10$, although if the number of samples in each fold is low (usually < 30 [because this allows for approximating the binomial distribution of the number of correctly classified samples in a fold by normal distribution]), other k are used (most commonly $k = 5$). An extreme value is $k = n$, in which all samples except one are used to train, and that one is also used to test. This method is called "leaving-one-out" [17]. Using 10-fold cross-validation, the results obtained are 57.6% accurate; the confusion matrix is listed in Table 37.1.

Table 37.1 Results obtained after applying the RF classifier to the multi-class data set Heart Disease (Cleveland)

	Class 1	Class 2	Class 3	Class 4	Class 5
Class 1	149	2	7	2	0
Class 2	10	12	6	7	0
Class 3	33	8	2	11	0
Class 4	6	11	11	6	1
Class 5	4	2	1	4	2

Table 37.2 Results obtained after applying the RF classifier to the binary version of the data set Heart Disease (Cleveland)

	Class 0 (no heart disease)	Class 1 (heart disease)
Class 0	137	23
Class 1	26	111

If the data set is converted to a binary one, then only the class absence of heart disease (class 1, 160 samples) and the presence of heart disease (classes 2 through 5 with 137 samples total) are taken into account; the accuracy increases to 83.5% using the same classifier and with the contingency table listed in Table 37.2.

Thus, it can be seen that multi-class classification is a more difficult problem for the algorithm because the number of samples available is not enough for a good generalization. Another problem mentioned, one that is quite common in medical data bases, is the existence of missing values that should be treated accordingly [45, 45]. In our case, because the missing data are only present in three samples, we opted for the simplest operation, that is, elimination. Another common problem in medical data sets [37] are incorrectness (the presence of noise, probably due to sensor errors), inexactness (the presence of redundant data, which can imply the need of more complex models that in fact would not be necessary), and sparseness (sometimes there might be few records available for certain studies).

- Unsupervised learning, in which the training data set is unlabeled {xi} and the aim is to unravel the underlying similarities, obtains a plausible compact description of the data. An objective is used to quantify the accuracy of the description. In the case of unsupervised learning, the aim is to model the distribution p(x). The likelihood of the model to generate the data is a popular measure of the accuracy of the description.

The most common unsupervised learning task is clustering, in which the aim is the construction of a function (f), which partitions the training data set into k clusters. Several algorithms can be applied for clustering, but typically they work by assessing the similarity between instances by assigning similar samples to the same cluster and dissimilar ones to different cluster. Using a simple and well-known cluster algorithm, k-means, with the binary Heart Disease (Cleveland) data set and not supplying the information related to the class (because it is assumed that the data set is unlabeled in this case), the error of the algorithm is 20.2%; the results are listed in Table 37.3.

Other unsupervised tasks also exist—such as association rules (which build rules associating items that occur together with a certain frequency, discover patterns in the data, or can alternatively use correlation between real-valued variables—that are popular in machine learning. Then, to end we can say that machine learning is the task of building a model from data that generalizes a decision against a performance measure.

Most times, learning pipelines must include some kind of preprocessing operations, for example, noise must be eliminated, data discretized, and so on. Feature selection is also an important operation to consider because it can help with the generalization of machine-learning algorithms, thus improving their performance and perhaps the interpretability of the obtained results (see Fig. 37.7).

Table 37.3 Results of the k-means algorithm for clustering the Heart Disease (Cleveland) data set (binary)

	Class 0 (no heart disease)	Class 1 (heart disease)
Class 0	126	34
Class 1	26	111

A Case Study

In this section we present a case study in which we applied machine-learning methods to classify HF subtypes based on the work by Alonso-Betanzos et al. [10]. HF is a relatively common cardiac syndrome known for its severe sequelae, including death. The diagnosis is often only evident from the combination of symptoms (e.g., fatigue, dyspnea, etc.) and signs (e.g., ankle edema), plus clinical investigations—including the determination of LV size and chamber filling pressure—and information derived from specific biomarkers.

HF is manifested in at least two subtypes. The current paradigm distinguishes them by using metric EF and constraint for end-diastolic volume. Approximately half of all HF patients, often including women and elderly, exhibit HFpEF. Thus, as life expectancies continue to increase in western societies, the prevalence of HFpEF will continue to grow. However, compared with "classical" HF with decreased ejection fraction (HFrEF), only a limited spectrum of treatment modalities seems to be effective for improving the morbidity and mortality rates in patients with HFpEF.

Traditionally, EF has been widely applied to assess the severity of cardiac problems. In the

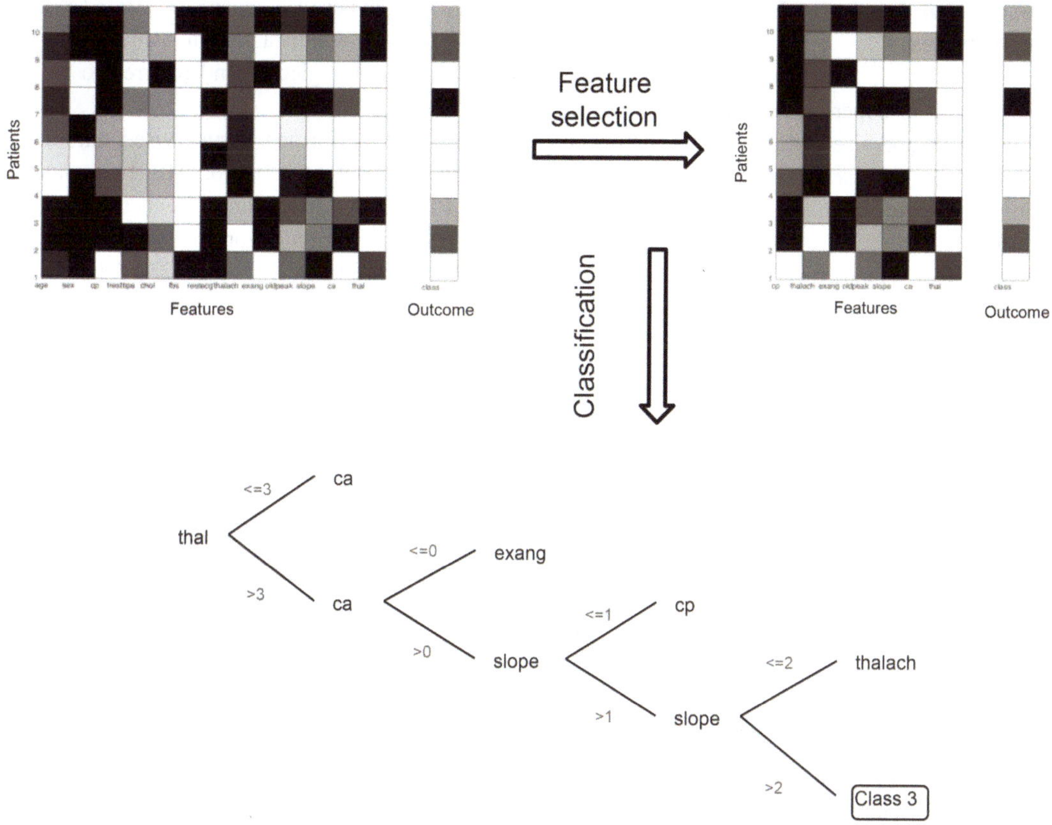

Fig. 37.7 Example of feature-selection and -classification process on the Heart Disease (Cleveland) data set. At the bottom, we can see a part of the C4.5 decision tree built for predicting class 3 (presence of heart disease). The variables used are Thal (Thalassemia), ca 8 (number of major vessels colored by fluoroscopy), exang (Exercise-Induced Angina), slope (Slope of the Peak Exercise ST Segment), cp (Chest Pain Type), and thalach (Maximum Heart Rate Achieved)

particular case of HF, EF is one of the many indicators to characterize the various aspects of the syndrome [47]. Typically, a low EF value corresponds with serious cardiac problems and a poor prognosis. Calculation of EF is performed by taking the ratio of two LV volume determinations during a cardiac cycle, namely, at the completion of filling and again at maximal contraction. Advised cut-off levels to distinguish HFrEF from HFpEF are clearly formulated, but they vary between 40% and 50%, which defines a linear divider. In addition, some studies opt for eliminating HF patients from consideration if 40 < EF < 50% ("gray zone" [currently often referred to as the "mid-range" phenotype]). Thus, there is a need to develop documented classification guidelines, solve gray-zone ambiguity, and formulate crisp delineation of the transition between phenotypes.

Subgroups of HF patients are located in at least two distinct regions on the basis of their end-systolic volume index and end-diastolic volume index; therefore, they are uniquely located within the LV-volume domain. In this case study, we present the application of machine-learning techniques to explore a more rational foundation for classifying two phenotypes of HF.

In summary, this case study addresses several relevant issues regarding the classification of HF patients: How can machine-learning models assist clinicians in the classification of major HF subtypes, the consequences of varying the cut-off values, and describe implications for borderline patients (in the gray zone).

Results for Applying Machine-Learning Techniques

The first set of experiments performed consisted of using unsupervised ML methods for the following three data sets (for more details, check Alonso-Betanzos et al. [10]):

- Data set 1: Data from real patients, a total of 48 instances where 35 belong to class HFpEF and 13 to class HFrEF.
- Data set 2: Data simulated with Monte Carlo, a total of 63 instances where 34 belong to class HFpEF and 29 to class HFrEF.
- Data set 3: Monte Carlo data generated as testing data, a total of 403 instances where 150 refer to class HFpEF; 137 belong to class HFrEF; and a third group (n = 116) still requires classification because on the basis of current guidelines they belong to neither HFpEF nor HFrEF. The third group is specifically introduced to challenge the universal validity of the current EF–EDVI paradigm, which favors a linear separator based on a fixed EF value.

Because we are using unsupervised algorithms, we focused on clustering, which consists of grouping a set of data in such a way that those belonging to the same group (called a "cluster") are more similar (in one sense or another [which is defined by the type of algorithm and its parameters]) to each other than to those in other clusters. To perform an unsupervised separation of the two major phenotypes of HF patients, we evaluated several different clustering algorithms, using different approaches all implemented in the Weka software tool [25]: K-means, Expectation Minimization (EM), and Sequential Information Bottleneck (sIB). As can be seen in Fig. 37.8, for Data sets 1 and 2 (real patients and simulated Monte Carlo, including only the two major types of patient subgroups), only the sIB algorithm tried to separate the samples using a similar approach as the current clinical guidelines. However, it can also be seen that the patients reclassified in an alternative manner (see squares in Fig. 37.8) and are all located within a region, which in some other studies is neglected and referred to as the "gray zone."

Then, we performed a supervised automatic classification of both major HF types using SVM PEGASOS, which implements a sequential minimal-optimization algorithm for training an SVM, also available in Weka [25]. The set of experiments carried out included different cut-off points (EF at 40%, 45%, 50%, and 55%) in the training data set (Data Set 1) to evaluate the consequences of adopting different criteria for defining major HF phenotypes and the ability of machine-learning methods to correctly classify the patients in each case. A summary of the results is listed in Table 37.4; more details can be found in Alonso-Betanzos et al. [10].

As can be seen, the results obtained after classifying the data with an SVM are quite satisfactory, with true-positive rates >0.90 in most of the cases. Moreover, we performed experiments to see how a machine-learning method would classify those patients belonging to the gray zone (i.e. the area where $40 < EF < 50\%$), which is not yet classified.

In Figs. 37.9 and 37.10, we see an example for cut-off 45%; the complete results can be checked in Alonso-Betanzos et al. [10]. In general, we can see that that the third group can largely be classified as HFpEF (although it varies when changing the cut-off). Interestingly, the separation does not follow the linear division as prescribed by the concept referring to a constant EF value for the cut-off. As seen, the points that are labeled differently seem to be located on the border between the main classes.

We can thus conclude that machine-learning models offer promise for making a computer-assisted distinction between the two major phenotypes of HF patients on the basis of ventricular-volume analysis. Moreover, selected machine-learning tools may assist during the classification of individual patients having measurements located in the (clinically often neglected) gray zone where $40 < EF < 50\%$.

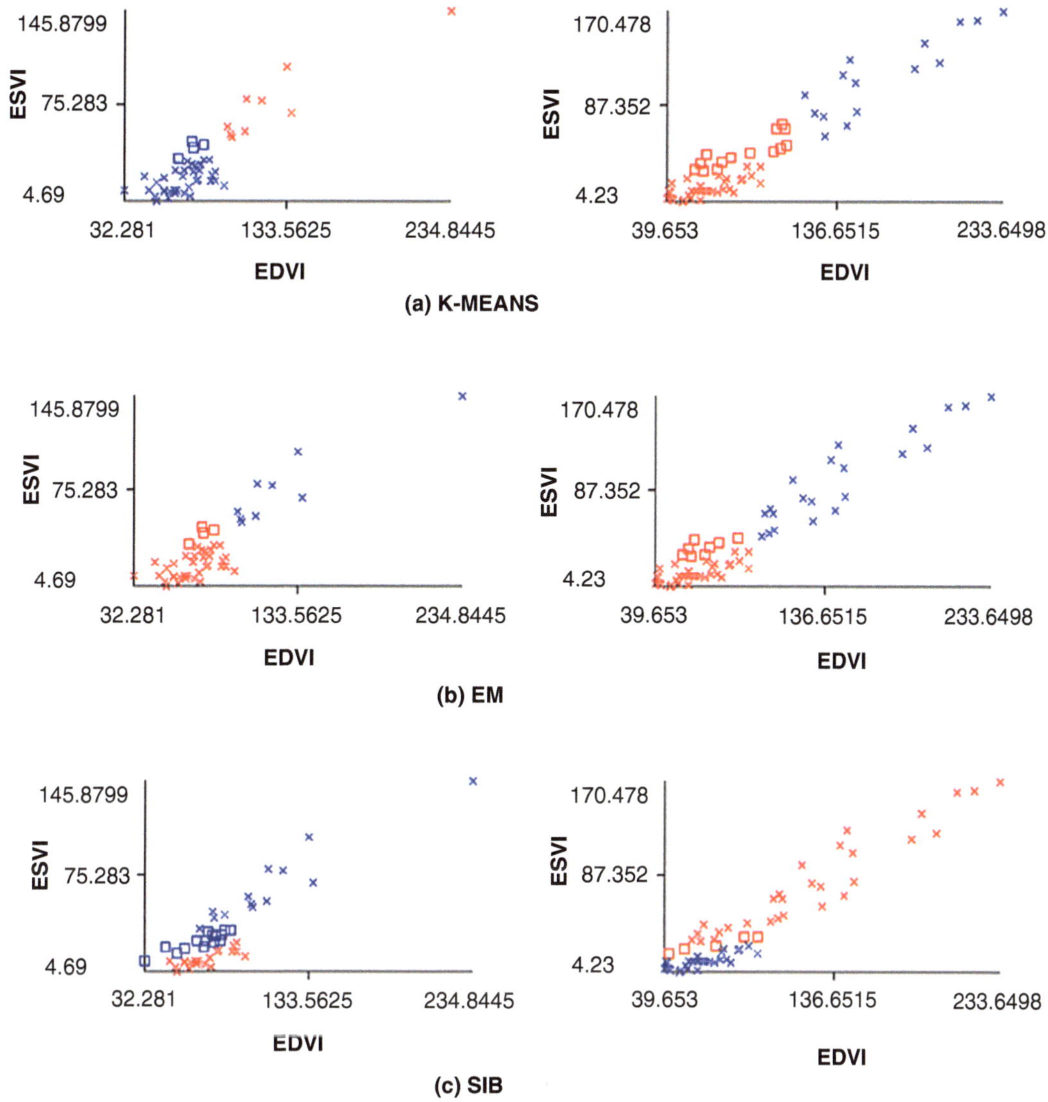

Fig. 37.8 Results of clustering analysis. Squares represent instances that are incorrectly assigned to a cluster. *EM* expectation minimization, *sIB* sequential-information bottleneck. (Image reprinted with permission from Alonso-Betanzos et al. [10])

Table 37.4 Results obtained after applying the RF classifier to the binary version of the data set Heart Disease (Cleveland)

EF CUT-OFF	40%	45%	50%	55%
TPR (HFpEF)	1 (169)	0.91 (161)	0.98 (147)	0.99 (136)
TPR (HFrEF)	0.87 (82)	0.96 (104)	0.97 (133)	0.98 (164)

TPR true-positive rate

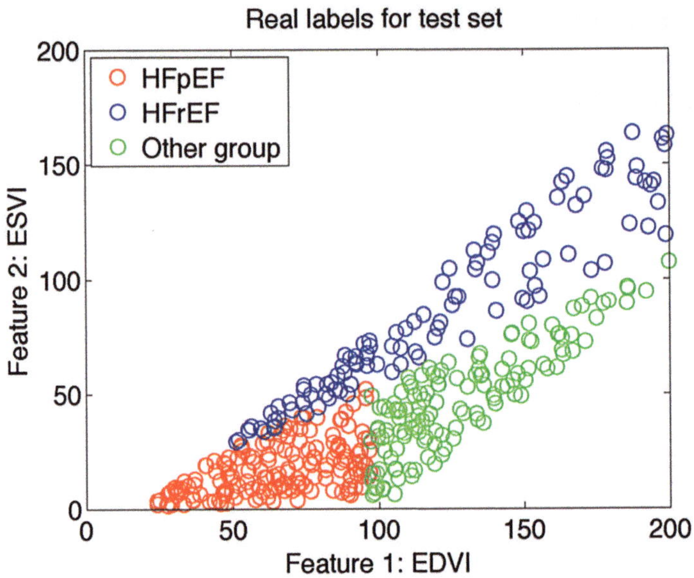

Fig. 37.9 Real labels for the test set ($n = 403$ [including the third group, which is newly assigned to either of the existing phenotypes]) for the 45% cut-off of the patient data as a training set (image reprinted with permission from Alonso-Betanzos et al. [10])

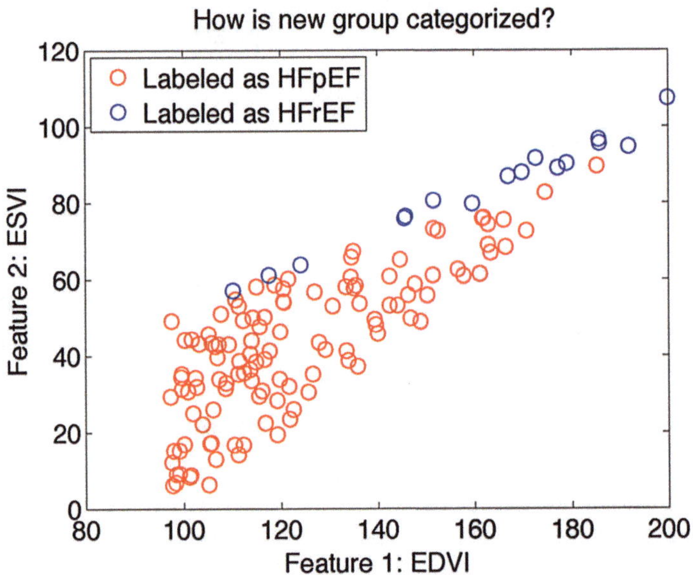

Fig. 37.10 Enlarged picture of the third group of data labels showing in detail that the algorithm applies a nonlinear division rather than a straight EF line (image reprinted with permission from Alonso-Betanzos et al. [10])

Future Directions

The need to apply machine-learning methods to the field of medicine has increased dramatically in recent years to face challenges brought by the advent of big data, for which it is necessary to cope with an unprecedented large number of features and samples. The increasingly decreased cost of storage technology has enabled us to store all kind of information about patients, with the aim of extracting useful and valuable information. However, these data are messy when they come out of the electronic health record of a patient and, even when they were collected from a study, the data might be of very different natures, which complicates the task of developing machine-learning algorithms.

Several research lines are open in this area. First is the problem of data distribution and data privacy. In some cases, information about patients is distributed across geographical and organizational boundaries (i.e. different hospitals), and it is not legal or affordable to gather it in a single location. In this case, it is necessary to develop distributed approaches for existing machine-learning methods that preserve privacy. It can be the case of a vertical distribution (each party has partial information about all the patients) or a horizontal distribution (each party involved in data sharing has information about all the variables but for different sets of patients). Although some approaches already exist that try to deal with this issue, there is still room for more works solving a problem that is especially important in the medical field. Another open question is the necessity of real-time processing in computer-assisted methods for the analysis of medical data. If a practitioner must wait a long time to obtain a recommendation from a computerized system, it is likely that he or she will stop using it. To avoid this, it is crucial to process and analyze data in real time, which can be performed either by accelerating the processing of the data (with feature selection or discretization methods) or by using online approaches (which are still relatively scarce in the literature). Anonymization, a process in which data sets are purged of personally identifying information, is another important challenge that has constituted an open research question for years. Although some recent attempts have been made to be able to use well-known privacy models, such as k-anonymity, in big data [48] and others, such as ε-differential privacy [49], there is still a long way to go because there are in fact very sophisticated methods [50] that can work backwards and re-identify individuals. Finally, another challenge is the visualization and interpretation of results. In recent years, several dimensionality-reduction techniques have been developed, aiming at a better visualization of the data. However, some of them have the limitation that the features being visualized are transformations of the original features, which greatly complicates the task of interpretation usually required by practitioners.

In conclusion, machine-learning models are much needed to help in clinical medicine. However, this enthusiasm does not generally match the level of actual activity in the field. It is very well to have machine-learning algorithms that are good at predicting and may help in clinical routine, but it is more complicated to employ them in the real world. We must make sure that they can be applied in a safe, responsible, and ethical way and—most of all—that people would accept being diagnosed by a machine rather than a person.

References

1. Hood L. A vision for personalized medicine. MIT Technology Review. Available at: https://www.technologyreview.com/s/417929/a-vision-forpersonalized-medicine/. Accessed 6 Apr 2017.
2. Deo RC. Machine learning in medicine. Circulation. 2015;132:1920–30.
3. Krittanawong C, et al. Future physicians in the era of precision cardiovascular medicine. Circulation. 2017;136:1572–4.
4. Shah SJ, Katz DH, Selvaraj S, Burke MA, Yancy CW, Gheorghiade M, Bonow RO, Huang CC, Deo RC. Phenomapping for novel classification of heart failure with preserved ejection fraction. Circulation. 2015;131:269–79.
5. Zhang X, Ambale-Venkatesh B, Bluemke DA, Cowan BR, Finn JP, Kadish AH, Lee DC, Lima JA, Hundley WG, Suinesiaputra A, Young AA, Medrano-Gracia P. Information maximizing component analysis of left ventricular remodeling due to myocardial infarction. J Transl Med. 2015 Nov 3;13 (1):343. https://doi.org/10.1186/s12967-015-0709-4.
6. Rumsfeld JS, Joynt KE, Maddox TM. Big data analytics to improve cardiovascular care: promise and challenges. Nat Rev Cardiol. 2016 Jun;13(6):350–9.
7. Tripoliti EE, Papadopoulos TG, Karanasiou GS, Naka KK, Fotiadis DI. Heart failure: diagnosis, severity estimation and prediction of adverse events through machine learning techniques. Comput Struct Biotechnol. 2017;15:26–47.
8. Kerkhof PLM, Alonso-Betanzos A, Moret-Bonillo V. Medical expert systems. In: Wiley encyclopedia of electrical and electronics engineering. Wiley; 2017.
9. Austin PC, Tu JV, Ho JE, Levy D, Lee DS. Using methods from the data mining and machine learning literature for disease classification and prediction: a case study examining classification of heart failure subtypes. J Clin Epidemiol. 2013;66:398–407.
10. Alonso-Betanzos A, Bolón-Canedo V, Heyndrickx GR, Kerkhof PLM. Exploring guidelines for classification of major heart failure subtypes by using machine learning. Clin Med Insights Cardiol. 2015;9(Suppl 1):57–71. https://doi.org/10.4137/CMC.S18744.
11. Narula S, Shameer K, Omar AMS, Dudley JT, Sengupta PP. Machine-learning algorithms to Automate morphological and functional assessments in 2D echocardiography. J Am Coll Cardiol. 2016 Nov;68(21):2287–95. https://doi.org/10.1016/j.jacc.2016.08.062.
12. Quinlan JR. Induction of decision trees. Mach Learn. 1986;1 (1):81–106.

13. Ahmad T, Testani JM, Desai NR. Can big data simplify the complexity of modern medicine? Prediction of right ventricular failure after left ventricular assist device support as a test case. JACC Heart Failure. 2016;4(9):722–4.
14. Motwani M, Dey D, Berman DS, Germano G, Achenbach S, Al-Mallah MH, Andreini D, Budoff MJ, Cademartiri F, Callister TQ, Chang HJ, Chinnaiyan K, Chow BJ, Cury RC, Delago A, Gomez M, Gransar H, Hadamitzky M, Hausleiter J, Hindoyan N, Feuchtner G, Kaufmann PA, Kim YJ, Leipsic J, Lin FY, Maffei E, Marques H, Pontone G, Raff G, Rubinshtein R, Shaw LJ, Stehli J, Villines TC, Dunning A, Min JK, Slomka PJ. Machine learning for prediction of all-cause mortality in patients with suspected coronary artery disease: a 5-year multicentre prospective registry analysis. Eur Heart J. 2017;38(7):500–7. https://doi.org/10.1093/eurheartj/ehw188.
15. Loghmanpour NA, Kormos RL, Kanwar MK, Teuteberg JJ, Murali S, Antaki JF. A Bayesian model to predict right ventricular failure following left ventricular assist device therapy. JACC Heart Failure. 2016;4(9):711–21.
16. Ng K, Steinhubl SR, de Filippi C, Dey S, Stewart WF. Early detection of heart failure using electronic health records: practical implications for time before diagnosis, data diversity, data quantity, and data density. Circ Cardiovasc Qual Outcomes. 2016;9:649–58.
17. Spertus JV, Normand ST, Wolf R, Cioffi M, Lovett A, Rose S. Assessing hospital performance after percutaneous coronary intervention using big data. Circ Cardiovasc Qual Outcomes. 2016;9:659–69.
18. Bourne PE, Bonazzi V, Dunn M, Green ED, Guyer M, Komatsoulis G, Larkin J, Russell B. The NIH big data to knowledge (BD2K) initiative. J Am Med Inform Assoc. 2015;22(6):1114. https://doi.org/10.1093/jamia/ocv136.
19. Shein E. Combating Cancer with data. Supercomputers will shift massive amounts of data in search of therapies that work. Commun ACM. 2017;60(5):10–2.
20. Erl T, Khattak W, Buhler P. Big data fundamentals. Concepts, drivers & techniques. Boston: Prentice-Hall; 2016.
21. Ankam V. Big data analytics. Birmingham: Packt Publishing; 2016.
22. Ramírez-Gallego S, García S, Mouriño-Talín H, Martínez-Rego D, Bolón-Canedo V, Alonso-Betanzos S, Benítez JM, Herrera F. Data discretization: taxonomy and big data challenge. Wiley Interdiscip Rev Data Min Knowl Disc. 2016;6 (1):5–21.
23. Yang Y, Webb GI. Discretization for naive-Bayes learning: managing discretization bias and variance. Mach Learn. 2009;74(1):39–74.
24. Quinlan JR. Induction of decision trees. Mach Learn. 1986;1 (1):81–106.
25. Frank E, Hall MA, Witten IH. The WEKA Workbench. Online appendix for "Data mining: practical machine learning tools and techniques". 4th ed. Cambridge, MA: Morgan Kaufmann; 2016.
26. Yang Y, Webb GI. Proportional k-interval discretization for naive-Bayes classifiers. In: European conference on machine learning. Berlin/Heidelberg: Springer; 2001. p. 564–575.
27. Irani KB. Multi-interval discretization of continuous-valued attributes for classification learning. 1993.
28. Ramírez-Gallego S, García S, Mourino-Talin H, Martínez-Rego D, Bolón-Canedo V, Alonso-Betanzos A, Herrera F. Distributed entropy minimization discretizer for big data analysis under apache spark. In Trustcom/BigDataSE/ISPA, 2015 IEEE. Vol. 2. IEEE; 2015, August. p. 33–40.
29. Zhai Y, Ong YS, Tsang IW. The emerging "Big Dimensionality". IEEE Comput Intell Mag. 2014;9(3):14–26.
30. Bolón-Canedo V, Sanchez-Marono N, Alonso-Betanzos A. Feature selection for high-dimensional data. Cham: Springer; 2015.
31. Hall MA. Correlation-based feature selection for machine learning. Doctoral dissertation, The University of Waikato. 1999.
32. Dash M, Liu H. Consistency-based search in feature selection. Artif Intell. 2003;151(1–2):155–76.
33. Kononenko I. Estimating attributes: analysis and extensions of RELIEF. In European conference on machine learning. Berlin/ Heidelberg: Springer; 1994, April. p. 171–82.
34. Peng H, Long F, Ding C. Feature selection based on mutual information criteria of max-dependency, max-relevance, and min-redundancy. IEEE Trans Pattern Anal Mach Intell. 2005;27(8):1226–38.
35. Guyon I, Weston J, Barnhill S, Vapnik V. Gene selection for cancer classification using support vector machines. Mach Learn. 2002;46(1):389–422.
36. Ramírez-Gallego S, Lastra I, Martínez-Rego D, Bolón-Canedo V, Benítez JM, Herrera F, Alonso-Betanzos A. Fast-mRMR: fast minimum redundancy maximum relevance algorithm for high-dimensional big data. Int J Intell Syst. 2016;0:1–19.
37. Bolón-Canedo V, Remeseiro B, Alonso-Betanzos A, Campilho A. Machine learning for medical applications. In: Proceedings of the European Symposium on Artificial Neural Networks (ESANN). 2016. p. 225–34.
38. Flach P. Machine learning. The art and science of algorithms that make sense of data. Cambridge: Cambridge University Press; 2012.
39. Murphy KP. Machine learning. A probabilistic perspective. Cambridge, MA: MIT Press; 2012.
40. Shalev-Schwartz S, Ben-David S. Understanding machine learning. Cambridge: Cambridge University Press; 2014.
41. Barber D. Bayesian reasoning and machine learning. Cambridge: Cambridge University Press; 2012.
42. Domingos P. The master algorithm. How the quest for the ultimate learning machine will remake our world. New York: Basic Books; 2015.
43. Bishop CM. Pattern recognition and machine learning. New York: Springer; 2006.
44. Fernández-Delgado M, Cernadas E, Barro S, Amorim D. Do we need hundreds of classifiers to solve real worls classification problems? J Mach Learn Res. 2014;15:3133–81.
45. Little RJA, Rubin DB. Statistical analysis with missing data. Chicester: Wiley; 2002.
46. Horton NJ, Kleiman KP. Much ado about nothing: a comparison of missing data methods and software to fit incomplete data regression model. The American Statiscian. 2007;61 (1):79–90.
47. Kerkhof PL. Characterizing heart failure in the ventricular volume domain. Clin Med Insights Cardiol. 2015;2015 (Suppl. 1):11.
48. Domingo-Ferrer J, Soria-Comas J. Anonymization in the time of big data. In: Domingo-Ferrer J, Pejić-Bach M, editors. Privacy in statistical databases. PSD 2016, Lecture notes in computer science, vol. 9867. Cham: Springer; 2016.
49. Domingo-Ferrer J, Soria-Comas J. From t-closeness to differential privacy and vice versa in data anonymization. Knowl-Based Syst. 2015;74:151–8. https://doi.org/10.1016/j.knosys.2014.11.011.
50. Sweeney L, Abu A, Winn J. Identifying participants in the Personal Genome Project by Name. Harvard University. Data Privacy Lab. White Paper 1021-1. April 24, 2013.

Genome-Wide Association Studies and Risk Scores for Coronary Artery Disease: Sex Biases

38

Sean G. Byars and Mike Inouye

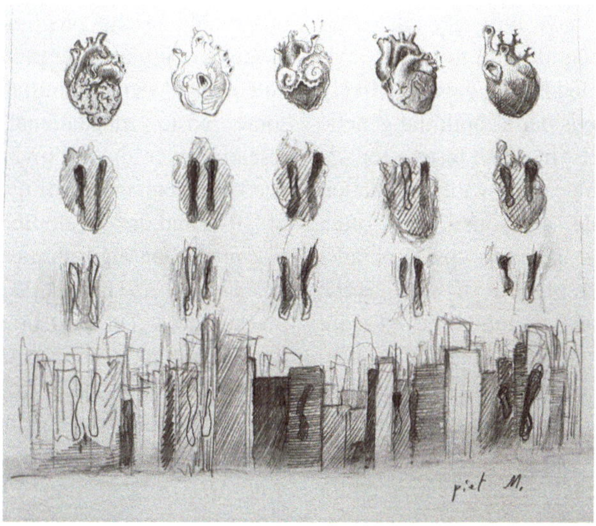

GWAS and risk scores. Art work by Piet Michiels, Leuven, Belgium

S. G. Byars (✉)
Melbourne Integrative Genomics, School of Biosciences, The University of Melbourne, Parkville, VIC, Australia

Department of Pathology, The University of Melbourne, Parkville, VIC, Australia
e-mail: sean.byars@unimelb.edu.au

M. Inouye
Melbourne Integrative Genomics, School of Biosciences, The University of Melbourne, Parkville, VIC, Australia

Department of Pathology, The University of Melbourne, Parkville, VIC, Australia

Systems Genomics Lab, Baker Heart and Diabetes Institute, Melbourne, VIC, Australia

Abstract

Phenotypic sex differences in coronary artery disease (CAD) and its risk factors have been apparent for many decades in basic and clinical research; however, whether these are also present at the gene level and thus influence genome-wide association and genetic risk prediction studies has often been ignored. From fundamental and medical standpoints, this is critically important to assess in order to fully understand the underlying genetic architecture that predisposes to CAD and better predict disease outcomes based on the interaction

P. L. M. Kerkhof, V. M. Miller (eds.), *Sex-Specific Analysis of Cardiovascular Function*,
Advances in Experimental Medicine and Biology 1065, https://doi.org/10.1007/978-3-319-77932-4_38

between genes, sex effects, and environment. In this chapter we aimed to (1) integrate the history and latest research from genome-wide association studies for CAD and clinical and genetic risk scores for prediction of CAD, (2) highlight sex-specific differences in these areas of research, and (3) discuss reasons why sex differences have often not been considered and, where present, why sex differences exist at genetic and phenotypic levels and how important they are for consideration in future research. While we find interesting examples of sex differences in effects of genetic variants on CAD, genome-wide association and genetic risk studies have typically not tested for sex-specific effects despite mounting evidence from diverse fields that these are likely very important to consider at both the genetic and phenotypic levels. In-depth testing for sex effects in large-scale genome-wide association studies that include autosomal and often excluded sex chromosomes alongside parallel improvements in resolution of sex-specific differences for risk factors and disease outcomes for CAD has the potential to substantially improve clinical and genetic risk prediction studies. Developing sex-tailored genetic risk scores as has been done recently for other disorders might be also warranted for CAD. In the era of precision medicine, this level of accuracy is essential for such a common and costly disease.

Keywords
GWAS · Coronary risk factors · Single nucleotide polymorphism · Sex-specific analysis · Genetic risk score · Women's Genome Health Study

History, Utility, and Sex Biases in Clinical Risk Scores for CAD

Historically, *cardiovascular disease (CVD)* risk assessment was initiated during the middle of the twentieth century spurred on by the spike in deaths in the USA due to *coronary artery disease (CAD)* that rose from ~10% of all deaths in 1900 to ~40% by 1960 [1]. This disease epidemic resulted in the *Framingham Heart Study (FHS)* being founded in 1947 by the US National Heart Institute in order to study and uncover causes of CVD in a relatively stable and well-defined community of mostly European ancestry [1]. Over a decade later in 1960, results from the FHS resulted in the initial concept of *coronary risk factors (CRFs*, see Fig. 38.1 for a comparison of their predictive performance in Framingham males and females), including, for example, high blood pressure, high cholesterol, history of smoking, *type 2 diabetes (T2D)*, age, and sex that are observable in the preclinical phase before the disease occurs and are predictive of future CVD outcomes. After this initial success in providing some basic translational disease indicators, research over the following 20 years provided deeper understanding of the multiplicative nature of CRFs and the development of a cardiovascular risk prediction algorithm called the *Framingham risk score (FRS)* pioneered again by Framingham investigators. The first Framingham risk prediction algorithm to predict CAD *(FRS-CAD)* was designed in 1998 to combine individual patient's risk factor information and produce a 10-year absolute risk score for angina, *myocardial infarction (MI)*, or death due to CAD in order to help clinicians decide whether therapy should be initiated [2]. In 2001, the Third Report from the Expert Panel on Detection, Evaluation, and Treatment of High Blood Cholesterol in Adults *(Adult Treatment Panel III)* recommended a modified 10-year Framingham risk score for CAD *(ATP III-FRS-CAD)* combined with a web-based calculator for use in patients exhibiting two or more risk factors to decide on thresholds for treatment of *low-density lipoprotein (LDL)* cholesterol [3, 4]. It was estimated that the use of these risk prediction scores increased the average life expectancy of CAD patients by 3 years in the USA [5].

While the FRS has often been the standard against which other subsequently developed CAD risk prediction tools are measured [2, 6–9], the general applicability of Framingham risk equations to modern populations has been

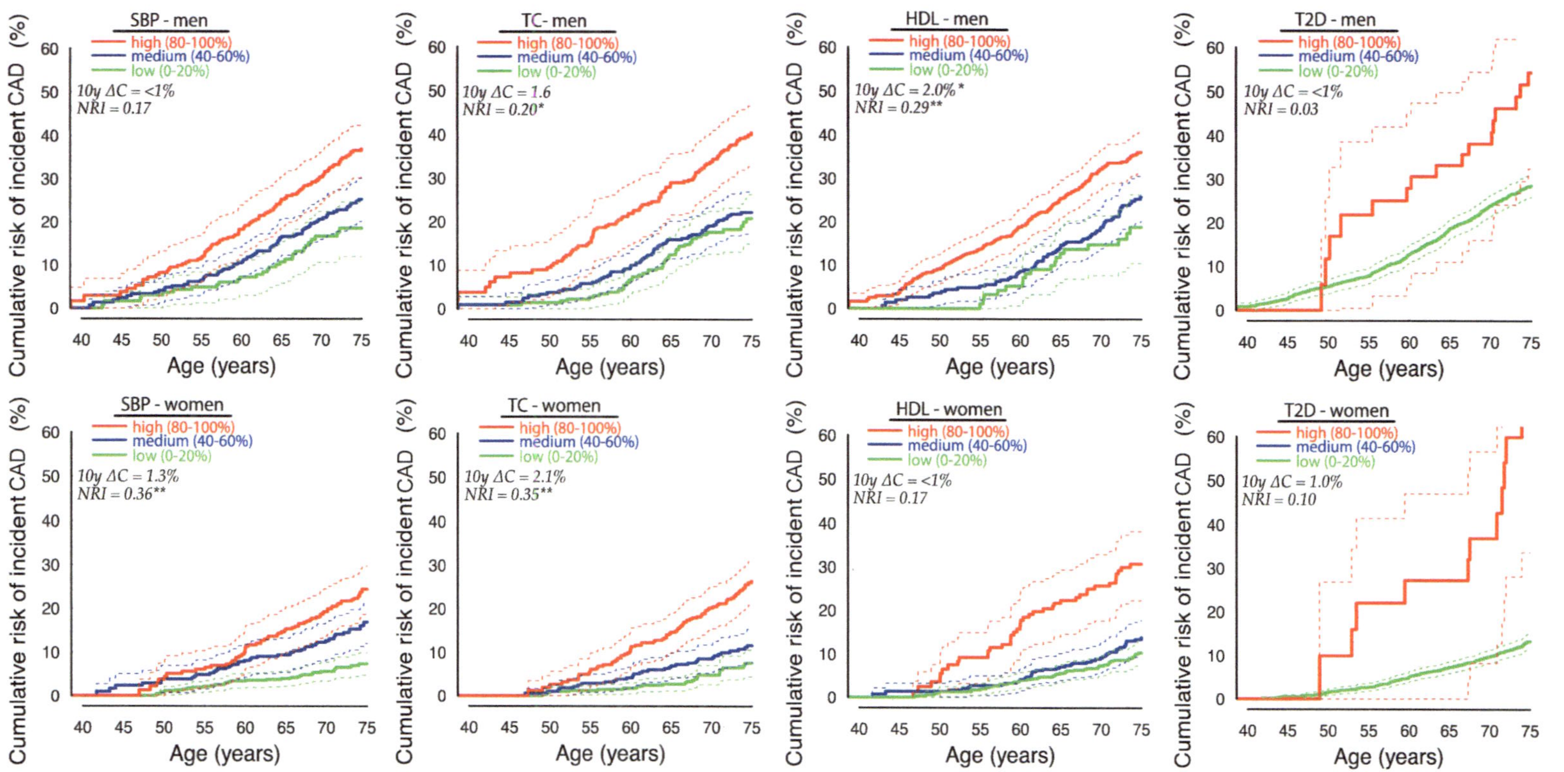

Fig. 38.1 Variation in lifetime coronary artery disease (CAD) risk depending on coronary risk factors (CRFs) in Framingham men and women. Each CRF was divided into quintiles (only highest, medium, lowest shown in Kaplan-Meier plots). Dashed lines are 95% CIs. Values shown in plots: Harrell's C-index difference (ΔC) and significance test (assessed using a correlated jackknife test) between two models (with/without each CRF); net reclassification improvement (NRI) and associated *p*-value. *SBP* systolic blood pressure, *TC* total cholesterol; HDL, high-density lipoprotein cholesterol; T2D, type 2 diabetes. *p*-values: $*p < 0.05$, $**p < 0.01$, $***p < 0.001$

debated over the last 10–15 years. This was partly because the FRSs are based on an ethnically and geographically limited, male-dominated sample when cardiovascular risk profiles were less well developed and also because many recent studies in diverse populations suggest that the FRS poorly classifies risk, particularly in women [10–15]. This is likely partly due to sex differences in progression and outcomes of CAD. For example, approximately 20% of all coronary events in women occur in the absence of major CRFs [16], and many women with CRFs do not experience coronary events [17]. As a consequence, the *Reynolds Risk Score (RRS)* for women was developed in 2007 that included C-reactive protein and parental history of premature CAD [18], and a revised Framingham risk score *(FRS-CVD)* that included increasingly common *atherosclerotic cardiovascular disease (ASCVD)* events (i.e., stroke) was developed in 2008 [13]. The RRS for women [18] tested 35 potential risk factors on 24,558 initially healthy US women, and the resulting accuracy of the clinical risk score was greatly improved demonstrating the benefits of a sex-specific algorithm. This was further supported by a study in 2008 where the RRS for men was tested and optimized in 10,724 US men, which also significantly improved global cardiovascular risk prediction [19]. While we largely discuss US-based risk scores here, other non-US-based consortia have also developed other risk prediction algorithms (all simply include sex as a covariate) including, for example, the German *Prospective Cardiovascular Munster Heart Study (PROCAM)* [20], the European *Systematic Coronary Risk Evaluation (SCORE)* system [21], and *QRISK2* [22].

Most recently, the *American Heart Association (AHA)* and the *American College of Cardiology (ACC)* developed a new ASCVD risk score *(AHA-ACC-ASCVD)* in 2013 (see Fig. 38.2 below for an indication of its performance in Framingham males and females) derived from four (including FHS data) racially and geographically diverse prospective cohort studies [23–26] that utilize the same traditional CRFs as the original FRSs while offering tailored equations for white and African-American men and women in order to improve and guide ASCVD risk-reducing therapy [26]. However, problems still reside in this most contemporary risk score as it overestimates risk in independent cohorts [27, 28]. Moreover, there is a certain level of misclassification of most clinical risk scores developed to date depending on sex. For example, DeFilippis et al. [29] compared performance of the AHA-ACC-ASCVD risk score with four other risk prediction equations (FRS-CAD, FRS-CVD, ATP III-FRS-CAD, RRS) utilizing the *Multi-Ethnic Study of Atherosclerosis (MESA)* study, which is a community-based, sex-balanced, multiethnic cohort consisting of 54% women, 42% European, 26% African-American, 20% Hispanic, and 12% Chinese. The authors found that the performance of the five risk scores to discriminate between those who did or did not have an ASCVD or CAD event was fairly similar. However, the FRS-CAD, FRS-CVD, ATP III-FRS-CAD, and AHA-ACC-ASCVD overestimated CAD and ASCVD risk by 37–154% in men. In comparison, the RRS estimated ASCVD and revascularization events in men relatively accurately (9% discordance). In women, the FRS-CVD accurately estimated risk (8% discordance), while the FRS-CAD, ATP III-FRS-CAD, and AHA-ACC-ASCVD overestimated risk (46–67% discordance) and the RRS underestimated (−21% discordance) the rate of events [29].

In summary, none of the most well-known ASCVD and CAD clinical risk prediction scores in the USA are highly accurate for both men and women [29]. More generally, CRFs underlying clinical risk scores provide only modest discrimination and cannot fully capture underlying risk [30]. This has clinical relevance for the current 10-year predicted risk threshold of 7.5% in the USA: over- or underestimation is likely to result in individuals with lower or higher risk receiving or not reaching the cutoff value for preventative treatment (statins), respectively [31–33]. Further testing and refinement of clinical risk scores, tailored to distinct populations and sex, is needed. While the RRS and AHA-ACC-ASCVD developed between 2007 and 2013 offer sex-specific clinical risk prediction algorithms, we now know that many more sex differences exist in the progression of CRFs and cardiovascular outcomes between men and women [34–37] (discussed

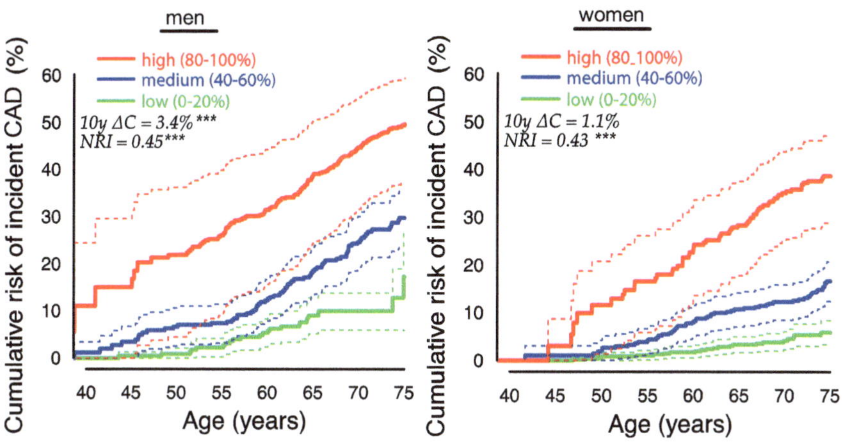

Fig. 38.2 Lifetime risk of coronary artery disease (CAD) depending on individual variation in the AHA-ACC-ASCVD clinical risk score in Framingham men and women. AHA-ACC-ASCVD scores were divided into quintiles (highest, medium, lowest shown in Kaplan-Meier plots). Dashed lines are 95% CIs. For Framingham males, integration of the AHA-ACC-ASCVD score into a simple CAD risk prediction model (first five principal components on genotypes) improved 10-year risk prediction (Harrell's C-index, C-index difference (ΔC) between two models assessed using a correlated jackknife test) by 3.4% ($p < 0.001$) with a net reclassification improvement (NRI) score of 0.45 ($p < 0.001$). For females, addition of the same score did not significantly improve 10-year C-index but did significantly improve the NRI (0.43, $p < 0.001$). *P*-values: $*p < 0.05$, $**p < 0.01$, $***p < 0.001$

below) that need to be accounted for in order to improve prediction of CAD and broader CVD. With the ever-increasing availability of large clinical and biobank-scale data as well as inclusion of genetic risk predictors (discussed below), these will lead to new more powerful clinical risk prediction equations.

GWAS and Sex-Specific Genetic Effects on CAD

Typical of common diseases, individual risk for CAD is modulated by the interplay between lifestyle and genetic factors [38]. Some of the first evidence that CAD had a genetic component was through early clinical observations [39], twin [40, 41] and cohort [42, 43] studies, which found increased CAD risk depending on whether there was direct or indirect family history of the disease and also estimated the heritability of CAD as ~50%. Genome-wide methods estimate CAD heritability between 40% and 50% [44]. Collectively these seminal studies laid the groundwork that helped drive the development of modern genetic tools to better define the underlying genetic architecture of CAD and begin to translate these findings into practice.

Historically, *linkage analysis studies* on families with a predisposition to early-onset CAD provided some initial insights into Mendelian or monogenic-like drivers of CAD. Linkage analysis can be an effective means of utilizing large multigenerational families to identify highly penetrant variants responsible for disease. Familial *hypercholesterolemia* defined by a heritable pattern of increased cholesterol and premature CAD was first described in 1938 [45], and in 1985 a deletion in the *LDLR* gene was found to be causative [46] demonstrating that a discrete mutation in a single gene predisposed to CAD. Subsequent family-based studies have identified further mutations in *APOB*, *PCSK9*, and *ABCG5/ABCG8* [47–50] that contribute to familial hypercholesterolemia, but beyond this condition that predisposes to CAD, the use of family studies to identify drivers of CAD has been limited. This was especially clear in linkage studies performed on families with more common presentations of CAD that failed to detect any susceptibility loci [51, 52]. Therefore, despite the tendency to cluster in families, these earlier studies highlighted the complex polygenic nature of CAD.

After the sequencing of the human genome in the late 1990s and early 2000s, which resulted in ~1.5 million *single nucleotide polymorphisms (SNPs)*, researchers used small-scale *candidate gene studies* (i.e., using previously identified potentially causative variants) to identify genes involved in disease; however, this underpowered approach with small sample sizes often resulted in false-positive findings [53, 54]. Many of the larger-scale candidate gene studies on CAD (or stroke) performed in the mid-late 2000s pooled data on both sexes and added sex as a covariate [55–60] without consideration for potential sex-specific effects. However, in 2002, a Japanese case-control study that tested variants predictive for MI found differences in the significance of those between males and females when they were analyzed separately [61], supporting the presence of sex-specific genetic risk factors for CAD. In 2008, a case-control study of over 14,000 Finnish individuals where authors tested for sex-specific effects on 46 candidate genes for CAD and CVD found that while some variants conferred risk in both sexes, others showed significant effects only in males (e.g., rs2069840 in *IL6* for CVD) and females (e.g., SNPs in *SELENOS* for stroke) with a larger number of sex-specific effects for women than men [62]. These studies showed some of the first evidence for sex differences in the effects of known variants in genes underlying CAD and CVD.

Subsequent further sequencing with the availability of cheaper genotyping chips (designed to capture common variation in populations) and careful documentation by the International HapMap Consortium of ~3.1 million SNPs in 270 individuals from four ethnic backgrounds in 2007 [63, 64] provided the necessary resources for initial *genome-wide association studies (GWAS)*. The first CAD GWAS was published in 2007, and during that time, three independent groups reported variants at the 9p21 locus with the risk allele significantly increasing risk of CAD in European populations by approximately 30% [65–67]. Following those initial discoveries, progressively larger sample sizes have been used to define many variants with relatively small effects that contribute significantly to CAD. Over the last 5 years, large international consortia have utilized massive sample sizes from mixed ethnic backgrounds in *GWAS meta-analyses* for CAD and have identified ~60 distinct genetic loci significantly linked to CAD [68–73]. Six more common variant loci were identified in 2017 [74], and *rare variant* (captured with whole exome/genome sequencing, i.e., not included on genotyping chips) *association studies* have identified at least nine genes in which mutations significantly alter CAD risk [75].

The majority of common CAD variants identified to date have a minor allele frequency of >5% in most worldwide populations, are located in nonprotein-coding regions of the genome highlighting their impact on CAD risk through gene regulation [73, 76], only cause modest increases in CAD risk (i.e., typically <20% change in risk per allele), collectively explain ~30–40% of CAD heritability [72, 73], and just under half (~40%) modulate CAD risk by influencing CRFs such as cholesterol, T2D, and hypertension [68, 72, 77] suggesting that the mechanisms whereby a majority of these loci mediate CAD risk remain unknown.

Since the first CAD GWAS was performed a decade ago (including previous candidate gene studies), researchers to date have typically not tested for sex-specific effects despite earlier findings (particularly in candidate gene studies) that these exist. Most studies have pooled men and women (adjusting for sex as a covariate) and typically account for population structure by including the first 5–10 principal components derived on genotypes from the sample. For example, one of the most recent GWAS meta-analyses of CAD [73] that used 60,801 CAD cases and 123,504 controls from a mix of males and females from different ancestral backgrounds (77% European, South (13% India and Pakistan) and East (6% China and Korea) Asian, Hispanic and African-American ~4%) included sex as a covariate. However, a slightly earlier study [72] published by the same consortium with a similar sample performed subgroup analyses by sex, and while they observed no higher risk trends for males vs females for any of the 46 genome-wide significant CAD risk loci identified in the main analysis (with sex added as a covariate), they did find one new locus (rs16986953 in gene desert, 1.3 Mb away from *APOB*) that reached genome-wide significance in males [72] but not females. Other non-CAD large-scale GWAS studies have revealed significant sex differences in contributing genetic variants [78–80] demonstrating that these are present in many polygenic traits and diseases.

A small subset of genome-wide CAD studies focusing on women exist but have often been limited by sample sizes with insufficient power to test for association [81]. For example, in 2012 Orozco et al. [82] tested for sex-specific associations in the Wellcome Trust Case Control Consortium of 399/1527 cases and 1492/1446 controls, for women and men, respectively, and found no CAD variants (previously identified from GWAS) were significant in women or men. More recent studies with larger sample sizes have found some convincing sex-specific genetic effects on CAD. Goodarzynejad et al. [83] in a sex-stratified study discovered a SNP in *SCARB1* (plasma membrane receptor for HDL) was associated with CAD in women but not men. A SNP in *CPS1* has also been found significantly associated with CAD in women but not men [84]. Liu et al. [78] found a SNP on 9p21 that showed a large ($p = 1.38 \times 10^{-8}$) male-specific association with CAD, but not females ($p = 0.167$). Taylor et al. [80] found sex-specific SNPs associated with risk factors (HDL, LDL) for CAD. This supports earlier findings from candidate gene studies that uncovered sex-specific genetic effects on CAD. Other recent GWAS have found autosomal variants contributing to complex traits in a sex-specific manner (e.g., adiposity [85]). Therefore, the investigation of sex-specific effects on CAD with larger better-powered studies is warranted. Resolving the full complement of shared and sex-specific CAD genetic loci will help provide a better understanding of the pathological pathways in progression and thus improve individually tailored treatments for this disease.

Notably, GWAS to date have almost exclusively focused on autosomal variation [86], even though the *X chromosome* is included on all current microarrays, and as a result, the role the X chromosome plays in CAD (and most complex diseases) remains largely unknown. This is mainly due to difficulty in accounting for analytical problems arising from the X's unique mode of inheritance and expression (i.e., dosage compensation in females), but recent analytical tools to perform XWAS [87] could begin to resolve this. The *Y chromosome* might also bring further understanding of sex-based differences in CVD development and outcomes [37]. For example, significant associations between Y chromosome variants and blood pressure have been found in different populations [88, 89], and some studies have suggested that blood pressure in men is largely determined by paternal (and not maternal) blood pressure status [90]. In males, the *Sry locus* of the Y chromosome contributes to hypertension through regulation of tyrosine hydroxylase, whereas in women, estrogen modulates the activity of this enzyme [36]. Therefore, modification of blood pressure, a key risk factor for CAD, is partly regulated by different genetic loci interacting with environmental stimuli that vary between males and females (i.e., changes in estrogen with pregnancy and menopause, discussed below). For an in-depth discussion of sex-specific differences in CVD related to the sex chromosomes, see Sampson et al. [37], Kling et al. [36], and Regitz-Zagrosek and Kararigas [35]. A recent study has also shown CVD is more prevalent among patients with pathogenic mtDNA mutations [91]. Given the seemingly incomplete autosomal nuclear genetic picture of CAD, variation in sex chromosomes and mtDNA could represent part of the missing heritability of CAD.

Genetic and Genomic Risk Prediction in CAD

While many genetic loci for CAD have now been identified, individually these are not clinically useful. For example, based on the most recent meta-analysis [73], the 9p21 locus remains the strongest genetic risk factor for CAD boasting an impressive significance of $p = 2.29 \times 10^{-98}$; however, the odds ratio (OR) for the effect allele is a modest 1.21 [92]. This represents the basis for recently developed *genetic risk scores (GRS)* that aim to improve CAD risk prediction. By combining GRS with or without CRFs (often referred to as modifiable risk factors because of the large environmental component to their variation), this should provide more accurate risk stratification of individuals [33, 93]. GRS are calculated essentially by counting the number of risk alleles (adjusted for effect size) inherited for each individual in a given population and provide a quantitative measure of genetic risk [94]. While family history might naturally serve as a substitute for genetic risk, individual risk loci and GRS predict CAD independent of family history, and the association between family history and CAD tends not to be attenuated after the inclusion of GRS [95]. This suggests that the proportion of heritability captured by current GRS is not currently high enough to erode the predictive power of family history and that family history of CAD partially reflects common environment effects.

The first attempt to include genetic markers for CVD prediction was in 2008 where variants associated with cholesterol were used to predict cardiovascular events [96] and extended later with nine significant independent risk variants for CAD [68]. These initial studies showed that individuals in the top GRS quintile had a twofold increased risk of CAD compared to those in the lowest GRS quintile [68] and paved the way for the exploration of various GRS based on different combinations of genetic loci for CAD in different populations [96–105]. For example, Mega et al. [106] utilized a GRS based on 27 CAD genetic risk variants *(GRS27)* and showed that genetic risk is independent of CRFs. Shortly after, Tada et al. [95] incorporated these same 27 with additional 23 CAD risk variants to form a 50-variant GRS *(GRS50)* and tested its performance on 23,595 individuals from the Malmo Diet and Cancer Study. They found their GRS50 further significantly improved CAD discrimination (i.e., C-index: measures how well a model discriminates individuals with(out) the outcome of interest; values range from 0.5 (performs no better than random) to 1.0 (perfectly distinguishes outcomes)) and reclassification (estimates improvement in performance of risk prediction model after additional risk variable is included). A recent study in 2016 by Abraham et al. [107] advanced the GRS by taking account of 49,310 small-effect SNPs predicted to underlie polygenic CAD to construct a genomic risk score *(GRS49k)*, and tested this on 12,676 Finnish men and women (Cox regression stratified by sex). The GRS49k confirmed the increased predictive value of including a large number of SNPs, by improving CAD risk prediction largely independently of CRFs (including family history) and significantly improved 10-year CAD risk prediction when it was combined with the ATP III-FRS-CAD or AHA-ACC-ASCVD clinical risk scores.

These studies show risk stratification provided by GRS is superior to clinical risk scores and the inclusion of many genetic risk variants significantly improves prediction of CAD outcomes. This could be particularly important in asymptomatic individuals without major risk factors. Because DNA is essentially stable across the lifetime of individuals, genetic risk can be obtained at birth. GRS could therefore be particularly useful for very early (i.e., young adults) intervention, which is especially important for a disease like CAD that begins to develop very early in life and is a result of lifetime modifiable exposures (diet, smoking, exercise) interacting on a background of low to high genomic risk. Adherence to a healthy lifestyle will be particularly important for young individuals with a high CAD GRS. For example, a recent analysis of high GRS individuals found a 46% attenuation of CAD risk in those with a favorable versus unfavorable lifestyle [38], and in another study a 50% reduction in CAD risk was achieved through statin therapy in those with high GRS [106]. Another recent study importantly showed the ability of GRS not only to predict incident CAD but also recurrent CAD independent of all CRFs and family history [106]. This shows that GRS hold great potential to improve both primary and secondary prevention strategies for CAD.

While the performance of GRS for CAD prediction is now being tested on less well-studied populations [98, 108–111], almost all GRS studies to date have not examined sex-specific effects in GRS. Of the few studies that have, the predictive performance and ability of GRS to stratify high- and low-risk individuals for CAD tends to perform much more poorly in women. For example, Paynter et al. [101] used a 101 SNP GRS based on GWAS loci significantly associated with CVD (or intermediate phenotypes) and found the GRS hazard ratio was initially weak (1.02, $p = 0.006$) but then not significantly associated with incident CVD in multivariable models in the *Women's Genome Health Study (WGHS)*. Compare this to some of the most recent GRS predicting CAD, where hazard ratios for GRS are typically highly significant ranging from 1.27 to 1.74 [102, 103, 107] in multivariable models. A recent study demonstrated that the ability of the GRS49k and CRFs to stratify high- and low-risk individuals for CAD performed much better in males than females (e.g., see Fig. 38.3 and Fig. S11–S13 in Abraham et al. [107]). This is also supported statistically by subsequent analyses on these cohorts that highlight male-female differences for key risk prediction performance indicators (see Fig. 38.3 for estimates).

Less successful prediction and stratification in females is perhaps not surprising in these studies given the combination of genetic and phenotypic biases that currently exist in CAD research. Current GRS are based on variants from GWAS that insufficiently capture sex-specific genetic effects on CVD and CAD (particularly females, discussed above), and clinical definitions for CAD outcomes (i.e., progression, presentation, age at onset) are male-dominated, resulting in lower specificity and accuracy of CAD outcomes captured and predicted in females. Testing a sex-specific GRS for CAD might be appropriate. Sex-tailored GRS have been constructed in other fields for conditions and diseases other than CAD. For example, female-specific GRS have been tested in relation to natural menopause [112], *polycystic ovary syndrome (PCOS)* [113], multiple sclerosis [114], and breast cancer [115], and a male-specific GRS has been tested for prostate cancer [116]. Indeed, developing sex-specific GRS can be useful when there are known differential sex differences in clinical phenotypes [117]. For CAD, the success of this will depend on better annotation of sex-specific variants identified in large GWAS that specifically search for sex-specific effects.

Improvement of Future Clinical and Genetic Risk Scores for CAD Through More Precise Definitions of Sex-Specific Effects

As is the case for many common human diseases, improving our understanding of the genetic architecture underlying CAD, which is essential for enhancing predictive performance of GRS, must also involve parallel improvements in the definitions of sex-specific differences in phenotypic presentations of CAD, which will reduce case/control misclassification and improve precision and risk stratification of males and females. The performance of clinical risk scores will also benefit from more refined CRFs that account for sex differences in their progression and interactions with environmental stimuli. As we discuss below, using exactly the same CAD and CRF definitions for males and females is not optimal in the era of precision medicine.

Firstly, because pathophysiology of CAD in men and women is quite different, grouping males and females and simply including sex as a covariate will reduce trait specificity and cause a level of bias in GWAS [118] from sex-based misclassification of CAD cases/controls. This bias is likely further compounded (especially for rarer genetic variants or variants with weaker yet significant effects) with the disproportionate ratio of men and women when they are pooled in GWAS. In general, analyzing sex differences in genetic associations for CAD risk is complicated because the age at onset is sex-dependent. For example, mortality from CAD tends to appear more commonly [119] and much earlier in males than females [120, 121]. Various factors have been proposed (e.g., psychological stressors, access and utilization for health care, inappropriate surveillance, and treatment guidelines for women) to contribute toward this mortality difference [36].

Progression of CAD shows some distinct differences by sex. For example, more severe structural/functional abnormalities in epicardial coronary arteries typify CAD progression in men, while more abnormal coronary reactivity, microvascular dysfunction, and plaque erosion/

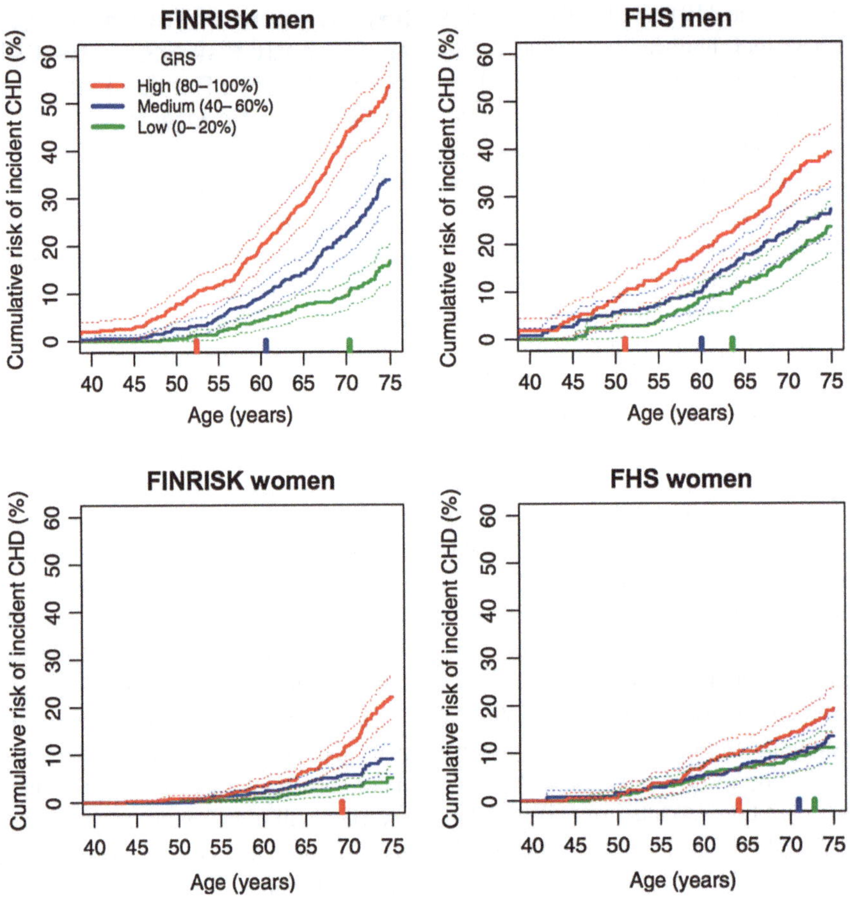

Fig. 38.3 Differences in the ability of the genomic risk score (GRS49k) to stratify men and women into high and low coronary artery disease (CAD) risk groups. Plots show Kaplan-Meier estimates for cumulative risk up to age 75 in 12,676 FINRISK and 3406 Framingham Heart Study (FHS) men and women. GRS were grouped into quintiles (only 0–20, 40–60, 80–100% shown), with vertical bars showing the age at which each group attained a cumulative CAD risk of 10%. Dashed lines are 95% CIs (Sourced from Abraham et al. [107]). For Framingham males, integration of the GRS with a simple CAD risk prediction model (first five principal components on genotypes) or complex model (first five PCs, total and HDL cholesterol, systolic blood pressure, T2D, smoking, treatment for high blood pressure, family history of CAD) improved 10-year risk prediction (Harrel's C-index, C-index difference between two models assessed using a correlated jackknife test) by 2.5% ($p = 0.02$) to 1.6% ($p = 0.05$), respectively. For Framingham females, integration of the GRS with simple and complex CAD risk prediction models did not improve 10-year risk prediction with C-index changing by 0.2% ($p = 0.78$) to 0.03% ($p = 0.86$), respectively. The same sample and methods as described in Abraham et al. [107] were used to obtain these estimates

distal microembolism typify CAD progression in women [122]. This has led to the recent suggestion that the definition of *ischemic heart disease (IHD)* might be more appropriate for women than CAD [81]. Heart failure in women is often more associated with *preserved ejection fraction (HFpEF)*, and if women suffer from left ventricular hypertrophy, they have a higher mortality risk compared to men [123]. Compared to women, men also present with more severe atherosclerosis in their coronary arteries. Correspondingly, MI typically occurs 10 years earlier in men [35]. Some recent studies suggest that most newly identified CAD GWAS loci mainly predict atherosclerosis [124]. Given that the pathophysiology of CAD can be roughly divided into three major processes (plaque formation (atherosclerosis), plaque rupture, thrombotic response to plague rupture [125]) and atherosclerosis tends to be more typical of male CAD progression [125], this suggests many newly identified CAD loci might be more predictive in males. This corresponds with differences of GRS to stratify high- and low-risk males and females for CAD outcomes [107] discussed in Sect. "Genetic and Genomic Risk Prediction in CAD" above.

Risk factors for CAD (i.e., cholesterol, BP, T2D) also show distinct differences between men and women [126–132] that often vary with age (see Fig. 38.4 for examples in Framingham males

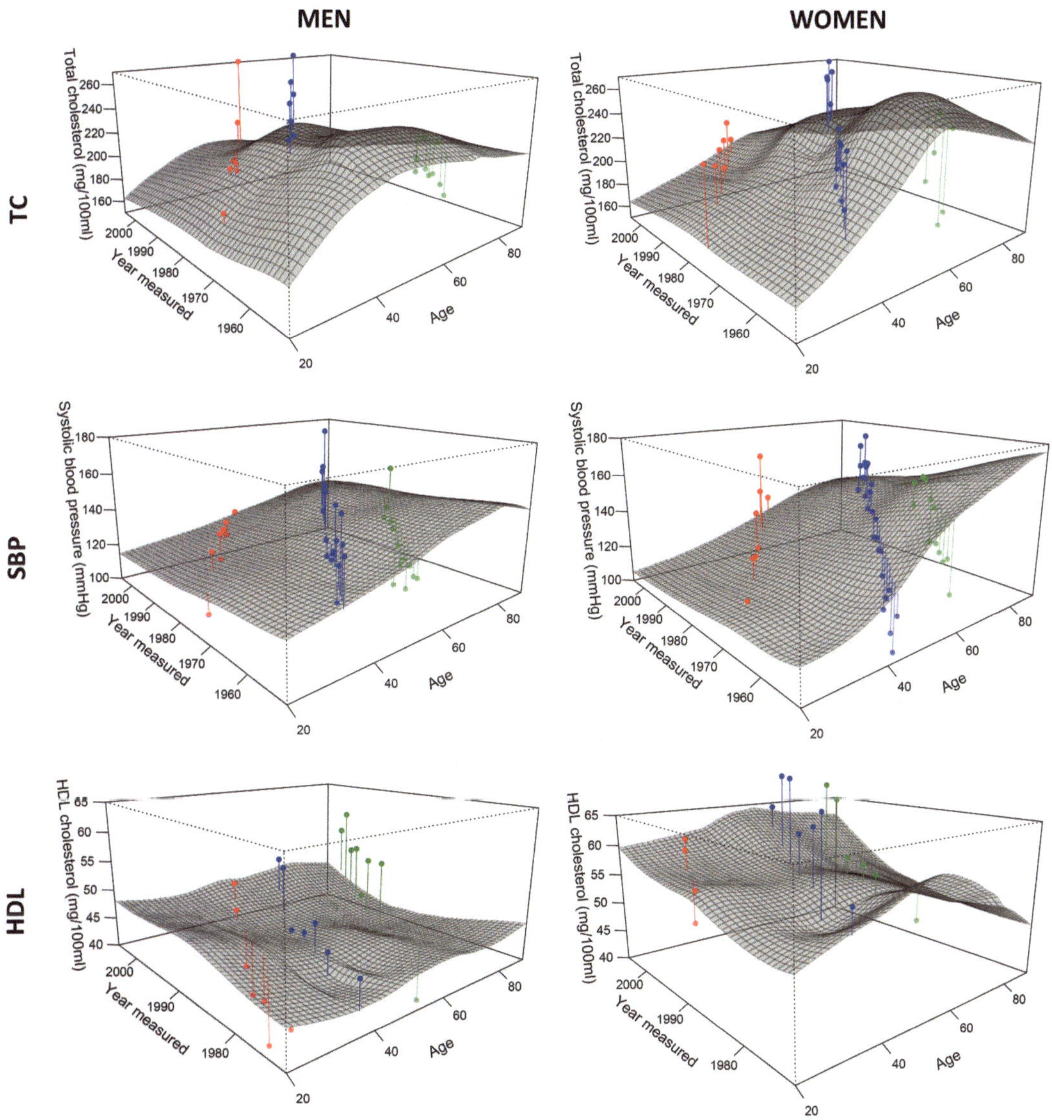

Fig. 38.4 Variation in coronary risk factors (CRFs) between the ages of 20 and 80, 1955–2005, for males and females in the Framingham Heart Study. Note the very different lifetime profiles of cholesterol for males and females. At around 60 years of age, total cholesterol (TC) tends to be much higher in females compared to males, and high-density lipoprotein (HDL) cholesterol levels in females tend to be generally higher in females compared to males. Plots show the 3-D surface accurately derived from many repeated individual measures (total cholesterol measured 37,317 times in 6605 males, 44,798 times in 7518 females; systolic blood pressure (SBP) measured 45,990 times in 6626 males, 58,690 in 7546 females; HDL cholesterol measured 17,207 times in 5334 males, 20,095 times in 6210 females) in a generalized additive model. Each plot shows example of individual repeated measures from three randomly chosen individuals

and females). For example, even though men often present with systemic hypertension at an earlier age, the incidence of pulmonary hypertension is greater in women than men [133–135]. Further, hypertension and T2D can be triggered by pregnancy and menopause [36]. For example, mothers experiencing pregnancy disorders such as pre-eclampsia and gestational diabetes subsequently have a much higher risk of hypertension and T2D [136–142]. Menopause also involves significant changes in estrogen levels, and estrogen modulates tyrosine hydroxylase, which is a key enzyme in the development of hypertension [36]. Correspondingly, premenopausal women have a reduced risk for hypertension and IHD compared to men, but this reverses after menopause [143, 144]. There is some evidence that women may be more protected from atherosclerosis before menopause [35], which is supported by the observation that women who have hormonal (estrogen) disturbances due to PCOS develop atherosclerosis and subsequent MI earlier in life [145, 146]. Men with a mutation in the estrogen receptor gene (*ESR1*) also have earlier CAD [147, 148].

In summary, more accurate classification of phenotypic sex differences in CAD risk factors, environmental stimuli that interact with these differently by sex, and sex differences in clinical presentations of CAD will in turn allow for further and more accurate identification of genes in female-focused and male-focused studies [149] and better predictive models that utilize GRS and CRFs. This will become increasingly important as we move further into the era of precision medicine.

Acknowledgments The Framingham Heart Study (FHS) dataset was obtained from dbGaP (phs000007), approved by the University of Melbourne Health Sciences Human Ethics Sub-Committee (HREC 1442186). The FHS is conducted and supported by the National Heart, Lung, and Blood Institute (NHLBI) in collaboration with Boston University (Contract No. N01-HC-25195 and HHSN268201500001I). This chapter was not prepared in collaboration with investigators of the Framingham Heart Study and does not necessarily reflect the opinions or views of the Framingham Heart Study, Boston University, or NHLBI. The funders had no role in study design, data collection and analysis, decision to publish, or preparation of the chapter.

References

1. Wong ND, Levy D. Legacy of the Framingham heart study: rationale, design, initial findings, and implications. Glob Heart. 2013;8:3–9.
2. Wilson PW, D'Agostino RB, Levy D, Belanger AM, Silbershatz H, et al. Prediction of coronary heart disease using risk factor categories. Circulation. 1998;97:1837–47.
3. National Cholesterol Education Program Expert Panel on Detection E, Treatment of High Blood Cholesterol in A. Third report of the National Cholesterol Education Program (NCEP) expert panel on detection, evaluation, and treatment of high blood cholesterol in adults (adult treatment panel III) final report. Circulation. 2002;106:3143–421.
4. Expert Panel on Detection, Evaluation, and Treatment of High Blood Cholesterol in Adults. Executive summary of the third report of the National Cholesterol Education Program (NCEP) Expert Panel on Detection. Evaluation, and treatment of high blood cholesterol in adults (adult treatment panel III). JAMA. 2001;285(19):2486–97.
5. Lenfant C. Shattuck lecture – clinical research to clinical practice – lost in translation? N Engl J Med. 2003;349:868–74.
6. Fletcher B, Berra K, Ades P, Braun LT, Burke LE, et al. Managing abnormal blood lipids: a collaborative approach. Circulation. 2005;112:3184–209.
7. Goldstein LB, Bushnell CD, Adams RJ, Appel LJ, Braun LT, et al. Guidelines for the primary prevention of stroke: a guideline for healthcare professionals from the American Heart Association/American Stroke Association. Stroke. 2011;42:517–84.
8. Mosca L, Benjamin EJ, Berra K, Bezanson JL, Dolor RJ, et al. Effectiveness-based guidelines for the prevention of cardiovascular disease in women – 2011 update: a guideline from the American Heart Association. Circulation. 2011;123:1243–62.
9. Pletcher MJ, Moran AE. Cardiovascular risk assessment. Med Clin North Am. 2017;101:673–88.
10. Bastuji-Garin S, Deverly A, Moyse D, Castaigne A, Mancia G, et al. The Framingham prediction rule is not valid in a European population of treated hypertensive patients. J Hypertens. 2002;20:1973–80.
11. Brindle P, Emberson J, Lampe F, Walker M, Whincup P, et al. Predictive accuracy of the Framingham coronary risk score in British men: prospective cohort study. BMJ. 2003;327:1267.
12. Cook NR, Paynter NP, Eaton CB, Manson JE, Martin LW, et al. Comparison of the Framingham and Reynolds risk scores for global cardiovascular risk prediction in the multiethnic Women's Health Initiative. Circulation. 2012;125:1748–56. S1741–1711.
13. D'Agostino RB Sr, Vasan RS, Pencina MJ, Wolf PA, Cobain M, et al. General cardiovascular risk profile for use in primary care: the Framingham heart study. Circulation. 2008;117:743–53.
14. Liu J, Hong Y, D'Agostino RB Sr, Wu Z, Wang W, et al. Predictive value for the Chinese population of the Framingham CHD risk assessment tool compared with the Chinese multi-provincial cohort study. JAMA. 2004;291:2591–9.
15. Ridker PM, Cook NR. Statins: new American guidelines for prevention of cardiovascular disease. Lancet. 2013;382:1762–5.
16. Khot UN, Khot MB, Bajzer CT, Sapp SK, Ohman EM, et al. Prevalence of conventional risk factors in patients with coronary heart disease. JAMA. 2003;290:898–904.

17. Greenland P, Knoll MD, Stamler J, Neaton JD, Dyer AR, et al. Major risk factors as antecedents of fatal and nonfatal coronary heart disease events. JAMA. 2003;290:891–7.
18. Ridker PM, Buring JE, Rifai N, Cook NR. Development and validation of improved algorithms for the assessment of global cardiovascular risk in women: the Reynolds risk score. JAMA. 2007;297:611–9.
19. Ridker PM, Paynter NP, Rifai N, Gaziano JM, Cook NR. C-reactive protein and parental history improve global cardiovascular risk prediction: the Reynolds Risk Score for men. Circulation. 2008;118:2243–51. 2244p following 2251.
20. Assmann G, Cullen P, Schulte H. Simple scoring scheme for calculating the risk of acute coronary events based on the 10-year follow-up of the prospective cardiovascular Munster (PROCAM) study. Circulation. 2002;105:310–5.
21. Conroy RM, Pyorala K, Fitzgerald AP, Sans S, Menotti A, et al. Estimation of ten-year risk of fatal cardiovascular disease in Europe: the SCORE project. Eur Heart J. 2003;24:987–1003.
22. Hippisley-Cox J, Coupland C, Vinogradova Y, Robson J, Minhas R, et al. Predicting cardiovascular risk in England and Wales: prospective derivation and validation of QRISK2. BMJ. 2008;336:1475–82.
23. The Atherosclerosis Risk in Communities (ARIC) Study: design and objectives. The ARIC investigators. Am J Epidemiol. 1989;129:687–702.
24. Fried LP, Borhani NO, Enright P, Furberg CD, Gardin JM, et al. The cardiovascular health study: design and rationale. Ann Epidemiol. 1991;1:263–76.
25. Friedman GD, Cutter GR, Donahue RP, Hughes GH, Hulley SB, et al. CARDIA: study design, recruitment, and some characteristics of the examined subjects. J Clin Epidemiol. 1988;41:1105–16.
26. Goff DC Jr, Lloyd-Jones DM, Bennett G, Coady S, D'Agostino RB, et al. 2013 ACC/AHA guideline on the assessment of cardiovascular risk: a report of the American College of Cardiology/American Heart Association task force on practice guidelines. Circulation. 2014;129:S49–73.
27. Kavousi M, Leening MJ, Nanchen D, Greenland P, Graham IM, et al. Comparison of application of the ACC/AHA guidelines, adult treatment panel III guidelines, and European Society of Cardiology guidelines for cardiovascular disease prevention in a European cohort. JAMA. 2014;311:1416–23.
28. Muntner P, Colantonio LD, Cushman M, Goff DC Jr, Howard G, et al. Validation of the atherosclerotic cardiovascular disease pooled cohort risk equations. JAMA. 2014;311:1406–15.
29. DeFilippis AP, Young R, Carrubba CJ, McEvoy JW, Budoff MJ, et al. An analysis of calibration and discrimination among multiple cardiovascular risk scores in a modern multiethnic cohort. Ann Intern Med. 2015;162:266–75.
30. Wang TJ, Gona P, Larson MG, Tofler GH, Levy D, et al. Multiple biomarkers for the prediction of first major cardiovascular events and death. N Engl J Med. 2006;355:2631–9.
31. Brindle P, Beswick A, Fahey T, Ebrahim S. Accuracy and impact of risk assessment in the primary prevention of cardiovascular disease: a systematic review. Heart. 2006;92:1752–9.
32. Stone NJ, Robinson JG, Lichtenstein AH, Bairey Merz CN, Blum CB, et al. 2013 ACC/AHA guideline on the treatment of blood cholesterol to reduce atherosclerotic cardiovascular risk in adults: a report of the American College of Cardiology/American Heart Association task force on practice guidelines. Circulation. 2014;129:S1–45.
33. Thanassoulis G, Vasan RS. Genetic cardiovascular risk prediction: will we get there? Circulation. 2010;122:2323–34.
34. Arain FA, Kuniyoshi FH, Abdalrhim AD, Miller VM. Sex/gender medicine. The biological basis for personalized care in cardiovascular medicine. Circ J. 2009;73:1774–82.
35. Regitz-Zagrosek V, Kararigas G. Mechanistic pathways of sex differences in cardiovascular disease. Physiol Rev. 2017;97:1–37.
36. Kling JM, Miller VM, Mulvagh SL. Transitions across a lifetime: unique cardiovascular physiology of women and relationship to cardiovascular disease risk. In: Spangenburg EE, editor. Integrative biology of Women's health. New York: Springer; 2013. p. 141–56.
37. Sampson AK, Jennings GL, Chin-Dusting JP. Y are males so difficult to understand?: a case where "X" does not mark the spot. Hypertension. 2012;59:525–31.
38. Khera AV, Emdin CA, Drake I, Natarajan P, Bick AG, et al. Genetic risk, adherence to a healthy lifestyle, and coronary disease. N Engl J Med. 2016;375:2349–58.
39. Gertler MM, Garn SM, White PD. Young candidates for coronary heart disease. J Am Med Assoc. 1951;147:621–5.
40. Marenberg ME, Risch N, Berkman LF, Floderus B, de Faire U. Genetic susceptibility to death from coronary heart disease in a study of twins. N Engl J Med. 1994;330:1041–6.
41. Zdravkovic S, Wienke A, Pedersen NL, Marenberg ME, Yashin AI, et al. Heritability of death from coronary heart disease: a 36-year follow-up of 20 966 Swedish twins. J Intern Med. 2002;252:247–54.
42. Lloyd-Jones DM, Nam BH, D'Agostino RB Sr, Levy D, Murabito JM, et al. Parental cardiovascular disease as a risk factor for cardiovascular disease in middle-aged adults: a prospective study of parents and offspring. JAMA. 2004;291:2204–11.
43. Murabito JM, Pencina MJ, Nam BH, D'Agostino RB Sr, Wang TJ, et al. Sibling cardiovascular disease as a risk factor for cardiovascular disease in middle-aged adults. JAMA. 2005;294:3117–23.
44. Won HH, Natarajan P, Dobbyn A, Jordan DM, Roussos P, et al. Disproportionate contributions of select genomic compartments and cell types to genetic risk for coronary artery disease. PLoS Genet. 2015;11:e1005622.
45. Müller C. Xanthomata, hypercholesterolemia, angina pectoris. J Intern Med. 1938;89:75–84.
46. Lehrman MA, Schneider WJ, Sudhof TC, Brown MS, Goldstein JL, et al. Mutation in LDL receptor: Alu-Alu recombination deletes exons encoding transmembrane and cytoplasmic domains. Science. 1985;227:140–6.
47. Abifadel M, Varret M, Rabes JP, Allard D, Ouguerram K, et al. Mutations in PCSK9 cause autosomal dominant hypercholesterolemia. Nat Genet. 2003;34:154–6.
48. Soria LF, Ludwig EH, Clarke HR, Vega GL, Grundy SM, et al. Association between a specific apolipoprotein B mutation and familial defective apolipoprotein B-100. Proc Natl Acad Sci USA. 1989;86:587–91.
49. Garcia CK, Wilund K, Arca M, Zuliani G, Fellin R, et al. Autosomal recessive hypercholesterolemia caused by mutations in a putative LDL receptor adaptor protein. Science. 2001;292:1394–8.
50. Berge KE, Tian H, Graf GA, Yu L, Grishin NV, et al. Accumulation of dietary cholesterol in sitosterolemia caused by mutations in adjacent ABC transporters. Science. 2000;290:1771–5.
51. Musunuru K, Kathiresan S. Genetics of coronary artery disease. Annu Rev Genomics Hum Genet. 2010;11:91–108.

52. Risch NJ. Searching for genetic determinants in the new millennium. Nature. 2000;405:847–56.
53. Tabor HK, Risch NJ, Myers RM. Candidate-gene approaches for studying complex genetic traits: practical considerations. Nat Rev Genet. 2002;3:391–7.
54. Wacholder S, Chanock S, Garcia-Closas M, El Ghormli L, Rothman N. Assessing the probability that a positive report is false: an approach for molecular epidemiology studies. J Natl Cancer Inst. 2004;96:434–42.
55. McCarthy JJ, Parker A, Salem R, Moliterno DJ, Wang Q, et al. Large scale association analysis for identification of genes underlying premature coronary heart disease: cumulative perspective from analysis of 111 candidate genes. J Med Genet. 2004;41:334–41.
56. Ozaki K, Ohnishi Y, Iida A, Sekine A, Yamada R, et al. Functional SNPs in the lymphotoxin-alpha gene that are associated with susceptibility to myocardial infarction. Nat Genet. 2002;32:650–4.
57. Pare G, Serre D, Brisson D, Anand SS, Montpetit A, et al. Genetic analysis of 103 candidate genes for coronary artery disease and associated phenotypes in a founder population reveals a new association between endothelin-1 and high-density lipoprotein cholesterol. Am J Hum Genet. 2007;80:673–82.
58. Topol EJ, McCarthy J, Gabriel S, Moliterno DJ, Rogers WJ, et al. Single nucleotide polymorphisms in multiple novel thrombospondin genes may be associated with familial premature myocardial infarction. Circulation. 2001;104:2641–4.
59. Yamada Y, Matsuo H, Segawa T, Watanabe S, Kato K, et al. Assessment of genetic risk for myocardial infarction. Thromb Haemost. 2006;96:220–7.
60. Yamada Y, Metoki N, Yoshida H, Satoh K, Ichihara S, et al. Genetic risk for ischemic and hemorrhagic stroke. Arterioscler Thromb Vasc Biol. 2006;26:1920–5.
61. Yamada Y, Izawa H, Ichihara S, Takatsu F, Ishihara H, et al. Prediction of the risk of myocardial infarction from polymorphisms in candidate genes. N Engl J Med. 2002;347:1916–23.
62. Silander K, Alanne M, Kristiansson K, Saarela O, Ripatti S, et al. Gender differences in genetic risk profiles for cardiovascular disease. PLoS One. 2008;3:e3615.
63. International HapMap C. The international HapMap Project. Nature. 2003;426:789–96.
64. International HapMap C, Frazer KA, Ballinger DG, Cox DR, Hinds DA, et al. A second generation human haplotype map of over 3.1 million SNPs. Nature. 2007;449:851–61.
65. Samani NJ, Erdmann J, Hall AS, Hengstenberg C, Mangino M, et al. Genomewide association analysis of coronary artery disease. N Engl J Med. 2007;357:443–53.
66. Helgadottir A, Thorleifsson G, Manolescu A, Gretarsdottir S, Blondal T, et al. A common variant on chromosome 9p21 affects the risk of myocardial infarction. Science. 2007;316:1491–3.
67. McPherson R, Pertsemlidis A, Kavaslar N, Stewart A, Roberts R, et al. A common allele on chromosome 9 associated with coronary heart disease. Science. 2007;316:1488–91.
68. Myocardial Infarction Genetics C, Kathiresan S, Voight BF, Purcell S, Musunuru K, et al. Genome-wide association of early-onset myocardial infarction with single nucleotide polymorphisms and copy number variants. Nat Genet. 2009;41:334–41.
69. Erdmann J, Grosshennig A, Braund PS, Konig IR, Hengstenberg C, et al. New susceptibility locus for coronary artery disease on chromosome 3q22.3. Nat Genet. 2009;41:280–2.
70. Coronary Artery Disease Genetics C. A genome-wide association study in Europeans and south Asians identifies five new loci for coronary artery disease. Nat Genet. 2011;43:339–44.
71. Consortium IKC. Large-scale gene-centric analysis identifies novel variants for coronary artery disease. PLoS Genet. 2011;7:e1002260.
72. Consortium CD, Deloukas P, Kanoni S, Willenborg C, Farrall M, et al. Large-scale association analysis identifies new risk loci for coronary artery disease. Nat Genet. 2013;45:25–33.
73. Nikpay M, Goel A, Won HH, Hall LM, Willenborg C, et al. A comprehensive 1,000 genomes-based genome-wide association meta-analysis of coronary artery disease. Nat Genet. 2015;47:1121–30.
74. Webb TR, Erdmann J, Stirrups KE, Stitziel NO, Masca NG, et al. Systematic evaluation of pleiotropy identifies 6 further loci associated with coronary artery disease. J Am Coll Cardiol. 2017;69:823–36.
75. Do R, Stitziel NO, Won HH, Jorgensen AB, Duga S, et al. Exome sequencing identifies rare LDLR and APOA5 alleles conferring risk for myocardial infarction. Nature. 2015;518:102–6.
76. Maurano MT, Humbert R, Rynes E, Thurman RE, Haugen E, et al. Systematic localization of common disease-associated variation in regulatory DNA. Science. 2012;337:1190–5.
77. Schunkert H, Konig IR, Kathiresan S, Reilly MP, Assimes TL, et al. Large-scale association analysis identifies 13 new susceptibility loci for coronary artery disease. Nat Genet. 2011;43:333–8.
78. Liu LY, Schaub MA, Sirota M, Butte AJ. Sex differences in disease risk from reported genome-wide association study findings. Hum Genet. 2012;131:353–64.
79. Morris AP, Voight BF, Teslovich TM, Ferreira T, Segre AV, et al. Large-scale association analysis provides insights into the genetic architecture and pathophysiology of type 2 diabetes. Nat Genet. 2012;44:981–90.
80. Taylor KC, Carty CL, Dumitrescu L, Buzkova P, Cole SA, et al. Investigation of gene-by-sex interactions for lipid traits in diverse populations from the population architecture using genomics and epidemiology study. BMC Genet. 2013;14:33.
81. Kavousi M, Bielak LF, Peyser PA. Genetic research and Women's heart disease: a primer. Curr Atheroscler Rep. 2016;18:67.
82. Orozco G, Ioannidis JP, Morris A, Zeggini E, Consortium D. Sex-specific differences in effect size estimates at established complex trait loci. Int J Epidemiol. 2012;41:1376–82.
83. Goodarzynejad H, Boroumand M, Behmanesh M, Ziaee S, Jalali A. The rs5888 single nucleotide polymorphism in scavenger receptor class B type 1 (SCARB1) gene and the risk of premature coronary artery disease: a case-control study. Lipids Health Dis. 2016;15:7.
84. Hartiala JA, Tang WH, Wang Z, Crow AL, Stewart AF, et al. Genome-wide association study and targeted metabolomics identifies sex-specific association of CPS1 with coronary artery disease. Nat Commun. 2016;7:10558.
85. Shungin D, Winkler TW, Croteau-Chonka DC, Ferreira T, Locke AE, et al. New genetic loci link adipose and insulin biology to body fat distribution. Nature. 2015;518:187–96.
86. Wise AL, Gyi L, Manolio TA. eXclusion: toward integrating the X chromosome in genome-wide association analyses. Am J Hum Genet. 2013;92:643–7.
87. Gao F, Chang D, Biddanda A, Ma L, Guo Y, et al. XWAS: a software toolset for genetic data analysis and association studies of the X chromosome. J Hered. 2015;106:666–71.

88. Charchar FJ, Tomaszewski M, Padmanabhan S, Lacka B, Upton MN, et al. The Y chromosome effect on blood pressure in two European populations. Hypertension. 2002;39:353–6.
89. Ellis JA, Stebbing M, Harrap SB. Association of the human Y chromosome with high blood pressure in the general population. Hypertension. 2000;36:731–3.
90. Uehara Y, Shin WS, Watanabe T, Osanai T, Miyazaki M, et al. A hypertensive father, but not hypertensive mother, determines blood pressure in normotensive male offspring through body mass index. J Hum Hypertens. 1998;12:441–5.
91. Bray AW, Ballinger SW. Mitochondrial DNA mutations and cardiovascular disease. Curr Opin Cardiol. 2017. https://doi.org/10.1097/HCO.0000000000000383.
92. Assimes TL, Roberts R. Genetics: implications for prevention and Management of Coronary Artery Disease. J Am Coll Cardiol. 2016;68:2797–818.
93. Manolio TA. Genomewide association studies and assessment of the risk of disease. N Engl J Med. 2010;363:166–76.
94. Chatterjee N, Shi J, Garcia-Closas M. Developing and evaluating polygenic risk prediction models for stratified disease prevention. Nat Rev Genet. 2016;17:392–406.
95. Tada H, Melander O, Louie JZ, Catanese JJ, Rowland CM, et al. Risk prediction by genetic risk scores for coronary heart disease is independent of self-reported family history. Eur Heart J. 2016;37:561–7.
96. Kathiresan S, Melander O, Anevski D, Guiducci C, Burtt NP, et al. Polymorphisms associated with cholesterol and risk of cardiovascular events. N Engl J Med. 2008;358:1240–9.
97. Anderson JL, Horne BD, Camp NJ, Muhlestein JB, Hopkins PN, et al. Joint effects of common genetic variants from multiple genes and pathways on the risk of premature coronary artery disease. Am Heart J. 2010;160:250–6. e253.
98. Qi L, Ma J, Qi Q, Hartiala J, Allayee H, et al. Genetic risk score and risk of myocardial infarction in Hispanics. Circulation. 2011;123:374–80.
99. Ripatti S, Tikkanen E, Orho-Melander M, Havulinna AS, Silander K, et al. A multilocus genetic risk score for coronary heart disease: case-control and prospective cohort analyses. Lancet. 2010;376:1393–400.
100. Thanassoulis G, Peloso GM, Pencina MJ, Hoffmann U, Fox CS, et al. A genetic risk score is associated with incident cardiovascular disease and coronary artery calcium: the Framingham heart study. Circ Cardiovasc Genet. 2012;5:113–21.
101. Paynter NP, Chasman DI, Pare G, Buring JE, Cook NR, et al. Association between a literature-based genetic risk score and cardiovascular events in women. JAMA. 2010;303:631–7.
102. Tikkanen E, Havulinna AS, Palotie A, Salomaa V, Ripatti S. Genetic risk prediction and a 2-stage risk screening strategy for coronary heart disease. Arterioscler Thromb Vasc Biol. 2013;33:2261–6.
103. Ganna A, Magnusson PK, Pedersen NL, de Faire U, Reilly M, et al. Multilocus genetic risk scores for coronary heart disease prediction. Arterioscler Thromb Vasc Biol. 2013;33:2267–72.
104. Hughes MF, Saarela O, Stritzke J, Kee F, Silander K, et al. Genetic markers enhance coronary risk prediction in men: the MORGAM prospective cohorts. PLoS One. 2012;7:e40922.
105. Weijmans M, de Bakker PI, van der Graaf Y, Asselbergs FW, Algra A, et al. Incremental value of a genetic risk score for the prediction of new vascular events in patients with clinically manifest vascular disease. Atherosclerosis. 2015;239:451–8.
106. Mega JL, Stitziel NO, Smith JG, Chasman DI, Caulfield MJ, et al. Genetic risk, coronary heart disease events, and the clinical benefit of statin therapy: an analysis of primary and secondary prevention trials. Lancet. 2015;385:2264–71.
107. Abraham G, Havulinna AS, Bhalala OG, Byars SG, De Livera AM, et al. Genomic prediction of coronary heart disease. Eur Heart J. 2016;37:3267–78.
108. Shahid SU, Shabana, Cooper JA, Beaney KE, Li K, et al. Genetic risk analysis of coronary artery disease in Pakistani subjects using a genetic risk score of 21 variants. Atherosclerosis. 2017;258:1–7.
109. Joseph PG, Pare G, Asma S, Engert JC, Yusuf S, et al. Impact of a genetic risk score on myocardial infarction risk across different ethnic populations. Can J Cardiol. 2016;32:1440–6.
110. Shanker J, Arvind P, Jambunathan S, Nair J, Kakkar V. Genetic analysis of the 9p21.3 CAD risk locus in Asian Indians. Thromb Haemost. 2014;111:960–9.
111. Sotos-Prieto M, Baylin A, Campos H, Qi L, Mattei J. Lifestyle cardiovascular risk score, genetic risk score, and myocardial infarction in Hispanic/Latino adults living in Costa Rica. J Am Heart Assoc. 2016;5(12):e004067.
112. Pasquale LR, Aschard H, Kang JH, Bailey JN, Lindstrom S, et al. Age at natural menopause genetic risk score in relation to age at natural menopause and primary open-angle glaucoma in a US-based sample. Menopause. 2017;24:150–6.
113. Lee H, Oh JY, Sung YA, Chung HW. A genetic risk score is associated with polycystic ovary syndrome-related traits. Hum Reprod. 2016;31:209–15.
114. Xia Z, Steele SU, Bakshi A, Clarkson SR, White CC, et al. Assessment of early evidence of multiple sclerosis in a prospective study of asymptomatic high-risk family members. JAMA Neurol. 2017;74:293–300.
115. Vachon CM, Pankratz VS, Scott CG, Haeberle L, Ziv E, et al. The contributions of breast density and common genetic variation to breast cancer risk. J Natl Cancer Inst. 2015;107:dju397.
116. Helfand BT, Kearns J, Conran C, Xu J. Clinical validity and utility of genetic risk scores in prostate cancer. Asian J Androl. 2016;18:509–14.
117. Carayol J, Schellenberg GD, Dombroski B, Genin E, Rousseau F, et al. Autism risk assessment in siblings of affected children using sex-specific genetic scores. Mol Autism. 2011;2:17.
118. Thornton-Wells TA, Moore JH, Haines JL. Dissecting trait heterogeneity: a comparison of three clustering methods applied to genotypic data. BMC Bioinformatics. 2006;7:204.
119. Go AS, Mozaffarian D, Roger VL, Benjamin EJ, Berry JD, et al. Heart disease and stroke statistics – 2013 update: a report from the American Heart Association. Circulation. 2013;127:e6–e245.
120. Leening MJ, Ferket BS, Steyerberg EW, Kavousi M, Deckers JW, et al. Sex differences in lifetime risk and first manifestation of cardiovascular disease: prospective population based cohort study. BMJ. 2014;349:g5992.
121. Mosca L, Barrett-Connor E, Wenger NK. Sex/gender differences in cardiovascular disease prevention: what a difference a decade makes. Circulation. 2011;124:2145–54.
122. Bairey Merz CN, Shaw LJ, Reis SE, Bittner V, Kelsey SF, et al. Insights from the NHLBI-sponsored Women's ischemia syndrome evaluation (WISE) study: part II: gender differences in presentation, diagnosis, and outcome with regard to gender-based pathophysiology of atherosclerosis and macrovascular and microvascular coronary disease. J Am Coll Cardiol. 2006;47:S21–9.
123. Regitz-Zagrosek V, Lehmkuhl E. Heart failure and its treatment in women. Role of hypertension, diabetes, and estrogen. Herz. 2005;30:356–67.

124. Salfati E, Nandkeolyar S, Fortmann SP, Sidney S, Hlatky MA, et al. Susceptibility loci for clinical coronary artery disease and subclinical coronary atherosclerosis throughout the life-course. Circ Cardiovasc Genet. 2015;8:803–11.
125. Assimes TL, Salfati EL, Del Gobbo LC. Leveraging information from genetic risk scores of coronary atherosclerosis. Curr Opin Lipidol. 2017;28:104–12.
126. Barrett-Connor E. Sex differences in coronary heart disease. Why are women so superior? The 1995 Ancel keys lecture. Circulation. 1997;95:252–64.
127. Jousilahti P, Vartiainen E, Tuomilehto J, Puska P. Sex, age, cardiovascular risk factors, and coronary heart disease: a prospective follow-up study of 14 786 middle-aged men and women in Finland. Circulation. 1999;99:1165–72.
128. Pilote L, Dasgupta K, Guru V, Humphries KH, McGrath J, et al. A comprehensive view of sex-specific issues related to cardiovascular disease. CMAJ. 2007;176:S1–44.
129. Tunstall-Pedoe H, Kuulasmaa K, Amouyel P, Arveiler D, Rajakangas AM, et al. Myocardial infarction and coronary deaths in the World Health Organization MONICA project. Registration procedures, event rates, and case-fatality rates in 38 populations from 21 countries in four continents. Circulation. 1994;90:583–612.
130. Yusuf S, Hawken S, Ounpuu S, Dans T, Avezum A, et al. Effect of potentially modifiable risk factors associated with myocardial infarction in 52 countries (the INTERHEART study): case-control study. Lancet. 2004;364:937–52.
131. Carroll MD, Lacher DA, Sorlie PD, Cleeman JI, Gordon DJ, et al. Trends in serum lipids and lipoproteins of adults, 1960-2002. JAMA. 2005;294:1773–81.
132. Mittendorfer B. Sexual dimorphism in human lipid metabolism. J Nutr. 2005;135:681–6.
133. Roger VL, Go AS, Lloyd-Jones DM, Benjamin EJ, Berry JD, et al. Heart disease and stroke statistics – 2012 update: a report from the American Heart Association. Circulation. 2012;125:e2–e220.
134. Townsend EA, Miller VM, Prakash YS. Sex differences and sex steroids in lung health and disease. Endocr Rev. 2012;33:1–47.
135. Nauser TD, Stites SW. Diagnosis and treatment of pulmonary hypertension. Am Fam Physician. 2001;63:1789–98.
136. Lykke JA, Langhoff-Roos J, Sibai BM, Funai EF, Triche EW, et al. Hypertensive pregnancy disorders and subsequent cardiovascular morbidity and type 2 diabetes mellitus in the mother. Hypertension. 2009;53:944–51.
137. Wu P, Haththotuwa R, Kwok CS, Babu A, Kotronias RA, et al. Preeclampsia and future cardiovascular health: a systematic review and meta-analysis. Circ Cardiovasc Qual Outcomes. 2017;10
138. Garovic VD, Bailey KR, Boerwinkle E, Hunt SC, Weder AB, et al. Hypertension in pregnancy as a risk factor for cardiovascular disease later in life. J Hypertens. 2010;28:826–33.
139. Funai EF, Friedlander Y, Paltiel O, Tiram E, Xue X, et al. Long-term mortality after preeclampsia. Epidemiology. 2005;16:206–15.
140. Brown DW, Dueker N, Jamieson DJ, Cole JW, Wozniak MA, et al. Preeclampsia and the risk of ischemic stroke among young women: results from the stroke prevention in young women study. Stroke. 2006;37:1055–9.
141. Vikse BE, Irgens LM, Leivestad T, Skjaerven R, Iversen BM. Preeclampsia and the risk of end-stage renal disease. N Engl J Med. 2008;359:800–9.
142. Kuklina EV, Ayala C, Callaghan WM. Hypertensive disorders and severe obstetric morbidity in the United States. Obstet Gynecol. 2009;113:1299–306.
143. Ostadal B, Ostadal P. Sex-based differences in cardiac ischaemic injury and protection: therapeutic implications. Br J Pharmacol. 2014;171:541–54.
144. Nwankwo TYS, Burt V, Gu Q. Hypertension among adults in the United States: National Health and nutrition examination survey, 2011–2012. NCHS Data Brief. 2013;133:1–8.
145. Legro RS. Polycystic ovary syndrome and cardiovascular disease: a premature association? Endocr Rev. 2003;24:302–12.
146. Vryonidou A, Papatheodorou A, Tavridou A, Terzi T, Loi V, et al. Association of hyperandrogenemic and metabolic phenotype with carotid intima-media thickness in young women with polycystic ovary syndrome. J Clin Endocrinol Metab. 2005;90:2740–6.
147. Puzianowska-Kuznicka M. ESR1 in myocardial infarction. Clin Chim Acta. 2012;413:81–7.
148. Sudhir K, Chou TM, Chatterjee K, Smith EP, Williams TC, et al. Premature coronary artery disease associated with a disruptive mutation in the estrogen receptor gene in a man. Circulation. 1997;96:3774–7.
149. Shaw LJ, Bugiardini R, Merz CN. Women and ischemic heart disease: evolving knowledge. J Am Coll Cardiol. 2009;54:1561–75.

Gendered Innovation in Health and Medicine

39

Londa Schiebinger and Ineke Klinge

Art work by Piet Michiels, Leuven, Belgium.

L. Schiebinger (✉)
History of Science, Stanford University, Stanford, CA, USA
e-mail: schieb@stanford.edu

I. Klinge
Horizon 2020 Advisory Group for Gender, European Commission, Brussels, Belgium

Abstract

Excellence in research requires careful attention to sex and gender analysis. The Gendered Innovations project, initiated in 2009, develops state-of-the-art methods of sex and gender analysis for basic and applied research.

P. L. M. Kerkhof, V. M. Miller (eds.), *Sex-Specific Analysis of Cardiovascular Function*,
Advances in Experimental Medicine and Biology 1065, https://doi.org/10.1007/978-3-319-77932-4_39

This chapter reviews recent developments in cardiovascular disease for (1) analyzing sex, (2) analyzing gender, and (3) policy initiatives.

Keywords

Gender medicine · Gendered innovations · Innovation in health · Cardiovascular disease · Matera Alliance · Biomedical research · Policy · Sex · Gender · Intersex · Drug development · Biomedical devices

Doing research wrong costs lives and money. For example, between 1997 and 2000, ten drugs were withdrawn from the US market because of life-threatening health effects. Eight of these posed "greater health risks for women than for men" [1]. Not only does developing a drug in the current market cost billions—but when drugs failed, they caused human suffering and death. We can't afford to get it wrong.

Doing research right can save lives and money. An analysis of the US Women's Health Initiative Hormone Therapy Trial, for example, found that for every $1 spent, $140 were returned to US taxpayers in healthcare savings. The study also saved lives: there were 76,000 fewer cases of cardiovascular disease, 126,000 fewer breast cancers, and 145,000 more quality-adjusted life years. While most of the results were positive, the analysis did find 263,000 more osteoporotic fractures [2].

It is crucially important to identify gender bias. But analysis cannot stop there: Gendered Innovations offers state-of-the-art methods of sex and gender analysis. Integrating these methods into basic and applied research produces excellence in science, health and medicine, and engineering research, policy, and practice. The methods of sex and gender analysis are one set of methods among many that a researcher will bring to a project.

According to the US National Institutes of Health (NIH), both sex (biological characteristics) and gender (sociocultural beliefs, attitudes, and behaviors) "play a role in how health and disease processes differ among individuals." While methods for analyzing sex are increasingly well understood [3], methods for analyzing gender within biomedicine are not.

This chapter reviews recent developments in cardiovascular disease for (1) analyzing sex, (2) analyzing gender, and (3) policy initiatives. Funding agencies expect these analyses to contribute to the pace of new discoveries, diminish errors of extrapolation between sexes, and mitigate adverse events in the development pipeline [4].

Background

Governments, universities, and, increasingly, corporations in the USA and Western Europe have taken three strategic approaches to gender equality over the past several decades [5]:

1. "Fix the numbers of women" focuses on increasing the numbers of women participating in science, health and medicine, and engineering.
2. "Fix the institutions" promotes gender equality in careers through structural change in research organizations, such as dual-career hiring and family-friendly policies.
3. "Fix the knowledge" or "gendered innovations" stimulates excellence in science and technology by integrating sex and gender analysis into research. This paper focuses on this third strategic approach. It's the newest and the most important for the future of innovation and discovery.

Gendered Innovations in Science, Health and Medicine, Engineering, and Environment was produced through a large interdisciplinary collaboration funded by Stanford University (2009–2017), the European Commission (2011–2013), and the US National Science Foundation (2012–2015) [6]. To match the global reach of science and technology, the project was developed through a series of international workshops bringing together over 80 natural scientists, biomedical researchers, engineers, and gender experts from across the USA, EU, and Canada. We've now expanded into South Korea, Taiwan, and Silicon Valley companies. This project (1) developed state-of-the-art methods of sex and gender analysis and (2) produced 26 case studies to demonstrate through concrete examples how gender analysis leads to discovery and innovation. The operative question is: Can we harness the creative power of sex, gender, or ethnicity analysis for discovery? Does considering gender, for example, add a valuable dimension to research? Does it take research in new directions?

Gendered Innovations formed the empirical basis for European Commission's Horizon 2020, its current funding framework. In December 2013, to better meet its grand societal challenges, the European Commission (EC) implemented policies to integrate gender/sex analysis in research and innovation (R&I). In the proposal template, under "concept and approach," applicants are asked "Where relevant, describe how sex and/or gender analysis is taken into account in the project's content" [7]. In 2015, the EC identified over 130 areas of science and technology where gender analysis could benefit research, including health and biomedicine, computer hardware and architecture, nanotechnology, oceanography, geosciences, organic chemistry, aeronautics, space medicine, biodiversity, ecology, and biophysics, among others.

Results have been good; overall 36% of EC Horizon 2020 grants funded in 2015 integrated sex or gender analysis into research. Proposals that integrate gender analysis into the design of research are more competitive [8].

The EC is currently the global leader in policy in this area, but they were not the first. In 2010, all 13 Canadian Institutes of Health Research (CIHR) required applicants to consider sex and gender in their research (Fig. 39.1). CIHR states that "the purpose of this tool is to give health researchers a framework for thinking through how gender and/or sex might be integrated into their research designs" [9]. Further, CIHR has produced three trainings to assist researchers to learn how to apply the tools of sex and gender analysis. These include, "Sex and Gender in Biomedical Research," "Sex and Gender in Primary Data Collection with Human Participants," and "Sex and Gender in Secondary Data Collected from Human Participants." These trainings are short, precise, and highly recommended (http://www.discoversexandgender.ca/).

Finally, January 1, 2016, the US National Institutes of Health (NIH) required that public-funded research "account for the possible role of sex as a biological variable in vertebrate animal and human studies" [10]. Further, NIH made available $10.1 million as a catalyst for considering sex as a fundamental variable in research (PA-15-034).

Gendered Innovations demonstrates that prospective discoveries lie at the interface of disciplinary boundaries [11]. Gendered Innovations has been highly successful in developing and promoting methods for sex and gender analysis to encourage robust, reliable, and reproducible science [12]. Gendered Innovations was presented at the United Nations in 2011, 2014, and 2016; the National Science Foundation in 2012 and 2017; the European Parliament in 2013 [13]; CNRS, Paris in 2013; the National Parliament, South Korea, in 2014; and elsewhere. The globally accessible peer-reviewed Gendered Innovations website has been translated into Chinese, German, Korean, Spanish, and Swedish [14].

Definition of Terms

The terms gender and sex are often conflated in the biomedical literature [15]. These, however, are distinct terms and must be used correctly. *Sex* is "a biological quality or classification of sexually-reproducing organisms, generally female, male, and/or intersex, according to functions that derive from the chromosomal complement, reproductive organs, and specific hormones that affect the expression of phenotypic traits" [16].

While we generally analyze sex as "male" and "female" (intersex represents approximately 1–2% of the population, though this is not well documented), it is important to recognize differences *within* groups of females and males/women and men. Take, for example, height. In the USA, women are shorter than men on average, but about 3% of women are taller than the

Organization	Policy to: FIX the Numbers	Policy to: FIX the Institutions	Policy to: FIX the Knowledge	Policy to Integrate Gender Analysis into Research
Canadian Institutes of Health Research (CIHR)	Yes	Yes	Yes	In 2010, CIHR required all grant applicants to respond to mandatory questions about "whether their research designs include gender and sex." August 2015 released online training for Sex and Gender in Biomedical Research.
European Commission Directorate-General for Research and Innovation	Yes	Yes	Yes	Since 2003, the European Commission has supported "questioning systematically whether, and in what sense, sex and gender are relevant in the objectives and in the methodology of projects." In 2013, these policies were reaffirmed and expanded in Horizon 2020, the Commission's current funding framework. The Commission states, "Integrating gender/sex analysis in research and innovation (R&I) content helps improve the scientific quality and societal relevance of the produced knowledge, technology and/or innovation. In March 2015, the EC Gender Advisory Group published an advice paper on preparing grants that integrate the gender dimension into research.
US National Institutes of Health (NIH)	Yes	Yes	Yes	Guidelines for considering sex as a biological variable in research were released June 2015. Policy implemented January 1, 2016. (Notice Number: NOT-OD-15-102.) These guidelines are mainstreamed in "Enhancing Reproducibility through Rigor and Transparency" (Notice Number: NOT-OD-15-103.)

Fig. 39.1 Gender policies of major granting agencies. (For more, see http://genderedinnovations.stanford.edu/sex-and-gender-analysis-policies-major-granting-agencies.html)

average man, and 6% of men are shorter than the average woman. The height difference between the average woman and man is less than the height difference between a 90th percentile woman and a 10th percentile woman, or the difference between 90th and 10th percentile men (Fig. 39.2).

Gender, by contrast, refers to culturally-defined norms and relations that shape our identities and behaviors as women, men, and gender-diverse individuals. More specifically, "gender" refers to sociocultural processes that interact with and thus may influence human biology (Fig. 39.3). Gender includes "gender norms" (spoken and unspoken rules—in family, social circles, workplace, institutional, or global culture—that influence individual attitudes and behaviors), "gender identity" (how individuals and groups perceive and present themselves), and "gender relations" (the social relations between individuals of different gender identities, such as the relations between a male patient and a female physician)—see Fig. 39.4.

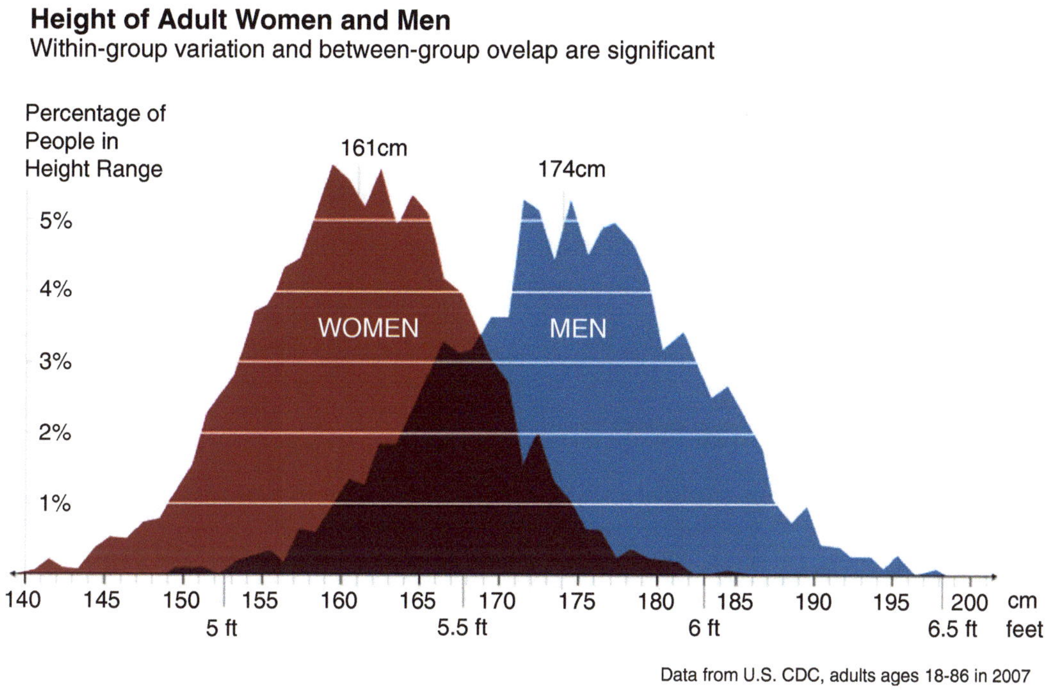

Fig. 39.2 Analyzing sex. (http://genderedinnovations.stanford.edu/methods/sex.html)

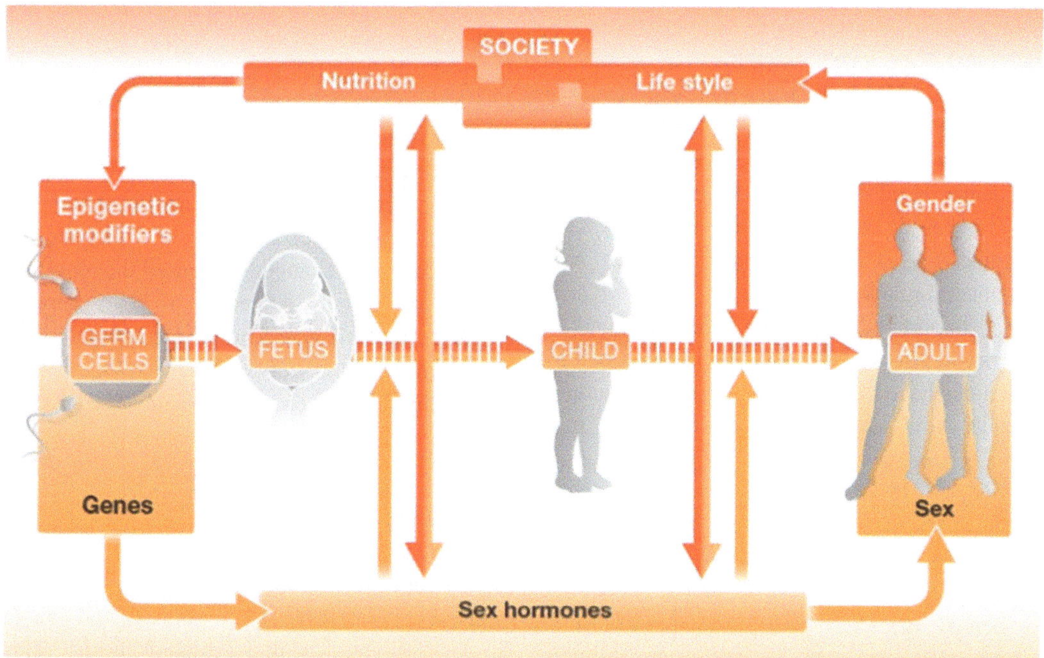

Fig. 39.3 Sex and gender interact throughout the human life cycle. Sex influences health directly or by modifying behavior; at the same time, gender behaviors may modify biological factors. Regitz-Zagrosek V. Sex and gender differences in health. Eur Mol Biol Organ Rep. 2012, 13 (7):596–603. Reproduced by kind permission

Analyzing Gender. Points to Keep in Mind:

Gender refers to cultural attitudes and behaviors traditionally ascribed to women, men, and gender diverse people.

1. Gender consists of:
   a. **Gender Norms** (spoken and unspoken cultural rules in the family, workplace, society, institutional or global culture that influence individual attitudes and behaviors).
   b. **Gender Identity** (how individuals and groups perceive and present themselves in relation to gender norms).
   c. **Gender Relations** (the power relations between individuals of different gender identities).
2. Gender attitudes and behaviors vary by culture, historical era, ethnicity, socioeconomic status, geographic location, and other factors. For example, gender norms may be very different on the US West coast vs. East coast, or in Italy vs. India.
3. Investigators should recognize within group variations (differences in attitudes and behaviors among women or among men) and between-group overlap (overlap in attitudes and behaviors between women and men).
4. It is important to consider factors intersecting with sex and gender (e.g., age, socio-economic status, or ethnicity).
5. There is no necessary relationship between gender characteristics and sex.

Fig. 39.4 Analyzing gender. Points to keep in mind

Analyzing Sex

Heart disease research in women offers one of the most developed examples of gendered innovations. Although heart disease is a major killer of women in developed countries, it has been defined primarily as a male disease, and "evidence-based" clinical standards have been created based on male pathophysiology and outcomes. As a result, women are often mis- and underdiagnosed [17].

An important innovation here has been to understand **sex** differences in the pathophysiology underlying heart disease. To take just one example, consider how underlying pathophysiology may differ between women and men [18]. Coronary angiography, the "gold standard" for diagnosing patients with chest pain, typically results in a diagnosis of obstructive coronary artery disease (CAD) in men (see chart below, right), but frequently fails to identify the cause in a large proportion of women [19]. As a result, many women with chest pain, but "normal" angiograms (Fig. 39.5, left panel), may be told that they have no significant disease and sent home.

New studies show, however, that the prognosis for these women is not benign: Women with a primary diagnosis of "non-specific chest pain" may suffer heart attack or stroke shortly after being discharged from hospitals [20]. This may also be true for some men. Large-scale randomized trials are needed to better understand the pathophysiology and optimal therapies for women and men with angina and "normal" angiograms.

After 20 years of research, sex analysis has prompted policy changes, increased the representation of women subjects in heart disease research, and enhanced knowledge about diagnosis and treatment in women and men alike. But, of course, much still needs to be done [21].

Analyzing Gender

Gender also plays a role in heart disease. A new Canadian study has analyzed both sex and gender in heart disease—with surprising results. This study seeks to predict outcomes after Acute Coronary Syndrome (ACS) in young adults, analyzed the independent effect of sex and gender on

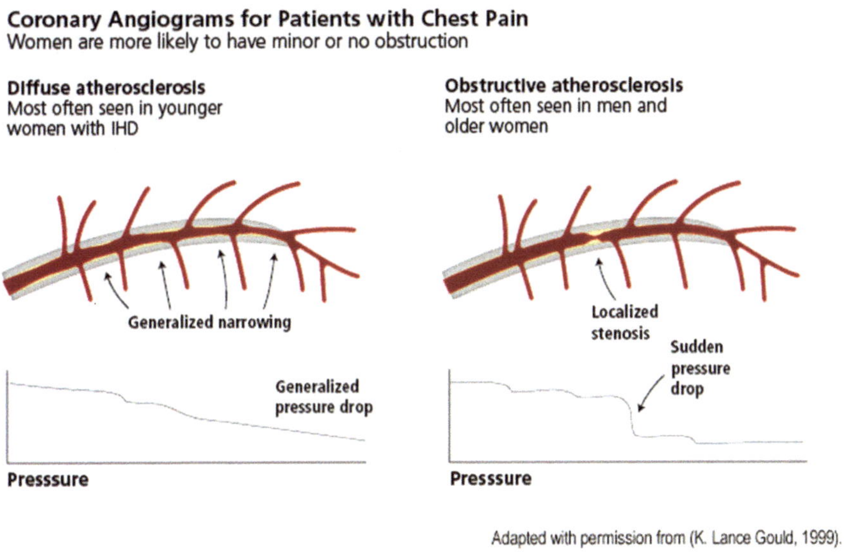

Fig. 39.5 Men and women have different patterns of heart disease. (Adapted with permission from K. Lance Gould 1999)

incidence of recurrence of ACS and death twelve months after diagnosis [22].

To do the study, the team reduced sex to a variable, which is straightforward. They found no difference between women and men. The principal investigators also reduced gender to a variable—which is highly interesting. If researchers do not find useful ways to measure gender, they miss important effects of gender on human health.

After testing 17 variables, the PIs found seven that were significant and developed a gender-related score from these:

1. Household primary earner
2. Personal income
3. Number of hours per week spent doing housework
4. Status of primary person responsible for doing housework
5. Level of stress at home
6. Bem Sex Role Inventory masculinity score
7. Bem Sex Role Inventory femininity score

These they controlled for sex, age, the Global Registry of Acute Coronary Events (GRACE) score, previous CVD events, and a number of traditional CV risk factors, such as smoking, obesity, hypertension, etc.

The results suggest that gender matters. Patients received a gender-related score of between 1 and 100. Patients with a higher "femininity" score were more likely to experience a recurrence of ACS—regardless whether they were women or men. That is to say, a man with a high femininity score was more likely to suffer a recurrence; and a woman with a high masculinity score was not.

Building on this radically new work, an international, interdisciplinary team centered at Stanford University has undertaken a large project to develop gender variables for health research. We argue that numerous extant measures of gender, such as the 1974 Bem Sex Role Inventory (used, e.g., in the Canadian study) [23], are outdated and likely to lead to erroneous conclusions. Further, we move beyond femininity and masculinity by reconceptualizing gender as key theory-based constructs or latent variables. Building on excellent work emerging in this area [24], this project aims to identify key gender-related variables along with relevant measures for each (and point to potential associations with health outcomes). Results will be available in early 2018 [25].

New Directions

Several exciting new directions are in play. Designers of biomedical devices are increasingly incorporating sex and gender. After the dramatic failure of Apple's HealthKit in 2014, designers are beginning to understand that not considering sex can lead to failure. Apple's HealthKit is an app that tracks numerous biological functions, including blood pressure, daily calories, blood alcohol content, percent body fat, respiratory rate, intake of sodium, magnesium, calcium, fiber, iodine, chromium, etc. It failed, however, to track women's menstrual cycles. Negative publicity forced Apple to release a corrected version of the app in 2015 at a high cost in terms of profit, poor publicity, and team morale.

Companies now understand that biometrics need to consider physiological differences between men and women (sometimes also age or ethnicity) to insure usability and safety for users. A tech company in Silicon Valley, for example, is designing a wearable to detect early signs of heart disease. Recognizing that men and women have different patterns of heart disease will enhance the usability and uptake of the device.

In 2016, the Matera Alliance was launched at a Gender Medicine congress in Matera, Italy [26]. It is comprised of individuals and organizations committed to increasing the applicability of drug development and testing to women across the lifespan. In a recent review in *Pharmacological Research*, two members of the Matera Alliance—Cara Tannenbaum and Danielle Day—summarized the current state of science for considering age and sex-related factors in the drug development pipeline, from cell culture and animal research through all phases of clinical trials in humans [27].

The article highlights the various pitfalls and problems in current drug development that takes the male as the norm, resulting in many adverse drug reactions in women. It offers a set of recommendations to improve standards for integrating age and sex into study design, analysis, and reporting of preclinical and clinical assessment of new molecular entities and biologics in adults. These recommendations provide guidance for drug development and testing across the spectrum of age and sex in adults.

In May 2017, at the Organization for the Study of Sex Differences (OSSD) meeting in Montreal, the Matera Alliance organized a roundtable with representatives of pharmaceutical industry to create and implement a multi-stakeholder roadmap that enables basic scientists, clinical trialists, regulatory agencies, payers, physicians, and patients to transform the way drugs and biologics are developed, tested, approved, and prescribed in a more tailored sex-specific manner (Fig. 39.6).

Integrating Sex and Gender into the Scientific Pipeline

Policy is one driver of innovation that can help encourage health and biomedical researchers integrate sex and gender analysis into the design of their research. Interlocking policies need to address granting agencies, editors of peer-reviewed journals, industry leaders, and educators.

At the beginning of the research pipelines, granting agencies can ask applicants to explain how sex and gender analysis is relevant their proposed research—as discussed above. The EC program, Horizon 2020, lists three objectives for gender equality: (1) to encourage gender balance in research teams, (2) to ensure gender balance in decision making, and (3) to integrate the gender dimension in research and innovation. To support this third initiative, EC included gender trainings among the eligible project costs to encourage researchers to further develop and share gender expertise in research.

Further, the EC founded the Advisory Group for Gender (AGG) in 2014. This group, chaired by Ineke Klinge as rapporteur for the Gendered Innovations project, is composed of 27 gender experts, one appointed from each of the advisory groups that comprise Horizon 2020 subject areas. Its mandate is to provide advice pertaining to all gender-equality activities under Horizon 2020 with special attention to the integration of the gender dimension in research and innovation.

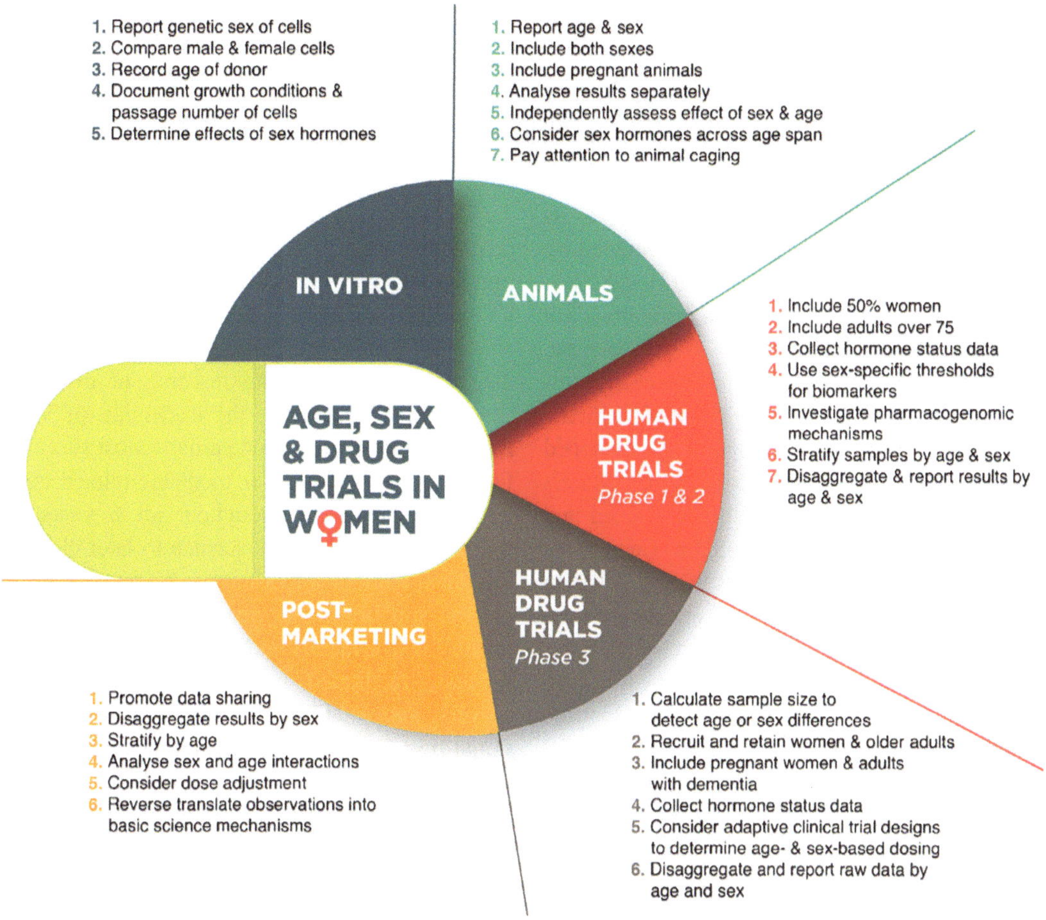

Fig. 39.6 Cycle of age- and sex-specific translational and posttranslational drug development. Tannenbaum C, Day D. Age and sex in drug development and testing for adults. Pharmacol Res. 2017; 121:83–93, esp. 90. doi:https://doi.org/10.1016/j.phrs.2017.04.027. Published by Elsevier Ltd. This is an open access article under the CC BY license (http://creativecommons.org/licenses/by/4.0/)

In 2015, the AGG produced "For a Better Integration of the Gender Dimension in the Horizon 2020 Work Programme 2016–2017." The AGG takes the basic position that "the gender dimension is part and parcel of research excellence. It enhances the societal relevance of the produced knowledge, technologies and innovations and contributes to the production of goods and services better suited to potential markets. To support this process, it is essential to devote resources to the gender aspects of research" [28]. The collaborative work of this advisory group resulted in integrating more gender-flagged topics in the EC's upcoming Work Programmes. They also clarified concepts and offered expert materials to program officers writing Work Programmes.

Following the EC's lead, a number of the EU member states have also established robust policies that integrate sex and gender analysis into calls for funding proposals, including Austria, Ireland, and Germany—both the Federal Ministry of Education and Research (BMBF) and the German Research Foundation (DFG) [29]. To take one example, in 2013 the Dutch Alliance for Gender and Health, a bottom-up multi- stakeholder Alliance, brought together policy makers, scientists, medical professionals, insurance companies, gender experts, and knowledge institutes around the theme gender and health, coordinated by Women Inc. A major outcome has been the Gender and Health Knowledge Agenda. Commissioned by the Dutch Ministry of Health, Welfare and Sport,

ZonMw (the Netherlands Organisation for Health Research and Development) has developed a Gender and Health Knowledge Program. This program is based on the previously published Knowledge Agenda and the subsequent Priority Study [30]. This grant program draws upon the Gendered Innovations methods and case studies as well as from materials and online trainings developed by the Canadian Institute for Gender and Health.

At the university level, research institutions can develop programs and policies to promote the gender dimension in research. One successful initiative is Stanford University's School of Medicine seed grants for medical researchers. Developed by Gendered Innovations co-director, Marcia Stefanick, who heads up Stanford's Women and Sex Differences in Medicine (WSDM) Center, this program offers Stanford medical PIs, between $20,000 and $35,000 to integrate sex and/or gender analysis into their current research. Examples of successful grants include "Investigating Sex Differences in Human Plasmacytoid Dentritic Cell Function and Regulation" and "Sex Differences in Hippocampal Physiology & Response to Circadian Dysfunction." The idea is to ready Stanford researchers to apply for large NIH grants that now require considering sex as a biological variable [31]. This initiative also contributes to integrating sex/gender analysis into the Stanford School of Medicine curriculum (see below). Once professors see the value of sex and gender analysis for their research, they are likely to add it to their lectures.

Importantly, at the end of the research pipeline, editorial boards of peer-reviewed journals can require sophisticated sex and gender analysis when selecting papers for publication. If sex and gender are not analyzed properly, an article is not excellent and worthy of publication. On the Gendered Innovations website we include a list of journals and their guidelines to authors and reviewers for sex and gender analysis [32]. The *Journal of the American College of Cardiology* and *American Journal of Physiology* were early adapters of these policies and guidelines. In December 2016, *Lancet* published proposed guidelines for reporting sex and gender in medical journals (Fig. 39.7). These guidelines have also been adopted by the International Committee of Medical Journal Editors. For biomedical and health research, it is now clear that sex and gender must be considered as health-relevant variables.

For the scientific pipeline, integrating knowledge of sex and gender into medical school curricula may, in fact, be a matter of life and death. Integrating sex and gender knowledge throughout the curriculum ensures adequate knowledge and skills for future physicians in etiology, pathogenesis, clinical presentation, diagnosis, treatment, and research of diseases. Charité in Berlin, under the leadership of Vera Regitz-Zagrosek, has been perhaps most successful in this regard. Gender medicine-related content was integrated throughout all 6 years of training from early basic science to later clinical modules. Factors important to the process were (1) the support of the dean and (2) a "change agent," i.e., a faculty member who oversaw the process, facilitated interactions with all key players, and established a supporting organizational framework. An evaluation of this program at Charité showed that sex and gender elements were integrated into 21% of lectures, 12% of seminars, and 8% of practical courses [33].

In the USA, Marjorie Jenkins of the US Food and Drug Administration (FDA)'s Office of Women's Health and Texas Tech along with a number of colleagues are working with the National Board of Medical Examiners to include questions about sex and gender medicine on the US Medical Licensing Exam. This is a major step forward. A Sex and Gender-Based Medical Education Summit was held at Mayo Clinic in 2015 [34].

In industry, too, it is important to integrate sex and gender into products and product design. Biomedical products and devices that incorporate the smartest aspects of sex and gender can open new markets. Products that meet the needs of complex and diverse user groups enhance global competitiveness and sustainability.

There is much work to be done. Researchers need to learn sophisticated methods of sex and gender analysis. Universities need to incorporate these methods into their curricula. Granting agencies need to ask applicants to explain how

Lancet: Proposed guidelines on reporting sex and gender in medical journals

1. Require correct usage of the terms "sex"and "gender."Using these terms precisely increases clarity, enables critical review, and facilitates meta-analysis.

2. Require the reporting of the sex and/or gender of the study group/participants, and the sex of animals or cells. If males and females were not studied in appropriate proportions (e.g. because the condition is sex-specific or because a convenience sample was used), these elements of study design should be justified in the Methods section, and considered in the Discussion section.

3. Consider analyzing data by sex and/or gender, where appropriate, or providing the raw data in the main manuscript, supplemental material, or in an accessible data repository. Report on the approach chosen for sex and gender analysis and comment on it in the Discussion section. In studies that are underpowered to detect sex (or gender) differences, access to data allows for use of those data in meta-analyses and systematic reviews.

4. Analyze the influence (or association) of sex and/or gender on the results of the study where appropriate, or indicate in the methods section why such analyses were not performed. Where those analyses were not performed, consider covering this topic in the Discussion section. Readers need toknow whether the results generalize to both sexes. Include "no difference"results as well as results that demonstrate differences.

5. If sex or gender analyses were performed post-hoc, indicate that these analyses should be interpreted cautiously. No-difference post-hoc analyses may be underpowered, leading to a false conclusion of no difference. By contrast, if many such analyses were performed, the additional comparisons may lead to spurious significance suggesting an erroneous conclusion of a sex-or gender-related difference where no such difference was in fact present. To minimize this likelihood, authors might consider making a statistical adjustment (such as a Bonferroni correction).

Schiebinger, L., Leopold, S. S., & Miller, V. M. (2016). Editorial policies for sex and gender analysis. *The Lancet*, *388* (10062), 2841-2842.

Fig. 39.7 Lancet proposed guidelines on reporting sex and gender in medical journals

sex and gender analysis is relevant their proposed research. Editorial boards of peer-reviewed journals need to require sophisticated sex and gender analysis when selecting papers for publication. Industry needs to incorporate the smartest aspects of gender in innovative products, processes, services, and infrastructures.

But eyes have been opened, and we cannot return to a world that ignores gender. Innovation is what makes the world tick. Gendered innovations spark creativity by offering new perspectives, posing new questions, and opening new areas to research. Can we afford to ignore such opportunities?

References

1. United States General Accounting Office. Drug safety: most drugs withdrawn in recent years had greater health risks for women. Washington, DC: Government Publishing Office; 2001.
2. Roth J, Etzioni R, Waters T, Pettinger M, Rossouw J, Anderson G, Chlebowski R, Manson J, Hlatky M, Johnson K. Economic return from the women's health initiative Estrogen Plus Progestin clinical trial: a modeling study. Ann Intern Med. 2014;160(9):594–602.
3. Cornelison TL, Clayton J. Considering sex as a biological variable in biomedical research. Gend Genome. 2017; 1 (2):89–93; Clayton J. Studying both sexes: a guiding principle for biomedicine. FASEB J. 2015;30:1–6.
4. Collins FS, Tabak LA. Policy: NIH plans to enhance reproducibility. Nature. 2014;505:612–3.
5. Schiebinger L. Has Feminism changed science? Cambridge, MA: Harvard University press; 1999.

6. Schiebinger L, Klinge I. Gendered Innovation in Health and Medicine. Gender: Zeitschrift für Geschlecht, Kultur, und Gesellschaft. 2015;2(2015):29–50.
7. European Commission. Fact sheet: gender equality in Horizon 2020. Brussels; 2013 December 9; European Commission. Vademecum on gender equality in Horizon 2020. Brussels: RTD-B7 Science with and for Society; 2016.
8. European Commission. European Commission. Horizon 2020. Monitoring Report 2015. Directorate-General for Research and Innovation. Brussels: European Commission, 53; 2016.
9. Canadian Institutes of Health Research (CIHR). Gender, sex and health research guide: a tool for CIHR applicants. Ottawa: Canadian Institutes of Health Research (CIHR); 2012. Enacted December 2010
10. Clayton J, Collins F. NIH to balance sex in cell and animal studies. Nature. 2014;509:282–3; Clayton JA. Studying both sexes: a guiding principle for biomedicine. FASEB J. 2016;30 (2):519–24; Tannenbaum C, Schwarz JM, Clayton JA, de Vries GJ, Sullivan C. Evaluating sex as a biological variable in preclinical research: the devil in the details. Biol Sex Diff. 2016;7(1):13.
11. Schiebinger L, Schraudner M. Interdisciplinary approaches to achieving gendered innovations in science, medicine, and engineering1. Interdiscip Sci Rev. 2011;36(2):154–67.
12. For method of sex and gender analysis, see http://genderedinnovations.stanford.edu/methods-sex-and-gender-analysis.html.
13. Schiebinger L, Klinge I. Gendered innovations: how gender analysis contributes to research. Luxembourg: Publications Office of the European Union, Foreword by EC Commissioner for Research, Innovation, and Science, Máire Geoghegan-Quinn; 2013
14. Schiebinger L, Klinge I, Paik HY, Sánchez de Madariaga I, Schraudner M, Stefanick M (eds). Gendered innovations in science, health & medicine, engineering, and environment; 2011–2017. genderedinnovations.stanford.edu.
15. Ristvedt SL. The evolution of gender. JAMA Psychiatry. 2014;71(1):13–14. Oertelt-Prigione S, Parol R, Krohn S, Preißner R, Regitz-Zagrosek V. Analysis of sex and gender-specific research reveals a common increase in publications and marked differences between disciplines. BMC Med. 2010;8(1):1.
16. http://genderedinnovations.stanford.edu/terms/sex.html.
17. Oertelt-Prigione S, Regitz-Zagrosek V (eds). Sex and gender aspects in clinical medicine. London: Springer; 2012; Regitz-Zagrosek V (ed). Sex and gender differences in pharmacology. London: Springer; 2012; Schenck-Gustafsson K, DeCola P, Pfaff D, Pisetsky D (eds). Handbook of clinical gender medicine. Basel: Karger; 2012.
18. Bairey Merz C, Mark S, Boyan B, Jacoba A, Shah P, Shaw L, Taylor D, Marbán E. Proceedings from the scientific symposium: sex differences in cardiovascular disease and implications for therapies. J Women's Health. 2010;19 (6):1059–72.
19. Bugiardini R, Bairey Merz C. Angina with "normal" coronary arteries: a changing philosophy. J Am Med Assoc. 2005;293 (4):477–84.
20. Robinson J, Wallace R, Limacher M, Ren H, Cochrane B, Wassertheil-Smoller S, Ockene J, Blanchette P, Ko M. Cardiovascular risk in women with non-specific chest pain (from the women's health initiative hormone Trials). Am J Cardiol. 2008;102(6):693–9.
21. GenCAD Partners. Gender in coronary artery disease. Brussels.: European Union; 2017.
22. Pelletier R, Ditto B, Pilote L. A composite measure of gender and its association with risk factors in patients with premature acute coronary syndrome. Psychosom Med. 2015;77(5):517–26. See also Schiebinger L, Stefanick ML. Gender matters in biological research and medical practice; 2016.
23. Bem SL. The measurement of psychological androgyny. J Consult Clin Psychol. 1974;42:155–62.
24. Bekker MHJ. Investigating gender within health research is more than sex disaggregation of data: a multi-facet gender and health model. Psychol Health Med. 2003;8(2):231–43; Nowatzki N, Grant KR. Sex is not enough: the need for gender-based analysis in health research. Health Care Women Int. 2011;32(4):263–77.
25. Nielsen M, Brooks C Cullen M, Einstein G, Ioannidis J, Klinge I, LeBlanc H, Neilands T, Paik H, Peragine D, Pilote L, Prochaska J, Ristvedt S, Stefanick M, Schiebinger L. Gender variables in health research; in progress.
26. www.materaalliance.org.
27. Tannenbaum C, Day D. Age and sex in drug development and testing for adults. Pharmacol Res. 2017;121:83–93.
28. http://ec.europa.eu/transparency/regexpert/index.cfm?do=groupDetail.groupDetailDoc&id=28824&no=1 AGG position paper 2016: For a better integration of the gender dimension in the Horizon 2020 Work Programme 2018-2020.
29. For details, see http://genderedinnovations.stanford.edu/sex-and-gender-analysis-policies-major-granting-agencies.html.
30. https://www.zonmw.nl/uploads/tx_vipublicaties/Programmatekst_Gender_en_Gezondheid.pdf.
31. http://med.stanford.edu/wsdm.html.
32. http://genderedinnovations.stanford.edu/sex-and-gender-analysis-policies-peer-reviewed-journals.html.
33. Ludwig S, et al. A successful strategy to integrate sex and gender medicine into a newly developed medical curriculum. J Women's Health. 2015;24(12):996–1005.
34. https://ce.mayo.edu/special-topics-in-health-care/node/3095.

Gendered Innovations in the Study of Cardiovascular Diseases

40

Evgeny Shlyakhto

Dress Red Day. Art Work by Piet Michiels, Leuven, Belgium

E. Shlyakhto (✉)
Russian Society of Cardiology, Saint-Petersburg, Russia

Almazov National Medical Research Centre, Saint-Petersburg, Russia
e-mail: e.shlyakhto@almazovcentre.ru

Abstract

The chapter summarizes modern knowledge on gender aspects of major cardiovascular diseases starting from basic research. It covers arterial hypertension along with treatment

P. L. M. Kerkhof, V. M. Miller (eds.), *Sex-Specific Analysis of Cardiovascular Function*, Advances in Experimental Medicine and Biology 1065, https://doi.org/10.1007/978-3-319-77932-4_40

aspects in males and females, as well as menopausal replacement therapy and testosterone use. Authors analyze both literature and original data on obesity and metabolic syndrome, coronary artery disease, arrhythmias including gender aspects of sudden cardiac death, cardiomyopathies, and diseases of aorta and aortic valve. A significant part deals with cardiac imaging and sex differences.

Keywords

Gendered innovations · Translational research · Basic science · Cardiovascular system · Sex-specific difference · Metabolic syndrome · Heart failure · Diabetic cardiomyopathy · Cardiac amyloidosis · Adiponectin · Bariatric procedure · Cardioprotection · Review

Introduction

Gendered Innovations is a scientific initiative that examines sex and gender as a variable for investigation, advancing science and technology across the globe. "Gendered Innovations" employs methods of sex and gender analysis to create new knowledge.

New technologies and medical innovations are rapidly increasing in cardiology, starting from genome editing and modulation of fetal programming and ending by development of new implantable devices and creating artificial organs. Significant epidemiological and clinical data has amassed over the years indicating important differences between women and men in the prevalence, course, and expression of various cardiovascular problems. However, considering gender in cardiovascular research and clinical practice is still quite rare, and most theories (especially biological models), as well as available treatments (particularly pharmacotherapy), are based almost exclusively on findings in male subjects (animal and/or human).

Taking into account all sex-specific aspects of cardiovascular function in health and disease, innovations should also be gender-oriented. Thus, new cardiovascular drugs can be personalized to gender-specific action; several male or female gender-specific markers of disease development can be targeted, including sex hormones and receptors. Any of new technology introduced for diagnosis and treatment, in particular drugs and invasive procedures, should be tested for its efficacy in different gender, including side effects and efficacy predictors. Gender discrepancies in epidemiology, biomarkers, cardiovascular structure, and function allow us to identify novel molecules and pathways than can be translated into new treatment strategies. There is now abundant evidence emphasizing the need to consider sex differences. Implementation of associated new technologies is summarized, along with a brief overview concerning upcoming innovations.

Gender Differences in Cardiovascular Disease: Basic and Translational Research

Molecular mechanisms of gender-based differences in incidence, clinical course, and pathogenesis of cardiovascular disease (CVD) remain poorly understood. Significant contribution to the understanding of sex differences in CVD is made by use of animal models of such common disorders as *arterial hypertension, myocardial ischemia-reperfusion, postischemic heart failure (HF)*, and *myocardial hypertrophy* due to pressure or volume overload. In general, the above pathologies are characterized by greater severity and earlier decompensation in male animals as compared to females. Male spontaneously hypertensive rats (SHRs) have higher systolic blood pressure (BP) than age-matched females [119]. In addition, male SHRs demonstrate earlier deterioration in left ventricular (LV) function progressing to chronic HF as compared to female animals. Higher BP values and more advanced target organ injury were found in male vs. female rats with DOCA-salt hypertension [6] and N-ω-nitro-L-arginine methyl ester (L-NAME)-induced hypertension [125]. It has been shown in the isolated perfused heart model that female

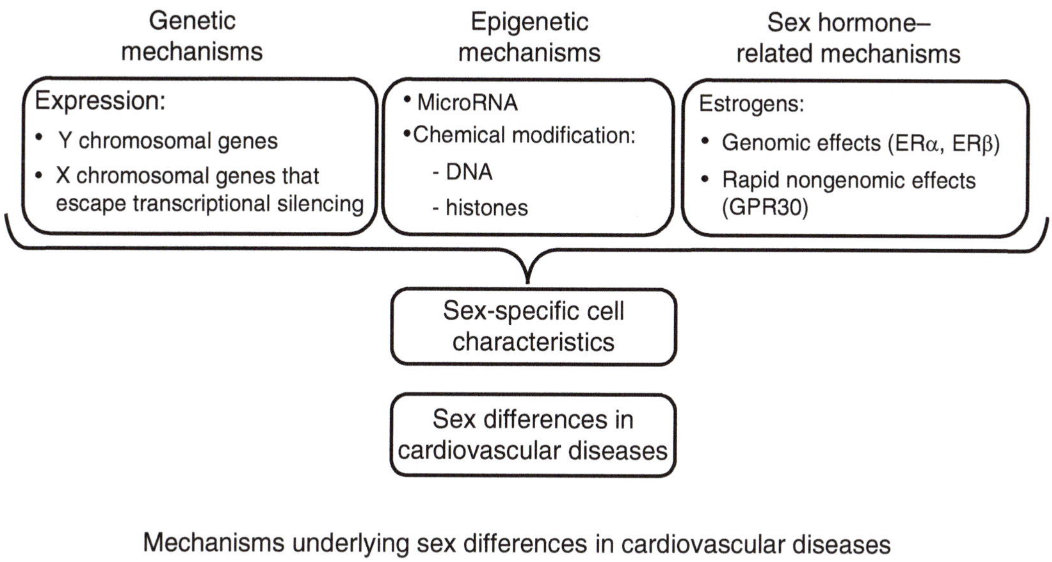

Fig. 40.1 Major mechanisms include genetic and epigenetic effects as well as sex hormone-induced events. The above mechanisms result in gender-specific cell characteristics which determine differential expression of cardiovascular disease in males and females (see text for more details)

rat hearts developed smaller infarcts and demonstrated better postischemic functional recovery after global *ischemia-reperfusion* in comparison with male hearts [5]. In the in vivo experiments, myocardial infarct size and the degree of LV remodeling have been shown to be significantly smaller in female rabbit [15], rat [59], and dog [61] versus male animals. Male mice compared with female mice demonstrated significantly higher rate of myocardial rupture and greater mortality within first 7 days after coronary ligation [22]. This difference can be explained by greater wall stress, more intensive inflammatory response, and activation of matrix metalloproteinase in male versus female animals [39]. Compared to females, poorer outcomes were also observed in males subjected to cardiac *pressure and volume overload*. Male rats with transverse aortic constriction (TAC) developed predominantly eccentric hypertrophy and more pronounced fibrosis as compared to females, which demonstrated concentric hypertrophy [33]. Chronic TAC-induced HF develops earlier in males than in females [143]. Administration of estradiol (E2) to ovariectomized females with TAC prevented the development of LV dysfunction and hypertrophy which provides evidence for protective effects of estrogens [112]. Female rats with atrioventricular shunt-induced volume overload demonstrated less severe HF, better LV function, and smaller LV dilatation [43].

The mechanisms of gender differences in CVD are under intense investigation (Fig. 40.1). Major mechanisms include (epi)genetic effects and *sex hormone-induced events. Genetic mechanisms* involve expression of Y chromosomal genes in males and expression of X chromosome genes that escape inactivation during transcriptional silencing of one X chromosome in females. It has been shown, for example, that the expression of certain variants of *Uty* gene which is localized in male-specific region of Y chromosome is associated with the level of HDL cholesterol in mice [134]. *Epigenetic mechanisms* include chemical modification of DNA and histones as well as gender-specific expression of certain microRNAs. Recent studies showed that the expression of microRNA-21, microRNA-24, microRNA-27a, microRNA-27b, microRNA-106a, and microRNA-106b is gender-dependent and is subjected to regulation by E2 and estrogen receptor (ER) β. Estrogen-dependent changes in the level of microRNA-21 resulted in varying intensity of myocardial fibrosis in pressure-overloaded mice [118]. Gender-dependent differences in CVD could be also attributed to the effects of sex hormones, among which the effects of E2 on ERα and ERβ are best studied. E2 exerts two types of effects, namely, *genomic effects* (estrogen receptor α- and estrogen receptor β-mediated gene expression changes) and *rapid non-genomic effects* which seem to be evoked by E2 binding with G-protein-coupled receptor GPR30. Non-genomic effects include activation of several kinases, such as ERK1/ERK2, Src, and JNK [80]. The same kinase pathways are known to be activated after myocardial conditioning. Significant progress in the elucidation of the role of certain ER in protective signaling has been achieved by using genetically engineered mouse models (GEMMs). GEMMs either lack ERα/ERβ or exhibit their systemic or tissue-specific overexpression [120]. Studies on ERα or ERβ knockout mice have shown that both types of the receptor might be involved in cardioprotection. Cardiomyocyte-specific hyperexpression of ERα has decreased myocardial fibrosis and remodeling only in female mice 2 weeks after MI which was accompanied by hyperphosphorylation of JNK [81]. In contrast, cardiomyocyte-specific hyperexpression of ERβ improved survival in both sexes in mice [127]. In special *four-core genotype (FCG) mouse*, gonadal sex and sex chromosome complement are uncoupled by means of moving testis-determining gene, *Sry*, from the Y chromosome to an autosome. As a consequence, FCG mouse offers the opportunity of studying differences in XX vs. XY mice that have the same sexual phenotype and same sex hormone levels, i.e., XX male vs. XY male and XX female vs. XY female. With this approach, the effects of sex hormones and sex chromosomes on the resulting phenotype can be assessed separately. However, recent experiments showed that both genes and hormones play a major role in determining the level of resistance to Coxsackievirus B3 and ischemia-reperfusion [75, 76, 122]. Cardiovascular effects of ERα/ERβ can be also elicited by *specific agonists*. For example, administration of ERβ agonist 2,3-bis(4-hydroxyphenyl)-propionitrile (DPN) to ovariectomized mice resulted in augmented functional recovery of the LV after ischemia-reperfusion in the isolated heart [100]. In the in vivo infarct model in rabbit, selective agonist of ERα 4,4′,4″-(4-propyl-(1H)-pyrazole-1,3,5-triyl) trisphenol, but not DNP, reduced infarct size [13]. The above mechanisms of gender influence on cardiovascular system result in *gender-specific cell characteristics*. For example, female mitochondria are more tolerant to oxidative stress and, in addition, faster restore ATP stores after episode of acute hypoxia. Female cardiomyocytes are more tolerant to apoptosis in ischemia-reperfusion because of increased expression of Bcl-2 and decreased expression of Bax.

In conclusion, regulatory bodies encourage the use of the animals of both sexes for research purposes [23]. Similarly, primary cells should be harvested from both male and female donors. Experimental studies in both sexes represent major tool for discovery of new molecular mechanisms of disease and therapeutic targets that might benefit both men and women.

Arterial Hypertension

Arterial hypertension is a well-known major cardiovascular risk factor with high prevalence. Almost two-third of postmenopausal women have elevated blood pressure [105]. It is widely accepted that menopause is an independent risk factor for development and progression of arterial hypertension [130, 131]. Although fewer women than men less than age 45 have hypertension, the genders share a similar prevalence between ages 45 and 64, and, older than the age of 64, the prevalence of hypertension in women surpasses that of men. Sex-specific factors in females are obesity, cystic ovarian syndrome, oral contraceptive drugs, and menopause resulting in increased salt sensitivity. New mechanism of blood pressure rise, such as oligonephronia [103, 105], and unbalanced renal medullar circulation, multiple aldosterone secretion disturbances, and increased large artery stiffness can also have sex-specific features demanding corresponding distinct treatment options.

Moreover, recent studies show that genetics of hypertension can be sex-specific, indicating that analyses that incorporate sex-dependent and epistatic effects could reconcile past inconsistencies and account for some of the missing heritability of blood pressure and are generally relevant to SNP association studies for any phenotype. Thus, Scurrah et al. in 2017, tested 88 tagSNPs in 2872 white individuals in 809 pedigrees from the Victorian Family Heart Study and showed sex-specific associations for 3 SNPs in men (rs2468523 and rs2478544 at AGT and rs11658531 at ACE) and only 1 SNP in women (rs12451328 at ACE) [109, 128]. The earlier PAMELA study of 3705 SNPs from 1550 patients showed that there are sexual dimorphisms in the association of blood pressure with the state of several SNPs [110].

It seems likely that salt sensitivity and its sex-specific prevalence can also be genetically based. Extensive research efforts have identified genes in the RAAS, ion and water channels, transporters, and exchangers, endothelial system, apelin-APJ system, sympathetic nervous system, intracellular messengers and kallikrein-kinin system, and many others related to this complex phenotype [65].

Novel methods of estimation salt sensitivity are in progress. Commonly used tests for diagnosis of salt-sensitive hypertension (SSH) are complex and time-consuming, so new methods are required. Many studies have demonstrated roles of miRNAs in hypertension, and a potential diagnostic value of miRNAs for human salt sensitivity has been reported. In a recent study, of Qi et al. [116], it was shown that hsa-miR-361-5p combined with dietary factors can be a good marker for diagnosis of salt sensitivity.

Antihypertensive Drugs and Gender

Data from large clinical trials and a meta-analysis offer strong evidence that the preventive effectiveness of the various drug classes does not differ by sex, and therefore the choice of the drug cannot be based on these criteria in postmenopausal women. There are currently no specific BP treatment goals for postmenopausal hypertension. However, there are studies indicating that diuretics are more effective in postmenopausal females, while beta-blockers have higher blood pressure lowering effects in males [144].

Neuromodulation in Hypertension Treatment: Gender Effect

There is again a growing evidence that essential hypertension is commonly neurogenic and is initiated and sustained by sympathetic nervous system overactivity. Potential mechanisms include increased central sympathetic outflow, altered norepinephrine (NE) neuronal reuptake, diminished arterial baroreflex dampening of sympathetic nerve traffic, and sympathetic neuromodulation by angiotensin II. During the last decade, novel invasive approaches for the management of resistant hypertension are rapidly spreading, including renal denervation (RDN), chronic baroreflex activation therapy (BAT), vagus stimulation, and others.

Although the first results from studies suggested important benefits regarding blood pressure (BP) control in resistant hypertension by the use of diverse systems of RDN in the setting of resistant hypertension, the Symplicity HTN-3, randomized sham-controlled trial, reduced the enthusiasm and led to a more critical approach toward this neuromodulation innovative therapy. There are preliminary data that the procedure is more effective in males and in younger females [137]. In several studies, BAT has been demonstrated to decrease office blood pressure and ABPM; it was equally effective in both sexes of all ages [50, 52, 142].

Hormone Replacement Therapy in Females

It is well known that hormone replacement therapy (HRT) is a solitary option to reliably eliminate climacteric disorders and to improve significantly the quality of life in postmenopausal women [94, 136]. The problem of HRT administration has become particularly relevant after HERS (the Heart and Estrogen-Progestin Replacement Study, [57]), HERS II [48], and WHI (Women's Health Initiative, [148], 2004) studies have been terminated showing some increase in acute cardiovascular events in females taking HRT compared to placebo groups. The reasons for such results have been widely discussed in the literature. The results of prospective clinical trials, as well as the data of some experiments, have shown beneficial effects of HRT on both total and cardiovascular mortality and level of left ventricle hypertrophy and left ventricle performance [49, 93, 124]. Secondary analyses of the controversial results of HRT clinical trials have found a number of concomitant factors contributing to HRT effects: age and time of treatment onset, comorbidities, way of administration, and doses of estrogens and progestins. By now, the majority of authors agree that administration of HRT to healthy females under age of 60 without underlying cardiovascular diseases and significant atherosclerosis in early menopause is safe and may decrease the risk of future cardiovascular disease [45, 53, 108, 136]. So far, the main indication for HRT is still the presence of climacteric syndrome. However, it is time to acknowledge that some consideration might be given to HRT as a prevention strategy. It should be started early after menopause but at least within 10 years after, and low doses at the beginning and non-oral estrogens are preferable, combined with progesterone 10 days per month for females with a uterus. Five to ten years of therapy at the beginning of menopause may have potentially valuable effect in many years. Thus, estradiol therapy was associated with less progression of subclinical atherosclerosis (measured as CIMT) than placebo when therapy was initiated within 6 years after menopause [54].

A new era is opening for personalized approaches, which enables prescribing according to a woman's genotype, for example, estrogen receptors. In preclinical studies, it was shown that 17β-estradiol (E2) inhibits the vascular injury response in young rodents, E2 inhibits TNF-α-induced inflammation in isolated rat aortic smooth muscle cells, and E2 has age-dependent effects on neointima formation and inflammation in vivo and in vitro.

Effects of HRT on blood pressure are mixed and may depend on age of initiation of therapy, type, dose, and duration of treatment as well as initial health status such as existing conventional risk factors (e.g., body mass index, blood pressure, lipid profile, fasting glucose, and smoking status) [149]. Current treatment guidelines do not recommend the use of HRT for stroke prevention or recurrence [46, 51, 85]. Current drug development strategies focus on targeted estrogen mimetics yielding benefit without unwanted side effects. That can better modulate specific ER-α-regulated pathways involved in CVD development. This could be achieved possibly by fusion peptides [41] or through novel selective estrogen receptor modulators (SERMs), which retain the beneficial metabolic effects of E2 in desired tissues, while exerting minimal or antagonistic effects on estrogen receptors in the breast and uterus. Future studies should focus on identifying the critical brain sites where estrogen receptors regulate body weight and signaling pathways that are required for estrogen action. Additionally, determining the functional role and molecular mechanisms of estrogen receptors' action in immune cells, the skeletal muscle may reveal additional pharmacological targets for therapeutic intervention.

Testosterone Therapy in Males

Over the last 10 years, there have been mounting concerns that testosterone therapy (TTh) may increase the risk for CV events, such as heart attacks and strokes [42, 140]. However, an increased CV risk was reported only in these two studies, whereas in all other reports, a neutral or beneficial effect of TTh was found. In particular, in 2017, the largest observational study published so far became available [21]. The study evaluated 44,335 male patients at Kaiser Permanente medical centers in Northern and Southern California who had been diagnosed with androgen deficiency. Of these, 8808 men were treated with TTh, whereas 35,527 were never dispensed with TTh. Among men with androgen deficiency, those who were dispensed with TTh prescriptions were associated with a lower risk of CV outcomes. Epidemiological studies have failed to find an association between endogenous TTh levels and risk of prostate cancer.

TTh is able to improve all aspects of sexual function independently of the pathogenetic origin of the disease. CV safety concerns related to TTh are essentially based on a limited number of observational and randomized controlled trials which present important methodological flaws. When HG is properly diagnosed and TTh correctly performed, no CV and prostate risks have been documented [24].

Obesity and Metabolic Syndrome

Molecular, Genetic, and Pathophysiology Gendered Innovations

Experimental data have revealed that X chromosome dosage (XX chromosome complement in females vs. XY in males) influences food intake, which affects adiposity and metabolic disorders. To optimize prevention and treatment strategies for obesity, it is important to understand sex differences in fat storage and their association with the development of insulin resistance [121].

Meta-analyses of 114 studies (up to 320,485 individuals) with genome-wide chip and/or Metabochip data by the Genetic Investigation of Anthropometric Trials (GIANT) Consortium—large-scale genome-wide interaction study—showed sex-specific effects or age-specific effects that differed between men and women. There were identified sex specificities of genetic effects of waist-hip ratio on 44 loci with sex-specific effects of which 28 showed larger effects in women than in men and five showed larger effects in men than in women. These results may provide further insights into the biology that underlies weight change with age or the sexually dimorphism of body shape [146].

A cross-sectional analysis was performed using data from the Dallas Heart Study, a multiethnic population-based study. Significant sex-based differences were observed in multiple biomarkers reflecting pathways of cardiovascular risk. Women have higher levels of leptin and D-dimer and lower levels of biomarkers reflecting endothelial dysfunction and inflammatory cell recruitment. Body composition and menopausal status have important influence on the sex-based differences in biomarkers and CVD between men and women [74].

Gender differences were revealed in serum leptin and adiponectin levels—in women with unhealthy abdominal obesity (abdominal obesity accompanied by metabolic syndrome), serum leptin concentration was higher, and serum adiponectin was lower than in healthy abdominal obesity women (abdominal obesity not accompanied by metabolic syndrome and cardiovascular diseases). There were no differences in serum leptin and adiponectin levels in males patients with healthy and unhealthy abdominal obesity [8]. According to our data, the risk of metabolic syndrome is associated with adiponectin levels in women but not in men (Fig. 40.2).

Neck circumference is independently associated with elevated muscle sympathetic nerve activity in overweight and obese men but not in women and thus may be relevant to cardiometabolic risk prediction [132].

Lower muscle mass is positively associated with the risk of metabolic syndrome in Taiwanese middle-aged and elderly females but not in males. Thus preventing loss of muscle mass at older ages might have the potential to reduce metabolic syndrome, and females should be more aware of their loss of muscle mass problem [107].

Diabetic (metabolic) cardiomyopathy is more prevalent in diabetic women than in diabetic men—in women, diabetes outweighs the incidence of heart disease by five times, whereas in men heart disease is two times as common in diabetes. Heart failure with preserved ejection fraction (HFpEF) is equally present with heart failure with reduced ejection fraction (HFrEF) in women, while in men HFrEF occurs more frequently. HFpEF has a strong link with metabolic syndrome, diabetes mellitus, and increase in glycolysis relative to mitochondrial oxidative metabolism including glucose oxidation. Peroxisome proliferator-activated receptor alpha (PPARα) may play a role in sex difference in metabolic cardiomyopathies, and PPARα expression in the heart is lower in female mice compared to male mice. PPARα and NOS are well-known regulators of metabolism and calcium handling and mediate development of heart failure. The role for these signaling pathways in regulating sex differences of HFrEF is proposed. As females

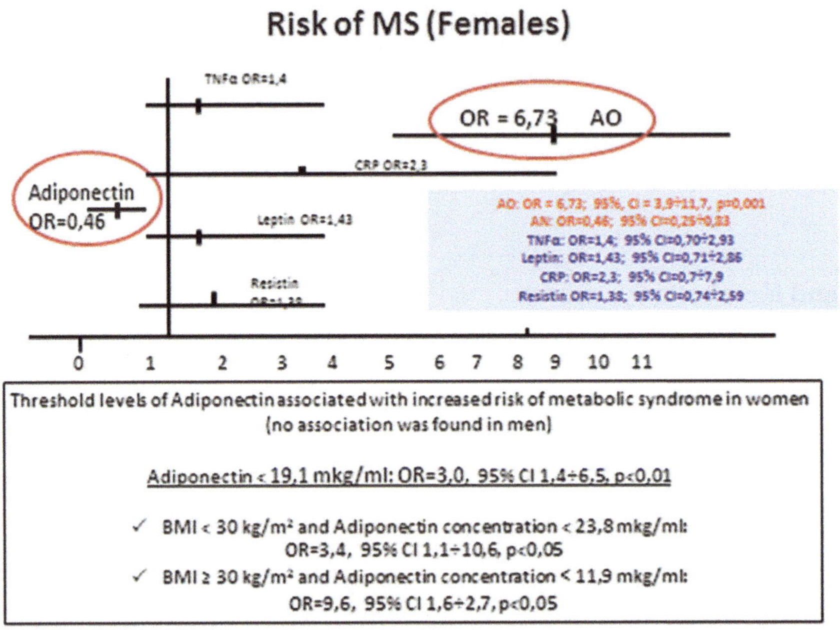

Fig. 40.2 Statistical analysis revealed that abdominal obesity (AO) is a strict predictor of metabolic syndrome (MetS) diagnosed by IDF criteria. AO is associated with more than sixfold increase of the risk of MetS. Additional MetS risk was associated with low adiponectin level. In females, adiponectin levels below 19 μg/mL led to threefold increase of MetS risk. Moreover, in BMI ≥30 kg/m2 and adiponectin below 12 μg/mL, MetS risk is augmented tenfold in women. So hypoadiponectinemia could be an additional MetS component and/or possible pathogenetic factor

have increased baseline NOS, they would be more susceptible to NOS uncoupling. Further studies are needed to test this hypothesis [96].

Gendered Innovations of Obesity and Metabolic Syndrome Therapy

Continuous positive airway pressure (CPAP) treatment relieves obstructive sleep apnea (OSA) symptoms, reduces CV risk factors, and improves functional status in both males and females, although women's pathophysiological consequences of CPAP therapy are different from men—improvement of endothelial dysfunction and chronic inflammation in females with OSA are less obvious than in males, but CPAP treatment brings down the cortisol level among women with OSA [62, 69]. CPAP therapy with high adherence has excellent therapeutic efficacy in long-term CPAP users. Adherence is both age- and gender-dependent, but the differences are small and not clinically relevant [147].

A decade analysis between 2002 and 2011 of trends and outcomes of 810,999 patients who underwent bariatric surgery demonstrated a small male fraction (19.3%) compared with the number of female patients (80.7%). Men undergoing bariatric surgery tend to have higher severity of illness, with higher risk-adjusted serious morbidity and mortality rates. Additional studies are necessary to examine barriers in obtaining treatment for obese men [151].

One half of all bariatric procedures are carried out on women of reproductive age. At least one in three of these women has menstrual dysfunction. Data from the National Bariatric Surgery Registry (NBSR) (15,222 women) and Health Survey for England (HSE, 2007–2013) (1073 women) were compared. Bariatric surgery improved factors that underlie fertility and pregnancy outcomes: in the postoperative period, the prevalence of type 2 diabetes fell by 54%, polycystic ovarian syndrome by 15%, and any menstrual dysfunction by 12%.

A prospective study is required to verify these effects [34].

Coronary Artery Disease

Numerous studies have shown that coronary artery disease occurs more frequently in men than in women (7.9 and 5.1% in men and women >20 years old, respectively) [47]. This trend reverses in older age (>75 years), possibly due to menopause and related hormonal changes in women. In a study on postmenopausal women, Tanko et al. showed that severity of osteoporosis—a markedly high disorder in older women—is associated with the risk of coronary event (s) independent from other risk factors. Anand et al. studied the progression of coronary artery calcification in type 2 diabetic patients and deduced that male sex was a risk factor for this disease. Their results were based on glycated hemoglobin (HbA_1c), lipid profile, high-sensitivity C-reactive protein (hs-CRP), interleukin(IL)-6, and plasma OPG levels, which showed that coronary artery calcification progressed only in patients with pre-existing conditions. In another study, sex difference in plaque characteristics was examined in patients with coronary stenotic lesions [98]. They measured vessel volume, plaque volume, and calcium content in the plaque in coronary lesions in elderly patients and found that women had greater amounts of calcium. This supports other reports that women have an increased rate of mortality following coronary interventions compared to men over 65 years [48, 84]. In addition, myocardial infarction and coronary artery bypass grafting surgery had more adverse outcomes in younger women than men of a similar age group [11, 141]. The rate of inhospital mortality following a bypass surgery was greater in women in all stratified age groups, but the risk of death decreased with age [139]. Another study reported that when other pre-existing conditions, such as diabetes, were not accounted for, the surgical mortality rate was greater in women [1]. They observed that female sex was at risk for wound infection and had longer hospital and post-surgery stays, but also had lower neurological complications and re-sternotomy bleeding.

Coronary artery anomalies are a diverse group of congenital disorders described as anomalous, abnormal, or atypical coronary arteries [4, 10]. Aydar et al. studied sex difference in different subtypes of coronary artery anomaly. The anomalous origin of the coronary arteries is the most common form of coronary artery anomaly [2]. They found that for a common type of anomalous origin, the circumflex and left anterior descending artery originating from separate ostia in the left aortic sinus had notably greater incidents in women than in men. On the other hand, the second type of circumflex that originated from right aortic sinus had similar frequency in both sexes. Likewise, the distribution of the second type of coronary artery anomaly, myocardial bridge, was similar between men and women, but the mean length was significantly greater in men. Finally, coronary artery fistulas had no significant difference in the origin as well as termination site of the fistulas between the sexes [4]. The frequency of small or large fistulas was also similar in men and women.

The challenges related to gendered innovations are still under investigation, as we could see a great diversity of the results of the trials evaluating gender effect on the course of cardiovascular diseases. To detect gender differences, a careful analysis of all risk factors of the disease development and progression at all stages of pathogenesis is mandatory. Careful planning of medical research with enrollment of female patients with all morphological, physiological, and social features typical for both genders also seems important. That would lead to more evident results and to the development of highly effective personalized medicine.

Innovations in Cardiac Imaging

Cardiovascular screening of women is a challenge. Using traditional risk factors, a majority of women are being stratified to a lower-risk group than men. Obviously, obstructive coronary artery disease (CAD) is more prevalent in men than in women. Nevertheless, mortality rates for cardiovascular disease are higher for women [95]. Cardiovascular risk in women is associated with significantly impaired coronary flow reserve (CFR) due to microvascular dysfunction and vessel remodeling, which may contribute to myocardial ischemia and adverse cardiac events. Recently, Taqueti et al. [135] have demonstrated that impaired CFR was similarly observed among symptomatic women and men, but in women low CFR (<1.6) was frequently associated with non-obstructive CAD, whereas in men severely obstructive CAD significantly contributed to diminished CFR ($P = 0.002$). Positron emission tomography (PET) with flow radiopharmaceuticals was used to apply for quantification of myocardial blood flow (MBF) for research. Nowadays cardiac PET is optimal and well-established tool for non-invasive measurement of MBF and CFR. Wide introduction of PET in clinical practice has changed a paradigm in the diagnosis and management of patients with CAD by shifted focus from an anatomical assessment coronary angiogram to a functional evaluation of microcirculation.

The PET Prognosis Multicenter Registry described in a total of 6307 patients that stress myocardial perfusion with ^{82}Rb-PET provided clinically meaningful risk stratification in both women and men. There was reported a proportional relationship between CAD mortality and myocardial perfusion defect size and extent at stress myocardial perfusion study, as assessed by percentage of involved myocardium [64]. The unadjusted 5-year CAD mortality ranged from 0.9% to 12.9% for women ($P < 0.0001$) and from 1.5% to 17.4% for men ($P < 0.0001$). The WISE study demonstrated a subgroup of symptomatic women with non-obstructive CAD that showed a microvascular dysfunction defined as CFR <2.5 in response to adenosine [86]. A majority of perfusion PET studies were traditionally performed with flow tracers ^{13}N-ammonia and ^{82}Rb. However, there are some disadvantages of the traditional radiopharmaceuticals such as the suboptimal relation between absolute blood flow and myocardial extraction for ^{82}Rb and requirement of nearest cyclotron for ^{13}N-ammonia due to short isotope half-live (9.9 min). Novel ^{18}F-labeled agents for PET myocardial perfusion imaging can take the full advantage of PET superior spatial resolution. Long period of ^{18}F half-live (110 min) doesn't require a cyclotron on site. ^{18}Flurpiridaz has demonstrated an excellent relationship to flow in animal models and in humans has shown a very good diagnostic accuracy for the detection of significant CAD [30, 99]. (^{18}F-Fluoropentyl) triphenylphosphonium salt (^{18}F-FPTP) is a new promising myocardial PET imaging tracer. It shows high accumulation in cardiomyocytes and rapid clearance from the liver, which made it appropriative for quantitative analysis of myocardial blood flow and CFR [66]. Impaired CFR seems to be represented as a target for novel therapeutic approach in women with non-obstructive CAD.

The combination of the nuclear medicine technique and computed tomography (CT) in hybrid scanners may yield complementary information concerning calcium score, severity of obstructive CAD, and myocardial blood flow, because CT coronary angiography is an excellent noninvasive imaging modality to rule out obstructive CAD. The large study by Kelkar et al. investigated long-term prognosis after measurement of coronary artery calcium scoring among low-intermediate risk women and men. In this study, women compared to men had a greater prevalence of coronary artery calcinosis and a higher 15-year mortality. It should be noted that women were older than men in the study. Nevertheless, these findings support the addition of measurement of coronary calcium scoring in women may improve the risk stratification beyond traditional risk factors. Makaryus et al. [82] showed clear differences between males and females regarding total vessel calcium scores: men tended to have higher average calcium scores in each coronary artery than women with a greater tendency to have multiple vessel involvement. The results of single-center cohort study of gender influence in coronary CT angiography demonstrated that women had a lower value of coronary calcium score than men and underwent less invasive coronary angiography and revascularization than men. This may reflect a gender-specific difference in the balance between coronary calcification and obstructive coronary heart disease [102]. In the study by Nakanishi et al., which included 13,092 asymptomatic patients without known cardiovascular disease, the relative risk according to coronary artery calcification category did not differ between genders. Nafakhi et al. studied the influence of gender on the correlations between coronary artery calcification with aortic root calcification and pericardial fat volume in 130 patients with intermediate pretest probability of CAD. The study demonstrated the pericardial fat volume was significantly associated with coronary artery calcification in female patients, while aortic root calcification showed a significant association with coronary artery calcification in male patients.

Gender-related differences in progression of atherosclerosis were demonstrated in an animal study [114]. It was revealed that male gender is aggravated of lipid deposition and vascular senescence in ApoE−/− mice of advanced age. Noninvasive diagnostic tools seem to be attractive for assessment of atherosclerotic lesions especially at high risk of acute complications. Molecular imaging approaches for the detection of vulnerable plaques would contribute to prevention of acute CAD. Nuclear techniques such as PET and SPECT enable to investigate the biological phenomena, which lead to plaque rupture: inflammation, macrophage infiltration, apoptosis, microcalcification, etc. Hybrid scanners PET/CT and PET/MR provide with information concerning biochemical process with precise anatomic details. Seo et al. described the use of novel tracer for target p32 proteins on macrophages ^{64}Cu-labeled dendritic form of a cyclic nine-amino acid peptide, named LyP-1, in atherosclerotic ApoE−/− mice. Significant uptake of the tracer was showed after 2 h of biodistribution in the aortic root and descending aorta. Foss et al. developed the labeled compound ^{125}Iiodo-DPA-713, for detection of macrophage infiltrates in the aorta of ApoE−/− mice. Specific nanobody cAbVCAM-1-5, labeled with ^{18}F, was successfully applied by Bala et al. for identification of atherosclerotic lesions in the aortic arches of ApoE−/− mice. There was revealed a significant correlation between intensity of the tracer uptake and level of expression of the VCAM-1 receptor in atherosclerotic plaques. Minimal expression of the VCAM-1 receptors was noted in non-inflamed tissue and high expression of the receptors on activated macrophages.

Other relevant biomarkers of vulnerable plaque are as follows: P-selectin [75, 76] labeled by ^{68}Ga; ^{18}F-RGD-galacto peptide, a ligand to $\alpha v\beta 3$ receptor [70], or ^{18}F-flotegatide, a ligand to integrin [133]; extracellular matrix proteins ^{64}Cu-labeled GPVI [9]; specific ^{99m}Tc-antibody to fibronectin [29]; and labeled markers of apoptosis such as ^{99m}Tc-AnxF568 and ^{124}I-Hypericin [27]. Recently, contrast agents PEGylated gold nanoparticles of CT have been developed for detection of macrophage-rich atherosclerotic plaques [117]. Translation of labeled biomarkers of the vulnerable plaque to clinical practice might offer some novel therapeutic approaches such as targeted delivery of drug [44].

Heart Rhythm Disturbances

Sex Differences in Atrial Fibrillation Substrate

Catheter ablation of symptomatic atrial fibrillation (AF) is indicated in patients with drug-refractory arrhythmia. The primary target of ablation strategy is pulmonary vein electrical disconnection, aiming to cease pathological ectopic firing inducing and sustaining AF. Along with the most frequent localization of AF triggers in the pulmonary veins, there are other sources harboring ectopic activity. Progression of AF to persistent forms is associated with extended atrial fibrosis (AF substrate) and autonomic nervous system dysregulation. Differences in atrial fibrillation (AF) triggers' localization have been demonstrated in male and female patients, while previous research has shown that non-pulmonary vein triggers are more frequently found in females [73]. Moreover, AF ablation is thought to be less effective in female patients, while this finding has not been confirmed consistently in all studies. At the same time, female gender is associated with a higher risk of complications associated with both radiofrequency and cryoballoon AF ablation [91, 111].

There is a lack of data regarding sex differences in AF substrate in the left atrium. In one study the authors analyzed pulmonary vein sleeves obtained from 166 patients (85 females, 101 patients with long-standing persistent AF, 65 patients with no history of AF) [77]. The authors showed that there are significant gender differences in the fibrotic remodeling in patients with persistent AF. A higher degree of fibrosis was found in pulmonary vein samples obtained from females with AF. These gender differences were mainly associated with different expression of fibrosis-related genes and proteins. Up-regulation of the TGFβ/Smad3 pathway was suggested to aggravate fibrosis remodeling.

In our recent postmortem analysis of 41 hearts, obtained from patients with structural heart diseases (23 males and 18 females, 15 patients with AF), density of atrial fibrosis and intramyocardial nerves and ganglia was differently distributed in men and women. The maximum ganglia density in men was found around the coronary sinus ostium in the right atrium and between the lower pulmonary veins in the left atrium. In women maximum ganglia density was found between upper and lower pulmonary veins. Contrarily, fatty tissue was more prominent around coronary sinus ostium in women. These differences in atrial fibrosis and intramyocardial innervation may contribute to different localization of atrial arrhythmogenic substrate and ectopic focuses in men and women and consequently lead to different efficacy of pulmonary vein isolation. Pharmacological provocation of non-pulmonary vein triggering activity during AF ablation procedures is advised by some groups.

Radiation Exposure Minimization During Catheter Ablation for Atrial Fibrillation and Other Cardiac Arrhythmias

Radiation exposure is a well-recognized risk factor for cancer. It is estimated that about 1/1000 individuals will develop cancer from an exposure similar to what provided by CT scan or atrial fibrillation (AF) ablation. According to the European Society of Cardiology Atrial Fibrillation Ablation Long-Term Registry data, CT scan is performed in 77.5% of patients during hospitalization for AF ablation [3]. Therefore, considering combination of CT and intraprocedural fluoroscopy in the majority of patients is even higher. Susceptibility to radiation varies according to patient age and sex, and women of fertile age are more sensitive. After a 64-slice CT coronary angiography, the lifetime risk of radiation-related cancer was reported to be 2.4–4.8 times greater in women [35].

A study performed on implantable cardioverter-defibrillator (ICD) recipients showed that there is a 10% excess risk of cancers during a median follow-up of 2.8 years, confined to tobacco-related cancers in patients with ischemic heart disease without difference between genders [113].

Nowadays, innovative electrophysiological mapping technologies have been developed that can already permit “near-zero” fluoroscopic exposure during catheter ablation procedures. The use of these technologies has been associated with a 96% reduction of the risk of developing a procedure-related cancer without compromising efficacy and safety [17]. These results are of significant importance for women of fertile age. Since electrophysiological procedures could be offered to some pregnant women with drug-refractory life-threatening arrhythmia, radiation minimization technologies should be considered [28].

Gender Differences in Periprocedural Sedation During Catheter Ablation for Atrial Fibrillation and Other Cardiac Arrhythmias

Periprocedural sedation is an important issue in electrophysiological interventions and device implantation. Thus, in a study by Ezzat and co-authors, excess of pain was found the principal source of disappointment after AF ablation, with a 56% prevalence among all study patients [37]. One study showed that women experience more pain than men during AF ablation, while the active analgesic strategy was with reduced discomfort [18]. In another study the authors demonstrated that female sex was the only variable significantly associated with early post-procedural pain following either catheter ablation or device implantation at multivariate analysis [12].

These data are in favor of more extensive analgesia in women during cardiac implantable electronic device and catheter ablation procedures. Regarding AF ablation procedure-related pain reduction, cryoballoon pulmonary vein isolation has been associated with less discomfort [18].

Sex Differences in the Noninvasive SCD Risk Stratification

It is known that women have lower incidence of sudden cardiac death (SCD) than men [71], but women have higher inhospital mortality after cardiac arrest than men [67]. SCD is more often caused by CAD in men than in women (80% and 45%, resp.) [63]. Men and women with nonischemic cardiomyopathy and hypertrophic cardiomyopathy may be at similar risk for arrhythmic SCD. Despite men having a more severe arrhythmogenic right ventricular dysplasia phenotype, the incidence of life-threatening ventricular arrhythmias was similar in both sexes [19]. However, women with long QT syndrome have greater risk of cardiac events, both in drug-inducible QT prolongations and genetically determined [55, 60, 123]. The adult type of catecholaminergic polymorphic ventricular tachycardia (VT) is diagnosed in women about 40 times more often. Perhaps this is due to special genetic mutations or hormonal effects [7, 123].

Risk Stratification Markers

LV Dysfunction Women are less likely than men to have LV dysfunction at the time of SCD. Compared with men, women were more likely to have normal LV function (48.9% vs. 35.2%), underscoring the difficulty in predicting the risk of SCD for women.

Autonomic nervous system function differs between men and women. Resting vagal tone as assessed by heart rate variability (HRV) is higher in women. Women are less likely to have exertion-related SCD. Not only does resting autonomic tone differ between men and women, but differences in autonomic reactivity between men and women are well documented. Despite the popular stereotypes of sex and emotion, women have less adrenergic response to induced mental stress [19].

The frequency of ventricular premature complexes is lower in the ovulation period [32]. Idiopathic VTs with the right ventricular outflow tract are more often registered in women [97].

Pregnancy predisposes to arrhythmias: different arrhythmias during pregnancy are registered in 1–4% of women without structural heart disease. Usually, these arrhythmias are benign and are associated with an imbalance of the autonomic neural system, electrolyte disturbances, and hormonal changes, and the presence of an arrhythmic anamnesis is the most significant factor in the risk stratification [14, 36].

Several ECG-based parameters, including fragmentation of QRS (fQRS) complex, heart rate turbulence (HRT), and microvolt T-wave alternans (mTWA) have been widely studied in the context of prediction of ventricular arrhythmias in different groups of patients. Gender differences of ECG markers are studied in patients with ventricular arrhythmias (VA), mostly in patients with CAD.

Pathological HRT has not been shown to be gender related in patients after myocardial infarction [129]. In only one study, it was identified that women with CAD were more likely to have higher levels of a turbulence onset parameter [129].

Backward, mTWA is considered a gender-related phenomenon in patients with CAD and congenital heart diseases. Pathological mTWA increased risk of arrhythmic events in men [25].

The prognostic value of fragmented QRS and other ECG abnormalities was confirmed in postmenopausal women with or without risk factors for CAD [126].

The group of patients with ventricular arrhythmia (VA) without structural heart disease is still less investigated. *In our center we studied ECG markers of electrical myocardial instability* (EMI) (VT, fQRS in sinus and premature ventricular complexes (PVC), turbulence onset (TO) and turbulence slope (TS), and mTWA) in patients with idiopathic VA and pregnant women without structural heart disease: 69 women aged 47 ± 9 years, 65 pregnant women aged 29.5 ± 5.5 years, and 62 men (mean age 44 ± 12 years). Abnormal ECG electrical myocardial instability markers were identified in all groups. There were differences between groups of men and women with idiopathic VA. In the group of males with idiopathic VA, there was the larger number of pathological fQRS and mTWA, which suggests that this group requires more focused follow-up. In the groups of women, it is possible to assume that female hormones have a protective effect, especially during pregnancy, although the younger age group of pregnant women should be considered too.

Traditional risk factors may play only a minor role in SCD in patients with idiopathic VA. Some investigators have suggested this protection to be caused by female endogenous estrogen [89]. The so-called estrogen-effect has been considered to have both long-term and rapid effects, both through an atheroprotective effect on serum lipid concentrations [16] and through a direct action on blood vessels. On the other hand, researchers have suggested that male sex hormones decrease life expectancy in men. A previous study on human eunuchs (castrated males) found that their lifespan was 14 years longer than the non-castrated males [92].

It is also known that autonomic nervous system function differs between men and women. Resting vagal tone as assessed by heart rate variability (HRV) is higher in women. Women are less likely to have exertion-related SCD [19]. Despite the popular stereotypes of sex and emotion, women have less adrenergic response to induced mental stress [19].

In the majority of cases, there appears to be a protective role of female sex hormones, although in some cases, hormonal effects can have an arrhythmogenic effect. Many studies have confirmed significant differences between men and women in arrhythmogenesis and in *the pathophysiological foundations of SCD*. This means that a gender approach is strongly indicated in risk stratification of SCD.

Thus, further work on the identification of gender differences will make it possible to expand the possibilities of modern preventive cardiology. We should try to move from identifying the main risk factors and predictors to selective use of screening diagnostic methods. It will help to develop effective preventive options that can improve the quality of life and reduce the risk of SCD, both in female and male patients.

Cardiomyopathies

The vast majority of dilated cardiomyopathy (DCM) cases is characterized by more severe clinical symptoms, lower life expectancy, and higher frequency of sudden cardiac death in males. In contrast, hypertrophic cardiomyopathy (HCM) in women is characterized by higher frequency and severity of congestive heart failure

Table 40.1 Sex-associated cardiomyopathies

Name	Predominant gender	Clinical characteristics
Dilated cardiomyopathy caused by DYS mutations	Males only	Disease manifestation in yearly teens. Rapid DCM progression with highly elevated creatine kinase level. Subclinical or manifested myopathic syndrome. No respiratory deficiency
Takotsubo cardiomyopathy	Females aged 55–75 (90%)	Clinical manifestation of acute MI after severe emotional stress. Rapid but reversal decline of contractile myocardial function, possibly, with obstruction of LVOT. Decreased density of myocardial ß1 adrenoceptors
Peripartum cardiomyopathy	Females in late gestation and several months after delivery	Rapid DCM progression often accompanied by myocardial microvascular dysfunction and preeclampsia. Familial history of DCM can be traced. Partly reversible at the beginning, but DCM progression can happen later in life with poor prognosis

(Table 40.1). This difference is most probably caused by a change in hormone status in teenagers with familial HCM and alteration in women during postmenopause.

In familial forms of HCM, risk of sudden cardiac death (SCD) as well as life-threatening ventricular arrhythmias is higher between 8 and 16 years of age [106]. Importantly, in males the average age of the time of the diagnosis being established is younger compared to females (38 ± 18 compared to 47 ± 23). In females with familial forms of HCM, the condition is independently associated with congestive heart failure progression (class III/IV NYHA) or with risk of death from heart failure or stroke [90]. In older groups of HCM females, the fraction of shortening is usually higher compared to males of the similar age [31].

In familial dilated cardiomyopathy, the disease manifestation usually occurs in the young and middle age groups. At the age below 50, the disease occurs two to three times more frequent in males compared to females, mostly due to dystrophic-associated dilated cardiomyopathy. In males with DCM compared with females, a frequency of reduced ejection fraction, end-stage congestive heart failure, and life threatening ventricular arrhythmias is higher. The morbidity in males with DCM is also higher than in females [90]. In contrast, the age-dependent regression of ß1 adrenoceptors is more prominent in women, giving more evidence for sex hormone-dependent signs of heart failure in women. This may explain higher frequency of takotsubo cardiomyopathy in women after menopause. The latter occurs predominantly in females (90%) age 55–75 after severe emotional stress and reflects the decreased density of ß1 adrenoceptors in combination with decreased estrogenic production and increased sensitivity of myocardial tissue to catecholamine stimulation [79].

Males with myocarditis have more severe clinical signs [90] and higher induction of matrix proteins, in particular collagen expression and matrix metalloproteases. Higher level of testosterone in men with myocarditis promotes myocardial inflammation and fibrosis, leading to DCM transformation and heart failure [38].

In alcoholic cardiomyopathy, the mortality is higher in males. It is particularly high in males and females of Afro-American origin. In females, alcoholic liver disease is found to be more often accompanied by alcoholic cardiomyopathy, as well as by brain and skeletal muscle damage in spite of the fact that the average amount of alcohol intake in females is only 60% of that in males [138]. Moreover, the average dose in females leading to cardiomyopathy is lower than in males [40, 90].

In familial transthyretin (TTR) amyloidosis, cardiac involvement occurs more often in males rather than in females. The same is valid for the degree of left ventricular hypertrophy. In women, the degree of myocardial hypertrophy depends on age and is higher in postmenopause. Males with TTR amyloidosis and cardiac involvement at the similar age have less degree of hypertrophy underlining the role of sex hormones in TTR amyloid cardiomyopathy. In women with late onset of amyloid cardiomyopathy and polyneuropathy, the survival rate after heart transplantation is higher compared to male patients who were transplanted. In contrast to familial TTR amyloidosis, wild-type (wt)-TTR amyloidosis occurs in males more often pointing to the role of sex hormones in liver synthesis of transthyretin. The slow rate of sex hormone declines in males leads to a higher frequency of wt-TTR amyloidosis and cardiomyopathy. Female sex hormones also have an impact on transthyretin synthesis in the liver. Thus, 15-ß-dihydrotestosterone has a stronger induction effect on transthyretin synthesis compared to 17ß-estradiol [90].

Several types of cardiomyopathy occur only in males or in females by their nature, for example, dystrophin-associated cardiomyopathy or peripartum cardiomyopathy (Table 40.1). The latter occurs before or shortly after delivery in the frame of unexpected and severe systolic dysfunction. This condition has complex etiology including immune response and microvascular dysfunction. However, the role of genetic predisposition and TNN and other sarcomeric gene variants in peripartum cardiomyopathy is emerging making the predisposition to this condition potentially suitable for predictive personalized medicine. The combination of genetic predisposition, viral genome, and endothelial dysfunction upon pregnancy leads to development of severe reduction of systolic function and peripartum cardiomyopathy [115].

Diseases of the Aorta and Aortic Valve

Aneurysm

It is known that the incidence of aortic dilatation is higher in men than in women and increases with age [104]. Further analysis has shown that compared with men, women were more likely to suffer perioperative cardiac complications during open surgical repair of the aorta but less susceptible to acute kidney failure [78].

Despite the detrimental effects of male sex hormones in experimental aneurysm of the abdominal aorta (AAA), it seems that lower testosterone levels in men are associated with aortic dilatation, and physiological levels of endogenous male sex hormones mediate protection [150]. However, further studies are required to establish more conclusive results. Currently, there are a few animal studies suggesting a protective role of estrogens. The use of exogenous female sex hormones in menopausal women has been found to be contradictory [56, 58, 72]. Finally, differential molecular mechanisms of sex hormones can provide potential therapeutic goals for AAA [83]. Davis et al. have shown that female mesenchymal stem cells (MSC) more significantly inhibit aneurysm progression and inflammation in the aortic wall. Furthermore, they demonstrated that priming male MSC with finasteride resulted in therapeutic and anti-inflammatory characteristics similar to those of female MSC. Perhaps the application of MSC from different gender sources will provide hope for their future use in biological methods of treating AAA [26].

Aortic Stenosis

There are no sex differences in the aortic stenosis occurrence in patients with a tricuspid aortic valve, while bicuspid aortic valve is three times more common among men than women. Despite the fact that the incidence of the bicuspid aortic valve is higher in men, in men the manifestation of the disease is more likely to occur with aortic insufficiency, while in women it is more often manifested with aortic stenosis [68]. Estrogen receptors regulate transcriptional pathways of proteins affecting myocardial hypertrophy or interrupt calcium channel expression on the cell membrane. It was shown that β-estrogen receptor mRNA is significantly increased in patients with aortic stenosis and more pronounced in women [101]. Study of healthy porcine valves identified 183 significantly different genes in male and female; many of those are relevant to valvular heart diseases [88]. In men valvular interstitial cells are more prone to calcification. β-Estradiol in women has a protective effect in heart valve diseases [20]. Furthermore inhibitory effect of the β-estradiol on the rat aortic valve interstitial cells proliferation was found only in female. It's possible that men and women have different density of estrogen receptors in valvular interstitial cells or different density of predisposed to osteogenic differentiation cells in the aortic valve [87].

Conclusions

This survey demonstrates how integrating gender into strategies for new technologies can lead to more streamlined utilization of research evidence and more tailored interventions.

In cardiology we can describe gendered innovations as the following:

1. Analyzing sex in clinical research that can lead to an understanding that diseases in women often have a different pathophysiology than in men.
2. New diagnostic and treatment techniques (can be still experimental)—some are more or less effective than old ones in women and in men, respectively.
3. Understanding sex differences in symptoms can lead to earlier and better diagnosis of CVD in men and women.

Table 40.2 Future opportunities to search for gendered innovations in cardiovascular medicine

Methodology	Major perspectives
Epidemiology	Gender disproportion in clinical studies and outcome studies should be avoided. Causes of discrepancies in cardiovascular outcome can be explained and new prevention strategies implemented
Biological mechanisms of sex differences	Meaning of inactivation of second X chromosome
	Gendered genetics and epigenetics
	Hormones and receptors, membrane and nuclear receptors to sex steroids, cardiovascular action, and role in atherosclerosis
	Nutritional genomics
Sex-specific factors that alter cardiovascular risk	Menopause and andropause
	Pregnancy
	POS, POF, etc.
Sex-specific diagnosis and treatment	Gendered visualization
	Stress-testing
	Personalized treatment according to differences in outcome during treatment
	Sex-specific drugs and devices
Reproductive history	Reproductive technologies and risk
	Pregnancy and CVD programming

4. Rethinking the cardioprotective effect of hormone replacement therapy.

We will have to focus more on sex in our future literature reviews, formulate scientific hypotheses that let us include sex as a contributing factor, plan experiments and collect data accordingly, and pay special attention to sex in data analysis.

To investigate the influence of sex on CVD development, one must therefore not only include samples from females and males but also design a methodological framework by which this influence can be analyzed. For example, to avoid sex being either overrepresented or introducing a bias in the analysis of results, similar numbers of male and female subjects must be included in every clinical study of new drugs, technologies, biomarkers, and devices (Table 40.2).

References

1. Al-Alao BS, Parissis H, McGovern E, Tolan M, Young VK. Gender influence in isolated coronary artery bypass graft surgery: a propensity match score analysis of early outcomes. Gen Thorac Cardiovasc Surg. 2012;60:417–24.
2. Angelini P. Coronary artery anomalies—current clinical issues: definitions, classification, incidence, clinical relevance, and treatment guidelines. Tex Heart Inst J. 2002;29:271–8.
3. Arbelo E, Brugada P, Blomstrom-Lundquist C, et al. Contemporary management of patients undergoing atrial fibrillation ablation: in-hospital and 1-year follow-up findings from the ESC-EHRA atrial fibrillation ablation long-term registry. Europace. 2017;38(17):1303–16.
4. Aydar Y, Yazici HU, Birdane A, Nasifov M, Nadir A, Ulus T, et al. Gender differences in the types and frequency of coronary artery anomalies. Tohoku J Exp Med. 2011;225:239–47.
5. Bae S, Zhang L. Gender differences in cardioprotection against ischemia/reperfusion injury in adult rat hearts: focus on Akt and protein kinase C signaling. J Pharmacol Exp Ther. 2005;315:1125–35.
6. Bayorh MA, et al. The role of gender in salt-induced hypertension. Clin Exp Hypertens. 2001;23:241–55.
7. Behere SP, Weindling SN. Catecholaminergic polymorphic ventricular tachycardia: an exciting new era. Ann Pediatr Cardiol. 2016;9(2):137–146. https://doi.org/10.4103/0974-2069.180645
8. Berezina A, Belyaeva O, Berkovich O, Baranova E, Bazhenova E, Brovin D, Grineva E, Shlyakhto E. Prevalence, risk factors, and genetic traits in metabolically healthy and unhealthy obese individuals. BioMed Res Int. 2015;2015:548734. https://doi.org/10.1155/2015/548734.
9. Bigalke B, Phinikaridou A, Andia ME, et al. Positron emission tomography/computed tomographic and magnetic resonance imaging in a murine model of progressive atherosclerosis using 64Cu-labeled glycoprotein VI-Fc. Circ Cardiovasc Imaging. 2013;6:957–64.
10. Blake HA, Manion WC, Mattingly TW, Baroldi G. Coronary artery anomalies. Circulation. 1964;30:927–40.
11. Blankstein R, Ward RP, Arnsdorf A, Jones B, Lou YB, Pine M. Female gender is an independent predictor of operative mortality after coronary artery bypass graft surgery: contemporary analysis of 31 Midwestern hospitals. Circulation. 2005;112:I323–7.
12. Bode K, Breithardt OA, Kreuzhuber M, Mende M, Sommer P, Richter S, et al. Patient discomfort following catheter ablation and rhythm device surgery. Europace. 2015;17:1129–35.

13. Booth EA, et al. Activation of estrogen receptor-alpha protects the in vivo rabbit heart from ischemia-reperfusion injury. Am J Physiol Heart Circ Physiol. 2005;289:H2039–H204.
14. Boriani G, Lorenzetti S, Cerbai E, Oreto G, et al. The effects of gender on electrical therapies for the heart: physiology, epidemiology, and access to therapies: a report from the XII Congress of the Italian Association on Arrhythmology and Cardiostimulation (AIAC). Europace. 2017;19:1418–26. https://doi.org/10.1093/europace/eux068.
15. Bouma W, et al. Sex-related resistance to myocardial ischemia-reperfusion injury is associated with high constitutive ARC expression. Am J Physiol Heart Circ Physiol. 2010;298:H1510–7.
16. Bush TL, Barrett-Connor E, Cowan LD, Criqui MH, Wallace RB, Suchindran C, et al. Cardiovascular mortality and noncontraceptive use of estrogen in women: results from the Lipid Research Clinics Program Follow-up Study. Circulation. 1987;75(6):1102–9.
17. Butter C, Schau T, Meyhoefer J, Neumann K, Minden HH, Engelhardt J. Radiation exposure of patient and physician during implantation and upgrade of cardiac resynchronization devices. Pacing Clin Electrophysiol. 2010;33:1003–12.
18. Carnlof C, Insulander P, Jensen-Urstad M. An active supply of analgesics during pulmonary vein isolation reduces pain. Scand Cardiovasc J. 2014;48:35–40.
19. Cha Y-M, Lloyd MA, Birgersdotter-Green UM. Arrhythmias in women: diagnosis and management. Oxford: Oxford University Press; 2014.
20. Chakrabarti S, Lekontseva O, Davidge ST. Estrogen is a modulator of vascular inflammation. IUBMB Life. 2008;60 (6):376–82.
21. Cheetham TC, An J, Jacobsen SJ, Niu F, Sidney S, Quesenberry CP, et al. Association of testosterone replacement with cardiovascular outcomes among men with androgen deficiency. JAMA Int Med. 2017;177:491–9.
22. Chen Q, et al. An association between gene expression and better survival in female mice following myocardial infarction. J Mol Cell Cardiol. 2010;49:801–11.
23. Consideration of Sex as a Biological Variable in NIH-funded Research. 2015. https://grants.nih.gov/grants/guide/notice-files/NOT-OD-15-102.html
24. Corona G, Sforza A, Maggi M. Testosterone replacement therapy: long-term safety and efficacy. World J Mens Health. 2017;35:e2.
25. Cygankiewicz I, Wranicz JK, et al. Clinical covariates of abnormal heart rate turbulence in coronary patients. Ann Noninvasive Electrocardiol. 2003;8(4):289–95.
26. Davis JP, Salmon M, Pope NH, Lu G, Su G, Sharma AK, Ailawadi G, Upchurch GR Jr. Attenuation of aortic aneurysms with stem cells from different genders. J Surg Res. 2015;199(1):249–58.
27. De Saint-Hubert M, Bauwens M, Deckers N, et al. In vivo molecular imaging of apoptosis and necrosis in atherosclerotic plaques using microSPECT-CT and microPET-CT imaging. Mol Imaging Biol. 2014;16:246–54.
28. Diemberger I, Marazzi R, Casella M, et al. The effects of gender on electrical therapies for the heart: procedural considerations, results and complications. Europace. 2017;19:1911–21. https://doi.org/10.1093/europace/eux034.
29. Dietrich T, Berndorff D, Heinrich T, et al. Targeted ED-B fibronectin SPECT in vivo imaging in experimental atherosclerosis. Q J Nucl Med Mol Imaging. 2015;59:228–37.
30. Dilsizian V, Taillefer R. Journey in evolution of nuclear cardiology: will there be another quantum leap with the F-18-labeled myocardial perfusion tracers? JACC Cardiovasc Imaging. 2012;5:1269–84.
31. Dimitrow PP, Czarnecka D, Kawecka-Jaszcz K, Dubiel JS. The influence of age on gender-specific differences in the left ventricular cavity size and contractility in patients with hypertrophic cardiomyopathy. Int J Cardiol. 2003;88 (1):11–6.
32. Dogan M, Yiginer O, Uz O, Kucuk U, et al. The effects of female sex hormones on ventricular premature beats and repolarization parameters in physiological menstrual cycle. Pacing Clin Electrophysiol. 2016;39:418–26.
33. Douglas PS, et al. Hypertrophic remodeling: gender differences in the early response to left ventricular pressure overload. J Am Coll Cardiol. 1998;32:1118–25.
34. Edson E, Whyte M, van Vlymen J, et al. Bariatric surgery in obese women of reproductive age improves conditions that underlie fertility and pregnancy outcomes: retrospective cohort study of National Bariatric Surgery Registry (NBSR). Obes Surg. 2016;26:2837–42.
35. Einstein AJ, Henzlova MJ, Rajagopalan S. Estimating risk of cancer associated with radiation exposure from 64-slice computed tomography coronary angiography. JAMA. 2007;298:317–23.
36. Enriquez AD, Economy KE, Tedrow UB. Contemporary management of arrhythmias during pregnancy. Circ Arrhythmia Electrophysiol. 2014;7:961–7.
37. Ezzat VA, Chew A, McCready JW, Lambiase PD, Chow AW, Lowe MD, et al. Catheter ablation of atrial fibrillation-patient satisfaction from a single-center UK experience. J Interv Cardiac Electrophysiol. 2013;37:291–303.
38. Fairweather D, Cooper LT, Blauwet LA. Sex and gender differences in myocarditis and dilated cardiomyopathy. Curr Probl Cardiol. 2013;38(1):7–46. https://doi.org/10.1016/j.cpcardiol.2012.07.003.
39. Fang L, et al. Differences in inflammation, MMP activation and collagen damage account for gender difference in murine cardiac rupture following myocardial infarction. J Mol Cell Cardiol. 2007;43:535–44.
40. Fernandez-Sola J, Estruch R, Nicolas JM, Pare JC, Sacanella E, Antunez E, Urbano-Marquez A. Comparison of alcoholic cardiomyopathy in women versus men. Am J Cardiol. 1997;80:481–5.
41. Finan B, Yang B, Ottaway N, Stemmer K, Muller TD, Yi CX, et al. Targeted estrogen delivery reverses the metabolic syndrome. Nat Med. 2012;18:1847–56.
42. Finkle WD, Greenland S, Ridgeway GK, Adams JL, Frasco MA, Cook MB, et al. Increased risk of non-fatal myocardial infarction following testosterone therapy prescription in men. PLoS One. 2014;9:e85805.
43. Gardner JD, et al. Gender differences in cardiac remodeling secondary to chronic volume overload. J Card Fail. 2002;8:101–7.
44. Gargiulo S, Gramanzini M, Mancini M. Molecular imaging of vulnerable atherosclerotic plaques in animal models. Int J Mol Sci. 2016;17:1511. https://doi.org/10.3390/ijms17091511.
45. Gebbie AE. Premature ovarian failure. Menopause Int. 2008;14:96.
46. Giordano S, Hage FG, Xing D, Chen Y-F, Allon S, Chen C-J, Oparil S. Estrogen and cardiovascular disease: is timing everything? Am J Med Sci. 2015;350:27–35.
47. Go AS, Mozaffarian D, Roger VL, Benjamin EJ, Berry JD, Blaha MJ, et al. Heart disease and stroke statistics-2014 update: a report from the American Heart Association. Circulation. 2014;129:e28–e292.

48. Grady D et al. Cardiovascular disease outcomes during 6.8 years of hormone therapy: heart and estrogen/progestin replacement study follow-up (HERS II). JAMA 2002;288 (1):49–57.
49. Grodstein F, Manson JE, Colditz GA, Willett WC, Speizer FE, Stampfer MJ. A prospective, observational study of postmenopausal hormone therapy and primary prevention of cardiovascular disease. Ann Int Med. 2000;133:933–41.
50. Halbach M, Hickethier T, Madershahian N, et al. Acute on/off effects and chronic blood pressure reduction after long-term baroreflex activation therapy in resistant hypertension. J Hypertens. 2015;33:1697–703.
51. Harvey R, Coffman KE, Miller VM. Women-specific factors to consider in risk, diagnosis and treatment of cardiovascular disease. Women's Health (Lond Engl). 2015;11:239–57.
52. Hevener AL, Clegg DJ, Mauvais-Jarvis F. Impaired estrogen receptor action in the pathogenesis of the metabolic syndrome. Mol Cell Endocrinol. 2015;418:306–21.
53. Hodis HN, Mack WJ, Azen SP, et al. For the Women's estrogen-progestin lipid lowering hormone atherosclerosis regression trial research group. N Engl J Med. 2003;349:535–45.
54. Hodis HN, Mack WJ, Henderson VW, Shoupe D, Budoff MJ, Hwang-Levine J, Li Y, Feng M, Dustin L, Kono N, Stanczyk FZ, Selzer RH, Azen SP, ELITE Research Group. Vascular effects of early versus late postmenopausal treatment with estradiol. N Engl J Med. 2016;374:1221–31.
55. Hreiche R, Morissette P, Turgeon J. Drug-induced long QT syndrome in women: review of current evidence and remaining gaps. Gend Med. 2008;5(2):124–35. doi:10.1016/j.genm.2008.05.005.
56. Hsia J, Criqui MH, Rodabough RJ, et al. Estrogen plus progestin and the risk of peripheral arterial disease: the Women's Health Initiative. Circulation. 2004;109:620e6.
57. Hulley S. Randomized trial of estrogen plus progestin for secondary prevention of coronary heart disease in postmenopausal women. JAMA. (1998);280(7):605.
58. Iribarren C, Darbinian JA, Go AS, et al. Traditional and novel risk factors for clinically diagnosed abdominal aortic aneurysm: the Kaiser multiphasic health checkup cohort study. Ann Epidemiol. 2007;17:669e78.
59. Jain M, et al. Influence of gender on the response to hemodynamic overload after myocardial infarction. Am J Physiol Heart Circ Physiol. 2002;283:H2544–50.
60. James AF, Choisy SC, Hancox JC. Recent advances in understanding sex differences in cardiac repolarization. Prog Biophys Mol Biol. 2007;94(3):265–319. Epub 2005 Jun 13
61. Johnson MS, et al. Sex differences in myocardial infarct size are abolished by sarcolemmal KATP channel blockade in rat. Am J Physiol Heart Circ Physiol. 2006;290:H2644–7.
62. Kallianos A, Panoutsopoulos A, Mermigkis C, et al. Sex differences of continuous positive airway pressure treatment on flow-mediated dilation in patients with obstructive sleep apnea syndrome. Clin Interv Aging. 2015;10:1361–7.
63. Kannel WB, Wilson PW, D'Agostino RB, Cobb J. Sudden coronary death in women. Am Heart J. 1998;136(2):205–12.
64. Kay J, Dorbala S, Goyal A, et al. Influence of sex on risk stratification with stress myocardial perfusion Rb-82 positron emission tomography: results from the PET Prognosis Multicenter Registry. J Am Coll Cardiol. 2013;62:1866–76.
65. Kelly TN, He J. Genomic epidemiology of blood pressure salt sensitivity. J Hypertens. 2012;30:861–73.
66. Kim JW, Seo S, Kim HS, et al. Comparative evaluation of the algorithms for parametric mapping of the novel myocardial PET imaging agent ^{18}F-FPTP. Ann Nucl Med. 2017;31:469–79. https://doi.org/10.1007/s12149-017-1171-6.
67. Kim LK, Looser P, Swaminathan RV, Horowitz J, et al. Sex-based disparities in incidence, treatment, and outcomes of cardiac arrest in the United States, 2003–2012. J Am Heart Assoc. 2016;5(6):e003704. https://doi.org/10.1161/JAHA.116.003704.
68. Kong WK, Regeer MV, Ng AC. Differences in phenotypes of bicuspid aortic valve and Aortopathy: insights from a large multicenter, international registry. Circ Cardiovasc Imaging. 2017;10(3):e005155.
69. Kritikou I, Basta M, Vgontzas AN, et al. Sleep apnoea and the hypothalamic-pituitary-adrenal axis in men and women: effects of continuous positive airway pressure. Eur Respir J. 2016;47:531–40.
70. Laitinen I, Saraste A, Weidl E, et al. Evaluation of αvβ3 integrin-targeted positron emission tomography tracer 18F-galacto-RGD for imaging of vascular inflammation in atherosclerotic mice. Circ Cardiovasc Imaging. 2009;2:331–8.
71. Lampert R, McPherson A, Clancy JF, et al. Gender differences in ventricular arrhythmia recurrence in patients with coronary artery disease and implantable cardioverter-defibrillators. J Am Coll Cardiol. 2004;43:2293–9.
72. Lederle FA, Larson JC, Margolis KL, et al. Abdominal aortic aneurysm events in the women's health initiative: cohort study. BMJ. 2008;337:a1724.
73. Lee SH, Tai CT, Hsieh MH, et al. Predictors of non-pulmonary vein ectopic beats initiating paroxysmal atrial fibrillation: implication for catheter ablation. J Am Coll Cardiol. 2005;46(6):1054–9.
74. Lew J, Sanghavi M, Ayers CR, et al. Sex-based differences in cardiometabolic biomarkers. Cicrulation. 2017;135:544–55.
75. Li J, et al. The number of X chromosomes influences protection from cardiac ischaemia/reperfusion injury in mice: one X is better than two. Cardiovasc Res. 2014a;102:375–84.
76. Li X, Bauer W, Israel I, et al. Targeting P-selectin by gallium-68-labeled fucoidan positron emission tomography for non-invasive characterization of vulnerable plaques: correlation with in vivo 17.6T MRI. Arterioscler Thromb Vasc Biol. 2014b;34:1661–7.
77. Li Z, Wang Z, Yin Z, Zhang Y, Xue X, Han J, et al. Gender differences in fibrosis remodeling in patients with - long-standing persistent atrial fibrillation. Oncotarget. 2017;8:53714–29. https://doi.org/10.18632/oncotarget.16342.
78. Liang NL, Genovese EA, Al-Khoury GE, Hager ES, Makaroun MS, Singh MJ. Effects of gender differences on short-term outcomes in patients with type B aortic dissection. Ann Vasc Surg. 2017;38:78–83. https://doi.org/10.1016/j.avsg.2016.06.006.
79. Lindenfeld J, Cleveland JC Jr, Kao DP, et al. Sex-related differences in age-associated downregulation of human ventricular myocardial β1-adrenergic receptors. J Heart Lung Transplant. 2016;35(3):352–61. https://doi.org/10.1016/j.healun.2015.10.021.
80. Mahmoodzadeh S, et al. 17beta-estradiol inhibits matrix metalloproteinase-2 transcription via MAP kinase in fibroblasts. Cardiovasc Res. 2010;85:719–28.
81. Mahmoodzadeh S, et al. Cardiomyocyte-specific estrogen receptor alpha increases angiogenesis, lymphangiogenesis and reduces fibrosis in the female mouse heart post-myocardial infarction. J Cell Sci Ther. 2014;5:153.
82. Makaryus AN, Sison C, Kohansieh M, et al. Implications of gender difference in coronary calcification as assessed by CT coronary angiography. Clin Med Insights Cardiol. 2014;8 (Suppl 4):51–5.
83. Makrygiannis G, Courtois A, Drion P, Defraigne JO, Kuivaniemi H, Sakalihasan N. Sex differences in abdominal

aortic aneurysm: the role of sex hormones. Vasc Surg. 2014;28(8):1946–58. https://doi.org/10.1016/j.avsg.2014.07.008.
84. Manfredini R, Fabbian F, Pala M, Tiseo R, De Giorgi A, Manfredini S, et al. Seasonal and weekly patterns of occurrence of acute cardiovascular diseases: does a gender difference exist? J Womens Health (Larchmt). 2011;20:1663–8.
85. Marjoribanks J, Farquhar C, Roberts H, Lethaby A, Lee J. Long-term hormone therapy for perimenopausal and postmenopausal women. Cochrane Database Syst Rev. 2017;1: CD004143.
86. Marroquin OC, Holubkov R, Edmundowicz D, et al. Heterogeneity of microvascular dysfunction in women with chest pain not attributable to coronary artery disease: implications for clinical practice. Am Heart J. 2003;145:628–35.
87. Masjedi S, Lei Y, Patel J, Ferdous Z. Sex-related differences in matrix remodeling and early osteogenic markers in aortic valvular interstitial cells. Heart Vessel. 2017;32(2):217–28.
88. McCoy CM, Nicholas DQ, Masters KS. Sex-related differences in gene expression by porcine aortic valvular interstitial cells. PLoS One. 2012;7(7):e39980.
89. Mendelsohn ME, Karas RH. The protective effects of estrogen on the cardiovascular system. N Engl J Med. 1999;340 (23):1801–11.
90. Meyer S, van der Meer P, van Tintelen JP, van den Berg MP. Sex differences in cardiomyopathies. Eur J Heart Fail. 2014;16:238–47. https://doi.org/10.1002/ejhf.15.
91. Mikhaylov EN, Lebedev DS, Pokushalov EA, Davtyan KV, Ivanitskii EA, Nechepurenko AA, et al. Outcomes of cryoballoon ablation in high- and low-volume atrial fibrillation ablation centres: a Russian pilot survey. Biomed Res Int. 2015;2015, Article ID 591603, 8 pages. doi:https://doi.org/10.1155/2015/591603.
92. Min K-J, Lee C-K, Park H-N. The lifespan of Korean eunuchs. Curr Biol. 2012;22(18):R792–3.
93. Mori T, Kai H, Kajimoto H, et al. Enhanced cardiac inflammation and fibrosis in ovariectomized hypertensive rats: a possible mechanism of diastolic dysfunction in postmenopausal women. Hypertens Res. 2011;34:496–502.
94. Mosca L, Appel LJ, Benjamin EJ. Evidence-based guidelines for cardiovascular disease prevention in women. ACC Curr J Rev. 2004;13(5):29–30.
95. Mozaffarian D, Benjamin EJ, Go AS, et al. American Heart Association Statistics Committee and Stroke Statistics Subcommittee. Heart disease and stroke statistics–2015 update: a report from the American Heart Association. Circulation. 2015;131:e29–322.
96. Murphy E, Amanakis G, Fillmore N, et al. Sex differences in metabolic cardiomyopathy. Cardiovasc Res. 2017;113:370–7.
97. Nakagawa M, Takahashi N, Nobe S, Ichinose M, et al. Gender differences in various types of idiopathic ventricular tachycardia. J Cardiovasc Electrophysiol. 2002;13:633–8.
98. Nakamura T, Ogita M, Ako J, Momomura S. Gender differences of plaque characteristics in elderly patients with stable angina pectoris: an intravascular ultrasonic radiofrequency data analysis. Int J Vasc Med. 2010;134:692.
99. Nekolla SG, Reder S, Saraste A, et al. Evaluation of the novel myocardial perfusion positron-emission tomography tracer 18F-BMS-747158-02: comparison to 13N-ammonia and validation with microspheres in a pig model. Circulation. 2009;119:2333–42.
100. Nikolic I, et al. Treatment with an estrogen receptor-beta-selective agonist is cardioprotective. J Mol Cell Cardiol. 2007;42:769–80.
101. Nordmeyer J, Eder S, Mahmoodzadeh Sh, Martus P, Fielitz J, Bass JN, Zurbrügg HR, , Roland Hetzer RP, Regitz-Zagrosek V. Upregulation of myocardial estrogen receptors in human aortic stenosis. Circulation 2004; 110: 3270–3275.
102. Nørgaard KS, Isaksen C, Buhl JS, et al. Single-centre cohort study of gender influence in coronary CT angiography in patients with a low to intermediate pretest probability of coronary heart disease. Open Heart. 2015;2:e000233. https://doi.org/10.1136/openhrt-2014-000233.
103. Nuyt AM, Alexander BT. Developmental programming and hypertension. Curr Opin Nephrol Hypertens. 2009;18:144–52.
104. Olsson C, Thelin S, Stahle E, Ekbom A, Granath F. Thoracic aortic aneurysm and dissection: increasing prevalence and improved outcomes reported in a nationwide population-based study of more than 14,000 cases from 1987 to 2002. Circulation. 2006;114:2611–8.
105. Oparil S, Miller AP. Gender and blood pressure. J Clin Hypertens. 2005;7:300–9.
106. Ostman-Smith I, Wettrell G, Keeton B, Holmgren D, Ergander U, Gould S, Bowker C, Verdicchio M. Age- and gender-specific mortality rates in childhood hypertrophic cardiomyopathy. Eur Heart J. 2008;29:1160–7. https://doi.org/10.1093/eurheartj/ehn122.
107. Ou YC, Chuang HH, Li WC, et al. Gender difference in the association between lower muscle mass and metabolic syndrome independent of insulin resistance in middle-aged and elderly Taiwanese population. Arch Gerontol Geriatr. 2017;72:12–8.
108. Ouyang P, Michos ED, Karas RH. Hormone replacement therapy and the cardiovascular system lessons learned and unanswered questions. J Am Coll Cardiol. 2006;47:1741–53.
109. Padmanabhan S, Newton-Cheh C, Dominiczak AF. Genetic basis of blood pressure and hypertension. Trends Genet. 2012;28:397–408.
110. Padmanabhan S, Menni C, Lee WK, et al. The effects of sex and method of blood pressure measurement on genetic associations with blood pressure in the PAMELA study. J Hypertens. 2010;28:465–77.
111. Patel D, Mohanty P, Di Biase L, Sanchez JE, Shaheen MH, Burkhardt JD, et al. Outcomes and complications of catheter ablation for atrial fibrillation in females. Heart Rhythm. 2010;7:167–72.
112. Patten RD, et al. 17 Beta-estradiol differentially affects left ventricular and cardiomyocyte hypertrophy following myocardial infarction and pressure overload. J Card Fail. 2008;14:245–53.
113. Pedersen SB, Nielsen JC, Botker HE, Farkas DK, Schmidt M, Sorensen HT. Implantable cardioverter-defibrillators and subsequent cancer risk: a nationwide population-based cohort study. Europace. 2015;17:902–8.
114. Pereira TMC, Nogueira BV, Lima LCF, et al. Cardiac and vascular changes in elderly atherosclerotic mice: the influence of gender. Lipids Health Dis. 2010;9:87–96.
115. Pinto YM, Elliott PM, Arbustini E, Adler Y, Anastasakis A, Böhm M, Duboc D, Gimeno J, de Groote P, Imazio M, Heymans S, Klingel K, Komajda M, Limongelli G, Linhart A, Mogensen J, Moon J, Pieper PG, Seferovic PM, Schueler S, Zamorano JL, Caforio AL, Charron P. Proposal for a revised definition of dilated cardiomyopathy, hypokinetic non-dilated cardiomyopathy, and its implications for clinical practice: a position statement of the ESC working group on myocardial and pericardial diseases. Eur Heart J. 2016;37(23):1850–8. https://doi.org/10.1093/eurheartj/ehv727.
116. Qi H, Liu Z, Liu B, Cao H, Sun W, Yan Y, Zhang L. MicroRNA screening and prediction model construction for diagnosis of salt-sensitive essential hypertension. Medicine (Baltimore). 2017;96:e6417.

117. Qin J, Peng C, Zhao B, Ye K, Yuan F, Peng Z, Yang X, Huang L, Jiang M, Zhao Q, et al. Noninvasive detection of macrophages in atherosclerotic lesions by computed tomography enhanced with PEGylated gold nanoparticles. Int J Nanomedicine. 2014;9:5575–90.
118. Queiros AM, et al. Sex- and estrogen-dependent regulation of a miRNA network in the healthy and hypertrophied heart. Int J Cardiol. 2013;169:331–8.
119. Reckelhoff JF, et al. Testosterone exacerbates hypertension and reduces pressure-natriuresis in male spontaneously hypertensive rats. Hypertension. 1998;31:435–9.
120. Regitz-Zagrosek V, Kararigas G. Mechanistic pathways of sex differences in cardiovascular disease. Physiol Rev. 2017;97:1–37.
121. Reue K. Sex differences in obesity: X chromosome dosage as a risk factor for increased food intake, adiposity and co-morbidities. Physiol Behav. 2017;176:174–82.
122. Robinson DP, et al. Sex chromosome complement contributes to sex differences in coxsackievirus B3 but not influenza A virus pathogenesis. Biol Sex Differ. 2011;2:8.
123. Rodriguez I, Kilborn MJ, Liu XK, Pezzullo JC, et al. Drug-induced qt prolongation in women during the menstrual cycle. JAMA. 2001;285:1322–6. https://doi.org/10.1001/jama.285.10.1322.
124. Rosano GMC, Vitale C, Silvestri A, Fini M. Metabolic and vascular effects of progestins in post-menopausal women. Implications for cardioprotection. Maturitas. 2003;46: S17–29.
125. Sainz J, et al. Role of sex, gonadectomy and sex hormones in the development of nitric oxide inhibition-induced hypertension. Exp Physiol. 2004;89:155–62.
126. Schröder K, Wegscheider K, Wenger NK, Vettorazzi E, Schröder R. Resting electrocardiogram predicts mortality in postmenopausal women with coronary heart disease or with risk factors for coronary heart disease. Eur J Prev Cardiol. 2014;21(6):749–57. https://doi.org/10.1177/2047487312454022. Epub 2012 Jun 29
127. Schuster I, et al. Cardiomyocyte-specific overexpression of oestrogen receptor beta improves survival and cardiac function after myocardial infarction in female and male mice. Clin Sci. 2016;130:365–76.
128. Scurrah K, Lamantia A, Ellis JA, Harrap SB. Familial analysis of epistatic and sex-dependent association of genes of the renin-angiotensin-aldosterone system and blood pressure. Circ Cardiovasc Genet. 2017;10(3):e001595.
129. Średniawa B, Mitręga KA, Cebula S, et al. Gender-dependent profile of heart rate turbulence parameters in patients after acute myocardial infarction treated invasively. Kardiol Pol. 2016;74(3):274–9.
130. Staessen JA, Ginnochio G, Thijs L, Fagard R. Conventional and ambulatory blood pressure and menopause in a prospective population study. J Hum Hypertens. 1997;11:507–14.
131. Staessen JA, Wang J, Bianchi G, Birkenhäger WH. Essential hypertension. Lancet. 2003;361:1629–38.
132. Straznicky NE, Grima MT, Sari CI, et al. Neck circumference is associated with muscle sympathetic nerve activity in overweight and obese men but not women. Front Physiol. 2017;8:203. https://doi.org/10.3389/fphys.2017.00203.
133. Su H, Gorodny N, Gomez LF, et al. Atherosclerotic plaque uptake of a novel integrin tracer 18F-Flotegatide in a mouse model of atherosclerosis. J Nucl Cardiol. 2014;21:553–62.
134. Suto J, Satou K. Effect of the Y chromosome on plasma high-density lipoprotein-cholesterol levels in Y-chromosome-consomic mouse strains. BMC Res Note. 2014;7:393.
135. Taqueti VR, Shaw LJ, Cook NR, et al. Excess cardiovascular risk in women relative to men referred for coronary angiography is associated with severely impaired coronary flow reserve, not obstructive disease. Circulation. 2017;135:566–77.
136. The 2017 hormone therapy position statement of The North American Menopause Society. Menopause. 2017; 24:728–53.
137. Tsioufis C, Dimitriadis K, Kordalis A. Renal denervation therapy: can it contribute to better blood pressure control in hypertension? Curr Vasc Pharmacol. 2017;16:66–9. https://doi.org/10.2174/1570161115666170426151649.
138. Urbano-Manquez A, Estruch R, Fernandez-Sola J, Nicolas JM, Pare JC, Rubin E. The greater risk of alcoholic cardiomyopathy fnd myopathy in women compared with men. JAMA. 1995;274(2):149–54.
139. Vaccarino V, Abramson JL, Veledar E, Weintraub WS. Sex differences in hospital mortality after coronary artery bypass surgery: evidence for a higher mortality in younger women. Circulation. 2002;105:1176–81.
140. Virgen R, O'Donnell CI, Barón AE, Grunwald GK, Maddox TM, Bradley SM, et al. Association of testosterone therapy with mortality, myocardial infarction, and stroke in men with low testosterone levels. JAMA. 2013;310:1829–36.
141. Wake R, Yoshiyama M. Gender differences in ischemic heart disease. Recent Pat Cardiovasc Drug Discov. 2009;4:234–40.
142. Wallbach M, Lehnig LY, Schroer C, et al. Effects of baroreflex activation therapy on ambulatory blood pressure in patients with resistant hypertension. Hypertension. 2016;67:701–9.
143. Weinberg EO, et al. Gender differences in molecular remodeling in pressure overload hypertrophy. J Am Coll Cardiol. 1999;34:264–73.
144. Weyer GW, Dunlap B, Shah SD. Hypertension in women: Evaluation and management. Obstet Gynecol Clin N Am. 2016;43:287–306.
145. Wilhelm-Bals A, Chehade H, Girardin E, Gonzalez E, Parvex P. [Premature kidneys and oligonephronia: long term repercussion]. [Article in French]. Rev Med Suisse. 2014;10:435-6–438-41.
146. Winkler TW, Justice AE, Graft M, et al. The influence of age and sex on genetic associations with adult body size and shape: a large-scale genome-wide interaction study. PLoS Genet. 2015;11(10):1–42.
147. Woehrle H, Graml A, Weinreich G. Age- and gender-dependent adherence with continuous positive airway pressure therapy. Sleep Med. 2011;12(10):1034–6.
148. Writing Group for the Women's Health Initiative Investigators. Risks and benefits of estrogen plus progestin in healthy postmenopausal women: principal results from the women's health initiative randomized controlled trial. JAMA: J Am Med Assoc. 288(3):321–33.
149. Yassin D-J, et al. Long-term testosterone treatment in elderly men with hypogonadism and erectile dysfunction reduces obesity parameters and improves metabolic syndrome and health-related quality of life. J Sex Med. 2014;11:1567–76.
150. Yeap BB, Hyde Z, Norman PE, et al. Associations of total testosterone, sex hormone-binding globulin, calculated free testosterone, and luteinizing hormone with prevalence of abdominal aortic aneurysm in older men. J Clin Endocrinol Metab. 2010;95:1123e30.
151. Young MT, Phelan MJ, Nguyen NT. A decade analysis of trends and outcomes of male vs female patients who underwent bariatric surgery. J Am Coll Surg. 2016;222 (3):226–31.

41 Sex- and Age-Related Reference Values in Cardiology, with Annotations and Guidelines for Interpretation

Peter L. M. Kerkhof, Richard A. Peace, and Peter W. Macfarlane

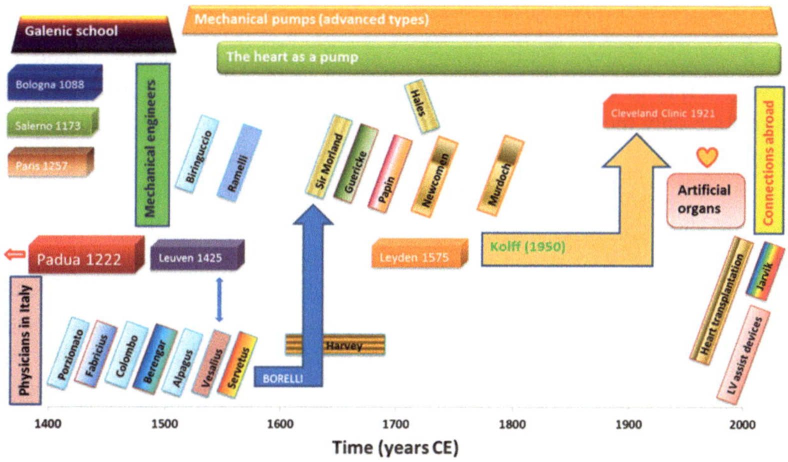

Time line indicating the transition from the Galenic school to the concept of a cardiac pump, based on the work of various scholars, both from Italy and those coming from abroad, including Vesalius (from Leuven), Servetus (from Spain, via Paris), and finally Harvey. Remarkably, this pattern was paralleled, if not preceded, by the perfection of mechanical pump systems, also starting in Italy. Dr. Kolff began designing an artificial kidney and later focused on the artificial heart. Interestingly, the reverend Hales (1733) studied both the volume of the heart and arterial pressure. The year listed at the location of various universities refers to the year of foundation. The mathematician Borelli (1608–1679) who conceived the body as a machine [37] seems to bridge the concept of the physiological and mechanical pump. Further details can be found in the Appendix.

P. L. M. Kerkhof (✉)
Department of Radiology and Nuclear Medicine, Amsterdam Cardiovascular Sciences, VU University Medical Center, Amsterdam, The Netherlands
e-mail: plm.kerkhof@VUmc.nl

R. A. Peace
Department of Nuclear Medicine, Royal Victoria Infirmary, Newcastle upon Tyne, UK

P. W. Macfarlane
Institute of Health and Wellbeing, Electrocardiology Group, University of Glasgow, Glasgow, Scotland, UK

Abstract

The definition of "abnormal" in clinical sciences is often based on so-called reference values which point to a range that experts by some sort of consensus consider as normal when looking at biological variables. Such a level is commonly calculated by taking (twice) the standard deviation from the mean, or considering certain percentiles. The suspicion or even confirmation of a disease is then

P. L. M. Kerkhof, V. M. Miller (eds.), *Sex-Specific Analysis of Cardiovascular Function*, Advances in Experimental Medicine and Biology 1065, https://doi.org/10.1007/978-3-319-77932-4_41

established by demonstrating that the value measured exceeds the upper or lower reference value. As is often the case, the measurement accuracy may depend on the conditions and specific method employed to collect and analyze data. This implies that, for example, data assessed by 2D echocardiography possibly differ from those obtained by MRI and therefore require modality-specific reference values. In this review we summarize reference values for the electrocardiogram, cardiac compartmental volumes, and arterial vessel size in males and females for various age groups. These values may further depend on other variables such as body size, physical training status, and ethnicity. Additional variables relevant for cardiology such as those referring to the microcirculation and biomarkers are only mentioned with reference to the pertinent literature. In general, the sex- and age-specific differences observed are often remarkable and warrant consideration in clinical practice and basic biomedical sciences.

Keywords
Cardiology · Reference values · Normal limits · Nomogram · Ejection fraction · Cutoff values · Heart rate · Heart rate variability · Electrocardiogram · Pediatric cardiology · Blood pressure · Pulsology · Ventricular size · Right ventricle · Left atrium · Right atrium · Aortic diameter · Coronary artery · Coronary-aorta index · Biomarkers · Surrogate index · Tissue Doppler imaging · Age-specific analysis · Sex-specific analysis · Aging effects · Giovanni Borelli · Review

Introduction

Age, sex, size, and ethnicity all matter.

Compiling a survey of reference values, regardless of the area of application, is a continuously evolving project. In the field of cardiology, not always a clear distinction has been made between sexes, although Löllgen [76] in the general guidelines for his handbook on cardiopulmonary diagnostics mentions the word "sex" twice, namely, as a codeterminant concerning *Muskelkraft* and *Leistungsfähigkeit.*

Investigational methods are improved over time, insights adjusted, or previously established approaches simply replaced by more advanced metrics. In the field of physiology the task of assembling reference values is particularly complex, because most metrics depend on sex and age, among other possible confounders such as the use of medication, time of the day, or level of physical training. The factors age and body mass are dominantly present in growing children, and for girls and boys alike. In adults certain variables may diverge as evidenced from cardiac sizes. Particularly in women the interpretation of measurements may require reference to their sex hormone levels which oscillate during childbearing ages and decline with menopause. For reference values regarding blood components, see Huxley and Kemp [46]. While the measurement methods employed to obtain an electrocardiogram (ECG) or to record blood pressure (BP) levels follow a rather standard protocol with national or international reference values, this is not always the case when cardiac dimensions are concerned. Apart from a diversity of techniques (such as echocardiography, MRI and SPECT, etc.) we face variability related to interobserver differences. A plethora of papers on cardiac dimensions can be found in recent review papers and meta-analyses, e.g., Dorosz et al. [27], Bhave and Lang [11], Wood et al. [125], Buccheri et al. [14], Kawel-Boehm et al. [53], and Kerkhof [55]. We list those as a source for further reference. The examples presented in this survey are by no means selected on the basis of alleged "superiority" but rather collected to reflect the diversity encountered in the literature with occasionally some attractiveness based on the scope (in terms of sex, age, ethnic groups, and normalization specifications). It is left to the reader to decide which publication best serves the individual needs, aided by the list of references.

Sex refers to a person's biological status and is typically categorized as male, female, or intersex. There are a number of indicators of biological sex, including sex chromosomes, gonads, internal reproductive organs, and external genitalia [3].

Gender refers to the attitudes, feelings, and behaviors that a given culture associates with a person's biological sex. Behavior that is compatible with cultural expectations is referred to as gender-normative; behaviors that are viewed as incompatible with these expectations constitute gender nonconformity [3].

Heart Rate

Resting heart rate (HR) is higher in smaller animals (up to 1200 beats per minute (bpm) in a hummingbird compared to 8 in a whale). Similarly, HR declines markedly during progressing childhood, without appreciable differences between boys and girls [102]. This trend is reasonably well described from infancy to adulthood for both sexes by an allometric relationship involving the inverse 0.25 power of body mass [73]. In healthy adults resting HR remains fairly constant (with slightly higher values in women), although maximum achievable values during exercise decline with age in both sexes.

Several epidemiological studies demonstrate the association between resting HR above 70 beats per minute and cardiovascular morbidity and mortality in adults [119]. Likewise, in animal research high frequency pacing is an established method to induce heart failure (HF). On the other side of the spectrum, HR reduction is suggested to be a mechanism explaining the prognostic benefit of beta-blockers (BB) after myocardial infarction (MI) or in HF patients. A prolonged diastolic phase supposedly enhances ventricular filling volume and thus increases stroke volume (SV). The SV vs HR diagram provides insight into the relative contribution of both components to cardiac output (CO) as illustrated elsewhere [62]. However, in view of the negative inotropic effect of BB, it remains unclear whether other effects besides HR reduction affect cardiovascular prognosis when using BB.

Throughout all species the life span correlates inversely with resting HR, but the relationship between HR and longevity is not entirely clear. In a Danish study involving 1233 twin pairs, the resting HR was independently associated with longevity, even when familial factors were controlled for. Heritability estimates were 0.27 (0.15–0.38) for males and 0.17 (0.06–0.28) for females. In multivariable models adjusting for age, sex, body mass index (BMI), diabetes, hypertension, pulmonary function, smoking, physical activity, and zygosity, resting HR was significantly associated with mortality [49]. In another study a 14% reduction of HR by daily administration of ivabradine in (male) mice prolonged life ($P = 0.01$) only by 6.2% compared to a placebo group, while excluding effects related to caloric restriction [38]. That study could not clarify underlying mechanisms and emphasized that generally mice do not die from cardiac disease, making cardioprotection by lowering HR in these animals of less importance.

Individual HR values in adults vary by time of the day, sex, body mass, physical condition, lifestyle, and many other factors, making it difficult to establish universal reference values. Bjerregaard [13] studied HR in 260 healthy subjects (age 40–79 years) during a full day. Mean 24 h HR varied from 53 to 95 beats per minute (bpm). On average males had a lower HR than matched female counterparts, as also documented in a recent study [62] based on a single HR determination in each participant. Older subjects had a lower HR than younger individuals. HR exhibits a circadian rhythm, with lowest readings around 4 a.m. and highest values showing almost a plateau starting in the morning at 8 o'clock. Peak HR during exercise declines gradually with age, approximately as given in the following expression: $HR_{max} = \{208–0.75 \text{ (age as years)}\}$; see Medicographia [83].

The study of HR variability (HRV), based on beat-by-beat oscillations in RR interval length, reveals information on underlying neural control mechanisms, referring to the instantaneous balance between parasympathetic and sympathetic innervation/activity. Sex-specific differences for HRV are well-known [66, 91]. The field is intensively explored (now >22,300 PubMed listed publications), including studies in both sexes regarding healthy individuals [94], during endurance sports with aging [123] and in patients, for example, with Brugada syndrome [16]. Sex-specific reference values of HRV during ordinary daily activity with age as covariable have recently been published [108]. BP variability (BPV) was also studied in combination with HRV [51].

Electrocardiogram

Apart from pulsology (i.e., the art of feeling and interpreting the pulse), no other area of cardiovascular investigations has thus far received as much attention as that devoted to electrophysiology of the heart [80]. One important practical advantage of recording the electrical activity resides in the fact that the (electrical) signals originating in the heart can be detected on the surface of the body. What remains is to solve the interpretation based on the so-called inverse problem, meaning the derivation of information about the heart (as a source) from body surface potentials.

The advantage of directly recording transmitted signals does not apply to the other two important measures, namely, pressure and volume. Only the *ictus cordis* (apex beat), parasternal heave, and systolic pressure can be estimated from the chest wall and periphery, respectively. Determination of diastolic pressure requires either an invasive procedure or the application of (often not very reliable) surrogate measures. Even more difficult is the determination of the phasic volume of the lumen of each cardiac compartment. The inherent limitations to accurately collect details on pressure and volume have further promoted investigations regarding the ECG. This process leads to refinements and the recognition of a wealth of information based on noninvasive recordings of electrical phenomena originating in the heart. Details on the ECG with attention to sex, age, and ethnicity have been described in detail [81]. In summary, Tables 41.1, 41.2, and 41.3 concern interval and duration of ECG measurements from Caucasian healthy males and females, similar measurements versus HR ranges, and miscellaneous measures including cardiac axes and the Lewis index, respectively.

Blood Pressure and Pulsology

The art of feeling the arterial pulse goes back to China in the period 2500 BC [8]. In the morning, and before taking any meal, the physician analyzed his patient, while following a strict protocol to carefully feel the pulse at three locations on both wrists, using index, middle, and ring fingers. Especially pulses originating from the liver, the heart being less relevant and regarded as the son, appeared informative, but interpretation always required consideration of the season [8]. In current times interpretation of the arterial pulse is still important, especially the more quantitative aspects including the mean value and augmentation index [7]. This section is limited to adults, as patterns of BP in children are discussed in the chapter by Dallaire and Sarkola [24].

Brachial or finger pressure levels are usually determined using noninvasive methods. The values refer to peripheral systolic (SBP) and diastolic BP (DBP). Age-dependent viscoelastic properties of the arterial vessel wall as well as reflections at branching sites modify the pressure waveform. This explains why the radial pressure wave is more peaked than the central aortic or carotid pressure wave. Central pressure can be derived from peripheral pressure recordings using a transfer function. Estimated central SBP (cSBP) and amplification (brachial SBP-cSBP) are noninvasively obtained measures expectedly offering prognostic information on cardiovascular disease (CVD). Until recently, no worldwide, multiple-device reference values were available. However, a multicenter study [43] established reference values and investigated how these values alter by presence of cardiovascular risk factors (CVRFs). Data on 45,436 untreated subjects without diabetes or overt CVD as

Table 41.1 Interval and duration of measurements (in milliseconds) from Caucasian normal adults

Normal Limits of the 12-Lead ECG in White Caucasians

The data in the following tables have been derived from a study of over 1450 apparently healthy Caucasians living in the west of Scotland. A few tables are based on smaller numbers. Further details can be found in [79] and in [81]. The results are presented as mean ± standard deviation together with the 96% range; that is, 2% of measurements have been excluded from each extreme of the range. P-, Q-, S-, T-wave amplitudes are presented as negative measurements.

Age group	Sex	PR Interval	QRS Duration	QT Interval	QT_c Interval Hodges[a]	QT_c Interval Bazett[b]	P-wave Duration	Heart rate (bpm)
18–29	Male	152.5 ± 23.0	96.4 ± 8.6	385.5 ± 28.9	403.6 ± 19.0	413.9 ± 23.1	103.0 ± 14.2	70 ± 12
		112–208	80–114	336–442	368–444	370–463	72–128	48–98
		$n = 265$	$n = 265$	$n = 265$	$n = 265$	$n = 265$	$n = 266$	$n = 266$
	Female	145.9 ± 19.7	87.7 ± 7.8	380.0 ± 27.8	411.6 ± 18.0	429.7 ± 22.9	99.0 ± 12.7	76 ± 12
		114–194	72–104	322–440	378–451	386–477	70–122	55–108
		$n = 317$	$n = 317$	$n = 317$	$n = 317$	$n = 317$	$n = 318$	$n = 318$
30–39	Male	155.7 ± 21.4	95.4 ± 9.8	385.5 ± 29.5	404.8 ± 19.4	416.0 ± 22.9	105.0 ± 12.3	71 ± 12
		116–206	78–114	326–448	366–448	375–468	78–130	52–99
		$n = 218$	$n = 218$	$n = 218$	$n = 218$	$n = 218$	$n = 221$	$n = 221$
	Female	145.7 ± 18.6	88.6 ± 7.3	386.6 ± 27.7	415.2 ± 16.9	432.6 ± 20.9	99.0 ± 11.6	77 ± 13
		114–184	76–106	330–438	384–445	395–473	72–122	57–105
		$n = 115$	$n = 115$	$n = 115$	$n = 115$	$n = 115$	$n = 118$	$n = 118$
40–49	Male	157.2 ± 21.8	94.4 ± 9.9	390.8 ± 29.3	409.2 ± 17.9	420.0 ± 21.9	106.0 ± 11.2	70 ± 12
		116–210	78–114	340–450	377–450	377–464	84–128	49–96
		$n = 119$	$n = 119$	$n = 119$	$n = 119$	$n = 119$	$n = 119$	$n = 119$
	Female	154.9 ± 20.4	89.4 ± 7.9	386.1 ± 27.0	415.2 ± 22.5	433.7 ± 28.4	104.0 ± 12.9	77 ± 12
		108–200	74–108	328–434	347–457	350–483	78–128	59–106
		$n = 72$	$n = 72$	$n = 72$	$n = 72$	$n = 72$	$n = 73$	$n = 73$
50 +	Male	161.5 ± 18.9	92.7 ± 9.3	385.5 ± 26.0	407.4 ± 17.5	420.9 ± 22.7	110.0 ± 10.5	73 ± 12
		120–196	74–112	320–434	374–444	380–475	86–134	54–100
		$n = 123$	$n = 123$	$n = 123$	$n = 123$	$n = 123$	$n = 125$	$n = 125$
	Female	155.6 ± 6.9	87.1 ± 8.7	390.7 ± 31.5	419.5 ± 22.7	438.2 ± 24.8	106.0 ± 9.5	77 ± 11
		122–196	68–104	336–488	376–486	392–506	88–126	59–104
		$n = 79$	$n = 79$	$n = 79$	$n = 79$	$n = 79$	$n = 80$	$n = 80$

[a] $QT_C = QT \pm 1.75\,(\text{rate} - 60)$
[b] $QT_C = QT\,(\text{rate}/60)^{1/2}$

Table 41.2 Interval measurements (in milliseconds) from Caucasian normals versus heart rate

Heart rate (bpm)	PR interval	QT interval	QT_c interval Hodges[a]	QT_c interval Bazett[b]
<50	145.067 ± 18.19	443.6 ± 21.62	420.5 ± 20.49	391.537 ± 17.70
	116–176	396–482	376.75–459.25	357.864–426.6
	$n = 15$	$n = 15$	$n = 15$	$n = 15$
50–59	154.753 ± 25.81	417.425 ± 21.27	410.197 ± 21.08	402.633 ± 21.28
	110–220	378–468	370–458.5	358.078–454.235
	$n = 146$	$n = 146$	$n = 146$	$n = 146$
60–69	154.769 ± 22.62	399.339 ± 21.51	408.173 ± 20.54	415.549 ± 21.38
	116–208	354–444	368.75–450.25	372.64–462.2
	$n = 372$	$n = 372$	$n = 372$	$n = 372$
70–79	151.196 ± 20.28	382.44 ± 19.82	407.516 ± 19.18	425.481 ± 21.51
	114–196	344–426	372.25–452	388.32–477.234
	$n = 408$	$n = 408$	$n = 408$	$n = 408$
80–89	150.51 ± 20.19	367.522 ± 17.36	409.076 ± 16.93	434.074 ± 20.06
	114–194	332–406	376.5–444.5	395.044–474.053
	$n = 247$	$n = 247$	$n = 247$	$n = 247$
90–99	151.368 ± 17.52	356.842 ± 17.59	415.958 ± 17.34	446.019 ± 21.80
	114–182	320–388	383.5–449.25	405.54–488.23
	$n = 114$	$n = 114$	$n = 114$	$n = 114$
≥100	151.459 ± 14.14	328.108 ± 16.25	408.372 ± 18.63	435.705 ± 23.93
	130–180	298–352	377.25–439.5	395.685–476.246
	$n = 37$	$n = 37$	$n = 37$	$n = 37$

[a] $QT_c = QT + 1.75(\text{rate} - 60)$
[b] $QT_c = QT\ (\text{rate}/60)^{1/2}$

collected from 50 centers were available. Of those, 18,183 (52% females) subjects made up the "normal population" (no CVRFs), and 27,253 subjects (48% females) having at least one CVRF were categorized as the "reference population." Values of cSBP and amplification were calculated as percentiles for both populations and for each sex stratified by brachial BP categories (as explained in Fig. 41.1.) and age decade. The study showed that amplification decreased with age and more so in males than in females. Sex was the most powerful factor associated with amplification being 6.6 mmHg (range 5.8–7.4) higher in males than in females. Amplification was marginally but significantly influenced by CVRFs, with smoking and dyslipidemia decreasing amplification, but more pronounced with increasing levels of blood glucose. Thus, amplification is significantly influenced by CVRFs but differently in men and women.

Cardiac Output (CO)

Ideally, ventricular function should be analyzed in the pressure-volume-time domain [71]. Due to the technical complexity, this goal is not yet a routine procedure in clinical practice. Therefore, often derived metrics such as SV, CO, or ejection fraction (EF) are used. A few studies have investigated if major hemodynamic variables are sex-specific or change during the menstrual cycle. The outcomes are summarized elsewhere [90] and generally indicate that HR, SV, CO, end-systolic volume (ESV), end-diastolic volume (EDV), and EF do not significantly fluctuate when early follicular, ovulatory, and mid-luteal phases are compared.

The CO is clearly associated with body size [88]. Therefore, CO as well as other volume-related variables are often indexed (i) by considering body surface area (BSA), here yielding

Table 41.3 Miscellaneous measures for sex- and age-specific ECG, including axes and Lewis index

Age group	Sex	QRS axis (degrees)	T axis (degrees)	QRS-T angle (frontal plane) (degrees)	Lewis index[a] (mV)	$SV_1 + RV_5$ (mV)	$SV_1 + RV_6$ (mV)	$SV_2 + RV_6$ (mV)
18–29	Male	57.5 ± 25.6	40.9 ± 16.7	16.6 ± 23.9	−0.117 ± 0.956	3.331 ± 0.881	3.905 ± 0.995	2.99 ± 0.79
		−10-91	2–69	−39-71	−1.831-2.048	1.685–5.252	2.049–6.372	1.59–4.85
		n = 265	*n* = 265	*n* = 265	*n* = 266	*n* = 264	*n* = 265	*n* = 265
	Female	51.2 ± 24.9	38.7 ± 15.0	12.5 ± 25.1	0.110 ± 0.736	2.442 ± 0.690	2.741 ± 0.727	2.32 ± 0.64
		−9-91	6–65	−46-59	−1.288-1.574	1.112–3.816	1.430–4.635	1.20–3.68
		n = 317	*n* = 317	*n* = 317	*n* = 317	*n* = 316	*n* = 317	*n* = 318
30–39	Male	46.5 ± 29.2	39.3 ± 17.2	7.2 ± 27.3	0.266 ± 0.899	3.007 ± 0.775	3.430 ± 0.875	2.69 ± 0.72
		−22-92	−1-70	−61-59	−1.476-1.962	1.549–4.680	1.846–5.257	1.31–4.14
		n = 220	*n* = 220	*n* = 220	*n* = 221	*n* = 220	*n* = 219	*n* = 220
	Female	49.6 ± 24.2	39.0 ± 15.4	10.6 ± 21.7	0.129 ± 0.626	2.454 ± 0.644	2.662 ± 0.834	2.29 ± 0.60
		−14-81	1–66	−47-61	−1.011-1.293	0.483–4.287	1.565–4.803	1.43–3.59
		n = 118	*n* = 118	*n* = 118	*n* = 118	*n* = 118	*n* = 115	*n* = 118
40–49	Male	37.7 ± 31.6	42.5 ± 17.3	−4.8 ± 26.3	0.491 ± 0.892	2.702 ± 0.763	2.842 ± 0.673	2.39 ± 0.69
		−37-85	5–71	−67-37	−1.060-2.395	1.420–4.227	1.634–4.170	1.22–3.77
		n = 117	*n* = 117	*n* = 117	*n* = 118	*n* = 118	*n* = 117	*n* = 118
	Female	36.2 ± 29.1	38.0 ± 15.7	−1.8 ± 27.9	0.521 ± 0.698	2.291 ± 0.634	2.418 ± 0.566	2.13 ± 0.57
		−53-85	5–67	−86-40	0.762–1.818	1.087–3.411	1.429–3.433	1.07–3.16
		n = 73	*n* = 73	*n* = 73	*n* = 73	*n* = 73	*n* = 73	*n* = 73
50+	Male	31.4 ± 30.6	44.4 ± 18.7	−13.0 ± 27.1	0.627 ± 0.742	2.709 ± 0.736	2.746 ± 0.703	2.35 ± 0.63
		−33-77	4–74	−82-40	0.703–2.120	1.612–4.558	1.546–4.144	1.34–3.78
		n = 125	*n* = 125	*n* = 125	*n* = 125	*n* = 125	*n* = 124	*n* = 125
	Female	26.9 ± 29.2	41.9 ± 15.9	−15.0 ± 27.8	0.714 ± 0.691	2.259 ± 0.567	2.316 ± 0.574	2.10 ± 0.55
		−36-73	4–72	−89-26	0.578–2.165	1.086–3.635	1.298–3.419	0.94–3.14
		n = 80	*n* = 80	*n* = 80	*n* = 80	*n* = 79	*n* = 80	*n* = 80

[a]Lewis index = (RI + SIII) − (SI + RIII)

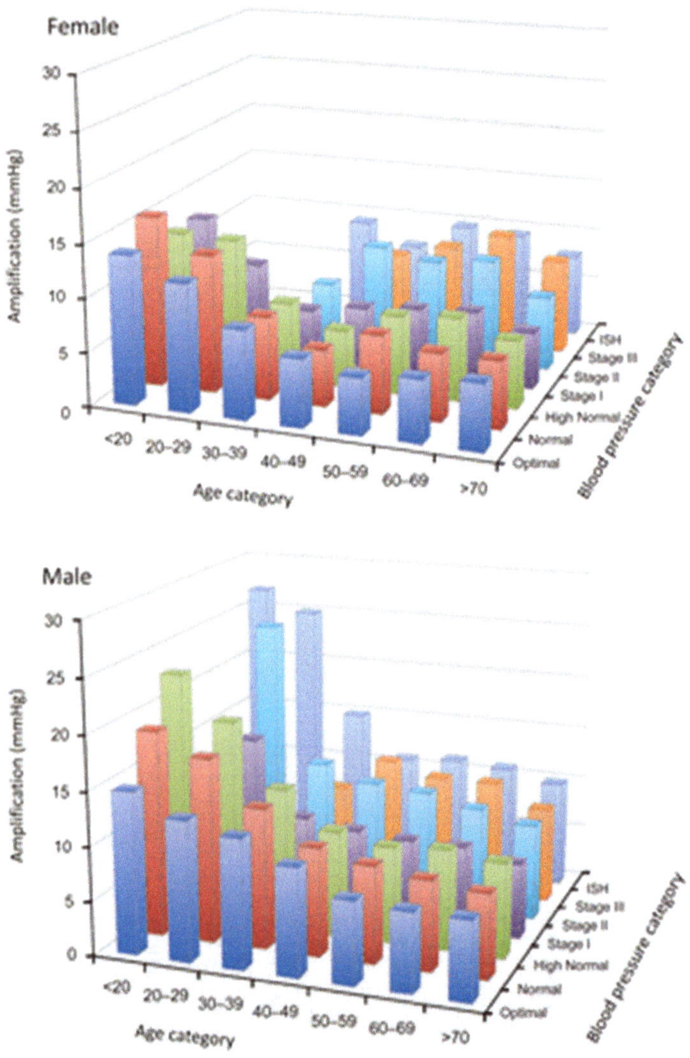

Fig. 41.1 Tridimensional bar graphs representing amplification (peripheral-central systolic blood pressures) according to sex (females upper panel, males below), age categories, and blood pressure categories. The value represented here is the median of the group. Some categories are not represented because there were less than 50 observations. ISH, isolated systolic hypertension. Stages 1, 2, and 3 refer to 50th percentile values of 143, 161, and 183 mmHg, respectively. (From Herbert et al. [43], reproduced with permission)

COi. Cattermole et al. [18] reported for healthy Chinese children ($N = 1197$), adolescents ($N = 460$), and adults ($N = 667$, aged 19–89 years) the normal ranges of cardiovascular variables measured using an ultrasonic CO monitor. In children there was no sex-related difference for HR, SVi, COi, mean arterial pressure (MAP), and peripheral resistance. Remaining data are published for three age groups, reportedly showing only minor differences between men and women, presented as centile curves but tabulated without sex stratification [18]. Obviously, COi increases in pregnant women [33].

Left Ventricular Mass (LVM)

An echo study involving 100 healthy participants (48 males) showed that LV mass (LVM, by Devereux formula), adjusted or not for BSA or body height, was significantly higher in males [112]. Other studies, e.g., using 2D echo, confirmed that the LVM (by Penn convention) in healthy men ($N = 503$) was larger ($P < 0.0001$), regardless of indexation for BSA, compared to 540 matched women [125]. Similarly, one study using multidetector computed tomography (MDCT) in 103 normotensive, nonobese adults

(43% women, age 51 ± 14 years) showed that LVM was significantly higher in men, even after indexation to body height [75].

A prospective study concerned longitudinal tracking of LVM over the adult life course (16 years), based on 4217 Framingham participants (mean age 45 years, 53% women). Women and participants with diabetes mellitus (DM) experienced a steeper increase in LVM with advancing age (compared with men and those without DM). Women also displayed greater increments in LVM with increasing BMI compared with men. Analyses evaluating short-term (4-year) changes in LVM identified the same key determinants (i.e., BMI, sex, SBP, antihypertensive treatment, and smoking) that influenced its long-term trajectory [74]. In a subsequent longitudinal analysis of 4320 participants (52% women) of the Framingham offspring cohort at echo examination cycles (1987–1991), (1991–1995), (1995–1998), and (2005–2008), sex-related differences in trends in LVM were observed. In men, age-adjusted LVM modestly increased (likely attributable to increasing body size) from first to last examination (192 to 198 g, P-trend = 0.0005), whereas in women it decreased from 147 to 140 g (P-trend <0.0001). Indexing LVM to BSA abolished the increasing trend in men (P-trend = 0.49), whereas the decreasing trend in women was maintained. Elucidation of the basis for these sex-related differences requires further study [50].

Age Effects A CMR study included 804 healthy Caucasian adults who were stratified by sex and age (45–54, 55–64, 65–74). LVM was significantly higher in males compared to females (mean ± SD of 53 ± 9 g/m^2 vs 42 ± 7 g/m^2). In older age groups, the LVM was lower in men but remained virtually unchanged in women [97]. In a relatively young group using MRI, 75 healthy subjects (age range 8–55, mean 28 years, 28 females) were studied, and no significant difference between males and females was found for the LVM to EDV ratio [78].

Exercise The LVM may increase with repeated physical activity and was measured by echocardiography in 1051 elite athletes (26% female, aged 18–40 years) and in 338 sedentary controls matched for age, sex, and body size. Normative ranges were calculated scaled to BSA, height, height(^2.7), and fat-free mass (FFM). The strongest correlation was found for FFM (R = 0.70). The LVM differed significantly (P < 0.05) between athletes of low, moderate, and high dynamic disciplines. LVM remained significantly increased in high dynamic athletes despite correction for body size. Sex differences for scaled LVM remained significantly increased in male athletes [99].

Myocardial Mass and Coronary Vasculature Scaling relationships between myocardial mass and the morphometry of coronary vasculature have been studied, where volume, diameter, and cumulative length of coronary arteries were reconstructed from casts and related to the perfused myocardial mass [21]. This aspect is discussed further in the section on coronary arteries.

Left Ventricular Size and Related Indexes

A classical paper highlights the importance of investigating the size of the heart by stating that normal heart muscle grows in order to maintain a constant relationship between systolic pressure and the ratio of wall thickness to radius [31]. The author claims that EF is a *double sensitive index* of cardiac failure, because it is affected by both SV and EDV. Nowadays we observe that half of all HF patients do not exhibit the anticipated decline of EF. This finding necessitates a reevaluation of the traditional model. Subsequent publications have added the relevance of the shape of the ventricle [122] and are reviewed elsewhere [55]. Size is still important but only within the context of age, sex, body mass [95, 98], ethnicity, and pathology (if present).

A systematic survey on LV volumes and EF assessed by 3D echocardiography can be found in Dorosz et al. [27] and Buccheri et al. [14], both based on meta-analysis. Similarly, Bhave and Lang [11] summarize research on the use of 3D echocardiography, also including LVM. Because 3D transthoracic echo (TTE) aids in identification of the true LV apex, it provides more accurate LV volumes than its 2D counterpart. As compared with a CMR standard, 3D TTE tends to slightly underestimate LV volumes, partly because identification of the true endocardial border is more difficult. Also, 3D TTE better assesses the extent of regional wall motion abnormalities, is superior for quantification of LVM, and offers applications such as speckle-tracking strain assessment, dyssynchrony analysis, and LV shape determination. The authors (as well as [70]) discuss in particular the publications by Aune et al. [6], Chahal et al. [19], Muraru et al. [85], and Fukuda et al. [34]. The latter concerned 410 healthy subjects (without history of cardiac disease and no risk factors) aged from 20 to 69 years (253 men) who had a 3D echo. The mean values in men and women were 50 ± 12 and 46 ± 9 mL/m^2 for LV EDVi, 19 ± 5 and 17 ± 4 mL/m^2 for ESVi, 61 ± 4% and 63 ± 4% for EF, and 64 ± 12 g/m^2 and 56 ± 11 g/m^2 for LVM. LV sizes decreased with age, whereas LVMi (i.e., LVM indexed for BSA) did not change according to Fukuda et al. [34].

For the range 20–80 years, mean ± 95% CI for EF, ESVi, EDVi, LVM and sphericity index measured by CMR (while including papillary muscle to LV mass) have been reported for males and females, based on data collected in a review by Kawel-Boehm et al. [53]. For LV and RV in both sexes, these authors found that ESVi and EDVi decline with age, while EF slightly increases.

Another CMR study included 804 healthy Caucasian adults who were stratified by sex and age (45–54, 55–64, 65–74 years). LV volumes were larger in males compared to females for absolute and indexed values. With advancing age, LV volumes were mostly smaller in both sexes. EF was significantly greater in females compared to males (mean ± SD of 61 ± 5% vs 58 ± 5%) and remained static with age for both sexes [97].

Using MDCT in 103 normotensive, nonobese adults (43% women, age 51 ± 14 years, 47% smokers, 28% non-Caucasian), it was shown that LVEDV was significantly larger in men, even after indexation to BSA [75]. However, EF was reported not to be different between men and women in this somewhat heterogeneous study group.

For cardiac patients with widely differing levels of HR, Fig. 41.2 shows a simple nomogram combining information on EF, EDVi, and ESVi.

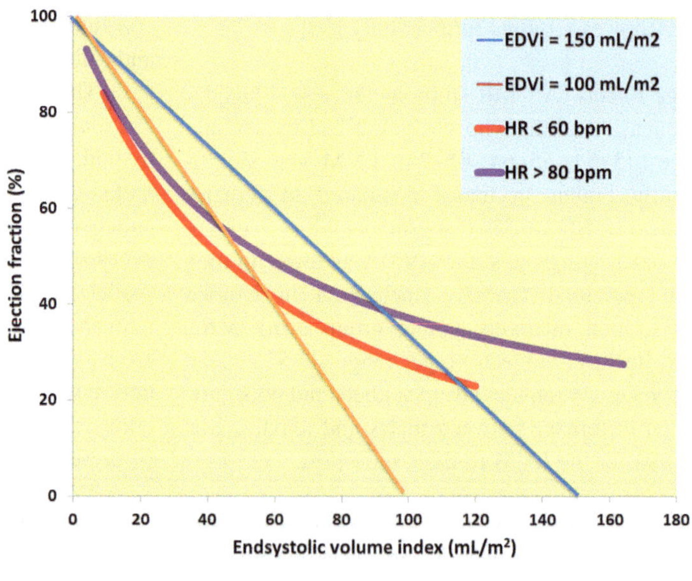

Fig. 41.2 Schematic design of a nomogram relating ejection fraction (%) to end-systolic volume index for fixed values (straight lines) of end-diastolic volume index (EDVi), on the basis of two nonlinear regression curves referring to cardiac patients with heart rate (HR) below 60 or above 80 beats per minute (bpm), as described elsewhere [61]

Examples are provided for EDVi values of 100 and 150 mL/m^2. However, any other fixed level of EDVi can be inscribed by taking the pertinent EDVi identical with ESVi (yielding EF = 0) along the abscissa. The figure applies to both men and women, although it must be kept in mind that generally average ESVi is smaller in females compared to matched males. When LV size is estimated on the basis of diameter (as was common practice in older studies), then LV diameter may be found to be larger in women, probably because of their more spherical LV shape.

There is evidence that following cardiac transplantation the LV size of the transplanted heart tends to adapt to the sex of the recipient and that ESV is significantly smaller in female recipients, irrespective of the sex of the donor [64].

Right Ventricular Mass (RVM)

Using MRI the RV in 75 healthy subjects (age range 8–55, mean 28 year, 28 females) was studied [78]. In contrast to findings for the LV, in the same study the ratio of RVM to RV EDV was significantly lower ($P < 0.05$) in males (0.33 ± 0.06 g/mL) compared to females (0.38 ± 0.06 g/mL).

Maceira et al. [79] studied 120 healthy subjects (60 women, 20–80 years) by CMR. Sex, BSA, and age were independent predictors of several RV parameters including RVM. The RV volumes and RVM were larger in males ($P < 0.01$). Normalized RVM and both absolute and normalized RV volumes decreased significantly with age, whereas RV EF increased. In that study increasing BSA values were associated with elevated RVM and larger RV volumes.

Right Ventricular Size and Related Indexes

Assessment of RV volume has traditionally been challenging, predominantly due to the complex anatomy of the RV, which does not resemble a simple geometrical shape. Therefore, cardiologists often rely on RV diameter or area, with associated fractional shortening (FS) and fractional area change (FAC) as alternatives for EF. Indeed, Aune et al. [5] warned that 3D echo may have limited value for the RV. Ostenfeld and Flachskampf [93] reviewed various studies on RV volume. Our Table 41.4 is an updated version, showing that average ESVi is smaller in women compared to men and that the associated average EF is higher. This implies that in this respect the sex-specific observations for the RV are similar to findings described for the LV. Interestingly, Rominger et al. [104] studied both LV and RV in the same subjects ($N = 377$, subdivided into 23 diagnostic groups), and the relationship between EF and ESVi superimposes for LV and RV, as described elsewhere [63]. In contrast, for post Fallot repair patients distinct regression curves were found [59] indicating a different pattern of volume regulation in these patients. Later studies, e.g., the one by Maffessanti et al. [82] using 3D echo in 507 healthy volunteers, similar as for the LV, also found for the RV a larger EF in women.

Effect of Age A CMR study by Petersen et al. [97] included 804 healthy Caucasian adults who were stratified by sex and age (45–54, 55–64, 65–74 years). RV volumes were significantly larger in males compared to females for absolute and indexed values, and were smaller with advancing age. RV EF was higher with increasing age in females only. Similarly, the impact of sex and age on RV volumes were analyzed by CMR in a longitudinally followed cohort free of pulmonary and cardiovascular disease, while RV trabeculations and papillary muscles were considered cavity volume. Women had higher RV EF than men (Table 41.4). Sex, age, height, BMI, and HR were found to account for most of the variability in RV volumes. Sex-related differences in volume persisted after indexation to height and to BSA (all $P < 0.001$ for both sexes). In summary: RV volumes increase with body size, are greater in men, and are smaller in older people [30]. A 3D echo with disk summation study [39] performed acceptably when compared with CMR for the RV in 71 healthy

Table 41.4 Right ventricular (RV) volume index in healthy adults (mean ± SD)

Authors	Year	Method	N	Sex	RVEDVi	RVESVi	RVEF	Age
Aune et al.	2009	3D echo	87	F	38 ± 10	15 ± 5	62 ± 10	29–80
Aune et al.	2009	3D echo	79	M	42 ± 11	17 ± 6	60 ± 11	29–80
Gopal et al.	2007	3D echo	36	F	65.4 ± 13.4	29.2 ± 10.7	56.2 ± 9.1	56 ± 14
Gopal et al.	2007	3D echo	35	M	74.7 ± 13.0	37.8 ± 7.4	48.9 ± 9.5	56 ± 14
Foppa et al.	2016	CMR	760	F	59 ± 10	19 ± 5	68 ± 6	64 ± 9
Foppa et al.	2016	CMR	576	M	70 ± 13	26 ± 8	64 ± 7	63 ± 9
Petersen et al.	2017	CMR	432	F	77 ± 13	33 ± 9	58 ± 6	45–74
Petersen et al.	2017	CMR	368	M	93 ± 17	43 ± 11	54 ± 6	45–74

RVEDVi right ventricular end-diastolic volume index, *RVESVi* right ventricular end-systolic index, *RVEF* right ventricular ejection fraction

volunteers (36 women), with similar differences for men and women (Table 41.4).

Ethnicity Another study [117] established MRI reference ranges for 500 subjects (268 women) free of clinical CVD. Except for EF, all unadjusted RV parameters were significantly greater in men than in women ($P < 0.001$). Even after adjusting for BSA and height, Chinese participants ($N = 70$) had significantly smaller RV volumes compared with Caucasians.

RV and a Regional Curvature Index RV remodeling involves changes in size, wall thickness, function, and shape. It has been suggested that regional curvature indices (rCI) may be useful for shape analysis [2]. Using 3D echo 6 regional values of rCI in 245 healthy subjects (age 42 ± 12 years) were established. Mean curvature was calculated for RV inflow tract (RVIT) and RV outflow tract (RVOT), besides free wall and septal regions for both apex and body. Septal curvature did not change significantly from end-diastole to end-systole, but RVOT and RVIT flattened. There were no sex-related differences in rCI. In subjects ≥55 years the RV free wall and RVOT were flatter, and the apex less pointed. These changes suggest that the RV is stiffer in older subjects, with less dynamic contraction of the RVIT and less bellows-like movement [2].

While summarizing current knowledge on the pathophysiology underlying RV failure, this part of the heart has most appropriately been termed "the dark side of the moon" [32]. However, newer imaging modalities are addressing some of the underexposed aspects, e.g. by introducing regional curvature indices mentioned above [2].

The lapidary question *What is a "normal" right ventricle?* has been posed by Kovalova et al. [67]. What makes their study particularly interesting is the fact these authors also evaluated the LV in the same individuals and compared the findings for men and women. Their fascinating results are presented and discussed later in the section on EF.

Left Atrium (LA)

Enlargement of the LA volume index (LAVi) is a risk factor for cardiovascular complications and death [106]. Recently, LA size has been recognized as a powerful imaging biomarker. Reference values can be found in the review by Kawel-Boehm et al. [53]. Using 3D echo in 410 healthy Japanese subjects (253 men, age 20–69 years), it was found that in both sexes LA maximum and minimum sizes slightly increase with age, particularly after indexation for BSA [34], whereas LV size decreased. Similarly, in 1480 healthy individuals (mean age 36.1 ± 15.5 years, range 20–80; 61% males) a comprehensive TTE showed that mean LAVi in the overall population was 29.5 ± 10.8 mL/m^2 (range: 26.1–41.8 mL/m^2). Again, distinct higher values were found in subjects ≥50 years as compared with those who were younger (33.4 ± 12.5 vs. 29.1 ± 13.5; $P < 0.001$). On univariate analysis, LA volume was significantly associated with age (R = 0.48, P < 0.0001), male gender (R = 0.28, P < 0.05), BSA (R = 0.51, P < 0.0001), LVEDV (R = 0.52, P < 0.0001), and LVMi (R = 0.31, P < 0.05), as reported by D'Andrea et al. [25].

A CMR study including 804 healthy Caucasian adults who were stratified by gender and age (45–54, 55–64, 65–74) showed that LA maximal volume was significantly larger in males compared to females, but only for absolute values, not for indexed values. LA EF was similar for both sexes [97].

Clinical correlates and prognostic value of (enlarged) LA volumes have also been studied in the elderly, because they form a subgroup at high risk for cardiovascular events. $LAVi_{max}$, $LAVi_{min}$, and emptying fraction (EmF) were measured by 3D echocardiography in a community-based cohort including 706 participants (71 ± 9 years of age; 59% women). $LAVi_{max}$ and $LAVi_{min}$ were not associated with age in a healthy subgroup ($N = 142$; 66 ± 9 years of age; 55% women) but progressively increased with age in the entire cohort ($P < 0.001$). In univariate analysis, $LAVi_{max}$, $LAVi_{min}$, and reservoir function parameters were significantly associated with outcome in terms of stroke and MI [106].

The relationship of demographics and cardiovascular risk factors to LA size was also explored in a more diverse study. LAVi was measured by CMR in 2576 asymptomatic participants of MESA (average 68.7 years, 53.0% women, white 42.2%, Chinese American 12.0%, black 24.5%, and Hispanic 21.2%). The mean LAVi was 36.5 ± 11.4 mL/m^2 in the entire cohort and 35.5 ± 10.1 mL/m^2 in subjects free of cardiovascular risk factors ($N = 283$). Multivariable adjusted analysis showed age ($\beta = 0.2$ mL/m^2 per year, P < 0.0001), male sex ($\beta = -4.2$ mL/m^2, P < 0.0001), obesity ($\beta = 1.3$ mL/m^2, P < 0.01), EDVi ($\beta = 0.4$ mL/m^2, P < 0.0001), Chinese American ($\beta = -2.6$ mL/m^2, P < 0.0001), and Hispanic ($\beta = 1.1$ mL/m^2, P < 0.05) ethnicities were associated with LAVi, but DM and smoking were not [128].

Right Atrium (RA)

2D speckle-tracking echocardiography (2D-STE) and 3DE were used to define normative reference values of RA volumes and longitudinal strain (LS) in 200 healthy volunteers (43 ± 15 years, range 18–75; 44% men). RA total, passive, and active *emptying fractions* (EmFs) were calculated. 3D echo volumes and EmFs were found to be larger than 2DE ones (all P < 0.0001). BSA-indexed 3D volumes were significantly larger in men than in women and all EmFs smaller. Aging was associated with a decrease in passive EmF (P < 0.0001) and an increase in active RA negative LS (P < 0.0001) in the study by Peluso et al. [96].

A CMR study included 804 healthy Caucasian adults who were stratified by sex and age (45–54, 55–64, 65–74). RA maximal volume was significantly larger in males for both absolute and indexed values, while RA EmF was significantly higher in females [97]. Further details can be found in the review by Kawel-Boehm et al. [53].

Myocardial Strain (Rate)

Analysis of longitudinal and circumferential strain using two-dimensional speckle-tracking imaging (2D-STI) is considered an important method for the assessment of real-time, quantitative global, and regional LV function. In a prospective study involving 119 healthy volunteers (age range 22–76 years, 60 women), the impact of age, sex, SV, and HR on LV longitudinal and circumferential strains was investigated for three myocardial layers, using short-axis (for circumferential strain) and apical (for longitudinal strain) views [114]. Longitudinal and circumferential three-layer strain was highest in the apex and lowest in the base. The mean global longitudinal strain in the endocardial layer (GLS-endo), in the mid-myocardial layer (GLS-mid) and in the epicardial layer (GLS-epi) were −24.3 ± 3.1%, −21.3 ± 2.9%, and −18.9 ± 2.8%, respectively. Sex, HR, and SV were independent predictors of GLS-endo, GLS-mid, and GLS-epi. The mean

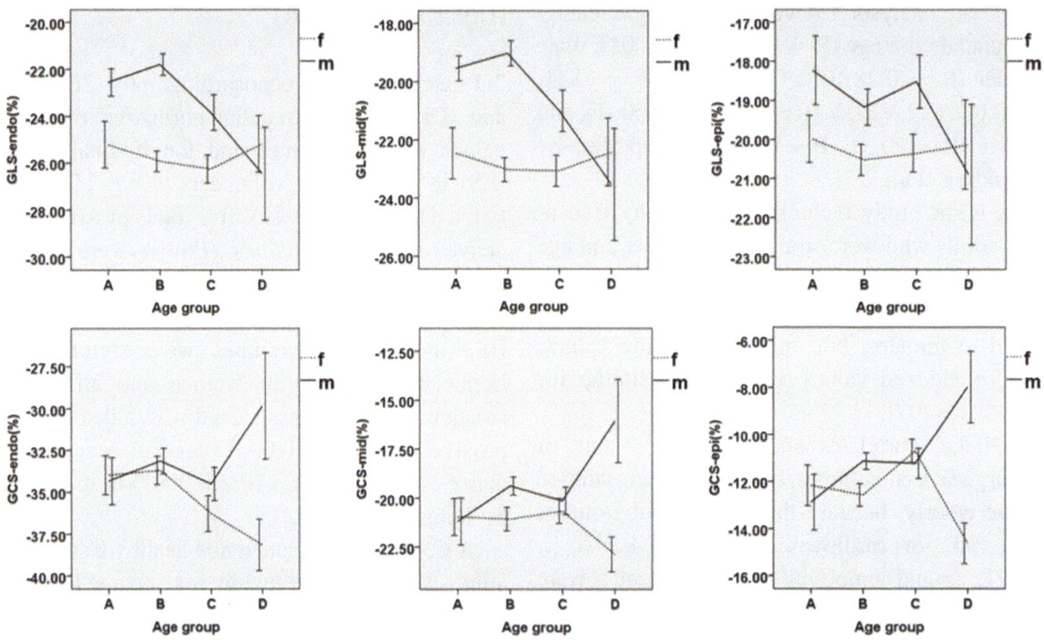

Fig. 41.3 Graphs illustrating left ventricular layer-specific global longitudinal strain (GLS) and global circumferential strain (GCS) changes across age groups in men and women. A = 22–40 year (11 males and 10 females); B = 41–50 year (31 males and 26 females); C = 51–60 year (15 males and 17 females); D = 61–76 year (2 males and 7 females). GLS and GCS seem to be independent of age and sex. (Reproduced from Shi et al. [114], with permission)

global circumferential strain in the endocardial layer (GCS-endo), in the mid-myocardial layer (GCS-mid), and in the epicardial layer (GCS-epi) were −34.3 ± 4.4%, −20.5 ± 3.0%, and −11.8 ± 2.7%, respectively. HR independently predicted GCS-endo, GCS-mid, and GCS-epi [114]. Figure 41.3 shows the findings for four adult age groups, illustrating the often diverging outcomes in both sexes. As a matter of fact, a recent study by Biering-Sørensen et al. [12] including 1296 participants undergoing a health examination with primary end point the composite of incident HF, acute MI, or CVD, indicated that after multivariable adjustment GLS was an independent predictor of outcomes in men but not in women.

A total of 440 (age, 45 ± 13 years) healthy subjects were analyzed by 3DE. Upper limits of LV EDVi and ESVi were larger in men (97 and 42 mL/m^2) than in women (82 and 35 mL/m^2; $P < 0.0001$). Conversely, lower limits of LV EF were higher in women than in men (51% vs. 50%; $P < 0.01$). Similarly, all strain components were higher in women than in men. Lower range was −18.6% in men and − 19.5% in women for longitudinal strain, −27.0% and − 27.6% for circumferential strain, −33.2% and − 34.4% for tangential strain, and 38.8% and 40.7% for radial strain, respectively. Among the strain components, longitudinal strain was the only one which did not change with age [10]. Sex-specific reference values for RV strain estimated by 2D echo have also been reported [86]. These findings emphasize the necessity to establish robust reference values in this relatively new area of evaluating cardiac function. The outcomes may be further complicated by size-dependent factors also reported. An animal study was designed to assess to what extent LV size influences strain and strain rate (SR) during controlled HR levels and different loading conditions. Longitudinal strain and SR were found to decrease with increasing LV dimensions in spite of unaltered contractility, showing that heart size influences strain and SR, which in addition are highly load-dependent parameters [105]. This observation has been confirmed in a meta-analysis [127] involving 2597 healthy subjects from 24 studies.

Ejection Fraction: A Measure Based on Physiology?

EF is the most widely applied metric to evaluate "systolic function." Mathematically, EF is "one" minus the ratio of two volume-related numbers, where the smaller one is compared to the larger one of each pair. Therefore, this ratio yields a number between 0 and 1 or between 0 and 100 if expressed as a percentage. A low value for EF is supposed to reflect serious cardiac illness. EF does not depend on any scaling or indexation (such as BSA), since it is a dimensionless ratio. This sounds like an advantage. However, the sole fact that it is a ratio implies that some information is missing. In practice, EF is a number obtained by dividing the outcome of two measurements that have the same dimension (i.e., mL or mL/m^2). If, for example, EF is 60% then EDV could be 100 mL while ESV is 40 mL, or EDV is 150 mL with an ESV of 60 mL. Yet, the two ventricles characterized by these {EDV, ESV} pairs, are different in many respects. The inverse relationship between EF and ESV as observed for patient data [54, 95] can also be found by using Monte Carlo simulation [56]. It is important to appreciate that EF has no strict physiological basis other than that ESV cannot be larger than EDV [56]. In a meta-analysis on LV reference values by 3D echo [14], including 2806 subjects from 13 studies, it was found that average LV EF was significantly higher in female subjects ($P = 0.003$). This finding is not surprising in view of the smaller ESVi values found in women and the inverse association for EF vs ESV(i) [60, 61, 95].

EF and LV volumes are considered of vital importance in clinical cardiology [27, 61]. In particular, LV EF is often considered a predictor of outcome in patients with chronic HF. Some treatments cause a small increase in EF and may, thereby, improve prognosis. However, HF patients ($N = 99$) with very low EF (all $\leq 20\%$) were censored at 3 years follow-up, and mortality appears to be 75%, while EF was found not to be a predictor ($P = 0.36$). The authors concluded that once the EF $\leq$ 20% this metric is no longer a predictor of mortality [89]. See within this context also Fig. 41.2 which features an asymptotic range where (extrapolated) EF does barely change, while rather ESV is the major determinant of ventricular function.

Interestingly, EF is higher in women, both in healthy individuals [15, 62], nearly normals [36], and in cardiac patients [1, 63], as well as specifically in HF patients [57, 120]. Remarkably, EF has a (supra)normal value in half of all patients with HF. A rationale for the universal applicability of EF has never been published. Despite the obscure origin, it has been demonstrated that EF often corresponds with improvement following therapeutic intervention and seems to be associated with prognosis. This is not totally surprising, as it has been documented that EF is primarily determined by ESV [54]. In turn, ESV is the major contributor to the numerical value of end-systolic elastance [65].

To derive age-, sex-, and ethnicity-specific reference values for LV volumes and EF using 3D echo, a total of 978 healthy European White and Asian Indian subjects were studied [19]. Indexed LV volumes were significantly smaller in female compared with male subjects and in Indian Asians compared with European whites. Upper limit of normal (mean $\pm$ 2 SD) reference values for ESVi and EDVi as follows: European white men, 29 mL/m^2 and 67 mL/m^2; Indian Asian men, 26 mL/m^2 and 59 mL/m^2; European white women, 24 mL/m^2 and 58 mL/m^2; Indian Asian women, 23 mL/m^2 and 55 mL/m^2, respectively.

Using a careful diagnostic pathway (including CMR) based on current best evidence in one study, 245 (47% females) out of 319 patients admitted to an HF clinic met the criteria and were phenotyped. Depending on the threshold selected (at 55, 50, 45, and 40% for EF), the ratio of HFrHF to HFpEF was 5.2, 3.2, 1.9, and

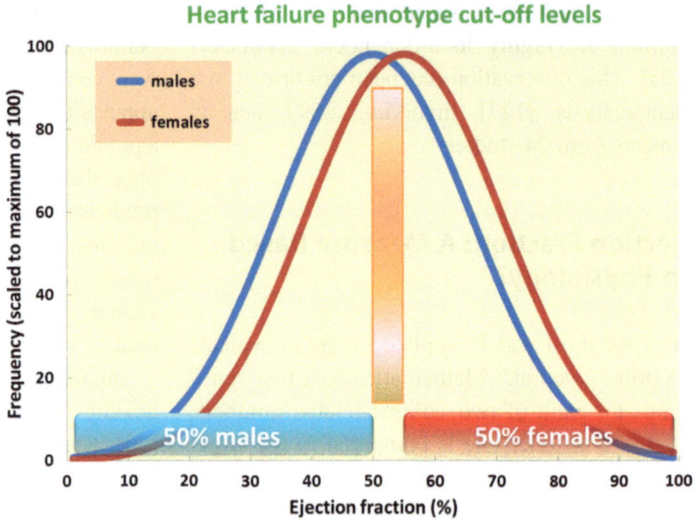

Fig. 41.4 Since women have on average a higher ejection fraction (EF, both in health and disease), it seems plausible to consider a higher cutoff value for heart failure (HF) phenotypes, e.g., at 55% rather than a single traditional value such as 50% for both sexes. In fact, several studies have proposed a higher EF cutoff value for women. In the situation suggested, the prevalence of HF with preserved EF is equal in men and women, given the distribution assumed as in this theoretical example

1.3 to one, respectively [121]. This study demonstrates that the estimated impact of the new HFpEF epidemic strongly depends on a single criterion. Unfortunately, this study did not stratify for sex, implying that further refinements may be required (Fig. 41.4). In addition, the imaging modalities employed yield varying outcomes as is often reported in the literature [125]. For example, in a comparative study concentrating on 52 HF patients it was found that EF measurements by M-mode echo, 2D echo, radionuclide ventriculography, and CMR are not interchangeable. As a consequence the conclusions and recommendations in HF studies should be interpreted in the context of locally available techniques [9].

In an interesting combined LV and RV 2D echo study concerning 91 healthy volunteers (age 17–62 years, 46 males), Kovalova et al. [67] found that the RV EF was lower than the LV EF, but the difference was more pronounced in men (50.0% ± 9.7 vs. 60.7% ± 8.4) than in women (58.0% ± 13.6 vs. 61.7% ± 9.4). These findings can be fully explained on the basis of the inverse EF-ESVi relationship described for both LV and RV [61], when keeping in mind that the SVi values observed in this investigation were similar in men and women. As a result the authors found two distinct distributions for EF which was most pronounced for the relatively larger RV in both sexes as determined by their method and also supported by their literature study (Fig. 41.5). In a review (ages 20–68 years) Kawel-Boehm et al. [53] reported average values for EDVi, being larger (91 and 80 mL/m^2) for the RV, than in the LV (81 and 76 mL/m^2) for males and females, respectively. Figure 41.5 compares well with the theoretical one composed for the LV in HF patients which for simplicity is assumed symmetrical for both sexes (Fig. 41.4).

Meta-analysis Comparing SPECT and CMR By considering only few subjects, many studies left uncertainty about the frequency of discrepancies between SPECT and CMR. Therefore, a meta-analysis of data on 164 subjects from 9 studies was performed. Correlation between both methods was good: EDV (R = 0.89), ESV (R = 0.92), and EF (R = 0.87). However, rates of discrepancies for individual subjects were considerable (37% for at least 30 mL in EDV; 35% for at least 20 mL in ESV; 52% for at least 5% in EF; and 23% for at least 10% in EF. The misclassification rate for the 40% EF cutoff was 11% [47].

Importance of Software In a gated SPECT study two software programs (Quantitative Gated SPECT (QGS) and Segami) were compared in a sex-specific study by Yamada et al. [126] to

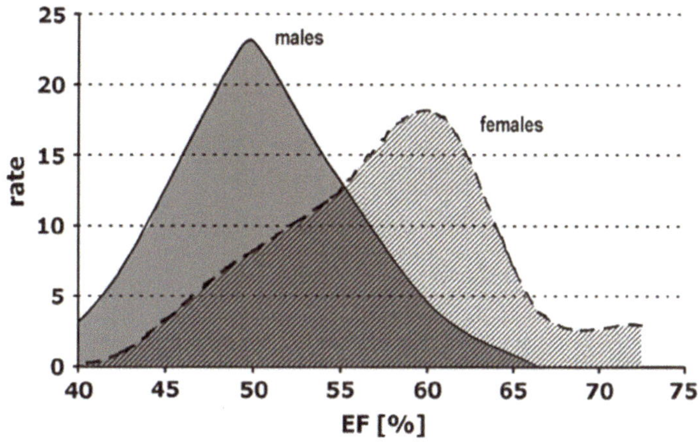

Fig. 41.5 Distribution of ejection fraction (EF) for the right ventricle in healthy men ($N = 46$) and women ($N = 45$), determined by 2D echo. (Reproduced, with permission, from Kovalova et al. [67])

evaluate the influence of age, weight, height, HR, SBP, DBP, BMI, and BSA in 181 healthy Brazilian individuals (20–80 years, 96 women). Previously it was demonstrated that QGS underestimates LV volume, which deviation is nonlinear for the range 37–101 mL. Yet, for both methods significant differences ($P \leq 0.001$) for EDVi, ESVi, and EF were found between men and women. Neither method showed a significant association between EF and age.

Exercise in Trained and Untrained Men It is believed that EF is a useful variable to evaluate LV function. However, changes of EF values in athletes during exercise are controversial [28], although baseline LV volumes are consistently increased. Unfortunately, these authors studied only males when comparing EF response to supine bicycle exercise in normals and athletes. Within this context it is relevant that we have shown that EF vs ESVi is described by a nonlinear relationship in both sexes, which for males is not different for baseline data compared to exercise [61]. Given the fact that the athletes reportedly had larger average ESVi and EDVi at rest, it may be expected that on average their EF changes as induced by exercise are smaller in magnitude compared to those observed in the control group. In the latter group the average ESVi decreased ($P < 0.001$) during exercise from 29 to 17 mL/m^2 and in the athletes from 50 to 40 mL/m^2 ($P < 0.001$), thus explaining the relatively minor EF changes ($P < 0.001$) in these male athletes.

Exercise test and EF in Cardiac Patients Sex-related differences were not only investigated in healthy individuals (e.g., [15, 28, 126]) but also in cardiac patients. In the study by Ababneh et al. [1] patients were referred for the evaluation of chest pain or dyspnea, often having cardiac risk factors, but all with normal exercise capacity corrected for age, no ECG signs of ischemia, normal perfusion scans, and no wall motion abnormalities. A total of 1513 patients (mean age 60 ± 12, range 24–92 years) who had normal results on Bruce exercise tests had either Tc-99 m sestamibi ($N = 884$) or Tl-201 ($N = 629$) injected at peak stress. The mean EF for gated Tc-99 m sestamibi scans did not differ from gated Tl-201 scans. However, when sex was considered, the mean EF for women was 66% ± 8% with Tc-99 m sestamibi gated SPECT ($N = 519$), higher than the mean EF for men (58% ± 8%, $n = 365$, $P < 0.0001$). Similarly, the mean EF for women studied with Tl-201 gated SPECT (67% ± 8%, $N = 326$) was higher than that of men (59% ± 7%, $N = 303$, $P < 0.0001$). Unfortunately, these investigators did not consider age and medication in the analysis of their cardiac patients. However, they demonstrated that the sex-related difference in mean EF found for normals in other studies persists in these cardiac patients, thus providing evidence for the concept developed in Fig. 41.4.

The EF Paradigm EF has played a central role in the design and interpretation of almost every

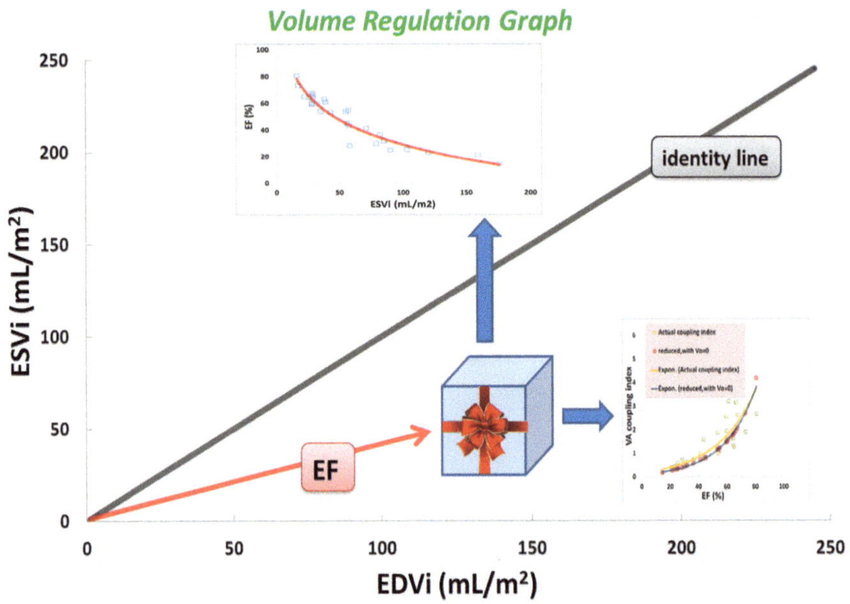

Fig. 41.6 Ventricular volume domain showing the relationship of end-systolic volume index (ESVi) versus end-diastolic volume index (EDVi). Actual data points are only located below the (black) identity line because ESVi cannot be larger than EDVi. All data points (in the two insets) refer to heart failure patients. Ejection fraction (EF) for any {EDVi, ESVi} is reflected by the slope of the (red) line with arrow head. The metric EF has two close family members, namely, ESVi itself (see upper inset) and the ventriculo-arterial (VA) coupling index (see inset to the right). For further explanation see text

clinical trial that involves cardiac function. Although cutoff values may differ from one study to another, well-documented sex-related differences regarding EF have barely been incorporated. Whenever considering future studies, sex-specific aspects may be the first step to take. Various major drug-related trials have been carefully analyzed by Fogel [29]. How predictive an increase in EF on therapy is of morbidity and mortality remains a matter of debate. His review outlines the utilization of EF (or lack thereof), in various congestive HF drug trials, and its relationship to symptomatic improvement and morbidity/mortality.

The VRG and Derived Metrics in Cardiac Disease

The volume regulation graph (VRG) relates ESVi to EDVi and applies to all four cardiac compartments, as described elsewhere [65]. Since EF = 100*(1 –ESVi/EDVi), any EF value can be readily identified in this representation, as shown in Fig. 41.6. An arbitrary EF value has been selected to illustrate that this metric associated with a given {EDVi, ESVI} pair (as represented by the surprise box) includes information on various other well-known variables. The upper inset figure shows that EF is inversely related to ESVi, with a high correlation for this group consisting of 28 HF patients, described by Asanoi et al. [4]. The inset to the right illustrates that EF is also highly correlated with the ventriculo-arterial coupling (VAC) index [63], not only when the volume intercept (Vo) is omitted but also when the actual value of Vo is considered. Thus, the information offered by the triad EF, ESVi, and VAC is nearly equivalent.

Aortic Diameter

Reference ranges for echocardiographic dimensions of the proximal aorta [111] have been based on 704 (mean age: 46.0 ± 13.5 years) healthy volunteers (394 women). Aortic dimensions were obtained in systole and diastole, following both the leading-edge to leading-edge and the inner-edge to inner-edge conventions. Diameters were measured at four levels: ventricular-arterial junction, sinuses of Valsalva, sinotubular junction, and proximal tubular ascending aorta. Measures of aortic root in the short-axis view following the orientation of each of the three sinuses were also performed. Men had significantly larger body sizes when compared with women and showed larger aortic dimensions independently of the measurement method used. Dimensions indexed by height and BSA are provided, as well as stratification by age ranges. In multivariable analysis, the independent predictors of aortic dimensions were age, gender, and height or BSA. Thus, age, sex, and body size are significant determinants of aortic dimensions [111].

Reference values for aortic root diameter (ARD) have been determined by Doppler TTE [124] in a cohort of 1043 Caucasian healthy volunteers (mean age 44.7 ± 15.9, range 16 to 92 years, 503 men [48%]). Measurements were made at end-diastole in parasternal long-axis views at four levels: (1) annulus, (2) sinuses of Valsalva, (3) sinotubular junction, and (4) proximal ascending aorta. The absolute ARDs were significantly greater in men at all four levels, whereas BSA-indexed diameters were greater in women ($P = 0.0001$). No significant sex differences were registered for sinuses of Valsalva and sinotubular junction to annulus diameter ratios ($P = 0.9$), whereas ascending aorta to annulus diameter ratio was higher in women ($P = 0.0001$). There was a linear correlation between ARDs (both absolute and indexed values), their ratios, and age in both sexes ($P = 0.0001$), as described by Vriz et al. [124].

Kawel-Boehm et al. [53] review varies studies reporting aortic diameter at well-defined anatomical locations in men and women for various age groups (20–80 years). Size appears to vary with BMI, and BSA-indexed readings seem to be slightly higher in women older than 30 years, both systolic and diastolic. Also, aortic diameters in men and women in various ethnic populations have been reported, e.g., in healthy Japanese individuals (age 20–80 years) by Daimon et al. [23].

Obviously, aortic diameters are increased in hypertensives. Effects of prolonged exercise are less clear. Aortic dissection is a highly morbid and mortal condition (in-hospital mortality over 25%), with dilation of the thoracic aorta as the primary risk factor, associated with sex, age, body size, hypertension, and inherited/congenital conditions. However, patients often present with significant aortic dilation in the absence of a clear risk factor or cause. Current consensus indicates that athletic participation has minimal impact on aortic size, implying that significant aortic dilation in an athlete is viewed as pathological. Also, it has been hypothesized that remodeling and dilation of the aorta may be stimulated by exercise-induced hemodynamic factors (including increased SV and elevated cSBP). However, all prior research has limitations because it has focused on athletes with an average age below 30 years [22].

Coronary Arteries

Coronary dominance is known to be influenced by postnatal workload changes of the ventricles, making serial assessments concerning size of special importance. Z-score equations for diameter reference values of two major coronary arteries adjusted for growth changes have been determined in 200 healthy, near-term neonates (106 boys) at birth, with follow-up at 1 and 6 months. Several regression models for the left (LCA) and right coronary artery (RCA) diameters were tested with weight, height, BSA, and aortic annulus (AoA) diameter. The coronary diameters at birth were strongly correlated with birth weight, height, BSA, and AoA diameter ($R > 0.8$, all $P < 0.01$). Only at 6 months the diameters in boys were larger than in girls, but this difference did not reach significance [52]. Growth in caliber of the coronary arteries is definite and progressive during postnatal life. As with other variables (including BP and HR) sex-specific differences occur only later in life. Further details on major studies in children can

Table 41.5 Coronary artery size (by IVUS) following heart transplantation (M/F)

	Initial area (mm^2)	Follow-up area (mm^2)	P
Female donors	15.7 (13.8–17.9)	17.0 (14.3–19.2)	0.49
Female recipients ($N = 16$)	15.5 (13.7–18.5)	15.9 (14.1–17.9)	0.63
Male recipients ($N = 21$)	16.1 (13.9–17.9)	17.9 (15.4–19.6)	**0.01**
Male donors	16.8 (15.4–18.7)	17.7 (15.2–19.7)	0.86
Female recipients ($N = 17$)	16.3 (15.1–17.8)	17.3 (14.6–19.5)	0.55
Male recipients ($N = 32$)	17.1 (15.4–19.5)	17.8 (15.4–19.8)	0.89

Data as the median value (interquartile range). *IVUS* intravascular ultrasound. Changes in vessel area of the proximal left anterior descending coronary artery. From Herity et al. [44], reproduced and slightly modified with permission

be found in the extensive review by Dallaire and Sarkola [24], and regarding adults in Taqueti [118].

In a prospective study reference ranges of the proximal LCA and RCA internal diameters (in diastole) in 390 healthy Asian children (ages of 2 months to 8 years, 55% boys) at predetermined anatomical sites were established. The LCA and RCA diameters correlated linearly with age, height, weight, BSA, as well as the AoA diameter, yielding $R > 0.8$, $P < 0.005$. The coronary-aorta index (i.e., coronary artery to AoA ratio) falls within a narrow range: LCA/AoA $= 0.15 \pm 0.02$ (range 0.09–0.21), RCA/AoA $= 0.13 \pm 0.02$ (range 0.09–0.20). This index may serve as a quick guide to detect coronary dilatation [116].

In healthy adults several studies have demonstrated that coronary vessel size is significantly larger in 50 males (compared to 25 women) for left main and LAD [113], as well as for LCA and RCA in 100 males vs 100 females [68], while Hiteshi et al. [45] found in 710 subjects (327 females) that neither body habitus nor LVM relate to the sex-related difference in various coronary artery diameters. Obviously, these differences between men and women have consequences for coronary artery bypass grafting (CABG), as discussed, e.g., by Dickerson et al. [26] and Hannan et al. [42].

Scaling relationships between myocardial mass and the morphometry of coronary vasculature have been studied in normal porcine hearts. Volume, diameter, and cumulative length of coronary arteries were reconstructed from casts and related to the perfused myocardial mass. It was shown that arterial volume is linearly related to regional myocardial mass, whereas the sum of coronary arterial branch lengths, vessel diameters, and volumetric flow show an approximately 3/4, 3/8, and 3/4 power-law relationship, respectively, in relation to myocardial mass. Fundamental design principles underlying the structure-function relations were suggested based on these scaling laws [21]. This publication triggered an editorial entitled: "Why are arteries the size they are?" by Santamore and Bove [109] bringing up questions such as what is driving arterial size: an energy cost factor, oxygen carrying capacity of blood, shear rate, or circumferential strain?

Remarkably, at follow-up in 86 heart transplant patients (median age 24 years) it was found that coronary artery size (expressed as - cross-sectional area) was significantly ($P = 0.01$) increased in male recipients of female hearts (Table 41.5), as reported by Herity et al. [44].

Biomarkers in Clinical Chemistry

Relevant data are presented elsewhere [84].

Microvasculature

Relevant data are presented elsewhere [46].

Fetal Development and Pediatric Cardiology

A cross-sectional ultrasound study [41] evaluated in 184 fetuses of normal pregnancies (19–42 weeks) the LV and RV volumes, SV,

COi, and EF. With advancing gestation, ventricular volumes, SV, and COi increased, whereas EF decreased. RV ESV was larger than LV ESV (0.50 vs 0.27 mL; $P < 0.001$), and a similar difference was found for diastole (1.20 vs 1.03 mL; $P < 0.001$). LV EF was greater than the RV EF (72.2 vs 62.4%; $P < 0.001$).

During childhood, most cardiovascular variables depend on body mass [24]. Starting at a reference range from 95 to 155 bpm during the first 2 years, HR in Chinese children was found to steadily decline and reach a lowest value of 65 at the age of 12 years [100]. Over the same period, SBP rises from a maximum of 95 to 135 mmHg, while average SVi sharply increases until the age of 5 years when a plateau at 55 mL/m^2 is reached. Limited information is available on cardiac size in children. A classical study reported on LV volume, LV mass, and LA volume [40]. SVi and, to a lesser extent, the COi generally increase from ages 1 to 5 years, but plateau or fall slightly thereafter [100]. Four studies on LV volume were conducted in children and young adolescents; one study provided data in an independent pediatric subgroup. In pediatric studies, EDV pooled mean value was 53.1 mL (95% CI, 38.1–68 mL), while for ESV, it was 19.8 mL (95% CI, 14.8–24.8 mL); LV EF mean value was 63.3% (95% CI, 61.6–65%). Significant heterogeneity and inconsistency were noted among studies. Age, SBP, and HR were identified as a source of between-studies variation for LV volumes [14]. In 60 healthy children (age 8–17 years, 30 girls), no difference has been found for EF in both LV and RV in three age groups on the basis of CMR [103]. Further details can be found in the review by Kawel-Boehm et al. [53], reporting on normal values for CMR in adults and children, including chamber size and function, aortic diameters, and distensibility. Remarkably, the MRI study by Sarikouch et al. [110] found that BSA-indexed LV EDV was 10–15% larger in boys (age 8–20 years). Published nomograms of pediatric echocardiographic measurements are often limited by insufficient sample size to assess the effects of age, sex, race, and ethnicity. Data collected from healthy nonobese children $\leq$18 years of age at 19 centers with a normal echocardiogram included age, sex, race, ethnicity, height, weight, echocardiographic images, and measurements performed at the Core Laboratory. Z-score models involved indexed parameters (X/BSAα) that were normally distributed without residual dependence on BSA. The models were tested for the effects of age, sex, race, and ethnicity. Raw measurements from models with and without these effects were compared, and < 5% difference was considered clinically insignificant because interobserver variability for echocardiographic measurements are reported as $\geq$5% difference. Of the 3566 subjects, 90% had measurable images [77]. Furthermore, meta-analysis based normal ranges of RV strain (rates) in children have been published [72]. Similar data for the LV have been reviewed elsewhere [17].

National High Blood Pressure Education Program [87] reports on reference values, see their Table 3 (BP levels for boys by age (1 through 17 years) and height percentile) and Table 4 (BP levels for girls by age (1 through 17 years) and height percentile), available via: http://pediatrics.aappublications.org/content/114/Supplement_2/555.long. To use the tables in a clinical setting, the height percentile is determined by using the newly revised CDC growth charts (www.cdc.gov/growthcharts). Central and brachial BP values for children have been presented in Avolio et al. [7]. BMI Tables for children can be found at: http://www.who.int/childgrowth/standards/bmi_for_age/en/

Metrics, Indexes, and (Doppler) Surrogates in Cardiology

The easiest and most common way to assess the RV is by two-dimensional (2D) transthoracic echocardiography measuring surrogate indexes, such as tricuspid annular plane systolic excursion (TAPSE) and fractional area change (FAC) [32]. TAPSE reflects longitudinal myocardial shortening, being the main component of RV contraction in normal hearts.

Choi et al. [20] found that mitral early diastolic (E) and late diastolic (A) velocity as well as E/A ratio were significantly higher in women compared to those in men. Conversely, mitral peak systolic and late diastolic annular velocities in both septal and lateral mitral annuli were significantly lower in women compared to those in men. However, there were no significant differences in both septal and lateral mitral early diastolic annular (e') velocity between men and women. In both men and women, mitral E velocity and its deceleration time as well as both E/A and E/e' ratio considerably increased with age. There were no significant differences in tricuspid inflow velocities and tricuspid lateral annular velocities between men and women except e' velocity, which was significantly higher in women compared to that in men. However, changes in both tricuspid inflow and lateral annular velocities according to age were similar to those in mitral velocities. Lancellotti et al. [69] have also described echo surrogates for LV end-diastolic pressure (EDP) in 159 patients (25% EF $<$ 50%, 64% NYHA $\geq$ II, and 53% with coronary artery disease) who underwent simultaneous echo evaluation (E/A, E/e', LA volume, tricuspid regurgitation jet velocity) and invasive measurement of EDP. Sixty-four patients had elevated LVEDP ($\geq$15 mmHg). The 2016 recommendations (sensitivity 75%, specificity 74%, positive predictive value 39%, negative predictive value 93%, AUC 0.78) identified slightly better patients with elevated invasive LVEDP ($\geq$ 15 mmHg) as compared with the 2009 recommendations. Reference values for variables from tissue Doppler imaging (TDI) according to sex, age aspects, and specific ethnic groups are summarized in Table 41.6.

How to Interpret Measurements?

Many routes are available to collect information on the heart. The next task is to interpret these data. Obviously, no single measurement can reveal all details about cardiac functioning, not even in the healthy heart. Various pieces of information each provide some insight. For example, the heart axis based on the ECG may suggest the presence of left ventricular hypertrophy. When this suspicion is signaled, additional measurements (e.g., by echocardiography) are required to confirm and quantify the abnormality. The various strategies involved can be viewed as an exercise to identify from a particular distance the full characteristics of a remote object, in Fig. 41.7 symbolically depicted by the large turquois circle. This item will be referred to as the "real picture" (RP). EF is a popular metric to assess so-called systolic function. Since EF refers to a ratio, it is impossible that this single number (theoretically between 0 and 100%) can offer a complete picture. Rather we must accept that EF is an incomplete index. All missing details can be lumped into a secondary index, which we will refer to as the EF-extra (EFe), depicted as the gray cylinder. The positive note comes from the fact that "any" additional variable (such as LAVi, strain, MAPSE, and many more not shown) will assist in more completely describing the partially unknown RP. The negative side stems from the fact that any candidate selected is only a small component of the solution. A smart approach would be to not repeat the shortcoming inherent to EF and thus avoid introducing any more ratios, such as the ventriculo-arterial coupling index [56].

Rather than collecting more tables or charts with reference values for current metrics to evaluate cardiac function, we explore their anticipated practical importance and interconnections. If we represent the full information needed to describe ventricular function as a central circular area (the turquoise RP mentioned earlier), then each known metric reveals part of the complete set, as reflected by the partial overlap of circles in Fig. 41.7. Appreciating additional sex-related differences in this figure (not shown) may imply specific therapeutic guidelines, as illustrated in resynchronization therapy for women [130].

LV systolic long-axis shortening is one determinant of volume change during the cardiac cycle and can be measured as mitral annular plane systolic excursion (MAPSE). The surrogate measure MAPSE is age dependent, but similar for men and women in comparable age groups

Table 41.6 Studies that provided reference values for variables Doppler and TDI according to sex and age groups in specific ethnic groups

Study	Population	n	Age, year	Male (%)	E	E/A (age, year)	Septal e'	Septal a'	Septal s'	E/e'	Changes with ageing				
											E	E/A	e'	a'	E/e'
Dalen et al. [a]	Norwegian	1266	M: 50.6 ± 13.7 F: 47.8 ± 13.6	48	M < F	M < F	M < F	M > F	M > F	M < F	↓	↓	↓	↑	↑
NORRE[b]	European white	449	M: 45.9 ± 14.0 F: 45.7 ± 13.4	44	M < F	M = F	M = F	M > F	M > F	M = F	↓	↓	↓	↑	↑
JAMP[c]	Japanese	700	M: 43.7 ± 14.5 F: 43.5 ± 14.5	55	M < F (< 50)	M < F (< 50)	M < F (< 50)	M > F (< 60)		M < F (>60)	↓	↓	↓	↑	↑
					M = F (> 50)	M = F (>50)	M = F (>50)	M = F (> 60)							
EMINCA[d]	Han Chinese	1394	M: 47.1 ± 16.2 F: 47.5 ± 15.8	49	M < F	M = F	M = F	M > F	M > F	M < F	↓	↓	↓	↑	↑

Reproduced from Ryu [107], with permission

The references mentioned are as follows:

[a]Dalen H et al. A. Reference values and distribution of conventional echocardiographic Doppler measures and longitudinal tissue Doppler velocities in a population free from cardiovascular disease. Circ Cardiovasc Imaging. 2010;3:614–622.

[b]Caballero L et al. Echocardiographic reference ranges for normal cardiac Doppler data: results from the NORRE Study. Eur Heart J Cardiovasc Imaging. 2015;16:1031–1041.

[c]Daimon M et al. Normal values of echocardiographic parameters in relation to age in a healthy Japanese population: the JAMP study. Circ J. 2008;72:1859–1866.

[d]Yao GH et al. Echocardiographic Measurements in Normal Chinese Adults (EMINCA) Study Investigators. Doppler echocardiographic measurements in normal Chinese adults (EMINCA): a prospective, nationwide, and multicentre study. Eur Heart J Cardiovasc Imaging. 2016;17:512–522

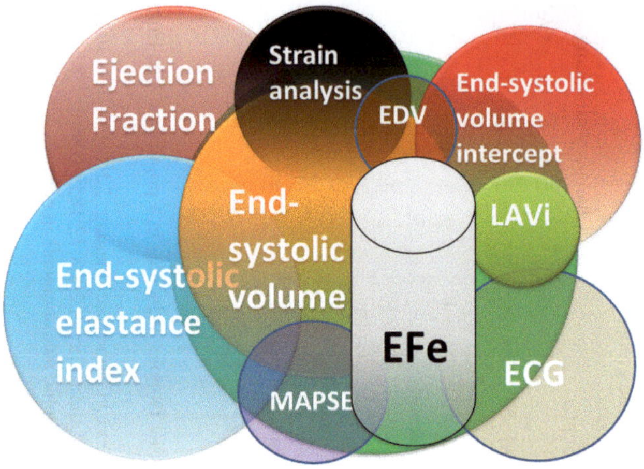

Fig. 41.7 Schematic overview of indicators of cardiac performance and the potential insight they provide. The complete "real picture" (RP) is represented by the underlying large turquoise-colored circular area. Various attempts have been made to characterize (part of) this incompletely understood territory. The sizes tentatively reflect the relative impact of each metric, and there is also some overlap meaning that two metrics partly tell the same story. One popular index is ejection fraction (EF) which unfortunately can never describe the full truth about RP, since it is based on a ratio. The information not embodied by EF is supposed to be contained in a cylinder (EFe, meaning EF-extra) [131]. The cylindrical shape symbolizes that this metric primarily has a repair function. In contrast, end-systolic volume appears to play a pivotal role [60]. *ECG* electrocardiogram [101], *EDV* end-diastolic volume, *LAVI* left atrial volume index, *MAPSE* mitral annular plane systolic excursion

[92]. In a recent study of 1266 subjects without evidence of heart disease, MAPSE was measured in all three standard apical views. Mean MAPSE was 1.58 ($\pm$ 0.25) cm, declining with age, with correlations between −0.50 and − 0.41. Again, MAPSE was reported to be sex independent, although values diverged for men and women above 60 years [115].

No publications exist about a possible sex-related difference regarding the end-systolic volume intercept. All other metrics in Fig. 41.7 that refer to mechanical variables are sex-dependent [63], with apparently an exception in case of MAPSE for healthy individuals <60 years (Table 3 in [115]). This complexity makes it an enormous challenge to identify a unique combination of metrics which in a concerted action could reliably describe cardiac function in the healthy individual. This task likely becomes formidable when addressing disease states.

Conclusions

Almost all volume-related metrics used to evaluate ventricular and atrial function show sex-specific reference values (Fig. 41.8). Age dependence has also been noted, and occasionally additional effects related to ethnicity can be discerned. Most of the distinctions referring to length, area, or volume persist after indexation for BSA. Thousands of publications report reference values for the size of each cardiac compartment, occasionally with conflicting outcomes, likely based on minor differences in study design and population characteristics. Furthermore, various degrees of interobserver, intra-observer, intervendor, and interphase variability are reported.

For a particular group of metrics, there is no meaning in indexation by whatever standard, namely, those indicators which consist of a

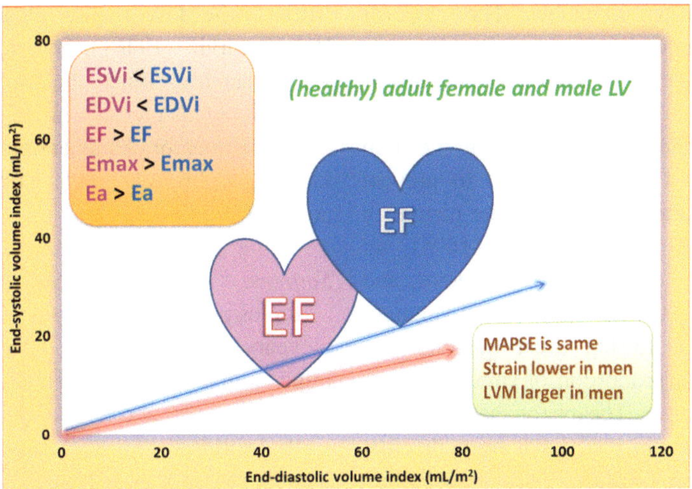

Fig. 41.8 Schematic overview of indicators of left ventricular (LV) function to highlight male/female differences. Ejection fraction (EF) in women is significantly higher compared to matched male counterparts, primarily because end-systolic volume index (ESVi) is smaller. Since end-diastolic volume index (EDVi) is linearly correlated with ESVi, the values for EDVi are in men usually larger than in women. The smaller ESVi in women implies a larger end systolic elastance (Emax) as well as a higher effective arterial elastance (Ea). *LVM* LV mass, *MAPSE* mitral annular plane systolic excursion. As emphasized in this survey, typical values may also depend on age and ethnicity (For further explanation, see Kerkhof et al. [62])

dimensionless ratio such as EF and VAC index. The EF and VAC are telling the same story if the end-systolic volume intercept is not considered in the calculation (Fig. 41.6). Similarly it should be appreciated that strain analysis may have limitations. As noted, a few metrics may only be meaningful over a limited trajectory and are not automatically applicable to all disease states. The examples of Emax and HFpEF have been discussed. Figure 41.8 summarizes the take-home message. Regarding volume-related metrics, it is generally not simple to establish reference values, because their estimates depend on imaging modality besides many other factors. Indeed, *age, sex, size, and ethnicity all matter*.

Appendix on Biological and Technical Pumps

Prototypes of technical pumps have been designed for a long time. To raise water for irrigation, Egyptians around 2000 BC invented the shadoof (or shaduf, also called a counterpoise lift) which is hand-operated and uses a long suspended rod with a bucket at one end and a weight at the other. Another invention, the Archimedean screw pump (200 BC), is actually still in use in our times. It may not be a surprise to learn that around the time that William Harvey (1578–1657) was dealing with the heart as a pump, engineers had already developed various types of mechanical pumps, including the water wheel [129] as described a century earlier by Vannoccio Biringuccio (c.1480–c.1539) in *De la Pirotechnia* (Venice 1540). In 1588 the sliding vane water pump technology was detailed by the Italian engineer Ramelli (1531–1600) in his book *The Diverse and Artifactitious Machines of Captain Agostino Ramelli*, while in 1593 the Frenchman Nicolas Grollier de Servière (1596–1689) created an early design for a gear pump, followed in 1636 by Pappenheim, a German engineer – walk of life largely unknown – who invented the double deep-toothed rotary gear pump, which is still used to lubricate engines. This gear pump made it possible to dispense with the reciprocating slide valves used by Ramelli. Pappenheim drove his machine by an overshot water wheel set in motion by a stream and was used to feed water fountains. The emperor Ferdinand II granted him a "privilege" – the equivalent of a patent – in respect of this invention. In 1650 Otto van Guericke (1602–1686) invented in Magdeburg the piston vacuum pump, which used leather washers to prevent leakage between the cylinder and the piston. In 1675 Sir Samuel Morland (1625–1695), an English academic, diplomat, spy, inventor, and mathematician, patented the packed plunger pump, capable of raising great quantities of water with far less proportion of strength than a chain or other pump. The piston had a leather seal. Morland's pump may have been the first use of a piston rod and stuffing box (packed in a cylinder) to displace water. In 1687 the French-born inventor Denis Papin (1647–1713), who spent much of his life in London and Marburg, developed the first true centrifugal pump, one with straight vanes used for local drainage. With the advent of steam engines, the design of efficient mechanical pumps received strong impetus, particularly in England with great names as Thomas Newcomen and William Murdoch [129].

It is interesting to observe that the first technical designs originated in Italy, also the cradle of the anatomical and physiological study of the cardiac pump (notably Borelli), and the subsequent spread toward northern Europe (Germany and France), to come to full glory in England at the time of the industrial revolution.

Giovanni Borelli (1608–1679) was an Italian iatromechanical physiologist, student of Galileo, and teacher of Malpighi. In his early 30s he was called to the University of Messina as professor of mathematics, later becoming "physician" to Christina of Sweden, who had chosen voluntary exile in Rome. Apart from devoting some of his time to astronomy, he also studied the heartbeat in the frog [35]. Later he wrote: The true action of the muscle of the heart is the contraction of its ventricles, and the compression and expression of the blood contained in them is carried out [like] ... a winepress [48]. The Leiden-born physician Willem Kolff (1911–2009) devoted most of his life to the development of an artificial kidney. In 1950 he immigrated to the USA and continued his research at the Cleveland Clinic, extending his work to the development of an artificial heart. Robert Jarvik (1946–), an American physician and researcher, joined the University of Utah's artificial organs program in 1971, then headed by Kolff, his mentor. As an entrepreneur he became known for his role in developing the Jarvik-7 artificial heart.

Sources: https://www.pumpsandsystems.com/topics/pumps/pumps/history-pumps-through-years

https://pumpsolutions.com.au/a-brief-history-of-water-pumps-and-how-they-have-affected-the-world/

https://pressurewasherify.com/blog/pumps-history/

References

1. Ababneh AA, Sciacca RR, Kim B, Bergmann SR. Normal limits for left ventricular ejection fraction and volumes estimated with gated myocardial perfusion imaging in patients with normal exercise test results: influence of tracer, gender, and acquisition camera. J Nucl Cardiol. 2000;7 (6):661–8.
2. Addetia K, Maffessanti F, Muraru D et al. Morphologic analysis of the normal right ventricle using three-dimensional echocardiography-derived curvature indices. J Am Soc Echocardiogr. 2018. pii: S0894-7317(17)30883-0. https://doi.org/10.1016/j.echo.2017.12.009.
3. American Psychological Association (APA). Definition of terms: sex and gender. 2012. https://www.apa.org/pi/lgbt/resources/sexuality-definitions.pdf. Accessed 5 Feb 2018.
4. Asanoi H, Sasayama S, Kameyama T. Ventriculoarterial coupling in normal and failing heart in humans. Circ Res. 1989;65:483–93.
5. Aune E, Baekkevar M, Rødevand O, Otterstad JE. The limited usefulness of real-time 3-dimensional echocardiography in obtaining normal reference ranges for right ventricular volumes. Cardiovasc Ultrasound. 2009;7:35. https://doi.org/10.1186/1476-7120-7-35.
6. Aune E, Baekkevar M, Rødevand O, Otterstad JE. Reference values for left ventricular volumes with real-time 3-dimensional echocardiography. Scand Cardiovasc J. 2010;44:24–30.
7. Avolio AP, Kuznetsova T, Heyndrickx GR, Kerkhof PLM, Li JJ-K. Arterial flow, pulse pressure and pulse wave velocity in men and women at various ages. In: Kerkhof PLM, Miller VM, editors. *Sex-specific analysis of cardiovascular function*. New York: Springer; 2018. https://doi.org/10.1007/978-3-319-77932-4_10.
8. Bedford DE. The ancient art of feeling the pulse. Br Heart J. 1951;13:423–37.
9. Bellenger NG, Burgess MI, Ray SG, Lahiri A, Coats AJ, Cleland JG, Pennell DJ. Comparison of left ventricular ejection fraction and volumes in heart failure by

echocardiography, radionuclide ventriculography and cardiovascular magnetic resonance; are they interchangeable? Eur Heart J. 2000;21(16):1387–96.
10. Bernard A, Addetia K, Dulgheru R, et al. 3D echocardiographic reference ranges for normal left ventricular volumes and strain: results from the EACVI NORRE study. Eur Heart J Cardiovasc Imaging. 2017;18(4):475–83. https://doi.org/10.1093/ehjci/jew284.
11. Bhave NM, Lang RM. Evaluation of left ventricular structure and function by three-dimensional echocardiography. Curr Opin Crit Care. 2013;19:387–96. https://doi.org/10.1097/MCC.0b013e328364d75e.
12. Biering-Sørensen T, Biering-Sørensen SR, et al. Global longitudinal strain by echocardiography predicts long-term risk of cardiovascular morbidity and mortality in a low-risk general population: the Copenhagen City Heart Study. Circ Cardiovasc Imaging. 2017;10(3):pii: e005521. https://doi.org/10.1161/CIRCIMAGING.116.005521.
13. Bjerregaard P. Mean 24 hour heart rate, minimal heart rate and pauses in healthy subjects 40–79 years of age. Eur Heart J. 1983;4:44–51.
14. Buccheri S, Costanzo L, Tamburino C, Monte I. Reference values for real time three-dimensional echocardiography-derived left ventricular volumes and ejection fraction: review and meta-analysis of currently available studies. Echocardiography. 2015;32:1841–50. https://doi.org/10.1111/echo.12972.
15. Buonanno C, Arbustini E, Rossi L, Dander B, Vassanelli C, Paris B, Poppi A. Left ventricular function in men and women. Another difference between sexes. Eur Heart J. 1982;3:525–8.
16. Calvo M, Le Rolle V, Romero D, Béhar N, Gomis P, Mabo P, Hernández A. Gender differences in the autonomic response to exercise testing in Brugada syndrome. In: Kerkhof PLM, Miller VM, editors. Sex-specific analysis of cardiovascular function. New York: Springer; 2018. https://doi.org/10.1007/978-3-319-77932-4_12
17. Cantinotti M, Kutty S, Giordano R, Assanta N, Murzi B, Crocetti M, Marotta M, Iervasi G. Review and status report of pediatric left ventricular systolic strain and strain rate nomograms. Heart Fail Rev. 2015;20(5):601–12.
18. Cattermole GN, Leung PY, Ho GY, Lau PW, Chan CP, Chan SS, Smith BE, Graham CA, Rainer TH. The normal ranges of cardiovascular parameters measured using the ultrasonic cardiac output monitor. Physiol Rep. 2017;5:pii: e13195. https://doi.org/10.14814/phy2.13195.
19. Chahal NS, Lim TK, Jain P, et al. Population-based reference values for 3D echocardiographic LV volumes and ejection fraction. JACC Cardiovasc Imaging. 2012;5:1191–7.
20. Choi JO, Shin MS, Kim MG, et al. Normal echocardiographic measurements in a Korean population study: part II. Doppler and Tissue Doppler Imaging. J Cardiovasc Ultrasound. 2016;24:144–52. https://doi.org/10.4250/jcu.2016.24.2.144.
21. Choy JS, Kassab GS. Scaling of myocardial mass to flow and morphometry of coronary arteries. J Appl Physiol (1985). 2008;104(5):1281–6. https://doi.org/10.1152/japplphysiol.01261.2007.
22. Churchill TW, Baggish AL. Is big truly bad? Aortic dilation in former national football league players. Circ Cardiovasc Imaging. 2017;10(11):pii: e007157. https://doi.org/10.1161/CIRCIMAGING.117.007157.
23. Daimon M, et al. Normal values of echocardiographic parameters in relation to age in a healthy Japanese population: the JAMP study. Circ J. 2008;72:1859–66.
24. Dallaire F, Sarkola T. Growth of cardiovascular structures from the fetus to the young adult. In: Kerkhof PLM, Miller VM, editors. Sex-specific analysis of cardiovascular function. New York: Springer; 2018. https://doi.org/10.1007/978-3-319-77932-4_22.
25. D'Andrea A, Riegler L, Rucco MA, et al. Left atrial volume index in healthy subjects: clinical and echocardiographic correlates. Echocardiography. 2013;30(9):1001–7.
26. Dickerson JA, Nagaraja HN, Raman SV. Gender-related differences in coronary artery dimensions: a volumetric analysis. Clin Cardiol. 2010 Feb;33(2):E44–9. https://doi.org/10.1002/clc.20509.
27. Dorosz JL, Lezotte DC, Weitzenkamp DA, Allen LA, Salcedo EE. Performance of 3-dimensional echocardiography in measuring left ventricular volumes and ejection fraction: a systematic review and meta-analysis. J Am Coll Cardiol. 2012;59:1799–808. https://doi.org/10.1016/j.jacc.2012.01.037.
28. Fisman EZ, Frank AG, Ben-Ari E, Kessler G, Pines A, Drory Y, Kellermann JJ. Altered left ventricular volume and ejection fraction responses to supine dynamic exercise in athletes. J Am Coll Cardiol. 1990;15:582–8.
29. Fogel MA. Use of ejection fraction (or lack thereof), morbidity/mortality and heart failure drug trials: a review. Int J Cardiol. 2002;84:119–32.
30. Foppa M, Arora G, Gona P, Ashrafi A, Salton CJ, Yeon SB, Blease SJ, Levy D, O'Donnell CJ, Manning WJ, Chuang ML. Right ventricular volumes and systolic function by cardiac magnetic resonance and the impact of sex, age, and obesity in a longitudinally followed cohort free of pulmonary and cardiovascular disease. Circ Cardiovasc Imaging. 2016;9:e003810. https://doi.org/10.1161/CIRCIMAGING.115.003810.
31. Ford LE. Heart size. Circ Res. 1976;39:297–303. https://doi.org/10.1161/01.RES.39.3.297.
32. Foschi M, Di Mauro M, Tancredi F, et al. The dark side of the moon: the right ventricle. J Cardiovasc Dev Dis. 2017;4(4): pii: E18. https://doi.org/10.3390/jcdd4040018.
33. Fu Q. Hemodynamic and electrocardiographic aspects of uncomplicated singleton pregnancy. In: Kerkhof PLM, Miller VM, editors. *Sex-specific analysis of cardiovascular function*. New York: Springer; 2018. https://doi.org/10.1007/978-3-319-77932-4_26.
34. Fukuda S, Watanabe H, Daimon M, et al. Normal values of real-time 3-dimensional echocardiographic parameters in a healthy Japanese population: the JAMP-3D study. Circ J. 2012;76:1177–81.
35. Fye WB. Giovanni Alfonso Borelli. Clin Cardiol. 1996;19:599–600.
36. Gebhard C, Buechel RR, Stähli BE, Gransar H, Achenbach S, Berman DS, et al. Impact of age and sex on left ventricular function determined by coronary computed tomographic angiography: results from the prospective multicentre CONFIRM study. Europ Heart J Imag. 2017;18:990–1000.
37. Gedeon A. Science and Technology in Medicine. New York: Springer; 2006.
38. Gent S, Kleinbongard P, Dammann P, Neuhäuser M, Heusch G. Heart rate reduction and longevity in mice. Basic Res Cardiol. 2015;110:2. https://doi.org/10.1007/s00395-014-0460-7.
39. Gopal AS, Chukwu EO, Iwuchukwu CJ, Katz AS, Toole RS, Schapiro W, Reichek N. Normal values of right ventricular size and function by real-time 3-dimensional echocardiography: comparison with cardiac magnetic resonance imaging. J

Am Soc Echocardiogr. 2007;20:445–55. https://doi.org/10.1016/j.echo.2006.10.027.
40. Graham TP Jr, Jarmakani MM, Canent RV Jr, Capp MP, Spach MS. Characterization of left heart volumes and mass in normal children and in infants with intrinsic myocardial disease. Circulation. 1968;38:826–37.
41. Hamill N, Yeo L, Romero R, Hassan SS, Myers SA, Mittal P, Kusanovic JP, et al. Fetal cardiac ventricular volume, cardiac output, and ejection fraction determined with 4-dimensional ultrasound using spatiotemporal image correlation and virtual organ computer-aided analysis. Am J Obstet Gynecol. 2011;205:76.e1–10. https://doi.org/10.1016/j.ajog.2011.02.028.
42. Hannan EL, Zhong Y, Wu C, et al. Comparison of 3-year outcomes for coronary artery bypass graft surgery and drug-eluting stents: does sex matter? Ann Thorac Surg. 2015;100(6):2227–36.
43. Herbert A, Cruickshank JK, Laurent S, Boutouyrie P. Reference values for arterial measurements collaboration. Establishing reference values for central blood pressure and its amplification in a general healthy population and according to cardiovascular risk factors. Eur Heart J. 2014;35:3122–33. https://doi.org/10.1093/eurheartj/ehu293.
44. Herity NA, Lo S, Lee DP, Ward MR, Filardo SD, Yock PG, Fitzgerald PJ, Hunt SA, Yeung AC. Effect of a change in gender on coronary arterial size: a longitudinal intravascular ultrasound study in transplanted hearts. J Am Coll Cardiol. 2003;41:1539–46.
45. Hiteshi AK, Li D, Gao Y, Chen A, Flores F, Mao SS, Budoff MJ. Gender differences in coronary artery diameter are not related to body habitus or left ventricular mass. Clin Cardiol. 2014;37:605–9.
46. Huxley VH, Kemp SS. Sex-specific characteristics of the microcirculation. In: Kerkhof PLM, Miller VM, editors. *Sex-specific analysis of cardiovascular function*. New York: Springer; 2018. https://doi.org/10.1007/978-3-319-77932-4_20.
47. Ioannidis JP, Trikalinos TA, Danias PG. Electrocardiogram-gated single-photon emission computed tomography versus cardiac magnetic resonance imaging for the assessment of left ventricular volumes and ejection fraction: a meta-analysis. J Am Coll Cardiol. 2002;39:2059–68.
48. JAMA. Giovanni Borelli (1608–1679) – Iatromechanical physiologist. JAMA 1966;195:485. doi:https://doi.org/10.1001/jama.1966.03100060125039
49. Jensen MT, Wod M, Galatius S, Hjelmborg JB, Jensen GB, Christensen K. Heritability of resting heart rate and association with mortality in middle-aged and elderly twins. Heart. 2018;104:30–6. https://doi.org/10.1136/heartjnl-2016-310986.
50. Kaess BM, Gona P, Larson MG, Cheng S, Aragam J, Kenchaiah S, Benjamin EJ, Vasan RS. Secular trends in echocardiographic left ventricular mass in the community: the Framingham heart study. Heart. 2013;99:1693–8. https://doi.org/10.1136/heartjnl-2013-304600.
51. Kappus RM, Ranadive SM, Yan H, et al. Sex differences in autonomic function following maximal exercise. Biol Sex Differ. 2015;6:28. https://doi.org/10.1186/s13293-015-0046-6.
52. Karagol BS, Orun UA, Zenciroglu A, Yuksel SP, Okumus N, Karademir S. The diameter of coronary arteries in healthy newborns at birth, 1 and 6 months of ages. J Matern Fetal Neonatal Med. 2012 Dec;25(12):2729–34. https://doi.org/10.3109/14767058.2012.718390.
53. Kawel-Boehm N, Maceira A, Valsangiacomo-Buechel ER, et al. Normal values for cardiovascular magnetic resonance in adults and children. J Cardiovasc Magn Reson. 2015;17:29. https://doi.org/10.1186/s12968-015-0111-7.
54. Kerkhof PLM. Importance of end-systolic volume for the evaluation of cardiac pump performance. In: Chazov EI, Smirnov VN, Oganov RG, editors. Cardiology, An International Perspective. New York: Plenum Press; 1984. p. 1339–52.
55. Kerkhof PLM. Characterizing heart failure in the ventricular volume domain. Clinical Medicine Insights: Cardiology. 2015;9(S1):11–31.
56. Kerkhof PLM, Kresh JY, Li JK-J, Heyndrickx GR. Left ventricular volume regulation in heart failure with preserved ejection fraction. Physiol Rep. 2013;1:e00007.
57. Kerkhof PLM, Heyndrickx GR, Li JJ-K. Hemodynamic determinants and ventriculo-arterial coupling are sex-associated in heart failure patients. Conf Proc IEEE Eng Med Biol Soc. 2016;2016:3286–9. https://doi.org/10.1109/EMBC.2016.7591430.
58. Kerkhof PLM, Yoo BW, Merillon JP, Peace RA, Handly N. Monte Carlo method applied to the evaluation of the relationship between ejection fraction and its constituent components. Conf Proc IEEE Eng Med Biol Soc. 2017a:1295–8. https://doi.org/10.1109/EMBC.2017.8037069.
59. Kerkhof PLM, Yoo BW, van de Ven PM, Handly N. Sex-specific aspects of left and right ventricular volume regulation in patients following tetralogy of Fallot repair. Conf Proc IEEE Eng Med Biol Soc. 2017b;2017:1303–6. https://doi.org/10.1109/EMBC.2017.8037071.
60. Kerkhof PL, Peace RA, Handly P. Importance of (measuring) the end-systolic volume index in predicting survival. Heart. 2018a;104:180. https://doi.org/10.1136/heartjnl-2017-312051.
61. Kerkhof PL, van de Ven PM, Yoo BW, Peace RA, Heyndrickx GR, Handly N. Ejection fraction as related to basic components in the left and right ventricular volume domains. Int J Cardiol. 2018b;255:105–10.
62. Kerkhof PLM, Kuznetsova T, Kresh JY, Handly N. Cardiophysiology illustrated by comparing ventricular volumes in healthy adult males and females, in: *Sex-specific analysis of cardiovascular function* (PLM Kerkhof & VM Miller, eds), New York, Springer 2018c. https://doi.org/10.1007/978-3-319-77932-4_8.
63. Kerkhof PLM, Peace RA, Heyndrickx GR, Meijboom LJ, Sprengers RW, Handly N. Heart function analysis in cardiac patients with focus on sex-specific aspects. In: Kerkhof PLM, Miller VM, editors. *Sex-specific analysis of cardiovascular function*. New York: Springer; 2018d. https://doi.org/10.1007/978-3-319-77932-4_23.
64. Kerkhof PLM, Parry G. Sex-specific aspects in cardiac transplantation evaluated by left ventricular size in male and female recipients. FASEB J. 2018. https://www.fasebj.org/doi/10.1096/fasebj.2018.32.1_supplement.901.1.
65. Kerkhof PLM, Kuznetsova T, Ali R, Handly N. Left ventricular volume analysis as a basic tool to describe cardiac function. Adv Physiol Educ. 2018f;42:130–9. https://doi.org/10.1152/advan.00140.2017.
66. Koenig J, Thayer JF. Sex differences in healthy human heart rate variability: a meta-analysis. Neurosci Biobehav Rev. 2016;64:288–310. https://doi.org/10.1016/j.neubiorev.2016.03.007.
67. Kovalova S, Necas J, Vespalec J. What is a "normal" right ventricle? Eur J Echocardiogr. 2006;7:293–7. https://doi.org/10.1016/j.euje.2005.06.010.
68. Kucher N, Lipp E, Schwerzmann M, Zimmerli M, Allemann Y, Seiler C. Gender differences in coronary artery

size per 100 g of left ventricular mass in a population without cardiac disease. Swiss Med Wkly. 2001;131:610–5.
69. Lancellotti P, Galderisi M, Edvardsen T, et al. Echo-Doppler estimation of left ventricular filling pressure: results of the multicentre EACVI Euro-Filling study. Eur Heart J Cardiovasc Imaging. 2017;18(9):961–8.
70. Lang RM, Badano LP, Mor-Avi V, et al. Recommendations for cardiac chamber quantification by echocardiography in adults: an update from the American Society of Echocardiography and the European Association of Cardiovascular Imaging. J Am Soc Echocardiogr. 2015;28(1):1–39.e14. https://doi.org/10.1016/j.echo.2014.10.003.
71. Langfy M. Cardiology in perspective. Gothenburg: Medicosport; 1976.
72. Levy PT, Sanchez Mejia AA, Machefsky A, Fowler S, Holland MR, Singh GK. Normal ranges of right ventricular systolic and diastolic strain measures in children: a systematic review and meta-analysis. J Am Soc Echocardiography. 2014;27:549–60.
73. Li JK-J. cardiovascular allometry: analysis, methodology, and clinical applications. In: Kerkhof PLM, Miller VM, editors. *Sex-specific analysis of cardiovascular function*. New York: Springer; 2018. https://doi.org/10.1007/978-3-319-77932-4_14.
74. Lieb W, Xanthakis V, Sullivan LM, Aragam J, Pencina MJ, Larson MG, Benjamin EJ, Vasan RS. Longitudinal tracking of left ventricular mass over the adult life course: clinical correlates of short- and long-term change in the Framingham offspring study. Circulation. 2009;119(24):3085–92.
75. Lin FY, Devereux RB, Roman MJ, et al. Cardiac chamber volumes, function, and mass as determined by 64-multidetector row computed tomography: mean values among healthy adults free of hypertension and obesity. JACC Cardiovasc Imaging. 2008;1:782–6. https://doi.org/10.1016/j.jcmg.2008.04.015.
76. Löllgen H. Kardiopulmonale Funktionsdiagnostik. Weh/Baden: Documenta Geigy; 1983.
77. Lopez L, Colan S, Stylianou M, et al. Relationship of echocardiographic z scores adjusted for body surface area to age, sex, race, and ethnicity. Circulation Cardiovasc Imag. 2017;10:e006979.
78. Lorenz CH, Walker ES, Morgan VL, Klein SS, Graham TP Jr. Normal human right and left ventricular mass, systolic function, and gender differences by cine magnetic resonance imaging. J Cardiovasc Magn Reson. 1999;1:7–21. https://doi.org/10.3109/10976649909080829.
79. Maceira AM, Prasad SK, Khan M, Pennell DJ. Reference right ventricular systolic and diastolic function normalized to age, gender and body surface area from steady-state free precession cardiovascular magnetic resonance. Eur Heart J. 2006;27:2879–88.
80. Macfarlane PW, van Oosterom A, Janse M, Kligfield P, Camm J, Pahlm O, editors. Comprehensive Electrocardiology. London: Springer; 2011.
81. Macfarlane PW. The influence of age and sex on the electrocardiogram. In: Kerkhof PLM, Miller VM, editors. *Sex-specific analysis of cardiovascular function*. New York: Springer; 2018. https://doi.org/10.1007/978-3-319-77932-4_6.
82. Maffessanti F, Muraru D, Esposito R, et al. Age-, body size-, and sex-specific reference values for right ventricular volumes and ejection fraction by three-dimensional echocardiography: a multicenter echocardiographic study in 507 healthy volunteers. Circ Cardiovasc Imaging. 2013;6:700–10.
83. Medicographia: heart rate modulation and exercise capacity. 2012; 34(4). www.medicographia.com
84. Mingels AMA, Kimenai DM. Sex related aspects of biomarkers in cardiac disease. In: Kerkhof PLM, Miller VM, editors. *Sex-specific analysis of cardiovascular function*. New York: Springer; 2018. https://doi.org/10.1007/978-3-319-77932-4_33.
85. Muraru D, Badano LP, Peluso D, et al. Comprehensive analysis of left ventricular geometry and function by three-dimensional echocardiography in healthy adults. J Am Soc Echocardiogr. 2013;26(6):618–28. https://doi.org/10.1016/j.echo.2013.03.014.
86. Muraru D, Onciul S, Peluso D, et al. Sex- and method-specific reference values for right ventricular strain by 2-dimensional speckle-tracking echocardiography. Circ Cardiovasc Imaging. 2016;9(2):e003866. https://doi.org/10.1161/CIRCIMAGING.115.003866.
87. National High Blood Pressure Education Program Working Group on High Blood Pressure in Children and Adolescents. The fourth report on the diagnosis, evaluation, and treatment of high blood pressure in children and adolescents. Pediatrics. 2004;114(2 Suppl 4th Report):555–76.
88. Nevill AM, Bate S, Holder RL. Modeling physiological and anthropometric variables known to vary with body size and other confounding variables. Am J Phys Anthropol. 2005;(Suppl 41):141–53.
89. Niebauer J, Clark AL, Anker SD, Coats AJ. Three year mortality in heart failure patients with very low left ventricular ejection fractions. Int J Cardiol. 1999;70(3):245–7.
90. Nio AQ, Stöhr EJ, Shave R. The female human heart at rest and during exercise: a review. Eur J Sport Sci. 2015;15(4):286–95. https://doi.org/10.1080/17461391.2014.936323.
91. O'Neal WT, Chen LY, Nazarian S, Soliman EZ. Reference ranges for short-term heart rate variability measures in individuals free of cardiovascular disease: the Multi-Ethnic Study of Atherosclerosis (MESA). J Electrocardiol. 2016;49:686–90. https://doi.org/10.1016/j.jelectrocard.2016.06.008.
92. Ochs MM, Fritz T, André F, et al. A comprehensive analysis of cardiac valve plane displacement in healthy adults: age-stratified normal values by cardiac magnetic resonance. Int J Cardiovasc Imaging. 2017 May;33(5):721–9. https://doi.org/10.1007/s10554-016-1058-y.
93. Ostenfeld EA, Flachskampf F. Assessment of right ventricular volumes and ejection fraction by echocardiography: from geometric approximations to realistic shapes. Echo Res Pract. 2015;2(1):R1–R11. https://doi.org/10.1530/ERP-14-0077.
94. Pagani M, Sala R, Malacarne M, Lucini D. Benchmarking heart rate variability to overcome sex-related bias. In: Kerkhof PLM, Miller VM, editors. *Sex-specific analysis of cardiovascular function*. New York: Springer; 2018. https://doi.org/10.1007/978-3-319-77932-4_13
95. Peace RA, Adams PC, Lloyd JJ. Effect of sex, age, and weight on ejection fraction and end-systolic volume reference limits in gated myocardial perfusion SPECT. J Nucl Cardiol. 2008;15:86–93.
96. Peluso D, Badano LP, Muraru D, et al. Right atrial size and function assessed with three-dimensional and speckle-tracking echocardiography in 200 healthy volunteers. Eur Heart J Cardiovasc Imaging. 2013;14(11):1106–14.
97. Petersen SE, Aung N, Sanghvi MM, et al. Reference ranges for cardiac structure and function using Cardiovascular Magnetic Resonance (CMR) in Caucasians from the UK Biobank population cohort. J Cardiovasc Magn Reson. 2017;19(1):18.
98. Pfaffenberger S, Bartko P, Graf A, et al. Size matters! Impact of age, sex, height, and weight on the normal heart size. Circ Cardiovasc Imag. 2013;6:1073–9.
99. Pressler A, Haller B, Scherr J, Heitkamp D, Esefeld K, Boscheri A, Wolfarth B, Halle M. Association of body

composition and left ventricular dimensions in elite athletes. Eur J Prev Cardiol. 2012;19(5):1194–204.

100. Rainer TH, Cattermole GN, Graham CA, Chan SS. Anthropometric and physiological measurements in healthy children. Hong Kong Med J. 2013 Dec;19(Suppl 9):26–9.
101. Rautaharju PM. The female electrocardiogram: special repolarization features, gender differences, and the risk of adverse cardiac events. Cham: Springer; 2015.
102. Rijnbeek PR, Witsenburg M, Schrama E, Hess J, Kors JA. New normal limits for the paediatric electrocardiogram. Eur Heart J. 2001 Apr;22(8):702–11.
103. Robbers-Visser D, Boersma E, Helbing WA. Normal biventricular function, volumes, and mass in children aged 8 to 17 years. J Magn Reson Imaging. 2009;29:552–9.
104. Rominger MB, Bachmann GF, Pabst W, Rau WS. Right ventricular volumes and ejection fraction with fast cine MR imaging in breath-hold technique: applicability, normal values from 52 volunteers, and evaluation of 325 adult cardiac patients. J Magn Reson Imaging. 1999;10:908–18.
105. Rösner A, Bijnens B, Hansen M, How OJ, Aarsaether E, Müller S, Sutherland GR, Myrmel T. Left ventricular size determines tissue Doppler-derived longitudinal strain and strain rate. Eur J Echocardiogr. 2009;10:271–7.
106. Russo C, Jin Z, Homma S, Rundek T, Elkind MSV, Sacco RL, Di Tullio MR. LA phasic volumes and reservoir function in the elderly by real-time 3D echocardiography: normal values, prognostic significance, and clinical correlates. JACC Cardiovasc Imaging. 2017;10(9):976–85. https://doi.org/10.1016/j.jcmg.2016.07.015.
107. Ryu DR. Normal reference values for Doppler echocardiography: influences of ageing, gender and ethnicity. J Cardiovasc Ultrasound. 2016;24:112–4. https://doi.org/10.4250/jcu.2016.24.2.112.
108. Sammito S, Böckelmann I. New reference values of heart rate variability during ordinary daily activity. Heart Rhythm. 2017;14(2):304–7. https://doi.org/10.1016/j.hrthm.2016.12.016.
109. Santamore WP, Bove AA. Why are arteries the size they are? J Appl Physiol (1985). 2008;104(5):1259. https://doi.org/10.1152/japplphysiol.90391.2008.
110. Sarikouch S, Peters B, Gutberlet M, Leismann B, Kelter-Kloepping A, Koerperich H, Kuehne T, Beerbaum P. Sex-specific pediatric percentiles for ventricular size and mass as reference values for cardiac MRI: assessment by steady-state free-precession and phase-contrast MRI flow. Circ Cardiovasc Imaging. 2010;3(1):65–76.
111. Saura D, Dulgheru R, Caballero L, et al. Two-dimensional transthoracic echocardiographic normal reference ranges for proximal aorta dimensions: results from the EACVI NORRE study. Eur Heart J Cardiovasc Imaging. 2017;18(2):167–79. https://doi.org/10.1093/ehjci/jew053.
112. Schvartzman PR, Fuchs FD, Mello AG, Coli M, Schvartzman M, Moreira LB. Normal values of echocardiographic measurements. A population-based study. Arq Bras Cardiol. 2000;75:107–14.
113. Sheifer SE, Canos MR, Weinfurt KP, et al. Sex differences in coronary artery size assessed by intravascular ultrasound. Am Heart J. 2000;139:649–53.
114. Shi J, Pan C, Kong D, Cheng L, Shu X. Left ventricular longitudinal and circumferential layer-specific myocardial strains and their determinants in healthy subjects. Echocardiography. 2016;33:510–8. https://doi.org/10.1111/echo.13132.
115. Støylen A, Mølmen HE, Dalen H. Relation between mitral annular plane systolic excursion and global longitudinal strain in normal subjects: the HUNT study. Echocardiography. 2018. https://doi.org/10.1111/echo.13825.
116. Tan TH, Wong KY, Cheng TK, Heng JT. Coronary normograms and the coronary-aorta index: objective determinants of coronary artery dilatation. Pediatr Cardiol. 2003;24(4):328–35.
117. Tandri H, Daya SK, Nasir K, Bomma C, Lima JA, Calkins H, Bluemke DA. Normal reference values for the adult right ventricle by magnetic resonance imaging. Am J Cardiol. 2006;98(12):1660–4.
118. Taqueti VR. Sex differences in the coronary system. In: Kerkhof PLM, Miller VM, editors. *Sex-specific analysis of cardiovascular function*. New York: Springer; 2018. https://doi.org/10.1007/978-3-319-77932-4_17.
119. Tardif JC. Heart rate as a treatable cardiovascular risk factor. Br Med Bull. 2009;90:71–84. https://doi.org/10.1093/bmb/ldp016.
120. Tasevska-Dinevska G, Kennedy LM, Cline-Iwarson A, Cline C, Erhardt L, Willenheimer R. Gender differences in variables related to B-natriuretic peptide, left ventricular ejection fraction and mass, and peak oxygen consumption, in patients with heart failure. Int J Cardiol. 2011;149:364–71. https://doi.org/10.1016/j.ijcard.2010.02.018.
121. Thompson A, Crilley J, Wilson D, Hungin APS, Fuat A, Murphy J. An epidemic of HFpEF? Heart. 2016;102(Suppl 6):A15–6.
122. Udelson JE. Left ventricular shape: the forgotten stepchild of remodeling parameters. JACC Heart Fail. 2017;5(3):179–81. https://doi.org/10.1016/j.jchf.2017.01.005.
123. van Beek JH, Kirkwood TB, Bassingthwaighte JB. Understanding the physiology of the ageing individual: computational modelling of changes in metabolism and endurance. Interface Focus. 2016;6:20150079. https://doi.org/10.1098/rsfs.2015.0079.
124. Vriz O, Aboyans V, D'Andrea A, et al. Normal values of aortic root dimensions in healthy adults. Am J Cardiol. 2014;114(6):921–7. https://doi.org/10.1016/j.amjcard.2014.06.028.
125. Wood PW, Choy JB, Nanda NC, Becher H. Left ventricular ejection fraction and volumes: it depends on the imaging method. Echocardiography. 2014;31:87–100.
126. Yamada AT, Campos Neto Gde C, Soares J Jr, Giorgi MC, Araújo F, Meneghetti JC, Mansur AJ. Gender differences in ventricular volumes and left ventricle ejection fraction estimated by myocardial perfusion imaging: comparison of Quantitative Gated SPECT (QGS) and Segami software programs. Arq Bras Cardiol. 2007;88:285–90.
127. Yingchoncharoen T, Agarwal S, Popović ZB, Marwick TH. Normal ranges of left ventricular strain: a meta-analysis. J Am Soc Echocardiogr. 2013;26(2):185–91. https://doi.org/10.1016/j.echo.2012.10.008.
128. Zemrak F, Ambale-Venkatesh B, Captur G, et al. Left atrial structure in relationship to age, sex, ethnicity, and cardiovascular risk factors: MESA (Multi-Ethnic Study of Atherosclerosis). Circ Cardiovasc Imaging. 2017;10(2):pii: e005379. https://doi.org/10.1161/CIRCIMAGING.116.005379.
129. Ziman J. The force of knowledge; the scientific dimension of society. Cambridge: Cambridge University Press; 1976.
130. Zusterzeel R. Safety and effectiveness of medical device therapy. In: Kerkhof PLM, Miller VM, editors. *Sex-specific analysis of cardiovascular function*. New York: Springer; 2018. https://doi.org/10.1007/978-3-319-77932-4_7.
131. Kerkhof PLM, Mérillon JP, Yoo BE, Peace RA, Parry G, Heyndrickx GR, Kuznetsova T, Meijboom LJ, Sprengers RW, Park HK, Handly, N. The Pythagorean theorem reveals the inherent companion of cardiac ejection fraction. Int J Cardiol. 2018. https://doi.org/10.1016/j.ijcard.2018.06.074

Index

A
Abdominal obesity (AO), 662
Activities of daily living (ADLs), 581, 582, 585
Acute coronary syndrome (ACS), 232, 648
Acute Decompensated Heart Failure National Registry (ADHERE), 535
Acute myocardial infarction (AMI)
  diagnosis, 547
  prognosis, 546
Acute rejection (AR), 383
Adipokines, 232, 473
Adiponectin, 661, 662
Adolescent
  blood pressure, 354
  ECG parameters, 352
  resting blood pressure, 353
β1-Adrenoceptors, 669
β-Adrenergic receptor antagonists, 143
Adrenocorticotropic hormone (ACTH), 227
Adult congenital heart disease (ACHD), 53
Advisory Group for Gender (AGG), 650
Aerobic fitness
  blood volume, 332
  cardiac output, 330
  cardiovascular control, 334
  exercise physiology, in humans, 332
  Frank-Starling mechanism, 331
  hemoglobin, 332
  lean body mass, 332
  scaling/normalization, 331
  sex-related differences, 331
  uptake and transport, of oxygen, 330
  VO_2peak, 330, 333
Afterload, 127, 136, 163, 164, 177, 213, 234, 303, 334, 351, 367, 415, 416, 460, 478–480, 514, 516, 520
Age differences, ECG, 95–97, 99
Age-specific analysis, v, vii, 1, 7, 34, 68, 93, 153, 161, 165, 201, 203, 394, 396, 406, 557, 651, 677, 683
Aging, 391, 394–404
  animal models, 392
  cardiac dimension, 362
  cardiac and vascular, 401
  cardiovascular diseases (*see* Cardiovascular diseases)
  CVD, 390
  fibroblast proliferation, 393
  mass-to-volume ratio, 375
  myocardial function
    atrial remodeling, 396–397
    diastolic, 396–397
    macroscopic and microscopic structure, 391
    microscopic and macroscopic changes, 394
    pacemaker function, cardiac conduction and heart rate, 394–395
    systolic, 395–396
  vascular function
    arterial structure, 397
    blood vessel stiffness, 397
    endothelial dysfunction, 397–399
    vascular contractility and reactivity, 399–400
  vasculature, 393–394
  ventricular dimensions, 370, 371
Aging effects, 685, 687
Aging heart, 247
Aging, in human hearts
  myocardium loss, 384
  ventricular myocardial mass, 384
Albrecht von Haller, 5
Alarmins, 230
Allograft vasculopathy, 383
Allometry, 7, 212–215, 350
  aortic diameter, 210
  Buckingham's Pi theorem, 214
  in comparative physiology, 210
  defined, 208
  deviations, from simple allometry, 211–212
  differential growth and growth in time, 211
  geometric similarity, 209
  hemodynamics
    blood pressure and flow, 212–213
    dimensional matrix, 214, 215
    peripheral resistance to flow, 213
  MLT system, 213, 214
  power law, 208
  proportional growth, 210–211
  in rheology and vascular mechanobiology, 215, 216
  sex differences, aging and hypertension, 216–218
  similarity criterion, 209
  simple allometry, 208

P. L. M. Kerkhof, V. M. Miller (eds.), *Sex-Specific Analysis of Cardiovascular Function*,
Advances in Experimental Medicine and Biology 1065, https://doi.org/10.1007/978-3-319-77932-4

American College of Cardiology (ACC), 109, 630
American College of Rheumatology (ACR), 501
American Heart Association (AHA), 109, 630
Anaerobic threshold (AT), 173–176, 178
Anamnesis, 5, 667
Androgen
 angiotensinogen synthesis, 142
 BP, 147
 chronic, 146
 and estrogens, 146
 MC4R, 148
 SHR, 145
 virilization, 147
 in women and female rats, 147
Androgen receptors (AR), 42, 43
Anemia, 19, 416, 490, 501, 502, 534–536, 559, 561
Aneurysm, 208, 219, 295, 303, 400, 401, 403, 670
Ang I-converting enzyme (ACE), 142
Ang II AT1 receptor (AT1R), 142
Angiotensin II, 425
Angiotensin-converting enzyme inhibitors (ACEIs), 140, 438, 536
Angiotensinogen synthesis, 142
Angiotensin receptor blockers (ARBs), 141, 537
Animal models
 aging, 392, 405
 collagen types I and III, 393
 endothelium, 394
 heterogeneity, 406
 myocyte hypertrophy, 393
 vasodilatory responses, 399
Ankylosing spondylitis (AS), 498, 499
Anonymization, 612, 613, 625
Anorexia religiosa, 24
Anthracycline, 482, 483
Anthracycline cardiomyopathy, 482
Anti-angiogenic factors, 457, 460
Antiques, 3–5
Aortic blood flow, 156
Aortic diameter, 694–695
Aortic pressure, 154
Aortic stenosis (AS), 51, 403, 670
Aortopathy, 51
Apelin-APJ system, 659
Apoptosis, 244
 AIF, 243
 Bcl-2, 243
 cell death, 243
 mitochondrion-toxic agents, 245
 PTP, 243
 ROS, 244, 245
Apple's HealthKit, 650
Aquaporins, 309
ARMin robot, 582, 583, 586
Aromatase, 517–518
Arrhythmogenesis, 183
Arterial compliance, 163, 166, 167, 302–303, 419–420
Arterial hypertension, 658–661
Arterial pressure, 6, 156–158
Arterial stiffness, 159, 397, 497
Arterial system, 292
Arterial-ventricular (AV) coupling, 368
Arterial wall, 296, 299–303
 composition, 293
 extracellular matrix components, 293
 gender-specific differences, in aortic stiffness, 293
 hemodynamic measurement (*see* Hemodynamic monitoring)
 to intraluminal blood pressure distention, 295
 pulse waveform analysis (*see* Pulse waveform analysis)
 stress–strain relations, 294
 structure, 292–293
 thickness, 293, 294
Arteries, 292–296
 aortic arch junction, 292
 arterial wall, 292, 293 (*see also* Arterial wall)
 brachial and radial, 292
 mechanical properties
  compliance and distensibility, 296
  pressure–strain elastic modulus, 294–296
  vascular stiffness, 293
  wall stress–strain relations, 294
 thoracic aorta, 292
Arteriitis temporalis, 502–503
Arterioles
 blood flow, 309
 capillary hydrostatic pressure, 309
 ET-1, 312
 porcine coronary, 313
 pressure-induced myogenic constriction, 311
 pulsatile pressure, 308
 sexual dimorphism, 311–313
 terminal, 309
Arthritis, 145, 262, 380, 490, 493, 498, 499, 501, 503–504, 530, 534
Artificial intelligence (AI), 608, 609, 617, 618
Atherosclerosis, 266, 492
 CCTA and PET, 268
 CFR, 268
 CMD, 269
 MACE, 268
 myocardial ischemia, 267
Atherosclerotic cardiovascular disease (ASCVD), 630, 631
Atopic dermatitis (AD), 504, 505
Atrial fibrillation (AF), 281–284, 400–402, 436, 666
 catheter ablation, 666–667
 causes, 437
 radiofrequency catheter ablation, 437
 sex differences, 665–666
Atrial Fibrillation Follow-up Investigation of Rhythm Management (AFFIRM) study, 283
Atrial kick, 368
Atrial natriuretic peptides (ANPs), 230, 553, 554
Atrial septal defect (ASD), 50

Atrioventricular septal defect (AVSD), 51
Augmentation index (AIx), 156, 161–163, 221, 292, 300, 497, 680
in central aortic, 162
in central arteries, 167
Augmented pressure, 301
Auscultatory method, 156, 221, 296
Auto-catheterization, 22
Autoimmune diseases, 491
Automated ECG analysis, 94, 96, 103, 105
Autonomic circulatory control
baroreflexes, 425
vasoconstrictor responsiveness, 425
vasomotor sympathetic activity, 422
Autonomic innervation, 193
Autonomic nervous system (ANS), 183, 186, 189, 667
Autonomic nervous system index (ANSI), 200, 201, 203, 204
Autonomic regulation, 194, 199
Autophagy, 245–247
Autopsy, 4, 17, 262, 437, 501, 596

B

Balloon angioplasty, 11
Bariatric procedure, 662
Baroreflex function, 425–426
Baroreflex sensitivity, 426
Barrier function
acute loss, 309
epithelial, 312
pathophysiology, 313
transient loss, 309
Barthel index, 280
Bayesian statistical model, 609
Beat-by-beat oscillations, 194
Berlin Heart Center, 383
Beta-blockers, 233
Bicycle face, 24
Big data, 598
Big-data analysis, 611, 612, 614–617, 621
collection, 613
data-analysis stage, 613
described, 609
digitalization process, 610
five "Vs", 611
validity and veracity, 611, 612
variety, 611
velocities, 611
volume, 611
identification, 612
image analysis, 610
and machine learning, 610
planning and careful execution, 612
to precision medicine, 609
preprocessing data, 613
preprocessing methods, 614
discretization, 614, 615
feature selection, 616, 617, 621
security issues, 612
typical life cycle, 612
value, 610
visualization, 613, 614
Big dimensionality, 616
Bioelectricity, 6
Biohumoral systems, 475
Biomarker, 590, 591, 598, 602, 688, 696
AMI, 561
cardiac, 561
cTn levels, 561
cTn, NT-proBNP and BNP, 561
Biomedical devices, 650
Bleomycin, 519
Blood-brain barrier (BBB), 314
Blood flow regulation, 309, 311
Bloodletting, 5, 7–9
Blood pressure (BP), 353, 354, 678, 680–682, 684
Blood volume expansion, 418
Body composition, 349, 350, 354
Body mass index (BMI), 146, 174, 175, 201–203, 261, 266, 362, 383, 461, 472, 474, 662, 679, 685, 687, 693, 695, 697
Body surface area (BSA), 348–350, 383, 391
Bone morphogenetic protein receptor-II (BMPR-II), 515–516, 518–520, 523
Borelli, 3, 6, 677, 678, 702
Brain-heart axis, 228
Brain injury
after neonatal, 580
upper extremity, 582
Braunwald, 11
Braveheart, 38
BridgeT project, 586
Brugada syndrome (BS), 183, 680
autonomic response, 183
description, 182
HRV analysis (*see* Heart rate variability (HRV))
multifactorial etiology, 182
B-type natriuretic peptide (BNP), 547, 553–555, 557, 558, 560
Buckingham's Pi theorem, 212–215
Buonanno, 18, 24, 130, 134

C

Calcium overload, 251, 252
Canadian Institutes of Health Research (CIHR), 645
Capillaries, 210, 265, 266, 308–314, 319, 321–323, 330, 478, 479, 482, 489, 494, 503, 507
Cardiac amyloidosis, 669
Cardiac apoptosis-related lncRNA (CARL), 38
Cardiac arrhythmias, 666–667
Cardiac computed tomography angiography (CCTA), 262
Cardiac contraction, 126
Cardiac electrophysiology, 422
Cardiac-hypertrophy-related factor (CHRF), 38
Cardiac innervation, 193
Cardiac magnetic resonance imaging (CMRI), 435
Cardiac output (CO), 124, 129, 330, 682–684
exercise physiology, in humans, 332

Cardiac output (CO) (*cont.*)
  lean body mass, 332
  oxygen uptake, 330
  VO_2peak, in men and women, 330
Cardiac patients, 372
  arterial pressure, heart rate, 362
  diabetes mellitus, diastolic HF, 372
  DM, 375
  EF, 373
  heterogeneous group, 375
  MRI-based RA diameters, 370
Cardiac rehabilitation
  core components, 566, 567
  coronary artery dissection, 572
  evidence, 574
  gender, 569–572
  gender gap, 568–569, 572–574
  indications, 566
  microvascular angina, 572
  outcomes, 566–568
  participation, 575
  referral, enrollment and adherence, 575
  spontaneous coronary artery, 572
Cardiac Rehabilitation Care Continuity through Automatic Referral Evaluation (CRCARE) study, 572
Cardiac remodeling
  chemotherapy, 482–483
  diabetes, 475
  hyperthyroidism, 480–481
  hypothyroidism, 478–480
  metabolic syndrome, 476–478
  oncology patients, 483–484
Cardiac resynchronization therapy, 597
Cardiac resynchronization therapy-pacemaker (CRT-P), 537
Cardiac transplantation, 7, 9, 15, 20, 23, 34, 135, 379–386, 506, 537, 570, 669, 696
  longitudinal studies, 385
  LV remodeling, 386
Cardiac troponin (cTn)
  cTnT and cTnI, 547, 548
  diagnosis, AMI, 549–551, 553
  healthy women and men, 548–549
  hs-cTnT and hs-cTnI, 552
  prognostic biomarker, 553, 554
  sex differences, 546
Cardiac troponin I (cTnI), 548, 550, 551, 553
Cardiac troponin T (cTnT), 548, 550, 551
Cardio-ankle vascular index (CAVI)
  vs. age, 161
  blood pressure, 158
  cardiovascular risk, 161
  conventional measurements, 157
Cardioembolic stroke, 280
Cardioembolism, 66
Cardiokine, 230
Cardiology
  clinical, 691
  metrics, indexes and Doppler surrogates, 697–699
  pediatric, 696–697
Cardiology-related discoveries, 10, 11
Cardiometabolic syndrome, 493
Cardiomyocyte, 533
  β-adrenergic stimulation, 125
  β-AS dynamic effects, 127
  cardiac muscle contraction, 125
  ECG, 125
  estrogen receptor activation, 125
  N2B expression, 127
  sex-specific differences, 126–129
  SR Ca^{2+} content, 125
  XB cycle, 127
Cardiomyocyte apoptosis, 534
Cardiomyocyte hypertrophy, 405
Cardiomyogenesis, 129
Cardiomyopathy, 438, 440, 668–670
Cardiomythology, 22, 24
Cardio-oncology, 566
  cardiac remodeling, 483–484
  chemotherapy, 482–483
  chemotherapy-related cardiotoxicity, 482
  mortality rate, 481
  radiation therapy, 482, 483
Cardioprotection, 188, 189, 658
Cardiopulmonary exercise testing, 352
Cardiopulmonary test, 171, 174
Cardiorespiratory fitness, 338–340
Cardiovascular adaptation, 330
  aerobic fitness (*see* Aerobic fitness)
  fitness, 339, 340
  to exercise training, left ventricular and myocardial, 334–336
  ventricular-vascular coupling, 338, 339
Cardiovascular allometry, 208–209
  See also Allometry
Cardiovascular development
  DNA methylation, 34
  lncRNAs role, 38
Cardiovascular dimensions
  abnormal, 354
  blood pressure, 353, 354
  cardiopulmonary exercise testing, 352
  challenges, in body size, 350
  determinants, in healthy child, 348, 349
  ECG, 352
  hemodynamics parameters, 351, 352
  prenatal growth, 349
  vascular normal and reference values, 350, 351
Cardiovascular disease (CVD), 20, 312, 656
  age-related changes, 406
  aneurysms, 403
  aortic stenosis, 403
  atrial fibrillation, 400–402
  bradycardia and pacemaker implantation, 400
  cardiac care, 546
  developments, 644
  DNA methylation, 34, 35
  FCG mice, in sex chromosomes, 40

frailty, 405–406
hazard ratios, men *vs.* women, 66
HFpEF, 402–403
HFrEF, 402
histone modifications, 35–37
ischemic heart disease, 404
isolated systolic hypertension, 403, 404
lncRNAs role, 38
manifestation, 62–63, 65, 67
miRNAs role, 37, 38
mortality, 62, 260
RCTs, 261
risk and expression, 404–405
risk stratification, 546
Cardiovascular medicine, 602
Cardiovascular mortality, 444
Cardiovascular research, 73, 74
epidemiology, 72
guidelines, 91
medical statistics (*see* Medical statistics)
research designs, 73
database, 73
prospective cohort study, 73
RCT, 73
results, 73
retrospective cohort study, 73
statistical analysis, 73
systematic review study, 73
research question and hypothesis, 72
Cardiovascular risk factors (CVRFs), 680
Cardiovascular system, 658
Carotid intima-media thickness (cIMT/CIMT), 466, 494, 496
Catechol-O-methyltransferase (COMT), 518
Catheter ablation, 666–667
Caucasian normal adults, 681
Caucasian normals *vs.* heart rate, 682
Cellular senescence, 247
Central aortic augmentation index, 162
Central artery stiffness, 398
Central elastic arteries, 392
Central hemodynamics, 159
Cerebral circulation, 280–281
Chemiluminescence, 144
Chest pain, 435
Child
blood pressure, 353
body size, 350
body surface area (BSA), 348
cardiopulmonary exercise testing, 352
ECG parameters, 352
LMS modeling technique, 351
Chinese database, 95
Chi-square test, 80, 83
Chronic baroreflex activation therapy, 659
Chronic inflammation, 497
Chronic kidney disease (CKD), 535
Chronic obstructive pulmonary disease (COPD), 494, 534, 536
Clinical cardiology, 19–26
Cluster analysis, 608, 620, 622, 623
Coarctation of aorta (COA), 51
Colombo, 3, 12, 13, 20, 25
Comorbidities
anemia and iron deficiency, 535
chronic kidney disease, 535
COPD, 536
depression, 536
diabetes mellitus, 534
frailty assessment, 536
heart failure, in community, 535
HFpEF and HFrEF, 534
hypertension, 535
Comparative hemodynamics, in stenosis and atherosclerosis, 221–222
Computer-assisted reporting of electrocardiograms (CARE), 94
Conduction system, 57, 105, 108, 109, 182, 183, 243, 351, 394, 395, 401, 406, 498, 499, 502
Confounding, 86, 88
Congenital heart disease (CHD)
adult and pediatric populations, 51
birth prevalence, 50, 51
complications, 54–55
cumulative survival curves, 56
functional class, 54
incidence, 50, 51
investigation, 50
male and female, 51–53
manifestation, 52–53
PAH, 53, 54
percentage, 55
preponderance, 52
prevalence, 53
prognosis, 55
treatment, 54–56
Congestive heart failure (CHF), 79, 548, 559
Conjugated equine estrogens (CEE), 146, 147
Connective tissue disease (CTD), 505
The COnsensus-based Standards for selection of health Measurement INstruments (COSMIN), 91
Consolatory reflections, 25–26
The CONsolidated Standards Of Reporting Trials (CONSORT), 91
Constraint-induced movement therapy (CIMT), 581
Continuous positive airway pressure (CPAP), 662
Contractility, 8, 127, 128, 136, 170, 178, 231, 247, 313, 331, 367, 396, 399, 402, 405, 415, 426, 479, 516, 690
Cornell criteria, 99–100
Coronary
anatomy and function, 265
artery calcium scores, 267
artery size, 266
atherosclerosis, 266
blood flow, 266
endothelial shear stress, 266
Coronary-aorta index, 696

Coronary arteries, 695–696
Coronary artery calcium scoring (CACS), 466
Coronary artery disease (CAD), 258, 263, 434, 531, 549, 663
  clinical and genetic risk scores, 635–638
  clinical risk, 628–631
  exercise, 375
  genetic and genomic risk, 633–636
  GWAS and sex-specific genetic effects, 631–633
Coronary artery dissection, 572
Coronary flow reserve (CFR), 266, 268, 269, 271, 435, 664
Coronary heart disease (CHD), 312
  cardiac rehabilitation, 567
  complication of, 67
  epicardial artery disease, 65
  etiological factor, 65
  fatal and nonfatal, 65
  mortality ratios, 62
  myocardial ischemia, 65
  secondary prevention, 566
  and stroke, 64
Coronary microvascular dysfunction (CMD), 259, 268, 269, 271, 272, 312, 435
  CMRI, 436
  diagnostic test, 435
  risk factors, 435
  treatment, 436
Coronary microvessels, 312
Coronary risk factors (CRFs), 628–630, 633–635, 637, 638
Coronary vasculature, 685
Corpus Hippocraticum, 5
Corticotropin-releasing hormone (CRH), 227
Cox proportional hazards model, 82–85
Cox regression model, 82–85
C-reactive protein (CRP), 490–492, 498, 499, 504
Creatine kinase muscle-brain (CKMB), 546, 547
C-type natriuretic peptide (CNP), 230
Cutoff values, heart, 692
Cyanotic congenital malformations, 10
Cyclic adenosine monophosphate (cAMP), 514
Cyclic guanosine monophosphate (cGMP), 514, 520
Cyclophosphamide, 483
Cytochrome P450 enzyme (CYP), 517, 518

D
D2:D4 (sex-specific ratio), 7, 19
Danish Osteoporosis Prevention Study (DOPS), 447
Dario Maestrini, 25
Databases, ECG
  Chinese database, 95
  Glasgow adult normal database, 94, 95
  Glasgow paediatric normal database, 95
  Indian database, 95
  Nigerian database, 95
Dehydroepiandrosterone (DHEA), 516, 517, 520, 523
De Medicina, 5
Demographic subgroups, 117
Depression, 536
Dermatology, 493, 505
Dermatomyositis, 493
Diabetes mellitus (DM), 534, 685
  cardiac computed tomography, 362
  comorbidity, 472, 473
  mortality risks, 478
  and ventricular function, 474–475
Diabetic cardiomyopathy, 661
Diagnostic criteria, 99, 100, 103
Diastolic BP (DBP), 680, 693
Diastolic pressure, 163–166
Diencephalon, 193
Dihydrobiopterin, 145
Dilated cardiomyopathy (DCM), 597, 668
Discretization, 614, 615
Disseminated intravascular coagulation (DIC), 308
DNA methylation
  in cardiovascular development, 34
  in cardiovascular disease, 34
  embryogenesis and aging, 35
  genome-wide, 34
DNA methyltransferase (DNMT) enzymes, 33
Donor organ, 383
Doppler echocardiography, 436
Doppler guide wire, 435
Doppler ultrasound, 420
Double outlet right ventricle (DORV), 51
Drp1 gene, 246
Dynamin, 248

E
Early *vs.* Late Intervention Trial with Estradiol (ELITE), 447
Effect modification, 86, 88–90
Eisenberg, 12
Ejection fraction (EF), 18, 384, 386, 396
  anginal complaints, 375
  beta-blockers, 366
  in cardiac patients, 693
  cardiotoxicity, 374
  end-systolic volume index (ESVi), 372
  and ESVi, 133, 369, 375
  LV volume, 374
  measurement, 691–694
  MESA, 374
  nomogram, 686
  paradigm, 693
  quantify indicator dilution curves, 366
  volume regulation graph, 134
  in women, 18, 67, 124, 128, 130, 132, 133, 136, 163, 363, 365, 370–373, 395, 396, 437, 482, 484, 530, 557, 620, 682, 688, 691–693, 701
Electrical myocardial instability (EMI), 667
Electrocardiogram (ECG), 6, 94, 96–100, 103–105, 422, 678, 680
  automated ECG interpretation, 94
  databases (*see* Databases, ECG)
  eyeballing scatter plots, 352

LVH
Cornell criteria, 99–100
Romhilt-Estes point score system, 100
S-L index, 97–99
normal limits, 94
orthogonal lead, 94
parameters, 352
P wave
amplitude, 96
duration, 96
QRS complex, amplitudes, 96, 97
QRS duration, 100
sex differences, 96, 97, 352
diagnostic criteria, 99
Glasgow database, 100
QRS complex, 96
ST elevation myocardial infarction, 103
ST-T segment
JT/J-Tpeak/Tpeak-Tend intervals, 104, 105
QT interval, 104
ST amplitude, 100, 103
T amplitude, 103, 104
Electronic health records (EHRs), 608–610
Emotion, 230
cardiovascular afferents, 231
in cardiovascular equilibrium, 228
stress and myocardial ischemia, 233
End-diastolic volume (EDV), 124, 125, 129, 130, 132–136, 367, 384, 385
End-diastolic volume index (EDVi), 57, 133, 686, 689, 690, 692–694, 701
Endoglin, 457, 460
Endorphins, 227
Endothelial function, 420
Endothelial nitric oxide synthase (eNOS), 286, 398
Endothelial progenitor cells (EPCs), 286
Endothelin, 147–148
Endothelin-1 (ET-1), 512, 513, 515, 519
Endothelium, 310
End-systolic elastance, 130, 367
End-systolic pressure (ESP), 130, 132
End-systolic volume (ESV), 124, 125, 128–136, 366, 367, 371, 384–386
End-systolic volume index (ESVi), 57, 132–134, 136, 369, 396, 694
Endurance training, 334–337, 339
E-patient, 23, 25
Epidemiology, 11, 12, 24, 49–57, 61–69, 72, 91, 146, 221, 259, 280, 354, 444, 514, 530, 580, 598, 602, 608, 656, 671, 679
Epigenetic regulation
DNA methylation, 33–35
histone modifications, 35–37
ncRNAs, 37, 38
sex hormones, AR and ER receptors, 42–43
Epigenetics, 39–42, 597
described, 32
mechanisms, dynamic and reversible, 33
modification, 32
phenotypic phenomena, 32
regulation, 32, 33
and sex differences
genomic imprinting, 39
X-escapee genes, 40, 41
X-inactivation, 39
Y-chromosome epigenetic modifiers, 41–42
transcription of genes, 32
Epistotherapy, 3
Epoxyeicosatrienoic acids (EETs), 518
ERα 4,4′,4″-(4-propyl-(1H)-pyrazole-1,3,5-triyl) trisphenol, 658
ERβ agonist 2,3-bis(4-hydroxyphenyl)-propionitrile (DPN), 658
Erectile dysfunction, 474
β-Estradiol, 670
17β-Estradiol, 441, 444, 447
Estrogen, 475, 476, 478, 492, 549, 555, 557
androgens, 146
AT1Rs, 142
CEE, 146
chronic, 146
and nitric oxide, 418
progesterone, 147
Estrogen modulates gene transcription, 441
Estrogen receptor (ERs), 42, 43
Estrogen receptor-alpha (ER-α), 533
Estrogen and Thromboembolism Risk (ESTHER), 447
Ethnicity, v, 41, 51–53, 84, 266, 324, 349, 351, 354, 374, 414, 426, 504, 558, 570, 591, 593, 630, 632, 645, 648, 650, 661, 678, 680, 685, 688, 689, 691, 695, 697, 699, 701
EuroHeart Failure survey II, 534
Excitatory–inhibitory balance, 201
Exercise
arteriovenous oxygen, 334
autonomic function, measures, 188
benefits, 340
cardiovascular adaptations, 336
endurance training, 334
exercise-induced arterial adaptations, 337
healthy older individuals, 336
HRV features, 187
incremental exercise, 185–188
in medicine and health care, 340
myokine, 341, 342
phases, testing, 185
physical exercise stress test, 184
regular exercise/exercise training, 334
sex differences, in left ventricular hypertrophy, 335
stroke volume, 334
structural and functional adaptations, 336
training-induced, in stroke volume, 336
ventricular-vascular coupling, 336–339
ventricular/vascular response, 338
VO_2peak, 334
Exercise testing, cardiopulmonary, 352

Exogenous hormones, 442
Exoskeletons, 582
Exosuits, 584
Extracellular regulated kinase (ERK) pathway, 522

F
Feature selection, 616
approaches, 616
classification, 616
consistency-based, 617
correlation-based, 617
description, 616
information gain, 617
learning algorithm, 616
minimum redundancy maximum relevance, 617
recursive feature elimination, for SVMs, 617
relief, 617
Female donors (FD), 381
Fetal growth restriction, 456, 458, 459
Fetus
body growth, 349
body size, 349
cardiac imaging, 350
echocardiography, 351
Fibrinolysis, 323
Fibroblast-like synoviocytes (FLS), 498
Fibrosis
and collagen accumulation, 393
hypertrophy, 400
LV wall thickness, 391
SAN, 406
Fibrotic disorders, 497
Fine particulate matter, 5
Finger length, 7
The Fire of Life, 7
Fission/fusion cycle, 249
Fluid homeostasis, 324
^{18}F-Fluorodeoxyglucose (FDG), 494, 495, 506, 507
(^{18}F-Fluoropentyl)triphenylphosphonium salt (^{18}F-FPTP), 664
^{18}Flurpiridaz, 664
Force-frequency relation (FFR)
described, 170
graded bicycle exercise, 171
LV-ejection performance, 178
LV pressure, measurement, 170
sex-specific differences, 170
Four core genotype (FCG) mouse model, 40, 41, 658
Fractional area change (FAC), 687, 697
Fragmentation of QRS (fQRS) complex, 667
Frailty assessment, 390, 404–406, 534, 536, 571
Frailty index (FI), 404
Framingham Heart Study, 155
Framingham risk score (FRS), 491, 500, 505

G
Galen's teachings, 12
Gangtrainer®, 583
Gender, 4, 590, 591, 601, 679
CHD, 51
diabetes and heart, 472–474
difference, 52
mortality, 55
volumetric assessment, 56
Gender differences, in periprocedural sedation, 666–667
Gendered innovations
age- and sex-specific translational, 651
analyzing gender, 648–649
analyzing sex, 648
biomedical devices, 650
biometrics, 650
in cardiology, 670, 671
cardiovascular disease, 644
in cardiovascular medicine, 671
definition, 645–647
European Commission's Horizon 2020, 645
gender policies, 646
molecular, genetic and pathophysiology, 661–662
obesity and metabolic syndrome therapy, 662–663
OSSD, 650
in pharmacological research, 650
posttranslational drug development, 651
research pipelines, 650–653
sex and gender, 656
the US market, 644
the US National Institutes of Health, 644
in USA and Western Europe, 644
the US Women's Health Initiative Hormone Therapy Trial, 644
Gender gaps, 568–569
Gender medicine
training, 652
the US Medical Licensing Exam, 652
Gender-specific cell characteristics, 657
Generation R Study, 426
Genetically engineered mouse models (GEMMs), 658
Genetic Investigation of Anthropometric Trials (GIANT) Consortium, 661
Genetic risk scores (GRS), 633, 635–638
Genome-wide association studies (GWAS), 593, 631–633, 635, 636
CVD, 635
and sex-specific genetic effects, 631–633
Genomics, 590–592, 594, 597, 598, 601
Geometrical taper, 218
German Health Examination Surveys (GHES), 140
Gestation, 414, 415, 418, 420, 423, 425, 426
Gestational hypertension, 456
Get With The Guidelines-Heart Failure (GWTG-HF) study, 111
Giant cell arteritis (GCA), 502–503
Giants of cardiology, 10
Glasgow adult normal database, 94, 95
Glasgow paediatric normal database, 95
Global (systemic) arterial compliance, 419
Global Burden of Disease Study, 62
Global circumferential strain (GCS), 364, 690
Global longitudinal strain (GLS), 364, 690

Global Registry of Acute Coronary Events (GRACE) score, 649
Glycocalyx, 309, 310, 314
G protein-coupled estrogen receptor (GPER), 521
Graded exercise, 171
  bicycle exercise, 174
  breath-by-breath cardiopulmonary test, 174
  hemodynamic and ventilatory responses, 170
  parameters (*see* Wave-intensity (WI) analysis)
  WI, 170
Growth
  and blood pressure, 354
  body composition, 349
  fetal body, 351
  prenatal growth, cardiovascular dimensions, 349–350

H
Habitual exercise, 340
Haemodynamic characteristics, healthy subjects, 158
Hales, 6, 7, 124, 296, 677
HapticWalker®, 583
Heart
  AF, 436
  anatomic structures, 227
  brain-heart axis, 228
  cardiovascular afferent activity, 229
  cholesterol, nutrition and sedentary lifestyle, 232–233
  hypertrophy and remodeling, 421
  as immune organ, 230, 232
  left ventricular function, 421–422
  morphologic types, neurons, 228
  natriuretic peptides, 230
  neuro-endocrine-immune symphony, 232
  physiological remodeling, 420–421
  psyche-brain-heart connection, 228
  as symbol, 2, 3
  woman's, 233–234
Heart-brain interaction, 228–230
Heart and Estrogen-Progestin Replacement Study (HERS), 659
Heart failure (HF), 244, 249, 251, 397, 437, 438, 497, 502
  cardiac (co)morbidity, 362
  cardiac structure and function, 533–534
  chronic systolic, 568
  clinical characteristics, 534
  comorbidities, 534–536
  DCM transformation, 669
  definition, 530
  diagnosis, 558–559
  diastolic, 479
  EF, 366
  etiology, 531–533
  HF-ACTION, 570
  HFpEF, 482, 483, 661
  HFrEF, 437, 661
  hospitalization, 570
  LVEF, 437
  management, 560
  pathophysiology, 531–533
  phenotypes, 372, 373
  phenotyping, 608, 621, 622
  postischemic, 656
  preserved ejection fraction, 472
  prevalence and incidence, 531
  prognosis, 531
  quality of life, 531
  severity, 668
  tailoring therapies and health care, 537
  treatment guidelines, 439
  treatments, 536–537
  ventricular arrhythmias, 669
  VRG, 372–374
Heart Failure–A Controlled Trial Investigating Outcomes of Exercise TraiNing (HF-ACTION), 570
Heart failure with preserved ejection fraction (HFpEF), 67, 259, 269, 272, 400, 402, 403, 473, 482, 483, 530, 531, 533, 559, 661
Heart failure with reduced ejection fraction (HFrEF), 400, 402, 530, 531, 661
Heart rate (HR), 129, 131, 330, 332–334, 366, 415, 679, 680
Heart rate turbulence (HRT), 667
Heart rate variability (HRV), 183, 668
  ANSI, 200
  asymptomatic males and females, 188
  asymptomatic men and women, 187, 188
  autonomic evaluation, 199
  autonomic function, measurement, 188
  in autonomic responses, 188
  baroreflex-mediated changes, 195, 196
  on benchmark population, sex and age differences, 201, 203
  binary logistic regression, 201
  in bioengineering, 192
  clinical applications, physiological and pathophysiological, 196
  complex dynamics, 201
  description, 183
  effort testing, 186, 187
  excitatory–inhibitory balance, 201
  factor analysis, 201, 203
  frequency-domain analyses, 183
  hormones, 188
  ICD implantation, 184
  LF/HF ratio, 194
  neural information, 194
  neural regulation, 197
  nomalized-unit evaluation, 198
  oscillations, spectral analysis, 199
  physiological regulatory mechanisms, 194
  practical value, 197
  proxies, 200, 201
  signal acquisition, physical stress test, 184, 185
  SPWVD, 186
  study population, 184
  subsystems, 183
  symptomatic and asymptomatic groups, 184
  time–frequency signal analysis, 185

Heart rate variability (HRV) (*cont.*)
  vagal activity, 183
  "vagal tone", 183
Heart research, 11
Heart Rhythm Society (HRS), 109
Heart transplantation, 7, 9, 15, 20, 23, 34, 135, 379–386, 506, 537, 570, 669, 696
Heliocentric paradigm, 20
Hemodynamic adaptations, 414
Hemodynamic coupling (k), 367
Hemodynamic monitoring
  sex differences, in different cultures, 304–305
  wall properties, to pulsatile pressure and flow, 297–299
  Windkessel model, 296–297
Hemodynamic variables, 363
Hemothorax, 111
Heroic art of healing, 4
HF-ACTION trial, 571
HF with mid-range (HFmrEF), 530, 531
Hidradenitis suppurativa (HS), 493, 505
Hildegard von Bingen, 16, 17, 19
Hippocrates of Kos, 4
Hippocrates of the Renaissance, 6
Hirsch (h)-index, 12
Histone acetyltransferases (HATs), 35
Histone deacetylases (HDACs), 35–37
Histone methyltransferases (HMTs), 35, 36
Histone modifications, 35–37
  in cardiovascular disease
    cardiomyocyte tri-methylation, H-3, 35
    G9a, 36
    gene expression, 35
    HDACs activity, 36, 37
    JMJD2A, 36
    to pathological cardiac hypertrophy, 36
  common histone modifications, 35
  dimers, 35
  epigenetic regulation, 33
  HATs, 35
History of cardiology, 2, 4–5
History of Catholic nuns, 19
History of medicine, 10
Homoscedasticity, 78, 85
Hormonal changes, 418
Hormonal effects, 419
Hormone replacement therapy (HRT), 659
Hormone treatment, 444
Horton-Brown syndrome, 502–503
Hot flashes, 440, 441
HR variability (HRV), 679
Human cardiac stem cells (hCSCs), 128
Human proBNP, 554
Hydrostatic pressure, 309, 311, 313, 314, 322, 323
Hydroxyeicosatetraenoic acids (HETEs), 518
20-Hydroxyeicosatetraenoic acids (20-HETEs), 143, 144
2-Hydroxyestradiol (2OHE2), 519, 520, 523
4-Hydroxyestradiol (4OHE2), 520
16α-Hydroxyestrone (16αOHE1), 518, 519, 523
17β Hydroxysteroid dehydrogenase 1 (17βHSD1), 517
17β Hydroxysteroid dehydrogenase 2 (17βHSD2), 517
5-Hydroxytryptamine (5-HT), 521, 522
Hyperinsulinemia, 313
Hypertension, 311, 324, 535, 593
  Global Burden of Disease Study, 140
  HIV patients, 145
  immune system, 146
  MC4R, 148
  mechanisms responsible, 141–142, 148
  obesity, 146
  in old male, 148
  prevalence, 140, 146
  RAS, 142–143
  SHR, 142, 143, 145, 149
  systolic BP, 141
  treatment, 149
  US adults, 140
  women, 140
  in young male, 148
Hypertrophy, 390–393, 396, 405, 406
Hypogonadism, 147
Hypoplastic left heart syndrome (HLHS), 52
Hypothalamic-pituitary-adrenal axis (HPAA), 227

I

Idiopathic pulmonary fibrosis (IPF), 497
Illustris medicus, 5
Immune-mediated inflammatory disease (IMID), 495, 503
Immune system, 145–146
Immune system-mediated hypertension, 146
Immunological difference, 383
Implantable cardioverter defibrillator (ICD), 537
  arrhythmic substrate, 112
  baseline characteristics, 111
  clinical indications, 109
  CRT, 108, 115
  evidence, 110, 112
  females and males, 111
  implantation, 184
  inappropriate therapies, 111
  inappropriate therapy, 111
  RV lead, 109
  S-ICD, 109, 112, 118
Incidence, 50, 51
Independent samples *t*-test, 76, 77
Indexing, 350, 351
Indian database, 95
Infirmaries, 19–26
Inflammasomes, 494
Inflammation imaging, 494–495
Inflammatory bowel disease (IBD), 494, 497, 498
Inflammatory joint disease (IJD), 498
Innervation, 183, 192, 231, 386, 497, 666, 680
International Society of Heart and Lung Transplantation (ISHLT), 382
Intersex, 645
Intima plus media thickness (IMT), 393
Intracerebroventricular (ICV), 143
Intuitive-emotional processes, 230

Iron deficiency, 535
Ischemic heart disease (IHD), 491, 492, 494
  autopsy evaluations, 262
  CAD, 259, 262
  CVD, 258
  epidemiology, 259, 260
  GDMT, 263
  management strategies, 262
  mortality, 265
  risk factors, 259–262
Ischemic preconditioning (IPC), 244

J
JT interval, 104, 105
Justification for the Use of Statins in Primary Prevention: An Intervention Trial Evaluating Rosuvastatin (JUPITER) trial, 599

K
Kallikrein-kinin system, 659
Kaplan-Meier survival rates, 111, 114
Krebs cycle, 242
Kronos Early Estrogen Prevention Study (KEEPS), 447

L
Lady's heart, 24
"The lady with the lamp", 22
Lambda mu sigma (LMS) modeling technique, 351
LA volume index (LAVI/LAVi), 369, 688
L-arginine, 420
Lean body mass, 349, 354, 355
Left atrium (LA), 688–689
Left bundle branch block (LBBB), 108, 109, 113–116, 400
Left ventricular (LV), 437
  dimensions, 367
  dysfunction, 667
  fast cine MRI, 369
  function, 422
  male/female differences, 701
  RA and LA, 370
  volume-related variables, 135
Left ventricular assist devices (LVAD), 439, 609
Left ventricular ejection fraction (LVEF), 108, 110, 112, 113, 116, 117, 558, 561
Left ventricular hypertrophy (LVH)
  Cornell criteria, 99–100
  Romhilt-Estes point score system, 100
  S-L index, 97–99
Left ventricular mass (LVM), 684–685
Left ventricular outflow tract obstruction (LVOTO), 439
Left ventricular remodeling, 421
Levene's test, 76–79, 87
Lewis index, 680, 683
Linear regression model, 77, 78
Locomotor training, 583
Logistic regression model, 81
Log-rank test, 82, 83
Long ncRNAs (lncRNAs), 37
  in cardiovascular development, 38
  in cardiovascular diseases, 38
Low-density lipoprotein (LDL), 628
Lucigenin chemiluminescence, 144
Lymphatics, 310

M
Machine learning, 608, 620–622
  AI–based methods, 608
  algorithms, 618
  and big-data analytics (*see* Big-data analysis)
  cardiovascular outcome, 609
  classification, 618, 619
  in clinical medicine, 625
  clustering, 618, 620
  computational tools, 608
  density estimation, 618
  description, 617
  dimensionality reduction, 618
  distribution of classes, in heart disease, 619
  feature-selection and -classification process, heart disease, 621
  HF
    diagnosis, 620
    ejection fraction (EF), 620
    experiments, 622
    subgroups, 621
    subtypes, 620
  learning pipelines, 620
  regression, 618
  RF classifier, 619
  supervised learning, 618
  types, 618
  unsupervised learning, 620
MADIT-CRT trial, 111
Major adverse cardiovascular events (MACE), 261, 268–270
Male donors/male recipients (MD/MR), 381
Male recipients (MR), 381
Mass, length and time (MLT) system, 213–215
Matera Alliance, 650
Maternal hemodynamic, 414, 459, 460, 463
  adaptations, 422, 426
  disrupted
    blood pressure regulation, 463
    in cardiac function, 460
    in kidney function, 460
    sensitivity, RAAS, 459
    vascular dysfunction, 460
    vascular function, 463
  preeclampsia, 459
  pregnancy, 459
Maternal venous system, 420
Mean arterial pressure (MAP), 158, 311, 312, 321, 415, 684
Mechanobiology, *see* Arteries
Medical device therapy, 108–117
  anatomy and physiology, 108
  cardiac device trials, 108

Medical device therapy (*cont.*)
clinical trials, 117
ICDs, 117 (*see also* Implantable cardioverter defibrillators (ICD))
LBBB with CRT, 113
male and female patients, implantable devices, 110
normal and LBBB activation, 109
physicians, 108
pulse generators, ICD, S-ICD and CRT(-D), 109
regulatory agencies, 118
sex differences (*see* Sex differences)
sex-specific safety
cardiac resynchronization therapy, 113–116
defibrillators, 109–112
medical device therapies, 117
Medical statistics
hypothesis test, 75
independent variable/determinant, 74
PICO model, 74
standard error (SE), 74
Medicine of the future, 608
Melanocortin-4 receptor (MC4R), 143
Menopausal hormone treatments (MHT), 441, 445
Menopause, 285–286, 434, 441, 442, 444, 447, 475
Menstruation, 7, 18, 314, 319, 341, 442, 556, 650, 662, 682
Mental stress, 230–233, 235
Mental stress-induced myocardial ischemia (MSIMI), 233
Mercury sphygmomanometer, 6
Merit Ptah, 16
Metabolic equivalent task (MET), 340
Metabolic syndrome (MetSy), 142, 146, 476, 661–663
and cardiac remodeling, 476–478
cardiovascular risk factors, 475
gender-specific cutoff criteria, 475
gender-specific manner, 475
hyperandrogenism, 475
metabolic syndrome (*see* Metabolic syndrome)
mortality, 476
obesity (*see* Obesity)
visceral fat, 476
Metabolomics, 598
2-Methoxyestradiol (2ME2), 519, 520, 523
Microchimerism, 285
Microcirculation, 310–311
anatomical prominence, 308
arterioles, 308–309
capillaries, 309
clinical chemistry reference values, 315
endothelium, 310
ethnicity, 324
hemodynamic reference values, 315, 321
heterogeneous, 324
lymphatics, 310
microvasculature, 308–310
microvessels, 308
myriad functions, 308
sex differences, 323
sexual dimorphisms (*see* Sexual dimorphisms)
sexually dimorphic constituents, 323
venules, 309–310
Microcirculatory dysfunction, 65
Microneurographic studies, 423
MicroRNAs, in cardiovascular disease, 37, 38
Microvascular network, 309, 310
Microvasculature, 696
Microvasculopathy, 383
Microvolt T-wave alternans (mTWA), 667
Midrange, 373
Midwifery, 14, 20, 21, 27
Migraine, 441
Migrainous infarct, 285
Miguel Serveto, 12
Military antishock trouser (MAST), 368
Military hospital, 20, 21
6-Min walk distance (6MWD), 518
Misdiagnosis, 65, 462
Mitochondrial biogenesis, 242
Mitochondrial biology
and apoptosis, 243–245
ATP, 242
autophagy, 245–247
biogenesis, 242
cardiac function, 242
cardiovascular diseases, 250
fusion and fission, 247–250
gender difference, 250–252
mtDNA, 242
and senescence, 247
TFAM, 242
Mitochondrial dysfunction, 243, 245–247, 251
Mitochondrial transcription factor A (TFAM), 242
Mitofusin 1 (MFN1), 248
Mitofusin 2 (MFN 2), 248
Mitogen-activated protein kinase (MAPK), 54, 519
Mitophagy, 246, 247, 249
Mitral annular plane systolic excursion (MAPSE), 698, 700
Modified Rankin Scale (mRS) score, 281
Molsidomine, 144
Monocrotaline, 516, 518–520
Morbidity, 62, 68
Morphea, 493
Mortality
age-adjusted CVD, 68
CVD, 62, 66, 67
and heart failure, 116
LBBB, 115
males with ICDs, 110
men-to-women mortality rate ratios, 64
SCD-HeFT trial, 110
Motor function
ADLs, 581
neurorehabilitation, 581–582
Motor training
lower extremity, 583–584
upper extremity, 582–583
Mount Olympus, 3

Multidetector computed tomography (MDCT), 684
Multi-Ethnic Study of Atherosclerosis (MESA), 374, 630
Multiplayer strategy, 585
Mutations of mtDNA, 242, 251
Mutterrecht, 17, 27
Mycophenolate mofetil (MMF), 145
Myocardial disease, 375
Myocardial infarction (MI), 628, 679
Myocardial mass, 685
Myocardial strain, 689–691
Myocytes
  aged female, 396
  aged male, 396
  atrial, 395
  ventricular, 392, 395
Myofilament, 125–127
Myogenic response, 311
Myokine, 341, 342
Myosin binding protein-C (MyBP-C), 125
Myosuit, 584

N
Nailfold videocapillaroscopy, 494, 503
National Bariatric Surgery Registry (NBSR), 662
National Cardiovascular Data Registry (NCDR) ICD Registry, 110
The National Cholesterol Education Program's Adult Treatment Panel III (NCEP-ATP III), 475
National Health Service (NHS), 592
National Institutes of Health (NIH), 645
The National Institute for Health and Care Excellence (NICE) guidance, 593
Natriuretic peptides, 230, 558–560
  age, 557–558
  blood circulation, 555
  healthy women and men, 555–557
  HF
    diagnosis, 558–559
    management, 560
  NT-proBNP, 555
  pregnancy, 557
  prevention, prognosis and risk stratification, 560–561
Neural cardiac regulation, 194
Neuro-endocrine-immune symphony, 232
Neuroendocrine regulation, 232
Neuromodulation, 659
Neurorehabilitation, 581–582
Neurotransmitters, 226
Neurovisceral system, 192
Neutrophil extracellular trap (NET), 497
Nigerian database, 95
Nitro-L-arginine methyl ester (L-NAME), 145, 656
Nomogram
  EF, EDVi and ESVi, 686
  ejection fraction, 686
  pediatric echocardiographic measurements, 697
Noncoding RNAs (ncRNAs)
  epigenetic regulators, 37
  lncRNAs, in cardiovascular development, 37, 38
  miRNAs, in cardiovascular disease, 37, 38
Noninvasive imaging, 435
Nonobstructive coronary artery disease, 259, 268, 272
Non-ST-elevation myocardial infarction (NSTEMI), 550
Normal limits, in white caucasians, 681
Normal ranges, 95
N-terminal pro-brain-type natriuretic peptide (NT-proBNP), 554–558, 560
Nuclear respiratory factors (NRF) 1 and 2, 242
Nurses, 15, 19, 27, 95

O
Obesity
  abdominal, 477
  androgen synthesis, 147
  cardiovascular disease, 473
  gendered innovations, metabolic syndrome therapy, 662–663
  HDL cholesterol reduction, 475
  and insulin resistance, 472
  and metabolic syndrome, 142, 146
  microvascular dysfunction, 473
  molecular, genetic and pathophysiology, 661–662
Obstructive sleep apnea (OSA), 441, 559, 662
Optical atrophy 1 (OPA1), 248, 249
Oral contraceptives, 442
Origins, Western clinical medicine, 3
Orthostatic intolerance, 322
Orthotopic heart transplantation (OHT), 383
  cardiac chamber dimensions, 384–386
  donor/recipient sex mismatch
    genetic and hormonal factors, 383
    immunologic factors, 383
    size mismatch, 383
  heart transplant, 380
  sex mismatch, cardiac transplants, 381–384
Oscillometric methods, 156
Ovariectomy, 142
Overall survival (OS), 81
Oxygen uptake, 330

P
Pacemaker, 109, 113
Paired antagonistic innervation, 192
Paracelsus, 6
Parasympathetic innervation, 192–194
Parasympathetic system, 183, 186
Parity, 284, 285, 349, 351
Patent ductus arteriosus (PDA), 50
Pathophysiology, HF, 533
Pattern recognition receptors (PRRs), 230
Peak oxygen uptake (VO_2peak), 330–334
Peak systolic pressure (PSP), 130
Pearson's chi-square test, 80
Pediatric cardiology, 348, 350
  fetal development, 696–697
Percussion technique, 6
Percutaneous coronary intervention (PCI), 103, 262
Pericardiocentesis, 111

Pericyte
  abluminal surface, 309
  barrier properties, 309
  venular capillary segments, 310
Peripartum cardiomyopathy, 440
Peripheral arterial diseases (PAD), 63
Peripheral BP components, 160
Peripheral-central systolic blood pressures, 684
Peripheral resistance, 311
Periprocedural sedation, 666–667
Permeability transition pore (PTP), 243
Peroxynitrate, 398
Peroxynitrite, 144
Personalized medicine, vi, 203, 302, 546, 561, 590–592, 596, 601, 607, 608, 656, 660, 663, 669–671
Pharmacology, 192, 193
Phlebotomy, 3, 8
Phosphodiesterases (PDEs), 514
Phospholamban (PLB), 125
Physiological anemia, 416
PICO model, 72, 74
Pilote, 12
Placental dysfunction, 457–459
Placentation, 457, 459
Plasma renin activity, 418
Plasma volume, 416
Pneumothorax, 111
Policy, gender, 646
Polycystic ovary syndrome (PCOS), 146, 147, 635, 638
Positron emission tomography (PET), 262, 493, 494, 506, 507
Postmenopausal women, 140, 143, 146, 147
Post-transplant mortality
  female donor sex, 382
  stratification, by UNOS, 382
Precision medicine, 149, 273, 537, 589–602, 608, 609, 628, 635
  big data, 598
  bioinformatics and modeling, 598, 599
  biomarkers and imaging, 598
  cardiovascular continuum, 596–599
  in cardiovascular diseases, 592
  cardiovascular medicine, 590
  challenges and barriers, 601
  domains, 594–596
  genetics and genomics, 597
  non-cardiovascular field, 592, 599–600
  omics, 597, 598
  patient factors, 599
  personalized medicine, 590, 592
  response, 595
  sex differences, 600–601
  stratified medicine, 590
  terminology, 590
  therapeutic approach, 590
  trial designs, 599
Preeclampsia, 284, 457–465
  classification, 456
  clinical features, 456
  complications, 457
  definition, 456
  diagnostic criteria, 457
  disrupted maternal hemodynamics (*see* Maternal hemodynamics)
  early- and late-onset, 456
  long-term complications, 461
    blood pressure regulation, 463
    cardiovascular disease, 461
    disrupted maternal hemodynamics, 463
    heart failure, 462
    hypertension, 461
    implications, for child born, 463
    mechanisms, 463–465
    metabolic disease, 462
    renal disease, 461–462
    vascular function, 463
  pathophysiology, 457
    factors, 457
    maternal factors, role, 459
    placental factors and maternal susceptibility, 457
    placenta, role, 458–459
    as a three-stage model, 457
  screening and prevention, 465–467
  therapy, 460–461
The Preferred Reporting Items for Systematic Reviews and Meta-Analyses (PRISMA), 91
Pregnancy, 314, 414
  cardiac electrophysiology, 422
  estrogen, 418
  glomerular filtration rate, 418
  hemodynamic adaptations, 416–418
  hemodynamic changes, 418
  maternal hemodynamic adaptations, 426
  microneurographic studies, 423
  natriuretic peptides, 557
  ovarian hyperstimulation, 418
  RAAS, 417
  relaxin, 418
  renal vasoconstricting systems, 418
  sex differences, 284
  vasoconstrictor responsiveness, 425
  vasomotor sympathetic activity, 422, 425
  venous distensibility, 420
Preload, 125, 127, 351, 367, 368, 415, 421, 475, 480
Premenopausal female donor, 384
Pressure–strain elastic modulus, 294–296
Pressure–volume loop, 129–131
Prevention And Treatment of Hypertension With Algorithm-based therapy (PATHWAY)-2 study, 593
Primary systemic vasodilation, 417
Prognosis, 55–57
  and beta-blockers, 679
  and biomarkers, 546
  in breast cancer, 482
  in Brugada syndrome, 182, 183
  in CAD and IHD, 258, 260
  and CFR, 268

in congenital heart disease, 50–57
and ejection fraction, 385, 620, 691
and exercise capacity, 570
in fibrotic disorder, 497
in heart failure, 531–534
in heart transplantation, 381, 537
and IHD in women, 271, 404
and machine learning, 609
and risk factors, 362
and stress test, 340
in sympathetic activation, 463
in TRIPOD, 91
and troponin, 553
Pro-inflammatory cytokines, 457
Prospective cardiovascular munster heart study (PROCAM), 630
Protein kinase A (PKA), 514
Proteomics, 594, 597, 598, 601
Psoriasis
in AD, 495
cases, 493
pathogenic immune pathways, 495
and psoriatic arthritis, 493, 503–504
Psychic factors, in heart disease, 230
Psycho-neuro-endocrine-immunology (PNEI), 226–227, 234
Psychosocial stress, 233
PTEN-induced putative kinase 1 (PINK1), 246
Puberty, 142, 161, 227, 349, 434, 442, 491, 505
Pulmonary arterial hypertension (PAH)
aromatase, 517–518
BMPR-II signalling pathway, 515–516
characterisation, 512
CHD, 54
clinical classification, 513
CYP1B1, 518–519
cytochrome P450 enzymes, 518
endothelin receptor antagonists, 512–513
estrogen, 520, 523
estrogen biosynthesis and metabolism, 517
16α-hydroxyestrone, 519
idiopathic and familial, 512
2-methoxyestradiol, 519
non-specific nature, 512
PDE-5 inhibitors, 514
predisposition, 53
prostacyclin analogues, 513–514
risk, 53
risk factor, 514–515
serotonin and sex differences, 521–523
sex hormones, 516–517, 520–521, 523
and sex-specific differences, 521
sGC, 514
Pulmonary artery endothelial cell (PAEC), 512
Pulmonary artery smooth muscle cell (PASMC), 512, 514, 519, 521, 522
Pulmonary circulation, 6
Arabic medical manuscripts, 13
bedside teaching and medical journals, 15, 16
blood circulation, discovery, 12
Galen's teachings, 12
gender models, female physicians and women writers, 16–18
history of, 12, 13
mechanism of, 13
medical women question, 16–19
pulsology, 12
women, in medicine, 13–15
Pulsatile pressure, 154
Pulseless electrical activity (PEA), 108
Pulse pressure (PP), 163–166
age, in cardiac patients, 165
cardiac patients, 165
diastolic arterial pressure, 164
systolic arterial pressure, 163
in women, 155
Pulse waveform analysis
augmentation index, 300–302
cyclic variationm, arterial compliance, 302, 303
pulse wave velocity, 299
transfer function, 302
wave separation, forward and reflected components, 299–301
Pulse wave velocity (PWV), 299, 337, 497
arterial stiffness, 157–158
Pulsology, 3, 5, 12, 680–682, 684

Q
QT interval, 104, 105

R
Race/ethnic differences, v, 41, 51–53, 84, 266, 324, 349, 351, 354, 374, 414, 426, 504, 558, 570, 591, 593, 630, 632, 645, 648, 650, 661, 678, 680, 685, 688, 689, 691, 695, 697, 699, 701
Radiofrequency catheter ablation, 437
Random forest (RF), 609, 619, 623
Randomized controlled trial (RCT), 73
Rankin scale, 280–283
Rarefaction, 311
Raynaud's phenomenon (RP), 491, 499, 500, 503
Reactive oxygen and/or nitrogen species (ROS/RNS), 144, 145
Reactive oxygen species (ROS), 242–245, 247, 251
Red blood cell, 416
Redox messengers, 244
Reference ranges, ECG, 95, 99
Glasgow database, 102
normal limits, 94
QRS duration, 101
S-L index, 99
T waves, in Glasgow database, 103
in V2–V4, for males and females, 102
Reference value
ARD, 695
blood components, 678
and body size-adjusted measurements, 350
cardiovascular dimensions, 350, 351, 355

Reference value (*cont.*)
for cardiovascular size and function, in children, 350
for cardiopulmonary exercise testing, 352
coronary arteries, 695
determination, 350
diastolic indices, in children, 351
3D echo, 691
ECG parameters, 352
ESVi and EDVi, 691
multiple-device, 680
National High Blood Pressure Education Program, 697
RA volumes, 689
sex-specific, 352, 680, 700
survey of, 678
TDI, 698, 699
24-hour ambulatory blood pressure, 354
Regression models
checking effect modification, 89
cholesterol levels and age, 86
Cox model, 82–85
linear regression model, 77, 78
logistic model, 81
Rehab gyms, 585
Relman, 25
Remodeling
cardiac, 36–38, 54, 226, 230, 233, 244, 266, 331, 336, 362, 364, 369, 372, 374, 385, 386, 390, 392, 402
left ventricular (LV), 362, 364, 367, 374
myocardial, 374
pattern, 372
Renaissance medicine, 13
Renal denervation (RDN), 659
Renal function, 460, 462
Renin-angiotensin aldosterone system (RAAS), 417, 459
Renin-angiotensin system (RAS), 142
Review, neuroendocrine, 232
Rheumatoid arthritis (RA), 490, 498
Right atrium (RA), 370, 689
Right ventricular (RV), 55, 131, 368, 688
fast cine MRI, 369
MAST, 368
Right ventricular hypertrophy (RVH), 100
Right ventricular mass (RVM), 687
Robot
research, 584–585
telerehabilitation, 585–586
Robot-assisted therapy, 586
Romhilt-Estes point score system, 100
Rosacea, 496
Ross, 11
Rufus of Ephesus, 5
RV inflow tract (RVIT), 688
RV outflow tract (RVOT), 688
RV systolic pressure (RVSP), 519

S
Salt-sensitive hypertension (SSH), 659
Sarcoidosis, 505–506
Scaling law, 209
SCD-HeFT trial, 110
Schaper, 12
School-based physical activity intervention, 340
Scleroderma, 491, 493, 494, 497, 499–500
Secondary prevention
coronary heart disease, 566
women and men, 571
Selective estrogen receptor modulators (SERMs), 660
Selective serotonin reuptake inhibitors (SSRIs), 441
Sepsis, 308, 322
Serotonin and sex differences, 521–523
Serum leptin, 661
Serveto, 2, 12, 20, 25, 26
Sex-associated cardiomyopathies, 668
Sex chromosomes
in epigenetic regulation, 33
hormones, AR and ER receptors, 42, 43
protein-coding genes, 41
X-chromosomes (XX), 39
X-escapee genes, 40
X-inactivation, process of, 39
X- and Y-chromosome (XY), 39
Y-chromosome effects, in CVD, 41
Sex differences, 75–90, 218–222
acute ischemic stroke, 280
acute stroke care, 281–282
in age-related changes, 394, 406
allometric analysis, 216–218
geometric taper, 218–220
in stenosis and atherosclerosis, 221, 222
vascular branching effects, 220
androgens, 147
in aneurysms, 403
in arterial wall properties, 296
atrial fibrillation, 283–284, 665–666
in blood pressure, 161–163, 313
capillary endothelial cells, 313
in cellular calcium regulation, 534
cellular-level, 393
CHD, 65
clinical trials, 287
confounding variables, 280
in continuous outcome, 75
independent samples *t*-test, 76, 77
linear regression model, 77, 78
mean and median, 75
non-normal outcome, 77
standardization, 78, 79
in CVD, 62, 67, 68, 397
in dichotomous outcome
logistic regression model, 81
Pearson's chi-square test, 80
prevalence, CHF, 79
RD and RR, 80
elastin, 403
in endothelial dysfunction, 398
endothelial cells, 286
FCG mouse model, 40
glucocorticoid action, 322

heart failure and cardiac electrophysiology, 108–109
HF, 67, 532, 533
hormone replacement therapy, 285–286
INSTRUCT trial, 281
JMJD2A, 36
LBBB, 400
in LV mass, 391
in MAP, 312
menopause, 285–286
microscopic changes, 393
in microvasculature, 404
migraine headache with aura, 285
mRS score, 281
in myocyte hypertrophy, 393
in noninvasive SCD, 667
in pacemaker cell, 393
parity and experimental stroke, 284–285
in pericyte function/dysfunction, 309
pregnancy, 284
pre-stroke, 280, 281
risk factors, 402
steroid hormones and their receptors, 42
stroke, 287, 586
in evaluation, 282–283
in incidence, 66
symptom presentation, 282
in survival outcome
Cox proportional hazards model, 82–85
log-rank test, 82, 83
OS, 81
survival function, 81
systolic function, 396
variables, in association, 86
confounding, 86–88
effect modification, 88, 89
multiple confounders/effect modifiers, 89, 90
in vascular reactivity, 400
vascular risk factors, 284
in vascular stiffness and augmentation index, 303 304
vessel size, 286–287
volume regulation, 314
X-encoded miRNAs, 41
X-escapee genes, 40
XX and XY mice, 41
XY* mouse model tests, 41
Sex disparities
pathophysiology, 580–581
stroke epidemiology, 580–581
treatment, 580–581
Sex hormone, 491, 501
AR and ER activity, 42, 43
aromatase, 517–518
CYP1B1, 518–519
DHEA, 520
as epigenetic cofactors, 39
in epigenetic regulation, 33
estrogen, 517, 518, 520
16α-hydroxyestrone, 519
2-methoxyestradiol, 519
in pulmonary hypertension, 516–517
pulmonary vasculature, 520
signalling and processing pathways, 514
vasodilatory property, 520
Sex hormone-binding globulin (SHBG), 557
Sex mismatch, 22, 135, 380–386
Sex ratio, 51, 52
Sex-related bias, 197, 201
Sex-specific analysis, 133, 227, 683
Sex-specific difference, 664
Sex-specific risk factors, 490
Sex steroids, 146–147
Sexual asymmetry, 4
Sexual dimorphisms
arterioles, 311–313
capillary function, 313–314
EC phenotype, 310–311
rarefaction, 311
vascular reactivity, 311–313
venules, 314–323
Sexualization, 32, 39
Shannon–Bers Ca^{2+} electrophysiological myocyte model, 127
Similarity criterion, 209
Single nucleotide polymorphisms (SNPs), 632, 633
Sinus tachycardia, 422
Sjögren's syndrome, 491, 493
Smoothed pseudo Wigner-Ville distribution (SPWVD), 185, 186
Sodium channels, 422
Sokolow-Lyon (S-L) index, 97–99
Soldiers heart, 24–25
Soltis–Saucerman model, 127
Soluble endoglin (sEng), 457, 460
Soluble fms-like tyrosine kinase-1 (sFlt-1), 457, 460, 464, 465
Sonnenblick, 11
Spontaneous coronary artery dissection (SCAD)
associated conditions, 436
coronary angiography, 436
diagnosis, 436
history, 436
mechanisms, 436
Spontaneously hypertensive rats (SHRs), 142, 656
Sprague-Dawley rats, 144
ST amplitudes, 103
The STandards for Reporting of Diagnostic Accuracy (STARD), 91
Stanford's Women and Sex Differences in Medicine (WSDM) Center, 652
Starling concept, 125
ST-elevation myocardial infarction (STEMI), 103, 550
Stiffness parameter, 172
Strain analysis
GCS, 364
GLS, 364
longitudinal strain, 374
rate, 375
Stratified medicine, 590, 599
The Strengthening The Reporting of OBservational studies in Epidemiology (STROBE), 91
Stress cardiomyopathy, 439
Stress response, 226, 227

Stroke
  and CHD, 62, 64
  coronary heart disease, 500
  CVD mortality, 66, 67
  and heart rate, 129
  HF, 67
  ischemic, 63
  SLE, 500
Stroke volume, 330, 366, 415
  acute exercise, 336
  index, 331
  lean body mass, 332
  maximal exercise, 336
  older women, 331
  submaximal and maximal exercise, 330
  VO_2peak, 333
  young women, 334, 336
Subcutaneous ICDs (S-ICDs), 109, 112, 117
Sudden cardiac death (SCD), 108, 112, 113, 117, 498, 667
Supervised learning, 618
Support vector machine (SVM), 609, 617, 622
Surrogate index/metric, 153, 158, 172, 349, 361, 504, 680, 697, 698
Swedish Primary Care Cardiovascular Database (SPCCD), 141
Sympathetic innervation, 192–194
Sympathetic nervous system (SNS), 142
Sympatho-vagal model, 198
Syncytiotrophoblast, 457, 459
Syndecan-1, 314
Systematic coronary risk evaluation (SCORE) system, 630
Systemic lupus erythematosus (SLE), 500–502
Systemic sclerosis (SySc), 490, 497, 499–500
Systolic blood pressure, 68
Systolic Blood Pressure Intervention Trial (SPRINT), 141, 599
Systolic and diastolic blood pressure, 364
Systolic elastance, 130
Systolic function, 395–396
Systolic Hypertension in the Elderly Program (SHEP), 166
Systolic pressure, 163–166, 338
Systolic pressure augmentation, 301

T
Takotsubo cardiomyopathy, 24, 25, 197, 232, 434, 439, 549, 668, 669
Tamponade, 111
Telerehabilitation, 585–586
Temporal arteritis, 502–503
Teresa of Avila, 19, 24
Testosterone, 475, 492
Testosterone therapy (TTh), 660–661
Tetrahydrobiopterin, 145
2,3′,4, 5′-Tetramethoxystilbene (TMS), 518
Tetralogy of Fallot (TOF), 51
Theatrum anatomicum, 6
Therapy device, 582
Thyroid dysfunction
  cardiac remodeling, 478–481
  hyperthyroid patients, 478
  hyperthyroidism, 478, 480–481
  hypothyroidism, 478–480
  stroke/myocardial infarction, 478
Time-frequency analysis, 183, 185–187
Tissue Doppler imaging (TDI), 698, 699
Toll-like receptors (TLRs), 230
Torsion
  natural pump, 375
  speckle tracking, 364
  TCR, 364
  torque mechanics, 364
  women, 364
Total peripheral resistance, 416
Transfusion history, 386
Transgender, 4
Translational research
  gender differences in CVD, 656–658
The Transparent Reporting of a multivariable prediction model for Individual Prognosis Or Diagnosis (TRIPOD), 91
Transposition of great arteries (TGA), 51
Transthoracic echo (TTE), 686
Transthyretin (TTR) amyloidosis, 669
Transverse aortic constriction (TAC), 657
Trastuzumab, 245, 373–374, 438, 482–484
Treatments
  ACEI therapy, 536
  DIG trial, 537
  HF, 536
  ICD/CRT-D-associated reduction, 537
  left ventricular assist devices, 537
  modalities, 532
Tricuspid annular plane systolic excursion (TAPSE), 520, 697
Troponin C (TnC), 547, 548
Trota of Salerno, 17
Tryptophan hydroxylase (TPH), 521
Two-dimensional speckle-tracking imaging (2D-STI), 689
Type 2 diabetes (T2D), 492

U
Ultrasonic measurements, 170, 171, 173
United Network for Organ Sharing (UNOS), 382
Unsupervised learning, 620
Upper paleolithic period, 124
US National Health and Nutrition Examination Survey (NHANES), 140
Uteroplacental perfusion, 457, 458
Uteroplacental vascular circuit, 418

V
Vascular endothelial dysfunction, 404
Vascular endothelial growth factor (VEGF), 286, 314, 440, 521
Vascular homeostasis, 308
Vascular mechanics, 293–296
Vascular remodelling, 516, 520, 522
Vascular smooth muscle (VSM), 308–313, 323
Vascular smooth muscle cells (VSMCs), 249
Vasoactive intestinal peptides (VIPs), 227
Vasoconstrictor, 399
Vasoconstrictor responsiveness, 425

Vasomotor sympathetic activity, 422, 425
Vasomotor symptom, 441, 447
Vatner, 12
Venous capacitance and compliance, 420
Ventricular arrhythmia (VA), 667
Ventricular hypertrophy (LVH), 548, 549, 559
Ventricular premature complexes, 667
Ventricular pressure, in males and females, 128–133
Ventricular size, 685–688
Ventricular-vascular coupling, 336–339
Ventricular volume, 6
  in males and females, 135
  regulation, 57, 131–134, 362–375, 694
  sex-mismatch cases, 135
Ventriculo-arterial coupling (VAC), 128, 130, 134, 169, 361, 375, 694, 698
  and EF, 134, 367
  ESVi, 132
  in women, 136
Venules, 309–310, 314–323
Victorian Family Heart Study, 659
Videocapillaroscopy, 507
VO_2peak (peak oxygen uptake), 330–334
Volume regulation graph (VRG), 57, 694
  heart disease, 371
  heart failure, 372–374
  LA and RA diameters, 375
  left ventricle (LV), 371
  linear borders, 374
  LV volumes, 370
  pressure, 375
  regression lines, 368
Von Willebrand factor, 283, 397
Vulnerable plaque, 506

W

Wave-intensity (WI)
  AT, 173
  body mass index (BMI), 174
  "characteristic impedance", 177
  data analysis, 174
  defined, 170, 171
  feature, 172
  FFR and other hemodynamic parameters, 170
  lung volume and forced vital capacity (FVC), 175
  "the method of characteristics", 171
  noninvasive measurements, 172
  pressure and velocity waves, 177
  protocol, 173, 174
  respiratory compensation (RC) point, 173
  sex-specific differences, 177
  stiffness parameter, 172
  subjects, 171
  24-h ambulatory blood-pressure monitoring, 176
  "water hammer formula", 171
Windkessel model, 166, 167, 296–298
Wistar rats, 146
Women
  acute coronary syndromes, 52
  PAH, 53
  TOF, 56
Women's Genome Health Study (WGHS), 635
Women's Health Initiative (WHI), 444, 447
Women's Ischemia Syndrome Evaluation (WISE) study, 435
Wound care, 2
Wounded soldiers, 6

X

X chromosome transcriptome, 506
X-escapee genes, 40, 41
X-inactivation center (Xic), 39
X-inactive-specific transcript (Xist), 39
XY* mouse model, 40, 41

Y

Young's modulus of elasticity, 294

Z

Z score, 78, 79, 350–352, 355, 695, 697

MIX
Papier aus verantwortungsvollen Quellen
Paper from responsible sources
FSC® C105338

If you have any concerns about our products,
you can contact us on
ProductSafety@springernature.com

In case Publisher is established outside the EU,
the EU authorized representative is:
Springer Nature Customer Service Center GmbH
Europaplatz 3, 69115 Heidelberg, Germany

Printed by Libri Plureos GmbH
in Hamburg, Germany